THE OXFOR

4E COGNITION

Albert Newen received his PhD in 1994 from the University of Bielefeld. He became associate professor in 2003 at Tübingen, changed to the Ruhr-University Bochum (RUB) in 2007 and was appointed to full professor in 2010. He is director of the interdisciplinary Center for Mind and Cognition. He received several research awards, including the Bennigsen-Foerder Award (North-Rhine Westfalia) as well as the award for "Philosophy in Psychiatry" from the society of psychiatry in Germany (DGPPN). He was visiting professor in Oxford, Stanford and Urbana-Champagne. His research combines philosophical theory formation with research in psychology, psychiatry and neurosciences.

Leon de Bruin (1979) obtained his PhD in philosophy from the University of Leiden in 2010 with an interdisciplinary study on social cognition. After his PhD, he worked as a postdoc at the Ruhr-University Bochum on the development of false belief understanding. He was appointed assistant professor of philosophy of mind at the Radboud University Nijmegen in 2012, and associate professor of philosophy of mind in 2017.

Shaun Gallagher is the Lillian and Morrie Moss Professor of Excellence in Philosophy at the University of Memphis. His areas of research include phenomenology and the cognitive sciences, especially topics related to embodiment, self, agency and intersubjectivity, hermeneutics, and the philosophy of time. Dr. Gallagher has a secondary research appointment at the University of Wollongong, Australia. He is Honorary Professor at the University of Tromsø, Norway. He has held visiting positions at the Cognition and Brain Sciences Unit, Cambridge University; the Center for Subjectivity Research, University of Copenhagen; the Centre de Recherche en Epistémelogie Appliquée (CREA), Paris; the Ecole Normale Supériure, Lyon; the Humboldt University in Berlin, and most recently at Keble College, University of Oxford.

THE OXFORD HANDBOOK OF

4E COGNITION

Edited by

ALBERT NEWEN

LEON DE BRUIN

and

SHAUN GALLAGHER

Great Clarendon Street, Oxford, OX2 6DP,
United Kingdom

Oxford University Press is a department of the University of Oxford.
It furthers the University's objective of excellence in research, scholarship,
and education by publishing worldwide. Oxford is a registered trade mark of
Oxford University Press in the UK and in certain other countries

© Oxford University Press 2018

The moral rights of the authors have been asserted

First published 2018
First published in paperback 2020
Impression: 1

All rights reserved. No part of this publication may be reproduced, stored in
a retrieval system, or transmitted, in any form or by any means, without the
prior permission in writing of Oxford University Press, or as expressly permitted
by law, by licence or under terms agreed with the appropriate reprographics
rights organization. Enquiries concerning reproduction outside the scope of the
above should be sent to the Rights Department, Oxford University Press, at the
address above

You must not circulate this work in any other form
and you must impose this same condition on any acquirer

Published in the United States of America by Oxford University Press
198 Madison Avenue, New York, NY 10016, United States of America

British Library Cataloguing in Publication Data
Data available

Library of Congress Cataloging in Publication Data
Data available

ISBN 978-0-19-873541-0 (Hbk.)
ISBN 978-0-19-886347-2 (Pbk.)

Printed and bound by
CPI Group (UK) Ltd, Croydon, CRO 4YY

Links to third party websites are provided by Oxford in good faith and
for information only. Oxford disclaims any responsibility for the materials
contained in any third party website referenced in this work.

Contents

PART X APPLICATIONS

List of Contributors

Ken Aizawa Department of Philosophy, Rutgers University, Newark, USA

Louise Barrett Psychology Department, University of Lethbridge, Canada

Harold Bekkering Donders Institute for Brain, Cognition and Behaviour, Radboud University Nijmegen, The Netherlands

Cameron Buckner Department of Philosophy, University of Houston, USA

Evan W. Carr Department Of Psychology, University of California, San Diego, USA

Anthony Chemero Departments of Philosophy and Psychology, University of Cincinnati, USA

Giovanna Colombetti Department of Sociology, Philosophy and Anthropology, University of Exeter, UK

Amy Cook Department of English, Stony Brook University, USA

Rick Dale Department of Communications, University of California, Los Angeles, USA

Leon de Bruin Department Of Philosophy, Radboud University Nijmegen, The Netherlands

Hanne De Jaegher IAS-Research Centre for Life, Mind, and Society, Department of Logic and Philosophy and Science, University of The Basque Country, Spain

Frédérique de Vignemont Institut Jean Nicod, Département d'études cognitives, ENS, EHESS, PSL Research University, CNRS, Paris, France

Damiaan Denys Academic Medical Center, Department of Psychiatry, University of Amsterdam, The Netherlands

Ezequiel A. Di Paolo Ikerbasque, Basque Foundation for Science, Spain

Tom Froese Institute of Applied Mathematics and Systems Research (IIMAS), National Autonomous University of Mexico (UNAM), Mexico City, Mexico

Shaun Gallagher Department of Philosophy, University of Memphis, USA

Vittorio Gallese Department of Medicine and Surgery, Unit of Neuroscience, University of Parma, Italy; Institute of Philosophy, School of Advanced Study, University of London, UK

Hans-Johann Glock Institute of Philosophy, University of Zurich, Switzerland

Mitchell Herschbach Department of Philosophy, California State University, Northridge, USA

R. Peter Hobson Tavistock Clinic and Institute of Child Health, University College London, UK

Matej Hoffmann Department of Cybernetics, Faculty of Electrical Engineering, Czech Technical University in Prague, Czech Republic

Jakob Hohwy Cognition & Philosophy Lab, Department of Philosophy, Faculty of Arts, Monash University, Melbourne, Australia

Daniel D. Hutto School of Humanities and Social Inquiry, Faculty of Law, Humanities and the Arts, University of Wollongong, Australia

Mark Johnson Department of Philosophy, University of Oregon, USA

Anne Kever Psychological Sciences Research Institute (IPSY), Université Catholique de Louvain, Louvain-La-Neuve, Belgium

Michael D. Kirchhoff Department of Philosophy, University of Wollongong, Australia

Julian Kiverstein Academic Medical Center, Department of Psychiatry, University of Amsterdam, The Netherlands

Joel Krueger Department of Sociology, Philosophy, and Anthropology, University of Exeter, UK

Maurice Lamb Department of Psychology, University of Cincinnati, USA

Ulf Liszkowski Department of Developmental Psychology, University of Hamburg, Germany

Lambros Malafouris Keble College, Oxford University, UK

Richard Menary Faculty of Arts, Macquarie University, Sydney, Australia

John Michael Department of Philosophy, Warwick University, UK

Barbara Gail Montero Philosophy Department, The City University of New York (CUNY), USA

Erik Myin Centre for Philosophical Psychology, Department of Philosophy, University of Antwerp, Belgium

Albert Newen Institute for Philosophy II, Ruhr University Bochum, Germany

Elisabeth Pacherie Institut Jean Nicod, Département d'études cognitives, ENS, EHESS, PSL Research University, CNRS, Paris, France

Rolf Pfeifer Osaka University, Japan, and Shanghai Jiao Tong University, China

Hannes Rakoczy Institute of Psychology, Department of Developmental Psychology, Georg-August-University Göttingen, Germany

Matthew Ratcliffe Department of Philosophy, University of Vienna, Austria

Vasudevi Reddy Centre For Situated Action and Communication, Department of Psychology, University of Portsmouth, UK

Erik Rietveld Department of Philosophy and Academic Medical Center, Department of Psychiatry, University of Amsterdam, The Netherlands

Andreas Roepstorff School of Culture and Society, Department of Anthropology, Aarhus University, Denmark

Mark Rowlands College of Arts & Sciences, Department of Philosophy, University of Miami, USA

Tobias Schlicht Institute for Philosophy II, Ruhr University Bochum, Germany

Marco F.H. Schmidt Department Psychology, Ludwig Maximilian University of Munich (LMU), Germany

Corrado Sinigaglia Department of Philosophy, University of Milan, Italy

Tobias Starzak Institute for Philosophy II, Ruhr University Bochum, Germany

Achim Stephan Institute of Cognitive Science, University of Osnabrueck, Germany

Kim Sterelny School of Philosophy, Research School of Social Sciences, Australian National University, Australia

Deborah Tollefsen Department of Philosophy, University of Memphis, USA

Michiel van Elk Department of Psychology, University of Amsterdam, The Netherlands

Maarten van Westen Academic Medical Center, Department of Psychiatry, University of Amsterdam, The Netherlands

Somogy Varga Department of Philosophy, University of Memphis, USA

Kai Vogeley Department of Psychiatry, University of Cologne, Germany

Gottfried Vosgerau Department of Philosophy, Heinrich Heine University, Düsseldorf, Germany

Sven Walter Institute of Cognitive Science, University of Osnabrueck, Germany

Arne M. Weber Department of Philosophy, Heinrich Heine University, Düsseldorf, Germany

Piotr Winkielman Department of Psychology, University of California, San Diego, USA

Dan Zahavi Center for Subjectivity Research, Department of Media, Cognition and Communication, University of Copenhagen, Denmark

Tadeusz Wiesław Zawidzki Department of Philosophy, Columbian College of Arts & Sciences, The George Washington University, USA

PART I

INTRODUCTION

CHAPTER 1

4E COGNITION

Historical Roots, Key Concepts, and Central Issues

ALBERT NEWEN, LEON DE BRUIN,
AND SHAUN GALLAGHER

Historical Roots of the Debate

THE debate about the role of the body in cognition has been ongoing since close to the beginnings of philosophy. In Plato's dialogue *The Phaedo*, for example, Socrates considers the idea, which he attributes to Anaxagoras, that one could explain his decision to remain in prison by a purely material or physical explanation in terms of bodily mechanisms. Socrates himself rejects this idea—surely, he thinks, there is something more to reason than just bodily processes. Aristotle, however, was motivated by the idea that Anaxagoras was not entirely wrong. While Aristotle did not accept the radical view of Anaxagoras, he considered that the body (with special reference to the hands) may play some role in what makes for human rationality. Such debates considering the role of the body for the mind can be traced through medieval texts authored by Neoplatonists, Aquinas, and others, and are given their modern formulations in thinkers such as Spinoza, La Mettrie, Condillac, and many others. Pragmatists, phenomenologists, and philosophers of mind wrestle with the same issues throughout the twentieth century. The more proximate background for the current debates about embodied cognition, however, is to be found in the disagreements between behaviorists and cognitivists. Continuing tensions within cognitivism, and the cognitive sciences more generally, brought on by contrasting functionalist and neurobiological accounts that tended to ignore the role of body and environment and focus on internalist explanations of brain function, set the stage for the emergence of contemporary views on embodied cognition.

In the 1990s, Varela, Thompson, and Rosch's (1991) *The Embodied Mind*, drawing on phenomenological and neurobiological resources, proposed an enactivist account of cognition that emphasized the role of the dynamical coupling of brain–body–environment. Around the same time, a paper by Flor und Hutchins (1991) introduced distributed cognition as a "new branch of cognitive science" for which the unit of analysis includes external structures, collectives, and artifacts organized as a system to perform a task. Hutchins's (1995) *Cognition in the Wild* was a direct influence on Clark and Chalmers's (1998) now-classic philosophical essay, "The Extended Mind." Throughout this time period, additional work inspired by Gibson's ecological approach to psychology contributed to a growing realization that cognition was not limited to processes in the head, but was embodied, embedded, extended, and enactive.

Although the concept of 4E cognition[1] brings these different approaches together under one heading and conceives of them as coherently opposed to the internalist, brain-centered views of cognitivism, there are continuing disagreements about a variety of issues within and among these embodied approaches. Is cognition embodied, embedded, extended, or enactive? The issues that continue to be debated concern the very nature of embodiment, the precise way that brain, body, and environment are coupled or integrated in cognition, and how much we can generalize from the observation of embodiment in one type of cognitive performance to others. Furthermore, there are questions about the role of representations and what it means to say that cognition is "constituted" by bodily and environmental processes.

Key Concepts

How to Individuate Cognitive Processes

Before introducing the key concepts in the debate, we first need to consider whether there are certain constraints that need to be taken into account in order to answer the question of what cognition is and how we should individuate cognitive processes.

If we take cognition as a *natural kind* (even if we do not know the underlying mechanisms) this would limit the nature of our investigation to a search for the relevant mechanism constituting it. But there is no consensus on this question: while Buckner (2015) argues that cognitive processes are indeed natural kinds, evidence about neural plasticity presents a strong challenge to this claim (Hübener and Bonhoeffer 2014). An alternative strategy for answering the question is to focus on *typical examples* (Newen 2015). This seems to be a promising strategy, but it is not without problems. One complication is that a selection of the typical examples is already biased by certain assumptions

[1] Mark Rowlands (2010, p. 3) attributes the 4E label to Shaun Gallagher who organized a conference on 4E cognition in 2007 at the University of Central Florida. The first use of that term, however, as far as we know, emerged in discussions at a workshop on the embodied mind at Cardiff University, in July 2006, which included the following participants: Shaun Gallagher, Richard Gray, Kathleen Lennon, Richard Menary, Søren Overgaard, Matthew Ratcliffe, Mark Rowlands, and Alessandra Tanesini. Richard Menary edited a special issue of the journal *Phenomenology and the Cognitive Sciences* on 4E cognition based on selected papers from the 2007 conference (Menary 2010).

concerning the nature of cognition. Thus, where traditional cognitive science focused primarily on playing chess and mastering the "Tower of Hanoi," i.e., tasks that are strongly rule-governed, proponents of 4E cognition appeal to experiments that involve spatial navigation, face-based recognition of emotion, and basic forms of social interaction. It is therefore paramount to get a clear view on the assumptions about cognition that are made by proponents of both positions.

4E Cognition and Traditional Cognitive Science

The foundation of traditional cognitive science used to be the representational and computational model of cognition (RCC). According to this model, cognition is a kind of information processing that consists in the syntactically driven manipulation of representational mental structures. In particular, cognitive processes were said to be (1) abstract, a-modal processes that mediate between modality-specific sensory inputs (perception) and motor outputs (action), and (2) computations over mental representations that are either symbolic (e.g., concepts in a "language of thought"; Fodor 1975) or sub-symbolic (e.g., activations in neural networks; Rumelhart et al. 1986). The RCC also involves a specific view of where cognition was supposed to take place—some kind of "contingent intracranialism" (Adams and Aizawa 2008). On this view, cognitive processes are, as far as their ontology is concerned, realized by brain processes only (at least in the case of humans and other animals), and as far as their explanation is concerned, understandable and explainable by focusing on brain processes only.

During the past couple of decades, these key elements of the RCC—the pivotal role of computation and representation in all cognitive processing and the pivotal role of a central processing unit in the brain as the sole relevant factor of cognitive processing—have come under pressure (Gallagher 2005; Walter 2014). Proponents of 4E cognition have argued against the assumption that cognition is an isolated and abstract, quasi-Cartesian affair in a central processing unit in a brain. This idea is typically associated with functionalism, which claims that cognitive phenomena are fully determined by their functional role and therefore form an autonomous level of analysis. According to proponents of 4E cognition, however, the cognitive phenomena that are studied by modern cognitive science, such as spatial navigation, action, perception, and understanding other's emotions, are in some sense all dependent on the morphological, biological, and physiological details of an agent's body, an appropriately structured natural, technological, or social environment, and the agent's active and embodied interaction with this environment. Even most of the phenomena studied by traditional cognitive science—such as language processing (e.g., Glenberg and Kaschak 2002), memory (Casasanto and Dijkstra 2010), visual-motor recalibration (Bhalla and Proffitt 1999) and perception-based distance estimation (Witt and Proffitt 2008)—are not abstract, modality-unspecific processes in a central processing area either, but essentially rely on the system's body and its dynamical and reciprocal real-time interaction with its environment.

Thus, by maintaining that cognition involves extracranial bodily processes, 4E approaches depart markedly from the RCC view that the brain is the sole basis of cognitive processes. But what precisely does it mean to say that cognition involves extracranial processes? First of all, the involvement of extracranial processes can be understood in a strong and a weak way. According to the strong reading, cognitive processes are partially *constituted* by extracranial processes, i.e., they are essentially based on them. By contrast, according to the weak reading, they are non-constitutionally related, i.e., only *causally dependent* upon extracranial processes. Furthermore, cognitive processes can count as extracranial in two ways. Extracranial processes can be *bodily* (involving a brain–body unit) or they can be *extrabodily* (involving a brain–body–environment unit).

Following this line of reasoning, we can distinguish between four different claims about embodied cognition:

a. A cognitive process is *strongly embodied by bodily processes* if it is partially constituted by (essentially based on) processes in the body that are not in the brain;
b. A cognitive process is *strongly embodied by extrabodily processes* if it is partially constituted by extrabodily processes;
c. A cognitive process is *weakly embodied by bodily processes* if it is not partially constituted by but only partially dependent upon extracranial processes (bodily processes outside of the brain);
d. A cognitive process is *weakly embodied by extrabodily processes* if it is not partially constituted by but only partially dependent upon extrabodily processes.

The last version of the claim (d) is identical with the property of *being embedded*, i.e., being causally dependent on extrabodily processes in the environment of the bodily system. Furthermore, *being extended* is a property of a cognitive process if it is at least partially constituted by extrabodily processes (b), i.e., if it *extends* into essentially involved extrabodily components or tools (Stephan et al. 2014; Walter 2014).

Many proponents of 4E cognition not only maintain that cognition involves extracranial processes, but also that cognition is *enacted* in the sense that it involves an active engagement in and with an agent's environment (Varela, Thompson, and Rosch 1991). We can distinguish between two versions of this claim:

e. A cognitive process is *strongly enacted* if it is partially constituted by the ability or disposition to act;
f. A cognitive process is *weakly enacted* if it is only partially dependent upon the ability or disposition to act.

It should be emphasized that proponents of 4E cognition differ greatly in terms of their commitments to these claims, and consequently in their interpretation of what it means for cognition to be embodied, embedded, extended, and enactive. One famous example of an enacted theory of cognition is Noë's (2004) theory of perception, according to which perception is not something passive that happens to us or in us but something we do: according to him, having a 3D-perceptual experience of an object includes having a specific disposition to act which he spells it out in terms of implicit knowledge of sensorimotor contingencies. It is part of the discussion whether this justifies a strong or only a weak enactment claim (Engel et al. 2013).

Constitution Versus Causal Dependency

As we saw earlier, the distinction between constitution and causal dependency plays an important role in the debate on embodied cognition. But what exactly grounds this distinction? Consider the example of cognitive processes involved in solving a simple math problem. It likely involves visual perception (if the problem is presented on paper), memory, language or symbol processing, etc. This means it would depend on a variety of elements and processes that include neuronal processes in the visual cortex, in motor areas, in language areas, the hippocampus, frontal areas, etc. In addition, as I read the problem I move my eyes, and likely my head. I posture my body so that my eyes are a certain distance from the text. I may gesture with my hands as I work out the solution. All of these factors can be involved even if I am solving the problem "in my head," without pencil and paper or other instruments. If I am involved in a competition to solve the problem, that stressful fact may have an effect on my cognitive performance. Can proponents of embodied cognition claim that not only the neuronal processes, but also eye movements, head movements, posture, use of pencil and paper, and perhaps even the competitive situation are all parts of the cognitive system that constitutes cognition in this case? When they make such claims, critics have accused them of the so-called coupling/constitution fallacy (Adams and Aizawa 2008; Rupert 2009), according to which the strong coupling between neural and extraneural processes, including bodily movement and use of pencil and paper, for example, does not suffice to make the non-neural processes constituents, rather than just causal or enabling conditions of the cognitive process. Quite generally, the question is whether, and if so, how, we are able to decide (either empirically, pragmatically, or a priori) whether a particular cognitive process is constituted by or merely dependent upon extracranial or extrabodily processes.

One strategy in this debate is to question whether the concept of constitution necessarily involves just non-causal, part-whole relations (e.g., Craver 2007), or in some cases requires diachronic and dynamical relations that depend on reciprocal causality (e.g., Kirchhoff 2014, 2015; Leuridan 2012). Another strategy is to take relevant features as constitutive of a cognitive process (e.g., an emotion or an episode of self-consciousness) if it is a characteristic feature of the phenomenon and part of a minimal pattern of integrated features sufficient to realize this phenomenon (e.g., Newen et al. 2015; Gallagher 2013). It may be that most of the features of mental phenomena are neither necessary nor sufficient but only characteristic. For example, a facial expression of fear is partially constitutive of fear although there are realizations of fear that do not involve the typical facial expression, e.g., in the case of a trained poker face (Newen et al. 2015). Issues about the relation of constitutive, causal, or background conditions are unresolved, and are still subject to ongoing debate in the embodied cognition literature.

Mental Representations

Another important question in the debate on embodied cognition concerns what role, if any, *mental representations* play in cognitive processing. The theoretical landscape is such that 4E approaches can and in fact do have supporters from both the computational/representational and the anti-computational/anti-representational camp. Dynamicists like Chemero (2009), for instance, defend a decidedly anti-computational/anti-representational version of embodied cognition (see also Barrett 2011), while Wilson's (1994) "wide computationalism" and Clark's (2008) "extended functionalism," according to which the mind is the joint product of intracranial processing, bodily input, and environmental scaffolding, are unequivocally computational/representational. In a similar vein, while some proponents of embodied cognition, for instance, in the area of vision research, explicitly try to supersede traditional computational/representational approaches (Gibson 1979; Noë 2004; Hutto and Myin 2013), others merely try to enrich them by integrating environmental resources (Ballard et al. 1997; Clark 2013). Thus, embodied approaches range from the computation/representation friendly variety (Alsmith and de Vignemont 2012; Prinz 2009) to accounts that are explicitly anti-computational and/or anti-representational (see Thelen et al. 2001; Brooks 1991; Pfeifer and Bongard 2006). This shows that the 4E approach as such does not presuppose a specific view on representation and computation.

An Overview of This Book

Since the volume is organized in nine additional parts, we will provide a short overview of the main questions that are treated in these parts.

Part 2: What is Cognition?

The second part of essays explores the concept of cognition specifically from the perspectives offered by 4E approaches to the mind. From a standard viewpoint, the debates around embodied approaches seem to turn the "what" question into the "where" question, so that the answer to the question about the nature of cognition is first of all about location: precisely where is cognition located? In this regard the line that demarcates between inside and outside plays an important role. From the perspective of the 4Es, however, the question of location is less critical; indeed, the distinction between inside and outside is downplayed, and the boundary line turns out to be a movable and permeable border. Thus, on the extended mind paradigm, if you happen to be using a piece of the environment to assist memory or to solve a problem, then in that case the mind extends into the environment; on the enactivist view, if there is a dynamical coupling to others or to tools in joint action, then there is no line that cuts the organism off from these other social and environmental factors. Cognition is affordance-based, where affordances are always relational (between the cognizing subject or some form of life and the possibilities offered by some entity or complex of entities), and where entity may be some physical part of the environment, another person who can provide information or opportunity, a social or cultural structure, or even something more abstract, such as a concept that, with some manipulation, offers a solution to a problem. Such approaches transform the question about cognition into questions about the nature of affordances, about whether cognition is extended or extensive, about what precisely we mean by coupling, about whether a dynamical systems approach can do without representations, and so forth.

Part 3: Modeling and Experimentation

How should we go about answering such questions? This question is taken up in Part 2, as well as in other parts of this volume. There is general agreement that a priori definitions or models of cognition are not helpful, and that we need to conduct experiments and consult the empirical literature. 4E approaches are part of cognitive science and as such offer models that need to be tested using a variety of methods drawn from different disciplines. This part draws on research in experimental psychology and neuroscience, developmental psychology, dynamical systems theory, predictive processing, and so on. Testable models are required, not only for the most basic forms of human cognition found in infancy, or in perceptual crossing experiments, but in the more complex instances of social interactions and cultural expression. One question here is whether one model (e.g., predictive processing or dynamical systems theory) can explain the broad varieties of cognitive events by itself, or whether we need an integration of different models for different forms of cognition. This pushes further to the question of whether such integration is possible and whether there is some consistency between predictive processing, dynamical systems theory, and the various interpretations of these models found in cognitivism, extended mind, enactivist, and ecological approaches. Equally critical are questions about whether experimental science remains business as usual, or whether the more holistic demands of 4E approaches—to account for not just brain processes, and not just bodily and affective processes, and not just environmental and social and cultural processes, but all of these as they function together to shape cognition—put pressure on what we can operationalize and test.

Part 4: Cognition, Action, and Perception

Traditional analyses of perception tend to focus on sensory processing as it happens in cortical areas that correspond to different sense modalities, and questions concerning cognitive penetration. 4E approaches, in contrast, place significant emphasis on embodied action and the idea that perception is action-oriented. Furthermore, it often challenges the orthodox view, found in Helmholtz and recent models of predictive coding, that perception is inferential. Gibson worked out a theory of direct (non-inferential) perception that was controversial from the start, but that nonetheless continues to be developed in recent work in ecological psychology. Putting direct perception together with the focus on action complicates the picture, which is complicated further if we think that object perception is not equivalent to social perception, and that direct social perception is involved in joint actions. These are issues explored in this part, but they are basic ones that tie directly into questions about intentionality, spatial perception, social cognition, evolution, culture, brain plasticity, and the nature of cognition in nonhuman animals and robots—all of which are explored in later parts. Importantly, it remains controversial whether the principles worked out for perception and action, sometimes referred to as "basic" cognition, scale up to apply to higher-order operations and cognition in general.

Part 5: Brain–Body–Environment Coupling and Basic Sensory Experiences

This part explores concepts of intentionality found in 4E approaches. The notion that perception is action-oriented leads to a consideration of a very basic motor intentionality—a concept that derives from phenomenology (e.g., Merleau-Ponty 2012), but that can also be found in pragmatists such as John Dewey. As Robert Brandom notes, citing Dewey, the "most fundamental kind of intentionality (in the sense of directedness toward objects) is the practical involvement with objects exhibited by a sentient creature dealing skillfully with its world" (2008, p. 178). This captures a form of intentionality that is built into skillful bodily movement in tandem with environmental demands. Indeed, one might argue that it is just this kind of intentionality that should be considered "non-derived" intentionality, which is seemingly the favorite candidate for the "mark of the mental." Alternatively, one might think that given the complexity of cognition, there is no one mark of the mental, but that one requires, perhaps, a pattern of factors to explain the varieties of cognitive practices. One issue at stake here is the very notion of embodiment as it defines embodied cognition. Whether embodiment is something that is reducible to neural representations, or requires some forms of complex coupling between brain, body, and environment, is one of the central issues that defines debates about cognition.

Part 6: Social Cognition

In many explanations of cognition, the concept of social cognition is regarded as a specialized topic. Although it is, in some regards, a specialized form of cognition that involves understanding other conspecifics, for some 4E approaches it also forms a more generalized constraint on cognition overall since most of what we consider human cognition originates in social interactions. Social cognition is itself a sophisticated form of cognition that spans a large spectrum of circumstances, from very basic embodied interactions that involve perception of and response to movement, posture, facial expression, gestures, and situated actions, to complex actions and joint actions within a large variety of everyday and specialized social and institutional frameworks. In this regard, social cognition may involve capacities for basic, empathic, embodied resonance processes, as well as more knowledge-based practices that involve conscious inference and familiarity with the person or group with whom one is engaged. If the various theories of embodied cognition have sometimes challenged the more standard theory-of-mind approaches to this topic, the overall suggestion of the papers in this part is that a pluralistic approach that includes a variety of capabilities and practices may be more appropriate in order to deal with the multiple forms of social cognition that need to be explained.

Part 7: Situated Affectivity

The concept of emotion, or more generally, affect, has come to play a larger role in mainstream analyses of cognition over the past 20 years. Cognition is not the narrow, hard, cold process of ratiocinative intellect that seems to fit so well with the computational model. Affect requires a more embodied and situated conception of cognition, and we need to recognize that it permeates cognitive processes, rather than occasionally penetrating them. One can trace the role of embodied affect from early infancy, through empathic processes, into sophisticated social situations that characterize adulthood. In this respect, it is not just emotion or the conscious feeling of emotion that is important; rather, non-conscious and wide-ranging affective processes that manifest in terms of hunger, fatigue, pain and pleasure, satiation and satisfaction can bias perception and thinking. My everyday intentionality, for example, is always conditioned by particular interests, and such interests are always modulated by a variety of affects, including emotions and moods. My anger makes me see things in specific ways; my joy leads me to ignore some of the negative factors in my environment; my fear moves me to act one way rather than another; my dark funk makes this rather than that matter. What I remember, what I perceive, how I respond to another person—all of these cognitive performances are pushed and pulled by affective factors, and these need to be accounted for in any account of cognition, whether it's framed in terms of predictive processing, dynamical systems theory, social and environmental situations or empathic resonance.

Part 8: Language and Learning

On some accounts, language is deeply rooted in bodily movements, not only for its material performance, but also for its semantic sense, and even to the extent that language transcends the body toward high cultural accomplishments (as Merleau-Ponty 2012 suggests), it remains tied to it. At the same time that "speech accomplishes thought" (again to borrow a phrase from Merleau-Ponty), it remains a form of action, and most frequently a form of communicative action. Communication is not all linguistic, strictly speaking, since there are significant aspects of nonverbal communication involved from the very beginning; but linguistic communication is required for establishing most of human social practices and the normativity that comes with those practices. In that sense, language is a bridge from very basic embodied practices to the most sophisticated practices and rituals of instituted and normative life that come along with standards of correctness, the senses of rightness and wrongness, and the practice of giving reasons in our everyday social engagements. The bridge goes both ways since what results from linguistic practice loops back to shape our bodily actions and affective life.

Part 9: Evolution and Culture

4E cognition, in contrast to anthropocentric views, which take cognition to be defined in representationalist terms, provides a perspective on cognitive evolution where principles of biological organization (or, for some, life-to-mind continuity) help us to understand cognition. This more biological approach points to the importance of the adaptive, flexible behavior of agents who operate in an ambiguous, precarious, and generally unstable worldly environment that they help to rearrange to reduce precariousness and increase stability. One problem, it seems, is that if we start our evolutionary story on this basis of continuity across the nonrepresentational aspects of life, this seems to lead to a significant gap between prelinguistic and linguistic cognition, if we take the latter to involve representation. Can we have evolutionary continuity that leads to a psychological discontinuity? The discontinuity, however, may not be about the advent of language and/or representation, but rather may be opened up by differences in embodiment introduced by evolutionary forces themselves, and corresponding differences in sociocultural practices, the use of artifacts, and the construction of affordance-based niches. These are predicated on a mix of material and social resources. Here, then, it is not just the biology of genes or organism that evolves, nor just the accompanying plastic changes of the brain that account for the rise of human cognition; it's the physical environment and what we can do with it in terms of moving *things* about to create a species-relative livable niche—where *things* are at first natural things, and then artifacts, and then later become things like words. We have a coevolution that involves corresponding changes in brain, body, tools, artifacts, language and cultural practices, and so on, and on, and on.

Part 10: Applications

In the last part, the essays examine the practical implications of the various theoretical insights to be found in the 4E literature. What can theories of embodied, embedded, extended, and enactive cognition tell us about psychopathology, animal cognition, robotic design, social and political institutions, or about the less practical but not less important aspects of aesthetic judgment, literature, and the arts? In all of these cases the central principles of the 4E approaches are relevant. When cognition or everyday communicative practices fail, as in psychopathology, we need to look not only at neuronal anomalies, but also at basic variances in embodied social interactions and the social structures that may themselves promote pathologies. When we attempt to understand nonhuman animals, we need to suspend our anthropocentric notions of cognition as primarily linguistic and representational and look more closely at the kinds of coupling and coping mechanisms that exist between body and environment. It may be that rethinking robotics from the bottom up (Brooks 1991) may have been one of the prime motivators for the development of 4E cognition, but it is also the case that robotic design can continue to learn from insights taken from the various aspects of biological self-organization, sensorimotor contingencies, evolutionary niche construction, affordance-based coping, social interaction, etc. that 4E theory has been advancing.

Can similar resources in the ecosystem of 4E theory help us explain juridical reasoning and how priming effects and biases generated in situated bodily processes can enter into such higher social-cognitive processing? Are such effects and biases strong enough or pervasive enough to challenge the legitimacy of a judicial system? Would a similar analysis tell us something important about a first responder's perception and response in a life-or-death situation? And can those same resources explain the generation of aesthetic experience by the imaginative drive of the humanities and arts? What is the nature of literature (or the theatrical play, or a film) if it enacts meaning or a world only when the reader (or audience) engages with it? What is the nature of that engagement if it is embodied, embedded, extended, enactive, and affective? These are questions that are clearly at the cutting edge of 4E research, not because they are recent applications of 4E principles, but because answers to these questions have the potential to loop back into theory and to challenge already formulated principles.

We are obviously in need of an improved theory of cognition. Why should we go for it now? The answer is a philosophical one that we can formulate by borrowing some famous words from the US President John F. Kennedy about reaching the moon: it is incumbent to resolve these issues, "not because they are easy, but because they are hard, because that goal will serve to organize and measure the best of our energies and skills, [and] because that challenge is one that we are willing to accept."

ACKNOWLEDGMENTS

We are grateful for the financial support of this edition delivered by the Anneliese-Maier Research Award given to Prof. Gallagher (from the Humboldt Foundation, Germany) and by the Research Training Group (DFG-Graduiertenkolleg "Situated Cognition," GRK 2185/1) directed by Prof. Newen. The editorial work was administratively supported by Yasmin Schwetz and Jonas van de Loo, whom we also want to thank.

REFERENCES

Adams, F. and Aizawa, K. (2008). *The bounds of cognition*. Malden: Blackwell.
Alsmith, A. and de Vignemont, F. (2012). Embodying the mind and representing the body. *Review of Philosophy and Psychology*, 3, 1–13.
Ballard, D., Hayhoe, M., Pook, P., and Rao, R. (1997). Deictic codes for the embodiment of cognition. *Behavioral and Brain Sciences*, 20, 723–67.
Barrett, L. (2011). *Beyond the brain*. Princeton, NJ: Princeton University Press.
Bhalla, M. and Proffitt, D. (1999). Visual-motor recalibration in geographical slant perception. *Journal of Experimental Psychology*, 25, 1076–96.
Brandom, R.B. (2008). *Between saying and doing: toward an analytic pragmatism*. Oxford: Oxford University Press.
Brooks, R.A. (1991). Intelligence without representation. *Artificial Intelligence*, 47(1–3), 139–59.
Buckner, C. (2015). A property cluster theory of cognition. *Philosophical Psychology*, 28(3), 307–36.
Casasanto, D. and Dijkstra, K. (2010). Motor action and emotional memory. *Cognition*, 115(1), 179–85.
Chemero, A. (2009). *Radical embodied cognitive science*. Cambridge, MA: MIT Press.
Clark, A. (2008). *Supersizing the mind*. Oxford: Oxford University Press.
Clark, A. (2013). Whatever next? Predictive brains, situated agents, and the future of cognitive science. *Behavioral and Brain Sciences*, 36(3), 181–204.
Clark, A. and Chalmers, D. (1998). The extended mind. *Analysis*, 58, 7–19.
Craver, C.F. (2007). *Explaining the brain: mechanisms and the mosaic unity of neuroscience*. Oxford: Oxford University Press.
Engel, A.K., Maye, A., Kurthen, M., and König, P. (2013). Where's the action? The pragmatic turn in cognitive science. *Trends in Cognitive Sciences*, 17(5), 202–9.
Flor, N. and Hutchins, E. (1991). Analyzing distributed cognition in software teams: a case study of team programming during perfective software maintenance. In: J. Koenemann-Belliveau, T.G. Moher, and S. Robertson (eds.), *Proceedings of the Fourth Annual Workshop on Empirical Studies of Programmers*. Norwood, NJ: Ablex Publishing, pp. 36–59.
Fodor, J. (1975). *The language of thought*. Cambridge, MA: MIT Press.
Gallagher, S. (2005). *How the body shapes the mind*. Oxford: Oxford University Press.
Gallagher, S. (2013). A pattern theory of self. *Frontiers in Human Neuroscience*, 7(443), 1–7. doi:10.3389/fnhum.2013.00443
Gibson, J. (1979). *The ecological approach to visual perception*. Boston: Lawrence Erlbaum.

Glenberg, A. and Kaschak, M. (2002). Grounding language in action. *Psychonomic Bulletin and Review*, 9, 558–65.
Hübener, M. and Bonhoeffer, T. (2014). Neuronal plasticity: beyond the critical period. *Cell*, 159(4), 727–37.
Hutchins, E. (1995). *Cognition in the wild*. Cambridge, MA: MIT Press.
Hutto, D. and Myin, E. (2013). *Radicalizing enactivism*. Cambridge, MA: MIT Press.
Kirchhoff, M. (2014). Extended cognition & constitution: re-evaluating the constitutive claim of extended cognition. *Philosophical Psychology*, 27, 258–83.
Kirchhoff, M. (2015). Extended cognition & the causal-constitutive fallacy: in search for a diachronic and dynamical conception of constitution. *Philosophy and Phenomenological Research*, 90(2), 320–60. doi:10.1111/phpr.12039
Leuridan, B. (2012). Three problems for the mutual manipulability account of constitutive relevance in mechanisms. *British Journal for the Philosophy of Science*, 63, 399–427.
Menary, R.A. (2010). Introduction to the special issue on 4E cognition. *Phenomenology and the Cognitive Sciences*, 9(4), 459–63.
Merleau-Ponty, M. (2012). *Phenomenology of perception*. London: Routledge.
Newen, A. (2015). Understanding others: the person model theory. In: T. Metzinger and J.M. Windt (eds.), *Open MIND*, 26, pp. 1–28.
Newen, A., Welpinghus, A., and Juckel, G. (2015). Emotion recognition as pattern recognition: the relevance of perception. *Mind & Language*, 30(2), 187–208.
Noë, A. (2004). *Action in perception*. Cambridge, MA: MIT Press.
Pfeifer, R. and Bongard, J. (2006). *How the body shapes the way we think*. Cambridge, MA: MIT Press.
Prinz, J. (2009). Is consciousness embodied? In: P. Robbins and M. Aydede (eds.), *The Cambridge handbook of situated cognition*. Cambridge: Cambridge University Press, pp. 419–37.
Rowlands, M. (2010). *The new science of the mind*. Cambridge, MA: MIT Press.
Rumelhart, D., McClelland, J., and the PDP Research Group (eds.) (1986). *Parallel distributed processing* (2 vols.). Cambridge, MA: MIT Press.
Rupert, R. (2009). *Cognitive systems and the extended mind*. Oxford: Oxford University Press.
Stephan, A., Walter, S., and Wilutzky, W. (2014). Emotions beyond brain and body. *Philosophical Psychology*, 27, 65–81.
Thelen, E., Schöner, G., Scheier, C., and Smith, L. (2001). The dynamics of embodiment. *Behavioral and Brain Sciences*, 24, 1–86.
Varela, F., Thompson, E., and Rosch, E. (1991). *The embodied mind*. Cambridge, MA: MIT Press.
Walter, S. (2014). Situated cognition: a field guide to some open conceptual and ontological issues. *Review of Philosophy and Psychology*. doi:10.1007/s13164-013-0167-y
Wilson, R. (1994). Wide computationalism. *Mind*, 103, 351–72.
Witt, J. and Proffitt, D. (2008). Action-specific influences on distance perception. *Journal of Experimental Psychology*, 34, 1479–92.

PART II

WHAT IS COGNITION?

CHAPTER 2

EXTENDED COGNITION

JULIAN KIVERSTEIN

INTRODUCTION

4E cognitive science is a broad church housing a number of theoretical perspectives that to varying degrees conflict with each other (Shapiro 2010). In this chapter I will argue that the debates within 4E cognitive science surrounding extended cognition boil down to competing ontological conceptions of cognitive processes. The embedded theory (henceforth EMT) and the family of extended theories of cognition (henceforth EXT) disagree about what it is for a state or process to *count* as cognitive. EMT holds that cognitive processes are deeply dependent on bodily interactions with the environment in ways that more traditionally minded cognitive scientists might find surprising. The strong dependence of some cognitive processes on bodily engagements with the world notwithstanding, EMT claims that cognitive processes are nevertheless wholly realized by systems and mechanisms located inside of the brain. Thus advocates of EMT continue to interpret the concept of cognition along more or less traditional lines (Adams and Aizawa 2008; Rupert 2009). That is to say, they think of cognitive processes as being constituted by computational, rule-based operations carried out on internal representational structures that carry information about the world.

EXT by contrast argues that bodily actions and the environmental resources that agents act upon can, under certain conditions, count as constituent parts of a cognitive process. Consider, for example, how thoroughly integrated mobile phones have become in those moments in our lives when we are left with our own thoughts. Chalmers describes how he uses his iPhone to daydream, "idly calling up words and images when my concentration slips" (Chalmers 2008, ix). Smartphones and other mobile technologies are so thoroughly interwoven in our everyday lives that according to EXT they might be now thought of as parts of our minds.

The debate between EMT and EXT is often taken to turn on the cognitive status or otherwise of bodily actions in which agents exploit the material and technological resources of their environments for cognitive purposes (see Rowlands 2010, ch. 3; Wheeler

2014). Does thinking always take place entirely inside the head of individuals? Does it sometimes constitutively depend upon an agent's coupled interactions with structures and resources found in the environment? I shall argue that to resolve this issue we need a mark of the cognitive (Adams and Aizawa 2008; Wheeler 2010; Rowlands 2009). We need a theory of what makes a state or process a state or process of a particular cognitive kind.

The mark of the cognitive consists of properties a system must possess if it is to count as cognitive. Not everyone is agreed that there is any such well-defined set of properties. Clark (2008) has argued, for instance, that the processes and mechanisms that fall under the category of the cognitive are too disunified for there to be any distinguishing properties they share in common. Yet there remains a question to be settled about whether instances of cognitive processes, which seem to work in very different ways, count as instances of the same kind of process. We might try to answer this question by comparison with prototypes of a given cognitive process such as learning, memory, categorization, decision-making, and so on. We might appeal to folk intuition as Clark and Chalmers (1998) propose. Either way, we are assuming a position on what makes a process count as a process of a particular cognitive kind. We might be drawing our standard from folk psychology or relying on some other standard to identify prototypical examples of cognition. In either case, we are relying at least tacitly on a mark of the cognitive.

The mark of the cognitive is also at the heart of a debate *within* EXT about the nature of extended cognition. One side in this debate makes the case for EXT on the basis of considerations drawn from functionalism in the philosophy of mind. I will label this position "extended functionalism" (abbreviated as FEX). FEX is in agreement with the cognitive science orthodoxy that cognitive processes are essentially computational in nature (Clark 2008; Wheeler 2011a). FEX departs from the cognitive science orthodoxy in arguing that some of the relevant computations take place in the world, through bodily actions on information-bearing structures located in the environment.[1]

The self-declared "radical" theorists of extended cognition (henceforth REX) propose an alternative explanatory framework to that of classical cognitive science drawn from dynamical systems theory and ecological psychology (Chemero 2009; Silberstein and Chemero 2012; Hutto and Myin 2013). REX claims that we find extended cognitive processes whenever the variables that describe one system are also the parameters that determine change in the other system, and vice versa. In such a system, it is only as a matter of explanatory convenience that we treat the agent and its environment as separately functioning systems. In reality the dynamics of the two systems are so tightly

[1] Whether or not a cognitive process counts as an extended cognitive process is to be settled on a case-by-case basis. We see this policy at work in the debate about the realizers of phenomenal conscious experience. Prominent defenders of extended cognition deny that the material vehicles of phenomenal experience ever extend into the world (see, e.g., Clark 2012; Wheeler 2015). These authors argue that the computational processing that forms the basis for phenomenal experience is firmly encased within the heads of individuals.

correlated and integrated that they are best thought of as forming a single extended brain–body–world system (Silberstein and Chemero 2012).

REX claims that basic forms of cognitive processes are essentially extended. Basic cognition is the type of cognition found in non-language-using creatures (Hutto and Myin 2013). It is nonrepresentational and unfolds over time through the skilled bodily engagements of agents with the affordances of the environment. The terminology of "extended" cognition is thus potentially misleading insofar as it seems to imply that cognitive processes have their home inside of the heads of individuals, and occasionally reach out into the world. The extendedness of cognitive processes is to be understood not only in a spatial sense as a claim about the location of the boundaries of the mind. It also refers to the relational character of basic cognitive processes. Basic cognition is relational in the sense of being constituted by an agent's skilled activity in relation to its environment (Hutto et al. 2014).

My argument will proceed in two stages. In the first stage (first and second sections), I outline the debate between EXT and EMT. In the first section I show how there is substantial agreement in both camps about how cognitive science is to proceed. Both sides agree that the best explanation of human problem-solving will often make reference to bodily actions carried out on externally located information-bearing structures. The debate is not about how to do cognitive science. It is instead, to repeat, a debate about the mark of the cognitive: the properties that make a state or process count as being of a particular cognitive kind. In the second section, I then turn to functionalist formulations of EXT that make appeal to what has come to be called the parity principle. The third section shows how after many twists and turns the debate has reached a deadlock, notwithstanding arguments to the contrary recently developed on either side of the EXT-EMT divide. I then turn my attention to REX and argue that EXT would fare better were it to drop its commitment to a representationalist mark of the cognitive.

The EMT and EXT Debate

Since Descartes's skeptical arguments in the *Meditations* it has seemed natural to many philosophers to think of mind and cognition as essentially inner phenomena. The mind partakes in causal transactions with the world by means of epistemically, more or less reliable sensory channels. The mind can likewise produce effects in the world by sending commands to the muscle systems in the body to move in particular ways. However, mental and cognitive processes such as perceiving, remembering, thinking, and reasoning take place within the minds of individuals. EMT departs from this Cartesian tradition in stressing the ways in which mental processes causally depend on the environment in which the agent is embedded in deep and surprising ways.[2] When,

[2] See, e.g., Rupert 2009.

for instance, we use a calculator to divide a bill in a restaurant, the calculator is an essential part of how we successfully compute a solution to an otherwise computationally challenging arithmetical problem. The calculator "scaffolds" mathematical thinking that is fully constituted and realized by causal mechanisms found within a person's brain.[3]

EXT is even more thoroughgoing in its rejection of the Cartesian legacy. EXT claims that cognitive processes can, under certain conditions, extend or spread across the boundary separating the agent's body from the rest of the world. Consider, by way of illustration, the expert bartender who lines up different glasses in a particular spatial order as he prepares a drinks order (Beach 1988).[4] This simple trick makes the bartender's task of remembering which drink to serve next far easier than it would otherwise be. Instead of needing to store all of this information and keep it in mind, some of the work of remembering is offloaded onto the environment in the line of glasses. The environment now functions as an external store of information, and performs the role of a stand-in for the drinks order. To work out which drink to serve next, the bartender need only look and reach for the next glass in the line. The information-bearing load on his working memory is thereby significantly lightened. Part of this work is delegated to the representational structure temporarily assembled in the world, which can then be used to control and guide action so as to bring the task at hand to successful completion.

The bartender's initial action of arranging the line of glasses is what David Kirsh has called an "epistemic action." It is an action that gives structure to the information processed by internal cognitive systems in ways that fit with the goals of the system. The result of this active structuring of information is that now the agent can couple with the external structure (the line of glasses), and through this coupling gather the information needed about the next drink to be served. When the bartender generates a plan to serve the next drink, he does so on the basis of his coupling with this external structure. It is

[3] Sutton et al. (2010) argue for a distinction between embedded and scaffolded theories of cognition. The former position, which they attribute to Adams and Aizawa, holds that cognition is fundamentally intracranial but may causally depend upon interactions with external resources located in the environment. Theories of scaffolded cognition by contrast argue that cognitive processes can unfold through couplings between heterogeneous internal and external resources. Distributed cognitive processes made up of heterogeneous elements, some inner and some outer, should figure among the processes that are investigated in cognitive science. Sutton and colleagues distinguish scaffolded cognition from extended cognition, arguing that the former comes in degrees and often falls short of satisfying the degree of integration required for some external resource to count as a constitutive part of a cognitive process. (See also Sterelny 2010.) In what follows I will for the most part treat embedded and scaffolded cognition as coextensive, but it should be noted that some cases of scaffolded cognition may also qualify as cases of extended cognition. Thus, the category of scaffolding cognition may cross-cut the distinction between extended and embedded cognition.

[4] See Kirsh (1995) for a classic treatment of more cases of problem-solving of this flavor in which actions are performed that structure the spatial environment in ways that simplify the reasoning the subject engages in to solve a problem. For more recent update, see Kirsh (2009).

on the basis of such a coupling that the information is generated necessary for successfully planning and accomplishing his task.[5]

It might be naturally objected that this is just another example of the scaffolding of internal cognitive processes by the environment. EXT and EMT can both agree that the line of glasses functions as an external store of information that is tightly integrated with inner perceptual, working memory, and attentional processes in such a way as to guide and control action. EXT claims that the external structure works together with inner cognitive processes to form a softly assembled cognitive system that brings about the bartender's behavior.[6] Describing the system as "softly assembled" marks a contrast with systems made up of component parts, each of which has a pre-specified and fixed function (Anderson et al. 2012). Softly assembled cognitive systems are characterized by an "interaction-dominant dynamics," which makes it difficult, if not impossible, to assign specific functions to specific component parts. For instance, each time the bartender looks to the row of the glasses, this delivers the systems responsible for planning the next action with just the information they need. The systems that are planning the bartender's actions are simultaneously influencing and being influenced by the perception-action systems that are sustaining the coupling with the environment. These systems stand in a relation of mutual and continuous causal influence on each other.

As already noted in my introduction, the dispute between EMT and EXT is typically taken to concern the cognitive status of external information-bearing structures and the bodily actions that are performed on those structures. EXT claims that environmentally located structures and resources, and the operations that are carried out on them form

[5] There is an important question in the literature about the nature of the extra conditions coupling with an external resource has to meet in order for some external resource to count as a part of a cognitive process. Everyone is agreed that causal coupling on its own is not sufficient. (See Adams and Aizawa 2008, 2010, on what they call the coupling-constitution fallacy.) Some degree of functional integration of the external resource is necessary. (See Menary 2007 for an account of cognitive integration.) What are the additional conditions that need to be satisfied in order for the external resource to be integrated in the right way? Clark and Chalmers (1998) make some tentative suggestions, specifying what have come to be called "conditions of glue and trust" (see also Wilson and Clark 2009). The glue conditions relate to the availability and accessibility of the information that an external resource provides. The trust conditions concern the reliability of this information and the degree to which the individual accepts it without question or critical scrutiny. Sterelny (2010) describes a spectrum of possible cases of cognitive integration of an external resource, identifying three key dimensions—trust, entrenchment, and individualization. For further discussion of these dimensions, see Sutton et al. (2010). Colombetti and Krueger (2015) make use of Sterelny's dimensional analysis of integration to argue for extended and scaffolded affectivity.

[6] Not everyone in the EMT camp would reject this claim. Adams and Aizawa (2008, ch. 7) make a distinction between extended cognitive systems and extended cognition. They allow that softly assembled systems like the ones I've just been describing count as examples of extended cognitive systems. They deny, however, that the extension of a cognitive system suffices for extended cognition on the basis that not every part of a cognitive process itself counts as cognitive. Rupert (2009), by contrast, does deny that softly assembled systems count as cognitive systems. They fail to meet Rupert's integration condition, according to which systems count as cognitive only when made up of persisting mechanisms, the integrated functioning of which is explanatory of intelligent behavior. My thanks to an anonymous reviewer for reminding me of the lack of consensus in the EMT camp about this issue.

constitutive parts of a cognitive process. EMT agrees that coupling to the environment can contribute in an ongoing and interactive way in the production of a cognitive phenomenon of interest. However, proponents of this theory argue that the contribution of such couplings to cognitive processes is best understood as causal, not as constitutive.

This disagreement notwithstanding, there is, however, much that EMT and EXT are agreed upon. Both theories agree that the *explanation* of how internal cognitive processes give rise to some behavior of interest will often need to advert to bodily actions on external, environmentally located structures. Embedded and extended theorists therefore agree that internal cognitive processes will often not be sufficient for *explaining* cognitive behaviors. Given the extent of this agreement about how to go about explaining many of our problem-solving behaviors, one might be forgiven for wondering what is really at stake in this debate.

Sprevak (2010) has argued, for instance, that the dispute is unlikely to make a difference to how cognitive scientists go about their everyday business. The two theories can both equally well accommodate the available experimental evidence.[7] Cognitive scientists could frame their theories either in embedded terms or in extended terms. It would make little or no difference when it comes to the explanatory value of the resulting theories. Thus EMT and EXT do not seem to be genuinely competing theories when judged from an empirical perspective by the experimental data each theory can explain. Moreover, it is far from clear that any explanatory advantage is really gained from labeling the construction and manipulation of environmental structures as parts of a "cognitive" process. If this is what the debate is about, it has all the hallmarks of being a merely semantic disagreement.

I agree with Sprevak, however, that the debate between EXT and EMT isn't about the best conceptual framework for interpreting findings in cognitive science. It is a debate in metaphysics about "what makes a state or process count as mental or non-mental" (Sprevak 2010, p. 361).[8] For instance, the two theories fundamentally disagree about the body and world and their role in cognitive processes. EXT casts the body in the role of a tool for mediating between neural processes and the intelligent use of the environment. The body of the agent is what Clark describes as "a bridging instrument" that enables "the emergence of new kinds of distributed information-processing organization" (Clark 2008, p. 207). EMT argues by contrast that it is the body and world as represented in the brain that plays a necessary part in problem-solving behavior.

In the next section I explain how EXT has been defended by appeal to the so-called parity principle by defenders of extended functionalism (FEX for short). FEX claims that bodily action on external information-bearing structures is one of the many ways in which the computational processes that underpin cognitive processes can be

[7] Also see Rupert (2009, ch. 5) for an argument to this effect. Clark echoes Sprevak's worry when he writes that the debate "though scientifically important, and able to be scientifically informed, looks increasingly unlikely to admit of straightforward scientific resolution" (Clark 2011, p. 454).

[8] This is a question whose answer I have said will come from providing a mark of the cognitive. I depart from Sprevak in framing the debate as being about cognition rather than the mental.

implemented. The human brain has a rich variety of modes of encoding and processing information, some of which involve constructing and acting on information-bearing structures located in the environment.[9] We shouldn't treat instances of problem-solving differently when the agent makes active use of resources located in the environment. We will see, however, that the appeal to parity and the equality of treatment for the inner and outer have been found to be less than persuasive by advocates of EMT. In the third section, I will argue this has led to stalemate in the debate, suggesting that EXT might be in need of new, more radical ideas.

The Parity-Based Defense of EXT

The parity principle was first formulated by Clark and Chalmers in a paper that initiated the debate about the extended mind as we know it today (Clark and Chalmers 1998). Here is how they formulated the parity principle:

> If as we confront some task, a part of the world functions as a process which, were it to go on in the head, we would have no hesitation in accepting as part of the cognitive process, then that part of the world is (for that time) part of the cognitive process. (p. 8)

Consider how the parity principle might apply to the bartender example discussed in the previous section. Instead of physically arranging the glasses in the world, suppose instead that the bartender visually imagines the same line of glasses. He then keeps in mind this visual image, accessing it when he needs to, until the order is completed. Now, most of us would, I guess, be willing to say that such a visual image would count as a part of the cognitive process that causes the bartender's behavior. However, if we say this of the visual image of the line of glasses, then surely we ought to say the same of the actually existing line of glasses in the world. The visual image is nothing but an inner reconstruction of the same physical structure in the world. All I have done in constructing this example is transpose a process that in the original example takes place partly in the world into one that instead takes place wholly inside the head. The parity principle says that if we count a process as cognitive when it takes place inside the head, we should also count it as cognitive when it extends into the world. A cognitive process ought to be counted as cognitive regardless of where (inside the head or out in the world) it takes place.

[9] The exact balance of internal and external resources recruited to solve a problem is negotiated on a case-by-case basis, in ways that are constrained by the task at hand. Whether the cognitive agent makes use of structures located in the environment in problem-solving will depend on the costs and benefits of doing so, as evaluated, for instance, in terms of energy expenditure, risks, and uncertainties (Clark 2008, ch. 7; Rowlands 2010). There are interesting connections here to Clark's recent work on predictive processing (Clark 2015).

Taken as a self-standing principle, the parity principle doesn't settle anything. It works as an argument for EXT only when taken in conjunction with some pre-existing conception of when a process counts as a cognitive process (Adams and Aizawa 2001; Rupert 2009; Wheeler 2011b; Walter and Kästner 2012). After all, it is the *cognitive* status of a process that partly takes place in the world that we are using the parity principle to try to settle. In order to apply the parity principle, we must therefore have some pre-existing standards for making judgments about which processes are cognitive and which are not. We must have some pre-existing philosophical theory of what makes a state or process count as a state or process of a particular cognitive kind.[10]

Clark and Chalmers answer this question in part on the basis of considerations drawn from commonsense psychology, and in part by reference to cognitive science. Consider once again Clark and Chalmers's infamous case of Otto who because of his Alzheimer's relies on a notebook to remember the location of the Museum of Modern Art (MOMA) in New York (Clark and Chalmers 1998). The notebook can do the work of storing information just as well as the brain can. It doesn't matter that information is encoded, stored, and recalled very differently in the Otto notebook system as compared with Inga, who recalls the location of MOMA using her biological memory. The entries in Otto's notebook causally interact with his perception, beliefs, desires, and behavior in many of the same sorts of ways as the memory states of Inga. The Otto notebook system and the declarative memory systems in Inga's brain can thus be seen as different physical realizations of functionally equivalent dispositional belief states. They are instances of the same type or kind of mental state because they play the same coarse-grained, action-guiding role.

Proponents of EMT have, however, queried this last move by pointing to significant fine-grained functional differences in how memory works in Inga and Otto. Otto, for instance, would likely show no difference when it comes to remembering items at the beginning or end of a list of items as compared with those that occur in the middle of a list (Adams and Aizawa 2008).[11] Functional differences like these, and there are many others, give us grounds for doubting that extended and inner cognitive processes really do count as functionally equivalent.

This objection is not based on the claim that fine-grained similarity of internal and extended cognitive processes is necessary for the application of the parity principle.[12] The objection is rather based on a claim about what makes a state or process count as an instance of a particular cognitive kind. EMT takes cognitive processes to have a nature that is determined by causally explanatory properties, identified by our best theories in

[10] We will eventually see how this question in turn depends on one's preferred mark of the cognitive. Adams and Aizawa (2001) also make this point (p. 46). Thus I am in effect repeating their claim here that the deployment of the parity principle to motivate EXT will depend on a prior commitment to a mark of the cognitive.

[11] For this line of argument, also see Rupert (2004).

[12] Criticisms of EMT along these lines can be found in Clark (2008, ch. 7), Sprevak (2009), and Wheeler (2010).

cognitive science. It is for this reason that they pay close attention to the details of the causal roles played by cognitive processes in humans.

Clark and Chalmers applied the parity principle according to standards that were in part based on common sense. They take ordinary folk to identify the causal properties that determine the nature of a given cognitive state in the types of explanations they give when making sense of one another as rational agents. Proponents of FEX more generally promote an attitude of maximal inclusiveness in the range of non-standard realizers we count as possible realizers of the mental. They encourage us to be as liberal as possible in the creatures we count as being minded like us (Clark 2008; Sprevak 2009; Wheeler 2010). AIs, robots, and other creatures of science fiction all count as having states that mediate between inputs and outputs in ways that are similar enough to the states appealed to in folk psychological explanations. All of these systems have states that guide action in roughly the same type of way as the states picked out in folk psychological explanations.

The debate between EMT and EXT (in its functionalist formulations) is at least in part a debate about which theories we appeal to in fixing the reference of our cognitive concepts. These theories tell us whether there is sufficient similarity between two instances of a cognitive process for both to count as tokens of the same type of process. Some have allowed for folk psychological explanation to do the work of fixing the reference of our cognitive concepts. Others have argued that folk psychology is irrelevant and have instead made appeal to scientific theories and explanations to identify causally explanatory properties. It is to these questions that I turn in the next section.

The Varieties of (Extended) Functionalism

Clark and Chalmers suggested we look to folk psychology to decide whether an inner (e.g., Inga) and extended (e.g., Otto) cognitive process count as tokens of the same type of cognitive process. Folk psychology, as has often been noted, doesn't guarantee an answer to this question that favors EXT. As Chalmers (2008) has noted, folk psychology gives us some reason to treat perception and action as marking the boundaries of the mind. The only way Otto has of retrieving information from his notebook is by means of perception and action. It might then be argued based on folk psychology, that Otto does not have any beliefs before he checks to see what is written in his notebook. The notebook functions as at best an environmental cue for the formation of internal mental representations. Sure, Otto depends on his notebook for the guidance of his behavior, but conceding this is consistent with his behavior being in part the outcome of beliefs located entirely inside of his head about the contents of his notebook. These are beliefs that Otto can form only by reading what is written in his notebook.[13] Folk psychology

[13] Clark (2008) calls this move "the Otto 2-step" and offers a brief rebuttal (p. 80).

therefore gives us some reason to treat the notebook as lying outside of the boundaries of the cognitive system.[14]

One might wonder in any case how much trust one should put in intuitions drawn from folk psychology. Folk psychology may invite us to count the states of Otto and Inga as states of the same kind by virtue of their action-guiding role (though there is room for debate on this point). Playing a similar role in guiding action is, however, not a scientifically illuminating causal property. We want to know more precisely what the action-guiding causal role is in virtue of which Otto and Inga count as sharing a state or process of the same kind. There is little reason to think folk psychology will prove a useful guide when it comes to identifying causally explanatory properties. Indeed the judgments that folk psychology yields may well encourage us to mistakenly lump together states and processes that empirical science distinguishes, and conversely to make distinctions where none are found in nature.

It is this type of reasoning that has led parties on both sides of the debate to look instead to cognitive science to identify the causally explanatory properties that make a state or process the type of state or process it is. What verdict would we reach if we individuated functional roles instead on the basis of our best empirical scientific theories? Do such theories allow for kinds of cognitive processes that sometimes extend into the world beyond the boundaries of the organism? Once again, there is no clear consensus.

Wheeler has argued in a series of papers for an affirmative answer to this question. He agrees that our empirical theories may well point to many functional differences between extended and internal memory processes. However, he asks us to imagine that cognitive psychologists found people whose internal memory processes exhibited the same functional differences as the Otto notebook system (Wheeler 2010). Wheeler thinks the psychologists would still classify the people in question as having declarative memory processes so long as they exhibited, for instance, "context-sensitive storage and retrieval of information" (Wheeler 2011b). He thus doubts that the existence of functional differences speak against treating extended and inner memory processes as processes of the same *generic* kind.

Why should we count extended and inner memory processes as instances of a generic kind of declarative memory? There is some debate within EXT about whether we must do so because generic cognitive processes share common underlying mechanisms. As we saw in the introduction, Clark argues that extended cognitive processes may prove to be fundamentally disunified, a motley of different mechanisms. The methodological moral of EXT for cognitive science would therefore be to let a thousand flowers blossom.

[14] Chalmers responds to this objection by arguing that folk psychological explanation is context-sensitive. Sometimes our explanatory interests concern an individual's large-scale behavior, in which case it makes sense to look at the larger system of agent together with environmental resource. On other occasions our explanatory interests might be more local, relating to Otto and the interactions with his notebook, in which case it makes sense to treat perception and action as marking the boundaries of the cognitive system.

Although Sutton (2010) presents himself as making the case for extended cognition on the basis of computational cognitive science, he can also be read as falling within this first camp. Sutton has argued for a second wave in EXT that stresses the functional differences between inner and outer problem-solving resources. Second-wave EXT argues that external structures like Otto's notebook complement the cognitive capacities of the biological brain, joining forces with them to deliver new hybrid cognitive systems that are part biological and part cultural.[15] A key claim of second-wave EXT is that the scientific study of intracranial, or inner cognitive processes is just one part of cognitive science. The scientific study of extended cognitive systems is, Sutton argues, also a central research question in cognitive science, often undertaken in collaboration with the social sciences.

The complementarity of heterogeneous inner and outer elements that is emphasized by the second wave seems to have the consequence that extended and intracranial cognitive systems are unlikely to be built on the basis of the same mechanistic principles. On the contrary, the point of stressing complementarity is that "in extended cognitive systems, external states and processes need not mimic or replicate the formats, dynamics, or functions of inner states and processes. Rather, different components of the overall (enduring or temporary) system can play quite different roles and have different properties" (Sutton 2010, p. 194).

The alternative empirical functionalist view defended by Wheeler requires that the underlying mechanisms that support generic memory must at least share in common some family resemblance (Wheeler 2011b).[16] Wheeler allows for a variety of mechanisms, some of which are dynamical, nonrepresentational, and noncomputational, others of which are computational and representational (Wheeler 2005; Wheeler 2011b). However, he argues that these mechanisms are nevertheless partially unified: either as computational and representational processes on the one hand, or as dynamical and nonrepresentational on the other.

Trouble lies in waiting for FEX in playing this empirical functionalist card. First, cognitive science licenses a number of different theory-loaded accounts of the cognitive, not all of which support EXT. Wheeler proposes the physical symbol systems hypothesis of Newell and Simon (1976) as an uncontroversial account of the nature of cognitive

[15] How does the second wave in EXT differ from EMT? Sutton et al. (2010) make a distinction between embedded and scaffolded cognition (discussed in fn. 5). Scaffolded cognition can allow for a spectrum of cases, some of which fit the description of extended cognitive processes in which cognitive processing is partially externally constituted.

[16] It is not entirely clear to me whether the second-wave EXT would disagree with Wheeler on this point. Indeed I suspect there may be no agreement on this point within the second-wave camp. Sutton and colleagues, for instance, stress that the interdisciplinary science of biotechnological minds is one that still works within the classical framework of cognitive science, the only difference being that the object of study is "cognitive and computational architectures whose bounds far exceed those of skin and skull." (Sutton 2010, p. 191, quoting Clark 2001, p. 138). Menary (2010) by contrast presents the second wave as being aligned with enactive cognitive science, viewing cognition as "constituted by our bodily activities in the world in conjunction with neural processes and vehicles" (p. 227).

processes that he thinks every party in the EXT-EMT debate ought to be able to accept.[17] We can think of Otto together with his notebook as forming "a sufficiently complex and suitably organized physical symbol system" (Wheeler 2011a, p. 236).[18] Both the Otto notebook system and Inga can thus be interpreted as physical symbol systems. While there are no doubt significant differences in the mechanisms that realize declarative memory in Otto and Inga, there is nevertheless a family resemblance. Both can be described as physical symbol systems.

Conceding this much doesn't settle the issue of whether to describe the notebook as making a cognitive as opposed to a merely causal contribution to Otto's memory processes. Take Robert Rupert as an example of a proponent of EMT;[19] he could no doubt agree with Wheeler's proposal to use Newell and Simon as a scientifically informed account of the cognitive. However, he is well known for his skepticism about the concept of generic memory as a scientifically well-formed category (Rupert 2004, 2009, 2013). Rupert argues that science would only treat Otto and Inga as instantiating a generic kind of declarative memory if memory in Otto and Inga was brought about by the same cluster of integrated and persisting mechanisms. The Otto notebook system arguably does form a cluster of integrated and persisting mechanisms. Thus at first glance it seems to pass one of Rupert's tests. However, before we celebrate the victory of EXT, we should note that the clusters of mechanisms that are the basis for remembering in Otto are very different to those of Inga. Rupert argues that we would only be warranted in attributing to Otto states of the same kind as Inga on the following condition. Our model of how Otto's behavior is produced would have to overlap significantly with our model of the mechanisms that bring about Inga's behavior. Rupert argues that there is no reason to think this will be the case.

This conclusion doesn't settle the empirical functionalist case in Rupert's favor. For it should be noted that any mechanistic system can be described at many levels of organization (Craver 2007). Wheeler could argue that extensive lower-level implementational differences may disguise from view significant higher-level functional similarities. However, such a reply doesn't take us far. Suppose a proponent of EMT were to concede, as I think they should, that declarative memory processes allow for information to be stored either internally (as in Inga) or externally (as in Otto).[20] Consistent with

[17] It should be noted, however, that the effect of this stipulation might be to exclude REX from the debate. For suppose we go along with Wheeler and agree that the physical symbol systems hypothesis does tell us what it is for a state or process to count as cognitive. Either it will follow that REX is wrong to claim that dynamical and ecological cognitive processes are nonrepresentational and noncomputational, or it will follow that the processes REX investigates are not cognitive at all.

[18] This quote is actually taken from Wheeler's discussion of a neural network model that is able to do pattern completion and recognition for external symbol systems, such as systems of mathematical notation or written symbols. I take it the conclusion Wheeler wants to draw for this particular example generalizes to the Otto notebook system.

[19] I choose Rupert here because he has written extensively against generic memory. Related arguments to those of Rupert can also be found in Adams and Aizawa (2008, ch. 4)

[20] Again see Sutton et al. (2010) for detailed arguments that this is actually the case based on empirical research concerned with transactional memory.

this concession, they could nevertheless insist against EXT that the contribution of the external elements is only causal, and not cognitive. Memory processes so conceived can take the form of *hybrid* processes made up of causally interacting elements, some of which are cognitive and some of which are noncognitive.[21] Such a version of EMT wouldn't yield a genuinely competing explanation to EXT (Sprevak 2010). It would agree with EXT that memory processes can take a variety of forms, some of which are wholly internal and others of which are environment-involving. Disagreement would persist over whether to count the environmental components as constituent parts of the memory process.

The upshot of all this is that empirical functionalism leaves us pretty much where we started. In order to bring science to bear on this debate, we must settle the philosophically prior issue of what makes a state or process count as a cognitive state or process of a particular kind. The empirical functionalist answers this question by looking to the causal properties of a state or process as identified by our best scientific theories. However, we have just seen that cognitive science might be taken to yield an answer to this question that doesn't decide between EXT and EMT. The causal properties in putative cases of extended cognition can be interpreted as hybrid: part cognitive, and part noncognitive. We thus have two possible descriptions. One favors EMT (the description of the system as a hybrid of cognitive and noncognitive elements); the other favors EXT. Science on its own doesn't seem to allow us to decide which is the better description. To settle the matter we need a mark of the cognitive: a philosophical theory of what it is that makes a state or process a cognitive state or process.

Should we conclude then that we have no alternative but to rely on folk psychology? This would seem to follow on the assumption that our philosophical intuitions about the nature of the mind have their basis in folk psychology. We've seen, however, that folk psychology will prove to be of only limited help in making the case for EXT. It can help us to form a pre-theoretical sense of what states and processes stand in need of explanation. However, if we assume it is a state or process' causally explanatory properties that make it the state or process that it is, folk psychology cannot help us with our original question. It is through scientific investigation that we will learn about causally explanatory properties, not by recourse to folk psychological intuition.

This all leaves us in a rather unsatisfying place. Scientific findings need to be given a philosophical interpretation if they are to settle the issue of which causally explanatory properties are constitutive of a given cognitive state or process. We need a theory that tells us which causal properties count as cognitive, and which do not. However, if a theory is to do this work for us it must be based on scientific findings that identify causally explanatory properties. Philosophical theorizing needs scientific grounding, but scientific theorizing needs philosophical interpretation. Neither empirical functionalism nor commonsense functionalism succeeds in providing us with the philosophical

[21] Recall Adams and Aizawa's (2008, ch. 7) distinction between extended cognition and extended cognitive systems (see fn. 6): elements can be a part of an extended cognitive system while not themselves counting as cognitive.

account of the cognitive we need to settle the debate between EXT and EMT. It is time to try something different.

Taking the Radical Option

Chemero (2009) writes that radical embodied cognitive science (henceforth RECS) is "a variety of extended cognitive science" (p. 31). He characterizes this branch of cognitive science as having its roots in the American pragmatist tradition of William James, John Dewey, George Mead, and Charles Peirce.[22] A theme that looms large in the work of these philosophers (and also in RECS) is the mutuality of animal and environment. Here is Dewey explaining the central idea:

> To see the organism in nature, the nervous system in the organism, the brain in the nervous system, the cortex in the brain is the answer to the problems which haunt philosophy. And when seen thus they will be seen to be in, not as marbles are in a box but as events are in history, in a moving, growing, never finished process. (Dewey 1958, p. 295)

To say that animal and environment stand in a relation of mutual dependence is to claim that animal and environment are interdependent in the sense of together forming a "moving, growing, never finished process." The connection of RECS to this older tradition in naturalistic thinking comes from the thesis that if we model the agent and environment as coupled dynamical systems, then it is only as a matter of convenience that we treat them as separate systems. Instead of describing how external environmental factors cause changes in the agent's behavior, we instead model how the whole agent-environment system as a single process changes over time.

The argument for this claim is based in part on the mathematics of dynamical systems theory. Dynamical systems theory models change over time in complex systems using the mathematics of differential equations. Examples of complex systems are the solar system, weather systems, the diffusion of ink in water, interaction of populations of predator and prey, and so on.[23] The key concept we will need for the arguments that follow is that of the "coupling" of the agent and environment. Two systems S_1 and S_2 are said to be coupled when the equations describing one system S_1 contain variables whose value is a function of the variables in the equations describing S_2, and vice versa. Thus take the example of dynamical systems description of rate of change in a population of predators and prey. The equations describing a population of predators will include variables for prey, and the equations describing change in the population of prey

[22] Here is not the place to enter into the historical details, but for excellent accounts, see Heft (2001) and Gallagher (2014).

[23] For useful entry points see Kelso (1995), Ward (2001), and Chemero (2009).

will include variables for predators.[24] For example, if the number of predators steadily increases, the number of prey will steadily decrease, thereby putting pressure on the predators. As the predators begin to die off, the prey can begin to recover, and this dynamic will continue until the two populations reach some sort of equilibrium, and the size of each population remains relatively stable.

Now if we apply the concept of coupling to agent–environment interactions, we get the following result. We have two equations: one describing the changes that take place in the agent, and the second describing the changes that take place in the environment. The variables in the respective equations describe how components of the agent and the environment change in relation to each other. For the agent, some of the components are located in the brain, others in the rest of its body relating, for instance, to its bodily movements or affective states. On the environment side, the components will in cases of extended cognition be the information-bearing structures that the agent makes use of in the performance of a cognitive task. The agent and environment are dynamically coupled in cases of extended cognition because the equation describing change in the environment contains variables whose values are determined by the changes taking place in the agent. Similarly, the equations describing change in the agent contain variables whose values are determined by change in the environment.

Two systems that are coupled resist decomposition into separately functioning systems. We cannot model the behavior of the system as the additive product or sum of the interactions of separate structures and components, some on the side of the agent, and others on the side of the environment. Such is the degree of continuous, integrated, and coordinated mutual influence between the two systems that we can't solve the equations describing the behavior of each system separately.[25] There are really two claims here that combine to yield the result that the agent–environment system is best described as a single system. First, the components that make up the agent system exhibit fluctuating rates of change that depend on components belonging to the environment system, and vice versa. Second, if we look at the behavior of the agent–environment as a whole, this behavior isn't the product of the behavior of each of the components. We must also look at the interaction of the components and the nonlinear effects that arise from those interactions due to the continuous causal influence of the components on each other.

[24] The Lotka–Volterra equations describe variation in population size in predators and prey using two equations. The first describes how the prey population changes over time as a function of growth minus the effect of predation. The second describes change in predator population as a function of size of prey population minus natural loss of predators.

[25] Van Orden et al. (2003); Silberstein and Chemero (2012); Anderson et al. (2012). The type of nonlinear ongoing causal influence between coupled systems or components we have described previously is sometimes described as "interaction dominance" (Anderson et al. 2012). Van Orden et al. (2003) show how 1/f scaling, also known as "pink noise," is a "signature" of interaction dominance. Pink noise has also been found in agent–environment interaction. For instance, Dotov et al. (2010) found 1/f scaling when subjects were playing a video game, controlling an object on a monitor using a mouse. When the mouse connection was temporarily disturbed, however, 1/f scaling decreased, indicating that "during normal operation, the computer mouse is part of the smoothly functioning interaction-dominant system engaged in the task" (Silberstein and Chemero 2012, p. 45).

Clark has also made extensive use of these types of considerations in arguing for extended cognition. He characterizes the interactions between internal and external resources as "highly complex nested and nonlinear." He continues:

> As a result, there may, in some cases, be no viable means of understanding the behavior and potential of the extended cognitive ensembles by piecemeal decomposition and additive reassembly. To understand the integrated operation of the extended thinking system created, for example, by combining pen, paper, graphics programs, and a trained mathematical brain, it may be quite insufficient to attempt to understand and then combine (!) the properties of pens, papers, graphics programs and brains. (Clark 2008, p. 116)

Clark retains a commitment to representational and computationalist explanation. This renders his appeal to dynamic coupling vulnerable to Rupert's objections (Rupert 2009, ch. 7). Rupert argues that in a range of cases in which dynamical systems theory is used to model cognitive behaviors, we do not find coupling. We don't find variables for environmental elements showing up in the equations describing the agent's behavior. Instead we find the environment causing variation "among a small number of dimensions (e.g., input units) of the organismic system" (2009, p. 136). It is the value of these internal organismic systems that then determine how the agent behaves, not the states of the external environment.

Rupert's objection depends for its success on his conception of cognitive processes as persisting and integrated sets of mechanisms that causally contribute in the production of a wide range of cognitive phenomena (Rupert 2009, ch. 3). Armed with such a conception of cognitive systems, he can argue that agent–environment interaction leads to changes in the agent's internal representational states. The agent is sensitive to such changes in its internal representational states, and on the basis of this sensitivity it adapts its behavioral outputs so as to accomplish its tasks. The agent's interaction with the environment is causally relevant only through the changes it brings about in the agent's internal representational states.

REX argues by contrast that the agent doesn't interact with the environment through the intermediary of internal representations. It is not only dynamical systems theory and the concept of coupling that does the argumentative work but also crucially ideas drawn from ecological psychology, or so I shall propose. Ecological psychologists show how the layout of the ecological niche of a given species of animal is rich with higher-order, structural invariants that specify affordances, and which the mobile animal is able to immediately and directly detect. Warren summarizes the idea well:

> The perceptual system simply becomes attuned to information that, within its niche, reliably specifies the environmental situation and enables the organism to act effectively. (Warren 2005, p. 358)

Perceptual systems function first and foremost to guide action. The perceiving animal is immediately and directly sensitive to higher-order invariants or patterns in sensory stimulation that specify affordances, the possibilities for action provided to an animal by its surrounding environment. Interaction with the environment produces patterns of energetic stimulation, which form the basis for directly and immediately detecting higher-order invariants that specify affordances. As an agent approaches the edge of a precipice there is an immediate shearing off of the texture of the ground of the supporting surface (Gibson 1969) and the perceiver immediately detects that here is a place that offers the potential to fall. This is a meaning that is carried in the light that reflects from this place. This type of informational regularity can be thought of as an ecological constraint under which the perceptual systems of animals evolved. Thus the tusk of the narwhal is "tuned to the salinity differentials that specify the freezing of the water's surface overhead," information that is critical to its survival (Warren 2005, p. 341).

With the ecological context in place, the argument from coupling looks a little different. It can be argued that the environment doesn't causally influence behavior only by means of internal representations. Interaction with the environment isn't only about the delivery of afferent stimuli that can be used by the brain to construct internal representations. It can be argued instead that the agent dynamically couples with information-bearing structures located in the environment. Agent and environment exert continuous and mutual causal influence on each other making it the case that agent and environment cannot be modeled as separate systems. They are instead best modeled as a single extended cognitive system.

Rupert, however, has another argument up his sleeve. He claims that even if we grant that the external environment can causally influence behavior, this still doesn't suffice for coupling. For the direction of causal influence is only one-way: from environment to agent and not the other way around (Rupert 2009, p. 136). The environment may make a difference to behavior sometimes, but the agent makes no difference to the environment. For example, when in a game of Tetris the subject rotates the zoid in order to see into which space it might fit, "the fundamental dynamics of the object are not changed: its evolution in state space from any given point remains the same as it was before the rotation" (Rupert 2009, p. 136).

What exactly does Rupert mean when he claim that the dynamics of the zoid remain the same before and after rotation? The player's rotation of a zoid is a now-classic example of an epistemic action. Players perform this action in order to recognize the shape of a zoid, and to verify whether a given orientation will help to fill a line or not. Rotation therefore influences the spatial path of the falling zoid and the place it comes to rest. The path the zoid traverses and its orientation when it finds its resting place all clearly influence play and the arrangement of the pieces on the board. Thus we have what looks like a case of coupling: the equations describing the game will include variables for the player's action of rotation. The equations describing the player's actions will include variables for the rotation of the zoid. Recognition of the shape is facilitated by the act of rotation.

Perhaps, however, Rupert has something more demanding in mind when he talks of coupling. He describes the coupling relation as holding when "an order parameter of one subsystem acts as control parameter of the other, and vice versa; as a result, one subsystem's evolution can change the very character of the evolution of the other" (Rupert 2009, p. 133). A control parameter is a value whose continuous quantitative change leads to qualitative change in the behavior of the system. An example is the temperature of a fluid: a difference in this parameter can change dramatically how the fluid changes over time from an initial state. An order parameter is a composite or macroscopic state of a system such as the convection patterns or Bénard rolls that can be seen in a viscous fluid, such as oil when it reaches a certain temperature. In the Tetris example, it is the player's action of rotating the zoid that is the candidate for the control parameter. It is tricky to say what would count as the order parameter of the game. Let us suppose that it is the overall configuration of the pieces on the board at a particular moment in the game. We can see why Rupert would think the order parameter so conceived isn't affected by rotation. The pieces do not suddenly change their position when one rotates.

Notice, however, that this is to describe the game at a single point in time. Kirsh's classic research on Tetris shows that rotation when done early enough in the game does have a sizable influence on the player's success. Without rotation the layers of the board would fill up much sooner. With rotation the player succeeds in filling more lines and thus playing longer. Suppose we are given the task of predicting how a given game is going to play out. We would need to take into account whether the player used the strategy of rotation, and when they chose to do so. Compare this with the earlier example of populations of predators and prey. The growth in the population of prey, for instance, is dependent on the effects of predation. Similarly, how fast the layers of the board fill up in part depends on the performance of the action of rotation.

Both of Rupert's arguments fail. An obvious further objection, however, appeals to the concept of hybrid cognitive systems used so effectively against EXT earlier in the chapter. Why doesn't an extended cognitive system count as a hybrid cognitive system composed of cognitive elements inside of the head of the agent, and noncognitive elements in the environment? (Adams and Aizawa 2008, ch. 7).

This objection assumes precisely the kind of decomposition that has been called into question earlier. The elements inside the head could realize cognitive processes only by representing what is outside the head. The debate between EXT and EMT (with its commitment to hybrid cognitive systems) thus turns on whether one takes representation to be the mark of the cognitive. I've argued this is a mark of the cognitive that EXT must reject.

Conclusion

The central claim of this chapter has been that to resolve the debate about extended cognition we will need to come up with a mark of the cognitive. We will need to say what makes a state or process count as a state or process of a particular cognitive kind. All sides in the EXT-EMT debate have supposed that we must answer this question by appeal to the causally explanatory properties of a state or process. However, of the causally explanatory properties, some may be only causally relevant and not constitutively relevant.[26] Externally located structures might be argued to fall in the category of causally relevant but not constitutively relevant properties. In order to come down on one side or the other in this debate, we will need to have some basis for deciding whether a causally explanatory property is constitutively relevant. This requires us to have a theory of which causal processes count as cognitive.

RECS may hold the key to breaking the stalemate that has been reached in the debate between EMT and EXT. Interaction with the environment cannot be argued to be of only causal relevance so long as agent and environment are exerting continuous mutual causal influence on each other. This mutual causal influence stands in the way of modeling agent and environment as separate, independently functioning systems.

Doesn't the success of this argument depend on the empirical functionalist claim that it is causal explanatory properties that make a state or process an instance of a particular cognitive kind? Empirical functionalism has traditionally been aligned with the computational and representational theories of mind, a connection that REX seeks to break. Empirical functionalists take the states that mediate between inputs and outputs to be representational states with an internal structure to which computational processes are sensitive. This commitment to computational and representational explanation is arguably essential to empirical functionalism. One of the main selling points of the computer theory of mind was supposed to be that it can make it intelligible how a causal and mechanistic process can also be sensitive to semantic properties of thinking (Crane 1995; Fodor 2000). The computer theory of mind was supposed to help us to understand how causal properties can constitute a cognitive process. REX, however, rejects the computer theory of mind, and stripped of this theoretical commitment, empirical functionalism has little to recommend it.

REX draws its mark of the cognitive from a variety of sources that spans the phenomenological and American naturalist tradition. REX takes extended cognitive systems to be perception-action systems on the basis of which the person or animal is adapted to its environment and so able to deal adequately with its affordances. REX is thus committed to a pragmatist interpretation of what cognition is, inspired by the mutual fit and complementarity of the animal and its environment. The argument I gave earlier for EXT stressed the importance of nonlinear causality and the dynamical properties of interaction-dominant systems. However, what makes all of this relevant to cognition in the end is the way in which dynamical properties of this kind relate to the mutuality of animal and its environment. It is this mutuality that grounds the mark of the cognitive needed to make a successful case for extended cognition.

[26] I borrow this distinction between causal relevance and constitutive relevance from Craver (2007).

References

Adams. F. and Aizawa, K. (2001). The bounds of cognition. *Philosophical Psychology*, 14(1), 43–64.

Adams, F. and Aizawa, K. (2008). *The bounds of cognition*. Oxford: Blackwell.

Adams, F. and Aizawa, K. (2010). Defending the bounds of cognition. In: R. Menary (ed.), *The extended mind*. Cambridge, MA: MIT Press, pp. 67–80.

Anderson, M.L., Richardson, M.J., and Chemero, A. (2012). Eroding the boundaries of cognition: implications of embodiment. *Topics in Cognitive Science*, 4(4), 717–30.

Beach, K. (1988). The role of external mnemonic symbols in acquiring an occupation. In: M.M. Gruneberg and R.N. Sykes (eds.), *Practical agency of memory*. New York: Wiley, pp. 342–6.

Chalmers, D. (2008). Forword. In: A. Clark *Supersizing the mind: embodiment, action and cognitive extension*. Oxford: Oxford University Press, pp. ix–xvi.

Chemero, A. (2009). *Radical embodied cognitive science*. Cambridge, MA: MIT Press.

Clark, A. (2001). Reasons, robots and the extended mind. *Mind and Language*, 16(2), 121–45.

Clark, A. (2008). *Supersizing the mind: embodiment, action and cognitive extension*. Oxford: Oxford University Press.

Clark, A. (2011). Finding the mind. *Philosophical Studies*, 152, 447–61.

Clark, A. (2012). Spreading the joy? Why the machinery of consciousness is (probably) still in the head. *Mind*, 118(472), 963–93.

Clark, A. (2015). *Surfing uncertainty: prediction, action and the embodied mind*. Oxford: Oxford University Press.

Clark, A. and Chalmers, D. (1998). The extended mind. *Analysis*, 58, 7–19.

Colombetti, G. and Krueger, J. (2015). Scaffoldings of the affective mind. *Philosophical Psychology*, 28(8), 1157–76.

Crane, T. (1995). *The mechanical mind: a philosophical introduction to minds, machines and mental representation*. London: Routledge, Taylor & Francis.

Craver, C. (2007). *Explaining the brain: mechanisms and the mosaic unity of neuroscience*. Oxford: Oxford University Press.

Dewey, J. (1958). *Experience and nature*. New York: Dover Books.

Dotov, D., Nie, L., and Chemero, A. (2010). A demonstration of the transition from readiness to hand to unreadiness to hand. *PLoS ONE*, 5, e9433. doi:10. 1371/journal.pone.0009433

Fodor, J. 2000. *The mind doesn't work that way: the scope and limits of computational psychology*. Cambridge, MA: MIT Press.

Gallagher, S. (2014). Pragmatic interventions into enactive and extended conceptions of cognition. *Philosophical Issues*, 24(1), 110–26.

Gibson, E.J. (1969). *Principles of perceptual learning and development*. New York: Appleton Century Crofts.

Heft, H. (2001). *Ecological psychology in context: James Gibson, Roger Barker and the legacy of William James's radical empiricism*. Mahwah, NJ: Lawrence Erlbaum Associates.

Hutto, D., Kirchhoff, M., and Myin, E. (2014). Extensive enactivism: why keep it all in? *Frontiers in Human Neuroscience*, 8, 706.

Hutto, D. and Myin, E. (2013). *Radicalising enactivism: basic minds without content*. Cambridge, MA: MIT Press.

Kelso, J.A.S. (1995). *Dynamic patterns: the self-organisation of brain and behaviour*. Cambridge, MA: MIT Press.

Kirsh, D. (1995). The intelligent use of space. *Artificial Intelligence*, 72(1–2), 31–68.

Kirsh, D. (2009). Problem solving and situated cognition. In: P. Robbins and M. Aydede (eds.), *The Cambridge handbook of situated cognition.* Cambridge: Cambridge University Press, pp. 264–306.

Menary, R. (2007). *Cognitive integration: mind and cognition unbounded.* Basingstoke, UK: Palgrave MacMillan.

Menary, R. (2010). Cognitive integration and the extended mind. In: R. Menary (ed.), *The extended mind.* Cambridge, MA: MIT Press, pp. 227–43.

Newell, A. and Simon, H. (1976). Computer science as empirical enquiry: symbols and search. *Communication of the Association for Computing Machinery*, 19(3), 113–26.

Rowlands, M. (2009). Extended cognition and the mark of the cognitive. *Philosophical Psychology*, 22(1), 1–19.

Rowlands, M. (2010). *The new science of the mind: from extended mind to embodied phenomenology.* Cambridge, MA: MIT Press.

Rupert, R. (2004). Challenges to the hypothesis of extended cognition. *Journal of Philosophy*, 101(8), 389–428.

Rupert, R. (2009). *Cognitive systems and the extended mind.* Oxford: Oxford University Press.

Rupert, R. (2013). Memory, natural kinds, and cognitive extensions; or, Martians don't remember and cognitive science is not about cognition. *Review of Philosophy and Psychology*, 4, 25–47.

Shapiro, L. (2010). *Embodied cognition.* London: Routledge, Taylor & Francis.

Silberstein, M. and Chemero, A. (2012). Complexity and extended phenomenological cognitive systems. *Topics in Cognitive Science*, 4(1), 35–50.

Sprevak, M. (2009). Extended cognition and functionalism. *Journal of Philosophy*, 106, 503–27.

Sprevak, M. (2010). Inference to the hypothesis of extended cognition. *Studies in History and Philosophy of Science*, 41, 353–62.

Sterelny, K. (2010). Minds: extended or scaffolded? *Phenomenology and the Cognitive Sciences*, 9(4), 465–81.

Sutton, J. (2010). Exograms and interdisciplinarity: history, the extended mind and the civilizing process. In: R. Menary (ed.), *The extended mind.* Cambridge, MA: MIT Press, pp. 189–225.

Sutton, J., Harris, C.B., Keil, P.G., and Barnier, A.J. (2010). The psychology of memory, extended cognition and socially distributed remembering. *Phenomenology and Cognitive Sciences*, 9(4), 521–60.

Van Orden, G.C., Holden, J.G., and Turvey, M.T. (2003). Self-organization of cognitive performance. *Journal of Experimental Psychology: General*, 132(3), 331–50.

Walter, S. and Kästner, L. (2012). The where and what of cognition: the untenability of cognitive agnosticism and the limits of the motley crew argument. *Cognitive Systems Research*, 13, 12–23.

Ward, L.M. (2001). *Dynamical cognitive science.* Cambridge, MA: MIT Press.

Warren, W.H. (2005). Direct perception: the view from here. *Philosophical Topics*, 33(1), 335–61.

Wheeler, M. (2005). *Reconstructing the cognitive world: the next step.* Cambridge, MA: MIT Press.

Wheeler, M. (2010). In defence of extended functionalism. In: R. Menary (ed.), *The extended mind.* Cambridge, MA: MIT Press, pp. 245–70.

Wheeler, M. (2011a). Embodied cognition and the extended mind. In: J. Garvey (ed.), *The Continuum companion to philosophy of mind.* London: Continuum Press, pp. 220–38.

Wheeler, M. (2011b). In search of clarity about parity. *Philosophical Studies*, 152(3), 417–25.

Wheeler, M. (2014). Revolution, reform, or business as usual? The future of embodied cognition. In: L. Shapiro (ed.), *The Routledge handbook of embodied cognition.* London: Routledge, Taylor & Francis, pp. 374–84.

Wheeler, M. (2015). Extended consciousness: an interim report. *The Southern Journal of Philosophy*, 53, 155–75.

Wilson, R.A. and Clark, A. (2009). How to situate cognition: letting nature take its course. In: P. Robbins and M. Aydede (eds.), *The Cambridge handbook of situated cognition.* Cambridge: Cambridge University Press, pp. 55–77.

CHAPTER 3

ECOLOGICAL-ENACTIVE COGNITION AS ENGAGING WITH A FIELD OF RELEVANT AFFORDANCES

The Skilled Intentionality Framework (SIF)

ERIK RIETVELD, DAMIAAN DENYS, AND MAARTEN VAN WESTEN

Introduction

The topic of this Oxford handbook is "4E cognition": cognition as embodied, embedded, enactive, and extended. However, one important "E" is missing: an E for *ecological*. In this chapter we will sketch an ecological-enactive approach to cognition that presents a framework for bringing together the embodied/enactive program (Chemero 2009; Thompson 2007) with the ecological program originally developed by James Gibson, in which affordances are central (e.g., Gibson 1979). We call this framework the skilled intentionality framework.

The skilled intentionality framework (SIF) is a philosophical approach to understanding the situated and affective embodied mind. It is a new conceptual framework for the field of 4E cognitive science that focuses on skilled action and builds upon an enriched notion of affordances, which we have recently argued for in *Ecological Psychology* (Rietveld and Kiverstein 2014). We define skilled intentionality as the selective engagement with multiple affordances simultaneously in a concrete situation (Rietveld, de Haan, and Denys 2013; Bruineberg and Rietveld 2014; Kiverstein and Rietveld 2015; Van Dijk and Rietveld 2017). The skilled intentionality framework clarifies how complementary insights on affordance responsiveness from philosophy/

phenomenology, ecological psychology, emotion psychology, and neurodynamics hang together in an intertwined way. The long-term ambition of the SIF research program is to understand the entire spectrum of skilled human action,[1] including social interaction, creativity, imagination, planning, and language use in terms of skilled intentionality.

By "affordances" we mean the possibilities for action provided to us by the environment (Gibson 1979; Chemero 2003, 2009; Michaels 2003; Reed 1996; Costall 1995; Heft 2001; Rietveld and Kiverstein 2014). Structuring and scaffolding our skilled activities, affordances are crucial for understanding the embodied mind. Grasping a glass, riding a bike, or improving an architectural design, for instance, can all be seen as a skilled individual's immediate responsiveness to affordances. An individual can respond to affordances thanks to abilities. In the relational approach to affordances developed in this chapter, the possession of the relevant ability is seen as necessary for being able to act on an affordance. Someone who does not have the ability to read English cannot be responsive to the possibility this sentence offers of being read. Humans typically acquire their abilities thanks to a history of interactions in sociocultural practices (Rietveld 2008a). For example, architects acquire their skills thanks to their being selected for education in specialized architecture academies, their traineeships in architecture firms, and repeated interactions with builders, other architects, and clients for the projects they realize.

Both humans and animals respond to affordances in a context-sensitive way. To an earthworm, for instance, a leaf affords plugging its burrow and thus regulating the humidity of its immediate surroundings. Such a context-sensitive engagement with this affordance is important for the condition of its skin (Darwin 1881). In a similar way, the environment offers all sorts of possibilities for humans, including possibilities for social interaction. For example, given a certain context, the sad face of a friend can invite a consoling gesture, a person waiting in a queue at a coffee machine can invite a conversation, and an extended hand can invite a handshake. Crucially, skilled responsiveness to affordances is not only encountered in everyday skilled activities, but also in activities that are traditionally characterized as "higher" cognition. Skills are crucial for knowledgeable action. For example, through her interaction with a patient, a skilled psychiatrist could intuitively, without explicit reflection, diagnose the patient with depression, based on pale complexion, red eyes, rigid and slow movements, disturbed language, pace of thinking, way of dressing, smell, and specific use of words. The prototypical example that we will use in this article to theorize the role of affordances in "higher cognition" is the design process of architects, which involves *both unreflective and reflective* episodes (Rietveld 2008a; Rietveld and Kiverstein 2014; Rietveld and Brouwers 2016; Van Dijk and Rietveld 2017).

Unlike, for example, Dreyfus's work on skilled action (Dreyfus 2002a, 2002b, 2006) or Hutto and Myin's (2012) early work on basic minds in enaction, the richer

[1] The words "action" and "skill" should be understood in a very broad sense in this chapter. For example, just like Alva Noë (2012) we even see perception as something we do, and can do more or less skillfully.

notion of affordances we have developed includes possibilities for long-term planning, possibilities for reflection, possibilities for creative imagination, possibilities for social interaction, and possibilities for language use (Rietveld and Kiverstein 2014; Rietveld and Brouwers 2016; Van Dijk and Rietveld 2017). The skilled intentionality framework (SIF) dissolves the dichotomy between "lower cognition" and "higher cognition" by interpreting affordances for the latter types of skilled activities as *just more affordances* available in our human ecological niche (left part of Figure 3.1) and responsiveness to them as just a manifestation of skilled intentionality in context. Moreover, a key aspect of so-called "higher" cognition regards the way in which persons are oriented toward the possible. The concept of skilled intentionality as multiple simultaneous states of *action readiness* for engagement with affordances entails orientation toward and preparation for possibilities for future action, which is a situated form of anticipation.

Skilled action is paradigmatic for embodied/enactive cognition (Rietveld 2008a, 2008c) and is investigated by different scientific traditions. The notions of affordances and affordance responsiveness are becoming central in various disciplines studying skilled action, including philosophy/phenomenology (Abramova and Slors 2015; Noë 2012; Kiverstein and Miller 2015; Van Dijk and Withagen 2016; Ramstead, Veissière, and Kirmayer 2016), sports/ecological psychology (Hristovski, Davids, and Araújo

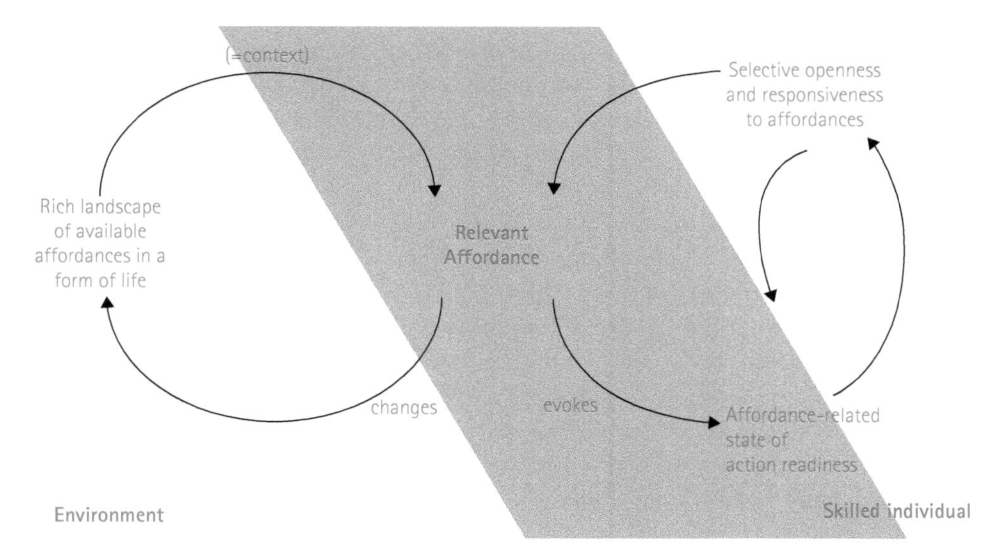

FIGURE 3.1. The skilled intentionality framework is a philosophical framework for understanding skilled action in context that integrates perspectives of various disciplines: ecological psychology (landscape of affordances), phenomenology (selective openness to and relevance of affordances), emotion psychology (states of action readiness, along the lines of Frijda 2007), and embodied neurodynamics (self-organizing affordance-related states of action readiness).

Adapted from J. Bruineberg and E. Rietveld, Self-organization, free energy minimization, and optimal grip on a field of affordances, *Frontiers in Human Neuroscience*, 8, e599, Figure 1, doi:10.3389/fnhum.2014.00599 © 2014 Bruineberg and Rietveld. This work is licensed under the Creative Commons Attribution License (CC BY). It is attributed to the authors Bruineberg and Rietveld.

2009; Chow et al. 2011; Withagen, Araújo, and De Poel 2017), affective science (Frijda, Ridderinkhof, and Rietveld 2014), and neuroscience (Friston et al. 2012; Schilbach et al. 2013; Dotov 2014; Dotov et al. 2010; Kirchhoff 2015; Jelic 2016; Pezzulo and Cisek 2016). For example, *affordance-related states of action readiness* are central to understanding both emotions (Frijda et al. 2014; cf. Frijda 2005) and the neurodynamics of skilled action (Bruineberg and Rietveld 2014). We will see later that these varying perspectives on skilled action can be understood as describing the same phenomenon of skilled intentionality from different yet complementary points of view. Ultimately, we will need all of them for a solid understanding of skilled action in context (for more on this integrative methodology based on complementarity of different scientific fields, see Klaassen, Rietveld, and Topal 2010; Rietveld 2008a, 2008c; Van Dijk and Rietveld 2017).

In short, the skilled intentionality framework (SIF) aspires to do justice to the complex phenomenon of embodied cognition as skilled engagement with multiple affordances by integrating perspectives at different levels of analysis: ecological psychology, phenomenology, emotion psychology, and neurosciences. The aim of this chapter is to summarize the distinctive Amsterdam SIF approach to skilled action in context, which can be characterized as *ecological-enactive* cognition.

In a series of papers, we have shown that the tendency toward a grip on multiple affordances simultaneously is something that is found at each of these levels of analysis and thus provides a way of conceptually bridging them (Bruineberg and Rietveld 2014; Rietveld and Kiverstein 2014; Rietveld and Brouwers 2016; Bruineberg, Kiverstein, and Rietveld 2016). It will be seen later how our concept of affordance-related states of action readiness in particular is able to facilitate crossings between these levels of skilled action in context.

SIF acknowledges that different fields of study like ecological psychology, phenomenology, affective science, and neurodynamics approach the same phenomenon over different time scales. For instance, what from a phenomenological perspective is described in philosophy as the experienced invitation of an affordance (i.e., a solicitation, Dreyfus and Kelly 2007; Rietveld 2008a), can be measured (and analyzed) as a state of "action readiness" in emotion psychology and (affective) neuroscience (Frijda 1986, 2007; Rietveld 2008b; Bruineberg and Rietveld 2014; Van Dijk and Rietveld 2017). In the next three sections, we will show how the notion of skilled intentionality returns in particular ways at various levels of analysis and time scales of the integrated individual-environment system (Figure 3.1).

The first section describes skilled intentionality at the ecological level (left part of Figure 3.1), i.e., the ecological niche that forms the context in which individuals are situated. We will discuss how our rich definition of affordances relates to different kinds of skilled activities, including social interaction, language use, and reflection. This situates the skilled individual in the context of a rich *landscape of affordances* that is shared with the other individuals inhabiting the same ecological niche.

The second section describes skilled intentionality at the phenomenological level of analysis (depicted in the middle of Figure 3.1). We discuss how a particular individual can be *selectively open* to this landscape, responding only to the *relevant* affordances in the particular situation. An individual can be solicited or drawn to act on relevant affordances and doing so will change their surroundings. This relevance of affordances relates to a disequilibrium within a self-organizing individual–environment system (the whole of Figure 3.1). We will explain later—using Merleau-Ponty's phenomenology of life—that such a disequilibrium is inherent to all living beings (Merleau-Ponty 1968/2003.). This disequilibrium develops dynamically as a result of material changes in the context/situation (left dynamic, Figure 3.1) and changes of states of the active individual (right dynamic, Figure 3.1). Crucially, skilled intentionality means reducing disequilibrium by moving toward an optimal grip on multiple relevant affordances simultaneously, that is, on a *field of relevant affordances*.[2]

The third section describes how at the embodied neurodynamic level, which is depicted in the right of Figure 3.1, skilled intentionality is understood as expressing a process of self-organization of multiple affordance-related states of action readiness. Due to the fact that we analyze the same self-organizing system from these different perspectives in the different sections, some amount of reiteration is inevitable.

SIF's Rich and Resourceful Landscape of Affordances and "Higher" Cognition

The SIF builds not just upon own work in different fields of embodied/enactive cognitive science (philosophy, emotion psychology, psychiatry, and radical embodied cognitive neuroscience), but also on decades of research on affordances in the tradition of ecological psychology (Gibson 1979; Heft 2001; Reed 1996; Chemero 2003; Withagen et al. 2012; Withagen et al. 2017). Starting from this latter tradition, we have argued that the first question to ask about an affordance is what the ecological niche is in which it is embedded or "nested" (Rietveld and Kiverstein 2014). This allows us to stay close to Gibson's idea of the primacy of the ecological niche for understanding the kind of animal one is interested in. In recent philosophical work we (Rietveld and Kiverstein 2014) have refined Chemero's (2003) definition of affordances, using Wittgenstein (1953), to show that affordances always have to be understood in the context of an ecological niche that implies the form of life of a certain kind of animal. Therefore, we define an affordance as a relation between (a) an aspect of the (sociomaterial) environment and (b) an ability *available in a "form of life"* (Wittgenstein 1953).

[2] Our synonym for the field of relevant affordances is the field of solicitations. Thanks to Shaun Gallagher for urging us to make this explicit.

A form of life is a kind of animal with a certain way of life and ecological niche. A form of life refers to a certain kind of practice: coordinated patterns of behavior of multiple individuals. The main reason we prefer the use of the Wittgensteinian notion of a form of life is because, at least in certain contexts, it is important to acknowledge the fact that within the human form of life there are many different sociocultural practices (e.g., communities of English language speakers, builders, academics, and architects, etc.). The notion "form of life" can refer both to a sociocultural practice and to a species (e.g., lions, earthworms, humans). The form of life of a certain kind of animal or a sociocultural practice is manifested in relatively stable patterns of behavior, generated by the coordinated activities of many individuals over time.[3] As such, a form of life is independent of any particular individual. A novice typically acquires his or her skill within an already existing form of life. Just like Wittgensteinian norms (Wittgenstein 1953; Rietveld 2008a), affordances continue to exist when an individual dies, because they are not related to a particular individual but to an entire practice, to a form of life (Rietveld and Kiverstein 2014; Van Dijk and Rietveld 2017). Affordances are just as deeply social as, for example, the norms of spelling are, because they are by definition related to (abilities available in) a practice in SIF.

The variety (cf. Roepstorff, Niewöhner, and Beck 2010; Roepstorff 2008) that is manifested in both relata of the definition of affordances, i.e., in both the sociomaterial environment and in available abilities in a form of life, allows us to see the human ecological niche as a rich and resourceful *landscape of affordances* (Rietveld and Kiverstein 2014). The variety in the environmental structure, which is one relatum of our definition, was outlined by Gibson (1979) already: different surfaces afford locomotion and support; substances afford nutrition and manufacture; objects afford many kinds of manipulation; animals afford each other all sorts of interactions (sexual, playful, fighting, cooperating, communicating, predatory, nurturing, etc.; see Gibson 1979).

Following an important development in the social sciences, we have suggested that in the human case the material environment is best understood as a *sociomaterial* environment (Mol 2002; Orlikowski 2007; Suchman 2007) because of "the intertwinement of the material and the social in practice" (Rietveld and Brouwers 2016; Rietveld and Kiverstein 2014; Van Dijk and Rietveld 2017). With the second relatum (of abilities available in a form of life) in our definition of affordances we go beyond Chemero's (2003, 2009) original and influential relational definition and are able to clarify how an affordance-based account of skilled action can do justice to the "whole spectrum of social significance" in the human form of life (Gibson 1979, pp. 127–8). The human form of life encapsulates many different sociocultural practices, which in turn entail and include many different abilities (and tools) (Wittgenstein 1953; Varela 1999). This move, which we have argued for elsewhere (Rietveld and Kiverstein 2014; Kiverstein and Rietveld

[3] Here we have to be very concise so we refer readers interested in the work that the notion of a form of life does to another paper in which we explained this relation between the form of life and affordances as Gibson conceived of them (Rietveld and Kiverstein 2014; see also the discussion of Wittgensteinian "blind rule-following" in Rietveld 2008a).

2015; Rietveld and Brouwers 2016; Van Dijk and Rietveld 2017), broadens the notion of affordances and, crucially, opens it up to include affordances for activities that people would traditionally classify as forms of "higher" cognition. For example, the ability to make correct epistemic judgments is part of the human form of life. So a particular sociomaterial aspect of the environment, say the letters typed here, can afford—in the context of our form of life—not just reading, copying, or photographing them, but also making a correct explicit color judgment.

The Fundamentally Social Character of the SIF's Landscape of Affordances

According to our Wittgensteinian definition of affordances, affordances are relative to the abilities available in a form of life (Rietveld and Kiverstein 2014). Because abilities thrive in particular social situations embedded in a sociocultural practice, it follows from our definition that the human landscape of affordances is thoroughly social.[4] Novices also acquire their abilities in these situations in practice (Rietveld 2008a). Examples of forms of life within the overarching human form of life are builders, English language users, concert pianists, and academics. A human individual typically belongs to multiple partially overlapping forms of life. The notion of a form of life is central in the Wittgensteinian account of *situated normativity* that we have developed to do justice to the normative aspect of embodied/enactive cognition (Rietveld 2008a). Sociocultural practices (i.e., forms of life) provide a frame for understanding the normative aspect of embodied/enactive cognition in a way that individual action or dyadic moments of social interactions fail to do (Rietveld 2008a). In other words, the forms of life provide the right level of analysis for understanding Wittgensteinian normativity and, as we saw above, affordances.

Situated normativity is crucial for understanding skilled "higher" cognition (e.g., of an architect correcting the design of a door) both in its linguistic and nonlinguistic forms (Klaassen, Rietveld, and Topal 2010; Kiverstein and Rietveld 2015; Rietveld and Kiverstein 2014).[5] This normative aspect of such skillful action is about distinguishing between correct and incorrect or better and worse in the context of a particular situation in a form of life. By placing Wittgenstein's notion of a form of life at the heart of SIF's definition of affordances, we give skilled intentionality the normativity that is necessary

[4] This has important consequences for the field of social-cognitive neuroscience (Schilbach et al. 2013) and thinking about the socially extended mind (Krueger 2013) that we do not have space to go into here.

[5] Separating the linguistic from the nonlinguistic might actually turn out to be artificial given the earlier-mentioned intertwinement of the social and the material in human practices, i.e., the sociomaterial nature of our environment.

for dealing with the whole spectrum of human social significance. Given the abilities available in our sociocultural practice, it is, for example, possible to state correctly that, independently of a particular individual's actual perception of it, but not independently of the form of life (i.e., our practice), the color of the letters on my computer screen is black. Or, to give an example that involves another sociocultural practice, it is possible to judge correctly that the word "black" in this context affords being translated into Dutch by using the word "zwart."

Recent work in embodied and enactive cognition has been right to emphasize the importance of social interactions and that social cognition fundamentally encompasses the bodily and affective aspects of these social interactions (e.g., Schilbach et al. 2013). Although social interactions are extremely important for understanding both our everyday life and possible disorders of it, it is key to take into account that they take place within a broader context. Crucially, it is not just the *moment* of interaction that is social, but, rather, our *whole landscape* of available affordances reflects the abilities that originate in our sociocultural practices (Rietveld, de Haan, and Denys 2013). This foundational character of the social follows from our definition of affordances as relations between aspects of the (sociomaterial) environment and abilities available in *a form of life, in a practice*.

On "Higher" Cognition: The Landscape of Affordances Includes Affordances for "Higher" Cognition

As has been noted rightly by Alva Noë (2012) in his criticism of Hubert Dreyfus's work, we should avoid "over-intellectualizing the intellect." In recent ethnographic work, we (Rietveld and Brouwers 2016) have shown how in architectural design, which is a typical form of "higher" cognition, architects tend toward a grip on affordances in their situation. The following fragment from that paper shows how this tendency dissolves the distinction between "lower" and "higher" cognition by making engagement with multiple affordances central to the way architects do, for example, problem-solving and long-term planning:

> Continuously adjusting their creations [in the design process] the architects seek insight into how they can advance the architectural art installation. They particularly do so through switching between different ways of visualizing the design, thus keeping the design "moving," as they, repeatedly discontent with a new result, over and over again evaluate the different ways in which the design could be made. . . . After spending several days optimizing the sculpture's rear wheel, AM and RR still experience discontent with its design and continue their search. They study the sketched design-possibilities for some moments before RR decides that he has to see the design in 3D: "I cannot see it well in this way, I want to see it in 3D." . . . They immediately switch from the design as visualized on paper to the design as visualized in 3D in the CAD computer program. . . . The process resembles a kind

> of situation-specific improvisation in which they "join forces" (Ingold 2013) with the available affordances. They experiment by actively manipulating aspects of the design, thus finding out what the design affords (cf. Charbonneau, 2013, p. 592) and which of these possibilities they *experience* as improvements of the overall design. In this manner they explore various adjustments. In the episode we highlight here RR is also unhappy with the 3D visualization as drawn in the CAD program. He concludes that it doesn't look good and that, in order to get insight into how this detail should be designed, they again need to visualize it differently—this time as a cardboard model. In such practices of switching between various visualizing forms the design evolves and takes shape. The architects move toward an optimal grip on their design. (Rietveld and Brouwers 2016, pp. 12–13)

Affordances in that real-life case include, for example, possibilities for making a sketch, making a 3D visualization, making an architectural model in cardboard, possibilities for reflection,[6] and elsewhere in that paper even possibilities for communicating with a physically distant collaborator (Rietveld and Brouwers 2016). This kind of ethnographic work situated in real-life practices fits in well with our Gibsonian and Wittgensteinian approach. It is also important because, as explained in the introduction, it is often assumed that the increasingly influential paradigm of embodied/enactive cognitive science (Chemero 2009; Thompson 2007; Varela, Thompson, and Rosch 1991) has sensible things to say about so-called "lower" cognition, such as grasping a glass or riding a bike, but not about "higher" cognition, such as using creative imagination, comforting a sad friend, or seeking the right word in writing a sentence. (We have made a first attempt to show how our approach can deal with these latter two linguistic cases in Klaassen, Rietveld, and Topal 2010.) Similarly, it is assumed that embodied/enactive cognition can deal only with the immediately present environment but not with the absent or the abstract, such as a plan for a new building or the concerns of an absent (or better: spatially distant) collaborator (Rietveld and Brouwers 2016; Clark and Toribio 1994; Clark 1999; Noë 2012; Degenaar and Myin 2014; Di Paolo, De Jaegher, and Rohde 2010; Van Dijk and Withagen 2016). In our skilled intentionality framework these problematic divides between "higher" and "lower" cognition dissolve, because we are able to understand human "higher" cognition along the same lines as skilled "lower" cognition: both are seen as forms of *skilled engagement with multiple affordances offered by the sociomaterial environment in the context of the human ecological niche.*

Our improved definition of affordances made this move possible. It follows from our definition of affordances that a given aspect of the sociomaterial environment can offer a broad range of affordances, dependent on the abilities available in the form of life. These abilities include linguistic abilities, such as, for example, the ability to point out things about the world with words, to orient someone's attention to an aspect of the environment, and to use words for naming things (in the context of a form of life). The ability to state things is very

[6] Reflecting is just one of the abilities available in our human form of life. See Section 4 of Rietveld (2013) for a short discussion of different kinds of possibilities for reflection.

important because you can do all sorts of things with it in all sorts of practices. However, the abilities in our form of life are obviously far more diverse than just linguistic abilities. The following example of a towel in a bathroom (cf. Wittgenstein 1969, pp. 510–11) makes this centrality of the whole spectrum of abilities clear, by showing how in the context of our form of life this aspect of the sociomaterial environment (i.e., the towel) offers multiple possibilities for action, such as:

(a) hanging it on a hook;
(b) getting perceptual access to aspects of the towel;
(c) grasping and taking hold of the towel;
(d) stating correctly, "that is a towel";
(e) drying my hands;
(f) judging correctly, "that the towel is gray";
(g) reflecting on sustainability of the material of which it is made, and many more.

So, in the context of our human form of life, just this one aspect of the sociomaterial environment offers many affordances. All affordances together contribute to the richness of *the landscape of affordances* of the human form of life in which individuals are situated (see the left part of Figure 3.1). However, with this towel example it is crucial to keep in mind that people typically act in the context of a sociomaterial practice. While engaged in a sociomaterial practice it is the landscape of affordances that forms the context of an agent's actions.

In our framework it is abilities acquired in such a form of life that allow individuals to engage with affordances adequately, and, crucially, this includes affordances for what others have called "higher" cognition. The possession of a skill allows an individual to coordinate actions with the sociomaterial practice in which the skill was acquired; to join forces with its affordances. Engaging with different affordances will require different abilities.

From the perspective of SIF, the possibility to perceive something is also afforded by an aspect of the sociomaterial environment. Gaining perceptual access to the world is a skilled activity (Noë 2012). Perceiving something[7] is just one of the many things we can do skillfully. Where Noë has argued that we use skills to get access to the world, our ethnographic observations suggest that skilled individuals tend toward an optimal grip on the landscape of affordances available in a form of life (Rietveld and Brouwers 2016). From a (complementary) phenomenological perspective, this is best characterized as tending toward a grip on a field of solicitations. Note that such an understanding radically undermines any separation between perception and action and makes responsiveness to (or coordination with) affordances a more basic notion than perception. The

[7] Also for Gibson things afford multiple activities including perceiving what they really are after one has acquired the right skills: "If the affordances of a thing are perceived correctly, we say that it looks like what it *is*. But we must, of course, *learn* to see what things really are—for example, that the innocent-looking leaf is really a nettle or that the helpful-sounding politician is really a demagogue. And this can be very difficult" (Gibson 1979, p. 142; see Rietveld and Kiverstein 2014).

phenomenon that we characterize as "responsiveness to an affordance for perceptual access to an aspect of the environment" offers an affordance-based way of talking about perception. This is useful for certain purposes, because in many situations states of action readiness related to affordances for perceptual access compete at a bodily level with states of action readiness related to affordances for doing other things. In our framework, perception is really just one of the many things people do, as in the towel example earlier where the possibility of drying one's hands is on equal footing with the possibility of getting perceptual access to aspects of the towel.

Moreover, SIF shifts the focus away from sensorimotor skills (which dominate embodied/enactive cognitive science at the moment) to *all skills* available in the human form of life. Once we possess the necessary skills, we can take hold of affordances for "higher" cognition, such as reflecting, judging, or naming something, in a similar way as we take hold of affordances for very mundane activities, such as drying our hands:

> If I say "Of course I know that that's a towel" I am making an utterance. . . . For me it is an immediate utterance. . . . It is just like directly *taking hold* of something, as I take hold of my towel without having doubts. And yet this direct taking-hold corresponds to a sureness, not to a knowing. But don't I take hold of a thing's name like that, too? (Wittgenstein, 1969, pp. 510–511, our italics)

Unlike the work of Dreyfus (2002b) and Hutto and Myin (2012, see Kiverstein and Rietveld 2015), the reach of skilled intentionality is not limited to nonlinguistic activities. A skilled speaker of language can just as easily engage with the affordance for stating correctly, "That is a towel" as with the affordance for drying her hands offered by the towel. SIF broadens the scope of human abilities beyond (nonlinguistic) sensorimotor skills.

Skilled intentionality is skilled responsiveness to the *rich* landscape of affordances. This landscape in which we situate the embodied mind includes, for example, possibilities for social interaction in practice (affordances related to the abilities of architects, conductors of orchestras, and psychiatrists, for instance), and possibilities for language use, as well as affordances for making correct explicit epistemic judgments (Rietveld and Kiverstein 2014). An important part of the SIF research program for the coming years is observing, describing, analyzing, and understanding these different affordances for forms of "higher" cognition in the context of different real-life sociocultural practices.

A Situated Individual's Selective Openness and Responsiveness to Relevant Affordances

The immense variety of affordances available in the landscape of affordances of a form of life raises the question how, in a given situation, an individual can be selectively open to this landscape. How and why is an individual selectively responsive to only the *relevant* affordances out of all these available possibilities for action? And how do affordances solicit a particular course of action in a given situation? We distinguish *affordances* from *solicitations* (Rietveld and Kiverstein 2014; Rietveld 2008a). Solicitations (Dreyfus and Kelly 2007) are the affordances that show up as relevant to a situated individual, and generate bodily states of action readiness. As argued earlier, affordances should be understood as flowing from the form of life as a whole rather than being merely an individual matter. The right level of analysis for affordances is the form of life, and for solicitations it is an individual in a concrete situation.

Our focus in this section will be on how relevance arises for the situated individual. We will first show how for living beings relevance originates from the tendency toward a relative equilibrium in the individual–environment system. Being an inviting or relevant possibility for action, a solicitation is the pre-reflective experiential equivalent of a bodily action readiness moving toward this optimum. With this operationalization, SIF calls attention to the close relation between skilled action and consciousness or lived experience (the invitational character of affordances). Next to "solicitation," this section will introduce two other phenomenological notions, which help the reader see why we understand skilled intentionality as coordination with multiple affordances simultaneously. While the landscape of affordances comprises the affordances available to a form of life, the *field of relevant affordances* reflects the multiplicity of inviting possibilities for action for an individual in a concrete situation. So the field of *relevant* affordances is a field of solicitations. From a phenomenological perspective, the situated individual's integrated responsiveness to multiple solicitations simultaneously can be characterized as a *tendency toward optimal grip* on a field of relevant affordances.

We will start by explaining the phenomenon of being drawn by one relevant affordance and then go on to discuss engagement with multiple relevant affordances.

Relevance and the Tendency Toward an Optimal Grip

Within the skilled intentionality framework we are careful not to presuppose goals, tasks, or aims of some mysterious origin as the source of relevance, but instead see the emergence of the soliciting character of affordances as the result of a process of self-organization. Merleau-Ponty's (1968/2003) philosophy of life helps us see that the environment always already solicits something to the active individual. Merleau-Ponty observes that, as complex biological systems, living organisms are always simultaneously "in a state of relative equilibrium and in a state of disequilibrium" (p. 149). Crucially, this inherent disequilibrium "inspires or motivates self-organized compensatory activity" (Merleau-Ponty 1968/2003, p. 149; Rietveld 2008c, ch. 7). This happens for example, when the organism repairs its tissue damage or restores its glucose level by eating (Rietveld 2008c; Kiverstein and Rietveld 2015). This inherent disequilibrium of the living animal (to the right in Figure 3.1) is the source of a lack that can never be compensated for and will always give rise to selective openness to the landscape of affordances and responsiveness to relevant affordances (middle of Figure 3.1) (Rietveld 2008a; Bruineberg and Rietveld 2014). Due to this lack, the material environment is always encountered as a world of value or significance, of affordances having affective allure. To use the words of enactive philosopher of emotions Giovanna Colombetti (2014), living beings have a "fundamental lack of indifference." Due to this source of primordial affectivity, all living beings are affective beings and there will always be a field of significant affordances soliciting the human being.

So, due to this inherent disequilibrium, this inevitable lack, humans and other living beings are concern-ful systems of possible actions and actually never manage to realize an optimal grip on their situation. They can only *tend toward* an optimal grip in the dynamic coupling of world and active body. (The need for the tendency toward an optimal grip will become clear later.) Our grip on the situation can only be a local optimum because our existence as a whole has "a problem," an absence, which is "not a lack of this or that" (Merleau-Ponty 1968/2003, pp. 155–6).

Solicitations are fundamentally related to the individual's need to re-establish this relative equilibrium.[8] We might say that a skilled individual can be "moved to improve" its situation by being responsive to solicitations (Rietveld 2008a). The inviting or soliciting character of affordances can be characterized phenomenologically by the idea that individuals are being "drawn" (Dreyfus and Kelly 2007) to affordances that they care about and are able to act on.[9] Such a description emphasizes the invitational or soliciting character of the environment. Merleau-Ponty describes this in the following example of the tendency toward an optimal grip:

> For each object, as for each picture in an art gallery, there is an optimal distance from which it *requires* to be seen. (Merleau-Ponty 1945/2002, p. 352, our italics)

Standing too close to a painting might make us, for example, lose grip of the overall composition, insofar as it impedes the "appearance" of the object. On the other hand, standing too far away may make the colors blend in such a way that we cannot grasp the texture of the brushstrokes.[10] Note the deliberate use of the word "grip," which brings a sense of actively maintaining oneself in relation to one's situation. In other words, there seems to be an optimum or equilibrium in the individual–environment relation that structures the individual's experience of (not) having grip. Accordingly, an individual's

[8] "The stability of the organism is a stability endlessly reconquered and compromised" (Merleau-Ponty 1968/2003, p. 150).

[9] For an affordance to stand out as relevant, the individual also has to possess the necessary ability or skill (see the first section). So not only what one cares about but also what one *can do* in the context of a practice is reflected by a solicitation (Rietveld 2008b; Merleau-Ponty 1945/2002). With the exception of some very basic innate abilities, these skills are acquired in sociocultural practice in our human case. What one is able to do also develops dynamically during the course of a day: when one is very tired one may not be able to pick up an available affordance.

[10] Note that both relevant affordances and grip vary with respect to the skills of the individual (an art historian will look in a different way than a child) as well as with the current concerns of the individual.

lived experience and the dynamically developing state of (dis)equilibrium of the living being can be seen as two sides of the same coin.

This notion of optimum can easily be misunderstood. We (Bruineberg and Rietveld 2014) have explained the tendency toward an optimal grip using empirical work from ecological dynamical systems theory, which will be described in the third section. One of these studies on boxing (Hristovski, Davids, and Araújo 2009) showed that in boxing there actually is an optimal distance from the heavy bag that is used in training. This optimum is a kind of relative equilibrium in the individual–environment relationship that allows the boxer to be ready to respond to multiple affordances simultaneously and rapidly switch from making one kind of punch (say a jab) to making another (a hook or an uppercut).[11] Our technical term for such an optimal position in which rapid switching is possible is the "metastable zone."

Metastability is a property of coupled dynamical systems in which over time the tendency to integrate and segregate coexist (Kelso 2012). Empirical work suggests that expert athletes make use of these metastable regimes to achieve functional performance outcomes (Seifert et al. 2014). Using these zones makes sense because there they are able best to join forces with the multiplicity of affordances (possible punches) that the situation affords. Crucially, we expect that the tendency toward an optimal grip can be formalized in terms of the tendency toward an "optimal metastable zone" (Bruineberg and Rietveld 2014) (see the third section). As will have become clear earlier, this optimal metastable zone can only be a relative equilibrium.

Earlier we also mentioned the link between affectivity and the inherent disequilibrium of living beings. The tendency toward optimal grip characterizes the internal relation between affectivity and adequate performance in a way that is well described in Dreyfus's work on skilled action:

> According to Merleau-Ponty, . . . absorbed, skillful coping . . . is experienced as a steady flow of skillful activity in response to one's sense of the situation. Part of that experience is a sense that when one's situation deviates from some *optimal* body-environment relationship, one's activity takes one closer to that optimum and thereby relieves the "tension" of the deviation. One does not need to know, nor can one normally express, what that optimum is. (Dreyfus 2002b, p. 378)

Disequilibrium, suboptimality, or a lack of adequate grip can be experienced as an affective tension that needs to be reduced (cf. Rietveld 2008a, 2008c). In an informative example, Wittgenstein (1978) describes this integrated engaged responsiveness and lived affective experience. A door is appreciated as too low in its current context by an

[11] Switching has been understood by Wheeler (2008) as intra-context sensitivity to relevance, which he explained dynamically. He distinguishes it from outer-context sensitivity to relevance. In earlier work we criticized this distinction by Wheeler and showed that in reality relevance sensitivity is actually related to the field of relevant affordances as a whole (Rietveld 2012).

expert architect. The dissatisfied architect immediately and skillfully joins forces with one of the affordances offered by this aspect of the material environment: with the solicitation to increase the height of the door. In working on improving the door, the architect expresses a basic form of normativity, in the sense that he distinguishes better from worse or correct from incorrect in the context of the particular situation. As mentioned earlier, we have called this normative aspect of being moved to improve in skilled action "situated normativity" (Rietveld 2008a). The architect's discontent—directed at the door in its context—and, related to that disequilibrium, the solicitation of the relevant affordance, shows how lived affective experience and context-sensitive performance are two sides of the same coin in skilled intentionality. Even without explicitly verbalizing a judgment or articulating any feelings, the architect's (facial/bodily) expression can show how he appreciates the situation. And, inversely, his action aimed at changing the design of the door is an expression of his discontent. Therefore affectivity is a central aspect of selective responsiveness to relevant affordances.[12] The notion of action readiness from emotion psychology can shed further light on this, as we will see in the next section.

Bodily Action Readiness Links Emotion and Ecological Psychology with Phenomenology

A core concept from the field of emotion psychology (Frijda 1986, 2007) is central in SIF: action readiness. The phenomenological observation that relevant affordances evoke or solicit bodily action readiness enables us to show how the perspectives of ecological psychology (Gibson 1979; Chemero 2009; Reed 1996; Heft 2001) and emotion psychology (Frijda 1986, 2007) converge: relevant affordances are bodily potentiating. The notion of action readiness was introduced by emotion psychologist Nico Frijda and identified as typical for a spectrum of emotions (Frijda 1986, 2007). States of action readiness characterize affective states in ways that reflect the strivings of organisms to modify their relation to the environment.

Action readiness is a bodily phenomenon in-between overt action and ability, a form of action preparation. States of action readiness can be observed, measured, and analyzed. Emotions, and states of action readiness, in particular, reflect a tendency of the individual to modify the relation between herself and the environment in a way that is in line with what matters to her. Relevant affordances move us, affect, and solicit us as they get us ready to act. Often they move us to improve our situation, as we have seen earlier. Affective tension and action readiness are two sides of the same coin. Affective tension is not necessarily felt phenomenologically.

[12] Note it is possible that it turns out, for example, that an important governmental regulation blocks the architect's plan. In that case the architect will typically experience discontent again and see other action possibilities that would allow him to deal with the situation.

For action control it is important that multiple states of action readiness can *self-organize* into a macrolevel pattern of preparation for action (Bruineberg and Rietveld 2014; Lewis and Todd 2005). It is this characteristic of states of action readiness that allows SIF to avoid presupposing goals of mysterious origin and make self-organization central instead. Frijda and colleagues write in this regard that "multiple states of action readiness may interact in generating action, by reinforcing or attenuating each other, thereby yielding . . . control" (Frijda, Ridderinkhof, and Rietveld 2014). Below we will show that due to this process of self-organization, multiple states of action readiness fuse in a way that is similar to mixed emotions like nostalgia (which, for example, might reflect both sadness and happiness).[13]

The tendency toward an optimal grip we pre-reflectively experience when a relevant affordance (i.e., a solicitation) shows up is related to the readiness of the affordance-related ability (Rietveld 2008a; Bruineberg and Rietveld 2014). Importantly, the notion of *relevant affordance-related states of action readiness* links the phenomenological level (pre-reflectively experienced solicitation) to the ecological level of analysis (relevant *affordance*-related action readiness). It makes explicit how the disequilibrium in the individual–environment relation makes a particular affordance stand out as soliciting and drives bodily action readiness (bottom right of Figure 3.1).

From Engagement with a Single Relevant Affordance to a Field of Multiple Relevant Affordances

With the exception of the boxing example, up until now, our discussion in this section has focused mostly on the soliciting character of one single affordance. However, skilled intentionality as we encounter it in our real-life practices implies responsiveness to *multiple* affordances *simultaneously*. The situated individual responds in an integrated way to what we call a *field of relevant affordances* (Rietveld 2012). This phenomenological notion describes how the affordances that a situated individual simultaneously responds to are related. When an expert boxer is training on a heavy bag, for example (Hristovski, Davids, and Araújo 2009), the field of relevant affordances reflects the integrated readiness for multiple kinds of punches (left or right jab, hook, and uppercut, for example), as well as drinking water. In a field of affordances we understand the various relevant affordances to provide the *context* for one another. Accordingly, the SIF provides a

[13] "Simultaneous [states of action readiness] can be expected to interact. They in fact have to interact, since they have to share output channels: action provisions, attention resources, logistic support resources, and so forth. The interactions are required for coordinating the multiple emotions' calls for action. Such coordinations lead to motive states, actions, and feelings that differ from those that would have become manifest when each emerged in isolation. Together, they result in mixed emotions or mixed feelings. . . . True mixed feelings are observed in nostalgia, consisting of pain moderated by the happiness that was, together with pleasure moderated by the regret that it had gone. . . . But what happens when multiple kinds of action readiness [interact depends] upon their relative strengths" (Frijda, Ridderinkhof, and Rietveld 2014, p. 5).

very simple, yet elegant understanding of situational context as the multiple relevant affordances that are in play and of context sensitivity as selective openness to a multiplicity of relevant affordances simultaneously. So in the SIF, context turns out to be "just more affordances" (Rietveld 2012).

A situated individual's field of relevant affordances should be distinguished from the landscape of affordances available in a form of life. The landscape of affordances is not dependent on the abilities of a particular individual, but on the abilities available in the form of life as a whole; in the entire ecological niche or sociocultural practice. This locates the landscape of affordances at the proper level of analysis for dealing with normativity, as mentioned earlier. Take, for example, the norms of spelling in a language: what today counts as a correct way to write a word is not dependent on any particular individual but on the community or practice as a whole. The landscape of affordances should thus be seen as independent of particular individuals.[14]

The structure of the situated individual's field of affordances is sketched in a very rudimentary and schematic way in Figure 3.2(left). The first dimension of this figure, namely the width of the field, reflects the amount of affordances the individual is simultaneously responsive to. The height of the columns (the second dimension) indicates the relevance or strength of attraction of the different solicitations. We say that these solicitations "stand out" as relevant (against the background of other affordances in the situation); they have affective allure (Rietveld 2008a). The last dimension, namely the depth of the field, reflects the anticipatory character of affordance responsiveness. This is the action preparation aspect of engagement with affordances. It regards one's readiness for what one can do next. For instance, while reading this chapter you might already experience pre-reflectively a sense of excited anticipation for tonight's dinner with your best friend. Observe, by way of contrast, Figure 3.2(right), which depicts the field of relevant affordances of a person suffering from depression. To this person every solicitation is equally unattractive. The scope of possibilities for action is diminished and at this moment it seems like there will be no improvement in the future. For example, the possibility of meeting up with a friend now or in the future is experienced as lacking affective allure. In other words, depression results in the deactivation of the soliciting field of relevant affordances that normally drives individuals toward an optimal grip on their situation (Rietveld, de Haan, and Denys 2013; de Haan et al. 2013; de Haan et al. 2015). On this basis it can be said that depression entails a breakdown of a key aspect of everyday skillful action.

The depth-dimension of the field of relevant affordances is crucial, because our current actions are often performed while reckoning with future possibilities for action that exist "on the horizon." For example, a study in ice climbing showed that the climbers anticipated not only the next step, but the entire route ahead (Seifert et al. 2014). Since action readiness is a situation-dependent bodily phenomenon in-between overt action and ability, it is a useful notion for understanding such action preparation or anticipation

[14] Or, more precisely, relatively independent, because an individual is herself also part of the sociomaterial environment and her activities contribute over time to maintaining the patterned practice of the form of life (see Van Dijk and Rietveld 2017).

FIGURE 3.2. A sketch of the field of relevant affordances at a certain point in time for a normal person (left) and a depressed person (right). The height of the columns refers to the relevance or strength of attractiveness of the different solicitations. The width, depicted as the number of columns that are placed next to each other horizontally, reflects the scope of affordances the individual is engaging with. The depth of the field reflects the temporal dimension, namely, the anticipatory character of engagement with relevant affordances. In other words, one is not only ready for the affordances one is engaging with now, but also for possibilities for action one might engage with in the future (just as the skilled boxer who is performing a right jab now is already poised for performing a left hook and right uppercut next). It is a dynamic field: dynamics in the landscape of affordances (e.g., in the material environment) and dynamics of the instability on the side of the individual can both lead to a restructuring of the field.

Reproduced from S. de Haan, E. Rietveld, M. Stokhof, and D. Denys, The phenomenology of deep brain stimulation-induced changes in OCD: an enactive affordance-based model, *Frontiers in Human Neuroscience*, 7, e653, Figure 1 (a and b), doi:10.3389/fnhum.2013.00653 © 2013 de Haan, Rietveld, Stokhof and Denys. This work is licensed under the Creative Commons Attribution License (CC BY). It is attributed to the authors de Haan, Rietveld, Stokhof and Denys.

(Rietveld 2008a; Bruineberg and Rietveld 2014). Anticipation of the trajectory of affordances ahead is about developing a bodily readiness for what you can do next.

There is an interesting link with work on spatial experience in enactivism (Jelic 2016; Rietveld 2016; Rietveld et al. 2015; Bruineberg and Rietveld 2014). Like the heavy bag for a boxer with the relevant punching skills, places we are familiar with generate a multiplicity of states of action readiness simultaneously. In this way, arriving at a particular *place* or "behavior setting" (Barker 1968; Heft 2001), such as a party, swimming pool, climbing wall, or construction site, pre-structures our field of relevant affordances readiness. For example, at a swimming pool we are ready for encountering people in bathing suits, but not at a construction site. We speak of "place-affordances" because places are aspects of the sociomaterial environment that offer possibilities for action and can generate a multiplicity of states of action readiness. Accordingly, places can put constraints on the structure of our field of relevant affordances over a somewhat longer time scale (see the third section).

The field of relevant affordances is a highly dynamic structure. Relevant affordances move the individual, but are also "consumed" in the process of acting on them when the individual–environment relation is changed and other affordances come to stand out as relevant (Bruineberg and Rietveld 2014). An example of this would be a boxing situation in which the state of the material environment changes rapidly due to the fluctuating movements of the heavy bag. Every now and then the boxer switches unreflectively between affordances; from jab to hook to uppercut and back. Crucially, as in the ice-climbing example, these switches are not independent of each other: the individual is responsive to the entire field of relevant affordances (Rietveld 2012). At a dance party, for example, I might quickly finish the conversation when I hear the first notes of a popular song, but I would refrain from dancing if my friend were to say, for example, that "something terrible happened." In the field of relevant affordances, the possibility of asking what happened shapes the context of the possibility to dance (Klaassen, Rietveld, and Topal 2010). To put it more generally, what is at the foreground and what is at the background shifts continuously (Rietveld, de Haan, and Denys 2013). This means that the field of relevant affordances depicted in Figure 3.2 represents a snapshot of the continuously changing field of relevant affordances.

The field of relevant affordances is a *dynamical* phenomenon, as mentioned earlier. Changes in this field of solicitations can originate in the individual and her actions, but also in the sociomaterial environment. Change often results from the individual's current concerns (i.e., needs, interests, and preferences), which are related to its field of solicitations. These current concerns in turn depend on the individual's inherent disequilibrium in the situation. For example, drinking another beer makes the possibility of going to the toilet more urgent.

A change in the landscape might also change the field of relevant affordances by putting constraints on what is possible and appropriate. An example of this would be the way in which the optimal metastable zone (of distance) for conversation with a friend at a party changes when the noise level in the room increases. When the volume of the music is turned up, I need to speak louder, but at a certain music volume I cannot make myself heard and it is appropriate to stand closer to my friend. So in this case, a change in the environment changed the fields of relevant affordances of both myself and my friend. It changed what is appropriate and what counts as optimal grip. This example illustrates again how we are skillfully attuned to the context, that is, to the available affordances.

In sum, at the level of the situated individual, skilled intentionality is characterized as an integrated response to the field of relevant affordances *as a whole*. Using ethnographic research methods, we investigated this phenomenon in a complex architectural design practice (Rietveld and Brouwers 2016). When tending toward an optimal grip on this integrated field of solicitations, the individual can improve the situation in line with what matters to him or her.

Tendency toward an Optimal Grip as Reduction of Disequilibrium in a Brain–Body–Environment System and the Friston Connection

In the previous section it was shown that the situated individual's field of relevant affordances is continuously restructured. Changes of the field of relevant affordances result from (a) the agent's actions that modify the environment, (b) from the agent-independent dynamics of the situated individual's physical environment, which is in flux (Ingold 2000, 2013), but follow also from (c) an ongoing dynamic within the individual's body and brain (e.g., Freeman 2000; Merleau-Ponty 1968/2003). In recent years, changes in body and brain gained a lot of attention in research on what is called "the anticipating brain" (see, e.g., Friston 2010, 2011; Allen and Friston, 2016; Allen and Gallagher 2016; Kiebel, Daunizeau, and Friston 2008; Cisek 2007; Cisek and Kalaska 2010; Pezzulo and Cisek 2016). These popular ideas in neurodynamics are contextualized in the SIF by connecting them to relevant findings from the fields of ecological psychology and phenomenology via the notion of states of action readiness (which we discussed in the previous sections) and by drawing on principles of the study of complex and dynamical systems (Bruineberg and Rietveld 2014; Kelso 2012; Friston 2011; Tschacher and Haken 2007).

Reduction of Disequilibrium as the Individual's Most Basic Concern

In the SIF the tendency toward an optimal grip on a field of relevant affordances is connected to the reduction of disequilibrium in the dynamical system "brain–body–landscape of affordances" (depicted schematically in Figure 3.3). The skilled individual is situated at a specific location in the landscape of affordances (say at a party or a swimming pool) and is selectively open and responsive to solicitations that reduce its state of *dis-equilibrium*. In a more technical paper (Bruineberg and Rietveld 2014), we characterized the reduction of disequilibrium within the brain–body–environment system as reduction of *dis-attunement* between the two dynamics depicted in Figure 3.3, namely, the internal dynamics (of multiple interacting and self-organizing affordance-related states of action readiness of the individual) and the external dynamics (of the dynamically changing landscape of affordances) (cf. Dotov 2014; Kirchhoff 2015; Malafouris 2014).

Doing so, we incorporated the work of Karl Friston on the so-called "free energy principle" (FEP) (Friston 2010, 2011) into the SIF.[15] Friston takes an important philosophical stance when he calls his FEP enactive. The SIF takes this very seriously by developing and integrating an ecological-enactive interpretation of FEP. On our (Bruineberg and Rietveld 2014; Bruineberg, Kiverstein, and Rietveld 2016) interpretation, Friston's free energy principle applied to living organisms is about improving the individual's grip on

[15] Although Friston's language might sound too cognitivist to some to be united with enactive/embodied cognition, we have shown that it is possible to give a more charitable ecological and enactive/embodied interpretation of his work (Bruineberg and Rietveld 2014; Bruineberg, Kiverstein, and Rietveld 2016; Allen and Friston 2016). It is good to keep in mind that together with Walter Freeman (e.g., 2000), Karl Friston (e.g., 1997) is one of the world's main pioneers of neurodynamics and the metastable brain (see Rietveld 2008b for how this links up with Varela's and Kelso's ideas).

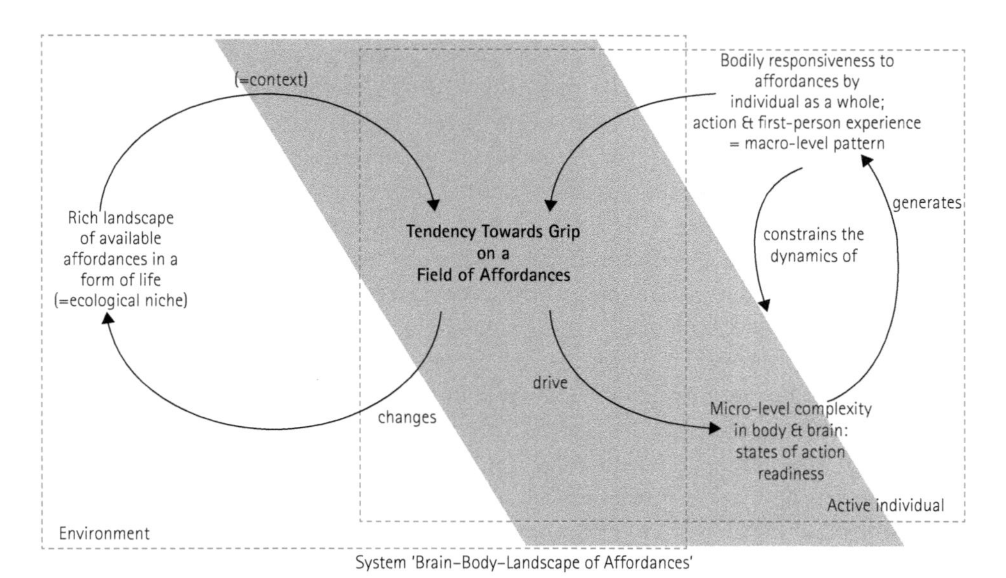

FIGURE 3.3. Sketch of the skilled intentionality framework.

Reproduced from J. Bruineberg and E. Rietveld, Self-organization, free energy minimization, and optimal grip on a field of affordances, *Frontiers in Human Neuroscience*, 8, e599, doi:10.3389/fnhum.2014.00599 © 2014 Bruineberg and Rietveld. This work is licensed under the Creative Commons Attribution License (CC BY). It is attributed to the authors Bruineberg and Rietveld.

the environment.[16] Crucially, Friston argues that you can predict the structure of the embodied brain from the structure of the environment (Friston 2011, 2013; Bruineberg, Kiverstein, and Rietveld, 2016).[17] It is precisely due to the fact that the structure of the individual's body reflects the structure of the ecological niche of the individual that she can stay attuned to it by being selectively open to affordances.

This fits in with the way in which the SIF explains how the field of solicitations and the individual coevolve during skilled action (and over a very long time scale during skill acquisition). According to the SIF, the multiple simultaneous states of action readiness that are generated are related to relevant affordances of the specific place and practice in which the individual is situated (see second section). Therefore, the SIF highlights the relevance of investigating the region of the landscape of affordances in which the individual is involved (e.g., by analyzing involvement in the sociomaterial practice over longer periods of time, see Rietveld and Brouwers 2016) and the structure of the resulting field of solicitations (including relevant place-affordances) to learn something about the activity of the brain and body, and vice versa (Bruineberg and Rietveld 2014).

[16] From this perspective, free energy can be seen a measure of the individual's grip in terms of attunement of internal and external dynamics.

[17] By means of example, think of the way the skills, muscular body, and style of movement of a dancer reflect the practice she participates in. Similarly, the brain has become adapted to this niche over time.

In line with the discussion of the tendency toward an optimal grip in the second section earlier, reduction of disequilibrium within the system brain–body–landscape of affordances is seen in our SIF approach as a basic and continuous concern that drives the individual's selective openness to relevant affordances. This can, for instance, be observed in the improvement of energy levels by eating or by sleeping, but also in more complex improvements of a person's situation in the context of sociocultural practices, such as a discontented architect reducing a disequilibrium by improving the design of a door (in its context). Making the door higher reduces the architect's discontent and the disequilibrium of the situation. Continuously, the individual's readinesses of skills and her behavior are geared toward a re-establishment of relative equilibrium (Rietveld 2008a, 2008c; Bruineberg and Rietveld 2014). By generating responsiveness to solicitations, which are the (pre-reflective) experience of states of action readiness for resolving a suboptimality or disequilibrium in the individual–environment relation, this basic concern for reduction of disequilibrium moves the skilled individual to improve his or her situation.

Although disequilibrium is continuously reduced by an individual's skilled intentionality, complete stability will never follow as long as the organism is living (Merleau-Ponty 1968/2003). We have stated in the second section that this is why we talk about a *tendency* toward an optimal grip. Any movement toward optimal grip on the situation can only bring relative equilibrium, but will not lead to a fully stable state of the system individual–environment. Crucially, it is in virtue of this intrinsic instability or disequilibrium that affordances get their relevance, multiple states of action readiness are generated, and an organism can respond flexibly to the environment and maintain its structural organization.

Interacting States of Action Readiness

Skilled intentionality, understood as the tendency toward optimal grip on a field of relevant affordances, typically describes the change of an individual's situation as responsiveness to multiple solicitations simultaneously (see the second section). Earlier we mentioned that multiple affordance-related states of action readiness interact to generate a coordinated engagement with multiple affordances simultaneously, which makes it possible to understand integration of different states of action readiness (Frijda, Ridderinkhof, and Rietveld 2014) as well as the capacity to switch rapidly from doing one thing to doing another (Hristovski et al. 2009; Rietveld 2012, 2008b, 2008c). In a sense, the SIF generalizes some of the insights gained in the fields of emotion psychology and ecological dynamical systems theory. Research on self-organization and so-called coupled pattern generators, which produce rhythmic patterns in robot locomotion (see Beer and Chiel 1993), provides a paradigm for understanding this interaction or coordination of states of action readiness (Bruineberg and Rietveld 2014). When a pattern generator oscillates at a particular frequency, it can influence the frequency of other coupled pattern generators. Crucially, slower dynamics on longer time scales enslave or entrain the faster oscillations (cf. Dotov 2014). This mechanism of enslavement is also hypothesized to be the mechanism that leads neuronal populations to synchronize transiently (Freeman 2000; Varela et al. 2001; Friston 1997; Bruineberg and Rietveld 2014). This fits with what we know from complex systems theory that describes how macrolevel patterns typically constrain the movements of the microlevel parts, while at the same time being generated by these parts (Tschacher and Haken 2007). We observe that these principles of self-organization hold for states of action readiness as well (Bruineberg and Rietveld 2014; Kiebel, Daunizeau, and Friston 2008), which can be seen at different levels of organization in brain and body and are central in the SIF. In the right part of Figure 3.3 we depicted how coordination of multiple microlevel patterns of action readiness generates a macrolevel pattern of action readiness which constrains the dynamics of these microlevel patterns. In this way, self-organization of multiple affordance-related states of action readiness generates a macrolevel pattern of selective openness by the individual to the field of solicitations as a whole. Continuously, the individual's readiness of skills and her behavior are geared toward a re-establishment of equilibrium in the system brain–body–landscape of affordances.

Dotov (2014) suggests what this might mean for understanding neural activity: brain activity at the microscopic level (e.g., neural activity evoked by the detection of a relevant affordance) contributes to behavior but these (microscopic) neural dynamics can perhaps best be understood as enslaved (Dotov 2014; Kelso 2012; Tschacher and Haken 2007) by the slower (macroscopic) dynamics of the larger dynamical system, that is, of what we call the system "individual-landscape of affordances." More research is needed to understand better how microscopic neural activity of certain brain areas is enslaved by the dynamics of the brain as a whole, which is in turn constrained by the dynamics of the macroscopic system "individual-landscape of affordances." This kind of research will benefit from keeping in mind that the brain is not only embedded in a body but also situated in a place (Heft 2001; Rietveld and Kiverstein 2014).

The relation between affordances and bodily (including neural) action preparation connects to our discussion on place-affordances and anticipation in the previous section. Behavior settings such as libraries, walls for ice climbing, and restaurants have a certain stability over a somewhat longer and slower time scale. We might say that place-affordances (e.g., the aspect of the sociomaterial environment that we call a library) generate patterns of action readiness over a longer time scale that can enslave or entrain faster affordance-related states of action readiness. As such, a place-affordance pre-structures which states of action readiness can be adopted, contributing to the situated individual's tendency toward an optimal grip on the situation embedded in the broader practice. In other words, this is a form of affordance responsiveness that unfolds over a somewhat longer time scale. States of action readiness related to place-affordances are high up in the hierarchical—or, better, nested—cascade of constraining states of action readiness. Similarly, anticipation of affordances on the horizon of the field of solicitations can influence our current affordance responsiveness (Van Dijk and Rietveld 2017). The action possibility to have dinner with one's best friend tonight can, for example, increase one's focus on the most relevant affordances so that one will finish working in time.

In the previous section, we explained that it is when we are well attuned to the dynamically changing landscape of affordances that we have the possibility to switch rapidly from doing one thing to doing another (Rietveld 2012). Being able to flexibly switch activities is described by the phenomenon of *hypergrip* on a field of relevant affordances (Bruineberg and Rietveld 2014). This notion of hypergrip is another expression for being in a (relatively) optimal metastable zone. We encountered a possible example of this earlier: for a skilled boxer, the zone of optimal metastable distance might solicit moving forward, because in this zone he or she is simultaneously ready for multiple relevant action opportunities and for flexibly switching between them in line with environmental fluctuations, like the sometimes very fast movements of a boxing bag (for another real-life example see Rietveld and Brouwers 2016). This phenomenon of tending toward an optimal metastable zone is potentially so important that it is worth taking a second look at the empirical study on optimal movement pattern variability in boxing (Hristovski, Davids, and Araújo 2009). At a critical distance of 0.6 (the distance to the punching bag scaled by arm length), boxers "could flexibly switch between any of the boxing action modes" (Chow et al. 2011; Hristovski, Davids, and Araújo 2009). At this distance the boxing bag "invited" (cf. Withagen et al. 2012, 2017) a wider variety of punches (left and right uppercuts, hooks, and jabs) than it did at other distances.

This boxing study and ethnographic observations of expertise in the practice of architecture indicate that we can describe optimal grip on a field of relevant affordances at a different level of analysis as optimal *metastable* attunement to the field of affordances (Bruineberg and Rietveld 2014; Rietveld and Brouwers 2016). Metastable attunement is a technical term for the ease with which a system can switch to another state (Kelso 2012; Bruineberg and Rietveld 2014) and the study of it in the brain–body–environment system as a whole provides a paradigm for understanding hypergrip. In the relative equilibrium of an optimal metastable zone, a self-organizing system as the "brain–body–landscape of affordances" can adopt a great number of states with only the slightest change (perturbation) in the environment (e.g., a random movement of the boxing bag) or the individual's internal state (the individual is also in motion; think, for example, of the many interacting patterns of bodily action readiness). Note that the affordances the individual has a readiness for are not endless, but limited to the relevant affordances given the agent's abilities and state of disequilibrium. A small disruption such as a random movement of the heavy bag can drive the system to settle on a new form of organization, which impacts the individual's phenomenology. The solicitation and related action readiness that gave rise to the movement are "consumed" in the process, making other solicitations stand out next (Bruineberg and Rietveld 2014). A new macrolevel pattern of selective openness to the landscape of affordances arises (right dynamic in Figure 3.3). As such, hypergrip on a field of relevant affordances is functional with respect to both the demands of the environment and the basic concern of the organism of tending toward an optimal grip on affordances in the situation.

Conclusion

The landscape of affordances in the human form of life is very rich and forms the context in which we should situate ecological-enactive cognition. In this chapter we have made skills for engaging with these affordances central to dissolve the distinction between "lower" and "higher" cognition. The long-term ambition of the Amsterdam SIF research program is to explore if the whole spectrum of things people do skillfully, including social interaction, language use, and other forms of "higher" cognition, can be understood in terms of skilled intentionality, which is the selective engagement with multiple affordances simultaneously. Both poles of our new Wittgensteinian interpretation of Gibson's (1979) notion of affordances (Rietveld and Kiverstein 2014; Chemero 2009), as relations between (a) aspects of the sociomaterial environment in flux and (b) abilities available in a "form of life" (Wittgenstein 1953), manifest an enormous variety. It is this definition that allows us to see the human ecological niche as a rich and resourceful landscape of affordances (Rietveld and Kiverstein 2014). The definition of affordances also makes it possible to deal with situated normativity because, just like Wittgensteinian normativity, affordances are always to be understood as related to a particular form of life. This practice-based normativity (Rietveld 2008a) is crucial for dealing with certain kinds of higher cognition, for example, the possibility of making correct epistemic judgments. In the human form of life the social dimension is implicated in a fundamental way as shaping and sustaining this landscape of affordances. The SIF approach shows that abilities are embedded in and acquired through participation in a sociocultural practice (Rietveld 2008a; Rietveld and Kiverstein 2014). The other relatum of an affordance, the environment, is also defined as sociomaterial from the start (Rietveld and Brouwers 2016; Van Dijk and Rietveld 2017). Moreover, the SIF approach distinguishes itself from more purely philosophical work in embodied/enactive cognition (e.g., Noë 2012) in that it is able to link complementary findings established in different scientific disciplines in one integrative conceptual framework. The SIF integrates the neurodynamic, the ecological/contextual, the affective, and the personal/phenomenological levels of analysis by showing how these perspectives on cognition describe different aspects of one self-organizing system that includes both the individual and its sociomaterial environment: the self-organizing system "brain–body–landscape of affordances." Re-establishment of equilibrium through reduction of disattunement between the internal dynamics (a hierarchy of interacting states of action readiness at multiple time scales, or, in other words, a nested cascade of constraining states of action readiness) and external dynamics (the dynamically changing landscape of affordances) is the individual's primary and ongoing concern. This primary and ongoing concern can phenomenologically be described as a tendency toward an optimal grip on the various relevant affordances encountered in a particular situation. In this process of skilled responsiveness to affordances, the sociomaterial environment is typically transformed as well. Moving toward an optimal grip on the field of solicitations implies reducing tension or discontent by engaging one's skills to join forces with multiple relevant affordances.

Acknowledgments

We would like to thank Jelle Bruineberg, Azille Coetzee, Ludger van Dijk, Marek McGann, Julian Kiverstein, Janno Martens, and an anonymous reviewer for feedback on an earlier version of this chapter. This research was financially supported by an ERC Starting Grant (number 679190, EU Horizon 2020) for Erik Rietveld's project AFFORDS-HIGHER.

References

Abramova, E. and Slors, M. (2015). Social cognition in simple action coordination: a case for direct perception. *Consciousness and Cognition*, 36, 519–31. Available at: http://linkinghub.elsevier.com/retrieve/pii/S1053810015000902

Allen, M. and Friston, K.J. (2016) From cognitivism to autopoiesis: towards a computational framework for the embodied mind. *Synthese*. doi 10.1007/s11229-016-1288-5

Allen, M. and Gallagher, S. (2016) Active inference, enactivism and the hermeneutics of social cognition. *Synthese*. doi:10.1007/s11229-016-1269-8

Barker, R.G. (1968). *Ecological psychology: concepts and methods for studying the environment in human behavior*. Stanford, CA: Stanford University Press.

Beer, R.D. and Chiel, H.J. (1993). Simulations of cockroach locomotion and escape. In: R.D. Beere, R.E. Ritzmann, and T. McKenna (eds.), *Biological neural networks in invertebrate neuroethology and robotics*. San Diego, CA: Academic Press Professional Inc., pp. 267–86.

Bruineberg, J., Kiverstein, J.D., and Rietveld, E. (2016) The anticipating brain is not a scientist: the free-energy principle from an ecological-enactive perspective. *Synthese*. doi:10.1007/s11229-016-1239-1

Bruineberg, J. and Rietveld, E. (2014). Self-organization, free energy minimization, and optimal grip on a field of affordances. *Frontiers in Human Neuroscience*, 8, 1–14. Available at: http://journal.frontiersin.org/Journal/10.3389/fnhum.2014.00599/abstract

Charbonneau, M. (2013). The cognitive life of mechanical molecular models. *Studies in History and Philosophy of Biological and Biomedical Sciences*, 44, 585–94.

Chemero, A. (2003). An outline of a theory of affordances. *Ecological Psychology*, 15(2), 181–95.

Chemero, A. (2009). *Radical embodied cognitive science*. Cambridge, MA: MIT Press.

Chow, J. Davids, K., Hristovski, R., Araújo, D., and Passos, P. (2011). Nonlinear pedagogy: learning design for self-organizing neurobiological systems. *New Ideas in Psychology*, 29(2), 189–200. doi:10.1016/j.newideapsych.2010.10.001

Cisek, P. (2007). Cortical mechanisms of action selection: the affordance competition hypothesis. *Philosophical transactions of the Royal Society of London. Series B, Biological sciences*, 362(1485), 1585–99.

Cisek, P. and Kalaska, J.F. (2010). Neural mechanisms for interacting with a world full of action choices. *Annual Review of Neuroscience*, 33, 269–98.

Clark, A. (1999). An embodied cognitive science? *Trends in Cognitive Sciences*, 3(9), 345–51.
Clark, A. and Toribio, J. (1994). Doing without representing? *Synthese*, 101(3), 401–31.
Colombetti, G. (2014). *The feeling body: affective science meets the enactive mind*. Cambridge, MA: MIT Press.
Costall, A. (1995). Socializing affordances. *Theory & Psychology*, 5(4), 467–81.
Darwin, C.R. (1881). *The formation of vegetable mould, through the actions of worms, with observations on their habits*. London: John Murray.
Degenaar, J. and Myin, E. (2014). Representation-hunger reconsidered. *Synthese*, 191, 3639–48.
de Haan, S., Rietveld, E., Stokhof, M., and Denys, D. (2013). The phenomenology of deep brain stimulation-induced changes in OCD: an enactive affordance-based model. *Frontiers in Human Neuroscience*, 7(653), 1–14.
de Haan, S., Rietveld, E., Stokhof, M., and Denys, D. (2015). Effects of deep brain stimulation on the lived experience of obsessive-compulsive disorder patients: in-depth interviews with 18 patients. *PLoS ONE*, 10(8), 1–29. doi:10.1371/journal.pone.0135524
Di Paolo, E., De Jaegher, H., and Rohde, M. (2010). Horizons for the enactive mind: values, social interaction, and play. In: J. Steward, O. Gapenne, and E. Di Paolo (eds.), *Enaction: toward a new paradigm for cognitive science*. Cambridge, MA: MIT Press. Available at: http://books.google.co.uk/books?id=UtFDJx-gysQC&printsec=frontcover&source=gbs_ge_summary_r&cad=0#v=onepage&q&f=false
Dotov, D.G. (2014). Putting reins on the brain: how the body and environment use it. *Frontiers in Human Neuroscience*, 8, 1–12.
Dotov, D.G., Nie, L., and Chemero, A. (2010). A demonstration of the transition from ready-to-hand to unready-to-hand. *PLoS ONE*, 5(3), e9433.
Dreyfus, H. and Kelly, S.D. (2007). Heterophenomenology: heavy-handed sleight-of-hand. *Phenomenology and the Cognitive Sciences*, 6(1–2), 45–55.
Dreyfus, H.L. (2002a). Intelligence without representation—Merleau-Ponty's critique of mental representation the relevance of phenomenology to scientific explanation. *Phenomenology and the Cognitive Sciences*, 1(4), 367–83. Available at: http://philpapers.org/rec/DREIWR-2
Dreyfus, H.L. (2002b). Refocusing the question: can there be skillful coping without propositional representations or brain respresentations? *Phenomenology and the Cognitive Sciences*, 1(4), 413–25.
Dreyfus, H.L. (2006). Overcoming the myth of the mental: how philosophers can profit from the phenomenology of everyday expertise. *Topoi*, 25, 43–9.
Freeman, W.J. (2000). *How brains make up their minds*. New York: Columbia University Press.
Frijda, N.H. (1986). *The emotions*. New York: Cambridge University Press.
Frijda, N.H. (2005). Dynamic appraisals: a paper with promises. *Behavioral and Brain Sciences*, 28(2), 205–6.
Frijda, N.H. (2007). *The laws of emotion*. Mahwah, NJ: Lawrence Erlbaum Associates, Inc.
Frijda, N.H., Ridderinkhof, K.R., and Rietveld, E. (2014). Impulsive action: emotional impulses and their control. *Frontiers in Psychology*, 5, 518. Available at: http://www.pubmedcentral.nih.gov/articlerender.fcgi?artid=4040919&tool=pmcentrez&rendertype=abstract
Friston, K. (2010). The free-energy principle: a unified brain theory? *Nature Reviews Neuroscience*, 11(2), 127–38. doi:10.1038/nrn2787
Friston, K. (2011). Embodied inference: or "I think therefore I am, if I am what I think." In: W. Tschacher and C. Bergomi (eds.), *The implications of embodiment (cognition and communication)*. Exeter, UK: Imprint Academic, pp. 89–125.

Friston, K. (2013). Active inference and free energy. *Behavioral and Brain Sciences*, 36(3), 212–3. doi:10.1017/S0140525X12002142

Friston, K.J. (1997). Transients, metastability, and neuronal dynamics. *NeuroImage*, 5(2), 164–71.

Friston, K.J., Shiner, T., FitzGerald, T., Galea, J.M., Adams, R., Brown, H., Dolan, R.J., Moran, R., Stephan, K.E., and Bestmann, S. (2012). Dopamine, affordance and active inference. *PLoS Computational Biology*, 8(1), e1002327.

Gibson, J.J. (1979). *The ecological approach to visual perception*. Boston: Houghton Mifflin.

Heft, H. (2001). *Ecological psychology in context: James Gibson, Roger Barker, and the legacy of William James's radical empiricism*. Hillsdale, NJ: Erlbaum.

Hristovski, R., Davids, K.W., and Araújo, D. (2009). Information for regulating action in sport: metastability and emergence of tactical solutions under ecological constraints. In: D. Araújo, H. Ripoll, and M. Raab (eds.), *Perspectives on cognition and action in sport*. Hauppauge: Nova Science, pp. 43–57.

Hutto, D. and Myin, E. (2012). *Radicalizing enactivism: basic minds without content*. Cambridge: Cambridge University Press.

Ingold, T. (2000). *The perception of the environment: essays on livlihood, dwelling and skill*. New York: Routledge.

Ingold, T. (2013). *Making: anthropology, archaeology, art and architecture*. New York: Routledge.

Jelic, A. (2016). The enactive approach to architectural experience: a neurophysiological perspective on embodiment, motivation, and affordances. *Frontiers in Psychology*, 7(481), 1–20.

Kelso, J.A.S. (2012). Multistability and metastability: understanding dynamic coordination in the brain. *Philosophical transactions of the Royal Society of London. Series B, Biological sciences*, 367(1591), 906–18.

Kiebel, S.J., Daunizeau, J., and Friston, K.J. (2008). A hierarchy of time-scales and the brain. *PLoS Computational Biology*, 4(11), e1000209.

Kirchhoff, M.D. (2015). Experiential fantasies, prediction, and enactive minds. *Journal of Consciousness Studies*, 22(3–4), 68–92.

Kiverstein, J. and Miller, M. (2015). The embodied brain: towards a radical embodied cognitive neuroscience. *Frontiers in Human Neuroscience*, 9, 1–11. Available at: http://journal.frontiersin.org/article/10.3389/fnhum.2015.00237

Kiverstein, J. and Rietveld, E. (2015). The primacy of skilled intentionality: on Hutto and Satne's the natural origins of content. *Philosophia*, 43(3), 701.

Klaassen, P., Rietveld, E., and Topal, J. (2010). Inviting complementary perspectives on situated normativity in everyday life. *Phenomenology and the Cognitive Sciences*, 9(1), 53–73.

Krueger, J. (2013). Ontogenesis of the socially extended mind. *Cognitive Systems Research*, 25–26, 40–6.

Lewis, M.D. and Todd, R.M., 2005. Getting emotional: a neural perspective on emotion, intention, and consciousness. *Journal of Consciousness Studies*, 12, 210–35.

Malafouris, L. (2014). Creative thinging: the feeling for clay. *Pragmatics & Cognition*, 22(1), 140–58.

Merleau-Ponty, M. (1945/2002). *Phenomenology of perception*. London: Routledge.

Merleau-Ponty, M. (1968/2003). *Nature: course notes from the College de France* (trans. R. Vallier). Evanston, IL: Northwestern University Press.

Michaels, C.F. (2003). Affordances: four points of debate. *Ecological Psychology*, 15(3), 135–48.

Mol, A. (2002). *The body multiple: ontology in medical practice*. Durham, NC: Duke University Press.

Noë, A. (2012). *Varieties of presence.* Cambridge, MA: Harvard University Press.
Orlikowski, W.J. (2007). Sociomaterial practices: exploring technology at work. *Organization Studies*, 28(09), 1435–48.
Pezzulo, G. and Cisek, P. (2016). Navigating the affordances landscape: feedback control as a process model of behavior and cognition. *Trends in Cognitive Sciences*, 20(6), 414–24. doi:10.1016/ j.tics.2016.03.013
Ramstead, M.J.D, Veissière, S.P., and Kirmayer, A.Y. (2016). Cultural affordances: scaffolding local worlds through shared intentionality and regimes of attention. *Frontiers in Psychology*, 26(7), 1090. doi:10.3389/fpsyg.2016.01090
Reed, E. (1996). *Encountering the world.* Oxford: Oxford University Press.
Rietveld, E. (2008a). Situated normativity: the normative aspect of embodied cognition in unreflective action. *Mind*, 117(468), 973–97.
Rietveld, E. (2008b). The skillful body as a concernful system of possible actions: phenomena and neurodynamics. *Theory & Psychology*, 18(3), 341–63.
Rietveld, E. (2008c). *Unreflective action: a philosophical contribution to integrative neuroscience.* Amsterdam: ILLC-Dissertation Series.
Rietveld, E. (2012). Context-switching and responsiveness to real relevance. In: J. Kiverstein and M. Wheeler (eds.), *Heidegger and cognitive science: new directions in cognitive science and philosophy.* Basingtoke, UK: Palgrave Macmillan, 105–35.
Rietveld, E. (2013). Affordances and unreflective freedom. In: D. Moran and R. Thybo Jensen (eds.), *Embodied subjectivity.* New York: Springer, pp. 1–43.
Rietveld, E. (2016). Situating the embodied mind in landscape of standing affordances for living without chairs: materializing a philosophical worldview. *Journal of Sports Medicine*, 46(7), 927–32. doi:10.1007/s40279-016-0520-2
Rietveld, E. and Brouwers, A.A. (2016). Optimal grip on affordances in architectural design practices: an ethnography. *Phenomenology and the Cognitive Sciences.* doi:10.1007/ s11097-016-9475-x
Rietveld, E., de Haan, S., and Denys, D. (2013). Social affordances in context: what is it that we are bodily responsive to? *Behavioral and Brain Sciences*, 36(4), 436. Available at: http://www.ncbi.nlm.nih.gov/pubmed/23883765
Rietveld, E. and Kiverstein, J. (2014). A rich landscape of affordances. *Ecological Psychology*, 36(4), 325–52.
Rietveld, E., Rietveld, R., Mackic, A., Van Waalwijk van Doorn, E., and Bervoets, B. (2015). The end of sitting: towards a landscape of standing affordances. *Harvard Design Magazine*, 40, 180–1.
Roepstorff, A. (2008). Things to think with: words and objects as material symbols. *Philosophical transactions of the Royal Society of London. Series B, Biological sciences*, 363(1499), 2049–54.
Roepstorff, A., Niewöhner, J., and Beck, S. (2010). Enculturing brains through patterned practices. *Neural Networks*, 23(8-9), 1051–9. doi:10.1016/j.neunet.2010.08.002
Schilbach, L., Timmermans, B., Reddy, V., Costall, A., Bente, G., Schlicht, T., and Vogeley, K. (2013). Toward a second-person neuroscience. *Behavioral and Brain Sciences*, 36(4), 393–414. Available at: http://www.ncbi.nlm.nih.gov/pubmed/23883742
Seifert, L., Wattebled, L., Herault, R., Poizat, G., Adé, D., Gal-Petitfaux, N., and Davids, K. (2014). Neurobiological degeneracy and affordance perception support functional intra-individual variability of inter-limb coordination during ice climbing. *PLoS ONE*, 9(2), e89865.

Suchman, L.A. (2007). *Human-machine reconfigurations: plans and situated actions.* Cambridge: Cambridge University Press.

Thompson, E. (2007). *Mind in life: biology, phenomenology, and the sciences of the mind.* Cambridge, MA: The Belknap Press of Harvard University Press.

Tschacher, W. and Haken, H. (2007). Intentionality in non-equilibrium systems? The functional aspects of self-organized pattern formation. *New Ideas in Psychology*, 25, 1–15.

Van Dijk, L. and Withagen, R. (2016). Temporalizing agency: moving beyond on- and offline cognition. *Theory & Psychology*, 26(1), 5–26.

Van Dijk, L. and Rietveld, E. (2017). Foregrounding sociomaterial practice in our understanding of affordances: the skilled intentionality framework. *Frontiers in Psychology.* doi:10.3389/fpsyg.2016.01969

Varela, F. (1999). *Ethical know-how: action, wisdom and cognition.* Stanford, CA: Stanford University Press.

Varela, F., Lachaux, J.P., Rodriguez, E., and Martinerie, J. (2001). The brainweb: phase synchronization and large-scale integration. *Nature Reviews Neuroscience*, 2, 229–39.

Varela, F., Thompson, E., and Rosch, E. (1991). *The embodied mind: cognitive science and human experience.* Cambridge, MA: MIT Press.

Wheeler, M. (2008). Cognition in context : phenomenology, situated robotics and the frame problem. *International Journal of Philosophical Studies*, 16(3), 323–49.

Withagen, R., Araújo, D., and de Poel, H.J. (2017) Inviting affordances and agency. *New Ideas in Psychology*, 45, 11–8.

Withagen, R., de Poel, H.J., Araújo, D., and Pepping, G.J. (2012). Affordances can invite behavior: reconsidering the relationship between affordances and agency. *New Ideas in Psychology*, 30(2), 250–8.

Wittgenstein, L. (1953). *Philosophical investigations.* Oxford: Blackwell.

Wittgenstein, L. (1969). *On certainty.* Oxford: Blackwell.

Wittgenstein, L. (1978). Lectures on aesthetics. In: *Lectures and conversations on aesthetics, psychology and religious belief.* Oxford: Blackwell, pp. 1–40.

CHAPTER 4

THE ENACTIVE CONCEPTION OF LIFE

EZEQUIEL A. DI PAOLO

A Project of the World

A widely known quote by Maurice Merleau-Ponty encapsulates the dialectical relation between subject and world in a succinct formula:

> The world is inseparable from the subject, but from a subject who is nothing but a project of the world; and the subject is inseparable from the world, but from a world that it itself projects. (Merleau-Ponty 1945, p. 454)

Meant to overcome the opposing pulls of idealism and empiricism, this statement contains in its formulation—if not in its meaning—a pregnant mystery.

In the context of this phrase, we are told that while the perceptual world may be posited by consciousness, this is only possible if there is underlying this consciousness an already oriented body engaged in transactions with the world. The perceiving subject is not an absolute world anchor, since all the existence bestowed on this world as a totality of meaning through sense-giving activity is inextricably entangled in the ways the world gives itself to the subject as a person, but also as an animal, as a living organism, and as a complex stream of material flows and potentialities. Another way to put it is that the subject is *in* the world but also *of* the world.

It is indeed a very enactive thing to say that the subject lays down the path in walking, that is, that the frames of signification are given by sense-making activity itself, which is by nature transactional and constrained within material and historical possibilities.

So far, so good. But what is the mystery in Merleau-Ponty's resounding formula? I find it in the idea of the subject as being a *project* of the world. Taken literally, it is a dissonant turn of speech as only those things are projects that are some*body*'s project.

And the world is not an entity that projects anything—if we decide to stand firm on non-teleological ground and avoid seeing nature as a whole as directed toward ends.

In what ways could then this be more than figurative language metaphorically referring to the transactional, materially constrained aspects of sense-making already mentioned? I will suggest in what follows that, as is often the case, a mysterious, yet beautiful, formulation invites a deeper truth. In the context of embodied perspectives on cognitive phenomena, this truth has been the concern of enactivist researchers. For them it has become clear that to ask questions about how the mind works is at the same time to ask questions about what is it about certain entities that they can be minds at all, and how can such entities emerge in a natural world. These two questions, which might be divorced in other areas of inquiry, are for the enactive perspective one single question with different facets. Hence the insistence on the part of some enactive thinkers on the need to understand life and mind as part of a continuity.

Differently put, I am talking about the difficult question: what is a body? This question, not always put in these explicit terms, is the platform on which enactive theory[1] is raised. It is, in my opinion, what differentiates the enactive approach from all other so-called embodied approaches: the thematization of bodies as a prerequisite for understanding *anything* about minds.[2] This is not a line of theorizing that emerges

[1] By enactive theory I refer here to the application of the enactive approach to specific scientific problems in psychology, neuroscience, cognitive science, AI, etc. Examples that explicitly use the label enactive include the development of a dynamical systems interpretation of sensorimotor contingencies theory (Buhrmann et al. 2013; Di Paolo et al. 2014; Di Paolo et al. 2017), nonrepresentational accounts of the phenomenology of the sense of agency (Buhrmann and Di Paolo 2017), neurophysiological models of multi-joint movements (Buhrmann and Di Paolo 2014), clarification of explanatory roles of social interaction in psychology (De Jaegher et al. 2010), hypotheses on social brain function (Di Paolo and De Jaegher 2012), hypotheses on sociocognitive development (e.g., Gallagher 2015), integrative, person-based approaches to autism (De Jaegher 2013), accounts of intentionality, action, and free will (Gallagher 2017), models of metabolism-based bacterial chemotaxis (Egbert et al. 2010), hypotheses on the interactive factors affecting imitation (Froese et al. 2012), organism-based theories of color vision (Thompson et al. 1992), general perspectives on brain function (Fuchs 2011, 2017; Gallagher et al. 2013), accounts of synesthesia (Froese 2014), accounts of neurobiological and embodied factors in prehistoric art and material culture (Froese et al. 2013; Malafouris 2007, 2013), and others.

[2] Unfortunately, the term "embodied" has become one of the most abused keywords in cognitive science. It is not necessary to produce a sophisticated critique of this term in order to see that whatever legitimate meaning it used to have has now been relentlessly diluted thanks to its adoption by brain-centered, individualistic, representational theories that are veiled versions of computationalism. Tenuous conceptual connections with the body do not make a classically disembodied approach any less disembodied. This is a sad state of affairs for which those truly interested in embodiment are partly to blame for often failing to specify the precise connection between their proposals and the body, and failing to describe what kind of bodies they have in mind. As a rule of thumb, any talk of bodily formatted representations belongs strictly to good old-fashioned computationalism, the corporeal adverb being superfluous. To point this out is only fair to researchers in the computational camp because their positions have never been naively unaware of bodily constraints when it came to concrete implementations, say, in classical robotics. The notion of embodiment must be revalorized. Intellectual honesty demands that any embodied theory should be able to provide precise answers two questions: What is its conception of bodies? What central role do bodies play in this theory different from the roles they play in traditional computationalism?

from scratch with the enactive approach (Varela et al. 1991; Thompson 2007), even though it saw one of its clearest formulations in Francisco Varela's later work (e.g., Varela 1997, 2000). The idea has roots in the earlier theory of autopoiesis (Maturana and Varela 1980), an attempt to give a systematic, generative, logical answer to the question: what is a living system? It also traces back to other notable precursors, as I will mention later.

If we take the project-of-the-world image at face value, then, albeit voided of any teleological implications, we get a hint of the kind of inquiry we are trying to circumscribe; ultimately one that offers important conceptual categories for any theorizing about cognitive phenomena. To ask about the meaning of this image, to ask how a medium projects itself into a subjects and objects, is to ask about the material conditions out of which pre-individual processes result in the individuation of living organisms, and the concomitant emergence of *their* world. It is also to demonstrate the intimate relation between these two moments, subject and world, as they co-emerge dialectically out of the same tensions found in pregnant materiality (see, e.g., Grosz 2011). It is also to ask in what ways these conditions relate to forms of psychic and collective individuation. Finally, it is to ask whether these material conditions provide only a background of enabling factors, which can then be assumed invariant across different instances of cognition, and therefore "safely ignored" for specific research projects, or whether, on the contrary, these conditions permeate all cognitive and social phenomena and make their understanding inescapable for any scientific project concerning the mind, no matter how specific.

Life and Mind Continuity

The enactive insistence on the continuity between life and mind has often been met either with impatience or misunderstanding. It is one of those situations where language can fail, giving the impression that one is talking nonsense or else saying something trivial and widely accepted. This means we must go back into it and attempt once again to clear the ground.

It is true: to say that there is continuity between life and mind could be seen as trivial and unimportant, especially in the context of widespread belief in the unity of science. After all, we do not know of any empirical instance of mental phenomena that does not also involve at least one living organism. The question is, does this knowledge matter for attempting to explain specific mental phenomena? Will attention to life-mind continuity have an influence in, say, theories of perception? Even to say that continuity implies that certain explanations used for understanding life will play important roles in any attempt to understand the mind could be met with shrugged shoulders. Do we not after all in disciplines like neuroscience, ethology, psychology, psychopharmacology, psychiatry, etc., already lean strongly on biological knowledge for support in explaining mental phenomena?

So one reaction is: we have been doing life-mind continuity all along. The other reaction is: this is nonsense. Defenders of the enactive approach may be at fault to some extent in this case. Synoptic formulas such as life = cognition have had their provocative initial impact diluted by their implausibility according to reasonable interpretations. Do such statements mean that life is coextensive with cognition? Are we performing a cognitive operation when, say, we digest our lunch? Are all currently living species equally "cognitive" since they are all equally alive? Is psychology reducible to biology? These and similar questions can tire the enactivist, but they are only fair if she is seen as standing behind a notion of continuity as the conflation of psychic and biological phenomena.

Between triviality and nonsense lies a deeper meaning of the continuity thesis. In the fewest possible words: mental phenomena constitutively demand explanations of individuality, agency, and subjectivity, and the principles and conceptual categories for these explanations are the same as those required by attempts to explain the phenomenon of life. Moreover, those conceptual categories and principles are not incidentally useful, but lie at the core of the question we have raised earlier: what are bodies?

Another way to put this is to say that a continuity thesis underlines the naturalistic project of the enactive approach. This is comparable to the same attitude adopted by John Dewey in his naturalistic theory of logic. According to him the primary postulate of such a theory is the "continuity of the lower (less complex) and the higher (more complex) activities and forms" (Dewey 1938, p. 30). Dewey maps the contour of the notion of continuity by making explicit what it excludes: a "complete rupture on one side and mere repetition of identities on the other; it precludes reduction of the 'higher' to the 'lower' just as it precludes complete breaks and gaps" (Dewey 1938, p. 30). Take the example of biological development; we cannot say in advance:

> that development proceeds by minute increments or by abrupt mutations; that it proceeds from the part to the whole by means of compounding of elements, or that it proceeds by differentiation of gross wholes into definite related parts. None of these possibilities are excluded as *hypotheses* to be tested by the results of investigation. What *is* excluded by the postulate of continuity is the appearance upon the scene of a totally new outside force as a cause of changes that occur. (Dewey 1938, p. 31)

To this we would add not so much an emphasis on "forces" outside the naturalistic framework but the rejection of the sudden appearance of fully independent novel levels of description—for instance, the realm of human normativity—without an account of how their emergence and relative autonomy is grounded on (understandable in terms of and interaction with) phenomena at other levels. This is as much a causal/historical point as it is ontological. The continuity thesis therefore proposes the need for a theoretical path that links living, mental, and social phenomena. The project, however, remains non-reductionistic for these three reasons: (1) it seeks explanations of emergent phenomena through theoretical and experimental investigations of, for example, self-organization and complex multi-scale interactions; (2) it replaces the notion of an independence of levels of inquiry (e.g., biology, psychology, sociology) with a notion of relative autonomy and postulates the conditions by which this autonomy can be tested;

and (3) it advances the possibility of various kinds of interactions between levels leading potentially to evolving forms of cross-level mutual dependence and transformations.

We have said it on other occasions: the enactive approach is best described as a non-reductive naturalism (Di Paolo et al. 2010) and the life-mind continuity is its core methodological, epistemological, and ontological attitude. Living and mental phenomena belong to intertwined branches of a same ontological tree (not that they must be equivalent or coextensive) and their study demands related epistemological tools.

Do the Sciences of the Mind Need an Account of Individuation and Agency?

Theories of cognition should be able to provide the operational conceptual categories with which to describe their objects of study and distinguish them from those outside their remit. They should be able to say in concrete terms what sort of system, event, or phenomenon counts as cognitive and in which cases it does not. Accounts that do not meet this mark are pre-scientific. This does not mean they cannot lead to interesting or important knowledge or even to practical solutions to problems or technological innovations. It only means that the bits of knowledge so generated are provisionally held together by intuition or tradition and not by an articulated theoretical framework.

The conceptual categories mentioned earlier—individuality, agency, and subjectivity—lie at a blind spot of functionalist approaches to cognition, whether classical or "embodied." Such approaches must assume these notions as given and unproblematic. Otherwise, they cannot work. Let us examine why.

The idea that it is possible to explain cognitive phenomena in terms of the commerce of functional, representational neural states, bits of information, vehicles and content, etc. implies that a certain stationarity[3] is needed in the permitted variations of states that the cognitive system may undergo. We can call this the informational frame within which functional states have well-defined roles. By definition of what it means to be an information-processing system, the cognitive machinery that processes information cannot therefore change in non-stationary and open-ended ways without at the same time limiting the range of applicability of a functional explanation.[4] One solution adopted for dealing with this problem, say, in theories of learning, is to assume that cognitive systems operate in at least two sufficiently distinct time scales: a fast time scale that corresponds to a settled functional system, and a slow time scale that corresponds to

[3] A process is said to be stationary if the probability distributions for its states do not change over time, that is, if it does not present transient trends that alter general statistical properties such as mean or standard deviation.

[4] It may still be possible for non-stationary changes to occur within a complex system such that certain regions of the system conserve a relative stationarity. Within these regions, at the appropriate time scales, it may be possible to perform valid functional analyses at a local level.

how this system changes its functional frame over time. The interaction between these time scales is hierarchical and essentially non-messy. They accord with Simon's (1962) postulates for near-decomposable systems. The whole framework remains functionalist in that changes in the frame occur in a way that is itself given by a more encompassing stationary frame, for example, plasticity rules in a neuronal network.

But here we face a problem: the question of what constitutes a cognitive system as an extended spatiotemporal entity (in essence, the issue of how it *becomes* a cognitive system) as well as the question of how a domain of significance is constituted in the here and now of a concrete situation (e.g., how activity is framed as appropriate to a context or a motivation)—two intimately connected questions in the enactive approach—demand answers in terms of transformative (frame-changing, frame-establishing) processes, i.e., they demand a non-stationary story. Given these constraints, only two options are open for functionalists: either (1) assume that the issue of becoming a cognitive system is a non-question, i.e., that nothing except convention distinguishes cognitive systems from any other system of functional relations, such as toasters, or (2) assume that it is interesting but irrelevant, i.e., that the answer to this question corresponds to a different science, such as biology, and that once given, it does not contaminate cognitive science and one can safely assume that it does not bear on the explanation of concrete cognition. Similar options are available for the second question, that of the emergence of concrete frames of significance.

Functionalism, even in its embodied versions, has (mostly tacitly) gone for either one of these options. The functionalist would be safe if there were nothing special about cognitive systems that would distinguish them from other systems that could be assumed to be stationary. Alternatively, even if during a period of construction the cognitive system does not verify the assumption of stationarity, this could be assumed to be a well-delimited period of transient transformations outside the remit of cognitive science, after which, the cognitive system can be safely be treated as stationary.

Why is stationarity at odds with an account of individuation? There is, first, an empirical answer, namely that such seems to be the nature of all known forms of cognitive systems: they grow, develop, adapt to unforeseen circumstances, and seem to have an open-ended (though not unconstrained) reserve of potentialities, which we have no reason to assume are all pre-given at birth, since potentialities are always relational with respect to an open-ended environment. As the enactive story unfolds, a stronger, conceptual answer emerges. It postulates that ongoing, open-ended, precarious processes are logically necessary for what makes a system cognitive. Like living systems, cognitive systems are identifiable as centers of activity and perspective. Cognition occurs when there is a cognizer that cognizes about something. This means that there is an entity that takes a stance, and from this stance relations between itself (the cognizer) and its world are inherently meaningful. But there cannot be any such relations unless the entity we call the cognizer is also an individuated entity. And as we will see, these relations cannot be meaningful unless individuation is an ongoing, open, precarious process; i.e., a non-stationary one. The possibility of unpredictable, frame-transforming changes is inherent to being a cognitive system, even in the particular circumstances where these changes are not actually occurring. Hence, to be a cognitive entity is to be a (generally) non-stationary organization in a (generally) non-stationary relation with the world. Since functionalism is limited to cases in which we can safely make the stationary approximation, it follows that it cannot account for fundamental aspects of cognition.

This is not merely an arcane conceptual issue. In many ways, its implications are always close to the surface in concrete research. When we study attention, volition, sense of agency, decision-making, value systems, learning, etc., all of these aspects are implicit. What makes a cognitive system one that can act purposefully, do so with spontaneity, have concerns about its ongoing well-being and activities as well as concern for others, decide correctly, recognize and solve problems, and so on? To try to answer how such acts are performed without understanding why they are carried out at all, why they are of any relevance for the cognitive agent in the first place, does not even amount to half the story. In the mind sciences, there can be no general account of "how it works" without also offering an explanation of "what is at stake and for whom," since these questions are inseparable. Otherwise, we are speaking of complex systems theory, not cognitive science.

Without a solid account of individuality, agency, and subjectivity, we have not even scratched the surface of a theory of the mind, and all the well-established results are provisional because we have no theory that specifies their range of validity, only intuitions and empirical data (which can only give instances of (non-)contradiction of an assumption, but not in themselves explicate the limits of its generalization).

Similar points can be made about the inexistence of a theory of agency in cognitive science, both traditional and "embodied." Again, nothing in functionalism, except external convention or convenience, enables us to theoretically distinguish between a system that is simply coupled to its environment, like the planets in the solar system are gravitationally coupled to the sun and each other, and a cognitive system that is an agent in a meaningful world. An agent does things as well as has things happen to it. Again, in practice, this lack is always complemented by some tacit commonsense assumptions when focusing on specific research concerns. We tend to assume that there is a clear difference between a person moving an arm of her own volition vs. having it moved by an experimenter. But do we have in principle ways to distinguish less obvious cases or to question whether accompanying the experimenter's movements and not opposing them is not also a volitional act?

In short, it seems that there are good reasons to bring to the surface some of the hidden assumptions of the prevalent functionalist framework in the sciences of mind—not only as a healthy exercise, but in order to offer a possible explanation of why certain questions have never been the center of cognitive science research, such as the question of cognitive becoming or the question of the constitution of agency. Enactive theory has, in addition, deeper reasons. These are the issues that permeate all aspects of cognition for this approach. However, this does not mean that it is not possible sometimes to assume that some of these theoretical worries will have limited impact in specific cases of interest. Whether this is a good epistemological move or not, however, necessitates a theoretically loaded framework to justify it, very much in the same way as the theory of relativity itself provides the justification of what conditions validate the applicability of Newtonian mechanics.

We then turn to reviewing these deeper reasons in the next sections.

Autopoiesis

The enactive view of life and mind derives from the theory of autopoiesis—if by derivation we mean the historical sense of progression of ideas and not the logical sense of entailment. In fact, much of what is predicated by enactivists, especially in relation to norms, agency, and social interaction, is different and even quite at odds with classical autopoietic theory (Maturana and Varela 1980; Maturana 2002). I will not rehearse the technical arguments but will highlight some of these differences as we proceed.

The theory of autopoiesis emerged in the 1970s as a response to prevailing views in biology, neuroscience, and psychology, which lacked deep scientific conceptions of organisms, agents, or persons. Autopoietic theory rejected notions of information processing, since they tend to conflate phenomena at the supra-, sub-, and organismic levels. Instead, it adopted an epistemology based on systems theory to postulate that the identifying feature of any system is not its conventional labels, nor its contingent spatiotemporal arrangements, but its organization. Define the organization of a system and you will achieve two things. First, you will define a class of systems that share this organization, such that different instances in this class can be said to belong together, regardless of how much they differ in terms of how their organization is actually instantiated. This seems obvious enough. However, any predicate based on a logical analysis of the properties of a given organization, will ipso facto apply to all instances of a class, which is a useful conceptual tool when we deal with complex systems. Second, given a particular instance of a class of systems, the conservation of its organization is what permits an observer to postulate its identity through time as that particular instance; its haecceity. Thus, while the observed system may change in structure, an observer can say that this is the same system provided its organization is unchanged. Conversely, a change of organization is sufficient to say that the same spatiotemporal arrangement of processes is no longer the same system that it was before. It has transformed into something else or it has disappeared.[5]

It is possible to raise some criticisms at this stage already, for instance, the apparent lack of material and temporal constraints underpinning the idea of a sustained

[5] Classical autopoietic theory works like mathematical set theory: while a given organization may be instantiated in different concrete structures, it is also the case that a particular concrete system may embody more than one organization, on condition that there is no contradiction between them. Thus *this* dog is a mammal, but also an animal and also a living system, each category implying a broader organization and class identity.

organization. Notice also the silence about history and about relations; strictly speaking, all that needs to be known about the identity of a system is given intrinsically and contemporaneously from within the system itself. It is not tendencies or relations or forms of becoming or potentialities that will take ontological precedence in the theory of autopoiesis, but conserved being as defined by organizational properties. The latter may themselves be relational but they belong to the system in that they are the sufficient constitutive relations that establish the class identity of the system. They are to be distinguished from relations *between* systems so defined.

Many conceptual categories that are historical or relational cannot be approached in this way: e.g., being the offspring of, being a twin, being capable of reproduction, belonging to a group, etc. Such relational, historical observations, according to autopoietic theory, belong to the cognitive domain of the observer, which is a quick way of dismissing them in order to focus only on set-theoretic, immanent, operational relations as they are actualized here and now. These, however, are no less observer-dependent since they always imply an epistemic grounding on the part of the observer. In practical terms, the choice of what counts as relevant variables, parameters, and constraints is always precisely that: a choice. Conceived at the level of quantum interactions or at a scale of attoseconds, a living cell does not reveal itself to an observer as autopoietic. Exactly the same epistemic condition holds for the relational/historical categories that autopoietic theorists feel uncomfortable with—once the domain of observation is chosen, their manifestation in this domain is not arbitrary nor is it a matter of convenience or convention.

These worries will resurface later. In any case, based on this systemic framework, to ask the question of how an organism, a living system, should be conceived of is to ask about its organization. More precisely, the central move in autopoietic theory is to propose the description of the organization of a class of systems such that living systems fall neatly into this class, and nonliving systems fall outside it. The proposed description is the definition of autopoiesis as a network of molecular processes undergoing material transformations (including production and destruction) with the following organizational conditions: (1) the network realizes the relations that give rise to the production of its constitutive processes, and (2) the processes constitute the network as a concrete unity in space and define its topological relations (see Maturana and Varela 1980, pp. 78–9; Varela 1979, p. 13).

We may call Condition 1 the *self-production* condition and Condition 2 the *self-distinction* condition.

"Autopoiesis is necessary and sufficient to characterize the organization of living systems" (Varela 1979, p. 17; Maturana and Varela 1980, p. 82, emphasis removed). It is a set-theoretic description of a self-producing and self-distinguishing system, an idea with precursors such as Kant's conception of life as the mutual production of parts and whole and Hegel's reworking of this idea (Kreines 2009; Michelini 2012), the work of physiologists in the Revolutionary period like Xavier Bichat (1800), and mid-twentieth-century philosophers like Hans Jonas (1966) and Georges Canguilhem (1965) (who already in 1951 used the term *autopoietic* to refer to the character of organic activity).[6] Autopoiesis is the idea that a living system is one that is constantly constructing itself and by this activity making itself distinct from its environment.

Introducing an operational description of the organization of living system is only the first step because, as we said earlier, anything that can be deduced logically from this definition will apply to any system that shares this organization. In this way, autopoietic theory challenges various widely held postulates, such as the very possibility of mental representations, and offers non-traditional conceptions of communication, sociality, evolution, and language (Maturana and Varela 1980; Maturana 2002). It would take too long to review these implications here.

The force of some of these challenges to tradition is inherited by the enactive approach, which nevertheless questions some of the starting assumptions and interpretations of autopoietic theory.

The Co-Definition Between Organism and Environment

Worries about autopoietic theory cannot hide the fact that the enactive approach is a historical offshoot of these ideas, in particular, as exemplified in the later work of Francisco Varela. Already moving beyond some of the limitations mentioned earlier, Varela conceived of autonomy as an idea wider than autopoiesis, applicable to other phenomena exhibiting some form of self-sustained identity through operational closure. In other words, the conditions of self-production and self-distinction could be applied in several different domains, not just the domain of molecular transformations (Varela 1979). Eventually, his contributions to work on immune networks cell assemblies in neuroscience, and his contact with various traditions exploring human experience (phenomenology, meditation, psychoanalysis) led him along a path of refinement and reformulation of some of the assumptions of autopoietic theory (e.g., Varela 1997, 2000; Varela et al. 1991).

There are some important differences between the enactive and the strictly autopoietic conceptions of life. These differences involve dropping some conceptual barriers such as the one separating constructive (internal) and relational (cognitive) phenomena. They also involve the introduction of precariousness as a constitutive element eliminating trivial interpretations of autonomy and necessitating a notion of adaptivity (Di Paolo 2005; Di Paolo and Thompson 2014). Perhaps the most crucial difference is an idea proposed by Varela explicitly toward the end of his life (Weber and Varela 2002): that autonomy as a

[6] "Only after a long series of obstacles surmounted and errors acknowledged did man come to suspect and recognize the autopoetic [sic] character of organic activity" (1965, p. 9). The text on experimentation in animal biology was part of a presentation given by Canguilhem in 1951 in Sèvres, France, and published as part of the collection *La Connaissance de la vie* in 1952 and re-edited in 1965.

property of living and cognitive systems is the grounding of the first layer of teleology and normativity according to which an organism can relate to the world in meaningful terms. In other words, adaptive autonomy, a non-mysterious property of some natural systems, is the condition of possibility, as well as the conceptual ground of sense-making, or simply, mind.

Instead of rehearsing this story, which has been told on other occasions (Di Paolo 2005, 2009; Thompson 2007), I would like here to present the enactive concept of life in a different form. I will approach it as the overcoming of dialectical tensions in the relation between organism and environment.

This relation is indeed where all the fascinating stuff happens, at least for those interested in cognition. The state-determined dynamics of the internal processes that constitute an autopoietic system have been used in the original literature to argue against the possibility of mental representations. In a nutshell, whatever impinges on an autopoietic system has only a triggering, not a formative effect on what subsequently goes on internally within the system. This is determined solely by the internal dynamics (Maturana and Varela 1980, pp. 81, 127). For this reason, all that is possible to say about the internal processes and the relations that the system enters into as a whole is that they have a relation of coherence, meaning that these conditions are not at odds which each other; otherwise the system would stop being autopoietic.[7] This is what is called structural

[7] Because it does not bear on the main thrust of this section, I mention only as an aside an overlooked problem with this kind of reasoning. The notion that the determining role in a state-determined system is played only by the states of that system and that an external perturbation at most triggers a particular chain of internal changes seems quite straightforward from an abstract perspective. And yet we should ask whether this is sufficient for claiming that the triggering perturbation is incapable of playing a formative/structuring role in the system's subsequent changes of state. It obviously could not create these changes all by itself, and the autopoietic argument against a *determining* effect of an external stimulus resulting in a representational token "shaped" or informed solely by the properties of this stimulus still stands (all that matters for this is that internal dynamics are state-*dependent*, not state-*determined*). However, once we consider complex systems far from equilibrium that are materially and energetically open to exchanges with the environment, we must also take into account that such systems can reach states with a variety of types of stability, including critical metastable states which are poised between a few lower energy, relatively more stable options. For some types of complex systems, the evolution toward such critical states is actually likely (Bak et al. 1987). These critical states by themselves, save for fluctuations, will not evolve into any of the lower energy options unless presented with an external trigger. This is precisely the meaning of information used by Gilbert Simondon (2005) in his analysis of individuation processes in nature. One of his recurrent examples is the process of crystallization. A supersaturated solution (the system) will remain in a metastable state until the process of crystallization is initiated by the presence of a seed (the perturbation). During this process there is a passage to a lower energy state "shaped" or in-*formed* by physical characteristics of the seed. Some compounds can potentially crystallize into more than one lattice shape—which lattice is actualized will depend on the seed. In this sense, a trigger is indeed a perturbation and does not in itself carry out the process of crystallization or the determination of the possible options for crystal structures. These virtual alternatives are defined by the system's critical state and the energy and material resources for formation of crystals are also provided by the system. Depending on the particular question of interest, however, it may make sense to speak of an *in-formational* process (Oyama 2000) that occurs at the *encounter* of seed and solution involving not abstract form being bestowed on formless matter (as content on vehicles), but material form (the seed) and being propagated (transduced) in interaction with both actually and virtually pre-formed matter (the critical-state solution).

coupling. The interpretation of the relation between internal and external phenomena that emerges from taking structural coupling as fully characterizing the organism-environment relation is a strong one:

> Systems as composite entities have a dual existence, namely, they exist as singularities that operate as simple unities in the domain in which they arise as totalities, and at the same time they exist as composite entities in the domain of the operation of their components. The relation between these two domains is not causal; these two domains do not intersect, nor do the phenomena which pertain to one occur in the other. (Maturana 2002, p. 12)

There is no possibility of naturalizing cognitive phenomena by using only the notion of structural coupling. It specifies only a minimal condition on the organism–environment relation: an organism is alive as long as it does not enter into destructive relations with the environment. But this minimal condition is insufficient to understand the inner links between life and meaning. In line with the principles of continuity discussed at the beginning, we must seek more specific organism–environment relations that allow—without contradicting structural coupling—a conclusion different from the dualism of non-intersecting domains.

Here it is worth quoting Varela at length in a passage written in 1996 for the preface to the second Spanish edition of the canonical text *De Máquinas y Seres Vivos*:

> One of the criticisms that could be made of this work (as well as of my 1979 book), is that the critique of representation as a guide for explaining cognitive phenomena is replaced by a weak alternative: externality as a mere perturbation of the activity generated by operational closure, and which the organism interprets, be it at the cellular, immune, or neural level. To replace the notion of *input-output* with that of structural coupling was an important step in the right direction because in this way we avoid the classical language trap of making the organism into an information-processing system. But it is a weak formulation because it does not propose a constructive alternative since it leaves interaction in the fog of being a mere perturbation. Often, the critical point has been made, that *autopoiesis*, as formulated in this book, leads to a solipsistic position. Because of what I have just said, I think this criticism has some merit. The temptation of a solipsistic reading of these ideas derives from this: the notion of perturbation during structural coupling does not adequately take into account the regularities that emerge from a *history* of interaction in which the cognitive domain is neither constituted internally (in a way that effectively leads to solipsism) nor externally (as in traditional representational thought). In these last few years I have developed an explicit alternative that avoids these two stumbling blocks, making of historical reciprocity the key of a *co-definition* between an "autonomous" system and its environment. It is what I propose to call the perspective of *enaction* in biology and cognitive science. (Varela 2000, pp. 447–8, original emphasis, author's translation)

It seems that the canonical picture of non-intersecting domains, one corresponding to the phenomena participating in self-construction, the other corresponding to relations entered into by the organism as a whole, is misleading. It is true that the lack of a determining relation between phenomena in one domain on phenomena on the other means that these domains are irreducible to each other. Thus, if we are in possession of all the relational facts between organism and environment, this knowledge will not suffice to predict with certainty how the internal dynamics of the organism will unfold. Conversely, knowledge about all the neural and physiological facts is insufficient for ascertaining the behavioral and perceptual facts likely to ensue in the relational domain. In each case there is a remainder of determination provided by each domain on itself. This is an important systemic lesson, but all it says is that one cannot reduce one domain to the other, not that the domains are non-intersecting, not that they cannot enable and constrain each other. A history of influences between phenomena in each domain may indeed lead to more intimate (partial) co-determinations than simply the minimal mutual tolerance of structural coupling.

Let us consider again the two conditions of the definition of autopoiesis (self-production and self-distinction), paying special attention to what they imply with respect to the organism–environment relation.

The self-production condition specifies that the network of component processes realizes the relations that give rise to the production (or regeneration) of the same processes. Realizing such relations in the real world implies establishing the conditions by which the flows of matter and energy present in the environment can be used in the regeneration of metabolic processes. Let us for a moment imagine what would be the ideal situation in which a form of life could realize these relations. This situation would correspond to the idealized circumstances in which every possible encounter of the organism with the external world produced a positive contribution to autopoiesis and none produced a negative contribution. In other words, if we take self-production on its own, the ideal condition would be one of total openness, such that every possible flow of matter and energy is taken advantage of. Unrealistic as this is, the case is that no relation with the environment would facilitate self-production more than this one if it were possible.

Consider now the self-distinction condition: the autopoietic system constitutes itself as a well-delimited unity with specific topological relations. What would be the relation with the environment that would most ideally realize self-distinction? One of total robustness to any environmental influence, i.e., perfectly shielded boundaries protecting the system. In this case, no possible interaction with the world could possibly put at risk the condition of being a distinct unity, simply because no interaction with the world would have any effect on the system.

There is a primordial tension to this definition of life insofar as the organism-environment relations that best satisfy each of its two conditions tend in exact opposite directions. The tension is well captured by the original split of the autopoiesis definition in two separate conditions. The organism must tend to be self-enclosed to assert its distinctiveness as an individual, but it must also tend to be open to sustain its self-production as a far-from-equilibrium system. In the classical literature, this tension is not further thematized. It is apparently resolved by an offhand clarification: operational closure does not imply material or energetic closure. But where in the world can we expect matter and energy to always flow in or out of a system in the "abstract," i.e., with zero influence on organizational/structural relations? This separation of function and flows is an abstraction aimed at in technological applications. Fuel is supposed to provide pure energy for the car engine and not alter its function. But we know that even in a system specifically designed to approximate this condition as much as possible, this is only an idealization (witness the effects of low octane fuels on uncontrolled ignition in the combustion chamber, leading to engine "knock" and eventually to serious engine damage). In biology especially, most of the matter that flows across the cellular membrane is pre-formed (high-energy compounds, proteins, plasmids, etc.)

It is problematic to say that matter and energy may flow freely across the boundary of the organism, because if this happened it would soon become a violation of the self-distinction requirement due to uncontrolled transformative effects. On the contrary, pre-formed active matter and energy can only flow *conditionally* across the organismic boundary.

Let us clarify how. The ideal organism–environment relation for each requirement in the definition of autopoiesis negates the other requirement. Given this pull of opposites for the organism–environment relation, there is one solution, which is the dialectical overcoming of this tension. A real-world autopoietic system would also need to be a dynamically adaptive one, which by necessity would be open to selected environmental flows (e.g., those that contribute to the condition of self-production) and closed to others (e.g., those that act against the condition of self-distinction). These options are presented schematically in Figure 4.1.

The overcoming of the primordial tension of autopoiesis takes us closer to the enactive conception of life. As we see, this conception is derived from autopoiesis, but provides an alternative interpretation of the possible relations between the constructive domain of production and regeneration, and the domain of organism–environment coupling. These irreducible domains are no longer interpreted as non-intersecting. On the contrary, assuming precarious, far-from-equilibrium conditions, the two requirements of autopoiesis demand opposing ideal situations; they "pull" in different directions as tendency and counter-tendency. This tension is managed over time and the internal and interactional domains relate in ways defined by the adaptive capabilities of the organism. Matter and energy flows do not provide abstract and form-less raw resources; they contribute to sustaining but also possibly modifying the way autopoiesis is realized. These effects are more salient as we observe a history of interactions.[8]

[8] Some of the transformative effects of environmental intercourse on processes of self-individuation are studied in origins of life and early evolution modeling. For instance, Froese, Virgo, and Ikegami (2012) explore possible routes to the origins of metabolism in a model of autocatalytic individuation in an excitable spatial medium. The authors show how reaction-diffusion self-individuated patterns become mobile, even in the absence of chemical gradients, when the spatially individuated pattern corresponding to one autocatalytic reaction "incorporates" another pattern formed by a different autocatalytic reaction. The resulting system becomes spatially asymmetric and spontaneously mobile. It is also well known that bacteria can exchange plasmids (small DNA molecules) horizontally, and even the consumption of certain chemicals can alter metabolic pathways and overall functionality,

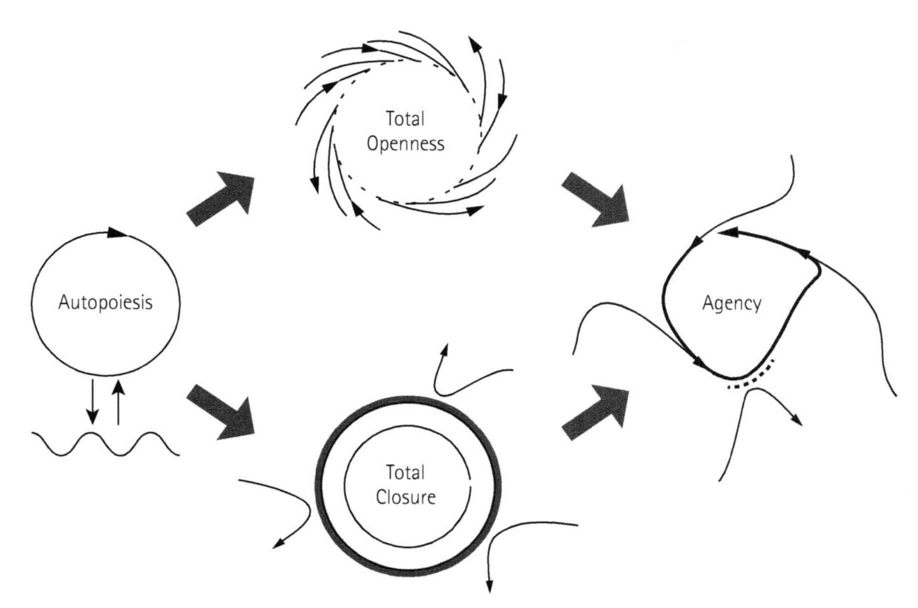

FIGURE 4.1. The primordial tension of life. An autopoietic system is represented on the left by a circle with an arrow closing on itself. The environment is represented as a wavy line and flows of pre-formed matter and energy in coupling with the autopoietic system are indicated with arrows. Under precarious conditions, each of the two requirements for autopoiesis (self-production and self-distinction) leads separately to opposing ideal relations with the environment. If all possible environmental flows could be used for self-production, this would entail an indistinct open system (top). In contrast, a system shielded from environmental flows would be optimal for self-distinction, thus resulting in full isolation (bottom). The organism would not be alive in either of these extreme cases (this is depicted by the open circles). The ideal organism-environment relation for each requirement negates the other requirement, which is why in neither extreme condition can both requirements be met simultaneously. On the right we see the dialectical overcoming of this primordial tension: an adaptive autopoietic system—an agent—able to distinguish and regulate flows that contribute to self-production and self-distinction and avoid flows that act against these conditions. The primordial tension is actively regulated over the time domain. Agency is therefore entailed in any material realization of autopoiesis.

Copyright © 2017 Ezequiel A. Di Paolo, Thomas Buhrmann, and Xabier E. Barandiaran, with permission. Originally published in *Sensorimotor Life: An Enactive Proposal*, Ezequiel Di Paolo, Thomas Buhrmann, and Xabier E. Barandiaran, Figure 5.6, p. 135, Oxford University Press, Oxford, 2017. https://global.oup.com/academic/product/sensorimotor-life-9780198786849). Reprinted here with permission of the authors.

The enactive conception of life expresses in systemic terms the characterization of metabolism offered by Hans Jonas (1966). In looking at the relation of life to matter, he describes it as a "dialectical relation of *needful freedom*" (Jonas 1966, p. 80). He does not specify the need for an adaptive regulation through which the organism evaluates its

for instance, in energy-taxis in bacteria. Egbert, Barandiaran, and Di Paolo (2012) coined the term "behavioral metabolution" to describe some of the evolutionary implications of these processes of transformation in metabolism and behavior through the incorporation of pre-formed matter.

coupling to specific environmental flows, but the need for active regulation is implied in his recognition of the primordial tension of life: materials are essential to the living organism but its identity is dynamic, not tied to the individuation of material constituents but emerging instead as the (risky) ongoing adventure of "riding" material changes "like a crest of a wave" and "as its own [the organism's] feat" (Jonas 1966). This helps us highlight a difference between classical autopoiesis and the enactive view of life. Unlike the former, life in the latter is inherently dynamic and inherently "at risk" because the overcoming of the primordial tension is an ongoing achievement (see also Froese and Stewart 2010, for whom the more static conception of life is connected to the influence of cybernetics on autopoietic theory). Life in the enactive approach is always in a transient, not just empirically, but constitutively as the only way of managing the primordial tension between its own opposing trends.

We should notice that the intrinsic dynamical and relational nature of the enactive conception of life as adaptive, precarious autopoiesis is problematic for functionalism. It is thanks to the ongoing and risky engagement and regulation of the coupling with the world that any identity is sustained and any interiority possible. There is no stopping at stable states in this view of life, no equilibrium, and due to the unpredictable nature of environmental relations, no guaranteed stationarity either. Functional analysis applied to life and mind can only be valid as a limiting case, not as general theory.[9]

Projecting a World

If we add to the minimal condition of structural coupling a condition of adaptive regulation of the organism–environment interaction, as seems necessary for the overcoming of the primordial tension of autopoiesis, then we are on the road toward articulating the historical co-definition between organism and environment that Varela speaks of.

First, it is important to notice that these moves do not imply abandoning the naturalistic approach, nor do they imply sneaking in teleology within a systemic description (as Villalobos 2013 worries). Quite the contrary, by introducing the notion of adaptivity, the enactive approach proposes a bona fide naturalization of teleology and normativity (Di Paolo 2005; Thompson 2007) as well as the notion of agency (Barandiaran et al. 2009). Naturalization would not be valid if we assumed teleology and normativity in the starting postulates, but would fail if these did not appear at some point in our story.

Second, it is also important to highlight that by co-definition we are not simply talking about historical correlation; i.e., the mere concatenation of state-dependent changes whereby the current state depends on the past history of internal changes that have occurred in the autopoietic system, which of course are historically related to

[9] For an enactive/organizational approach to biological function, see the work of Alvaro Moreno and colleagues (e.g., Moreno 2010; Nunes-Neto et al. 2014; Saborido et al. 2011). We should also mention that regulation could act with the effect that some subsystems are able to adaptively sustain a given function or repair it when partially lost.

perturbation events and their internal compensation (e.g., Maturana and Varela 1980, p. 102). The notion of co-definition entails stricter conditions.

To see these points more clearly we need to consider adaptivity in more detail. Adaptivity (Di Paolo 2005, p. 438) is operationally defined as a system's capacity, in some circumstances, to regulate[10] its states and its relation to the environment with the result that, if the states are sufficiently close to the boundary of viability:[11]

1. Tendencies are distinguished and acted upon depending on whether the states will approach or recede from the boundary and, as a consequence,
2. Tendencies of the first kind are moved closer to or transformed into tendencies of the second, and so future states are prevented from reaching the boundary with an outward velocity.

Here we should notice that adaptivity relies on a broader dynamical systems ontology than the one used by autopoietic theory. The latter leans strongly on the notion of a state-determined system according to which the state of any variable in the system is determined only by the previous state of this and other variables in the same system, not by anything external to it. Any external interaction enters into the system as a perturbation, to which the system responds by an internally determined compensatory change. This notion is compatible with the enactive approach but its application differs in two crucial ways. Firstly, while state-*dependence* is always valid, state-*determination* as such is only valid in mathematically autonomous systems, i.e., systems that among other things are not subject to time-dependent couplings. This is not necessarily always the condition in which a living system finds itself. Coupling with other systems introduces all kinds of time-dependent and quasi-regular forms of perturbation (e.g., the rhythms of daylight, the rhythms of the seasons) as well as the emergence in interaction with other organisms of stable patterns of relational dynamics underdetermined by

[10] Here, before the adaptivity requirements are mentioned, it would have been more appropriate to use the term "modulate" instead of "regulate" in the original definition. The difference is subtle. To modulate a system is to alter the conditions of its operation, for instance, by altering parameters or relations to other systems. To regulate a system is to modulate it according to some norm. It is correct to use regulation when speaking of adaptive systems and modulation when speaking in more general terms.

[11] There are different ways to measure viability according to the organism or model in question. In a model of mobile self-producing protocells, Barandiaran and Egbert (2014) distinguish not only viable and nonviable states in the space of essential variables that affect the system, but they also define precarious regions in this space as those states in which the system is not yet dead but in the absence of any environmental changes (e.g., external intervention) the trajectory of its states will inevitably cross the viability boundary. In a related model studying plastic transformation in self-individuated protocells, Agmon, Gates, and Beer (2015) measure the viability of a given configuration as the average number of perturbations it would take for the protocell to disintegrate. In this model, moreover, the authors demonstrate that adaptive transitions can occur in the absence of strict monitoring and regulation of changes in viability, but as an emergent aspect of transformations that tend to increase viability. These two models show two possible examples of how viability space can be modeled and measured.

the participants themselves, which constrain their individual operation (De Jaegher and Di Paolo 2007). The second difference is that enactive theory uses notions of dynamical landscapes and different kinds of stability to operationalize concepts such as tendencies and dispositions in dynamical terms (see Di Paolo 2015). In short, there is an acceptance of the concrete reality of potentialities, virtualities, dispositions, etc., which are articulated in operational concepts such as critical metastable states, gradients, flows, and dynamical landscapes.

A system able to adaptively regulate itself and its coupling with the environment is by necessity a system in which parameters and conditions of operation change over time. This being so, the notion of co-definition is not one whereby systems' *states* become historically correlated, but more strongly, one in which systems' *operating conditions* do. We are then likely to find between organism and environment a relation of mutual shaping, not just a relation of correlated states. In other words, adaptivity operates within the constraints of structural coupling (the organism conserves its autopoietic organization) but it can introduce through a history of interaction additional conditions and coherences between the transformation of both states and conditions of operation in organism and environment. These coherences are not just in the cognitive realm of the observer as autopoietic theory says, but operate as constraints and facilitative conditions that shape the structure of the organism and of the environment (including other organisms) through time. This is how the enactive approach can articulate operationally the notion of *co-definition* described by Varela in the passage quoted earlier.

Thus, adaptive interactions contribute to shaping the organism in fundamental ways. Durable structural/functional effects on the organism related to habit, history of use, and training are too numerous to mention. To name just one striking example, consider the case of underwater vision in Southeast Asian children. Moken children living in the Burma archipelago along the West coast of Thailand routinely dive in search of shells and clams without the use of visual aids. Their underwater acuity is roughly twice as good as that of European children. On land both groups do not differ in acuity or accommodative power. They differ, however, in pupil size: a Moken child's being significantly smaller, which helps to improve acuity underwater (Gislén et al. 2003). The difference is demonstrably associated with the regular performance of underwater visual activities: pupil size reduction and improved underwater acuity can also occur in Swedish children after only one month of training in recognizing visual patterns underwater (Gislén et al. 2006).

Many other clear examples of adaptive historical co-structuring involve social and collective systems. Consider horizontal gene transfer in bacteria whereby plasmids are exchanged that may be incorporated into chromosomes and alter bacterial resistance to antibiotics (Koonin et al. 2001). Or consider the epigenetic effects of mother rats licking and grooming their young offspring. If a rat mother fails to engage in grooming and licking of her pups during the first couple of weeks after birth, those pups will grow to have deficient regulation of acute stress responses (Liu et al. 1997). Licking and other forms of tactile stimulation promote the formation of glucocorticoid receptors (GR) in the pups. This increased expression of GR associated with different brain regions, including the hippocampus, closes a negative feedback loop that helps in the regulation of short-lived stress responses. Insufficient GR make this feedback loop fail and lead to sustained states of high anxiety (which in adult mothers are likely to promote poor contact with pups in the next generation).

The question is whether it is possible to provide explanations of these effects using the notion of structural coupling. The answer is no. The condition of non-lethality of interactions is necessary but insufficient to explain the regularity of these effects. We could simply imagine as viable, for instance, that pupil size in Moken children remains unaffected by frequent underwater diving. Instead, there is in these examples an undeniably adaptive shaping at play, with cumulative historical effects that lead to differentiation in the realization of self-production and self-distinction. An enacted relation with the environment leaves both external and internal historical traces that coherently alter both the domain of organismic construction and the domain of external interactions because each domain constrains and enables the other.

The intimate relation that develops historically between agent and environment is one therefore of *mutual shaping*, which is also manifested on the environmental side. Autopoietic theory puts most of the emphasis on one arc of the organism–environment coupling, the perturbations that impinge on the autopoietic system. But the enactive approach—one could say this is one of the defining features, the one that gives the approach its name—emphasizes the role of action and involvement in the world. The emphasis is on the whole organism–environment coupling (sensorimotor or otherwise).

It is in the environmental consequences of living activity that the organism objectifies its sense-making, both for itself and for other organisms. We see this clearly in the wide variety of examples of niche construction in biology (Odling-Smee et al. 2003), in particular as they involve spatial constraints that enhance historical relations through the path-dependence brought forth by locality (Silver and Di Paolo 2006). Both the increasingly self-differentiated subjective world of the sense-maker and the increasingly objectified properties of the environment partially lose their mutual externality as organisms act (and eventually labor) in a transforming relation of productive activity (see also McGann 2014 for discussion about the need to elaborate an enactive theory of collectively shaped environments).

A history of mutual structuring can also lead to novel forms of extended autonomy. Such is the case of extended physiological circuits in hermatypic coral, the trapping of air bubbles that allow insects and spiders to swim underwater, or the construction of sound amplifying burrows by mole crickets (to name a few of the fascinating examples discussed in Turner 2000).

The picture of mutual co-definition between organisms and environment is even more compelling when we consider life as originating in communities from the very beginning, an issue that we have not discussed here and would deserve a more thorough separate treatment.[12] If environments can be the source of structuring powers, which the organism can to some degree adaptively select to be open or closed to, this is a fortiori the case if we take account of the collective nature of life. Here not only do we find organisms interacting with structuring/structured flows of active matter and energy available in the inorganic world but with objectified biological and historical products, sedimented practices and acts that play the role of signals, symbiotic relations, and even whole other organisms. Historicity is fueled by numbers, as there are clear material and temporal limitations to the effects that a single organism may have in transforming its world.

Conclusion

Returning to Merleau-Ponty's formulation, which we deliberately pushed in a direction that was latent in its language, we can say that a subject projects a world and is itself projected by it, in virtue of how sense-making is constituted by adaptive, historically shaped, organism–environment relations. It is on this world that subjects depend for their continuous existence both as experiencing beings but also, more fundamentally, as living bodies. Varela's long quote echoes this view by speaking of a relation of co-definition.

Our examination of the enactive conception of life, however, emphasizes two aspects that are not explicit either in Varela's or in Merleau-Ponty's formulations. The first is that the constitutive precarious conditions of all life, without which it would not exist as such, demand an ongoing process of organismic individuation which is primordially at odds with itself and as a consequence can only surpass its own tensions dialectically by adaptively regulating its relations with the environment from which it emerges. Sense-making is precisely the opening into the temporal/historical dimension in which viability is made possible by time-managing otherwise unsolvable contradictions. At the fundamental level, enactive bodies are constantly buying time. The adaptive relation, the basis of all forms of cognition—insofar as cognition implies time-oriented subjects

[12] The question of whether some form of fundamental collectivity is implied in the enactive conception of life is an important one, but remains so far unresolved. It seems, at first, plausible that this is not the case. However, this view is based only on the apparent conceivability of the emergence of a singular organism without entering into the detailed conditions of feasibility of such an emergence. It may well be the case that it is impossible to conceive of life arising singularly once these conditions are taken into account. For instance, studies in synthetic biology are beginning to pay attention to the effects of collective protocell interactions on the formation of prebiotic lipid vesicles (e.g., Shirt-Ediss, Ruiz-Mirazo, Mavelli, and Solé 2014). The very conditions for life to exist might imply a constitutive collectivity (like vortices that emerge in a zero angular momentum fluid, or poles of a magnet, it may be the case that adaptive operational closure cannot emerge in the singular). But this is an open question. Suffice it to say the empirical fact that all known forms of life are collective has clear implications for the history of organism–environment co-definition that we are examining here.

capable of caring about impending things and events in the world—is material at its core, as one would expect from a non-dualistic philosophy of nature.

The processes of mutual co-definition between organism and environment, or mutual projection between subject and world, acquire true historical power in their collective dimension. Organisms self-differentiate and produce shared worlds through common paths of interaction. The collective potentiates the structuring powers but also amplifies contradictions like the primordial tension of life, which is manifested as a primordial tension of participatory sense-making (Cuffari et al. 2015; Di Paolo et al. 2018). Life and mind never fully lose their constitutive spontaneity due to the inherent need to always keep active. This is the second enactive emphasis—one that deserves further development—the importance to conceive of the phenomena of life and mind as plural from the start. It may turn out to be that the true protagonists of Varela's relation of co-definition are collectivities of organisms and their common environment, as much as the true protagonists of Merleau-Ponty's formulation are a community of people and their shared history.

Acknowledgments

Thanks to Tom Froese, Hanne De Jaegher, and Marek McGann for helpful comments on previous versions of this chapter. This work is supported by the project "Interidentidad: Identity in Interaction" (MINECO, FFI2014-52173-P).

References

Agmon, E., Gates, A.J., and Beer, R.D. (2015). Ontogeny and adaptivity in a model protocell. In: P. Andrews, L. Caves, R. Doursat, S. Hickinbotham, F. Polack, S. Stepney et al. (eds.), *Proceedings of the European Conference on Artificial Life*. Cambridge, MA: MIT Press, pp. 216–23.

Barandiaran, X. and Egbert, M. (2014). Norm-establishing and norm-following in autonomous agency. *Artificial Life*, 20, 5–28.

Barandiaran, X., Rohde, M., and Di Paolo, E.A. (2009) Defining agency: individuality, normativity, asymmetry and spatio-temporality in action. *Adaptive Behavior*, 17, 367–86.

Bak, P., Tang, C., and Wiesenfeld, K. (1987). Self-organized criticality: an explanation of 1/f noise. *Physical Review Letters*, 59, 381–4.

Bichat, X. (1800). *Recherche physiologiques sur la vie et la mort*. Paris: Brosson, Gabon & Cie.

Buhrmann, T. and Di Paolo, E.A. (2014). Spinal circuits can accommodate interaction torques during multijoint limb movements. *Frontiers in Computational Neuroscience*, 8, 144.

Buhrmann, T. and Di Paolo, E.A. (2017). The sense of agency—a phenomenological consequence of enacting sensorimotor schemes, *Phenomenology and the Cognitive Sciences*, 16, 207–36.

Buhrmann T., Di Paolo E.A., and Barandiaran X. (2013). A dynamical systems account of sensorimotor contingencies. *Frontiers in Psychology*, 4, 285.

Canguilhem, G. (1965/2008). *Knowledge of life*. New York: Fordham University Press.
Cuffari, E., Di Paolo, E.A., and De Jaegher, H. (2015). From participatory sense-making to language: there and back again. *Phenomenology and the Cognitive Sciences*, 14, 1089–1125.
De Jaegher, H. (2013). Embodiment and sense-making in autism. *Frontiers in Integrative Neuroscience*, 7, 15.
De Jaegher, H. and Di Paolo, E.A. (2007) Participatory sense-making: an enactive approach to social cognition. *Phenomenology and the Cognitive Sciences*, 6, 485–507.
De Jaegher, H., Di Paolo, E.A., and Gallagher, S. (2010). Can social interaction constitute social cognition? *Trends in Cognitive Sciences*, 14, 441–7.
Dewey, J. (1938/1991). *Logic: the theory of inquiry*. Carbonsdale and Edwardsville: Southern Illinois University Press.
Di Paolo, E.A. (2005). Autopoiesis, adaptivity, teleology, agency. *Phenomenology and the Cognitive Sciences*, 4, 429–52.
Di Paolo, E.A. (2009). Extended life. *Topoi*, 28, 9–21.
Di Paolo, E.A. (2015). Interactive time-travel: on the intersubjective retro-modulation of intentions. *Journal of Consciousness Studies*, 22, 49–74.
Di Paolo E.A., Barandiaran X.E., Beaton M., and Buhrmann T. (2014). Learning to perceive in the sensorimotor approach: Piaget's theory of equilibration interpreted dynamically. *Frontiers in Human Neuroscience*, 8, 551.
Di Paolo, E.A., Buhrmann, T., and Barandiaran, X.E. (2017). *Sensorimotor life: an enactive proposal*. Oxford: Oxford University Press.
Di Paolo, E.A., Cuffari, E., and De Jaegher, H. (2018). *Linguistic bodies: the continuity between life and language*. Cambridge, MA: MIT Press.
Di Paolo, E.A. and De Jaegher, H. (2012). The interactive brain hypothesis. *Frontiers in Human Neuroscience*, 6, 163.
Di Paolo, E.A., Rohde, M., and De Jaegher, H. (2010). Horizons for the enactive mind: values, social interaction and play. In: J. Stewart, O. Gapenne, and E.A. Di Paolo (eds.), *Enaction: toward a new paradigm of cognitive science*. Cambridge, MA: MIT Press, pp. 33–87.
Di Paolo, E.A. and Thompson, E. (2014). The enactive approach. In: L. Shapiro (ed.), *The Routledge handbook of embodied cognition*. London: Routledge, pp. 68–78.
Egbert, M., Barandiaran, X., and Di Paolo, E.A. (2010). A minimal model of metabolism-based chemotaxis. *PLoS Computational Biology*, 6(12), e1001004.
Egbert, M., Barandiaran, X., and Di Paolo, E.A. (2012). Behavioral metabolution: the adaptive and evolutionary potential of metabolism-based chemotaxis. *Artificial Life*, 18, 1–25.
Froese, T. (2014). Steps toward an enactive account of synesthesia. *Cognitive Neuroscience*, 5, 126–7.
Froese, T., Lenay, C., and Ikegami, T. (2012). Imitation by social interaction? Analysis of a minimal agent-based model of the correspondence problem. *Frontiers in Human Neuroscience*, 6, 202.
Froese, T. and Stewart, J. (2010). Life after Ashby: ultrastability and the autopoietic foundations of biological individuality. *Cybernetics and Human Knowing*, 17, 7–50.
Froese, T., Virgo, N., and Ikegami, T. (2014). Motility at the origin of life: its characterization and a model. *Artificial Life*, 20, 55–76.
Froese, T., Woodward, A., and Ikegami, T. (2013). Turing instabilities in biology, culture, and consciousness? On the enactive origins of symbolic material culture. *Adaptive Behavior*, 21, 199–214.
Fuchs, T. (2011). The brain—a mediating organ. *Journal of Consciousness Studies*, 18, 196–221.

Fuchs, T. (2017). *Ecology of the brain: the phenomenology and biology of the embodied mind.* New York: Oxford University Press.

Gallagher, S. (2015). The problem with 3-year olds. *Journal of Consciousness Studies*, 22, 160–82.

Gallagher, S. (2017). *Enactivist interventions: rethinking the mind.* New York: Oxford University Press.

Gallagher, S., Hutto, D., Slaby, J., and Cole, J. (2013). The brain as part of an enactive system. *Behavioral and Brain Sciences*, 36, 421–2.

Gislén, A., Dacke, M., Kröger, R.H.H., Abrahamson, M. Nilsson, D.-E., and Warrant, E.J. (2003). Superior underwater vision in a human population of sea-gypsies. *Current Biology*, 13, 833–6.

Gislén, A., Warrant, E.J., Dacke, M., and Kröger, R.H.H. (2006). Visual training improves underwater vision in children. *Vision Research*, 46, 3443–50.

Grosz, E. (2011). Matter, life, and other variations. *Philosophy Today*, 55, 17–27.

Jonas, H. (1966). *The phenomenon of life: toward a philosophical biology.* Evanston, IL: Northwestern University Press.

Koonin, E.V., Makarova, K.S., and Aravind, L. (2001). Horizontal gene transfer in prokaryotes: quantification and classification. *Annual Review of Microbiology*, 55, 709–42.

Kreines, J. (2009). The logic of life: Hegel's philosophical defense of teleological explanation in biology. In: F. Beiser (ed.), *The Cambridge companion to Hegel and nineteenth-century philosophy.* Cambridge: Cambridge University Press, pp. 344–77.

Liu, D., Tannenbaum, B., Caldji, C., Francis, D., Freedman, A., Sharma, S. et al. (1997). Maternal care, hippocampal glucocorticoid receptor gene expression and hypothalamic-pituitary-adrenal responses to stress. *Science*, 277, 1659–62.

Malafouris, L. (2007). Before and beyond representation: towards an enactive conception of the palaeolithic image. In: C. Renfrew and I. Morley (eds.), *Image and imagination: a global history of figurative representation.* Cambridge: McDonald Institute for Archaeological Research, pp. 289–302.

Malafouris, L. (2013). *How things shape the mind: a theory of material engagement.* Cambridge, MA: MIT Press.

Maturana, H. (2002). Autopoiesis, structural coupling and cognition: a history of these and other notions in the biology of cognition. *Cybernetics and Human Knowing*, 9, 5–34.

Maturana, H. and Varela, F.J. (1980). *Autopoiesis and cognition: the realization of the living.* Dordrecht, Holland: D. Reidel Publishing.

McGann, M. (2014). Enacting a social ecology: radically embodied intersubjectivity. *Frontiers in Psychology*, 5, 1321.

Merleau-Ponty, M. (1945/2012). *Phenomenology of perception* (trans. D.A. Landes). London: Routledge.

Michelini, F. (2012). Hegel's notion of natural purpose. *Studies in History and Philosophy of Biological and Biomedical Sciences* 43, 133–9.

Moreno, A. (2010). Modularity and function in early prebiotic evolution. *Origins of Life and Evolution of the Biosphere*, 40, 475–7.

Nunes-Neto, N. Moreno, A., and El Hani, C.N. (2014). Function in ecology: an organizational approach. *Biology and Philosophy*, 29, 123–41.

Odling-Smee, J.F., Laland, K.N., and Feldman, M.W. (2003). *Niche construction: the neglected process in evolution.* New Jersey: Princeton University Press.

Oyama, S. (2000). *The ontogeny of information: developmental systems and evolution.* Durham: Duke University Press.

Saborido, C., Mossio, M., and Moreno, A. (2011). Biological organization and cross-generation functions. *British Journal for the Philosophy of Science*, 62, 583–606.

Shirt-Ediss, B., Ruiz-Mirazo, K., Mavelli, F., and Solé, R.V. (2014). Modelling lipid competition dynamics in heterogeneous protocell populations. *Scientific Reports*, 4, 5675.

Silver, M. and Di Paolo, E.A. (2006). Spatial effects favour the evolution of niche construction. *Theoretical Population Biology*, 70, 387–400.

Simon, H.A. (1962). The architecture of complexity. *Proceedings of the American Philosophical Society*, 106, 467–82.

Simondon, G. (2005). *L'individuation à la lumière des notions de forme et d'information.* Grenoble: Millon.

Thompson, E. (2007). *Mind in life: biology, phenomenology, and the sciences of mind.* Cambridge, MA: Harvard University Press

Thompson, E., Palacios, A., and Varela, F.J. (1992). Ways of coloring: comparative color vision as a case study for cognitive science. *Behavioral and Brain Sciences*, 15, 1–26.

Turner, J.S. (2000). *The extended organism: the physiology of animal-built structures.* Cambridge, MA: Harvard University Press.

Varela, F.J. (1979). *Principles of biological autonomy*. New York: North Holland.

Varela, F.J. (1997). Patterns of life: intertwining identity and cognition. *Brain and Cognition*, 34, 72–87.

Varela, F.J. (2000). *El Fenómeno de la Vida*. Santiago de Chile: J.C. Sáez.

Varela, F.J., Thompson, E., and Rosch, E. (1991). *The embodied mind: cognitive science and human experience*. Cambridge, MA: MIT Press.

Villalobos, M. (2013). Enactive cognitive science: revisionism or revolution? *Adaptive Behavior*, 21, 159–67.

Weber, A. and Varela, F.J. (2002). Life after Kant: natural purposes and the autopoietic foundations of biological individuality. *Phenomenology and the Cognitive Sciences*, 1, 97–125.

CHAPTER 5

GOING RADICAL

DANIEL D. HUTTO AND ERIK MYIN

> Make not impossible.
> That which but seems unlike: 'tis not impossible.
>
> Isabella, *Measure for Measure*, Act V, Scene I.

E is the letter, if not the word, in today's sciences of the mind. E-approaches to the mind—those that focus on embodied, enactive, extended, embedded, and ecological aspects of mind—are now a staple and familiar feature of the cognitive science landscape. Many productive scientific research programs are trying to understand the significance of E-factors for the full range of cognitive phenomena, with new proposals about perceiving, imagining, remembering, decision-making, reasoning, and language appearing apace (Wilson and Foglia 2016).

Some hold that these developments mark the arrival of a new paradigm for thinking about mind and cognition, one radically different from cognitive science as we know it (e.g., Thompson 2007; Chemero 2009; Di Paolo 2009; Bruineberg and Rietveld 2014).[1] Others maintain that accommodating E-factors, while important, requires either only very modest twists or, at most, some crucial but still limited revisions to the framework of otherwise business-as-usual cognitive science (Goldman 2012, 2014; Gallese 2014; Clark 2008; Wheeler 2010). By conservative lights, radicals vastly exaggerate the theoretical significance of the so-called E-turn. Moderates hold that whatever changes are required will fall short of reconceiving cognition. Who is right? Are we, in fact, witnessing a revolution in thinking about thinking?

[1] Big claims have been made on behalf of E-approaches. They are said to be "gradually supplanting" their traditional cognitivist competitors (Cappuccio and Froese 2014, p. 3; Stewart et al. 2010). Some go further, attesting that E-approaches have now "matured and become a viable alternative" to traditional representational-cum-computational approaches to cognition, yielding theoretical and methodological advances that "avoid or successfully address many of the fundamental problems" faced by their rivals (Froese and Ziemke 2009, p. 466; see also Froese and Di Paolo 2011).

Revolution in Mind

To assess the scale and magnitude of the theoretical changes that may need to be wrought in order to properly accommodate E-findings, it is important to be clear about which traditional assumptions are being challenged and are at stake. Doing so requires getting clear about the central tenets of classic cognitivism, which has enjoyed the status of default view in the sciences of the mind since the cognitive revolution of the 1950s.

Contemporary cognitivism takes it to be axiomatic that "the mind represents and computes" (Branquinho 2001, p. xv). In doing so it endorses an intellectualist vision of minds that made its debut in early modern times, making representationalism and computationalism the two main pillars of cognitivism.[2]

Today's cognitivism tends to make a further specific assumption about the mechanisms responsible for intelligent activity—viz., that cognitive processes that give rise to such activity take the form of brain-based computations over internal mental contents. In making this particular assumption about the character of representational-computational mechanisms, contemporary cognitivism in its most familiar guise endorses an I-conception of mind that is methodologically and metaphysically committed to individualism, intellectualism, and internalism. Thus, in its standard format, contemporary cognitivism rarely questions the idea that cognition only goes on in the intellectual insides of individuals.

There are well-known limitations to this cognitivist package when it comes to modeling the fluid and plastic nature of cognition. These are especially conspicuous when it comes to trying to account for the intelligence of fast-paced, on-the-fly skilled performance in terms of classic reasoning processes involving the manipulation in-the-head amodal symbols and propositions (Sutton et al. 2011). Even assuming that such reasoning may be tacit, that style of processing is deemed too slow, rigid, and abstractly formatted to properly account for the dynamically updated character of real-time intelligent activity. In this light, the field has sought to characterize the contextualized sensitivity of such intelligence and knowledge to our embodied situations in ways other than positing "a clunky set of internalised propositions" (Sutton and McIlwain 2015, p. 100; see also Dreyfus 2014).

Radically enactive and embodied accounts of cognition, REC, ask us to rethink—root and branch—old-school conceptions of cognition, demanding that we revise our views of the mind's core work and how it gets it done. Like sensorimotor enactivism (O'Regan

[2] Brook (2007), who identifies the historical roots of these ideas, reminds us that "Descartes conceived of the materials of thinking as representations in the contemporary sense. And Hobbes was the first to clearly articulate the idea that thinking is operations performed on representations. Here we have two of the dominating ideas of all subsequent cognitive thought: the mind contains and is a system for manipulating representations" (p. 5). Chomsky (2007), a principal architect of the most recent cognitive revolution, recognizes it as a "second" cognitive revolution that built upon the previous cognitive revolution of the early modern era; he is also clear about this historical debt.

and Noë 2001; Noë 2004), REC understands cognition as something that organisms do. Cognition is a kind of embodied activity that is out in the open, not a behind-the-scenes driver of what would otherwise be mere movement. REC understands basic forms of cognition as extensive and dynamically loopy processes that are responsive to information in the form of environmental variables spanning multiple temporal and spatial scales (Chemero 2009; Bruineberg and Rietveld 2014).

Crucially, REC conceives of cognition in terms of unfolding, world-relating processes rather than in terms of content-bearing states and their interactions. Cognitive processes, unlike states, have reach and unfold over time. Importantly, unlike a state or event, a process is "something which goes on through time and can change as it does so" (Steward 2016, p. 76).[3]

When it comes to explaining intelligent activity, REC advocates the equal partner principle. Accordingly, appeals to neural factors in such explanations stand on the same footing as appeals to bodily and environmental factors. On this score, REC disagrees with conservative views of cognition that hold that only neural factors (or an extended set of neural-plus-other-specified-factors as per extended functionalism) are properly cognitive factors. Such theories hold that only truly cognitive factors, however exactly they are demarcated, make a special contribution to explaining intelligent activity and imbue it with mind.

An example of a RECish explanation that honors the equal partner principle is found in the constraints-led approach to skill acquisition, which draws heavily on Gibson's (1979) ecological psychology. The constraints-led approach assumes a tight fit between animals and their environment; perception is understood as being fundamentally bound up with and for action. Perceiving is a matter of getting a grip on the world as opposed to representing it. The constraints-led approach augments these core ideas from ecological psychology by calling on dynamical systems theory (Davids et al. 2008; Chow et al. 2011, 2015). Combining these two frameworks, as in Chemero (2009), has opened the way for fruitfully investigating complex self-organizing responsiveness of learners acquiring skills in embodied activities.

Putting all of this together, from a constraints-led approach perspective the processes involved in the mastery of embodied skills can be characterized by a number of interacting variables and the continuous, temporal, and interdependent changes in their unfolding patterns can be expressed in terms of laws captured by mathematical equations. Importantly, although it is possible to focus on the dynamics of the parts of the wider system for various purposes, the basic unit of analysis for ecological dynamics is the nonlinearly coupled organism–environment system.

Crucially, from the vantage point of a constraints-led approach, individuals are understood as situated dynamical systems that are simultaneously open to influence and

[3] In making this important adjustment REC invites questions about cognition that only make sense when it is understood along processual lines, such as "Why, for example, did the process unfold in this way rather than that? Why has it not stopped? What is sustaining it and keeping it going? What is responsible for any regular patterns we may observe in its progression? And so on" (Steward 2016, p. 78).

intervention on multiple scales. Training is focused on the self-organizing antics of such dynamical systems, which are always open to reconfigurations enabling them to self-organize quickly and flexibly to the contextual demands (Kelso 1995). Therefore trainers working within this paradigm selectively modify specific bodily, environmental, and task constraints—for example, changing the size of playing fields, adjusting distances between players, fatiguing players—to shape and control the emergence of skills and expertise over time (Newell 1986; Hutto and Sánchez-García 2015).

Those who favor REC seek to explain skilled performance in terms of embodied activity that involves dynamic processes that span brain, body, and environment. Accordingly, cognitive processes are not, for example, conceived of as mechanisms that exist only inside individuals. Instead they are identified with nothing short of bouts of extensive, embodied activity that takes the form of more or less successful organism-environment couplings. Likewise, embodied skills are acquired and emerge as a consequence of a history of interactions between learners and their embedding environments in ontogeny and phylogeny. Through sustained context-sensitive active engagements with worldly offerings, organisms are changed in ways so as to be able to, in Clark's (2015) apt formulation, get "a grip on the patterns that matter for the interactions that matter" (p. 5).

In sum, through adjusting and attuning to the world over time, in complex and nested ways, organisms enactively evolve their most fundamental cognitive capacities. Their tendencies for interaction—their patterns of sensitivity and responsiveness—alter across many and various spatial and temporal scales at once. Attention is a pivotal driver of these processes of adjusted sensitivity and refined responsiveness. To which features of their environments organisms attend and, more fundamentally, how they attend in full-bodied active ways shifts over time, and this leaves in its wake a variety of structural changes. How organisms respond to a range of features thereby constrains their further interactions and makes its mark in the form of adaptive changes—neural, bodily, and environmental—altering organismic capacities and dispositions.

Through these processes of organism–environment adjustment the weights of neural connections change and are recalibrated. But relevant changes can also be located in the environment, such as when modification to the spatial arrangement of objects or created artifacts alters the tendencies of how organisms interact with their worlds. In this way, "the technologies, objects, rituals, and practices of the material/social world are intimately intertwined with the many different ways in which we might remember a time in our past" (Campbell 2006, p. 363).

As just noted, REC acknowledges that cognitive capacities at least in part depend on structural changes inside of organisms. But REC is studiously austere in the way it understands the character and basis of such changes—in particular, it steers clear of casting them in information processing and representational terms. Does REC's restraint on this front put it at odds with standard explanations of how, say, neural changes in memory are wrought by the most basic forms of learning? The simple answer is: yes and no.

Breaking with Tradition

Can an adequate account be given in REC terms of the neural, biochemical, and genetic changes that contribute to the most elementary form of learning in memory? At first glance, there seems no difficulty here if we attend only to the big lessons of the current state of the art. Consider Eric Kandel's Nobel Prize–winning research on learning and memory. In seeking to understand how learning produces changes in the human brain, Kandel focuses on simple animal models that he assumes obey the same basic principles of neuronal organization that apply to humans. This choice is justified to the extent that "elementary forms of learning are common to all animals with an evolved nervous system" (Kandel 2001, p. 1030, 2009).[4]

The full story of the relevant neural changes connected with learning, even in the simplest of creatures, is enormously complicated. Telling it would involve a foray into a great many tricky details, not only of neuroscience but also biochemistry and genetics. For our purposes it suffices to highlight two major take-home lessons from Kandel's groundbreaking work. The first is that the special memory-enabling capacities of the hippocampus are not due to the intrinsic properties of its neurons, since all nerve cells have similar properties, but from "the pattern of functional interconnections of these cells, and how those interconnections are affected by learning" (Kandel 2001, p. 1030). The second is that "learning results from changes in the strength of the synaptic connections between precisely interconnected cells . . . [such that] experience alters the strength and effectiveness of these preexisting chemical connections" (Kandel 2001, p. 1032). On the face of it, these lessons are entirely congenial to REC.

However, there's a catch. For in framing his work, Kandel tells us that "To tackle that problem we needed to know how sensory information about a learning task reaches the hippocampus and how information processed by the hippocampus influences behavioral output" (Kandel 2001, p. 1030). This is perhaps to be expected given that Kandel set out with the express aim to contribute to a new synthesis that would combine "the mentalistic psychology of memory storage with the biology of neuronal signaling" (p. 1030). Such an ambition is hardly surprising in today's climate. Scientific research on memory in general is rife with the free-and-easy use of the language of "memory traces," of "encoded and retrieved information," and of "the storage and retrieval of information and representations."

Despite the popularity of these familiar metaphors, close inspection of how they operate in science reveals them to have serious limitations—limitations that make them

[4] As Rosenberg (2014) emphasizes in discussing Kandel's research, "The difference between humans and rats and sea slugs is of course proportionately larger numbers of neurons in more complicated circuits are involved. . . . [Yet] what is going on in all three cases is just input-output wiring and rewiring" (p. 27).

prime candidates for theoretical explication or elimination (see Roediger 1980). For example, as De Brigard (2014) observes:

> "Storing" is a rather misleading term. What seems to occur *when we encode information* is the strengthening of neural connections due to the co-activation of different regions of the brain, particularly in the sensory cortices, the medial temporal lobe, the superior parietal cortex, and the lateral prefrontal cortex. *During encoding*, each of these regions performs a different function depending on the moment in which *the information gets processed*. A memory trace is the dispositional property these regions have to re-activate, when triggered by the right cue, in roughly the same pattern of activation they underwent during *encoding*. (De Brigard 2014, p. 169, emphases added)

REC assumes that De Brigard's analysis in the previous quotation is mostly correct, sans its commitment to the highlighted claims that information is encoded and processed. REC seeks to explain basic forms of learning and memory entirely in terms of re-enacted know-how.

Following Kandel's lead and focusing solely on the simpler kinds of procedural memory widespread in the animal kingdom, remembering can be understood as the capacity to re-enact embodied procedures, often prompted and supported by external items. Memory of this sort is a kind of knowing what to do in familiar circumstances. It is surely not necessary to posit stored mental contents in order to explain the dispositional basis of such capacities (Ramsey 2007, ch. 5). The brain's underlying contribution to such capacities "turns out to be just a matter of either organizing extant synaptic circuits in new wiring patterns or switching on genes in neurons that produce new synapses. . . . The brain does everything without thinking *about* anything at all" (Rosenberg 2014, pp. 26–7, emphasis in original). Importantly, purely embodied know-how is not grounded in or mediated by any kind of knowledge; rather it can be understood as the overall responsiveness of a complex system that is laid down through habit and past experience (Barandiaran and Di Paolo 2014).

This REC account of basic memory can be told, without gaps, so long as no appeal is made to the encoding and processing of information or representations. Insisting on this crucial edit to the standard picture is pivotal for understanding what motivates the REC framework. According to the standard cognitivist account, information is supposed to be picked up via the senses through multiple channels, encoded and then further processed and integrated in various ways, allowing for its later retrieval.

The crucial question is: does an information-processing story add anything of explanatory value to the radically reductionist account of memory offered by Kandel—that given in terms of how experience modulates neural connections and weights? If not, then the information-processing gloss is explanatorily superfluous. But if the information-processing account does add explanatory value, then what exact additional contribution does it make, and how precisely do its explanations work?

There is a deep theoretical problem—which Hutto and Myin (2013) dub "the hard problem of content"—that gives us reason to be skeptical of the standard informational processing story. The problem arises from the fact that the notion of information that can be most easily called upon to do serious explanatory work and provide the details of the information-processing story is that of information-as-covariance. According to that notion of information, a state of affairs is said to carry information about another just in case it lawfully covaries with that other state of affairs to some specified degree. The go-to example is the age of a tree covarying with the number of its rings. Information in this sense is perfectly objective and utterly ubiquitous—it literally litters the streets. Moreover, this notion of information has impeccable naturalistic credentials: it is used in many sciences and thus can clearly serve the needs of a cognitive science with explanatorily naturalistic ambitions.

In trying to fill out the standard information-processing story, cognitivists face a dilemma. In dealing with the first horn, they can try to give a naturalistically respectable explanation of information encoding and processing by appeal to the notion of information-as-covariance. Yet if that is the only notion of information at play in cognitivist theorizing then it is difficult to understand what it could possibly mean for information to be literally encoded. How can relations that hold between covarying states of affairs be literally "extracted" and picked up from the environment so as to be "encoded" within minds?

Perhaps it will be said that what should be focused on here is not the medium but the message. Sometimes the information-processing story is told in quasi-communicative terms of signaling and receiving messages. Yet how seriously should we take these analogies and the talk of encoding and decoding "messages"?

Again there are grounds for caution. Despite its widespread popularity, attempts to seriously explicate the nature of neural or mental "codes" and their alleged encoded content are few and far between. Goldman gives a frank appraisal of the current situation: "There is no generally accepted treatment of what it is to be . . . a mental code, and little if anything has been written about the criteria of sameness or difference for such codes. Nonetheless, it's a very appealing idea, to which many cognitive scientists subscribe" (Goldman 2012, p. 73).

Moreover, how should we understand the nature and source of the putative contentful messages? The notion of information-as-covariance is surely not able to help us to understand how sense perception supplies the mind with messages—informational contents—that can be encoded and decoded, conveyed and communicated.[5] Once again we lack any respectable scientific account of how to understand the idea that cognition is literally a matter of trafficking in such informational contents.

At this juncture it would be natural for cognitivists to try instead to deal with the second horn of the dilemma. They might try to call on some other naturalistically

[5] Friends of cognitivism readily acknowledge that "no philosopher has ever claimed that covariation by itself constitutes or confers content" (Shapiro 2014a, p. 216; see also Matthen 2014, pp. 124–5; Miłkowski 2015, p. 78). Yet they fail to draw the full consequences from this admission.

respectable notion of information that will enable them to tell their encoding and processing stories in full detail.[6] Telling a different tale requires identifying an alternative notion of information with sound naturalistic credentials that can do the precise, additional explanatory work required in order to make good on the standard information-processing story. As things stand, it is at best unclear which if any candidate notion of information has the right characteristics to play such a role.

In the light of this analysis, it becomes clear that the "storage" metaphor is not the only, or even the most, problematic card in the cognitivist deck, pace De Brigard. All of the familiar metaphors relating to the way cognitivists talk about the processing of information—certainly, any that rely on picturing information as some kind of commodity or abstract contentful message—generate equally deep and serious scientific mysteries. Such mysteries need dispelling, one way or another—they want explaining or explaining away.

Less Can Be More

Famous cases in the history of science teach us that sometimes less is more. The need for and adequacy of explanantia are re-assessed during conceptual revolutions in science as and when the nature of explananda is reconceived. The evolution of our thinking about what is required for explaining motion provides a shining example. Before a correct understanding of inertia was achieved in classical physics, rest was thought to be the natural state of bodies. Because of this assumption the initiation and continued motion of objects were treated as primary explananda. Explanations of the forces that moved objects from rest and kept them in motion were thought to be always necessary even in the basic cases. In Aristotelian and medieval frameworks, respectively, such explanations were given in terms of the medium or an impetus. Such proposals were plagued by internal theoretical problems. They also impeded progress by providing apparently compelling reasons to reject Copernican heliocentrism (Dijksterhuis 1961). It is only with a major and hard-won shift in theoretical perspective, in which unaccelerated linear motion came to be understood as the natural state of objects, that it became clear that the problematic explanations previously demanded for it were entirely hollow and unnecessary. This is a clear instance in which, by reconceiving what needed to be explained, and removing the demand for distracting and misguided explanations, thinking was liberated and barriers to progress were removed.

Like those involved in the early debates about the basis of object motion, cognitivists are prepared to overlook deep theoretical problems in their proposals on the grounds that they believe the type of account they demand is the only kind that can meet the special explanatory needs of the case. Consequently, they hold that nonrepresentationalist

[6] See Piccinini 2015, p. 30, for details of other possible ways of understanding information.

accounts of intelligent behavior only appear credible if one systematically underestimates what those explanations require.

Conservatives hold that once the relevant explanatory needs are made transparent, it becomes evident that only the explanatory posits of the sort cognitivists supply can properly account for "the stunning successes" of cognitive science (Shapiro 2014 p. 214; see also Burge 2010). Prima facie, this assessment looks plausible enough given the current state of the art: "Examination of research on memory, attention, and problem-solving . . . leaves little doubt that cognitive scientists are neck deep in representational commitments" (Shapiro 2014, p. 218).

The master intuition behind these assessments, concerning the special features of the explanandum, is that cognizers manage to concretely represent abstract properties. Apparently, it is this explanatory need that rules out the possibility that intelligent behavior might be explained by contentless structural changes that are selected for through the history of an individual's interactions with worldly offerings. Something more is still needed, so the persistent intuition says, and that something more must be a content derived from experiential encounters and carried by a discrete representational vehicle that modulates behavior.

The driving assumption is that abstract properties must be literally derived from a diverse array of perceptual inputs such that flexible responsiveness to such properties is controlled by discrete items that "stand in" for them. The key is that abstract properties need to be "re-coded into simple, useable objects" (Colombo 2014a, p. 16).

Clark and Toribio (1994) is the locus classicus for a contemporary version of the argument that the representation of abstract properties is needed to explain at least some cognitive tasks (see Degenaar and Myin 2014). Summing up the conclusion of that argument, positing representations is needed to understand how we manage to think about "states of affairs that are unified at some rather abstract level, but whose physical correlates have little in common" (Clark 1997, p. 167). To cognize in such ways requires:

> that all the various superficially different processes are first assimilated to a common inner state or process such that further processing can then be defined over the inner correlate: an inner item, pattern, or process whose content then corresponds to the abstract property. (Clark 1997, p. 167)[7]

Such distilled representational items are assumed to modulate a host of behavioral responses, as Matthen (2014) makes clear with his example of a dog that has learned to expect its food at 5 p.m., and as a consequence has a multiplicity of dispositions—sitting by its bowl, bothering its mistress, whining, and so on—the manifestation of which depends on further contextual factors, such as whether its mistress is present or

[7] In the quoted passage, Clark (1997) talks about reasoning, but Clark and Toribio (1994) make it clear that such reasoning is to be understood in a very broad sense. For example, keeping track of and behaviorally anticipating something not present is taken to be an instance of "reasoning about the absent" (see Clark and Toribio 1994, p. 419).

absent. Matthen holds that "a single learned association controls all of these behaviors. This argues for discrete representation. The whole organism *cannot* just be rewired so that it incorporates all of these context-dependent behaviors" (p. 121, emphasis added).

By this reasoning it can come to seem, by sheer intuitive insistence, that representations of some sort must feature in the best explanations of how cognizers "track abstract properties across situations" (Colombo 2014a, p. 16).[8]

The putative work that discrete, concrete representations mediating perception and action do is twofold; they are meant to distill what is encountered through learning and thereby modulate complex possible behaviors. Those who frame matters in this way will feel an irresistible need to posit such unifying, unitary representations. It can seem, given familiar lines of reasoning, that there must be concrete mediators of perception and behavior in the form of discrete representations that bear abstract contents.

The conclusion seems unavoidable. Crucially, however, this proposal raises precisely the worries about the information-processing story outlined earlier. The attempt to explain how information processing and contentful representations add explanatory value in such cases encounters the insuperable problems already mentioned. There are seemingly intractable problems when it comes to explaining in a naturalistically illuminating way how information or content can be literally distilled from environmental interactions into concrete representational vehicles without assuming some unexplained leap from concrete interactions to an abstract representation. Overcoming those problems in a straight way requires providing the relevant details that make clear how abstract properties can be made concrete.

Is there another option? As Drefyus (2014) intimates, "In spite of the authority and influence of Plato and 2,000 years of philosophy, . . . one must be prepared to abandon the traditional view" (Dreyfus 2014, p. 30). In other words, we can escape the hard problem of content by letting go of the idea that learning from specific cases requires abstracting and interiorizing contents.

A Radical REConceiving

Conceptual revolutions are rare, to be sure. Even so, REC has all the hallmarks of being bona fide revolutionary given that it presses for "the replacement of a whole system of concepts and rules by a new system" (Thagard 1992, p. 6). Like other major revolutions in thought, REC does not force us to jump from one conceptual branch to another while staying within the same familiar tree—it demands that we switch to a new tree altogether.

[8] Elsewhere we suggest that it amounts to adopting a totemic theory of mind, since totems too are concrete figures that represent what spiritually binds together otherwise disparate things (Hutto and Myin 2014).

In doing away with the idea that content is a defining feature of basic cognition, REC undermines the most foundational notions of representational cognitive science. Going radical the REC way is to abandon the information-processing and representationalist views of cognition in favor of a purely embodied know-how.

It follows that "some enactivists do not mean by 'cognition' what traditionalists have meant by 'cognition'" (Aizawa 2014, p. 40). Aizawa (2014) draws the important conclusion from this in stressing that "by adopting a new conception of cognition . . . [revolutionaries] have detached themselves from the traditions of cognitive science" (Aizawa 2014, p. 40).

Moreover, Aizawa crucially observes that REC does "not so much solve traditional problems, as merely walks away from them" (Aizawa 2014, p. 22). That is exactly right. Certain conceptual problems do not warrant straight solutions; they warrant dissolution by rethinking the background framework commitments or assumptions that bring them into being and make them seem, at once, both intractable and yet unavoidable.[9]

Walking away is precisely what REC recommends when it comes to dealing with framework-dependent questions such as how to understand how information can be literally acquired, stored, and processed in brains. Yet while this is so, Aizawa (2014) overstates the case in claiming that this means RECers "use different tools to study different issues" (p. 21). It is clearly true that REC proposes a different framework for thinking about cognition and different sets of tools for studying cognitive phenomena than employed by cognitivists. Nevertheless, RECers target the very same phenomena of interest to cognitivists—e.g., perceiving, imagining, remembering—albeit in doing so they significantly REConceive the nature of those phenomena.

In sum, REC avoids theoretical difficulties by making a shift from conceiving the fundamental job of cognition as "accurately representing an environment to continuously engaging that environment with a body so as to stabilize appropriate coordinated patterns of behavior" (Beer 2000, p. 97). REC differs from its close conservative cousin, sensorimotor enactivism, in making a firm move away from all forms of representational thinking about basic cognition. RECers deny that all cognition, and in particular its root forms, involve informational processing and contentful representations.

Importantly, REC's decision to go radical is not unmotivated. In providing a diagnosis of the deep theoretical difficulties associated with the hard problem of content, REC—at least as advanced in Hutto and Myin (2013)—supplies something of special importance to those who are attracted to making a break with cognitivism.

According to Aizawa (2014, 2015), most defenders of REC do not do enough to motivate its acceptance. He finds a dearth of arguments. Aizawa (2015) readily admits that "if one understands 'cognitive processes' as behavioral processes, then of course, 'cognitive

[9] In the surrounding discussion, Aizawa (2014) hits the nail on the head again: "One might dissolve the problem or abandon the problem, if one rejects the traditional concept, but one cannot solve it" (Aizawa 2014, p. 40). See Hutto (2013a) for further discussion of how deeply held framework assumptions can, at once, generate impossible conceptual problems as well as apparently irresistible explanatory needs.

processes' are typically realized in the brain, body, and world" (p. 762). But those who make this claim typically provide no argument in support of the identification.

Aizawa (2015) observes, for example, that Maturana offers only speculative pronouncements in "an oracular tone" (p. 760) and that Chemero too gives "no reason in support of his revisionary interpretation. . . . There really is no argument there" (Aizawa 2015, p. 764). In a nutshell, Aizawa (2014) is correct to say that "one *cannot argue* that cognition is embodied and extended, by observing that behavior is embodied or extended" (p. 40, emphasis added).

Of course, the "no arguments for REC" claim is exaggerated. For example, one can find plenty of arguments through the years in the collected work of Dreyfus (2014), such as the infamous frame or infinite regress arguments invoked there. Nor is it true that all defenders of REC have failed to note "that cognition has generally been proposed to be a cause of cognition . . . [and failed to provide] reasons or evidence against it" (Aizawa 2015, p. 759). Concerns about the causal impotency of content are well known and have been used by RECers to provoke a shift in our thinking (see, e.g., Hutto 2013c).

Still, as arguments go, the hard problem of content provides the basis for an especially tough one. If not by endorsing REC, how else might cognitivists handle it?

Handling the Hard Problem

Taking the radical REC line is motivated by a desire to provide a complete and gapless naturalistic account of cognition, right here, right now. REC predicts that, when combined with other E-resources, appeal to scientifically respectable contentless notions of information-as-covariance and the norms of biological functionality offer all that is needed for understanding basic minds.

All explanatory naturalists competing to understand basic cognition must face up to the hard problem of content, or HPC, one way or another. As noted, answering the HPC would require explaining how it is possible to get from informational foundations that are non-contentful to a theory of mental content using only the resources of a respectable explanatory naturalism. Adequate explanations of this kind have systematically eluded us.

Unlike phenomenal consciousness, it used to be said that mental content neither posed any "deep metaphysical enigmas" (Chalmers 1996, p. 24) nor "any deep philosophical difficulty" (Strawson 1994, p. 44). For the longest time such was the orthodox view of analytic philosophers of mind. Indeed, in the 1980s and '90s not only did it seem possible that a workable naturalized theory of mental content might be in the cards, it looked to be just around the corner. In those days the race was on to be the first to cross the finish line. Hence, as Kriegel (2011) observes, "A generation ago . . . finding a place for intentionality in the natural order—'naturalizing intentionality'—consumed more intellectual energy than virtually any other issue in philosophy" (p. 3). Nowadays, however, that research program has all but petered out.

Of course, a putting-down of tools is what we should expect if someone has already crossed the finish line—namely, if a workable naturalized theory of content were currently in hand. Some believe this is just what happened. For example, Miłkowski (2015) tells us that the HPC "has already been solved" (p. 74); that it was "solved a couple of dozen years ago" (p. 74); that it was "already solved by Dretske (and Millikan, and Fodor, and Bickhard, by the way)" (p. 78); it was solved "at least in principle . . . [by] . . . various accounts . . . [that] usually recruit a similar solution" (p. 83). Like the French soldiers who taunt Arthur's knights in *Monty Python and the Holy Grail*, some defenders of Castle Cognitivism clearly assert there is no need to seek a solution to the HPC because "we've already got one."

What was the solution? Obviously, it required relying on something more than the notion of information-as-covariance. Miłkowski (2015) holds that other scientifically respectable notions of information were successfully called upon for understanding cognition, namely, information-as-control and information-as-structural-similarity (2015, p. 76). Such proposals, or something near enough, are found in O'Brien and Opie's (2015) understanding of mental representation in terms of structural similarity. Those authors argue that the content of an analog representational vehicle is just the structural resemblance holding between that vehicle and its object. Yet as Miłkowski (2015) acknowledges, rightly in our view, even if these richer notions of information and information processing do play a role in understanding cognition, they do not constitute content; rather, they are "are necessary but not sufficient for representation" (p. 82). So, on their own, appeal to such notions cannot be the solution to the HPC.

The crowning move that allegedly solved the HPC was the invocation of the notion of teleological function by the likes of Dretske and Millikan (Miłkowski 2015, p. 83). That move is ultimately what accounts for the special kind of normativity needed to understand content, and it does so by appealing to a biological notion of function (see Millikan 2005). Or rather, as Miłkowski (2015) puts it, quite ambiguously, something like this move is what allegedly already solved the HPC, "one way or another" (p. 84).

Despite Miłkowski's (2015) confidence, there are serious reasons to doubt that teleosemantics did the required work. There is widespread consensus among those with remarkably diverse philosophical predilections and agendas that teleosemantics fundamentally fails to deliver an adequate theory of content. The big problem is that even if it is allowed that biological function entails some kind of normativity—e.g., such that it implies that organismic responses can be misaligned with respect to certain features of the world that they target—the kind of normativity supplied falls a good distance short of what is required to explain how an organism comes to have mental contents with specified truth conditions.

This verdict on the shortcomings of teleosemantic accounts is repeatedly voiced. We are warned that "Evolution won't give you more intentionality than you pack into it" (Putnam 1992, p. 33); that there is a crucial distinction between "functioning properly (under the proper conditions) as an information carrier and getting things right (objective correctness or truth)" (Haugeland 1998, p. 309); that "natural selection does not care about truth; it cares about reproductive success" (Stich 1990, p. 62).

In sum, the problem with teleosemantics, as discussed at length in Hutto and Myin (2013, ch. 4), is that it fails to account for intensionality (with an s). As such, it is in no position to account for truth-conditional content. Put otherwise, "No amount of environmental appropriateness of a neural state or its effects is fine-grained enough to give unique propositional content to the neural state" (Rosenberg 2014, p. 26).

Nor do these worries evaporate if we swap a notion of mental content cast in terms of truth conditions for a weaker notion of content cast in terms of accuracy or veridicality conditions. The same basic problem recurs: "Evolution does not care about veridicality. It does not select for veridicality per se" (Burge 2010, p. 303). Fundamentally, as Burge diagnoses it, what's wrong with any attempted teleosemantic solution to the HPC is that there is "a root mismatch between representational error and failure of biological function" (Burge 2010, p. 301).

Returning to Kriegel's (2011) assessment, it would seem, in this light, that the true explanation for the cessation of attempts to produce the required naturalistic theory of content is gloomier and far from triumphalist. In fact, "The naturalizing intentionality research program bears all the hallmarks of a degenerating research program. . . . [It] has run up against principled obstacles it seems unable to surmount. Far from being technical, the problems just mentioned are fatal" (Kriegel 2011, pp. 3–4).

Another way to avoid the HPC is simply to adopt a pluralist and metaphysically noncommittal, anti-realist take on the nature of scientific explanation. According to those who advocate going this way:

> Truth or existence is not a necessary condition for theoretical posits like representations to be legitimate epistemological/methodological tools. . . . So, even if . . . cognitive systems are not representational systems at all . . . representationalism can still be successfully defended, and "representation-talk" can still be justifiably preserved. (Colombo 2014b, p. 271)

Accordingly, even if a naturalistic account of mental content is never forthcoming—even if the HPC is never solved—"little hangs on the matter" (Colombo 2014b, p. 271). Indeed, nothing does. This can be seen from the fact that the anti-realist approach is so liberal it allows that—despite their logical incompatibility—cognitivitist and REC explanations can both valuably illuminate the very same phenomena, at the same time.

Sprevak's (2013) observations about fictionalism reveal why anti-realist ways of avoiding the HPC are generally unattractive. Sprevak notes that going fictionalist is a way of avoiding having to make the unpalatable choice of either dealing with the HPC or going radical, à la REC, and thus revising the theory and practice of cognitive science. In this respect going fictionalist, prima facie, "offers a neat way out" (Sprevak 2013, p. 540).[10]

[10] The fictionalist approaches discussed here must not be confused with the anodyne view that talk of content may have practical or instrumental value talk even if we don't take it seriously in terms of metaphysics. For "one benefit of this talk is that it provides us with a way of referring to the (real) neural causes of behaviour" (Sprevak 2013, p. 555). Thus, even if attribution of content to cognitive states is systematically false, it can still provide a way of labeling mental states "and hence of keeping track of them" (Sprevak 2013, p. 555).

But there is heavy price to pay. Fictionalists, like all anti-realists, break the links between truth, existence, and explanation in ways that make it unclear just what kind of explanatory value is yielded by citing theoretical posits. Fictionalists, for example, must hold that a revealing explanation "can be provided by a fiction just as well as by truth" (Sprevak 2013, p. 556). Fair enough. Still, defenders of this family of views owe us an account of the precise nature of these explanatory offerings.

Some possibilities can be ruled out in advance. In general, anti-realist explanations cannot be causal in character. For example, if mental content does not really exist it cannot feature in causal explanations because only real entities can be causes. Although deciding on the correct analysis of causation and causal explanation remains a philosophically vexed matter, one thing everyone can agree on is that for something to be a cause it must exist. This presents a particular problem for fictionalism about content, because it seems that contents must exist and make a real difference if they are to ground relevant fictions about content. If fictionalists try to deny the grounding relation, they are left with the problem of explaining how there can be fictions at all. This problem is quite serious on the assumption that fictions must be contentful (see Sprevak 2013, p. 553).[11]

The foregoing analysis reveals that anti-realism will only provide a tenable way of dealing with the HPC if we are provided with an independently compelling account of how anti-realist explanations explain.[12]

RECers agree with their realist-minded cognitivist opponents that questions of metaphysics matter in science. Thus, in rejecting an overly liberal pluralism, Fodor states: "There are lots of issues that a sufficiently shameless philosopher of mind can contrive to have both ways but not the issue between [REC] pragmatists and [cognitivist] Cartesians" (Fodor 2008, p. 12).

So how do realist defenders of cognitivism hope to handle the HPC? Some admit that the problem is real and serious, but strike an optimistic attitude about the prospects for closing its explanatory gap in due course, or they simply deny that closing that epistemic gap is necessary. Optimistic realists assume the HPC will solve itself; that the metaphysics will come out in the wash or that the naturalness of content can be taken for granted.

Thus in discussing this line of argument, Shapiro (2014) outlines two options—one more agnostic than the other—maintaining that:

> Perhaps Hutto and Myin are correct that *no* extant theory of content succeeds, but this fact does not preclude future success.
>
> (1) Why not take the challenges to naturalized theories of content that they pose to be a call for further work on naturalizing content, or further reflection on what it means for content to be natural? . . .

[11] Sainsbury (2010) questions the general logic of fictionalism along these very lines. In a nutshell, Sainsbury's complaint is that in trying to make metaphysical savings by assuming that the ontology of some region of discourse is fictional, "No progress towards nominalism has been made. . . . Mathematics is a fiction, a story. But what is a story?" (p. 2)

[12] McDowell (1994/1998) claims that explanations involving subpersonal contents have an "enormous capacity for illumination" despite being "irreducibly metaphorical" (p. 349). But he fails to make it clear how this can be so (see Hutto 2013b for a critique).

(2) Perhaps one should look at struggles to naturalize content as misguided from the start, and see the tremendous gains of cognitive science as themselves sufficient to establish that content—whatever it is—is natural. (Shapiro 2014, p. 218)

The first option requires the HPC to be solved in due course. The assumption is that mental representations exist and their contentful properties are what really and truly explain the successes of cognitive science, and we will come to know more about how this all works someday. Assuming the metaphysics will one day be shown to be all in order, they say, "Find the right architecture of mind, and the naturalizing strategy will follow. . . . This is why . . . arguments about naturalizing content almost completely irrelevant at this stage of the inquiry" (Matthen 2014, p. 126).[13]

This may seem all well and good. But until a solution to the HPC is in hand, we can't know whether the metaphysics of content is all in order. Consequently, we can't know now that contentful properties do the relevant explaining. The history of science is littered with cases of theoretical posits that proved to be nonexistent but were deemed to successfully explain certain phenomena. Fictionalists, as we have seen, try to account for this by breaking the assumed links between existence and explanation; but committed realists cannot go fictionalist at this juncture. Consequently, no realist is currently in the epistemically privileged position to justifiably claim that contentful properties are what best explain the successes of cognitive science.

This analysis reveals that current cognitivist theorizing incurs massive debts against the future. What we can know, right now, is that the claimed explanatory power of cognitivist theories is in hock: any current explanatory power they may possibly have is mortgaged against future theoretical developments that may not come about. Optimistic cognitivist realists are placing a bet—a bet on which bookmakers would be forgiven for offering only long odds, given the history of failure in attempts to naturalize content.

The second option does not require the HPC to be solved, ever. Instead it takes it as an article of metaphysical faith that there is no fundamental tension about how content can be natural even if we never manage to figure out how it could be so. Going this way would be to adopt a kind of metaphysical mysterianism about content. The trouble with the mysterian line is that such realism is difficult to rationally motivate. How can we be confident that mental content plays a part in the stunning successes of cognitive science if we are—*forever*—debarred from understanding how content could play such a part in securing those successes?

[13] Despite making this claim, it is unclear to what extent Matthen (2014) actually endorses an optimistic realism as opposed to some version of anti-realism. He vacillates in what he says on this matter. On the one hand, he speaks of a multiplicity of modules that communicate by sending and receiving signals, remarking that "on natural assumptions, these communications have a semantics" (p. 123). On the other hand, he also remarks, "I am not suggesting that this kind of agentive talk should be taken literally. My point is that it provides a design perspective on the machine without you cannot comprehend the setup" (p. 122).

Another option is out-and-out, utterly barren content eliminativism. Flatly denying all intentionality is, for true nihilists, the only rational conclusion to be drawn from "all the unsuccessful programs of research [that failed] . . . to provide a non-circular, let alone a naturalistic account of content" (Rosenberg 2015, p. 456).

The scorched earth approach of such thoroughgoing eliminativism generates its own deep questions. Indeed, it leaves us trading one mystery for another. Anyone who claims that cognition is entirely a matter of contentless computations—anyone who allows that content falls out of the equation entirely, and offers no successor notion—will be unable to explain how organisms relate to and connect with the specific worldly offerings of their environments. Any theory of this extremely austere sort will be woefully ill-placed to explain the array of findings that give us reason to think that cognitive activity is deeply influenced by E-factors.

This may explain why there have been so few advocates of content-free or computation-only accounts of cognition (Stich 1983; Piccinini 2008). Even computationalists who are skeptical about the existence of a deep metaphysical link between computation and content tend to hold a more subtle position. They allow that even if computations are not essentially individuated by semantic properties—even if computations have a wholly non-semantic and entirely mechanistic nature—they can still be sensitive to semantic properties (Rescorla 2012, 2014; Piccinini 2015).

Why so? The reason this is the preferred view is clear enough. Such theorists feel compelled to assume that "There are . . . semantic properties that relate many computing systems to their environment—they relate internal representations to their referents—and are encoded in the vehicles in such a way that they make a difference to the system's computation, as when an embedded system computes over vehicles that represent its environment" (Piccinini 2015, p. 32). We have been at pains to show that paying for that assumption requires facing up to the HPC in one or the other ways described earlier.

REC offers a different way out. It avoids the HPC by promoting a revolutionary shift in standard thinking about mind and cognition. REC's rejection of the I-conception of mind, but also the twin representational and computational pillars of cognitivism, is motivated by the avoidance of deep theoretical mysteries. But avoiding those mysteries requires more than just tinkering at the edges of the cognitivist thinking about the basic character and architecture of mind or shifting our views on how cognitive processes function, interact, and unfold. REC asks us to REConceive and RECast our understanding of what cognition is, of how it works and of what it does. It asks us to fundamentally adjust how we think about minds.

REC questions whether, on close inspection, there is any need to posit any kind of content in order for the sciences of the mind to do their fundamental explanatory work. On the positive side, REC recommends getting by with something less—a replacement, contentless notion of intentionality (see Hutto 2008, ch. 3, 2011; Hutto and Myin 2013, 2017; Hutto and Satne 2015 for more on this teleosemiotic replacement notion).

Consequently, REC's strategy must not be confused with more extreme eliminativist ways of dealing with the HPC. REC does not propose simply biting the bullet and surrendering the idea that content is needed when it comes to understanding cognition without offering anything in its place.

Conclusion

Importantly, as with other major conceptual revolutions, should we succeed in radicalizing our conception of cognition and avoid what has been a theoretical hindrance in the old framework, we can still retain what is of value from that tradition by RECtifying it. RECtifying our thinking about the basic nature of cognition in this way will assist in unifying what is best in cognitivism and other E-approaches within a single framework. The hope is that pursuing such a program of work will yield many hard-to-predict philosophical, scientific, and practical fruits down the line.

Some, like Shapiro (2014), doubt that surrendering cognitivist thinking about basic minds is a live option for cognitive science. He advises against letting go of notions of informational processing and representational content. For him, REC goes too far: he doubts that its thoroughgoing radicalism will take our thinking about cognition "to the next step" (p. 215). The jury is still out on that question. But given its revolutionary ambitions, Shapiro is certainly correct to say that REC is not simply "a more ferocious breed" of E-account, but "a different animal altogether" (p. 215).

Yet for all its ferocious radicality, it is important to recognize that REC is a package deal that seeks to accentuate the positive and eliminate the negative without messing with what's in-between. In this respect, its avoidance of the HPC is not as extreme as nihilistic eliminativist theories that do so by giving up any idea of intentionality altogether.

References

Aizawa, K. (2014). The enactivist revolution. *Avant*, 5(2), 19–42.

Aizawa, K. (2015). What is this cognition that is supposed to be embodied? *Philosophical Psychology*, 28(6), 755–75.

Barandiaran, X. and Di Paolo, E. (2014). A genealogical map of the concept of habit. *Frontiers in Human Neuroscience*, 8, 522. doi:10.3389/fnhum.2014.00522

Branquinho J. (2001). *The foundations of cognitive science*. Oxford: Oxford University Press.

Brook, A. (2007). Introduction. In: A. Brook (ed.), *The prehistory of cognitive science*. Basingstoke: Palgrave Macmillan, pp. 38–66.

Bruineberg, J. and Rietveld, E. (2014). Self-organization, free energy minimization, and optimal grip on a field of affordances. *Frontiers in Human Neuroscience*, 8, 1–14.

Burge T. (2010). *The origins of objectivity*. Oxford: Oxford University Press.

Campbell, S. (2006). Our faithfulness to the past: reconstructing memory value. *Philosophical Psychology*, 19(3), 361–80.

Cappuccio, M. and Froese, T. (eds.) (2014). *Enactive cognition at the edge of sense-making.* London: Palgrave Macmillan.

Chalmers, D. (1996). *The conscious mind.* Oxford: Oxford University Press.

Chemero A. (2009). *Radical embodied cognitive science.* Cambridge, MA: MIT Press.

Chomsky N. (2007). Language and thought: Descartes and some reflections on venerable themes. In: A. Brook (ed.), *The prehistory of cognitive science.* Basingstoke, UK: Palgrave Macmillan, pp. 38–66.

Chow, J.-Y., Davids, K., Button, C., and Renshaw, I. (2015). *Nonlinear pedagogy in skill acquisition: an introduction.* London: Routledge.

Chow, J.-Y., Davids, K., Hristovski, R., Araújo, D., and Passos, P. (2011). Nonlinear pedagogy: learning design for self-organizing neurobiological systems. *New Ideas in Psychology,* 29, 189–200.

Clark, A. (1997). *Being there: putting brain, body and world together again.* Cambridge, MA: MIT Press.

Clark A. (2008). Pressing the flesh: a tension in the study of the embodied, embedded mind? *Philosophy and Phenomenological Research,* 76, 37–59.

Clark, A. (2015). Embodied prediction. In: T. Metzinger and J.M. Windt (eds.), *Open MIND,* 7(T). Frankfurt am Main: MIND Group, pp. 1–21. doi:10.15502/9783958570115

Clark, A. and Toribio, J. (1994). Doing without representing? *Synthese,* 101(3), 401–31.

Colombo, M. (2014a). Explaining social norm compliance: a plea for neural representations. *Phenomenology and the Cognitive Sciences,* 13(2), 217–38.

Colombo, M. (2014b). Neural representationalism, the hard problem of content and vitiated verdicts: a reply to Hutto and Myin. *Phenomenology and the Cognitive Sciences,* 13(2), 257–74.

Davids, K., Button, C., and Bennett, S. (2008). *Dynamics of skill acquisition.* Champaign, IL: Human Kinetics.

De Brigard, F. (2014). Is memory for remembering? Recollection as a form of episodic hypothetical thinking. *Synthese,* 191(2), 1–31.

Degenaar, J. and Myin, E. (2014). Representation-hunger reconsidered. *Synthese,* 191(15), 3639–48.

Dijksterhuis, E. (1961). *The mechanization of the world picture.* New York: Oxford University Press (trans. C. Dikshoorn of Dijksterhuis, E. (1950). *De Mechanisering van het* wereldbeeld. Amsterdam: Meulenhoff).

Di Paolo, E. (2009). Extended life. *Topoi,* 28(1), 9–21.

Dreyfus, H.L. (2014). *Skillful coping: essays on the phenomenology of everyday perception and action* (ed. Mark Wrathall). Oxford: Oxford University Press.

Fodor, J.A. (2008). *LOT 2.* Cambridge, MA: MIT Press.

Froese, T. and Di Paolo, E.A. (2011). The enactive approach: theoretical sketches from cell to society. *Pragmatics and Cognition,* 19(1), 1–36. doi:10.1075/pc.19.1.01fro

Froese, T. and Ziemke, T. (2009). Enactive artificial intelligence: investigating the systemic organization of life and mind. *Artificial Intelligence,* 173, 466–500.

Gallese V. (2014). Bodily selves in relation: embodied simulation as second-person perspective on intersubjectivity. *Philosophical transactions of the Royal Society of London. Series B, Biological sciences* 369, 20130177. doi:10.1098/rstb.2013.0177

Gibson, J.J. (1979). *The ecological approach to visual perception.* Boston: Houghton Mifflin.

Goldman A.I. (2012). A moderate approach to embodied cognitive science. *Review of Philosophy and Psychology,* 3(1), 71–88.

Goldman, A.I. (2014). The bodily formats approach to embodied cognition. In: U. Kriegel (ed.), *Current controversies in the philosophy of mind*. London: Routledge, pp. 91–108.

Haugeland, J. (1998). *Having thought: essays in the metaphysics of mind*. Cambridge, MA: Harvard University Press.

Hutto, D.D. (2008). *Folk psychological narratives: the sociocultural basis of understanding reasons*. Cambridge, MA: MIT Press.

Hutto, D.D. (2011). Elementary mind minding, enactivist-style. In: A. Seemann, (ed.), *Joint attention: new developments in philosophy, psychology, and neuroscience*. Cambridge, MA: MIT Press, pp. 307–41.

Hutto, D.D. (2013a). Enactivism: from a Wittgensteinian point of view. *American Philosophical Quarterly*, 50(3), 281–302.

Hutto, D.D. (2013b). Fictionalism about folk psychology. *The Monist*, 96(4), 585–607.

Hutto, D.D. (2013c). Exorcising action oriented representations: ridding cognitive science of its Nazgûl. *Adaptive Behavior*, 21(1), 142–50.

Hutto, D.D. and Myin, E. (2013). *Radicalizing enactivism: basic minds without content*. Cambridge, MA: MIT Press.

Hutto, D.D. and Myin, E. (2014). Neural representations not needed: no more pleas, please. *Phenomenology and the Cognitive Sciences*, 13(2), 241–56.

Hutto, D.D. and Myin, E. (2017). *Evolving enactivism: basic minds meet content*. Cambridge, MA: MIT Press.

Hutto, D. D. and Sánchez-García, R. (2015). Choking RECtified: embodied expertise beyond Dreyfus. *Phenomenology and the Cognitive Sciences*, 14(2), 309–31.

Hutto, D.D. and Satne, G. (2015). The natural origins of content. *Philosophia*, 43(3), 521–36.

Kandel, E. (2001). The molecular biology of memory storage: a dialogue between genes and synapses. *Science*, 294, 1030–8.

Kandel, E. (2009). The biology of memory: a forty-year perspective. *Journal of Neuroscience*, 29, 12748–56.

Kelso, S. (1995). *Dynamic patterns*. Cambridge, MA: MIT Press.

Kriegel, U. (2011). *The sources of intentionality*. Oxford: Oxford University Press.

Matthen, M. (2014). Debunking enactivism: a critical notice of Hutto and Myin's radicalizing enactivism. *Canadian Journal of Philosophy*, 44(1), 118–28.

McDowell, J. (1994/1998). The content of perceptual experience. *The Philosophical Quarterly*, 44(175), 190–205 (Reprinted in 1998 in *Mind, value and reality*. Cambridge, MA: Harvard University Press, pp. 341–58. Page references from reprinted version).

Miłkowski, M. (2015). The hard problem of content: solved (long ago). *Studies in Logic, Grammar and Rhetoric*, 41(1), 73–88.

Millikan, R.G. (2005). *Language: a biological model*. Oxford: Oxford University Press.

Newell, K.M. (1986). Constraints on the development of coordination. In: M.G. Wade and H.T.A. Whiting (eds.), *Motor development in children: aspects of coordination and controls*. Amsterdam: Martinus Nijhoff Publishers, pp. 341–61.

Noë, A. (2004). *Action in perception*. Cambridge, MA: MIT Press.

O'Regan, J.K. and Noë, A. (2001). A sensorimotor account of vision and visual consciousness. *Behavioural and Brain Sciences*, 24, 939–1031.

Piccinini, G. (2008). Compuation without representation. *Philosophical Studies*, 137(2), 205–41.

Piccinini, G. (2015). *Physical computation: a mechanistic account*. Oxford: Oxford University Press.

Putnam, H. (1992). *Renewing philosophy*. Cambridge, MA: Harvard University Press.

Ramsey W.M. (2007). *Representation reconsidered*. Cambridge: Cambridge University Press.
Rescorla, M. (2012). Are computational transitions sensitive to semantics? *Australasian Journal of Philosophy*, 90(4), 703–21.
Rescorla, M. (2014). The causal relevance of content to computation. *Philosophy and Phenomenological Research*, 88(1), 173–208.
Roediger, H.L. (1980). Memory metaphors in cognitive psychology. *Memory and Cognition*, 8(3), 231–46.
Rosenberg, A. (2014). Disenchanted naturalism. In: B. Bashour and H. Muller (eds.), *Contemporary philosophical naturalism and its implications*. London: Routledge, pp. 17–37.
Rosenberg, A. (2015). The genealogy of content or the future of an illusion. *Philosophia*, 43, 537–47.
Sainsbury, R.M. (2010). *Fiction and fictionalism*. London: Routledge.
Shapiro, L. (2014). Radicalizing enactivism: basic minds without content (Review). *Mind*, 123(489), 213–20.
Sprevak, M. (2013). Fictionalism about neural representations. *The Monist*, 96, 539–560.
Steward, H. (2016) Making the agent reappear: how processes might help. In: R. Altshuler and M. Sigrist (eds.), *Time and the philosophy of action*. London: Routledge, pp. 67–84.
Stewart, J., Gapenne, O., and Di Paolo, E. (eds.) (2010). *Enaction: toward a new paradigm for cognitive science*. Cambridge, MA: The MIT Press.
Stich, S. (1983). *From folk psychology to cognitive science*. Cambridge, MA: MIT Press.
Stich, S. (1990). *The fragmentation of reason: Preface to a pragmatic theory of cognitive evaluation*. Cambridge, MA: MIT Press.
Strawson, G. (1994). *Mental reality*. Cambridge, MA: MIT Press.
Sutton, J. and McIlwain, D.J.F. (2015). Breadth and depth of knowledge in expert versus novice athletes. In: J. Baker and D. Farrow (eds.), *The Routledge handbook of sport expertise*. London: Routledge, pp. 95–105.
Sutton, J., McIlwain, D., Christensen, W., and Geeves, A. (2011). Applying intelligence to the reflexes: embodied skills and habits between Dreyfus and Descartes. *Journal of the British Society for Phenomenology*, 42(1), 78–103.
Thagard, P. (1992). *Conceptual revolutions*. Princeton, NJ: Princeton University Press.
Thompson, E. (2007). *Mind in life: biology, phenomenology and the sciences of the mind*. Cambridge, MA: Harvard University Press.
Wheeler, M. (2010). In defence of extended functionalism. In: R. Menary (ed.), *The extended mind*. Cambridge, MA: MIT Press, pp. 245–70.
Wilson, R.A. and Foglia, L. (2016). Embodied cognition. In: E.N. Zalta (ed.), *The Stanford encyclopedia of philosophy* (spring 2016 ed.). Retrieved from http://plato.stanford.edu/archives/spr2016/entries/embodied-cognition/

CHAPTER 6

CRITICAL NOTE

So, What Again is 4E Cognition?

KEN AIZAWA

In principle, one might expect each of the papers in this section, "What is Cognition?," to present some 4E answer to what cognition is. Perhaps this would be a definition of "cognition," or a theory of what cognition is, or a conceptual framework that articulates what the concept of cognition is. Nevertheless, in the chapters in this section, as in the 4E literature more generally, the question of what cognition is does not come to the forefront. Moreover, even when the question is taken seriously, the answers do not seem to be worked out in much detail.

Before turning to the 4E treatments of cognition, it is worthwhile reviewing the way the question is handled by cognitivism, wherein the general outlines of an account are relatively familiar. (Hutto and Myin also see merit in beginning with cognitivism.) This provides a model of an answer one might give to what cognition is.

To begin with, cognitivists draw a distinction between cognition and behavior. While one might wonder exactly what behavior is, just as one might wonder what exactly cognition is, there is a broad consensus that behavior is the product of endogenous and exogenous factors. So, consider solving the famous Tower of Hanoi puzzle. The puzzle consists of three upright dowels with multiple doughnut-shaped disks of different sizes stacked on one of the three rods. The puzzle is to move all of the disks from one dowel to another subject to three simple rules:

1. Each disk must be moved individually.
2. Only the top disk of a stack can be moved.
3. No disk can be placed on top of a smaller disk.

Various factors, endogenous and exogenous, influence how individuals behave when solving this problem; in other words, many factors in and around the organism influence performance on this task. Exogenous factors, such as the lighting, the size of the disks, and background noise influence performance. To take extreme examples, it can be harder to solve the problem in the dark, and very large, heavy disks can be hard to manipulate. Endogenous factors, such as fatigue, attention, and (most importantly) cognition can also influence performance. So, while there may be problems with characterizing cognition and behavior, it is typically accepted that cognition is among the endogenous factors influencing behavior.

In general, cognitive psychologists study cognition by developing tests that they hope will reveal features of the underlying cognitive processes. (See Shapiro in press, for a nice account of a cognitivist study of memory.) Cognitivism is, in the first instance, a theory regarding one endogenous factor influencing behavior. There are, of course, differences of opinion among mainstream cognitivists about the character of these endogenous processes, but, in the main, cognitivists are committed to the empirical hypothesis that cognitive processes are inferential, computational processes over representations.[1] In addition, it is common to suppose that these representations have non-derived content; that is, they do not bear their contents based on social conventions, but based on some set of natural (i.e., non-semantic) conditions. This is a "bare bones" sort of cognitivism articulated for the 4E literature in Adams and Aizawa (2008), but cognitivists often include other hypotheses. So, for example, it is sometimes proposed that cognition involves a "language of thought," a system of syntactically and semantically combinatorial representations (see, e.g., Fodor and Pylyshyn 1988; Aizawa 2003). And sometimes there is thought to be a distinction between perceptual processing, on the one hand, and ("higher") cognitive processing, on the other (see, e.g., Fodor 1983; Pylyshyn 2003). And cognitivism often embraces the existence of something answering to the concept of "innate knowledge" (see, again, Fodor 1983).

Cognitivists generally take the existence of inferences and representations as empirical hypotheses, since the hypotheses are meant to explain certain psychological phenomena. Among the phenomena thought to be explained by inferences and representations is the ability to acquire a natural language. The proposal is that the so-called "primary linguistic data" to which young children are exposed do not fully specify the details of the grammar that children ultimately acquire. Children must use the sentences, and indeed non-sentences, to which they are exposed to infer the grammar of the language to which they have been exposed. English speakers must figure out that "Kangaroos have not piloted a spaceship to the sun" is a grammatically correct English sentence, even though they may never have heard it before, while "want food," is not a grammatically correct English sentence, while they have probably never heard it before and have probably never been told that it is not grammatical.

Essentially every part of the cognitivist view has been challenged in the 4E literature. So, for example, Clark (2005), doubts the very idea of non-derived content. Hutto and Myin (2013, 2018), rehash the consensus view that the naturalized semantics project of the 1980s and 1990s did not succeed. (For a contrary view, see, e.g., Ryder 2004 and

[1] Hutto and Myin (2018) correctly pick out these elements of cognitivism. Interestingly, while Rupert (2009) seems to be very much in touch with contemporary cognitive psychology, he does not seem to embrace the view that cognitive processes are inferential processes over representations.

Miłkowski 2015.) Gibson (1979) doubted the need for mental inference in perception. Despite the challenges, there is at least a reasonably widespread understanding of what cognitivists think cognition is.

Turn now from the question of what cognition is to the question of where in the world it might be found. In principle, one might think—as do most cognitivists—that cognition, as they conceive it, is localized in the brain. Alternatively, one might think—as do some advocates of what is sometimes called "extended functionalism"—that cognition, as cognitivists conceive it, is (sometimes) localized in the brain, body, and world. So, for example, Rob Wilson's "wide computationalism" proposes that computational cognition might be realized in the brain, body, and world, and, in one collaborative paper with Wilson, Andy Clark appears to embrace this conception of cognition (cf. Wilson 1994; Wilson and Clark 2009, but see also Clark 2015, Wheeler 2010, and Kiverstein, 2018, p. 19). So, to suppose that cognition is a form of computation over representations in no way begs the question about where in the world cognitive processing is to be found. Nevertheless, as we shall now see, none of the contributors to this section of the book adopt the cognitivist conception of cognition.

Di Paolo

In his contribution to this volume, Di Paolo (2018) embraces the idea that there is some need for a "mark of the cognitive." He writes:

> Theories of cognition should be able to provide the operational conceptual categories with which to describe their objects of study and distinguish them from those outside their remit. They should be able to say in concrete terms what sort of system, event, or phenomenon counts as cognitive and in which cases it does not. Accounts that do not meet this mark are pre-scientific. (p. 75)

One might, of course, think it is too much to say that lacking an account of the mark of the cognitive renders a theory pre-scientific or that what is needed are "operational conceptual categories," but for present purposes all that matters is that Di Paolo accepts the importance to enactivism of the mark of the cognitive question.

Move on, then, to Di Paolo's enactivist account of cognition. He proposes that "ongoing, open-ended, precarious processes are logically necessary elements in what makes a system cognitive" (p. 76). A logically necessary element would appear to be less than an operational conceptual category or enough to distinguish cognition from noncognition, but we can nevertheless consider what has been proposed. So, we have a proposal regarding a feature of cognition, but why think it is true? Why are ongoing, open-ended precarious processes logically necessary elements of cognitive systems? Di Paolo begins with an "empirical answer": it is "the nature of all known forms of cognitive systems . . . [to] grow, develop, adapt to unforeseen circumstances, and seem to have an open-ended (though not unconstrained) reserve of potentialities, which we have no reason to assume are all pre-given at birth" (p. 76). But isn't it in the nature of all known forms of cognitive system to be massive? So, why isn't being massive a logically necessary element in what makes a system cognitive? Moreover, suppose it is in the nature of cognitive systems to have ongoing and open-ended precarious processes. How does Di Paolo get from this empirical hypothesis to the view that it is *logically necessary* that cognitive systems have ongoing and open-ended precarious processes? He hints that it has to do with how the enactive story unfolds (p. 76), but does not elaborate. But perhaps Di Paolo does not really mean "logically necessary." Maybe that was a bit of an exaggeration. Or maybe Di Paolo had some other notion of necessity in mind.

Or maybe Di Paolo did not mean this at all. Later, he proposes a somewhat different formulation: "To be a cognitive entity is to be a (generally) non-stationary organization in a (generally) non-stationary relation with the world" (p. 77). (The notion of ongoing and open-ended, precarious processes no longer appears.) But aren't there noncognitive organisms that have a (generally) non-stationary organization in a (generally) non-stationary relation to the world? One might think so. Moreover, Di Paolo seems to assume that not all life-forms are cognitive systems: "Synoptic formulas such as life = cognition have had their provocative initial impact diluted by their implausibility according to reasonable interpretations. Do such statements mean that life is coextensive with cognition? Are we performing a cognitive operation when, say, we digest our lunch? Are all currently living species equally "cognitive" since they are all equally alive?" (p. 74). So, this seems to open Di Paolo's view to the challenging question: why are there no noncognitive organisms that have a (generally) non-stationary organization in a (generally) non-stationary relation to the world? Indeed, Di Paolo's theory seems to fail to meet a standard he apparently sets for himself: "Theories of cognition should be able to provide the operational conceptual categories with which to describe their objects of study and distinguish them from those outside their remit" (p. 75).

Kiverstein

Kiverstein's chapter discusses non-enactivist approaches to embodied cognition. Kiverstein (2018) rightly recognizes that some 4E theorists embrace a vaguely mainstream version of computational cognition. These are the "wide computationalists" or "extended functionalists" mentioned earlier. He distinguishes this type of 4E theory from so-called "radical embodied cognitive science" (RECS) articulated by, for example, Chemero (2009), and Hutto and Myin (2013). Kiverstein also does a valuable job of picking up on a strand of RECS thinking that does not seem to be much in evidence in Hutto and Myin 2018. Kiverstein takes it that radical embodied cognitive science:

> claims that we find extended cognitive processes whenever the variables that describe one system are also the parameters that determine change in the other system,

> and vice versa. In such a system, it is only as a matter of explanatory convenience that we treat the agent and its environment as separately functioning systems. In reality the dynamics of the two systems are so tightly correlated and integrated that they are best thought of as forming a single extended brain–body–world system. (2018, pp. 20–1; cf. Chemero 2009, Box 2.1, pp. 31–2)

Part of what makes Kiverstein's discussion valuable is that, while he is generally sympathetic to the 4E project, he picks up on a line of thought that one might find uncharitable in the mouth of a critic.

So a critic might well suspect that the foregoing sketch must be incomplete. Presumably there is a difference between those reciprocally influenced dynamical systems that are cognitive and those that are not cognitive. To make the point more concrete, turn to the well-known example that Kiverstein uses, the Lotka–Volterra model of the population dynamics of a single-predator and single prey (see also Shapiro 2013). This is a system in which the parameters mutually influence each other. Is that supposed to be an instance of extended cognition? Presumably not. Or what about coupled pendulums? Are they extended cognitive systems? Presumably not. But if they are, is this supposed to be the result of a terminological stipulation? Or is there an argument for the view that all reciprocally coupled dynamical systems are instances of extended cognition?

Assume, for the sake of argument, that a single-predator, single-prey system is not a cognitive system. Suppose that something has been left out, that some further condition is needed. The options for RECS seem to be limited. The "single extended brain–body–world system" narratives typically involve an agent in continuous reciprocal causal (CRC) relations with something in the environment. Such narratives often suppose that somehow the mutual influence or the continuous reciprocity can finesse the coupling-constitution fallacy. (See, for example, Chemero 2009; Palermos 2014; for discussion of the coupling-constitution fallacy, see Adams and Aizawa 2008, ch. 6, 7.) But this seems to presuppose that there is something about an agent that makes the agent a "source" of cognition, so that when the agent couples to, say, a notebook, in the right way through continuous reciprocal causation, cognition extends. But this would seem to require a mark of the cognitive that an agent alone (or perhaps the agent's brain?) could satisfy. But this leaves us where we started in searching for a REC theory of what cognition is. What is it about the agent that makes it a cognitive agent? Moreover, attempting to say how an agent or an agent's brain might realize cognitive processes does not appear to be consonant with other elements of the RECS approach. Note that Chemero claims that "radical embodied cognitive science can explain cognition as the unfolding of a brain-body-environment system" (cf. Chemero 2009, p. 43). Invoking an argument from causal coupling to a constitutive claim about a cognitive system seems to go against the grain of other elements of the RECS approach.

The tensions between the CRC idea and the idea of cognition as brain–body–world unfolding might well give advocates of RECS the further idea that they should abandon the arguments based on continuous reciprocal causation. A further reason to think

CRC should go is that it does not appear to do the work that Chemero, Kiverstein, and others seem to want it to do. Return to the Lotka–Volterra model which is supposed to give us a single unified system, a system that is not decomposable into other systems. Suppose, for the sake of argument, that it is. Nevertheless, as Kiverstein notes, there are components and these components undergo processes. Predators starve or eat prey. Prey die. Right? The Lotka–Volterra model does not show that there are no such processes in components. Right? So, suppose we have a dynamical systems model of Otto and his notebook. Let this be a unitary system. Why couldn't Otto's brain be a component with cognitive processes? Maybe there is an answer to this question, but it is a question that has not even been asked in Chemero (2009) or Hutto and Myin (2013).

It is important not to shoot the messenger here. Kiverstein rightly appreciates the importance of having an account of what cognition is: "The central claim of this chapter has been that to resolve the debate about extended cognition we will need to come up with a mark of the cognitive. We will need to say what makes a state or process count as a state or process of a particular cognitive kind" (p. 36). Nevertheless, as part of following through on this, Kiverstein has what seems to be the unenviable task of presenting the RECS theory of cognition. Kiverstein accurately, I believe, reports on one portion of what he finds and tries to present it sympathetically, but cannot paper over the difficulties. There are apparent gaps in the RECS use of dynamical systems and CRC as an account of cognition.

Rietveld, Denys, and van Westen

Rietveld, Denys, and van Westen (2018) do not say much about cognition (just that their framework does not mark a distinction between "higher" and "lower" cognition) and they say nothing about what they take cognition to be. Instead, they articulate what they call the skilled intentionality framework (SIF) as a way of understanding skilled human action: "The SIF research program is to understand the entire spectrum of skilled human action, including social interaction, creativity, imagination, planning, and language use in terms of skilled intentionality" (p. 42). In the very broadest of terms, the approach takes skilled human action to be a matter of tending toward an optimal grip on one of many possible affordances.

The focus of SIF framework apparently differs from that of cognitivism, on the one hand, and of Chemero, Hutto, and Myin's REC, on the other. SIF is concerned with skilled human action. By contrast, cognitivism is a theory of one of the putative endogenous factors that contribute to skilled human action. So, return to the example of solving the Tower of Hanoi problem. Consider how cognitivists view this task. Newcomers to the task lack the skilled actions required to solve the problem. As they become skilled, however, there are many endogenous changes that mark the acquisition of this skill. Some of these may be perceptual, some motor, some attentional, and some cognitive. Participants might better perceive the sizes of the disks, may become better able to handle the disks, might become better at attending to the size and positions of the disks, and might learn particular combinations of moves that enable the participant to come closer to a solution. For cognitivists, one of the central concerns is to disentangle and characterize any number of endogenous and exogenous factors that contribute to skilled action. By contrast, Rietveld, Denys, and van Westen show little interest in endogenous factors.

The contrasts between SIF and cognitivism are probably evident, but the differences between SIF and RECS, as described in earlier publications by Chemero and Hutto and Myin, may be less so. Aizawa (2015, following Shapiro 2013), makes the case that Chemero's embodied cognition is a proposal to study what is more commonly known as behavior. Chemero (2009) conjectures that "cognition is the ongoing, active maintenance of a robust animal-environment system, achieved by closely co-ordinated perception and action" (p. 212, fn. 8) and that "radical embodied cognitive science can explain cognition as the unfolding of a brain-body-environment system, and not as mental gymnastics" (p. 43). These appear to be roundabout ways of saying that cognition is what is commonly understood to be behavior. This seems to be the natural way of understanding what Chemero is claiming, although it is surprising that Chemero makes such a bold proposal with little commentary and without any argumentation (see Aizawa 2015; Hutto and Myin 2018, p. 106, apparently agree that Chemero provides no reason in support of this view). Hutto and Myin (2013) propose to equate "basic cognition with concrete spatiotemporally extended patterns of dynamic interaction between organisms and their environments" (p. 5). This again appears to be a convoluted way of saying that (basic) cognition is behavior; hence that radical embodied cognitive science should study behavior.

The study of behavior, however, is not (in any obvious sense) what Rietveld, Denys, and van Westen appear to have in mind. They are not concerned with just any sort of behavior; they are concerned with an apparent special case of behavior, namely, skilled human action. The SIF does not try to understand how I do the tango, but it might try to understand how I ride a bike. So, Rietveld, Denys, and van Westen apparently part company with Chemero (2009) and Hutto and Myin (2013).

Hutto and Myin

Hutto and Myin (2018) begin their contribution with the idea of contrasting cognition as they conceive it from cognition as cognitivists conceive it. They correctly note that cognitivists are committed to (typically, brain-bound) computations over representations. Unfortunately, they fail to note the distinction between cognition and behavior. This is a distinction accepted by cognitivists, behaviorists, and many dynamical systems theorists. For Hutto and Myin (2013), (basic) cognition and behavior, at least at times, appear to be the same thing. As noted earlier, they propose the equation of "basic cognition with concrete spatiotemporally extended patterns of dynamic interaction between organisms and their environments." Moreover, their contribution to this volume suggests something similar: "REC understands cognition as something that organisms do. Cognition is a kind of embodied activity that is out in the open, not a behind-the-scenes driver of what would otherwise be mere movement" (p. 97). This sounds like the view that cognition is behavior.

One might well reject the distinction between cognition and behavior, as Hutto and Myin seem to do, but it is important to recognize this distinction as being part of the cognitivist view so as to properly represent cognitivism and what is at issue between cognitivists and RECers. So, for example, Hutto and Myin write, "Today's cognitivism tends to make a further specific assumption about the mechanisms responsible for intelligent activity—viz., that cognitive processes that give rise to such activity take the form of brain-based computations over internal mental contents" (p. 96). If we take solving the Tower of Hanoi problem as an example of intelligent activity, then the cognitivist view is that cognitive processes are not *the* mechanisms responsible. There are perceptual, motor, bodily, and environment mechanisms as well. By cognitivist lights, there is a form of continuous reciprocal causation between the disks on the dowels and the next move the subject makes. The subject views the puzzle, then moves a disk, then thinks some more, then moves another disk. The configuration of the puzzle disks is among the mechanisms that are responsible for the subject's intelligent activity. So, if one wishes to explain intelligent activity, then of course one needs to mention environmental factors. As another example, Hutto and Myin claim that "When it comes to explaining intelligent activity REC advocates the equal partner principle. Accordingly, appeals to neural factors in such explanations stand on the same footing as appeals to bodily and environmental factors" (p. 97). But when it comes to intelligent activity, cognitivism can accept *some* sort of equal partner principle. It depends on what "equal" and "same footing" come to. The environmental factors are equal in being just as much causally efficacious as the cerebral factors, but they are not equal, by cognitivist lights, in being just as cognitive as the cerebral factors. What REC evidently wants to conclude is that the environment is an equal *cognitive* partner. One would have that view if, per Hutto and Myin's REC, cognition were behavior. This makes it all the more pressing to ask why we should think that cognition is behavior.

One more thing. Suppose that cognition is behavior. Clearly behavior is embodied. Clearly one does not solve the Tower of Hanoi problem entirely in one's mind. Clearly the behavior that is solving the problem involves physically moving the disks between the dowels. Few if any cognitive scientists deny this. (Perhaps Cartesian substantival dualists would deny this on the grounds that the cognitive part of the behavior is not realized in the brain.) But given the consensus that behavior is embodied, the remaining question that should be up for debate is whether cognition is behavior. Why think cognition is behavior?

There is not space here to run through an extensive discussion of this issue, so a simple report on the state of play may have to suffice. Why should we think that cognition is behavior? Aizawa (2014) noted that certain texts familiar to enactivists and RECers—e.g., Maturana and Varela (1980), and Chemero (2009)—propose that cognition is behavior, but that these texts provide no arguments for the view. Hutto and Myin (2018) reply by suggesting that we should consult Dreyfus (2014). Okay, but does this mean that there really are no arguments in Maturana and Varela (1980), Chemero (2009), or, indeed, Hutto and Myin (2013, 2018) for the view that cognition is behavior? That seems to be the implication of what they write, but can that be so? Can there really be whole books with a central presupposition that cognition is behavior but provide no reason for the view?

Conclusion

Unfortunately, the foregoing critical comments leave much of the substance of the papers in this section untouched. One (boring) reason for this is simply the lack of space. One cannot review and comment on the tens of thousands of words on the 4E frameworks (think, for example, of Di Paolo's account of the concept of life, Rievelt, Denys, and van Westen's discussion of affordances, or Hutto and Myin's case for anti-representationalism) in the space of a few thousand words. The more telling reason, however, is that so little of the foregoing papers have to do with the issue of what cognition is. Di Paolo and Kiverstein advance what appear to be implausible or incomplete theories of what cognition is. Rievelt, Denys, and van Westen have hardly anything to say about cognition per se. Hutto and Myin just seem to want to call behavior "cognition." Moreover, they seem uninterested in presenting reasons for thinking that cognition is behavior.

References

Adams, F. and Aizawa, K. (2008). *The bounds of cognition*. Malden, MA: Blackwell Publishers.

Aizawa, K. (2003). *The systematicity arguments*. Boston: Kluwer Academic.

Aizawa, K. (2014). The enactivist revolution. *Avant*, 5(2), 19–42. doi:10.12849/50202014.0109.0002

Aizawa, K. (2015). What is this cognition that is supposed to be embodied? *Philosophical Psychology*, 28(6), 755–75.

Chemero, T. (2009). *Radical embodied cognitive science*. Cambridge, MA: MIT Press.

Clark, A. (2005). Intrinsic content, active memory and the extended mind. *Analysis*, 65(1), 1–11.

Clark, A. (2015). *Surfing uncertainty: prediction, action, and the embodied mind*. New York, NY: Oxford University Press.

Di Paolo, E.A. (2018). The enactive conception of life. In: this volume, pp. 71–94.

Dreyfus, H.L. (2014). *Skillful coping: essays on the phenomenology of everyday perception and action* (ed. M. Wrathall). Oxford: Oxford University Press.

Fodor, J. (1983). *The modularity of mind: an essay on faculty psychology*. Cambridge, MA: MIT Press.

Fodor, J. and Pylyshyn, Z. (1988). Connectionism and cognitive architecture: A critical analysis. *Cognition*, 28(1–2), 3–71.

Gibson, J.J. (1979). *The ecological approach to visual perception*. Hillsdale, NJ: Lawrence Erlbaum.

Hutto, D.D. and Myin, E. (2013). *Radicalizing enactivism*. Cambridge, MA: MIT Press.

Hutto, D.D. and Myin, E. (2018). Going radical. In: this volume, pp. 95–116.

Kiverstein, J. (2018). Extended cognition. In: this volume, pp. 19–40.

Maturana, H.R. and Varela, F.J. (1980). *Autopoiesis and cognition: the realization of the living*. Dordrecht: Springer.

Miłkowski, M. (2015). The hard problem of content: Solved (long ago). *Studies in Logic, Grammar and Rhetoric*, 41(1), 73–88.

Palermos, S.O. (2014). Loops, constitution, and cognitive extension. *Cognitive Systems Research*, 27, 25–41.

Pylyshyn, Z.W. (2003). *Seeing and visualizing: it's not what you think*. Cambridge, MA: MIT Press.

Rietveld, E., Denys, D., and van Westen, M. (2018). Ecological-enactive cognition as engaging with a field of relevant affordances: the skilled intentionality framework (SIF). In: this volume, pp. 41–70.

Rupert, R. (2009). *Cognitive systems and the extended mind*. Cambridge, MA: MIT Press.

Ryder, D. (2004). SINBaD neurosemantics: a theory of mental representation. *Mind and Language*, 19(2), 211–40.

Shapiro, L. (2013) Dynamics and cognition. *Minds & Machines*, 23(3), 353075.

Shapiro, L. (in press). Mechanism or bust? Explanation in psychology. *The British Journal for the Philosophy of Science*.

Wheeler, M. (2010). In defense of extended functionalism. In: R. Menary (ed.), *The extended mind*. Cambridge, MA: MIT Press, pp. 245–70.

Wilson, R. (1994). Wide computationalism. *Mind*, 103(411), 351–72.

Wilson, R. and Clark, A. (2009). How to situate cognition: letting nature take its course. In: M. Aydede and P. Robbins (eds.), *The Cambridge handbook of situated cognition*. Cambridge: Cambridge University Press, pp. 55–77.

PART III

MODELING AND EXPERIMENTATION

CHAPTER 7

THE PREDICTIVE PROCESSING HYPOTHESIS

JAKOB HOHWY

Introduction

A millennium ago the great polymath Ibn al Haytham (Alhazen) (ca. 1030; 1989) developed the view that "many visible properties are perceived by judgment and inference" (II.3.16). He knew that there are optical distortions and omissions of the image hitting the eye, which without inference would make perception as we know it impossible (Lindberg 1976; Hatfield 2002). Al Haytham was aware it is counterintuitive to say perception depends on typically intellectual activities of judgment and inference and so remarks that "the shape and size of a body . . . and such like properties of visible objects are in most cases perceived extremely quickly, and because of this speed one is not aware of having perceived them by inference and judgment" (II.3.26).

Since al Haytham, many in optics, psychology, neuroscience, and philosophy have advocated the role of inference in perception, and have insisted too that this inference is somehow unconscious (for review, see Hatfield 2002). With characteristic clarity, Hermann von Helmholtz coined the phrase *unconscious perceptual inference* and said that the "psychical activities" leading to perception:

> are in general not conscious, but rather unconscious. In their outcomes they are like inferences insofar as we from the observed effect on our senses arrive at an idea of the cause of this effect. This is so even though we always in fact only have direct access to the events at the nerves, that is, we sense the effects, never the external objects. (Helmholtz 1867, p. 430)

The starting point for this inferential view is the conviction that perception can be explained only if a particular, fundamental problem of perception is solved, namely, how the brain can construct our familiar perceptual experience on the basis only of the imperfect data delivered to the senses, and without ever having unfettered access to the true hidden causes of that input. This type of problem is also at the heart of massive scientific endeavors in contemporary artificial intelligence and machine learning.

Recently, the notion of unconscious perceptual inference has been embedded in a vast probabilistic theoretical framework covering cognitive science, theoretical neurobiology, and machine learning. The basic idea is that unconscious perceptual inference is a matter of Bayesian inference, such that the brain in some manner follows Bayes's rule and thereby can overcome the problem of perception. The most comprehensive, ambitious, and fascinating of these probabilistic theories build on the notion of *prediction error minimization* (PEM) (this notion arose in machine learning research, with versions of it going back to 1950s; for recent philosophical overviews, see Clark 2013; Hohwy 2013).

Several aspects of unconscious perceptual inference are anathema to many versions of enactive, embedded, embodied, and extended (4E) cognition. If perception is a matter of Bayesian inference, then perception seems a very passive, intellectualist, neurocentric phenomenon of receiving sensory input and performing inferential operations on them in order to build internal representations. This process is divorced from action and active interaction with the environment; it appears insensitive to the situation in which the system is embedded; it leaves no foundational role for the body in cognitive and perceptual processes; and it makes perceptual processes a matter of what happens behind the sensory veil with no possibility of extension to mental states beyond the brain, let alone the body (4E cognition is now a vast and varied area of research; the types of approaches that stress anti-representational and anti-inferential elements are, for example, Varela et al. 1991; Clark 1997; Noë 2004; Gallagher 2005; Thompson 2007; Clark 2008; Hutto and Myin 2013).

The tension between perceptual inference and 4E cognition matters because both are influential attempts at explaining the same range of phenomena. Having noticed the initial tension between them, there are three main options: (1) perceptual inference and 4E cognition are incompatible as foundational accounts of perception and cognition, which means one must be false (Anderson and Chemero 2013; Barrett 2015); this option appears unattractive because key aspects of both seem believable and important. The next two options are more discursive: (2) perceptual inference and 4E cognition should be considered compatible, but only because perceptual inference, rightly understood, is not a matter of neurocentric, representationalist inference but yields just the kinds of processes necessary for 4E cognition (Clark 2013, 2015, 2016). (3) Perceptual inference and 4E cognition should be considered compatible, but only because 4E cognition, rightly understood, is nothing but representation and inference (Hohwy 2016b). Options 2 and 3 deflate perceptual inference and 4E cognition, respectively, that is, they achieve reconciliation by recasting one of the sides of the debate in terms of the other.

This chapter aims to show that Option 3 is reasonable. Perceptual inference, in the shape of PEM, is tremendously resourceful and can therefore encompass phenomena highlighted in debates on 4E cognition. Reconciliation with somewhat deflated 4E notions is achieved without compromising PEM's representationalist and inferentialist essence. This advances the debate about 4E cognition because, in the context of PEM, inference and representation are both shown to have several surprising aspects, such that, perhaps, 4E cognition need not abhor these notions altogether.

The chapter first explains PEM and lays out its specific notion of inference. Then action is subsumed under PEM's inferential scheme, and the role of representation in perception and action is explained. Finally, select aspects of 4E cognition are incorporated into the PEM fold.

Predictive Processing and Inference

In many approaches to unconscious perceptual inference, the notion of inference is left unspecified; as Helmholtz says, our psychical activities are "like" inference. Here, the notion of inference captures the idea that the perceptual and cognitive systems need to draw conclusions about the true hidden causes of sensory input vicariously, working only from the incomplete information given in the sensory input.

On modern approaches, this is given shape in terms of Bayesian inference. This yields a concrete sense of "inference" where Bayes's rule is used to update internal models of the causes of the input in the light of new evidence. A Bayesian system will arrive at new probabilistically optimal "conclusions" about the hidden causes by weighting its prior expectations about the causes against the likelihood that the current evidence was caused by those causes (there are useful textbook sources on machine learning such as Bishop 2007; philosophical reviews such as Rescorla 2015; see also recent treatments of hierarchical Bayes and volatility such as Payzan-LeNestour and Bossaerts 2011; Mathys et al. 2014).

Consider a series of sensory samples, for example, auditory inputs drawn from a sound source. The question for the perceiver is where the sound source is located (somewhere on a 180° space in front of the perceiver). Assume the samples are normally distributed and that the true source is 80°. Before any samples come in, the perceiver expects—predicts—samples to be distributed around 90°. The first sample comes in indicating 77°, and thereby suggests a prediction error of 13°. Which probabilistic inference should the perceiver make? Inferring that the source is at 77° would disregard prior knowledge and lead to a model overfitted to noise. Ignoring the prediction error would prevent perceptual learning altogether. So the right weight to assign to the prediction error in updating the prior belief of 90° ought to reflect an optimal, rational balance between the prior and the likelihood, and this is indeed what Bayes's rule delivers. So probabilistic inference should be determined by Bayes's rule. In other words, the *learning rate* in Bayesian inference is determined by how much is already known and how much is being learned by the current evidence, reflected in the likelihood. (In this toy example, I set aside the question how the perceiver knows not to add the weighted prediction error to 90°, moving toward 103° and away from 80°; notice that if the system does this, then prediction error will tend to grow over time).

The correct weights to give to the prior and the prediction error can be considered transparently through the variance of their probability distributions. The more the variance, the less the weight. A strong prior will have little variance and should be weighted highly, and a precise input, which fits well the expected values of the model in question, should be weighted highly. The inverse of the variance is called the *precision*, and it is a mathematically expedient convention to operate with precisions in discussions of inference: the learning rate in Bayesian inference therefore depends on the precisions of the priors and prediction errors. As will become apparent later, precisions are important to PEM and its ability to engage 4E-type issues.

So far, only one inferential step is described. For subsequent samples, Bayes's rule should also be applied, but for the old inferred posterior as the new prior. Since there is an optimal mix of prior and likelihood, the model will converge on the true mean (80°) in the long run. Critically, in this process, the average prediction error is minimized over the long run. Even for quite noisy samples (imprecise distributions, or probability density functions), a Bayesian inference system will eventually settle on an expectation for the mean that keeps prediction error low. This can be turned around such that, subject to a number of assumptions about the shape of the probability distributions and the context in which they are considered, a system that minimizes prediction error in the long run will *approximate* Bayesian inference.

The heart of PEM is then the idea that a system need not explicitly know or calculate Bayes's rule to approximate Bayesian inference. All the system needs is the ability to minimize prediction error in the long run. This is the sense in which unconscious perceptual inference is inference: internal models are refined through prediction error minimization such that Bayesian inference is approximated. The notion of inference is therefore nothing to do with propositional logic or deduction, nor with overly intellectual application of theorems of probability theory.

It would be misguided to withdraw the label "inference" from unconscious perceptual inference, or from PEM, just because it is an approximation to Bayes, or because the process is not an explicit application of a mathematical formalism by the brain. If the inferential aspect is not kept in focus, then it would appear to be a coincidence, or somehow an optional aspect of perceptual and cognitive processes that conform to what Bayes's rule dictate. Put differently, anyone who subscribes to the notion of predictive processing must also accept the inferential aspect. If it is thrown out, then the "prediction error minimization" part becomes a meaningless, unconstrained notion.

PEM thus says that perceivers harbor internal models that give rise to precision-weighted predictions of what the sensory input should be, and that these predictions can be compared to the actual sensory input. The ensuing prediction error guides the updates of the internal model such that prediction error in the long run is minimized and Bayesian inference approximated.

However, this description of PEM is still too sparse. In any given situation, a PEM system will not know how much or how little to weight prediction error even if it can assess the precisions of the prior and of the current prediction error. In essence, a system that operates with only those precisions will be assuming the world is more simple and persistent than it really is. For example, different sensory modalities have different precisions in different contexts, and without prior knowledge of these precisions, the system can make no informed decisions about how to weight prediction error. For example, similarly sized prediction errors in the auditory and visual modalities should not be weighted the same, since the precisions of each should be expected to be different. Therefore a PEM system would need to have and shape expectations about the precisions as well as the means of probability distributions. The need for such *expected precisions* is also driven by the occurrence of multiple interacting causes of sensory input within and across sensory modalities. In the example of the location of the auditory source, variability in the sensory sampling might be due to a new cause interfering with the original sound source (e.g., a moving screen intermittently obscures the location of the sound). If the system does not have robust expectations for the precision of the sound source, then it will be unable to make the right inferences about the input (i.e., is it one cause with varying precisions or is it two interacting causes that gives rise to the nonlinear evolution in the auditory sensory input?).

A PEM system must model expectations of precisions, and this part of the PEM system itself needs to be Bayes-optimal. Models will harbor priors for precisions; they will predict precisions and generate precision prediction errors. Moreover, they will need to do this across all the hidden causes modeled such that their interactions can be taken into account. This calls for a *hierarchical* structure where the occurrence of various causes over many different time scales can impact on the predictions of the sensory input received at any given time. For example, the interaction of relatively slow time scale regularities (e.g., the trains driving past your house two or three times an hour) need to influence the predictions of faster time scale regularities (e.g., the words heard in a conversation in your lounge room), and vice versa.

A PEM system that operates in a complex environment, with levels of uncertainty that depend on the current state of the world and many interacting causes at many different time scales, will thus build up a vast internal model with many interacting, hierarchically ordered levels, which all pass messages to each other in an attempt to minimize average prediction error over the long term.

Consider finally what happens over time to the models harbored in the brain, on the basis of which predictions are made and prediction errors minimized. The parameters of these models will be shaped by the Bayesian inferential process to mirror the causes of the sensory input. In the example earlier, by minimizing prediction error over time for the location of the cause of auditory input, the model will revise its initial false belief that the location is at 90°, and come to expect it to be at its true position of 80°. Further, by minimizing precision prediction error, the model may be able to anticipate interacting causes, such as a moving screen intermittently blocking the sound. This means that, by approximating Bayesian inference, the models of a PEM system must *represent* its world.

Here, the notion of representation is not just a matter of receptor covariance, where the states of neural populations covary with the occurrence of certain environmental causes. The hierarchical model is highly structured, and performs operations over the parameters. For example, there will be model selection. In our example, the system might ask whether there is another cause interacting with the sound source, or if the signal itself is becoming noisier. In addition, there are convolutions of separate expected signals generated on the basis of the models; for example, when a cat and a fence are detected, the expected sensory signals from both hidden causes are convolved into one stream by the brain to take the occlusion of the cat by the fence into account. As will become clear, the representational aspects of PEM are critical when it comes to incorporating action, too.

The representational nature of a PEM system is not optional. The ability to minimize prediction error over time depends on building better and better representations of the causes of its sensory input. This is encapsulated in the very notion of model revision in Bayesian inference. (There is extensive discussion of what it takes for perception to be representational; for examples of relevance to Bayesian inference, see Ramsey 2007; Orlandi 2013, 2014; Gładziejewski 2015; Ramsey 2015.)

So far, it appears that predictive processing is inferential and representational in a specific Bayesian sense. Traditionally, 4E approaches have rejected both notions. Next, PEM will be shown to have explanatory reach into 4E cognition too.

PEM and Action

A representationalist and inferentialist account of cognition and perception may appear divorced from the concerns and activities of a real, embodied agent operating in its environment. Thus enactive and embodied accounts have de-emphasized classic representationalist understandings of cognition and perception and with it much semblance to inference (there are many versions and much discussion of embodiment; see, e.g., Brooks 1991; Noë 2004; Gallagher 2005; Alsmith and Vignemont 2012; Hutto and Myin 2013; Orlandi 2014).

Perhaps the basic sentiment could be summed up in the strong intuition that embodied action is not inference, and yet the body and its actions are crucial to gain any kind of understanding of perception and cognition. PEM can, however, easily cast action as a kind of inference—as *active* inference (Friston, Samothrakis, et al. 2012).

Recall that any system that minimizes prediction error over time will approximate Bayesian inference; that is, such a system will be inferential in the Bayesian sense that it increases the evidence for its internal model. Using the example from earlier again, by minimizing prediction error the system could accumulate evidence for the model that represents the sound source as located at 80°. In that case, the internal model is revised from the initial 90° to the new estimate of 80°.

It is trivial to observe that the perceiver could also have minimized prediction error by turning the head 10° to the left and thereby have accumulated evidence for the prediction that the sound source is located at 90°. Prediction error can be minimized both through passive updating of the internal model and through active changes to the sensory input. Action, such as turning one's head, can therefore minimize prediction error.

Since, as argued earlier, minimizing prediction error is inference, and action is inference. There is then no hindrance to incorporating action into an inferentialist framework.

In active inference, representations are central to guiding action. This is because action only occurs when a hypothesis—in this case a representation of a state that is yet to occur—has accumulated sufficient evidence relative to other hypotheses to become the target of PEM. This yields two aspects that are sometimes seen as hallmarks of representations: they are action-guiding and they are somehow detached from what they stand for (for discussion and review, see Orlandi 2014). Active inference therefore has a good claim to be both inferential and representational.

For perceptual inference, precisions were shown to be critical. Without precisions, the PEM system would not be able to minimize error in a world with state-dependent uncertainty and interacting causes. The same holds for active inference. Without any notion of how levels of prediction error tend to shift over many interacting time scales, the system would pick the action that minimizes most error here and now—for example, by entering and remaining in a dark room (for discussion, see Friston, Thornton, et al. 2012). This would be analogous to overfitting, and would come at the cost of increasing prediction error over the longer term. For example, even though the perceiver might minimize prediction error by forcing the sound to come at the 90° midline, this might make it difficult to ascertain the true source of a potentially moving cause such as the trajectory of a mosquito buzzing about (since direction detection is harder over the midline due to minimal interaural time difference). This calls for even more hierarchical model-building, namely, in terms of the precisions expected in the evolution of the prediction error landscape as a result of the agent's active intervention in the world. These self-involving, modeled regularities are, however, not fundamentally different from the regularities involved in perceptual inference. They simply concern the sensory input the agent should expect to result from the interaction of one particular cause in the world—the agent itself—with all the other causes of sensory input (for discussion of self-models, see, e.g., Synofzik et al. 2008; Metzinger 2009).

There is thus room for a notion of action within PEM. But this possibility alone does not imply that a PEM system is likely to actually *be* an agent. If the system is endowed with a body such that it could act, then the imperative for minimization of prediction error will make actual action highly likely.

If the system has accumulated strong evidence for, say, an association between two sounds, it may still be unable to distinguish several hypotheses, for example, whether the sounds are related as cause and effect or if they are effects of some common cause. It is standard in the causal inference literature that intervention is required to acquire evidence for or against these hypotheses (Pearl 2000; Woodward 2003). For example, if variation in one sound persists even if the other sound is actively switched off, then that is evidence the latter sound is not the cause of the first. The necessity of action is generalized in the observation from earlier that the system needs to learn differences in precisions and patterns of interactions among causes, such as occlusions and other causal relations that change the sensory input in nonlinear ways. Such learning thus requires action. The price of not engaging the body plant to intervene in the environment is that prediction error will tend to increase since predictions will be unable to distinguish between several different hypotheses. A PEM system that can act will therefore be best served to actually act.

This simple account of agency has profound consequences. It will be a learnable pattern in nature that inaction will tend to increase prediction error in the longer term (due to the inaccuracy of the hypotheses the system can accumulate evidence for by using only passive inference). Conversely, the system can learn that action tends to allow minimization of prediction error at reasonable time scales. Overall, this teaches the system that, on balance, its model will accumulate more precise evidence through action than through inaction. This will bias it to minimize prediction error through active inference. Of course, a system that only ever acts on the basis of unchanging models will never be able to learn new patterns, which is detrimental in a changing world. Therefore action must be interspersed with perceptual inference where models are updated, before new action takes place.

The mechanism by which this switching between perception and action takes place is best conceived in terms of precision optimization. Recall that the PEM system will build up expectations for precisions, which are crucial for dealing with state-dependent noise in a world with interacting causes. The role of expected precisions in inference is to optimally adjust weights for expected sensory input: input that is expected to be precise is favored in Bayesian inference whereas input that is expected to be imprecise is not favored. Mechanistically, this calls for a neuronal gating mechanism that inhibits or excites sensory inputs according to their expected precisions. This gating mechanism serves as a kind of probabilistic searchlight and thus plays the functional role of *attention* (Feldman and Friston 2010; Brown et al. 2011; Hohwy 2012, 2016a).

As the system gates its sensory input according to where it expects the most precise sensory input will occur, across several time scales, it may switch between perception and action. For example, if more precision is expected by the agent having its hand at the position of the coffee cup rather than at the current position at the laptop, then it will begin gating the current sensory input, which suggests the hand is at the laptop. This in turn allows the coffee hypothesis to gain relative weight over the laptop hypothesis, and the prediction error generated by that hypothesis can easily by minimized by moving the hand. Since the gain is high on this prediction error, the new hypothesis quickly accumulates evidence for its truth, and the hand will find itself at the coffee cup (for more on the dynamics of action and perception in relation to temporal phenomenology, see Hohwy et al. 2015; for the formal background, see Friston, Trujillo-Barreto et al. 2008).

Embodied, Embedded, and Inferential and Representational

When all the elements described in the last section are combined, a wholly inferential conception of agency begins to take shape. If action and agency are moments of PEM, then desires are just beliefs (or priors) about states that happen to be future, with a focus on their anticipated levels of prediction error, and where reward is the absence of prediction error. This suggests a neat continuity with perceptual inference, which also relies on priors and the imperative to minimize prediction error.

The idea that action is driven by PEM relative to a model does raise a question about the content of the model relative to which error is minimized. This model is what defines what we would normally describe as the agent's desires. In the wider PEM framework—which, as shall be described later, relies on notions of *free energy minimization*—the expected states that anchor active inference relate to set points in terms of the organism's homeostasis. This immediately evokes an evolutionary perspective, where expected bodily states are central to behavior. Apart from the specific evolutionary aspects, this suggests an *embodiment* perspective, because all aspects of perception and cognition then have a foundation in bodily states, and movement and purposeful behavior have a foundation in the environment. This element of embodiment makes it more likely that contact can be made between probabilistic theories of perception and action and embodied cognition approaches (such as, e.g., Varela et al. 1991; Gallagher 2005; Thompson 2007; for recent treatments that relate to PEM, see Bruineberg and Rietveld 2014; Fazelpour and Thompson 2015).

However, even this foundational embodiment is conceived probabilistically in PEM. A set of expectations for bodily states (relating to homeostasis) is essentially a model. In probabilistic terms, this model gives the probability of finding the organism in some subset of the overall set of states it could be in. The model is specified in terms of internal states, as signaled in interoception, but is tied to the overall setting of the organism in a subset of environmental states. The expected states defined in interoceptive terms would, in real organisms traversing actual environments, be mirrored in the expected states described in environmental terms, or in terms of their sensory input or exteroception. For example, fish are most likely to find their sensory organs impinged upon from watery states and this is associated strongly with the homeostatic needs specified in their model. In general, within this probabilistic reading of the foundational embodiment of a PEM organism, there is thus a tight coupling between the interoceptive and exteroceptive prediction error landscapes for any PEM system.

Not only does PEM provide a notion of embodiment, it also speaks to elements of embedded or situated cognition (see van Gelder 1995; Clark 1997; Aydede and Robbins 2009). With the tight coupling of the organism's expected states in terms of interoception and exteroception, perception and cognition cannot be separated from bodily or environmental aspects of the PEM system.

Crucially, this reading of embodiment and embedding leads directly to inferential processing and PEM. The model specifies the probability of finding the organism in any one of all the possible states. To know this model directly would require the agent averaging over all possible states and ascertaining the occurrence of itself in them. This is not possible for a finite organism to learn directly. Instead, the organism must essentially guess what its expected states are and minimize the ensuing error through perceptual and active inference. In slightly more formal terms, the organism needs to minimize surprise; that is, it needs to avoid finding itself in states that are surprising given its model. The sum of prediction error is always equal to or larger than the surprise, so minimizing prediction error will implicitly minimize surprise. This bound on surprise is also known in probabilistic terms as the free energy, and so this challenging idea is enshrined in the so-called free energy principle (Friston 2010).

When viewed in this larger context of the free energy principle, promising notions of embodied and embedded cognition present themselves. More research is needed on the extent to which they capture facets of the wide-ranging and heterogeneous 4E body of research. However, for the conception of embodiment and embedding mooted here, an inferential conception is inescapable.

Hierarchical Inference for a Changing World

In much 4E research there is a focus on fluid interactions with the world, characterized by non-inferential, nonrepresentational, "quick and dirty" processing. This picture is set up to contrast with inferential, representational, "slow and clean" processing (Clark 1997, 2013, 2015). Often, this kind of quick and dirty, situated cognition is discussed in terms of *affordances*: salient elements of the environment that are in some sense perceived directly and are immediately action-guiding. Affordances in quick and dirty processing are thought to evade the computational bottleneck that a traditional representational system would have trying to passively encode the entire sensory input presented at any given time. For some types of action and at some stages of learning, performance is rather plodding and sluggish, but there is an important insight in how the notion of situated cognition highlights the fluid swiftness with which organisms can perform some complex actions in their environment.

In a PEM system there is no bottleneck problem in the first place, however. There is never an issue of starting from scratch and encoding an entire natural scene in order to be able to perceive it. Hierarchical Bayesian inference is based on prior learning, which over time has shaped priors at many levels. Given priors, the sensory input is no longer something that needs to be encoded here and now. Instead the sensory input is, functionally speaking, the *feedback* to the forward predictive signal generated by the brain's internal model (Friston 2005). The model predicts what will happen and gets confirmation or disconfirmation on these predictions from the sensory input. There is thus no encoding of the entire sensory input in each perceptual instance. This means the PEM system has no need to resort to quick and dirty processing tricks to overcome a computational bottleneck. Instead, the system relies on slow and clean learning in order to facilitate swift and fluid perception of and interaction with the world. This learning is "slow" because is relies on meticulous accumulation of evidence for hypotheses at multiple time scales. It is "clean" because the learning slots into a hierarchy with clearly defined, general functional roles for time scales, for predictions of values, and for predictions of precisions.

The difference between swift and fluid processing and plodding and sluggish processing can easily be accommodated within a PEM system. Affordances are just causes of sensory input that, on the basis of prior learning, are strongly expected to give rise to high precision prediction error. To maintain Bayes optimality, the system gates sensory input accordingly, and strongly focuses both perceptual and active inference on these affordances. In this setting, PEM happens quickly, since highly precise distributions are easier to deal with computationally than imprecise ones. This means that the agent in question will obtain its expected states swiftly and fluidly.

Typically, the 4E preference for quick and dirty processing and affordances comes with a rejection of rich representational states (Clark 2008, 2015). The point is that such representations cannot come about due to the bottleneck problem. Moreover, the appeal to affordance-based quick and dirty processing is thought to obviate the need for rich internal representations altogether as the world's affordances in some sense are its own representation (Brooks 1991).

On the PEM-based account of swift and fluid processing, internal representations are, however, necessary. Over time, multilayered representations are constructed and shaped, and Bayesian model selection picks the model with the best evidence as the representation of the world relative to which prediction error is minimized in active inference (this kind of approach is developed in more detail for PEM in Seth 2014, 2015). Again, we get the result that PEM has the resources to speak to typical 4E discussions, but that it happens on the basis of representation and inference.

It could be that the brain builds rich representations as it learns about the world, and then gradually substitutes these much sparser and representation-poor, purpose-made representations that more directly tie in with and engage the environment. One argument here derives from Occam's razor, in the sense that there are simplicity gains from opting for a simple over a complex, rich model (Clark 2015). However, simplicity is not something additional to inference. Complex models are to be avoided because they are overfitted and thereby incur a prediction error cost in the longer run. How rich or simple a model should be is thus fully given by PEM in the first place.

In fact, there is reason to think the PEM account is preferable to the affordance-based account. It is true that swift and fluid processing is a salient and impressive aspect of human cognition. But so is the flexible way we shift between contexts, projects, beliefs, and actions. We might engage in attentive, fluid, and swift interaction for a period of time, but other beliefs and concerns always creep in and make it imperative to shift to another behavior. On the affordance-based account, it is not readily explained how the agent might disengage from a given set of affordances; the focus is at best on how representation-rich learning is needed before swift and fluid processing is possible, rather than the role of rich representation during swift and fluid processing. The agent seems tightly knitted to its environment, and it is not clear how the agent can step back and reconsider its current course of action.

In contrast, flexible cognition is a central motivation for adopting PEM's hierarchical Bayesian inference in the first place. Active inference is driven by the most probable hypothesis at any given time. The system will have built up expectations not just for what the most likely causes of sensory input might be but also for the typical evolution of prediction error precision. In particular, there will be accumulated evidence that any given hypothesis under which prediction error is minimized at a certain time will have a limited life span—in essence, the system will know that it lives in a changing world where precise evidence for any given hypothesis will soon begin to be hard to find. For example, as the agent fluidly and swiftly catches baseballs, it will know that the sun will soon set and make the visual input imprecise. It will therefore begin accumulating evidence for the next hypothesis (e.g., "I am eating dinner") under which evidence will soon begin to be accumulated and prediction error minimized.

This speaks to a crucial balance, which a PEM system must obtain. As prediction error is minimized in active inference, the hypothesis relative to which error is minimized is held stable. This means that, as prediction error is minimized, the world can in fact change "behind the scenes" to such an extent that it would eventually be better to abandon the current hypothesis and adopt a new one. Anticipating such change in the environment matters greatly to the agent because it should never engage in any behavior, no matter how swift and fluid, for so long that when it ceases the behavior, the world has changed in other respects and predictive error will be very large. A PEM agent therefore will be inclined to believe that the current state of affairs will change, and therefore the agent will intersperse active inference with perceptual inference, where the internal model is checked and the size of the overall prediction error is adjusted and tightened up before a new hypothesis is selected for active inference (see Hohwy 2013; Hohwy et al. 2015).

A hierarchical system operating with slow and clean processing can thus economically explain both swift and fluid, affordance-based cognition as well as flexible cognition. This is an important point to make in the context of PEM's affinity to 4E cognition. The motivation for PEM is, in the end, the simple observation that we live in a changing world. Our world presents many different causes of our sensory input, and these causes interact with each other to create nonlinearities in the input; moreover, these interactions happen concurrently at many different time scales (e.g., "The setting sun makes the balls hard to see, but this time of the year the janitor often turns on the floodlights at the far pitch"). This complexity is what creates the need for hierarchical Bayesian inference in the first place: a rich internal model that keeps track of all these contingencies and can mix the various causes in the right way to anticipate the sensory input. This has a 4E-type ring to it: the cognitive system is the way it is because the agent's world and body are the way they are. In particular, PEM is not the best solution for non-ecological, lab-style model environments where typically context and interactions between hidden causes are kept to a minimum. In other words, a machine learning researcher who never tests their system against the real world will have little impetus to build a PEM system. On 4E approaches, there is also a strong focus on real-world settings, but the response is typically to tie the agent very closely to its environment. This, however, makes it harder to see how the real world, and also that fact that the real world is a changing place, can be taken into consideration. PEM, in contrast, makes room for the changing world by retracting farther away from the world, into a vast internal model that seeks to represent the full richness of the world and the way it changes over many time scales. On the PEM conception of the agent's place in the world, cognition is not a matter of being closely in tune with and driven by the sensory input. Rather, cognition is a matter of having richly represented expectations for the world and the body and seeking confirming feedback on those expectations through the senses.

The Mind and Things Without It

Both perception and action are inferential and representational. The PEM system's process of minimizing prediction error implies that the sensory input is explained away on the basis of the evolving hypotheses of an internal model. The more the system can minimize its prediction error, the more it will accumulate evidence for its own model. This is a trivial observation: if I can minimize prediction error for my theory that my hamster has escaped, the more evidence I have for that theory. If we consider the PEM system an agent, then it acquires evidence for its own existence through its activities (Friston 2010). Borrowing a term from philosophy of science, the PEM system can thus be said to be *self-evidencing* (Hempel 1965; Hohwy 2016b).

A self-evidencing system creates a sensory boundary between itself (i.e., the model) and the causes of its sensory input. This again is a trivial consequence of self-evidencing: there is something that garners evidence and then there is what the evidence is evidence of. Or again, in both perceptual and active inference there is something doing the inference and something being inferred. This boundary can also be described in terms of causal nets, where a set of inner states (i.e., brain states) can be said to have a "Markov blanket" (Pearl 1988) consisting of the inner states' parents (i.e., the sensory states) and their children and other parents of the children (i.e., the active states driving active inference) (Friston 2013; Hohwy 2015, 2017; causal Bayes nets must be acyclic, but brains have recurrent (cyclic) states; there are technical ways, such as dynamical Bayes nets, to deal with such problems). The activity of the states within a Markov blanket is wholly determined by the states of the blanket. In principle, nothing about the environmental states beyond the blanket need be known to know what the system is doing. By extension, in principle, only the states of the sensory organs need be known to know everything the mind does.

PEM then comes with a principled way of drawing a boundary between the mind and the outside world. If a particular state is part of what is doing the inference, then it must be within the sensory boundary, as a part of what approximates inference about outside causes of sensory input. This may relate to the vigorous debate about *extended cognition* (Clark and Chalmers 1998; Clark 2008), which is the last member of 4E cognition to discuss.

Extended cognition is the idea that some objects, such as notebooks and smartphones, play such an integrated, memory-like function in the mental economy of some agents that, by parity of reasoning, they should be considered part of the agent's mental states even though they reside outside the central nervous system. There is much discussion of this idea (see, e.g., Menary 2007; Adams and Aizawa 2008; Anderson et al. 2012; Spaulding 2012). PEM brings with it a new way of thinking about the role of such external objects. On the one hand, these objects are inferred (e.g., on the basis of the sensory input from the notebook) and as such they are outside the mental states of the system. On the other hand, if the extended cognition hypothesis is correct, they are within the sensory boundary, forming part of the inner states behind a Markov blanket inferring the hidden causes beyond it.

Interpreting purported cases of extended cognition according to PEM thus leaves two main options. There might be contradiction, since something cannot be both within and beyond the same boundary at the same time. Or, there might be multiple coexisting sensory boundaries. The second option is very interesting and very likely to be true, since Markov blankets occur easily. There is an associated cost, however: we have identified the inner states (or the model) with the agent, and if there are multiple Markov blankets then there are multiple agents coexisting at the same time. Though this may be true in a weak sense of agent, it is explanatorily messy. When asking which agent is acting, there would then be a multitude of correct answers, depending on how many nested Markov blankets are involved in the same action. This speaks in favor of using inference to the best explanation to identify the agent whose relatively invariant involvement accounts for most of observed behavior over time. It seems likely this more pragmatically identified agent would be the agent as specified by the model harbored just in the nervous system. This is the agent relative to which prediction error is minimized over the longer time scale, which as we saw is central to understanding predictive processing accounts in the first place (for discussion, see Hohwy 2016b). Bringing this discussion back to extended cognition, the pragmatic method of identifying the agent suggests that there is no extended cognition, since the special objects in question are beyond the one Markov blanket. The more lax way of identifying agents suggests that extended cognition ambiguous, since the special objects are beyond some blankets and within others.

The existence of the sensory boundary or Markov blanket implies that perception and agency are confined to the inner states of the PEM system (wherever the boundary or boundaries of the system are located). Those inner states will mirror the states outside the boundary: the inner states will, through PEM, come to represent the worldly causes of the sensory input impinging at the system's periphery. Conversely, through active inference, the outside states will come to conform to the expectations harbored in the internal states.

There is then an intriguing duality to this sensory boundary between mind and world. On the one hand, the boundary is epistemic (cf. self-evidencing): the worldly causes can only be known vicariously, through inference on sensory input. On the other hand, the boundary is characterized in causal terms (cf. Markov blanket): there is a dynamic coupling between mind and world, enabled through both perception and action.

This duality summarizes well why PEM is a good fit for many of the issues in 4E debates: PEM is able to throw light on embodied agents dynamically interacting with the environment in which they are embedded. This good fit with 4E cognition is, however, made possible precisely because PEM is inferential and representational.

References

Adams, F. and Aizawa, K. (2008). *The bounds of cognition*. Oxford: Blackwell.

Alsmith, A.J.T. and Vignemont, F. (2012). Embodying the mind and representing the body. *Review of Philosophy and Psychology*, 3(1), 1–13.

Anderson, M. and Chemero, A. (2013). The problem with brain GUTs: conflation of different senses of "prediction" threatens metaphysical disaster. *Behavioral and Brain Sciences*, 36, 204–5.

Anderson, M.L., Richardson, M.J., and Chemero, A. (2012). Eroding the boundaries of cognition: implications of embodiment. *Topics in Cognitive Science*, 4(4), 717–30.

Aydede, M. and Robbins, P. (2009). *The Cambridge handbook of situated cognition*. New York: Cambridge University Press

Barrett, L. (2015). A better kind of continuity. *The Southern Journal of Philosophy*, 53, 28–49.

Bishop, C.M. (2007). *Pattern recognition and machine learning*. Dordrecht: Springer.

Brooks, R.A. (1991). Intelligence without representation. *Artificial Intelligence*, 47(1–3), 139–59.

Brown, H., Friston, K.J., and Bestmann, S. (2011). Active inference, attention and motor preparation. *Frontiers in Psychology*, 2, 218.

Bruineberg, J. and Rietveld, E. (2014). Self-organization, free energy minimization, and optimal grip on a field of affordances. *Frontiers in Human Neuroscience*, 8, 599.

Clark, A. (1997). *Being there*. Cambridge, MA: MIT Press.

Clark, A. (2008). *Supersizing the mind: embodiment, action, and cognitive extension*. New York: Oxford University Press.

Clark, A. (2013). Whatever next? Predictive brains, situated agents, and the future of cognitive science. *Behavioral and Brain Sciences*, 36(3), 181–204.

Clark, A. (2015). Radical predictive processing. *The Southern Journal of Philosophy*, 53, 3–27.

Clark, A. (2016). *Surfing uncertainty*. New York: Oxford University Press.

Clark, A. and Chalmers, D. (1998). The extended mind. *Analysis* 58(1), 7–19.

Fazelpour, S. and Thompson, E. (2015). The Kantian brain: brain dynamics from a neurophenomenological perspective. *Current Opinion in Neurobiology*, 31, 223–9.

Feldman, H. and Friston, K. (2010). Attention, uncertainty and free-energy. *Frontiers in Human Neuroscience*, 4, 215.

Friston, K. (2010). The free-energy principle: a unified brain theory? *Nature Reviews Neuroscience*, 11(2), 127–38.

Friston, K. (2013). Life as we know it. *Journal of The Royal Society Interface*, 10, 20130475.

Friston, K., Samothrakis, S., and Montague, R. (2012). Active inference and agency: optimal control without cost functions. *Biological Cybernetics*, 106(8), 523–41.

Friston, K., Thornton, C., and Clark, A. (2012). Free-energy minimization and the dark room problem. *Frontiers in Psychology*, 3, 130.

Friston, K.J. (2005). A theory of cortical responses. *Philosophical transactions of the Royal Society of London. Series B, Biological sciences*, 369(1456), 815–36.

Friston, K.J., Trujillo-Barreto, N., and Daunizeau, J. (2008). DEM: a variational treatment of dynamic systems. *NeuroImage*, 41(3), 849–85.

Gallagher, S. (2005). *How the body shapes the mind*. Oxford: Oxford University Press.

Gładziejewski, P. (2015). Predictive coding and representationalism. *Synthese*, 193(2), 559–82.

Hatfield, G. (2002). Perception as unconscious inference. In: D. Heyer and R. Mausfeld (eds.), *Perception and the physical world*. Chichester: John Wiley & Sons, Ltd., pp. 113–43.

Helmholtz, H.v. (1867). *Handbuch der Physiologishen Optik*. Leipzig: Leopold Voss.

Hempel, C.G. (1965). *Aspects of scientific explanation and other essays in the philosophy of science*. New York: Free Press.

Hohwy, J. (2012). Attention and conscious perception in the hypothesis testing brain. *Frontiers in Psychology*, 3, 96.

Hohwy, J. (2013). *The predictive mind*. Oxford: Oxford University Press.

Hohwy, J. (2015). The neural organ explains the mind. In: T. Metzinger and J.M. Windt (eds.), *Open MIND*. Frankfurt am Main: MIND Group, pp. 1–23.

Hohwy, J. (2016a). Prediction, agency, and body ownership. In: A.K. Engel, K.J. Friston, and D. Kragic (eds.), *The pragmatic turn: toward action-oriented view in cognitive science*. Cambridge, MA: MIT Press.

Hohwy, J. (2016b). The self-evidencing brain. *Noûs*, 50(2), 259–85.

Hohwy, J. (2017). How to entrain your evil demon. In: T.K. Metzinger and W. Wiese (eds.), *Philosophy and predictive processing*. Frankfurt am Main: MIND Group. doi:10.15502/9783958573048

Hohwy, J., Paton, B., and Palmer, C. (2015). Distrusting the present. *Phenomenology and the Cognitive Sciences*, 15(3), 315–35.

Hutto, D. and Myin, E. (2013). *Radicalizing enactivism: basic minds without content*. Cambridge, MA: MIT Press.

Lindberg, D.C. (1976). *Theories of vision: from Al Kindi to Kepler*. Chicago: University of Chicago Press.

Mathys, C.D., Lomakina, E.I., Daunizeau, J., Iglesias, S., Brodersen, K.H., Friston, K.J. et al. (2014). Uncertainty in perception and the hierarchical Gaussian filter. *Frontiers in Human Neuroscience*, 8, 825.

Menary, R. (2007). *Cognitive integration: mind and cognition unbounded*. Basingstoke, UK: Palgrave Macmillan.

Metzinger, T. (2009). *The ego tunnel*. New York: Basic Books.

Noë, A. (2004). *Action in perception*. Cambridge, MA: MIT Press.

Orlandi, N. (2013). Embedded seeing: vision in the natural world. *Noûs*, 47(4), 727–47.

Orlandi, N. (2014). *The innocent eye: why vision is not a cognitive process*. Oxford: Oxford University Press.

Payzan-LeNestour, E. and Bossaerts, P. (2011). Risk, unexpected uncertainty, and estimation uncertainty: Bayesian learning in unstable settings. *PLoS Comput Biol*, 7(1), e1001048.

Pearl, J. (1988). *Probabilistic reasoning in intelligent systems: networks of plausible inference*. San Fransisco: Morgan Kaufmann Publishers.

Pearl, J. (2000). *Causality*. Cambridge: Cambridge University Press.

Ramsey, W. (2007). *Representation reconsidered*. Cambridge: Cambridge University Press.

Ramsey, W. (2015). Must cognition be representational? *Synthese*, 1–18.

Rescorla, M. (2015). Bayesian perceptual psychology. In: M. Matthen (ed.), *The Oxford handbook of the philosophy of perception*. Oxford: Oxford University Press, pp. 694–716.

Seth, A.K. (2014). A predictive processing theory of sensorimotor contingencies: explaining the puzzle of perceptual presence and its absence in synesthesia. *Cognitive Neuroscience*, 5(2), 97–118.

Seth, A.K. (2015). The cybernetic Bayesian brain. In: T.K. Metzinger and J.M. Windt (eds.), *Open MIND*. Frankfurt am Main: MIND Group, pp. 1–24.

Spaulding, S. (2012). Overextended cognition. *Philosophical Psychology*, 25(4), 469–90.

Synofzik, M., Vosgerau, G., and Newen, A. (2008). I move, therefore I am: a new theoretical framework to investigate agency and ownership. *Consciousness and Cognition*, 17(2), 411–24.

Thompson, E. (2007). *Mind in life: biology, phenomenology, and the sciences of mind*. Harvard: Harvard University Press.

van Gelder, T. (1995). What might cognition be, if not computation? *Journal of Philosophy*, 91, 345–81.

Varela, F., Thompson, E., and Rosch, E. (1991). *The embodied mind*. Cambridge, MA: MIT Press.

Woodward, J. (2003). *Making things happen*. New York: Oxford University Press.

CHAPTER 8

INTERACTING IN THE OPEN

Where Dynamical Systems Become Extended and Embodied

MAURICE LAMB AND ANTHONY CHEMERO

INTRODUCTION

IN 1951, a Russian chemist, Boris Pavlovich Belousov, discovered a chemical reaction in which the color of the reaction solution oscillated from amber to colorless with a stable frequency. In 1951, and again in 1955, Belousov attempted to publish his discovery of this homogeneous clock reaction in major Russian chemistry journals and was rejected both times (Pechenkin 2009). Unfortunately for Belousov, clock reactions in homogeneous solutions, where the reactants are in the same phase, were not thought possible at the time of his discovery and his results were rejected out of hand. Belousov's reaction was rediscovered in the early 1960s by Anatol Markovich Zhabotinsky. Later, the Belousov-Zhabotinsky reaction, as it is now known, became a paradigmatic example of non-equilibrium thermodynamic systems. Despite its later significance and striking visual appearance, Belousov's discovery was originally rejected because it failed to fit the general scientific consensus regarding what was required of chemical clock reactions. In fact, his results were rejected because they were thought to be impossible. As it turns out, this presumed impossibility was due to a prevailing treatment of chemical reactions as closed systems at or near thermodynamic equilibrium (Burger and Field 1985). Such systems cannot exhibit more orderly behavior over time, as Belousov's reaction did. Closed systems tend toward disorder and remain disordered. A form of the mistakenly narrow view held by Belousov's original reviewers, that the system of interest must be closed and near equilibrium, also commonly appears in cognitive science today.

Cognitive systems are open, far-from-equilibrium systems. Treating them as closed systems at equilibrium makes them appear as mysterious and impossible as Belousov's reaction did to his original reviewers. In philosophy this apparent mystery is the basis for explanatory gaps.[1] Belousov's reaction appeared problematic at first because it initially became more orderly, something physical systems aren't supposed to do when they are at or near equilibrium. The "problem" with cognitive systems is that they are manifestly ordered, and apparently tend toward increased order over some period of human growth and development. In the context of extended and embodied cognition, the order of cognitive systems manifests as the product of soft assemblies of many components defined primarily by their interactions across scales. Beginning with an assumption that the order apparent in cognitive systems is specified by program-like structures makes cognition appear as impossible as Belousov's reaction did to his initial reviewers. In fact, cognition itself becomes mysterious on the assumption that cognitive systems are primarily self-contained systems. This mystery is perpetrated when philosophers or scientists imagine that brains continue to be functioning cognitive systems when isolated in a vat or imagine that cognitive practices must only occur within a small irregular spheroid of bone, blood, and skin.

Dynamical systems theory in cognitive science is a set of methods, theories, and models based on the same tools that dissolved the mystery of Belousov's far-from-equilibrium reaction in chemistry. Belousov's discovery came about in the context of a science that lacked the tools for characterizing it and hence lacked the means of explaining it (Burger and Field 1985; Winfree 1984). The same could be said for much of the history of human attempts to explain cognitive systems. In this chapter, we will focus on how dynamical systems theory provides a basis for understanding the order of cognitive systems, and consequently why, given dynamical systems theory, we should expect embodied and extended approaches to cognition to be successful.

What Do We Mean by Dynamical Systems Theory?

Dynamical systems theory is not a unified approach within cognitive science. Dynamical systems theory may refer to any number of theoretical orientations (Beer 1995, 2000; Chemero 2000, 2009; Haken 1983; Kelso 1995; Schöner and Spencer 2015; Sporns 2011; Zednik 2011). As a result, we focus narrowly on a set of hypotheses that appear in many forms of dynamical systems theories and that we believe provide a strong theoretical basis for some claims made by 4E approaches to cognition. This means that some dynamical systems approaches, perhaps even the reader's preferred approach, will not fit the framework we lay out in this chapter. Henceforth, we will use the term "dynamical systems" to refer to entities, theories, and approaches that depend on two

[1] This is a strong claim. We consider appearance of unexplained order in non-dynamical systems approaches to the world to be the primary reason why dynamical systems theories are necessary for adequate explanation of the physical world.

proposed hypotheses which we will refer to as the interaction hypothesis and the openness hypothesis.

Dynamical systems approaches of interest in the context of 4E cognition assume that features of cognitive systems are not well defined independent of one another and that cognitive systems are not well defined in isolation from one another and from the environment. These assumptions are at the core of the interaction and openness hypotheses, respectively. According to the interaction hypothesis, any state or behavior within a cognitive system is adequately characterized only in relation to other states and behaviors within the cognitive system. Interaction is not limited to momentary state features and relations. The states and behaviors of entities are also only well defined as temporally extended entities with relations extending over time. That is, a precise copy of all of the cognitive system states that constitute an individual at this moment is not identical to that individual's cognitive system if it only persists for an instant. Every cognitive system has a particular history and a particular temporal trajectory; without these, an exact copy of a set of current cognitive states isn't any particular cognitive system, and depending on its temporal trajectory, may not even count as a cognitive system. Understanding the messy picture that the interaction hypothesis entails is a central aim of many of the methods used in dynamical systems theory.

Moreover, we can treat the interaction hypothesis as a principle for delimiting systems. Relative to a given phenomenon of interest, a system is all and only those elements that depend, to a specified extent, upon one another. In other words, everything is not related to everything else; we do not live in either a perfectly random or perfectly symmetrical universe. As a result, some things are more dependent on one another than they are on other things. How much and in what ways a collection of elements must depend on one another in order to count as a system is partly grounded in how the system of interest is defined, along with empirically observable facts about entities in the collection. The openness hypothesis, then, is grounded in the fact that variations in dependence relations give rise to system boundaries and, thus, different systems. According to the openness hypothesis, the states and behaviors of any cognitive system are only persistent when they are not at equilibrium with other systems that they interact with. Openness is introduced on the assumption that dependence is not absolute; systems are bounded, but their boundaries are semi-permeable. For our current purposes, dynamical systems theory is any set of research interests, methods, theories, and hypotheses related to and in support of the interaction and openness hypotheses.[2]

Interaction and openness are empirical hypotheses, not metaphysical ones. Thus, the hypotheses do not entail that there cannot be entities whose states and behaviors are not interactive and open. Any metaphysician should be able to disprove metaphysically strong versions of interaction and openness in their sleep. Neo-mechanists with strong

[2] We should note here that the systems that enactive cognitive scientists call *autopoietic* are interactive and open (Varela, Thompson, and Rosch 1993). But not all interactive and open systems are autopoietic. Dynamical systems theory in cognitive science is applicable to both autopoietic and non-autopoietic systems.

ontological inclinations sometimes seem to assume such a proof, which entails that scientists should seek to identify just those cognitive components that they take to be isolatable and relatively independent (e.g., Machamer, Darden, and Craver 2000). As empirical hypotheses, we claim only that interaction and openness are true of cognitive systems as they occur in the context of the observable universe. Whether or not interaction and openness are metaphysically necessary for cognition is another question, one which neither dynamical systems theory nor contemporary science attempts to answer. As a result, dynamical systems theory, as we are framing it, applies to cognitive systems to the extent that interaction and openness are required for cognitive systems to occur as they have been observed or as we expect to observe them in our universe. Note that the current position, i.e., that all observed cognitive systems adhere to both the interaction and openness hypotheses, is not an extreme position. Consider that every cognitive system that has been observed is a biological system, and has been observed in a far-from-equilibrium context. For example, if we force an observed biological cognitive system to approach equilibrium, for example, by cutting off its supply of oxygen, we expect to extinguish the features that make the cognitive system cognitive—not because oxygen is constitutive of cognitive systems, but because prolonged isolation from oxygen reduces the metabolic resources available to the cognitive system, resulting in the breakdown of cells, leading to a host of potential cognitive impairments and, eventually, death. Without oxygen, the system's cognitive elements become less dependent upon one another due to the decrease in available metabolic resources. The cognitive system has become closed relative to an atmospheric system necessary for its far-from-equilibrium persistence. As the elements that make up this doomed cognitive system become increasingly independent, the system becomes less capable of producing distinctively cognitive phenomena.

We realize that characterizing the dynamical approach in cognitive science in terms of openness and interaction, rather than mathematics, is atypical. The mathematical methods of dynamical systems theory are extremely important to understanding its overall significance and application in various sciences. However, by shifting our discussion to higher-level theoretical elements involved in the dynamical systems approach to cognition, we hope to make the discussion and implications clear to an audience less interested in the technical details. For those readers inclined toward more technical discussions, we suggest any of the following resources: Haken (1983, 2000); Kaplan (1995); Kelso (1995); Strogatz (1994). Our aim in this chapter is to make the connections between embodied and extended cognitive science and dynamical systems plain. In doing so we also aim to demonstrate that there are empirically testable consequences of embodied and extended cognition.

The Interaction Hypothesis

The interaction hypothesis is more than just the claim that an entity's states and behaviors are the result of deterministic chains of interrelated causes. According to the interaction hypothesis, the states and behaviors of any entity in a cognitive system are highly dependent on the states and behaviors of some other entity or set of entities. A state or behavior that is strongly dependent on some other set of states and behaviors cannot be adequately characterized without also saying something about the states and behaviors that it is dependent upon. In such a case, while it is possible to simplify a characterization of a system by leaving off descriptions of dependence relations, doing so does not make the dependence relations any less significant. For example, in an extremely simple coupled system:

$$x_{n+1} = x_n + (y_n + z_n)^2 \quad \text{[Eq. 1]}$$

$$y_{n+1} = y_n + (x_n + z_n)^2 \quad \text{[Eq. 2]}$$

$$z_{n+1} = z_n + (x_n + y_n)^2 \quad \text{[Eq. 3]}$$

The coupling involved in this simple mathematical system entails that no equation can be solved independent of the others. Their solutions are interdependent. In this system x, y, and z represent a set of dimensions or degrees of freedom required to characterize the states and evolution of a system. If this were a mathematical model of a real system, in order to adequately characterize the collective state of the system in terms of its three state dimensions, identify the results of a manipulation of a current collective state, or predict a future collective state, we would need to observe or guess the states of all three system variables, x_n, y_n, and z_n. In actual scientific practice, dependence relations within cognitive systems are typically considerably more complicated than these. However, no matter how complicated a system is, in order to predict, manipulate, or understand a system and its behaviors, we must first observe or guess the initial conditions of the states of the system. As a system is observed over time, any state changes will be simultaneously amplified and dampened by the other states of the system. What matters for the interaction hypothesis isn't that one cannot, in principle, map out the states and trajectories of this system by solving for the individual variables given some initial conditions. Rather, for a system like this one, the individual components do not matter more than their interactions. If we focus on Eq. 1 and observe the state of x_n while approximating a solution for $(y_n + z_n)^2$, we do not thereby identify an intrinsic characteristic state of the system for x_{n+1}. Unless the values of y_n and z_n remain sufficiently small relative to that of x_n, the state x_{n+1} is not independent of y_n and z_n. According to the interaction hypothesis, cognitive systems exhibit interaction relations like Eq.1–3. The strength of the interactions among elements in a cognitive system provides the basis for the appearance of both multi-scaled phenomena and order in cognitive systems.

According to the interaction hypothesis, the components that make up a cognitive system exhibit some level of correlation—this follows from the fact that the parts interact with and are typically strongly dependent upon one another. In this case, the level of correlation among a system's components is the spatial or temporal extent over which the behavior of a system's components are quantitatively similar. Thus, to the extent that components in a system can be individuated, the individuated components will still behave in qualitatively and quantitatively similar ways. Individuation is possible when components act significantly independent of one another, where significance is determined in the context of scientific method, theory, and practice. Often, this determination depends on the aims and interests of a research program or group of researchers (Lamb 2015).

Here, it may be useful to introduce a simple example from outside the cognitive sciences. In laser systems at low energy levels, laser-active photons emit light in random directions. Low energy lasers have an extremely short temporal correlation length (Haken 1983). An observation of a laser-active atom at one moment doesn't allow for an accurate prediction or significant manipulation of the system at a later time, say, any observation of the system over 10^{-100} seconds later. However, an increase in energy pumped into the system results in an increase in the temporal correlation length. The laser-active atoms not only emit light in an increasingly similar direction, but they do so over a longer time period, e.g., approximately one second. In this ordered, high-energy state, predictions can be made regarding the behavior of the system and its constituting atoms over a one-second time period and manipulations on the system will have ramifications, i.e., will be difference-making, for the next second of the system's behavior.

We should expect temporal correlations of this sort whenever the parts of a system are mutually coupled to one another, i.e., when they are dependent upon one another (Chemero 2009; Haken 1983; Kelso 1995). We can see this even in the simple system described by Eq. 1–3. Suppose that the component or quantity that is measured by the variable x changes its value. That change will affect the values of y and z, which, in turn, affect the value of x, which, in turn, will affect the values of y and z, and so on . . . Because of this, even minor fluctuations in x (or y or z) will percolate through the system and affect its behavior for long periods of time. This percolation is the source of the long-term temporal correlation among the parts of a highly interactive system. This sort of temporal correlation is ubiquitous in biological systems, which, like lasers, are far-from-equilibrium. Human gait and heart beats, for example, exhibit long-term temporal correlation, so that small—even apparently random—fluctuations in the timing of foot falls or ventricular contractions have measurable effects minutes later (Haken, Kelso, and Bunz 1985; Hausdorff et al. 1996; Ivanov 2003; Ivanov et al. 2001; Van Orden, Holden, and Turvey 2003).

Thus, according to a dynamical systems theory approach in cognitive science, the states and behaviors of components of cognitive systems interact with one another in the way that x, y, and z or the chambers of the heart do. They exhibit long-term temporal or spatial correlations and cannot be adequately characterized independently of one another. Of particular relevance for 4E cognitive science is the spatial extent of the components of cognitive systems that exhibit dependence.

The Openness Hypothesis

The openness hypothesis is a natural extension and refinement of the interaction hypothesis. It is an extension because it is an application of interaction between systems as opposed to within systems. All known instances of cognitive systems occur in contexts where multiple cognitive and noncognitive systems interact. According to openness, cognitive systems only persist in the context of other systems. Openness is a refinement because it includes the requirement that related systems are not at equilibrium with one another. This non-equilibrium requirement is not trivial. Far-from-equilibrium systems make the appearance of system ordering without prior or built in order, i.e., self-organization, possible.

Openness requires at least three systems that are significantly different along at least one independent degree of freedom. In the most intuitive example, a pot of water on a stove can be characterized in terms of the temperatures of a burner, the pot of water, and the surrounding air. While there are many similarities and differences between these three systems, the only difference that matters for our discussion of openness is their temperatures. When the burner is on, it is at a higher temperature than the pot of water and the surrounding air. However, the temperatures of each are highly interdependent. When there is a temperature gradient from hot burner to cooler water to cold air, the system may be treated as open.[3] Depending on the temperature gradient across the system, the water molecules will act more or less like a system, i.e., they will be more or less dependent upon one another. At a certain temperature gradient, the spatial and temporal dependence relations among water molecules exponentially extends, resulting in a qualitative and quantitative shift in system behavior. Random and mostly independent molecule motions become structured over time and space, resulting in the appearance of persistent rotating cells. The laser system in the previous section is similar; the energy pumped into the system is supplied from an external source. Thus, the increasing temporal dependence of the laser-active photons is the result of the openness between the laser system and an external energy source. While these examples may seem trivial, it has been shown over and over that this type of self-organizing process can be the basis for the order and complexity that characterize biological systems (Boerlijst and Hogeweg 1991; Camazine et al. 2001; Deacon, Srivastava, and Bacigalupi 2014; Haken 1975, 1983, 2000; Kelso 1995). The systems of dependence that enable heart beats and gait patterns can self-organize because they are open to energy and information from their

[3] Because we are isolating our view to just these three systems in a particular context, it appears that the dependence relations run only in one direction, but this is an artifact of our selection of contexts. In very specific contexts, if the air were cold enough, and artificially made colder, the dependence relations could appear reversed. This is likely an artifact of some human (though more likely a Western/patriarchal) preference for categorizing things in terms of active and passive.

surrounds (Haken et al. 1985; Hausdorff et al. 1996; Ivanov 2003; Ivanov et al. 2001; Van Orden et al. 2003).

Both interaction and openness are required to account for the fact that there is life in our universe and that life can be accounted for naturalistically, without appeal to magic (i.e., inexplicable order). Living things in our universe involve complex sets of interaction relations extended over a variety of spatial and temporal scales. These interaction relations embody the order and information that make up both the biological and cognitive life of humans. Order and information are varying correlation relations that can be exploited by systems to perceive and act. Moreover, dependency over time and space account for the processes of persistence, perception, action, and reproduction central to understanding life and cogitation. However, interaction is not sufficient for these processes to occur and persist. Given the second law of thermodynamics, information in a closed system must either be quantitatively unchanging or decreasing. Information and order only persist and increase in far-from-equilibrium open systems for as long as a sufficient gradient is maintained. Earth, with its atmosphere, sun, hot core, ecosystems, books, computers, language, and so on is a complicated system of many open systems of energy and information. In this context, neither gradual death nor magic needs to be assumed to explain the existence and persistence of cognitive systems. The continual transfer of energy into, and out of, the system results in increasing variety and complexity of interactions across multiple scales. If we ignore the interactions and openness of this system, life and cognition become either impossible or magical. Belousov's contemporaries rejected even the possibility of his findings because they could not accept magic in chemistry. They missed the third option; Belousov's reaction was open and far-from-equilibrium, in the same way that many of our terrestrial systems are. The appearance of human life may seem miraculous, resulting as if by magic. But appeal to magic ignores the local openness, i.e., the local spatial and historical information and energy gradients that make up our terrestrial context. It is these energy gradients that eliminate the need for magic.[4]

According to the dynamical systems approaches we are describing, cognitive systems are self-organizing systems of components whose behavior and states are highly dependent upon each other. These systems of components can self-organize, non-magically, because they are open to energy and information from their surrounds. This is more than just an interesting backstory: it means that the conclusions of many 4E approaches to cognition are not only possible but they are plausible and testable.

Practical Effects: Interaction, Openness, Embodiment, Extension

With dynamical systems theory understood in terms of interaction and openness, we can see why dynamical systems theory is useful for embodied and extended approaches

[4] And also happens to keep us from just simply ending up dead, i.e., everything about us becomes completely independent.

to cognition. First, interaction doesn't limit cognitive systems to the boundaries that humans are most likely to notice. System boundaries are defined in terms of interactions relative to a phenomenon of interest. For cognitive systems these interactions can be measured only once a phenomenon is defined. This means that assumptions about the boundaries of cognitive phenomena can be tested by measuring interactions among proposed elements in a cognitive system. Moreover, given openness, the boundaries of a cognitive system may be variable depending on the energy and information available to the cognitive system. The internal interactions of an open system are not fixed, and are subject to change over time without explicit direction. Given the interaction and openness hypotheses, any of the forms of embodied and extended cognition become just as theoretically plausible as more traditional approaches to cognition. However, it remains to be shown which approaches to cognition are correct given the available empirical evidence in cognitive science. Note that interaction and openness are already well established, given the empirical and theoretical evidence in physics, chemistry, and mathematics, along with many other disciplines (Alligood 1997; Fisher 1983; Gleick 1988; Haken 1983; Hilborn 1994; Kaplan 1995; Kelso 1995; Lorenz 1972; Strogatz 1994). As a result, embodied and extended cognition cannot be rejected merely by appeal to established scientific theory, much less intuitions about boundaries. In the remainder of this chapter we will look at how some of these methods have been employed in the study of system boundaries.

Testable Theories of Boundedness

The interaction hypothesis indicates a way to identify and measure differently scaled systems that aren't entirely determined by human convention or human biological and socially evolved categorization abilities. This opens the door for making various 4E hypotheses testable, contrary to the claims of some cognitive scientists/philosophers (Adams and Aizawa 2001, 2009). Humans quite regularly identify the boundaries of things with varying degrees of ease, e.g., the edges of a square, the boundaries of a country, the beginning and end of a day. For many ordinary purposes our ability to identify boundaries is sufficient. However, cognitive scientists should not only rely on their intuitions regarding the spatial or temporal boundaries of cognitive systems. As discussed, systems with many strongly interacting components, like cognitive systems, will exhibit correlations that can be measured. The spatial and temporal extent of these various correlations map onto the spatial and temporal extent of both the cognitive system and its features (Lamb 2015). The threshold for counting interactions as strong enough is determined by both the extent to which the systems components are dominated by their interactions and the aims and interests of cognitive researchers (Dotov, Nie, and Chemero 2010). Regarding the latter, if the aim is to affect some change in an individual's cognitive behavior, then a threshold that includes correlated things that don't affect that behavior at all would be inadequate. The same is true for a threshold that leaves out things that have a significant effect on the behavior. However, as stated before, the amount of effect required to include an element in a system is partially determined by the aims of the researcher. Note that this is a shift from having the practitioner determine the *boundaries* of a system to determining a *threshold* for what is included in a system based on the effects of potential system elements. Often the default assumption in contemporary cognitive science has been that the spatial and temporal boundaries of cognitive systems are identical to the spatial and temporal length of the brain or nervous system, hence the current trend to append the term "neuro" to anything related to human behavior and cognition. Thus, the belief is that there is no need to determine what the system is, as it has been a priori determined. However, this assumption should be tested, and the various 4E approaches to cognition introduce alternative theories to test.

If cognition is embodied, then in at least some cases the collection of highly dependent elements that make up a cognitive system includes parts of the body outside the nervous system. That this is the case seems to us to be well established empirically (Anderson, Richardson, and Chemero 2012; Chemero 2009; Clark 2006; Gibson 1979; Haken et al. 1985). We also find Thompson and Cosmelli's (2011) thought experiment about brains in vats to be very compelling. Thompson and Cosmelli wonder what would actually be required to sustain an experiencing brain in a vat, which should be a possible cognitive system if cognition is not embodied. First, they note that the human brain's activity is largely spontaneous and endogenously controlled, and that maintaining this activity requires tremendous resources, including metabolic and hormonal processes from the body. Moreover, among the many activities of the brain is the regulation of these very metabolic and hormonal processes, including acting so as to replenish many of the raw materials for these processes. They conclude that any brain in a vat that had experiences would require a vat that is functionally equivalent to a human body in a real environment. Thus, there is no isolated or disembodied cognitive system. If the endogenous and spontaneous activity that the brain requires is non-magical, then its activities are most likely the result of the self-organizing processes of an open system. Moreover, the self-regulating metabolic and hormonal processes are highly dependent on one another. This entails at least the possibility that these processes or some aspects of them are among the set of dependent elements that make up a cognitive system. Although we believe that arguments in the cognitive sciences should not depend upon thought experiments, Thompson and Cosmelli's thought experiment illustrates how significant interaction and openness are when considering actual, as opposed to metaphysically possible, cognitive systems.

Thompson and Cosmelli's thought experiment has recently become a genuine experiment. In summer of 2015, Rene Anand presented his research on a functioning human brain that he developed in vitro from pluripotent stem cells, i.e., a real brain in a vat (Thomson 2015). There were no ethical consequences of this, Anand explained to the press, because "We don't have any sensory stimuli entering the brain. This brain is not thinking in any way." Put in the terminology we are using here, the cognitive system includes a brain and body that are strongly interacting with one another, and are open to an environment with which they are not at equilibrium. A brain that is not in this relation to a body and environment is not part of a cognitive system.

Thought experiments and in vitro brains aside, there are many actual experiments that show that cognitive systems include aspects of the brain and body. We will limit our discussion here to a single now-classic study by Kelso et al. (1998). Kelso and colleagues measured the neural activity of human participants who were moving their right index fingers in four different patterns, at varying rates, along with a metronome. The right index finger movement patterns were flexing the finger in rhythm with the metronome, flexing in syncopation with the metronome, extending the finger in rhythm with metronome, and extending in syncopation with the metronome. They found very strong correlations (with values approaching 1) between metronome rate and activity in the left sensorimotor cortex that were independent of the type of right index finger movement. That is, all finger movements (flexing and extending, syncopated and in rhythm) at the same rate came along with the same pattern of activity in the left sensorimotor cortex. In this case, parts of the brain, the broader nervous system, and parts of the musculoskeletal system make up a temporary, self-organizing system of interacting parts, each of which alters the dynamics of the others. This embodied system can self-organize because it is open to the surrounding environment, both materially (oxygen, energy, etc.) and informationally (experimental instructions, metronome, etc.). The embodied correlated system has a temporal limit in that it lasts only as long as the participant is engaged in the task; once the participant stops engaging in the task, the parts no longer correlate with one another in the same way.

If cognition is extended, the collection of interacting components that make up cognitive systems at least sometimes includes parts outside an individual biological body. That cognitive systems are extended in this way is empirically less well established than that they are embodied, but two types of extended cognitive systems have been demonstrated experimentally: multi-agent extended cognitive systems and human-tool extended cognitive systems. With respect to multi-agent extended systems, Richardson and colleagues have shown in multiple settings that humans engaged in a cooperative task while seated in rocking chairs form a system of dependent parts (Frank and Richardson 2010; Richardson et al. 2012; Richardson et al. 2007). In particular, when two participants are engaged in a cooperative task, they tend to rock in their chairs in phase with each other (Richardson et al. 2007). In this case, the correlated elements include parts of two brains, two nervous systems, and two musculoskeletal systems, along with two rocking chairs. As is the case with the brain areas and body parts in the Kelso et al. experiment discussed earlier, these parts are no longer correlated in the same way when they are not engaged in the cooperative task: the rocking patterns of participants in rocking chairs who are not cooperating are not correlated. The systems come together as one through typically unnoticed informational coupling that occurs among cooperating individuals. When working together, the coupling among cognitive and biomechanical elements identified as the individual participants dominates their individual behaviors. Thus, their unintentional rocking patterns become driven as much by their partner as by their own biomechanical and cognitive processes. The result is the emergence of a new system extending across the participants during the task. The task changes their dependency relations and the systems that they are open to. The result is a change in boundaries of the systems that they participate in and that drive their behaviors. Notably the boundaries of this system are defined by both the spatial and temporal correlations among components engaged in the task, and by the fact that we are defining the system relative to a task to be completed. If we change the task, the system boundaries will be defined differently. As a result, there are potentially as many definitions of cognitive systems as there are distinct cognitive phenomena.

Dotov, Nie, and Chemero (2010) have shown that humans and tools also form extended cognitive systems (see also Dotov and Chemero 2014; Dotov et al. 2017). Participants in these experiments played a simple video game, controlling an object on a monitor using a mouse. For a six-second period of each trial, the connection between the mouse and the object on the monitor was disrupted, so that the mouse movements no longer controlled the object on the monitor. Dotov et al. (2017) measured the movement dynamics at the hand–mouse interface and found that the participant-plus-tool formed a single dependent system during regular game play, but not during the mouse disruption. That is, during smooth playing of the video game, the human-plus-video-game were a strongly coupled system of components, dependent upon one another but open to the environment. When the coupling among this extended system's components was disrupted, the extent of the correlations broke down, resulting in a change in the system's boundaries. Note that the correlations used to identify the system boundaries were not the obvious ones. The very tiny fluctuations in the hand movements, typically referred to as noise in the data, exhibited relatively long-range temporal correlations when the mouse acted normally, and these long-range temporal correlations broke down when the mouse cursor's movements were disrupted.

When Lines are Drawn

Understanding dynamical systems in terms of interaction and openness does not result in the kind of winner-takes-all approach to cognitive science that is often argued for. The boundaries of systems are determined by complex interactions of entities in open contexts and by the phenomena that one seeks to attribute to a system or systems. As a result, it may be the case that for some—or possibly many—cognitive phenomena, the cognitive systems that produce those phenomena are strictly bounded to neural activations in a relatively well-defined region of space. On the other hand, it may be that relatively few cognitive phenomena are the result of cognitive processes confined to the skull. The proposed dynamical systems approach thus diverges from other accounts of cognition by allowing for a plurality of cognitive systems with many different forms, often existing simultaneously. This may be a tough pill to swallow. It entails that sometimes our intuitions of what counts as an individual, a person, or a cognitive agent are incorrect relative to a given set of phenomena. It also means that while what one typically refers to as oneself is a particular temporally and spatially extended and bounded thing, it is, at the same time, a part of larger and smaller cognitive systems. If this point seems counterintuitive, then it should be possible to prove it incorrect either by (1) showing empirically that for any phenomenon properly considered cognitive, the entities that account for that phenomenon are only highly interacting brain-bound entities and are only self-organized when treated as open to systems outside brain boundaries, or by (2) showing that for any phenomenon that is accounted for by a system that extends beyond the bounds of a brain, the phenomenon may not be counted as properly cognitive. The significant contribution of a dynamical systems approach to cognition is that it provides a basis for determining the validity of 4E cognition, both empirically and analytically.

Not only does the current dynamical systems approach not make any a priori commitments to the boundaries of cognitive systems, it also doesn't presuppose a particular best explanatory approach to cognitive systems. It may be the case that for many cognitive phenomena the best explanations are provided in terms of discrete-but-causally-related mechanisms. However, even if this is the case, each of these mechanisms is a system that only exists and persists as a well-bounded system of strongly interacting entities that are open in a non-equilibrium context. The spatial and temporal extents of the dimensions that define the individual mechanisms are the result of their dynamics as defined in this chapter. In this case, dynamical approaches best account for the occurrence of the mechanistic phenomena, and the mechanistic approach best accounts for the appearance of the cognitive phenomena. Variations in the dynamics that do not extinguish the mechanism do not make a difference to the cognitive phenomena, but variations in the mechanism, due to something other than complexity or openness, do make a difference to the cognitive phenomena. This point is purely hypothetical. Our aim is only to clarify that dynamical systems, as defined here, is not in any way a polemical position. It is a framework for good empirical and philosophical work that does not depend on built-in order or magic.

Conclusion

We have characterized dynamical systems research in cognitive science so as to make its applicability to embodied and extended cognition apparent. Dynamical cognitive science takes cognitive systems to be strongly interacting collections of parts that are highly correlated along defining dimensions or degrees of freedom and not at equilibrium with their surroundings. Being not at equilibrium enables the flow of energy and information, and allows the collection of dependent parts to form self-organizing patterns. Questions related to embodied and extended cognition typically concern the boundaries of particular dynamical systems. We can put this in terms of which collections of things in the neighborhood of cognition form the highly correlated, interacting system that makes that cognition possible, and which other things in the neighborhood form the context in which the cognition exists. While defining these neighborhoods is not a trivial task, it should not be done in a way that makes cognition magic or impossible (de Oliveira and Chemero 2015). According to dynamical systems theory, given the identification of some cognitive phenomenon, we should expect cognitive systems exhibiting that phenomenon to consist of many strongly interacting components. Moreover, the cognitive systems should not be isolated from other systems. Given these two claims, dynamical systems theory entails that some forms of 4E cognition are at least possible, and perhaps likely.

References

Adams, F. and Aizawa, K. (2001). The bounds of cognition. *Philosophical Psychology*, 14(1), 43–64.

Adams, F. and Aizawa, K. (2009). Why the mind is still in the head. In: P. Robbins and M. Aydede (eds.), *The Cambridge handbook of situated cognition*. Cambridge: Cambridge University Press, pp. 78–95.

Alligood, K.T. (1997). *Chaos: an introduction to dynamical systems*. New York: Springer.

Anderson, M.L., Richardson, M.J., and Chemero, A. (2012). Eroding the boundaries of cognition: implications of embodiment. *Topics in Cognitive Science*, 4(4), 717–30. doi:10.1111/j.1756-8765.2012.01211.x

Beer, R.D. (1995). Computational and dynamical languages for autonomous agents. In: R.F. Port and T. van Gelder (eds.), *Mind as motion: explorations in the dynamics of cognition*. Cambridge, MA: MIT Press, pp. 121–47.

Beer, R.D. (2000). Dynamical approaches to cognitive science. *Trends in Cognitive Sciences*, 4(3), 91–9.

Boerlijst, M. and Hogeweg, H. (1991). Self-structuring and selection: spiral waves as a substrate for prebiotic evolution. In: C.G. Langton, C. Taylor, J.D. Farmer, and S. Rasmussen (eds.), *Artificial life* (vol. 2). Redwood City, CA: Addison-Wesley, pp. 255–76.

Burger, M. and Field, R. (eds.) (1985). *Oscillations and traveling waves in chemical systems*. New York: Wiley.

Camazine, S., Franks, N.R., Sneyd, J., Bonabeau, E., Deneubourg, J.-L., and Theraula, G. (2001). *Self-organization in biological systems*. Princeton, NJ: Princeton University Press.

Chemero, A. (2000). Anti-representationalism and the dynamical stance. *Philosophy of Science*, 67(4), 625–47. doi:10.1086/392858

Chemero, A. (2009). *Radical embodied cognitive science*. Cambridge, MA: MIT Press.

Clark, A. (2006). Language, embodiment, and the cognitive niche. *Trends in Cognitive Sciences*, 10(8), 370–4.

Deacon, T.W., Srivastava, A., and Bacigalupi, J.A. (2014). The transition from constraint to regulation at the origin of life. *Frontiers in Bioscience (Landmark Ed.)*, 19, 945–57.

de Oliveira, G.S. and Chemero, A. (2015). Against smallism and localism. *Studies in Logic, Grammar and Rhetoric*, 41(1), 9–23.

Dotov, D.G. and Chemero, A. (2014). Breaking the perception-action cycle: experimental phenomenology of non-sense and its implications for theories of perception and movement science. In: M. Cappucio and T. Froese (eds.), *Enactive cognition at the edge of sense-making: making sense of non-sense*. Basingstoke: Palgrave Macmillan, pp. 37–60.

Dotov, D.G., Nie, L., and Chemero, A. (2010). A demonstration of the transition from ready-to-hand to unready-to-hand. *PLoS One*, 5(3), e9433.
Dotov, D.G., Nie, L, Wojcik K., Jinks, A., Yu, X., and Chemero, A. (2017). Cognitive and movement measures reflect the transition to presence-at-hand. *New Ideas in Psychology*, 45, 1–10.
Fisher, M.E. (1983). Scaling, university and renormalization group theory. In: F.J.W. Hahne (ed.), *Critical phenomena*. Berlin: Springer, pp. 1–139.
Frank, T.D. and Richardson, M.J. (2010). On a test statistic for the Kuramoto order parameter of synchronization: an illustration for group synchronization during rocking chairs. *Physica D: Nonlinear Phenomena*, 239(23), 2084–92.
Gibson, J.J. (1979). *The ecological approach to visual perception*. Boston: Houghton Mifflin.
Gleick, J. (1988). *Chaos: making a new science*. New York, NY: Penguin.
Haken, H. (1975). Cooperative phenomena in systems far from thermal equilibrium and in nonphysical systems. *Reviews of Modern Physics*, 47(1), 67–121.
Haken, H. (1983). *Synergetics: an introduction: nonequilibrium phase transitions and self-organization in physics, chemistry, and biology* (3rd rev. and enl. ed). Berlin: Springer.
Haken, H. (2000). *Information and self-organization: a macroscopic approach to complex systems* (2nd enl. ed). Berlin: Springer.
Haken, H., Kelso, J.A.S., and Bunz, H. (1985). A theoretical model of phase transitions in human hand movements. *Biological Cybernetics*, 51(5), 347–56.
Hausdorff, J.M., Purdon, P.L., Peng, C.K., Ladin, Z.V.I., Wei, J.Y., and Goldberger, A.L. (1996). Fractal dynamics of human gait: stability of long-range correlations in stride interval fluctuations. *Journal of Applied Physiology*, 80(5), 1448–57.
Hilborn, R.C. (1994). *Chaos and nonlinear dynamics: an introduction for scientists and engineers*. New York: Oxford University Press.
Ivanov, P.C. (2003). Long-range dependence in heartbeat dynamics. In: G. Rangarajan and M. Ding (eds.), *Processes with long-range correlations: theory and applications*. Berlin: Springer, pp. 339–72. Retrieved from http://link.springer.com/chapter/10.1007/3-540-44832-2_19
Ivanov, P.C., Nunes Amaral, L.A., Goldberger, A.L., Havlin, S., Rosenblum, M.G., Stanley, H.E. et al. (2001). From 1/f noise to multifractal cascades in heartbeat dynamics. *Chaos (Woodbury, NY)*, 11(3), 641–52. doi:10.1063/1.1395631
Kaplan, D. (1995). *Understanding nonlinear dynamics*. New York: Springer.
Kelso, J.A.S. (1995). *Dynamic patterns: the self-organization of brain and behavior*. Cambridge, MA: MIT Press.
Kelso, J.A.S., Fuchs, A., Lancaster, R., Holroyd, T., Cheyne, D., and Weinberg, H. (1998). Dynamic cortical activity in the human brain reveals motor equivalence. *Nature*, 392(6678), 814–8. doi:10.1038/33922
Lamb, M. (2015). *Characteristics of non-reductive explanations in complex dynamical systems research* [dissertation]. Cincinnati: University of Cincinnati. Retrieved from https://etd.ohiolink.edu/ap/10?0::NO:10:P10_ACCESSION_NUM:ucin1427982269
Lorenz, E.N. (1972). Predictability: does the flap of a butterfly's wings in Brazil set off a tornado in Texas? *139th Annual Meeting of the American Association for the Advancement of Science*. Washington, DC, December 29.
Machamer, P., Darden, L., and Craver, C.F. (2000). Thinking about mechanisms. *Philosophy of Science*, 1–25.
Pechenkin, A. (2009). BP Belousov and his reaction. *Journal of Biosciences*, 34(3), 365–71.

Richardson, M.J., Garcia, R.L., Frank, T.D., Gergor, M., and Marsh, K.L. (2012). Measuring group synchrony: a cluster-phase method for analyzing multivariate movement time-series. *Frontiers in Physiology*, 3, 405. doi:10.3389/fphys.2012.00405

Richardson, M.J., Marsh, K.L., Isenhower, R.W., Goodman, J.R., and Schmidt, R.C. (2007). Rocking together: dynamics of intentional and unintentional interpersonal coordination. *Human Movement Science*, 26(6), 867–91.

Schöner, G. and Spencer, J. P. (eds.) (2015). *Dynamic thinking: a primer on dynamic field theory*. New York: Oxford University Press.

Sporns, O. (2011). *Networks of the brain*. Cambridge, MA: MIT Press.

Strogatz, S.H. (1994). *Nonlinear dynamics and chaos: with applications to physics, biology, chemistry, and engineering*. Boulder, CO: Westview Press.

Thompson, E. and Cosmelli, D. (2011). Brain in a vat or body in a world? Brainbound versus enactive views of experience. *Philosophical Topics*, 39(1), 163–80.

Thomson, H. (2015, August 18). First almost fully-formed human brain grown in lab, researchers claim. *The Guardian*. Retrieved from http://www.theguardian.com/science/2015/aug/18/first-almost-fully-formed-human-brain-grown-in-lab-researchers-claim

Van Orden, G.C., Holden, J.G., and Turvey, M.T. (2003). Self-organization of cognitive performance. *Journal of Experimental Psychology: General*, 132(3), 331–50. doi:10.1037/0096-3445.132.3.331

Varela, F.J., Thompson, E., and Rosch, E. (1993). *The embodied mind: cognitive science and human Experience*. Cambridge, MA: MIT Press.

Winfree, A.T. (1984). The prehistory of the Belousov-Zhabotinsky oscillator. *Journal of Chemical Education*, 61(8), 661–3.

Zednik, C. (2011). The nature of dynamical explanation. *Philosophy of Science*, 78(2), 238–63. doi:10.1086/659221

CHAPTER 9

SEARCHING FOR THE CONDITIONS OF GENUINE INTERSUBJECTIVITY

From Agent-Based Models to Perceptual Crossing Experiments

TOM FROESE

Introduction

Asocial Assumptions at the Origins of Social Cognition Research

The existence of other subjects apart from oneself has long been a challenging topic in analytic and phenomenological philosophy, the so-called problem of other minds (for recent reviews of the literature, see Hyslop 2016; Overgaard 2012). The same is true of cognitive science: for if my first-person experience is only realized by processes within my individual body, or even just in my brain, my experience of the existence of other minds must also result from some of those internal processes. In particular, this leads to the mainstream hypothesis that my experience of another person must therefore be constituted by a mental representation created by processes in my brain, for example, by inference and/or simulation. This standard working hypothesis is often referred to as *methodological individualism*, a phrase that has been adapted from sociology to signify the sufficiency of individual processes for explaining social cognition even in the context of interaction with others (Boden 2006).

The details of this representationalist-internalist response to the problem of other minds, also known as *theory of mind*, vary substantially. But most accounts implicitly agree that whether other minds exist or not—and most if not all in the end assume that other minds do exist—is actually beside the point, at least for scientific explanatory purposes. Somewhat paradoxically, the aim is instead to explain social cognition in terms of one individual's isolated mind, such that the role played by other people is relegated to being an instrumental trigger for or external cause of social cognition mechanisms in that individual's brain. Hence, these theory of mind accounts also agree that social interaction between people cannot be *constitutive* of social cognition, i.e., the mechanisms of social cognition cannot be directly realized in terms of one's interaction with others.

To be fair, it is becoming widely accepted that social interaction may call forth special forms of shared or we-intentionality. But even so, such an "irreducibly collective mode" (p. 160) is generally taken to be realized within one individual's head (Gallotti and Frith 2013). In other words, even when it is accepted that jointly acting according to a "we intend to do" is qualitatively distinct from and cannot be reduced to a mere sum of several "I intend to dos," this concession of collective intentionality does not challenge methodological individualism, as was famously emphasized by Searle:

> Of course I take it in such cases that my collective intentionality, is in fact shared; I take it in such cases that I am not simply acting alone. But I could have all the intentionality I do have even if I am radically mistaken, even if the apparent presence and cooperation of other people is an illusion, even if I am suffering a total hallucination, even if I am a brain in a vat. Collective intentionality in my head can make a purported reference to other members of a collective independently of the question whether or not there actually are such members. (Searle 1990, p. 407)

For Searle, and according to methodological individualism more generally, it is sufficient that my mind/brain behaves *as if* there are other agents with minds of their own in my environment, but whether they really exist makes no difference for explaining the phenomenon of collective intentionality. That is, even though traditional accounts, as epitomized by Searle's proposal, generally accept that other minds exist in addition to our own, their methodological individualism means that these accounts are similarly consistent with the extreme possibility of solipsism. They are essentially neutral on this important ontological question (whereas this kind of Cartesian skepticism is incompatible with the alternative claim that interaction with others can be constitutive of collective intentionality and of certain forms of social cognition). Of course, no one takes the possibility of solipsism seriously. Yet this problematic "as-if" assumption about the existence of other minds tends to motivate the type of explanations offered by theory of mind approaches: what ultimately matters is that my intentional stance is useful when I am dealing with certain complex "objects" in my environment by allowing me to better explain and predict their future movements. Dennett has formulated a canonical version of this type of explanation:

> Here is how it works: first you decide to treat the object whose behavior is to be predicted as a rational agent; then you figure out what beliefs that agent ought to

> have, given its place in the world and its purpose. Then you figure out what desires it ought to have, on the same considerations, and finally you predict that this rational agent will act to further its goals in the light of its beliefs. A little practical reasoning from the chosen set of beliefs and desires will in most instances yield a decision about what the agent ought to do; that is what you predict the agent will do. (Dennett 1987, p. 17)

Readers not trained in analytic philosophy or classical cognitive science may be forgiven for thinking how utterly strange and far removed from our normal lives these claims are. Searle's total hallucination and Dennett's autistic detachment certainly do not do justice to the complex phenomenology of normal human sociality in which we directly experience each other as persons in their own right (Ratcliffe 2007). It is beyond the scope of this chapter to uncover the reasons for this rather strange state of affairs in cognitive science. Suffice it to note that these approaches follow rather straightforwardly from the assumption of mind–body dualism, which entails that we can perceive nothing but surface behavior whose causes are not directly given (Reddy 2008). Moreover, we can understand the motivation for theory of mind accounts when we consider that traditionally the function of perception has been limited to making internal representations of physical objects located in the external environment. For if the mind is isolated inside the head, and if only the other's object-body is in principle accessible by means of my perception, then it must require extra cognitive work to ascertain others' mental states (i.e., a theory of mind).

This assumption, often known as the assumption of hidden minds, has been criticized on phenomenological grounds (e.g., Gallagher 2012; but see Bohl and Gangopadhyay 2014). The problem is that an encounter of others as opaque objects whose behavior must be inferred and predicted may describe what it is like to suffer from certain kinds of psychopathology (Froese, Stanghellini, and Bertelli 2013), but it is not our normal condition with most of the people we meet every day. It would therefore be preferable to develop an approach that is more in line with actual experience, that is, which agrees with the fact that we always already find ourselves in a world shared with others, while at the same time staying scientifically grounded in formal models and empirical results.

Phenomenological Clarification of Social Cognition

As long noted by phenomenological philosophy, we have the direct perceptual experience that others exist, and we often share experiences with others while we engage in affiliative social interaction (León and Zahavi 2016). Moreover, the latter types of intersubjective encounters are more adequately described as two first-person perspectives temporarily becoming integrated into a second-person perspective (Zahavi 2016): you and I are having a laugh, we are eating a meal together, we are shaking hands, etc. We share an awareness of jointly being in one and the same unfolding situation. I will refer to these forms of being related by participating in each other's experience as *genuine intersubjectivity* in order to highlight that what I am interested in is that the other person as such plays a constitutive role in the social phenomenon to be explained. This term also serves to thereby distinguish it from theory of mind approaches that are at least willing to accept that the phenomenology of these intimate forms of intersubjectivity is presenting us with shared moments of experience, but that nevertheless insist on explaining it within the constraints of methodological individualism.

Although in what follows I will not explicitly argue against the other foundational assumption of theory of mind approaches, i.e., the cognitive unconscious, I will demonstrate that it is actually possible to investigate the conditions of genuine intersubjectivity without postulating any subpersonal explanations that appeal to representational concepts. Indeed, I will not even make use of belief-desire psychology at the personal level because, arguably, for the kind of social interactions we will be considering, mutually engaging in embodied practices is sufficient for adequate social understanding (Kiverstein 2011). We will consider the most minimal form of understanding that can be considered as social, namely, the understanding that one is dealing with another subject.

This phenomenological tradition of emphasizing the varied possibilities of direct perceptual and embodied interactive understanding of other minds continues to be developed by contemporary phenomenological and enactive approaches to cognitive science (Gallagher 2008; Froese and Leavens 2014; Froese and Gallagher 2012; Krueger 2012). More recently, even among mainstream approaches there is growing acceptance that perceptual experience does not merely present us with meaningless surface behavior, but we can directly perceive aspects of other minds (Wiltshire et al. 2015). The precise nature of the mechanisms that enable the direct perception of other minds, and the extent of this phenomenon, is currently hotly debated (Michael and de Bruin 2015).

The problem is that it is one thing to agree that social encounters, like sharing a meal, are qualitatively experienced as social events, meaning that we can directly perceive the participation of another subject. But it is quite another to argue that the other person genuinely takes part in the constitution of that experience (De Jaegher, Di Paolo, and Adolphs 2016). In other words, it is possible to accept the general validity of the phenomenological description, while insisting that it is still best explained by representationalist-internalist theory of mind mechanisms at the subpersonal level of one individual agent (Spaulding 2010). For example, the discovery of mirror neurons lent itself to an account of direct perception in terms of simulation theory (Gallese 2005). But even quite traditional "theory" theory of mind approaches to direct social perception are possible (Carruthers 2015), with some of them featuring new ways of thinking about the mechanisms of unconscious inferences (Friston and Frith 2015).

From an Instrumental to a Constitutive Role of Interaction

Thus, despite these phenomenological concessions from the mainstream, what still requires more systematic scientific assessment is whether an intersubjective experience at the personal level may not actually be better explained by positing an equally intersubjective mechanism also at the unconscious subpersonal level, i.e., a mechanism that involves internal and relational activities from more than one person. In other words, it remains to be seen to what extent we can replace the mainstream restriction to subpersonal mechanisms consisting only of one individual's neural activity with a much broader perspective on subpersonal mechanisms consisting of all kinds of activity crisscrossing the brains, bodies, and environment of two or more people.

This alternative possibility is being taken seriously by a growing number of researchers who adopt a more encompassing view of subpersonal mechanisms, for example, in terms of interpersonal synergies (Chemero 2016), autonomous dynamics of mutual interaction (De Jaegher 2009), and inter-brain coordination (Di Paolo and De Jaegher 2012). What these recent developments demonstrate is that, at least conceptually and methodologically, there is no longer any reason to limit our explanatory toolkits to fit methodological individualism. But are there actually any phenomena at the personal level that force us to move beyond this individualist restriction at the subpersonal level? That is, can we find empirical evidence that is consistent with genuine intersubjectivity such that there is co-constitution of one and the same shared moment of experience? Can we thereby ground our experience of being with others?

Fortunately, that finding this evidence is a promising possibility is suggested by an essential characteristic of second-person interaction, namely, reciprocity (de Bruin, van Elk, and Newen 2012). As an example let us consider in detail the reciprocity of emotional interaction, as illustrated in Figure 9.1.

Following Froese and Fuchs (2012), let us assume that in this illustration person A undergoes an emotion, e.g., anger, which is manifested in typical bodily (facial,

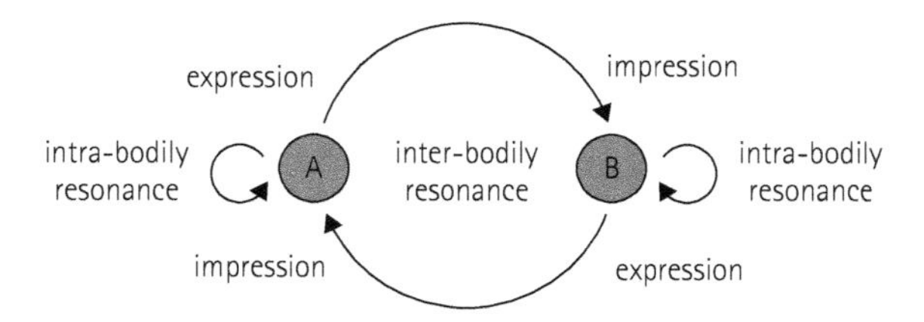

FIGURE 9.1. Illustration of pre-reflective intra- and inter-bodily affective resonance.

Reproduced from *Phenomenology and the Cognitive Sciences*, 11(2), pp. 205–35, The extended body: a case study in the neurophenomenology of social interaction, Tom Froese and Thomas Fuchs, doi:10.1007/s11097-012-9254-2 © Springer Science+Business Media B.V. 2012. With permission of Springer.

gestural, interoceptive, adrenergic, circulatory, etc.) changes. His pre-reflectively experienced lived body thus functions as a felt resonance board for the emotion: person A feels the anger as the tension in his face, as the sharpness of his voice, the arousal in his body, etc. These proprio- and interoceptive bodily feelings may be termed intra-bodily resonance. However, this intra-bodily resonance is an external expression at the same time: the anger becomes visible and is directly perceived as such by A's partner B.[1] What is more, A's expression will produce an impression in B by triggering corresponding or complementary bodily feelings. Thus, A's sharpness of voice might induce in B an unpleasant tension, a tendency to withdraw, etc. Person B not only sees the anger immediately in A's face and gesture, but also senses it with his body, through his own intra-bodily resonance. However, it does not stay like this, for the bodily reaction caused in B in turn becomes an expression perceived by A, causing another impression; it will affect his bodily reaction, change his expression, however slightly, and so forth.

Froese and Fuchs (2012) argue that there is a circular interplay of expressions and reactions running in split seconds and constantly modifying each subject's bodily state, in a process that becomes highly autonomous and is not directly controlled by either of them. They have become parts of a dynamic sensorimotor and inter-affective system that connects both bodies by reciprocal movements and reactions leading to inter-bodily resonance. Both subjects experience a feeling of being connected with the other in a dynamic way that may be termed mutual incorporation (Fuchs and De Jaegher 2009). Each lived body reaches out, as it were, to be extended by the other, dynamically forming an extended body. This is accompanied by a concrete, holistic impression of the interaction partner and a feeling for the atmosphere of the shared situation: you and I are having a tense argument that is out of control.

Note that, although this is an example of a confrontational situation rather than a co-operative one, according to the phenomenological tradition of intersubjectivity, it is still a case of interdependent empathic understanding, where empathy is defined as a basic and intuitive experience of the embodied and expressive mind of the other (Zahavi 2016). To be sure, this kind of second-person perspective is not always sufficient to constitute an integrated "we" in which two subjects share a specific experience, such as two parents identifying with each other by sharing the joy of seeing their child's first steps. In contrast, in the example described by Froese and Fuchs, the interaction resulted in interdependent yet divergent and individual-specific emotions, e.g., anger in person A and corresponding shock in person B. The emergence of a "we," in which such "I-you" interdependence is transformed into a new, qualitatively shared intersubjective perspective, may require further integration via collective concerns and values.

No appeal to theorizing and/or simulation is necessary in describing this kind of second-person interaction, at least not at the phenomenological level, although a subpersonal explanation adhering to representationalist-internalist mechanisms is

[1] There is a growing literature about the conditions for the direct perception of emotions, which I will not enter into here (see, e.g., Stout 2012).

probably always conceivable—for example, by means of fully duplicating the interaction between the other person and myself as a theory/simulation in my head. On the other hand, why duplicate anything that is there in the situation already? In the general case of perception it has long been argued that internal models are not necessary if the world itself can play the required role, so it is worth considering whether we could extend this idea to social scenarios. If it is possible to take advantage of the situational complexity of social interaction and thereby externalize some of the brain's social-cognitive load, it is unlikely that evolution would have let such an opportunity go to waste (De Jaegher et al. 2016).[2] In other words, it is also possible that we could explain such interactive direct social understanding as being grounded in a pre-reflectively lived coupling, an inter-bodily reciprocity that has created a "mixture of myself and the other" (Merleau-Ponty 1960/1964, p. 155). This is an empirically verifiable hypothesis: if we go beyond the representationalist-internalist constraints of traditional cognitive science, might it be possible to find examples of such a mixture of self and other at the level of extended subpersonal mechanisms underlying shared experience?

The mirror neuron system might be the mainstream's most suitable starting point because it is activated no matter whether the action is executed by the self or perceived as something done by the other. Indeed, a time series of mirror neuron firing could therefore involve an indistinguishable mixture of self- and other-generated activity. But this self–other neutrality is also a problem because genuine second-person interaction actually depends on the maintenance of a distinction between you and me (Zahavi 2016), and it has long been recognized that the mirror neuron system by itself is not sufficient for explaining self–other distinction (Jeannerod and Pacherie 2004). And, more importantly, mirror neuron activation does not depend on reciprocal interaction; one individual's passive observation of another's action is sufficient as a trigger.

Instead we are looking for a mechanism that is reciprocally distributed across two subjects while they are still retaining their distinct individuality. Thus, a phenomenological analysis of intersubjectivity can lead to a very different conceptualization of how we come to know other minds, relegating reasoning by analogy or simulation to a derivative role dependent on direct perception, while instead emphasizing the more fundamental role of embodied co-determination of autonomous systems (Fuchs and De Jaegher 2009). In the words of Merleau-Ponty:

> Between my consciousness and my body as I experience it, between this phenomenal body of mine and that of another as I see it from the outside, there exists an internal relation which causes the other to appear as the completion of the system. (Merleau-Ponty 1945/2002, p. 410)

[2] A range of evolutionary robotics experiments, in which embodied agents are evolved to solve a task in an environment populated with other agents, supports this conjecture. Invariably, the evolutionary algorithm encounters solutions to the task whereby the mechanisms generating an agent's performance cannot be separated from its ongoing interaction with other agents (see, e.g., Di Paolo, Rohde, and Iizuka 2008; Froese, Iizuka, and Ikegami 2013).

On this view, one's engagement with another person is not something external to the process of perceiving them as such; instead it is constitutive of an irreducible second-person perspective of "me and you" that reciprocally integrates the first-person perspectives of both subjects. Even more radically, it is plausible that such embodied interaction is our primary social capacity, developmentally and in adult life (Reddy 2008), whereas the first- and third-person perspectives, as appealed to by simulation and theory versions of theory of mind, respectively, are the basis of derivative modes giving rise to more complex forms of social understanding (Fuchs 2013; Merleau-Ponty 1960/1964; Gallagher 2012). Roughly, on this view, it seems plausible that pre-reflective embodied interaction without much self–other distinction is the most fundamental social mechanism that enables more explicit forms of self-awareness, which gives rise to a more fully articulated second-order perspective, which then enables more advanced social interaction skills to be deployed, and so forth in a dialectical spiral. At some point this history of social interaction will enable the emergence of more individual forms of social cognition that have been the focus of theory of mind approaches, namely, those involving detached reflection and pretense.

The upshot is that there is nothing standing in the way of the possibility of a cognitive science of genuine intersubjectivity. Moreover, if we do not specifically look for the involvement of the other when trying to account for social processes, it is unlikely that we will find such involvement, thereby inadvertently turning methodological individualism into something like a self-fulfilling prophecy. It is time for a more comprehensive research program to be established in order to better elucidate the potential mechanisms of genuine intersubjectivity.

Toward a Science of Genuine Intersubjectivity

The enactive approach to social cognition is in a promising position to meet the requirements (De Jaegher and Di Paolo 2007; Froese and Di Paolo 2011; Froese and Gallagher 2012; Fuchs and De Jaegher 2009). Its novel research program enables us to better integrate both empirical and phenomenological findings in order to determine to what extent embodied interaction can indeed give rise to a socially extended mind in the strong sense of genuine intersubjectivity. Importantly, it argues that social interaction dynamics can be constitutive of social cognition (De Jaegher, Di Paolo, and Gallagher 2010; Di Paolo and De Jaegher 2012), which is in line with phenomenological analyses of intersubjectivity, for example, of how the second-person perspective is constituted by the reciprocal interlocking of two first-person perspectives (León and Zahavi 2016). The enactive approach, in agreement with the phenomenological tradition, therefore marks a clear break with theory of mind approaches. Yet it can also

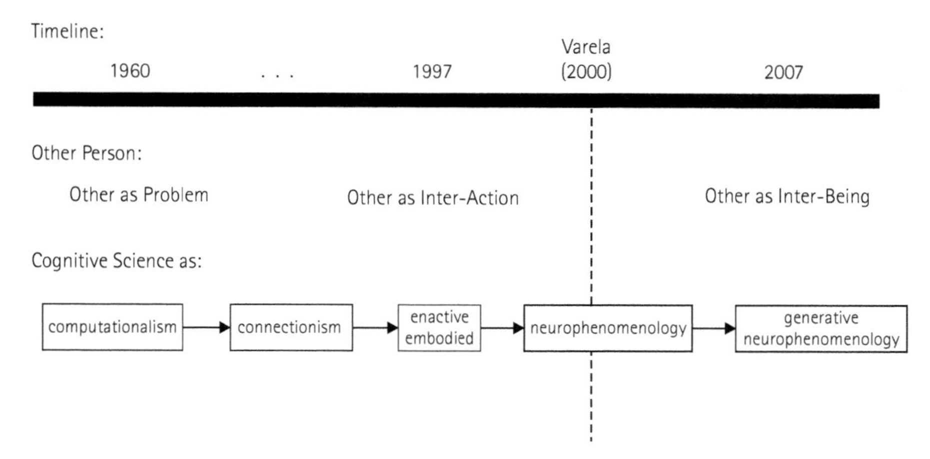

FIGURE 9.2. Steps to a science of inter-being. According to Varela (2000), the successive stages in the history of cognitive science can be seen as corresponding to a transformation in the scientific understanding of the other. He projected that in the 2000s there will arise a cognitive science of inter-being, what we have called the enactive approach to genuine intersubjectivity.

Reproduced from *The Psychology of Awakening: Buddhism, Science, and Our Day-to-Day Lives* by Gay Watson. Published by Rider. Reprinted by permission of The Random House Group Limited. © 1999, The Random House Group Limited.

be seen as a development arising from within the history of cognitive science, as illustrated in Figure 9.2.

As we have already seen, for classical cognitive science, here represented by the paradigms of computationalism and connectionism, the other's mind is mostly a problem: the self is an isolated agent who must figure out how to survive in a world populated by potentially mindless bodies. For embodied cognitive science and static neurophenomenology, i.e., approaches that acknowledge the role of embodied interaction and direct perception without changing their underlying presuppositions about the limits of constitution, the other is mainly conceived of in terms of an affordance for interaction; the appearance of the other is a given fact but external links must still be built between two independent minds. At this intermediate stage, which is the mainstream's state of the art (Wiltshire et al. 2015), there is still no room for Merleau-Ponty's radical concept of an internal relation that allows the other to be the completion of a larger, integrated system.

Yet Varela already foresaw the seeds of a future stage of cognitive science in which genuine intersubjectivity, what he calls inter-being, i.e., an interaction in which we co-constitute a shared awareness in contrast to an interaction that remains external input to independent individuals, may be thematized without contradiction: "The other and I are common ground, a joint tissue which is tangibly present in empathy and affect, which offer a possible level of analysis if we avail ourselves of the means to do so" (Varela 2000, p. 87). However, while Varela was optimistic about this transformation of social cognition research, he was also cautious about the speed of acceptance of the alternative view given how ingrained methodological individualism and skepticism of phenomenology are in the modern worldview:

> These are heavily inertial assumptions that will move as slowly as continents. The natural attitude of the scientist and the public today is to see the mind as a distinct, brain-encased self. Breaking that illusion from within science seems, today, not a complete impossibility—some cracks are opening for a science of interbeing. (Varela 2000, p. 87)

Indeed, such cracks have been appearing since Varela published his projected timeline, with contributions picking up speed and recognition since a decade ago.

Methodology

We can group these contributions into three major pillars that together form the overall enactive research program: phenomenology, theory, and experiment, as illustrated in Figure 9.3. As we have seen, much progress has already been made in terms of the first two pillars. Classical and contemporary phenomenological philosophy provide descriptions of intersubjectivity, including empathy, direct social perception, second-person perspective, intercorporeity, etc. (Gallagher and Zahavi 2008). The theory has been developed by the enactive approach to social cognition, which emphasizes the constitutive role of social interaction in terms of a variety of concepts, such as self-other co-determination (Thompson, 2001), participatory sense-making (De Jaegher and Di Paolo 2007), mutual incorporation (Fuchs and De Jaegher 2009), self–other contingencies (McGann and De Jaegher 2009), and the extended body (Froese and Fuchs 2012). This is not the place to compare and contrast all of these different concepts. What they share in common is the aim to highlight that the embodied mind is intersubjectively constituted at its most fundamental levels.

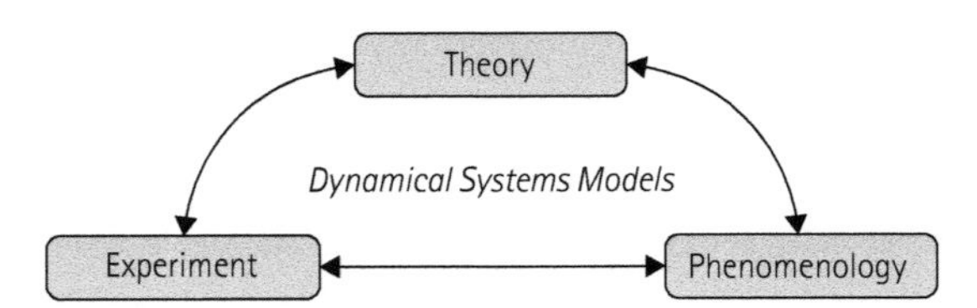

FIGURE 9.3. The three pillars of the enactive research program. The overall aim is to integrate theory, experiment, and phenomenology into one coherent framework by mutually informing and constraining interdependencies. Ideally, the three pillars should be formally integrated by means of dynamical systems theory. When the experiments are conducted in the field of neuroscience, this research program is known as neurophenomenology (e.g., Varela 1999).

Reproduced from *Phenomenology and the Cognitive Sciences*, 11(2), pp. 205–35, The extended body: a case study in the neurophenomenology of social interaction, Tom Froese and Thomas Fuchs, doi:10.1007/s11097-012-9254-2 © Springer Science+Business Media B.V. 2012. With permission of Springer.

However, there continues to be a lamentable disconnect between this growth in phenomenological and theoretical research on the one hand and the relative lack of experiments on genuine intersubjectivity on the other. To be fair, there are some compelling psychological studies on the potentially constitutive role of real-time embodied interaction in social cognition (Auvray and Rohde 2012; Dumas, Kelso, and Nadel 2014; De Jaegher et al. 2010). And there is a growing sophistication of hyperscanning methods in social-cognitive neuroscience, which have revealed various ways in which brains mutually influence each other during online social interaction (Dumas, Laroche, and Lehmann 2014; Chatel-Goldman et al. 2013; Schilbach et al. 2013; Hari et al. 2015). But so far there exists not a single study aimed at specifically verifying if genuine intersubjectivity is possible by integrating all of the methodological elements shown in Figure 9.3.

One important factor is that many of the studies on social interaction whose findings might be interpreted from the perspective of genuine intersubjectivity are conceived and analyzed from within the strict confines of methodological individualism (Chatel-Goldman et al. 2013). Accordingly, no evidence of social interaction could ever amount to evidence of Varela's inter-being if the latter is not even conceivable in principle. Another important factor is that none of these studies included a systematic assessment of the experience that participants undergo during their interaction. Here the enactive approach collides with yet another deep-seated prejudice found in experimental cognitive science, namely, that subjective reports are not trusted as reliable data. This premise is also slowly changing, partly because of the concerted push by phenomenological and enactive approaches for lived experience to be taken seriously (Froese, Gould, and Barrett 2011), and partly because the burgeoning field of consciousness science is in need of more reliable methods of measurement (Sandberg et al. 2010). In general, attitudes toward consciousness are changing, even if important practical difficulties remain.

For the enactive approach to social cognition the more fundamental of these two problematic factors is the continued insistence on methodological individualism as the gold standard of acceptability in the sciences of the mind. For as we have seen, even if it has become more accepted that we have the experience of directly perceiving other-minded beings and sharing experiences with them, this insight by itself is not sufficiently compelling for most cognitive scientists to expand their traditionally internalist scope of subpersonal explanations so as to consider social interaction dynamics in themselves as a part of the necessary causal or even constitutive mechanisms. To change this conservative attitude, it would help if it could be demonstrated that in principle there is nothing mysterious about the theoretical shift from treating an interaction as an external affordance for *independent* individuals to conceiving their interaction as an opportunity for the establishment of an internal relation between *interdependent* individuals.

Agent-Based Modeling as Proof of Concept

Due to this conceptual requirement of demonstrating an in-principle possibility, dynamical systems theory has been playing a somewhat different role in the enactive approach to social cognition, compared to how Varela (1999) had envisioned its place in his neurophenomenology of time consciousness. For him a key advantage of explanations in terms of dynamical system theory, in contrast to the representational terminology of the cognitive unconscious, is its neutrality with respect to scale and domain. The theory can account for and integrate activity in brain, body, and environment, including interactions with others, by abstractly modeling the relevant dynamics (Kelso 1995). Moreover, it therefore does not matter whether the phenomena to be explained are subjective or objective; activity in both domains is unfolding in time and can therefore be integrated into one explanatory framework.

Finally, and here we briefly return to the mainstream's problematic assumption of the cognitive unconscious, using dynamical systems theory to explain what is going on at the subpersonal level has another double advantage: (1) it makes these accounts formally continuous with the standard mathematics of the rest of the natural sciences, in contrast to the fundamental break introduced by the mainstream's appeal to homuncular or representational mechanisms; and (2) it thereby helps to refocus the study of subjectivity to where it belongs—in the concrete lifeworld of the personal level—rather than projecting its abstractions into the hypothetical domain of the cognitive unconscious.

These advantages are retained in the study of genuine intersubjectivity, but the aim of formally integrating concrete subjective and objective descriptions into one dynamical account faces considerable difficulties at the moment. Currently we do not have sufficiently fine-grained descriptions of the real-time interaction dynamics in its objective and especially its subjective dimensions. Instead we make use of another key advantage of the dynamical approach, namely, that it lends itself to creating and analyzing minimal agent-based models of cognition (Beer 2000). These simulation models can also be related to the three pillars of the enactive research program (Figure 9.3), e.g., by helping to formalize theoretical claims and phenomenological insights and by providing fresh perspectives on psychological and neuroscientific studies (Froese and Gallagher 2012; Rohde 2010).

One of the most basic insights of the dynamical approach is that all behavior is a relational phenomenon belonging to the whole brain–body–environment system (Beer 2000). Accordingly, all behavior is extended and must be explained in an appropriately distributed manner. Social behavior is no exception, as confirmed by a series of modeling studies, which lead us to conceive of social interaction as a relational property of one integrated brain–body–environment–body–brain system (Froese, Iizuka et al. 2013). This idea is illustrated in Figure 9.4.

What this dynamical perspective to social interaction implies is that changes to neural, bodily, environmental, or intersubjective conditions can lead to changes at level of the system as a whole, which in turn changes the conditions of the components. In

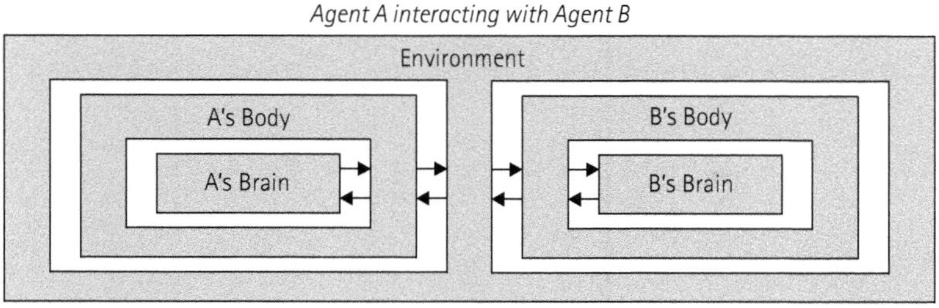

FIGURE 9.4. A dynamical perspective on the interaction between two situated, embodied agents. An agent's nervous system (abbreviated as "brain"), body, and environment are conceptualized as nonlinear dynamical systems that are parametrically coupled, thus forming an irreducible whole. When Agent A is interacting with Agent B, their mutual coupling constitutes a whole brain–body–environment–body–brain system.

Reproduced from Tom Froese, Hiroyuki Iizuka, and Takashi Ikegami, From synthetic modeling of social interaction to dynamic theories of brain–body–environment–body–brain systems, *Behavioral and Brain Sciences*, 36(4), pp. 420–1, doi:10.1017/S0140525X12001902 © Cambridge University Press 2013. Reproduced with permission.

this way social interaction is constitutive of social cognition: an individual agent's social behavior depends on the coupling of all the subsystems, including the other agent, and cannot be attributed to any one component in isolation. Importantly, what this means is that the brain is not an isolated black box that must try to represent what is happening in the outside world in order to predict what will happen next and to infer what it ought to do in response (Clark 2013; Dennett 1987). Rather, the brain is an important, even essential, part of the whole intersubjective situation and is thereby directly participating in the social interaction (Gallagher, Hutto, Slaby, and Cole 2013). We can illustrate this implication in terms of a minimalist agent-based model. This model suffers from obvious practical limitations, such as reducing an agent's situatedness and embodiment to nothing but being a moving object in an empty space, as well as from in-principle limitations, especially the lack of an experiential dimension. Nevertheless, it is also precisely because of the model's minimalism that we stand a chance to understand its emergent behaviors in some detail.

Froese and Fuchs (2012) describe a computer simulation in which two agents are situated in an open-ended 1D space along which they can move left- and rightwards. One agent is oriented upward and the other downward. The agents are equipped with a touch sensor, which becomes active when the bodies of the agents are overlapping and remains turned off otherwise. The artificial brain of each agent is a standard continuous-time recurrent neural network that consists of three fully interconnected neurons. Activity of all the neurons is modulated by the status of the touch sensor, and two dedicated motor neurons modulate the velocity of movement, thereby forming a closed sensorimotor loop.

The parameters of the neural network (weights, time constants, biases) and of the way in which it is coupled with the body and environment (sensor and motor gains) are automatically shaped by an evolutionary algorithm, which selects pairs of structurally

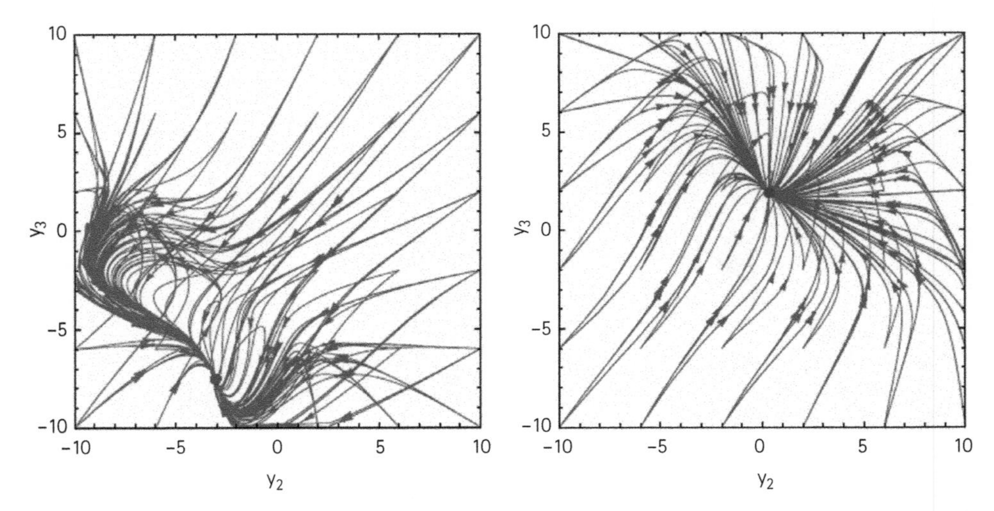

FIGURE 9.5. Representative flow structure of a section of the activation space of an agent's continuous-time recurrent neural network. Notice how the structure of this system changes depending on whether the contact sensor is fixed to off (left) or on (right).

Reproduced from *Phenomenology and the Cognitive Sciences*, 11(2), pp. 205–35, The extended body: a case study in the neurophenomenology of social interaction, Tom Froese and Thomas Fuchs, doi:10.1007/s11097-012-9254-2 © Springer Science+Business Media B.V. 2012. With permission of Springer.

identical agents according to how well they coordinate their behavior. The task consists of finding each other, negotiating a shared direction of movement, and jointly moving toward that direction while repeatedly making contact. Once adequate social interaction has evolved it is possible to analyze the state space of an isolated neural network, as shown in Figure 9.5.

To facilitate the analysis, we focus only on the activation vector of a subsystem consisting of the two motor neurons y_2 and y_3, whose outputs modulate left- and rightward components of overall velocity, respectively. To isolate the agent from its environment, we fix its sensor to be continuously off or on. In both conditions flows of activation eventually converge on a single stable fixed-point attractor. However, the position of this equilibrium point and the structure of its basin of attraction differ depending on the state of the sensor. In other words, for this neural network making contact does not merely serve as an external input to be processed by a fixed subpersonal architecture. Rather, contact has the effect of reorganizing the dynamical structure of the neural network's motor subsystem, such that the contact itself cannot be considered as separate from the way in which the agent's behavior is realized. But this point holds equally for all forms of sensorimotor interaction, not just for the special case when the neural activation is generated via coordinated interaction with another agent (see also Chemero 2016). What happens with the state of the motor subsystem during conditions of social interaction is qualitatively different, as can be seen in Figure 9.6.

It turns out that the motor subsystem behaves very differently when both agents coregulate their movements (top row of Figure 9.6), compared to when an agent is isolated

FIGURE 9.6. Trajectories in neural activation space for representative trials of the "live" interaction condition (top row) and "playback" condition (bottom row), while heading leftward (left column) or rightward (right column). Activations of Agent A (black line) and Agent B (gray line) always start near the origin and then progress outwards with time. The locations of the normally separately existing equilibrium points, i.e., one for each sensor status, are superimposed simultaneously for ease of reference (solid dots).

Reproduced from *Phenomenology and the Cognitive Sciences*, 11(2), pp. 205–35, The extended body: a case study in the neurophenomenology of social interaction, Tom Froese and Thomas Fuchs, doi:10.1007/s11097-012-9254-2 © Springer Science+Business Media B.V. 2012. With permission of Springer.

by fixing its input (Figure 9.5). During reciprocal interaction the agents' activations are maintained as one of two far-from-equilibrium transients, which can take place in two distinct regions of state space depending on whether the agents are jointly moving leftward or rightward. These self-organized transients of neural activity are the cause and effect of the co-constitution of a robust yet flexible coordination of behaviors. In other words, reciprocal sensorimotor interaction between the two agents spontaneously transforms the dynamical organization of their motor subsystems, permitting them to jointly generate more complex patterns of behavior than would be possible for either agent in isolation.

Despite the model's simplicity (or rather thanks to it), this finding might be a good starting point for formalizing Merleau-Ponty's claim that our perceptions of others "arouse in us a reorganization of motor conduct, without our already having learned the gestures in question" (1960/1964, p. 145). Moreover, it also provides us with a clue to formalizing his claim that when perceiving others "there exists an internal relation which causes the other to appear as the completion of the system" (1945/2002, p. 410).

This is because the maintenance of the coordinated behavior depends on the active participation of the other agent. When we prevent the possibility of a reciprocal modulation of neural dynamics, for example, when we re-initialize a trial but replace the other agent with a playback of its movements that were recorded during a previous run of that trial, the complex style of transient neural dynamics can no longer be maintained by the remaining active agent (Figure 9.6, bottom row).

This minimal agent-based model therefore serves as a formal proof of concept that an extended subpersonal mechanism that is compatible with the concept of genuine intersubjectivity, i.e., such as when we reciprocally participate in the interactive realization of each other's socially contingent behavior, is possible in principle. We are therefore in a suitable position to verify that this possibility can also be empirically confirmed in terms of actual psychological experiments of social interaction, in particular those that also take into account the experiential dimension of the participants.

Perceptual Crossing Experiments

Dynamical systems theory casts doubt on the foundational assumptions of mainstream cognitive science by questioning the necessity of its internalist-representationalist framework. In particular, conceiving of cognitive systems as nonlinearly coupled organism-environment systems "removes the pressure to treat one portion of the system as *representing* other portions of the system" (Silberstein and Chemero 2012, p. 40). In addition, it leads us to expect that "the processes crucial for consciousness cut across brain-body-world divisions, rather than being brain-bound neural events" (Thompson and Varela 2001, p. 418; but see Clark 2009). Moreover, given the proof of concept provided by the agent-based model (and many others similar to it), it suggests that there is nothing standing in the way of generalizing these implications to a multi-agent system nonlinearly integrated via social interaction, that is, in which the world of each agent includes other agents. The next step is therefore to empirically verify that the processes crucial for experiences involving others can cut across self–other divisions, as required by genuine intersubjectivity.

Methodologically, this means that we are looking for an experimental paradigm that enables us to systematically investigate the extent to which second-person interaction dynamics allow subjects to participate in each other's experience, while still retaining the tractability of the minimalist agent-based models. One suitable way is to mediate participants' sensorimotor loops via minimal human–computer interfaces that are designed to distill the full complexity of human interaction capacities to their essential features (Rohde 2010). For the study of embodied social interaction, a design that fulfills these requirements is the so-called perceptual crossing paradigm (Auvray and Rohde 2012), which was first popularized by Auvray, Lenay, and Stewart (2009).

Auvray, Lenay, and Stewart (2009) set out to investigate whether recognition of another's presence can be brought about on the basis of embodied interaction in a minimalist virtual environment. No other interaction between the participants is

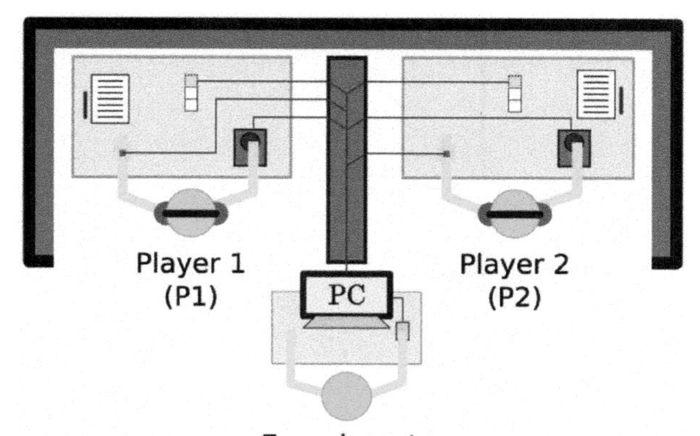

FIGURE 9.7. Experimental setup of a perceptual crossing experiment. The two players can only engage with each other via a human–computer interface that reduces their scope for embodied interaction to a bare minimum of translational movement and binary tactile sensation.

Reproduced from Tom Froese, Hiroyuki Iizuka, and Takashi Ikegami, Embodied social interaction constitutes social cognition in pairs of humans: A minimalist virtual reality experiment, *Scientific Reports*, 4(3672), Figure 1, doi:10.1038/srep03672 © 2014 Tom Froese, Hiroyuki Iizuka, and Takashi Ikegami. This work is licensed under the Creative Commons Attribution 3.0 Unported License. It is attributed to the authors Tom Froese, Hiroyuki Iizuka, and Takashi Ikegami.

possible other than binary tactile feedback. The basic experimental setup, with minor modifications introduced by Froese, Iizuka, and Ikegami (2014), is shown in Figure 9.7. Each participant's interface consists of two parts: a trackball that controls the linear displacement of their virtual avatar, and a handheld haptic feedback device that vibrates at constant frequency for as long as the avatar overlaps with another virtual object, and remains off otherwise. Participants are separated by a wall and wear noise-canceling headphones. Three small lights on each desk signal the start, halftime (30 seconds), and completion of each one-minute trial.

Players are virtually embodied as "avatars" in an invisible 1D space that wraps around, as illustrated in Figure 9.8. A player can encounter three types of object that have equal sizes and give rise to the same constant haptic feedback upon contact: a static object and two moving objects, one of which is the other player's virtual avatar and the other is an irresponsive object that directly copies the other's movements. Participants are informed about these three types of object, although they are not told that the irresponsive mobile object is an immediate copy of the other's movements. The point of this "shadow" object is to determine if players are sensitive to the reciprocity of interaction as such, i.e., to its social contingency, rather than to just to the presence of external motion alone. The task given to each player is to click whenever they judge themselves to be interacting with the other player's avatar. No explicit feedback regarding the correctness of a participant's click is provided during the experiment, and a click is unperceivable to the other participant.

Auvray, Lenay, and Stewart (2009) discovered that even under these minimalist and ambiguous conditions, most clicks happened during interaction with the other

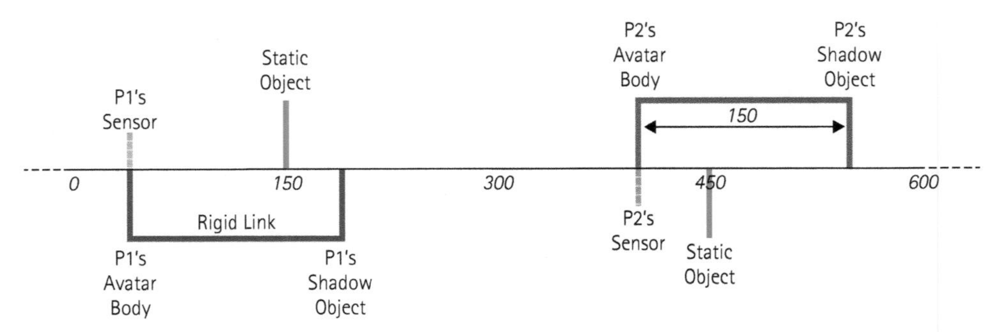

FIGURE 9.8. Virtual environment of a perceptual crossing experiment. Each avatar consists of a binary contact sensor and a body object. Unbeknownst to the players, a "shadow" object is attached to each avatar body at a fixed distance of 150 units. There are also two static objects, although only one is noticeable to each player. All objects are 4 units long and can therefore only be distinguished interactively in terms of their different affordances for engagement.

Reproduced from Tom Froese, Hiroyuki Iizuka, and Takashi Ikegami, Embodied social interaction constitutes social cognition in pairs of humans: A minimalist virtual reality experiment, Scientific Reports, 4(3672), Figure 2, doi:10.1038/srep03672 © 2014 Tom Froese, Hiroyuki Iizuka, and Takashi Ikegami. This work is licensed under the Creative Commons Attribution 3.0 Unported License. It is attributed to the authors Tom Froese, Hiroyuki Iizuka, and Takashi Ikegami.

participant. The study was therefore a success. However, a closer look at the results challenged the most straightforward explanation, i.e., that participants clicked correctly because they were sensitive to social contingency. Participants were indeed less likely to click following tactile stimulation by a static object, but there was no significant difference between the probability of a click following stimulation by the other's avatar or by their shadow object. In other words, this suggests that individuals could not properly distinguish between the two moving objects from their own perspective.

So how is it possible that the experiment was a success? It turns out that while participants were not more likely to respond to contact with the other player, it was nevertheless more likely that they were in fact in contact with them than with the shadow object—thereby guaranteeing that most clicks following a stimulation by a moving object were in fact correct. And the reason why they tended to spend more time with each other was an emergent property of the situation: mutual interaction was relatively more stable than interaction with the other's shadow, given that in the latter situation the other player would still be searching for the partner and thereby dragging their shadow along with them. A one-sided interaction is more unstable and difficult to maintain, as we already saw from the spontaneous breakdown of interaction during the playback condition of the agent-based model.

This first perceptual crossing study was welcomed by many enactivists as an experimental demonstration of the claim that social interaction can constitute social cognition, at least under some conditions (De Jaegher et al. 2010). Yet this radical interpretation of the results has received extensive criticism and may not be tenable in its original form, at least not if we take cognition to be a form of sense-making, that is, as a process of which an essential component is its being meaningful for the agent.

Importantly, the experiment provided no evidence that social cognition was actually shaped by the social interaction. To the contrary, participants were equally likely to click no matter whether the moving object that had caused the stimulation afforded a reciprocally co-regulated interaction or not. The relatively enhanced stability of the reciprocal interaction dynamics may have permitted the external self-organization of the objective correctness of the clicks, but that notable achievement is arguably not sufficient to talk about constitution of cognition (Overgaard and Michael 2015). For the social interaction to be essentially constitutive, rather than just contextually enabling, we expect something to be qualitatively different for the participants during reciprocal interaction that marks it as a social situation, whether by transformation of the pre-reflectively felt meaning, perceptual experience, or judgment of the situation, which should have been reflected in their clicking behavior (for an alternative interpretation of constitution, see De Jaegher et al. 2016).

Moreover, the null result of a lack of co-constitution of a genuine second-person perspective should actually be expected according to the enactive approach. As Froese and Di Paolo (2011) have argued, just as sense-making in the individual case requires the regulation of interaction rather than just interaction per se, sense-making in the intersubjective case requires the co-regulation of reciprocal interaction rather than just individually regulated interaction that coincidentally happened to be with another subject. Co-regulation signifies that the conditions of normativity are distributed across the agents. For example, the success of a handshake cannot be reduced to one person alone; the other person has to make a corresponding gesture. And the original experiment's task of agency detection does not require such co-regulation: we can imagine an extreme case in which a participant does not move at all and simply clicks when receiving consecutive stimulations, which is indicative of the searching activity of the other player.

Froese, Iizuka, and Ikegami (2014) set out to experimentally verify Froese and Di Paolo's argument by changing the instructions given to the players at the start of the perceptual crossing experiment, and by collecting subjective reports about the experience of the interaction preceding each click. Crucially, they told pairs of participants that they formed a team and that they needed to help each other inside the virtual environment to recognize each other's presence because their combined clicking performance would be ranked against other teams. The aim of these prosocial instructions was to encourage the formation of a "we."

This variation of the perceptual crossing paradigm resulted in several interesting findings. First, in contrast to the study by Auvray et al. (2009), participants were significantly more likely to click on the other's avatar than on the other's shadow, which provides compelling empirical support for the claim that social interaction can be constitutive of social cognition, even when adopting stricter criteria such as that the interaction process must also give rise to qualitative changes at the personal level. Second, in trials when participants jointly clicked correctly on each other, these clicks were more likely to lead to reports of a significantly clearer experience of the presence of the other subject, as measured by ratings using an adapted perceptual awareness scale (Sandberg et al. 2010). Third, clicks in trials with joint success and associated with clear social perception were preceded by increased co-regulation of behavior, as measured by the extent of turn-taking between players. Turn-taking is interesting because its normativity is structured like a handshake: you cannot successfully turn-take by yourself. Fourth, the delays between jointly successful clicks tended to be small, with the most common delay range being within two seconds of each other. This is suggestively close to the one-second time scale that Varela (1999) took to be indicative of the time taken by large-scale neural integration during the emergence of a coherent cognitive act, at least for the case of one individual. Could a similar mechanism of neural integration be at work when the process involves more than one brain? This is an intriguing hypothesis. However, to further support this possibility of large-scale inter-brain integration, we would need to conduct a perceptual crossing study that includes a double-EEG analysis of the interacting players.

Conclusions

Overall, enactive theory, models, and experiments converge on the claim that co-regulated social interaction constitutes genuine intersubjectivity, such that it is indeed possible for us to mutually participate in the generation of our intentions, emotions, and experiences more generally. This is a desirable outcome because it means that cognitive science can be made consistent with the folk psychological intuition that the presence of others makes a difference to how we behave, think, and live. Others transform how the world shows up for us.

In contrast, according to the internalist starting point adopted by mainstream social cognition research, all aspects of sociality must ultimately be consistent with an overall ontology and metaphysics in which isolated individuals (or even just their brains) are the sole repositories of intentionality, whether individual or collective. If the mainstream were to allow even the mere possibility of forms of cognition whose underlying mechanisms are not reducible to what is going on within isolated individuals, then methodological individualism would have to be replaced with a framework that is at least potentially capable of interindividual explanations. This would have far-ranging repercussions for how we understand ourselves as well as our place in this world.

Searle is to be commended for making explicit the constraints imposed on our understanding by mainstream cognitive science:

(1) It must be consistent with the fact that society consists of nothing but individuals. Since society consists entirely of individuals, there cannot be a group mind or group consciousness. All consciousness is in individual minds, in individual brains.
(2) It must be consistent with the fact that the structure of any individual's intentionality has to be independent of the fact of whether or not he is getting things right.... One way to put this constraint is to say that the account must be consistent

> with the fact that all intentionality, whether collective or individual, could be had by a brain in a vat or by a set of brains in vats. (Searle 1990, pp. 406–7)

Searle recognizes that these constraints resemble methodological individualism and methodological solipsism, respectively, although he prefers to construe them "as just commonsensical, pre-theoretical requirements." Yet it should be evident by now that these requirements are neither commonsensical nor pre-theoretical; they are deeply rooted in the internalist commitments of mainstream tradition of cognitive science. To be fair, many proponents of this tradition probably pursue their work without considering the possibility of solipsism. They prefer to treat others as real subjects in their own right, albeit with the caveat that they are still forced to claim that their existence is epiphenomenal from the perspective of scientific accounts of social cognition. To be able to scientifically recognize that the presence of others does make a difference in a meaningful way, we must first overcome the constraints identified by Searle.

More importantly, as argued in this chapter, it is possible to break free from these constraints while staying within the remit of natural science. Rejecting them does not lead to accepting the existence of a mysterious group mind or consciousness (Boden 2006). Rather, it allows us to acknowledge and properly investigate the role of social interaction dynamics in constituting individual and intersubjective capacities. Once this turn toward the constitutive role of embodiment and interaction is undertaken, the brain-in-a-vat thought experiment loses its intuitive force (Thompson and Cosmelli 2011). And accepting that we sometimes get things wrong does not entail that we are never right about experiencing ourselves to be in direct relation with others and the world around us (Beaton 2013).

The upshot is that the enactive approach is bringing about a shift in mainstream ontology and metaphysics toward a worldview that is more consistent with the phenomenological approach to social reality: reciprocal interaction with other people plays a constitutive role in implicitly and explicitly shaping our lives and minds. We genuinely have relationships with others; we are not independent but interdependent. These insights return cognitive science closer to actual human experience, and thereby allow us to better appreciate and address the growing challenges of living in an increasingly interconnected world.

ACKNOWLEDGMENTS

Tom Froese thanks Guillaume Dumas and Jonas Chatel-Goldman for the many fruitful discussions. Thanks also to Ezequiel Di Paolo for his attentive comments on an earlier version of this manuscript. Detailed feedback provided by an anonymous reviewer helped to further fine-tune the arguments. This work was supported by SEP-CONACyT project #221341 CB-2013-01 ("Análisis de redes complejas en sistemas biológicos y sociales").

References

Auvray, M., Lenay, C., and Stewart, J. (2009). Perceptual interactions in a minimalist virtual environment. *New Ideas in Psychology*, 27(1), 32–47.

Auvray, M. and Rohde, M. (2012). Perceptual crossing: the simplest online paradigm. *Frontiers in Human Neuroscience*, 6(181). doi:10.3389/fnhum.2012.00181

Beaton, M. (2013). Phenomenology and embodied action. *Constructivist Foundations*, 8(3), 298–313.

Beer, R.D. (2000). Dynamical approaches to cognitive science. *Trends in Cognitive Sciences*, 4(3), 91–9.

Boden, M.A. (2006). Of islands and interactions. *Journal of Consciousness Studies*, 13(5), 53–63.

Bohl, V. and Gangopadhyay, N. (2014). Theory of mind and the unobservability of other minds. *Philosophical Explorations*, 17(2), 203–22.

Carruthers, P. (2015). Perceiving mental states. *Consciousness and Cognition*, 36, 498–507.

Chatel-Goldman, J., Schwartz, J.-L., Jutten, C., and Congedo, M. (2013). Non-local mind from the perspective of social cognition. *Frontiers in Human Neuroscience*, 7(107). doi:10.3389/fnhum.2013.00107

Chemero, A. (2016). Sensorimotor empathy. *Journal of Consciousness Studies*, 23(5-6), 138–52.

Clark, A. (2009). Spreading the joy? Why the machinery of consciousness is (probably) still in the head. *Mind*, 118(472), 963–93.

Clark, A. (2013). Whatever next? Predictive brains, situated agents, and the future of cognitive science. *Behavioral and Brain Sciences*, 36(3), 181–204.

de Bruin, L., van Elk, M., and Newen, A. (2012). Reconceptualizing second-person interaction. *Frontiers in Human Neuroscience*, 6(151). doi:10.3389/fnhum.2012.00151

De Jaegher, H. (2009). Social understanding through direct perception? Yes, by interacting. *Consciousness and Cognition*, 18, 535–42.

De Jaegher, H. and Di Paolo, E.A. (2007). Participatory sense-making: an enactive approach to social cognition. *Phenomenology and the Cognitive Sciences*, 6(4), 485–507.

De Jaegher, H., Di Paolo, E.A., and Adolphs, R. (2016). What does the interactive brain hypothesis mean for social neuroscience? A dialogue. *Philosophical transactions of the Royal Society of London. Series B, Biological sciences*, 371(1693), 20150379. doi:10.1098/rstb.2015.0379

De Jaegher, H., Di Paolo, E.A., and Gallagher, S. (2010). Can social interaction constitute social cognition? *Trends in Cognitive Sciences*, 14(10), 441–7.

Dennett, D.C. (1987). *The intentional stance*. Cambridge, MA: MIT Press.

Di Paolo, E.A. and De Jaegher, H. (2012). The interactive brain hypothesis. *Frontiers in Human Neuroscience*, 6(163). doi:10.3389/fnhum.2012.00163

Di Paolo, E.A., Rohde, M., and Iizuka, H. (2008). Sensitivity to social contingency or stability of interaction? Modelling the dynamics of perceptual crossing. *New Ideas in Psychology*, 26(2), 278–94.

Dumas, G., Kelso, J.A.S., and Nadel, J. (2014). Tackling the social cognition paradox through multi-scale approaches. *Frontiers in Psychology*, 5, 882. doi:10.3389/fpsyg.2014.00882

Dumas, G., Laroche, J., and Lehmann, A. (2014). Your body, my body, our coupling moves our bodies. *Frontiers in Human Neuroscience*, 8(1004). doi:10.3389/fnhum.2014.01004

Friston, K.J. and Frith, C.D. (2015). A duet for one. *Consciousness and Cognition*, 36, 390–405.

Froese, T. and Di Paolo, E.A. (2011). The enactive approach: theoretical sketches from cell to society. *Pragmatics & Cognition*, 19(1), 1–36.

Froese, T. and Fuchs, T. (2012). The extended body: a case study in the neurophenomenology of social interaction. *Phenomenology and the Cognitive Sciences*, 11(2), 205–35.
Froese, T. and Gallagher, S. (2012). Getting interaction theory (IT) together: integrating developmental, phenomenological, enactive, and dynamical approaches to social interaction. *Interaction Studies*, 13(3), 436–68.
Froese, T., Gould, C., and Barrett, A. (2011). Re-viewing from within: a commentary on first- and second-person methods in the science of consciousness. *Constructivist Foundations*, 6(2), 254–69.
Froese, T., Iizuka, H., and Ikegami, T. (2013). From synthetic modeling of social interaction to dynamic theories of brain-body-environment-body-brain systems. *Behavioral and Brain Sciences*, 36(4), 420–1.
Froese, T., Iizuka, H., and Ikegami, T. (2014). Embodied social interaction constitutes social cognition in pairs of humans: a minimalist virtual reality experiment. *Scientific Reports*, 4(3672). doi:10.1038/srep03672
Froese, T. and Leavens, D.A. (2014). The direct perception hypothesis: perceiving the intention of another's action hinders its precise imitation. *Frontiers in Psychology*, 5, 65. doi:10.3389/fpsyg.2014.00065
Froese, T., Stanghellini, G., and Bertelli, M.O. (2013). Is it normal to be a principal mindreader? Revising theories of social cognition on the basis of schizophrenia and high functioning autism-spectrum disorders. *Research in Developmental Disabilities*, 34(5), 1376–87.
Fuchs, T. (1996). Leibliche Kommunikation und ihre Störungen. *Zeitschrift für klinische Psychologie, Psychopathologie und Psychotherapie*, 44, 415–28.
Fuchs, T. (2013). The phenomenology and development of social perspectives. *Phenomenology and the Cognitive Sciences*, 12, 655–83.
Fuchs, T. and De Jaegher, H. (2009). Enacting intersubjectivity: participatory sense-making and mutual incorporation. *Phenomenology and the Cognitive Sciences*, 8(4), 465–86.
Gallagher, S. (2008). Direct perception in the intersubjective context. *Consciousness and Cognition*, 17(2), 535–43.
Gallagher, S. (2012). In defense of phenomenological approaches to social cognition: interacting with the critics. *Review of Philosophy and Psychology*, 3(2), 187–212.
Gallagher, S., Hutto, D.D., Slaby, J., and Cole, J. (2013). The brain as part of an enactive system. *Behavioral and Brain Sciences*, 36(4), 421–2.
Gallagher, S. and Zahavi, D. (2008). *The phenomenological mind: an introduction to philosophy of mind and cognitive science*. London, UK: Routledge.
Gallese, V. (2005). Embodied simulation: from neurons to phenomenal experience. *Phenomenology and the Cognitive Sciences*, 4(1), 23–48.
Gallotti, M. and Frith, C.D. (2013). Social cognition in the we-mode. *Trends in Cognitive Sciences*, 17(4), 160–5.
Hari, R., Henriksson, L., Malinen, S., and Parkkonen, L. (2015). Centrality of social interaction in human brain function. *Neuron*, 88(1), 181–93.
Hyslop, A. (2016). Other minds. In: E.N. Zalta (ed.), *The Stanford encyclopedia of philosophy* (spring 2016 ed.). https://plato.stanford.edu/entries/other-minds/
Jeannerod, M. and Pacherie, E. (2004). Agency, simulation and self-identification. *Mind & Language*, 19(2), 113–46.
Kelso, J.A.S. (1995). *Dynamic patterns: the self-organization of brain and behavior*. Cambridge, MA: The MIT Press.
Kiverstein, J. (2011). Social understanding without mentalizing. *Philosophical Topics*, 39(1), 41–65.

Krueger, J. (2012). Seeing mind in action. *Phenomenology and the Cognitive Sciences*, 11(2), 149–73.

León, F. and Zahavi, D. (2016). Phenomenology of experiential sharing: the contribution of Schutz and Walther. In: A. Salice and H.B. Schmid (eds.), *The Phenomenological Approach to Social Reality*. Switzerland: Springer, pp. 219–34.

McGann, M. and De Jaegher, H. (2009). Self-other contingencies: enacting social perception. *Phenomenology and the Cognitive Sciences*, 8(4), 417–37.

Merleau-Ponty, M. (1945/2002). *Phenomenology of perception* (trans. C. Smith). Milton Park, UK: Routledge.

Merleau-Ponty, M. ([1960] 1964). The child's relations with others. In: J.M. Edie (ed.), *The primacy of perception*. Evanston, IL: Northwestern University Press, pp. 96–158.

Michael, J. and de Bruin, L. (2015). How direct is social perception? *Consciousness and Cognition*, 36, 373–5.

Overgaard, S. (2012). Other people. In: D. Zahavi (ed.), *The Oxford handbook of contemporary phenomenology*. Oxford: Oxford University Press, pp. 460–79.

Overgaard, S. and Michael, J. (2015). The interactive turn in social cognition research: a critique. *Philosophical Psychology*, 28(2), 160–83. doi:10.1080/09515089.2013.827109

Ratcliffe, M. (2007). *Rethinking commonsense psychology: a critique of folk psychology, theory of mind and simulation*. New York: Palgrave Macmillan.

Reddy, V. (2008). *How infants know minds*. Cambridge, MA: Harvard University Press.

Rohde, M. (2010). *Enaction, embodiment, evolutionary robotics: simulation models for a post-cognitivist science of mind*. Amsterdam, Netherlands: Atlantis Press.

Sandberg, K., Timmermans, B., Overgaard, M., and Cleeremans, A. (2010). Measuring consciousness: is one measure better than the other? *Consciousness and Cognition*, 19(4), 1069–78.

Schilbach, L., Timmermans, B., Reddy, V., Costall, A., Bente, G., Schlicht, T. et al. (2013). Toward a second-person neuroscience. *Behavioral and Brain Sciences*, 36(4), 393–462.

Searle, J.R. (1990). Collective intentions and actions. In: P. Cohen, J. Morgan, and M.E. Pollack (eds.), *Intentions in communication*. Cambridge, MA: MIT Press, pp. 401–16.

Silberstein, M. and Chemero, A. (2012). Complexity and extended phenomenological-cognitive systems. *Topics in Cognitive Science*, 4(1), 35–50.

Spaulding, S. (2010). Embodied cognition and mindreading. *Mind & Language*, 25(1), 119–40.

Stout, R. (2012). What someone's behavior must be like if we are to be aware of their emotions in it. *Phenomenology and the Cognitive Sciences*, 11(2), 135–48.

Thompson, E. (2001). Empathy and consciousness. *Journal of Consciousness Studies*, 8(5-7), 1–32.

Thompson, E. and Cosmelli, D. (2011). Brain in a vat or body in a world? Brainbound versus enactive views. *Philosophical Topics*, 39(1), 163–80.

Thompson, E. and Varela, F.J. (2001). Radical embodiment: neural dynamics and consciousness. *Trends in Cognitive Sciences*, 5(10), 418–25.

Varela, F.J. (1999). Present-time consciousness. *Journal of Consciousness Studies*, 6(2-3), 111–40.

Varela, F.J. (2000). Steps to a science of inter-being: unfolding the Dharma implicit in modern cognitive science. In: G. Watson, S. Batchelor, and G. Claxton (eds.), *The psychology of awakening: Buddhism, science, and our day-to-day lives*. Boston, MA: Weiser Books, pp. 71–89.

Wiltshire, T.J., Lobato, E.J.C., McConnell, D.S., and Fiore, S.M. (2015). Prospects for direct social perception: a multi-theoretical integration to further the science of social cognition. *Frontiers in Human Neuroscience*, 8(1007). doi:10.3389/fnhum.2014.01007

Zahavi, D. (2016). You, me, and we: the sharing of emotional experiences. *Journal of Consciousness Studies*, 22(1–2), 84–101.

CHAPTER 10

COGNITIVE INTEGRATION

How Culture Transforms Us and Extends Our Cognitive Capabilities

RICHARD MENARY

Introduction

Cognitive integration is a contribution to the embodied, embedded, and extended cognition movement in philosophy and cognitive science and the extended synthesis movement in evolutionary biology—particularly cultural evolution and niche construction. It is a framework for understanding and studying cognition and the mind that draws on several sources: empirical research in embodied cognition, arguments for extended cognition, distributed cognition, niche construction and cultural inheritance, developmental psychology, social learning, and cognitive neuroscience.

Its uniqueness rests in its ability to account for a range of cognitive phenomena diachronically across both ontogenetic and phylogenetic time scales as well as synchronically on faster time scales incorporating real-time online bodily interactions with the local environment. Furthermore, it does so by going beyond straightforwardly causal and dynamical descriptions of the phenomena in question to include normative social practices that govern and coordinate the brain–body–environment interactions that form the core of the cognitive integration (CI) framework. The three main pillars of the CI framework are interaction, cognitive practices, and the transformation of cognitive abilities.

In this chapter I will outline the framework in terms of its core commitments, provide motivations for these core commitments, and discuss core examples of CI at work. I shall also respond to several recent criticisms of the CI framework. In the first section I provide an initial sketch of the core commitments of the CI framework. I also outline some of the main criteria for CI across several dimensions. In the second section I provide some of the primary motivations for CI, both ontogenetic and phylogenetic. In the final section I explore the relationship CI has to other 4E approaches and deal with some of the recent objections that have been raised against it.

What Is Cognitive Integration? An Initial Sketch

The CI framework is a sustained attempt to explain how human cognition has been influenced by cultural evolution. It takes as fundamental the thesis that humans are active cognitive agents who think by interacting with and manipulating their physical and social environments and that phylogenetically earlier forms of cognition are both built upon and transformed by more recent cultural innovations.

Human cognition is fundamentally interactive. Thinking, perceiving, and acting can all be understood in terms of interaction with the environment and with one another via the environment. This is not to say that there cannot be any thought without environmental interaction, but interactive (or online) thought is primary; non-interactive (or offline) thought is secondary. What I mean by this is that online cognition is a phylogenetically earlier form of cognition than offline cognition. Offline cognition builds upon existing cognitive circuitry for online cognition. This is a plausible approach given that the earliest forms of cognition were for sensorimotor interactions with the environment—for perceiving and moving, hunting, avoiding, seeking a mate, and so on.[1] For example, Trestman proposes that the Cambrian explosion involved the spread of organisms with complex sensorimotor mechanisms: "The appearance of spatially savvy, visually guided predators with complex active bodies radically changed the array of selection pressures for animals in general, across many phyla, putting a premium on mechanical defenses, crypsis, chemical defense, remote sensing, mobility, selective movement, and choice and construction of safe micro-habitats, driving the widespread adaptive radiation that makes up the CE" (Trestman 2013, p. 89).

Indeed, in some cases what is traditionally thought of as primarily offline cognition—mathematical cognition, planning, problem-solving, and remembering—is really a case of the integration of online and offline cognition. A typical case is the reuse of neural circuitry for finger gnosis for numerical cognition, where neural regions for body-schema mapping of finger position overlaps with the capacity for identifying numerical quantities (Pinel et al. 2004; Venkatraman, Ansari, and Chee 2005; Anderson and Penner-Wilger 2012). These are central cases of cognitive integration: cases in which sensorimotor capacities are built upon by enculturated capacities for creating, maintaining, and manipulating complex systems of representation and communication.

[1] Although I am not denying that the evolution of cognition is a complex matter involving the coevolution of many different cognitive traits.

CI explains this as being the result of dual component transformations: our learning histories transform both our online and offline capacities. Our offline capacities for reasoning about the present and future—e.g., problem-solving and planning, imagining, inner speech—are still capable of functioning without their online components. This might lead us to suspect that we don't really need the online capacities; this would be a mistake. It is much more credible to think that both online and offline capacities mostly function together when completing cognitive tasks. To say either that there is no inner cognitive life—cognition is online all the time—or that there is only inner cognitive life (a kind of cognitive solipsism) are both positions that are too extreme to be credible.

Given that CI begins with interaction and online cognition, how does it deal with charges of ableism, or cases of extreme offline cognition such as locked-in syndrome? CI may appear to be ableist, but I don't think that it is. There are many different and various ways to interact with the environment and they don't all include able-bodied action. Impairments to senses or limbs do not necessarily preclude other routes to environmental interaction. Look, for example, at the phenomenon of echolocation in the visually impaired (Thaler 2011; Downey 2016). Some visually impaired people have discovered that they can use a form of echolocation, reflected sound waves, to perceive objects in their local environment. They do so by producing clicks with their tongues and listening for the echo. Remarkably, a number of blind people are able to determine size and distance of an object and to use the information to navigate their environment.

It might seem that a major objection to CI is locked-in syndrome. In the majority of these cases, patients had already lived a life of interacting with the environment. As such, their cognitive capacities have already been transformed by a life of learning, training, and interacting with the environment. Since these interactions (and their associated capacities) are not activated during locked-in syndrome, there must be more to the capacities than interaction alone. This is entirely right, but it doesn't follow that CI is therefore false. CI does not claim that interactions are necessary for all cases of cognition, just that they are sufficient for most. Consequently, it seems likely that there are some inner echoes of our fully engaged cognitive lives still active in locked-in syndrome.

An evolutionary argument can also be made for CI. Humans are adapted to interact with their physical and social environments. The phylogeny of cognition begins with evolved cognitive circuits that interact with environmental variables, primarily via sensorimotor operations. CI does not deny that there are complex internal processes, nor does it deny that there are representations of the environment and of hand and limb position in space. But notice the direction of fit: these cognitive circuits are adapted primarily for sensorimotor interaction with the environment. Cognition begins in exploration, interaction, and manipulation. It does not begin in the evolution of complex language-like representations with syntactic structure. Rather, the growing complexity of cognitive capacities evolves gradually and is a response to environmental complexity and hostility (Godfrey-Smith 1996; Sterelny 2003).

Growing complexity in the social and physical environment drives the evolution of cognitive complexity (Godfrey-Smith 1996; Menary 2017). Drivers of cognitive complexity include environmental complexity, social complexity, social learning in specialized learning environments, hostility and arms races, and intra- and interspecies competition for resources. Complexity and hostility partly explain the expansion of modern human cognitive capacities in the early to late Paleolithic, but they don't entirely explain human uniqueness. The uniqueness of modern human cognitive capacities is also explained by cultural inheritance, high-fidelity social learning, cultural/social intelligence, and a high degree of plasticity and redundancy in neural systems (Sterelny 2012; Menary 2014, 2015a). The coevolution of these factors produces the human syndrome[2] for creating vast and complex social systems and systems of representation that allow us to narrate, theorize, and determine the ethical and legal norms by which we shall live (Menary and Gillett 2017). It is at this point that CI really comes to the party.

These cultural and representational systems allow us to measure and quantify as well as explain physical and social phenomena (which we are partly responsible for). CI helps to explain why we innovated these complex systems of representation and interaction. It also explains how cognitive systems evolve through the creation and deployment of representational systems and the (cognitive) practices for "creating, maintaining and manipulating" those systems (Menary 2012, p. 148).

The main mechanism for the spread of representational systems and cognitive practices is social learning. Modern humans have evolved the capacity for high-fidelity learning over long developmental periods (and in some cases long into adulthood). "The evolution of accumulating social learning was one central causal factor in the evolution of human uniqueness" (Sterelny 2012, p. 23). One way of thinking about human uniqueness is in terms of phenotypic plasticity—the capacity of a phenotype to adaptively change in response to changes in the environment. In the human case, this has become a specialization for social learning, with a high degree of neural plasticity. Development takes place in specialized learning environments, and so is, partly, exogenous. Development is scaffolded, but the scaffolding becomes part of the mature system (Menary 2007, 2015).

Some of our initial interactions with the environment are based in our developmental need to explore the physical and social environments and to engage in exploratory play (Gopnik et al. 1999; Schultz and Bonawitz 2007; Menary 2016). However, these early exploratory interactions soon become scaffolded, through learning, by the acquisition of cognitive practices. These normative practices regulate the interactions by making them more specific and targeted at cognitive tasks, or bringing about cognitive ends—such as the completion of problem-solving, memory, and logical/mathematical tasks. The specific way in which the interactions can be understood is in terms of embodied engagements, or manipulations. For example, the capacity for tool use and early developing language can be recruited by cognitive practices for writing to produce cognitive ends such as writing lists to order a sequence of actions or remember in which order to perform them, completing simple or complex equations, making an argument, etc. The acquisition of cognitive practices leads to the transformation of our cognitive

[2] I'll have more to say about this in the second section.

abilities—for example, acquiring a novel ability by transforming an existing capacity (Menary 2007a, 2015a).

I now go on to explain the concept of a cognitive practice.

Cognitive Practices

Cognitive practices (CPs) are cultural practices that are cognitive in nature. The practices are acquired through learning and training, transforming our genetically endowed biological capacities and allowing us to think in ways that our un-enculturated brains and bodies do not. As a species of cultural practice, cognitive practices are patterns of action spread out across a population (Roepstorff et al. 2010; Hutchins 2011; Menary 2007a, 2013a, 2015a). As such they are normative patterned practices[3] (Menary 2016). As we have already seen, CI should be understood as a thesis about the enculturation of human cognition. It is a thesis about how phylogenetically earlier forms of cognition are built upon by more recent cultural innovations (e.g., systems of symbolic representation). The resulting integrated cognitive system is a multilayered system with heterogeneous components, dynamically interwoven into a cooperative of processes and states. The coordination dynamics of the system are, at least in part, understood in terms of the physical dynamics of brain–body–niche interactions in real time; however, they are also to be understood in terms of CPs that govern and determine those interactions (over time). To properly understand the role of CPs, we have to distinguish between their operation at both social/population levels and individual, even subpersonal, levels.

"A patterned practice approach assumes that regular, patterned activities shape the human mind and body through embodiment, and internalization. Vice versa, enacting practices shape and re-shape norms, processes, institutions, and forms of sociality" (Roepstorff, Niewhonner, and Beck 2010, p. 1052). Consequently, they should be understood primarily as public systems of activity and/or representation that are susceptible to innovative alteration, expansion, and even contraction over time. They are transmitted horizontally across generational groups and vertically from one generation to the next. At the individual level they are acquired, most often by learning and training, and they manifest themselves as changes in the ways in which individuals think, but also the ways that they act and the ways in which they interact with other members of their social group(s) and the local environment. CPs, therefore, operate at different levels (groups and individuals) and over different time scales (intergenerationally and in the here-and-now) (Menary 2013).

How do individuals embody CPs? They do so by transforming body schemas or motor programs (Menary 2007a, 2010b; Farne et al. 2007)—such as the case of finger gnosis and numerical cognition. Motor programs are acquired through learning and training, but existing programs may also be extended during training. Learning to catch,

[3] Normative in a weak sense, because they specify a cognitively right and wrong way to act.

write, type, or flake a hand-ax are examples of acquired motor programs. Cognition or thought is accomplished through the coordination of body and environment and is, therefore, governed both by body schemas and by CPs.

After this initial sketch, I move on to providing a dimensional analysis of integration bringing together some of the central criteria for integrated cognitive systems.

What is an Integrated Cognitive System?

Providing a definition of cognition in terms of necessary and sufficient conditions has not been a great success, primarily because most of the main proposals rely on definitions that are themselves contested or incomplete. For example, Adams and Aizawa (2001, 2008) have long advocated a definition of cognition in terms of representations with underived/original content. However, the concept of content, let alone underived content, is contested and incomplete (Menary 2010c; Hutto and Myin 2013). Furthermore, such definitions may be too narrow or restrictive, or they may rule out more exotic cases of cognition—for example, artificial cognition and cognition in organisms without complex nervous systems (Brooks 1999). Alternatively, other proposed criteria may be too broad: information processing may be done by too many entities that we would be loath to count in the cognitive club. Similarly, many devices compute, but are not thereby cognitive.

I won't spend time here rebutting these claims any further than I already have, save to say that nature is complex, and the very idea of natural content is itself complicated. The needs of simple animal targeting systems are unlikely to be exact replicas of sentential representations with propositional content (Sterelny 2003), but we also hope for an account of how to get from simpler tracking systems to representations with structured propositional contents. Although we might group these representations together as falling under one or another of the available naturalistic theories of content (Millikan 1993; Dretske 1988; Fodor 1990), the nature of the contents, their conditions of determination and, indeed, whether they have truth conditions or express complex propositions, will vary significantly across representational types[4] and the mechanisms that produce and consume them.[5]

We can, however, provide the main criteria for CI. This will at best provide sufficiency and should not be interpreted in terms of jointly necessary and sufficient conditions. This is because the criteria are really dimensions along which various cases can be analyzed. Therefore, the criteria provide sufficiency for cognition, but they are not necessary for it. This allows that cognition could vary along these dimensions, across

[4] Elsewhere I have provided at least a basic account of the fundamental structure of representations (Menary 2007a, 2009, 2013b, 2016) that can provide the basis of a continuity account of representation; I don't propose to go through the details here.

[5] This is an issue that recurs in the final section where we look at recent objections to CI.

different cases (Sutton 2006; Menary 2010b; Heersmink 2015). Some cases may score too low on the dimensional analysis and other cases may meet a clear threshold. Still further cases may be borderline. I think that the dimensional analysis is helpful and accurately represents the complexity of the phenomena in question. It may not satisfy those seeking a simple set of necessary and sufficient criteria, but then it is unlikely that they will ever reach satisfaction in this regard.

The Dimensions of Integration and the Cognitive Threshold

The cognitive threshold is not a precise scoring system. It is not supposed to be a definitive test for cognition or for integration. Rather it is a heuristic, or rough guide to analyzing cases for their degrees of integration. What is required for integration? We might provide a dimensional analysis looking at how cases score on a dimensional framework:[6] does the case score highly on interaction? Is it governed by cognitive practices? What kind of cognitive norms for manipulation does it involve (what are the specifics of the interactions)? Does it involve learning? If so, does it result in the transformation of an existing capacity or the acquisition of a novel ability? If the case scores low on most of these criteria, then perhaps it is a case of mere cognitive offloading or merely embedded cognition. The higher the score, the higher the degree of integration. Many of the cases I will look at in the next section score highly on these dimensions of integration. Before turning to those, I present a visual guide to the dimensions of integration (Figure 10.1).

Any case of CI will score highly across these dimensions of interaction, cognitive practices, and transformation. One cannot have a case of CI without these core dimensions. Therefore, the main criteria for cognitive integration are the following:

- coordinated interaction
- cognitive practices (regulative of interactions)
- normative manipulations/embodied engagements (specifics of interactions)
- cognitive transformation (acquisition of novel cognitive ability, transformation of existing capacity)

[6] Richard Heersmink (2015) has also provided a dimensional analysis of integration, but while his dimensions partly overlap with mine, they also differ in important respects. For example, Heersmink has an informational dimension, which I do not, because I don't think that integration, or cognition, should be understood in terms of informational transfer. His conception of transformation is equivalent to mere internalization, but I think that transformation is not the creation of internal symbols, representations, and the like—it is the capacity to recognize, create, and manipulate public representations, and this happens via learning the relevant cognitive practices and norms for their creation and manipulation.

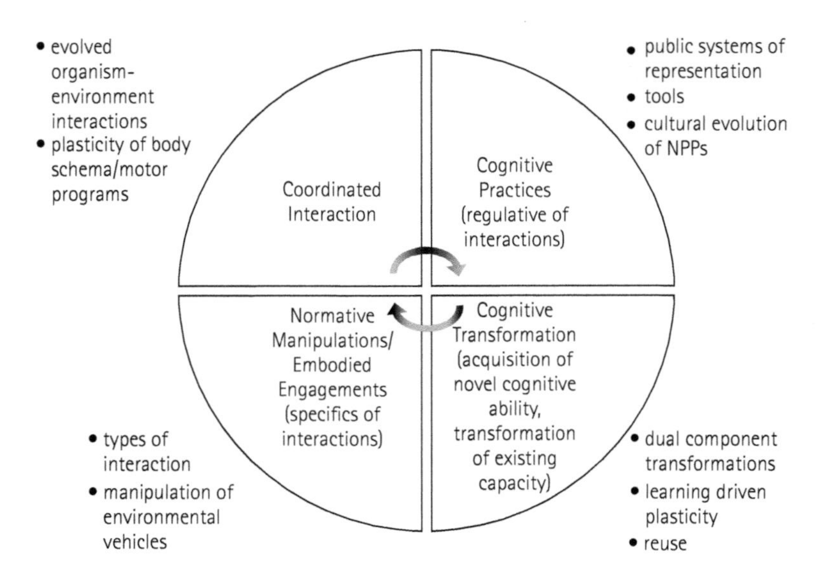

FIGURE 10.1. A visual guide to the dimensions of integration.

Consequently, we would expect to find cases of integration to score very highly on coordinated interactions, as opposed to mere causal influence of the environment on the body—such as ultraviolet radiation effecting alterations in skin pigmentation (see Figure 10.2).

The details of coordinated interactions are spelled out in terms of normative cognitive practices and the details of these practices are spelled out in terms of varieties of manipulation. The manipulation thesis (Rowlands 1999; Menary 2007a) concerns our embodied engagements with the world, but it is not simply a causal relation. Bodily manipulations are also normative; they are embodied practices developed through learning and training (in ontogeny). Below I outline six different classes of bodily manipulation of the environment with the general label of cognitive practices, and then provide a table that organizes them and provides examples of the different kinds of manipulation (for examples, see Table 10.1).[7]

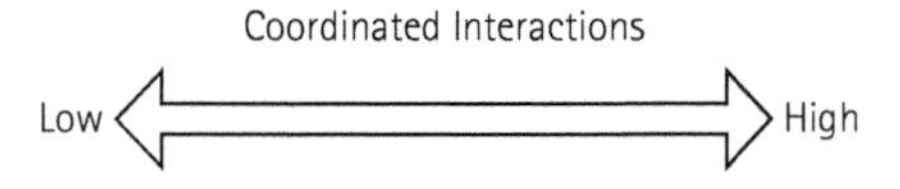

FIGURE 10.2. Coordinated interactions.

[7] This is a revision of previous formulations where cognitive practices were applied to a more restricted range of manipulations. See also Menary (2016) and Menary and Gillett (2017) for versions of the revision.

Table 10.1 Table of Manipulations

Types of Cognitive Practice

The table below has two sections: The first section briefly summarizes what each of the four types of cognitive practice are (epistemic tools and representational systems are amalgamated); the second section lists examples of each of these. Additionally, examples marked with * indicate blending of multiple types of practices.

Biological Interactions	Corrective Practices	Epistemic Practices	Epistemic Tools and Representational Systems
"Organism-environment relations" (Menary 2010b, p. 564); sensorimotor contingencies and animate perception dependent on and altered by context of the environment and state of the organism—in the case of humans, developing these requires training and expertise and also norms in some instances (Menary 2007a, p. 82).	Use of language/ exogenous props to direct/structure practical actions toward achieving tasks (channeling behavior) (Menary 2007a, p. 84). See Menary (2010b, p. 564) for differentiation of corrective practices from epistemic practices by stating that the former do not involve direct manipulation of the physical environment.	Using the environment as its own representation. Movements not just in physical space but in an abstract problem space (Menary 2007a, pp. 84, 89, 91). Kirsh and Maglio (1994, p. 514) propose that EP entail cognitive offloading in three ways: -reduce memory load -reduce time complexity -reduce error. Kirsh (1995b) proposes that EA as manipulations of spatial arrangements -simplify choice -simplify perception -simplify internal computation.	Manipulations of external representations according to learned cognitive norms (Menary 2007a, p. 84; see p. 137 for types of cognitive norms—updated to 5 in 2010b, pp. 570–71): 1 purposive 2 corrective 3 manipulative 4 interpretative 5 creative. See Menary (2007a, p. 139) for a list of properties associated with normative phenomena: non-moral "ought-ness."
Examples			
Saccades/animate perception (see Menary 2007a, pp. 126–7 for summary of Clark 1999 on Yarbus 1967, Ballard 1991, and Churchland, Ramachandran, and Sejnowski 1994). Repertoires of body schema and motor programs as embodied engagements (Menary 2007a, pp. 78–80, 85–6).	Instructional nudges (Menary 2007a, p. 83). Children's use of egocentric language to direct actions and keep task oriented moving forward (Vygotsky 1978; see Menary 2007a, pp. 179–85). Sutton's (2007) sporting example from cricket (see Menary 2007a, pp. 81–3).	Tetris Zoids (Kirsh and Maglio 1994). Kirsh's (1995a,b) five examples: - jigsaw sorting (p. 60)* - alphabetizing bookshelves (p. 65) - Tetris (p. 61) - cryptography - geometry (p. 65).	Rumelhart, Smolensky, and Hinton's PDP (1986, pp. 44–8) example of long-multiplication (see Menary 2007a, p. 167 and many others for taking this further). Zhang and Norman's (1994) Tower of Hanoi experiment (see Menary 2007a, pp. 146–9).

(continued)

Table 10.1 Continued

Attuned affordances (p. 80). Extended phenotypes (p. 84, and see Dawkins 1982). Sensorimotor contingencies (p. 84, and see O'Reagen and Noë 2001). Tool use in macaques (Menary 2010b, p. 574 cites this. Also see Maravita and Iriki 2004, and Iriki and Taoka 2012).	Realigning one's body in space to improve perception (Menary 2010b, p. 564). Kendall and Hollon's (1981) four categories of the regulatory use of speech for problem-solving (see Menary 2010b, p. 568).	These are summarized by Menary 2007a, p. 93, who also observes that the last two blend into the use of public representational systems).* Another blending case is Beach's (1988) study of how expert bartenders remember long orders through the learned use of items in their environment (see Menary 2007a, pp. 94–5); blended because it involves norms.*	Zhang and Norman's (1995) analysis of numeral systems and why Hindu-Arabic numerals are better than Roman—related to the PDP example (see Menary 2007a, p. 146). Linguistic systematicity (Menary 2007a, pp. 149–71). Mathematics in general (Menary 2010b, pp. 575–7 and 2015a; and see Landy et al. 2007, 2014 and Aydin et al. 2007. Also see Menary and Kirchhoff 2014 on mathematical expertise) Reading and writing (Menary 2007b; 2010b, p. 571, 2014, and see Dehaene 2009). Use of a memory note books by memory-impaired patients (Menary 2012, and see Sohlberg and Mateer 1989). Maps, lists, diaries, and plans (Menary 2010b, p. 570). See Menary (2017) on a wide range of "epistemic tracking tools."

© Alexander James Gillett, 2018

1. Biological interactions or direct sensorimotor interactions with the environment; an obvious example is sensorimotor contingencies (O'Regan and Noë 2001).
2. Corrective practices are a form of exploratory inference and are clearly present early in cognitive development. In the self-corrective cases, we create linguistic structures that control and direct future action. We might narrate the sequence of actions to ourselves in our heads or out loud, or we might write them down. The linguistic sequencing might be exclusively for our own use, or it might be designed for use by others. We might structure a sequence of actions collectively as well as individually, by jointly creating linguistic control structures.

3. Epistemic practices: experts would often perform actions that did not directly result in a pragmatic goal. The actions simplify cognitive processing. Other examples include the epistemic probing of an environment and epistemic diligence, and maintaining the quality of information stored in the environment (Menary 2012). Diligent cognitive agents have cognitive abilities for inspecting, testing, and correcting the informational structure of the environment and/or the beliefs they have formed about it. Therefore, they are not dependent upon the favorability of their epistemic environments.
4. Epistemic tools: many tools aid in the completion of cognitive tasks, from rulers to calculators, pen and paper to computers (Menary 2017). Manipulating the tools as part of our completion of cognitive tasks is something that we learn, often as part of a problem-solving task.
5. Representational systems: behaviorally modern humans display an incredible facility for innovating new forms of representational systems. Humans also display a general capacity for learning how to create, maintain, and deploy representations. Alphabets, numerals, diagrams, and many other forms of representation are often deployed as part of the processing cycle that leads directly to the completion of a cognitive task (Menary 2015a).
6. Blended interactions are complex cognitive tasks that may involve combinations of practices in cycles of cognitive processing.

Motivating Cognitive Integration

Introduction

The payoff from CI is twofold: the ability to do things that we otherwise cannot do and the transformation of existing abilities, making us smarter and better at difficult and demanding cognitive tasks. Humans are not that good at mathematics without years of training and teaching. Indeed, we just don't do symbolic mathematics spontaneously and naturally in the head. Without systems of mathematical representation and practice, we just can't do complicated mathematical cognition (Menary 2015a, De Cruz and De Smedt 2013). It's true that we have an intuitive sense of quantity and continuity and it is also true that we can develop rudimentary counting systems without symbolic mathematical systems, but that does not take us very far at all. A key motivating support for CI is that modern human minds have seen a vast and rapid expansion of external representational systems for abstract and theoretical thinking and that rather than simply internalize these systems (or think of them as simply the externalization of existing internal representational structures), we have redeployed existing sensorimotor resources for manipulating the environment to serve as a medium by which we can manipulate environmentally located representational vehicles. The next step of the argument is to claim that these manipulations are part of the cycles of cognitive processing required to complete cognitive tasks (see the previous section). In this section, I want to further motivate the key claim that we don't simply internalize or offload representational complexity; this is because of what I have elsewhere called a process of dual component transformation (Menary 2010b). In the second part of this section, I will look at the evolutionary grounds for CI.

Ontogeny: Representational Complexity and Dual Component Transformations

Symbolic[8] forms of representation for writing and mathematics were innovated less than 10,000 years ago and probably a lot more recently even than that (Ifrah 2000). There is evidence of proto-writing and mathematical systems that predates these periods, but even the oldest pictorial representations are fairly recent, at perhaps 30,000–50,000 years ago (Conrad 2006). The spread of external representations is recent when compared to the long history of hominin tool use, which stretches back at least 2.6–2.8 million years to *Homo habilis* (Stout et al. 2008). It is likely that the advent of external representations was a result of growing social complexity and the recent innovation of spoken language (probably 100,000–70,000 years ago).

Fully symbolic writing systems and number systems are very recent indeed, with many mathematical notations being only hundreds of years old and computer languages only decades. The spread of numeracy and literacy has also been very recent in the last couple of hundred years, with large swathes of the global population still illiterate and innumerate. It is highly unlikely, therefore, that our capacity for writing and symbolic math is innate, or that there are any specialized modules or core systems for writing, reading, and symbolic mathematics. Rather, it seems, these things must be learned. Consequently, there is overwhelming evidence that symbolic representations are not simply the result of externalizing pre-existing symbolic representations.

Furthermore, we don't simply internalize symbol systems after learning and training either. The argument for this latter position is complicated, but I will present a simpler version of the argument for the sake of brevity. The simple version of the argument runs as follows:

> Standard internalization requires the re-representation of external symbols as neural representations, perhaps imagistic in form. Symbols get mapped to neural populations and those populations become tuned to the recognition of, for example, Arabic numerals or letters. Phonetic knowledge and knowledge of arithmetic facts

[8] Here I'm talking about genuine abstract symbols as a conventional system of representation. Tally systems, especially with one-to-one mappings, are considerably older, perhaps as old as 70,000 years (see d'Erricho et al. 2001).

> then get mapped to these "representations." So far so straightforward, but the next step in the argument is crucial: once the symbols are internalized, cognitive operations can be performed on them without the need for the external "scaffolds." However, this ignores several important facts about the development of symbolic cognition and the transformed capacity that is the result of development. The developmental fact concerns the abilities that the neophyte gains after learning and teaching. For the neophyte learns to recognize symbols, numerals, and letters by manipulating them, drawing them, reciting them, and learning how to place them on a page[9] and to order them both physically/visually and verbally. Consequently, the child does not simply internalize symbols that are presented to it; the child actively creates symbols and learns to order and manipulate them in physical space on a page or a screen (this is why writing is still important), or in some cultures by manipulating beads on an abacus. This learning process results in a dual transformation of not just our ability to recognize symbols (what is usually referred to as an internal representation) but also our ability to create and physically order symbols on a physical medium (such as the page).

To put it slightly differently, our existing capacities for pattern recognition are reoriented (or redeployed) to recognizing symbols (as a process of specialization) and our existing sensorimotor abilities for manipulation and tool use are reoriented to the creation and ordering of symbols—writing (Dehaene 2009; Menary 2014). There is a dual route to symbolic cognition, an active creative one that depends upon existing sensorimotor abilities for tool use and a recognitional one that depends upon existing abilities for (higher-order) pattern recognition.

Why though don't we jettison these dual routes in maturity and simply rely upon internal resources? Because symbolic cognition never dispenses with the dual routes. This is, of course, partly due to the fact that we continue to write out our symbolic forms of thought as sentences or equations. Symbol systems evolved as a public medium of knowledge, representation, and communication, allowing for the dissemination of knowledge and thought both horizontally across social groups and vertically between generations (Donald 1991). However, rather than simply thinking of writing as a public record of inner thought, the peculiar nature of the writing process, the act of writing, is partly constitutive of symbolic thought.

Some of the strongest evidence for this claim comes from mathematical cognition, where there is evidence that mathematicians rely upon the regular spacing and the FOIL (first, outer, inner, last) ordering of symbols in algebraic equations to be able to complete mathematical operations (Landy and Goldstone 2007; Landy et al. 2014). Further evidence can be solicited from the fact that, in general, even expert mathematicians use the same creative loop through stable written symbol strings in the process of solving complex mathematical problems (De Cruz and De Smedt 2013; Menary and Kirchhoff 2014; Menary and Gillett 2017). One might argue that the expert simply offloads the products of internal cognition onto the environment, by simply writing out equations

[9] And, increasingly, on a touch capacitive screen.

that have already been completed in the head. This idea is just too simplistic to be credible, because mathematicians are involved in a continuous interplay between the public symbols on the page (or whiteboard) and the transformed processes for imagining solutions to problems.

I turn now to the evolutionary case.

Phylogeny: The Evolutionary Case for ICs and the Evolutionary Continuity Condition

There is a good evolutionary case for ICS, which coheres with the developmental case given in the last subsection. To begin with, it is obvious that public symbol systems are very recent cultural innovations with a high degree of cultural variation (Coltheart 2014; Downey 2014; Menary 2015a). However, many of our core cognitive abilities are phylogenetically much older than these recent innovations. A puzzle emerges: how do cognitive resources that were not evolved for symbolic cognition come to be able to create and manipulate public systems of symbolic representation? The answer lies in the convergent evolution of phenotypic plasticity, neural plasticity, cultural inheritance, and language that sprung from the need to adapt to variable environments and growing social complexity. Cognitively speaking, modern humans have minds that are prepared for symbolic cultures. What this means is that we reached a stage of cognitive evolution such that the need for symbol systems did not require an evolutionary leap to new specialized systems. Rather, existing systems could be reoriented, by learning and training, to perform new functions required for symbolic thought (Heyes 2012; Anderson 2010; Menary 2014). This could only be the case if modern humans were already capable of social learning and their brains were plastic enough that learning could drive the necessary functional changes in the brain.

Phylogenetically, humans are already adapted to their environments by interacting with them, often by creating and manipulating tools and artifacts, but also practices, norms, and forms of social interaction. This results in physical and sociocultural niche construction and the heritability of physical artifacts, practices, social norms, and knowledge (perhaps in narrative form prior to writing). Tool use is ancient in the hominin lineage; it follows that hominins have been interacting with and manipulating their environments for millions of years (Stout et al. 2008). Although modern humans have arrived only very recently in evolutionary terms, they were already adapted to manipulating the environment (Menary 2007a). Cognitively modern humans were ready to innovate public symbol systems in the right kind of social environment. In the previous section I outlined the different varieties of manipulation, from basic sensorimotor interactions that are continuous with much of the animal world to tool use and symbol use, which are more obviously associated with human cognition. In the remainder of this section I will say a few words about the importance of phylogenetically older forms of sensorimotor interaction with the environment for the evolution of symbolic manipulations.

Evolutionary continuity is one of the key background conditions for CI (Menary 2009, 2015a). The central claim is one of continuous evolutionary development of the biological basis of cognition. However, CI is fully committed to the extended synthesis (Pigliucci and Muller 2010), which introduces the importance of extra-genetic channels of inheritance, including ecological inheritance, epigenetic inheritance, and the role of a developmental niche in assembling phenotypic traits. Although modern humans are adapted to sensorimotor interactions, or manipulations of the environment, these capacities can be redeployed in development to the creation and manipulation of symbolic representations. The importance of earlier sensorimotor capacities in the evolution of more recent capacities to manipulate symbols can be summarized in the following way (see Menary 2013, p. 27):

1. Organisms are reciprocally coupled to their environmental niches, resulting in an organism–environment system (Odling-Smee, Laland, and Feldman 2003).
2. As an organism–environment system, the organism is predisposed to manipulate its environmental niche, or in some cases create it. This is an adaptation of the organism.
3. An organism's manipulations of its environment, while part of its phylogenetic history, can, in many cases, be fine-tuned and calibrated through learning or reinforcement as part of its ontogenetic history.
4. That these manipulations are adaptations gives them a basic kind of normativity. This normativity allows for the beginnings of an organism's sensitivity to salient environmental variables. It allows for the possibility of intentional directedness.
5. Humans are predisposed to manipulate their environment, but the fine-tuning and calibration of these manipulations in ontogeny is not part of their phylogenetic history. The role of culture in providing systems of representation and methods for their manipulation must be learned and practiced before fluent bodily manipulations of public representations become part of a human's cognitive repertoire.
6. The phylogenetic history of *Homo sapiens* illustrates how we move on a continuum from biological manipulations as adaptations in our hominin forebears to more complicated forms of tool use and imitation, through to language and the development of public representational systems. They all involve manipulations of the environment and eventually result in a culture that is a repository of knowledge, skills, and representational systems that is passed on to later generations via learning and development.

The account of the evolution of ICS is significant because it argues for a continuity between biological sensorimotor interactions (online cognition) and higher, specifically human, cognitive capacities that are the result of development in richly structured cultural niches (Menary 2009, 2015a). A central strand in the argument for cognitive integration is a thesis of continuity: interactive abilities, such as an infant's capacity to explore its environment and interact with objects, are continuous with cognitive manipulative abilities, such as my capacity to think by constructing, maintaining, and manipulating representational systems. The continuity thesis at the heart of cognitive integration requires that there is no deep metaphysical discontinuity between the mind and the world. Consequently, to be a cognitive integrationist is to be a thoroughgoing naturalist. In the final section I deal with some recent objections to CI.

CI's Relations to other Approaches and Defending CI Against Objections

Introduction

A number of objections have been raised against CI and I deal with a selection of them here. Firstly, I differentiate CI from a Clark and Chalmers (1998) style "artifact extension" model of the extended mind. Then I look at Sterelny's (2010) objection to an artifact extension version of the extended mind. I show that Sterelny's own position is quite close to CI and that his objections to Clark's formulation of the extended mind do not apply to CI. Then I reject Rupert's (2009, 2010) arguments from persisting integrated cognitive systems and his objection to a developmental argument for CI. Finally I rebut two arguments that CI is overly representational and, therefore, at least consistent with internalism and representationalism about the mind.

Artifact Extension, Functionalism, and Extended Predictive Processing vs. CI

Artifact extension (AE) depends upon functionalism: the artifact must function in a way similar enough to the original cognitive function that it can be considered to be doing the cognitive work (Clark and Chalmers 1998; Wheeler 2010). There is a weak and a strong version of AE—the weak version is simply cognitive offloading. If the mathematical problem is too hard, delegate it to the calculator. The stronger version requires integration of the artifact into the cognitive processing routines of the individual (Wilson and Clark 2009; Pritchard 2010). However, now it matters what integration entails, one way of thinking is just that an artifact becomes so tightly grafted into the physical individual that it is no longer distinct from them. We can think of this in terms of implants; for example, neural or otherwise that might be linked to input and output from neural circuitry.[10] The other is some form of causal coupling. In the most recent formation, Clark

[10] Indeed this is already happening, and is one of the famous examples from Clark and Chalmers's (1998) original paper.

has updated causal coupling in terms of active inference (Clark 2013, 2016). Artifacts now become part of the cycle of active bodily movement that reduces the error in sensory signals from which the system must make predictions about the hidden variables in the environment (Clark 2016; Hohwy 2013). Consequently artifacts now become part of the processing for prediction error just because they reduce error in the statistical models of the environment generated by the brain. As Hohwy puts it, active inference and artifacts play a secondary role in optimizing the precision estimations to the sensory signals, thereby reducing error (Hohwy 2013). The main thing to notice is that the functional role of artifacts has now been shifted to service the purported core function of the brain—to always reduce error (Friston 2010) in line with the free energy principle, where reducing free energy is equivalent to reducing error: "any self-organising system that is at equilibrium with its environment must minimise its free energy" (Friston 2010, p. 127).

The alternative is CI: artifacts only become important insofar as we can manipulate them via cognitive practices acquired through learning. The primary function of cognitive practices and artifacts is not to optimize precision estimations. The free energy principle is not an ultimate explanatory principle that can be used to explain why there are practices and artifacts, as I have argued elsewhere (Menary 2015b, 2017). Rather, predictive processing may be a useful way of understanding the proximal mechanisms of the brain, but the ultimate explanations of neural organization, artifacts, and practices will be evolutionary. CI explains the existence of cognitive artifacts and practices in terms of cultural evolution and social learning. As we have already seen, there is a wealth of evidence in support of this model and, while it is at least compatible with predictive processing at a neural level (see Fabry 2015, 2017 for details), it rejects the free energy principle as an ultimate explanation of the enculturation of cognition. I would be delighted to discover that I have misunderstood Clark and that he thinks that artifacts and practices are not explicable simply in terms of error minimization, but the evidence suggests otherwise.

Sterelny's Objections

Sterelny's version of embedded cognition is far more sympathetic to what CI is trying to achieve. Sterelny's main critique is aimed at AE and they are similar to some of the worries raised about AE by CI. In differentiating his own scaffolded/niche constructionist position from AE, he reminds us that standard AE cases are not cases that fit well with cultural inheritance and niche construction: "Otto's solution to the problem of memory loss was not built by cumulative improvement over the generations" (Sterelny 2010, p. 469). We should think of these cases as examples of adaptive plasticity of a phenotype. Sterelny's point is that the next generation inherits the language, tools, and norms shaped by previous generations and which they themselves will shape before passing them on. However, the HEC (hypothesis of extended cognition) cases involve individually entrenched resources that are not candidates for shaping the cognitive capacities of future generations (we could imagine cases in which they are, but rarely).

Otto's notebook is just too individually entrenched to be passed on to the next generation. It is individually suited to Otto, not to Inga. The notebook is not part of Otto's mind but is simply a scaffold for his damaged memory system.

However, Sterelny does hold that the scaffolding of the cognitive niche can have a deeply transformative effect on our cognitive capacities (2010, p. 472). CI takes transformation to be linked to cognitive practices and interaction, where practices, norms, representational systems, and learning environments form the scaffolding. This is quite different to the AE position that Sterelny is criticizing. We can see why by formulating the following dilemma:

What's doing the work?

A. The individually entrenched artifacts that extend cognition, or
B. The practices for manipulating them.
   If A, then Sterelny's critique applies.
   If B, then AE is not so important.

Clark might respond by denying both Sterelny's and my criticisms, because he endorses the idea of incremental cognitive niche construction: "Under certain conditions, non-organic props and aids, many of which are either culturally inherited tools or structures manipulated by culturally transmitted practices, might themselves count as proper parts of extended cognitive processes" (Wheeler and Clark 2008, p. 3566).

This line of thinking appears to be consistent with CI. However, Clark is also on record denying that cognitive practices could be central to his account:

> It is crucial to the story I am telling that the biological brain adapts, selects, and alters, its own internal routines so as more and more fluently to exploit the reliable presence of all those specific, culturally selected, tuned, and delivered, resources. For it is only in that way that we achieve the kind of complex temporally nuanced dovetailing (between the neural and the rest) that warrants treating a temporary ensemble, Metamorpho-like, as a new, genuine, cognitive whole. (Clark 2011, p. 459)

I think that it is possible to nudge Sterelny toward CI. The first move is to ask whether it is true for Sterelny that transformation results in full internalization. Surely not. Dual component transformations are quite compatible with Sterelny's scaffolded cognition; indeed CI and scaffolding are very close in most respects. However, I think that it is fair to say that Sterelny's scaffolded view does not sufficiently take into account the role of cognitive practices and manipulations. When one combines scaffolding with cognitive practices and manipulations, one has something much closer to CI.

The other point of agreement concerns the role of cultural inheritance and social learning. Sterelny has been at the forefront of making a case for the influence of cultural evolution on cognition (2003, 2010, 2012). Cognitive scaffolding, what he sometimes refers to as "cognitive capital" (2012, pp. 13–14), is inherited and has a profound influence on cognitive capacity via social learning in an extended developmental period. The only point of departure appears to be CI's more detailed account of cognitive transformations.

Rupert's Criticism of Persistent Cognitive Systems

Rupert (2009, 2010) provides an "embedded cognition" alternative to AE, and is another competitor with CI. However, the motivations for endorsing an embedded view over CI are usually aimed at a version of AE and don't, directly, address the arguments in favor of CI. The main criticism concerns the supposed fleeting nature of extended cognitive systems. Rupert argues that the development of extended cognitive systems would cause problems, given that the system would have to be formed and dissolved anew each time there was an interaction with an external artifact (2010, p. 330). Perhaps this kind of objection has some purchase against AE,[11] but not against CI. The manipulations of the environment are not something that is formed and dissolved anew each time, but are enduring capacities of the organism, developed after learning.

Let's break the argument down:

1. "We want to understand how and why the capacities and abilities of individual persisting systems change over time, eventually taking a stable form" (Rupert 2010, p. 330).
2. However, extended systems are only fleeting (systems only while they are causally coupled).
3. Cognitive science is interested in persisting (and integrated) cognitive systems.
4. Therefore fleeting coupled systems are not the kind of systems of interest to cognitive science.

Rupert's argument is odd. Premise 2 is false, for the reasons given earlier. Although the enacting of patterned practices is part of the cycle of processing in ICS, the patterned practices and our embodiment of them are not fleeting, but stable and persisting. Because ICSs are structured by normative patterned practices, they do not solely depend upon fleeting causal interactions with the environment. Normative patterned practices are always there!

Rupert also criticizes an evolutionary approach to integration, via developmental systems theory. Developmental systems theory (Oyama, Griffiths, and Gray 2000) proposes that we think of the entire developmental system, including nongenetic lines of inheritance, as being reconstructed "from one generation to the next via numerous interdependent causal pathways" (Griffiths and Stotz 2000, p. 35). In the cognitive case Griffiths

[11] But it seems to me these are not arguments against extended predictive process because of the role of active inference in minimizing prediction error. Active inferences are prompted by errors in prediction; they produce movement and action to make the sensory signals more precise. Active inferences have an important role to play in the overall cycle of processing in the system.

and Stotz argue that: "Our cognitive processes assume and exploit the presence of a highly structured environment, including languages and other systems of representation. With the aid of this external scaffolding, we can accomplish tasks beyond the capacity of the 'naked brain'" (2000, p. 39). Rupert takes developmental systems theory to support AE in a fairly straightforward way: "According to developmental systems theory, evolutionary forces often operate on transorganismic, or extended, biological systems. Assuming that cognitive traits were selected for, it is no surprise that such traits should be exhibited by transorganismic systems" (Rupert 2009, p. 113). However, the biological ground for an extended system applies only when all parts of the system share a common fate.

Rupert lays out the following criticism of the case for AE via developmental systems theory:

> Imagine an extended cognitive system including an artist and her sketchpad (Rupert 2009, p. 116). The artist and the sketchpad do not share a common fate—the survival of the artist is not dependent upon the sketchpad and vice versa. Therefore, the artist and sketchpad are not an extended cognitive system (on biological grounds). A second argument would run as follows: humans create mathematical systems in order to solve mathematical problems. Mathematical systems do not create their users. There is an asymmetry between the creator of the mathematical system and the system itself (which does not create the user). Therefore, humans and mathematical systems cannot be part of the same system. On these grounds Rupert denies the route from developmental systems theory to AE.

However, neither of Rupert's arguments works against CI. Rupert says human and sketchpad do not share a common fate—because the human would survive without sketchpad. Rather than applying this to a case of AE, let's see if it works against CI. What would happen if we took away the cognitive-environmental niche (e.g., public language)? If we took away all of our cognitive practices and, thereby, removed the capacity to create, maintain, and manipulate it? One likely outcome is that modern humans would be under severe selective pressure. It seems clear that humans and their cultural surrounds are under shared selective pressure and this is no surprise if our cultural, and cognitive, environments are inherited and make a profound difference to the cognitive abilities that we have.

The second argument would be true of examples of niche construction. It assumes a one-way relationship between humans and their environments, but this is precisely what niche construction denies. Humans construct their environments which then have selective and developmental effects on the population over time. Rupert's argument depends upon an asymmetric relationship where an organism creates its niche; the niche does not create the organism (Premise 2 earlier). However, this second premise at best depends upon an equivocation over the word "create"; at worst it is false, since organism and environment coevolve over time, with phenotypes being "constructed" in inherited developmental niches (Stotz 2010). The organism and niche do coevolve over time. This makes Rupert's second premise false; our capacity for mathematical cognition is "created" in development by the interaction between the developing child and public systems of mathematics and systems of pedagogy for acquiring them in development.

The Representational Double-Act

Does CI commit a fallacy of doubling up on representations, both internal and external, where the doubling up leads to overdetermination and, consequently, no need for the external representations? Both Steiner (2015) and Hutto and Myin (2013) claim that it does, which is ironic since Brogaard (2014) says that CI is committed to the exact opposite claim—that there are no intrinsic representations in the brain! Steiner and Hutto and Myin are wrong about CI, but for interesting reasons—I shan't address Brogaard's argument here, but the previous sections of this chapter should put paid to the notion.

First to Steiner's (2015) argument. Steiner's argument runs like this:

1. To apprehend the contents of external representations that are manipulated for cognizing, the cognitive agent must internally represent the content of the external representation.
2. The causal roles of internal and external representations are *too similar*, external representations replicate or "double up" on the causal role of internal representations, leading to a causal overdetermination problem.
3. Faced with an overdetermination problem we should dispense, on grounds of parsimony, with the "double" of the causal role, in this case the external representation.
4. Therefore: external representations cannot play the role in cognition that AE and CI claim that they do.

Steiner is right that if internal representations and external representations do the same work, then why should we count them as being part of the same cognitive system, or cycle of processing? The problem with this argument should be clear enough to see. There is no overdetermination of the roles of representations in CI, primarily because CI is not committed to Premise 1. Given dual component transformations and the role of cognitive practices, CI cannot be committed to the claim that understanding public systems of representation is an achievement based upon the creation of inner representations of those systems. The sensorimotor manipulations of mathematical notations, for example, are partly constitutive of our understanding of mathematical practice. Those manipulations are governed by the public systems of mathematical practice and not by inner representations of such. If anything, the reverse of Premise 1 is a consequence of CI. If there are internal representations of mathematical notations, their contents are due to and dependent upon the contents of the public system of mathematical practice.[12] I have made this point in a number of places (Menary 2006, 2010c, 2015a).

[12] This is what I take Brogaard to be objecting to.

Steiner's aim seems to be to discredit a representationalist version of AE, to force a concession that a commitment to inner representations for "understanding" external representations causes an overdetermination problem for AE. However, like Procrustes, Steiner has had to stretch CI to fit the representationalist commitments that it does not endorse and hence the argument fails to hit its target. However, Steiner's strategy of pushing representationalist AE on the role of "internal" and "external" representations reveals some interesting differences between AE and CI. Steiner completely ignores the roles of dual component transformations and cognitive practices, except for one curious comment which I shall quote in full here:

> "Occasional nonrepresentationalism," as we might call it, about the use of external representations, does not currently figure in the representationalist versions of HEC [hypothesis of extended cognition]. But it could be proposed, for instance, in the form of a coordination-dynamics style of explanation. According to such an explanation, in some circumstances, the manipulation of external representations and their contents is only enacted from a set of normative and nonrepresentational skills (a nonrepresentationalist radicalization of Menary, 2007a). (Steiner 2015, pp. 14–15)

CI already says this, so it is hardly radical. However, the norms and practices by which the sensorimotor abilities function are not themselves lacking in content and which can themselves be communicated by public representations—as algorithms, text, written examples of notation, speech, and so on.

One shouldn't confuse the capacity to recognize symbols with the capacity to represent them, but let us add that one should not further confuse this with the capacity to create and manipulate public symbols.

Steiner is right to argue that if internal and external representations are doing the same job in any cycle of cognitive processing, there is really no need for external representations. However, he is simply wrong to suggest that CI has this commitment. The manipulation of public representations is given a specific role in CI, which Steiner fails to engage with.

CI Just Isn't Radical Enough

Hutto and Myin (2013) go after CI directly. They have some positive things to say about CI: "Menary (2007a, 2010a) has developed this insight into a rich and nuanced position that provides a new framework for thinking about the EMH [extended mind hypothesis]" (Hutto and Myin 2013, p. 148). However, they go on to equate CI with a quasi-internalist position in a manner reminiscent of Steiner. Their main bone of contention with CI is that it is committed to there being internal representations and that this opens up an objection based upon two commitments of psychological individualism, which they term "content-involving cognition" (CIC) and the "default internal mind assumption" (DIM). Their definition of representational content is straightforward: a representation that makes a claim about the world and has a truth condition, or condition of satisfaction (Hutto and Myin 2013). As they say, "Anything that deserves to be called content has special properties—e.g., truth, reference, implication—that make it logically distinct from, and not reducible to, mere covariance relations holding between states of affairs" (Hutto and Myin 2013, p. 67). CIC is the assumption that cognition involves contentful representations all the way down. CIC, in their view, "sponsors" DIM, and DIM is the antithesis of CI.

Why though do they think that CI is committed to, or at least consistent with CIC? Their "evidence" is shaky; it is based upon a single quote from my 2007 book (Menary 2007a, pp. 58–9). However, when put in context, the charge of CIC is easily dispelled, for in the same book, I make it clear that a cognitive vehicle need not be contentful: "This is a very general definition which does not tie us to a process or vehicle having to be internal or external, or whether vehicles must be representations or have a particular kind of content" (Menary 2007a, p. 15). Indeed, it need not be a representation at all and certainly not a representation with truth-conditional content. As I have already argued in this chapter, there is a clear line of argument in CI that cognition moves from cases of non-contentful sensorimotor cognition to complex representational cognition required for high-level problem-solving and prospective thinking, with many gradations in-between. Hutto and Myin try to wedge CI into a position where it is consistent with CIC and DIM and from which it cannot escape internalist objections.

However, CIC and DIM are already ruled out by CI, first because CI begins with non-contentful sensorimotor interactions and second because it takes cognition to be adapted to interacting with the environment. In Chapter 5 of Menary (2007a), I make this very clear and have argued for it again in the second section of this chapter. I shan't repeat that argument here, but simply remind the reader of where they can easily find it.

Indeed, here there is some complementarity between enactivism and CI. Cognition begins in sensorimotor couplings with environmental variables that form the basic mechanisms out of which more complex and eventually fully contentful thought arises. However, I don't think that a simple claim that content and vehicles exist—"but they are associated with linguistic symbols" (Hutto and Myin 2013, p. 151)—can be justified. It is tantamount to stating a dichotomy between biology and culture.[13] Social relations are biological; if norms are social, then by definition they are within the ambit of biology. The position that Hutto and Myin take here leads to some painful contortions, for example, when they try to describe what offline cognition looks like on their view, they produce the following: "Rather, what is gained is an ability to perform operations that previously required the manipulation of external symbols but have now become possible in the absence of external symbols" (Hutto and Myin 2013, p. 152).

[13] This is something that Hutto and Myin apparently deny in their forthcoming book. Indeed, they appear to recognize the importance of evolutionary continuity, although it remains to be seen whether they can maintain the distinction between basic and non-basic cognition and a commitment to evolutionary continuity (also see Menary 2015a, p. 3 for similar criticisms).

If our capacities for offline symbolic thinking are not transformed by our acquisition of the online capacities to manipulate them, then it becomes a mystery as to how we are able to do so. For a full account of how this transformation works in the case of mathematics, see Menary (2015a).

Conclusion

The last 20 years have seen an explosion of interest in extended cognition in cognitive science and the extended synthesis in biology. The CI framework provides a way of combining the best insights from both these traditions into a coherent account of cognition. In this chapter, I've given a detailed account of how the framework works and how we can understand the core dimensions of integration, and I have responded to critics. CI continues to be put to work and is being extended into new domains, including predictive processing (Fabry 2015, 2017), social cognition (Menary, Brown and Mostafavi in progress), and even the metaphysics of mind (Kirchhoff 2014).

Acknowledgments

The research for this article was supported by the Australian Research Council, Future Fellowship FT 130100960. Thanks to Alexander James Gillett for collating the citations included in Table 10.1 and feedback on an earlier draft.

References

Adams, F. and Aizawa, K. (2001). The bounds of cognition. *Philosophical Psychology*, 14, 43–64. doi:10.1080/09515080120033571

Adams, F. and Aizawa, K. (2008). *The bounds of cognition*. Oxford: Blackwell.

Anderson, M.L. (2010). Neural reuse: a fundamental organizational principle of the brain. *Behavioral and Brain Sciences*, 33, 245–64. doi:10.1017/S0140525X10000853

Anderson, M.L. and Penner-Wilger, M. (2012) Neural reuse in the evolution and development of the brain: evidence for developmental homology? *Developmental Psychobiology*, 55(1), 42–51. doi:10.1002/dev.21055

Aydin, K., Ucar, A., Oguz, K.K., Okur, O.O., Agayev, A., Unal, Z. et al. (2007). Increased gray matter density in the parietal cortex of mathematicians: a voxel-based morphometry study. *American Journal of Neuroradiology*, 28, 1859–64. doi:10.3174/ajnr.A0696Ballard

Ballard, D.H. Animate vision. (1991). *Artificial Intelligence Journal*, 48, 57–86.

Ballard, D., Hayhoe, M., and Pelz, J. (1995). Memory representations in natural tasks. *Journal of Cognitive Neuroscience*, 7(1), 66–80. doi:10.1162/jocn.1995.7.1.66

Beach, K. (1988). The role of external mnemonic symbols in acquiring an occupation. In: M.M. Gruneberg and R.N. Sykes (eds.), *Practical aspects of memory.* New York: Wiley, pp. 342–6.

Brogaard, B. (2014). A partial defense of extended knowledge. *Philosophical Issues*, 24, 39–62. doi:10.1111/phis.12025

Brooks, R.A. (1999). *Cambrian intelligence: the early history of the new AI.* Cambridge, MA: MIT Press.

Churchland, P.S., Ramachandran, V., and Sejnowski, T. (1994). A critique of pure vision. In: C. Koch and J. Davis (eds.), *Large scale neuronal theories of the brain.* Cambridge, MA: MIT Press, pp. 23–60.

Clark, A. (1999). Where brain, body, and world collide. *Journal of Cognitive Systems Research*, 1, 5–17. doi:10.1016/S1389-0417(99)00002-9

Clark, A. (2011). Finding the mind. *Philosophical Studies*, 152(3), 447–61. doi:10.1007/s11098-010-9598-9

Clark, A. (2013). Whatever next? Predictive brains, situated agents, and the future of cognitive science. *Behavioral and Brain Sciences*, 36(3), 181–204. doi:10.1017/S0140525X12000477

Clark, A. (2016). *Surfing uncertainty: prediction, action, and the embodied mind.* New York: Oxford University Press.

Clark, A. and Chalmers, D. (1998). The extended mind. *Analysis*, 58(1), 7–19. doi:10.1080/09515080903538917

Coltheart, M. (2014). The neuronal recycling hypothesis for reading and the question of reading universals. *Mind and Language*, 29(3), 255–69. doi:10.1111/mila.12049

Conard, N. (2006). An overview of the patterns of behavioural change in Africa and Eurasia during the Middle and Late Pleistocene. In: F. d'Errico and L. Blackwell (eds.), *From tools to symbols from early hominids to humans.* Johannesburg: Wits University Press, pp. 294–332.

Dawkins, R. (1982). *The extended phenotype.* Oxford: Oxford University Press.

De Cruz, H. and De Smedt, J. (2013). Mathematical symbols as epistemic actions. *Synthese*, 190, 3–19. doi:10.1007/s11229-010-9837-9

Dehaene, S. (2009). *Reading in the brain: the science and evolution of a human invention.* New York: Viking.

d'Errico, F., Henshilwood, C., and Nilssen, P. (2001). An engraved bone fragment from c. 70,000-year-old Middle Stone Age levels at Blombos Cave, South Africa: implications for the origin of symbolism and language. *Antiquity*, 75, 309–18. doi:10.1017/S0003598X00060968

Donald, M. (1991). *Origins of the modern mind—three stages in the evolution of culture and cognition.* Cambridge, MA: Harvard University Press.

Downey, G. (2014). All forms of writing. *Mind and Language*, 29(3), 304–19. doi:10.1111/mila.12052

Downey, G. (2016). Sensory enculturation and neuroanthropology: the case of human echolocation. In: J. Chiao, S. Li, R. Seligman, and R. Turner (eds.), *Oxford handbook of cultural neuroscience, vol. 1: Cultural neuroscience and health.* Oxford: Oxford University Press, pp. 41–57. doi:10.1093/oxfordhb/9780199357376.013.23

Dretske, F.I. (1988). *Explaining behavior.* Cambridge, MA: MIT Press.

Fabry, R. (2015). Enriching the notion of enculturation: cognitive integration, predictive processing, and the case of reading acquisition. In: T. Metzinger and J.M. Windt (eds.), *Open MIND*, 25(C), 1–23. doi:10.15502/9783958571143

Fabry, R. (2017). Betwixt and between: the enculturated predictive processing approach to cognition. *Synthese.* doi:10.1007/s11229-017-1334-y

Farne, A., Serino, A., and Làdavas, E. (2007). Dynamic size-change of peri-hand space following tool-use: determinants and spatial characteristics revealed through cross-modal extinction. *Cortex*, 43, 436–43. doi:10.1016/S0010-9452(08)70468-4
Fodor, J. (1990). *A theory of content and other essays*. Cambridge, MA: MIT Press.
Friston, K. (2010). The free-energy principle: a unified brain theory? *Nature Reviews Neuroscience*, 11, 127–38. doi:10.1038/nrn2787
Godfrey-Smith, P. (1996). *Complexity and the function of mind in nature*. Cambridge: Cambridge University Press.
Gopnik, A., Meltzoff, A., and Kuhl, P. (1999). *How babies think: the science of childhood*. London: Weidenfeld & Nicholson.
Griffiths, P.E. and Stotz, K. (2000). How the mind grows: a developmental perspective on the biology of cognition. *Synthese*, 122, 29–51. doi:10.1023/A:1005215909498
Heersmink, R. (2015). Dimensions of integration in embedded and extended cognitive systems. *Phenomenology & Cognitive Science*, 14, 577–98. doi:10.1007/s11097-014-9355-1
Heyes, C. (2012). Grist and mills: on the cultural origins of cultural learning. *Philosophical transactions of the Royal Society of London. Series B, Biological sciences*, 367, 2181–91. doi:10.1098/rstb.2012.0120
Hohwy, J. (2013). *The predictive brain*. Oxford: Oxford University Press.
Hutchins, E. (2011). Enculturating the supersized mind. *Philosophical Studies*, 152, 437–46. doi:10.1007/s11098-010-9599-8
Hutto, D.D. and Myin, E. (2013). *Radicalizing enactivism: basic minds without content*. Cambridge, MA: MIT Press.
Ifrah, G. (2000). *Universal history of numbers. From prehistory to the invention of the computer* (trans. D. Bellos, E.F. Hardin, S. Wood, and I. Monk). New York: John Wiley & Sons.
Iriki, A. and Taoka, M. (2012). Triadic (ecological, neural, cognitive) niche construction: a scenario of human brain evolution extrapolating tool use and language from the control of reaching actions. *Philosophical transactions of the Royal Society of London. Series B, Biological sciences*, 367, 10–23. doi:10.1098/rstb.2011.0190
Kendall, P.C. and Hollon, S.D. (1981). Assessing self-referent speech: methods in measurement of self-statements. In: P.C. Kendall and S.D. Hollon (eds.), *Assessment strategies for cognitive-behavioral interventions*. New York: Academic, pp. 85–118. doi:10.1016/B978-0-12-404460-9.50010-0
Kirchhoff, M.D. (2014). Extended cognition & constitution: re-evaluating the constitutive claim of extended cognition. *Philosophical Psychology*, 27(2), 258–83. doi:10.1080/09515089.2012.724394
Kirsh, D. (1995a). Complementary strategies: why we use our hands when we think. In: J.D. Moore and J.F. Lehman (eds.), *Proceedings of the Seventeenth Annual Conference of the Cognitive Science Society*. Mahwah, NJ: Erlbaum, pp. 212–17.
Kirsh, D. (1995b). The intelligent use of space. *Artificial Intelligence*, 73, 31–68. doi:10.1016/0004-3702(94)00017-U
Kirsh, D. and Maglio, P. (1994). On distinguishing epistemic from pragmatic actions. *Cognitive Science*, 18, 513–49. doi:10.1207/s15516709cog1804_1
Landy, D., Allen, C., and Zednik, C. (2014). A perceptual account of symbolic reasoning. *Frontiers in Psychology*, 5, 1–10. doi:10.3389/fpsyg.2014.00275
Landy, D. and Goldstone, R.L. (2007). How abstract is symbolic thought? *Journal of Experimental Psychology*, 33(4), 720–33. doi:10.1037/0278-7393.33.4.720

Maravita, A. and Iriki, A. (2004). Tools for the body (schema). *Trends in Cognitive Sciences*, 8(2), 79–86. doi:10.1016/j.tics.2003.12.008

Menary, R. (2006). Attacking the bounds of cognition. *Philosophical Psychology*, 19(3), 329–44. doi:10.1080/09515080600690557

Menary, R. (2007a). *Cognitive integration: mind and cognition unbounded.* Basingstoke, UK: Palgrave Macmillan.

Menary, R. (2007b). Writing as thinking. *Language Sciences*, 29, 621–32. doi:10.1016/j.langsci.2007.01.005

Menary, R. (2009). Intentionality, cognitive integration and the continuity thesis. *Topoi*, 28, 31–43. doi:10.1007/s11245-008-9044-1

Menary, R. (2010a). Cognitive integration and the extended mind. In: R. Menary (ed.), *The extended mind.* Cambridge, MA: MIT Press, pp. 227–44.

Menary, R. (2010b). Dimensions of mind. *Phenomenology and Cognitive Science*, 9, 561–78. doi:10.1007/s11097-010-9186-7

Menary, R. (2010c). The holy grail of cognitivism: a response to Adams and Aizawa. *Phenomenology and the Cognitive Sciences*, 9(4), 605–18. doi:10.1007/s11097-010-9185-8

Menary, R. (2012). Cognitive practices and cognitive character. *Philosophical Explorations*, 15(2), 147–64. doi:10.1080/13869795.2012.677851

Menary, R. (2013a). Cognitive integration, enculturated cognition and the socially extended mind. *Cognitive Systems Research*, 25–26, 26–34. doi:10.1016/j.cogsys.2013.05.002

Menary, R. (2013b). The enculturated hand. In: Z. Radman (ed.), *The hand, an organ of the mind: what the manual tells the mental.* London: MIT Press, pp. 349–67.

Menary, R. (2014). Neural plasticity, neuronal recycling and niche construction. *Mind and Language*, 29(3), 286–303. doi:10.1111/mila.12051

Menary, R. (2015a). Mathematical cognition: a case of enculturation. In: T. Metzinger and J.M. Windt (eds.), *Open MIND*, 25(T), 1–20. doi:10.15502/9783958570818

Menary, R. (2015b). What? Now. Predictive coding and enculturation: a reply to Regina E. Fabry. In: T. Metzinger and J.M. Windt (eds.), *Open MIND*, 25(R), 1–8.

Menary, R. (2016). Pragmatism and the pragmatic turn in cognitive science. In: A.K. Engel, K.J. Friston, and D. Kragic (eds.), *The pragmatic turn: toward action-oriented views in cognitive science. Strüngmann Forum Reports* (vol. 18). Cambridge, MA: MIT Press, pp. 217–37.

Menary, R. (2017). Keeping track *with* things. In: A. Carter, A. Clark, J. Kallestrup, S.O. Palermos, and D. Pritchard (eds.), *Extended knowledge*. Oxford: Oxford University Press.

Menary, R. and Gillett, A.J. (2017). Embodying culture: integrated cognitive systems and cultural evolution. In: J. Kiverstein (ed.), *The Routledge handbook of philosophy of the social mind.* New York: Routledge, pp. 72–87.

Menary, R. and Kirchhoff, M. (2014). Cognitive transformations and extended expertise. *Educational Philosophy and Theory*, 46(6), 610–23. doi:10.1080/00131857.2013.779209

Millikan, R. (1993). *White queen psychology and other essays for Alice.* Cambridge, MA: Bradford Books/MIT Press.

Odling-Smee, F.J., Laland, K.N., and Feldman, M.F. (2003). *Niche construction: the neglected process in evolution*. Princeton, NJ: Princeton University Press.

O'Regan, J.K. and Noë, A. (2001). A sensorimotor account of vision and visual consciousness. *Behavioural and Brain Sciences*, 24(5), 939–73. doi:10.1017/S0140525X01000115

Oyama, S., Griffiths, P.E., and Gray, R.D. (2000). *Cycles of contingency: developmental systems and evolution*. Cambridge, MA: MIT Press.

Pigliucci, M. and Muller, G.B. (eds.) (2010). *Evolution: the extended synthesis.* Cambridge, MA: MIT Press.

Pinel, P., Piazza, M., Le Bihan, D., and Dehaene, S. (2004). Distributed and overlapping cerebral representations of number, size, and luminance during comparative judgments. *Neuron*, 41(6), 983–93. doi:10.1016/s0896-6273(04)00107-2

Pritchard, D. (2010). Cognitive ability and the extended cognition thesis. *Synthese*, 175 (S1), 133–51. doi:10.1007/s11229-010-9738-y

Roepstorff, A., Niewöhner, J., and Beck, S. (2010). Enculturing brains through patterned practices. *Neural Networks*, 23, 1051–9. doi:10.1016/j.neunet.2010.08.002

Rowlands, M. (1999). *The body in mind: understanding cognitive processes.* Cambridge: Cambridge University Press.

Rumelhart, D.E., Smolensky, P., and Hinton, G.E. (1986). Schemata and sequential thought processes in PDP models. In: J. McClelland and D.E. Rumelhart (eds.), *Parallel distributed processing: explorations in the microstructure of cognition* (vol. 2). Cambridge, MA: MIT Press, pp. 7–57.

Rupert, R. (2009). *Cognitive systems and the extended mind.* Oxford: Oxford University Press.

Rupert, R. (2010). Representation in extended cognitive systems: does the scaffolding of language extend the mind? In: R. Menary (ed.), *The extended mind.* Cambridge, MA: MIT Press, pp. 325–53.

Schulz, L.E. and Bonawitz, E.B. (2007). Serious fun: preschoolers play more when evidence is confounded. *Developmental Psychology*, 43(4), 1045–50. doi:10.1037/0012-1649.43.4.1045

Sohlberg, M.M. and Mateer, C.A. (1989). *Introduction to cognitive rehabilitation: theory and practice.* New York: Guilford Press.

Steiner, P. (2015). A problem for representationalist versions of extended cognition. *Philosophical Psychology*, 28(2), 184–202. doi:10.1080/09515089.2013.811482

Sterelny, K. (2003). *Thought in a hostile world: the evolution of human cognition.* Oxford: Blackwell Publishing.

Sterelny, K. (2010). Minds: extended or scaffolded? *Phenomenology and the Cognitive Sciences*, 9(4), 465–81. doi:10.1007/s11097-010-9174-y

Sterelny, K. (2012). *The evolved apprentice: how evolution made humans unique.* Cambridge, MA: MIT Press.

Stotz, K. (2010). Human nature and cognitive-developmental niche construction. *Phenomenology and the Cognitive Science*, 9, 483–501. doi:10.1007/s11097-010-9178-7

Stout, D., Toth, N., Schick, K., and Chaminade, T. (2008). Neural correlates of Early Stone Age toolmaking: technology, language and cognition in human evolution. *Philosophical transactions of the Royal Society of London. Series B, Biological sciences*, 363, 1939–49.

Sutton, J. (2006). Distributed cognition—domains and dimensions. *Pragmatics & Cognition*, 14(2), 235–47. doi:10.1075/pc.14.2.05sut

Sutton, J (2007). Batting, habit, and memory: the embodied mind and the nature of skill. *Sport in Society*, 10(5), 763–86. doi:10.1080/17430430701442462

Thaler, L., Arnott, S.R., and Goodale, M.A. (2011). Neural correlates of natural human echolocation in early and late blind echolocation experts. *PLoS One.* doi:10.1371/journal.pone.0020162

Trestman. M. (2013). The Cambrian explosion and the origins of embodied cognition. *Biology & Theory*, 8, 80–92. doi:10.1007/s13752-013-0102-6

Venkatraman, V., Ansari, D., and Chee, M.W.L. (2005). Neural correlates of symbolic and non-symbolic arithmetic. *Neuropsychologia*, 43, 744–53. doi:10.1016/j.neuropsychologia.2004.08.005

Vygotsky, L. (1978). *Mind in society*. Cambridge, MA: Harvard Press.

Wheeler, M. (2010). In defense of extended functionalism. In: R. Menary (ed.), *The extended mind*. Cambridge, MA: MIT Press, pp. 245–70.

Wheeler, M. and Clark, A. (2008). Culture, embodiment and genes: unravelling the triple helix. *Philosophical transactions of the Royal Society of London. Series B, Biological sciences*, 363, 3563–75. doi:10.1098/rstb.2008.0135

Wilson, R. and Clark, A. (2009). How to situate cognition: letting nature take its course. In: P. Robbins and M. Aydede (eds.), *The Cambridge handbook of situated cognition*. New York: Cambridge University Press, pp. 55–77.

Yarbus, A. (1967). Eye movements during perception of complex objects. In: L.A. Riggs (ed.), *Eye movements and vision*. New York: Plenum Press, pp. 171–96.

Zhang, J. and Norman, D.A. (1994). Representations in distributed cognitive tasks. *Cognitive Science*, 18, 87–122. doi:10.1016/0364-0213(94)90021-3

Zhang, J. and Norman, D.A. (1995). A representational analysis of numeration systems. *Cognition*, 57, 271–95. doi:10.1016/0010-0277(95)00674-3

CHAPTER 11

CRITICAL NOTE

Cognitive Systems and the Dynamics of Representing-in-the-World

TOBIAS SCHLICHT

GIVEN its young history, cognitive science is still searching for its proper research paradigm that may help it "to enter upon the secure path of a science," as Kant (1998, p. B ix) once wrote with respect to the state of metaphysics at the time. For a long time, the philosophical foundation of cognitive science was characterized in terms of the notions of representation and computation (e.g., Pylyshyn 1984). In this "classical" approach, cognitive states or structures were supposed to be representational in the sense that they stand for a particular part of the world (with various degrees of abstraction), and they were supposed to be computational in the sense that cognition was constituted by formal rule-governed manipulations of these representations. Since such cognitive processes were supposed to be realized in the brains of cognitive systems (i.e., animals), the focus has been on neuroscience to discover the biological mechanisms enabling cognition, leading to the second paradigm of connectionist approaches suggesting a different architecture underlying cognition (Smolensky 1988). In the last few decades, though, proponents of 4E approaches to cognitive phenomena have questioned many of these assumptions to various degrees, introducing the third paradigm of dynamic and enactive approaches to cognition. Radical enactivism (Varela et al. 1991; Hutto and Myin 2013; Chemero 2009; Gallagher 2005) conceives of cognition as embodied, embedded, enactive, and extended (thus 4E) and rejects representations in explanations of basic mental phenomena like action and perception. Detailed characterizations of what these features of cognition mean can be found in various contributions to this volume. A more moderate variety of enactivism attempts to develop 4E cognition while retaining the notion of mental representation (Clark 2008; Wheeler 2005). But while some of the claims made by 4E proponents are very widely accepted, others are very controversial. For example, no cognitive scientist rejects the claim that (human) cognition is embodied in the sense that it depends on functioning brains situated in organisms. But claims about radical anti-representational enactivism as well as about extended cognition are much more controversial, since they break with traditional types of explanation.

Despite my reservations with regard to radical 4E views, the very rich and interesting papers in this section are filled with important and original ideas, contributing in various ways to the search for the best paradigm for cognitive science. But although many writers make general claims regarding "cognition" in general, it is a matter of dispute whether any one of the paradigms they propose can adequately address all mental phenomena, from basic perception via imagination and memory to propositional thought and language use. Maybe cognitive scientists will one day have to accept different paradigms to explain different phenomena.

The papers in this section break with traditional distinctions and boundaries. Some central distinctions in "classical" cognitive science include the allegedly clear boundaries between perception and cognition, between (intentional) action and (non-intentional) behavior, between causation and constitution, and between cognitive systems and the environment. Not only 4E approaches, but also recent developments like the predictive processing framework (Hohwy 2018) cast doubt on these distinctions from various perspectives. The distinction between perception and cognition is blurred in predictive processing; the distinction among action, perception, and behavior is blurred by radical enactivist approaches to basic mental phenomena (Hutto and Myin 2013); the distinction between causation and constitution is questioned in dynamical approaches (Froese 2018; Lamb and Chemero 2018); and the clear demarcation of a cognitive system from its environment is questioned in the literature on dynamical systems, the extended mind, and (to some extent) cognitive integration (Menary 2018; Lamb and Chemero 2018).

For reasons of space, not all fascinating details mentioned in the target papers can be recapitulated here. This commentary is structured according to the following discussion points: (1) the compatibility of the recently popular predictive processing framework with 4E approaches; (2) the possible constraints that phenomenological characterizations of cognitive phenomena put on modeling and experimentation, illustrated by the debate on social cognition; (3) the putative replacement of the classical representationalist approach to cognition by a rigorously dynamic and embodied approach to cognition; and (4) the contrast between brain-bound and extended or integrated cognitive systems.

Predictive Processing and 4E Cognition

Hohwy's predictive coding framework may be the most traditional of the four proposals since it is compatible with classical understandings of cognition in terms of computation and representation and with conservative 4E proposals that retain mental representations. Lots of work has already demonstrated the fruitfulness of this perspective for explanations of cognitive phenomena, and future work will certainly illuminate the relation to 4E accounts in much more detail. Menary's cognitive integration framework also seems compatible with classical cognitive science in the sense that (a) acts of integration and transformation presuppose a well-defined cognitive system in the first place, and (b) acts of cognitive integration do not replace representational mechanisms but enhance them and allow for further flexibility by way of integration. Froese as well as Lamb and Chemero depart heavily (and intentionally) from classical cognitive science, although these departures are not without problems, as will be discussed later in this commentary.

In a number of influential writings, Hohwy (2013, 2018) and Clark (2013, 2016) defend the view that the brain is a prediction machine. This idea, going back more than a century at least to Helmholtz (but partly even to Kant), means that perception is achieved by unconscious inferences. The "delicious" (Dennett 2017, p. 167) basic idea of Bayesian predictive coding is that the brain is in the business of constantly generating expectations and hypotheses (on many levels of complexity) about the possible causes of its sensory input. Such top-down expectations are generated by a model in the brain and (after being processed downstream) subsequently compared to actual bottom-up input, yielding error messages of various magnitudes (being processed upward). Neural processing based on sensory input is thus reinterpreted as the processing of prediction error enforcing an update of the model or the hypothesis that issued the top-down expectations in the first place. This process continues until the model (more or less) matches the feedback and can thus count as accurate with respect to the state of the world. Furthermore, this process is "Bayesian" because it is supposed to obey Bayes's rule, according to which these hypotheses express possibilities in probabilistic terms. The overarching goal (and unifying principle) of the brain's cognitive activities is to minimize prediction error, or "surprise" (hence PEM for prediction error minimization). Viewed from this perspective, perception, action, attention, and other cognitive phenomena are all strategies of "doing the same thing" (Hohwy 2013, p. 2).

Since Hohwy's explanation of the details of the framework is so competent and accessible, I will use the space available for evaluating his goal of showing that PEM can be reconciled with 4E cognition. In order to evaluate this claim, it is important to know that various formulations of the basic idea of Bayesian predictive processing, e.g., by Friston (2010), Clark (2016), and others, differ in important details. Thus, what is at issue here is Hohwy's development of the framework. He emphasizes that PEM retains two crucial features from classical cognitive science: it is representational and inferential. Accordingly, PEM can only be compatible with the moderate 4E approaches mentioned earlier. Indeed, Hohwy intends to show that "perceptual inference and 4E cognition should be considered compatible, but only because 4E cognition, rightly understood, is nothing but representation and inference" (Hohwy 2018, p. 2). Given that such a moderate enactivism retains representations, it is not, for example, compatible with Hutto and Myin's (2013) radical enactivism, because Hutto (2005, p. 389) takes "the recent enactivist turn in cognitive science" to be motivated by a "rejection of the very idea that we can make sense of the basis of everyday skills in terms of the manipulation of underlying tacit representations."

With respect to the oppositions mentioned at the outset, PEM remains firmly classical in its representational and inferential character. It is representational since the brain's generative model of the world plays a central role in the account, issuing expectations and predictions. Sensory input does not shape perception directly; it is better conceived as "feedback to the queries issued by the brain" (Hohwy 2013, p. 2). Our expectations "drive what we perceive" and, consequently, "the hypothesis that is selected determines perceptual content" (Hohwy 2013, p. 37). What we experience is not the world, "it *is* the predictions of the currently best hypothesis about the world" (Hohwy 2013, p. 48). In a similar vein, Frith (2007, p. 111) captures this with his slogan that "our perception of the world is a fantasy that coincides with reality."

Considering the contributions to this section with respect to the issue of compatibility with predictive processing, Hohwy's representationalist account of the predictive processing framework is compatible only with Menary's cognitive integration theory, since the other contributors eschew representations. Froese (2018) echoes roboticists like Rodney Brooks (1991) and enactivists such as O'Regan and Noë (2001) who declared that the brain can simply rely on the availability of the world outside rather than being forced to represent it, and writes: "In the general case of perception it has long been argued that internal models are not necessary if the world itself can play the required role" (Froese 2018, p. 169). He adds that "the brain is not an isolated black box that must try to represent what is happening in the outside world in order to predict what will happen next and to infer what it ought to do in response" (p. 175). Froese cites Clark (2013) in support, but in the cited paper, Clark defends a predictive processing perspective starting explicitly by characterizing the "task of the brain," which "can seem impossible: it must discover information about the likely causes of impinging signals without any form of direct access to their source" (Clark 2013, p. 183). As Clark (2015) has also made clear, he intends that framework to be understood as dealing heavily in the computation of mental representations on all levels of the complex perceptual hierarchy. Those enactive accounts that are heavily influenced by Gibson's ecological approach (such as Noë 2004; Hutto and Myin 2013; Chemero 2009; Froese 2018), which was hostile to mental representations and models, seem incompatible with predictive processing. (Clark 2016 attempts to develop a compatible solution.) Gibson claimed that perception is direct, and of the world, and of affordances, i.e., possibilities for action. According to Gallagher (2005, 2008), one prominent defender of 4E cognition, we do not have to infer, say, that someone else is happy; we can see her happiness directly in her face, just like we see directly that a chair affords sitting in it. Although, phenomenologically, proponents of each side can agree that our access to the world seems to be direct, PEM holds that the mechanism in the brain enabling such access is indirect and inferential. Radical 4E accounts rely on Gibson in the sense that they suggest his notion of affordance, together with the notion of coupling, does all the explanatory work. For example, Hutto and Myin (2013, p. 19) write that "experiencing organisms are set up to be set off by certain worldly offerings." This suggests a rigidity of the "quick and dirty processing" (Hohwy 2018, p. 138) in the coupling of an environment's offerings and an organism's actions that completely ignores this flexibility in our behavior, the possibility to reflect on affordances and deliberate actions, and to switch swiftly "between contexts, projects, beliefs, and actions" (Hohwy, p. 139). When I am standing at the buffet of the "all you can eat" restaurant, although I may be set up to automatically pick up the gustatory affordances, I can also pause and contemplate my recent diet plans. By relying on a representational generative model, Hohwy argues that PEM can much better account for the flexibility of human cognition and action (For a comparative discussion of Hohwy and Clark see Dolega 2017).[1]

How does the PEM framework accommodate affordances, i.e., possibilities for action? Hohwy stresses that action is conceived as "active inference," i.e., "action only occurs when a hypothesis—in this case a representation of a state that is yet to occur—has accumulated sufficient evidence relative to other hypotheses to become the target of PEM" (p. 135). That is, instead of accumulating sensory input in order to update a given hypothesis by perceptual inference, here a given hypothesis is held constant, issuing actions resulting in the state of the world that is envisaged by the hypothesis. Both 4E accounts and predictive coding accounts claim a close relationship between action and perception. But in contrast to Gibson and his followers, who claim that action is for perception in the sense that we directly see the affordances as objective features of the environment, it is the generative model in the Bayesian predictive coding framework that can be seen as a "method for generating affordances galore," as Dennett puts it: "We expect solid objects to have backs that come into view as we walk around them; we expect doors to open, stairs to afford climbing, and cups to hold liquid" (Dennett 2017, p. 168). In line with Millikan (1995, 2004) and Siegel (2014), PEM yields a representational analysis of affordances (see Martens and Schlicht 2017). This is in stark contrast to the radical treatments of affordances in most 4E theories. I suspect that this is why Froese is openly opposed to the predictive processing framework put forward by Hohwy.

[1] A more general problem for the claim that we can perceive affordances *directly* without the need to postulate representations is posed by vision science (Palmer 1999). If we cannot posit affordances on the bottom of the visual hierarchy, then we perceive them by perceiving objects and surfaces in the environment. But vision science informs us about the complexity of visual processing of basic features that have to be integrated (bound) to objects in space. We do not perceive colors, shapes, and textures in isolation, but the brain apparently combines them via integration mechanisms to bounded objects. Pathological cases inform us that lesions can lead to isolated and more or less well-defined impairments regarding the processing of such features. Yet, these basic features by themselves do not afford anything for the cognitive system. Therefore, perceiving affordances may *seem* direct, but the underlying cognitive neural processing is very complex and usually described making use of representations. I can only sketch this problem here.

Phenomenology and Cognitive Mechanisms

Unlike Hohwy, Froese starts on the phenomenological rather than the neurobiological level or the level of cognitive architecture. His contribution thus partly concerns the question whether a description of a cognitive phenomenon on the phenomenological level puts constraints on the nature and structure of the possible mechanisms and processes enabling or underlying the cognitive phenomenon in question. Froese focuses on social understanding and makes a case for the claim that by breaking free from any representationalist commitments it can—much better than its classical rivals—integrate the personal-level phenomenology with the multi-scale dynamics occurring within and between interacting subjects. I want to comment on two aspects of this claim: first, the relation between features of our phenomenal experience and the cognitive mechanisms underlying such phenomenology; second, the phenomenological adequacy and pervasiveness of Froese's description of genuine intersubjectivity.

Froese's alternative characterizes "genuine intersubjectivity" in terms of an interactive scene in which each agent participates in the other's experience such that "the other person as such plays a constitutive role in the social phenomenon to be explained" (p. 166). On the basis of this allegedly more adequate phenomenological description of social understanding, Froese calls for a more interactive investigation of the mechanisms underlying social cognition that respects the constitutive role of the other's performance for my understanding of her mental states. Theoretical advances in enactive approaches must now be matched, Froese argues, by new experimental paradigms that overcome the methodological individualism implicit in traditional approaches. He is not satisfied with various concessions from defenders of more traditional, individualist approaches to social cognition that "social encounters, like sharing a meal, are qualitatively experienced as social events," because this concession, in his view, is typically combined with an insistence on methodological individualism. He is right that it is a further step to claim that since "the other person genuinely takes part in the constitution of that experience," an adequate explanation of this phenomenal experience should posit "an equally *intersubjective mechanism* also at the unconscious subpersonal level, i.e., a mechanism that involves internal and relational activities from more than one person" (p. 167, my emphasis). This mechanism should then encompass, in addition to an individual's neural activity, "all kinds of activity crisscrossing the brains, bodies, and environment of two or more people" (p. 167).

Let's grant—for the sake of argument—that in genuinely social encounters characterized by interaction and reciprocity, it feels exactly like Froese describes: it seems as if the other participates in my first-person experience leading to "mutual incorporation" (p. 168). What follows, if anything, for the nature and structure of the subpersonal cognitive processes and mechanisms enabling this phenomenal experience? Must there be, or should we expect, a strict isomorphism between the structural features of our experience as it presents itself to us from the first-person perspective, and the neural, bodily, and environmental structures that play a role in bringing about this experience? If we follow Froese, then the demand is not only that there be an isomorphism between my phenomenal experience (presenting you, among other things) and neural events and processes in my brain. For this would only amount to the methodological individualism that Froese rejects. Froese and other enactivists demand a complex dynamic relationship between processes in the two (or more) brains of the agents involved in the interaction *and* bodily and environmental processes that are involved, and he requires that this dynamic relationship mirrors the phenomenal experiences of both agents (see also de Jaegher, Di Paolo, Gallagher 2010).

Frankly, I do not see why this should be the case. First of all, it is hard to make predictions about the shape of naturalist theories that eventually will explain mental phenomena. But it cannot be taken as a general constraint on theories in cognitive neuroscience, for example, that the putative neural mechanisms mirror the structure of the phenomenal experience that they are supposed to explain. Indeed, such an isomorphism should be quite rare. This can be seen quite easily if we look at examples. Take the unity of conscious experience. At any given moment, we enjoy a rich experiential state that integrates bodily sensations, emotions, thoughts, and memories, etc. into one unified state; these mental phenomena do not occur in a serial fashion but together. Neuroscientists are looking for mechanisms that could achieve this kind of unified conscious experience (cf. Crick and Koch 1990). If Froese were right, neuroscientists would, for example, be forced to find a unique spot in the brain where all the information is coming together, where this unified conscious experience "happens," a "Cartesian theater" as it were (Dennett 1991). But neuroscientists agree that finding such a place is very unlikely. Instead, what they find is a multitude of possible integrative processes like, for example, 40 Hz-oscillations (Crick and Koch 1990; Engel et al. 1999). The structural features of such putative mechanisms do not in any sense match the structural features of conscious experience, and no one has objected that 40 Hz-oscillations *could not* be a candidate for yielding a unified conscious experience just because they are not unified themselves.[2]

Take a second example: our self-awareness as a single, identical, and enduring subject of experience, a "haven of stability" (Damasio 1999) and invariance, persisting over time in the face of an ever-changing stream of conscious experience. This feature of self-consciousness is the source of all our Cartesian intuitions regarding the existence and nature of a self, existing separately from (or "in") the brain or organism. An intuitive force that Dennett (2017, p. 16) aptly calls "Cartesian gravity" seduces us into

[2] Note that Chalmers's (1996) notion of functional isomorphism does not put any constraints on the implementation of this functional description.

thinking that there simply *must* either be a point in the brain where such a little (wo) man is watching all the events on the screen of consciousness, or, if there is not such a place, then we should conclude that the self and self-consciousness simply cannot be explained naturalistically. Of course, no one takes this reasoning seriously. In his introduction to *The Intentional Stance*, Dennett (1987, p. 6) predicted, 30 years ago, that "whatever the true theory of the mind turns out to be, it will overturn some of our prior convictions. . . . Any theory that makes progress is bound to be *initially* counterintuitive." The take-home message is that we should be open to the possibility that whatever scientists may discover about the brain need not match our experience on the phenomenal level. This open-mindedness of course entails the possibility that enactivists may be correct in calling for a truly dynamic theory that grants a more than causal role to bodily and environmental features in the production of phenomenal experience (Noë 2004; Lamb and Chemero 2018). But the phenomenal features of our conscious experience do not put any constraints on scientific theories of conscious experience, including the experience of genuine intersubjectivity.[3] Let me now evaluate the adequacy of Froese's phenomenological claims.

[3] Froese also calls for a more rigorous, phenomenologically informed experimental investigation of genuine intersubjectivity. He criticizes an allegedly "deep-seated prejudice found in experimental cognitive science, namely, that subjective reports are not trusted as reliable data" (p. 173). This is simply not true. In the science of consciousness, for example, most cognitive scientists even make it a condition of the investigation of consciousness that participants in experiments report their experience and investigators take such reports at face value. Stanislas Dehaene (2014, p. 9) *defines* consciousness via cognitive access, i.e., the ability to access and *report* experiences to others: "When we are fully awake and attentive, sometimes we can see an object and describe our perception to others, but sometimes we cannot—perhaps the object was too faint, or it was flashed too briefly to be visible. In the first case, we are said to enjoy conscious access, and in the second we are not." Although various philosophers and scientists criticize Dehaene's restriction (Block 2007; Tsuchiya et al. 2016), many others share his methodology. For decades, Daniel Dennett has pushed his favorite method of *heterophenomenology* to investigate consciousness. Similarly, it purports to take reports seriously but emphasizes that the experimenter must still "interpret" them from the intentional stance (Dennett 2007, p. 251). Reports play a central role in investigating cognitive phenomena. But not many cognitive scientists take very seriously—as Froese demands they should—Varela's stronger notion of "inter-being," the equivalent of Froese's notion of genuine intersubjectivity. Froese correctly notes that the much-discussed experiment by Auvray et al. (2009) does not amount to a demonstration of a constitutive role for interaction in social understanding (Overgaard and Michael 2015), although it was applauded as such by De Jaegher et al. (2010). Even the more recent hyperscanning methods used by some neuroscientists still focus on the neural mechanisms alone (Schilbach et al. 2013), while not being able to take into account bodily and environmental dynamics. It is interesting to note that Froese himself calls for experimental setups that take into account the dynamics of brain, body, and environment but only suggests a concrete experiment that retains a focus on brain dynamics. In their variation on Auvray's setup, Froese et al. primed participants by *telling them that they formed a team* playing against other teams, and thereby encouraged "the formation of a 'we'" (p. 181). Not surprisingly, when participants were debriefed after the experiment, they reported a "significantly clearer experience of the presence of the other subject" (p. 181). By taking these reports seriously, the experimenters are only taking themselves seriously.

Phenomenological Adequacy and Social Interaction

Froese's chapter also nicely illustrates the recent clash of representationalist and radically enactive paradigms of cognition, with respect to social cognition. As mentioned earlier, 4E defenders seem to offer notions like "affordance" and "coupling" as replacements of notions like "computation" and "representation." Froese criticizes traditional approaches to social understanding from the standpoint of a radically enactive, embodied approach to cognition. In particular, he claims that classical approaches, like theory-theory and simulation theory (which postulate extra-perceptual unconscious cognitive processes like inference or simulation that enable access to someone else's mental states), presuppose methodological individualism, which results in taking the very existence of the other person as a problematic assumption. He claims that his phenomenological description of *genuine* intersubjectivity is much more adequate. Let's examine this claim more closely. I should say at the outset that in what follows my commentary on Froese's paper is much more critical than my commentary on Hohwy's framework in the previous section.

In his discussion of what Froese and Fuchs (2012) call "inter-bodily resonance," Froese claims that if person A undergoes an emotion, anger, say, then A can feel her anger in terms of typical bodily changes. At the same time, A's anger is visible because part of her emotional state is a typical bodily expression. Thus, person B can directly see the emotional expression in A's face, for example. Anyone participating in the debate can grant this because the facial expression is part of the emotion (Barlassina and Newen 2014). But Froese claims that during a process of *mutual influence*, A and B "become parts of a dynamic sensorimotor and inter-affective system that connects both bodies by reciprocal movements and reactions" such that "each lived body reaches out . . . to be extended by the other, dynamically forming an extended body" (p. 168). Not everyone must share this bold intuition, and it is questionable that it is phenomenologically adequate because by mutually affecting each other, A and B need not experience "mutual incorporation." Why should we move from causation to constitution? Why should we, as Adams and Aizawa (2008, p. 91) put it, "move from an observation of a causal dependency between cognition and the body and the environment to the conclusion that cognition extends into the body and environment"? Even Froese describes the situation in purely causal terms when he writes that A's expression will "produce an impression in B" or "induce in B an unpleasant tension," in turn "causing another impression" (p. 168), so nothing seems to force us to subscribe to the constitutional claim. Anyway, A's emotional state will only *sometimes* induce a duplicate emotion in B, resulting in, say, *joint* happiness or sadness; more often, a different emotional state like sympathy or empathy will be induced in B by A's sadness. In such cases, A and B will continue to experience the clear difference between them rather than experience a mutual incorporation of the other. Even if we grant that, in special cases, the dynamical reciprocal relationship between A and B may yield an experience as described by Froese, it is a strong claim that *only* such cases would count as instances of *genuine* intersubjectivity whereas all the other cases don't. The emotional experiences of A and B can be explained in terms of a complex causal relationship where brain and bodily processes in A and B give rise to respective conscious feelings in both agents. There is nothing in here that cannot be accommodated by classical accounts.

One central theme is Froese's emphasis on *interaction* between two or more agents that has apparently been neglected by defenders of theory-theory and simulation theory. This criticism is not new. In recent years, many contributors to the debate have argued in this direction (e.g., Gallagher 2001; De Jaegher and Di Paolo 2007). What about Froese's claim that traditional theories (like TT and ST) are "consistent with the extreme possibility of solipsism" (p. 164) because the very existence of the other person is in question if it is not so incorporated. His objection is that in such accounts the other is only "given in" or "presented by" one's own first-person experience. Froese mentions Searle and Dennett in this context, two philosophers with radically different positions regarding mental phenomena. Froese accuses Dennett for the ontological neutrality of his intentional stance theory. According to Dennett, the intentional stance provides us with a useful (sometimes the only) yet risky heuristic to predict and explain the behavior of a system (or phenomenon more generally), in particular the behavior of human beings. We treat other people as intentional systems by presupposing that they are more or less rational and that they use information in order to execute their functions, which means that their behavior is guided by representations and rationality. One advantage of the intentional stance is that it can be applied to any system if it is too complex for us to come up with an explanation from the physical or design stances (Dennett 1987). Dennett's favorite example is the chess computer whose next moves are best predicted from the intentional stance, presupposing a *desire* to win the game and an *intention* to make the best move. This advantage comes at the risk of getting it completely wrong, since we may assume specific beliefs and desires although the system either harbors none at all or other beliefs and desires with different contents. Dennett argues that there is no fact of the matter whether the system entertains belief A or belief B, since many different attributions of content-involving states may yield correct predictions of behavior: "She will open the fridge because she believes/desires that . . ." But the intentional stance remains ontologically neutral only with respect to a commitment to specific mental states. The ontological neutrality does not extend toward the existence of the system, agent, or person in question; their existence is taken for granted.

Froese may think that the problem of solipsism arises because of a commitment to the authority of the first-person perspective and one's individual subjective conscious experience. This charge is strange if one is familiar with Dennett's philosophy. It may be different in the case of Searle, who is definitely an internalist. But anyway, it is easy to see that if this is a problem, then Froese is equally affected by it. He claims that the enactive view does not face this (alleged) problem because (and only because?)

enactivism assigns a constitutive role to the other in my attempt to understand their state of mind: "An individual agent's social behavior depends on the coupling of all the subsystems, including the other agent, and cannot be attributed to any one component in isolation" (Froese, p. 175). But Froese defines the relevant second-person perspective "as two first-person perspectives temporarily becoming integrated" (p. 165): "you and I are having a laugh, we are eating a meal together, we are shaking hands, etc. We share an awareness of jointly being in one and the same unfolding situation." Froese uses such examples in support of his notion of "participation in the other's experience." But in Froese's account, the first-person perspective enjoys the same systematic priority as in Searle's account. Initially, the other person is presented in my first-person experience just like in cases of "mere" observation. If the existence of the other posed a problem for traditional accounts, then it would equally present an analogous problem for Froese's enactivism. B shows up in A's conscious experience, while A shows up in B's. Later in his contribution, Froese seems to contradict this way of thinking about the relation between first-person and second-person perspective when he attempts to reverse their relationship: relying on work by Merleau-Ponty (2002) and Reddy (2008), he suggests that "embodied interaction is our primary social capacity, developmentally and in adult life, whereas the first- and third-person perspectives . . . are the basis of derivative modes giving rise to more complex forms of social understanding" (p. 170). How can the first-person perspective be derivative from the second-person perspective if the latter is defined in terms of a "reciprocal interlocking of two first-person-perspectives" (p. 170)? His statement is ambiguous between assigning an "irreducible" fundamental role to the second-person perspective (a) only for social encounters or (b) even beyond social encounters, for subjectivity. The latter seems to be conceptually impossible if a second-person perspective is defined as an integration of two first-person perspectives.

Second-person perspectives do not appear out of the blue; they have to be established, negotiated, and maintained by two or more agents (with first-person perspectives each) and must meet important constraints in order to provide a situation that is different from mere (individualistic) observation. Other enactivists like De Jaegher and Di Paolo (2007, p. 490) make clear that the relevant social interaction between the agents must be "coordinated," emphasizing the *active* and *joint* character of social understanding based on the *coupling* between two agents. They argue that the interaction process gains a certain autonomy only at a certain level of complexity. Agents exchanging body heat in a full subway does not count as coupling of the relevant kind. They agree with Froese's anti-individualism when they argue that "in such situations of interactive engagement, it is not individual cognizing and behavior that sufficiently determines the relevant phenomena: both social acts and meanings are constituted socially and during the interactive encounter. . . . The interactive constitution of social acts and meanings is a joint cognitive process that necessitates, but is underdetermined by, individual cognition" (De Jaegher, Di Paolo, and Adolphs 2016, p. 8). Interaction is supposed to be part of what produces social understanding such that it happens "between individuals" (De Jaegher, Di Paolo, and Gallagher 2010, p. 446). Crucially, De Jaegher et al. emphasize that the individual agents do not lose their autonomy in the process of interaction.

But given the emphasis on the coordinated character of interaction, all these radical enactivist approaches face the same problem (cf. Martens and Schlicht 2017): while they intend to *explain* social cognition in the sense of how meaning is generated jointly, they *presuppose* meaningful social cognition (in the individual sense) because the coupling and interaction have to be achieved in the first place and this cognitive achievement cannot itself be explained in terms of interactive mechanisms. Consider the shy student at the bar who sees a girl at the other end of the room, potentially the love of his life. If neither he nor she is going to make a move, no coupling or interaction is ever going to take place. But making this move presupposes that either student consider the other as a potential candidate for interaction in the first place. Searle (1990) argued that *any* kind of social interaction or collective intentionality presupposes this individual cognitive act. One must see the other as a minded creature, or as a second-order intentional system, as Dennett (1976) put it, i.e., if the student ends up interacting with the girl, then this is because he presupposed—individually—that she does not only hold beliefs and desires (as an intentional system), but even beliefs about other people's beliefs (as a second-order intentional system). Otherwise she would not be able to reciprocate the relevant communicative acts, gestures, and bodily signals, which could then sustain the encounter and elevate the interaction process to the relevant level. At the point where interaction can play the autonomous constitutive role envisaged by Froese and by De Jaegher, Di Paolo, and Gallagher (2010), the coupled agents must be engaged in a complex joint activity of exchanging information on various levels. And such a joint activity can only get off the ground via an initial process of social cognition—if not full-blown mindreading—simply understood as entertaining the idea that the other can reciprocally relate to myself.

One way of accounting for the preconditions of coupling and social interaction is to refer, like Martens and Schlicht (2017), to core cognitive systems (Spelke 2000; Carey 2009). On this approach, young infants come equipped with a limited set of innate special-purpose representational systems that encode information about objects, agents, number, and geometrical shape on the basis of a computation of different sets of cues or signature limits. For example, even young infants seem to be able to perceptually process relevant cues signaling the presence of an agent, e.g., biological motion, self-propelled motion, goal-directedness, animacy, etc., while they process other cues to represent objects (e.g., inertness, cohesion, and continuity). Based on their studies, Spelke and Carey suggest that this difference in representation remains constant in adulthood. These representations could provide infants and adults with perceptually basic ways of approaching objects and agents in different ways, leading to different kinds of interaction. We simply do not approach trees and houses in the way we approach other people or, more generally, agents. Only agents are considered as candidates for social interaction and the core cognitive mechanisms hypothesis can account for this fact. Although there may be other accounts that can explain this, it is clear that individual cognitive acts will play a crucial enabling role to get any social interaction off the ground. And it may well be that a representationalist alternative will be superior to radical enactive accounts like Froese's.

Integrating two first-person perspectives does not provide us with a situation that is radically different in terms of the cognitive presuppositions from an observational starting point. In particular, it is questionable that Froese is correct in describing the situation *preceding* the interaction in a way that the other "participates" in my own first-person experience any more than trees and houses "participate" in my perceptual encounters with them. For example, from the predictive processing perspective, the brain's generative model will yield expectations both about the physical and the social environment as far as causes of sensory inputs are concerned. If we take the core systems account seriously, then the expectations will differ dramatically regarding physical objects and agents given their respective behavior. Frith (2007, p. 175) notes:

> Communication, when we confront each other face-to-face, is not a one-way process from me to you. The way you respond to me alters the way I respond to you. This is a communication loop. But in addition it is not just me who is trying to predict what you will say next on the basis of my model of your idea. You also have a model of my idea in your mind. You are also trying to predict what I will say next. You also will alter what you say to indicate that your model of my meaning is not quite working to predict what I am going to say. This is the big difference from my interactions with the physical world. The physical world is utterly indifferent to my attempts to interpret it. (Frith 2007, p. 175)

Frith allows interaction to play a causal role because "individuals engaged in joint action have a broader understanding of the behavior of their partners, and thus of options available for action, by representing aspects of the interactive scene in the we-mode" (Gallotti and Frith 2013). Similarly, Butterfill (2013) emphasizes that interaction can generally provide a richer knowledge base about someone else's state of mind than mere observation. But he recognizes, at the same time, the need for at least "minimal mindreading" in order to get the interaction off the ground.[4]

Cognitive Systems

Along with many 4E proponents, Menary, Froese, and Lamb and Chemero raise the issue of the boundaries and identification criteria of cognitive systems. Typically, 4E defenders agree in their criticism of classical cognitive science for its focus on methodological individualism, in particular on the brain as the locus of where cognition happens or where the physiological basis of cognitive processes should be identified. Cognitive processes are associated with cognitive systems so it is important to know what we

[4] As far as the representational commitments are concerned, Menary's (2018) cognitive integration approach is much less radical in its departure from classical cognitive science than Lamb and Chemero's. It should rather be seen as an important addition and enrichment with different emphases. See below for details.

should regard as cognitive systems. The central objection put forward by defenders of extended cognition is that drawing the boundary (of the supervenience base of cognitive processes) at the skull, or even at the skin, seems arbitrary in light of various examples of extensive tool use (e.g., Clark 2008) or joint and distributed cognition (e.g., Hutchins 1995). Examples from this area, many of which are mentioned in the table provided by Menary (2018), seem to suggest that the vehicles of cognitive processes extend way beyond the organism's body. Phenomena like the rubber hand illusion (Botvinick and Cohen 1998) demonstrate the flexibility of our body schema, and extensive tool use allegedly forces us to reconsider the boundaries of a cognitive system, extending it out into the world to include extensively used tools such as laptops, smartphones, and other gadgets. This section discusses how the various contributions address this issue.

According to dynamical systems theory, as presented by Lamb and Chemero (2018), what constitutes a cognitive system may vary in different circumstances, leaving us with a "messy picture" (p. 149) of the boundaries of cognitive systems. Dynamical systems theory is characterized as a "set of methods, theories, and models" that yields arguments in favor of embodied and extended approaches to cognition because it holds that all features (parts) of a cognitive system are interdependent and that different cognitive systems as such are interdependent. The central features of cognitive systems, in Lamb and Chemero's version, are thus "interaction" and "openness." According to the former, "any state or behavior within a cognitive system is adequately characterized only in relation to other states and behaviors within the cognitive system" (p. 149). This includes relations among the elements over time. But the dependencies are not fixed; they differ in intimacy and strength and are subject to variations that give rise to boundaries and demarcate different systems from each other. According to the openness hypothesis, "the states and behaviors of any cognitive system are only persistent when they are not at equilibrium with other systems that they interact with" (p. 149). Autopoietic systems (Maturana and Varela 1980; Thompson 2007) count as a subclass of dynamical systems, as Lamb and Chemero note. Froese (2018) subscribes to this kind of enactivism, claiming "where there is life there is mind" (Kirchhoff and Froese 2017, p. 1). Attempting to identify the criteria for cognitive systems with reference to biological features such as metabolism, etc. is certainly a move that seems justified given that paradigmatic cognitive systems are biological systems. Yet the scope of the cognitive and the scope of the biological may not coincide, and it is difficult to decide whether it is already time to put all our money on the autopoietic approach. While it may be difficult for the opposition to point out exactly which (higher) animals may be cognitive systems and which (lower) animals are not and why this is so, the autopoietic approach may identify too many cognitive systems, since it implies that even a single cell has basic mental capacities (a little bit of mind, so to speak), simply by sharing the same organizational properties of mind. It is not clear to me why such an overly inclusive view of cognition should be a useful research strategy for cognitive science because it is difficult to single out a mark of the cognitive among the living (see Schlicht 2011 for discussion). If cognitive science has a *subjectum*, a proper realm of investigation, then this must be sufficiently different from the subject matter of biology or any other science.

Given Froese's opposition to predictive coding, it is ironic that Kirchhoff and Froese (2017) motivate their autopoietic view with the "free energy principle" (Clark 2013; Friston 2010; Hohwy 2015), according to which all organisms must minimize their free energy in order to maintain their identity and existence. Kirchhoff and Froese present it as "an overarching rationale for brain functioning," sharing this enthusiasm with Hohwy (2018) and Friston (2010). The predictive processing account, at least according to Hohwy, assigns a prominent role to the brain by claiming that prediction error minimization is the primary task of the *brain*, so that the account does not automatically yield a claim about *all* living systems. *Embrained autopoietic systems* may provide us with a useful criterion for identifying cognitive systems, but it must be motivated independently (see below). My point is that if the free energy principle is primarily invoked in order to explain *brain* function, being associated with active inference, then it remains a major task for defenders of autopoietic cognition to explain how this principle can be applied to organisms lacking brains *and* ground an explanation of cognition at the same time. The dilemma, in my view, is that if the free energy principle extends to all organisms (and possibly beyond since it is associated with thermodynamics), why should we think that minimizing free energy has anything to do with cognition? If it does not so extend, but is supposed to provide the key to an explanation of cognition, then why should we believe in the claim that where there is life there is mind? I think more work is needed to clarify this issue.

Returning to the dynamical systems approach, the identity and boundary of a cognitive system is determined "relative to a given phenomenon of interest" (Lamb and Chemero 2018, p. 149) and constituted by:

> all and only those elements that depend, to a specified extent, upon one another. . . . Everything is not related to everything else; we do not live in either a perfectly random or perfectly symmetrical universe. As a result, some things are more dependent on one another than they are on other things. How much and in what way a collection of elements must depend on one another in order to count as a system is partly grounded in how the system of interest is defined and empirically observable facts about the entities in the collection. (p. 149)

Variations in dependence give rise to new delineations of systems. As Lamb and Chemero suggest correctly, this seems to echo the claim made by proponents of extended cognition that the boundaries of skull and skin are not always the correct boundaries to be drawn (Clark and Chalmers 1998). Lamb and Chemero suggest that such boundaries should be testable, claiming that experiments could help reveal the boundary and identity of a given cognitive system at a time. But if it is largely a matter of "the aims and interests of cognitive researchers" (Lamb and Chemero, p. 155) to determine which interactions are *considered* strong enough, then it seems unlikely that experimental investigations of the sort suggested by Lamb and Chemero will tell us *definitely* what constitutes a cognitive system and where to distinguish one cognitive system from another one, since it is difficult to see how there can be an *objective* answer that is independent of such aims and interests. According to Lamb and Chemero, the interaction and openness of cognitive systems imply that "the continual transfer of energy into, and out of, the system results in increasing variety and complexity of interactions across multiple scales" (p. 154). Such variety indeed yields a messy picture of the boundaries of cognitive systems because what may count as one system now may not count as one system from another viewpoint or at a later stage. I wonder whether this implication of the dynamical systems approach so formulated—if exclusive—makes it a useful paradigm in cognitive science. While dynamical systems theory may add important insights into the complexity of the causal interactions of a given cognitive system (if coupled with more traditional approaches), on its own it does not seem to retain a robust enough notion of a persisting and stable "core" cognitive system. If the main objection to classical cognitive science is that the boundaries of skull and skin are supposedly arbitrary, then Lamb and Chemero's framework invites the same objection, namely, that their messy way of making the boundaries dependent on (and variable according to) the aims and interests of researchers is equally if not more arbitrary. At this point, Rupert's (2010) worry is important:

> Humans categorize, perceive, remember, use language, reason, make sense of the actions of others—these and more are all persisting abilities of persisting systems; they do not consist in the activities of relatively short-lived coupled systems. . . . The importance of systems that persist and cohere, even through change, is especially clear in developmental psychology: we want to know how *that* system—that single developing human—came to be the way it is and how a similar course of development happens, on average, for the relatively homogeneous multitude of such persisting human systems. We want to understand how and why the capacities and abilities of individual persisting systems change over time, eventually taking a stable form. (Rupert 2010, p. 330)

Ever since Tolman's (1948) experiments with rats, psychologists and other cognitive scientists have had well-defined cognitive systems at their disposal, namely animals, people, and, more recently, robots. The advantage of considering these entities as cognitive systems is that this assumption yields questions about cognitive development of *these* systems in comparison to others and with respect to different cognitive capacities. Any comparative investigation in developmental psychology, say, presupposes a working assumption about the developing system in question. In order to tell whether *a given system* cognitively developed in this or that direction, we must assure to have the same system in view. What counts as a human being or a rat is neither a matter of convention nor a matter of varying research perspectives. If one rejects the relatively clear boundaries of higher animals, there seem to be two possibilities: either we go smaller or go bigger. Advocating the autopoietic approach (Froese 2018) amounts to going smaller, since even single cells are candidate cognitive systems (Thompson 2007). The versions of the extended mind thesis, including Menary's (2018) cognitive integration approach, are examples for the second option, going bigger. Lamb and Chemero's proposal seems to allow for all three possibilities since different researchers at different times or with varying interests may consider different entities that then cannot be comparatively studied. Whether and in what way the predictive processing framework is compatible with (or committed to) either the classical computational paradigm in cognitive science or the more recent 4E paradigm is currently under intensive discussion (Clark 2015; Hutto 2017).

So let's have a look at Menary's and Lamb and Chemero's proposal. What about cognitive integration? Menary's (2018) approach of considering integrated cognitive systems is presented as a variant of the extended mind hypothesis, motivated by its payoff to explain "our ability to do things we otherwise cannot do and the transformation of existing abilities, making us smarter and better at difficult and demanding cognitive tasks" (p. 197). By integrating systems of mathematical representation and the performance of the relevant practices, we can enhance our set of cognitive skills. By using much more basic sensorimotor skills in order to manipulate our environment in a process Menary calls "niche construction," we formed the basis of all our higher-order cultural and representational developments in human history. The approach diverts from radical 4E accounts by accepting mental representations and thus seems to belong to the group of "conservative" enactivist views (cf. Hutto and Myin 2013 for this label). In this process, the manipulated environment also plays an active role by shaping and even constituting our cognitive skills. This process cannot only be traced back to our ancestors; it still has to be considered in explanations of our present cognitive skills. Menary suggests that the complex cycles of cognitive processing cannot be easily decoupled from human beings performing the cognitive tasks at hand. Rather, we are reciprocally coupled to our niche in complex ways.

But Menary does not define what an integrated cognitive system is; instead, he suggests several criteria for cognitive integration that are formulated as dimensions along which any given system can score either highly or not. Among such criteria are "coordinated interaction," "cognitive practices," "normative manipulations," and "cognitive transformations." The most basic cognitive practices are bodily interactions based on the exploitation of sensorimotor contingencies (O'Regan and Noë 2001) and individual or collective sequencing and coordination of action. More sophisticated practices include the inspection and correction of the informational structures of the environment and the manipulation of tools to solve demanding problems. We do not need a complete list. As these examples make clear, there is no need to go beyond the descriptions offered, which clearly presuppose *someone*, some previously bounded cognitive system—presumably a human being, animal, or robot—carrying out all these manipulations, performing and integrating the relevant practices (for some further thoughts about identifying the cognitive system/subject of experience with the organism, see Schlicht 2017).

Obviously, Menary is right in drawing attention to how learning new practices from others or acquiring them through testing, etc. *shapes* and *changes* one's cognitive repertoire; but this does not force us to claim that the cognitive system has been *extended* in any significant way; nor does it force us to posit a *further* and *distinct* extended or coupled cognitive system that hasn't existed before. We can make do—in line with Rupert's reasoning (as discussed by Menary)—with the cognitive system we started with, the human being, animal, or robot. Drawing the boundary at skull and/or skin is not as arbitrary as suggested by some proponents of 4E approaches. With respect to the skull, psychopathology provides plenty of (important) evidence about the specific and central role the brain plays in enabling and maintaining cognition and consciousness. Damasio (1999) describes a rich set of pathological phenomena based on brain lesions diminishing varieties of cognition and consciousness. In general, such research demonstrates the asymmetry between the brain's role for the production of these phenomena and the body's and environment's roles. Nevertheless, the body does not only embed the brain but provides important cognitive tools itself. Thus, defenders of 4E cognition rightly stress that any comprehensive theory of cognition must adequately describe the role the body and also (tools found in) the environment play in the dynamic production of cognition. But the brain is central: without a functioning and controlling brain, bodily capacities and integrated tools are of no use, since no cognition will get off the ground, while the brain can still enable quite an amount of thinking in case of a damaged body. Despite its emphasis on integration of tools into the cognitive system, Menary's approach to the extended mind debate makes a clear concession to more classical approaches to cognition by presupposing a cognitive system that initially starts the process of niche construction or tool use and integration.

Defenders of extended cognition must find a way of delineating cognitive systems that both justifies the move beyond the skin, i.e., the boundary that is set by the biological animal, and also guards them against the cognitive bloat objection. This objection points out the danger of including too much into the cognitive system. For example, by analogy we may be forced to consider the sun as part of my cognitive system since its light makes it easier for me to perceive my environment. This is a consequence that 4E proponents should want to avoid since it makes it difficult if not impossible to investigate cognition at all (Sprevak 2009).

Determining the boundaries of cognitive systems and situating them within the dynamic causal relations to their immediate surroundings overlaps with the problem of determining causal and constitutive factors of cognitive processes. The proposals by Lamb and Chemero and Froese are prone to what Adams and Aizawa (2001) called the coupling-constitution fallacy early on in the debate, i.e., the mistake of considering causal factors as constitutive elements. But I would like to draw attention to the fact that Lamb and Chemero formulate their whole account in terms of elements of a cognitive system being "highly interdependent" (p. 153) such that they will "exhibit correlations that can be measured" (p. 155). As is well known, experimental data about (strong or weak) correlations underdetermine theoretical conclusions as to whether they should be *interpreted* as indicating (a) a mere correlation, (b) a causal relation, or (c) a constitutive relation. This problem is particularly salient in the search for the neural correlate of consciousness (Metzinger 2000; Tsuchiya et al. 2015) because the empirically available evidence is equally compatible with a causal account of neural processes and consciousness (Searle 1992), with an identity theory (Papineau 2002), and even with a common cause interpretation (Chalmers 2002).

Consider an example from outside the cognitive domain: in order to be able to drive my car, I depend on gas and oil and, possibly, roads, etc. But it is very useful, for various (practical, judiciary, economical, and other) purposes, to continue to demarcate my car very strictly from its environment and from all kinds of enabling devices and larger systems in which it may be embedded. A proponent of "extended cars" could object that drawing the boundary of my car at the steel and windows is just as arbitrary as drawing it at the engine. Of course, we could *decide* that the demarcation of my car at the steel and windows should be traded in for a more extended view. But we have to draw the boundary *somewhere*, and just like the cognitive bloat objection to extended cognition, it is not clear where to draw the boundary instead. My car needs gas and oil as well as a road to drive on, but it does not make much sense to treat oil and gas or even gas stations and the oil industry (and why stop there?) as other than causal contributors to the driving of the car. Analogously, what's the explanatory advantage in treating the water we drink or the sunlight we enjoy as other than causal contributors to the maintenance of an organism's identity and cognitive functioning? When Rupert (2009) emphasizes the importance of cognitive systems being persistent and integrated, it also applies to this example. For the time being, cognitive scientists are well advised to consider the animal as the cognitive system (see Schlicht 2017). Whether animals without brains are capable of cognition at all is an empirical issue, and it remains a difficult task—in light of the dominating functionalism—to determine where to draw the boundaries in the animal kingdom. It seems, however, that endowing organisms down to the single cell with cognitive capacities is not very useful when we want to leave open the possibility that cognitive science differs from biology.

Conclusion and Outlook

For reasons of space, I have highlighted those points I find problematic or where I see tensions and problems that should be addressed by defenders of the respective accounts. Hohwy's predictive coding framework may be the most traditional of the four proposals since it is compatible with classical understandings of cognition in terms of computation and representation, being compatible only with conservative 4E proposals that retain mental representations. Lots of work has already demonstrated the fruitfulness of this perspective for explanations of cognitive phenomena, and future work will certainly illuminate the relation to 4E accounts in much more detail. Menary's cognitive integration framework also seems compatible with classical cognitive science in the sense that (a) acts of integration and transformation presuppose a well-defined cognitive system in the first place, and (b) acts of cognitive integration do not replace representational mechanisms but enhance them and allow for further flexibility by way of integration. Froese as well as Lamb and Chemero depart heavily (and intentionally) from classical cognitive science, although these departures are not without problems. As I have argued, Froese's account of genuine (enactive) intersubjectivity (in terms of social interaction and mutual incorporation) presupposes preceding individual cognitive acts of (minimal) social understanding, just like the second-person perspective is defined as two first-person perspectives becoming temporally integrated. Furthermore, Froese's claim that phenomenological descriptions should constrain the range and nature of possible cognitive mechanisms is misguided. Of course, cognitive scientists are well advised to use subjective reports in their investigations of cognitive phenomena, but they already do so. And the purported mechanisms invoked in models of problem-solving processes need not match the structure suggested in such reports in order to count as possible explanations of them. What I take to be particularly problematic about Lamb and Chemero's proposal is (a) that it makes the boundaries of cognitive systems in principle a matter of research interests, and (b) that their radical dynamical perspective leaves us without a clear criterion for the diachronic identity of cognitive systems, which is needed for developmental investigations of such systems in cognitive science. In particular, with regard to the more radical 4E accounts, such as Froese's and Lamb and Chemero's, it is difficult to see how these accounts scale up to account for "representation-hungry" (Clark and Toribio 1994) phenomena on the basis of coupling and affordances alone. Nevertheless, there is a kernel of truth in all of these interesting proposals, but what I take to be true about them—as outlined in these contributions—can probably be integrated by classical accounts, i.e., conservative enactivism. What's radical about them will most probably not be retained in future paradigms for cognitive science. But that's an empirical question and it will be exciting to see where future investigations lead us with regard to finding a paradigm (or paradigms) for cognitive science.

References

Adams, F. and Aizawa, K. (2008). *The bounds of cognition*. Oxford: Wiley-Blackwell.

Auvray, M., Lenay, C., & Stewart, J. (2009). Perceptual interactions in a minimalist virtual environment. *New Ideas in Psychology*, 27(1), 32–47.

Barlassina, L. and Newen, A. (2014). The role of bodily perception in emotion: in defense of an impure somatic theory. *Philosophy and Phenomenological Research*, 89, 637–78.

Block, N. (2007). Consciousness, accessibility, and the mesh between psychology and neuroscience. *Behavioral and Brain Sciences*, 30, 481–548.

Botvinick, M. and Cohen, J. (1998). Rubber hands "feel" touch that eyes see. *Nature*, 391(6669), 756.

Brooks, R. (1991). Intelligence without representation. *Artificial Intelligence*, 47, 139–59.

Butterfill, S.A. (2013). Interacting mindreaders. *Philosophical Studies*, 165(3), 841–63.

Carey, S. (2009). *The origins of concepts*. New York: Oxford University Press.

Chalmers, D.J. (2002). Consciousness and its place in nature. In: D.J. Chalmers (ed.), *Philosophy of mind: classical and contemporary readings*. New York: Oxford University Press, pp. 247–72.

Chemero, A. (2009). *Radical embodied cognitive science*. Cambridge, MA: MIT Press.

Clark, A. (1997). *Being there: putting brain, body, and world together again*. Cambridge, MA: MIT Press.

Clark, A. (2008). *Supersizing the mind.* New York: Oxford University Press.
Clark, A. (2013). Whatever next? Predictive brains, situated agents, and the future of cognitive science. *Behavioral and Brain Sciences*, 36, 181–253.
Clark, A. (2015). Radical predictive processing. *Southern Journal of Philosophy*, 53, 3–27.
Clark, A. (2016). *Surfing uncertainty.* Oxford: Oxford University Press.
Clark, A. and Chalmers, D.J. (1998). The extended mind. *Analysis*, 58, 7–19.
Clark, A. and Toribio, J. (1994). Doing without representing? *Synthese*, 101, 401–31.
Crick, F.H. and Koch, C. (1990). Toward a neurobiological theory of consciousness. In: N. Block et al. 1997 (eds.), *The nature of consciousness.* Cambridge, MA: MIT Press, pp. 277–92.
Damasio, A.R. (1999). *The feeling of what happens: body and emotion in the making of consciousness.* New York: Harcourt Brace and Co.
Dehaene, S. (2014). *Consciousness and the brain: deciphering how the brain codes our thoughts.* New York: Viking.
De Jaegher, H. and Di Paolo, E. (2007). Participatory sense-making. *Phenomenology and the Cognitive Sciences*, 6(4), 485–507.
De Jaegher, H., Di Paolo, E. A., and Adolphs, R. (2016). What does the interactive brain hypothesis mean for social neuroscience? A dialogue. *Philosophical transactions of the Royal Society of London. Series B, Biological sciences*, 371(1693), 20150379. doi:10.1098/rstb.2015.0379
De Jaegher, H., DiPaolo, E., and Gallagher, S. (2010). Can social interaction constitute social cognition? *Trends in Cognitive Sciences*, 14(10), 441–7.
Dennett, D.C. (1976). Conditions of personhood. In: A. Oksenberg-Rorty (ed.), *The identities of persons.* Berkeley: University of California Press, pp. 175–96.
Dennett, D.C. (1987). *The intentional stance.* Cambridge, MA: MIT Press.
Dennett, D.C. (1991). *Consciousness explained.* London: Penguin.
Dennett, D.C. (2007). Heterophenomenology reconsidered. *Phenomenology and the Cognitive Sciences*, 6, 247–70.
Dennett, D.C. (2017). *From bacteria to Bach: the evolution of minds.* New York: Norton & Company.
Dolega, K. (2017). Moderate Predictive Processing. In T. Metzinger & W. Wiese (Eds.). Philosophy and Predictive Processing: 10. Frankfurt am Main: MIND Group. doi: 10.15502/9783958573116.
Engel, A. K., Fries, P., König, P., Brecht, M., and Singer, W. (1999). Temporal binding, binocular rivalry, and consciousness. *Consciousness and Cognition*, 8, 128–51.
Friston, K.J. (2010). The free-energy principle: a unified brain theory? *Nature Reviews Neuroscience*, 11, 127–38.
Frith, C.D. (2007). *Making up the mind: how the brain creates our mental world.* Oxford: Blackwell.
Froese, T. (2018). Searching for the conditions of genuine intersubjectivity: from agent-based models to perceptual crossing experiments. In: this volume, pp. 163–86.
Froese, T. and Fuchs, T. (2012). The extended body: a case study in the neurophenomenology of social interaction. *Phenomenology and the Cognitive Sciences*, *11*(2), 205–35.
Gallagher, S. (2001). The practice of mind: theory, simulation or primary interaction? *Journal of Consciousness Studies*, 8(5–7), 83–108.
Gallagher, S. (2005). *How the body shapes the mind.* Oxford: Oxford University Press.
Gallagher, S. (2008). Direct perception in the interactive context. *Consciousness and Cognition*, 17, 535–43.

Gallotti, M. and Frith, C. (2013). Social cognition in the we-mode. *Trends in Cognitive Sciences*, 17(4), 160–5.
Gibson, J.J. (1979). *The ecological approach to visual perception*. Boston, MA: Houghton Mifflin.
Hohwy, J. (2013). *The predictive mind*. Oxford: Oxford University Press.
Hohwy, J. (2015). The neural organ explains the mind. In: T. Metzinger and J. Windt (eds.), *Open mind*. Cambridge, MA: MIT Press, pp. 702–23.
Hohwy, J. (2018). The predictive processing hypothesis. In: this volume, pp. 129–46.
Hutchins, E. (1995). *Cognition in the wild*. Cambridge, MA: MIT Press.
Hutto, D.D. (2005). Knowing *what*? Radical vs. conservative enactivism. *Phenomenology and the Cognitive Sciences*, 4, 389–405.
Hutto, D.D. (2017). Getting into predictive processing's great guessing game: bootstrap heaven or hell? *Synthese*. doi:10.1007/s11229-017-1385-0.
Hutto, D.D. and Myin, E. (2013). *Radicalizing enactivism*. Cambridge, MA: MIT Press.
Kant, I. (1998). *Critique of pure reason* (trans. and eds. P. Guyer and A.W. Wood). Cambridge: Cambridge University Press.
Kirchhoff, M. and Froese, T. (2017). Where there is life there is mind: in support of a strong life-mind continuity thesis. *Entropy*, 19, 169. doi:10.3390/e19040169
Lamb, M. and Chemero, A. (2018). Interacting in the open: where dynamical systems become extended and embodied. In: this volume, pp. 147–62.
Martens, J. and Schlicht, T. (in press). Individualism vs. interactionism about social cognition. *Phenomenology and the Cognitive Sciences*.
Maturana, H.R. and Varela, F.G. (1980). *Autopoiesis and cognition: the realization of the living*. Boston Studies in the Philosophy of Science, vol. 42. Dordrecht: D. Reidel.
Menary, R. (2018). Cognitive integration: how culture transforms us and extends our cognitive capabilities. In: this volume, pp. 187–216.
Merleau-Ponty, M. (1945/2002). *Phenomenology of perception* (trans. C. Smith). Oxford, UK: Routledge.
Metzinger, T. (ed.) (2000). *Neural correlates of consciousness*. Cambridge, MA: MIT Press.
Millikan, R.G. (1995). Pushmi-pullyu representations. *Philosophical Perspectives: AI, Connectionism, and Philosophical Psychology*, 9, 185–200.
Millikan, R.G. (2004). *Varieties of meaning*. Cambridge, MA: MIT Press.
Noë, A. (2004). *Action in perception*. Cambridge, MA: MIT Press.
O'Regan, J.K. and Noë, A. (2001). A sensorimotor account of vision and visual consciousness. *Behavioral and Brain Sciences*, 24, 883–917.
Overgaard, S. and Michael, J. (2015). The interactive turn in social cognition research: a critique. *Philosophical Psychology*, 28(2), 160–83.
Palmer, S. (1999). *Vision science: photons to phenomenology*. Cambridge, MA: MIT Press.
Papineau, D. (2002). *Thinking about consciousness*. Oxford: Oxford University Press.
Pylyshyn, Z. (1984). *Computation and cognition*. Cambridge, MA: MIT Press.
Reddy, V. (2008). *How infants know minds*. Cambridge, MA: Harvard University Press.
Rupert, R. (2009). *Cognitive systems and the extended mind*. Oxford: Oxford University Press.
Rupert, R. (2010). Representation in extended cognitive systems. In: R. Menary (ed.), *The extended mind*. Cambridge, MA: MIT Press, pp. 325–54.
Schilbach, L., Timmermans, B., Reddy, V., Costall, A., Bente, G., Schlicht, T. et al. (2013). Toward a second-person neuroscience. *Behavioral and Brain Sciences*, 36(4), 393–414.
Schlicht, T. (ed.) (2011). *Journal of Consciousness Studies* [Special issue: *Consciousness and life*], 18(5,6).

Schlicht, T. (2017). Experiencing organisms: from mineness to subject of experience. *Philosophical Studies*. doi:10.1007/s11098-017-0968-4

Searle, J. R. (1990). Collective intentions and actions. In: P.R.C.J. Morgan and M. Pollack (eds.), *Intentions in communication*. Cambridge, MA: MIT Press, pp. 401–15.

Searle, J.R. (1992). *The rediscovery of the mind*. Cambridge, MA: MIT Press.

Siegel, S. (2014). Affordances and the content of perception. In: B. Brogaard (ed.), *Does perception have content?* Oxford: Oxford University Press, pp. 51–75.

Smolensky, P. (1988). On the proper treatment of connectionism. *Behavioral and Brain Sciences*, 11, 1–74.

Spelke, E. (2000). Core knowledge. *American Psychologist*, 55, 1233–43.

Sprevak, M. (2009). Extended cognition and functionalism. *Journal of Philosophy*, 106(9), 503–27.

Thompson, E. (2007). *Mind in life: biology, phenomenology, and the sciences of mind*. Cambridge, MA: Harvard University Press.

Tolman, E. (1948). Cognitive maps in rats and men. *Psychological Review*, 55, 189–208.

Tsuchiya, N., Wilke, M., Frässle, S., and Lamme, V.A.F. (2015). No-report paradigms: extracting the true neural correlates of consciousness. *Trends in Cognitive Sciences* 19(12), 757–70.

Varela, F., Thompson, E., and Rosch, E. (1991). *The embodied mind*. Cambridge, MA: MIT Press.

Wheeler, M. (2005). *Reconstructing the cognitive world: the next step*. Cambridge, MA: MIT Press.

PART IV

COGNITION, ACTION, AND PERCEPTION

CHAPTER 12

THE BODY IN ACTION

Predictive Processing and the Embodiment Thesis

MICHAEL D. KIRCHHOFF

Introduction

The ceramist working at a potter's wheel is involved in a genuinely dynamic process. If one focuses on the wet clay, it becomes apparent that it slowly morphs into shape, while spinning over fast enough time scales that its shape appears to be constantly coming into being and dissipating again. Moreover, the dynamic loops between the potter's body—her hands, arms, eyes, and so on—and the wet clay itself seem so interwoven and fluid that the dynamic profile of the overall process is constantly forming and re-forming over short periods of time. There is, or so it would seem, what Hurley terms a *dynamic singularity*: "a continuous and complex dynamic system . . . with feedback loops that . . . have external as well as internal orbits" (Hurley 1998, p. 333).

So, "trying to separate brain, body, and material culture in the above nexus of mediated activity, is like trying to construct a pot keeping your hands clean from the mud" (Malafouris 2010, p. 65). It cannot be done. A different reason for thinking that dynamic sensorimotor profiles and the embedding environment matter fundamentally to minds is that like the piece of wet clay, thrown onto the wheel of the potter, minds are subject to "continuous re-shaping, re-wiring and re-modelling" (Malafouris 2010, p. 55) in virtue of the plasticity of brain, body, and cultural practices.

Call this view the *embodiment thesis* (ET, hereafter). Varela, Thompson, and Rosch (1991) are arguably the pioneers of this view of embodiment in the relation to mind. For them, cognition is conceived of not as a detached reconstruction of the world but as a suite of dynamic processes enabling embodied activity (see also Engel et al. 2013). On such a view, action does not play a secondary role to perception and cognition. Instead, action is what enables perception and cognition.

ET finds a natural home in Hurley's dynamical singularity thesis (1998) and in Malafouris's material engagement theory (2007). And they are not alone. Nudging this way of thinking about embodiment and its relation to mind even further are arguments concerning the dynamical nature of mind (Kelso 1995; Thelen and Smith 1994), enactive considerations about perception as embodied activity (Hutto and Myin 2013; Noë 2004; Thompson 2007), and the extended mind hypothesis (Clark 2008; Menary 2007; Rowlands 2018).[1]

Those who defend ET often take it to imply some or all of the following set of claims about the nature of mind:

1. *The constitutive thesis*: Minds or cognitive systems are extensive in the sense that they are realized in patterns of sensorimotor activity nonlinearly coupled with the embedding environment.
2. *The nonrepresentational thesis*: The sensorimotor profile of organisms is sufficient for at least some kinds of cognitive activities, thus replacing the need for organisms to construct complex internal mental representations of the outside environment.
3. *The cognitive-affective inseparability thesis*: Affect, cognition, and sensorimotor contingencies are inseparable in the sense that patterns of affectivity are part and parcel of perception, action, and thinking.
4. *The metaplasticity thesis*: It is not only the brain that is plastic. The body in action is plastic, and the entire organism is situated in a plastic network of processes spanning brain, body, and world.

Together, these four theses put embodied activity at the core of a unified and dynamic singularity—one poised on the brink of instability such that "it can switch flexibly and quickly" and so be in a position "to anticipate the [immediate] future" rather than simply having to react it on the basis of constructing—on-the-fly—representations of it (Kelso 1995, p. 26; Kirchhoff 2015a).

But there is a complication. Despite the initial appeal of ET, new work on predictive processing models in neurobiology and philosophy threatens to dampen the plausibility of all four of these theses. That is, from the perspective of predictive processing (PP), it is by no means obvious that one must go this way about the role of the body in the sciences of mind and brain. PP states that brains are forever trying to minimize a prediction error quantity—viz., to minimize a mismatch between its predictions of the sensory input it expects and the sensory input it actually receives (Hohwy 2013). In its most ambitious form, PP is the view that cognition is nothing over and above minimizing prediction error on the basis of the brain's generative model (Friston 2009). This has implications for how to think of embodiment in relation to cognition.

On what I shall dub "internalist PP," embodiment plays a nontrivial role. However, it plays merely a *causal* role in the minimization of prediction error. Bodily activity can be

[1] I am aware that these views differ in their commitments about the nature of minds. It is not my intention to survey their points of continuity and discontinuity here. Instead I focus only on ET, while drawing out points for discussion between an interpretation of ET and its relation to predictive processing models in theoretical neuroscience.

understood as importantly modifying sensory signals in such a way that they match the brain's expectations. Nevertheless, the body stops short of playing a *constitutive* role in prediction error minimization. That is, if all that matters to the brain is error minimization, and if this process is directly determined by brain states, then bodily and worldly elements—however causally significant—fail to qualify as constitutive parts of prediction error minimization. As Hohwy states: "The parts of our own bodies that are not functionally sensory organs are beyond the boundary [of a cognitive system], so cognitive states are not extended into the body—there is no embodied extension. Likewise, things in the environment are outside the [cognitive system], as are other people and their mental states" (2014, p. 11).

This chapter investigates the extent to which it is possible to accept PP while also taking seriously the theoretical commitments of ET. It will be argued that PP and ET are not mutually exclusive. Instead, it is possible to show that PP and ET can come fruitfully together on all four issues mentioned earlier. The outcome of this is that it avoids obscuring from view that PP and ET unite on a number of important points—in particular, committing themselves to understand the basic conditions unifying life and mind (Kirchhoff 2016); seeking an explanation of cognition as rooted in action (Bruineberg et al. in press; Clark 2016); and as driving a shift away from orthodox cognitive science to a view of minds as realized in dynamics breaking across brain, body, and world (Kirchhoff 2015a, 2015b).

Overview

This chapter has four sections. The second section targets the constitutive thesis of ET. It will be argued that there are reasons to prefer, at least in certain circumstances, a wide reading of the realizers of prediction error minimization. The third section addresses the nonrepresentational thesis of ET. It will be argued that there is no need to posit the existence of internal representations to explain basic forms of prediction error minimization, on the one hand, and that there is no need to posit the existence of representations to explain prediction error minimization, if widely realized, on the other. The fourth section considers the cognitive-affective inseparability thesis of ET. It will be argued that there is no division of cognition and affectivity in PP. That is, cognition and affectivity can be shown to be dual aspects of the same underlying strategy for prediction error minimization. Finally, the fifth section attends to the metaplasticity thesis of ET. It will be argued that PP can accommodate the issue of plasticity, thus giving us an account of cognition as a "constitutive intertwining between *neural and cultural plasticity*" (Malafouris 2010, p. 56; italics in original).

The Constitutive Thesis

Over the last few decades, some theorists (in philosophy and cognitive science) have expressed deep skepticism about internalism vis-à-vis the mind–world relation. They advocate that the realizers of mentality are wider than the brain, comprising, at least in certain circumstances, bodily patterns of sensorimotor activity and the embedding environment. This is the constitutive thesis of ET.

By far the most influential objection to the constitutive thesis of ET is the familiar causal-constitutive fallacy invoked in the debate over the extent of mind. That is, there are some who would claim that arguments for ET, just as arguments for the extended mind hypothesis, conflate claims about causation with claims about constitution. Therefore (or so the objection goes), such arguments commit the causal-constitutive fallacy. It is commonly thought that causation and constitution are wholly independent relations. That is, facts about causation do not imply any facts about constitution. Because issues such as continuous reciprocal causation—evident in the dynamic singularity thesis (Hurley 1998)—are at the core of arguments for ET, those who think that ET commits the causal-constitutive fallacy are arguing that defenders of ET make an unjustifiable inference from observations about causal dependence to claims about constitutive dependence.[2]

Arguments for internalist PP share a strong structural resemblance to arguments against ET based on the causal-constitutive fallacy. Hohwy's (2013, 2014) arguments for internalist PP can be cast as arguments against ET on the basis of failing to distinguish clearly between what causally contributes to prediction error minimization and what *neurobiologically* constitutes prediction error minimization. In other words, although prediction error minimization depends causally on bodily activity and the embedding environment, it is directly constituted in physical brain states. As Hohwy says about PP, it is "a good companion for neurocentrism" (2014, p. 17). One can develop this claim neuroscientifically, which suggests that the constitutive boundary of mind stops at the dorsal horn of the spinal cord (Hohwy 2014, p. 18). Or one can argue for this claim on philosophical grounds. Here I focus on the philosophical argument for internalist PP.

The philosophical argument for internalist PP assumes that the brain is akin to a black box. It is epistemically secluded from what is outside the box. Here body and world are modeled as existing outside the box. Hence, for the brain to know anything about bodily and worldly states, it must generate inferences about those states. It must do so in virtue of the fact that such external states are hidden from view, yet are what causes the sensory input that the brain receives. Clark (2013) puts this "black box" view of the brain as follows: "All that it 'knows,' in any direct sense, are the ways its own states (e.g., spike trains) flow and alter. In that (restricted) sense, all the system has direct access to is its own states. The world itself is thus off-limits" (2013, p. 183). Hohwy's argument for this view takes the following structure:

- Premise 1: If PP is correct, then the brain is self-evidencing.
- Premise 2: If the brain is self-evidencing, then the mind is secluded from the body and surrounding environment.

[2] For a detailed discussion of the causal-constitution distinction and its implications for the extended cognition debate, see Kirchhoff (2015b).

- Premise 3: PP is correct.
- Conclusion: Therefore, the brain is self-evidencing and the mind is skull-bound, concealed from body and world.

This is the self-evidencing brain hypothesis. It implies that the brain is forever seeking to meet sensory input with its own top-down generated predictions. Crucially, for the argument, predictions are modeled on a plausible candidate for inference to the best explanation. In this context, this amounts to approximate Bayesian inference. Now, in inference to the best explanation, when a hypothesis *hi* best explains some evidence *ei*, the latter is understood as providing evidence for the former on the condition that *hi* accounts for *ei*. Thus, *hi* can be understood as self-evidencing. As Hohwy puts it:

> When *hi* is self-evidencing, there is an explanatory-evidentiary circle (EE-circle) where *hi* explains *ei* and *ei* in turn is evidence for *hi*. In Bayesian terms, the internal generative models—when inverted—generate predictions (hypotheses) that explain away prediction error (the sensory evidence), thus maximizing its evidence. (2014, p. 6)

Prediction error minimization is therefore self-evidencing (or so the argument goes). If this is correct, then neither bodily states nor worldly states are constitutively part of prediction error minimization. If all that really matters is what unfolds on the inside the EE-circle—in the relation between sensory input and internal model—then there is no good reason to insist that bodily and worldly states play anything else than a causal role for prediction error minimization. Of course, this does not rule out the body in perception. This follows "from the notion of active inference where hypotheses are tested in selective sensory sampling" (Hohwy 2014, p. 16). Active inference is what explains action. In the context of active inference, the brain generates predictions, which, in turn, control bodily movement in order to fulfill the predictions. Yet active inference unfolds internal to the EE-circle. As Hohwy states, "The role of the body is real and substantial, but only in the sense that the body is represented in the model, as a parameter useful for minimizing prediction error" (2014, p. 17). Thus, internalist PP implies a view of mind as constituted uniquely in the brain.

Two things are worth emphasizing.

First, from the point of view of ET, internalist PP significantly downplays the role of action, delegating embodied activity to the back seat in favor of higher-order predictive processing. There are essentially two ways for an agent to minimize prediction error. One is perception. It consists in changing an internal model such that it best fits the incoming sensory stream. The other is action. It consists of eliciting a change in the sensory signals so as to fit the internal model. Yet for the perspective of ET, embodied activity is what matters. For only via action can agents adapt in a swift and flexible fashion to changes in the environment. What internalist PP seems to miss or overlook is that unless an agent is able to engage in active inference, it cannot minimize prediction error. Like ET, proponents of PP often appeal to a tight coupling between perception and action (Clark 2016; Friston 2011; Hurley 1998; Noë 2004). But closer inspection reveals that perception and action are not only tightly coupled, but that action is essential for perception.[3] On the assumption that this is true, it puts PP into contact (contra internalist PP) with work on sensorimotor contingency theory (O'Regan and Noë 2001; Noë 2004). The core ideas of sensorimotor contingency theory are: (a) that perception is *embodied* because perceiving is a skillful activity in which both action and perception are constitutively intertwined; and (b) that perception is *counterfactual*. It is counterfactual because perception involves sensorimotor contingencies, i.e., patterns of regularities relating sensory input to movement and change. Thus, for proponents of sensorimotor contingency theory, to perceive is not simply to sense something. Rather, perceiving consists in sensitivity to modulations of one's sensations during action. This suggests that perceiving involves more than hierarchical generative models for predicting the likely causes of sensory input. That is, it involves sensitivity to how sensory signals would change on the basis of possible actions.[4]

Second, internalist PP is right to emphasize neurally encoded generative models. That is, such models might very well be the beginning of an account of minds as rooted in the minimization of prediction error, but they need not be the ends to that account. To see this, consider that for Friston and colleagues, generative models can be seen as having wide realizers, comprising parts of the body in addition to neural realizers. As they say:

> We must here understand "model" in the most inclusive sense, as combining interpretive dispositions, morphology, and neural architecture, and as implying a highly tuned "fit" between the active, embodied organism and the embedded environment. (Friston, Thornton, and Clark 2012, p. 6)

Failing to include this wider notion of "model," focusing only on neural parameters, is what paves the way for internalist PP. However, it is possible to suggest on the basis of this much more embodied notion of a model that minimization of prediction error is not restricted to the brain alone, but comprises the entire body plant (morphology, action capacities, and so on). It is in virtue of this embodied model that Friston goes on to claim that an "agent does not *have* a model of its world—it *is* a model" (2013, p. 213). Crucially, from the point of view of an embodied model, self-evidencing must be understood with respect to an organism's embedding environment. In other words, it is not possible to understand what counts as optimizing model evidence for some organism by abstracting away from the organism's local environment.

[3] To say that action is essential for perception such that action and perception turn out to be inseparable from each other is a contentious issue. See Block's (2005) excellent discussion of sensorimotor contingency theory on this. Nevertheless, if this is not the right track, it follows that action—embodied activity—is part of the minimal realization base of prediction error minimization.

[4] Of course, sensorimotor accounts could be interpreted as being compatible with internalism, not externalism. My point here is not to infer externalism about PP from its connection with sensorimotor contingency theories. Rather, it is merely to set up the idea of a constitutive link between action and perception.

On the assumption that this is correct, then the primary role of minimizing prediction error is not to generate inferences from within something akin to a black box but rather to coordinate and maintain the right balance "between the complexities of the agent's sensory, motor, and neural systems" (Pfeifer and Bongard 2006, p. 123). Clark (2015) discusses an example of this: what he refers to as the "outfielder's problem." It consists in running to catch a fly ball in baseball. As Clark states, "The solution . . . involves running in a way that seems to keep the ball movement in a constant speed through the visual field. As long as the fielder's own movements cancel any apparent changes in the ball's optical acceleration, she will end up in the location where the ball is going to hit the ground" (2015, p. 11). Interestingly, in a stationary position, players are unable to predict the location of where the ball will land (or they do very poorly). One implication of this is that perception is not enough. To successfully cancel out prediction error in catching a fly ball necessitates bodily movement. As Clark reports, "They are unable to predict the landing spot because OAC [optical acceleration cancellation] is a strategy that works by means of moment-by-moment self-correction that crucially involves the agent's own movement" (2015, p. 11). If this is correct, it yields a view of PP that is continuous with the constitutive thesis of ET, given that in at least certain circumstances, prediction error minimization breaks across neural and bodily realizers, enabling appropriate forms of sensorimotor coordination with respect to a task environment (see also Riley et al. 2012).

The Nonrepresentational Thesis

ET advances the view that dynamic sensorimotor profiles are at the heart of mind. If this is correct, then the following picture of the ceramicist working at her wheel is within reach: she (or her brain) need not be continuously seeking to represent the world from behind some epistemic veil of experience, forever trying construct and reconstruct representations of bodily and worldly aspects to succeed in shaping the seemingly ever-shifting clay material.

One might try to justify this claim along the following lines. Suppose that cognitive systems are reciprocally coupled brain–body–environment systems. If this is correct, then there is no need for one part of this system to represent another part. In other words, "If the animal-environment system is just one system, the animal portion of the system need not represent the environment portion of the system to maintain its connection with it" (Silberstein and Chemero 2012, p. 40). The reason underlying this claim is that there is no separation between the embodied individual and its local, embedding environment that needs to be "bridged by representations" (Silberstein and Chemero 2012, p. 40).

The central idea is that minds are not primarily for thinking, traditionally conceived, *but for action*—"for getting things done in the world in real time [primarily]" (Wilson and Foglia 2015, p. 5). Interestingly, in a recent blog post on his *Surfing Uncertainty*, Clark expresses a similar point for PP. For Clark, internalist PP "makes action play second fiddle to something like representational fidelity. But a moment's reflection ought to convince us that it is action—not perception—that real-world systems really need to get right." The main idea here is that perception per se is not worthwhile unless it yields some form of adaptively fruitful embodied activity. In fact, as Clark goes on, "In an often hostile, time-pressured world . . . [representational] fidelity needs to be traded against speed" (http://philosophyofbrains.com/2015/12/14/surfing-uncertainty-prediction-action-and-the-embodied-mind.aspx). Further support for this view of minds comes again from the example of running to catch a fly ball in baseball. As we saw, the solution, optical acceleration cancellation, highlights the importance of embodied activity. As Clark notes, "The point is simply that the canny use of data available in the optic flow enables the catcher to sidestep the need to create a rich inner model to calculate the forward trajectory of the ball" (2008, p. 16).

Elaborating on this, note that prediction error is a measure of surprise about future states, given a model. Because of this, "It can be expressed as *accuracy* minus *complexity*" (Hobson and Friston 2014, p. 19; italics in original). That is, organisms can be understood as continuously trying to increase the accuracy (precision) of their predictions, while, at the same time, seeking to minimize the complexity involved in doing so. Synergies, breaking across the embodied agent, are a nontrivial way by which organisms can minimize complexity. A synergy is an assembly (typically short-lived) of processes enslaved to act as a single coherent and functional unit (Kelso 1995). Specifically, synergies "are defined as compensatory, low dimensional relations in the dynamic activities of neuromuscular components (Kelso 2009), not as static representational structures such as motor programs" (Riley et al. 2012, p. 23). As Riley et al. further note, "Synergetics . . . allows control problems—conventional time-sinks of computation [and representation]—to be offloaded to the dynamical organization of the body" (2012, p. 22).

An example of this is a central pattern generator (CPG). CPGs are assembled when sensory and mechanical stimuli induce specific neurotransmitters, resulting in a functional organization of network connectivity. For example, "A mollusc slowly treading water changes abruptly, recruiting additional interneurons to enable rapid escape from a predator" (Riley et al. 2012, p. 24). Synergies orchestrate a circuit of processes enslaved to act as a single coherent and functional unit, "setting the balance between top-down and bottom-up models of influence" (Clark 2015, p. 16).

Another example of this is postural control. According to Riley et al. (2012), to be able to maintain balance, "the body's center of mass (CM) cannot exceed the limits of the base of support (usually, the boundaries of the feet). Movement must be anticipated in remote preflexes and ultrafast compensation across whole-body synergies, lest we tip over" (2012, p. 24). The solution to this lies in the dynamics of postural sway, "the ceaseless, arrhythmic, low-amplitude changes in CM position" (2012, p. 24). We all have the capacity to increase the accuracy of the sensory input we expect to meet by minimizing complexity, viz., by creating for ourselves an indefinite variety of synergies "to suit changes in postural demands" (Riley et al. 2012, p. 25).

No doubt there is self-evidencing here. By reducing complexity through an on-the-fly assembly of synergies, agents are able to maximize model evidence—maximizing the solution best suited to get the job done. But this need not entail any strict separation between agent and environment that can only be bridged by internal representations.

Further support for this claim comes from revisiting the idea that expected states have low surprise—i.e., low prediction error. Prediction error is an upper bound on surprise. Here surprise refers to a discrepancy between an organism's predictions of sensory input and the input it actually receives (Friston et al. 2012). Surprise is not a psychological notion. It is a technical measure of improbability or negative log evidence for a model. Here a fish out of water would be in a state of high surprise, while a fish in its aquatic milieu will be in a state of low surprise. Thus, surprise is relative to a model. This is the driving assumption behind the idea that an organism is a model of its environment, as we saw in the previous section. In this sense, each organism embodies "an optimal [though only approximate] model of its econiche" (Friston 2011, p. 89).

This puts pressure on internalist PP. Let us assume that self-evidencing creates an EE-circle (or a Markov blanket, as such a circle is formally known). Does it follow that an agent must represent the world in order to maintain her connection with it? Not necessarily. To see this, consider how Friston defines the interdependency between internal and external states making up an EE-circle:

> The dependencies induced by Markov blankets create a *circular causality* that is reminiscent of *the action-perception cycle*. The circular causality here means that external states cause changes in internal states, via sensory states, while the internal states couple back to the external states through active states—such that *internal and external states cause each other in a reciprocal fashion*. This circular causality may be a fundamental and ubiquitous causal architecture for self-organization. (Friston 2013, pp. 2–3, italics added)

For Friston, then, minimization of prediction error is embedded in a dynamical nexus comprising internal and external states, precisely because internal and external states cause one another in a circular and reciprocal fashion. Prima facie, at least, this should remove any temptation of conceiving of internal and external states as wholly and fundamentally separate. Indeed, it is consistent with the claim that there is no separation between the individual and her environment that must be bridged by representations. Clark seems to agree, as he says:

> Whatever the use of . . . "model" means in these low-level free energy minimization accounts, they do not seem to imply the presence of inner models or content-bearing states of the kinds imagined in traditional cognitive science. Instead, what are picked out seem to be physical processes defined over states that do not bear content at all—neither richly reconstructive nor of any more "action-oriented" kind. (2016, p. 14)

So, if it is not representations that maintain a connection between brain, body, and world, what it is? A strong candidate is synergetics. This fits snugly with the idea that action precedes perception (at least causally), for synergies enable systems to be poised on the brink of instability such that they can be in a position "to anticipate the [immediate] future" rather than simply having to react to it on the basis of constructing—on-the-fly—representations of it (Kelso 1995, p. 26; Kirchhoff 2015a)

The Cognitive-Affective Inseparability Thesis

The body in action, as we have seen, plays not merely a causal but a constitutive role in minimizing prediction error, and it does so in a way that frees agents from having to rely on the construction of internal representations to maintain their connection with the embedding environment. But there is more so say about the body in action. That is, it meets the world neither in some cold-blooded nor in some dispassionate sense. Instead, bodies are configured such that perceiving and acting are essentially affective as well. This is what I call the cognitive-affective inseparability thesis of ET: affect, cognition, and sensorimotor contingencies are inseparable given that patterns of affectivity are part and parcel of perception, action, and cognition (Colombetti 2013; Gallagher et al. 2013).

Internalist PP, however, would have us draw a different conclusion. On this particular interpretation of PP, the capacity to evaluate a situation or more generally understand the meaning of a situation is based on a neurally realized inferential process aiming to minimize prediction error. This yields a particular view of the body—one where the body plays second fiddle to the process of finding the hypothesis that best explains the incoming sensory input. On the one hand, prediction error minimization of this sort depends on the body's function to both generate and transmit information concerning exteroceptive, interoceptive, and proprioceptive stimuli. On the other hand, it also depends on the body to carry out and thus fulfill its predictions. In this sense, the body is reduced to a mere mechanical and functional vessel for a more abstract and neurally realized form of prediction error minimization. Hence, the brain's ability to generate a prediction that meets the sensory signals received is effectively what tells the body what to do. It follows that for internalist PP, affectivity does not have an embodied dimension in any constitutive sense. Instead, it is entirely a product of inference to the best explanation carried out in some Cartesian theater. In this sense, internalist PP treats affectivity as a product of hypothesis testing, not as part of the body itself. Hohwy expresses this view: "The upshot is a very deflated conception of the role of body, objects and action (including of self and of other people). All this is just representations that we harbor because they best explain away the sensory input" (http://philosophyofbrains.com/2014/06/28/prediction-error-minimization-and-embodiment.aspx).

Yet it is possible to give a less intellectualist account of affectivity, even from the point of view of PP. That is, it is possible to show that PP and the cognitive-affective inseparability thesis can join forces against the internalist form of PP. By way of example, let us consider a very simple creature, such as the bacterium *Escherichia coli* (*E. coli*).

E. coli move around by using flagella in search of substances required to maintain their homeostasis. Homeostasis refers to the ability of the body to maintain a stable internal environment in the face of changing external conditions. *E. coli* are capable of initiating active inference to maintain homeostasis. Expecting high glucose levels, *E. coli* will direct their motion to domains with high concentrations of lactose whenever their expectations of high glucose levels are not meet—that is, whenever interoceptive sensory signals result in high surprise relative to their generative model. In the enactivist tradition (Di Paolo 2005; Thompson 2007; Varela et al. 1991), it is argued that this basic form of metabolic organization yields a meaningful perspective on the environment rooted in a simple form of normativity, given that "*more* of it is better than *less*" (Di Paolo 2009, p. 13; italics in original). Indeed, for Di Paolo, such normativity is a uniquely "relational fact impossible to appreciate unless we have an organism present *for which* the effects of the chemical encounter on the processes of self-construction and self-distinction make sense as being good or bad" (2009, p. 13). This emphasis on homeostasis and its link to the enactivist notion of sense-making is akin to the theory of emotion put forth by Damasio (1999), for whom emotion and affect are essentially an embodied process of self-maintenance targeting homeostasis. As Colombetti notes, emotion and affect "thus conceived also provides action-guiding values, drives and preferences" (2010, p. 150).

Crucially, for our purposes, this is compatible with the assumption that an organism is a close-to-optimal model of its environment. One can understand "goodness" as well as "badness" as relative to the *E. coli* being a model of its local niche. That is, expected states are valuable precisely because it is those states that enable an agent to maximize self-evidence with respect to the environment. In the context of active inference, embodied activity, prior expectations, and affect can thus be understood as coming together. In other words, attending to the idea that organisms are models of their environment invites the idea that affectivity is a feature of the entire embodied organism in its embedding environment. If this is correct, then PP—in this more embodied and enactive sense—invites us to understand affectivity as anything but a byproduct of either perception or active inference. Instead, it suggests that affectivity and perception, say, are simultaneous effects of the same underlying strategy for minimizing prediction error.

A specific model of PP that targets this relationship between embodied affectivity and minimizing prediction error is Barrett and Bar's affective prediction hypothesis (2009). According to this hypothesis, "responses signaling an object's salience, relevance or value do not occur as a separate step after the object is identified. Instead, affective responses support vision from the very moment that visual stimulation begins" (2009, p. 1325) If this is on the right track, then in embodied activity one does not simply perceive the world in a cold and detached manner. Instead, to perceive is already to be in an affective state such that bodily changes are part and parcel of experience. In the terms of PP, an organism's model when perceiving an object, person, or situation is already configured in accordance with prior expectations involving not merely internal neural circuitry but an entire body adjustment.

This brings PP into close contact with yet another central tenet of enactivism. This is the core idea expressed in the claim that organisms "enact their world" via a process of sense-making. Di Paolo defines sense-making as follows: it is "the instauration of a natural perspective from which encounters in the world are intrinsically meaningful for the organism following the norm established by the continuing process of self-production" (2005, pp. 429–30). Crucially, the claim that PP turns on minimization of prediction error need not entail that it is incompatible with sense-making. As Clark (2015) points out, the constructive aspect of PP "corresponds rather closely to the [enactivist] notion of 'enacting a world' " (2015, p. 18). In the context of active inference, organisms can be understood as actively constructing what they experience via embodied activity. Here the active body is at the core of experience. Indeed, as Clark further states, "In this way, different organisms and individuals may selectively sample in ways that both actively construct and continuously confirm the existence of different 'worlds.' It is in this sense that, as Friston, Adams, and Montague (2012, p. 22) comment, our implicit and explicit models might be said to 'create their own data' " (2015, p. 20).

In this sense, PP can be understood as consistent with what Colombetti (2010) terms the "bodily cognitive-emotional form of understanding" (2010, p. 147). That is, on the assumption that affectivity is an essential component of perception, and if sense-making is inherently affective, then for organisms to enact their world in prediction error minimization is also for organisms to enact it affectively.

The Metaplasticity Thesis

This brings us to the final thesis of ET to be considered in this chapter—what I call the metaplasticity thesis. The specific idea is that minds in action are plastic in ways akin to the morphing dynamics between the potter's body, her arms, eyes, and so on, and the wet clay itself, such that minds are subject to "continuous re-shaping, re-wiring and remodelling" (Malafouris 2010, p. 55) in virtue of the plasticity of brain, body, and cultural practices.

The notion of metaplasticity was originally introduced in neuroscience, referring to neural development and plasticity—to how macroscopic changes in neural attributes come about by virtue of synaptic plasticity and subsequent modification (Zhang and Linden 2003). It captures the idea that developmental processes in the brain are activity dependent (Engel et al. 2013). In this sense, plasticity refers to the brain's ability to adjust to changes in the "internal and external milieu" (Jacobson 1991, p. 199) As Engel et al. (2013) report in the context of skilled musicians, they "often show functional and structural changes in their sensorimotor systems that result from action-dependent plasticity" (p. 203). Generally speaking, these "studies demonstrate that appropriate action, which allows exercise of relevant sensorimotor contingencies, is necessary throughout life to stabilize the functional architecture of the respective circuits" (p. 203).

An excellent example of metaplasticity is the work, by Iriki and colleagues, on tool use in Japanese macaques, and how such tool use induces both structural and functional changes in the brain (Maravita and Iriki 2004). Extended exposure to a scaffolded learning environment—the use of rakes—leads to structural and functional modification of the so-called body schema—a dynamic ensemble of processes underlying the control and awareness of body shape and posture. After two weeks of extensive training, the macaques had acquired the skill of using a rake to pull food items closer. Iriki and colleagues measured neuronal activity in these monkeys and found "pre-motor, parietal and putaminal neurons that respond both to somatosensory information from a given body region (i.e., the somatosensory Receptive Field; sRF) and to visual information from the space (visual Receptive Field; vRF) adjacent to it" (Maravita and Iriki 2004, p. 79). These neurons are called bimodal neurons. For example, some neurons respond to somatosensory stimuli (touches to the hand) and to visual stimuli near the hand. In a number of experiments, Iriki and colleagues found that in some of the bimodal neurons, whose original vRFs were associated with stimuli near the hand, had "expanded to include the entire length of the tool" (Maravita and Iriki 2004, p. 79). Other bimodal neurons, whose original sRFs were associated with stimuli around the shoulder and neck areas, had expanded to cover "space reached by the arm" (Maravita and Iriki 2004, p. 80). As they conclude, such "vRF expansions may constitute the neural substrate of use-dependent assimilation of the tool into the body schema, suggested by classical neurology" (2004, p. 80). Such studies illustrate body schemas, dynamic neural processes that control body shape and posture in terms of capabilities for action. And this to the extent that "embodied activity *enacts or brings forth new systemic wholes*" (Clark 2008, p. 39, italics in original; see Menary 2014 for an excellent discussion of these findings and their broader implications).

Metaplasticity is not restricted to the brain. In the context of material engagement theory, "the term is used in a much broader sense to characterize the emergent properties of the enactive constitutive intertwining between *neural and cultural plasticity*" (Malafouris 2010, p. 56; see also Malafouris 2009; Malafouris and Renfrew 2008). According to Malafouris, this implies that "differences and variations in life and learning experiences caused by social, environmental and cultural factors can cause individuals of the same genotype to have different neural, cognitive, and behavioral outcomes" (2010, p. 53).

This extensive sense of metaplasticity can be captured from the perspective of PP. I am not claiming that PP cannot accommodate neural plasticity.[5] Here I merely restrict my attention to PP and the metaplasticity thesis of ET. A good place to begin is by noticing that for human agents, embodied activity unfolds in cultural practices. Such practices exhibit regularities. That is, they are patterned in particular ways. As Roepstorff, Niewöhner, and Beck note, "From the inside of a practice, certain models of expectancy come to be established, and the patterns, which over time emerge from these practices, guide perception as well as action" (2010, p. 1056).

PP can be understood within this patterned practices framework. Indeed, according to Roepstorff, from the point of view of PP, "these practices may help to establish priors

[5] See Rabinovich et al. (2012) for discussion on PP and neural plasticity.

or even hyperpriors, sets of expectations that shape perception and guide action" (2013, p. 225). And, as Clark adds, "Such a perspective, by highlighting situated practice, very naturally encompasses various forms of longer-term material and social environmental structuring. . . . At multiple time scales . . . we thus stack the dice so that we can more easily minimize costly prediction errors in an endlessly empowering cascade of contexts" (2013, p. 195).

Indeed, the regularities embedded in cultural practices are likely to play a role in so-called precision estimation—minimizing uncertainty about error signals. Interestingly, precision is not merely causally relevant for minimizing prediction error. Precision is constitutively relevant given that any PP model lacking the capacity for precision estimation would be unable to determine the uncertainty or reliability of any given error signal. Hence, precision estimation "provides a powerful means of sculpting the larger patterns of 'effective connectivity' that modify the internal flow of influence and information according to task and context" (Clark 2016, pp. 146–7). In other words, precision estimation provides a means by which to shape, constrain, and transform patterns of inference and action. Yet precision weighting need not be a purely brain-bound process. Specifically, the process of shaping and constraining inference and action can be seen to break across neural and cultural parameters. If this is correct, then the patterned regularities that are embedded in cultural practices may also play a role in shaping and constraining action and inference. Hence, they may play a role in precision estimation.

In the specific landscape of PP, one way to explore this is to consider situations where the error signal is too extreme to be easily minimized by action—that is, situations in which there is very high negative log evidence for a model. One such case is culture shock in which subjects report experiencing distress and alienation. PP would predict that culture shock is the result of major discrepancies between onboard generative models and the embedding environment. A case in point is 13-year-old Eva Hoffman, who, with her parents, left Poland in 1959 for Vancouver, Canada. Although she had her immediate family by her side, everything else in her experiential world had changed. As she explains, "The country of my childhood lives within me with a primacy that is a form of love. . . . It has fed me language, perceptions, sounds. . . . It has given me the colors and the furrows of reality, my first loves" (Hoffman, 1989, pp. 74–5; quoted in Wexler 2008, p. 175). Indeed, after only three nights in Vancouver, upon wakening from a dream, she wonders: "What has happened to me in this new world? I don't know. I don't see what I've seen, don't comprehend what's in front of me. I'm not filled with language anymore, and I have only a memory of fullness to anguish me with the knowledge that, in this dark and empty state, I don't really exist" (Hoffman 1989, p. 180; quoted in Wexler 2008, p. 175). Indeed, when "she attempts to take in her new environment the requisite internal structures are lacking or the old structures are obstructing" (Wexler 2008, p. 176).

The first thing to note is that priors—tuning inner anticipatory patterns toward a unique configuration—may be acquired, calibrating them to particular regularities in an individual's embedding environment. In the context of PP, the metaplasticity thesis implies that any short-term explanation of prediction error minimization is premised on long-term minimization of prediction error. If this is correct, then any explanation of culture shock, when reviewed through the lens of the metaplasticity thesis, must make reference to the role of cultural practices in shaping, on average and over time, prior expectations encoded in generative models. To assume otherwise would be to commit what Hurley coins the "internal end point error," viz., that after maturation, neural mechanisms alone explain "how experience works, as well as what it is like" (Hurley 2010, p. 142). Hence, what best explains the experience of distress and alienation in culture shock is not best understood by appeal to purely neural parameters alone. It is better captured in what Hurley calls a "characteristic extended dynamic." That is, any explanation in terms of neural parameters must be given with reference to cultural practices. After all, embodied activities unfold neither merely in a brain nor in a vacuum. They unfold in practices.

Conclusion

This chapter has explored four points of overlap between the embodiment thesis and predictive processing models in neuroscience. First, it has been argued that both the embodiment thesis and predictive processing can be understood as pushing for the idea that extraneural parts of the body and embedding environment may play, in the right circumstances, a constitutive role with respect to mind. Second, rather than casting minds in representational terms, both frameworks drive the idea that minds are essentially for action. It has been argued that this view of minds need not be taken to imply that minds must represent in order to maintain a connection between agent and world. Third, predictive processing and the embodiment thesis underpin the idea that cognition and affect are inseparable. In the context of predictive processing, it has been argued that cognition and affectivity are dual aspects of the same basic strategy for minimizing prediction error. Finally, the capacity for brains to reorganize in terms structural, functional, and affective connectivity is well established. I have taken this idea and extended it to entire brain–body–environment systems. I have done so by appeal to the notion of metaplasticity (coined by Malafouris 2010).

Acknowledgments

Kirchhoff's work was supported by an Australian Research Council Discovery Project "Minds in Skilled Performance" (DP170102987), a John Templeton Foundation grant "Probabilitizing Consciousness: Implications and New Directions," and by a John Templeton Foundation Academic Cross-Training Fellowship at the Wellcome Trust Center for Neuroimaging at University College London (ID#60708). The opinions expressed in this publication are those of the author and do not necessarily reflect the views of the John Templeton Foundation.

References

Barrett, L.F. and Bar, M. (2009). See it with feeling: affective predictions during object perception. *Philosophical transactions of the Royal Society of London. Series B, Biological sciences*, 364, 1325–34.

Clark, A. (2008). *Supersizing the mind*. Oxford: Oxford University Press.

Clark, A. (2013). Whatever next? Predictive brains, situated agents, and the future of cognitive science. *Behavioral and Brain Sciences*, 36, 181–253.

Clark, A. (2015). Radical predictive processing. *The Southern Journal of Philosophy*, 53 (Spindel Supplement), 1–25.

Clark, A. (2016). *Surfing uncertainty*. Oxford: Oxford University Press.

Colombetti, G. (2010). Enaction, sense-making and emotion. In: J. Stewart, O. Gapenne, and E. Di Paolo (eds.), *Enaction: toward a new paradigm for cognitive science*. Cambridge, MA: The MIT Press, pp. 145–64.

Colombetti, G. (2013). *The feeling body*. Cambridge, MA: The MIT Press.

Damasio, A. (1999). *The feeling of what happens: body and emotion in the making of consciousness*. New York: Harcourt Brace.

Di Paolo, E. (2005). Autopoiesis, adaptivity, teleology, agency. *Phenomenology and the Cognitive Sciences*, 4, 97–125.

Di Paolo, E. (2009). Extended life. *Topoi*, 28, 9–21.

Engel, A.K., Maye, A., Kurthen, M., and König, P. (2013). Where's the action? The pragmatic turn in cognitive science. *Trends in Cognitive Sciences*, 17(5), 202–9.

Friston, K. (2009). The free-energy principle: a rough guide to the brain? *Trends in Cognitive Sciences*, 13(7), 293–301.

Friston, K. (2011). Embodied inference: or "I think therefore I am, if I am what I think." In: W. Tschacher and C. Bergomi (eds.), *The implications of embodiment (cognition and communication)*. Exeter, UK: Imprint Academic, pp. 89–125.

Friston, K. (2013). Life as we know it. *Journal of the Royal Society Interface*, 10, 20130475. doi:10.1098/rsif.2013.0475

Friston, K., Adams, R., and Montague, R. (2012). What is value-accumulated reward or evidence? *Frontiers in Neurorobotics*, 6, 11. doi:10.3389/fnbot.2012.00011

Friston, K., Thornton, C., and Clark, A. (2012). Free-energy minimization and the dark-room problem. *Frontiers in Psychology*, 3(130), 1–7.

Gallagher, S., Hutto, D.D., Slaby, J., and Cole, J. (2013). The brain as part of an enactive system. *Behavioral and Brain Sciences*, 36(4), 421–2.

Hobson, A. and Friston, K. (2014). Consciousness, dreams, and inference. *Journal of Consciousness Studies*, 21(1–2), 6–32

Hoffman, E. (1989). *Lost in translation*. New York: Penguin Books.

Hohwy, J. (2013). *The predictive mind*. Oxford: Oxford University Press.

Hohwy, J. (2014). The self-evidencing brain. *Noûs*, 1–27. doi:10.1111/nous. 12062

Hurley, S.L. (1998). *Consciousness in action*. Cambridge, MA: Harvard University Press.

Hurley, S.L. (2010). The varieties of externalism. In R. Menary (ed.), *The extended mind*. Cambridge, MA: The MIT Press, pp. 101–54.

Hutto, D.D. and Myin, E. (2013). *Radicalizing enactivism: basic minds without content*. Cambridge, MA: The MIT Press.

Jacobson, M. (1991). *Developmental neurobiology* (3rd ed.). New York: Plenum Press.
Kelso, S. (1995). *Dynamic patterns*. Cambridge, MA: MIT Press.
Kelso, S. (2009). Synergies: atoms of brain and behavior. In: D. Sternad (ed.), *Progress in motor control: a multidisciplinary perspective*. New York: Springer, pp. 83–91.
Kirchhoff, M.D. (2015a). Experiential fantasies, prediction, and enactive minds. *Journal of Consciousness Studies*, 22(3–4), 68–92.
Kirchhoff, M.D. (2015b). Species of realization and the free energy principle. *Australasian Journal of Philosophy*, 93(4), 706–23.
Kirchhoff, M.D. (2016). Autopoiesis, free energy, and the life-mind continuity thesis. *Synthese*. doi:10.1007/s11229-016-1100-6
Malafouris, L. (2007). Before and beyond representation: toward an enactive conception of the Palaeolithic image. In: C. Renfrew and I. Morley (eds.), *Image and imagination: a global history of figurative representation*. Cambridge: McDonald Institute Monographs, pp. 289–302.
Malafouris, L. (2009). “Neuroarchaeology”: exploring the links between neural and cultural plasticity. *Progress in Brain Research*, 178, 251–9.
Malafouris, L. (2010). Metaplasticity and the human becoming: principles of neuroarchaeology. *Journal of Anthropological Sciences*, 88, 49–72.
Malafouris, L. and Renfrew C. (2008). Steps to a “neuroarchaeology” of mind: an introduction. *Cambridge Archaeological Journal*, 18, 381–5.
Maravita, A. and Iriki, A. (2004). Tools for the body (schema). *Trends in Cognitive Sciences*, 8, 79–86.
Menary, R. (2007). *Cognitive integration: mind and cognition unbounded*. Basingstoke: Palgrave Macmillan.
Menary, R. (2014). Neural plasticity, neuronal recycling and niche construction. *Mind & Language*, 29(3), 286–303.
Noë, A. (2004). *Action in perception*. Cambridge, MA: MIT Press.
O’Regan, J.K. and Noë, A. (2001). A sensorimotor account of vision and visual consciousness. *Behavioral and Brain Sciences*, 25(4), 883–975.
Pfeifer, R. and Bongard, J. (2006). *How the body shapes the way we think: a new view of intelligence*. Cambridge, MA: MIT Press.
Rabinovich, M.I., Friston, K.J., and Varona, P. (2012). *Principles of brain dynamics: global state interactions*. Cambridge, MA: MIT Press.
Riley, M.A., Shockley, K., and Van Orden, G. (2012). Learning from the body about the mind. *Topics in Cognitive Science*, 4, 21–34.
Roepstorff, A. (2013). Interactively human: sharing time, constructing materiality. *Behavioral and Brain Sciences*, 36(3), 224–5.
Roepstorff, A., Niewöhner, J., and Beck, S. (2010). Enculturating brains through patterned practices. *Neural Networks*, 23(8–9), 1051–9.
Rowlands, M. (2018). Disclosing the world: intentionality and 4E cognition. In: this volume, pp. 335–52.
Silberstein, M. and Chemero, A. (2012). Complexity and extended phenomenological-cognitive systems. *Topics in Cognitive Science*, 4, 35–50.
Thelen, E. and Smith, L. (1994). *A dynamic systems approach to the development of cognition and action*. Cambridge, MA: The MIT Press.

Thompson, E. (2007). *Mind in life: biology, phenomenology and the sciences of mind.* Cambridge, MA: MIT Press.

Varela, F., Thompson, E., and Rosch, E. (1991). *The embodied mind.* Cambridge, MA: MIT Press.

Wexler, B. (2008) *Brain and culture: neurobiology, ideology, and social change.* Cambridge, MA: MIT Press.

Wilson, R. and Foglia, L. (2015). Embodied cognition. In: E.N. Zalta (ed.), *Stanford encyclopedia of philosophy.* Retrieved from https://plato.stanford.edu/archives/fall2015/entries/embodied-cognition/

Zhang W. and Linden D.J. (2003). The other side of the engram: experience-driven changes in neuronal intrinsic excitability. *Nature Reviews Neuroscience*, 4, 885–900.

CHAPTER 13

JOINT ACTION AND 4E COGNITION

DEBORAH TOLLEFSEN AND RICK DALE

WE engage in various tasks every day. We brush our teeth, drive our cars, drink our coffee, butter our bread. Even the most simple of these tasks requires complex cognitive mechanisms—the coordination of mind and body across time and space. Cognitive science is in the business of explaining how we do those tasks. But many of the tasks we engage in each day involve others, and this, in turn, involves coordination of bodies and minds over time. These actions, like the individual actions of which they are composed, involve complex cognitive and physical processes. Over the past 30 years there has been an increased interest in studying joint action across a number of different disciplines, including psychology, sociology, cognitive anthropology, cognitive science, and philosophy. In this chapter we canvas recent philosophical and empirical research on joint action. Along the way we highlight embedded, embodied, extended, and enactive approaches (4E cognitive approaches) and how these approaches challenge more orthodox approaches to the issue of what joint action is and how it comes about and is sustained.

We propose an ecumenical approach to the study of joint action. Joint action takes many forms and is executed in a variety of contexts. The cognitive science of joint action will have to integrate both high-level and low-level approaches across a variety of disciplines, including experimental psychology, neuroscience, and philosophy. The goal should *not* be to hanker for a single theory about joint action, simply because a single unitary account is unlikely to capture the nuanced and complex nature of joint action. Instead, we argue that we should seek a better understanding of how various accounts coalesce into a tapestry of explanatory tools. This tapestry permits a more flexible perspective on joint action, and embraces that "joint action" is a graded concept with instances that span different cognitive processes and tasks, across both human and nonhuman agents. This integrated perspective would include but also go beyond cognition, into shared processes in a shared environment, and so an ecumenical account is better viewed as an "ecognitive" account. It is both fully ecological and fully cognitive.

Defining Joint Action

There are different views in the literature on what joint action is. Some use the term "joint action" to refer to interactions in which individuals coordinate in space and time that bring about a change in their environment (Sebanz, Bekkering, and Knoblich 2006). According to this definition joint action is broad and, like other prominent theories, would encompass social interactions such as successfully walking past each other on a crowded street without bumping into one another, or successfully driving through a busy city (De Jaegher and Di Paolo 2007). Individual human beings in these cases coordinate their bodies in space and time and bring about change in their environment. Other theories emphasize the intentional, or *goal structure* that underlies interaction. Others (Tomasello 2014; Searle 1995; Bratman 2014; Tuomela 2013; Fiebich and Gallagher 2013) distinguish mere social interaction from joint action. Joint actions are those where individuals coordinate in space in time in order to bring out a specific goal. That is, they intend to do the action together *and they share a goal or intention to do so*. Two people who successfully pass each other on the sidewalk might be said to coordinate their bodies in space and time and bring about a change in the environment (after all, they have changed location as a result of their coordination), but they don't do it *together* and so don't, according to this conception, engage in joint action. These and other theories involve diverse commitments to varying dimensions, and while some theories emphasize perceptuomotor coordination as critical to joint action, others see different key ingredients, such as intentions or goals.

It is important to note here that what one thinks joint action is will often determine the sorts of cognitive processes and states that one thinks are relevant to explaining the phenomenon. If, for instance, one thinks joint action essentially involves higher-level forms of cognition such as shared intention, then this will determine the sorts of empirical approaches one takes toward the study of joint action. Experiments designed to reveal joint commitments, shared goals, and shared intentions will dominate. On the other hand, if one adopts a more perceptuomotor conception of joint action, basic forms of social interaction (joint attention, norm conformism, alignment, and mimicry) will play a much larger role in developing empirical approaches. The question that frames much of the literature on joint action is this: are lower-level cognitive processes sufficient for joint action, or is joint action something that necessarily involves high-level cognitive states such as belief and intention? These two levels have been identified as critical dimensions of joint action, and integrating them may facilitate a more flexible conception of it (cf. Fiebich, Nguyen, and Schwarzkopf 2015; Tollefsen and Dale 2012).

Philosophical Accounts of Joint Action

A good starting place for a chapter on joint action is the philosophical accounts offered over the past few decades. Philosophers of action are interested in providing an account of human action where action is defined in opposition to mere behavior. Actions are things that human beings *do* as opposed to things that merely happen to us. The link between action, control, and responsibility is a tight one for philosophers. Actions are those for which we can be held responsible and over which we have some control. Causal theorists have argued that what distinguishes actions from mere happenings is that actions are caused by antecedent mental states. When I go to the fridge to retrieve a beer it is because of a mental state, an intention, which causes my body to move in ways that get me to the fridge.

Philosophical accounts of joint action have taken the causal theory of action as their starting point. To understand the actions of multiple people, we need to identify the internal, individual, mental states responsible for causing bodies to move. If intention shapes and informs individual actions, then joint action must also be shaped and informed by intentions. Philosophers have argued, however, that we have to consider the ways in which the intentions involved in joint action are shared or *joint*.

Numerous positions have elaborated on this basic tenet. Michael Bratman (2014) has argued that joint action involves a state of affairs in which each individual intends "that we J," and these individual intentions are interdependent and had under conditions of mutual common knowledge. Raimo Tuomela (2013) has argued that joint action involves individual we-attitudes, individual intentions that are had in the we-mode and that these intentions are interdependent and had under conditions of mutual common knowledge. John Searle (1995) has argued similarly. Individual intentions to do some action will not explain the joint action. The intentions of the individual must be collective. According to Searle, when individuals act together they do so via we-intentions, and, in a similar account, Gilbert (2013) emphasizes the role of joint commitments. These are the intentions of the individual agents in the group, not some supra-agent.[1]

In each case, these theories begin with what is happening within the individual mind and focus on finding a representational state that can function to bring about the action. Though these theorists emphasize the need for individual intentional states to be interrelated, and so recognize the role of the other in bringing about the joint action, they are very much working within an orthodox theory of mind. Mental states are internal, representational states that direct behavior in specific ways. Philosophical

[1] A reviewer incisively remarked that Searle also sees the we-mode as so central to social *brains* that the perennial "brain in a vat" could engage in we-intentions with other brains in a vat (whether in the same vat or in a different vat is up for discussion, we suppose).

accounts also represent the paradigm of "intellectualist" approaches to joint action. They attempt to identify high-level representational states, such as intention, that are causally responsible for the behavior of individuals and that bring about the movement and coordination of bodies in space and time. Finally, philosophical theories also represent a paradigm of narrowing the conception of joint action. Joint action is defined in terms of shared intentions or shared goals. Consider Searle's well-known example of the distinction between two types of groups: the random collection of people at a park who run for shelter when it starts to rain and the dance troop that behaves in exactly the same manner. The difference in these cases, according to Searle, is that the individual actions of the dance troop are guided by we-intentions. The strangers' actions are not. According to most philosophical accounts of joint action, the difference between collective behavior and joint action is to be found either within the individual in the form of we-intentions or between the individuals in terms of joint commitments or Bratmanian shared intentions. Without this representational basis (we are doing this together), it is mere collective behavior—not action.

There have been a number of objections raised in the literature to philosophical approaches. First, the focus in these philosophical accounts has also been on identifying prior intention and has focused almost exclusively on pre-planned actions. But as Christopher Kutz (2000) and others (Tollefsen and Dale 2012) have argued, many joint actions occur on the fly and do not involve pre-planning. Consider the way that musicians play together and, in some cases, improvise. It is implausible to think that a jazz trio is formulating shared intentions as it plays together. Perhaps they formed the prior shared intention to "play together" but this very general intention doesn't seem to help us understand how they are capable of coordinating their minds and bodies over time.

Second, these accounts have been focused on higher-level cognitive representations (intentions, commitments) and the relation between them. But this approach seems to demand much from participants in a joint action. For instance, if Bratman is correct and joint action requires that we each form an intention "that we J" and we each know that we have formed that intention (it requires common knowledge), this requires sophisticated mindreading capacities. Yet young children and animals don't have such capacities, so it becomes difficult to understand joint action in young children and animals. Indeed, it seems to exclude them from the class of critters that can perform joint actions. This seems unjustified (Tollefsen 2005; Tollefsen and Dale 2012).

Third, although these accounts may enrich our understanding of what joint action is, they don't really help us to understand how joint action unfolds over time. Even if joint action does involve Searlean we-intentions, it would be nice to understand the ways in which we-intentions might interact with lower-level states such as motor representations in order to bring about the coordination of mind and body. Tollefsen and Dale (2012) have called this the execution problem. Philosophical theories remain skeletal. A cognitive science of joint action will require filling in the details.

Finally, there is a real need to operationalize theories of joint action. The philosophical literature on joint action has burgeoned over the past decade and a great deal of effort has gone toward developing counterexamples to proposed theories. This has led to ever more complex theories. What would help arbitrate between existing theories is if these theories could be operationalized in a way that would generate empirical results. But in order to operationalize these theories, we need to know more about the underlying mechanisms involved in acting together. Once we identify these mechanisms we can manipulate them, along with social context and physical environment, in order to understand when and where shared intentions, joint commitments, we-intentions, etc. come on board, if indeed they do.

Shared Task Representations

Sebanz and Knoblich (2009) have argued that although a shared intention to perform some task might be central to its performance, individuals also have to have shared task representations. When we engage in joint tasks we must understand the other's action *as part of* a larger action. I need to predict not just what my partner will do but how what she will do will fit with what I must do in order to reach our common goal. This suggests that what we need is a shared representation of the task at hand. The representation is still something private and in the head of an individual, but it is shared because the content is the same between individuals.

The ability to form two-person systems with shared constraints is exemplified in experiments by Richardson and colleagues (2007). They asked pairs of participants to lift wooden boards off a conveyer belt. The boards varied in length and appeared on the conveyor in ascending or descending length. The participants were instructed to lift the planks only by touching the ends. This meant that larger boards would need to be lifted together by the pair with one person at each end. Richardson and colleagues found that participants were taking each other's arm span into account as they completed the task. The transition from lifting individually to lifting jointly varied as a function of a pair's mean arm span. Pairs that had a larger mean arm span switched to joint lifting at a longer board length than pairs with a smaller mean arm span. Individual's action planning took into account the capacities of those with whom they were in collaboration.[2]

In a series of experiments, Sebanz et al. (2003, 2006) and Atmaca et al. (2008) have shown that knowledge of another's task affects the ways in which one will plan and perform their own action. They compared the performance of an individual on a solo task with the performance of two individuals when the same task was distributed across the dyad. The aim was to see whether a standard response found in individual task performance appears when individuals perform that task together. When a single person performs a task that requires repeatedly choosing between two actions (a left key press

[2] It is important to note that Richardson, Marsh, and Baron (2007) may dispute a representational account of this task; however, they would certainly embrace the notion that we directly perceive social affordances, and so the theoretical import is quite similar.

and right key press, for instance) when confronted with a stimulus (press the left key when you see the blue ring on the finger, press the right key when you see the red ring on the finger), they can be easily confounded by adding features that are irrelevant to the task. For instance, when an arrow randomly appears and points in a direction opposite the response required, participants find the task more difficult and perform less well. Sebanz and colleagues hypothesized that if participants in the joint task simply formed representations of their part of the task and ignored the other's part of the task, then when the other participant in the joint action was faced with an irrelevant feature it should not affect the other's performance. The experiments showed that participants experienced a similar conflict. This "joint Simon effect" suggests that participants didn't just form representations of the other as an agent but shared a task representation. They formed a representation of the task as a whole and their part within the task. Experiments by Wenke et al. (2011) have suggested a slightly different interpretation of the representations formed during a joint task. They argued that participants in a joint task don't co-represent what another must do but instead co-represent that another is jointly responsible for the task and when it is their turn.

Recently, some have argued that the joint Simon effect may be driven more by non-social attentional saliency effects (Dolk et al. 2013) and have questioned whether there is really a need to posit an automatic and dedicated social process. In a series of experiments, Dolk et al. (2011, 2013) de-socialized the joint Simon task environment and showed that a significant joint Simon effect could be obtained by sitting next to a passive observer of the alternative response button that is associated with the attention-distracting event; then they showed that the joint Simon effect occurred even when the passive observer left the room. This suggests that what is generating the effect is the attention-distracting event and not a representation of the other's role in the task. Non-social events seem to induce the joint Simon effect as well. Dolk et al. (2013) were able to induce significant joint Simon effects by placing a Japanese waving cat and a ticking metronome in the experimental environment.

Embedded, Embodied, and Enactive Approaches to Joint Action

In recent years an embodied and embedded approach has emerged and become influential in cognitive science (cf. Clark 1997; Barsalou 2008). This approach stresses that high-level cognition is a function of basic perception and action processes and that it emerges out of the interaction of the organism with its physical and social environment (Gibson 1979; Thelen and Smith 1994; van Gelder and Port 1995; Van Orden et al. 2003). The insights of embedded and embodied cognition extended to joint action and social interaction more broadly have resulted in a number of important findings. The core idea of this approach is that there are basic perceptual and motor processes that are sufficient for bringing about a number of basic forms of social interaction and enable more complex forms of joint action involving high-level mental states such as shared intention. In some cases, those advocating this approach will argue that these basic forms of social interaction are sufficient for joint action. This represents a "bottom-up" approach to joint action and can be contrasted with the intellectualist approach of philosophical theories.

For instance, Marsh and colleagues (2006) have argued that in order to fully understand joint action, we need to adopt an embedded and embodied approach. The idea of a task representation that guides individual behavior within the context of joint action assumes that what is primary in cognition is the mind/brain's ability to form internal representations that guide action. A more embedded and embodied approach emphasizes the role the environment (including the body and others) plays in shaping behavior. Some such approaches even go as far as to deny that there are inner representations guiding joint action. These ecological approaches view joint action from the dynamical systems perspective. Rather than seeing joint action as the result of individual cognitive systems that manage to coordinate bodies as a result of processing internal representations, they view groups as dynamical systems. The tools of dynamics allow us to investigate when a collection of entities is exhibiting behavior at a higher level, sometimes referred to as a "complex system" (Richardson, Dale, and Marsh 2014). A complex system is not simply a collection of randomly interacting entities (like a contained gas). Instead, elements that form a system interact in ways that lead to different kinds of aggregate behaviors that can be observed and measured at a higher level of organization (from multicellular organisms, to ant nests, to human interactions). These tools thus project human joint action into a broader landscape of system patterns, and joint action—and social interaction as a subset—are just human examples of such integrated systems.

Some of the empirical research on social dynamics is grounded in research on interpersonal synchrony. Findings in the area of interpersonal dynamics over the past 20 years have identified the ways in which certain behaviors become entrained or aligned—that is, where there is dynamic matching between behavioral and/or cognitive states of two or more people. For instance, when two people sit in rocking chairs next to each other and begin to rock, their movements become effortlessly synchronized such that within several minutes they are rocking at the same time; more subtle movement, such as postural sway, is also a paradigmatic example of a behavior that becomes synchronized across individuals (Fowler et al. 2008; Schmidt and Richardson, 2008; Shockley et al. 2009; van Ulzen et al. 2008). The ecological approach argues that human behavior is responsive to a "social pull" that fundamentally shapes individual and joint action (Shockley 2009). Marsh and colleagues represent a "bottom-up" and anti-intellectualist approach to joint action. Joint agency arises out of the complex interaction of individual agents embedded in an environment.

Sebanz and Knoblich (2009) argue that this ecological approach needs to be integrated with ideomotor theories, social cognition theories, and joint intentionality theories in order to be able to explain a wide range of joint action types (for additional discussion, see Vesper et al. 2010). Mechanisms of entrainment and simultaneous affordance alone don't allow us to understand how participants in a joint action can perform different actions. The focus on behavioral matching or "synchrony" misses the fact that often participants to a joint action have to do something radically different from their partner and thus make a distinction between self and other. Joint action requires an understanding of the other's role in the joint action and some shared goal or intention. Knoblich and Sebanz argue that the self–other distinction cannot be explained by a purely ecological perspective. An elegant example of the importance of more goal-driven adaptation between interacting persons is in Fusaroli et al. (2012), who find that participants do not simply "indiscriminately align" in their language when carrying out a task together, but rather do so in circumscribed goal-based ways.

Reflection on conversation might help to bring the intellectualist and non-intellectualist together. What research on conversation over past few decades reveals is that joint action is mediated by a number of lower-level processes. Although the sort of shared intentions and goals identified by philosophers of joint action may be at play in conversation, there are clearly lower-level processes that aid in the establishment and maintenance of linguistic interaction. Pickering and Garrod (2004; see also Garrod and Pickering 2009) argue that the joint activity of conversation is most successful when cognitive processes align across various levels of linguistic organization, from words and sentence structure to the use of figurative language, such as irony (Roche et al. 2012). Eye movements also become aligned when subjects are having a conversation about an item in the environment. For example, Richardson and Dale (2005) have demonstrated that during interaction, there is a tight coupling of visual attention. As people discuss a work of art, their eye movements become distinctly aligned in time. Indeed, it seems to be the case that the better the alignment, the better the participants are understanding each other or, in other words, the better they succeed in fulfilling the shared goal or intention of communicating with one another (among many: Brennan and Clark 1996; Richardson and Dale 2005; see also Dale, Kirkham, and Richardson 2011; Ireland et al., 2011; Ramseyer and Tschacher 2011; Tanenhaus and Brown-Schmidt 2008; Kavanagh and Winkielman 2016). This alignment of attention, or joint attention, in conversation suggests that joint action, in general, involves aspects of an alignment "system" (Tollefsen and Dale 2012) that helps start and sustain joint action over time. Both top-down and bottom-up processes might be involved in complex forms of joint action (cf. Louwerse et al. 2012).

We have so far focused on cognitive and perceptuomotor variables, but the relevant theoretical variables go beyond the individuals in an interaction—they may also involve the surrounding materials and institutions. For example, recent work by Gallagher and Ransom (2016) brings joint action and material engagement theory together. Lambos Malafouris's (2013) material engagement theory (MET) is akin to other nonrepresentationalist, enactivist approaches. The focus of his work, however, is on giving an archaeology of the mind. It attempts to provide an evolutionary framework that explains how the contemporary human mind emerged out of a dynamic interaction between physical, neurological, cultural, and material forces. According to MET, the mind depends on interactions between brain, body, and artifacts. In particular,

Malafouris highlights the ways in which artifacts made possible types of cognition that were not otherwise possible, and argues that the interaction between individual and artifact constitutes new cognitive capacities. Like the extended mind thesis, MET makes a strong metaphysical claim about the constitution of cognitive processes. Further, it makes a strong claim about agency. According to MET, agency should not be considered a property solely of humans, but a property of the process that involves both humans and things. He appeals to Bruno Latour's (2005) actor-network theory (ANT) to develop this idea. According to ANT, actions occur in networks that include people and things. Intentions are not formed independently of engagement with the world but via engagement with material reality. Malafouris suggests that the speed bump that causes one to slow down should be conceived as part of a network that includes a human agent. It is the network and not the human that acts—that slows down (Malafouris 2013, p. 124).

Gallagher and Ransom (2016) suggest that we look at joint action from the perspective of MET. They contrast this with accounts of joint action that focus on identifying the internal, individual, goal, or task representations (Sebanz et al., 2006). Consider, for instance, research by Richardson, Marsh, and Baron (2007) and Marsh et al. (2006) that shows that when subjects are lifting boards together they seem to take into account the material lengths of the boards. One approach to explaining this phenomenon would be to posit beliefs (or task representations) about the lengths of the boards in the minds of agents and predictions about how one's partner will perform. For Gallagher and Ransom, it isn't beliefs about the boards that modulate the interaction, but the boards themselves. The MET approach also emphasizes the role that artifacts play in modulating joint actions. They argue that MET can provide a better understanding of more complex forms of joint action as well. They offer the Occupy movement as an example. The coordinated activities of this group were made possible by the complex interaction of individuals, social media, technology, geography, and cultural practices.

If you adopt the ecological and MET approach advocated by Gallagher and Ransom, then the search for internal representational states that bring about joint action will seem mistaken. These are essentially anti-representationalist positions. Although not all enactivists deny the existence of representations, the explanation of human action is to be found essentially in the dynamic interplay between individuals and environments and not in internal representations, and this insight is taken to hold for joint action as well. This has the result of blurring the line between joint actions in the narrow sense and various other types of social interaction and collective behavior. Recall it was precisely because individuals formed representations of the task as something they will do together—a joint or shared intention—that allowed Searle to distinguish between the random collection of strangers and the dance troupe. The MET approach also ends up blurring the line between individual, environment, and other. According to MET, agency isn't something individual human beings do. That is, it is not something somehow contained within an individual agent but arises out of a dynamic interaction between individual and environment. Joint action, then, is not simply two agents coming together and interacting. Rather, joint agency arises out of the complex interaction of individuals and their environments. Environments supply conditions and constraints for interaction, prompts or salient information that individuals may grasp onto as they shape each other's interactive structure and engage in "participatory sense-making" (De Jaegher and Di Paolo 2007). The basic question here is whether we should see action or agency as distributed across individuals, environments, and artifacts, or somehow contained within the parameters of an individual. Does joint action require a rethinking of the bounds of cognition and agency?

Joint Action and Extended Mind

The idea that the mind is bounded by skin and bone has been significantly challenged over the past 30 years. In 1995, Edwin Hutchins published *Cognition in the Wild*, a detailed study of navigation on a naval vessel. Hutchins's work extends cognitive models such as constraint satisfaction models to groups, and his work has been instrumental in developing the field of distributive cognition. At about the same time, philosophers Andy Clark and David Chalmers published their article "The Extended Mind" (1998), which built on work in cognitive science and psychology on external representations (Kirsh and Maglio 1994; Zhang and Norman 1994) and provided a philosophical argument for the idea that the mind is not bounded by skin and bones but extends to encompass aspects of the environment. Clark and Chalmers introduced the *parity principle*: "If, as we confront some task, a part of the world functions as a process which, were it done in the head, we would have no hesitation in recognizing as part of the cognitive process, then that part of the world (so we claim) is part of the cognitive process" (1998, p. 8). Clark and Chalmers use a thought experiment to support their view. Consider Otto, an individual suffering from memory loss, who relies on his notebook in a substantive way to live in the world. Clark and Chalmers argue that Otto's notebook is an extension of his biological memory because it plays the same functional role as his biological memory. In this example, the mind extends to a single artifact. Distributed cognition, as developed by Hutchins, involves a distribution of cognition across artifacts, environments, and people. In "From Extended Mind to Collective Mind" (2006), Deborah Tollefsen discuses a difference between "solipsistic systems" and "collective systems." Clark and Chalmers (1998) concentrated on solipsistic systems insofar as their focus was on a single agent acting in the world, while those who focus on the operation of cognitive systems across multiple agents are concerned with collective systems.

Building on thought experiments introduced by Clark and Chalmers (1998), Tollefsen (2005) offers a thought experiment in defense of group cognition. Instead of Otto and Inga, Tollefsen considers the case of absent-minded philosophy professor Olaf and his partner Inga. Olaf often relies on Inga for a variety of cognitive tasks, such as remembering his meeting schedule or travel directions, much akin to how Otto uses his notebook. The interaction between Olaf and Inga is also guided in a closely analogous way to the situation between Otto and his notebook, including Olaf's trust in the information Inga provides and the fact that he often defers to her for help throughout the day. As a result, Tollefsen (2006) suggests that "if [Clark and Chalmers] are correct, the mind not only extended to encompass non-biological artifacts . . . but it also occasionally forms collective systems that support cognition and belief" (p. 143). Moreover, she argues that this collective reading is more resilient to counterarguments than solipsistic cases.

Some have read Tollefsen's extension of the extended mind to group cognition as a case of the social parity principle (Theiner 2008; Ludwig 2015). According to this principle, "If, in confronting some task, a group collectively functions in a process which, were it done in the head, would be accepted as a cognitive process, then that group is performing that cognitive process" (Ludwig 2015, p. 197). Importantly, Tollefsen points out that the possibility of group cognition rests on substantive two-way interaction between individuals. Without such interaction, Olaf's mind merely extends to Inga's. This two-way interaction provides for cognitive integration, an interdependence of cognitive functions and the basis for thinking of the group as a system in itself rather than simply a collection of individual cognitive systems.

Tollefsen further develops her example of Olaf and Inga by considering cases where their cognitive systems are interdependent and integrated. She appeals to research on transactive memory systems (TMS) (Wegner et al. 1991; Wegner 1995) as an example of substantive interaction between individuals. In "transactive memory," one individual's memory is shown to function alongside that of another person or persons, as a kind of socially extended memory system. A transactive memory system involves the complex interaction of individual memory systems. These individual memory systems are not merely external storage devices; both the process of remembering and storage of memories is done in an interactive and dialogical manner. Unlike a notebook, which does not itself have the characteristic of memory identified by mainstream memory research (learning time and access time), a group of individual memory systems can, it seems, become unified and produce a memory product (recall an event, say) that is richer than that recalled by an individual. Consider the case of a department subcommittee charged with revising the department's policies and procedures. Faculty member A might offer some information that he remembers about the history of the department; faculty member B might offer information about what the department chair requires of the committee; faculty member C might remember something said by the provost regarding policymaking. As a result, the committee might recall a great deal more information relevant to their task than any individual alone, and we can imagine that such recall is done in a collaborative and dialogical process. Information may have been stored individual minds or heads but the retrieval was done through conversation.

Transactive memory systems as described by Wegner go through the same stages that occur at the individual level: encoding, storage, and retrieval. Encoding at the collective level occurs when members discuss information and determine the location and form in which it will be stored within the group. Retrieval involves identification of the location of the information. Retrieval is transactive when the person holding an item internally is not the person who asked to retrieve it. In our case above, a faculty member may have promoted retrieval of information from another faculty member by asking questions and offering alternative hypotheses that the committee member needed to consider. Theiner (2013) has argued that transactive memory is an example of an emergent group-level property.

In *Macrocognition: A Theory of Distributed Minds and Collective Intentionality* (2014), Bryce Huebner argues that a successful account of group cognition should combine intentional systems theory with a cognitive architecture of a particular kind (p. 96). The type of cognitive architecture he proposes is a functionally decomposable system, wherein a variety of integrated subsystems run various computations, usually in a highly distributed manner and over multiple representational formats. This architecture, in turn, allows a system to carry out tasks necessary for its survival, including many fluid and flexible interactions with its surrounding environment. He discusses several potential cases of group minds—from beehives and stock markets to TMS and research in high-energy physics. Huebner offers a scale to situate these different putative cases from not yet minimally minded to maximally minded entities.

In *Cognitive Systems and the Extended Mind* (2009) and in more recent articles (e.g., Rupert 2011), Rupert defines a cognitive state in terms of its role in a larger system, such that a state is cognitive if and only if "it is the state of one or more of an integrated set of mechanisms that contribute distinctively (i.e., not as background conditions) to the production of a wide range of cognitive phenomena, across a variety of conditions, working together in overlapping subsets" (2011, p. 637). In short, a state is cognitive if it functions within a cognitive system, where a cognitive system involves various mechanisms and processes that "contribute causally to the production of a wide range of cognitive phenomena, across a variety of conditions, working together in an overlapping way with a variety of other mechanisms of similar standing" (2009, p. 41).

Cognitive systems are persistent and display a set of capacities that persist across different contexts. This persistence is best explained, according to Rupert, by the fact that they are realized in a physically bounded organism. This systems-based approach is then used by Rupert to rule out (or at least cast serious doubt on) artifacts like cellphones or calculators as part of the cognitive system and activities involving these artifacts as cognitive processes, because such things are not integrated in the ways that, say, vision, linguistic processes, and short-term memory are typically integrated. The systems objection can be used to rule out-group cognition as well, since group cognition is often context-dependent, intermittent, and transitory.

Tollefsen, Dale, and Paxton (2013) have argued that the conditions for an integrated system could be met by small task groups, particularly those that work together over time. Two individuals, when coordinating together, may form a kind of dynamic unit of analysis itself. Defining such a unit of analysis on a quantitative level has been successful in some domains. For example, using recurrence analysis methods, Dale, Kirkham, and Richardson (2011) have shown that perceptuomotor systems of two people interacting come to form a tightly coupled unit. Using both eye movements and computer mouse movements, they found that two people solving a particular puzzle take on particular roles of "parts" in a dyadic whole—in the sense that their behaviors are becoming so weaved in time that they should be regarded as causally coupled.

Stronger evidence for this comes from research by Riley et al. (2012) and Ramenzoni (2008), who show that we can consider two people in a joint action as a system because quantitative analysis is most efficient using fewer dimensions. They compared this to a situation in which two people were not in joint action. The idea is that two people are coming to form a "synergy"—their behaviors are more than the sum of their parts; they show responsive behavioral dynamics to each other, and so on. By using principal component analysis, they reveal that joint action induces the two-person system to "occupy fewer dimensions" than when acting independently. This means even at a quantitative level it may be possible to characterize two persons as a genuine integrated system.

Joint Action and Development

As we have seen, one of the objections raised to philosophical accounts of joint action is that they over-intellectualize joint action and in doing so risk ruling out children as participants (Tollefsen 2005). In *A Natural History of Human Thinking*, Michael Tomasello (2014) argues that joint action is distinctive of the human race, and that very young children exhibit the cognitive and motivational capacities distinctive of joint action. Joint action, according to Tomasello, involves the formation of joint goals. Although chimpanzees and other animals engage in complex group behaviors, they do not have joint goals. Rather, according to Tomasello, a "leaner" interpretation of the behavior is called for (Tomasello and Carpenter 2005). Consider the hunting of monkeys by chimpanzees (Boesch and Boesch 1989; Watts and Mitani 2002). The chimps seem to coordinate their actions and surround and capture the monkey. According to Tomasello and colleagues, a better interpretation of the phenomenon is to see each individual chimpanzee as pursuing the monkey on its own, taking into account the behavior and intentions of the others. To the extent that there is cooperative behavior, it is because each chimpanzee would prefer to have the monkey caught than to let it get away. According to Tomasello, the chimpanzees are engaged in co-action, not joint action. This captures the distinction between social interaction and joint action. When two people coordinate their bodies in a way as to not walk into each other on a sidewalk, this is co-action, not joint action, as it does not involve a shared goal. He extends this interpretation to all cases of nonhuman coordination.

Young children, on the other hand, exhibit the ability to form joint goals with others at a very young age—as early as 14 months. Warneken et al. (2006; Warneken and Tomasello 2007) ran a series of experiments in which young children play a game with an adult that requires that the child and the adult coordinate their actions (for instance, in one case, the object of the game is to retrieve a toy, and it requires that each person operate one side of the experimental apparatus). At some point during the game the adult stops playing (without any reason). The children find this very frustrating and exhibit "re-engagement" behavior to try to get the adult to participate again in the game. According to Warneken et al. (2014) and Tomasello (2014), this behavior reveals that the child has an expectation that the adult will contribute to the game and some rudimentary shared or joint goal—a commitment to do something together. The evidence for joint goals in young children becomes stronger at the age of three as they seem to exhibit a greater commitment to joint actions—even putting the joint action before their own personal gratification—and when rewards were involved, they often divided the spoils up equally (Hamann,Warneken, and Tomasello 2012; Hamann et al. 2011). No such behavior is found in nonhuman animals. Tomasello, like Margaret Gilbert (2013), believes that joint action essentially involves a normative element. When individuals commit to doing things together it brings about expectations, obligations, and entitlements that were not previously in place.

Tomasello's approach is clearly "intellectualist"—identifying the mechanisms that guide joint action in children, with high-level representations of goals and an appreciation for the entitlements and obligations generated by commitments. Although Tomasello admits that very young children may be engaging in less robust forms of joint action, he maintains that at three years of age children understand what others intend and the commitments others have made to engage in joint action (however, see work by Fantasia et al. 2014, suggesting such processes may begin even in the first few months of life). Such knowledge clearly presupposes social-cognitive capacities. Tomasello and colleagues (2005) are also working with a narrow conception of joint action. Joint action involves shared goals and intentions, and because they start with this narrow conception, nonhuman animals are only capable of co-action, not joint action.

Enactivist theories of social cognition provide a framework for thinking about the socia-cognitive capacities required for joint action that avoid over-intellectualization. Engaging in joint action requires some understanding of the other. Without such an understanding the coordination of minds and bodies could not happen. This understanding has been understood as a "theory of mind" and the standard approaches (theory-theory and simulation theory) share the view that this capacity is essentially the capacity to read minds and to see individuals as motivated by beliefs and desires. Social understanding is essentially mentalistic. Recent *enactive* approaches to social cognition (Hutto 2011, 2017; Reddy and Morris 2004; Ratcliffe 2007; Gallagher and Zahavi 2008; Fuchs and De Jaegher 2009) argue that the "mindreading" approach to social cognition is not able to capture the very practical, embodied nature of what goes on in our everyday social interactions. They argue that most dealings with others should be explained in terms of *second-person* embodied practices that involve various capabilities—imitation, intentionality detection, gaze following, social referencing, etc.—but do not depend on mindreading. These forms of social cognition have informed various accounts of joint action by providing the more basic forms of social interaction from which joint action emerges.

For instance, Fiebich and Gallagher (2013) have argued that we can come to understand join action by starting with primary intersubjectivity. Primary intersubjectivity refers to the set of sensorimotor abilities that provide us with a way of understanding another person's movements, gestures, facial expressions, eye direction, and goal-directed actions. These capacities allow us to engage in joint attention. Once we are able to engage in joint attention we are able to develop more complicated understanding of others. This represents secondary intersubjectivity (Trevarthen 1979, 1998; Trevarthen and Hubley 1978) and allows us to engage in more complex interactions, including joint actions and intentional joint attention (the formation of shared goals to attend to something together). Importantly, primary intersubjectivity is understood as an embodied, embedded skill rather than along theory-theory or simulation lines, which tend to offer an over-intellectualized account of our basic social interactions. This approach does not draw a sharp distinction between animal joint action and human joint action.

Moving Forward: An Ecumenical Approach

As we have seen, we can identify at least two broad approaches to the study of joint action: the intellectualist approach, which attempts to identify higher-order mental states such as goals and intentions in order to explain joint action, and a bottom-up approach, which focuses on lower-level processes that give rise to and sustain joint action. Embedded, embodied, and enactive approaches represent a more bottom-up approach. Philosophical accounts of joint action represent the paradigm of intellectualist approaches. We might also characterize the debate in terms of those who look for internal processes of the individual versus those who take a more external approach that focuses on the dynamic interaction between individuals and the environmental affordances that shape and inform joint actions. Philosophical accounts and those that focus on shared task representations represent the internal approach. The embedded, embodied, and enactive approaches provide us with external approaches to explaining joint action.

The present debate is of course ongoing. It seems to us that both internal processes along with coordinative external constraints are critical ingredients in theories of joint action. The *potential* that humans have to resonate with others and artifacts by its very existence requires particular kinds of cognitive capacities to make it happen. Put simply, people can dance but squirrels cannot, at least not well. So, an important future agenda is to integrate the individual and the group levels of description in a broader understanding of joint action. For example, the role of predictive processes may be critical to how individuals navigate conversational domains, along with alignment among *internal* processes like comprehension and production (Pickering and Garrod 2014). In another development, some have proposed that partners in interaction may process "single bits" of information—such as whether someone can simply see something that you can see, too—and this may help shape implicit behaviors or inferences, while being tractable in terms of memory and processing constraints on individual humans (Brennan et al. 2010). There may be a network of constraints at lower and higher levels that bring about joint actions such as conversation (Dale et al. 2014; Fusaroli et al. 2014).

One recent approach to solving this problem is to discern the brain mechanisms that may be involved in these joint processes. This of course highlights the individual level of analysis, but does so in contexts that are explicitly interactive and sometimes quite naturalistic, such as free-flowing interaction (Silbert et al. 2014). The underlying mechanisms of "brain-to-brain coupling" may open new avenues of investigation about these interactive capacities (Hasson et al. 2012). By understanding these capacities, we could better understand the joint actions that are possible and how they are instantiated. Such an understanding must, of course, be developed in the context of an appropriately embodied and embedded investigation; otherwise this neural investigation may summon similar concerns as the production tradition in psycholinguistics. With advances in imaging techniques, this synergy of the internal and external in natural contexts may be possible (e.g., Xu et al. 2014).

The study of joint action must integrate both high-level and low-level approaches and synthesize findings from neuroscience, experimental psychology, cognitive science, and philosophy. Drawing hard-and-fast boundaries between actions that are joint and mere collective behavior risks turning substantive debates into verbal disputes. What seems clear is that social interaction takes a variety of different forms, some more sophisticated than others. Understanding how the more sophisticated forms arise from less sophisticated forms will involve empirical research rather than armchair speculation and should be motivated not by prior theoretical commitments to either representationalism or nonrepresentationalism but by a commitment to understanding the social world.

References

Atmaca S., Sebanz N., Prinz W., and Knoblich G. (2008). Action co-representation: the joint SNARC effect. *Social Neuroscience*, 3, 1–11.

Barsalou, L.W. (2008). Grounded cognition. *Annual Review of Psychology*, 59, 617–45.

Boesch, C. and Boesch, H. (1989). Hunting behavior of wild chimpanzees in the Tai National Park. *American Journal of Physical Anthropology*, 78, 547–73.

Bratman, M.E. (2014). *Shared agency: a planning theory of acting together.* New York: Oxford University Press.

Brennan, S.E. and Clark, H.H. (1996). Conceptual pacts and lexical choice in conversation. *Journal of Experimental Psychology: Learning, Memory, and Cognition*, 22(6), 1482–93.

Brennan, S. E., Galati, A., and Kuhlen, A.K. (2010). Two minds, one dialog: coordinating speaking and understanding. In: B.H. Ross (ed.), *The psychology of learning and motivation* (vol. 53). Burlington, VT: Academic Press, pp. 301–44.

Clark, A. (1997). *Being there: putting brain, body, and world together again.* Cambridge, MA: MIT Press.

Clark, A. and Chalmers, D.J. (1998). The extended mind. *Analysis*, 58(1), 7–19.

Dale, R., Fusaroli, R., Duran, N., and Richardson, D.C. (2014). The self-organization of human interaction. *Psychology of Learning and Motivation*, 59, 43–95.

Dale, R., Kirkham, N.Z., and Richardson, D.C. (2011). The dynamics of reference and shared visual attention. *Frontiers in Psychology*, 2, 355. doi:10.3389/fpsyg.2011.00355

De Jaegher, H. and Di Paolo, E. (2007). Participatory sense-making. *Phenomenology and the Cognitive Sciences*, 6(4), 485–507.
Dolk T., Hommel B., Colzato L.S., Schütz-Bosbach S., Prinz W., and Liepelt R. (2011). How "social" is the social Simon effect? *Frontiers in Psychology*, 2, 84.
Dolk, T., Hommel, B., Prinz, W., and Liepelt, R. (2013). The (not so) social Simon effect: a referential coding account. *Journal of Experimental Psychology: Human Perception and Performance*, 39(5), 1248–60.
Fantasia, V., Fasulo, A., Costall, A., and López, B. (2014). Changing the game: exploring infants' participation in early play routines. *Frontiers in Psychology*, 5, 522.
Fiebich, A. and Gallagher, S. (2013). Joint attention in joint action. *Philosophical Psychology*, 26(4), 571–87.
Fiebich, A., Nguyen, N., and Schwarzkopf, S. (2015). Cooperation with robots? A two-dimensional approach. In: C. Misselhorn (ed.), *Collective agency and cooperation in natural and artificial systems: explanation, implementation and* simulation. Cham, Switzerland: Springer International Publishing, pp. 25–43.
Fowler C.A., Richardson, M.J., Marsh K.L., and Shockley K.D. (2008). Language use, coordination, and the emergence of cooperative action. In: A. Fuchs and V.K. Jirsa (eds.), *Coordination: neural, behavioral and social dynamics*. Berlin and Heidelberg: Springer, pp. 261–79.
Fuchs, T. and De Jaegher, H. (2009). Enactive intersubjectivity: participatory sense-making and mutual incorporation. *Phenomenology and the Cognitive Sciences* 8(4), 465–86.
Fusaroli, R., Bahrami, B., Olsen, K., Roepstorff, A., Rees, G., Frith, C. et al. (2012). Coming to terms quantifying the benefits of linguistic coordination. *Psychological Science*, 23(8), 931–9. doi:10.1177/0956797612436816
Fusaroli, R., Rączaszek-Leonardi, J., and Tylén, K. (2014). Dialog as interpersonal synergy. *New Ideas in Psychology*, 32, 147–57.
Gallagher, S. and Ransom, T. 2016. Artifacting minds: material engagement theory and joint action. In: C. Tewes (ed.), *Embodiment in evolution and culture*. Berlin: de Gruyter, pp. 337–51.
Gallagher, S. and Zahavi, D. (2008). *The phenomenological mind: an introduction to philosophy of mind and cognitive science*. New York: Routledge.
Garrod, S. and Pickering, M.J. (2009) Joint action, interactive alignment and dialogue. *Topics in Cognitive Science*, 1(2), 292–304.
Gibson, J.J. (1979). *The ecological approach to visual perception*. Boston: Houghton Mifflin.
Gilbert, M. (2013). *Joint commitment: how we make the social world*. New York: Oxford University Press.
Hamann, K., Warneken, F., Greenberg, J., and Tomasello, M. (2011). Collaboration encourages equal sharing in children but not chimpanzees. *Nature*, 476, 328–31.
Hamann, K., Warneken, F., and Tomasello, M. (2012). Children's developing commitment to joint goals. *Child Development*, 83(1), 137–45.
Hasson, U., Ghazanfar, A.A., Galantucci, B., Garrod, S., and Keysers, C. (2012). Brain-to-brain coupling: a mechanism for creating and sharing a social world. *Trends in Cognitive Sciences*, 16(2), 114–21.
Huebner, B. (2014). *Macrocognition: a theory of distributed minds and collective intentionality*. Oxford, UK: Oxford University Press.
Hutchins, E. (1995). *Cognition in the wild*. Cambridge, MA: MIT Press.

Hutto, D.D. (2011). Elementary mind minding, enactivist-style. In: A. Seemann (ed.), *Joint attention: new developments in philosophy, psychology, and neuroscience*. Cambridge, MA: MIT Press.

Hutto, D.D. (2017). Basic social cognition without mindreading: minding minds without attributing contents. *Synthese*, 194(3), 827–46.

Ireland, M.E., Slatcher, R.B., Eastwick, P.W., Scissors, L.E., Finkel, E.J., and Pennebaker, J.W. (2011). Language style matching predicts relationship initiation and stability. *Psychological Science*, 22(1), 39–44.

Kavanagh, L.C. and Winkielman, P. (2016). The functionality of spontaneous mimicry and its influences on affiliation: an implicit socialization account. *Frontiers in Psychology*, 7, 458.

Kirsh, D. and Maglio, P. (1994). On distinguishing epistemic from pragmatic action. *Cognitive Science*, 18, 513–49.

Kutz, C. (2000). Acting together. *Philosophy and Phenomenological Research*, 61(1), 1–31.

Latour, B. (2005). *Reassembling the social: an introduction to actor-network-theory*. Oxford: Oxford University Press.

Louwerse, M.M., Dale, R., Bard, E.G., and Jeuniaux, P. (2012). Behavior matching in multimodal communication is synchronized. *Cognitive Science*, 36(8), 1404–26.

Ludwig, K. (2015). Is distributed cognition group level cognition? *Journal of Social Ontology*, 1, 189–224.

Malafouris, L. (2013). *How things shape the mind: a theory of material engagement*. Cambridge, MA: MIT Press.

Marsh, K.L., Richardson, M.J., Baron, R.M., and Schmidt, R.C. (2006). Contrasting approaches to perceiving and acting with others. *Ecological Psychology*, 18, 1–38.

Pickering, M.J. and Garrod, S. (2004). Toward a mechanistic psychology of dialogue. *Behavioral and Brain Sciences*, 27(2), 169–90.

Pickering, M.J. and Garrod, S. (2014). Neural integration of language production and comprehension. *Proceedings of the National Academy of Sciences*, 111(43), 15291–2.

Ramenzoni, V. (2008). *Effects of joint task performance on interpersonal postural coordination* [electronic thesis or dissertation]. Retrieved from https://etd.ohiolink.edu/

Ramseyer, F. and Tschacher, W. (2011). Nonverbal synchrony in psychotherapy: coordinated body movement reflects relationship quality and outcome. *Journal of Consulting and Clinical Psychology*, 79(3), 284–95.

Ratcliffe, M. (2007). *Rethinking commonsense psychology: a critique of folk psychology, theory of mind and simulation*. New York: Palgrave Macmillan.

Reddy, V. and Morris, P. (2004). Participants don't need theories: knowing minds in engagement. *Theory & Psychology*, 14(5), 647–65. doi:10.1177/0959354304046177

Richardson, D.C and Dale, R. (2005). Looking to understand: the coupling between speakers' and listeners' eye movements and its relationship to discourse comprehension. *Cognitive Science*, 29, 1045–60.

Richardson, M.J., Dale, R., and Marsh, K. (2014). Complex dynamical systems in social and personality psychology: theory, modeling and analysis. In: H.T. Reis and C.M. Judd (eds.), *Handbook of research methods in social and personality psychology*. New York: Cambridge University Press, pp. 253–82.

Richardson, M.J., Marsh, K.L., and Baron, R.M. (2007). Judging and actualizing intrapersonal and interpersonal affordances. *Journal of Experimental Psychology: Human Perception and Performance*, 33(4), 845–59.

Riley, M.A., Shockley, K., and Van Orden, G. (2012). Learning from the body about the mind. *Topics in Cognitive Science*, 4, 21–34.

Roche, J., Dale, R.C., and Caucci, G. (2012). Doubling up on double meanings: pragmatic alignment. *Language and Cognitive Processing*, 27, 1–24. doi:10.1080/01690965.2010.509929

Rupert, R.D. (2009). *Cognitive systems and the extended mind*. Oxford: Oxford University Press.

Rupert, R.D. (2011). Empirical arguments for group minds: a critical appraisal. *Philosophy Compass*, 6(9), 630–9.

Schmidt R.C. and Richardson, M.J. (2008). Dynamics of interpersonal coordination. In: A. Fuchs and V.K. Jirsa (eds.), *Coordination: neural, behavioral and social dynamics*. Berlin and Heidelberg: Springer, pp. 281–308.

Searle, J. (1995). *The construction of social reality*. New York: Free Press.

Sebanz, N., Bekkering, H., and Knoblich, G. (2006). Joint action: bodies and minds moving together. *Trends in Cognitive Sciences*, 10(2), 70–6.

Sebanz, N. and Knoblich, G. (2009). Prediction in joint action: what, when, and where. *Topics in Cognitive Science*, 1(2), 353–67.

Sebanz, N., Knoblich, G., and Prinz, W. (2003). Representing others' actions: just like one's own? *Cognition*, 88(3), B11-B21.

Shockley, K., Richardson, D.C., and Dale, R. (2009). Conversation and coordinative structures. *Topics in Cognitive Science*, 1: 305–19. doi:10.1111/j.1756-8765.2009.01021.x

Silbert, L.J., Honey, C.J., Simony, E., Poeppel, D., and Hasson, U. (2014). Coupled neural systems underlie the production and comprehension of naturalistic narrative speech. *Proceedings of the National Academy of Sciences*, 111(43), E4687–E4696.

Tanenhaus, M.K. and Brown-Schmidt, S. (2008). Language processing in the natural world. *Philosophical transactions of the Royal Society of London. Series B, Biological sciences*, 363, 1105–22. doi:10.1098/rstb.2007.2162

Theiner, G. (2008). *From extended minds to group minds: rethinking the boundaries of the mental* [dissertation]. Bloomington: Indiana University.

Theiner, G. (2013). Transactive memory systems: a mechanistic analysis of emergent group memory. *Review of Philosophy and Psychology*, 4(1), 65–89.

Thelen, E. and Smith, L. (1994). *A dynamical systems approach to the development of cognition and action*. Cambridge, MA: MIT Press.

Tomasello, M. (2014). *A natural history of human thinking*. Cambridge: Harvard University Press.

Tomasello, M. and Carpenter, M. (2005). The emergence of social cognition in three young chimpanzees. *Monographs of the Society for Research in Child Development*, 70(1).

Tomasello, M., Carpenter, M., Call, J., Behne, T., and Moll, H. (2005). Understanding and sharing intentions: the origins of cultural cognition. *Behavioral and Brain Sciences*, 28(5), 675–91.

Tollefsen, D. (2005). Let's pretend: joint action and young children. *The Journal of the Philosophy of Social Science*, 35(1), 75–97.

Tollefsen, D. (2006). From extended mind to collective mind. *Cognitive Systems Research*, 7, 140–50.

Tollefsen, D. and Dale, R. (2012). Naturalizing joint action: a process-based approach. *Philosophical Psychology*, 25(3), 385–407.

Tollefsen, D., Dale, R., and Paxton, A. (2013). Alignment, transactive memory, and collective cognitive systems. *Review of Philosophy and Psychology*, 4(1), 49–64.

Trevarthen, C. (1979). Communiaction and cooperation in early infancy: a description of primary intersubjectivity. In: M. Bullowa (ed.), *Before speech: the beginning of human communication*. Cambridge: Cambridge University Press, pp. 321–48.

Trevarthen, C. (1998). The concept and foundations of infant intersubjectivity. In: S. Bråten (ed.), *Studies in emotion and social interaction, 2nd series: intersubjective communication and emotion in early ontogeny*. New York: Cambridge University Press, pp. 15–46.

Trevarthen, C. and Hubley, P. (1978). Secondary intersubjectivity: confidence, confiding, and acts of meaning in the first year. In: J. Lock (ed.), *Action, gesture and symbol*. London: Academic Press, pp. 183–229.

Tuomela, R. (2013). *Social ontology: collective intentionality and group agents*. New York: Oxford University Press.

van Gelder, T. and Port, R.F. (1995). Explorations in the dynamics of cognition. In: T. van Gelder and R. Port (eds.), *Mind as motion*. Cambridge, MA: MIT Press.

Van Orden, G. Holden, J., and Turvey, M. (2003). Self-organization of cognitive performance. *Journal of Experimental Psychology: General*, 132(3), 331–50.

van Ulzen, N., Lamoth, C., Daffertshofer, A., Semin, G., and Beek, P. (2008). Characteristics of instructed and uninstructed interpersonal coordination while walking side-by-side. *Neuroscience Letters*, 432(2), 88–93.

Vesper, C., Butterfill, S., Knoblich, G., and Sebanz, N. (2010). A minimal architecture for joint action. *Neural Networks*, 23(8), 998–1003.

Warneken, F., Chen, F., and Tomasello, M. (2006). Cooperative activities in young children and chimpanzees. *Child Development*, 77, 640–63. doi:10.1111/j.1467

Warneken, F., Steinwender, J., Hamann, K., and Tomasello, M. (2014). Young children's planning in a collaborative problem-solving task. *Cognitive Development*, 31, 48–58.

Warneken, F. and Tomasello, M. (2007), Helping and cooperation at 14 months of age. *Infancy*, 11, 271–94. doi:10.1111/j.1532-7078.2007.tb00227.x

Watts, D. and Mitani, J.C. (2002). Hunting behavior of chimpanzees in Ngogo, Kibale National Park, Uganda. *International Journal of Primatology*, 23, 1–28.

Wegner, D.M. (1995). A computer network model of human transactive memory. *Social Cognition*, 13(3), 1–21.

Wegner, D.M., Erber, R., and Raymond, P. (1991). Transactive memory in close relationships. *Journal of Personality and Social Psychology*, 61(6), 923–29.

Wenke, D., Atmaca, S., Holländer, A., Liepelt, R., Baess, P., and Prinz, W. (2011). What is shared in joint action? Issues of co-representation, response conflict, and agent identification. *Review of Philosophy and Psychology*, 2(2), 147–72.

Xu, Y., Tong, Y., Liu, S., Chow, H.M., AbdulSabur, N.Y., Mattay, G.S. et al. (2014). Denoising the speaking brain: toward a robust technique for correcting artifact-contaminated fMRI data under severe motion. *NeuroImage*, 103, 33–47.

Zhang, J. and Norman, D.A. (1994), Representations in distributed cognitive tasks. *Cognitive Science*, 18, 87–122.

CHAPTER 14

PERCEPTION, EXPLORATION, AND THE PRIMACY OF TOUCH

MATTHEW RATCLIFFE

Introduction

Philosophical discussions of perception tend to focus on vision. However, there is the concern that what applies to vision may not apply to perception more generally, and that other senses can offer different insights into its nature.[1] A stronger claim, made by Alva Noë, is that many philosophical treatments of perception not only overemphasize vision; they are also rooted in a mistaken photographic model of vision. Touch, he suggests, offers a better starting point from which to understand the nature of perception:

> Think of a blind person tip-tapping his or her way around a cluttered space, perceiving that space by touch, not all at once, but through time, by skillful probing and movement. This is, or at least ought to be, our paradigm of what perceiving is. The world makes itself available to the perceiver through physical movement and interaction. (Noë 2004, p. 1)

According to Noë, all perception is touch-like, insofar as it involves an active, exploratory process, rather than the passive receipt of information. Perception relies upon "knowledge" of "sensorimotor contingencies." In other words, it incorporates a tacit, non-conceptual appreciation of how various kinds of motor activity affect patterns of sensory stimulation.[2] For current purposes, I will assume that something along these lines is right: that most or even all perceptual achievements utilize stable relationships

[1] See, for example, O'Callaghan (2007).
[2] See also O'Regan and Noë (2001).

between kinds of activity and kinds of sensory change, and that active, tactual perception exemplifies this particularly well.[3]

On one interpretation of the claim that all perception is touch-like, the different perceptual modalities all share certain features, but these features are especially salient in the case of touch. So a good approach is to start by reflecting on the structure of touch and then generalize to the other senses. If this is right, then one could just as accurately state that touch is vision-like, at least when vision is characterized properly. Indeed, Gibson (1962, p. 477) observes that exploratory touch with the fingers is akin to eye movements, and goes so far as to say that "active touch can be termed *tactile scanning*, by analogy with ocular scanning." On the other hand, even if all sensory perception utilizes sensorimotor contingencies, this does not rule out the possibility that touch remains distinctive in some respect. Perhaps touch is an effective exemplar for enactivism because it depends upon bodily activity to a greater *extent* than the other senses and/or in a distinctive *way*. With this in mind, we might consider various claims to the effect that touch is the most fundamental sense, the only sense that is essential to the having of a body, to the capacity to interact with one's surroundings, to animal being (e.g., Jonas 1954; O'Shaughnessy 1989; Fulkerson 2014).

This chapter addresses the question of whether and how the relationship between touch, the body, and bodily activity might be distinctive. I begin by pointing out a complication for any claim that "touch is or involves *p*": it is unclear what touch is, and any generalizations concerning the nature of touch are susceptible to counterexamples, involving either cases of touch that do not involve *p* or non-tactual forms of perception that do. Even so, I argue that most uncontroversial instances of touch incorporate bodily activity in a way that is consistent with the sensorimotor contingency view. Following this, I consider a recent defense of the claim that tactual perception utilizes bodily exploration in a distinctive way and to a greater extent than the other externally directed senses. I raise some objections, and suggest that there are insufficient grounds for maintaining that touch is somehow "more enactive" than other senses. However, this leaves open the possibility that touch relates to the body in a unique way, in virtue of some other characteristic. Hence I go on to address other formulations of the claim that touch is both more fundamental than the other senses and more intimately bound up with our bodily nature. All of these turn out to be problematic. I conclude that the most plausible case for the primacy of touch involves an appeal to its diversity, rather than to any particular characteristic of touch.

[3] I do not go so far as to accept the view that sensorimotor contingencies are *sufficient* for tactual perception or for any other form of perception. The weaker claim I endorse here is that they are a necessary but not sufficient constituent of most—but perhaps not all—perceptual processes. My discussion does not require endorsement of various stronger claims made by O'Regan and Noë (2001) to the effect that their approach can accommodate all those aspects of perception that others have sought to account for in terms of representations, images, qualia, and so forth (although I confess that I would very much like to dispense with all talk of representations and qualia). In order to keep things manageable, I focus on this specific formulation of enactivism. However, many of the points I make are of relevance to other kinds of enactivist approach as well.

The Nature of Touch

Generalizations concerning what touch or "tactual perception" does or does not involve are invariably problematic, as it is unclear what counts as an instance of touch, what does not, and why. Philosophers have proposed various criteria for individuating the senses, most of which further develop one or more of the criteria proposed by Grice (1966): (1) what is perceived via a sense (e.g., color, rather than sound); (2) the distinctive introspectible quality of the sensory experience; (3) the kinds of stimuli detected; (4) the composition of sense organs and their connections to the brain.[4] More recently, Keeley (2002) has distinguished between the tasks of identifying commonsense criteria (which include but are not limited to the relevant phenomenology) and identifying those (non-phenomenological) criteria that are applicable in scientific inquiry. Turning to the latter, he proposes that each sense is receptive to a specific stimulus type (to be construed in terms of the kinds of energy identified by physics) and involves distinctive sense organs. In addition, the organism discriminates stimuli on the basis of a given form of energy without relying on additional means of detection, via mechanisms that are evolutionarily and developmentally dedicated to detecting those specific stimuli.[5]

Perhaps some combination of these or other criteria will succeed in individuating certain senses, but any such inventory is problematic for touch. There is no clearly bounded organ of touch. It utilizes exterior and interior surfaces of the organism, hair, various different types of sensory receptor, and, as I will make clear in the next section, a sense of bodily position and movement. Furthermore, touch is not receptive to a specific stimulus type. It is not only a "pressure sense." Certain tactual achievements in humans and other organisms depend more specifically upon the detection of vibration, which is not just pressure or change in pressure but a temporally extended, structured pattern of pressure changes (Katz 1925/1989). It can be added that tactual perception is inseparable from perception of hot and cold, and therefore reliant on at least two distinct forms of energy. Katz (1925/1989, p. 165) observes that temperature contributes to recognition of "almost every material," a point that applies equally to texture perception. For instance, cold surfaces generally feel smoother than warmer ones (Taylor, Lederman, and Gibson 1973, p. 260). The view that touch is inextricable from perception of hot/cold is also supported by neurobiological findings.[6] Some sensory neurons are receptive to heat

[4] In what follows, I refer to (1) in terms of the "content" of tactual experience. I use the term "content" in a noncommittal way, to mean whatever it is that one perceives through touch—a texture, an object, an occurrence. Although I do not deny that one could perceive something without experiencing it, my discussion of content is concerned more specifically with perceptual experience: I take the content of tactual perception to be whatever it is that we experience through touch. This use of the term "content" does not imply any commitment to the existence of perceptual "representations."

[5] For several other proposals for individuating the senses, see the essays collected in Macpherson (2011).

[6] I say "hot and cold" rather than "temperature" because different sensory receptors are responsive to heat and cold.

and to certain mechanical stimuli, as well as chemical irritants (Schepers and Ringkamp 2010, p. 181; Lumpkin and Caterina 2007).[7] Of course, one could respond by appealing to cases where hot/cold perception is absent but other aspects of touch remain. However, this does not give us sufficient grounds for maintaining that hot/cold perception is a separate sense, given that selective absences also arise within other established perceptual modalities. For example, loss of color vision does not amount to a complete loss of vision.

Even if perception of hot/cold were to be excluded from the sense of touch, its reliance on pressure/vibration does not serve to distinguish it from all other senses, as hearing relies on pressure changes in the inner ear.[8] Of course, one perceives sounds through hearing, rather than pressure changes. The sensory content of audition is thus distinct from the proximal stimulus upon which it depends. But this applies equally to certain instances of touch; what is perceived through touch need not be "physical contact with something" or "properties of the entity that one is in contact with." There are many plausible examples of distance touch.[9] In using a tool, the most salient content of tactual experience is often not the tool itself but something that is manipulated or investigated with the tool. And many other tactual experiences involve perceiving something *through* a medium of whatever thickness. When you touch something while wearing a glove, you perceive through the glove, rather than perceiving only the material that is in direct contact with your skin. This applies equally to passive touch; one can perceive bumps on the road while sitting on a moving bus. It might be objected that such cases involve perceiving both pressure/vibration and, via that, something else. However, I doubt that this is any more phenomenologically accurate for touch and pressure/vibration than it is for hearing and pressure/vibration or for vision and light. It is uncontroversial that the content of visual perception does not include photons hitting the retina and that the content of hearing does not include pressure changes in the ear. There are no grounds for imposing such a view on those tactual experiences for which it is no more phenomenologically plausible. A further problem for the pressure-sense view is posed by tactual experiences of absence or lack of pressure. For instance, Katz (1925/1989, p. 61) offers the example of running one's hand across a brush and perceiving not just the bristles themselves but also the space or "tactual ground" between the bristles.

[7] There is also the question of whether pain should be distinguished from tactual perception. Fulkerson (2014, p. 49) argues that pain is distinct from touch, on the basis that only the latter facilitates perception of entities outside of one's own body. For instance, when you have an injection, the pain "is not experienced as a property of the needle." But one could respond by suggesting that, although the pain is not itself experienced as a property of the needle, it does include an awareness of its being externally caused, of one's being hurt *by* something that is currently in contact with one's body. It also includes some appreciation of the nature of the stimulus, whether it is forcefully or non-forcefully applied, sharp, or blunt. However, although I think it is a mistake to separate hot/cold from touch, I am less convinced in the case of pain. At most, I would be inclined to say that some forms of pain are integral to some tactual experiences, or vice versa.

[8] See Gray (2013) for a recent defence of the view that the sense of touch should be distinguished from perception of heat and cold.

[9] See also Ratcliffe (2012) and Fulkerson (2014) for recent arguments in support of "distance touch."

Other examples include exploring a sculpture with one's hands and *feeling* empty spaces, experiencing a part of one's body as exposed when one's clothes become untucked, and feeling the absence of a wedding ring on one's hand, after taking it off to go swimming. So there is no straightforward relationship between the content of tactual perception and the detection of contact or pressure.

Such examples also render problematic any appeal to a distinctive content and/or an experiential quale or "what-it-is-likeness" of touch (where the latter requires acceptance of the position that a perceptual experience is not exhausted by its content). The contents of tactual perception are notably diverse. They include objects, textures, things coming into contact with the body, things that are not in contact with one's body, the absence of something, static properties such as shape, and dynamic properties such as patterns of movement. To add to this, it is far from clear what the limits of tactual perceptual content are. When one strokes the hair of one's child as he drifts into sleep, does one perceive a texture, a number of fine strands, hair, a roughly spherical entity with hair attached, a warm, animate body, a person, or a particular person?[10] Some descriptions suggest that the contents of tactual experience can be both extremely diverse and highly detailed. Consider this description that Helen Keller offers of her unusually refined sense of touch:

> The thousand soft voices of the earth have truly found their way to me—the small rustle in the tufts of grass, the silky swish of leaves, the buzz of insects, the hum of bees in blossoms I have plucked, the flutter of a bird's wings after his bath, and the slender rippling vibration of water running over pebbles. . . . I have endlessly varied, instructive contacts with all the world, with life, with the atmosphere whose radiant activity enfolds us all. The thrilling energy of the all-encasing air is warm and rapturous. Heat waves and sound waves play upon my face in infinite variety and combination, until I am able to surmise what must be the myriad sounds that my senseless ears have not heard. (Keller 1908/2003, pp. 40–41)

As for a distinctive experiential quality, it is thoroughly unclear what this might consist of, given the diversity of tactual experience. Perhaps, one might suggest, touch implicates the body in a distinctive way, generating a tactual "feel" that is not exhausted by perceptual content. But touch involves a wide range of different bodily experiences, involving differing degrees of bodily awareness. Sometimes, one's body or part of it is a conspicuous aspect of the experience but, in other cases, the body is arguably no more conspicuous than it is in visual perception (Ratcliffe 2012; Fulkerson 2014).[11] More

[10] To further complicate things, interpersonal touch is often communicative, and patterns of communicative tactual interaction with another person can feed into the task of perceiving that person through touch. It is unclear whether or how the communicative and perceptual functions of touch are to be separated.

[11] As Katz (1925/1989) puts it, the "object" and "subject" poles of touch vary in salience, while in the other senses they are more stable. Martin (1992, p. 204) conceives of a tactual experience as a "single state of mind, which can be attended to in different ways." In other words, there is a single, unified experience, with both bodily and non-bodily aspects that can differ in their relative prominence. While I think this

generally, touch has all manner of "feels," regardless of whether or not some of these feels can, in the way that Grice suggests, be distinguished from perceptual content.

For these same reasons, an appeal to sensorimotor contingencies will not suffice to individuate touch. O'Regan and Noë (2001, p. 941) maintain that the senses are to be distinguished by the characteristic types of sensorimotor pattern they utilize: "what *does* differentiate vision from, say, audition or touch, is the *structure of the rules* governing the sensory changes produced by various motor actions, that is, what we call the *sensorimotor contingencies* governing visual exploration." However, tactual perception employs the whole body, along with its many and varied capacities for activity. It therefore encompasses a wide range of exploratory strategies, which can involve any number of sensorimotor patterns. Consider the differences between exploring the location and intensity of a heat source from a distance, discriminating textures (which, for some types of texture, can be achieved either through passive receptivity to something moving across one's skin or through active tactual exploration) and perceiving the shape of an object (which depends more specifically upon active, exploratory touch). Of course, one could say that all of the sensorimotor associations implicated in touch can be distinguished from the "rules" that govern, for instance, visual perception. Then again, they also differ from each other, and so the question arises as to why they should be grouped together as a single sense, while those that feature in vision are excluded.

One might respond that tactual discrimination of objects and properties *does* utilize bodily activity in a sense-specific way. Touch not only involves moving in order to perceive; one also manipulates entities in a number of structured ways, so as to obtain specific types of sensory information. An account along these lines is proposed by Fulkerson (2014), who suggests that what distinguishes a unitary sense from a combination of senses working together is that a sense "binds" various "features" into unified perceptual "objects." Tactual perception, he says, achieves this in a unique fashion; it is the only sense that involves *exploratory binding*. I will further discuss Fulkerson's proposal in the next section, in addressing whether some forms of touch are "more enactive" than other forms of sensory perception. But, for now, I want to note that this approach does not succeed in distinguishing touch from other forms of perception. Fulkerson (2014, p. 33) proposes that "haptic perception is unified in virtue of the fact that all of its constituent physiological systems work together to assign sensory features to the same set of objects." However, this clearly does not apply to all uncontroversial cases of touch. Active and passive touch both involve variably sophisticated degrees of discrimination, and do not always facilitate unified perception of an "object." So the claim has to be that some cases of touch involve p, rather than that all cases of touch involve p. The question therefore arises as to why those instances or types of touch that do not involve p still qualify as "touch." Their identification as such rests upon an implicit

is right, it is not specific enough. One might take the line that the *content* of tactual experience has two different aspects. However, although one's body and parts of it certainly can feature as contents of tactual perception, it would be wrong to restrict the role of the body to perceptual content. The body is also that *through* which we perceive things tactually; it occupies the role of perceiver (Ratcliffe 2012).

or explicit understanding of what touch is, which is independent of and presupposed by the exploratory feature-binding account.

Touch, Movement, and Exploration

To some extent, the problem of individuating touch can be circumvented by addressing what a wide range of uncontroversial cases involve, or by making claims about a specific subset of tactual perceptions. For the most part, it is clear that touch cannot be extricated from a sense of bodily position and movement. "Tactile" perception is sometimes distinguished from "tactual" or "haptic" perception, where the former relies only upon sensory receptors in the skin, while the latter also includes proprioception and kinesthesia. It is clear, I think, that the sense of touch should be construed as tactual rather than just tactile. It is uncontroversial that we rely upon patterns of bodily activity in order to perceive various different properties through touch. Certain textures can be perceived either by moving against a surface or by that surface moving against one's body. Here, only a weak claim applies: some *token* perceptions of that texture depend upon bodily activity, although a texture of that *type* can also be perceived in other ways. However, other *types* of property, such as the shape of a three-dimensional object, are (other than in a few simple cases) only perceivable through exploratory touch. Now, the mere observation that touch depends upon bodily activity is compatible with a model where "tactile" and "bodily" information are independently obtained and only subsequently combined so as to facilitate more sophisticated perceptual discriminations. If that were right, it would be debatable whether touch is to be identified with the initial tactile achievement or with the haptic/tactual achievement that follows. But it is not right, given that self-initiated movements also shape sensory processes and regulate incoming sensory signals. Consider the example of self-tickling. Most people are unable to tickle themselves, as the perception that arises due to self-stimulation is somehow different, less intense. This difference between self- and other-induced perceptions is not attributable to information about self-produced bodily movement modifying an already given pattern of sensory activity. Rather, the initiation of movement attenuates peripheral sensory signals, and it is not that all incoming sensory signals are affected in the same way; initiation of specific actions modifies "precise sensory signals" (Blakemore, Wolpert, and Frith 2000). The point is not specific to self-touch and applies to touch more generally: sensory processes are regulated in a non-conscious way (and perhaps a conscious way too) by anticipated and actual bodily activities. Indeed, there is evidence to suggest that the same perceptual changes can be brought about either through changes in the pattern of sensory stimulation or through bodily activity; the two variables are "in principle fully interchangeable" (Saig et al. 2012, p. 14030).[12]

[12] What is perhaps not so clear is *when* actual and anticipated motor activities lead to sensory attenuation. On phenomenological grounds, it is arguable that the converse can equally occur. If you

It might be objected that this applies only to active touch. Furthermore, passive touch—which is not reliant upon bodily activity—can similarly facilitate sophisticated and fine-grained perceptions.[13] Fulkerson (2014) offers the examples of feeling a bug crawl along your arm and feeling your pet cat jump upon and then walk across the bed you are lying in. It is debatable whether such percepts are exclusively tactual in origin though. Try pressing a coin against your palm with your eyes open, and then again with your eyes closed. In the first case, it might seem that you perceive a clearly bounded circular shape through touch, unaided by vision, but the perception is less clearly defined when your eyes are closed. However, regardless of how discriminating passive touch might turn out to be, I do not think it poses a problem for the view that touch is inextricable from a sense of one's body and what it is doing. Both passive and active forms of touch involve proprioception; percepts generally include some sense of bodily location. As for movement, if one lacked all sense of the distinction between self-initiated movement and being acted upon by an external force, it is not that active touch would be lost and passive touch retained. Passive touch involves the sense that a percept is *not* attributable to self-initiated movement or, where something unexpected brushes against your hand as you reach for an object, that it is not an anticipated consequence of one's movements. In the absence of any ability to distinguish self-initiated from other kinds of movement, one would be just as unable to experience passive touch. Any remaining experience would be indeterminate, neither active nor passive. The discriminative power of both active and passive touch relies upon a sense of what, if anything, one is doing relative to a perceived entity and whether or not any movements are self-initiated. In both cases, the ability to perceive would be dramatically curtailed by a severing of the link between sensory stimulation and information concerning the presence or absence of self-produced movement.

Hence it would seem that the structure of tactual perception is consistent with the sensorimotor contingency view. What one perceives is dependent upon how changing patterns of stimulation relate to bodily activity (or to a lack of bodily activity), rather than the fusion of two prior percepts. The same pattern of sensory stimulation could result in two quite different percepts, depending on whether and how one is moving, and whether a movement is self-initiated. Nevertheless, it is questionable whether this

stick your hand into a hole, expecting to encounter something slimy and unpleasant, the subsequent tactual encounter with gunge remains a conspicuous one, perhaps more so than if it had not been anticipated it at all. In contrast, if you hold out your hand to receive some coins, the perception of their falling onto your palm is less salient than when something is dropped onto it in the absence of any expectation (an example that straddles the distinction between active and passive touch). Hence conscious anticipation is sometimes associated with attenuation and sometimes not, in a way that is closely related to the affective content of tactual experience. It is also unclear whether attenuation is specific to active touch. It seems plausible to maintain that two instances of passive touch involving the same patterns of sensory stimulation could differ considerably, depending on whether or not something is anticipated and whether or not it is welcome.

[13] Hence it is unhelpful to state, as O'Regan and Noë (2001, p. 939) do, that perceiving is a "way of acting"; this obscures the differences between active and passive forms of touch.

applies to all instances of touch. Take the case of heat perception. When one reaches out with one's hand to explore the proximity and temperature of a nearby heat source, the relevant information is indeed extracted from how patterns of stimulation change in response to bodily movements. What, though, should we say about the simple feeling that "I am hot" or "It is hot in this room"? Even if heat perception is ruled out as an instance of touch, many other tactual experiences similarly involve immersion in a diffuse medium. Take the example of swimming underwater and, as one does so, exploring the textures of marine plants with one's hands. Here, it seems right to say that one retains some kind of tactual awareness of the water surrounding one's body. But the water is a tactual background in the context of which one has localized tactual experiences of the plants, rather than something one "encounters" as an object of active or passive touch. And it is not clear that all-over bodily experiences of immersion in a medium can be satisfactorily accounted for in terms of sensorimotor couplings. Such media might be better conceived of as contexts within which sensorimotor couplings operate, media that are—to some extent—compensated for by sensorimotor processes.

We should also consider the possibility that touch is somehow "more enactive" than other senses. So far as I know, the clearest and most sophisticated statement of that view is Fulkerson's (2014) account of exploratory feature-binding. Fulkerson maintains that the senses bind "features" into "objects," and that this distinguishes them from multisensory phenomena, which involve associations between already bound percepts. Drawing on work by Susan Lederman and colleagues, Fulkerson notes that active touch involves a range of "exploratory procedures," each of which is specific to certain kinds of property (see, e.g., Klatsky and Lederman 2002). Some such procedures, he proposes, facilitate "exploratory binding." While the other senses involve feature-binding, they do not rely upon manipulating the environment in the way that touch does. Hence at least some tactual achievements are distinctive, as perceptual coherence is achieved by acting upon things in specific, temporally extended, and often complicated ways. We might therefore say that touch is the most enactive of the senses.[14]

My first worry about this proposal is that the nature of "features" and "objects" is not made sufficiently clear. Fulkerson does not limit his use of the term "object" to three-dimensional, temporally enduring entities, and neither is he referring simply to "objects of perception," meaning "whatever it is that one perceives"; unbound features and objects can equally be "objects of perception." Within the category of "objects," Fulkerson includes spatiotemporal objects, sounds, and also smells: "In vision, a set of unique features—including color, shape, texture, and motion—are all predicated, or bound, to visual objects. In addition, a range of auditory features are assigned to individual auditory objects, typically thought to be sounds. In olfaction, features or

[14] Fulkerson's discussion of touch is not principally concerned with enactivism, and he expresses reservations about the general applicability of a sensorimotor contingency approach. However, those reservations principally concern whether the approach can supply an account of perception that allows us to dispense with any appeal to representation or imagery, rather than whether sensorimotor contingencies are at play at all.

qualities are predicated to odors" (2014, p. 35). However, it is not clear to me that the unified "objects" of hearing are restricted to "sounds." As a motorbike roars past, does one simply hear a roaring sound or does one hear the "sound of something moving past" or, even more specifically, the "sound of a passing motorbike"? If we at least sometimes hear "the sound of *p*," then it is arguable that sounds are features, of a kind that are sometimes bound to objects by audition and sometimes not. It is similarly unclear why smells should be objects. Suppose I encounter a "cheesy smell." It is not that I experience a particular object, which just happens to have cheesiness as one of its contingent properties. If the cheesiness is subtracted from the smell, then I do not experience that smell at all. The smell is inseparable from its cheesiness; it is a "smell *of* cheesiness." This is quite different from perceiving a red cup, where the cup could equally have been blue rather than red. But let us suppose, for the sake of argument, that smells can be decomposed into features. Their status as objects would still be in question. Consider textures, which Fulkerson takes to be features of tactual objects, rather than objects in their own right. As noted by Taylor, Lederman, and Gibson (1973, p. 252), touch is receptive to properties such as "temperature, hardness, roughness, elasticity, stickiness, slipperiness, rubberiness," and so forth, which are "together perceived as texture." So, if texture is a feature rather than an object, then it is a feature assembled out of other features. If features can have features, then the possibility of decomposing a smell into component features does not suffice to make it an object rather than a feature.

Without a clearer account of what features and objects actually are, it is difficult to rule out various other candidates for exploratory feature-binding, which are either not exclusively tactual or not tactual at all. Take the experience of moving a piece of chocolate around in one's mouth, manipulating it with one's tongue, and sampling its taste, texture, and smell. This seems to involve a unified perceptual object, to which taste, texture, and smell are equally bound. Much the same can be said of the relationship between certain smells and what emits them. Suppose I smell something nearby and explore my surroundings to seek out the source. I eventually pick up a rotten apple from the floor and sniff it, while moving it toward and then away from my nose. At some point during this process, I come to experience the smell as a property or "feature" of the apple that I see, rather than a separate olfactory percept that is associated with a visual percept. Exploratory strategies are similarly implicated in auditory perception. In shaking a ball to discern its contents and repeatedly hearing a ringing sound, one experiences the sound *of* the ball or of something inside it. Both smelling the apple and hearing the ball involve assigning properties to objects by means of exploratory procedures, and it is unclear that the resulting percepts are any less unified than the objects of touch. One might respond that the two were experienced as separate to begin with and therefore remain separable. However, this applies equally to tactual exploration. Simple, brief movements seldom suffice to unify percepts in the way that Fulkerson describes, and the features that exploratory procedures bind together may start off as separate tactual percepts. Hence, given the way in which Fulkerson distinguishes features from objects, it is unclear that feature-binding in touch differs from various multisensory achievements.

Of course, one could maintain that combinations of other senses only *occasionally* bind features into objects, but again that applies to touch as well. According to Fulkerson, the unified "objects" of exploratory touch are three-dimensional objects in the familiar sense of the term (although it is debatable how fine-grained our tactual object-recognition capacities are). However, even in humans, touch is not principally an object-recognition sense. For the most part, it involves the detection of "features," such as textures. It can be added that tactual object recognition via touch may rely upon other senses as well. Cardini et al. (2010) report that whether someone feels a stimulus touch one or both of her cheeks can be influenced by showing her photographs of a face being touched in one or the other way, an effect that is most pronounced when the face she sees is her own. And Katz (1925/1989) observes that the more general ability to make fine-grained textural discriminations often relies on vision, even where it might seem that differences are detected through touch: "The dissimilarity of the visual impressions can make the tactual impressions appear different to some extent, even when the hand alone notices no differences, and one thus must speak of the strong effects of visual associations" (pp. 136–7). He further suggests that, where visual input is absent, it may remain the case that visual imagery is generated, and that this increases the specificity of experienced content. However, if it is at least accepted that not *all* tactual object recognition is multisensory, perhaps it is plausible to maintain that touch is distinctive in being the only *unitary* sense to employ exploratory feature-binding. Then again, in the absence of other criteria to specify why touch indeed is a unitary sense, exploratory feature-binding could just as well be taken to cast doubt upon its unity. And, even if it is granted that touch is a single sense while all other cases of exploratory feature-binding are multisensory, it would remain the case that exploratory feature-binding is not unique to touch (unless, that is, the concepts of "feature" and "object" can be refined so as to exclude all the multisensory cases while retaining the tactual ones).

Hence I am skeptical of the possibility that touch involves exploratory activity in a way that is *qualitatively* different from other forms of sensory perception. Nevertheless, it is plausible to maintain that there is a *quantitative* difference. Indeed, touch may well encompass *more* exploratory strategies than the other senses combined, or perhaps a *more diverse range* of strategies (although, to fully develop this position, one would need to supply criteria for distinguishing and comparing types of exploratory strategy). But, even if this is so, it is wholly symptomatic of the fact that the "organ" of touch encompasses the motor capacities of the entire body. The other senses rely on more localized structures that occupy fixed bodily locations, thus constraining the kinds of sensorimotor pattern they can take advantage of. As Gibson (1968, p. 42) remarks, "the eyeball is 'all of a piece', but it is an unusual sense organ," while the "organ of touch is dispersed over the whole body." It can be added that the hands alone have a degree of dexterity and flexibility that facilitates all manner of specialized exploratory procedures.[15] Even so, none of this implies a qualitatively distinct relationship between

[15] See Radman (2013) for several relevant discussions of the hand and its capacities.

perception and bodily activity. It is just that other senses rely on fewer and perhaps more neatly circumscribed sensorimotor patterns than touch.

The Primacy of Touch

Might there remain something distinctive about touch, something that I have so far failed to acknowledge? Independently of recent enactivist literature, various authors have claimed that touch is more intimately associated with the body than other senses, and uniquely indispensable to our engagement with the world. It is thus said to be our primary, most primordial, or most fundamental sense. Several subtly different primacy claims are to be distinguished, most of which are questionable and/or insufficiently clear. Sometimes, emphasis is placed upon the unique contribution of touch to our appreciation of force, contact, pressure, or causation, something that is essential to our sense of being situated in a world as one body among many. For instance, Jonas (1954, p. 507, p. 516) details the many virtues of sight, but adds that the "objectivity" gained via sight comes at the cost of "causal connection," something that does not feature in visual perceptual content. He adds that touch has a kind of primacy among the senses, as it is the only sense "in which the perception of quality is normally blended with the experience of force." As such, it is the sense through which "the original encounter with reality as reality takes place"; "the true test of reality." Of course, if one conceives of vision as an active, exploratory process, the disanalogy no longer holds; seeing does not involve a self-contained subject passively surveying a self-contained object in the way Jonas describes. Even so, it seems plausible to suggest that vision does not incorporate force perception. Then again, vision does encompass certain experiences of acting and at the same time being acted upon, which could equally be construed in terms of an "encounter with reality." In interpersonal contexts, one's gaze upon another is perceived to have a certain effect, and vice versa. We can add that tactual experiences of interaction and reciprocity are not invariably reliant upon physical contact with the relevant object, or upon manipulating it. It is unclear why those forms of touch that do involve force perception should take pride of place over other ways of encountering and interacting with reality, unless it is simply assumed from the outset that force is central to our grasp of the real.

Closely related to the "force" view is the claim that our concept of causation originates in touch, a view that Jonas also seems to subscribe to. More recently, Wolff and Shepard (2013) have offered a detailed statement and defense of this position, maintaining both that our concept of causation originates in an appreciation of force, and that we obtain our appreciation of force through touch.[16] It is questionable, however, whether our

[16] Interestingly, Wolff and Shepard (2013) suggest that the evidence supports a patient- rather than agent-oriented account of force perception. Hence, on their view, the primacy of touch relates to our being acted upon by our surroundings, more so than our acting upon them.

concept of causation originates exclusively in an experience of force. Plenty of other experiences involve a sense of interaction and reciprocity: I shut my eyes to avoid the light of the sun; I hear a loud noise that hurts my ears; and so forth. And do we not detect causation as we enter a hot room and gradually become uncomfortably hot ourselves? Furthermore, perceptions of force such as being shoved hard by somebody are not exclusively tactual in nature, and neither is it clear that they are *primarily* tactual. One equally hears a noise, experiences a sudden change in the pattern of visual sensation, and loses one's sense of balance. In the absence of changes in the pattern of visual stimulation that are consistent with being suddenly moved by an outside agency, would one still have the same, unambiguous experience of being acted upon? It is also debatable what our concept of causation actually consists of, whether it is homogeneous, and whether all typical people operate with the same concept. So I am doubtful of the claim that an appreciation of causation depends exclusively or primarily on a subset of tactual experiences that involve force perception.

Moreover, I question the assumption that tactual experiences, of the kind that do involve physical contact with an object of perception, incorporate the simple perception of "force" or—as others have suggested—"pressure."[17] I think that "perception of pressure through touch" is an abstraction from a diverse range of tactual experiences that are much richer in structure, most of which do not incorporate an isolable experience of "pressure." Many tactual experiences of physical contact are affectively charged. One feels the pleasantness of a caress, rather than first feeling the caress and then having an emotional response to it. This observation is consistent with the finding that certain parts of the body have sensory receptors that are specifically dedicated to pleasant touch (e.g., Olausson et al. 2010; Farmer and Tsakiris 2013). However, tactual experiences are not simply more or less pleasant than one another. Consider the experience of squeezing a large, warm piece of rubber, an experience that might well be pleasant in some way. Now suppose that the patterns of pressure and tactual stimulation are identical to those involved in squeezing one's naked spouse. The two perceptual experiences would still be quite different in character, and in a way that is not simply attributable to one's being more pleasant than the other. A range of different affective tendencies and patterns of anticipation serve to shape the relevant experiences, and the fact that both involve physical pressure gives us no grounds for maintaining that both actually include the same "experience of pressure." The fact that a perceptual experience depends upon an object of perception exerting physical pressure on one's body does not imply that the content of the experience includes pressure per se. I grant that certain perceptual contents may approximate a decontextualized, affectless perception of pressure more closely than others. But on what grounds should we regard these as more fundamental to our grasp of reality than all those other tactual experiences that together encompass various different *significant* relationships and interactions with the surrounding world? As noted earlier, it is

[17] The terms "force" and "pressure" do not mean the same thing. However, for current purposes they are interchangeable, as the concerns I raise apply to both.

also arguable that touch is a vibration sense, rather than just a pressure sense. This further distances us from the view that the kind of reciprocity with the environment that it facilitates is most centrally a matter of feeling and exerting pressure.

Other primacy claims place more emphasis on the inextricability of touch from the having of a human or animal body. For instance, O'Shaughnessy (1989, pp. 37–8) states that touch is "essential to the animal condition as such, something which seems to follow from the fact that tactile-sense is broad enough to overlap with the sheer capacity for physical action on the part of its owner. . . . In touch we directly appeal to the tactile properties of our own bodies in investigating the self-same tactile properties of other bodies."[18] However, the intuitive appeal of such claims rests upon a failure to distinguish three things: physical contact; a sense that utilizes physical contact; and a sense of physical contact. It is trivially true that being a part of the physical world involves coming into physical contact with other things. But it does not follow from this that the perceptual capacities of an animal body *must* be sensitive to physical contact. Martin (1993, p. 211) acknowledges the possibility of a "hypothetical jellyfish" that has bodily sensations but does not distinguish its own body from its surroundings. We can add that such a creature need not have sensory receptors on or near its external surfaces. Even so, it could still have an internally located perceptual system of sorts, one that responded to different kinds of stimuli with different motor activities. Candidate stimuli include magnetic fields, acidity, temperature, and chemical signatures that indicate the presence or absence of nutrition sources. So having a body and interacting with one's environment do not require perception of physical contact or even a sense that utilizes physical content with organismic boundaries. One might respond that hypothetical jellyfish perception of this kind still counts a form of touch. However, if one were to accept that, one's inclusion criteria would have to be so permissive as to include every other kind of actual and conceivable sensory achievement. "Touch" would become synonymous with "any way of sensing any property of one's environment." So the animality claim is either false or it is trivially true and uninformative.

Even if we restrict ourselves to the case of human touch, there are insufficient grounds for thinking that touch has a kind of priority, in virtue of its facilitating a sense of confrontation between one's body and another body. Again, it is important to stress the distinction between relying on physical contact in order to sense something and having a sense *of* physical contact. (By analogy, vision is reliant on photons coming into contact with the retina, but does not include a sense of something coming into contact with the retina.) In fact, many tactual experiences do not involve a clear sense of something as distinct from one's body. Ihde (1983, pp. 96–7) offers the example of sitting on a couch and having a diffuse, non-localized, all-over feeling of comfort, which involves a blurring of the boundaries between one's body and its proximal environment. More generally, touch contributes to the sense of being immersed in a medium

[18] The view that touch is essential to animality can be traced back to Aristotle. See Freeland (1992) for a good discussion.

of one or another kind, something that need not involve clear differentiation between one's body and its surroundings. This, Ihde suggests, is illustrated by the contrast with those experiences that do involve a pronounced sense of the distinction between one's body and its surroundings. Immediately after one has jumped into a cold lake, one experiences a contrast between oneself and a surrounding medium that is more usually lacking.[19] Another example, discussed by Gibson (1962, p. 480), is the "feeling of cutaneous contact with the earth" as we stand and walk. This is clearly a tactual experience of sorts, but one that is distinct from both active and passive touch; it is "a means of registering the stable environment with reference to which *both* the movements of one's body and the motions of objects occur." Hence some forms of touch constitute or at least contribute to a wider phenomenological context, within which more localized passive and active contact experiences arise (Ratcliffe 2013). Why should a subset of these localized experiences, rather than a variable sense of being "immersed" or "grounded," be most fundamental to our grasp of body, world, and/or the relationship between them? So far as I can see, no argument has been offered. Furthermore, it would be problematic to maintain that touch is primordial because it constitutes a variably differentiated perceptual ground and, at the same time, because it facilitates localized experiences of contact and difference. These are quite different kinds of perceptual achievement, and the acceptance of both as fundamental would suggest that the "perceptual encounter between body and world" is multifaceted, rather than grounded in some singular accomplishment that is unique to a form of touch. Once that much is conceded, a diverse range of other tactual and non-tactual perceptual achievements could just as well be added to the list.[20]

The diversity of touch also poses problems for the view that touch is somehow constitutive of bodily experience in a way that other senses are not. Katz (1925/1989, pp. 126–7) considers the experience of self-touching and remarks that "the mutual perception of two body parts having the same sensory organs is unique; there is nothing like it outside of the skin sense. . . . The moving part of the body feels the motionless part as object." He also addresses how the "subject" and "object" poles can reverse during self-touch; as one explores one's own body tactually, different parts of the body alternate between the roles of perceiver and perceived. In a similar vein, Husserl (1952/1989, §2, ch. 3) maintains

[19] In this example, the experience of contrast is largely attributable to perception of cold. Hence one might seek to further support the view that touch involves undifferentiating immersion in a medium by distinguishing tactual perception from perception of hot/cold. I have argued that hot/cold perception is more plausibly regarded as integral to touch. Even so, this remains compatible with the view that tactual perception only occasionally involves a salient contrast between one's body as a whole and one's surroundings.

[20] One might further speculate that the sense of coming into contact with something distinct from oneself and of the boundary between the two is partly attributable to vision. If tactual experiences involve varying degrees of differentiation and—as suggested earlier—visual input shapes tactual perception, it could be that, in some of those cases where there is a sharp distinction between perceiver and perceived, that distinction is partly attributable to seeing two clearly distinct, clearly bounded entities come into contact with each other.

that touch (and only touch) is indispensable to the experience of having a body; the body (*Leib*) is "constituted originarily only in tactuality" (p. 158).[21] However, the question again arises as to which tactual achievements are fundamental to human bodily experience and which are not. Rather than appealing to the case of one hand touching the other, where the two hands are experienced as distinct and as occupying reversible perceptual roles, we could just as well prioritize experiences such as rubbing one's palms together and experiencing a unified tactual percept rather than separate, albeit interdependent perceptions of each hand (Ratcliffe 2013). So my general worry about all these proposals is that they take the form "a certain kind of touch, *p*, is primary in virtue of *x*," when one could just as well say "a certain form of touch *q*, is primary in virtue of *y*." No independent grounds are supplied for construing our bodily experience or our relationship with the world as principally a matter of *x* and not of *y*. It is either asserted or assumed, often amounting to nothing more than a hazy intuition. Furthermore, in some cases, it is not clear whether *x* is even integral to the relevant tactual achievements, as in claims concerning force or pressure.

Hence the sheer diversity of touch renders general priority claims on behalf of touch either unclear or implausible, and there are also insufficient grounds for giving pride of place to any one subset of tactual achievements. Nevertheless, I think this adds up to a compelling reason for accepting a different formulation of the view that touch is more fundamental to human life (and, indeed, to most animal life) than the other senses. In short, touch accommodates so much that if *every one* of those sensory achievements encompassed by it were absent, a human life would be unsustainable.[22] This does not apply to the other established senses. Taylor, Lederman, and Gibson (1973, p. 270) suggest that:

> The "reality" of the touch sensation may possibly be related to the multimodal nature of touch. A thing seen and heard is more "real" than is the disembodied voice of a singer heard on a stereo system. A thing touched may be at once sensed as a vibrating object, a warm one, and a hard one. . . . By exploring freely, one obtains a succession of independent chunks of information about the object, such as could only in very unlikely circumstances have been produced by anything other than a real object.

I think this is the right approach to take. If touch is the "test of reality," it is because touch encompasses so many different ways of encountering an entity, which relate to each other and to the deliverances of the other senses in a cohesive way. Similarly, if touch is central to our appreciation of force, pressure, or causation, it is because touch

[21] Merleau-Ponty's various discussions of the perceptual reversibility between subject and object, and of its profound phenomenological implications, are indebted to both Katz and Husserl (e.g., Merleau-Ponty 1945/1962, 1968).

[22] This is not to deny that people can and sometimes do suffer very substantial impairments of the sense of touch. See, for example, Cole (1991, 2004) for interesting discussions. However, I know of no recorded case where a living person who is able to communicate has suffered complete loss of all aspects of the sense of touch throughout the entire body.

provides us with a wide range of perceptual encounters to abstract from, rather than a singular experience of collision. And, if touch is inextricable from bodily experience to an extent that the other senses are not, it is because touch includes many different ways of experiencing one's body, in relation to itself, to the surrounding world, and to other people. The ubiquity and indispensability of touch becomes even more apparent once it is acknowledged that what we perceive through one sense includes an appreciation of what *could* be perceived by means of other senses. A visually perceived cup looks graspable; a surface looks smooth to the touch (Husserl 1952/1989; Merleau-Ponty 1945/1962; Noë 2004; Ratcliffe 2008, 2015). Tactual possibilities permeate all experience. For instance, they are integral to the visual appreciation of something as here, now, in close proximity to oneself. If a visually perceived entity did not offer various kinds of tactual possibility, it would appear strangely distant, not quite there, somehow lacking (Ratcliffe 2013). Furthermore, touch not only gives us multiple actual and potential entry points onto a given object—it offers multiple ways of engaging with a wider, cohesive situation, in the context of which more localized perceptions arise. It is not that one nudges an object and—through some localized "reality quale," feeling of pressure, force, or solidity, or anything else that might be claimed to characterize the feel or content of tactual perception—mysteriously comes to grasp the reality of things. Such encounters arise within a wider intersensory situation, involving both a variably differentiated sense of immersion in one's environment and a coherent framework of actual and anticipated sensory experiences, many of which are tactual in nature.

So, in a way, touch is the "test of reality," but only because it encompasses so much and therefore has an unfair advantage over the other senses. It is not the case that some form of touch involves a singularly intimate relationship with the world, perhaps in virtue of its being "embodied" or "enactive" in a distinctive manner. Given the heterogeneity of touch, the question arises as to whether it should be regarded as a single sense at all, and Taylor, Lederman, and Gibson (1973) attribute its diversity to its multimodality. One option is to adopt a "wastebasket" model of touch: touch includes everything that is left over, once we have classified those sensory achievements that are notably circumscribed with respect to sense organs, stimuli, contents, and associated bodily activities. But I think that would be premature. Even if no criteria suffice to isolate all instances of tactual perception and only instances of tactual perception, it is plausible to maintain that what we call "touch" has a degree of unity comparable to the other established senses (Ratcliffe 2012). So any attempt to dismember touch might turn out to be just as applicable to them (Fulkerson 2014). Perhaps the most pressing issue to address concerns the nature and extent of multimodality. It could be that intersensory commerce and interdependence are so extensive that the senses should not be "individuated" at all. By analogy, "touch" itself encompasses various different kinds of sensory achievement, which are not usually categorized as separate senses. Given this, I will phrase my conclusion in a way that allows for an agnostic stance toward the "sense of touch": *what we currently refer to as touch* has a kind of primacy over the other senses, in virtue of its accommodating so much, but it would be a mistake to further insist that touch has this primacy in virtue of any singular characteristic.

Acknowledgments

Thanks to Matthew Fulkerson and an anonymous referee for commenting on an earlier version of this chapter.

References

Blakemore, S.-J., Wolpert, D., and Frith, C. (2000). Why can't you tickle yourself? *Neuroreport*, 11(11), r11–16.

Cardini, F., Costantini, M., Galati, G., Romani, G.L., Làdavas, E., and Serino, A. (2010). Viewing one's own face being touched modulates tactile perception: an fMRI study. *Journal of Cognitive Neuroscience*, 23, 503–13.

Cole, J. (1991). *Pride and the daily marathon*. London: Gerald Duckworth.

Cole, J. (2004). *Still lives: narratives of spinal cord injury*. Cambridge, MA: MIT Press.

Farmer, H. and Tsakiris, M. (2013). Touching hands: a neurocognitive review of intersubjective touch. In: Z. Radman (ed.), *The hand, an organ of the mind*. Cambridge, MA: MIT Press, pp. 103–30.

Freeland, C. (1992). Aristotle on the sense of touch. In: M.C. Nussbaum and A.O. Rorty (eds.), *Essays on Aristotle's De Anima*. Oxford: Oxford University Press, pp. 227–48.

Fulkerson, M. (2014). *The first sense: a philosophical study of human touch*. Cambridge, MA: MIT Press.

Gibson, J.J. (1962). Observations on active touch. *Psychological Review*, 69, 477–91.

Gibson, J.J. (1968). *The senses considered as perceptual systems*. London: George Allen & Unwin Ltd.

Gray, R. (2013). What do our experiences of heat and cold represent? *Philosophical Studies*, 166, s131–s151.

Grice, H.P. (1966). *Some remarks about the senses*. In: R.J. Butler (ed.), *Analytic philosophy: first series*. Oxford: Blackwell, pp. 133–53.

Husserl, E. (1952/1989). *Ideas pertaining to a pure phenomenology and to a phenomenological philosophy: second book* (trans. R. Rojcewicz and A. Schuwer). Dordrecht: Kluwer.

Ihde, D. (1983). *Sense and significance*. Atlantic Highlands, NJ: Humanities Press.

Jonas, H. (1954). The nobility of sight. *Philosophy and Phenomenological Research*, 14, 507–19.

Katz, D. (1925/1989). *The world of touch* (trans. L.E. Krueger). London: Lawrence Erlbaum Associates.

Keeley, B.L. (2002). Making sense of the senses: individuating modalities in other animals. *The Journal of Philosophy*, 99, 5–28.

Keller, H. (1908/2003). *The world I live in*. New York: New York Review of Books.

Kladsky, R.L. and Lederman, S.J. (2002). Haptic perception. In: L. Nadel (ed.), *Encyclopedia of cognitive science*. Basingstoke, UK: Palgrave Macmillan, pp. 508–12.

Lumpkin, E.A. and Caterina, M.J. (2007). Mechanisms of sensory transduction in the skin. *Nature*, 445, 858–65.

Macpherson, F. (ed.) (2011). *The senses: classic and contemporary philosophical perspectives*. Oxford: Oxford University Press.

Martin, M. (1992). Sight and touch. In T. Crane (ed.), *The contents of experience*. Cambridge: Cambridge University Press, pp. 196–215.

Martin, M. (1993). Sense modalities and spatial properties. In: N. Eilan, R. McCarthy, and B. Brewer (eds.), *Spatial representation: problems in philosophy and psychology*. Oxford: Blackwell, pp. 206–18.

Merleau-Ponty, M. (1945/1962). *Phenomenology of perception* (trans. C. Smith). London: Routledge.

Merleau-Ponty, M. (1968). *The visible and the invisible* (trans. A. Lingis). Evanston, IL: Northwestern University Press.

Noë, A, (2004). *Action in perception*. Cambridge, MA: MIT Press.

O'Callaghan, C. (2007). *Sounds*. Oxford: Oxford University Press.

Olausson, H., Wessberg, J., Morrison, I., McGlone, F., and Vallbo, A. (2010). The neurophysiology of unmyelinated tactile afferents. *Neuroscience and Biobehavioral Reviews*, 34(2), 185–91.

O'Regan, J.K. and Noë, A. (2001). A sensorimotor account of vision and visual consciousness. *Behavioral and Brain Sciences*, 24, 939–1031.

O'Shaughnessy, B. (1989). The sense of touch. *Australasian Journal of Philosophy*, 67, 37–58.

Radman, Z. (ed.) (2013). *The hand, an organ of the mind*. Cambridge, MA: MIT Press

Ratcliffe, M. (2008). *Feelings of being: phenomenology, psychiatry, and the sense of reality*. Oxford: Oxford University Press.

Ratcliffe, M. (2012). What is touch? *Australasian Journal of Philosophy*, 90, 413–32.

Ratcliffe, M. (2013). Touch and the sense of reality. In Z. Radman (ed.), *The hand, an organ of the mind*. Cambridge, MA: MIT Press, pp. 131–57.

Ratcliffe, M. (2015). *Experiences of depression: a study in phenomenology*. Oxford: Oxford University Press.

Saig, A, Gordon, G., Assa, E., Arieli, A., and Ahissar, E. (2012). Motor-sensory confluence in tactile perception. *The Journal of Neuroscience*, 32, 14022–32.

Schepers, R.J. and Ringkamp, M. (2010). Thermoreceptors and thermosensitive afferents. *Neuroscience and Biobehavioral Reviews*, 34, 177–84.

Taylor, M.M., Lederman, S.J., and Gibson, R.H. (1973). Tactual perception of texture. In: E.C. Carterette and M.P. Friedman (eds.), *Handbook of perception, vol. 3: the biology of perceptual systems*. New York: Academic Press, pp. 251–73.

Wolff, P. and Shepard, J. (2013). Causation, touch, and the perception of force. In: B.H. Ross (ed.), *The psychology of learning and motivation* (vol. 58). London: Academic Press, pp. 167–202.

CHAPTER 15

DIRECT SOCIAL PERCEPTION

JOEL KRUEGER

INTRODUCTION

DEFENDERS of a direct perception approach to other minds—what I'll here refer to as direct social perception (DSP)—argue that we directly perceive mental states.[1] I see my partner's sadness in her slumped shoulders, furrowed brow, and quiet speech; my niece's joy in her unrestrained laughter and sprightly gait; a friend's desire for a beer as he opens the refrigerator door and reaches for a bottle. In these and other cases, emotions and desires aren't hidden away *behind* behavior. They are concretely embodied *in* behavior. And when I see this behavior, I see these mental states directly, without inferential mediation. I see mind in action.

DSP challenges a dominant supposition about the hiddenness of mental states in philosophy of mind and cognitive science. Call this supposition the "unobservability principle" (UP). According to UP, we can't see mental states because, whatever their ontology, mental states are intracranial phenomena. As such, they are perceptually inaccessible to everyone but their owner. Accordingly, because we can't perceive other minds, we need to use an indirect method based on inference or simulation to reach them.

We can see just how widely UP is assumed by looking at a few representative quotes. Consider the following:

> One of the most important powers of the human mind is to conceive of and think about itself and other minds. Because the mental states of others (and indeed ourselves) are completely hidden from the senses, they can only ever be inferred. (Leslie 1987, p. 139)

[1] I have elsewhere referred to this idea as the "direct perception" account of other minds (Krueger 2012). Here, I'll follow Spaulding (2015) and instead speak of direct *social* perception to specify that not just any object but rather other people—specifically, their mental states—are the objects of our direct perception.

> Mental states, and the minds that possess them, are necessarily unobservable constructs that must be inferred by observers rather than perceived directly. (Johnson 2000, p. 22)
>
> People do not have direct information about others' mental states and must therefore base their inferences on whatever information about others' mental states they do have access to. This requires a leap from observable behavior to unobservable mental states that is so common and routine that people often seem unaware that they are making a leap. (Epley and Waytz 2009, p. 499)

Some even argue that the inaccessibility of other minds renders questions of whether a given entity is minded (e.g., a machine, animal, vegetative patient) unanswerable (Gray and Schein 2012, p. 407).

Within philosophy and cognitive science, UP generates at least two distinct questions. First, it generates the *epistemological* question of how, in the absence of perceptual verification, we can have knowledge of or justified belief in other minds. Second, it generates the *empirical* question of what sort of mechanisms enable us to attribute minds to others.[2] While the former question has a philosophical heritage within the Western canon stretching back at least to Mill—although Indian Buddhist philosophers were concerned with this question significantly earlier than that (Inami 2001)—the latter question has, for the past several decades, shaped ongoing debates about social cognition and theory of mind in cognitive science.

In what follows, I have two main objectives: first, to develop a version of DSP that draws upon both phenomenological literature and empirical work in cognitive science; second, to consider whether DSP can stand up to a number of objections. Ultimately I suggest that it can. After providing some historical context for DSP, I consider some different ways of interpreting the idea. I argue for a particular reading of DSP—one which recognizes that some mental states are partially constituted by expressive behavior—and, after considering some supporting empirical evidence, conclude by responding to a number of possible objections.

DSP in its Historical Context

DSP is not a new thesis. Many phenomenologists defend some version of it (Gallagher 2008; Gallagher and Zahavi 2008, ch. 9). Husserl, for example, tells us that we perceptually encounter another's "lived experiencing . . . completely without mediation and without consciousness of any impressional or imaginative picturing" (Husserl 2006, p. 84).[3]

[2] See Overgaard (2013) for a discussion of why it is important these two questions be kept distinct despite a persistent tendency in the literature to conflate them.

[3] While it seems Husserl ultimately endorses DSP, there is some degree of uncertainty in his account. See Zahavi (2014, pp. 125–32).

Scheler is more explicit. He argues that difficulties surrounding other minds are mostly self-created insofar as they are based on an unquestioned acceptance of UP (Scheler 1954, p. 238). But Scheler rejects UP and develops an alternative "perceptual theory of other minds," as he terms it. He insists—in perhaps the canonical phenomenological statement of DSP—that:

> we certainly believe ourselves to be directly acquainted with another person's joy in his laughter, with his sorrow and pain in his tears, with his shame in his blushing, with his entreaty in his outstretched hands, with his love in his look of affection, with his rage in the gnashing of his teeth, with his threats in the clenching of his fist, and with the tenor of this thoughts in the sound of his words. If anyone tells me that this is not "perception" [of the emotion itself], for it cannot be so, in view of the fact that a perception is simply a "complex of physical sensations," and that there is certainly no sensation of another person's mind nor any stimulus from such a source, I would beg him to turn aside from such questionable theories and address himself to the phenomenological facts. (Scheler 1954, p. 260)

Merleau-Ponty defends this idea in a number of places. For example, he writes:

> I perceive the grief or anger of the other in his conduct, in his face or his hands, without recourse to any "inner" experience of suffering or anger, and because grief and anger are variations of belonging to the world, undivided between the body and consciousness, and equally applicable to the other's conduct, visible in his phenomenal body, as in my own conduct as it is presented to me. (Merleau-Ponty 2002, p. 415)

We find similar ideas in other phenomenologists as well, such as Levinas's characterization of our experiential encounter with the face of the other (Levinas 1999; cf. Krueger 2008; Overgaard 2006) and Stein's account of empathy (Stein 1989; cf. Jardine and Szanto 2017).

But DSP isn't just found in the phenomenological tradition. Nathalie Duddington argues that "our knowledge of other minds is as direct and immediate as our knowledge of physical things" (Duddington 1918, p. 147). A consequence of Dewey's embodied view of cognition is that emotions can be proper objects of perception (Dewey 2008; cf. Krueger 2014). The Japanese philosopher Tetsurō Watsuji—working at roughly the same time as Merleau-Ponty and Sartre—draws upon both phenomenology and Zen Buddhism to develop a model of intersubjectivity consonant with DSP (Watsuji 1996; cf. Krueger 2013; McCarthy 2011). Wittgenstein tells us that:

> "We *see* emotion."—As opposed to what?—We do not see facial contortions and *make the inference* that he is feeling joy, grief, boredom. We describe a face as sad, radiant, bored, even when we are unable to give any other description of the features.—Grief, one would like to say, is personified in the face. (Wittgenstein 1980, §570; cf. Overgaard 2006)

And DSP receives support from more recent analytic philosophers, too: Austin (1979), Dretske (1973), Green (2010), McDowell (1982), McNeill (2012), Newen et al. (2015), Pickard (2003), Stout (2010), and Smith (2010) endorse some version of the thesis.

Before looking at DSP in more detail, we should briefly note a methodological distinction. Contemporary analytic defenders tend to focus on the *mechanisms* by which we come to be aware of others' mental states, such as the nature of visual perception or the various processes constitutive of our perceptual capacities. Framed thus, the debate then concerns how best to understand these mechanisms (e.g., inference, simulation, etc.) and how they factor into an epistemic explanation of our knowledge of others' mental states. Defenders working from a phenomenological perspective, however, are more concerned with the question of what mental states must be like in order for DSP to be plausible. This *ontological* orientation leads them to focus on the relation between mental phenomena and embodiment. I turn to this ontological question now.

DSP and the Embodied Mind

In what follows, I focus on a phenomenological approach to DSP for two reasons: first, as we've seen, phenomenology has a long history of rejecting UP and offers ample theoretical resources for exploring this issue. Second, since contemporary phenomenological approaches to DSP focus on the relation between mental phenomena and embodiment, they regularly draw upon 4E approaches to mind in philosophy and cognitive science. Their interdisciplinary perspective is therefore of particular relevance to this volume.

Phenomenologists claim that some mental phenomena are directly given within expressive behavior. As Scheler puts it, we see mental states because we perceptually encounter others as a psychophysical "expressive unity" (*Ausdruckseinheit*) (Scheler 1954, pp. 281, 261). More recently, Gallagher and Zahavi (2008)—largely responsible for reinvigorating current interest in phenomenological approaches to other minds—argue that "in seeing actions and expressive movements of other persons, one already sees their meaning. No inference to a hidden set of mental states is necessary. Expressive behavior is saturated with the meaning of the mind; it reveals the mind to us (Gallagher and Zahavi 2008, p. 185). Similarly, Thompson tells us that "we experience the other directly as a person, that is, as an intentional and mental being whose bodily gestures and actions are expressive of his or her experiences and states of mind" (Thompson 2005, p. 264; cf. Ratcliffe 2007).

But these formulations need refinement. There are several ways of understanding how behavior expresses mental phenomena, and not all of them are consistent with DSP (Krueger and Overgaard 2012).

DSP and the Co-Presence Thesis

One way of characterizing the relation between mind and behavior is to characterize the former as perceptually *co-present* within the expressive dynamics of the latter. Joel Smith (2010) develops a nuanced defense of this view. Drawing upon a functionalist view of mental properties as well as Husserl's analysis of the anticipatory structure of perception, Smith begins by observing that often what we *experience* outstrips what we actually *see*. When we see an apple, for example, we only see, strictly speaking, the part of the apple facing us; its occluded sides remain hidden. But it's a fact about perceptual consciousness that we nevertheless *experience* the apple as a three-dimensional object. Phenomenologically, its hidden sides are perceptually co-present along with the side facing us.

This basic structural feature of our perceptual consciousness, Husserl argues, follows from the fact that we are necessarily embodied subjects situated in a specific place and time. He observes:

> Of necessity a physical thing can be given only "one-sidedly.". . . A physical thing is necessarily given in mere "modes of appearance" in which necessarily a core of "what is actually presented" is apprehended as being surrounded by a horizon of "co-givenness." (Husserl 1998, p. 94)

For Smith and Husserl, hidden sides of objects are co-present because perception is a temporally extended process. Perception is structured by anticipatory appresentations of what we *would* see if we were to crane our neck for a different view, move around the object, or pick it up. As Husserl puts the idea, in perceiving objects as three-dimensional, we "recognize that a hidden intentional 'if-then' relation is at work here: the exhibitings must occur in a systematic order; it is in this way that [the occluded sides] are indicated in advance, in expectation, in the course of a harmonious perception" (Husserl 1970, pp. 161–2). These "if-then" relations or "sensorimotor contingencies" (Noë 2004) specify how movements will bring new sides of objects into view. They are part of the content of our experience, ensuring that occluded sides are perceptually co-present with visible sides.

How does this relate to other minds? Smith argues that, analogously, although we only ever *see* another's behavior, we nevertheless *experience* associated mental phenomena as experientially co-present: "Just as the rear aspect of the book is visually present without being visually presented, so another's misery is visually present even though only their frown is visually presented" (Smith 2010, p. 739). This is so, Smith continues, because my perception of another's mentality can be fulfilled by "the co-presented and presented taking part in a harmonious experience" (Smith 2010, p. 741). I can be said to experience another's anger or happiness, say, by perceiving ongoing patterns of behavior that continually confirm this anger or happiness: scowling and fist-shaking, or smiles and laughter. In this way, co-presented mental properties are experientially confirmed in ongoing presentations of another's "changing but incessantly *harmonious behavior*" the way that occluded sides of objects are co-present in my experience of the sides facing me (quoted in Smith 2010, p. 739). For Smith, this view is a genuine perceptual account of other minds that simultaneously respects the intuition that aspects of others' thoughts, feelings, and intentions remain private and hidden from view.

While a compelling view, it's not clear the co-presence thesis actually fits with DSP. This is because it still appears to tacitly confirm UP. For, what we see in others are not actually features of their mentality but rather features of their *behavior*. To be clear, Smith insists that we don't see "*mere* behavior" devoid of social significance (Smith 2010, p. 742)—we experience another's behavior as meaningful, as articulating their beliefs, desires, intentions, and emotions—but our experience of this behavior nevertheless only grants *indirect* access to their mentality. Even if we anticipate that the functional profile of James's anger will be "harmoniously" confirmed in the temporally extended dynamics of his flushed cheeks, contorted facial expressions, shaking fists, and brisk movements, we still only make direct perceptual contact with these bits of behavior. The full functional profile of James's anger remains beyond the scope of visual perception. According to the co-presence thesis, then, James's mental properties are not actually rendered visible in his behavior—at least in the way DSP appears to require. The co-presence thesis thus appears to be a *weakly perceptual* account of other minds (McNeill 2012). Mental properties are experientially present without, strictly speaking, actually being *seen*.[4]

DSP and the Constitution Thesis

A second way of characterizing the relation between mental states and behavior is not in terms of co-presence but rather *constitution*. This strategy involves rejecting the supposition of an ontological gap between mind and behavior. Stated positively, the idea is that "mind can be equally and unambiguously instantiated in experience and behavior" (Pickard 2003, p. 89). Again, DSP claims that overt actions such as smiling, scowling, shaking one's fists, gesturing while speaking, counting on one's fingers, or reaching for a beer grant direct perceptual access to other minds. Some mental states are concretely embodied within the expressive behavior we see. A "constitutive" sense of bodily expression is thus the idea that certain bodily actions are expressive of mind in that they actually constitute proper parts of some mental phenomena. To see these actions is thus to literally see part of another's mind—the external public-facing part—and not simply the subsequent causal effect of some internal mental state. Put otherwise, this rendering of DSP argues that minds are *hybrid* entities: they consist of both internal (neural, physiological, and phenomenal) and external (behavioral, environmental) parts and processes, integrated into a unified whole. While we can't see *all* of the constitutive parts

[4] For a longer discussion of this point, see Krueger (2012) and Gallagher (2016).

of another's hybrid mind, according to this view, we nevertheless do have direct perceptual access to the externally realized parts. Consequently, from a phenomenological perspective, there is no problem of other minds because the foundational supposition motivating the problem in the first place—UP—is denied.

Put this plainly, the constitution thesis might seem implausible—and probably rather philosophically unsophisticated. For one thing, it's clear that there are all sorts of mental phenomena we *can't* see. Additionally, one might worry that there is very little, if anything, in common between mind and behavior. So it's implausible to say that seeing the latter is to see the former. However, such an assessment is too hasty. First, there are several empirical streams of research that appear to support this hybrid view of mind. Second, a version of DSP that make use of this hybrid account appears capable of withstanding a number of objections. The view thus warrants a more careful consideration.[5]

Empirical Support

For the sake of space, I will focus on work supporting the idea that emotions may have visible parts.[6] It seems likely that, on one hand, emotions consist of various internal components: e.g., neurophysiological states and processes (Damasio 1999; LeDoux 1996; Prinz 2004), as well as cognitive components like evaluative judgments or appraisals of the objects and events toward which emotions are intentionally directed (Nussbaum 2001; Solomon 2004). On the other hand, however, there is evidence that emotions are not *exhausted* by these internal components. Some emotions may be partially composed of behavioral expressions—facial expressions, gestures, whole-body expressions, patterns of behavioral entrainment and sensorimotor coupling, etc.—that others can see. In other words, these bodily expressions, the emotion's public-facing profile, may be a constitutive part of the emotion.

How this is so can be made clearer by looking at cases where an emotion's visible bodily expression is compromised or altogether missing. People with Moebius syndrome—a congenital form of bilateral facial paralysis—cannot facially express emotion; they also commonly exhibit other motor impairments that further hinder their expressive capacities (Briegel 2006; Cole and Spalding 2009; Krueger and Henriksen 2016). As a result, many describe phenomenologically diminished emotional lives. For example, one person reports that "I sort of think happy or think sad, not really saying or recognizing actually feeling happy or feeling sad," and that the phenomenal qualities of his emotions "are there but they are probably reduced" (Cole 1999, p. 308). Another claims not to have had emotion as a child. She only learned to express and thus *feel* her

[5] A different version of direct perception of emotions is developed and defended in Newen et al. 2015.

[6] For a discussion of how other mental states like thoughts and intentions may similarly have visible parts, see Krueger (2012, pp. 157–62). For other arguments that we directly perceive intentions, see Pacherie (2005) and Proust (2003).

emotions after consciously mimicking others' expressions she observed while on holiday in Spain. In her own words, she started "using the whole body to express [her] feelings" (Cole and Spalding 2009, p. 154). People with Moebius syndrome often adopt alternative forms of bodily expression—exaggerated prosody, gestures, vocalizations, painting, dancing, playing a musical instrument—that enable them to express, recalibrate, and shape the phenomenal character of their emotional experience (Bogart and Matsumoto 2010). Without the ability to spontaneously express their emotions via various motor and behavioral channels, however, part of the emotion appears to be missing.

Not all facial paralysis is congenital.[7] A similar effect is observed in individuals who've voluntarily received Botox injections, which inhibits facial expressions (Baumeister et al. 2016; Davis et al. 2010; Havas et al. 2010). We also find this effect in cases of acquired facial paralysis, such as Bell's palsy. One individual suggestively describes entering into an "emotional limbo" while the paralysis was at its strongest; however, as he gradually regained facial animation over several months, the phenomenology of his emotions returned to its previous level (Cole 1998). There is also evidence that expressive components beyond the face are parts of certain emotions. For instance, individuals who've suffered severe spinal cord injuries and lack the ability to bodily express emotions using gesture, postural adjustments, or other whole-body movements report less intense feelings of high-arousal emotions like fear, anger, or sexual arousal (Chwalisz et al. 1988; Hohmann 1966; cf. Laird 2007, pp. 74–6; Mack et al. 2005). Many other studies suggest that manipulating facial expressions, postures, and gestures generates emotion-specific autonomic activity and produces a corresponding change in emotional phenomenology (e.g., Davis et al. 2009; Laird 2007; Niedenthal 2007).

These studies appear to support the idea that mental states like emotions are partially constituted by expressive behavior. The bodily or facial expression of some emotions is part of the physical vehicle by which we realize that emotion. Removing an aspect of this vehicle thus removes part of the emotion itself—much like removing spark plugs from a car engine removes its capacity to realize locomotion—and the experience of the emotion is altered accordingly. But this evidence does not suggest that emotions are *identical* with their behavioral expression. There are still internal neural, physiological, and phenomenal parts of emotions not exhausted by their behavioral manifestation or the subpersonal "affect programs" that underwrite them (more on this later). When I am genuinely happy and smile broadly, for example, my happiness is not simply *in* the physical features of my publically observable smile, or *in* the complex neural and physiological processes that enable me to perform such a smile. Both components are needed for the realization of my happiness. How these internal and external components integrate their respective functions in realizing emotional processes is an open empirical question that need not concern us here (see Laird 2007 and Niedenthal 2007 for overviews; see

[7] The following examples assist in responding to the objection that, since their facial paralysis is congenital, people with Moebius syndrome have no benchmark against which to measure "proper" emotional phenomenology and thus aren't in a position to make reliable judgments about its purported diminishment.

also Colombetti 2014 and Maiese 2011). The point is that they *do*. And the external parts of this process are publically available, ripe for seeing.

Objections

I now consider some common objections to this version of DSP. This list is not exhaustive; nor do I think that the discussion earlier is sufficient to establish the truth of DSP. Rather, my intention in what follows is simply to show that DSP has the resources to answer these objections.

The Behaviorism Objection

Pierre Jacob (2011) objects that DSP entails a kind of crude behaviorism. According to Jacob, another's bodily expressions either constitute their emotional states or they do not. If they do *not*, then we do not directly perceive another's mental states, only their behavior, and we've made no advance beyond inferentialist approaches to other minds that affirm UP. If they *do*, however, then DSP entails a crude reductive behaviorism that, among other faults, jettisons the first-person dimension phenomenologists claim to be most interested in.

This objection can be dealt with fairly easily since it rests on a mischaracterization of DSP. The relevant notion of "constitution" (i.e., in the claim that mental states are constituted by behavior) can be taken in a strong or weak sense. Taken in the strong sense, "constitutes" means "amounts to" or "equals." If phenomenologists mean to say that mental states amount to or equal behavior—e.g., John's anger simply *is* his frowning, fist-clenching, etc., and nothing more—this interpretation would commit them to a kind of crude behaviorism.

But phenomenologists don't endorse this strong sense of constitution. A second, weaker interpretation of "constitution" is available. On this interpretation, "constitutes" means "is a part of"—the way, for example, the tip of an iceberg constitutes a proper part of the iceberg (without, of course, constituting the *entire* iceberg). Tips are a constitutive part of icebergs, but icebergs are not wholly constituted by their tips. Analogously, we see others' emotions by seeing the external public-facing "tips," as it were. But this weaker sense of constitution doesn't entail crude behaviorism. Although certain expressive dynamics constitute an external "tip" of some emotional processes, the view doesn't thereby imply that we perceive *all* of the relevant mental phenomena. Once again, mental states such as emotions are structurally complex; they are *hybrid*, consisting of functionally integrated internal and external components.

So, saying that we perceive external components of some mental states is consistent with there being other aspects or components, such as their physiological signature or first-person dimension, that are not directly perceived by others. Something like this is what phenomenologists seem to have in mind when they speak of the irreducible *alterity* of the Other (e.g., Levinas 1999). In other words, phenomenologists are quite explicit about the fact that another's subjectivity is simultaneously both *immanent* (i.e., concretely embodied in their expressive behavior) as well as *transcendent* (i.e., partially beyond the reach of my perceptual capacities) (Taipale 2015). This reading of DSP—informed by a hybrid account of embodied mentality—thus appears to offer a way through the Scylla of Cartesian internalism and Charybdis of behaviorism Jacob's criticism rests on.[8]

The Absent Behavior Objection

Ken Aizawa offers a related objection. Following Jacob, he begins by observing that DSP claims cognition is a type of behavior. But Aizawa objects to this idea—the view, as he puts it, "that cognitive properties are properties of the brain-body complex, rather than properties of the brain" (Aizawa 2017, p. 4277) Aizawa appeals to two empirical cases that purport to show a distinction between cognition and behavior: experiments with neuromuscular blockade and locked-in syndrome. Aizawa argues that these cases are examples of cognitive processing continuing in the absence of behavior. Accordingly, they challenge any view identifying cognition with behavior insofar as the former can function without the latter.

A neuromuscular blockade operates by paralyzing patients while allowing them to retain conscious awareness. They can be useful for surgeons, for instance, who need to manipulate patients on the operating table without encountering muscular resistance. However, administering paralytics can potentially create scenarios where it's difficult to tell if a patient has been properly anesthetized. As a result, there have been occasions where patients have undergone surgery, fully aware and in great pain, but unable to report their suffering due to their induced paralysis (Osterman et al. 2001). Similarly, patients suffering from locked-in syndrome—a condition resulting from damage to the brainstem, which leads to a near-total paralysis except for vertical eye movement and blinking—appear to be fully conscious, with largely unaffected cognitive functioning (Schnakers et al. 2009; León-Carrión et al. 2002; see also Bauby 1997). Again, Aizawa argues that these cases challenge DSP since they appear to provide cases in which cognitive processing remains untouched in the absence of overt neuromuscular activity.[9]

DSP can offer several points in response. First, Aizawa may be too quick in assuming that cognitive processing remains untouched in the absence of overt neuromuscular activity. For instance, some of the phenomenological descriptions we

[8] See Krueger and Overgaard (2012) for further discussion of this objection.

[9] While I disagree with Aizawa's objection, I very much agree with his observation that the question of whether cognition is a type of behavior is more than a mere terminological dispute. As these cases indicate, the question has significant practical implications and is therefore worthy of careful empirical consideration (Aizawa 2017, p. 11).

find in Jean-Dominique Bauby's (1997) first-person account of living with locked-in syndrome—conveyed to his nurse via coded eye blinks—suggest otherwise (e.g., his experiences take on an unstable, shifting, or dream-like quality). Additionally, there is evidence that individuals who spend extended periods of time in solitary confinement—with severe restrictions on movement and environmental interaction—experience hallucinations and alterations of consciousness (depersonalization, derealization) (Guenther 2013). It may be that the phenomenal character of consciousness is altered via the prolonged restriction or total absence of neuromuscular activity.[10]

More substantively, Aizawa—like Jacob—seems to attribute a strong sense of constitution to DSP. But, as indicated previously, this is a crude behaviorism phenomenologically motivated defenders of DSP don't endorse. Again, the claim is not that cognition *amounts to* or *is* behavior—in the sense of being wholly reducible to it—but rather that behavior is an external part of some cognitive process involving dynamically integrated "brain–body complexes," to use Aizawa's expression. Acknowledging the role that the latter plays in driving some cognitive processes does not entail rejecting or disregarding the crucial role of the former.

In fact, this weaker sense of constitution would actually predict that some cognitive processing continues in the absence of overt neuromuscular activity since internal components of various cognitive processes may retain some level of functioning even when decoupled from external behavioral components. For example, we can think all manner of thoughts without any behavioral indication of these thoughts; and we can have emotional experiences or become sexually aroused while asleep. Our inner life continues to hum along in the absence of overt neuromuscular activity each night (although there is often *some* overt behavior going on, such as periodic limb movements and postural adjustments, or eye movements during REM sleep). Likewise, we can—at least potentially—perform mathematical operations entirely in our head, without using a pencil and paper, gesturing, or counting on our fingers. But the salient point is that using a pencil and paper, or gesturing while working through a problem, *amplifies* our computational facility; these artifacts and actions grant access to modes of thought that would otherwise remain very difficult or perhaps impossible to achieve without their ongoing input (Goldin-Meadow et al. 2001; Rumelhart and McClelland 1986).

Similarly, in the case of emotions, behavioral components may amplify the character or intensity of the emotion by contributing information and self-regulatory resources that open up new modes of affective experience. It seems that musculoskeletal feedback

[10] Aizawa might respond that long-term changes in a situation—e.g., being locked up in solitary confinement, which leads to an impoverished perceptual environment and diminished sensorimotor feedback—may indicate important relearning conditions that causally impact the functioning of the cognitive system as a whole. But it doesn't thereby follow that these changes pick out *constitutive* relations between features of the environment and, e.g., phenomenal experience. This worry highlights a lack of clarity with respect to the role temporal dynamics play in establishing putative cases of extended cognitive systems. This is a complicated point; I cannot do it justice here. The reader is encouraged to look at Clark's (2008) treatment of this question from the perspective of dynamic systems theory (e.g., pp. 24–9). My thanks to Albert Newen for raising this point.

from grimacing, for instance, can amplify the felt intensity of pain (Kleck et al. 1976). And the point, then, is that in some cases, we need these external resources to realize certain mental states, such as emotions, in their full capacity. Like Jacob's objection, Aizawa's objection only works if we attribute a crude behaviorism to DSP that isn't actually representative of the view.

The Part-Whole Objection

Although William McNeill (2012) is sympathetic to the idea that we can see mental states, he is critical of versions of DSP that rely on a hybrid notion of the embodied mind.[11] McNeill argues that even if we grant that some mental states are partially embodied in visible behavior, this fact alone is not sufficient for thinking we actually see the mental state. We might see some trees, for instance, and the trees we see might be part of a wood. But just because we see some of the trees making up the wood, it doesn't thereby follow that we see the wood (McNeill 2012, p. 583). Similarly, we might see cards without seeing the deck of which they are a part, or ships without seeing the whole fleet. McNeill argues that, analogously, we might see a part of John's anger in his frowning without actually seeing the anger of which it is a part. Seeing parts is not sufficient for seeing the whole.

As Overgaard (2014) observes, it's not clear this example works. There are different types of part-whole relations, each with their own logical structure, that McNeill's example overlooks. Specifically, McNeill's tree-wood example picks out a *member-collection* relation while DSP's behavior-mental state claim picks out a *component-integral object* relation (Winston et al. 1987). So, it may be that seeing particular members (trees, playing cards, ships) is in some cases insufficient for seeing the collection of which they are a part (forest, deck, fleet). But the logic of the component-integral object relation works differently. In these cases, certain components of "integral objects"—i.e., objects exhibiting a patterned organization bearing specific structural and functional relations to their components—may be so essential, structurally and functionally, that seeing a component *is* in fact sufficient to see the object of which it is a part.

For instance, when I show a visitor significant parts of the University of Exeter campus (the Forum, Reed Hall, Peter Chalk Centre, Northcote House, and Amory Building), they can rightfully claim to have seen the University of Exeter—even though there are many other parts I did not show them. Of course, just showing them a pebble I picked up from outside my office building would not be sufficient for them to see the University of Exeter. That pebble is not a significant part of the university. However, this particular collection of buildings *is* an integral part of the University of Exeter's identity. To see this collection of buildings is thus to see the University of Exeter. Similarly, as the empirical

[11] McNeill (2012) offers a number of careful objections to DSP that go well beyond the scope of what I can discuss here. See Overgaard (2014) for in-depth responses to these objections.

evidence canvassed earlier suggests, certain behavioral components are so structurally and functionally integral to the emotion that, in their absence, the emotion is profoundly compromised or actually disappears altogether. Accordingly, it seems as if the embodied component is significant enough to the structural and functional integrity of the emotion that to see the component is sufficient to see the emotion.

The Asymmetry of Access Objection

Another objection stems from the observation of a stark asymmetry between self-experience and other-experience. When I experience an emotion, say, I feel it immediately and know it directly *as mine*. However, I lack this sort of first-person access to the others' experiential lives. This stark asymmetry is what generates the epistemological problem of other minds (Hyslop 2015; see also de Vignemont 2010). But in claiming we enjoy direct access to other minds, DSP appears to deny or overlook this asymmetry. And by overlooking the fact that we don't have the same kind of access to the minds of others that we have to our own, DSP rests on a highly implausible claim and is therefore probably false.

In response, we can first note that DSP acknowledges—and actually insists on—the asymmetry between self-experience and other-experience. From a phenomenological perspective, it's clear this asymmetry exists and is an unavoidable feature of human experience; what's unclear is how one would go about challenging it. Husserl, for instance, argues this asymmetry is phenomenologically constitutional for intersubjectivity. Without it, I would be incapable of experiencing another's mind *as other* but would instead experience it as "merely a moment of my own essence, and ultimately he himself and I myself would be the same" (Husserl 1999, p. 109). Similarly, Merleau-Ponty observes that "the grief and anger of another have never quite the same significance for him as they have for me. For him these situations are lived through, for me they are displayed" (Merleau-Ponty 2002, p. 415).

The salient point is that while the asymmetry of self-experience and other-experience clearly entail different modes of access to minds, it doesn't follow that these different modes of access necessarily entail a difference in terms of *directness* (Overgaard 2013). On DSP, John's anger is visible, embodied within his expressive behavior. I know John's anger by seeing it directly. In my own case, I know my anger by feeling it directly. So, while my mode of access to these two instances of anger clearly differs—seeing vs. feeling—both nevertheless involve a direct awareness of the anger itself. Importantly, knowing John's anger by seeing it is only less direct than knowing my anger by feeling it if introspective access to one's mental state is taken to constitute the gold standard of what directness amounts to (Zahavi 2014, p. 165). But why assume this? Why can't minds be experienced in more than one way—and with equal directness? As Zahavi notes, "Arguably there is no more direct way of knowing that another is in pain than seeing him writhe in pain" (Zahavi 2014, p. 165). As should be clear, DSP can still accommodate the basic asymmetry between self- and other-experience while still consistently denying that either is less direct than the other.

The Irrelevance Objection

I conclude with a final objection. Some critics (Herschbach 2008; Spaulding 2010) argue that DSP—in contrast to alternative theories of social cognition in cognitive science, such as theory theory (TT) or simulation theory (ST)—has such a limited explanatory scope that it's irrelevant for developing an empirically grounded account of how we understand other minds. Clearly there are many instances when social perception is limited: while interacting with someone in less than ideal conditions, such as looking at them from far away or in poor lighting, or when trying to interpret ambiguous behavior or complex mental states like ulterior motives, irony, jealousy, or the like—states that don't necessarily have a discernible behavioral signature. In these cases, we clearly need extra-perceptual mechanisms (e.g., folk psychology, simulation, or some combination of the two). In short: direct social perception can only reach a very small part of the complex topography of our social life.

DSP need not entirely disagree with the spirit of this objection. In the hands of even its most ardent contemporary defenders, DSP is not offered as a comprehensive theory of social cognition (Gallagher 2008; Zahavi 2011). Rather, DSP acknowledges that perception is one tool in our social toolkit. It may arguably be a very important tool—but arguing for this point is not inconsistent with conceding that we sometimes draw upon other tools to navigate the complexities of our social life.

Still, the critic (e.g., Spaulding 2010) might press the point and argue that DSP targets a different *explanandum* than do other approaches, and ultimately what it accounts for (our perceptual encounter with others' mental states) has limited explanatory scope. In contrast to DSP, approaches such as TT and ST offer characterizations of the *subpersonal* cognitive or neural mechanisms causally responsible for our understanding of other minds. But these mechanisms, like physiological mechanisms enabling digestion or respiration, lie beneath the reach of our awareness. We can investigate the structure and functioning of these various mechanisms without worrying about consciousness. Moreover, since personal-level experiences (which DSP is concerned with) ultimately emerge from causally antecedent subpersonal processes, the latter ought to be given explanatory precedence when it comes to developing a theory of other minds. In other words, it's at *this* level that our explanations of how we experience and know other minds should "bottom out."

DSP can say several things in response. First, it's true that DSP is not looking to offer characterizations of subpersonal mechanisms responsible for our ability to see others as minded, insofar as these mechanisms work outside of consciousness. Phenomenology is first and foremost concerned with *experience*. So Spaulding, a consistently helpful critic of DSP, is in this case simply wrong when she suggests that DSP defenders think the role of phenomenology "is to dictate the nature of operative sub-personal processes" (Spaulding 2010, p. 131).

Yet the DSP defender can still grant Spaulding's point that, generally speaking, alternative approaches such as TT and ST are in fact working at a different level of explanation while still asserting DSP's relevance. For it seems odd to suggest that the phenomenology of certain cognitive processes *has no relevance whatsoever* when it comes to understanding the nature of those same processes (see Spaulding 2010, p. 131). Social cognition theorists should not accept this sort of unquestioned reductionism. And phenomenologists likewise ought not to accept the implication that theirs is a purely descriptive project of taxonomy and classification devoid of causal explanatory potency.

Consider phenomenological psychopathology. Phenomenological approaches have enabled us to get a clearer grip on the experiential dimension of the patient's disorder—how schizophrenia, for instance, is lived through from the first-person perspective—as well as the structures or modes of consciousness that allow the disorder to manifest the way that it does. These descriptions allow researchers to better understand the character of a given disorder and make important diagnostic distinctions based on this character. But they can also contribute to causal explanations, too.

For example, charting symptom progression in schizophrenia is not simply a matter of isolating subpersonal neurological abnormalities. This is because phenomenological features of the patient's subjective life exacerbate the experiential fragmentation distinctive of the schizophrenic illness; they "provide both the motivation and the field of possibility for the progressive symptomatic developments" (Parnas and Sass 2008, p. 270). Certain causal explanations require both neurobiological *and* phenomenological elements (McClamrock 1995). Even in cases where phenomenology need not be part of specific causal explanations, it can still provide helpful diagnostic clues about where to look for the relevant subpersonal mechanisms. Neuroscientific work on empathy and mirror neurons—under the broader rubric of ST—appeals centrally to individuals' perceptual experience of others' mental states and intentional actions. Both what they see and how they see it (i.e., their phenomenology) constrains the target explanandum at the subpersonal level (Gallese 2001).

Finally, Marchi and Newen (2015) suggests another strategy for responding to this objection. They survey extensive evidence suggesting that our perceptual experiences of others' emotional expressions are sensitive to, and modified by, sociocultural background knowledge (beliefs, values, images, etc.). This view—conjoined with a liberal-content view of perception (Newen 2016), according to which we can develop the skills to perceive more than mere colors, shapes, and edges, but also rich content like causal relations, intentions, agency, natural and artificial kinds, and social phenomena—suggests that, contra the irrelevance objection, perception needs to remain a central part of our considerations of social cognition. This is because perception—including social perception—is a kind of practical expertise that can be developed, shaped, and refined in various ways. And thus we can (at least potentially) develop the skills necessary to perceive much more of others' mental life than this objection concedes. To jettison perception from our consideration of social cognition thus seems premature.

In sum, DSP has an important role to play in ongoing debates about our ability to perceive and engage with other minds. It is both philosophically defensible and empirically supported. To be clear, I have not considered whether DSP ought to be *the* primary approach to other minds or, rather, whether it might supplement alternative approaches (e.g., TT and ST). That is a discussion for another time.[12]

References

Aizawa, K. (2017). Cognition and behavior. *Synthese*, 194(11), 4269–88.

Austin, J.L. (1979). Other minds. In: J.O. Urmson and G.J. Warnock (eds.), *Philosophical papers*. Oxford: Oxford University Press, pp. 76–116.

Bauby, J.-D. (1997). *The diving bell and the butterfly*. London: Harper Perennial.

Baumeister, J.-C., G. Papa, and F. Foroni.(2016). Deeper than skin deep—the effect of botulinum toxin-A on emotion processing. *Toxicon*, 118, 86–90.

Bogart, K. and Matsumoto, D. (2010). Living with Moebius syndrome: adjustment, social competence, and satisfaction with life. *The Cleft Palate-Craniofacial Journal*, 47(2), 134–42.

Briegel, W. (2006). Neuropsychiatric findings of Möbius sequence—a review. *Clinical Genetics*, 70(2), 91–7.

Chwalisz, K., Diener, E., and Gallagher, D. (1988). Autonomic arousal feedback and emotional experience: evidence from the spinal cord injured. *Journal of Personality and Social Psychology*, 54(5), 820–8.

Clark, A. (2008). *Supersizing the mind: embodiment, action, and cognitive extension*. Oxford: Oxford University Press.

Cole, J. (1998). *About face*. Cambridge, MA: MIT Press.

Cole, J. (1999). On "being faceless": selfhood and facial embodiment. In: S. Gallagher and J. Shear (eds.), *Models of the self*. Exeter: Imprint Academic, pp. 301–18.

Cole, J. and Spalding, H. (2009). *The invisible smile: living without facial expression*. Oxford: Oxford University Press.

Colombetti, G. (2014). *The feeling body: affective science meets the enactive mind*. Cambridge, MA: MIT Press.

Damasio, A. (1999). *The feeling of what happens: body and emotion in the making of consciousness*. New York: Harcourt Brace.

Davis, J.I., Senghas, A., Brandt, F., and Ochsner, K.N. (2010). The effects of Botox injections on emotional experience. *Emotion*, 10(3), 433–40.

Davis, J.I., Senghas, A., and Ochsner, K.N. (2009). How does facial feedback modulate emotional experience? *Journal of Research in Personality*, 43(5), 822–9.

de Vignemont, F. (2010). Knowing other people's mental states as if they were one's own. In: D. Schmicking and S. Gallagher (eds.), *Handbook of phenomenology and cognitive science*. New York: Springer, pp. 283–99.

Dewey, J. (2008). The theory of emotion. In: J.A. Boydston (ed.), *The early works, 1882–1898* (vol. 4). Carbondale: Southern Illinois University Press.

Dretske, F. (1973). Perception and other minds. *Noûs*, 7(1), 34–44.

Duddington, N.A. (1918). Our knowledge of other minds. *Proceedings of the Aristotelian Society*, 19, 147–78.

Epley, N. and Waytz, A. (2009). Mind perception. In: S.T. Fiske, D.T. Gilbert, and G. Lindzey (eds.), *The handbook of social psychology* (5th ed.). New York: Wiley, pp. 498–541.

[12] For a recent collection of articles devoted to this and related issues, see Michael and de Bruin (2015).

Gallagher, S. (2008). Direct perception in the intersubjective context. *Consciousness and Cognition*, 17(2), 535–43.
Gallagher, S. (2016). The minds of others. In: D.O. Dahlstrom, A. Elpidorou, and W. Hopp (eds.), *Philosophy of mind and phenomenology: conceptual and empirical issues*. New York: Routledge.
Gallagher, S. and Zahavi, D. (2008). *The phenomenological mind: an introduction to philosophy of mind and cognitive science*. New York: Routledge.
Gallese, V. (2001). The "shared manifold" hypothesis: from mirror neurons to empathy. *Journal of Consciousness Studies*, 8(5-7), 33–50.
Goldin-Meadow, S., Nusbaum, H., Kelly, S.D., and Wagner, S. (2001). Explaining math: gesturing lightens the load. *Psychological Science*, 12(6), 516–22.
Gray, K. and Schein, C. (2012). Two minds vs. two philosophies: mind perception defines morality and dissolves the debate between deontology and utilitarianism. *Review of Philosophy and Psychology*, 3(3), 405–23.
Green, M. (2010). Perceiving emotions. *Aristotelian Society Supplementary Volume*, 84(1), 45–61.
Guenther, L. (2013). *Solitary confinement: social death and its afterlives* (1st ed.). Minneapolis: University of Minnesota Press.
Havas, D.A., Glenberg, A.M., Gutowski, K.A., Lucarelli, M.J., and Davidson, R.J. (2010). Cosmetic use of botulinum toxin-A affects processing of emotional language. *Psychological Science*, 21(7), 895–900.
Herschbach, M. (2008). Folk psychological and phenomenological accounts of social perception. *Philosophical Explorations: An International Journal for the Philosophy of Mind and Action*, 11(3), 223–35.
Hohmann, G.W. (1966). Some effects of spinal cord lesions on experienced emotional feelings. *Psychophysiology*, 3(2), 143–56.
Husserl, E. (1970). *The crisis of european sciences and transcendental philosophy: an introduction to phenomenological philosophy* (trans. D. Carr). Evanston, IL: Northwestern University Press.
Husserl, E. (1998). *Ideas pertaining to a pure phenomenology and to a phenomenological philosophy—first book: general introduction to a pure phenomenology* (trans. F. Kersten). Dordrecht: Kluwer Academic Publishers.
Husserl, E. (1999). *Cartesian meditations: an introduction to phenomenology* (trans. D. Cairns). Boston: Kluwer Academic Publishers.
Husserl, E. (2006). *The basic problems of phenomenology: from the lectures, winter semester, 1910–1911* (trans. I. Farin and J.G. Hart). Dordrecht: Springer.
Hyslop, A. (2015). Other minds. In: E.N. Zalta (ed.), *The Stanford encyclopedia of philosophy* (fall 2015). Retrieved from http://plato.stanford.edu/archives/fall2015/entries/other-minds/
Inami, M. (2001). The problem of other minds in the Buddhist epistemological tradition. *Journal of Indian Philosophy*, 29(4), 465–83.
Jacob, P. (2011). The direct-perception model of empathy: a critique. *Review of Philosophy and Psychology*, 2(3), 519–40.
Jardine, J. and Szanto, T. (2017). Empathy in the phenomenological tradition. In: H. Maibom (ed.), *Routledge handbook of philosophy of empathy*. London and New York: Routledge, pp. 86–97.
Johnson, S.C. (2000). The recognition of mentalis tic agents in infancy. *Trends in Cognitive Sciences*, 4(1), 22–8.

Kleck, R.E., Vaughan, R.C., Cartwright-Smith, J., Vaughan, K.B., Colby, C.Z., and Lanzetta, J.T. (1976). Effects of being observed on expressive, subjective, and physiological responses to painful stimuli. *Journal of Personality and Social Psychology*, 34(6), 1211–8.

Krueger, J. (2008). Levinasian reflections on somaticity and the ethical self. *Inquiry: An Interdisciplinary Journal of Philosophy*, 51(6), 603–26.

Krueger, J. (2012). Seeing mind in action. *Phenomenology and the Cognitive Sciences*, 11(2), 149–73.

Krueger, J. (2013). Watsuji's phenomenology of embodiment and social space. *Philosophy East and West*, 63(2), 127–52.

Krueger, J. (2014). Dewey's rejection of the emotion/expression distinction. In: T. Solymosi and J.R. Shook (eds.), *Neuroscience, neurophilosophy, and pragmatism: brains at work in the world*. New York: Palgrave Macmillan, pp. 140–61.

Krueger, J. and Henriksen, M.G. (2016). Embodiment and affectivity in Moebius syndrome and schizophrenia: a phenomenological analysis. In: J.E. Hackett and J.A. Simmons (eds.), *Phenomenology for the 21st century*. New York: Palgrave Macmillan, pp. 249–67.

Krueger, J. and Overgaard, S. (2012). Seeing subjectivity: defending a perceptual account of other minds. *ProtoSociology: Consciousness and Subjectivity*, 47, 239–62.

Laird, J.D. (2007). *Feelings: the perception of self*. Oxford: Oxford University Press.

LeDoux, J.E. (1996). *The emotional brain*. New York: Simon and Shuster.

León-Carrión, J., van Eeckhout, P., Domínguez-Morales, Mdel R., and Pérez-Santamaría, F.J. (2002). Survey: the locked-in syndrome: a syndrome looking for a therapy. *Brain Injury*, 16(7), 571–82.

Leslie, A.M. (1987). Pretense and representation: the origins of "theory of mind." *Psychological Review*, 94, 412–26.

Levinas, E. (1999). *Alterity and transcendence* (trans. M.B. Smith). London: The Athlone Press.

Mack, H., Birbaumer, N., Kaps, H.P., Badke, A., and Kaiser, J. (2005). Motion and emotion: Emotion processing in quadriplegic patients and athletes. *Zeitschrift Für Medizinische Psychologie*, 14(4), 159–66.

Maiese, M. (2011). *Embodiment, emotion, and cognition*. New York: Palgrave Macmillan.

Marchi, F. and Newen, A. (2015). Cognitive penetrability and emotion recognition in human facial expressions. *Frontiers in Psychology*, 6(828), 1–12.

McCarthy, E. (2011). *Ethics embodied: rethinking selfhood through continental, Japanese, and feminist philosophies*. Lanham, MD: Lexington Books.

McClamrock, R. (1995). *Existential cognition: computational minds in the world*. Chicago: Chicago University Press.

McDowell, J. (1982). Criteria, defeasability, and knowledge. *Proceedings of the British Academy*, 68, 455–79.

McNeill, W.E.S. (2012). Embodiment and the perceptual hypothesis. *The Philosophical Quarterly*, 62(248), 569–91.

Merleau-Ponty, M. (2002). *Phenomenology of perception*. Routledge: New York.

Michael, J. and de Bruin, L. (2015). How direct is social perception? *Consciousness and Cognition (Special Issue)*, 36, 373–5.

Newen, A. (2016). Defending the liberal-content view of perceptual experience: direct social perception of emotions and person impressions. *Synthese*. doi:10.1007/s11229-016-1030-3

Newen, A., Welpinghus, A., and Juckel, G. (2015). Emotion recognition as pattern recognition: the relevance of perception. *Mind & Language*, 30(2), 187–208.

Niedenthal, P.M. (2007). Embodying emotion. *Science*, 316(5827), 1002–5.

Noë, A. (2004). *Action in perception*. Cambridge, MA: MIT Press.
Nussbaum, M. (2001). *Upheavals of thought: the intelligence of emotions*. Cambridge: Cambridge University Press.
Osterman, J.E., Hopper, J., Heran, W.J., Keane, T.M., and van der Kolk, B.A. (2001). Awareness under anesthesia and the development of posttraumatic stress disorder. *General Hospital Psychiatry*, 23(4), 198–204.
Overgaard, S. (2006). The problem of other minds: Wittgenstein's phenomenological perspective. *Phenomenology and the Cognitive Sciences*, 5(1), 53–73.
Overgaard, S. (2007). *Wittgenstein and other minds: rethinking subjectivity and intersubjectivity with Wittgenstein, Levinas, and Husserl*. London: Routledge.
Overgaard, S. (2013). Other People. In: D. Zahavi (ed.), *The Oxford handbook of contemporary phenomenology*. New York: Oxford: Oxford University Press, pp. 460–79.
Overgaard, S. (2014). McNeill on embodied perception theory. *The Philosophical Quarterly*, 64(254), 135–43.
Pacherie, E. (2005). Perceiving intentions. In: J. Sàágua (ed.), *A Explicação da Interpretação Humana*. Lisbon: Edições Colibri, pp. 401–14.
Parnas, J. and Sass, L.A. (2008). Varieties of "phenomenology": on description, understanding, and explanation in psychiatry. In: K.S. Kendler and J. Parnas (eds.), *Philosophical issues in psychiatry: explanation, phenomenology, and nosology*. Baltimore, MD: The Johns Hopkins University Press, pp. 239–78.
Pickard, H. (2003). Emotions and the problem of other minds. In: A. Hatzimoysis (ed.), *Philosophy and the emotions*. Cambridge: Cambridge University Press, pp. 87–104.
Prinz, J.J. (2004). *Gut reactions: a perceptual theory of emotions*. Oxford: Oxford University Press.
Proust, J. (2003). "Perceiving intentions." In: J. Roessler and N. Eilan (eds.), *Agency and self-awareness: issues in philosophy and psychology*. Oxford: Oxford University Press, pp. 296–320.
Ratcliffe, M. (2007). *Rethinking commonsense psychology: a critique of folk psychology, theory of mind, and simulation*. New York: Palgrave Macmillan.
Rumelhart, D.E. and McClelland, J.L. (eds.) (1986). *Parallel distributed processing: explorations in the microstructure of cognition*. Cambridge: MIT Press/Bradford Press.
Scheler, M. (1954). *The nature of sympathy* (trans. P. Heath). London: Routledge and Kegan Paul.
Schnakers, C., Vanhaudenhuyse, A., Giacino, J., Ventura, M., Boly, M., Majerus, S. et al. (2009). Diagnostic accuracy of the vegetative and minimally conscious state: clinical consensus versus standardized neurobehavioral assessment. *BMC Neurology*, 9, 35.
Smith, J. (2010). Seeing other people. *Philosophy and Phenomenological Research*, 81(3), 731–48.
Solomon, R. (2004). Emotions, thoughts, and feelings: emotions and engagements with the world. In: R. Solomon (ed.), *Thinking about feeling: contemporary philosophers on emotions*. Oxford: Oxford University Press, pp. 76–88.
Spaulding, S. (2010). Embodied cognition and mindreading. *Mind & Language*, 25(1), 119–40.
Spaulding, S. (2015). On direct social perception. *Consciousness and Cognition*, 36, 472–82.
Stein, E. (1989). *On the problem of empathy* (trans. W. Stein). Washington, DC: ICS Publications.
Stout, R. (2010). Seeing the anger in someone's face. *Aristotelian Society Supplementary Volume*, 84(1), 29–43.
Taipale, J. (2015). Empathy and the melodic unity of the other. *Human Studies*, 38(4), 463–79.

Thompson, E. (2005). Empathy and human experience. In: J. Proctor (ed.), *Science, religion, and the human experience*. New York: Oxford University Press, pp. 261–85.
Watsuji, T. (1996). *Watsuji Tetsurō's Rinrigaku: ethics in Japan* (trans. Y. Seisaku and R.E. Carter). Albany: SUNY Press.
Winston, M.E., Chaffin, R., and Herrmann, D. (1987). A taxonomy of part-whole relations. *Cognitive Science*, 11(4), 417–44.
Wittgenstein, L. (1980). *Remarks on the philosophy of psychology* (vol. 1; G.H. von Wright and H. Nyman, eds.; trans. C.G. Luckhardt and M.A.E. Aue). Oxford: Blackwell.
Zahavi, D. (2011). Empathy and direct social perception: a phenomenological proposal. *Review of Philosophy and Psychology* 2(3), 541–58.
Zahavi, D. (2014). *Self and other: exploring subjectivity, empathy, and shame*. Corby, UK: Oxford University Press.

CHAPTER 16

CRITICAL NOTE

Cognition, Action, and Self-Control from the 4E Perspective

SVEN WALTER

This section on Cognition, Action, and Perception comprises four (somewhat heterogeneous) chapters. Roughly, two are about action, the other two about perception.

In "The Body in Action: Predictive Processing and the Embodiment Thesis," Michael Kirchhoff (2018) argues that predictive processing approaches to cognition (e.g., Clark 2016; Hohwy 2013) that regard cognitive processing as essentially a matter of the brain's minimizing prediction errors (see also Hohwy 2018 and de Bruin 2018) are not committed to some kind of "intracranialism," but compatible with four characteristic tenets of 4E approaches to cognition: (1) cognitive systems or processes are realized by patterns of sensorimotor activity nonlinearly coupled with the environment; (2) an organism's sensorimotor profile renders elaborate internal representations of the environment dispensable; (3) affectivity is not just a byproduct of perception or active inference, but part and parcel of perception, action, and cognition; and (4) plasticity is a feature not of the brain alone, but of the complex network of processes spanning the brain, the body in action, and the environment. In "Joint Action and 4E Cognition," Deborah Tollefsen and Rick Dale (2018) argue for an ecumenical approach to joint action (e.g., Huebner 2014; Tollefsen 2015) that integrates high-level, "intellectualist" accounts that explain joint action in terms of abstract, higher-order mental states with lower-level, "bottom-up" accounts that regard joint agency as arising out of basic perceptual and motor processes underlying the complex interaction of environmentally embedded individuals (see also Reddy 2018).

In "Perception, Exploration, and the Primacy of Touch," Matthew Ratcliffe (2018) considers several reasons why one might take touch to be the fundamental or "first" among the senses (e.g., Fulkerson 2014) that appeal to the idea that only touch involves the body, bodily activity, or a particular relationship to the world (or does so in a special way or to a special degree). He finds all such considerations wanting, concluding that if touch is indeed primary, it is not because of any particular (4E) characteristic, but because of its sheer diversity: touch accommodates so much that it is practically indispensable. Joel Krueger's (2018) "Direct Social Perception" defends an uncompromising (partial) solution to the traditional "other minds problem." Drawing on his previous work on embodied and extended emotions (e.g., Colombetti and Krueger 2015; Krueger 2014; Krueger and Szanto 2016; see also Carr et al. 2018 and Colombetti 2018), Krueger argues for a direct account of social perception according to which instead of having to infer what is going on in others' minds, we can (sometimes) directly perceive their behaviorally embodied mental states: we can, for instance, literally *see* their joy in their laughter, their grief in theirs tears, etc. (see also Gallese and Sinigaglia 2018 and De Jaegher 2018).

Below, I offer some critical thoughts on some philosophical issues touched upon in these chapters. In doing so, I by no means want to downplay the value of these contributions to a fascinating field that, despite the impressive advances made over the past two decades, is arguably still in its infancy. I highlight these issues mostly because, apart from revealing some problematic aspects of the arguments presented therein, they illustrate a general concern I have about some prominent debates in the context of 4E approaches to cognition: that at some times we are so excited that we (finally!) can bring philosophy in close touch with empirical results that we forget our core business as philosophers—the argument—while at other times we can't stop overdoing it with our philosophical concept-mongery and thereby fail to see important lessons empirical results have to teach us. In addition, I want to draw attention to a topic that one might have expected to be covered in a handbook on 4E cognition, in particular, in the section on Cognition, Action, and Perception, but that isn't addressed. A cognitive system that is to react to a given perceptual input with an appropriate behavioral output must—among many other things, but crucially—make sure it actually does what it resolved to do (e.g., Holton 2009) or what it thinks it ought to do, all things considered (e.g., Mele 2012). It must, in other words, be capable of *self-control*, i.e., it must be able to causally influence the motivational strength of its desires in the light of its higher-order volitions, values, preferences, etc. While it has traditionally been taken for granted that the (typical, best, or maybe only) way to firm one's "mental resolve" is to consciously summon thoughts that bolster one's long-term values, recent research suggests that self-control, far from being solely a matter of one's internal "willpower," is an embodied and embedded capacity of a cognitive system that can be enhanced by exploiting the system's body and environment in pretty much the sense that (other kinds of) cognitive processes are usually taken to be embodied and environmentally "scaffolded." But first some remarks on some of the issues raised in the other papers.

In July 2015, sections of one of the wings of Malaysia Airlines flight MH370 were found on Reunion Island. Even after these parts had been retrieved, though, the search for the airplane wreck went on. Krueger's main argument seems to rely on a principle that, if generally true, would seem to render the continuation of the search unintelligible—for by having seen parts of the wreck, the search parties had already seen the whole wreck. According to Krueger, the view that affective states are partially embodied in their visible behavior allows for a non-inferential solution to the traditional "other minds problem": since visible behavior is a *constitutive part* of affective states, we can literally see others' mental states by observing their visible behavior. I won't quarrel with the claim that the behavioral aspect of affective states is a constitutive part of them (after all, the claim that the body substantially contributes to an agent's affective life is far less controversial than the claim that cognition is embodied—except perhaps for a short period of radical cognitivism, emotions have never been taken to be a purely abstract, "fleshless" affair to begin with; see Stephan et al. 2014; see also Stephan 2018). This, however, does not provide us with a non-inferential solution to the other minds problem.

The claim that we can literally see mental states because they are embodied in directly observable behavior rests on an inference that, as suggested earlier, is questionable, to say the least. Bill Murray is a constitutive part of the cast of *Lost in Translation*, but by seeing Bill Murray, you don't see the cast of *Lost in Translation*. The famous scene where Ilsa says "Play it once, Sam" is a constitutive part of *Casablanca*, but you haven't seen Casablanca just because you saw that scene. When the Costa Concordia capsized in January 2012 after having hit a submerged rock formation, captain Francesco Schettino was probably to blame for a lot of things; but to accuse him of having seen the submerged rock just because he could see the end of the formation far away on the shore line would have been absurd. In these cases, we see x and x is a constitutive part of y, but we clearly do not see y. Krueger tries to dissipate this "part-whole objection" by distinguishing between a "member-collection relation" (that holds between, say, a wood and the individual trees) and a "component-integral object relation" where "certain components of 'integral objects' . . . may be so essential, structurally and functionally, that seeing a component *is* in fact sufficient to see the object of which it is a part" (p. 312). Much of course depends upon exactly is meant by a "structural and functional" relation between an integral object and its constitutive parts. But for all I can tell, my computer's CPU is as much an essential structural and functional component of my computer as Bill Murray's character was an essential structural and functional component of *Lost in Translation*; and yet you have not seen my computer just because I show you a picture of its CPU, nor have you seen *Lost in Translation* just because you saw Bill Murray.

But suppose for the sake of argument there were a legitimate sense in which by seeing, say, someone's laughter (the constitutive part), we saw her joy (the integral object). Would that show that we enjoy non-inferential access to her mental state? No. Suppose Anne's joy is partially embodied in her laughter, and suppose that by perceiving the latter we perceive the former. This alone doesn't get us any closer to knowing which mental state Anne is in. For consider Anna, a super-actor displaying the very same laughter without, however, feeling the least (let alone the same) joy. On the basis of our seeing Anne's laughter, we cannot tell whether she really is in joy (and we thus see her joy) or just perfectly faking it, like Anna (and we thus see no joy). Far from having immediate access to Anne's mental state, we have to rely on a defeasible inference that we will be prepared to give up if it turns out that other components of joy are missing (action tendencies, for instance, neurophysiological processes, etc.). Krueger's bodily constitution approach doesn't therefore render social perception direct. Compared to an intracranial approach that regards behavior as a bodily *effect* of brain-bound emotions, it (at best!) replaces one inference with another. According to the latter, we see Anne's laughter and infer that she is in joy, since her joy is plausibly (though defeasibly) a cause of the laughter. According to the former, we see Anne's laughter and infer that she is in joy since her laughter is plausibly (though defeasibly) a constituent of joy. Embodied approaches to emotions may be superior to intracranial approaches. But to think that they render social perception more direct or the other minds problem more tractable is, I believe, premature. Irrespective of all empirical considerations regarding emotions, establishing this claim requires a careful *philosophical* discussion of the points just raised (and others). Krueger certainly knows this, and he will have a riposte and be able to finesse his account of direct social perception to accommodate the above worries. I just mention this as one example where discussions in the field of 4E cognition could benefit from *more* philosophy; let us now turn to some debates where, in a sense, *less* philosophy wouldn't hurt.

Predictive processing (PP) approaches regard cognition as a matter of minimizing the mismatch between the sensory input the brain expects based on its generative model and the sensory input it actually receives. It thus seems natural to adopt a position that Kirchhoff dubs "internalist PP": body and environment *causally contribute* to modifying the brain's input in such a way that prediction errors are minimized, but they are not themselves *constitutive* of cognition, which is an entirely intracranial affair. According to Kirchhoff, however, predictive processing approaches are actually compatible with four central tenets of what he calls the "embodiment thesis" (ET), and in particular with the claim that cognitive processes are constituted by patterns of sensorimotor activity nonlinearly coupled with the environment. He acknowledges that the debate between "internalist PP" and ET-friendly versions closely resembles the debate about the notorious "coupling/constitution fallacy" (e.g., Adams and Aizawa 2001, 2008). One might thus have expected him to be overcautious in avoiding the fallacy. Strikingly, though, when it comes to the argument, Kirchhoff points out that "what internalist PP seems to miss or overlook is that unless an agent is able to engage in active inference, it cannot minimize prediction error" (p. 247). This, however, is not true. No advocate of internalist PP would (let alone has to) deny that an agent can minimize prediction error only by engaging in active inference. Quite generally, intracranialists can happily endorse counterfactuals like "If an agent's body and environment were different than they actually are, her cognitive life would be different as well." All they insist on is that what makes them true is that body and environment causally contribute to the agent's cognitive life instead of being constitutive of it. Kirchhoff goes on to claim that because "action is essential for perception . . . it follows that action—embodied activity—is part of the minimal realization base of prediction error minimization" (p. 248, fn. 3). This is just the coupling/constitution fallacy pure and proper: not every *x* that is essential for *y* is thereby *eo ipso* a part of *y*. The same holds when Kirchhoff claims that predictive processing "must be understood with respect to an organism's embedding environment" because "it is not possible to understand what counts as optimizing model evidence for some organism by abstracting away from the organism's local environment" (p. 248).

The fact that it is impossible to understand the ant's celestial system of navigation by abstracting away from the sunlight (e.g., Wehner and Müller 2006) does not make the sun part of what finds its way back to the anthill—it merely causally enables the ant to find its way back.

I am not belaboring this in order to criticize Kirchhoff's attempt to show what an ET-friendly version of a predictive processing approach could look like (to which I am sympathetic). My point is more general. Undoubtedly, dependence and constitution are different relations. And undoubtedly, there are cases where we have clear and unequivocal (and apparently well-warranted) intuitions which of the two is at issue: the CPU is a constituent of my computer, while the power plant that produces the electricity or the wall socket is only something upon which its functioning causally depends. In other cases, however, intuitions are much less clear: are the power cord or a pen drive constituents or merely external causal contributors? What if the pen drive is the only bootable device? Likewise for the cognitive case: is error minimization constituted by body and environment or merely causally dependent upon them? Are the memories of Clark and Chalmers's (1998) Alzheimer's patient Otto constituted by the entries of his notebook or only causally dependent upon them? The heated and persistent arguments about the "coupling/constitution fallacy" presuppose that somewhere among these borderline cases a sharp boundary can be drawn. But as I have argued at length elsewhere (Walter 2014), it is hard to see by which means (either empirical or a priori) we could ever decide whether a given cognitive process is constituted by or merely causally dependent upon extracranial processes (see also Baumgartner and Wilutzky 2017). As the (all too brief) discussion earlier indicates, any interpretation of a given system (such as the error minimization complex of brain, body, and environment) as either causal or constitutive will be theory-driven and could easily reach the opposite conclusion if driven by a different theory. Causation and constitution, though theoretically distinct, seem to be *indiscernible in practice* in the sense that there is no principled criterion, apart from purely pragmatic considerations, that could settle whether the relation between a given intracranial and extracranial process is one of dependence or constitution—which, strikingly, Kirchhoff seems to accept, indeed argue for, at other places (e.g., Kirchhoff 2014; for a detailed discussion of several candidates that all fail to do the job, see Walter 2014). If this is so, however, then arguing about the pros and cons of dependence vs. constitution is not a very interesting project. What *is* interesting is whether, say, predictive processing provides us with an adequate account of the phenomena to be explained. This is important, not only for philosophy, but for cognitive science in general. The question, however, whether predictive processing is constituted by or only causally dependent upon body and environment isn't. It is empirically underdetermined and not resolvable a priori. And, in any case, since both constitution and causation entail the same counterfactual dependencies among brain, body, and environment, it hasn't any real (empirical) impact. The same holds for another issue raised in this section, viz., the debate between "extended" and "distributed" approaches to cognition.

Drawing on previous work by Tollefsen (2006), Tollefsen and Dale distinguish between "solipsistic" and "collective" systems. The former are individual agents like

Otto whose cognitive states "extend" into their non-social (natural or technological) environment. The latter are complex systems "distributed" over multiple interacting agents utilizing natural and technological resources; for example, when the navigation of a vessel is accomplished by the concerted effort of the technologically equipped and socially organized crew (e.g., Hutchins 1995). Intuitively, the difference between solipsistic and collective systems, between extended and distributed (or collective) cognitive processes, seems clear enough: while extended cognitive processes have a solipsistic system as the "cognitive core" from which they *ex*tend into the environment, distributed cognitive processes are "spread out" over a collective system, no individual member of which can be singled out as the "cognitive core." It is *Otto* who, by relying on his notebook, has dispositional beliefs about the location of the MOMA, but it is not a single nautical officer who, by relying on the rest of the crew and technological equipment, navigates the vessel, but the *crew as a collective*.

Note, however, a curious consequence of this seemly straightforward distinction. If Otto relies on his notebook, he is a solipsistic system and *he himself* is the bearer of an *extended belief*; if, however, he instead relies (in the appropriate way) on his wife, he is part of an integrated system, and it is this *collective system* that has a *distributed belief*. According to Tollefsen and Dale, what accounts for this peculiar difference is the fact that in the latter case there is a "two-way interaction" (p. 271) that makes the Otto-cum-wife system a single collective system rather than a complex of interacting individual systems. But this seems to undermine the very motivation for adopting an extracranial perspective on cognition or mentality: as Clark and Chalmers (1998) stress, what motivates such an approach is that it allows for a straightforward *explanation of Otto's behavior*: Otto goes to 53rd Street because he wants to see an art exhibition at the MOMA and has the (extended) belief that it is on 53rd Street. But if Otto relies (in a two-way interaction) on his wife instead of his notebook so that the bearer of the (distributed) belief is not Otto, but the collective Otto-cum-wife system, we lose exactly this explanation. We can then no longer say that Otto went to 53rd Street because *he* had the (extended) belief that MOMA is there; all we can say is that Otto went to 53rd Street because the collective system of which he is a part believed that MOMA is there, but this doesn't seem to explain Otto's behavior qua individual.

One solution would be to ascribe cognitive processes always to individuals and regard other agents as extraorganismic cognitive resources inter pares, so that it does not matter whether Otto relies on his notebook or his wife. Wilson's (2004, 2005) social manifestation thesis, for instance, maintains that although some cognitive processes require an appropriately organized and technologically equipped social collective, the bearer of these processes is an individual member of the collective, not the collective itself. However, there are clear cases in which the bearer must be the distributed system because there simply is no single "cognitive core": a bill is not passed by one senator embedded in a collective of senators, but by the senate as whole, and while it is true that the blind rage of a lynch mob or a mass riot is possible only in the context of the mob or the mass, the rage itself, emerging as it does from the dynamical, top-down influenced interplay among the members, is clearly not a feature of any single member, but of the collective (see also Stephan 2018).

It thus seems as if we have to ascribe a belief to Otto even when he is relying on his wife instead of his notebook, and reserve distributed beliefs for truly collective systems that have no clearly detectable individual "cognitive core." If this is so, then there are not only solipsistic systems like the Otto-cum-notebook system and collective systems like ship crews or senates, there are also systems where some cognitive process requires an appropriately organized social collective, but where it is still one individual member and not the collective that is the bearer of that process. But where—and on exactly what grounds—should we draw the boundary between cognitive processes that are ascribed to individuals interacting with, among other things, other individuals and cognitive processes that are ascribed to groups of interacting individuals that together form an integrated collective? When is the interaction intricate enough to ascribe a cognitive process to a group rather than to one member that plays a particularly salient role? The point, again, is that if there is ever going to be an answer to this question, it is going to be theory-driven, not empirically motivated. And all it will do is tell us which complex systems should, given the theory in question, be called "extended," which "distributed" and which possibly require yet another label. It will, however, not advance our understanding of the functioning of these systems by an inch. The latter (I hope we all agree) is what matters; the former is just word-mongering. In an area of philosophy that often pats itself on the back for finally having brought philosophy in close contact with the empirical sciences, this is something we should remind ourselves of from time to time: many aspects of the debates between intracranialists and 4E afficonados, between advocates of internalist PP and more 4E-friendly versions, between extended and distributed approaches to cognition, between extended and embedded approaches, and so on and so forth, are, from the point of view of the empirical cognitive sciences, as detached from reality as many other traditional philosophical disputes. They are good old-fashioned metaphysics (after all, it's hard to altogether shed the philosopher in us). That isn't necessarily bad. But maybe we should, after almost 20 years of arguments about the coupling/constitution fallacy and the relationships between the different 4E(+D) approaches to cognition, slowly get over what is effectively nothing but a disagreement about nomenclature and focus on all the fascinating phenomena the field holds for us. Thankfully, many of these phenomena are discussed extensively in this volume. Since at least one of them, however, isn't, I will close with some (rather sketchy) remarks about it.

The behavior of an agent is based, at least sometimes (or ideally), on processes of deliberation, decision-making, intention formation, and self-control. Almost every second of our waking life is full with incentives (affordances; see Rietveld et al. 2018) for doing something: the World Wide Web is (almost constantly) there to be browsed, our emails are (almost constantly) there to be checked, the chocolate cake is there to be eaten, the wine to be drunk, the petunias to be watered, and (last but not least) our partner to be kissed. If we are to accomplish our long-term goals (finish this paper, lose weight, stay sober), we have to resist many if not most of these short-term temptations. In terms of Frankfurt's (1971) hierarchical model: we have to seek a "mesh" between our second-order volitions and our first-order desires. This keeps us busy about a fifth of our waking life. Hofmann et al. (2012) asked subjects at random times during the day to report whether they were currently suppressing an urge for food, sleep, distraction, nicotine, sex, sports, etc. On about half of these occasions the subjects reported having felt an urge, 42 percent of which they reported to have actively resisted (p. 1325). Unfortunately, both body and environment can hamper our attempts at self-control. On the bodily side, purely internal willpower seems to be a muscle-like bodily resource (dependent actually on the blood sugar level; Gaillot et al. 2007) that, if used for too long, will be depleted (e.g., Baumeister and Tierney 2011; Baumeister, Muraven, and Tice 2000), and with detrimental effects: subjects who have to exercise self-control on a marginal task (like resisting to eat chocolate chip cookies or holding their hands in ice water) show a greater tendency to impulsively spend money (Vohs and Faber 2007), be aggressive (Stucke and Baumeister 2006), drink alcohol before driving (Muraven et al. 2002), or show sexually inappropriate behavior (Gailliot and Baumeister 2007; for a review, see Baumeister et al. 2006). On the environmental side, recent work on "situationism" in social psychology (e.g., Miller 2017) has detailed many surprising ways in which the situation can unconsciously affect us in a way that limits and reduces our capacity to bring our actions in line with our "real self," thereby preventing a mesh between how we are and how we want to be, or at least making it more difficult to achieve (e.g., Walter 2016, chs. 11, 12). How, then, can we control ourselves?

Glossing over many details, the standard answer is that making our decisions and actions accord with our best evaluative judgments about what we should do is a matter of achieving causal control over the motivational strength of our mental states. There are different accounts of self-control (for an overview, see Henden 2008), but they all share the conviction that we gain control by means of something *internal*, either by forming an extra intrinsic *desire* to act in accordance with what we take ourselves to have most reason to do (e.g., Mele 1987), or by means of certain cognitive skills we possess (e.g., Kennett 2001), or by relying on a volitional faculty of "will" or "willpower" (e.g., Searle 2001). The problem with these internal strategies, however, is that they let us down. As even the most mediocre self-help book will point out, and as the research on "ego depletion" described earlier confirms, as soon as we start to rely on willpower, the battle is virtually lost—in particular, if situationism is right and we often aren't even aware that we are acting against our better judgment. Relying on our purely internal "strength of will" is like paddling upstream against the current: it is exhausting; the chances of success are dim; and we shouldn't do it if we can avoid it.

As said earlier, all extant accounts of self-control agree that achieving self-control is a matter of causally influencing the motivational strength of one's desires, values, preferences, etc. What they disagree about is *what kind of mental state this requires*. What they seem to overlook, though, is the possibility that it may not be a mental state at all by which we bolster the efficacy of our desires, values, preferences, etc. Just as we make our cognitive life easier by outsourcing part of our cognitive work to our morphology or an appropriately structured environment, we can make our volitional life easier by taking advantage of our bodily makeup and an active structuring of the environment (for an early statement of this idea, see McGeer and Pettit 2002). For instance, something so simple as firming one's muscles can increase one's ability to withstand pain, overcome food temptation, or consume unpleasant medicines (Hung and Labroo 2011), and physical activity has a positive impact on the agent's discounting delayed rewards (Sofis et al. 2017). The idea that body-based strategies of self-control might supplement and even be superior to internal strategies has already successfully reverberated into accounts of consumer behavior for developing health-related advocacy and communication campaigns (Petit et al. 2016) and into the treatment of eating disorders (Cook-Cottone 2015). Moreover, psychologists are currently developing "situational accounts of self-control" according to which we are often able to minimize the intrapsychic struggle typically associated with self-control by simply manipulating the environment to our advantage: if you want to stay sober, don't meet your friends at the pub and bank on your willpower to resist the temptation: do not go to the pub to begin with; if you are on a diet, don't study the dessert menu and count on your willpower to resist the temptation: do not ask for the dessert menu to begin with (e.g., Duckworth et al. 2016).

In the heyday of first-generation science, it may have seemed obvious that the faster, the better, or "the more" a cognitive system can "think," the better its performance. As it turned out, this is wrong: in many circumstances, those who perform the best are those who make sure they have to do as little "thinking" as possible. The same, it seems, holds for self-control. It is not those who have the most willpower that are most capable of self-control, but those who make sure they have to exercise as little willpower as possible (by, like Ulysses, "tying themselves to the mast"; see also Vierkant 2015). All of this, of course, has been sketchy, and exactly how it can be fruitfully integrated with philosophical accounts of self-control and autonomy remains to be seen. What is clear, however, is that self-control, far from being only an intracranial affair, is as much a capacity that can be enhanced by exploiting the system's bodily makeup and the active structuring of its environment as many (other) cognitive processes. Andy Clark once said: "Our brains make the world smart so that we can be dumb in peace" (1997, p. 180). They do even more. They make the world firm, so that we can be tempted in peace.

Acknowledgments

The author is a member of and was supported by the bilocal DFG Research Training Group ("Graduiertenkolleg") on Situated Cognition, based at the Ruhr University Bochum and Osnabrück University (2017–2021).

References

Adams, F. and Aizawa, K. (2001). The bounds of cognition. *Philosophical Psychology*, 14, 43–64.
Adams, F. and Aizawa, K. (2008). *The bounds of cognition*. Malden: Blackwell.

Baumeister, R., Gailliot, M., DeWall, C.N., and Oaten, M. (2006). Self-regulation and personality. *Journal of Personality*, 74, 1773–1801.

Baumeister, R., Muraven, M., and Tice, D. (2000). Ego-depletion: a resource model of volition, self-regulation, and controlled processing. *Social Cognition*, 18, 130–50.

Baumeister, R. and Tierney, J. (2011). *Willpower*. New York: Penguin Books.

Baumgartner, M. and Wilutzky, W. (2017). Is it possible to experimentally determine the extension of cognition? *Philosophical Psychology*, 30, 1104–25.

Carr, E.W., Kever, A., and Winkielman, P. (2018). Embodiment of emotion and its situated nature. In: this volume, pp. 529–52.

Clark, A. (1997). *Being there: putting brain, body, and world together again*. Cambridge, MA: MIT Press.

Clark, A. (2016). *Surfing uncertainty*. Oxford: Oxford University Press.

Clark, A. and Chalmers, D. (1998). The extended mind. *Analysis*, 58, 7–19.

Colombetti, G. (2018). Enacting affectivity. In: this volume, pp. 571–8.

Colombetti, G. and Krueger, J. (2015). Scaffoldings of the affective mind. *Philosophical Psychology*, 28, 1157–76.

Cook-Cottone, C. (2015). Incorporating positive body image into the treatment of eating disorders: a model for attunement and mindful self-care. *Body Image*, 14, 158–67.

de Bruin, L. (2018). False-belief understanding, 4E cognition, and the predictive processing paradigm. In: this volume, pp. 493–512.

De Jaegher, H. (2018). The intersubjective turn. In: this volume, pp. 453–68.

Duckworth, A., Szabó Gendler, T., and Gross, J. (2016). Situational strategies for self-control. *Perspectives on Psychological Science*, 11, 35–55.

Frankfurt, H. (1971). Freedom of the will and the concept of a person. *Journal of Philosophy*, 68, 5–20.

Fulkerson, M. (2014). *The first sense: a philosophical study of human touch*. Cambridge, MA: MIT Press.

Gallese, V. and Sinigaglia, C. Embodied resonance. (2018). In: this volume, pp. 417–32.

Gailliot, M., and Baumeister, R. (2007). Self-regulation and sexual restraint. *Personality and Social Psychology Bulletin*, 33, 173–86.

Gailliot, M., Baumeister, R., DeWall, C.N., Maner, J., Plant, A., Tice, D. et al. (2007). Self-control relies on glucose as a limited energy source. *Journal of Personality and Social Psychology*, 92, 325–36.

Henden, E. (2008). What is self-control? *Philosophical Psychology*, 21, 69–90.

Hofmann, W., Baumeister, R., Förster, G., and Vohs, K. (2012). Everyday temptations: an experience sampling study of desire, conflict, and self-control. *Journal of Personality and Social Psychology*, 102, 1318–35.

Hohwy, J. (2013). *The predictive mind*. Oxford: Oxford University Press.

Hohwy, J. (2018). The predictive processing hypothesis. In: this volume, pp. 129–46.

Holton, R. (2009). *Willing, wanting, waiting*. Oxford: Clarendon Press.

Huebner, B. (2014). *Macrocognition: a theory of distributed minds and collective intentionality*. Oxford: Oxford University Press.

Hung, I. and Laproo, A. (2011). From firm muscles to firm willpower: understanding the role of embodied cognition in self-regulation. *Journal of Consumer Research*, 37, 1046–64.

Hutchins, E. (1995). *Cognition in the wild*. Cambridge, MA: MIT Press.

Kennett, J. (2001). *Agency and responsibility, a common-sense moral psychology*. Oxford: Oxford University Press.

Kirchhoff, M. (2014). Extended cognition and constitution: re-evaluating the constitutive claim of extended cognition. *Philosophical Psychology*, 27, 258–83.

Kirchhoff, M.D. (2018). The body in action: predictive processing and the embodiment thesis. In: this volume, pp. 243–6.

Krueger (2014). Varieties of extended emotions. *Phenomenology and the Cognitive Sciences*, 13, 533–55.

Krueger, J. (2018). Direct social perception. In: this volume, pp. 301–2.

Krueger, J. and Szanto, T. (2016). Extended emotions. *Philosophy Compass*, 11, 863–78.

McGerr, V. and Pettit, P. (2002). The self-regulating mind. *Language and Communication*, 22, 281–99.

Mele, A. (1987). *Irrationality: an essay on akrasia, self-deception, and self-control.* Oxford: Oxford University Press.

Mele, A. (2012). *Backsliding: understanding weakness of will.* Oxford: Oxford University Press.

Miller, C. (2017). Situationism and free will. In: K. Timpe, M. Griffith, and N. Levy (eds.), *The Routledge companion to free will.* New York: Routledge, pp. 407–22.

Muraven, M., Collins, L., and Nienhaus, K. (2002). Self-control and alcohol restraint. *Psychology of Addictive Behaviors* 16, 113–20.

Petit, O., Basso, F., Merunka, D., Spence, C., Cheok, A.D., and Oullier, O. (2016). Pleasure and the control of food intake: an embodied cognition approach to consumer self-regulation. *Psychology & Marketing*, 33, 608–19.

Ratcliffe, M. (2018). Perception, exploration, and the primacy of touch. In: this volume, pp. 281–3.

Reddy, V. (2018). Why engagement? A second-person take on social cognition. In: this volume, pp. 433–52.

Rietveld, E., Denys, D., and van Westen, M. (2018). Ecological-enactive cognition as engaging with a field of relevant affordances: the skilled intentionality framework (SIF). In: this volume, pp. 41–70.

Searle, J. (2001). *Rationality in action.* Cambridge, MA: MIT Press.

Sofis, M., Carrillo, A., and Jarmolowicz, D. (2017). Maintained physical activity induced changes in delay discounting. *Behavior Modification*, 41, 499–528.

Stephan, A. (2018). Critical note: 3E's are sufficient, but don't forget the D. In: this volume, pp. 607–19.

Stephan, A., Wilutzky, W., and Walter, S. (2014). Emotions beyond brain and body. *Philosophical Psychology*, 27, 65–81.

Stucke, T. and Baumeister, R. (2006). Ego depletion and aggressive behavior. *European Journal of Social Psychology*, 36, 1–13.

Tollefsen, D. (2006). From extended mind to collective mind. *Cognitive Systems Research*, 7, 140–50.

Tollefsen, D. (2015). *Groups as agents.* Cambridge: Polity Press.

Tollefsen, D. and Dale, R. (2018). Joint action and 4E cognition. In: this volume, pp. 261–8.

Vierkant, T. (2015). Is willpower just another way of tying oneself to the mast? *Review of Philosophy and Psychology*, 6, 779–90.

Vohs, K., and Faber, R. (2007). Spent resources: self-regulatory resource availability affects impulse buying. *Journal of Consumer Research*, 33, 537–47.

Walter, S. (2014). Situated cognition: some open conceptual and ontological issues. *Review of Philosophy and Psychology*, 5, 241–63.

Walter, S. (2016). *Illusion freier Wille?* Stuttgart: Metzler.

Wehner, R. and Müller, M. (2006). The significance of direct sunlight and polarized skylight in the ant's celestial system of navigation. *Proceedings of the National Academy of Sciences*, 103, 12575–9.
Wilson, R. (2004). *Boundaries of the mind*. Cambridge: Cambridge University Press.
Wilson, R. (2005). Collective memory, group minds, and the extended mind thesis. *Cognitive Processing*, 6, 227–36.

PART V

BRAIN–BODY–ENVIRONMENT COUPLING AND BASIC SENSORY EXPERIENCES

CHAPTER 17

DISCLOSING THE WORLD

Intentionality and 4E Cognition

MARK ROWLANDS

The View

There is a view of cognitive processes that I've been peddling for more years than I care to remember (Rowlands 1995, 1999, 2002, 2003, 2006, 2010, 2011, 2013, 2015a, 2015b). The view consists of three related claims:

1. *Some* (not all, by any means, but some) cognitive processes are *partly* (not completely, obviously) made up of processes whereby an individual operates on (typically, manipulates, transforms, and/or exploits) structures in its environment.
2. The structures carry information that is relevant to the cognitive task in which the individual is engaged.
3. The processes are ones whose function is to make information *available*, either to the subject of those processes or to further processing operations.

Consider, for example, Kirsh and Maglio's (1994) account of what they call *epistemic action*: action whose function is to change the nature of the cognitive task to be accomplished. Two jigsaw pieces will carry information about their relative fit with each other. This information is present in the pieces, but is unlikely to be available without further action on our part. The further action in question involves physically manipulating the pieces, bringing them into proximity, turning them relative to each other, etc. This action transforms the information contained in the pieces from information that is merely present to information that is available to the subject: the pieces can now be judged to fit, or not fit, with each other, and the basis of this judgment is information that was always there but has now been made available through these actions.

I have argued (2010) that this making available of information is the essence of cognition: what cognition is really all about. Sometimes—as in the jigsaw case—the information that is made available is entirely present in the structures on which cognitive processes are performed. In such cases, cognition involves simple transformation of information that is *present* to information that is *available*: the former is identical with the latter. Often, however, the information that is present in the structures will form only a *part* of the information that is subsequently made available. In such cases, the information that is present is supplemented or embellished with further information—information that is effectively inserted by operations performed on the relevant structure. But whether it consists in simple transformation of information that is present into information that is available, or whether the information made available exceeds that originally present in the information-bearing structure, the idea of making information available—whether to the organism or to subsequent processing operations—lies at the heart of the concept of cognition. When they are embodied in the right sort of system—I shall talk more about this later—processes that have this function are cognitive ones. And some actions performed on the world can have such a function. That is why they are cognitive.

Claims 1–3 stake out a recognizable position on the spectrum of views that constitute what has become known as 4E cognition—cognition conceived of as extended, embodied, enacted, and/or embedded. First, the claims entail that cognitive processes are *extended*: the manipulative, transformative, or exploitative operations performed on environmental structures are processes that occur outside the body of the cognizing organism. The result is an extrabodily extension of cognitive processes but not of cognitive states. I'm not keen on identifying sentences in a notebook with cognitive states such as beliefs, for example (2010). Second, if the environment is construed broadly enough to include the bodily but non-neural—and that is the way I have construed it—then this characterization entails that some mental processes are embodied, in at least one clear sense of that multiply ambiguous term. Third, the claims also either entail, or are at least consonant with, a central enactivist theme of mental processes being a transaction between individuals and their environments. Fourth, the claims are compatible with, but stronger than, the claim that mental processes are often embedded in environmental scaffolding. The embedding claim is one of *causation*: the ability of an individual to engage in a cognitive process or complete a cognitive task is often causally facilitated by his, her, or its reliance on external information-bearing structures. The claim I defend is one of *constitution* rather than causation: some mental processes are partly constituted by, or composed of, actions performed on the world.

A view is one thing, motivating and/or defending it quite another. In recent years, my favored argument is based on a general *picture* of intentionality. A picture is not an analysis. I do not offer necessary, sufficient, or necessary and sufficient conditions for an item to count as intentional. What I do offer is a picture of intentionality—a sketch of the sort of thing intentional directedness is, one that highlights its more abstract and general features. This picture has its roots in the work of Frege, and also the early phenomenologists (Rowlands 2010, 2015a, 2015b). In this chapter, however, I shall not focus on the history of the picture, but aim at its further articulation.

The Picture

Intentionality is directedness toward the world. To understand intentionality is to understand what it is for an intentional act to be directed toward an object. The notion of an *object* should be understood in the broad sense to incorporate whatever an intentional act is directed toward. What sort of item this is depends, to a considerable extent, on one's preferred conception of intentional directedness. Possible contenders include objects simpliciter, or other (arguably) concrete particulars such as events and processes, but also (more plausibly, I think) facts or states of affairs. I shall remain neutral on the question of the proper objects of intentional directedness, and my use of the term "object" should be understood in a sense sufficiently broad to incorporate all of these candidates. Whatever is the object of intentional directedness, one thing is clear: if you are trying to understand intentional directedness, you will look in vain to the objects of that directedness. To understand intentional directedness is to understand what *makes* something an intentional object: to identify that *by means of which* or *in virtue of which* an item becomes an object of directedness.

In developing this basic idea I employed a model of intentionality that is—despite some lean years ca. 1970–2000—still sufficiently widely accepted to be dubbed the standard model. According to this, intentionality has a tripartite structure, comprising *act, object,* and *mode of presentation.* The mode of presentation connects act and object. Typically, the mode of presentation is understood in this way: the act has a content, perhaps (or perhaps not) expressible in the form of a description, and the mode of presentation is that in virtue of which the object, in whatever way is deemed appropriate, "fits" the content (for example, satisfies that description). I chose the standard model not because it is, itself, essential to the case I want to develop, but rather because it houses a crucial ambiguity. It is this ambiguity that is important. The ambiguity pertains to the concept of a mode of presentation.

The language I employed in exploiting this ambiguity might be unfamiliar to many. This is because it provides an extremely abstract description of a phenomenon of a certain sort. However, the phenomenon itself, at least in its concrete forms, is more familiar than one might realize. Consider one specific instance of this general phenomenon: certain tribulations encountered by Frege when he tried to articulate his notion of *sense.*

As several commentators have noted, there is a pronounced tension in Frege's account of sense. I shall argue that this tension forces us to distinguish two distinct concepts of sense. One of these I shall label "empirical" and the other "transcendental." But, first, consider the tension. This results from Frege's desire to attribute two distinct types of feature or function to *senses* or *thoughts* (*gedanken*—the senses of declarative sentences). On the one hand, Frege claims that senses can be objects of mental acts in a way akin to that in which physical objects can be the objects of mental acts (Harnish 2000). Physical objects can be perceived; senses or thoughts (that is, the sense of a declarative sentence) can be *apprehended.* Moreover, when a thought is apprehended, Frege (1918/1994) claims, "something in (the thinker's) consciousness must be *aimed at* the thought." In one of its guises, therefore, a sense is an intentional object of an act of apprehension. This is what I shall call the *empirical* version of sense.

However, according to Frege, a sense also has the role of fixing reference. Although senses can be objects of reference, that is not their only, or even typical, role. In its second guise, the function of sense is to direct the speaker's or hearer's thinking not to the sense itself but to the object picked out by that sense. In this case, senses do not figure as intentional objects of mental acts, but as items *in virtue of which* a mental act can have an object. In their second role, senses are *determinants* of reference: they are what fix reference rather than objects of reference. This is what I shall call the *transcendental* version of sense.

We might frame the distinction in terms of a metaphor employed by Frege, in the service of explaining the concept of apprehension:

> The expression "apprehend" is as metaphorical as "content of consciousness." . . . What I hold in my hand can certainly be regarded as the content of my hand but is all the same the content of my hand in a quite different way from the bones and muscle of which it is made and their tensions, and is much more extraneous to it than they are. (1918/1994, p. 35)

Understood in the first way—the empirical construal—a sense is akin to an object held in the hand. However, understood in the second way—the transcendental construal—sense is akin to that *in virtue of which* this object can be held in the hand.

These functions of sense are not merely distinct. More importantly, when sense is playing the empirical role of object of apprehension, it cannot also play the transcendental role of determiner of reference. This inability to play both roles simultaneously manifests itself as a certain *non-eliminability* that attaches to sense in its transcendental, reference-determining role. In its empirical guise, a sense is an object of apprehension: an intentional object of a mental act. But the second, transcendental, characterization of sense tells us that whenever there is an intentional object of a mental act, there is also a sense that fixes reference to this object. If we combine these characterizations, therefore, it seems we must conclude that whenever sense exists as an intentional object of a mental act of apprehension, there must, in that act, be another sense that allows it to exist in this way. And if this latter sense were also to exist as an intentional object of a mental act, there would have to be yet another sense that allowed it to do so. Sense in its, transcendental, reference-determining guise, therefore, has a non-eliminable status within any intentional act. In any intentional act, there is always a sense that is not, and in that act cannot be, an intentional object.

The picture of intentionality I defend involves generalizing this ambiguity beyond the specifically Fregean context. If an object satisfies or "fits" the content of a mental act, this will be because the object possesses certain *aspects*: ones picked out by that content. For example, on a (non-obligatory) description-theoretic construal of content, the description applies to whatever object satisfies it, and the object will satisfy the description in virtue of possessing certain aspects. Aspects are not to be identified with objective *properties* of objects. Aspects are objects of awareness in an intentional rather than objective sense. Aspects are the ways objects are presented to subjects—and, therefore, at least partially dependent on subjects—and to any aspect there may or may not correspond an objective property.

Since an object will fit the content of an act only in virtue of possessing certain aspects, and since the mode of presentation is that which links act and object, this invites an almost irresistible identification: we identify the *mode of presentation* of the object with that object's *aspects*. This identification is, however, problematic to the extent it presupposes the empirical understanding a mode of presentation, and, therefore, overlooks the possibility of a transcendental alternative, understood along the lines suggested by Frege's second conception of sense.

Aspects are intentional objects of awareness—and, therefore, empirical items in the sense explained earlier. I can attend not only to the apple, but also to its size, color, and luster. Indeed, typically, I attend to the former by attending to the latter. Thus, if we (1) identify modes of presentation with aspects, and (2) adhere to the standard model's claim that an intentional object is determined only via a mode of presentation, it follows that (3) whenever there is a mode of presentation—an aspect—there must be another mode of presentation that fixes reference to it. If a mode of presentation is identified with an aspect, and an aspect is an intentional object, then the standard model commits us to the existence of another mode of presentation in virtue of which the original mode of presentation can be an intentional object. This latter—transcendental—mode of presentation, necessarily, cannot be identical with an aspect. In short, intentional objects require modes of presentation. If modes of presentation are themselves intentional objects—and they would be if we identify them with aspects—there must be another mode of presentation that allows them to be as such. Therefore, we must distinguish two different ways of understanding a mode of presentation:

1. *Empirical modes of presentation (aspects)*: Often, indeed typically, the notion of a mode of presentation is understood as the way objects appear to subjects. If the apple appears green and shiny, then greenness and shininess is the mode of presentation of the apple. In this sense, the mode of presentation is an intentional object—it is the sort of thing of which I can become aware if my attention is suitably engaged. I can attend not only to the apple, but also to its greenness and shininess. An empirical mode of presentation is an intentional object. As such, it is identical with an *aspect* (or aspects) of an object.
2. *Transcendental modes of presentation*: In any intentional act, there must be more than an empirical mode of presentation. There must also be a *transcendental* mode of presentation. The reason is that the mode of presentation is supposedly what determines the intentional object of a mental act. So, if the object of an intentional act is an empirical mode of presentation (for example, the greenness and shininess of the apple), there must be another mode of presentation—a *transcendental* mode of presentation—in virtue of which this empirical mode of presentation is

> an intentional object. The transcendental mode of presentation is that component of the intentional act that permits the object to appear under empirical modes of presentation (or aspects).

The directedness of an intentional act toward the world, then, consists in its transcendental mode of presentation. The transcendental mode of presentation is the intentional core of an act. That is, *the directedness of an intentional act is that in virtue of which objects appear to a subject under aspects* (or empirical modes of presentation).

Intentionality as Disclosure

The general idea of an intentional act as that in virtue of which objects appear under aspects can be captured under the rubric *disclosure* (or *revelation*—I shall use these terms interchangeably). An intentional act is one that discloses an object (to a subject), where:

> A transcendental mode of presentation, TMP, discloses an object, O, to subject S *iff* TMP is that in virtue of which O falls under empirical mode of presentation EMP for S.

I understand the expression "in virtue of" in the right-hand clause as expressing a *sufficiency* claim. However, there are at least two relevant types of sufficiency, yielding two forms of disclosure. On the one hand there is logical sufficiency, and this yields what I shall call *constitutive disclosure*. What it is like to have an experience—one example of a TMP—constitutively discloses an object in that it provides a logically sufficient condition for the world (in some other guise) to appear in a certain way. Suppose I have an experience characterized by a certain *what-it-is-like-ness*. This particular what-it-is-like-ness is logically sufficient for the world to appear in a certain way—a localized shiny green. If I have an experience characterized by this *what-it-is-like-ness* then (a localized part of) the world cannot fail to appear shiny and green. It may be that my experience is illusory. If so, there is no apple that appears shiny and green, but some other object does so. Or it may be that I am hallucinating. If this case, there is no object that appears shiny and green. However, it is still true that a certain region of the visual world appears this way.

Constitutive disclosure takes the form of a logically sufficient condition for the world to appear a certain way. Contrast this sort of disclosure with that implicated in another account of vision—one cast at the level of neural or computational processes. At the neural level, we might describe various processes beginning in the retina and culminating in the visual cortex. At the computational level, we might describe various information processing operations that progressively convert a pattern of activation values distributed over the retina into a visual representation. Both neural and computational processes of this sort can qualify as TMPs. If, however, one were sympathetic to certain intuitions located in the broad "explanatory gap" genre, then one would deny that these sorts of processes are logically sufficient for the experience of greenness. But one could still accept that these are physically or causally sufficient for this experience. Disclosure that occurs by way of causal or physical sufficiency I shall call *causal disclosure*.

If we accept the explanatory gap intuitions, then we are committed to this general rule. *Constitutive disclosure* is disclosure by way of *content*. *Causal disclosure* is disclosure by way of *vehicles* of content. Nothing in my argument turns on acceptance of these intuitions. My focus will be on the vehicles of content and the corresponding idea of causal disclosure, since these are most germane to the issue of 4E cognition.

From Disclosure to 4E Cognition

Intentional directedness, I have argued, is that which discloses objects—causally or constitutively, depending on whether we are dealing with the content or vehicle—as falling under aspects or empirical modes of presentation. Intentional directedness is, therefore, a form of disclosing activity, broadly understood. Given this picture of intentionality, the argument for 4E cognition runs as follows:

1. Intentional directedness is disclosing activity.
2. Disclosing activity often—not always, certainly not necessarily—straddles neural processes, bodily processes, and things that a subject does to and with its environment.
3. Therefore, intentional processes often—not always, certainly not necessarily—straddle neural processes, bodily processes, and things that a subject does to and with its environment.

I have defended the first premise. In this section, the focus switches to Premise 2.

The argument for Premise 2 is perhaps best introduced through a well-known example of Merleau-Ponty: a blind person's cane. The cane can, of course, be an (empirical) object of awareness. The blind person might concentrate on how it feels, its weight, texture, and so on. But this, in ordinary contexts of use, would be rare. Rather, its role is, typically, a transcendental one. Understood transcendentally, the cane is something in virtue of which the world is disclosed to the blind person as falling under certain aspects or empirical modes of presentation. In virtue of activity that involves the cane, an object may be disclosed to the blind person as being "in front" of him or her, as "near," "further away," "to the left," and so on. Merleau-Ponty is at pains to emphasize the phenomenology of the resulting perception of the world. The blind person does not experience aspects of the objects he encounters as occurring in the cane, even though this is part of the material basis of his perception of these aspects. Still less does he experience them as occurring in the fingers that grip the cane, or in the sensory cortex that processes the information. He or she experiences these objects and their aspects as being located in the world. Phenomenologically—in terms of what it is like to have it—the consciousness of the blind person passes all the way through the cane out to the world. The crucial question, however, is: why should the phenomenology be like this?

The answer lies in the fact that when the blind person uses the cane in this way, the cane is not an *object* of disclosure but a *vehicle* of disclosure. It is part of an activity in virtue of which objects in the world are disclosed as falling under certain aspects or empirical modes of presentation. Where does this disclosing activity take place? No precise answer is available, but the activity straddles neural processes, extraneural bodily processes, and things the blind person does in and to the world around him or her. Not much disclosing of the world will get done without neural processes, to be sure. But in this case at least, the disclosing activity encompasses things going on in the blind person's body and also things he or she does in the surrounding environment. Intentional directedness is disclosing activity and so intentional directedness, in this case, straddles all of these things.

Indeed, this revealing activity does not stop short of the world. In employing the cane, the blind person ceases to experience the cane as such—ceases to experience it as an object of awareness. As revealing activity, his experience travels all the way through the cane to the object itself. That is why his experience can be a disclosing of the aspects of those objects. The experience's phenomenology passes all the way through the cane to the world because the underlying disclosing activity—the activity that constitutes the experience's intentional directedness—does not stop short of the world.

Consider, next, the case of an unimpaired visual subject. Suppose I am asked, à la Yarbus (1967), to look at a picture and identify certain information contained in it. For example, suppose I am asked to determine the approximate ages of the people in the picture. To accomplish this task, my eyes engage in a certain saccadic scan path. When I am asked a different question—for example, "What were the people in the picture doing prior to the arrival of the visitor standing in the door?"—the scan path my eyes follow will be very different. These scan paths are part of the causal disclosure of the world. The world is disclosed as containing an object—a painted figure—that falls under a given empirical mode of presentation: for example, as being a depiction of someone roughly 30 years old, or as containing figures who had been talking together prior to the arrival of the visitor. As such, the saccadic eye movements are disclosing activity: part of the means by which an object in the world is revealed to me as falling under an empirical mode of presentation. The saccadic eye movements are, therefore, among the vehicles of intentional directedness.

These are examples of perception. But the same general model can be applied to non-perceptual cases. Suppose I am thinking about an object—say, an apple—and I am thinking about its being unusually green and shiny. Greenness and shininess are empirical modes of presentation—aspects—of the apple. These are empirical items: I am thinking about them, therefore they qualify as intentional objects of my act of thinking. The transcendental mode of presentation of my thought about the apple, therefore, is that aspect of the act of thinking in virtue of which the apple is presented to me, in thought, as green and shiny. Cognition, no less than perception, reveals objects as falling under empirical modes of presentation. Both these objects and their empirical modes of presentation are objects of intentional directedness. The intentionality of both perception and cognition is precisely that in virtue of which one type of intentional object (an object simpliciter) is disclosed as possessing or falling under another type of intentional object (an aspect or empirical mode of presentation). The intentional directedness of both perception and cognition is this non-eliminable activity in virtue of which this sort of disclosure takes place.

The argument for 4E cognition, then, follows the same path. The revealing activity essential to cognition often straddles internal and external processes. Consider, for example, Clark and Chalmers's (1998) case of Otto. Otto is in the early stages of Alzheimer's and writes down information that he thinks might prove useful in a notebook. One day, having learned of an exhibition at the Museum of Modern Art, he leafs through his book and finds a sentence: "The Museum of Modern Art is on East 53rd Street." This sentence combines with his desire to see the exhibition, and the result is behavior: Otto makes his way to East 53rd Street.

According to a common interpretation of Clark and Chalmers's argument, the sentence, "The Museum of Modern Art is on East 53rd Street," should be identified with a belief of Otto. The sentences in his notebook are identical with a subset of his beliefs. I do not endorse this claim (and am not entirely sure that Clark and Chalmers do either). Such a view identifies a cognitive state with an external structure. In contrast, the view I endorse is formulated in terms of processes. Otto's manipulation of his notebook can form part—not all—of a mental process: the process of remembering. This is because manipulation of the book—flipping through the pages, reading the inscribed sentences—is a form of disclosing activity. The activity of manipulating the book in this way is part of the means whereby, in the case of memory, Otto's intentional directedness toward the world is brought about. The manipulation of the book is, in part, that in virtue of which a certain object in the world—a particular museum—is disclosed to Otto as falling under a specific empirical mode of presentation: that of being located on 53rd Street. Otto's disclosure of the world, in this case, straddles processes occurring in his brain (he must be able to read and recognize the sentences he sees on the pages), extracranial bodily processes (the movements of his fingers, his eyes, and so on), and things he does to the book (flipping through the pages, etc.). This is disclosing activity and, in this case, it is activity that straddles all these things.

Information Processing: The Present and the Available

A picture of intentionality is one thing. But any picture can be interpreted in a variety of ways. Someone might accept that intentional directedness is revealing or disclosing activity. They might also accept that this activity can straddle processes occurring both inside and outside a subject's brain. Nevertheless, they might insist that there is something special about the revealing activity that goes on inside the brain: something distinctive about intracranial revealing activity that makes it alone worthy of labels such as "mental" or "cognitive." The remainder of this chapter will argue against this.

Consider a classical—and I mean *very* classical—account of a cognitive process: David Marr's (1982) account of vision. The "very classical" status of this account is important for my purposes. Marr's account was instrumental in shaping how cognitive scientists came to think of cognitive processes. It is the implicated conception of cognitive processes that is important, rather than the details or defensibility of his account.

Visual perception, Marr tells us, begins with the retinal image. This structure consists in a distribution of light intensities across the retina. It contains comparatively little information, and is notoriously ambiguous. Therefore, if it is to be of much use, it needs to be *processed*. This processing is carried out in several stages. The first wave of processing transforms the retinal image into the raw primal sketch. This information-bearing structure is, then, available for a second wave of processing—the result of which is the full primal sketch. This new information-bearing structure is, then, available for a third wave of processing, the result of which is the 2½D sketch, and so on. The details of each structure and the type of processing to which each is subject need not detain us. What is of interest is the general vision of cognition implicated in this account. This vision comprises two essential elements:

1. Information-bearing structures.
2. Operations performed on those structures.

Each successive structure carries information, at least in a conditional sense. Subsequent to the retinal image, each structure is formed, in effect, by the brain's "guess" as to what sort of thing in the world might have caused a structure of this sort. The brain does this by applying various *principles* to each successive structure. For example, the transformation of the raw primal sketch into the full primal sketch is effected by the application of various grouping principles—proximity, similarity, common fate, good continuation, closure, and so on—to the raw primal sketch. If these principles are correctly applied in a particular case—if the brain "guesses" right—then the full primal sketch will contain more information about the visual environment than its predecessor.

In addition to information-bearing structures, there are operations performed on them. It is a truism that information-processing operations are ones performed on information-bearing structures. But what is often overlooked is what an information-processing operation actually is. Marr's model is useful because it makes this clear. Vision, on Marr's account, is an *input-output* function. The input is retinal image and the output is a 3D object representation. To understand vision is to understand how to get from input to output. To this end, the overall process is broken down into smaller input-output functions: the retinal image is the input and the raw primal sketch the output, and the latter in turn provides input for the processes that will produce another output, the full primal sketch, and so on. To understand vision is to understand how to get from each input to each output, and this requires understanding the processes that effect such transformations.

Important for our purposes, however, are not the details of the processes but their overall function, understood in a relatively abstract way. At the heart of the picture of cognitive sketched by Marr is the idea of making information available. Processing operations performed on the retinal image result in the raw primal sketch. This carries new information, information that was not contained in the retinal image. This new information is now available for further processing, and the result is the full primal sketch. This, also, carries new information that was not present in the raw primal sketch, and this is now available for further processing, and so on. These operations can culminate either at the personal or subpersonal level. In the latter case, information is made available to the cognizing subject, rather than to further subpersonal processing operations. But whether it culminates at the personal or subpersonal level, the dominant theme is clear: the function of information processing is to make information *available*.

More precisely, the function of information processing is to make information available, either to the subject or to subsequent processing operations. In its simplest form, this would consist in transforming information from that which is merely present in a structure to that which is available. In such cases, the information that is present in the structure is the same as the information that is made available for further processing. The Kirsh and Maglio case of the jigsaw pieces is of this sort. Marr's model, however, is not of this simple form. Rather, the operations performed on each information-bearing structure augment the information contained therein. The information that is made available to further processing, therefore, is not the same as the information contained in the structure processed: it contains, or involves, the original information that was present in the structure together with additional information added by the operations performed on that structure. Nevertheless, the result of these operations performed on the structure is information that is *available* for further processing operations (or to the organism if these operations have reached completion).

The concept of *availability* can be understood in terms of the distinction between information that must be *constructed* and information that can be *detected*. The information contained in the raw primal sketch is *available* to subsequent processing operations in the sense that it can be *detected* (and thereby used) by those operations. No further constructive operations are required for the processes that subsequently operate on the raw primal sketch to do their specific work. If this condition is met, the information is available to detection by the subsequent processing operations.

Like the earlier treatment of intentionality, the preceding discussion is intended to provide a picture, rather than an analysis, of cognition: an adumbration of the sort of thing cognition is. The two pictures are intimately related. Not all systems that exemplify the sort of profile described earlier—one in which operations are performed on information-bearing structures in order to make information available—will qualify as cognitive systems. Cognition occurs when processes of the form outlined earlier are embodied in an intentional system. An intentional system is a system to which there can be such a thing as disclosure, in the sense explained earlier. Humans and many other conscious creatures are the most obvious example of such systems, but delineating the class of world-disclosing entities is beyond the scope of this chapter.

External Information Processing

If the preceding section is correct, it provides us with a general, abstract vision of the nature of the disclosing activity occurring inside the brain of a cognizing organism. The function of this activity is to transform information that is present in structures, perhaps augmenting it in the process, to information that is available—whether to the organism or to subsequent processing operations. This, at an extremely abstract level, is what cognition is. But, if so, there is no reason to suppose that cognition is restricted to what is going on inside the brains of cognizing organisms. This general schema is also applicable to the processes occurring outside the head when an organism successfully engages in cognitive tasks.

Consider, for example, Gibson's (1979) ecological alternative to Marr's model. Central to this account is the concept of the *optic array*. The visual environment is filled with rays of light traveling between the surfaces of objects. At any point in space, light will converge from all directions. Therefore, at each point there is what can be regarded as a densely nested set of solid visual angles, composed of inhomogeneities in the intensity of light. Thus we can imagine the observer as a point surrounded by a sphere divided into various segments. The intensity of the light and the mixture of the wavelengths vary from one segment to another. This spatial pattern of light is the optic array.

The optic array is an *external information-bearing structure*. It is external in the obvious sense that it is located outside the skin of the perceiver. It is information-bearing in virtue of the fact that the structure of the array is determined by the nature, location, and orientation of the surfaces from which light has been reflected. The boundaries between the segments of the array, since they mark a change in intensity and distribution of wavelength, provide information about the three-dimensional structure of the visual environment. At a finer level of detail, each segment will be subdivided in a way determined by the texture of surface from which the light is reflected, and so carries information about further properties of the surfaces of objects.

The second defining element of the classical conception of cognition consists in operations whose function is to make available—either to the cognizing subject or to further processing operations—information contained in relevant structures. This element is also present in Gibson's model. Gibson argued that perception is inextricably bound up with action—a theme that enactivist accounts would later come to accord a prominent role. Perceiving organisms actively explore their environment. The optic array does not merely impinge on passive observers. When an observer moves, the entire optic array is transformed, and such transformations contain information about the layout, shapes, and orientations of objects in the world.

By moving, and thus effecting transformations in the optic array, perceiving organisms can identify and appropriate what Gibson called the *invariant* information contained in the array. This is information that can be made available to an observer not by any one static array but only in the transformation of one optic array into another. Acting on the array, and thereby transforming it is, therefore, a means of changing the status of information in the array from that of the (merely) present to that of the available.

The same general picture emerges if we return to Clark and Chalmers's case of Otto. The notebook is a store of information that is, obviously, external to Otto's skull. That is, information is present in the book, but that is all. When Otto acts on the book—rifles through the pages and scans the various sentences—he transforms the information from information that is merely present (i.e., contained in the book) to information that is available to him. This is precisely the function of the operations Otto performs on the external information-bearing structure that is the book.

I could go on—at length—but word constraints militate against. The essential features of the classical conception of cognition seem to be:

1. Information-bearing structures: structures that carry information relevant to the accomplishing of a given cognitive task, and,
2. Operations performed on those structures, where,
3. The function of these operations it to make available, sometimes via augmentation, the information contained in those structures.

At this level of abstraction, essentially the same schema is applicable to processes occurring outside the skin or skull of cognizing individuals. If our concern is with information processing, the distinction between neural and extraneural is not one that carries any overriding theoretical significance.

Original Intentionality: Inside and Outside the Skull

One might accept that the same sorts of operations are performed on internal and external structures and, nevertheless, maintain that there is a crucial difference between inner and outer. This is because the *structures* on which the operations are performed are different. The most common, and important, version of this objection is based on the idea of *original intentionality*: some internal (i.e., neural) structures have it, but no external ones do (Adams 2010; Adams and Aizawa 2008; Fodor 2009).

The distinction between *original* and *derived* intentionality is widely, although not universally, accepted (for a dissenting voice, see Dennett 1988). Words and sentences are the clearest examples of items that have only derived intentionality. They are intentional in the sense that they are about things—objects or state of affairs. But, in themselves, they are merely patterns of shapes on a contrastive background. As such, they can mean anything—or nothing. In order to have meaning, they must be interpreted. This interpretation is achieved through mental acts, mediated by linguistic conventions. Thus, the intentionality of these linguistic items derives from the intentionality of mental acts: it is *derived* intentionality. The intentionality of mental acts, on the other hand, is generally assumed to be quite different. Thoughts, beliefs, desires, and emotions are all about things. But their intentionality does not derive from other mental acts. It is, accordingly, *original* intentionality.

This original intentionality objection is particularly forceful when directed at versions of 4E cognition that identify cognitive *states* with external *structures*. Beliefs have original intentionality. The sentences in Otto's notebook do not. Therefore, how can these sentences be identical with beliefs? (Fodor 2009). My version of 4E cognition, however, involves no such identification. It is cognitive *processes* that are (partly) constituted by manipulation and transformation of external, information-bearing structures. And while the external information-bearing structures do not have original intentionality, the processes of manipulating and transforming them certainly do (Menary 2006). Otto's manipulation of his notebook, for example, involves perceptual and linguistic processing—Otto has to read the sentence—and this will involve states with original intentionality.

Nevertheless, this response may seem less than satisfactory. While the overall process might involve original intentionality, this always attaches to segments of the process that occur inside the brain of the cognizing organism. Nothing outside the brain has original intentionality (Adams 2010). Therefore, nothing outside the head should be regarded as mental. And since cognition is a species of the mental, this means that nothing going on outside the head can be regarded as cognition.

To this objection, two responses are available. The first, which I have pursued elsewhere (Rowlands 2011), is to argue that many crucial segments of cognitive processes occurring inside the head do not have original intentionality, and this in no way precludes their counting as cognitive. Here, however, I shall pursue a different strategy.

Why should one be so confident that nothing outside the head possesses original intentionality? Words and sentences are, indeed, merely shapes inscribed on a page. But mental representations are merely patterns of activity inscribed in the brain—electrical and chemical activity, both underwritten by biological construction. Why suppose that original intentionality can belong *only* to the latter sort of inscription?

The puzzle deepens when we consider standard naturalistic criteria of intentionality. Despite the lack of general consensus on how to naturalize intentionality, there are certain commonly accepted criteria of representation that feature prominently in most attempts at naturalization. That is, there are certain conditions that an item must satisfy in order to qualify as capable of representing—or being about—a state of affairs distinct from it. The underlying approach is naturalistic if these conditions presuppose no intentional concepts. The following conditions are common:

1. *Informational*: An item *r* is about some state of affairs *s* if and only if it carries information about *s*.
2. *Teleological: r* is about *s* if and only if it has the proper function either of tracking *s* or of enabling an organism (the representational consumer) to achieve some task in virtue of tracking *s*.
3. *Misrepresentation: r* can be about *s* only if it is capable of misrepresenting *s*.
4. *Combinatorial: r* can be about *s* only if *r* is part of a wider representational framework.
5. *Decouplability: r* can be about *s* only if *r* is decouplable from *s*.

Not every theory need adopt all these constraints. For example, informational theories can be combined with teleological theories—informational-come-teleological theories, as we might think of them (Dretske 1986). However, there are also theories that adopt the informational constraint but eschew the teleological constraint (Fodor 1990). And there are theories that embrace the teleological constraint but reject the informational constraint (Millikan 1984). Nevertheless, these disagreements aside, one can conclude that if an item were to satisfy all of these constraints, then, from the perspective of a naturalistic account of intentionality, there would be a strong case for regarding it as intentional. The issue of whether original intentionality can be found outside the brain, then, amounts to the issue of whether non-neural items can satisfy these conditions.

With this in mind, consider an example of what I shall call a *deed* (Rowlands 2006, 2012). You are trying to catch a ball in a high velocity sport such as cricket. You have less than half a second before the ball reaches you. To compound difficulties, the ball is moving toward you at a tricky height: lower chest height. To complete the catch you have an awkward decision to make—whether to point your fingers up or down.

Pointing your fingers up or down is, in this case, an example of a *deed*. So too are all the additional online, feedback-modulated adjustments of the fingers that you will have to make in order to successfully receive the ball. In general, deeds are hierarchically structured: deeds have other deeds as components, and the order of performance is a function of the task. Deeds occupy a middle ground in the gamut of human behavior:

1. They are things we do rather than things that happen to us. In catching the ball it is not as if we discover our hands and fingers moving of their own accord—that would be a very alien and unnerving experience.
2. They are distinct from sub-intentional acts in O'Shaughnessy's (1980) sense. They are not at all like random tongue-waggings or toe-tappings where, as O'Shaughnessy puts it, reason plays neither a positive nor negative role in their genesis. Deeds are things we do precisely because we have general antecedent intentions we wish to satisfy.
3. They fall short of intentional action in the strict sense. Like actions, intentional states may play a role in explaining the status of a deed as something we do. I move my fingers because, ultimately, I have a general antecedent intention to catch the ball. However, unlike actions, general antecedent intentions are not sufficient to

> individuate deeds. An entire array of feedback-modulated adjustments may go into satisfying one general antecedent intention.

Actions, in the standard sense beloved of philosophers, are individuated by way of associated intentions, volitions, tryings, etc.—depending on one's preferred view of action. In any given case of behavior, which actions are present, and how many actions there are, depends on the antecedent intention, volition, or trying. Thus, while it is common to suppose that actions are intentional, it is also generally assumed that their intentional status derives from the intentionality of their associated intentions, volitions, or tryings.

No such individuative relationship holds between intentional states and deeds. Many deeds can be involved in satisfying a single intention, volition, or trying, and these intentional states, therefore, cannot individuate deeds in the way an intention, volition, or trying individuates an action. Thus, if (1) deeds could be shown to be intentional, there would (2) be no reason to suppose this is inherited from the intentionality of an associated intention, volition, or trying. They would be intentional in an original sense—or, at least, there would be no reason to suppose otherwise.

It was the burden of an earlier work of mine (Rowlands 2006) to show that some deeds—and in particular the sorts of deeds involved in manipulating and transforming external information-bearing structures—can satisfy all the naturalistic conditions of intentionality listed earlier. That was a book-length treatment, and each condition was allotted an entire chapter. The following is an unacceptably, but unavoidably, brief synopsis of the argument.

Consider, again, the simple binary deed introduced earlier. This can satisfy most, arguably all, of the listed naturalistic constraints on intentionality. The position of the fingers carries information about the trajectory of the ball; at least it does so to no lesser extent than traditional mental representations carry information about things extrinsic to them. It has the proper function either of tracking the trajectory of the ball or of enabling the organism—the catcher/consumer—to do something in virtue of tracking this trajectory. The proper function of fingers pointed up is either to indicate a trajectory of a certain sort or to enable the catcher to make a successful catch in virtue of such indication. It can certainly misrepresent: if the fingers are pointed up when they should have been pointed down, the catcher will receive a painful knock on the heel of the hand. It can form part of a larger combinatorial system. For example, pointing the fingers up then makes possible the more subtle modulation of fingers necessary to receive the ball. It makes no sense to modulate your fingers in this way unless and until the fingers are pointed up. It can also, I argued, be decoupled from the ball, in at least one sense of that multiply ambiguous idea.

Constraints of space do not permit me to develop these claims in any satisfactory way. In place of argument I shall instead, in the spirit of Wittgenstein, merely offer an invitation to think about the issue in a certain way. We should always remember that intentionality is a *relational*, not an *intrinsic* feature of things that have it. We might be tempted to suppose that something magical happens when we venture into the brain: that what we find in there is quite unlike anything else. Perhaps it is. But it is not what we find in there that is crucial. It is the relations in which these inner items stand to things outside them that determine whether or not these items are intentional. If we assume a naturalistic account of intentionality, then the relations in question are the ones identified in the previous criteria, or some subset thereof. Is there any reason for thinking that only neural inscriptions can enter into these sorts of relations? Apart from some rudimentary, and wholly inconclusive, remarks about the nature of signs, it is not clear if any real thought has been devoted to this question. We are accustomed to thinking that intentionality must emanate from the inside out: that its real locus is to be found inside the head, and everything outside must be intentional in only a derived sense. But this picture of intentionality is unsupported by satisfactory argument or evidence.

Conclusion

Intentionality is disclosing activity. Disclosing activity often—not always, not necessarily—straddles processes occurring in the brain, in the wider non-neural body, and in things that creatures do to and with the world around them. This, in my view, is the ultimate justification for thinking of cognition in (broadly) 4E terms. Disclosing activity can take place in the head, or it can take place (in part) outside the head. Processes occurring inside the head and outside it are fundamentally akin in this respect: they both comprise operations performed on information-bearing structures, where the function of these operations is to make information available, to the subject or subsequent processing operations, either by transforming the status of information in the structures from present to available, or by also augmenting that information. And no decisive reason has yet been given for supposing that the structures involved are fundamentally different. If this is correct, there is no reason to suppose that cognition is bound by skin or skull.

Acknowledgments

I would like to thank Tobias Starzak for some very helpful comments on an earlier version of this chapter.

References

Adams, F. (2010). Why we still need a mark of the cognitive. *Cognitive Systems Research*, 11, 324–31.
Adams, F. and Aizawa, K. (2008). *The bounds of cognition*. New York: Wiley-Blackwell.
Clark, A. and Chalmers, D. (1998). The extended mind. *Analysis*, 58, 7–19.

Dennett, D. (1988). Evolution, error and intentionality. In: *The intentional stance*. Cambridge, MA: MIT Press, pp. 278–99.
Dretske, F. (1986). Misrepresentation. In: R. Bogdan (ed.), *Belief*. Oxford: Oxford University Press, pp. 17–36.
Fodor, J. (1990). *A theory of content, and other essays*. Cambridge, MA: MIT Press.
Fodor, J. (2009). Where is my mind? *London Review of Books*, 31(3), 13–5.
Frege, G. (1918/1994). The thought: a logical inquiry. In: R. Harnish (ed.), *Basic topics in the philosophy of language*. Englewood Cliffs, NJ: Prentice Hall, pp. 517–35.
Gibson, J. (1979). *The ecological approach to visual perception*. Boston: Houghton Mifflin.
Harnish, R. (2000). Grasping modes of presentation: Frege vs Fodor and Schweizer. *Acta Analytica*, 15, 19–46.
Kirsh, D. and Maglio, P. (1994). On distinguishing epistemic from pragmatic action. *Cognitive Science*, 18, 513–49.
Marr, D. (1982). *Vision*. San Francisco: W.H. Freeman.
Menary, R. (2006). Attacking the bounds of cognition. *Philosophical Psychology*, 19, 329–44.
Millikan, R. (1984). *Language, thought and other biological categories*. Cambridge, MA: MIT Press.
O'Shaughnessy, B. (1980). *The will* (2 vols.). Cambridge: Cambridge University Press.
Rowlands, M. (1995). Against methodological solipsism: the ecological approach. *Philosophical Psychology*, 8, 5–24.
Rowlands, M. (1999). *The body in mind: understanding cognitive processes*. Cambridge: Cambridge University Press.
Rowlands, M. (2002). Two dogmas of consciousness. In: A: Noë (ed.), *Is the visual world an illusion?* Special ed. of *Journal of Consciousness Studies*, 9, 158–80.
Rowlands, M. (2003). *Externalism: putting mind and world back together again*. London: Acumen.
Rowlands, M. (2006). *Body language: representation in action*. Cambridge, MA: MIT Press.
Rowlands, M. (2010). *The new science of the mind: from extended mind to embodied phenomenology*. Cambridge, MA: MIT Press.
Rowlands, M. (2011). Intentionality and embodied cognition. *Philosophical Topics*, 39(1), 81–97.
Rowlands, M. (2012). Representing without representations. *Avant*, 3(1), 133–44.
Rowlands, M. (2013). Enactivism intentionality and content. *American Philosophical Quarterly*, 50(3), 303–16.
Rowlands, M. (2015a). Bringing philosophy back: 4E cognition and the argument from phenomenology. In: D. Dahlstrom, A. Elpidorou, and W. Hopp (eds.), *Philosophy of mind and phenomenology: conceptual and empirical approaches*. New York: Routledge, pp. 310–26.
Rowlands, M. (2015b). Consciousness unbound. *Journal of Consciousness Studies*, 22(3–4), 34–51.
Yarbus, A. (1967). *Eye movements and vision*. New York: Plenum Press.

CHAPTER 18

BUILDING A STRONGER CONCEPT OF EMBODIMENT

SHAUN GALLAGHER

EMBODIED cognition has been reframing our understanding of mind, brain, perception, action, and cognition more generally, since at least the 1990s. Even prior to the publication of *The Embodied Mind* by Varela, Thompson, and Rosch (1991), and Rodney Brooks's "Intelligence without Representation" (1991), there had been ongoing work on embodied cognition in phenomenological philosophy from the time of Merleau-Ponty (1945) onward (e.g., De Waelhens 1950; Gallagher 1986; Henry 1965; Young 1980). In philosophy, in a growing number of publications since that time, some of the deeper resources found in phenomenology, pragmatism, and ecological psychology have been gaining a foothold in the cognitive sciences (see Gallagher 2014).

As we see in the present volume, however, there is no consensus theory of embodied cognition (EC), and debates continue about the best way to understand this notion. The alternatives range from conservative models that remain close to cognitivist conceptions of the mind, to more moderate and radical camps that argue we need to rethink our basic assumptions about the way the brain and the mind work. Most recent debates have been focused on the pragmatic and action-oriented perspectives of ecological, enactive, and extended conceptions, which either minimize reliance on the notion of representation or eschew it altogether (e.g., Chemero 2011; Clark 2008; Hutto and Myin 2013; Thompson 2007).

Lawrence Shapiro (2014) worries that, unlike chemists and biologists, researchers in the area of EC are unable to reach consensus about how to answer questions concerning domain of investigation, what the central concepts in that domain are, and why embodied cognition is an improvement over older paradigms. This should not be a worry, however, because embodied cognition is not a science like chemistry or biology, even if it is something like cognitive science, which is Shapiro's third example. At best, EC is a research program *within* cognitive science; and in some respects it is more like a philosophical framework for research in cognitive science. Although Shapiro suggests that cognitive science does have good answers to the questions he poses, I'm not sure that's true. First, his answer to the question of which concepts are necessary for doing cognitive science, namely, as he suggests, "information, representations and algorithms" (p. 74), are precisely some of the concepts that are under current debate, in part because of challenges from EC. And second, cognitive science is interdisciplinary in a way that chemistry and biology (as Shapiro portrays them) are not. If you ask such questions of someone working in artificial intelligence, you will not necessarily get the same answers as when you ask a cognitive neuroscientist, a philosopher of mind, or a cognitive anthropologist. Any of these researchers may also be working on EC. In fact the kinds of disparate research topics that Shapiro lists under the heading of EC—e.g., motor behavior, robot navigation, action-sentence compatibility, the role of metaphor in concept acquisition, and various questions about perception—are in fact topics that are addressed in cognitive science. Indeed, cognitive science and the field of EC include a bit of chemistry (since hormones and neurotransmitters have some effect on cognition) and more than a bit of (neuro- and even extraneural) biology. Given that they also include philosophy, one should expect that there will be ongoing debates and dissensus in almost every corner.

Ecological, enactive, and extended approaches to cognition propose different ways in which the body plays a role in shaping cognition. One idea is that the non-neural body processes information both prior to and subsequent to central manipulations (e.g., Chiel and Beer 1997); another is that representations are minimal and action-oriented (Clark 1997; Wheeler 2005); still another is that the body itself plays a representational role (Rowlands 2006). Enactivist approaches suggest that sensorimotor contingencies (O'Regan and Noë 2001), bodily affects (Colembetti 2013; Gallagher and Bower 2014), and posture and movement (Gallagher 2005) all enter into cognition in a nonrepresentational way. For some the idea that the body is dynamically coupled to the environment is important (Thompson 2007; Di Paolo 2005); for others the idea that action affordances are body- and skill-relative (Chemero 2011) is essential. All of these ideas help to shift the ground away from orthodox cognitive science.

In these different proposals there are serious disagreements about the nature and role of representation, the precise nature of body–environment coupling, the role of the brain, the nature of affordances, and so on. One seeming point of agreement, however, is that the body as such is an important factor in cognition. There seems only one clear exception to this minimal consensus: specifically, what Alsmith and Vignemont (2012) call "weak," in contrast to "strong" embodiment. Strong embodiment endorses a significant explanatory role for the (non-neural) body itself in cognition; weak embodiment gives the significant explanatory role to what are variously called body or body-formatted (neural) representations (e.g., Glenberg 2010; Goldman 2012, 2014; Goldman and de Vignemont 2009).

Weak embodied theorists still retain the term "embodied," but, in fact, for them the body per se is not necessarily involved in the real action of cognition. Rather, the real action occurs in the brain. Indeed, the body, in this version of embodied cognition, is the "body in the brain" (a phrase that has appeared frequently since Berlucchi and Aglioti 1997 introduced it; e.g., Berlucchi and Aglioti 2010; Kammers 2008; Tsakiris 2010). So, what precisely is this concept of weak or minimal embodiment, and how does it work? Alvin Goldman, building on the work of a variety of researchers in psychology, neuroscience, linguistics, and philosophy, has provided the most developed account of this approach so far. In this chapter, I'll focus on Goldman's model of weak embodiment to see how it measures up to stronger models of EC.

Let me make one prefatory note. The perspective I'll be concerned with is large. By this I mean that the claims made about the various approaches to EC are not focused on a particular problem or particular type of problem in cognitive science. Rather, the claims being made by the various theorists I'll consider are generally overarching claims about the right way to approach any problem about cognition. It may be that given the current state of science, and a particular problem to solve, one of the approaches may have a better account on offer than any of the others. Each of the theories to be considered may have distinctive explanatory value for solving specialized problems. I acknowledge that in trying to solve a particular problem, a theorist is not attempting to solve all problems or to give the entire story of cognition. For the most part, however, I'll be concerned with the larger claims about the best way to conceive of the overall system.

Weak Embodiment

Goldman (2012, p. 85) proposes a "unifying and comprehensive" account of embodied cognition that is not only "the most promising one for promoting an embodied approach to social cognition" (Goldman and Vignemont 2009, p. 155), but to cognition more generally. On this account, however, actual bodies play a marginal and perhaps even trivial role in cognition. Rather, body-formatted (or B-formatted) representations in the brain do most of the work.

Goldman rules out aspects of anatomy, sensorimotor contingencies, and environmental couplings as relevant to cognition, and he makes it clear that all important aspects of cognition can be found in the brain: "The brain is the seat of most, if not all, mental events" (Goldman and Vignemont 2009, p. 154). The processes involving B-formatted neural representations are purely internal to the brain and in some sense disembodied. As Shapiro (2014) suggests, they could just as well be thought to occupy a well-equipped vat. Goldman introduces one careful qualification to this sort of claim, namely, that the contents of such representations require the brain to be embodied since it is "possible (indeed, likely)" that the contents will depend on what the representations "causally interact with. . . . Envatted brain states would not have the same contents as brain states of ordinary embodied brains" (2014, p. 104). The body, in its own peculiar way, is thus somewhat better than a vat for delivering the right kind of information to the brain.

What precisely are B-formatted (or B-coded) representations? Goldman (2012; Goldman and Vignemont 2009) admits that it is not clear what a format or mental code is, although they are frequently referenced in cognitive science. Such codes are classically thought to be language-like (propositional), with distinctive vocabularies and syntactical procedures or rules. Visual representations have a visual format, and other modalities have their own distinctive formats (Jackendoff 1992). B-formats, however, are non-propositional, and specifically they "represent states of the subject's own body, indeed, represent them from an internal perspective" (Goldman 2012, p. 73). Jesse Prinz (2009, p. 420) suggests that "such representations and processes come in two forms: there are representations and processes that represent or respond to the body, such as a perception of bodily movement, and there are representations and processes that affect the body, such as motor commands." In Goldman and Vignemont (2009), mirror neurons were given as an example of B-formatted representations, and the claim was that B-formatted representations likely had limited application, primarily in the realm of social cognition. Somatosensory, affective, and interoceptive representations may also be B-formatted, "associated with the physiological conditions of the body, such as pain, temperature, itch, muscular and visceral sensations, vasomotor activity, hunger and thirst" (Goldman and Vignemont 2009, p. 156). In subsequent papers, however, Goldman expands their role, and accordingly provides a wider set of examples.

B-formatted representations may *originally* have an interoceptive or motor task such that the content of the representation in some way references the body; they may involve proprioceptive and kinesthetic information about one's own muscles, joints, and limb positions, for example. Thus, *primarily*, B-formatted representations are interoceptive or motoric representations "of one's own bodily states and activities" (Goldman 2012, p. 71). Importantly, however, Goldman considers such information, which may originate peripherally, to be B-formatted only when represented centrally: "for example, codes associated with activations in somatosensory cortex and motor cortex" (2012, p. 74), but also the "interoceptive cortex" (Craig 2002) registering "pain, tickle, temperature, itch, muscular and visceral sensations, sensual touch, and other feelings from (and about) the body" (2012, p. 74). Exteroceptive information about one's body, however, comes by way of vision, touch, and so forth, and these modalities involve their own, non-B-formatted representations.

The terms "originally" and "primarily" play an important role in these explanations since the notion of B-formatting is only the first part of a two-part theory. If we stayed with only this first definition of B-formatting, we would have a quite limited explanatory scope, allowing explanation of cognitive operations that concern only the body per se—operations that take the body as an interoceptive object, or that involve body-specific processes. To expand the scope of application, Goldman adopts Michael Anderson's (2010) "massive redeployment hypothesis," i.e., the idea that, within an evolutionary time frame, neural circuits originally established for one use can be reused or redeployed for other purposes while still maintaining their original function. So, for example, mirror neurons start out as motor neurons involved in motor control; but they get exapted in the course of evolution for purposes of social cognition and now are also activated when one agent sees another agent act. Any cognitive task that employs a

B-formatted representation, in either its original function or its exapted/derived function is, on this definition, a form of embodied cognition.

Goldman points to another example of this reuse principle in linguistics. Pulvermüller's (2005) language-grounding hypothesis shows that language comprehension involves the activation of action-related cortical areas. For example, when a subject hears the word *lick*, one finds activation in a sensorimotor area that involves the tongue; action words like *pick* and *kick* activate cortical areas that involve hand and foot, respectively. Language comprehension thus reflects the reuse of interoceptive, B-formatted motor representations. This suggests that "higher-order thought is grounded in low-level representations of motor action" (Goldman 2014, p. 97). From here it is a short run to the type of work done by Glenberg (2010) and Barsalou (1999), or by Lakoff and Johnson (1999), showing how, by simulation or metaphor, respectively, one can explain the embodied roots of abstract thought. Moreover, to the extent that memory involves activation of motor control circuits (Casasanto and Dijkstra 2010), or counting involves activation of motor areas related to the hand (Andres et al. 2007), these cognitive activities should be considered as instances of embodied cognition.

Although Goldman at first seems to separate perception from this kind of embodied cognition, indicating that vision, touch, and so forth, involve their own, non-B-formatted representations, he does concede that B-formatted representations play some role in perception. He leans heavily on the work of Proffitt, who shows that bodily states (fatigue, physical fitness), anticipation of bodily effort, and even perception of one's own body can modulate perceptual estimations of distance, slope, and size of objects in the environment (Bhalla and Proffitt 1999; Linkenauger, Ramenzoni, and Proffitt 2010; Witt and Proffitt 2008). On Goldman's interpretation such bodily factors are monitored by the brain, which generates information that is integrated with B-formatted representations (e.g., internal motor simulations) that, in turn, inform perceptual judgments about distance, slope, size, etc. In the case of judging distance to a target, for example, "subjects made distance judgments from an 'actional' perspective" (2012, p. 83). If the subject is fatigued or if task A will take more effort than task B, with respect to the target, the distance will seem longer.

> The subject tries to reenact the cognitive activity that would accompany the motor activity in question—without actually setting any effectors in motion. During this series of steps—or perhaps at the end—the energetic or physiological states of the system are monitored. Distance judgments are arrived at partly as a function of the detected levels of these states. (2012, p. 83)

As Goldman acknowledges, Proffitt does not provide details about the internal mechanism; and as we should acknowledge, neither does Goldman's description. The agent simulates the action in question; in the simulation the agent's brain detects levels of energy required for the task, and this somehow informs distance judgment. Details aside, Goldman's point is that this type of process exemplifies neural reuse where B-formatted representations are redeployed for various cognitive tasks.

A Check-Up for Weak Embodiment

As Goldman shows, there are neuroscientific and behavioral studies that are consistent with his version of EC. Besides the research already mentioned, the work on mirror neurons and canonical neurons is such a good fit that Vittorio Gallese has now defined his notion of "embodied simulation" in terms of B-formatted representations and the reuse hypothesis (Gallese 2014; Gallese and Sinigaglia 2011). Gallese also favors Anderson's systematic, evolutionary account of "neural reuse" and the notion of exaptation (2014, p. 6). Thus, for Gallese, embodied simulation "and the underpinning [mirror mechanisms,] by means of neural reuse can constitutively account for the representation of the motor goals of others' actions by reusing one's own bodily formatted motor representations, as well as of others' emotions and sensations by reusing one's own visceromotor and sensorimotor representations" (2014, p. 7).[1]

The very same behavioral and neuroscientific studies that Goldman appeals to in order to support the notion of B-formatted cognition, however, also support the more radical notions of embodied cognition that suggest the body itself, or more precisely, the body as it is coupled to the environment, and not just B-formatted representations, plays a constitutive role in cognition. To be clear, it is the very same scientific data that is appealed to in all cases; the debate is about the interpretation of the data. Scientists rightly insist on data; but the real theoretical action is in their interpretation. Are activations of mirror and canonical neurons simulations that representationally repeat the world in an internal model? Or are they part of a dynamical response that involves an adjustment of the larger organism–environment system? If the latter, then we need to rethink what cognition is, what the mind is, and how the brain actually works. I think these are some of the real challenges posed by EC, and not only does Goldman's model not go far enough in that direction, it creates a detour back toward classical cognitivism. Alsmith and Vignemont (2012, p. 2) nicely formulate the issue as:

> whether positing body representations actually undermines the explanatory role of the body, in the same manner in which positing representations of the world has been thought to undermine the explanatory role of the environment (cf. Brooks 1991). Alternatively, perhaps certain types of representation are so closely dependent upon the non-neural body (i.e., the body besides the brain), that their involvement in a cognitive task implicates the non-neural body itself.

Goldman clearly rejects this latter alternative. The B-formatted internalist version of EC clearly does exclude a role for the body itself as well as for the

[1] Gallese distances himself from any simple conception of weak EC by acknowledging that the body as a whole plays a role in cognition, and that links between body and the functional properties characterizing the multimodal motor neurons that control its behavior are central (Gallese 2017).

environment.[2] Nonetheless, Goldman considers his proposal about B-formats to involve a real change from classic cognitivism. Indeed, as he points out, some critics would say it departs too radically from cognitivist models insofar as it gives sensorimotor processes a central role to play in what, according to classic cognitivism, should be disembodied, abstract processes (e.g., Mahon and Caramazza 2008; see Goldman 2014). In important respects, however, as Goldman also acknowledges, weak EC is quite consistent with the internalist and representationalist tendencies of classic cognitivism, and thereby inconsistent with other more radical accounts of EC—"radical" at least insofar as they do not exclude the body and environment as important factors in cognition. As Alsmith and Vignemont put it, ideas found in weak EC "are but a hair's breadth away from the (also familiar) neurocentric idea that cognitive states are exclusively realized in neural hardware" (2012, p. 5). I'm not sure that they are even that far away.

Not only is weak EC contrary to most accounts of EC, I think Goldman's explication of it leads to some problems about how to understand one of its major suppositions. The following statement, from Randall Beer, which expresses a central proposition shared by most of the other accounts of EC, one that is seemingly missing from the weak EC account, captures part of my worry: "Given that bodies and nervous systems coevolve with their environments, and only the behavior of complete animals is subjected to selection, the need for . . . a tightly coupled perspective should not be surprising" (Beer 2000, p. 97). Specifically, I want to consider weak EC's use of the reuse hypothesis in light of its evolutionary understanding.

As we've seen, Goldman acknowledges that the notion of reuse involves the evolutionary concept of exaptation which, as Gallese explains, "refers to the shift in the course of evolution of a given trait or mechanism, which is later on reused to serve new purposes and functions" (2014, p. 6). Neural reuse means that neural areas evolve in a way that allows them to be activated for tasks that were not the original tasks associated with such areas. Anderson (2010) clearly states this as an evolutionary concept and he distinguishes it from a similar phenomenon that has been proposed as a developmental concept, Dehaene's (2005) "neuronal recycling" hypothesis, which explains ontogenetic changes in the visual system (specifically in the "visual word form area") when the person learns to read. Anderson also distinguishes neural reuse from normal neural plasticity that may happen in any individual brain.

Anderson (2010) and Goldman (2012) mention Broca's area as a good example of neural reuse. The area homologous to what is Broca's area in the human has, in the monkey, a motor function, and Broca's area continues to involve this original function—movement preparation, action sequencing, and action imitation. But, of course, in the human, the area in question has evolved linguistic and action recognition functions. This is a clear example, framed in evolutionary terms.

[2] Goldman is not alone. Barsalou's notion of grounded cognition, for example, refers to the idea that cognition operates on reactivation of motor areas but "can indeed proceed independently of the specific body that encoded the sensorimotor experience" (2008, p. 619; see Pezzulo et al. 2011).

To the extent that the reuse hypothesis involves the evolutionary concept of exaptation, however, it is not a principle that applies at the level of token activations or representations in individual brains. In the case of a token activation of a specific brain area, the brain is not *reusing* for cognitive purposes neurons (or representations) that have a primary body-related purpose, although that is sometimes how Goldman seems to describe it. For example, he describes "reusing or redeploying B-formats to execute a fundamentally non-bodily cognitive task" (2012, p. 83). One can object to this on a narrow technical point. The reuse hypothesis explains how B-formats have come about across an evolutionary time scale, but reuse doesn't explain the actual deployment of B-formats. It's more correct to say "Cognition (token) C is a specimen of embodied cognition if and only if C *uses* some (internal) bodily format to help execute a cognitive task (whatever the task may be)" (2014, p. 103, emphasis added). Goldman's gloss on this statement, however, suggests that "uses" really means "reuses":

> Instead of saying that it suffices for a cognition to qualify as embodied that it reuse a B-code, the test for embodiment (via reuse) should require both that the cognition reuse a B-code *and* that it reuse the B-code in question *because* this B-code has the function of representing certain bodily states. This might indeed be a helpful addition. (2014, p. 107)

Likewise Goldman references the mirror system where activation of mirror neurons "as part of a representation of what another person is doing, or planning to do, [thereby involving] a different cognitive task than the fundamental [i.e., primary] one . . . constitutes a redeployment [reuse] of the motoric format in a novel, cognitively interpersonal, task" (2012, p. 79). Of course this may be a mere point of terminological divergence, and Goldman is entitled to his own usage (or reusage) of this concept. Nonetheless, even if we set this technical-terminological issue aside and focus on reuse in its phylogenetic or ontogenetic meaning, we find that this idea undermines the purely internalist account of cognition favored by weak EC.

On the evolutionary time scale, what accounts for this reuse possibility is the fact that the human brain has evolved specifically with the human body, which through the course of evolution gained the upright posture, leading to a restructuring of its skull and jaw to allow larger brain development and speech, along with many other morphological changes in the body. On the time scale of evolution, reuse has everything to do with the body—including its morphological features, which are dismissed as trivial by Goldman and Vignemont (2009, p. 154). Brain evolution does not happen in vitro or in a vat.

Not only the body, but also physical, social, and cultural environments are important factors both evolutionarily and, following Dehaene, developmentally for any understanding of neural reuse or neuronal recycling. The role of the cultural environment, relevant especially in developmental contexts, for example, remains unstated, but implicit even in Goldman's discussion of neural reuse as it applies to Pulvermüller's work in neural linguistics. Citing this as one of the best examples of neural reuse, however, comes along with the obvious relation between language and culture. That is, if we think of the reuse hypothesis in these terms, important parts of the story, even those parts that involve neural plasticity (via association, and Hebbian learning) in individuals, involve cultural learning (Overmann 2016). Even if Pulvermüller is right,[3] it should be clear that action words like *pick* and *kick* will not activate cortical areas that involve hand and foot, respectively, if we are scanning French speakers or speakers of Urdu.

The role of culture and context (including bodily and environmental factors), however, applies equally in token cases of cognition. Naumann (2016), for example, suggests that motor simulations related to word processing in the context of a sentence are more specific than the meaning represented by the abstract verb outside of a sentence. For example, the simulation of "Bill grasped the needle" will be different from "Bill grasped the barbell," since the shape of the hand grasp will be different. Importantly, the simulation will take shape not only if one knows what a barbell is, or what a needle is, but to some significant extent it will depend on the history of one's use of such items, and one's skill level. One can certainly ask about the difference in neural attunement between a novice and an expert seamstress or weightlifting body builder. Likewise, what we mean by "needle" can differ from one context to another (a sewing needle, a compass needle, a hypodermic needle, a tall structure in Seattle), and with respect to how we plan to use it, the reach and grasp will in each case be different due to very basic body-schematic and intentional factors. In any case, things like barbells and needles are not entities that find appropriate explanation in the evolutionary framework. If there is relevant semantic somatopic function in this regard, it's a matter of plasticity and cultural learning—which means it's a matter of *metaplasticity* where not only neurons undergo plastic changes, but bodily and cultural practices do too (Malafouris 2010; Overmann 2016).

That we are led directly to consider the role of body and environmental context, including in some cases cultural context, through considerations about the notion of reuse (whether understood in evolutionary or developmental, or as applied to token examples of cognition, as Goldman proposes) means that one would have to back off any strong statement of weak EC.

Body Building

The brain is the way it is, and operates the way it operates, because it evolved along with the body. The brain is part of the body, and has always been part of the body. If

[3] There is some evidence against this type of semantic somatotopy. Not all neuroimaging studies have found increased activity for action-verbs in the motor system (Bedny and Caramazza 2011); activations found in premotor areas reflect no necessary overlap or match of neural areas correlated with motor action and neural areas that activate for the language task (Willems et al. 2009); Postle et al. (2008) showed the same general activation in premotor cortex in response to action-verbs related to leg, arm, and hand, with no somatotopical differentiation, suggesting that such activation may simply be related to general language use (Bedny and Caramazza 2011). For fuller discussion, see Naumann (2016).

the human body were different—if humans had not attained the upright posture, for example, or did not evolve with hands, sensory and motor systems would be different (more attuned to the olfactory than to vision, for example), and cognition would not function in the specific way it functions now. Indeed, we would likely have to redefine what we mean by rationality.

This is not to deny the reuse hypothesis, but to keep it within the right framework. It also suggests that the body itself places important constraints on how reuse works. Indeed, there is a good clue about cognition in the reuse hypothesis. Specifically, it's likely that original use continues to govern the possibility of the reuse and how it functions. If the reuse hypothesis suggests that our perceptual-motor systems were originally and primarily designed for *action*, and specifically, only the kind of actions that our bodily form made possible—locomotion on two feet, reaching to grab near objects of certain sizes, etc.—then this primacy of action, and the specific characteristics of action affordances delimited by a human body with hands and feet, etc., arguably carry through to the reuse of these systems in different cognitive contexts of planning, language use, social interaction, etc. Thus, for example, mirror neuron activation differentiates between actions performed by others who are nearby—within reachable, peripersonal space—and actions performed by others who are not nearby (see Caggiano et al. 2009), specifying types of social affordances and possible interactions. The brain attunes itself to what the body and what the environment afford. This understanding of reuse would support a more enactivist version of EC. In contrast, to discount the actual physical body and its various properties and capabilities, or the physical and social environment (which is, at the same time, to discount evolution, development, brain plasticity, and the very real constraints of physical existence), and to replace all of this with B-formatted representations, is to offer an oversimplified, "sanitized"[4] cartoon of cognition.

A more radical EC response to the B-formatted version would, of course, emphasize the role of the body itself, as well as the environment, in individual instances of cognition. In a token act of cognition in the individual human, the brain does what it does in tandem with the body. When the bodily system is fatigued or hungry, which is something more than a neural state, these conditions influence brain function; the body regulates the brain as much as the brain regulates the body. Parts of the brain, e.g., the hypothalamus, operate on homeostatic principles rather than anything that can be construed as representational principles. Homeostatic regulation happens via mutual (largely chemical) influences between parts of the endocrine system and related processes in the autonomic system. Low glucose levels (hypoglycemia, which is a biological condition not caused by the brain) may mean slower or weaker brain function, or some things turning off, or at the extreme, brain death. In cases of hypoglycemia, perception is not modulated because the brain *represents* hunger and fatigue, but because the perceptual system (brain and body) is chemically (materially) affected by the actual

[4] This is Goldman and Vignemont's term: "Both proponents and opponents of EC might criticize our interpretations on the grounds of being excessively tame or 'sanitized.' There is some justice to this charge, but we regard sanitized variants as scientifically and philosophically fruitful" (2009, p. 155).

hunger and fatigue. There are real physical connections here in the complex chemistry of the body–brain system in its coupling with the environment. My hunger may not affect my perception so much if a sexually attractive person walks into the room; although this has something to do with processes in the hypothalamus, it is not the result of a confused hypothalamic B-representation.

Accordingly, embodiment as it relates to cognition is not just about anatomical structure, body parts, sensorimotor contingencies, and action capabilities. The body's *affective* life regulates brain function via hormonal and neurotransmitter levels. Together with peripheral and autonomic systems—including heart function (Garfinkel et al. 2013) and respiration (Liu et al. 2014)—*affect* shapes cognitive function. There are numerous relevant examples in the literature. Consider the 2011 study by Danziger et al. providing evidence that hunger can distort higher-order cognitive processes.[5] The study shows that whether a judge is hungry or satiated may play a significant role in judicial decisions. Favorable rulings drop from ≈65 percent to near zero from early morning to just before lunch, and rise again to ≈65 percent just after lunch. It's difficult to fit this kind of fact into a B-formatted representation. Perner and Ogden (1988) suggest that hunger is a "nonrepresentational internal state." But even if there were some kind of neural representation of hunger (in the orbital frontal cortex, or hypothalamus, for example; Tataranni et al. 1999), it's not at all clear how a B-representation of hunger would somehow get reused as a B-representation of harsher legal sentences.[6]

Moreover, to fully understand the cognitive event that constitutes the judge's legal ruling, one has to consider not only nonrepresentational and unruly processes complicated by autonomic and endocrine responses, but also social and institutional environments and practices, some of which would lead the judge to feel justified, and to think that she has reached a just decision (Gallagher 2013). Affective phenomena such as these, related in part to body chemistry and in part to environmental conditions, are good examples of how cognition can be the result of complex brain–body–environment couplings that simply can't be captured by the concept of B-formatted representations.

Conclusion

There are substantive differences in theoretical approaches between weak EC approaches, like Goldman's B-formatted conception of the body in the brain, and

[5] A similar example involves changes in the perception of sexual attractiveness due to testosterone depletion (hypogonadism) or across menstrual phase (Alexander and Sherwin 1991; Johnston et al. 2001).

[6] Mahon and Caramazza (2008) suggest that justice is an abstract disembodied concept; Lakoff and Johnson (1999) claim it involves an embodied image-schema (having to do with balance), and in Goldman's terms that would count as a B-formatted representation. No matter how we think of the roots of this concept, the actual cognitive process involved in administering it, which is something Andy Clark might call a representation-hungry problem, seems to be short-circuited by nonrepresentational hunger, and then supported when homeostasis is reinstated. In either case, cognition involves a real body.

theories that include the full body in its dynamical gestalt-like relations with its physical, social, and cultural environments. One might think that we could opt for a hybrid theory that would combine a B-formatted explanation of how the brain works internally with the externalist, ecological, enactive, and extended conceptions that would allow for all of the extraneural elements involved in cognition. It's difficult, however, to see how we might mesh the extraneural roles of peripheral, autonomic, and ecological aspects of embodiment, along with the physical, social, and cultural aspects of the environment, with a sanitized version of neural representations. The more radical theories force us to rethink the way the brain works. They suggest that the brain—attuned by evolutionary pressures and by the individual's personal (social and cultural) experiences, in its interwoven relations with body and environment—or in other words, the system as a whole, with its own structural features that enable specific perception-action loops, that in turn shape the structure and functioning of the nervous system—*responds to* the world rather than *represents* it. Specifically, it responds not by representing, but through a dynamical participation in a large range of messy adjustments and readjustments that involve internal homeostasis, external appropriation and accommodation, and larger sets of normative practices.[7] I've been suggesting that not just the brain, not just the body with its different systems, not just the physical and social environment—but all of these together play important roles in cognition in ways that are irreducible to B-representations. All of these factors would simply undermine the explanatory role of B-representations, and mess up the sanitized picture presented by weak embodied cognition.

Acknowledgments

The author acknowledges the support of the Humboldt Foundation's Anneliese Maier Research Award, and a Senior Research Visiting Fellowship at Keble College, Oxford. He also thanks Rob Rupert for his helpful comments on an earlier draft.

References

Alexander, G.M. and Sherwin, B.B. (1991). The association between testosterone, sexual arousal, and selective attention for erotic stimuli in men. *Hormones and Behavior*, 25(3), 367–81.

[7] It's beyond the scope of this chapter to address more positive accounts of how a human might accomplish the higher-order tasks that some consider representation-hungry. These involve debates about representational vs. nonrepresentational frameworks. For enactivist accounts of some of these issues, see Gallagher (2017) and Hutto (2015).

Alsmith, A.J.T. and de Vignemont, F. (2012). Embodying the mind and representing the body. *Review of Philosophy and Psychology*, 3(1), 1–13.
Anderson, M.L. (2010). Neural reuse: a fundamental reorganizing principle of the brain. *Behavioral and Brain Sciences*, 33, 245–66.
Andres, M., Seron, X., and Olivier, E. (2007). Contribution of hand motor circuits counting. *Journal of Cognitive Neuroscience*, 19, 563–76.
Barsalou, L.W. (1999). Perceptual symbol systems. *Behavioral and Brain Sciences*, 22, 577–660.
Barsalou, L.W. (2008). Grounding cognition. *Annual Review of Psychology*, 59, 617–45.
Bedney, M. and Caramazza, A. (2011). Perception, action and word meanings in the human brain: the case from action verbs. *Annals of the New York Academy of Sciences*, 1224, 81–95.
Beer, R. (2000). Dynamical approaches to cognitive science. *Trends in Cognitive Sciences*, 4, 91–9.
Berlucchi G. and Aglioti S.M. (1997). The body in the brain: neural bases of corporeal awareness. *Trends in Neurosciences*, 20, 560–4.
Berlucchi, G. and Aglioti, S.M. (2010). The body in the brain revisited. *Experimental Brain Research*, 200, 25–35.
Bhalla, M. and Proffitt, D.R. (1999). Visual-motor recalibration in geographical slant perception. *Journal of Experimental Psychology: Human Perception and Performance*, 25, 1076–96.
Brooks, R.A. (1991). Intelligence without representation. *Artificial Intelligence*, 47(1–3), 139–59.
Caggiano, V., Fogassi, L., Rizzolatti, G., Thier, P., and Casile, A. (2009). Mirror neurons differentially encode the peripersonal and extrapersonal space of monkeys. *Science*, 324(5925), 403–6.
Casasanto, D. and Dijkstra, K. (2010). Motor action and emotional memory. *Cognition*, 115(1), 179–85.
Chemero, A. (2011). *Radical embodied cognitive science*. Cambridge, MA: MIT Press.
Chiel, H.J. and Beer, R.D. (1997). The brain has a body: adaptive behavior emerges from interactions of nervous system, body and environment. *Trends in Neurosciences*, 20, 553–7.
Clark, A. (1997). *Being there: putting brain, body, and world together again*. Cambridge, MA: MIT Press.
Clark, A. (2008). *Supersizing the mind: reflections on embodiment, action, and cognitive extension*. Oxford: Oxford University Press.
Colombetti, G. (2013). *The feeling body: affective science meets the enactive mind*. Cambridge, MA: MIT Press.
Craig, A.D. (2002). How do you feel? Interoception: the sense of the physiological condition of the body. *Nature Reviews Neuroscience*, 3, 655–66.
Danziger, S., Levav, J., and Avnaim-Pesso, L. (2011). Extraneous factors in judicial decisions. *Proceedings of the National Academy of Sciences*, 108(17), 6889–92.
Dehaene, S. (2005). Evolution of human cortical circuits for reading and arithmetic: the "neuronal recycling" hypothesis. In: S. Dehaene, J.-R. Duhamel, M.D. Hauser, and G. Rizzolatti (eds.), *From monkey brain to human brain*. Cambridge, MA: MIT Press, pp. 131–57.
De Waelhens, A. (1950). La phénoménologie du corps. *Revue Philosophique de Louvain*, 48(19), 371–97.
Di Paolo, E.A. (2005). Autopoiesis, adaptivity, teleology, agency. *Phenomenology and the Cognitive Sciences*, 4(4), 429–52.
Gallagher, S. (1986). Body image and body schema: a conceptual clarification. *Journal of Mind and Behavior*, 7(4), 541–54.
Gallagher, S. (2005). *How the body shapes the mind*. Oxford: Oxford University Press.

Gallagher, S. (2013). The socially extended mind. *Cognitive Systems Research*, 25–26, 4–12.
Gallagher, S. (2014). Pragmatic interventions into enactive and extended conceptions of cognition. *Noûs—Philosophical Issues*, 24, 110–26.
Gallagher, S. (2017). *Enactivist interventions: rethinking the mind.* Oxford: Oxford University Press.
Gallagher, S. and Bower, M. (2014). Making enactivism even more embodied? *AVANT: Trends in Interdisciplinary Studies* (Poland), 5(2), 232–47.
Gallese, V. (2014). Bodily selves in relation: embodied simulation as second-person perspective on intersubjectivity. *Philosophical transactions of the Royal Society of London. Series B, Biological sciences*, 369, 2013–177. doi:10.1098/rstb.2013.0177
Gallese, V. (2017). Neoteny and social cognition: a neuroscientific perspective on embodiment. In: C. Durt, T. Fuchs, and C. Tewes (eds.), *Embodiment, enaction and culture.* Cambridge, MA: MIT Press, pp. 309–32.
Gallese, V. and Sinigaglia, C. (2011). What is so special about embodied simulation? *Trends in Cognitive Sciences*, 15(11), 512–9.
Garfinkel, S., Minati, L., and Critchley, H. (2013). Fear in your heart: cardiac modulation of fear perception and fear intensity. Poster presented at the British Neuroscience Association Festival of Neuroscience, London, April 8.
Glenberg, A.M. (2010). Embodiment as a unifying perspective for psychology. *Wiley Interdisciplinary Reviews: Cognitive Science*, 1(4), 586–96.
Goldman, A.I. (2012). A moderate approach to embodied cognitive science. *Review of Philosophy and Psychology*, 3(1), 71–88.
Goldman, A.I. (2014). The bodily formats approach to embodied cognition. In: U. Kriegel (ed.), *Current controversies in philosophy of mind.* New York and London: Routledge, pp. 91–108.
Goldman, A. and de Vignemont, F. (2009). Is social cognition embodied? *Trends in Cognitive Sciences*, 13(4), 154–9.
Henry, M. (1965). *Philosophie et phénoménologie du corps.* Paris: PUF.
Hutto, D.D. (2015). Overly enactive imagination? Radically re-imagining imagining. *The Southern Journal of Philosophy*, 53(S1), 68–89.
Hutto, D. and Myin. E. (2013). *Radicalizing enactivism: basic minds without content.* Cambridge, MA: MIT Press.
Jackendoff, R. (1992). *Languages of the mind: essays on mental representation.* Cambridge, MA: MIT Press.
Johnston, V.S., Hagel, R., Franklin, M., Fink, B., and Grammer, K. (2001). Male facial attractiveness: evidence for hormone-mediated adaptive design. *Evolution and Human Behavior*, 22(4), 251–67.
Kammers, M.P. (2008) *Bodies in the brain: more than the weighted sum of their parts.* Enschede: Gildeprint.
Lakoff, G. and Johnson, M. (1999). *Philosophy in the flesh: the embodied mind and its challenge to Western thought.* New York: Basic Books.
Leder, D. (1990). *The absent body.* Chicago: University of Chicago Press.
Linkenauger, S.A., Ramenzoni, V., and Proffitt, D.R. (2010). Illusory shrinkage and growth: body-based rescaling affects the perception of size. *Psychological Science*, 21(9), 1318–25.
Liu, L., Papanicolaou, A.C., and Heck, D.H. (2014). *Visual reaction time modulated by respiration.* Working paper. Memphis: University of Tennessee Medical Center.

Mahon, B.Z. and Caramazza, A. (2008). A critical look at the embodied cognition hypothesis and a new proposal for grounding conceptual content. *Journal of Physiology (Paris)*, 102(1): 59–70.

Malafouris, L. (2010). *How things shape the mind*. Cambridge, MA: MIT Press.

Merleau-Ponty, M. (1945). *Phénoménologie de la perception*. Paris: Gallimard.

Naumann, R. (2016). Dynamics in the brain and dynamic frame theory for action verbs. In: L. Ströbel (ed.), *Proceedings of the International Conference: Sensory Motor Concepts in Language & Cognition*. Düsseldorf, Germany: Düsseldorf University Press, pp. 109–30.

O'Regan, J.K. and Noë, A. (2001). A sensorimotor account of vision and visual consciousness. *Behavioral and Brain Sciences*, 24(5), 939–73.

Overmann, K. A. 2016. Beyond writing: the development of literacy in the Ancient Near East. *Cambridge Archaeological Journal* 26(2), 285–303.

Perner, J. and Ogden, J.E. (1988). Knowledge for hunger: children's problem with representation in imputing mental states. *Cognition*, 29(1), 47–61.

Pezzulo, G., Barsalou, L.W., Cangelosi, A., Fischer, M.H., McRae, K., and Spivey, M.J. (2011). The mechanics of embodiment: a dialog on embodiment and computational modeling. In: A. Borghi and D. Pecher (eds.), *Embodied and grounded cognition*. Frontiers E-books, p. 196.

Postle, N., McMahon, K.L., Ashton, R., Meredith, M., and de Zubicaray, G.I. (2008). Action word meaning representations in cytoarchitectonically defined primary and premotor cortex. *NeuroImage*, 43, 634–44.

Prinz, J. (2009). Is consciousness embodied? In: P. Robbins and M. Ayede (eds.), *The Cambridge handbook of situated cognition*. Cambridge: Cambridge University Press, pp. 419–36.

Pulvermüllerü, F. (2005). Brain mechanisms linking language and action. *Nature Reviews Neuroscience*, 6, 576–82.

Rowlands, M. (2006). *Body language*. Cambridge, MA: MIT Press.

Shapiro, L. (2014). When is cognition embodied? In: U. Kriegel (ed.), *Current controversies in philosophy of mind*. New York and London: Routledge, pp. 73–90.

Tataranni, P.A., Gautier, J.F., Chen, K., Uecker, A., Bandy, D., Salbe, A.D. et al. (1999). Neuroanatomical correlates of hunger and satiation in humans using positron emission tomography. *Proceedings of the National Academy of Sciences*, 96(8), 4569–74.

Thompson, E. (2007). *Mind in life: biology, phenomenology and the sciences of mind*. Cambridge, MA: Harvard University Press.

Tsakiris, M. (2010). My body in the brain: a neurocognitive model of body-ownership. *Neuropsychologia*, 48(3), 703–12.

Varela, F.J., Thompson, E., and Rosch, E. (1991). *The embodied mind: cognitive science and human experience*. Cambridge, MA: MIT Press.

Wheeler, M. (2005). *Reconstructing the cognitive world: the next step*. Cambridge, MA: MIT Press.

Willems, R.M. et al. (2009). Neural dissociations between action verb understanding and motor imagery. *Journal of Cognitive Neuroscience*, 22, 2387–400.

Witt, J.K. and Proffitt, D.R. (2008). Action-specific influences on distance perception: a role for motor simulation. *Journal of Experimental Psychology: Human Experimental Psychology*, 34, 1479–92.

Young, I.M. (1980). Throwing like a girl: a phenomenology of feminine body comportment motility and spatiality. *Human Studies*, 3(1), 137–56.

CHAPTER 19

MOTOR INTENTIONALITY

ELISABETH PACHERIE

INTRODUCTION

IN his famous book *Phenomenology of Perception*, first published in 1945, the French philosopher Maurice Merleau-Ponty coined the phrase "motor intentionality," using it to refer to the form of intentionality exemplified by purposive, skillful, unreflective bodily activities, as opposed to the more cognitive, conceptual, and representational forms of intentionality typical of conscious intentions (Merleau-Ponty 1945). He introduced this notion in his long discussion of the case of Schneider, a soldier in the German army who suffered serious brain injuries during World War I and displayed a large number of neuropsychological impairments. Merleau-Ponty used Schneider's case to highlight the contrast between motor and cognitive intentionality but also to emphasize their generally smooth interplay in normal agents. In what follows, I will explore this contrast and interplay. How should we characterize motor intentionality? Is it best described, as Merleau-Ponty would have it, as a form of nonrepresentational intentionality? If not, how do motor representations differ from the representations involved in conscious intentions? How can motor intentionality and more cognitive forms of intentionality be integrated?

In the second section, I take Merleau-Ponty's discussion of Schneider's case as my starting point. In the third section, I consider more recent conceptual and empirical work that can help not only elucidate the distinction between motor and cognitive intentionality but also shed light on the challenges raised by their interplay. In the fourth section, I defend a representational stance of motor intentionality and discuss the format and contents of motor representation. Finally, the fifth section will discuss the interplay between motor and cognitive intentionality and the problem of explaining how our motor behavior can be responsive to our intentions.

Merleau-Ponty on Motor Intentionality

In *Phenomenology of Perception*, Merleau-Ponty (1945) used the case of Schneider to motivate the need to posit motor intentionality as a basic form of intentionality. Schneider, a soldier in the German army in World War I, suffered serious brain injuries when wounded by the explosion of a mine. He became a patient of the psychologist Adhémar Gelb and the neurologist Kurt Goldstein, who in their case reports described the large array of neuropsychological impairments he displayed, including alexia, form agnosia, loss of movement vision, loss of visual imagery, tactile agnosia, loss of body schema, loss of position sense, acalculia, and loss of abstract reasoning (Goldstein and Gelb 1918; Goldstein 1923).[1] Merleau-Ponty was especially interested in Schneider's pattern of performance in different motor tasks, as described by Gelb and Goldstein. Schneider presented a dissociation between a preserved ability to perform what Gelb and Goldstein termed "concrete movements" and an impaired ability to perform "abstract movements." In their terminology, concrete movements correspond to habitual movements performed in everyday life and abstract movements are isolated, arbitrary movements not relevant to any actual situation, such as moving arms and legs to order, or bending and straightening a finger. For instance, Schneider could grasp his nose with his hand but not point to it; nor could he interrupt his grasping movement midway on order or touch his nose with a ruler. He could perform habitual actions with speed and precision, like taking a match out of a box and lighting a lamp, but was at a loss when asked to perform an abstract, arbitrary movement, like drawing a circle in the air with his arm. Finally, he could perform or pantomime habitual movements on order, but only by placing himself mentally in the actual situation to which they corresponded and then executing them in perfect detail.

Taking the dissociation between Schneider's inability to point to his nose and his preserved ability to grasp his nose as evidence in support of a distinction between cognitive and motor intentionality, Merleau-Ponty wrote:

[1] The validity of Schneider's case and the exact nature of his impairments have been a matter of debate. Goldenberg (2003) argues that Gelb and Goldstein's minds "were clouded by the enthusiasm of proving the truth of an all-embracing theory of the human mind and its reaction to brain damage" (2003, p. 292), leading them to embellish their description of the case, while comforted in their enthusiasm by a patient eager to please. Others, however, have pointed out that aspects of Schneider's behavior that raised Goldenberg's suspicions, such as the compensation of visual form agnosia by kinesthetic mediation, are modes of compensation spontaneously used by patients with similar deficits (Farah 2004; Marotta and Behrmann 2004). It is also a matter of debate whether Schneider's case should be classified as an example of apperceptive visual agnosia or rather of integrative agnosia (Marotta and Behrmann 2004). Importantly for present purposes, even if doubts are likely to persist regarding the validity of all of Goldstein and Gelb's claims about their patient, the dissociation between identification and localization that is the focus of Merleau-Ponty's discussion is now well documented, as will be discussed in Section 3, and has been found not just for the visual modality but also for the tactile modality (Paillard et al. 1983).

> It must therefore be concluded that "grasping" or "touching," even for the body, is different from "pointing." From the outset the grasping movement is magically at its completion; it can begin only by anticipating its end, since to disallow taking hold is sufficient to inhibit the action. And it has to be admitted that a point on my body can be present to me as one to be taken hold of without being given in this anticipated grasp as a point to be indicated. But how is this possible? If I know where my nose is when it is a question of holding it, how can I not know where it is when it is a matter of pointing to it? It is probably because knowledge of where something is can be understood in a number of ways. (Merleau-Ponty 2002, p. 119)

Merleau-Ponty proposed that this dissociation points to the existence of different ways of knowing or understanding locations in space. Pointing to one's nose demands that one be able to form a representation of the positions of one's nose and hand in objective space. In contrast, grasping one's nose involves a practical understanding of bodily space, "where the patient is conscious of his bodily space as the matrix of his habitual action, but not as an objective setting" (Merleau-Ponty 2002, p. 119). Merleau-Ponty also emphasized the independence of this practical understanding from an objective understanding of bodily space. He wrote:

> A patient of the kind discussed above, when stung by a mosquito, does not need to look for the place where he has been stung. He finds it straight away, because for him there is no question of locating it in relation to axes of co-ordinates in objective space, but of reaching with his phenomenal hand a certain painful spot on his phenomenal body, and because between the hand as a scratching potentiality and the place stung as a spot to be scratched a directly experienced relationship is presented in the natural system of one's own body. (Merleau-Ponty 2002, p. 121)

Importantly, this practical understanding is not confined to one's bodily space narrowly conceived and to actions directed at one's body. This system also encompasses the surrounding space and the familiar objects it contains, offering themselves as poles of action in relation to the body's potentialities. Thus, according to Merleau-Ponty, "In the action of the hand which is raised towards an object is contained a reference to the object, not as an object represented, but as that highly specific thing towards which we project ourselves, near which we are, in anticipation, and which we haunt" (Merleau-Ponty 2002, p. 159)

Here, Merleau-Ponty appears to take the dissociation between different types of motor tasks in Schneider's case as evidence for the existence of a way of being directed toward one's body and toward objects in one's surroundings that functions independently of conceptual representations of their locations in objective space. He seems to claim both that motor intentionality is preserved in pure form in Schneider and also, more generally, that motor intentionality is our normal way of relating to our body and surroundings and what enables our unreflective, skillful goal-directed activities.

There is, however, another line of argumentation that runs simultaneously in Merleau-Ponty's long discussion of Schneider's case and leads to a conflicting conclusion. Schneider is unable to draw a circle in the air in the normal way:

> Asked to trace a square or a circle in the air, he first "finds" his arm, then lifts it in front of him as a normal subject would do to find a wall in the dark and finally he makes a few rough movements in a straight line or describing various curves, and if one of these happens to be circular he promptly completes the circle. (Merleau-Ponty 2002, p. 126)
>
> Since the patient can move, he doesn't lack motility and since he can recognize when the movements he makes happen to be circular, he doesn't lack a representation of the movement. Here, Merleau-Ponty concludes that what he lacks is "something which is an anticipation of, or arrival at, the objective and is ensured by the body itself as a motor power, a 'motor project' (*Bewegungsentwurf*), a 'motor intentionality' in the absence of which the order remains a dead letter" (Merleau-Ponty 2002, p. 127)

As several authors have pointed out and as Jensen (2009) discusses in detail, these two lines of reasoning suggest there is at best an ambiguity and at worst an inconsistency in Merleau-Ponty's interpretation of Schneider. On the one hand, he appears to claim that pure motor intentionality is preserved in Schneider' case, but, on the other hand, he also appears to take his inability to convert the thought of a movement into actual movement as evidence of an impairment of motor intentionality. How can motor intentionality be claimed both to be preserved and to be impaired in the same person? Unless they are qualified, the two claims are clearly inconsistent. However, they might be reconciled if we consider that for Merleau-Ponty, motor intentionality is both (a) a basic form of intentionality, distinct from, and capable of functioning independently of, more abstract, conceptual, objective representational forms of intentionality and (b) a form of intentionality that also insures the transition between more abstract forms of intentionality (e.g., thoughts about movement) and actual movements. Merleau-Ponty (2002, pp. 127–8) contrasts concrete movement as centripetal and having as background the world as given and abstract movement as centrifugal and as constructing its own background and projecting it, or throwing it out, on the world. Importantly, he takes motor intentionality to be what makes possible both abstract and concrete movements. In a way, then, motor intentionality itself has both a "centripetal" dimension, where, as stated in claim (a), it can operate independently of more abstract forms of intentionality, and a "centrifugal" dimension where it serves a function of projection of abstract movements into the world, in accordance with claim (b). Thus, if we understand Merleau-Ponty as suggesting that, in Schneider's case, the centripetal dimension of motor intentionality is preserved, while its centrifugal dimension is impaired, the threat of inconsistency might be avoided. This also means that an account of motor intentionality should aim at elucidating not just what distinguishes motor intentionality from more cognitive forms of intentionality but also how motor intentionality relates to these more cognitive forms of intentionality. In particular, such an account should try to spell out what exactly the function of projection ascribed to motor intentionality by Merleau-Ponty involves and thus move beyond his own largely metaphorical description of this function.

Motor Intentionality as a Basic Form of Intentionality: Empirical Evidence

According to Merleau-Ponty, motor intentionality constitutes a basic form of intentionality, distinct from more cognitive forms of intentionality and capable of functioning independently of them. Findings from several lines of empirical research in cognitive science and neuroscience appear to support the distinction and dissociability of motor intentionality and other forms of intentionality.[2] In particular, a large body of empirical evidence ranging from electrophysiological studies of macaque monkey brains to neuropsychological studies of patients with brain damage and behavioral studies in healthy humans support a dual model of visual processing, with a visuomotor system subserving the visual guidance of actions directed at objects in the environment (vision-for-action) and a visual perceptual system subserving the construction of visual percepts and conscious object perception (vision-for-perception).

In the early 1980s, neuroanatomists and physiologists established the existence of two separate cortical pathways, ventral and dorsal, subserving different functions in the visual cortex of primates (Ungerleider and Mishkin 1982). In-depth studies of patients with lesions in either the dorsal or the ventral pathways provided evidence that processing in the ventral pathway supports vision-for-perception while processing in the dorsal pathway supports vision-for-action. The most famous and widely discussed evidence is probably Milner and Goodale's analysis of patient D.F. (Milner and Goodale 1995). As a consequence of carbon monoxide poisoning, D.F. suffered important lesions of the ventral pathway. As a result, she had visual form agnosia. D.F. is described by Milner and Goodale as unable to recognize everyday objects, to visually identify simple shapes, or to tell whether two visual shapes are the same or different. Yet her visuomotor abilities appeared intact. She could reach out and pick up objects with remarkable accuracy, shaping her hand optimally for the grip. When asked to post a card through a slit, she oriented the card correctly, despite being at chance when asked to report the orientation of the slit. In contrast to D.F., patient A.T., studied by Jeannerod and colleagues (Jeannerod et al. 1994), had a lesion of the dorsal stream and suffered from optic ataxia. A.T.'s perception of the shape, size, and orientation of objects was normal, but her grasping movements directed at objects were systematically incorrect. The coexistence in D.F. of impaired conscious visual perception and object recognition and of preserved visuomotor abilities and the inverse dissociation found in A.T. suggest that visuomotor representations need not be derived from conscious visual perceptions but can be built independently. These dissociations also suggest that conscious visual representations cannot be directly derived from intact visuomotor representations.

[2] See Jacob and Jeannerod (2003) for a detailed discussion of this evidence and an assessment of its significance.

Finally, psychophysical experiments in healthy human adults have also shown a dissociation between the processing responsible for accurate visuomotor processing for pointing or grasping and the processing responsible for perceptual awareness. For instance, Bridgeman and colleagues (Bridgeman et al. 1979) conducted a series of series of experiments that exploited the phenomenon of saccadic suppression. During saccades, i.e., rapid eye movements, vision is partially suppressed and changes in the positions of objects in the visual field are not consciously perceived. Bridgeman and colleagues instructed the participants to point to a target that had just been displaced and extinguished. On some of the trials, the displacement occurred during saccades, preventing the participants from perceiving the target displacement. Bridgeman and colleagues found that the accuracy of pointing was not affected by conscious detection or failure to detect the target displacement. In a later set of experiments, Bridgeman and colleagues (Bridgeman et al. 1981) used the dot in frame illusion, where a stationary dot set against a large undifferentiated background moving in one direction appears to be moving in the opposite direction. They found again that although perceptual judgments of the position of the dot were affected by the dot's apparent motion, pointing accuracy wasn't. These experiments suggest that visual awareness of the position and motion of a target and visually guided pointing at a target are largely independent processes. Similarly, size-contrast illusions have been shown to affect conscious perception and judgment but not grasping performance. The Titchener illusion (also known as the Ebbinghaus illusion) is a display consisting of two circles of equal size, one surrounded by a ring of smaller circles, the other surrounded by larger circles. As a result, the circle surrounded by smaller circles is perceived as larger than the other central circle. Aglioti et al. (1995) used a three-dimensional version of the illusion using plastic disks and had their participants make a perceptual judgment and pick up one of the two central disks. A grasping movement involves a progressive opening of the grip where the fingers stretch up to a maximum aperture, followed by a closure of the grip until it matches object size. Maximum grip aperture occurs at about 60 percent to 70 percent of the duration of the movement and is reliably correlated with the object's size (Jeannerod 1981). Aglioti and colleagues used this property of the motor grasping pattern as an index of the computation of the object size made by the visuomotor system. They found that while perceptual judgments about object size were affected by the illusion, the grip wasn't and remained correlated with the object's actual size.[3]

Similar findings regarding pointing and grasping have been reported for a variety of other visual illusions including the Müller-Lyer illusion (Daprati and Gentilucci 1997), the Ponzo illusion (Jackson and Shaw 2000), the Kanizsa compression illusion (Bruno and

[3] The design of these experiments has raised certain methodological criticisms. For instance, Franz et al. (2000) argued that their results might be due to an asymmetry between the perceptual and the motor task (the perceptual task requiring the subjects to compare two discs, whereas in the motor task they could focus their attention on a single disc), and as such provided no evidence for a dissociation between perception and action. However, Haffenden and Goodale (1998) obtained similar results in a modified version of the task where this asymmetry was not present and the motor task and perceptual task were matched. For discussions of the methodological issues concerning illusion studies and of the degree to which they support the dual visual system hypothesis, see Jacob and Jeannerod (2003) and Briscoe (2008, 2014).

Bernardis 2002), and the hollow-face illusion (Króliczak et al. 2006). In each case, there is a divergence between what subjects consciously see and their visually guided behavior, suggesting that the spatial information used for visually guided action and the (illusory) spatial content of conscious visual experience might be processed relatively independently.

It is important to note, however, that our understanding of the visual pathways has evolved considerably since Milner and Goodale (1995) proposed their dual-system model. Substantial evidence has accrued that the anatomical and functional separation between the dorsal and ventral pathways is far from complete, casting doubt of the validity of a simple dissociation between vision-for-perception and vision-for-action and suggesting instead a more complex organization of visual processing. Thus, Rizzolatti and Matelli (2003) have described two anatomically segregated subcircuits of the dorsal stream, a dorso-dorsal pathway and a ventro-dorsal pathway. It has been proposed that the dorso-dorsal pathway is concerned with immediate visuomotor control and the ventro-dorsal pathway with the long-term storage of the particular skilled actions associated with familiar objects, with lesions to one or the other pathways leading to different neuropsychological impairments (Binkofski and Buxbaum 2013; Pisella et al. 2006).

In addition, neuroanatomical studies have uncovered many connections between the dorsal substreams and the ventral stream, indicating that these streams are able to communicate with each other in a bidirectional way and suggesting that the ventro-dorsal substream may constitute an interface between the ventral and the dorsal streams of visual information processing. Similarly, brain imaging studies indicate that the dorsal and ventral streams are often jointly involved in grasping, notably in situations involving delayed or pantomimed grasping—situations when information about the object from pictorial cues or memory is needed to control the grasping movement—and tool use, when conceptual knowledge needs to be accessed to allow for the selection of the most appropriate grasp (for reviews, see Cloutman 2013; Grafton 2010).

Thus, on the one hand, evidence of dissociations between visuomotor processing and visual perception processing appears to support Merleau-Ponty's contention that motor intentionality constitutes a basic form of intentionality, distinct from more cognitive forms of intentionality and capable of functioning independently of them. On the other hand, evidence of substantial crosstalk between streams appears consistent with his further contention that motor intentionality insures the transition between more abstract forms of intentionality (e.g., thoughts about movement) and actual movements. Before I consider the challenges raised by the interfacing of motor intentionality and more cognitive forms of intentionality, let me try to offer first a fuller characterization of motor intentionality.

Motor Representations

Merleau-Ponty characterizes motor intentionality as nonrepresentational, whereas cognitive scientists are generally happy to talk of the dorsal pathway as computing sensorimotor representations. Is this just a matter of terminological sloppiness on the part of cognitive scientists or is instead Merleau-Ponty's use of the term "representation" highly loaded and perhaps overly restrictive?

For something to qualify as a representation in Merleau-Ponty's sense, it must have propositional, conceptual content, and represent an object or a situation in an objective or detached fashion. However, many cognitive scientists and philosophers currently operate with a less demanding notion of representation. For instance, according to the account proposed by Bermúdez (1998), for a state to qualify as representational, the following criteria should be met: (1) the state should have correctness conditions and allow for the possibility of misrepresentation; (2) it should be compositionally structured; (3) it should admit of cognitive integration; and (4) it should play a role in the explanation of behavior that cannot be accounted for in terms of invariant relations between sensory input and behavioral output. This characterization leaves it open whether a representation has conceptual content or not, whether its content is objective or detached or not, and whether its format is propositional. Importantly, both cognitive integration and compositionality are graded notions. So, one way of drawing the distinction between conceptual and non-conceptual representations would be to say that conceptual representations must satisfy more stringent criteria of full cognitive integration and full compositionality. Indeed, Bermúdez suggests that the distinction between conceptual and non-conceptual content may in part be a matter of degree of compositionality and cognitive integration.[4]

Format and Content of Motor Representations

Several authors (Butterfill and Sinigaglia 2014; Jacob and Jeannerod 2003; Pacherie 2000, 2011) have argued that sensorimotor representations, like perceptual representations, have non-conceptual content, but also that this non-conceptual content is of a different kind from the non-conceptual content of perception. According to these authors, a motor representation represents the goal of an action in a specific non-conceptual format. This representation of the goal of an action (say, reaching for an object) is not just a representation of the target object toward which the action is directed; it also includes a representation of the final state of the acting body when that object has been reached. In simple, object-oriented actions (i.e., when an object is the target of an action), the visual attributes of this object are represented in a specific, "pragmatic" mode used for the selection of appropriate movements and distinct from other modes of representation used for other

[4] Bermúdez (1998) also argues, perhaps more contentiously (see Levine 2001), that there is a constitutive link between a capacity for conceptual thought and a capacity for genuine inference, where having a capacity for genuine inference is linked to an ability to appreciate the rational grounds for, and thus to justify, one's inferences, and that capacity for justification requires language mastery.

aspects of object-oriented behavior (categorization, recognition, etc.). In that sense, pragmatic representations are not as informationally rich as perceptual representations, since they represent objects attributes only to the extent that they are relevant to the selection of motor patterns. Jeannerod (1997) suggests that the function of these representations "falls between" a sensory function (extracting from the environment attributes of objects or situations relevant to a given action) and a motor one (encoding certain aspects of that action). In other words, these representations should be viewed as relational, with the body and the target object functioning as the terms of the relation. What they represent are neither states of the body per se nor states of the environment per se, but rather relations between body and goal. To use a different formulation, we could say that the goal is given under a specific mode of presentation; it is represented in terms of the motor patterns that it affords to the agent.

Another important aspect of motor representations is their dynamical character: they do not just represent relations between body and goal, they represent dynamic relations between them. This characteristic is linked to their role in the guidance and control of the action as it unfolds. In order for a motor representation to guide an action, it must anticipate the future states of the environment and of the acting body itself; in order to control it, it must allow for adjustments during execution. In recent decades, theories of motor control have emphasized the role of internal forward or predictive models. These models capture the causal relationships between motor acts and their sensory consequences and can be used by the motor system to estimate the effects of the motor commands sent to the effectors, compare these predicted effects with sensory feedback, and make adjustments if needed (for full descriptions of these models, see Desmurget and Grafton 2000; Wolpert, Ghahramani, and Jordan 1995; Wolpert and Kawato 1998). The content of motor representations is thus dynamical in the sense both that it gets elaborated over time—it becomes more determinate through feedback—and that the motor representation is itself responsible for making available the information that will make the content more determinate. For instance, to adjust one's grip on an object, one needs accurate information about its weight, compliance, and surface texture, and sensory feedback will be needed to adjust initial estimates, but for sensory feedback to become available one needs to grasp the object in the first place.

Are Motor "Representations" Really Representations?

One may agree that motor intentionality operates along the lines just described, but still be skeptical that the concept of representation plays an explanatory role here and contend instead that motor intentionality is better characterized nonrepresentationally in terms of dynamic systems of self-organizing continuous reciprocal causation between sensorimotor processes and the environment (e.g., Dreyfus 2000; Gallagher 2008).

In the remainder of this section, I argue that motor "representations" meet the criteria for representationality set out by Bermúdez.

The first criterion for a state to count as representational is that it have correctness conditions. One important characteristic of motor representations is their Janus-faced structure, their function falling between a sensory function and a motor one. A motor representation represents a situation as affording a certain goal, and it does so by representing the motoric means by which the goal is to be achieved. For instance, it represents an object as reachable by representing how the reaching is to be effected. As a result, the classical distinction between states with a mind-to-world direction of fit, and states with a world-to-mind direction of fit (e.g., Searle 1983), while useful as a way of contrasting states such as beliefs and desires, does not easily apply to motor representations. Rather, motor representations may be seen as akin to what Millikan (1995) calls "pushmi-pullyu" representations (or PPRs), that is, hybrid representations with a dual direction of fit. PPRs, according to Millikan, are not simply conjunctions of a descriptive plus a directive representation; rather they are more primitive and computationally less demanding than either purely descriptive or purely directive representations. If we accept that motor representations have this hybrid character, this should be reflected in their correctness conditions. A motor representation of an object as to be reached by such and such motoric means would have dual correctness conditions. For it to be correct it would have to be the case both that the object in question is indeed reachable by these motoric means and that it actually be reached by these motoric means. This characterization of the correctness conditions of motor representations also makes sense of the idea that the success of an action does not just depend on the fact that a certain outcome is achieved, but also on the specific way in which the outcome is achieved. For a given motor representation to be correct, it is not sufficient that it causes some series of changes in the relations between body and world, where the last element in the series corresponds to some desired outcome—the changes must also conform to a certain dynamical pattern.

Motor representations also have structure and exhibit some form of compositionality, thus meeting the compositionality criterion of Bermúdez. They have identifiable constituent units (e.g., reaching, grasping, rotating, lifting, transporting, releasing) that can be combined in various ways. Different actions will involve different combinations of these and other categories of units, and, at a higher level of organization, more complex actions will in turn involve combinations of relatively simple actions such as putting an object in a container. For instance, this action could be a recurrent element in the complex action of packing my suitcase before a trip.

In addition, motor representations do not just have a lexicon; they also have what may be called a "grammar" for assembling the constituent units into a coherent pattern. There are spatial, temporal, and motor (kinematic and biomechanical) constraints on the coordination of action that must be reflected in this grammar. The coordination of reaching and grasping, some aspects of which were already briefly mentioned in the previous section, may serve as an illustration. First, the combination of reaching and grasping units must obey certain spatial constraints. Reaching is mostly achieved by the proximal joints of the arm and makes use of an egocentric or body-centered system of representation of locations. Grasping, on the other hand, is a function of the intrinsic shape and size of the target object; it involves a transformation of visual information encoded in allocentric, object-centered coordinates into motor information encoded in the system of coordinates used to define the prehension space. Yet reaching and grasping must be spatially compatible. In particular, reaching must take into account not just the location of the object but also its orientation, so that the final position of the arm is compatible with the correct position of the hand and fingers for grasping the object. Second, reaching and grasping must also be temporally coordinated. As we already mentioned, their temporal coordination goes beyond mere succession. The fingers begin to shape during transportation of the hand to the object location. Maximum grip aperture occurs at about 60 percent to 70 percent of the duration of the movement and is reliably correlated with the object's size. Third, a motor representation normally codes for transitive movements, where the goal of the action determines the global organization of the motor sequence. For instance, the type of grip chosen for a given object is a function not just of its shape and size but also of the intended activity. For instance, the same object may be held with a precision grip or with a power grip depending on whether I intend to put it in a large box or to insert it in a tight-fitting container. Similarly, the same cup will be seized in different ways depending on whether one wants to carry it to one's lips or turn it upside down. Finally, the biomechanical constraints and the kinematic rules governing the motor system are also reflected in motor representations. Bodily movements as represented in motor representations respect the isochrony principle (the tangential velocity of movements is scaled to their amplitude), Fitt's law (the time required to rapidly move to a target area is a function of the ratio between the distance to the target and the width of the target), and the two-third power law between curvature and velocity.

Motor representations also admit of cognitive integration (Bermúdez's third criterion). As we have just seen, how an object is grasped is a function not just of its size, shape, and orientation, but also of what we intend to do with it. In addition, how we interact with an object also depends on its function, where the function may not be visually salient. Thus, motor representations will be influenced by knowledge of function. More generally, our motor interactions with an object will often be determined not only by sensory information immediately available to the agent but also by stored beliefs and knowledge regarding certain attributes and properties of the object (for instance, I may know from previous experience that this pot is heavier than it looks). Motor representations also connect up with our motivational states. We do not blindly respond to all the solicitations for action that the environment provides. Which motor representations are formed and acted upon is not just a function of environmental saliencies; it can be determined in part by the agent's motivational states, higher-order goals, intentions, and emotional states (Pacherie 2002). Motor representations are thus cognitively penetrable to a certain extent and can be influenced by information coming from other sources.

Moreover, it may be argued that the cognitive integration of motor representations is not just a matter of motor representations being influenced by other cognitive states. The influence can also work in the other direction. In particular, there is evidence that motor representations may be activated not just when we prepare to act but also when we observe others acting. In the last two decades neurological studies have yielded a set of important results on mirroring processes. In a series of single-neuron recording experiments on macaque monkeys investigating the functional properties of neurons in area F5, Rizzolatti and his colleagues discovered so-called mirror neurons, i.e., sensorimotor neurons that fire both during the execution of purposeful, goal-related actions by the monkey and when the monkey observes similar actions performed by another agent (for reviews, see Rizzolatti and Craighero 2004; Rizzolatti and Sinigaglia 2008). In addition, a large body of neuroimaging experiments have investigated the neural networks engaged during action generation and during action observation in humans, revealing the existence of an important overlap in the cerebral areas activated in these two conditions (for reviews, see Grèzes and Decety 2001; Jeannerod 2006). The existence of such a mirror system in humans is also supported by behavioral experiments on motor interference, where observation of a movement is shown to degrade the performance of a concurrently executed incongruent movement (Brass, Bekkering, and Prinz 2001; Kilner, Paulignan, and Blakemore 2003).

These results have been interpreted as support for the existence of a process of motor simulation or motor resonance whereby the observation of an action activates in the observer a motor representation of the action that matches the motor representation activated in the brain of the agent. Once a match is established, it enables the observer to apply predictive models in his or her motor system to interpret observed movements and to infer their goal. Thus, motor representations may contribute at least certain premises to cognitive systems engaged in the interpretation of intentional behavior.[5]

Finally, the existence of a bidirectional link between the processing of linguistic items pertaining to action concepts and the activation of motor representations is also well documented. Thus, passively reading action verbs has been found to somatotopically activate areas of the motor and premotor cortex associated with the relevant body parts needed to carry out the specified actions (Hauk et al. 2004). For example, the different patterns of activation found in the motor cortex when reading the words "kick," "pick," or "lick" overlap significantly with the actual activation that takes place when carrying out these actions with the relevant effectors of foot, hand, and mouth, respectively. Conversely, stimulation of the motor system has been found to affect the linguistic processing of action concepts. For instance, one study found that applying TMS to hand and foot areas of the motor cortex improved the recognition of hand-related ("pick") and foot-related ("kick") action verbs, respectively, in lexical decision tasks (Pulvermüller et al. 2005; see also Kiefer and Pulvermüller 2012).

The last criterion to be considered is explanatory usefulness. For motor representations to be vindicated, it must also be demonstrated that a purely mechanical explanation of the motor behavior would not do. According to Bermúdez, the need for

[5] The extent to which mirroring processes can provide an understanding of others' actions and intentions has given rise to an intense debate, with some theorists seeing these processes as the fundamental neural basis of human social cognition (e.g., Gallese 2007), while others hold more deflationary views (e.g., Jacob 2008).

explanations appealing to contentful states arises in situations where the behavior to be explained cannot be accounted for in terms of invariant relations between sensory input and behavioral output. Our discussion of the influence of cognitive and motivational factors on the construction of motor representations should make it clear that the motor behavior they are meant to explain could not be explained in terms of a lawful correlation between sensory stimulus and behavioral response. For instance, the same sensory stimulus (a horizontal bar in front of the agent) will be responded to with either an overhand or an underhand grip depending on what the agent intends to do. A mechanistic explanation may perhaps be enough to account for reflexes, but the movements we want to explain are relationally characterized movements—movements related to a certain goal—and, as Bermúdez (1998, p. 86) suggests, for such movements we need intentional explanations.

The Interplay of Motor and Cognitive Intentionality

As we saw in the second section, there are two lines of argument in Merleau-Ponty's discussion of Schneider's case. In what Jensen (2009) calls the argument from concrete behavior, Merleau-Ponty appears to claim that motor intentionality is preserved in Schneider's case and this preservation is what enables him to perform concrete movements. In contrast, in the argument from abstract behavior, he seems to claim that motor intentionality is impaired in Schneider's case and that this impairment is what explains his inability to perform abstract movements. I suggested earlier that these two claims may be reconciled if we consider that for Merleau-Ponty motor intentionality has both a centripetal dimension, where it can operate independently of more abstract forms of intentionality, and a centrifugal dimension, where it functions as a bridge between abstract, cognitive forms of intentionality (e.g., thoughts about movement) and actual movements. If we can understand him as claiming that, in Schneider's case, the centripetal dimension of motor intentionality is preserved, while its centrifugal dimension is impaired, the threat of inconsistency might be avoided.

But this means that an account of motor intentionality should aim at elucidating not just what distinguishes motor intentionality from more cognitive forms of intentionality but also how motor intentionality relates to these more cognitive forms of intentionality.

As Jensen (2009) points out, Merleau-Ponty's argument from abstract movement targets intellectualist models of action, according to which intentional bodily actions can be analyzed in terms of two independent components: a conscious intention, representing the goal of the action and possibly the movements to be performed, and the physical movements themselves caused by the intentions. Schneider's capacity to perform physical movements is intact, and he can form representations of abstract movements, such as drawing a circle in the air, since he can recognize when the movements he makes happen to be circular, yet he cannot perform abstract movements in the normal way. Schneider's inability to perform abstract movements shows that this analysis is unsatisfactory. What remains a mystery and, for Merleau-Ponty, is doomed to remain one as long as we stay within an intellectualist framework, is "by what magical process the representation of a movement causes precisely that movement to be made by the body" (Merleau-Ponty 2002, p. 160, n. 94). We thus need to appeal to motor intentionality to make bodily agency intelligible.

What we have said about motor intentionality up to this point is not enough to dissolve the mystery. Motor intentionality, understood as a basic form of intentionality, may well explain how bodily movements can be exercises of agency, can be purposive, and can be imbued with meaning—and this independently of more abstract, conceptual, representational forms of intentionality. But we still lack an explanation of how motor intentionality and more cognitive forms of intentionality can be integrated and how our motor behavior can be responsive to our intentions. Merleau-Ponty claims that "the normal function which makes abstract movement possible is one of 'projection'" (Merleau-Ponty 2002, p. 128). Unless we can explain how this projection operates and how motor and cognitive intentionality are integrated, we are left with a projection process that appears no less magical than the process by which, in intellectualist accounts, the representation of a movement causes precisely that movement to be produced by the body.

Several attempts have been made to address this issue. If a full explanation of human agency as integrated rational and bodily agency needs to appeal to both propositional attitude states like beliefs, desires, and intentions qua propositional attitudes and motor representations, we need to explain how intentions and motor representations can be coordinated and pull in the same direction. This problem is what Butterfill and Sinigaglia (2014) call the interface problem.

Butterfill and Sinigaglia (2014) argue that intentions and motor representations have distinct but complementary roles in explaining the purposiveness of actions and have distinct representational formats adapted to the function they serve. Intentions, understood in the standard way, are propositional attitudes with a characteristic role in practical reasoning, and as such are subject to norms of rationality. We need to appeal to intentions and related propositional attitudes if we are to account for human agency as the agency of beings who do things for reasons. The main functions associated with motor representations involve selecting the motor patterns needed to perform an action in a given situation and guiding and controlling their execution. As we saw in the previous section, to serve these functions motor representations must have a proprietary representational format, distinct from the format of intentions. Butterfill and Sinigaglia characterize the interface problem as the problem of explaining how it is that intentions and motor representations, having as they do different representational formats, are able to coordinate such that the action outcomes that they specify "non-accidentally match."

Several approaches to the interface problem may be considered. What Butterfill and Sinigaglia (2014) call the common cause approach proposes that intentions and motor representations coordinate in virtue of sharing a common cause that triggers them both.

The idea here is that a sensory state of the agent (e.g., a perception of a coffee mug) or an environmental stimulus (e.g., a coffee mug) triggers both an intention and a motor representation with aligned contents relating to the grasping of the mug. An advantage of this solution is that the difference in formats between these two representations does not raise any difficulties, since it is not in virtue of a causal interaction between them that they align. However, as Butterfill and Sinigaglia note, this is unlikely to provide a full solution to the interface. Neither intentions nor motor representations are always triggered by environmental causes. Intentions are often the result of deliberation or planning, and motor representations are frequently keyed to intentions rather than stimuli in the environment or an agent's sensory states.

Wayne Wu (2011, 2015) develops another approach that appeals to intention-guided attention. Wu takes himself to be solving a slightly different problem, namely, what he calls the many-many problem. This is the problem that an agent faces of selecting, out of many potential target objects for action and out of many potential actions on a target object, a specific action on a specific target object. On Wu's view, intentions help an agent solve the many-many problem by serving as structural causes that constrain the space of possible solutions. They do so in two distinctive ways, both centrally involving the deployment of concepts in their content. First, intention-guided attention identifies the object or objects to be acted upon from among competing objects. Thus, if one's intention deploys the concept of, say, FORK, this thereby directs attention to the appropriate object in the agent's perceptual field. Second, intentions activate appropriate motor representations. For example, an intention to GRAB one's fork will guide the agent in attending to the spatial properties of the fork in appropriate ways, and activate motor representations constitutive of grabbing.

The first part of Wu's solution to the many-many problem may be seen as a sophisticated variant of the common cause approach: the intention is not caused by an object in the environment, but it directs attention to the relevant object, which in turn triggers a motor representation. However, this is only a partial solution to the many-many problem, merely reducing it to a one-many problem. Hence, the second role assigned to intentions, where the action concept deployed in the content of an intention activates a motor representation appropriate to this action. However, from the point of view of the interface problem, the second part of Wu's solution is problematic, as it appears to presuppose the existence of a connection between action concepts and motor representations rather than explaining it.

According to Butterfill and Sinigaglia (2014), the solution to the interface problem involves recognizing that the contents of intentions can be partially determined by the contents of motor representations and explaining what form this content-determining relation takes. Their explanation appeals to demonstrative and deferential action concepts: the idea is that our intentions sometimes deploy demonstrative concepts that defer to motor representations specifying certain action outcomes, and thereby refer to those action outcomes, without any need for translation. Thus, on this proposal, we can consider the content of an intention to be "Do that!" and the demonstrative "that" would defer to a motor representation referring to the relevant action. As Butterfill and

Sinigaglia put it, "These demonstrative concepts would be concepts of actions not of motor representations, but they would succeed in being concepts of actions by deferring to motor representations. For any such concept, it is a motor representation which ultimately determines what it is a concept of" (Butterfill and Sinigaglia 2014, p. 134).

Mylopoulos and Pacherie (2016) have pointed out several disanalogies between ordinary instances of demonstrative reference and Butterfill and Sinigaglia's proposed demonstrative deference in intention that raise important difficulties for their view. Mylopoulos and Pacherie develop an alternative solution to the interface problem. Like Butterfill and Sinigaglia's approach, this solution recognizes that the intention concepts deployed in the contents of intention can be partially determined by the contents of motor representations. However, they explain this content-determining relation by appealing to the notions of executable action concepts and motor schemas rather than to demonstrative deference. They propose that in order to properly interface with motor representations, intentions must have as constituents executable action concepts, where to have an executable concept for a given type of action one must have a motor schema for actions of that type. Motor schemas are more abstract and enduring representations than motor representations. They store knowledge about the invariant aspects and the general form of an action and are implicated in the production and control of action. On the one hand, they can be acquired through processes of probabilistic inductive generalization from motor representations or from already extant schemas. On the other hand, the activation of a motor schema once learned will yield a motor representation, when the information needed to specify its parameters is provided, typically via attentional processes. Motor schemas would thus be what bridge the gap between intentions and motor representations, ensuring proper, content-preserving coordination between them.

Concluding Remarks

An account of motor intentionality should aim at elucidating not just what distinguishes motor intentionality from more cognitive forms of intentionality but also how motor intentionality relates to these more cognitive forms of intentionality. While a wealth of conceptual and empirical work has helped sharpen our understanding of the distinctiveness of motor intentionality, our understanding of how motor and cognitive intentionality are integrated remains much more tentative, despite some recent attempts to address this issue. It remains to be debated as well whether, as Merleau-Ponty claimed, motor intentionality should be understood as nonrepresentational or whether the notion of representation he worked with was too loaded and restrictive, opening the possibility that the contrast between cognitive and motor intentionality should be understood not as a contrast between representational and nonrepresentational intentionality but rather as a contrast between conceptual and non-conceptual forms of intentionality. My own leanings are, as is probably already clear, toward the latter position. I favor a representational stance in part because, as I argued in the fourth section, motor "representations" appear to meet sufficiently robust criteria for representationality, in part also (exhibiting here—again!—my own limitations and prejudices) because this representational stance provides in my view a more promising starting point for understanding the interplay of motor and cognitive intentionality.

Acknowledgments

This work was supported by grants ANR-10-LABX-0087 IEC and ANR-10-IDEX-0001-02 PSL*.

References

Aglioti, S., DeSouza, J.F.X., and Goodale, M.A. (1995). Size-contrast illusions deceive the eye but not the hand. *Current Biology*, 5, 679–85.

Bermúdez, J.L. (1998). *The paradox of self-consciousness*. Cambridge, MA: MIT Press.

Binkofski, F. and Buxbaum, L.J. (2013). Two action systems in the human brain. *Brain and Language*, 127(2), 222–9.

Brass, M., Bekkering, H., and Prinz, W. (2001). Movement observation affects movement execution in a simple response task. *Acta Psychologica*, 106, 3–22.

Bridgeman, B., Kirch, M., and Sperling, A. (1981). Segregation of cognitive and motor aspects of visual function using induced motion. *Perception & Psychophysics*, 29(4), 336–42.

Bridgeman, B., Lewis, S., Heit, G., and Nagle, M. (1979). Relation between cognitive and motor-oriented systems of visual position perception. *Journal of Experimental Psychology: Human Perception and Performance*, 5(4), 692–700.

Briscoe, R. (2008). Another look at the two visual systems hypothesis: the argument from illusion studies. *Journal of Consciousness Studies*, 15, 35–62.

Bruno, N. and Bernardis, P. (2002). Dissociating perception and action in Kanizsa's compression illusion. *Psychonomic Bulletin & Review*, 9(4), 723–30.

Butterfill, S. and Sinigaglia, C. (2014). Intention and motor representation in purposive action. *Philosophy and Phenomenological Research*, 88(1), 119–45.

Cloutman, L.L. (2013). Interaction between dorsal and ventral processing streams: where, when and how? *Brain and Language*, 127(2), 251–63.

Daprati, E. and Gentilucci, M. (1997). Grasping an illusion. *Neuropsychologia*, 35, 1577–82.

Desmurget, M. and Grafton, S. (2000). Forward modeling allows feedback control for fast reaching movements. *Trends in Cognitive Sciences*, 4, 423–31.

Dreyfus, H.L. (2000). A Merleau-Pontyian critique of Husserl's and Searle's representationalist accounts of action. *Proceedings of the Aristotelian Society*, 100, 287–302.

Farah, M. (2004). *Visual agnosia*. Cambridge, MA: MIT.

Franz, V.H., Gegenfurtner, K.R., Bülthoff, H.H., and Fahle, M. (2000). Grasping visual illusions: no evidence for a dissociation between perception and action. *Psychological Science*, 11(1), 20–5.

Gallagher, S. (2008). Are minimal representations still representations? *International Journal of Philosophical Studies*, 16(3), 351–69.

Gallese, V. (2007). Before and below "theory of mind": embodied simulation and the neural correlates of social cognition. *Philosophical transactions of the Royal Society of London. Series B, Biological sciences*, 362, 659–69.

Goldenberg, G. (2003). Goldstein and Gelb's case Schn.: a classic case in neuropsychology? In: C. Code, C.W. Wallesch, Y. Joanette, and A.R. Lecours (eds.), *Classic cases in neuropsychology* (vol. 2). Hove, UK: Psychology Press, pp. 281–300.

Goldstein, K. (1923). Über die Abhängigkeit der Bewegungen von optischen Vorgängen. Bewegungsstörungen bei Seelenblinden. *Monatsschrift für Psychiatrie und Neurologie*, 54 (Festschrift Liepmann), 141–94.

Goldstein, K. and Gelb, A. (1918). Psychologische Analysen hirnpathologischer Falle auf Grund von Untersuchungen Hirnverletzer. *Zeitschrift für die gesamte Neurologie und Psychiatrie*, 41, 1–142.

Grafton, S.T. (2010). The cognitive neuroscience of prehension: recent developments. *Experimental Brain Research*, 204(4), 475–91.

Grèzes, J. and Decety, J. (2001). Functional anatomy of execution, mental simulation, observation and verb generation of actions: a meta-analysis. *Human Brain Mapping*, 12, 1–19.

Haffenden, A.M. and Goodale, M.A. (1998). The effect of pictorial illusion on prehension and perception. *Journal of Cognitive Neuroscience*, 10(1), 122–36.

Hauk, O., Johnsrude, I., and Pulvermüller, F. (2004). Somatotopic representation of action words in human motor and premotor cortex. *Neuron*, 41(2), 301–7.

Jackson, S.R. and Shaw, A. (2000). The Ponzo illusion affects grip-force but not grip-aperture scaling during prehension movements. *Journal of Experimental Psychology: Human Perception and Performance*, 26, 418–23.

Jacob, P. (2008) What do mirror neurons contribute to human social cognition? *Mind and Language*, 23(2), 190–223.

Jacob, P. and Jeannerod, M. (2003). *Ways of seeing: the scope and limits of visual cognition*. Oxford: Oxford University Press.

Jeannerod, M. (1981). Intersegmental coordination during reaching at natural visual objects. *Attention and Performance*, 9, 153–68.

Jeannerod, M. (1997). *The cognitive neuroscience of action*. Oxford: Blackwell.

Jeannerod, M. (2006). *Motor cognition*. Oxford: Oxford University Press.

Jeannerod, M., Decety, J., and Michel, F. (1994). Impairment of grasping movements following a bilateral posterior parietal lesion. *Neuropsychologia*, 32, 369–80.

Jensen, R.T. (2009) Motor intentionality and the case of Schneider. *Phenomenology and the Cognitive Sciences*, 8(3), 371–88.

Kiefer, M. and Pulvermüller, F. (2012). Conceptual representations in mind and brain: theoretical developments, current evidence and future directions. *Cortex*, 48, 805–25.

Kilner, J.M., Paulignan, Y., and Blakemore, S.J. (2003). An interference effect of observed biological movement on action. *Current Biology*, 13, 522–5.

Króliczak, G., Heard, P., Goodale, M.A., and Gregory, R.L. (2006). Dissociation of perception and action unmasked by the hollow-face illusion. *Brain Research*, 1080, 9–16.

Levine, J. (2001). The self, and what it is like to be one: reviews of Bermúdez and Weiskrantz. *Mind and Language*, 16 (1), 108–19.

Marotta, J.J. and Behrmann, M. (2004). Patient Schn: has Goldstein and Gelb's case withstood the test of time? *Neuropsychologia*, 42, 633–8

Merleau-Ponty, M. (1945). *Phénoménologie de la perception*. Paris: Gallimard.
Merleau-Ponty, M. (2002). *Phenomenology of perception* (trans. C. Smith). London: Routledge and Kegan Paul.
Millikan, R.G. (1995). Pushmi-pullyu representations. *Philosophical Perspectives*, 9, 185–200.
Milner, A.D. and Goodale, M.A. (1995) *The visual brain in action*. Oxford: Oxford University Press.
Mylopoulos, M. and Pacherie, E. (2016). Intentions and motor representations: the interface challenge. *Review of Philosophy and Psychology*. doi:10.1007/s13164-016-0311-6
Pacherie, E. (2000). The content of intentions. *Mind and Language*, 15, 400–32.
Pacherie, E. (2002). The role of emotions in the explanation of action. *European Review of Philosophy*, 5, 55–90.
Pacherie, E. (2011). Nonconceptual representations for action and the limits of intentional control. *Social Psychology*, 42(1), 67–73.
Pisella, L., Binkofski, F., Lasek, K., Toni, I., and Rossetti, Y. (2006). No double-dissociation between optic ataxia and visual agnosia: multiple sub-streams for multiple visuo-manual integrations. *Neuropsychologia*, 44(13), 2734–48.
Pulvermüller, F., Hauk, O., Nikulin, V.V., and Ilmoniemi, R.J. (2005). Functional links between motor language systems. *European Journal of Neuroscience*, 21, 793–7.
Rizzolatti, G. and Craighero, L. (2004) The mirror-neuron system. *Annual Review of Neuroscience*, 27, 169–92.
Rizzolatti, G. and Matelli, M. (2003). Two different streams form the dorsal visual system: anatomy and functions. *Experimental Brain Research*, 153(2), 146–57.
Rizzolatti, G. and Sinigaglia, C. (2008). *Mirrors in the brain: how our minds share actions and emotions*. Oxford: Oxford University Press.
Searle, J. (1983). *Intentionality*. Cambridge: Cambridge University Press.
Ungerleider, L. and Mishkin, M. (1982) Two cortical visual systems. In: D.J. Ingle, M.A. Goodale, and R.J.W Mansfield (eds.), *Analysis of visual behavior*. Cambridge: MIT Press, pp. 549–86.
Wolpert, D.M., Ghahramani, Z., and Jordan, M.I. (1995). An internal model for sensorimotor integration. *Science*, 269(5232), 1880–2.
Wolpert, D.M. and Kawato, M. (1998). Multiple paired forward and inverse models for motor control. *Neural Networks*, 11, 1317–29.
Wu, W. (2011). Confronting many-many problems: attention and agentive control. *Noûs*, 45(1), 50–76.
Wu, W. (2015). Experts and deviants: the story of agentive control. *Philosophy and Phenomenological Research*, 90(3). doi:10.1111/phpr.12170

CHAPTER 20

THE EXTENDED BODY HYPOTHESIS

Referred Sensations from Tools to Peripersonal Space

FRÉDÉRIQUE DE VIGNEMONT

> Stellarc: We can't continue designing technology for the body because that technology begins to usurp and outperform the body. Perhaps it's now time to design the body to match its machines.
>
> (Atzori and Woolford 1995)

THE possibility of replacing a defective part of one's body is not without raising a number of metaphysical and psychological questions. Is the patient who has received the transplant of another individual's body part still the same person before and after the graft? Does it depend on the type of body part (hand, face, internal organ)? To what degree is she able to appropriate the new body part? These questions seem to reach another level of complexity with the Australian artist Stellarc's redesigning of his own body. Grafts made for medical purposes still follow our genetic blueprint, that is, how evolution designed our body. Grafts made for artistic or technological purposes, on the other hand, no longer respect it. Since the 1960s, Stellarc has tested how far the body can be extended and reshaped. Some of his works might be seen as mere fantasy, like the graft of an ear on the back of his hand. But others might be conceived as improvements of the body, like the addition of a third arm that moved under his control. With technological progress, such bodily improvements risk becoming more and more crucial if one does not want the body to become obsolete, so to speak. It is thus especially important to assess the implications of bodily extension. In this chapter I will leave aside metaphysical and ethical issues in order to focus on phenomenological ones. I will first analyze the case of tool use as a prototypical example of bodily extension and show what effect it has on bodily awareness. I will then consider whether one can extend bodily awareness even though the body itself is not extended.

Stretching the Body

We do not need Stellarc and his futuristic performances to analyze the consequences of bodily extension. We actually extend our body a hundred times a day by constantly using tools, from toothbrush to knife and pen. By tool, I do not mean any kind of object, but only unattached external objects that one actively manipulates—and not simply holds—for functional purpose (Beck 1980). Tools extend our motor, sensory, and spatial abilities. In other words, we can do more and farther away. One may also say that we *feel* farther away. But to what extent are the referred sensations that we feel to be located in tools similar to the sensations that we feel in our own body? We shall see that although more and more findings can be taken as evidence for the embodiment of tools, this embodiment has limits.

Tool Use

The lower animals keep all their limbs at home in their own bodies, but many of humans' are loose, and lie about detached, now here and now there, in various parts of the world—some being kept always handy for contingent use, and others being occasionally hundreds of miles away. A machine is merely a supplementary limb; this is the be-all and end-all of machinery. We do not use our own limbs other than as machines; and a leg is only a much better wooden leg than anyone can manufacture. Observe a man digging with a spade; his right forearm has become artificially lengthened, and his hand has become a joint. (Butler 1872, p. 267)

In his utopia *Erewhon*, Samuel Butler denies any significant difference between tools and hands. Over a century later, empirical research seems to confirm his view, showing that tools are processed in many ways as hands. However, one must distinguish among different claims concerning the embodiment of tools. More specifically, one can ask the following two questions:

1. Do tools stretch our space of action or do they also stretch the space of our body?
2. If they stretch the space of our body, are they integrated only at the unconscious level of sensorimotor body representations or do they also modify our bodily awareness?

Let us start with the first question. There is little doubt that tools can enlarge the range of motor opportunities. This is precisely their function. One way to phrase it is to say that they extend what is known as *peripersonal space*, which can be defined as the space "within which it [the body] can act" (Maravita et al. 2003, p. 531).[1] For example, in a seminal study, Iriki and colleagues (1996) trained monkeys to use a rake to reach food placed outside their peripersonal space and recorded their neural activity. They found that some of the neurons that displayed no visual response to food at this far location before tool use began to display visual responses after tool use. A few minutes after tool use was interrupted, the visual receptive fields shrank back to their original size. Roughly speaking, what was far from the body was perceived as close thanks to tool use.

Does that entail that tools are represented as parts of the body? In other words, does peripersonal space extend farther because of the extension of the body? Not necessarily. One must distinguish within peripersonal space its internal boundaries (where bodily space ends and peripersonal space starts) and its external ones (how far peripersonal space stretches). Interestingly, the latter can be displaced without the former being modified. For instance, if one uses a remotely controlled device (like in telemedicine), the space surrounding the device is most probably processed as being peripersonal, although it seems highly unlikely that one represents the whole distance between one's body and the remote device as parts of one's body (Cardinali, Brozzoli, and Farnè 2009a). The situation is different in the case of a tool over which one has direct control: both the internal and the external boundaries are pushed forward because the tool, which is in continuity with the body, is processed as a part of it. This is confirmed by the following study (Cardinali et al. 2009b). Participants repetitively used a long mechanical grabber to grasp various objects. After their training session, they were subsequently retested with their hand alone without the grabber. The kinematics of their movements was then significantly modified when reaching to grasp compared to before their training session. More specifically, they planned their movements as if their arm were longer than before using the grabber. The effect of extension was generalized to other movements, such as pointing on top of objects, although they were never performed with the grabber. This clearly shows that the grabber was incorporated in the sensorimotor representation of the body.

This result may not appear surprising insofar as we already know that amputees can experience phantom limbs, and thus that the perceived boundaries of one's body do not always coincide with its biological boundaries. But can one really compare tools and phantom limbs? There is indeed one difference, which might make the embodiment of tools more difficult. Phantom limbs do not always involve the modification of body representations; it is rather the reverse. It is because body representations have not been correctly updated after amputation that amputees still feel their missing limb. By contrast, to embody tools, one must modify body representations. And one does so all the time, each time one uses a tool, and each time one drops it. Unlike phantom limbs, tool embodiment thus shows extensive plasticity of body representations, at least at the sensorimotor level.[2]

[1] I shall come back to the notion of peripersonal space in the last section.

[2] One might believe that there is a second major difference between phantom limbs and tools, namely, their shape. However, there is no clear-cut distinction at this level. On the one hand, phantom limbs do not always respect human anatomy. For instance, amputees can feel the presence of a hand attached at the level of their elbow. On the other hand, tools such as a grabber can look like a forefinger and a thumb in pincer grip.

We can now turn to our second question. Since Milner and Goodale (1995) proposed their perception-action model, it is well accepted that what is true at the sensorimotor level may not be true at the conscious level. Perception and action indeed require different transformations of sensory signals and obey different rules. Consequently, the sensorimotor embodiment of tools does not necessarily entail alterations in bodily awareness. Since tools are designed to improve our ability to act, it may well be that they have no effect outside the realm of action. The evidence, however, does not go in that direction: there are perceptual consequences of tool use.

Let us first consider the experience of the limb holding the tool. Butler said: "Observe a man digging with a spade; his right forearm has become artificially lengthened." We have seen that this is true at the motor level but it is also true at the perceptual level (Cardinali et al. 2009b). After using the grabber, participants were asked to localize their elbow, their wrist, and their fingertip. The results showed that they mislocalized their body parts, as if their arm were longer (larger distance than before tool use between the fingertip and the elbow). This seems to indicate that the tool has been included in the perceptual representation of one's body too. If this is the case, then one should expect that one could feel sensations in the tool. One way to test this prediction is to consider classic tactile illusions and see whether they can be found when the tactile stimulation is applied on the tool instead of the skin. Consider the following well-known effect: when one closes one's eyes and crosses one's hands, one takes more time and is less accurate in judging which hand was touched first (Yamamoto and Kitazawa 2001a). This difficulty can be explained by the conflict between two distinct spatial frames of reference of tactile experiences: the bodily frame (e.g., on the right hand) and the egocentric frame (e.g., on the left). What happens now when one holds two sticks that are crossed with one's hands uncrossed, and the two sticks are vibrated one after the other? If the vibration were felt in the hands holding the sticks, there should be no conflict (e.g., the vibration on the right hand is on the right), and one should have no difficulty judging which stick was vibrated first. However, this is not what was found. Participants had the same difficulties with their sticks crossed and their hands uncrossed as with their hands crossed (Yamamoto and Kitazawa 2001b). This indicates that the vibration was experienced as being located at the tip of the sticks (which were crossed), rather than on the hands holding them (which were uncrossed).

Still one may question whether the participants felt the sensations in the sticks *in the same way* as they felt sensations in their hands. For instance, one might claim that they only indirectly felt the vibration at the end of the sticks in virtue of directly feeling the vibration on their palms. It is true that insofar as tactile receptors are on the skin, and not on the tool, referred sensations must involve subpersonal mechanisms of projecting sensations to the tool and recruit different brain processing than non-referred sensations (Limanowski and Blankenburg, 2016). But this is not the same as to say that the participants only indirectly felt sensations in the sticks. According to Dretske (1995), perception is indirect if some of the information about the perceived object or event is not embedded in the information about the more proximal object: I can indirectly hear that the postman is coming on the basis of hearing that the dog is barking and of knowing that the dog barks each time that the postman comes. By contrast, perception is direct if all the information about the perceived object or event is embedded in the information about the more proximal object: I can directly hear that the postman is coming on the basis of hearing his voice. The crucial question thus is whether additional information is required for one to feel sensations in tools. To answer it, let us consider the most famous example of referred sensations, namely, the blind man and his white cane. Arguably, the blind man primarily feels the resistance of the floor rather than the resistance of the cane in his palm, which is less phenomenologically salient. Furthermore, information about the bumps on the floor is embedded in the information about the pressure on the palm; it just needs to be extracted and conceptually structured. No further knowledge is required: there is no need to first categorize the specific pressure that one feels in one's hand, and then infer on the basis of past associations that there is a bump on the floor. Arguably, the first time that the blind man holds his white cane, he can immediately feel the obstacles on the floor at the end of his cane. He might not be able to correctly categorize what he feels, but this can be the same when using his own fingers to recognize objects. What is important is that the first time he uses his cane, he immediately feels the world in a certain way at the end of his cane (the resistance of the floor, its volume, etc.). By contrast, the first time one hears that the dog is barking, one cannot hear that the postman is coming. Along with many others, I thus want to argue that we *directly* feel touch on tools (Lotze 1888; Martin 1993; O'Shaughnessy 2003; de Vignemont, 2018; Vesey 1961).

To summarize, I have argued that tools can extend (1) peripersonal space, (2) sensorimotor body representations, and (3) tactile experiences. Do I then agree with Butler? Is a machine merely a supplementary limb? I will now highlight some major differences between limbs and tools.

Pain, Itches, and Tickles

I have just argued that one can feel *tactile* sensations in tools but bodily awareness cannot be simply reduced to these sensations. There is a whole range of other types of bodily experiences, including pains, itches, tickles, and so forth. What is then interesting is that one cannot feel them as being located in tools. You cannot tickle your pen. Nor can you feel the urge to scratch the tip of your fork. As for pain, obviously you wince when your car bumps into another car, but you do not feel pain as being localized in the trunk. More generally, if a tool is damaged, one may feel annoyed, or even really upset if the tool is important, but one does not feel hurt. The fact is that one uses tools in harmful situations in which one would not use one's own limbs.

What is the origin of this fundamental limit of referred sensations in tools? We need first to rule out a possible explanation: it is not because pain cannot be experienced beyond the biological boundaries of the body. Unfortunately, patients with phantom limbs can feel excruciating pain there. Nor can we can appeal to a purely mechanical explanation. It is true that only some specific types of physical stimuli can be transmitted from the body part that holds the tool to the tool itself, such as vibration. However, we know that physical discontinuity does not preclude referred sensations in external objects, as in the rubber hand illusion (Botvinick and Cohen 1998). In this illusion, participants see a rubber hand in front of them, while their own hand is hidden from sight. After synchronous stroking of the biological and the rubber hands, participants report feeling as if they were touched on the rubber hand. Yet the rubber hand is not in physical contact with their body. Hence, the laws of physical transmission cannot be the full story.

One thus needs to analyze what distinguishes touch from pains, tickles, and itches. According to Armstrong (1962), they correspond to two distinct types of bodily sensation, which he calls transitive and intransitive sensations. In transitive sensations, one can easily draw the distinction between the sensation itself and the object of the sensation (the sensation of the pen in my hand, for instance). It has a clear exteroceptive dimension (about the pen) in addition to its bodily dimension (about the skin). By contrast, intransitive sensations have only a bodily component. The feather tickles me, but my tickling sensation is not about it. It is only about my body. Likewise, when my painful experience is caused by an external painful stimulus, it represents the body part in pain, and not the stimulus that caused the pain. This is not to say that one cannot localize pain in the external world. The localization of intransitive sensations can actually be quite accurate, and this is important to guide appropriate behaviors (like scratching or withdrawing). Still, it is primarily encoded relative to the body in a somatotopic reference frame (Haggard et al. 2013; Mancini et al. 2015). The content of the painful experience is filled in by the body.

Now the function of tools is to act on the external world. That is why we need to experience the external world at their end. This is also why, I argue, we can experience only transitive sensations there. For example, when the blind man uses his white cane, it is to explore his environment. Hence, when he feels the obstacles on the floor at the tip of his cane, he primarily experiences the exteroceptive component of tactile sensations. By contrast, when a noxious or a ticklish stimulus is applied at the end of the tools, it is hard to see what one could experience in tools given the lack of an exteroceptive component. In a nutshell, the function of touch is to acquire knowledge about the properties of *external objects*, whereas the function of pain is to prevent the *body* from damage. This difference in focus, I suggest, can account for the absence of referred intransitive sensations in tools.

There is another peculiarity of intransitive sensations. Pains, tickles, and itches all have an intrinsic affective dimension.[3] They feel pleasant or unpleasant, but they are not neutral, and they all play a motivational role. One withdraws the hand in pain, scratches the itchy leg, or wiggles and jiggles. Since one does not feel these affectively loaded sensations in tools, one may say that tools are spatially and motorically embodied, but not *affectively embodied*. This is not to say that we cannot have a deep attachment for some tools, or that we do not protect them when we can. I may not care about my fork,

[3] Touch can have an affective dimension too, but stroking a tool hardly makes sense.

which can be easily replaced, but I definitely care about my Montblanc pen that was given to me for my PhD. Still, even then do I care about it in the same way and to the same extent as I care about my body? Most probably not, because I do not feel pain in my pen. One may claim that it is actually important for tools *not* to be affectively embodied. If they were, then we would not be able to use them to stoke the hot embers of a fire, or to stir a pot of boiling soup (Povinelli et al. 2010).

The Sense of Bodily Ownership: The Body by Default

There is a further reason for which it is important for tools not to be affectively embodied. If they were completely embodied, if they were indeed only supplementary limbs, then we would have no template of our "habitual body," to borrow Merleau-Ponty's phrase. It is indeed important to remember that the plasticity of body representations goes both ways. We are able to incorporate tools hundreds of times a day, and a large part of the experimental investigation of body representations has put emphasis on their capacity to stretch. However, one should not forget that each time we drop a tool, we need to go back to our original body representation. We thus automatically recalibrate the correct size of our limbs a couple of minutes after tool use. How do we achieve this recalibration? Given the number of times we drop off tools in everyday life, it does not seem parsimonious to assume that each time we recompute the size of our limbs, as if it had never been computed before. If tools were fully embodied, we would risk losing track of our biological body. On the contrary, the merely partial embodiment of tools, and, more specifically, the lack of affective embodiment, entails that there is a representation of the body that is not altered by tool use, and thus more stable than the others. It is thanks to this representation that one generally experiences what Williams James (1890, p. 242) called the "the same old body always there." Another way to put it is to say that it represents the body by default. It can then be used to recalibrate the other types of body representations after they have been temporarily altered. Consequently, the motor system can incorporate the tool for the time of its use with no cost because there is a reference standard of the body that does not vary.

One can then note a last limit to the embodiment of tools. Most of the time, one does not experience a sense of ownership for tools. One does not feel the fork that one uses for lunch as part of one's body. Interestingly, when amputees describe the failure to appropriate their prosthesis, they often explain it by claiming that the prosthesis is just a tool. For example, a patient reported the following:

> Using a prosthetic is not a natural thing, because a prosthetic is not a substitute leg, *it is a tool* which may or may not do some of the things that a leg might have done. (Murray 2004, p. 971, my emphasis)

Here we are very far from Butler. Even if tools can be represented to a large extent in the same way as the parts of one's body, one still does not *experience* them as such. It is then tempting to relate the lack of ownership to the lack of affective embodiment. In a nutshell, one cannot feel pain in tools; thus, tools are not affectively embodied; thus, one does not feel ownership for tools. The relationship between pain, affective embodiment, and the sense of bodily ownership needs to be further developed (de Vignemont, 2017). Still, one can already contrast the case of tools with the case of the rubber hand illusion. This illusion indeed involves not only referred sensations, but also illusory ownership. Participants report feeling as if the rubber hand were part of their own body. What is then interesting is that participants can feel pain in the rubber hand. In one study, both the rubber hand and the subject's hand were synchronously stroked, as in the classic setup (Valenzuela-Moguillansky et al. 2011). The difference was that immediately after the stroking, they both received a painful stimulation. Subjects were then asked whether they felt pain *in the rubber hand*. The authors found a correlation between the intensity of the pain felt in the rubber hand and their sense of ownership of the rubber hand. Arguably, unlike tools, the rubber hand can be affectively embodied, and this explains the illusory experience of ownership.

We have just seen the limits of the embodiment of tools. In the absence of intransitive sensations, one does not embody tools at the affective level, although one does so at the sensorimotor and perceptual levels. Consequently, there is a type of body representation, which is affectively loaded, that is immune to the influence of tool use and that can carry information about the body by default, that is, the body to protect.

Stretching Bodily Awareness

Bodily awareness can be stretched, to some extent at least. So far I have explained that one can feel sensations in tools because tools are embodied at the perceptual and the sensorimotor levels. But is such embodiment necessary? Or is it possible to feel sensations in objects that bear no relationship with the body (i.e., exosomesthesia)? The localization of transitive sensations is not constrained by the biological limits of the body, but is it constrained by the limits of the body as they are mentally represented? I will now consider two series of cases: (1) referred sensations in non-bodily shaped objects that are not tools and (2) sensations in peripersonal space. Although some studies claim to be "explaining away the body" (Hohwy and Paton 2010) or claim that sensations can be "hopping out of the body" (Miyazaki et al. 2010), I will argue that none of these cases qualifies as exosomesthesia.

Beyond Tools

Could one feel sensations anywhere, maybe as far as the moon, as suggested by Armel and Ramachandran?

> If you looked through a telescope at the moon and used an optical trick to stroke and touch it in synchrony with your hand, would you "project" the sensations to the moon? (2003, p. 1500)

There are actually two distinct questions raised here. First, can the moon be embodied? Secondly, can you feel sensations in the moon? The first question concerns embodiment and inquires about the constraints that lay upon it: can you incorporate any object, even if it is not bodily shaped and not in continuity with the body and if you have no control over it? However, it might be that one can feel sensations in the moon although the moon is not embodied. What we really want to know is thus the constraints for referred sensations per se: does the object need to be embodied for one to feel sensations in it? Consider the following study by Hohwy and Paton (2010). They used the classic setup of the rubber hand illusion, previously described, but the difference was that the rubber hand was suddenly swapped with a small white cardboard box. This did not preclude participants from reporting sensations on the box. Does that show that one can dispense with embodiment? It is worth noting here that the experimenter could not induce the illusion for the box if no prior classic rubber hand illusion occurred before. Once normally elicited by a rubber hand, visual capture of touch was not disturbed by the intrusion of an object. One possible interpretation of this result is that the transition from the rubber hand to the box was perceived as a visual distortion of the hand (something like "My hand looks like a box"), to which body representations adjusted. This would show the flexibility of body representations, instead of showing that they played no role. Alternatively, as suggested by the authors themselves, it might have been that the box was perceived as hiding the hand (something like "My hand is in the box"). If this is the right interpretation, it is then not even clear that one can talk of referred sensations. It is rather that participants felt sensations on their hand, which they localized in the box. This hardly qualifies as exosomesthesia.

Let us now consider another bodily illusion, the cutaneous rabbit illusion, which was induced on a non-bodily shaped object (Miyazaki et al. 2010). We know that repeated rapid tactile stimulation at the wrist, then near the elbow, can create the illusion of touches at intervening locations along the arm, as if a rabbit hopped along it. In Miyazaki and colleagues' (2010) version of the illusion, participants lifted up a stick between their two index fingers until it was in contact with the system that delivered mechanical pulses on the fingers via the stick. They received a series of tactile stimulations on their left index finger, then on their right index finger. Participants then reported feeling touches between the two fingers, that is, on the stick that they were holding. The authors concluded that tactile sensations could "hop out of the body." However, tactile sensations experienced on the stick are not more surprising than sensations experienced on tools. Actually, one may even say that the stick is a kind of tool that the participants manipulated to interact with the stimulating device.

These studies show that embodiment is not constrained by bodily resemblance, but we already knew this thanks to tool embodiment. In this sense, *the body* can be said to be explained away. However, this is not to say that *embodiment* is explained away. The object is still integrated in body representations. A possibly more convincing case can be found in older reports, and in particular in a study by von Békésy:

> But if the observer was permitted to see the movements of the loudspeaker in the room and coordinate them with the sensations on his arms, after some training he began to project the skin sensations out into the room. (1959, p. 14)

Von Békésy's report seems to indicate that one can feel tactile sensations in external objects with no spatial contiguity and no resemblance with the body. Can one then explain referred sensations in terms of embodiment of the loudspeaker? This has been tried, but I think with little success (Martin 1993; Smith 2002). The problem is not that this strategy involves assuming puzzling distortions of body representations. The problem is the lack of independent reason to assume that the loudspeaker is incorporated. One cannot appeal to the fact that participants felt sensations in it for risk of circularity. And there seems to be no other plausible justification for the claim that the loudspeaker was incorporated. Participants had never interacted with it. Moreover, it cannot be explained as a kind of rubber hand illusion. The rubber hand illusion indeed involves *fusion* between a visual event and a tactile event (for example, seeing the stroking of the paintbrush). By contrast, what one may perceive here is a relationship of *causality* between the movement of the loudspeaker and the sensation. Causality involves a relationship between two distinct events, not a fusion of them. Finally, one does not even need to argue for the embodiment of the loudspeaker because it is not clear that the observer *directly* felt sensations there. In the previous section, I ruled out the interpretation of referred sensations in tools in terms of indirect perception, but this interpretation seems more plausible in von Békésy's case. As von Békésy described, the observer had to learn to project the sensation. Put another way, he only *indirectly* localized tactile sensations on the loudspeaker. Consequently, this last case shows that indirect referred sensations in external objects do not require the embodiment of the objects, but one is not entitled to draw conclusions about direct referred sensations on only this basis.

Peripersonal Sensations

So far, I have only considered cases in which one feels sensations in *objects*, whether they are tools, boxes, sticks, or loudspeakers. Although these entities bear little resemblance to body parts, at least they are like body parts in one respect: they are material objects. As such, they could conceivably be represented as parts or extensions of the body. There are, however, other reported cases in which one feels sensations as being located in a specific *empty* region of space. It may then seem that bodily sensations can stretch beyond what is embodied. It is difficult indeed to see how an empty region of space can be embodied. However, we shall see that there are regions of external space that are encoded in a bodily frame of reference, namely, peripersonal space.

Let us first reconsider Hohwy and Paton' s (2010) study. In one condition they stroked a discrete volume of empty space five centimeters above the rubber hand in synchrony with the biological hand. Interestingly, participants reported that they still felt sensations on their own skin, and not above it. A subject, for instance, described it as follows: "It's a magnetic field impacting on my arm" (p. 8). The point here is that even when the stimulations are not on the body, the subjects can still experience them on their body. There is, however, another version of the rubber hand illusion, called the invisible hand illusion (Guterstam et al., 2013). In this study, the experimenter synchronously stroked the participant's hidden hand and a discrete volume of empty space above the table in direct view of the participant. This time, participants localized their sensations of touch at the empty location. Von Békésy (1967) also reports a similar type of referred sensation. By placing two vibrators slightly out of phase with each other on two spread fingers or on the outspread thighs, healthy subjects described feeling the vibration in the region of empty space between the fingers or the legs. Finally, similar reports are also found in patient studies. An amputated patient described feeling a sensation "in space distal to the [phantom]-finger-tips" when his stump was stimulated (Cronholm 1951, p. 190). Another patient "mislocalized the stimulus to the left hand into space near that hand" (Shapiro et al. 1952, p. 484).

How should we interpret these puzzling cases? Are these referred sensations completely disconnected from the body? Can one feel sensations not on the moon itself, but simply up in the sky? The reply that I want to offer is negative. It is crucial to note that in all these cases, referred sensations are localized *close to the body*, that is, in peripersonal space. The name of peripersonal space finds its origin in a seminal study by Rizzolatti and colleagues (1981), who described bimodal neurons activated both by tactile stimuli and by visual stimuli close to the body. On the basis of this and many other related findings, it has been argued that the zone that surrounds the body (up to 30 cm) is represented differently from far space. We have seen earlier its motor properties, but it also displays specific sensory properties. For instance, in humans, a cross-modal congruency effect is found for stimuli presented in peripersonal space (Spence et al. 2004). Participants are asked to perform a speeded discrimination of the location of a vibrotactile stimulus presented either on the index finger or the thumb, while trying to ignore visual distractors presented simultaneously at either congruent or incongruent positions. Crucially, incongruent visual distractors interfere with the tactile discrimination (i.e., participants are both slower and less accurate) only when visual stimuli are close to the body. Visuo-tactile interference happens because both visual and tactile experiences share a common spatial frame of reference, which is centered on body parts (Kennett et al. 2002). A similar effect can be found in the neuropsychological syndrome of tactile extinction. After right-hemisphere lesions, some patients have no difficulty in processing an isolated tactile stimulus on the left side of their body. However, when they are simultaneously touched on the right hand, they are no longer aware of the touch on their left hand. Interestingly, the same is true when they see a visual stimulus *near* their right hand: the visual stimulus on the right side "extinguishes" the tactile stimulus on the left side so that they fail to detect the touch (di Pellegrino et al. 1997).

One way to interpret the influence of visual experiences on tactile experiences is to say that the perceptual system anticipates the contact of the seen stimulus on one's body (Hyvärinen and Poranen 1974). Recent theories have highlighted the importance of prediction in cognitive systems (Hohwy 2013). Expectations about upcoming sensory events can be used to prepare sensory systems and allow for enhanced processing of the forthcoming event (Engel et al. 2001). Specifically, the sight of objects moving toward one's body can generate an expectation of a tactile event. The tactile expectation then influences the experience of the actual tactile stimulus. For example, it was found that merely seeing the experimenter's hand *approaching* a rubber hand could induce sensations in the rubber hand (Ferri et al. 2013). What is interesting is why such an expectation is generated. One explanation is that the perceptual system expects the body to move. Peripersonal space is then the space where the body could be in a soon future, a gray zone between one's body and the external world.

Interestingly, it has been repeatedly shown that the rubber hand illusion works only if the rubber hand is placed in peripersonal space (Lloyd 2007; Preston 2013). Roughly speaking, what is in peripersonal space could be part of one's body. In the invisible hand illusion, the region of space that is stroked is also within the limits of peripersonal space. Actually, participants reported that it seemed as if they had an "invisible hand." Hence, it was not as if they perceived the empty space as being empty. They perceived it as being occupied by a hand that they could not see. To some extent, referred sensations in this illusion can be compared to sensations in phantom limbs, as suggested by the authors themselves (Guterstam et al. 2013). The other types of peripersonal sensations (i.e., bodily sensations felt in peripersonal space) may be less easily explained in terms of embodiment. Nonetheless, they can be understood only within a bodily frame of reference.

We can now offer the following explanation of peripersonal sensations (i.e., bodily sensations felt in peripersonal space). When an object or event enters peripersonal space, it is automatically encoded in relation to bodily boundaries as fixed by body representations. Under normal conditions, the perceptual system then generates tactile expectations, which can in turn generate tactile sensations, which are localized on the body. This involves a remapping of what occurs in peripersonal space onto the surface of the body. In illusory or pathological conditions, I suggest that this remapping can be disrupted. In the invisible hand illusion, sensations are still localized within bodily space, but the body is taken to be at a different location from where it is actually. In pathological conditions, the remapping simply fails to occur and sensations remain localized within peripersonal space. Peripersonal sensations are thus only the consequences of the exceptional disruption of the normal process of remapping in tactile expectation.

Conclusion

According to the extended mind hypothesis, "There is nothing sacred about skull and skin" for cognitive abilities (Clark and Chalmers 1998). Can we say the same for bodily awareness? Here I have considered several versions of what we can call the extended body hypothesis. According to a weak version, bodily awareness is not limited by the biological boundaries of our body. In light of tool embodiment, I highlighted the malleability of embodiment but also showed that there are important limitations to the sensations that we can feel in tools. Thanks to these limits, we cannot forget who we are and what we must protect in priority. I then considered a stronger version of the extended body hypothesis, according to which bodily awareness is not even constrained by the *apparent* boundaries of the body. In favor of this version, I have described how one can have peripersonal sensations in which one feels sensations outside apparent bodily boundaries. However, even peripersonal sensations are localized relative to a bodily frame of reference. Thus, there still seems to be something sacred about our apparent skull and skin for bodily awareness.

References

Armel K.C. and Ramachandran V.S. (2003). Projecting sensations to external objects: evidence from skin conductance response. *Proceedings of the Royal Society B: Biological Sciences*, 270(1523), 1499–1506.

Armstrong, D. (1962). *Bodily sensations*. London: Routledge and Paul.

Atzori, P. and Woolford, K. (1995). Extended-body: interview with Stelarc. *Ctheory. net*, June 9.

Beck, B.B. (1980). *Animal tool behavior: the use and manufacture of tools*. New York: Garland Press.

Botvinick, M. and Cohen, J. (1998). Rubber hands "feel" touch that eyes see. *Nature*, 391, 756.

Butler, S. (1872). *Erewhon*. London: Penguin Classics.

Cardinali, L., Brozzoli, C., and Farnè, A. (2009a). Peripersonal space and body schema: two labels for the same concept? *Brain Topography*, 21(3–4), 252–60.

Cardinali, L., Frassinetti, F., Brozzoli, C., Urquizar, C., Roy, A.C., and Farnè, A. (2009b). Tool-use induces morphological updating of the body schema. *Current Biology*, 19(12), R478–9.

Clark, A. and Chalmers, D. (1998). The extended mind. *Analysis*, 58(1), 7–19.

Cronholm, B. (1951). Phantom limbs in amputees: a study of changes in the integration of centripetal impulses with special reference to referred sensations. *Acta Psychiatrica Neurologica Scandanavia Supplement*, 72, 1–310.

de Vignemont, F. (2017). Pain and touch. *The Monist*, 100(4), 465–77.

de Vignemont, F. (2018). *Mind the body*. Oxford: Oxford University Press.

di Pellegrino, G., Làdavas, E., and Farnè, A. (1997). Seeing where your hands are. *Nature*, 388, 730.

Dretske, F. (1995). *Naturalizing the mind*. Cambridge, MA: MIT Press.

Engel, A.K., Fries, P., and Singer, W. (2001). Dynamic predictions: oscillations and synchrony in top-down processing. *Nature Reviews Neuroscience*, 2(10), 704–16.

Ferri, F., Chiarelli, A.M., Merla, A., Gallese, V., and Costantini, M. (2013). The body beyond the body: expectation of a sensory event is enough to induce ownership over a fake hand. *Proceedings of the Royal Society B: Biological Sciences*, 280(1765), 20131140. doi:10.1098/rspb.2013.1140

Guterstam, A., Gentile, G., and Ehrsson, H.H. (2013). The invisible hand illusion: multisensory integration leads to the embodiment of a discrete volume of empty space. *Journal of Cognitive Neuroscience*, 25(7), 1078–99.

Haggard, P., Iannetti, G.D., and Longo, M.R. (2013). Spatial sensory organization and body representation in pain perception. *Current Biology*, 23(4), R164–76.
Hohwy, J. (2013). *The predictive mind*. Oxford: Oxford University Press.
Hohwy, J. and Paton, B. (2010). Explaining away the body: experiences of supernaturally caused touch and touch on non-hand objects within the rubber hand illusion. *PLoS One*, 5(2):e9416.
Hyvärinen, J. and Poranen, A. (1974). Function of the parietal associative area 7 as revealed from cellular discharges in alert monkeys. *Brain*, 97(4), 673–92.
Iriki, A., Tanaka, M., and Iwamura, Y. (1996). Coding of modified body schema during tool use by macaque postcentral neurones. *Neuroreport*, 7(14), 2325–30.
James, W. (1890), *The principles of psychology* (vol. 1). New York: Holt.
Kennett, S., Spence, C., and Driver, J. (2002), Visuo-tactile links in covert exogenous spatial attention remap across changes in unseen hand posture. *Perception and Psychophysics*, 64(7), 1083–94.
Limanowski, J. and Blankenburg, F. (2016). That's not quite me: limb ownership encoding in the brain. *Social Cognitive and Affective Neuroscience*, 11(7), 1130–40.
Lloyd, D.M. (2007). Spatial limits on referred touch to an alien limb may reflect boundaries of visuo-tactile peripersonal space surrounding the hand. *Brain and Cognition*, 64(1),104–9.
Lotze, H. (1888). *Microcosmus: an essay concerning man and his relation to the world*. New York: Scribner &Welford.
Mancini, F., Steinitz, H., Steckelmacher, J., Iannetti, G., and Haggard, G. (2015). Poor judgment of distance between nociceptive stimuli. *Cognition*, 143, 41–7.
Maravita, A., Spence, C., and Driver, J. (2003). Multisensory integration and the body schema: close to hand and within reach. *Current Biology*, 13(13), R531–9.
Martin, M.G.F. (1993). Sense modalities and spatial properties. In: N. Eilan, R. McCarty, and B. Brewer (eds.), *Spatial representations*. Oxford: Oxford University Press.
Merleau-Ponty, M. (1945), *Phénoménologie de la perception*. Paris: Gallimard.
Milner, A.D. and Goodale, M.A. (1995). *The visual brain in action*. Oxford: Oxford University Press.
Miyazaki, M., Hirashima, M., and Nozaki, D. (2010). The "cutaneous rabbit" hopping out of the body. *Journal of Neuroscience*, 305, 1856–60.
Murray, C.D. (2004). An interpretative phenomenological analysis of the embodiment of artificial limbs. *Disability and Rehabilitation*, 26, 963–73.
O'Shaughnessy, B. (2003). *Consciousness and the world*. Oxford: Oxford University Press.
Povinelli, D.J., Reaux, J.E., and Frey, S.H. (2010). Chimpanzees' context-dependent tool use provides evidence for separable representations of hand and tool even during active use within peripersonal space. *Neuropsychologia*, 48, 243–7.
Preston, C. (2013). The role of distance from the body and distance from the real hand in ownership and disownership during the rubber hand illusion. *Acta Psychologica* (Amsterdam), 142(2), 177–83.
Rizzolatti, G., Scandolara, C., Matelli, M., and Gentilucci, M. (1981). Afferent properties of periarcuate neurons in macaque monkeys. II. Visual responses. *Behavioural Brain Research*, 2(2), 147–63.
Shapiro, M.F., Fink, M., and Bender, M.B. (1952). Exosomesthesia or displacement of cutaneous sensation into extrapersonal space. *AMA Archives of Neurology and Psychiatry*, 684, 481–90.
Smith, A.D. (2002). *The problem of perception*. Cambridge, MA: Harvard University Press.

Spence, C., Pavani, F., and Driver, J. (2004). Spatial constraints on visual-tactile cross-modal distractor congruency effects. *Cognitive, Affective, and Behavioral Neuroscience*, 4(2), 148–69.

Valenzuela-Moguillansky, C., Bouhassira, D., and O'Regan, K. (2011). Role of body awareness. *Journal of Consciousness Studies*, 18(9–10), 110–42.

Vesey, G.N.A. (1961). The location of bodily sensations. *Mind*, 70(277), 25–35.

Von Békésy, G. (1959). Similarities between hearing and skin sensations. *Psychological Review*, 661, 1–22.

Von Békésy, G. (1967). *Sensory inhibition*. Princeton, NJ: Princeton University Press.

Yamamoto, S. and Kitazawa, S. (2001a). Reversal of subjective temporal order due to arm crossing. *Nature Neuroscience*, 4(7), 759–65.

Yamamoto, S. and Kitazawa, S. (2001b). Sensation at the tips of invisible tools. *Nature Neuroscience*, 4(10), 979–80.

CHAPTER 21

CRITICAL NOTE

Brain–Body–Environment Couplings. What Do they Teach us about Cognition?

ARNE M. WEBER AND GOTTFRIED VOSGERAU

Introduction

As a matter of fact, we are not brains in vats, and cognition is not just the juggling of mental entities that are detached from the world. Indeed, there are many loops and couplings between brains, bodies, and environments. This fact, which was somewhat underemphasized by "classical" strands of cognitive science, has been brought into focus in 4E cognition. Although this shift of focus has produced a number of valuable insights about the nature and function of cognition, its theoretical status is still under debate (as Gallagher notices in response to the worries of Shapiro 2014; Gallagher, p. 354). Thus, in this critical note on the four views on brain–body–environment couplings, we will concentrate on the following question: What do such couplings teach us about cognition? As we will show, the answers to this question depend heavily on the explanatory aim one has in mind. This can be demonstrated by comparing the accounts of Gallagher, Pacherie, Rowlands, and de Vignemont.

Let us start with an analogy. Galileo's laws of falling objects state that everything falls with the same speed (in a vacuum). This is obviously wrong since a feather does not fall with the same speed as the cannonball under many conditions (i.e., a *ceteris paribus* clause is tacitly understood). This is because of factors that are not taken into account by Galileo's laws. Indeed, there are many factors interacting with falling objects, not just air. For example, if you attach the feather to the cannonball, they will fall with the same speed. If the feather is attached to a bird, it will eventually not fall down at all but rather "fall up." In other words: there are relevant feather-body couplings and feather-environment couplings that heavily influence the falling behavior of the feather. We can even construct cases that involve a social dimension: e.g., if Lieut. Smith commands recruit

Jones to blow the feather upwards, it will not fall down; however, if Jones commands Smith to blow, the feather will fall down nevertheless. Does all of this show that Galileo is fundamentally wrong? Certainly not. It just shows that real-world scenarios are complex. And explaining every bit of a real-world scenario involves a lot of different laws describing very different factors.

If we are clear what phenomenon we want to explain, we can abstract away from different factors and so develop explanations that are very restricted in application and only really account for idealized cases. However, scientific method stipulates that this reduction of complexity will give us insight into the basic mechanisms at work, which, in turn, will ultimately allow us to formulate very complex models that approximate real world scenarios with great precision. Thus, an explanation about why feathers stay in the air when attached to flying birds cannot be explained by Galileo's laws. And as long as we are clear about this, we can make perspicuous the explanatory role that the laws (and the factors mentioned by the laws) play: they explain certain aspects of the target phenomena, but not everything that is to be explained about that phenomena.

So, how does all of this relate to the question of how best to explain brain–body–environment couplings? It may be that our fundamental terms to describe the phenomena are not well chosen in the first place. Perhaps—as in the case of falling objects—the phenomena to be explained are so complex that we need to develop multifaceted theories involving a lot of different laws and mechanisms. In any case, we have to be clear about our explanatory target and the explanatory role of our theoretical posits. We will now discuss the four different accounts in light of the question what exactly is to be learned about cognition from coupling effects.

Gallagher—Universal but Messy Couplings

We agree with Gallagher's diagnosis that it is unclear how to understand the notion of embodied cognition. There are various more or less radical accounts available, which only have in common the fact that they are challenging the classical cognitivist's approach. The missing conceptual specifications of their presuppositions lead to further problems. In addition to the variety of accounts available, we also have to face certain difficulties regarding the overall enterprise of an interdisciplinary project as we find it in cognitive science. These further complications result especially from the absence of a common view or theoretical framework among the 4E approaches that would allow for a transfer or comparison of results between different disciplines.

We do not agree, however, with Gallagher's solution: to expand the conception of embodiment without specifying the explanatory target. Gallagher argues that we should broaden the notion of embodiment by citing an empirical study relating the metabolism of judges with their more or less harsh sentences at court. A second example provided by Gallagher is Pulvermüller's seminal "kick-pick-lick" study that would not work with French-speaking subjects. Gallagher argues from such examples that bodily, environmental, and cultural factors play an inevitable explanatory role for cognition. And this contradicts the classical cognitivist position, which holds that the brain and the central nervous system are primary influences on cognitive processing. Gallagher's conclusion is that all cognitive explananda result from couplings between brain, body, and environment: a cognitive system should generally be understood as something that is coupled during a "dynamical participation in a large range of messy adjustments and readjustments that involve internal homeostasis, external appropriation and accommodation, and larger sets of normative practices" (Gallagher, p. 364).

To take the analogy from above: Gallagher seems to conclude from the non-falling of feathers-attached-to-birds that Galileo's fundamental notions of mass and gravity are completely mistaken and that we should rather think of falling objects as constituting messy adjustments and readjustments. In other words, by shifting the explanatory target to virtually everything (the effect of metabolism on the judges' sentences and the effect of learned languages on brain activation), we are left with a system too complex to be explained by simple mechanisms or principles. Indeed, it is not clear why Gallagher does not include examples such as the effect of oxygen levels of surrounding air on the measurements of intelligence or the effect of the presence of terrorists threatening the judges on the sentences of those judges. It seems that he should include all of these cases and more. That is why the notion of embodiment loses its explanatory power: it turns into a name for something messy.

To focus on Gallagher's main objective, his examples for a homeostasis of a cognitive system in its environmental context are not only used to broaden the notion of embodiment. This notion should also make it possible to get rid of representations as an explanatory postulate in cognitive science. Therefore, he argues against the claim that the most fundamental representations are body-formatted representations. The conception of representations as body-formatted was originally introduced by Goldman to explain the specific effect of body representations on cognitive processing. Because it can be shown that the same body representation has similar effects independent of what the body really is like, it is inferred that it is not the body itself that plays the main role in cognition, but is rather neural representations. In opposition to Goldman, Gallagher's broader conception of embodiment refers not only to the anatomy of an organism and sensorimotor contingencies, but also to environmental couplings. His main argument is that bodily factors can have an influence on cognitive processing, while the exact effects that he wants to explain remain underspecified.

According to Gallagher, the empirical results of Pulvermüller's work cannot be interpreted as evidence for a reuse of neural body-formatted representations, since Pulvermüller's results are restricted to the community of English speakers. The brain of a French speaker would not react to English words by reusing the originally body-formatted representations. This is because languages are a matter of "cultural learning," such that reuse of a B-formatted representation can only be understood against this background. And cultural learning requires some kind of "metaplasticity where not only neurons undergo plastic changes, but bodily and cultural practices do too" (Gallagher p. 361). A metaplasticity that also incorporates the environmental and cultural context is the novel aspect added by Gallagher to establish his broader conception of embodiment.

But it seems to be a rather empirical question about learning a language if an English speaker who learns the French word *piétiner* will reuse her somatotopic representation in the brain while reading the translation in a dictionary or hearing a teacher saying "*piétiner* means 'kick' in French." In contrast to Gallagher's view regarding Hebbian learning—e.g., learning that requires an association with an already acquired neural basis—we find no compelling argument that the cultural context has primacy over the postulation of body-formatted representations. Again, the explanatory target seems to fluctuate: while Goldman tries to explain how a certain concept is represented in the brain by associating linguistic forms (*kick* or *piétiner*) with B-formatted representations (thus "reusing" them), Gallagher shifts the focus to the question of why it is *kick* rather than *piétiner* that is related to this kind of brain activity in some speakers. It is very unlikely that we will be able to attain one answer to both questions.

We do not want to deny that neural plasticity is sensitive to the environmental context. However, this can only answer some questions, while the answer to other questions might still call for the postulation of representations. In general, the appropriate scientific method for explaining messy systems is not to find a nice label for their messiness (e.g., "embodiment" or "coupling"), but to break the messy system down into not-so-messy components that might be explained very differently from one another and that might not occur in real-life situations in isolation.

Pacherie—Historical Remarks Based on Former Remarks

Pacherie refers to bodily boundaries when she asks if there are any representations involved in the case of motor intentionality. Further, she refers to a historical interpretation of a pathological case to give her own explanation of the interplay between representational elements and motor abilities. Therefore, the explanatory target of Pacherie is a certain kind of movement ("concrete movements"), and her specific question is whether postulating representations is necessary for their explanation. By critically assessing Merleau-Ponty's account—who denies that any representations are involved in motor intentionality—Pacherie diagnoses inconsistencies in his reasoning when he tries to explain certain motor deficits. She presents a more detailed view in order to explain the same deficits of motor abilities on the basis of an argument for representationalism. Pacherie argues that "the content of motor representations is . . . dynamical in the sense both that it gets elaborated over time—it becomes more determinate through feedback—and that the motor representation is itself responsible for making available the information that will make the content more determinate" (Pacherie, p. 377).

Here, then, we find a more restricted view on the dynamics between body and environment compared to Gallagher's, since only bodily movements and its feedback are taken into account. Pacherie claims that these dynamics can be determinate or can specify the content of motor representations during the progress of a cognitive system. We would like to add that their format could still be understood in Goldman's sense of B-formatted representations.

Merleau-Ponty discussed the pathological case of a person named Schneider, who had a selective impairment of actions that do not serve any purpose in everyday life, like drawing a circle in the air with his arm. These actions are called abstract movements and require relevant concepts; e.g., the concept of a circle. In contrast, habitual or concrete movements performed in everyday life are rather a kind of unreflected activity, which does not involve any conceptual and representational abilities. Because concrete movements were not impaired in the case of Schneider, Merleau-Ponty concludes that objective, abstract, and conceptual understanding is independent of a practical understanding. Pacherie's interpretation of Merleau-Ponty has it that practical understanding should be understood as nonrepresentational, because it can be explained without postulating conceptual representations of objects in objective space and can thus be sufficiently explained with reference to the acting body in its environment.

Pacherie suspects that Merleau-Ponty's interpretation of Schneider's abilities and the supposed dissociation between different ways of understanding locations in space with reference to bodily actions is inconsistent. He refers to Schneider's inability to transform an intention to act into an actual movement as evidence of an impairment of motor intentionality, but, at the same time, he claims that Schneider's motor intentionality is not impaired. It seems more plausible to Pacherie to assume that there is an interdependence between the unreflected sensorimotor level and the reflexive level of the conceptual representation when it comes to intentionality. Pacherie presents a very useful solution of the so-called interface problem regarding the connection between the sensorimotor and the conceptual level: concepts for intentional actions are partially determined by the contents of motor representations. After having argued that certain representations are involved during actions—contrary to Merleau-Ponty—she proposes that the content has been acquired in a format of fundamental motor schemas. The action concepts were acquired by generalization from motor representations or already extant schemas, and their reactivation is led by attentional processes. Thus, motor schemas can be thought of as a connection between intentions and motor representations. Even if the contribution of Pacherie is mainly a historical critique, she points out that general problems mentioned about the explanatory use of representations apply also to the claim that the body is a basis for acquiring and constituting higher-order cognition.

Unfortunately, Pacherie's paper only discusses a historical statement given by Merleau-Ponty on the basis of criteria for representationality recently given by Bermúdez. Thus, although the explanatory target is well specified and the solution of the interface problem is well taken, it is an open question what explanatory role the notion of representation fulfills: why should we classify things according to Bermúdez's criteria if the alleged features of representations do not play any explanatory role? Is the argument between Pacherie and Merleau-Ponty only a fight about words after all?

Rowlands—Interesting, but not Relevant

Rowlands is concerned with a system's active operations on the environment, particularly in relation to explaining intentionality as a cognitive phenomenon. According to him, intentionality is constituted in a specific way by processes and factors external to those that can be found in the brain and body. Rowlands argues that intentionality has to be understood as a "disclosing activity [that] often—not always, not necessarily—straddles processes occurring in the brain, in the wider non-neural body, and in things that creatures do to and with the world around them" (Rowlands p. 351). For illustration, he refers to Merleau-Ponty's example of a blind person using a cane. By using the cane, the person actively detects information that is only available through this specific operation, like tapping with the cane from the left to the right. Rowlands further analyses this kind of directedness toward the world by distinguishing between an empirical and a transcendental mode of presentation. The experience of certain aspects of objects like the imperviousness of an obstacle is made available by using the cane falls under Rowlands's category of an empirical mode of presentation, whereby the imperviousness is understood as the intentional object. The transcendental mode of presentation concerns that part of the intentional act that permits the object to appear under certain empirical modes of presentation. Following the example, the activity of using the cane should be understood as the transcendental mode of presentation that permits the blind person to get in contact with the external world.

Rowlands's explanation of intentionality is about the information that is actively made available through certain operations to further processing and is not about the information already contained in the structure of the environment, the body, or the brain. On the contrary, words and sentences—or thoughts, beliefs, and desires—are examples that have only derived intentionality, while intentionality as an activity should be seen as original and not dependent on other mental states. Thus, intentional directedness has to be understood not only as constituted by neural or bodily processes, but also by an activity disclosing the world, like using the cane to detect available information. More generally speaking, and independent of the restricted example of the blind person using a cane, we can transfer Rowlands's definition of an empirical and transcendental mode of presentation to explain intentionality as a disclosing activity by tool use.

Unfortunately, while Rowlands's explanatory target and his explanatory strategy are clear, it is not so clear how his points relate to an understanding of cognition. It is plausible that there are processes that make information about the external world available. However, the overall goal of cognitive science is to explain *flexible* behavior by postulating internal mechanisms and procedures (which makes cognitive science go beyond behaviorism in the psychological sense; cf. Neisser 1967 and many others). This flexibility cannot be sufficiently explained by reference to an actively operating system disclosing external aspects. Every artificial scanner can actively explore information about its environment, but it cannot flexibly react to the inputs—as opposed to humans and other animals. Flexible outputs are not only generated by an activity disclosing the world. In other words: what has been the focus of cognitive science is the difference between scanners and humans; i.e., the explanation of the flexible behavior of the latter. The intentionality-as-making-information-available-actively account of Rowlands does not seem to contribute to our understanding of this difference, and hence it is not clear that it contributes to an understanding of cognition at all (although it certainly contributes to an understanding of one very important precondition of cognition).

De Vignemont: Direct and Indirect Explanatory Roles

Frédérique de Vignemont starts with a representational stance that she has defended elsewhere. Against this background, she has the specific aim to explain bodily awareness in the light of the coupling of brain, body, and environment. She discusses if and how tools can be seen as an extension beyond the brain and body relevant to bodily sensations. Like Rowlands, she refers to the example of a blind person's usage of a cane to acquire "information about the bumps on the floor [that] is embedded in the information about the pressure on the palm" (de Vignemont p. 393). Discussing empirical results quite similar to this example—like being stimulated via vibration of crossed sticks being held in the hands—she argues that these examples cannot show that tools should count as extensions of the body, since they cannot be said to be objects of bodily sensations. Consequently, she argues for a more restricted view compared to Gallagher's or Rowlands's view regarding the possible external, non-bodily, or non-neural influences.

It is obvious that tools can extend our space of action, but it is rather unclear, as de Vignemont points out, if our body and the awareness of the body are also extended. Therefore, de Vignemont asks to what extent sensations are located in the tools as compared to sensations located in the body while using the tools. Tools are not only integrated on an unconscious level of sensorimotor processing, but also have an impact on the awareness of one's own body. She claims that although tools can extend the reach of peripersonal space—because one can feel tactile sensations in tools—it is not the case that bodily awareness can be said to be extended, since bodily awareness cannot be reduced to tactile sensations. The reason is that there are two kinds of perceptual consequences: the experience of the limb that holds the tool on the one side and of the tool itself on the other side. Since the blind person feels the cane in the palm, de Vignemont holds that the tactile experience mediated by the cane is only indirect.

Moreover, she investigates whether or not one can have extended bodily awareness without extended bodies by investigating exosomesthesia. After having examined several cases of referred sensations in non-bodily shaped objects that are not tools and other cases, de Vignemont concludes that body representations, rather than bodies, play the decisive explanatory role for bodily awareness.

To sum up, de Vignemont has a very clear explanatory target (bodily awareness) and is very clear about the explanatory role that bodies and environments play. In this way, she shows clearly what we can learn from brain–body–environment couplings for one specific phenomenon; namely, that bodies and environments do not play a direct role.

Conclusion

Our comparison of the four accounts of Gallagher, Pacherie, Rowlands, and de Vignemont has shown that the questions being addressed by each differ greatly. While Gallagher looks for a simple explanation for a complex phenomenon including all facets of human behavior (or at least for a large part of the facets of human behavior, while it is not at all clear which facets exactly), the other three have much more specific explanatory targets. Pacherie and de Vignemont concentrate on two phenomena that fall quite clearly into the realm of cognition broadly conceived; namely, the relation between motor intentionality and cognitive intentionality on the one hand, and body awareness on the other. Rowlands focuses on the explanation of intentionality understood as a process of making information available. Even though the availability of information is doubtlessly a key prerequisite for cognition, it is not clear that it is part of cognition, for the reason that noncognitive systems seem to do the same thing. So, even though the explanatory goal of Rowlands is clear, his discussion is probably not teaching us anything interesting about cognition. Again, this is not to say that it does not teach us anything interesting at all—it is just not relevant to our conception of cognition.

The account of Pacherie is, in our view, clearly concerned with cognition and it has a clear explanatory target. However, the explanation that Pacherie gives is in terms of representations. This is fine with regard to the goal to explain the interaction between the different levels of intentionality that she successfully explains. However, it is not clear why this should be done in terms of representations. The reason why this is not clear is that she uses a definition by Bermúdez without showing the explanatory role the single criterion plays in her account. Thus, we are left unclear about what features of representations we really need to presuppose for her account. And if it turns out that we need only two of the four criteria, her whole account does not clearly speak in favor of representationalism. Thus, if her goal is to convince us that even 4E cognition is in need of representations, she should convince us that all of the features that turn things into representations play a specific explanatory role. In our view, this goal is reached by de Vignemont, who shows that for the explanation of body awareness we need to introduce the notion of body representations. Otherwise, we cannot explain certain phenomena. In effect, this argument leads to the insight that in the case of body awareness, the body does not play a direct role for cognition but only an indirect one.

Taken together, the four accounts show that we have to be careful when specifying the kinds of explanations we are giving in order to know to which question we are contributing. It seems to be a general problem of the field of 4E cognition that this is often not done (cf. the introduction of Gallagher, p. 354). Recently, cognitive scientists were confronted with a huge variety of partly overlapping, partly contradictory theories that differ broadly regarding the degree to which cognition is thought to be embodied, embedded, enacted, extended, or coupled with the environment. These accounts use similar terms for different mechanisms and different terms for similar mechanisms.

In the absence of a systematic workup and comparison, this array of theories leads to a general confusion within and between different disciplines. Unfortunately, while 4E cognition is still debated, due to the diversity of theories and approaches, even the precise meaning of underlying terms—such as "embodied cognition" or "embodiment"—lacks univocality. The result is that we have no unitary research perspective, so far. To achieve some progress in this regard and to guide further research, we want to elucidate the theoretical presuppositions of a certain understanding of "embodied" cognition; namely, "grounded" cognition. We suggest that "grounded" should be interpreted in terms of the conditions of acquisition or constitution of a given ability by bodily abilities. Interpreted in terms of acquisition conditions, "grounded" means that ability A is grounded in ability B if B is necessary to acquire A. For example, once a cognitive system has acquired A on the basis of B, B can be lost without disturbance of A. Understood in terms of constitution conditions, "grounded" means that a cognitive ability A is grounded in a bodily ability B if B is necessary to possess ability A. Compared to the acquisition condition, this means that whenever B is lost, A is lost as well (or at least severely impaired).

During the discussion of the other four accounts, we have argued that one has to specify the explanatory target as well. By focusing particularly on the relation between motor control and action-related cognitive processes, like perceiving an action or thinking about an action, we investigated the grounding relation in a direct manner, while remaining neutral on other further higher-order cognitive processes. When we take a closer look at the introduced constitution condition, action-related cognition may be entirely or partially constituted by motor processes, or it might not be constituted by motor processes at all. By applying our conceptual specification to explain certain cognitive deficits found in subjects suffering from motor deficits, we showed elsewhere that motor features contribute to linguistic concept application and comprehension and that motor abilities are constitutive of processes within the domain of action cognition. However, we did not observe a complete breakdown of action-related cognition in the case of pathologies concerning the motor system. Thus, impairments of both action cognition and action perception by deficits in motor control mechanisms provide reasons to prefer a moderate thesis about only a partial constitution of cognition by bodily abilities. More abstract action cognitive processes might have become independent of motor abilities after acquisition in the course of phylo- and ontogenetic development (cf. Weber and Vosgerau 2012).

The further advantage of this conceptually specified methodology is that distinct approaches to 4E cognition can be empirically tested against one another. In this vein, we provided a systematic comparison of current models and prospective theories that deal with the overall relation between cognition, perception, and motor control mechanisms. Within this framework, the crucial distinction is drawn between constitution relation and acquisition relation, which offers a systematic classification tool for already existing and future theories. This comprehensive meta-theoretical framework not only offers systematic insights into current models and theories, but also defines all different hypothetical possibilities of the relations between the domains of cognitive, perceptual, and motor abilities (cf. Gentsch et al. 2016).

For an illustration, you can think of accounts like the theory of event coding, internal models, or simulation theories. Following the theory of event coding, there is a common representational code for action and perception structuring action planning and action perception. Neither motor control is grounded in perception nor the other way round, but both are linked through additional commonly coded representations. Therefore, from a constitution point of view, this theory cannot be considered as a genuine embodied or grounded cognition theory. Theories about internal models postulate neural processes that extrapolate the commands of the motor system in order to estimate and anticipate the outcome of a motor command. Because exteroceptive predictions are not identical with motor commands but are constituted by the latter, predictions and related action perception and cognition cannot be exhaustively grounded in motor commands. More than motor commands is required to produce these predictions, namely, sensory feedback, context cues, and context estimates. Following our methodology about an embodiment understood as a bodily constitution to generate a certain behavior, motor abilities are only partially constitutive of action-related perceptual and cognitive capacities. According to the family of simulation theories, cognition is essentially a simulation-like reuse of the brain's modal systems, such as the sensorimotor system. For example, the mirror neuron theory assumes that a reference to the neural system used for representing motor acts performed by others serves to explain higher-order abilities like mindreading and social understanding. If simulation is essentially seen as a reuse of sensorimotor modalities, simulation theories postulate a full constitution of action-related cognition by the motor system.

It is an interdisciplinary enterprise to explain behavior. Advocates of 4E cognition open the view to consider more explanatory factors than classical cognitivists did. Surely, we have to deal with complex couplings in-between, but the nature of these relations has to be conceptually defined prior to investigation. Likewise, we have to specify the explanatory target to avoid dealing with omnidirectional, but unclear, theses about a coupling between brain, body, and environment.

References

de Vignemont, F. (2018). The extended body hypothesis: referred sensations from tools to peripersonal space. In: this volume, pp. 389–404.

Gallagher, S. (2018). Building a stronger concept of embodiment. In: this volume, pp. 353–68.

Gentsch, A., Weber, A., Synofzik, M., Vosgerau, G., and Schütz-Bosbach, S. (2016). Toward a common framework of grounded action cognition: relating motor control perception and cognition. *Cognition*, 146, 81–9.

Neisser, U. (1967). *Cognitive psychology*. Englewood Cliffs, NJ: Prentice-Hall.

Pacherie, E. (2018). Motor intentionality. In: this volume, pp. 369–88.

Rowlands, M. (2018). Disclosing the world: intentionality and 4E cognition. In: this volume, pp. 335–52.

Shapiro, L. (2014). When is cognition embodied? In: U. Kriegel (ed.), *Current controversies in philosophy of mind*. New York and London: Routledge, pp. 73–90.

Weber, A. and Vosgerau, G. (2012). Grounding action representations. *Review of Philosophy and Psychology*, 3, 53–69.

PART VI

SOCIAL COGNITION

CHAPTER 22

EMBODIED RESONANCE

VITTORIO GALLESE AND CORRADO SINIGAGLIA

Introduction

Mental simulation was claimed to provide a distinctive way of gaining knowledge about others' actions and thoughts since the late 1980s (Gordon 1986; Heal 1986; Goldman 1989). A decade later, the discovery of mirror neurons in macaque monkeys and the evidence of mirror brain areas in humans presented a new angle on this claim (Gallese and Goldman 1998), suggesting also an embodied approach to simulation (Gallese 2003, 2005).

The present chapter aims at introducing and discussing such an embodied approach and its role in basic social cognition. To this aim, we shall start by characterizing the distinctive features of embodied simulation (ES). Although ES has been proposed to account for not only action mirroring, but also emotion and sensation mirroring (Gallese 2003; Gallese, Keysers, and Rizzolatti 2004; Gallese and Sinigaglia 2011; Gallese 2014) and even language processing (Gallese 2008; Glenberg and Gallese 2012), we shall confine ourselves here to its motor aspects.

There is substantial evidence that ES may critically contribute to understanding others' actions, or so we shall argue. In doing this, we shall also explore the conjecture that ES might involve a common ground for action execution and observation not only at the functional but also at the phenomenological level.

Mirror Neurons in a Nutshell

Mirror neurons were originally recorded from the most anterior region (area F5) of the ventral premotor cortex (PMV) of the macaque monkey (di Pellegrino et al. 1992; Gallese et al. 1996; Rizzolatti et al. 1996; Rizzolatti, Fogassi, and Gallese 2001). These neurons respond to specific actions such as grasping, tearing, holding, or manipulating a given item both when the monkey is so acting and also when it isn't acting other than in observing someone else (e.g., an experimenter or another monkey) performing those actions. Neurons with mirror properties were later discovered in a sector of the posterior parietal cortex reciprocally connected with area F5 (Bonini et al. 2010; Fogassi et al. 2005; Gallese et al. 2002), in the primary motor cortex (area F1) (Dushanova and Donoghue 2010; Tkach et al. 2007), and in the dorsal premotor cortex (PMD) (Cisek and Kalaska 2004). More recently, pyramidal tract neurons (PTNs) originating from both F5 and F1 were reported to respond to action observation (Kraskov et al. 2009; Vigneswaran et al. 2013). While most PTNs increased their discharge during action observation, some of them showed a discharge suppression, where this inhibitory effect might prevent the observer from executing the mirrored movements. Finally, mirror-like neurons were also found in the lateral intraparietal area (LIP), which is involved in controlling eye movements. These neurons signal both when the monkey looks in a given direction and when it observes another monkey looking in the same direction, thus facilitating shared attention (Shepherd et al. 2009).

Executing and observing actions have been shown to share a consistent pattern of cortical activations also in humans, with the main nodes located in the PMV, the caudal sector of the inferior frontal gyrus, and the inferior parietal lobule (Caspers et al. 2010; for a review, see Rizzolatti et al. 2014). Additional mirror-like activations have been reported in the PMD and the superior parietal lobule as well as in the middle cingulate and in the somatosensory areas (Gazzola and Keysers 2009). The human cortical areas with mirror properties are somatotopically organized; the same premotor and parietal regions normally activated when people perform mouth-, hand-, and foot-related actions are also activated when they observe the same motor actions executed by another individual (Buccino et al. 2001; see also Cattaneo and Rizzolatti 2009).

Since their discovery, mirror neurons have been claimed to encode *action goals* rather than muscle contractions or joint displacements (Rizzolatti, Fogassi, and Gallese 2001). This claim has been jointly corroborated by two complementary lines of evidence. On the one side, some studies demonstrated that premotor and parietal mirror neurons responded to the outcome to which the observed action was directed, say, the grasping of a piece of food, even when the action was performed by using different effectors—for instance, hand or mouth (Ferrari et al. 2005; Gallese et al. 1996) or—tools (normal or reverse pliers) requiring opposite sequences of movements, that is, closing or opening the fingers (Rochat et al. 2010). On the other hand, premotor and parietal mirror neurons have been shown to change their discharge profile when similar bodily movements (typically reaching arm movements) are directed to different action goals, such as the grasping of a piece of food for eating instead of for placing it (Bonini et al. 2010; Fogassi et al. 2005). Strikingly, mirror goal encoding has been also demonstrated to be independent from specific sensory modalities (Umiltà et al. 2001; Kohler et al. 2002). Similar results in mirror goal encoding have been obtained in humans (for a review, see Rizzolatti et al. 2014).

Mirror Neurons and Embodied Simulation

Because of their functional properties, mirror neurons have early on been associated with simulation. In a seminal paper, Goldman and Gallese (1998) conjectured that mirror neurons provide "a primitive version, or possibly a precursor in phylogeny, of a simulation heuristic" that might underlie human understanding of others' actions and thoughts (Goldman and Gallese 1998, p. 498). In a nutshell, the conjecture was that when observing another's action, mirror neuron activity recruits a motor representation similar to that which would occur if the observer herself were planning to execute that action. This would allow the observer to "retrodict" the goal to which the observed action was directed. Such goal retrodiction would involve nothing but a "primitive use of simulation," which doesn't go back to mental states such as beliefs and desires. Nevertheless, this primitive simulation heuristic would be critical in understanding others' actions, playing also a pivotal role in the development of full-blown mindreading human abilities (Goldman and Gallese 1998).

Subsequently, the attempt to better understand the link between mirror neurons and simulation lead to the proposal of ES (Gallese 2003, 2007). This proposal differs from the previous one on two main points, at least. The first concerns the notion of mental simulation. The second is about its embodied nature. In this section we will focus on the first point, whereas the next section will be entirely devoted to the second one.

Two different views on the core meaning of "mental simulation" are currently being proposed: simulation as resemblance and simulation as reuse. According to the first view, a mental state or process simulates another mental state or process just in case it copies, reproduces, or resembles the second state or process and in doing so performs a function (Goldman 2006; see also Gordon 1986; Heal 1986; Goldman 1989). The notion of simulation as resemblance seems to fit the standard story of simulation-type mindreading. The simulator supposedly forms pretend mental states matching, as closely as possible, initial mental states of the target, and uses her own decision-making system to generate pretend mental states which match the target's states as closely as possible (Goldman 2006, 2009).

According to the alternative view, simulation as reuse, there is mental simulation just in case the same mental state or process that is used for one purpose is reused for another purpose (Hurley 2008; Gallese 2009a, 2009b, 2011, 2014; Gallese and Sinigaglia 2011). The main argument of the reuse view is that, on almost any story, all simulation-type mindreading requires any resemblance of the mental states or processes between the simulator and the target to arise from the reuse of the simulator's own mental states or processes. At bottom it is mental reuse, not resemblance, that drives mindreading (Hurley 2008).

ES theory endorses a reuse notion of mental simulation. This notion seems to capture the characteristic feature of mirror neuron activity better than the notion of resemblance does. Indeed, there is evidence that mirror neuron motor and visual responses are in many cases far from being highly similar. The fact that neurons (or brain areas) have mirror properties does not rule out the possibility that their motor and sensory responses have different degrees of congruence—quite the contrary (Rizzolatti, Fogassi, and Gallese 2001). What is common to all mirror neuron activities is that the brain resources typically used for one purpose are reused for another purpose. To illustrate, consider the case of parieto-frontal cortical networks involved in performing hand grasping action: there is compelling evidence that parietal and premotor neurons typically represent motorically a goal (such as the grasping of a piece of food) or a hierarchy of goals (such as the grasping of a piece of food for bringing it to the mouth or for placing it in a container) when grasping actions are to be performed. Parietal and premotor neurons with mirror properties do nothing but reuse the same resources to represent motorically that goal, or that hierarchy of goals, when grasping actions are observed rather than performed (Rizzolatti, Fogassi, and Gallese 2001; Rizzolatti and Sinigaglia 2010).

Embodied Simulation and Bodily Formatted Representations

In the action domain, ES concerns not only action mirroring but also motor imagery and object perception. Imagining acting has been shown to recruit motor processes and representations as if the imaginer were actually acting (for a review, see Jeannerod 2001). The same holds for object perception: viewing a manipulable object such as a mug or a bottle elicits motor processes and representations as if the viewer were actually acting upon the object, and this happens even when she has no intention to act at all (Craighero et al. 1999, 2002; Jeannerod et al. 1995; Murata et al. 1997; Raos et al. 2006). What is common to mirroring another's action, imagining acting, and perceiving objects is that in all these cases processes and representations, which are typically involved in planning and executing actions, are used for another purpose, that is, for simulating those actions without any overt motor performance. But what makes all these action simulations instances of ES is that all of them not only rely on reuse, but also involve representations that are bodily in format (Gallese and Sinigaglia 2011).

Cognitive scientists and philosophers distinguish the content of a representation from its format, because each can be varied independently of the other (Goldman and de Vignemont 2009). To illustrate, consider a paper map, where two different routes are represented by two distinct lines. The difference between the lines is a difference in content. By contrast, imagine that the two routes are now represented by a distinct series of verbal instructions. Again, the difference between the instructions is a difference in content. But the lines and the verbal instructions differ in format, representing the two routes cartographically and propositionally, respectively. The format of a representation constrains its possible contents. For instance, a map cannot represent what is represented by a sentence such as "There could not be a mountain whose summit is inaccessible" (Butterfill and Sinigaglia 2014).

Claiming that ES involves representations that are bodily in format means that these representations share some characteristic performance profiles, which are different from those associated with other representational formats, starting from the propositional one. There is a large amount of evidence that many postural, temporal, and spatial constraints characteristic of planning and executing bodily actions are also effective when people just imagine acting rather than actually act (Jeannerod 2001). This is also the case in object perception. Indeed, it has been shown that the view of manipulable objects affords action representations provided that the objects fall into viewers' action space (Costantini et al. 2010; Cardellicchio et al. 2011; Bonini et al. 2014). Even more interestingly for our purposes, action mirroring has been also demonstrated to be spatially constrained. Two single cell studies showed that most of the recorded F5 mirror neurons selectively responded to the sight of grasping actions according to whether the actions were performed within or outside the monkey's peripersonal space. Strikingly, F5 mirror neuron spatial selectivity did not always reflect the distance between the monkey and the objects targeted by the experimenter's actions; rather it depended in many cases on the actual possibility of the monkey reaching for and acting upon the objects which others' actions were directed to (Caggiano et al. 2009; Bonini et al. 2014). No similar constraints have been reported as always effective when processing mere propositional representations such as beliefs and desires, and there is no plausible reason to assume that they should be.

Characterizing the embodied nature of ES by appealing to bodily formatted representations might seem like nothing but a snatcher version of embodiment. After all, one might object that this characterization of ES focuses on mental representations, leaving the body out of it (see Gallagher 2015a, 2015b). However, the characterization of ES in terms of bodily formatted representations does not rule out the possibility that body states and activities affect action mirroring and contribute to others' action processing. Quite the contrary. What makes a representation a bodily one is the relation to the body and its spatial, temporal, and biomechanical constraints. This relation makes the body something more than a representational content, by identifying a specific class of representations in virtue of their format. And it is because of its bodily format that ES can be shaped and constrained by body states and activities (Costantini, Ambrosini, and Sinigaglia 2012a, 2012b). Something similar holds also for motor imagery and object perception. For instance, the view of manipulable objects has been shown to afford ES provided that the objects fall into a bodily action space, regardless of whether it surrounds the observer's body or the body of another agent (Bonini et al. 2014; Cardellicchio et al. 2012; Costantini, Committeri, G., and Sinigaglia, C. 2011).

There is no need to say that this does not imply that embodied cognition should be interpreted in terms of bodily formatted representations only. ES theory aims at providing neither a global view on embodiment nor a general account of all embodied dimensions of social cognition. More modestly, it means to capture action mirroring and related phenomena, which may also include space representation and language

(Gallese and Sinigaglia 2011), within a theoretically unitary framework. This makes critical the notion of representation, provided it comes in different formats, like the bodily one. If this equates with a "moderate approach to embodied cognitive science" (see Goldman 2012), then ES approach is moderate too.

Embodied Simulation and Action Understanding

Appealing to the notion of reuse and its bodily format also helps in accounting for how mirror-driven ES might have a role in action understanding. A main claim of ES theory is that an observer might be able to identify others' action goals by reusing the processes and representations that would occur if the observer herself were planning to execute those actions (Gallese 2003, 2005; Gallese and Sinigaglia 2011). This has as a consequence that goal ascription—that is, the process of identifying another's action goal—might depend on how this action goal is motorically processed and represented.

Now there is evidence that the richer one's motor expertise, the greater her sensitivity to another's action and the better her ability to capture its goal. Several studies demonstrated that mirror-driven ES strongly correlated with motor rather than visual expertise (Aglioti et al. 2008; Calvo-Merino et al. 2006), so that the ability to judge the goal of another's action can be improved by practicing this action, even in absence of any visual feedback (Casile and Giese 2006). Furthermore, it has been shown that both temporary and permanent lesions of mirror cortical areas might result in deficits in another's action goal identification (for an exhaustive review, see Rizzolatti et al. 2014). For instance, a rTMS study demonstrated that a temporary lesion of PMV might selectively impair an observer's ability to proactively gaze at the target of an another's action (Costantini et al. 2014). More recently, it has been shown that delivering continuous theta-burst stimulation (cTBS) over the hand area of PMV made participants less accurate in identifying hand action goals, while after receiving cTBS on the lip area of PMV, participants were less accurate in identifying mouth action goals (Michael et al. 2014). Furthermore, slight manipulations of one's own motor processes and representations have been also found to impact on goal identification. For instance, it has been shown that people are significantly slower in gazing at the target of another's action when they are not in the condition to perform those actions (Costantini, Ambrosini, and Sinigaglia 2012a). Finally, there is compelling evidence that performing an action can impact the observer's ability not only to proactively gaze at another's action target (Costantini, M., Ambrosini, E., and Sinigaglia, C. 2012b) but also to judge what another individual is doing (Cattaneo et al. 2011). By using a habituation paradigm, a TMS study demonstrated that the blindfolded repeated motor performance of an action such as pushing or pulling an object induced in participants a strong perceptual aftereffect when judging whether other people were actually pushing or pulling an object. Crucially, the aftereffect disappeared after delivering TMS over participants' PMV, thus suggesting a causal role for PMV in judging others' action goals. Similar results have been obtained also for communicative actions (de la Rosa et al. 2016).

Positing that mirror-driven ES might be critically involved in action understanding does not amount to positing either that mirror-driven ES should be involved whenever action understanding occurs or that it should underpin any kind of action understanding (Gallese and Sinigaglia 2011). Fully understanding an action is a multilevel process, which may involve, at least, identifying which outcome is the goal of the observed action, representing others' mental states (e.g., beliefs, desires, etc.) that might provide reasons explaining why the action happened, and realizing how those reasons are linked to others' minds and their behavior (Sinigaglia 2013; Sinigaglia and Butterfill 2015b). There is evidence that mirror-driven ES might play a critical role in goal ascription, while reasoning about others' mental states might activate a putative "mindreading network" formed by the mesial frontal cortex, anterior cingulate cortex, and the temporoparietal junction (Brass et al. 2007). Although it is still far from clear how this network actually works and whether and to what extent its activation is mindreading specific (Ammaniti and Gallese 2014; Gallese 2014), a compelling issue for future research is whether and how mirror-driven ES might contribute, at least in part, to reading others' minds, as well as whether and how the mindreading network might downstream modulate ES.

Embodied Simulation and its Role in Social Cognition

ES theory posits that one observer might be able to identify others' action goals by reusing the processes and representations that would occur if the observer herself were planning to execute those actions. Because of its bodily format, such a goal identification would constitute a building block of social cognition, at least at the basic level (Gallese et al. 2009; Goldman and de Vignemont 2009).

An important source of data congruent with this role of ES in social cognition comes from developmental psychology. There is overwhelming evidence that infants succeed in predicting and capturing others' action goals by capitalizing on their own motor processes and representations from very early in life (for a review, see Woodward and Gerson 2014). For instance, it has been shown that six-month-old infants are sensitive to others' action goals, provided that they belong to their motor repertoire (Woodward 1998). Infants have been demonstrated to be sensitive to the action goals of others even at three months of age, but only when previously motorically trained to perform those actions (Sommerville et al. 2005). In the same vein, 10-month-old infants have been found to identify the goal of action sequences to the extent that they can perform them (Sommerville and Woodward 2005). Similarly, infants have been reported to proactively gaze at the target of an observed placing action only to the extent they can perform it (Falck-Ytter et al. 2006). Analogous results have been obtained in the case of grasping actions: 6- to 10-month-old infants were able to proactively gaze at the target of an observed grasping action just by capitalizing on motor cues such as the observed grip of the pre-shaping hand, and their ability in proactive gaze strictly correlated with their own grasping ability (Ambrosini et al. 2013).

Taken together, these findings clearly indicate that once the processes and representations that enable one to perform a given action are on board, the step to a basic form of social cognition such as identifying another's action goals need not involve something radically different from ES. This is not to say that ES alone accounts for all basic forms of social cognition. And this does not rule out the possibility for ES to be modulated or even reshaped by social interactions.

Consider acting together. There is evidence that when so acting, individuals might process and represent motorically their own actions differently from when they are performing those actions alone (Loehr et al. 2013; Vesper et al. 2013). For instance, when clinking glasses together or playing a piano duet together, people have been shown to process and represent motorically not only their own action (e.g., raising one glass or playing a chord) but also the other's action as being part of the same whole action (Loehr et al. 2013; Kourtis et al. 2013). Now there is also evidence that that action mirroring might be different in the case of individual and joint actions (Tsai et al. 2011). This difference does not imply something fundamentally different from ES. Rather, it requires us to refine the role of ES, distinguishing between individual and collective goals.

This applies even to more basic social interactions that do not require any common goal to be implemented (Gallese et al. 2009). Again, this is not to say that ES underpins all basic social interactions, or that social interactions, even when based on ES, are just matter of ES. But the emphasis on the role of social interactions in basic social cognition is fully consistent with the notion that basic social cognition might, at least in part, hinge on the same bodily resources recruited by ES. Thus, far from being a lonely, solipsistic activity (see Gallagher 2015b), ES reveals how close and intertwined individual and social cognition might be (Gallese 2016). Processes and representations primarily built for planning and executing actions can be reused also for representing those actions when performed by others. This could explain not only how individual motor processes and representations facilitate social interaction, but also how social interaction might impact on one's own way of processing and representing action motorically.

Embodied Simulation and Action Experience

Our overall aim in the present chapter was to introduce and discuss ES theory as inspired by early and more recent findings on the functional properties of mirror neurons and mirror brain areas as well as on their specific role in action goal encoding. A main claim was that mirror-driven ES might critically be involved in identifying another's action goal. The evidence we have reviewed indicates that an individual might be able to identify another's action goal and even perceptually judge it by capitalizing on the processes and representations that would occur if the individual herself were planning to execute this action—where these processes and representations are bodily in format. This format accounts for why ES and its contribution to action understanding might depend on an individual's body states.

A first consequence of ES theory is that knowledge of another's action goal does not necessarily require knowledge of particular contents of her mental states such as beliefs, desires, and intentions. Because an individual can represent in a bodily format not only her own but also others' action goals, and because these bodily formatted representations can allow her to identify these action goals when pursued by others and even to perceptually judge them, there is a route to knowledge of the action goals of others that can be taken independently of gaining knowledge of the contents of their mental states (Gallese et al. 2009; Gallese and Sinigaglia 2011; Gallese 2014, 2016; see also Sinigaglia and Butterfill 2015a).

But there is also another consequence of ES account of mirroring action. It has been proposed that ES does not concern the functional level only, but also the phenomenological one (Gallese 2001). Because bodily formatted representations occurred both when acting and also when just observing someone else acting, and because these representations might influence what people are experiencing in both cases, it has been conjectured that there could be aspects of phenomenal character common to experiences of one's own and others' actions (Gallese 2006; Sinigaglia and Butterfill 2015a).

The conjecture is not completely new. More than 20 years ago, Marc Jeannerod pointed out that imagining acting is so close to actually acting, involving many of the same representations and processes almost up to the actual muscle contractions, that the experience of imagining acting would share a core phenomenal aspect with the experience of actually performing the imagined action (Jeannerod 1994; see also Jeannerod 2001). In spite of their differences, motor imagery and action mirroring are both instances of ES, being both shaped by bodily formatted processes and representations that are similar to those involved in planning and executing action. As the experience of imagining acting could have some phenomenal aspects in common with the experience of actually performing the imagined action, so observing someone else performing a given action could have some phenomenal aspects in common with the experience of actually performing this action—or this is the conjecture.

Although further research is needed, there are some seminal studies providing evidence for the notion that bodily formatted processes and representation might shape experiences involved in both performing a particular action and in observing that action. For instance, it has been shown that bodily formatted processes and representations can influence perceptual experience not only in performing but also in observing action (Repp and Knoblich 2009). Indeed, expert and non-expert pianists were asked to perform a sequence of key presses and to observe someone else perform the same sequence. The key presses produced an ambiguous tone pair, that is, a pair with the property that the first tone is sometimes perceived as lower in pitch than the second whereas at other times it is perceived as higher in pitch. The results showed that, for the expert pianists only, the direction of the key presses influenced the perceived direction of the change in pitch, and this regardless of whether they were performing or observing the action.

Even more interestingly for our purposes, the influence of bodily formatted processes and representations on perceptual experience may concern also the observed action, and not their effects only. As already mentioned, manipulating the bodily formatted processes and representations involved in the performance of a given action can influence the perceptual experience of this action also when performed by someone else (Cattaneo et al. 2011). It is worth noting that this influence is content-respecting, varying what is perceptually experienced according to the variation of what is bodily processed and represented. Indeed, not only repeatedly pushing or pulling an object induced a strong perceptual aftereffect when observing other people performing a pushing or a pulling action, but the perceptual aftereffect disappeared once the bodily formatted processes and representations involved in pushing and pulling performance were altered by TMS deliverance.

In the same vein, congenitally aplasic individuals with and without phantom limb experience have been found to differently report what they were perceptually experiencing when observing a sequence of pictures representing a rotating hand. The hand could appear to rotate either clockwise, through the shorter but biomechanically implausible path, or anticlockwise, through the longer but biomechanically plausible path. The result showed that the individual with phantom limb experiences reported having experiences indicating that the hand rotates through the longer, biomechanically plausible path (providing, that is, that the interval between presentation of the two pictures is not too brief), whereas the other aplasic individuals only had experiences as if the hand were moving along the shorter path, violating biomechanical constraints. This was because the aplasic individual with a phantom limb was able to simulate in a bodily format the observed hand rotation, whereas this was not the case for the aplasic individuals without a phantom limb (Funk et al. 2005).

Taken together, these findings indicate, at least seminally, that in some respects what an individual experiences when others act is similar to what she experiences when she herself acts. But what does she experience when her action experiences are shaped by bodily formatted processes and representations? One possible view is that such action experiences are all experiences of bodily configurations, of joint displacements, and of effects characteristic of particular actions. So ES can influence experience but there is nothing characteristically motor in the phenomenology of experiences of action. A more radical view, which we tend to favor, is that some actions can be experientially present, in some way. For instance, they can be experientially present in something like the way that some physical objects can be experientially present (Sinigaglia and Butterfill 2015a, 2015b). There is evidence that experiential presence of physical objects may depend on a system of object indexes (Kahneman et al. 1992), which is not tied to a specific modality (Jordan et al. 2010) and may enable experiences of physical objects even while they are temporarily fully occluded. According to this more radical view, ES would stand to experiences of actions in something like the way that the object indexes stand to experiences of objects. Action experiences would not be only experiences of bodily configurations and joint displacements—they would include experiences of goal-directed actions.

The conjecture we have explored about how ES might provide a common phenomenal ground for experiencing one's own and other actions is consistent with either view about what is experienced. However, our hope is that future research will be directed to investigate whether ES may involve a characteristic phenomenology of action experience, enabling us to experience not just bodily configurations, joint displacements, and the sensory effects of an action, but also the action itself.

References

Aglioti, S.M., Cesari, P., Romani, M., and Urgesi, C. (2008). Action anticipation and motor resonance in elite basketball players. *Nature Neuroscience*, 11(9), 109–16.

Ambrosini, E., Reddy, V., de Looper, A., Costantini, M., López, B., and Sinigaglia, C. (2013). Looking ahead: anticipatory gaze and motor ability in infancy. *PLoS ONE*, 8(7), e67916.

Ammaniti, M. and Gallese, V. (2014). *The birth of intersubjectivity: psychodynamics, neurobiology and the self*. New York: W.W. Norton and Company.

Bonini, L., Maranesi, M., Livi, A., Fogassi, L., and Rizzolatti, G. (2014). Space-dependent representation of object and other's action in monkey ventral premotor grasping neurons. *Journal of Neuroscience*, 34 (11), 4108–19.

Bonini, L., Rozzi, S., Serventi, F.U., Simone, L., Ferrari, P.F., and Fogassi, L. (2010). Ventral premotor and inferior parietal cortices make distinct contribution to action organization and intention understanding. *Cerebral Cortex*, 20(6), 1372–85.

Brass, M., Schmitt, R.M., Spengler, S., and Gergely, G. (2007). Investigating action understanding: inferential processes versus action simulation. *Current Biology*, 17, 2117–21.

Buccino, G., Binkofski, F., Fink, G.R., Fadiga, L., Fogassi, L., Gallese, V. et al. (2001). Action observation activates premotor and parietal areas in a somatotopic manner: an fMRI study. *European Journal of Neuroscience*, 13, 400–4.

Butterfill, S. and Sinigaglia, C. (2014). Intention and motor representation in purposive action. *Philosophy and Phenomenological Research*, 88(1), 119–45.

Caggiano, V., Fogassi, L., Rizzolatti, G., Thier, P., and Casile, A. (2009). Mirror neurons differentially encode the peripersonal and extrapersonal space of monkeys. *Science*, 324(5925), 403–6.

Calvo-Merino, B., Grèzes, J., Glaser, D.E., Passingham, R.E., and Haggard, P. (2006). Seeing or doing? Influence of visual and motor familiarity in action observation. *Current Biology*, 16(19), 1905–10.

Cardellicchio, P., Sinigaglia, C., and Costantini, M. (2011). The space of affordances: a TMS study. *Neuropsychologia*, 49(5), 1369–72.

Cardellicchio, P., Sinigaglia, C., and Costantini, M. (2012). Grasping affordances with the other's hand. *Social Cognitive and Affective Neuroscience*, 8(4), 455–9.

Casile, A. and Giese, M.A. (2006). Nonvisual motor training influences biological motion perception. *Current Biology*, 16, 69–74.
Caspers, S., Zilles, K., Laird, A.R., and Eickhoff, S.B. (2010). ALE meta-analysis of action observation and imitation in the human brain. *NeuroImage*, 50, 1148–67.
Cattaneo, L., Barchiesi, G., Tabarelli, D., Arfeller, C., Sato, M., and Glenberg, A.M. (2011). One's motor performance predictably modulates the understanding of others' actions through adaptation of premotor visuo-motor neurons. *Social Cognitive and Affective Neuroscience*, 6, 301–10.
Cattaneo, L. and Rizzolatti, G. (2009). The mirror neuron system. *Archives of Neurology*, 5, 557–60.
Cisek, P. and Kalaska, J.F. (2004). Neural correlates of mental rehearsal in dorsal premotor cortex. *Nature*, 431, 993–6.
Craighero, L., Bello, A., Fadiga, L., and Rizzolatti, G. (2002). Hand action preparation influences the responses to hand pictures. *Neuropsychologia*, 40(5), 492–502.
Craighero, L., Fadiga, L., Rizzolatti, G., and Umiltà, C. (1999). Action for perception: a motor-visual attentional effect. *Journal of Experimental Psychology. Human Perception and Performance*, 25(6), 1673–92.
Costantini, M., Ambrosini, E., and Sinigaglia, C. (2012a). Tie my hands, tie my eyes. *Journal of Experimental Psychology. Human Perception and Performance*, 38(2), 263–6.
Costantini, M., Ambrosini, E., and Sinigaglia, C. (2012b). Does how I look at what are you doing depend on what I doing? *Acta Psychologica*, 141, 199–204.
Costantini, M., Ambrosini, E., Cardellicchio, P., and Sinigaglia, C. (2014). How your hand drives my eyes. *Social Cognitive and Affective Neuroscience*, 9(5), 705–11.
Costantini, M., Committeri, G., and Sinigaglia, C. (2011). Ready to your and my hands: mapping the action space of others. *PLoS ONE*, 4, e17923.
Costantini, M., Tieri, G., Sinigaglia, C., and Committeri, G. (2010). Where does an object trigger an action? An investigation about affordances in space. *Experimental Brain Research*, 207(1–2), 95–103.
Cross, E.S., Hamilton, A.F., and Grafton, S.T. (2006). Building a motor simulation de novo: observation of dance by dancers. *NeuroImage*, 31(3), 1257–67.
de la Rosa, S., Ferstl, Y., and Bülthoff H.H. (2016). Visual adaptation dominates bimodal visual-motor action adaptation. *Scientific Report*, 6, 23829.
di Pellegrino, G., Fadiga, L., Fogassi, L., Gallese, V., and Rizzolatti, G. (1992). Understanding motor events: a neurophysiological study. *Experimental Brain Research*, 91, 176–80.
Dushanova, J. and Donoghue, J. (2010). Neurons in primary motor cortex engaged during action observation. *European Journal of Neuroscience*, 31(2), 386–98.
Falck-Ytter, T., Gredeback. G., and von Hofsten, C. (2006). Infant predict other people's action goals. *Nature Neuroscience*, 9(7), 878–9.
Ferrari, P.F., Rozzi, S., and Fogassi, L. (2005). Mirror neurons responding to observation of actions made with tools in monkey ventral premotor cortex. *Journal of Cognitive Neuroscience*, 17, 212–26.
Fogassi, L., Ferrari, P.F., Gesierich, B., Rozzi, S., Chersi, F., and Rizzolatti, G. (2005). Parietal lobe: from action organization to intention understanding. *Science*, 308(5722), 662–7.
Funk, M., Shiffrar, M., and Brugger, P. (2005). Hand movement observation by individuals born without hands: phantom limb experience constrains visual limb perception. *Experimental Brain Research*, 164(3), 341–6.

Gallagher, S. (2015a). Invasion of the body snatchers: how embodied cognition is being disembodied. *The Philosophers' Magazine*, 96–102.

Gallagher, S. (2015b). Reuse and body-formatted representations in simulation theory. *Cognitive Systems Research*, 34, 35–43.

Gallese, V. (2001). The "shared manifold" hypothesis: from mirror neurons to empathy. *Journal of Consciousness Studies*, 8(5-7), 33–50.

Gallese, V. (2003). The manifold nature of interpersonal relations: the quest for a common mechanism. *Philosophical transactions of the Royal Society of London. Series B, Biological sciences*, 358, 517–28.

Gallese, V. (2005). Embodied simulation: from neurons to phenomenal experience. *Phenomenology and the Cognitive Sciences*, 4, 23–48.

Gallese, V. (2006). Intentional attunement: a neurophysiological perspective on social cognition and its disruption in autism. *Cognitive Brain Research*, 1079, 15–24.

Gallese, V. (2007). Before and below "theory of mind": embodied simulation and the neural correlates of social cognition. *Philosophical transactions of the Royal Society of London. Series B, Biological sciences*, 362, 659–69.

Gallese, V. (2008). Mirror neurons and the social nature of language: the neural exploitation hypothesis. *Social Neuroscience*, 3, 317–33.

Gallese, V. (2009a). Mirror neurons, embodied simulation, and the neural basis of social identification. *Psychoanalytic Dialogues*, 19, 519–36.

Gallese, V. (2009b). We-ness: embodied simulation and psychoanalysis [reply to commentaries]. *Psychoanalytic Dialogues*, 19, 580–4.

Gallese, V. (2011). Neuroscience and phenomenology. *Phenomenology & Mind*, 1, 33–48.

Gallese, V. (2014). Bodily selves in relation: embodied simulation as second-person perspective on intersubjectivity. *Philosophical transactions of the Royal Society of London. Series B, Biological sciences*, 369(1644), 20130177.

Gallese, V. (2016). Finding the body in the brain: from simulation theory to embodied simulation. In: H. Kornblith and B. McLaughlin (eds.), *Alvin Goldman and his critics*. New York: Blackwell.

Gallese, V., Fadiga, L., Fogassi, L., and Rizzolatti, G. (1996). Action recognition in the premotor cortex. *Brain*, 119, 593–609.

Gallese, V., Fogassi, L., Fadiga, L., and Rizzolatti, G. (2002). Action representation and the inferior parietal lobule. In: W. Prinz and B. Hommel (eds.), *Attention and performance* XIX. Oxford: Oxford University Press, pp. 247–66.

Gallese, V. and Goldman, A. (1998). Mirror neurons and the simulation theory of mind-reading. *Trends in Cognitive Sciences*, 2, 493–551.

Gallese, V., Keysers, C., and Rizzolatti, G. (2004). A unifying view of the basis of social cognition. *Trends in Cognitive Sciences*, 8(9), 396–403.

Gallese, V., Rochat, M., Cossu, G., and Sinigaglia, C. (2009). Motor cognition and its role in the phylogeny and ontogeny of action understanding. *Developmental Psychology*, 45(1), 103–13.

Gallese, V. and Sinigaglia, C. (2011). What is so special about embodied simulation? *Trends in Cognitive Sciences*, 15, 512–19.

Gazzola, V. and Keysers, C. (2009). The observation and execution of actions share motor and somatosensory voxels in all tested subjects: single-subject analyses of unsmoothed fMRI data. *Cerebral Cortex*, 19, 1239–55.

Glenberg, A. and Gallese, V. (2012). Action-based language: a theory of language acquisition production and comprehension. *Cortex*, 48, 905–22.

Goldman, A. (1989) Interpretation psychologized. *Mind and Language*, 4, 161–85.

Goldman, A.I. (2006). *Simulating minds: the philosophy, psychology, and neuroscience of mindreading*. Oxford: Oxford University Press.

Goldman, A.I. (2009). Mirroring, simulating, and mindreading. *Mind and Language*, 24, 235–52.

Goldman, A.I. (2012). A moderate approach to embodied cognitive science. *Review of Philosophy and Psychology*, 3, 71–88.

Goldman, A.I. and Gallese, V. (1998). Mirror neurons and the simulation theory of mind-reading. *Trends in Cognitive Sciences*, 2, 493–501.

Goldman, A.I. and de Vignemont, F. (2009). Is social cognition embodied? *Trends in Cognitive Sciences*, 13, 154–9.

Gordon, R. (1986). Folk psychology as simulation. *Mind and Language*, 1, 158–71.

Heal, J. (1986). Replication and functionalism. In: J. Butterfield (ed.), *Language, mind and logic*. Cambridge: Cambridge University Press, pp. 135–50.

Hurley, S. (2008). Understanding simulation. *Philosophy and Phenomenological Research*, 77, 755–74.

Jeannerod, M. (1994). The representing brain: neural correlates of motor intention and imagery. *Behavioral and Brain Sciences*, 17, 187–245.

Jeannerod, M. (2001). Neural simulation of action: a unifying mechanism for motor cognition. *NeuroImage*, 14, 103–9.

Jeannerod, M., Arbib, M.A., Rizzolatti, G., and Sakata, H. (1995). Grasping objects: the cortical mechanisms of visuomotor transformation. *Trends in Neurosciences*, 18, 314–20.

Jordan, K.E., Clark, K., and Mitroff, S.R. (2010). See an object, hear an object file: object correspondence transcends sensory modality. *Visual Cognition*, 18(4), 492–503.

Kahneman, D., Treisman, A., and Gibbs, B.J. (1992). The reviewing of object files: object-specific integration of information. *Cognitive Psychology*, 24, 175–219.

Kohler, E., Keysers, C., Umiltà, M.A., Fogassi, L., Gallese, V., and Rizzolatti, G. (2002). Hearing sounds, understanding actions: action representation in mirror neurons. *Science*, 297, 846–8.

Kourtis, D., Sebanz, N., and Knoblich, G. (2013). Predictive representation of other's people actions in joint action planning. *Social Neuroscience*, 8(1), 31–42.

Kraskov, A., Dancause, N., Quallo, M.M., Shepherd, S., and Lemon, R.N. (2009). Corticospinal neurons in macaque ventral premotor cortex with mirror properties: a potential mechanism for action suppression? *Neuron*, 64, 922–30.

Loehr, J.D., Kourtis, D., Vesper, C., Sebanz, N., Knoblich, G. (2013). Monitoring individual and joint action outcomes in duet music performance. *Journal of Cognitive Neuroscience*, 25(7), 1049–61.

Michael, J., Sandberg, K., Skewes, J., Wolf, T., Blicher, J., Overgaard, M. et al. (2014). Continuous theta-burst stimulation demonstrates a causal role of premotor homunculus in action understanding. *Psychological Science*, 25(4), 963–72.

Murata, A., Fadiga, L., Fogassi, L., Gallese, V., Raos, V., and Rizzolatti, G. (1997). Object representation in the ventral premotor cortex (area F5) of the monkey. *Journal of Neurophysiology*, 78(4), 2226–30.

Raos, V., Umiltà, M.A., Murata, A., Fogassi, L., Gallese, V. (2006). Functional properties of grasping-related neurons in the ventral premotor area F5 of the macaque monkey. *Journal of Neurophysiology*, 95 (2), 709–29.

Repp, B.H. and Knoblich, G. (2009). Performed or observed keyboard actions affect pianists' judgements of relative pitch. *The Quarterly Journal of Experimental Psychology*, 62(11), 2156–70.

Rizzolatti, G., Cattaneo, L., Fabbri-Destro, M., and Rozzi, S. (2014) Cortical mechanisms underlying the organization of goal-directed actions and mirror neuron-based action understanding. *Physiological Review*, 94(2), 655–706.

Rizzolatti, G., Fadiga, L., Gallese, V., and Fogassi, L. (1996). Premotor cortex and the recognition of motor actions. *Cognitive Brain Research*, 3(2), 131–41.

Rizzolatti, G., Fogassi, L., and Gallese, V. (2001). Neurophysiological mechanisms underlying the understanding and imitation of action. *Nature Review of Neuroscience*, 6, 889–901.

Rizzolatti, G. and Sinigaglia, C. (2010). The functional role of the parieto-frontal mirror circuit: interpretations and misinterpretations. *Nature Review Neuroscience*, 11, 264–74.

Rochat, M.J., Caruana, F., Jezzini, A., Escola, L., Intskirveli, I., Grammont, F. et al. (2010). Responses of mirror neurons in area F5 to hand and tool grasping observation. *Experimental Brain Research*, 204, 605–16.

Shepherd, S.V., Klein, J.T., Deaner, R.O., and Platt, M.L. (2009). Mirroring of attention by neurons in macaque parietal cortex. *Proceedings of the National Academy of Sciences*, 106(23), 9489–94.

Sinigaglia, C. (2013). What type of action understanding is subserved by mirror neurons? *Neuroscience Letters*, 540(12), 59–61.

Sinigaglia, C. and Butterfill, S. (2015a). A puzzle about the relations between thought, experience, and the motoric. *Synthese*, 192(6), 1923–36.

Sinigaglia, C. and Butterfill, S. (2015b). Motor representation in goal ascription. In: M.H. Fischer and Y. Coello (eds.), *Conceptual and interactive embodiment: foundations of embodied cognition* (vol. 2). Oxford: Routledge, pp. 149–64.

Sommerville, J.A. and Woodward A. (2005). Pulling out the intentional structure of action: the relation between action processing and action production in infancy. *Cognition*, 95(1), 1–30.

Sommerville, J.A., Woodward A., and Needham A. (2005). Action experience alters 3-month-old infants' perception of others' actions. *Cognition*, 96(1), 1–11.

Tkach, D., Reimer, J., and Hatsopoulos, N.G. (2007). Congruent activity during action and action observation in motor cortex. *Journal of Neuroscience*, 27(48), 13241–50.

Tsai, J.C.C., Sebanz, N., and Knoblich, G. (2011). The GROOP effect: groups mimic groups action. *Cognition*, 118(1), 135–40.

Umiltà, M.A., Kohler, E., Gallese, V., Fogassi, L., Fadiga, L., Keysers, C. et al. (2001). I know what you are doing: a neurophysiological study. *Neuron*, 31(1), 155–65.

Vesper, C., van der Wel, R.P.R.D., Knoblich, G., and Sebanz, N. (2013). Are you ready to jump? Predictive mechanisms in interpersonal coordination. *Journal of Experimental Psychology. Human Perception and Performance*, 39(1), 48–61.

Vigneswaran, G., Philipp, R., Lemon, R.N., and Kraskov, A. (2013). M1 corticospinal mirror neurons and their role in movement suppression during action observation. *Current Biology*, 23, 236–43.

Woodward, A.L. (1998). Infants selectively encode the goal object of an actor's reach. *Cognition*, 69, 1–34.
Woodward, A.L. and Gerson, S.A. (2014). Mirroring and the development of action understanding. *Philosophical transactions of the Royal Society of London. Series B, Biological sciences*, 369(1644), 20130181.

CHAPTER 23

WHY ENGAGEMENT?

A Second-Person Take on Social Cognition

VASUDEVI REDDY

THERE has, in recent years, been a second move in developmental psychology to place the social at the center of social cognition. The first occurred in the 1960s with Bruner's invention of a LASS to constrain Chomsky's LAD and, in the 1970s both with Donaldson's invocation of a "human sense" to modulate Piaget's abstract logic and with the "discovery" of infant sociality by Snow, Bruner, Stern, Trevarthen, Tronick, Shotter, and others. In broad terms, these were attempts to turn away from the individualism, internalism, and acontextuality of aspects of Piagetian and Chomskyan approaches. The current move, visible not only in developmental psychology but also in neuroscience and cognitive psychology, has a narrower focus on social cognition, but less clear targets against which it may be reacting. There are several terms in the recent literature that capture it—"interaction," the "second-person," "engagement," "child-directedness," "ostensive cues," "shared engagement," "joint engagement," the "interactive brain" and so on. However, these terms differ from each other in fundamental ways, leaving their role in social cognition unclear, unspecified, and contradictory. Part of the problem lies in confusions about what is meant by each of the terms. But a deeper problem may lie in the unwitting continuation of precisely the internalism and acontextuality that the earlier move in the 1960s and '70s was attempting to displace. All current theoretical contenders in infant social cognition would claim sociality as foundational—but often very differently so; and the challenges that characterized the first move have most often fallen by the wayside.

In this chapter I will attempt to tease apart some of the meanings of the terms and to argue that we do indeed need to be conscious of the principles of these challenges; that we need to see the social in social cognition not merely as supporting or permitting, but also as fundamentally constituting it (in the developmental and genealogical sense; Brinck 2016; Brinck personal communication). The typical development of social cognition, I will argue, originates in (and is sustained throughout adulthood by—this is another paper) second-person engagements that irresistibly involve the infant, changing not only the infant cognizer's capacity to cognize, but also that which develops to be cognized.[1] The emotional involvement of persons, in particular those most salient of emotional involvements that occur in second-person engagements where the infant is directly addressed or responded to by another, becomes the crucible of cognition. It may be time to reclaim the term cognition and once again reframe it as a phenomenon involving the person in relation.

The significance of emotional involvement is hardly debatable in our everyday lives—it provides meaning to existence, leading to better quality of life and longer mortality; it enhances school and academic performance in childhood (Furrer and Skinner 2003); it is an explicit strategic objective in advertising (Heath 2009); it is crucial, according to some, in the therapeutic process (Maroda 1998); it is one of the central factors driving autobiographical memory (Holland and Kensinger 2010); and it holds, according to innumerable writings over centuries, transformative power for individual as well as group living. I use the term engagement because it captures these emotional qualities better than the term "interaction," but it needs clarification. In the following sections I ask, first, how we can conceptualize engagement (in terms of its structure, its contexts, and its manifestation) and differentiate second- and third-person relations. Current explanations of social cognition, although emphasizing joint engagement and direct interactions, don't quite grasp the nettle of emotional involvement in their explanations. Lastly, I discuss evidence for the developmental significance of the You and of second-person engagements using evidence of attentional and intentional engagements in the first year, whose presence (as well as absence in atypical development) can only be explained through recognizing mutuality and emotional involvement.

What is Engagement?

This crucial representative image of engagement—emotional involvement—implies that "mere" interaction without acknowledgment of, or response to, the person-ness of the other—e.g., "merely" paying for petrol at a counter or showing the guard your ticket on a train or absent-mindedly saying "thank you" to someone ahead of you who holds a door open—does not typify engagement. Interaction must contain something else: a smile, a joke, gratitude, surprise, dislike, attraction, interest—for us to say that we are engaged. It must *involve* us emotionally. But this implies a categorical definition which may not be justified, and raises questions about other assumed dichotomies.

[1] If the infant herself is changed by engagements, then what the infant needs to understand about the social is also different: not only are new potential engagements afforded, but the new aspects of social being created in the engagements arise to be understood.

Is Engagement a Continuum or a Category?

Is there any interaction which really does *not* involve us emotionally? It could be argued that at least at some minimal level, all our interactions must involve emotional connection; involvement of *any* kind is emotional. To be aware of the attendant behind a counter looking at you expectantly, to perceive the guard's outstretched hand, to be aware that someone at the door is waiting till you get there requires some degree of interest and awareness in you. And awareness implies *some* level of affectivity (in Stern's sense of vitality affects rather than categorical affects; Stern 1985). Even if one excludes the most basic interactions as being too minimally affective, it leaves the vast range of interactions as a spread of varying degrees and kinds of emotional involvement. This means of course that while the criterion of emotional involvement may be a good representation of engagement, it can be misleading unless we see engagement as a continuum rather than as an either/or category.

Is Engagement Singular or Multiple?

Once again, the answer to this opens inconvenient complexities. Like "being," engagement not only occurs with different degrees of involvement, but also in different domains. We could be relishing the warmth of a hot shower while absorbed in solving a stubborn puzzle. We could be driving on a motorway but be reliving a painful event. We could be in the middle of a tragedy, weeping, while at the same time thinking through its practical implications as if it were a mundane logistical problem. Speaking of engagement as if it was one thing, even if at one moment, is inevitably limited. We only do so, in academic discourse, because to grasp and measure the full complexity of multiple engagements seems a Herculean task.

Does Engagement Occur not only with Persons but also with Objects?

There are many reasons for answering yes and *not* venerating the person–object distinction. First, the distinction is by no means absolute: persons are objects too, and we run the risk of another kind of dualism by ignoring this. Second, there are many ways in which persons personalize the material world: artifacts are structured by the intelligence, the bodies, and the desires of persons; and the object world is introduced *to* persons by a cultured and person-ed reality (Costall 1997; Rodriguez and Moro 2008; Rossmanith et al. 2014). Third, even newborn infants are interested in engaging with objects, with intense attention, whole-body movements, and rough swiping by the hands in their vicinity (von Hofsten 1982), and are interested in the effects of their actions on the world, seeking to explore visual-kinesthetic matching and the production of visual "sights" contingent on their actions (Van der Meer et al. 1995). The emotionality involved in attraction to perceived objects and in wishing to "grasp" them in some way is strong and moving evidence of an openness to engage even with the material world. No doubt engagements with objects are limited (Brazelton, et al. 1974; Legerstee 1994), and do not involve the rich mutuality of engagement with animate beings. But they are engagements nonetheless.

Can there be Engagement without Action?

One could regard looking at and "feeling" a response to what one sees as action of sorts. And of course one can be engaged, intensely, with only attention and responsive feelings—as to a movie or with a puzzle (albeit with imagined actions). But doing something manifest in the world—reaching out to what one sees, or turning away, or vocalizing to call—can lead not only to a change of state inside oneself, but to a change in the world, allowing responses and responses to responses, expanding, curtailing, or coloring the potential of the relation you are engaged in. The most impactful and consequential engagements, then, are those that involve action. Emotional involvement without action—as in intense feelings while watching a movie—is indeed engagement, but does not allow the kinds of mutuality, unscripted developments, and consequences that are possible in active engagements. But action too, like personhood and emotional involvement, is inevitably a continuum.

So what is the Difference between Second-Person and Third-Person Relations?

It is clear that engagement can occur in many different kinds of interaction—thinking out an argument with yourself, frowning in a one-to-one encounter with another person, crying in sympathy with a character in a movie or on the news, joining in with the joyful mood of a group of friends, taking part in witty banter with a colleague while feeding off the amusement of the audience. And it is clear that the term engagement is loaded with complexities of level, of types of relation, of intensity, of activity, and of modality. Within this quagmire of overlapping terms and complex phenomena, is engagement not only still worth talking about, but worth talking about in relation to another distinction—that of second- and third-person relations? How would this help us understand the development of social cognition in human infants?

What could we mean by a second-person and a third-person relation (for the sake of simplicity talking only of relations between persons)? Although simply operationalizable in structural terms—as the difference between dyadic interactions and interactions in which one person is the third party observing another dyad interact—the substance of the distinction (and its relevance for cognition) comes from an emotional source. It derives from Martin Buber's (1957) attempt to differentiate two different modes of knowing—the I-You and the I-It. At its heart, a second-person relation involves the experience of being addressed by another, of being seen as a You by another person, and of the mutuality that is generated in seeing the other as a You in turn. A third-person relation involves a more detached, observational, stance in which one sees the other as a He or She rather than as a You.

There are different contexts in which a second- or third-person stance can occur: one can be in a dyadic interaction but be thinking about the interlocutor in detached analytic terms ("He has done it again" or "Why does she act as if I am an idiot?" and so on); or one can be in a group or three-person situation or even in a movie primarily as an observer, but still see the other(s) as if they were speaking to oneself and feel involved with responsive sympathy or hate or anger or adoration toward them. The key difference between relating in the second person and relating in the third person is not one of the structure of the situation, but one of the openness or closed-ness with which one faces (and is faced by) the other. Closed-ness to the other—even in dyadic engagements—can occur in many ways: through categorizing or objectifying the other, through adopting an analytic stance, through having another agenda or concern. You can see the other through the filter of a label, a group category, or a dismissive analysis (she is just a student, he is an immigrant, she is autistic); you can depersonalize the other, literally objectifying them as a means to another end (in pornography, for instance, or in malicious teasing to entertain an audience); you can approach them with another agenda (I must convince him to buy this), which stops you from "hearing" them or from your own genuine expressiveness; you can approach the crying of a baby with an immediate search for expert advice (Should I go to her or let her cry a little more?) which could (momentarily) stop you hearing and relating to the crying *as* crying; you can be discussing a student's poor performance with him while worrying about what your head of department wants you to say to him. The extent to which one is open to and present to the other and to which the other is open to and present to you is what marks out a second-person relation. Of course this too consists of multiple levels and continua.

There are arguments against making this distinction at all (for understanding developing social cognition). Some oppose any relevant distinction between second- and third-person interactions (what one might call the "no real difference" objection): saying that in both cases the child or adult observes an *other* person (not the self) displaying actions and expressions, making the second-person situation only another kind of third-person situation (Barresi and Moore 1996). This objection is premised on the assumption of a gap between interactants that can only be bridged intellectually (hence the other is always a third person, as it were) even when in emotional engagement with you. A different objection (what one might call the "graded difference" objection) comes from the observation that since the distinction between active engagement and passive observation appears to be gradual rather than absolute, it "undermines the claims about the developmental primacy and phenomenological pervasiveness of 2P interactions" (de Bruin et al. 2012). But this only follows if the claims are themselves categorical, and they need not be; the fact of their graded distinction does not negate the fact of the different effects from different parts of the grading. Another argument (what one could call the "can't have two separate theories" objection), primarily in the context of understanding belief-desire reasoning (Schoenherr 2016), questions the plausibility of the distinction and its relevance for social cognition on two grounds. First, the boundaries between second- and third-person contexts in the real world are seen as simply not clear-cut enough to merit such a sharp theoretical distinction. Interactional and observational stances toward others are often so rapidly changing in a single social situation that "it would seem surprising if distinct theories were to apply to both contexts." Second, it is argued, many of the inferences that can be drawn in second-person interactive situations can also be drawn when observing others. The rapidity of shifting stances and the complexity of their intertwining is certainly true, and it is entirely sensible to assume that any theoretical and inferential grasp of minds will not differentiate between—or exist in two separate compartments for—the two types of interaction. Both these objections, however, relate to distinguishing second-person theories or inferences from third-person theories or inferences. My argument is not that second- and third-person relations build parallel theories of other minds, or indeed that they do not influence each other, but that they yield different *experiences* of others. And in the pre-theoretical infant, the experience of mutuality and emotional connection in second-person relations is developmentally crucial (and is crucial throughout life, although probably in different ways).

Both types of experiences, second-person involvements and third-person observations, must influence each other and both may be necessary even for stable pre-inferential perceptions of other minds. Indeed, in phenomena such as teasing and humor, even infants seem to show a rapid alternation between stances (Reddy 2008). But being addressed as a You and addressing the other as a You arouses emotional responses differently from watching someone else be addressed, and engenders—even if briefly—a mutuality and suspension of separateness. The other becomes a person to you, someone who knocks you off balance or enters your consciousness in a more fundamental way than when you are largely untouched by the other, or just watching them. Even though the boundaries between them may be thin and permeable and their occurrence fleeting and dynamic, the difference between the two relations—of being addressed or being "heard" versus being observed or watched or analyzed, of addressing or "hearing" versus watching, observing or analyzing—is undeniable. We cannot talk of a single kind of "other"; how you experience an other depends on the type of engagement you have with them. And in early infancy the most salient access to others (and I mean this in the non-Cartesian sense of "others" as persons or as beings with minds) is in engagements where the other addresses, is addressed by, and responds to the infant, which involve the infant perceptually and emotionally.

A second answer to the question of whether it is worth valuing the distinction is this: it matters on the ground. However messy the definitional distinctions, being addressed and being heard makes a difference to one's experience of being at any moment, and most crucially, it makes a difference to development. If we have evidence that such engagements matter, the phenomenon demands recognition, study, and intervention where it is endangered. Many current developmental theories, however, don't take it into account. One recent review (Schneidman and Woodward 2016) challenges the importance of child-directedness in learning, finding little evidence for its superiority in learning words but some evidence for its importance in learning to act on cultural artifacts. The authors suggest that any advantage comes not from child-directedness per se but from greater likelihood of attention to child-directed as compared to observed input, or from other properties accompanying it that support learning. Paradoxically, this suggestion supports a difference between second-person and third-person situations: while they may not always aid learning, child-directed acts are salient to children—they obtain more attention and might be accompanied by other experiences not present in merely observational situations. When we are talking about engaging with other minds, salience, in terms of attention and emotional relevance, is central.

Evidence for the Power and Primacy of Second-Person Relations

In adults, being directly addressed by another person leads to different neural processing and to enhanced sensory awareness. Hearing one's name being called leads to activation of the same brain areas as does seeing someone look at one, independent of arousal (Kampe, Frith, and Frith 2003). Direct gaze, smiles, and attracting the other's attention show distinct and localized neural activation processes (Schilbach et al. 2006, 2008). There also appears to be a processing advantage of "being addressed": even in fairly unnatural laboratory experiments where personal pronouns are presented through headphones, the second-person pronoun—"you"—is processed preferentially across different ERPs, showing up significantly earlier than the processing of first- or third-person pronouns, arousing enhanced self-related processing and reduced external processing (Herbert, Blume, and Northoff 2015). This suggests an attentional advantage—leading to greater internal sensory processing—by hearing an address. Being looked at (by eyes in a still photograph) arouses greater self-awareness: if preceded by a photograph of a face with direct gaze, participants show greater accuracy in self-reports of arousal to affectively arousing images than if they are preceded by a photograph of a face with averted gaze (Baltazar et al. 2014). If these patterns are also the case in infancy, early second-person engagements might change and enhance the awareness of other and of self, allow an awareness of both self and other as persons and also allow recognition of the marking of a shared world by the other.

What is necessary for second-person relations in infancy? Three things: infants need to be *open to engagement* with others (i.e., have interest in and an ability to act toward them); infants need to have *others who recognize them as persons* (i.e., persons who address them as a You and respond to their addresses); and infants need to be *able to recognize the recognition* of the other *and respond to* it (i.e., they need to recognize the other's actions as relevant responses connected with themselves and their actions and be able to pursue mutually responsive engagements).

Infant Openness to Engagement with Others

There is considerable evidence of interest in others and openness to their initiatives from birth and the first weeks and months of infancy. A specific predisposition for looking at human faces and face-like stimuli is present at birth, and allows not just greater duration of looking but also tracking across space (Goren et al. 1975; Morton and Johnson 1991). Direct gaze to the infant is not only preferred within at least a few days of birth (Farroni et al. 2002), but leads to greater accuracy of directional saccades if it precedes gaze directed elsewhere (Farroni et al. 2004). Recent findings from EEG and NIRS studies show that direct addresses result in different neurological effects even in infancy. Direct gaze results in different gamma band oscillation in four-month-old infants (Grossman et al. 2007), and both direct gaze and being addressed by their own name are correlated in five-month-old infants (Grossman, Parise, and Friederici 2010), suggesting a selective attention to direct communications. By five months mutual gaze leads to enhanced word learning (Parise et al. 2011). Being called by their own name leads to enhanced attention to objects (Parise, Friederici, and Striano 2010), and by six months mutual gaze leads to enhanced gaze-following (Senju and Csibra 2008). Similarly, a greater interest in sounds in the human voice range, in human voices, in female voices, and in the mother's voice is evident at birth, and may be a result of auditory experiences in utero (De Casper and Fifer 1980). Studies of neonatal imitation, regardless of the controversies surrounding their status as imitation, show powerfully that neonates within minutes of birth look with intense and focused interest at the facial and manual actions others direct toward them, and attempt to respond with actions themselves (Kougioumutzaki 1998; Meltzoff and Moore 1977). Such responsiveness suggests that the neonate's openness to others may, at least in healthy and neurologically intact neonates, be free of distrust and distress (Brazelton 1986), and is expressed in sustained and effortful attention.

Having Others who Recognize them as Persons

The availability of "others" who are keen to respond to the infant's actions and interests may be a bit of a circumstantial lottery, and varies in form with culture, but is certainly common enough for this to form part of the normal social environment that infants meet at birth. The simplest operational measures we can think of to pin down the recognition of personhood involve direct dialogic addresses rather than treating the infant as an object (for instance, by not just performing necessary caretaking or medical procedures interpersonally), affective attunement (by matching or complementing the other's affect, showing recognition that the other is feeling something and trying to tune into that mood or that rhythm), and sensitivity to changing initiatives and interests of the infant (by listening to and responding to them even if they are unexpected). In early encounters, and increasingly over the next weeks, the infant is often directly addressed by others—in face-to-face attempts to engage the infant (Trevarthen 1977), in responsive tactile engagements (Kaertner et al. 2010), and perhaps even in vocal responsiveness. Parents vary in their modes and degrees of affective attunement—sometimes described as variations in affective mirroring (Legerstee and Varghese 2001) or contingent smiling (McQuaid et al. 2009)—and are inevitably affected by their own state of depressiveness, leading to altered patterns of communication and different communicative expectancies in their infants (Field 1984; Murray et al. 1996). Having one's communicative initiatives and emotionality responded to by others not only gives the infant feedback about their relevance to the other, but potentially confirms the infant as an expressive emotional being. The influence of this feedback and confirmation, however, depends on what the infant recognizes of it.

The Ability to Recognize Others' Recognition

The recognition of the contingency of others' responses to their own actions is evident from at least two months of age, but possibly also from shortly after birth. We know that infants can detect contingencies between their own actions and their effects by four weeks of age (Van der Meer et al. 1995), and even in the neonatal period, as revealed by rapid instrumental conditioning (Siqueland and Lipsitt 1966) and by neonates' ability to detect temporal mismatches between actions felt on their face and actions seen on a face on a monitor (Filipetti et al. 2015). By two and three months of age, contingency detection is advanced and can be seen in different ways: in differentiating live versus videotaped "partners" (i.e., live versus replayed versions of their own faces in a double video setup, Reddy et al. 2007) and live versus videotaped body movements (Bahrick and Watson 1985). More than the detection of temporal contingency, however, infant distress at the lack of response to their initiatives is crucial evidence of the recognition of an absence of recognition at some level. Increased frowning, looking away, actual distress, or signs of helplessness as indicated by reduced attempts to engage can result from different types of unresponsiveness: the cessation of responses within a good engagement with a familiar person (still-face studies: Cohn and Tronick 1983; Markova and Legerstee 2006; Nagy 2008), the mismatch between action and response (Murray and Trevarthen 1985; Nadel et al. 1999), reduced contingent maternal responsiveness (Field 1984; Murray and Cooper 1997; Legerstee and Varghese 2001; McQuaid et al. 2009), and so on. Anecdotal evidence from nursing care about the reduction in distress at invasive medical procedures if there is first an interpersonal address to the infant adds to the implications of these experimental studies, suggesting that the quality of contingent interactions and the extent to which they respect the communicative potential of the infant is important in the immediate and the longer term. The infant does seem to recognize when she is recognized as a conversational being from very early in life.

Part of the difficulty in arguing this case—that the infant is indeed recognizing the other's actions *as recognition*—comes from different meta-theoretical assumptions about the nature and availability of mind. Where the task for the infant is seen as being an inferential construction of hidden mental states, this claim becomes very difficult to sustain. But the considerable recent evidence for the embodiment and perceptual availability of many mental states (Robertson and Johnson 2009; Becchio et al. 2010, 2013) have made the opacity of mind position rather difficult to sustain. Minds are embodied; mind *is* the way the body expresses attention, intention, and emotion, and while conceptualizing mind is a complex and developmentally late achievement, perceiving it is a simple achievement within engagement (see discussions in Leudar and Costall 2006; Reddy 2008; Schilbach et al. 2013).

In what follows I take two domains of social-cognitive awareness—of others' attention and of others' intentions—with two examples of second-person engagements in each domain: one showing infant responses to attention or intentions directed to the infant, and the other showing responses to attention or intentions directed to the infant's actions. I use these to argue two points: (1) These early engagements are crucial steps along typical developmental paths in awareness of attention and intention. Not recognizing their impact results, I argue, in inadequate theorizing. (2) These engagements show a profound mutuality of emotional involvement. Both infant and other are responding to each other, in the moment, as persons, and these responses open up new ways of being and new possibilities for understanding. The engagements not only *reveal* infant awareness but also *create* new things to be aware of, for both infant and adult.

Developments in Joint Attentional Engagements

Infants don't just detect others' attention toward them; they respond emotionally to it (for the moment, given patchy evidence about other modalities, I discuss only visual attention). And throughout early development the range of emotional responses to attention—positive, negative, indifferent, and ambivalent—is broadly similar, even while these infant emotional responses are increasingly elicited with age by more diverse and complex attentional acts by others. From birth, gaze to self is interesting, leading to longer durations of looks and more frequent looks (Farroni et al. 2002); and this very same interest and perception of relevance can lead to distress if the infant cannot disengage from it (Brazelton 1986). By the second month the onset of others' gaze evokes smiling and positive affect (Wolff 1987). But affective responses to attention can be more complex. Coy smiles in the two-month-old suggest a recognition of others' attention *as* attention (but only when directed to self), and clowning and showing off in the eight-month-old suggest complex elicitation of attention, not just to self, but to specific actions by the self.

Positive Shyness or Coy Smiles to Gaze to Self at Three and Four Months

Coy smiles in response to others' smiling gaze directed to them start to appear in the third month of life in typically developing infants, consisting of a pattern typically associated with embarrassed smiles in adults (Asendorpf 1990): intense smile, with gaze, head aversion, or arm raising occurring within the peak of the smile, and with frequent return of gaze. They occur more frequently at the start of an interaction than later on; and although they initially occur with familiar adults, by four months of age they can be seen prominently with strangers (Reddy 2000; Colonnesi et al. 2013). Positive shyness or coy smiles do not, however, occur in children with autism despite developmental age-appropriate achievement of mirror self-recognition (Hobson et al. 2006; Reddy et al. 2010). This phenomenon has two theoretical implications. First, it challenges the claim that emotional reactions akin to embarrassment cannot occur until the development of a concept of self in the second year (Lewis 1995). The pattern of behavior evident in these coy smiles is similar to adult expressions of embarrassed smiles (see Reddy 2005 for a discussion of similarities and differences over age). Importantly, it is dissociated from mirror self-recognition in autism and precedes rather than follows it in typical development, suggesting that self-consciousness begins as an affective response to attention to self, and that self-conscious affectivity might help to constitute, rather than solely be derivative of, a concept of self (Izard and Hyson 1986; Hobson 1990; Reddy 2008).

Second, the emotional responses to others' attention that begin from birth and shortly after show, continuities over the first year, long before the infant can engage in the traditionally conceived triadic joint attention (Bates et al. 1976; Tomasello 1999). The range of emotions, originally elicited only by attention directed to self, expand and start to be elicited by more complex situations and "stimuli," increasing in subtlety of expression and control but without dramatic changes in categorical affect (Reddy 2003, 2005, 2011). These continuities must be explained. They could, unconvincingly, be dismissed as mere pseudo-emotional responses (Reddy and Morris 2004), awaiting a developmental watershed such as the so-called nine-month-revolution before being considered as responses to *attention*. Alternatively, they could be seen as emotional responses to attention to self that are crucial first steps in an expanding realization of what attention is and can be.

Clowning and Showing-Off from Seven or Eight Months of Age

Soon after the middle of the first year, infants start to pick up on others' emotional responses to infant actions (Reddy 1991, 2001; Mireault et al. 2011). Often an accidental discovery that a certain facial expression, sound, or movement of the body leads to adult laughter or general positive attention can lead to the infant repeating that expression, sound, or movement (sometimes for weeks) to re-elicit the response (Reddy 1991, 2005). The occurrence of clowning is fundamentally mutual—it depends on the presence of an "other" who finds an action or expression funny, and it depends on the infant's pleasure in the other's amusement, the infant's awareness that the amusement is linked to the action or expression, and the infant's ability to re-elicit it by repeating the action or expression. Such emotional engagements show deficits in preschool children with autism (but not in developmental age-matched children with Down syndrome), with either very limited or formulaic evidence of clowning and showing-off (Reddy, Williams, and Vaughan 2002). This phenomenon reveals an expanding grasp of attentionality (the infant now knows that others attend not only to her, but to her actions) *before* the infant can engage in triadic joint attention—such as pointing to external objects—toward the end of the first year. Joint engagement with the infant's action or expression as a "shared object" is evident much earlier than the joint engagement at 14 months (Moll et al. 2007); the simpler mutuality of these engagements, with adult emotional responses and infant attempts to re-elicit it interdependent on each other, may be crucial for allowing infants further access to the nature of others' attentionality.

Both positive shyness and clowning/showing-off illustrate the powerful way in which people—both infants and adults—are emotionally moved by others' attention and actively seek to move each other. The crucial point about these engagements is not their age of occurrence—earlier than hitherto thought—but their developmental sequencing. First, infants become emotionally moved by others' attention to self. Some months after that they become emotionally moved by others' attention to their actions. Only after that do they become able to grasp others' attention to distal objects in space. This sequence is chronological (in typical development) as well as potentially causal (e.g., in children on the autistic spectrum). Given this continuity of emotional response to attention and this sequence, the argument that attention as a psychological phenomenon is "discovered" at nine months is untenable.

Developments in Joint Intentional Engagements

Infants don't just observe others' intentional actions toward distal objects; they also, and often powerfully from the moment of birth, experience intentional actions directed toward them. It would be strange to imagine that infant awareness of others' intentions was independent of infant experience of such actions directed to them. And if there is evidence that infants not only recognize infant-directed actions, but anticipate them with appropriate responses, then it would be strange indeed to ignore these responses in theorizing about action understanding.

Anticipatory Adjustments to Being Picked Up at Two and Three Months

Kanner discussed at some length as long ago as 1943, reports from several parents of school-age children with autism, that their children did not show anticipatory adjustments to being picked up. Such a lack of motor preparedness or anticipation of actions toward the self has been subsequently reported in feeding and other kinds of actions in children with autism (Brisson et al. 2012). Typically developing infants, however, from at least around two months of age, show anticipatory adjustments of their bodies when familiar adults are approaching to pick them up, even before actual contact is made. Three types of body adjustment occur, differently in different infants: raising and extending or tucking up the legs, opening out or raising the arms, raising the chin or turning the head and neck. The adjustments are present at two months but become smoother by three months (Reddy, Markova, and Wallot 2013). By at least three months, these adjustments cease if the adult delays a few seconds in picking the infant up (Fantasia et al. 2016). This phenomenon does not easily fit current explanations of social cognition. Given that the infant at two months is not yet capable of reaching and grasping in this manner, their experience of their own actions cannot be the explanation of the adjustments to the adult's approach. The arms directed to the infant could, at a stretch, be interpreted as ostensive cues, but if so, they are communicative in themselves rather than cues to information about the world, and these adjustments appear to emerge during the first weeks of life (Wallot et al., in preparation), suggesting that the arms approaching are not hardwired signals. The key point here is that the infant at two months, *before* she grasps intentional actions directed to other objects, is appropriately responsive to intentional actions directed to herself. Such engagements are often emotionally significant to the infant, and involve unthinking mutuality (sensitivity and ongoing adjustments by each person to actions as they unfold); they must form the basis of later, more complex awareness of intentions.

Compliant Responses to Others' Directives in the Second Half of First Year

Although infant compliance with others' verbal directives is clear by the end of the first year, adults issue commands and requests to infants from many months earlier. From the middle of the first year, but varying between family and culture, adults tend to increase the frequency of directives, embedding them in frequently repeated pragmatic formats (mostly involving display of positive actions and skills rather than prohibitions). And infants from around seven or eight months, also varying with their experience of the directives, start complying with the directives (Reddy et al. 2013). What is very clear in these directive episodes is the infant's interest in joining with the adult's intentions and interests; the adult's directive is interesting to the infant *because* it comes from the adult to the infant. And the infant's (initial) willingness to engage—however tentatively—with their directives encourages adults in constantly upping the ante and putting forward more and more complex invitations. The gradual increase in frequency of compliant responses from 6.5 to 12.5 months in both cultural groups in this study suggests that these engagements are part of a continual and expanding process of understanding intentions, challenging the claim that communicative intentions are complex things understood only in the second year of life (Tomasello 1999). The infant-directedness of the directives—as an "ostensive cue" to the purpose of the utterance—may indeed be evident before the content of the directive becomes clear, as some argue (Csibra and Gergely 2009). But compliance with different directive contents does not occur all at once and it is likely that it is the repetitive pragmatic formats with partial and increasing fulfillment of the intent that clarify their meaning.

Intention awareness would seem to be inseparable from intentional engagements. The infant gets drawn in to respond to others' intentions by having them be directed toward herself in the first place and then toward her actions. Infant appropriate responses to others' intentional actions follow the same chronological scheme as with attention—going from grasping intentions directed to self, to intentions directed to actions, and only later to intentions directed to objects in space. Adult sensitivity to infant capabilities and motivations is crucial in such engagements; both parties to the interactions change and adjust as they unfold, creating paths characteristic of each relationship.

Concluding Points

Most theorists of social cognition would take as given that in some sense, interaction or engagement is necessary for social cognition. But what is meant by this "necessity" differs significantly among them. The work of Csibra and Gergely (2009) in studying the role of ostensive cues in the development of learning through communication has been an important step forward in the field and is sometimes identified as an example of a second-person theory. They have argued that the direct address—looking at the infant, calling their name, or using infant-directed speech—is crucial in activating built-in attentional biases in the infant and in allowing the infant to see what follows as generalizable knowledge. However, their use of the term "ostension," deriving from their central focus on understanding the occurrence of communicative learning about the world (not especially about the other), takes a different slant from the second-person position outlined here. Their use of a cue-based explanation is substantially different from the emotional involvement–based explanation I offer here. The evidence—of the primacy of infant awareness of attention and intentions directed to the infant, and of the expansion of this awareness to attention and intentions directed to the infant's actions, and only then to awareness of attention and intentions directed to objects in space—extends their findings to the early emergence of other-awareness (not just about others as cues leading to an awareness of the world).

Similarly, the work of Michael Tomasello and his colleagues has been central to almost all debates about the emergence of social cognition and is also sometimes identified as involving a second-person position. Their recent focus on joint engagement—for example, showing that understanding what another person knows about an object depends initially on actual joint engagement with that adult and the object, and only subsequently relies on simple observation of the adult's engagement with the object (Moll, Carpenter, and Tomasello 2007)—suggests an intriguingly crucial role for engagement in social cognition. However, even though the implications of the position have very early roots in dyadic interaction, the evidence used is largely limited to triadic interactions occurring after nine months and the so-called nine-month sociocognitive revolution in understanding of intentionality (Tomasello 1999; Tomasello et al. 2005). The recognition of early second-person engagements allows a fuller recognition of what it means to be addressed, to *be seen* as a You by another person (and not only to seeing the other as a You with an object). The mutuality of seeing the other as a You and being seen as a You at the same time needs us to go a step further—or earlier—for understanding the second-personal roots of social cognition.

To sum up, to offer an adequate explanation of social cognition we need explanations that do not cast the infant in the role of an epistemic observer or analyst, that do not begin to explain social cognition until late in infancy, and that do not conceptualize engagement merely in terms of cues to the perception of the object world. To understand the origins of social cognition, we need to understand direct infant engagements with the social—their occurrence in early addresses by the world, the reasons for their occurrence, and the manner in which they seem to be sustained in a delicate mutuality of emotional involvement. That which is mutual not only involves a more messy developmental experience, but also involves a changing field; we seem to be poised at the beginning of an explanation but have not gotten very far.

Some might argue that the proper domain of study to explain social cognition is the individual brain—that is, that "the social neuroscientist does not also need to be studying the stock market" (Adolphs in De Jaegher et al. 2016). However, the presence of the stock market must change not only the content, but also the form of the cognitions. The shape and potential of cognition is not fixed independent of the material it cognizes. The material itself changes the shape and scope of cognition. Engagements are not just grist for a cognitive mill to grind; they create the very material that we seek to understand.

Like all those other terms that we use—mind, culture, emotion, love—engagement is a vague and multifaceted term. One could argue that this is an essential vagueness—an indeterminacy that encompasses the possibility of as yet unknown manifestations. But it captures something crucial about our social lives. The ability to connect with the emotions of others and to recognize their recognition of our own is at the heart of development and may involve a dynamic process of "identifying with" (Hobson 2002, 2007) and an openness to dialogue where the emergence cannot be predetermined. Similarly, in attempting to study sociocognitive development from birth, the scientist's recognition of the subjective orientations of infant and adult in engagement are crucial for understanding what it is that prompts infants to try to understand others, and what it is that adults are doing to invite and indeed allow it. As much as mutual engagement between infants and adults is needed for developing social cognition, mutual engagement between developmental scientists and infants is needed to develop an adequate theory of social cognition.

References

Asendorpf, J. (1990). The expression of shyness and embarrassment. In: R. Crozier (ed.), *Shyness and embasarrassment: perspectives from social psychology*. New York: Cambridge University Press, pp. 87–118.

Bahrick, L. and Watson, J. (1985). Detection of intermodal proprioceptive-visual contingency as a potential basis for self-perception in infancy. *Developmental Psychology*, 21, 963–73.

Baltazar, M., Hazem, N., Vilarem, E., Beaucousin, V., Picq, J.L., and Conty, L. (2014). Eye contact elicits bodily self-awareness in human adults. *Cognition*, 133(1), 120–7.

Barresi, J. and Moore, C. (1996). Intentional relations and social understanding. *Behavioral and Brain Sciences*, 19(1), 107–22.

Becchio, C., Del Giudice, M., Dal Monte, O., Latini-Corazzini, L., and Pia, L. (2013). In your place: neuropsychological evidence for altercentric remapping in embodied perspective taking. *Social Cognitive and Affective Neuroscience*, 8, 165–70.

Becchio, C., Sartori, L., and Castiello, U. (2010) Toward you: the social side of actions. *Current Directions in Psychological Science*, 19(3), 183–8.

Brazelton, B. (1986). *Affective development in infancy*. Westport, CT: Ablex.

Brinck, I. (2008). The role of intersubjectivity for the development of intentional communication. In: J. Zlatev, T. Racine, C. Sinha, and E. Itkonen (eds.), *The shared mind: perspectives on intersubjectivity*. Amsterdam: John Benjamins, pp. 115–40.

Brisson, J., Warreyn, P., Serres, J., Foussier, S., and Adrien, J.-L. (2012). Motor anticipation failure in infants with autism: a retrospective analysis of feeding situations. *Autism: The International Journal of Research and Practice*, 16, 420–9. doi:10.1177/1362361311423385

Cohn, J.F. and Tronick, E.Z. (1983). Three-month-old infants' reaction to simulated maternal depression. *Child Development*, 54(1), 185–93.

Colonnesi, C., Boegels, S., de Vente, W., and Majdandzic, M. (2013). What coy smiles say about positive shyness in early infancy. *Infancy*, 18(2), 202–20.

Costall, A. (1997). The meaning of things. *Social Analysis: The International Journal of Social and Cultural Practice*, 41(1), 76–85.

Csibra, G. and Gergely, G. (2009). Natural pedagogy. *Trends in Cognitive Sciences*, 13(4), 148–53.

de Bruin, L., van Elk, M., and Newen, A. (2012). Reconceptualizing second-person interaction. *Frontiers of Human Neuroscience*, 6, 151.

Fantasia, V., Markova, G., Fasulo, A., Costal, A., and Reddy, V. (2016). Not just being lifted: infants are sensitive to delay during pick up routine. *Frontiers in Psychology*, 6, 2065.
Farroni, T., Csibra, G., Simion, F., and Johnson, M.H. (2002). Eye contact detection in human from birth. *PNAS*, 99(14), 9602–5. doi:10.1073/pnas.152159999
Farroni, T., Massaccesi, S., Pividori, D., Simion, F., and Johnson, M. (2004). Gaze following in newborns. *Infancy*, 5(1), 39–60.
Field, T. (1984). Early interactions between infants and their depressed mothers. *Infant Behavior and Development*, 7(4), 517–22.
Filippetti, M., Oriolo, G., Johnson, M., and Farroni, T. (2015). Newborn body perception: sensitivity to spatial congruency. *Infancy*, 20(4), 455–65.
Furrer, C. and Skinner, E. (2003). Sense of relatedness as a factor in children's academic engagement and performance. *Journal of Educational Psychology*, 95(1), 148–62.
Goren, C., Sarty, M., and Wu, P. (1975). Visual following and pattern discrimination of face-like stimuli by newborn infants. *Pediatrics*, 56(4), 544–9.
Grossman, T., Johnson, M., Farroni, T., and Csibra, G. (2007). Social perception in the infant brain: gamma oscillatory activity in response to eye gaze. *Social Cognitive and Affective Neurocience*. doi:10.1093/scan/nsm025
Grossmann, T., Parise, E., and Friederici, A.D. (2010). The detection of communicative signals directed at the self in infant prefrontal cortex. *Frontiers in Human Neuroscience*, 4, 201. doi:10.3389/fnhum.2010.00201
Heath, R. (2009). Emotional engagement: how television builds big brands at low attention. *Journal of Advertising Research*, 49 (1), 62–73.
Herbert, C., Blume, C., and Northoff, G. (2015) Can we distinguish an "I" and "ME" during listening?—an event-related EEG study on the processing of first and second person personal and possessive pronouns. *Self and Identity*, 15(2). doi:10.1080/15298868.2015.1085893
Hobson, R. (1990). On the origins of self and the case of autism. *Development and Psychopathology*, 2(2), 163–81.
Hobson, R.P. (2002). *The cradle of thought: exploring the origins of thinking*. Oxford: Macmillan.
Hobson, R.P. (2007). Communicative depth: soundings from developmental psychopathology. *Infant Behavior & Development*, 30, 267–77.
Hobson, R. P., Chambi, G., Lee, A., and Meyer, J. (2006). Foundations for self-awareness: an exploitation through autism. *Monographs of the Society for Research in Child Development*, 71(2), 1–166.
Holland, A. and Kensinger, E. (2010). Emotion and autobiographical memory. *Physics of Life Reviews*, 7(1), 88–131.
Izard, C.E. and Hyson, M.C. (1986) Shyness as a discrete emotion. In: W.H. Jones, J.M. Cheek, and S.R. Briggs (eds.), *Shyness: emotions, personality, and psychotherapy*. Boston: Springer.
Kaertner, J., Keller, H., and Yovsi, R. (2010) Mother-infant interaction during the first 3 months: the emergence of culture-specific contingency patterns. *Child Development*, 81(2), 540–54.
Kampe, K., Frith, C., and Frith, U. (2003). "Hey John": signals conveying communicative intention toward the self activate brain regions associated with "mentalizing," regardless of modality. *The Journal of Neuroscience*, 23(12), 5258–63.
Kanner, L. (1943). Autistic disturbances of affective contact. *Nervous Child*, 2, 217–50.
Legerstee, M. (1994). Patterns of 4-month-old infant responses to hidden silent and sounding people and objects. *Early Development and Parenting*, 3(2), 71–80.

Legerstee, M. and Varghese, J. (2001). The role of maternal affect mirroring on social expectancies in three-month-old infants. *Child Development*, 72(5), 1301–13.

Leudar, I. and Costall, A. (2006). *Against theory of mind*. London: Palgrave Macmillan.

Lewis, M. (1995). Embarrassment: the emotion of self-exposure and evaluation. In: J.P. Tangney and K.W. Fischer (eds.), *Self-conscious emotions: the psychology of shame, guilt, embarrassment, and pride*. New York: Guilford Press, pp. 198–218.

Markova, G. and Legerstee, M. (2006). Contingency, imitation, and affect sharing: foundations of infants' social awareness. *Developmental Psychology*, 42(1), 132–41. doi:10.1037/0012-1649.42.1.132

Maroda, K. (1998). *Seduction, surrender and transformation: emotional engagement in the analytic process*. Hillsdale, NJ: The Analytic Press.

McQuaid, N., Bibok, M., and Carpendale, J. (2009). Relation between maternal contingent responsiveness and infant social expectations. *Infancy*, 14, 390–401. doi:10.1080/15250000902839955

Meltzoff, A. and Moore, M.K. (1977). Imitation of facial and manual gestures by human neonates. *Science*, 198, 75–8.

Mireault, G., Poutre, M., Sargent-Hier, M., Dias, C., Perdue, B., and Myrick, A. (2011). Humour perception and creation between parents and 3- to 6-month-old infants. *Infant and Child Development*, 21(4), 33–47. doi:10.1002/icd.757

Moll, H., Carpenter, M., and Tomasello, M. (2007). Fourteen-month-olds know what others experience only in joint engagement, *Developmental Science*, 10(6), 826–35.

Morton, J. and Johnson, M. (1991). CONSPEC and CONLERN: a two-process theory of infant face recognition. *Psychological Review*, 98(2), 164–81.

Murray, L. and Cooper, P. (1997). Effects of postnatal depression on infant development. *Archives of Disease in Childhood*, 77, 99–101. doi:10.1136/adc.77.2.99

Murray, L., Fiori-Cowley, A., Hooper, R., and Cooper, P. (1996). The effect of postnatal depression and associated adversity on early mother-infant interactions and later infant outcome. *Child Development*, 67(5), 2512–26.

Murray, L. and Trevarthen, C. (1985). Emotional regulation of interactions between two-month-olds and their mothers. In: T. Field and N. Fox (eds.), *Social perception in infants*. Norwood, NJ: Ablex, pp. 101–25.

Nadel, J., Carchon, I., Kervella, C., Marcelli, D., and Reserbat-Plantey, D. (1999). Expectancies for social contingency in 2-month-olds. *Developmental Science*, 2(2), 164–73.

Nagy, E. (2008). Innate intersubjectivity: newborns' sensitivity to communication disturbance. *Developmental Psychology*, 44(6), 1779 –84.

Parise, E., Friederici, A., and Striano, T. (2010). "Did you call me?" Five month-old infants' own name guides their attention. *PLoS One* 5(12), e14208.

Parise, E., Handl, A., Palumbo, L., and Friederici, A. (2011). Influence of eye gaze on spoken word processing: an ERP study with infants. *Child Development*, 82(3), 842–53.

Reddy, V. (1991). Playing with others' expectations: teasing and mucking about in the first year. In: A. Whiten (ed.), *Natural theories of mind*. Oxford: Blackwell, pp. 143–58.

Reddy, V. (2001). Infant clowns: interpersonal creation of humour in infancy. *Enfance*, 53, 247–56.

Reddy, V. (2003). On being the object of attention: implications for self-other consciousness. *Trends in Cognitive Sciences*, 7(9), 397–402. doi:10.1016/S1364-6613(03)00191-8

Reddy, V. (2008). *How infants know minds*. Cambridge, MA: Harvard University Press.

Reddy, V. (2011). A gaze at grips with me. In: A. Seemann (ed.), *Joint attention: new developments in psychology, philosophy of mind, and social neuroscience*. Cambridge, MA: MIT Press, pp. 137–57.

Reddy, V., Chisholm, V., Forrester, D., Conforti, M., and Maniatopoulou, D. (2007). Facing the perfect contingency: interactions with the self at 2 and 3 months. *Infant Behavior and Development*, 30(2), 195–212.

Reddy, V., Liebal, K., Hicks, K., Jonnalagadda, S., and Chintalapura, B. (2013). The emergent practice of infant compliance: an exploration two cultures. *Developmental Psychology*, 49(9), 1754–62.

Reddy, V., Markova, G., and Wallot, S. (2013). Anticipatory adjustments to being picked up in infancy. *PLoS One*, 8(6), e65289. doi:10.1371/journal.pone.0065289

Reddy, V. and Morris, P. (2004). Participants don't need theories. *Theory and Psychology*, 14(5), 647–65.

Reddy, V., Williams, E., Costantini, C., and Lang, B. (2010). Engaging with the self: mirror behaviour in autism, Down syndrome and typical development. *Autism*, 14(5), 531–46.

Reddy, V., Williams, E., and Vaughan, A. (2002). Sharing humour and laughter in autism and Down's syndrome. *British Journal of Psychology*, 93(2), 219–42.

Robertson, S. and Johnson, S. (2009). Embodied infant attention. *Developmental Science*, 12(2), 297–304.

Rodriguez, C. and Moro, C. (2008). Coming to agreement: object use by infants and adults. In: J. Zlatev, T. Racine, C. Sinha, and E. Itkonen (eds.), *The shared mind: perspectives on intersubjectivity*. Amsterdam: John Benjamins, pp. 89–114.

Rossmanith, N., Costall, A., Reichelt, A.F., López, B., López, B., and Reddy, V. (2014). Jointly structuring triadic spaces of meaning and action: book sharing from 3 months on. *Frontiers in Psychology*, 5, 1390. doi:10.3389/fpsyg.2014.01390

Schilbach, L., Eickhoff, S., Mojzisch, A., and Vogeley, K. (2008). What's in a smile? Neural correlates of facial embodiment during social interaction. *Social Neuroscience*, 3(1), 37–50.

Schilbach, L., Timmermans, B., Reddy, V., Costall, A., Bente, G., Schlicht, T. et al. (2013). Toward a second-person neuroscience. *Behavioral and Brain Sciences*, 36, 393–414. doi:10.1017/S0140525X12000660

Schilbach, L., Wohlschlaeger. A., Kraemer, N., Newen, A., Shah, J., Fink, G. et al. (2006). Being with virtual others: neural correlates of social interaction. *Neuropsychologia*, 44, 718–30.

Schneidman, L. and Woodward, A. (2016). Are child-directed interactions the cradle of social learning? *Psychological Bulletin*, 142(1), 1–17.

Schoenherr, J. and Westra, E. (in press). Beyond "interaction": how to understand social effects on social cognition. *British Journal for the Philosophy of Science*.

Senju, A. and Csibra, G. (2008). Gaze following in human infants depends on communicative signals. *Current Biology*, 18(9), 668–71.

Siqueland, E. and Lipsitt, L. (1966). Conditioned head-turning inn human newborns. *Journal of Experimental Child Psychology*, 3(4), 356–76.

Stern, D. (1985). *The interpersonal world of the infant*. New York: Basic Books, Inc.

Tomasello, M. (1999). Having intentions, understanding intentions and understanding communicative intentions. In: P. Zelazo, J. Astington, and D. Olson (eds.), *Developing theories of intention*. Mahwah, NJ: Lawrence Erlbaum, pp. 63–76.

Tomasello, M., Carpenter, M., Call, J., Behne, T., and Moll, H. (2005). Understanding and sharing intentions: the origins of cultural cognition, *Behavioural and Brain Sciences*, 28, 675–735.

Trevarthen, C. (1977). Descriptive analysis of infant communicative behavior. In: H.R. Schaffer (ed.), *Studies in mother-infant interaction*. London: Academic Press, pp. 227–70.

Van der Meer, A., Van der Weel, F.R., and Lee, D.N. (1995). The functional significance of arm movements in neonates. *Science*, 267(5198), 693–5.

Von Hofsten, C. (1982) Eye-hand coordination in the newborn. *Developmental Psychology*, 18(3), 450–61.

Wolff, P.H. (1987). *The development of behavioral states and the expression of emotions in early infancy: new proposals for investigation*. Chicago: University of Chicago Press.

CHAPTER 24

THE INTERSUBJECTIVE TURN

HANNE DE JAEGHER

From Interaction to Intersubjectivity

There is a lot of talk these days of an "interactive turn" in the study of social understanding. But the interactive turn in the embedded, embodied, and especially the enactive sciences of mind is more of an *intersubjective* turn. If by intersubjective (the two elements here separated for a moment, to emphasize each aspect) we understand that which happens between subjects, then several aspects of this are being increasingly investigated today, social interaction being only one of them.

True, there is a heightened attention to interaction processes (e.g., Hari et al. 2015), but there is also great interest in what it is like to connect with others, i.e., in the experience of interacting. Work like that of Shaun Gallagher, Peter Hobson, Vasu Reddy, Evan Thompson, Colwyn Trevarthen, and others, shows us that the science of social understanding needs to recognize connecting, which connotes both interacting and its personal, subjective aspects. The impulse for this movement toward an intersubjective turn comes in part from the phenomenological insights that animate many of the criticisms of and improvements upon cognitivist social cognition research (e.g., Ratcliffe 2007; Gallagher 2001, 2012; Zahavi 2001; Szanto and Moran 2015). What phenomenologists diagnose as missing are especially the human, personal aspects, and these center on subjectivity and interacting (Gill 2015; Satne and Roepstorff 2015).

Subjectivity and interaction have been repeatedly overlooked, but also again and again called back into psychology and cognitive science throughout their history. Often these appeals have only been heard with half an ear, or the enthusiasm quickly dissipated within the contingencies of the times (see, e.g., Bruner 1990). Many have thought and written about experiencing and interacting. Among others, in psychology there is the work of Vygotsky, Asch, and Donaldson; in sociology, of Mead, Sacks, and Goffman; in phenomenology, of Gurwitsch, Schutz, Stein, Husserl, and Merleau-Ponty; and in psychotherapy there is the work of Stern, Benjamin, and Beebe. Thinking such as theirs is now once again being re-discovered, -applied, -interpreted, -operationalized. There is renewed excitement in doing so, in part due to the increasing sophistication of technologies for studying interaction processes. Combining this with phenomenological insights, we are once more in a position to progress further toward a broader understanding of intersubjectivity than the dualist "prediction and explanation of another person's mental states" (whereby it is assumed that we cannot perceive the mental, and thus have to infer or simulate it on the basis of outward, in themselves meaningless, stimuli).

What is intersubjectivity? Developmental psychologist Vasu Reddy defines it as "engagement between subjectivities" (Reddy 2008, p. 23). To unfold this a little further, a definition from outside the strictly cognitive sciences is helpful. Feminist psychoanalyst Jessica Benjamin describes intersubjectivity as "a relationship in which each person experiences the other as a 'like subject,' another mind who can be 'felt with,' yet has a distinct separate center of feeling and perception" (Benjamin 2004, p. 5). Several elements come to the fore here: relating and interacting; experiencing the other as someone to whom things matter, like they do for me; someone whom I can connect with, through feeling, yet who is distinct from me and has their own perspective. Relating with and understanding each other and the world together is then not in the first place a question of figuring out each other's mental states. It is a rather a matter of participating in the creation and transformation of meaning together, between persons who each have their limited but inherently meaningful, evolving perspectives on the world, each other, and themselves, through acting and interacting.

Such an understanding of intersubjectivity is naturally compatible with embodied, embedded, and enactive cognitive science (Varela, Thompson, and Rosch 1991; Thompson 2007; Di Paolo, Rohde, and De Jaegher 2010). The enactive perspective in particular allows us to understand both how things are inherently meaningful to subjects, and the role of (embodied) interaction processes in establishing and modulating meaning together with others. In this chapter, I put forward a framework that I think contains the necessary elements for a comprehensive understanding of intersubjectivity. I show this frame at work in some empirical examples, and finish by indicating where further research is still needed.

A Framework for Intersubjectivity

A comprehensive approach to studying intersubjectivity should, first, account for social interaction processes *and* do justice to subjectivity in its bodily, experiential, existential, and historico-sociocultural complexity—and should do so, I would like to say, all at once. Of course, this is impossible; no investigation can capture all these elements at the same time. But explaining intersubjectivity (like mind, agency, subjectivity) on an enactivist approach entails a division into different "units of study" than the ones we know from classical psychology and cognitive science. The latter investigate perception, action, emotion, memory, but often with no clear view of why or how they hang together, only the vague hope that their interconnections will one day be clear. The enactive approach, in contrast, starts always from the contextualized coherence of meaningful behavior (McGann, De Jaegher, and Di Paolo 2013). On this basis, it goes in search of the "difference which makes a difference," as Bateson said (2000, p. 459). For finding these differences that matter (for attempts, see Barandiaran, Di Paolo, and Rohde 2009; Bührmann and Di Paolo 2017), basic principles need to be firmly grounded (Chiel and Beer, 1997).

Second, an integrative intersubjectivity framework has to connect physiological, neural, interactional, linguistic, and societal aspects and levels of explanation. This requires concepts and methodologies that span several different disciplines. Third, a comprehensive framework should encourage applications, and dialogue with experts in other sectors, such as teachers and therapists. Fourth, it should recognize and make explicit the values underlying it, so that it can be critically aware of how it influences and is influenced by societal institutions and norms. And finally, because its subject matter is the ways in which people understand and deal with each other, it should be prepared to deal with ethical questions and dimensions.

Are the foundations for such a comprehensive theoretical framework there? Let us take a closer look at each aspect in turn.

Interacting and Subjectivity

First, what about the intricate connection between interaction processes and subjectivity? To account for this would mean to show the interdependence between the very organization of interaction processes and of the individuals—and their experiences—taking part in them. Such an account is provided by the theory of participatory sense-making (De Jaegher and Di Paolo 2007). Perhaps the most radical aspect of this theory is its proposal that social interactions can take on a life of their own, or that they can self-organize in a particular, well-defined sense. A social interaction exists when two conditions hold: (1) a relational dynamic emerges and maintains itself for a while, while (2) not destroying the autonomy of the individuals involved in it, though it could decrease or increase their autonomy (De Jaegher and Di Paolo 2007; De Jaegher, Di Paolo, and Gallagher 2010; Froese and Di Paolo 2011; Gallagher 2017). This overturns cognitivist individualism, because it poses the possibility that social interaction processes, *as such*, can modulate individual intentions (De Jaegher and Di Paolo 2007; De Jaegher, Di Paolo, and Gallagher 2010). I will describe examples of this later.

On such a conception, interactions form and transform individuals and their intentions, just as individuals form and transform interactions. To think of interaction processes as effective factors in intersubjectivity means to understand them not just as contextual or enabling, but also as potentially constitutive of social understanding. A contextual factor is something that merely affects a particular phenomenon and modulates its properties; an enabling factor is necessary for a phenomenon to occur (if it is missing, the phenomenon does not happen); and a constitutive factor is also necessary for, and moreover part of what makes the phenomenon what it is (De Jaegher, Di Paolo, and Gallagher 2010). If interaction processes can (in part) constitute social understanding, then social understanding is not reducible to individual processes. Moreover, social understanding is, in some sense, not possible without social interaction, whether it be through live interaction, or through interactions as they developmentally determine the capacity for social understanding, including in the form of readiness-to-interact in current situations (Di Paolo and De Jaegher 2012).

What about subjectivity? Individuals on this account are characterized anew, in terms of both the individual and the interactive contributions to their self-organization. Enactivists understand cognizers as sense-makers, who enact and engage with their environment in terms of its significance and valence on the basis of their autonomy (Thompson and Stapleton 2009). Sense-makers are autonomous in the sense that they self-organize under precarious circumstances. This makes a sense-maker sensitive to what is beneficial and what is pernicious for its self-maintenance, and capable of adapting to its circumstances in the service of self-organization (Di Paolo 2005). Sense-making is thus always a relation of significance, a set of adaptations based in the needs and constraints of a precarious bodily being, living in specific circumstances. Subjects are, in consequence, experientially and existentially sensitive (Thompson and Stapleton 2009; Colombetti 2013; de Haan 2015). Things literally *matter* to sense-makers, all the way from making a difference to their metabolic self-maintenance to affecting their societal roles and existential concerns. This is the sense in which we speak of subjectivity.

On this view, social understanding is understood as the coordination of intentional activities in and through interaction, "whereby individual sense-making processes are affected and new domains of social sense-making can be generated that were not available to each individual on her own" (De Jaegher and Di Paolo 2007, p. 497). Individuals can thus *participate in each other's sense-making*. While this idea was first formulated to capture ongoing embodied interactions, it turns out to also be useful for describing social understanding in a more general sense (see, e.g., Gallagher 2009). One of the implications of this for fields like social neuroscience and psychology is a rethinking of the idea of social skills and social agency. Social skills, on this account, are understood as partly constituted in interactions and interactional histories between people (McGann and De Jaegher 2009). Social agency is the development of increasingly sophisticated ways to deal with the tensions between individual and interactional autonomies, leading to complex capacities such as self-control and languaging—i.e., genres of participation (Cuffari et al. 2015). All this happens in a field of sociocultural normativities that situated individuals cannot but variously incorporate, transform, and transmit through their interactions (e.g., Young 2005).

The "narrow corridor situation" illustrates this picture. When you encounter someone walking toward you from the other direction in a narrow corridor, you may end up stepping left and right in front of each other a few times before the brief spell of this emerging coordination can be broken. Here, an interaction emerges and maintains itself for a while, over and against the wishes of the individuals. The processes at work here can be individual, interactive, or both. The picture quickly turns complicated: while this kind of encounter illustrates that autonomous interaction dynamics can override individual intentions (you are briefly delayed in reaching your office), at the same time, some of these unintentional disruptions may in fact be desired (perhaps you had wanted to ask your colleague about something anyway, and this chance entanglement opens up the opportunity to do so). Moreover, this kind of interaction, however local it seems, is shot through with social norms and roles (remember Goffman's analysis of face-work or interaction ritual, 1955; 1967). Finally, these various self- and interactive regulations happen at either or both personal and subpersonal levels. You can report on aspects of this encounter (your interaction in words and overt meanings), while other aspects remain under the level of awareness (heart rate, breathing, and neural synchronization, galvanic skin response, and so on).

Phenomena, Levels of Explanation, and Interdisciplinarity

How should we understand the connections between the interactional, physiological, neural, coordinative, linguistic, and sociocultural levels at play in intersubjective encounters?

Enaction's phenomenologically and biologically sensitive logic of how autonomy, emergence, sense-making, embodiment, experience, and the sociocultural fit together can help with this (Di Paolo, Rohde, and De Jaegher 2010; Froese and Di Paolo 2011; Cuffari and Jensen 2014). One of the threads running through enactive research is the elaboration of a conceptual toolbox that can be used for doing interdisciplinary research in the sense not merely of dialoguing between disciplines, but of intervening in one discipline in the ways of another one, and building novel empirical paradigms that span disciplinary boundaries (Callard and Fitzgerald 2015). One way to generate hypotheses about the contributions of various elements is to try to determine which factors are contextual to a phenomenon, which ones are enabling, and which are constitutive of it (an illustration of how this can be done is given in De Jaegher, Di Paolo, and Gallagher 2010). The distinction between contextual, enabling, and constitutive factors is a meta-conceptual tool that can help study how particular physiological, coordinative, neural, linguistic, and sociocultural elements are at play in different aspects of intersubjectivity. Other tools are the concepts central to enactivism itself, such as self-organization, autonomy, social interaction, and sense-making. They can be operationalized in different ways.

Take the idea that interaction processes can constitute social understanding (De Jaegher, Di Paolo, and Gallagher 2010). This can be investigated in several ways. A first approach can consist in figuring out which elements of self-regulation, including neurological and physiological processes, coordinate across individuals, and how (for instance, in development, Trevarthen and Aitken 2001). Aspects of this self-organization or coordination may be explained by individual factors, where the interaction is considered no more than a context for the synchronizing effect. For instance, the interaction may be considered an input for a contingency perception system (Gergely and Watson 1999). But social interaction processes may themselves enable or even constitute the interpersonal synchrony, and, therefore, part of the social understanding that takes place (see, respectively, Di Paolo, Rohde, and Iizuka 2008, and De Jaegher, Di Paolo, and Gallagher 2010).

This can be tested, for instance, in minimalistic virtual world experiments. In the perceptual crossing paradigm (Auvray et al. 2009; Auvray and Rohde 2012), participants interact with each other through moving their computer mouses (which move their cursors in a virtual world they do not see), and feedback via tactile stimuli received on the fingertip each time they encounter an object. Objects in this world are the avatar of the other participant, and other moving and fixed objects. The catch of this experiment is that each object, no matter what kind, generates exactly the same stimulus on the fingertip. Notwithstanding this, participants can tell the difference between objects steered by the other person and other moving objects. This performance can be explained only by taking into account the interactive dynamics that emerge between the participants when they encounter each other's avatars, thus showing a constitutive role for interaction dynamics in the performance of this social task (De Jaegher, Di Paolo, and Gallagher 2010; De Jaegher, Di Paolo, and Adophs 2016). This kind of experiment can also show how interaction dynamics make a difference to experiencing another's presence (Froese, Iizuka, and Ikegami 2014), recognizing another's intentionality, imitation (for both of these, see Lenay and Stewart 2012), and to mutual awareness of a shared object (Deschamps et al. 2016).

That real-world social interactions can affect individual physiology can be shown in studies of various kinds of physiological coordination (e.g., Fusaroli et al. 2015). For instance, during a fire-walking ritual, audience members' heartbeats synchronize with that of the fire-walker, but only those in the audience who are his relatives—not those who are unrelated (Konvalinka et al. 2011). Such research shows the complex interactions between different levels of organization: physiological, personal, interactive, and cultural.

Work on music pedagogy is also interesting in this regard, because it investigates how self-regulation, interaction dynamics, and the experience of connecting relate to societal and cultural views on living and learning. Laroche and Kaddouch, for instance, show how an improvisation-based piano teaching method fosters and expands pupils' embodied musical personality (Laroche and Kaddouch 2014, 2015). Lessons start with the student performing a pattern that they enjoy playing—a phrase that is, in the parlance of the approach, "embodied with ease" and within the student's "spontaneous achievement zones" (Laroche and Kaddouch 2014, p. 35). Through the teacher's musical engagement with the student's pattern, first supporting and later challenging it by intentionally breaking down the musical interaction, the student is coaxed out of her comfort zone. This opens up the unfolding, now problematic, musical situation to possible recoveries, through the development of patterns that are novel to the student, yet fit the newly created intersubjective demand. These patterns, Laroche and Kaddouch hypothesize, are closer to the student's "core tastes," or what the student might want her musical expression to be. In this way, learning to play the piano involves spontaneous, interactively generated "new 'means of meaning'" (p. 36). Van der Schyff (2015), similarly, explores how enactive, embodied ideas of life and meaning transform the cultural ways in which music is understood. He suggests that an enactive approach to music's "bio-cultural nature" opens up the rational, individualistic Western view to an understanding of music as an "active ecological phenomenon" that allows a "deeper awareness of the primordial, empathic, and interpenetrative experience of being-in-the-world" (van der Schyff 2015, p. 12).

Several other related areas of research link interaction, personal experience, and sociocultural normativity. In ecological psychology, for instance, the study of interaction dynamics goes together with various ideas about how these dynamics, and language, are infused with and infuse personal and societal meaning (Hodges et al. 2012; Zlatev 2012; McGann 2014). Cowley, Moodley, and Fiori-Cowley (2014) show how, in interactions between mother and infant, culturally specific expectations play out and are learned in the mutual regulation of affect and interaction between them. A particular hand-waving gesture by a Zulu mother from South Africa is taken up by her baby as a request to be silent now. Cowley and colleagues argue that this is culturally specific because the particular fast-paced gesturing of the mother would be experienced as frightening or overwhelming in another culture, and yet the baby becomes quiet, not afraid or distressed. This becoming quiet or "thula" is an important element of the culture this dyad lives in, and is learned early on through lively bodily exchanges.

What is exciting about this kind of work is that, in different ways, it attempts to connect what happens at physiological levels with experienced significance at the personal and subjective level (i.e., experience that is meaningful to a particular subject, now, in this context), the interactive-dynamic, and the wider socio-historico-cultural levels. This brings us to the next two points: the necessary connection between research on social understanding and the wider socio-historico-cultural context, and ethical concerns.

Applications and Dialogue with Field Experts

Intersubjectivity concerns everyone on a daily basis, and a philosophical and scientific approach to it is at the same time relevant for and benefits from contact with real-world issues. This relates to another basic tenet of enaction: as a philosophical and scientific approach to subjectivity—to mind, agency, interpersonal understanding—it does not shy away from, but rather aims to do justice to the lived intricacies of the phenomena it studies. The guiding principle here is the concern that is always involved in subjectivity: enaction investigates what matters to sense-makers. This means researching "what we experience during the ongoing adventure of establishing, losing, and re-establishing meaningful relations between ourselves and the world" by "articulating in operational terms what these meaningful relations consist of, as well as what it means to establish or lose them" (Bührmann and Di Paolo 2017, p. 207).

Illustrative of enaction's potential in this is the fact that its theorizing on social understanding has been taken up in many different sectors. It is used, for instance, to help understand psychopathologies (Fuchs 2010), such as autism (De Jaegher 2013), schizophrenia (Kyselo 2016), and addiction (Zautra 2015). In general, these can be understood as relationships between specifically different kinds of embodied interacting with and experiencing the world. This gives a comprehensive, integrative picture of each pathology as a particular complex of meaningful interactions with the world, with their own specific logic, and implications for their own ways of sense-making. For instance, particularities of autistic embodiment include jerky instead of fluid postural adjustments to visual and auditory stimuli, visual motion integration deficits, hypo- and hypersensitivities, and motor coordination differences (Gepner and Mestre 2002a; Gepner and Mestre 2002b; Rogers and Ozonoff 2005; Fournier et al. 2010; Torres, Brincker, Isenhower et al. 2013). That social interactions are different in autism is well known but the enactive approach specifically grounds them in embodiment and connects them to ways of understanding each other (De Jaegher 2013). In this way, connections between embodiment, social relating, and, for instance, specifically autistic forms of thinking such as rigidity and literalness, can be elucidated (Hobson 2002). This has implications for therapy: if there is a strong bodily anchorage of certain specific disturbances, specific bodily forms of therapy will have high remedial benefits (Martin et al. 2016). It also has implications for psychopathology research, which itself should perhaps be seen as an intersubjective endeavor (Galbusera and Fellin 2014).

On the other hand, understandings originating in such real-world issues also enhance and further enactive theory and inquiry. In this, phenomenology, with its systematic investigation of the structures of experience, plays a crucial role in refining theoretical insights and frameworks (Varela 1996; Gallagher 2003). For instance, a phenomenological focus reveals a strong relationship between embodiment, feelings of loss of self, and hyperreflectivity in schizophrenia (de Haan and Fuchs 2010), and this leads to the insight that existential questions can and should be taken up by enactivist theory as well (de Haan 2015).

Similar cross-fertilizations exist for issues in clinical interactions and clinical reasoning (Øberg et al. 2014; Øberg et al. 2015), augmented communication in cerebral palsy (Auer and Hörmeyer 2015), martial arts (Light 2014), literature and narrative (Caraciollo 2014; Popova 2015), classroom interactions and learning (Towers and Martin 2015; Maiese 2017), and psychotherapy (Galbusera and Fuchs 2014; Röhricht et al. 2014).

Critical Engagements and the Ethical Dimension

Theories of the mind and of intersubjectivity, by the very nature of their subject matter, have societal relevance (see, for instance, recent work on solitary confinement that strongly leans on the lessons of embodied intersubjectivity: Guenther 2013; Gallagher 2014). In part, this goes beyond applying ideas on the one hand, or fine-tuning research through contact with real-world issues on the other; it veers into the realm of ethics. The scientific understanding of the mind has implications that researchers may not always be aware of, but that can profoundly influence how people are treated, and how they see themselves and their place in the world. It is important to keep this in mind, in particular when researching subjectivity and intersubjectivity.

The point can be illustrated with reference to the relationship between cognitive research on dementia and dementia care practices. Cognitive approaches often start from deficits in individual capacities thought to underlie social skills, for instance, emotion categorization (Lindquist et al. 2014). When there is diminished capacity for categorizing emotions, the conclusion could be drawn that people with dementia are socially impaired. But this understanding ignores the subjective, personal, and interactive aspects that I have been highlighting here—which are known to be a better basis for understanding the disorder and providing care (Kitwood 1997; Perrin and May 2000; Zeiler 2014). Similar points can be made for other disorders, for instance, autism (Donnellan, Hill, and Leary 2013; De Jaegher 2013).

There is an ethical issue in switching from an objective description of cognition (e.g., on the basis of a computer metaphor, but also from some embodied perspectives) to looking at the personal and the subjective life of cognizers. Many healthcare professionals, and scientists too, do conceptualize and organize their work—perhaps intuitively—from what we could call a humane perspective, with personalized, interpersonal care at the center of it. The problem is that this does not have a firm theoretical backing. Enaction can provide this, because it starts from the question of what matters to the individual and what processes constitute her as such. This starting point goes beyond merely finding that the body and social interactions are important—understanding disorders (and also any other issues that modulate participation in society) now becomes fundamentally about significance and values for the person, and how they play out in various kinds of engagements and their breakdowns. As Colombetti and Torrance (2009) argue, adopting such an ethical attitude based on the research perspective we use to understand embodied minds leads to a shift in how we understand responsibility: full responsibility for a situation is no longer understood as lying exclusively with individuals. Rather, responsibility shifts and becomes more or less possible or potent, always within a particular intersubjective situation. It thus needs to be assessed differently, and likely on a case-by-case basis. The shift in perspective can also disarm mechanisms behind discrimination and stigmatization by reframing ways in which people who are different are recognized in society (Urban 2015). Interactions themselves can have ethical qualities (they can be fair or just, discriminatory or not), for instance, in how they enable or constrain individual participation and autonomy, and align with or against sociocultural customs and practices (Di Paolo, Cuffari, and De Jaegher 2018).

Where Next?

It seems that a new approach—an intersubjective science of social understanding—is not only coming into view, but already starting to push past the cognitivist folklore. The theory of participatory sense-making was proposed precisely to contribute to this purpose: to bring together such different elements and levels, across disciplines, around the notions of interaction, autonomy, subjectivity, sense-making, affect, and embodiment, to better understand intersubjectivity in its various dimensions.

Of course, a lot more work is needed on all the issues discussed in the second section, but also on some that I have not been able to give much attention here. One entire such realm is the interplay between societal normativity and (inter)subjectivity. Interactions and human experience are always embedded in societal, cultural, economic, and historical contexts, which shape and are shaped by the persons engaging in them, in part through their interactions with each other.

Another question is how to bring coherence to the investigations. The main contribution of the enactive framework to this is its starting point in these two questions:

- What matters to a sense-maker? What is at stake for him or her in the particular situation? Which level of identity is being maintained and organized?
- How does this relate to how she or he moves and is moved, acts and perceives, in (social) interactions with the world?

One thing that will help bring coherence, and that is only beginning to be explored, is a better understanding of the role of the experience of interacting (Høffding and Martiny 2016; Kimmel and Preuschl 2016; De Jaegher et al. 2017). An intuition that *what it is like to interact matters* underlies much research on social understanding. But in order for this intuition to truly inform and engage with the extensive research on interaction dynamics we need systematic, hands-on ways to grasp interactive experience's structure and significance. This will inevitably also involve using our experience as researchers in our own investigations of intersubjectivity.

Once we learn to take experience into account not simply as another source of data, but also our own experience as embodied, embedded researchers, and the ethical dimensions implicated, we will truly begin to better understand embodied intersubjectivity.

Acknowledgments

Thank you Ezequiel Di Paolo, Shaun Gallagher, Michael Kimmel, participants of a draft session on academia.edu, and an anonymous reviewer for insightful comments.

References

Auer, P. and Hörmeyer, I. (2015). Achieving intersubjectivity in augmented and alternative communication (AAC): intercorporeal, embodied and disembodied practices. *InLiSt—Interaction and Linguistic Structures*, 55. http://www.inlist.uni-bayreuth.de/issues/55/inlist55.pdf

Auvray, M., Lenay, C., and Stewart, J. (2009). Perceptual interactions in a minimalist virtual environment. *New Ideas in Psychology*, 27(1), 32–47.

Auvray, M. and Rohde, M. (2012). Perceptual crossing: the simplest online paradigm. *Frontiers in Human Neuroscience*, 6, 181. doi:10.3389/fnhum.2012.00181

Barandiaran, X.E., Di Paolo, E.A., and Rohde, M. (2009). Defining agency: individuality, normativity, asymmetry and spatio-temporality in action. *Adaptive Behavior*, 17, 367–86.

Bateson, G. (2000). *Steps to an ecology of mind: collected essays in anthropology, psychiatry, evolution, and epistemology*. Chicago: The University of Chicago Press.

Benjamin, J. (2004). Beyond doer and done to: an intersubjective view of thirdness. *The Psychoanalytic Quarterly*, 73(1), 5–46. doi:10.1002/j.2167-4086.2004.tb00151.x

Bruner, J. (1990). *Acts of meaning*. Cambridge, MA: Harvard University Press.

Bührmann, T. and Di Paolo, E.A. (2017). The sense of agency—a phenomenological consequence of enacting sensorimotor schemes. *Phenomenology and the Cognitive Sciences*, 1–30. doi:10.1007/s11097-015-9446-7

Callard, F. and Fitzgerald, D. (2015). *Rethinking interdisciplinarity across the social sciences and neurosciences*. London: Palgrave MacMillan.

Caracciolo, M. (2014). *The experientiality of narrative: an enactivist approach*. Berlin: De Gruyter.

Chiel, H.J. and Beer, R.D. (1997). The brain has a body. adaptive behavior emerges from interactions of nervous system, body and environment. *Trends in Neurosciences*, 20, 553–7.

Colombetti, G. (2013). *The feeling body: affective science meets the enactive mind*. Cambridge, MA: MIT Press.

Colombetti, G. and Torrance, S. (2009). Emotion and ethics: an inter-(en)active approach. *Phenomenology and the Cognitive Sciences*, 8, 505–26.

Cowley, S.J., Moodley, S., and Fiori-Cowley, A. (2004). Grounding signs of culture: primary intersubjectivity in social semiosis. *Mind, Culture and Activity*, 11, 109–32.

Cuffari, E., Di Paolo, E.A., and De Jaegher, H. (2015). From participatory sense-making to language: there and back again. *Phenomenology and the Cognitive Sciences*, 14(4), 1089–125. doi:10.1007/s11097-014-9404-9

Cuffari, E.C. and Jensen, T.W. (2014). Living bodies: co-enacting experience. In: C. Müller, A. Cienki, E. Fricke et al. (eds.), *Body-language-communication: an international handbook on multimodality in human interaction* (vol. 2). Berlin/Boston: Mouton de Gruyter, pp. 2016–25.

de Haan, S.E. (2015). *An enactive approach to psychiatry* [unpublished PhD thesis]. Heidelberg: University of Heidelberg.

de Haan, S. and Fuchs, T. (2010). The ghost in the machine: disembodiment in schizophrenia—two case studies. *Psychopathology*, 43, 327–33.

De Jaegher, H. (2013). Embodiment and sense-making in autism. *Frontiers in Integrative Neuroscience*, 7, 15. doi:10.3389/fnint.2013.00015

De Jaegher, H. and Di Paolo, E.A. (2007). Participatory sense-making: an enactive approach to social cognition. *Phenomenology and the Cognitive Sciences*, 6(4), 485–507. doi:10.1007/s11097-007-9076-9

De Jaegher, H., Di Paolo, E.A., and Adolphs, R. (2016). What does the interactive brain hypothesis mean for social neuroscience? A dialogue. *Philosophical transactions of the Royal Society of London. Series B, Biological sciences*, 371(1693), 20150379. doi:10.1098/rstb.2015.0379

De Jaegher, H., Di Paolo, E.A., and Gallagher, S. (2010). Can social interaction constitute social cognition? *Trends in Cognitive Sciences*, 14(10), 441–7. doi:10.1016/j.tics.2010.06.009

De Jaegher, H., Pieper, B., Clénin, D., and Fuchs, T. (2017). Grasping intersubjectivity: an invitation to embody social interaction research. *Phenomenology and the Cognitive Sciences*, 16(3), 491–523. doi:10.1007/s11097-016-9469-8

Deschamps, L., Lenay, C., Rovira, K., Le Bihan, G., and Aubert, D. (2016). Joint perception of a shared object: a minimalist perceptual crossing experiment. *Frontiers in Psychology: Cognitive Science*, 7, 1059. doi:10.3389/fpsyg.2016.01059

Di Paolo, E.A. (2005). Autopoiesis, adaptivity, teleology, agency. *Phenomenology and the Cognitive Sciences*, 4(4), 97–125.

Di Paolo, E.A. (2016). Participatory object perception. *Journal of Consciousness Studies*, 23(5–6), 228–58.

Di Paolo, E.A., Cuffari, E.C., and De Jaegher, H. (2018). *Linguistic bodies. The continuity between life and language.* Cambridge, MA: MIT Press.

Di Paolo, E.A., and De Jaegher, H. (2012). The interactive brain hypothesis. *Frontiers in Human Neuroscience*, 6, 163. doi:10.3389/fnhum.2012.00163

Di Paolo, E.A., Rohde, M., and De Jaegher, H. (2010). Horizons for the enactive mind: values, social interaction, and play. In: J. Stewart, O. Gapenne, and E.A. Di Paolo (eds.), *Enaction: toward a new paradigm for cognitive science*. Cambridge, MA: MIT Press, pp. 33–87.

Di Paolo, E.A., Rohde, M., and Iizuka, H. (2008). Sensitivity to social contingency or stability of interaction? Modelling the dynamics of perceptual crossing. *New Ideas in Psychology*, 26(2), 278–94.

Donnellan, A., Hill, D.A., and Leary, M.R. (2013). Rethinking autism: implications of sensory and movement differences for understanding and support. *Frontiers in Integrative Neuroscience*, 6. doi:10.3389/fnint.2012.00124

Fournier, K., Hass, C., Naik, S., Lodha, N., and Cauraugh, J. (2010). Motor coordination in autism spectrum disorders: a synthesis and meta-analysis. *Journal of Autism and Developmental Disorders*, 40, 1227–40. doi:10.1007/s10803-010-0981-3

Froese, T. and Di Paolo, E.A. (2011). The enactive approach: theoretical sketches from cell to society. *Pragmatics & Cognition*, 19(1), 1–36.

Froese, T., Iizuka, H., and Ikegami, T. (2014). Using minimal human-computer interfaces for studying the interactive development of social awareness. *Frontiers in Psychology*, 5, 1061. doi:10.3389/fpsyg.2014.01061

Fuchs, T. (2010). Phenomenology and psychopathology. In: S. Gallagher and D. Schmicking (eds.), *Handbook of phenomenology and cognitive science*. Dordrecht: Springer, pp. 546–73.

Fusaroli, R., Bjørndahl, J.S., Roepstoff, A., and Tylén, K. (2015). A heart for interaction: physiological entrainment and behavioral coordination in a collective, creative construction task. *arXiv:1504.05750*. [physics.soc-ph].

Galbusera, L. and Fellin, L. (2014). The intersubjective endeavour of psychopathology research: methodological reflections on a second-person perspective approach. *Frontiers in Cognitive Science: Psychology*, 5, 1150.

Galbusera, L. and Fuchs, T. (2014). Embodied understanding: discovering the body from cognitive science to psychotherapy. *In Mind Italia*, V, 1–6.

Gallagher, S. (2001). The practice of mind: theory, simulation or primary interaction? *Journal of Consciousness Studies*, 8(5-7), 83–108.

Gallagher, S. (2003). Phenomenology and experimental design: toward a phenomenologically enlightened experimental science. *Journal of Consciousness Studies*, 10, 85–99.

Gallagher, S. (2009). Two problems of intersubjectivity. *Journal of Consciousness Studies*, 16(6-8), 298–308.
Gallagher, S. (2012). In defense of phenomenological approaches to social cognition: interacting with the critics. *Review of Philosophy and Psychology*, 3(2), 187–212. doi:10.1007/s13164-011-0080-1
Gallagher, S. (2014). The cruel and unusual phenomenology of solitary confinement. *Frontiers in Psychology*, 5, 585. doi:10.3389/fpsyg.2014.00585
Gallagher, S. (2017). Social interaction, autonomy and recognition. In: L. Dolezal, L. and D. Petherbridge (eds.), *Body/self/other: the phenomenology of social encounters*. London: Routledge.
Gepner, B. and Mestre, D. (2002a). Postural reactivity to fast visual motion differentiates autistic from children with Asperger syndrome. *Journal of Autism and Developmental Disorders*, 32, 231–8.
Gepner, B. and Mestre, D. (2002b). Rapid visual-motion integration deficit in autism. *Trends in Cognitive Sciences*, 6, 455.
Gergely, G. and Watson, J.S. (1999). Early socio-emotional development: contingency perception and the social-biofeedback model. In: P. Rochat (ed.), *Early social cognition*. Hillsdale, NJ: Erlbaum, pp. 101–37.
Gill, S. (2015). *Tacit engagement: beyond interaction*. Heidelberg: Springer.
Goffman, E. (1955). On face-work: an analysis of ritual elements in social interaction. *Psychiatry*, 18, 213–31.
Goffman, E. (1967). *Interaction ritual: essays on face-to-face behavior*. Garden City, NY: Doubleday.
Guenther, L. (2013). *Solitary confinement: social death and its afterlives*. Minneapolis: Minnesota University Press.
Hari, R., Henriksson, L., Malinen, S., and Parkkonen, L. (2015). Centrality of social interaction in human brain function. *Neuron*, 88, 181–93. doi:10.1016/j.neuron.2015.09.022
Hobson, R.P. (2002). *The cradle of thought*. London: Macmillan.
Hodges, B.H., Steffensen, S.V., and Martin, J.E. (2012). Caring, conversing, and realizing values: new directions in language studies. *Language Sciences*, 34, 499–506.
Høffding, S. and Martiny, K. (2016). Framing a phenomenological interview: what, why and how. *Phenomenology and the Cognitive Sciences*, 15(4), 539–64. doi:10.1007/s11097-015-9433-z
Kimmel, M. and Preuschl, E. (2016). Dynamic coordination patterns in *tango argentino*: a cross-fertilization of subjective explication methods and motion capture. In: J.-P. Laumond and N. Abe (eds.), *Dance notations and robot motion*. New York: Springer International Publishing, pp. 209–35.
Kitwood, T. (1997). *Dementia reconsidered: the person comes first*. Philadelphia: Open University Press.
Konvalinka, I., Xygalatas, D., Bulbulia, J., Schjødt, U., Jegindø, E.-M., Wallot, S. et al. (2011). Synchronized arousal between performers and related spectators in a fire-walking ritual. *Proceedings of the National Academy of Sciences*, 108(20), 8514–19. doi:10.1073/pnas.1016955108
Kyselo, M. (2016). The enactive approach and disorders of the self—the case of schizophrenia. *Phenomenology and the Cognitive Sciences*, 15(4), 591–616. doi:10.1007/s11097-015-9441-z

Laroche, J. and Kaddouch, I. (2014). Enacting teaching and learning in the interaction process: "keys" for developing skills in piano lessons through four-hand improvisations. *Journal of Pedagogy*, 5(1), 24–47.

Laroche, J. and Kaddouch, I. (2015). Spontaneous preferences and core tastes: embodied musical personality and dynamics of interaction in a pedagogical method of improvisation. *Frontiers in Psychology*, 6, 522. doi:10.3389/fpsyg.2015.00522

Lenay, C. and Stewart, J. (2012). Minimalist approach to perceptual interactions. *Frontiers in Human Neuroscience*, 6, 98.

Light, R.L. (2014). Mushin and learning in and beyond budo. *Ido Movement for Culture. Journal of Martial Arts Anthropology*, 14(3), 42–8.

Lindquist, K.A., Gendron, M., Barrett, L.F., and Dickerson, B.C. (2014). Emotion perception, but not affect perception, is impaired with semantic memory loss. *Emotion*, 14, 375–87. doi:10.1037/a0035293

Maiese, M. (2017). Transformative learning, enactivism, and affectivity. *Studies in Philosophy and Education*, 36(2), 197–216. doi:10.1007/s11217-015-9506-z

Martin, L.A.L., Koch, S.C., Hirjak, D., and Fuchs, T. (2016). Overcoming disembodiment: the effect of movement therapy on negative symptoms in schizophrenia—a multicenter randomized controlled trial. *Frontiers in Psychology*, 7, 483. doi:10.3389/fpsyg.2016.00483

McGann, M. (2014). Enacting a social ecology: radically embodied intersubjectivity. *Frontiers in Psychology*, 5, 1321. doi:10.3389/fpsyg.2014.01321

McGann, M. and De Jaegher, H. (2009). Self-other contingencies: enacting social perception. *Phenomenology and the Cognitive Sciences*, 8(4), 417–37. doi:10.1007/s11097-009-9141-7

McGann, M., De Jaegher, H., and Di Paolo, E.A. (2013). Enaction and psychology. *Review of General Psychology*, 17, 203–9. doi:10.1037/a0032935

Øberg, G.K., Blanchard, Y., and Obstfelder, A. (2014). Therapeutic encounters with preterm infants: interaction, posture and movement. *Physiotherapy Theory and Practice*, 30(1), 1–5. doi:10.3109/09593985.2013.806621

Øberg, G.K., Normann, B., and Gallagher, S. (2015). Embodied-enactive clinical reasoning in physical therapy. *Physiotherapy Theory and Practice*, 31(4), 244–52. doi:10.3109/09593985.2014.1002873

Perrin, T. and May, H. (2000). *Wellbeing in dementia: an occupational approach for therapists and carers*. Edinburgh: Churchill Livingstone.

Popova, Y.B. (2015). *Stories, meaning, and experience: narrativity and enaction*. New York and London: Routledge.

Ratcliffe, M. (2007). *Rethinking commonsense psychology: a critique of folk psychology, theory of mind and simulation*. New York: Palgrave Macmillan.

Reddy, V. (2008). *How infants know minds*. Cambridge, MA: Harvard University Press.

Rogers, S.J. and Ozonoff, S. (2005). Annotation: what do we know about sensory dysfunction in autism? A critical review of the empirical evidence. *Journal of Child Psychology and Psychiatry*, 46, 1255–68. doi:10.1111/j.1469-7610.2005.01431.x

Röhricht, F., Gallagher, S., Geuter, U., and Hutto, D.D. (2014). Embodied cognition and body psychotherapy: the construction of new therapeutic environments. *Sensoria: A Journal of Mind, Brain & Culture*, 10(1), 11–20.

Satne, G. and Roepstoff, A. (2015). Introduction: from interacting agents to engaging persons. *Journal of Consciousness Studies*, 22(1–2), 9–23.

Szanto, T. and Moran, D. (2015). Phenomenological discoveries concerning the "we": mapping the terrain. In: T. Szanto and D. Moran (eds.), *Phenomenology of sociality: discovering the "we."* London: Routledge.

Thompson, E. (2007). *Mind in life: biology, phenomenology, and the sciences of mind.* Cambridge, MA: Harvard University Press.

Thompson, E. and Stapleton, M. (2009). Making sense of sense-making: reflections on enactive and extended mind theories. *Topoi*, 28(1), 23–30.

Torres, E.B., Brincker, M., Isenhower, R.W., Yanovich, P., Stigler, K.A., Numberger, J.I. et al. (2013). Autism: the micro-movement perspective. *Frontiers in Integrative Neuroscience*, 7. doi:10.3389/fnint.2013.00032

Towers, J. and Martin, L.C. (2015). Enactivism and the study of collectivity. *ZDM Mathematics Education*, 47(2), 247–56. doi:10.1007/s11858-014-0643-6

Trevarthen, C. and Aitken, K.J. (2001). Infant intersubjectivity: research, theory, and clinical applications. *Journal of Child Psychology and Psychiatry*, 42(1), 3–48.

Urban, P. (2015). Enacting care. *Ethics and Social Welfare*, 9, 216–22. doi:10.1080/17496535.2015.1022356

van der Schyff, D. (2015). Music as a manifestation of life: exploring enactivism and the "Eastern perspective" for music education. *Frontiers in Psychology*, 6, 345. doi:10.3389/fpsyg.2015.00345

Varela, F.J. (1996). Neurophenomenology: a methodological remedy for the hard problem. *Journal of Consciousness Studies*, 3, 330–49.

Varela, F.J., Thompson, E., and Rosch, E. (1991). *The embodied mind: cognitive science and human experience* (6th ed.). Cambridge, MA: MIT Press.

Young, I.M. (2005). Throwing like a girl: a phenomenology of feminine body comportment, motility, spatiality. In: I.M. Young (ed.), *On female body experience: "Throwing Like a Girl" and other essays*. Oxford: Oxford University Press, pp. 27–45.

Zahavi, D. (2001). Beyond empathy: phenomenological approaches to intersubjectivity. *Journal of Consciousness Studies*, 8(5-7), 151–67.

Zautra, N. (2015). Embodiment, interaction, and experience: toward a comprehensive model in addiction science. *Philosophy of Science*, 82, 1023–34.

Zeiler, K. (2014). A philosophical defense of the idea that we can hold each other in personhood: intercorporeal personhood in dementia care. *Medicine, Health Care and Philosophy*, 17, 131–41.

Zlatev, J. (2012). Prologue: Bodily motion, emotion and mind science. In: A. Foolen, U. Lüdtke, T.P. Racine, and J. Zlatev (eds.), *Moving ourselves, moving others: motion and emotion in intersubjectivity, consciousness and language*. Amsterdam and Philadelphia: John Benjamins, pp. 1–26.

CHAPTER 25

THE PERSON MODEL THEORY AND THE QUESTION OF SITUATEDNESS OF SOCIAL UNDERSTANDING

ALBERT NEWEN

Background, Motivation, and Outline

In becoming a focus of debate in the 1990s, the key question of "How do we understand others?" received two different answers: the core proposal of theory-theory (TT) is that one's understanding of another essentially relies on a folk psychological *theory*, where some take the position that the relevant folk psychology is innate (e.g., Baron-Cohen 1995), while others claim that it is acquired (Gopnik 1993). In contrast, simulation theory (ST) holds that we understand others by means of *simulation*, where simulation can take place at two levels, referred to as low-level and high-level simulation (Goldman 2006).

This chapter has two aims: first, I argue that neither of the traditional leading accounts of understanding others is adequate but that we need to go in the direction of what I call the "person model theory." Second, an important question is which types of embodiment or enactment (or further aspects of 4E) are systematically relevant for social understanding according to the person model theory? I argue that there are clear cases of embodiment of social understanding for all humans, while extendedness and/or enactment seem to be only clearly implemented in early infancy; I contend that we may only be able to find examples concerning the latter 4E features which are rather *specific cases* within the large varieties of social understanding. Furthermore, 4E features of being embodied, enacted, extended, or embedded can only be ascribed to an implementation, a token of a specific type, which makes the 4E features intensely context-dependent.

Theory-Theory, Simulation Theory, and the Challenge of Early Intuitive Understanding of Others

Both the TT and the ST have at least one central shortcoming. TT actually combines two claims, namely (a) that the corpus of knowledge we rely on when understanding others has the status of a folk psychological theory, and (b) that our epistemic strategy consists in the usage of theory-based inferences. Both claims are problematic: a key counterexample is face-based recognition of emotion (as, e.g., described in Goldman 2006). An elaborate case concerns the study of disgust: it has been shown that experiencing as well as observing disgust are dependent on certain mirror neurons that are activated in both cases (Wicker et al. 2003). Since mirror neuron activation is a very basic brain process shared with nonhuman animals and it is sufficient to stimulate these mirror neurons to experience gustatory disgust, the registration of gustatory disgust thus appears to be a rather basic process. It is thus implausible to suppose that a folk psychological theory is always involved in early abilities like registration of gustatory disgust. Defenders of TT may point out that early cognitive abilities cluster in the manner of core cognitive abilities (Carey 2009) such that different sensitivities always cluster into theory-like structures of core cognition. Although it is convincing that during ontogeny different feature-based sensitivities will be integrated, thereby improving the baby's functional capacities, it is implausible that such feature-based sensitivities must cluster right from the start. Given the processing evidence from the observation of mirror neurons registering disgust (or pain) in others, it seems to be a relatively "isolated" ability. Given the primitiveness and stability of the reaction, this seems also to hold for basic abilities like neonate imitation (Meltzoff and Moore 1977) or the reaction of babies to a caregiver: if a mother stops reacting intuitively through normal facial expressions and gestures, and instead reacts with a "still face," the baby quickly starts to cry (Bertin and Striano 2006; Nagy 2008). The baby is irritated by the unexpected pattern of reaction. Babies only a few months old already expect a typical behavioral interaction pattern from the caregiver. Even if the innate or early acquired abilities like neonate imitation or the interaction pattern with a caregiver are rich, it is not fruitful to describe them as anchored in a theory: a theory is constituted by a minimal package of systematically interconnected beliefs, and these early cases of intuitive understanding are not constituted by beliefs but rather by imitation abilities or by interaction patterns; nor is it clear that these abilities are integrated into a cluster of core cognition right from the beginning, or, if such a cluster of core cognition is described, it remains open whether it in any fruitful way deserves to be called a "theory." Any plausible account of such "early" theories must rely on the problematic intellectualist presupposition of an *implicit* network of beliefs and inferences being actually in place almost from birth onwards instead of presupposing much more basic cognitive mechanisms of association, imitation, and simple transformations at this early age (de Bruin and Newen 2012).

This debate cannot be settled here but it leaves the burden of proof with the intellectualist position speaking of implicit beliefs and inferences (1) without clarifying what this could mean at an age of a few months when babies have or acquire some basic abilities and (2) without delivering any striking evidence to presuppose such mental states (given that is an empirical question still open after the clarification of the meaning). Thus, the TT approach does not adequately account for many early abilities of intuitive understanding (or at least we lack convincing arguments to account for a cluster of abilities as a theory leaving a heavy burden of proof on representatives of TT). However, the approach clearly has merits, since some representatives offer a helpful description of the later unfolding of children's abilities, especially from age three or four onwards (see Gopnik 1993).

Recent versions of ST have tried to account for the challenge of infants' early intuitive understanding by distinguishing between high-level and low-level "mindreading" (Goldman 2006). High-level mindreading consists of an *explicit simulation* of beliefs, desires, and related actions of others on the basis of one's own experiences, i.e., the observer puts him- or herself into the other person's shoes. Let me only mention one main limitation of explicit simulation here, namely, that it cannot account for cases of understanding other persons' suffering from mental disorders like persecution mania, since I am not able to simulate this radically different mindset, but nevertheless I am able to develop quite some understanding of these persons on the basis of adequate rules of thumb provided by an expert, or simply by accepting and registering usual rituals or dispositions (for further criticism of high-level mindreading, see Newen and Schlicht 2009). The main progress of recent versions of ST consists of describing low-level mindreading, which is supposed to account for face-based recognition of emotion and the other early abilities of intuitive understanding. In the following it is argued that the presupposition of low-level mindreading is still too intellectualist to do justice to early intuitive understanding of others. Goldman's claim is that low-level mindreading, as in the case of recognizing disgust as a state of the other, is a process of simulation that includes (1) a simulation in the sense of a *replication,* i.e., of being in the same state as the observed person (which is true for the mirror neuron activation in the case of disgust), (2) a recognition of oneself as being disgusted, and (3) a projection of this registered state onto the other person, normally by forming the belief that the observed person experiences disgust. For several reasons, however, this strategy for integrating early intuitive understanding into the framework of ST is inadequate: although Goldman (2006) claims that all these steps are "comparatively simple, primitive, automatic, and largely below the level of consciousness" (Goldman 2006, p. 13), he still thinks that this involves an act of self-recognition as described in (2) and a projection as described in (3). This analysis over-intellectualizes the basic types of understanding of others. Abilities like neonate imitation, normal reactions in parent–child interactions (and a registration of their breakdowns), as well as understanding of goal-directed actions of others, do not essentially involve steps of self-recognition and projection but can be accounted for by much simpler abilities of direct perception and situational activations of habitual patterns of interaction.

Before unfolding and defending a more parsimonious explanatory strategy, let me highlight a second deficit of ST: even if the "simulation strategy" could be described without over-intellectualization, ST cannot account for all those basic forms of social understanding that do not consist in a replication of the same mental but—as often in the case of basic smooth parent–child interaction—in complementary interactions, e.g., mother gives a cookie, child takes it; child throws a ball, mother catches it; or child points to an object and expects mother to hand it over. There are many basic communicative interactions in everyday life, especially in games with children, that are complementary in nature and involve such quick interaction (throwing and catching a ball), that it is not plausible it should involve a "simulation" of, e.g., the other's intention to throw in order to do the catching. Second-person understanding is often complementary not simulative in nature (de Bruin et al. 2012). In cases like throwing and catching, furthermore, it is sufficient just to *perceive* the throwing (not to simulate it) in order to interact adequately by catching the ball. Given that TT cannot account for implicit understanding and ST cannot account for understanding complementary interactions, some proposals of hybrid theories combining TT and ST (including Goldman 2006) do not solve the problem. This has motivated the development of *interaction* theories.

Interaction Theory: Merits and Limits

Interaction theory (IT) was first developed by Gallagher (2001) and in the meantime different versions have been developed by De Jaegher et al. (2010) and Froese (2018). The shared ideas of all variants of interaction theory are the following two: (1) the standard case of understanding others is not one of simply observing others, but of being involved in interaction with them. Whereas TT and ST focus on an observational stance ("off-line understanding"), IT argues that the dominant form of understanding others takes place while we are interacting with others in a situation ("online understanding"), or, to put it differently: online interaction is a constituent of the central type of understanding others. And to this it adds (2) the standard epistemic access in such a case of online understanding is *direct perception*.

Let me summarize the main arguments in favor of IT. This will provide a platform to describe the limits of IT, and so motivate the adoption of the *person model* theory that I set out later. We turn first to the role of online understanding as appealed to in (1). The capacity for basic nonlinguistic social understanding (including the cases mentioned earlier) is constituted by second-person interaction in concrete situations. Thus, to account for those social-cognitive abilities, it is sufficient to presuppose a form of online understanding, which also seems to be the only form of understanding available in early human life. But what happens when we grow up? Does this remain the only form of understanding? Does it remain the dominant one? Even defenders of IT accept that it is not the *only* form of understanding, e.g., Gallagher and Hutto (2008), for example, argue that during ontogenetic development children start with (1) direct perception in online interaction, then develop (2) contextual understanding since they "begin to see that another's movements and expressions often depend on meaningful and pragmatic contexts" (Gallagher and Hutto 2008, pp. 5–6), and then (3) come to understand others by relying on narratives (Hutto 2008). There is much evidence for these additional strategies of social understanding, and it is plausible that these forms of understanding remain intact at later stages, such that as adults we do not rely simply on one strategy but on a multiplicity. Given the evidence that a simulation strategy is used in at least some situations (Goldman 2006), as well as the strong evidence for inference-based interpretation of others relying on explicit mindreading, which is in favor of TT (Wimmer and Perner 1983; Perner and Ruffman 2005), we need to accept that humans rely on a multiplicity of strategies, which also includes offline strategies like simulation and inference-based explicit mindreading: thus the most plausible account accepts a multiplicity of online and offline cognitive strategies of social understanding (for a detailed argument, see Newen 2015, §3.4). For an evaluation of the merits and limits of IT, I want to show (1) that online understanding by direct perception is an important strategy of understanding; (2) that in contrast to IT we systematically rely on a *multiplicity of strategies of understanding* including online and offline understanding while (at least some versions of) IT only accounts for basic forms for online understanding.

The Relevance of Online Understanding by Direct Perception

An important case of understanding others is emotion recognition. In this section I argue that emotion recognition is often based on direct perception, and that direct perception is a core component of online understanding; but it will not follow that direct perception is only or dominantly involved in online understanding, since it is also the perfect tool for offline understanding in the sense of understanding by observation without involvement in interaction.

Basic emotions are evolutionarily old, shared with some (social) nonhuman species, and develop early in human life. By way of evidence for the latter claim, we may note that the first full-blown emotional response for joy emerges at 2–3 months, anger at 4–6 months (according to Stenberg et al. 1983; seven months according to Holodynski and Friedlmeier 1999), fear at 7–9 months (Lewis 2000; Schaffner 1974), and sadness at 3–7 months. In an earlier article on classifying emotions (Zinck and Newen 2008), we presented further evidence for the existence of basic emotions in line with Ekman (1999) and Damasio (1999). Independent from the question of how many basic emotions we may need to distinguish, they do clearly exist, they unfold early, and they can be recognized without any demanding cognitive processes being involved—although it may not be clear exactly when this happens in ontogeny. This is also supported by studies that demonstrate that we can easily perceive basic emotions while looking at a point-light walker (Atkinson et al. 2004). Basic emotions have an important adaptive function, which consists of the indication of fundamental challenges: fear indicates and is produced by danger, sadness by a loss or a separation from positive conditions (e.g., separation from a parent), anger by frustration of expectancies or registration of inhibitions, while joy indicates and is produced by self-efficiency and social acceptance. The evolutionary importance of basic emotions is obvious, and this supports the observation that emotion recognition takes place very early in life and does not involve demanding cognitive abilities (Griffiths 1997; Ekman 1999; Zinck and Newen 2008). Thus it is very plausible (1) that recognition of basic emotions does not depend on any special epistemic strategy but can rely simply on direct perception, and (2) that any species living in social groups depends on recognizing the emotions of other group members.

There may be two worries that perception cannot be the relevant epistemic process: (1) because the content of perception is always sparse, including only features like color, shape, or spatial organization, and thus even basic emotions would be too rich in content to be perceived. Or (2), if one allows that the content of perception can be rich enough to include emotions in principle, the worry may then be that they cannot be perceived because they are *internal* states or processes of the other person and thus can only be inferred on the basis of behavioral data. Neither reason is convincing: as for (1), there is now a whole battery of arguments available that defend the rich or liberal-content view of perceptual experience: arguments developed in the recent literature claim that we can directly perceive causal relations (Butterfill 2009; Siegel 2009), intentions (Pacherie 2005, Gallagher 2008, Becchio et al. 2012), and actions and agency (Gao et al. 2009; Rutherford and Kuhlmeier 2013), as well as natural and artificial kinds (Bayne 2009). Furthermore, several authors argue that emotions can be directly perceived for reasons drawn from phenomenology (Zahavi 2011) or from cognitive science (Newen 2016). Last but not least, against the worry expressed in argument (2) that emotions are purely internal entities, there are well-developed and empirically grounded accounts on which emotions are individuated as patterns of characteristic features (Izard 1993; Newen et al. 2015) including, e.g., the facial expression, gestures, or body postures typical for a basic emotion. Direct perception can therefore be defended as a plausible non-demanding cognitive strategy of early social understanding since it is available in early infancy. Thus, we should allow direct perception to be a basic strategy deployed in understanding others, e.g., in understanding others' emotions, their goal-directed actions, etc. Direct perception is clearly a key route to understanding others: it is involved in online understanding of others, and this point is well taken by IT; but the same cognitive strategy—once available—can also be used for offline understanding, in scenarios in which one observes the facial expressions of two persons interacting without being involved.

Limits of the Explanatory Strategy of it

A principal limitation of any approach offering a systematic theory of understanding others in the form of IT is the danger of simply redescribing the phenomenon in question so that it appears to be adequately characterized by our preferred philosophical theory:

1. The phenomenon: a great many phenomena of understanding others are online interactions in everyday life.
2. The theoretical approach of IT: the central constituent of the phenomena of understanding others is online interaction in everyday life.

If IT holds (2) to be a central claim, then a great deal of weight rests on the explication of the status of a "central constituent." The thesis must say more about online understanding than simply describe it as involved in the majority of cases of understanding others, which would be identical with the description of the relevant phenomena in (1). It must at least be shown that online interaction is the dominant form of understanding others—which would still leave us only with a systematic ordering of the phenomena—or it must be shown in more detail which component or components of online interaction are centrally involved in all (or most of the) cases of understanding others.

IT has thus far not sufficiently addressed these issues. Concerning the suggested dominance of online understanding, I argue that there is better evidence for a pluralistic view that allows strong context-dependence of multiple epistemic strategies. This argument is part of the constructive alternative called the person model theory (see the fourth section). If we are searching for an underlying central component that is constitutively involved in social interaction, there are two candidates to be considered: either (1) the epistemic access of direct perception is seen as the central component (Gallagher 2008), this being an important tool correctly highlighted by IT, but which can also be used for offline understanding (see earlier); or (2) we consider the proposal of "participatory sense-making" as the constituent component involved in any social understanding. The latter is an important feature for *some* specific types of understanding others, but cannot be generalized. To show this limitation, let me first discuss this suggestion by starting with the definition given by De Jaegher and Di Paolo (2007):

> This is what we call *participatory sense-making*: the coordination of intentional activity in interaction, whereby individual sense-making processes are affected and new domains of social sense-making can be generated that were not available to each individual on her own. (p. 497)

The core claim of IT is that the process of coordination in interaction is constitutive in some cases of social understanding (while it may be only contextual or enabling in others). Let us discuss a clear case of being constitutive such that the function of coordination in online interaction is to create or activate a special joint action based on mutual social understanding; e.g., if two persons want to dance together for the first time, they start with coordinative behavior and develop a mutual understanding in the form of expecting the other's moves. If it works out, the resulting dance can be seen as implementing a new kind of coordination, which establishes a new mutual understanding. This case of joint activity illustrates the core idea and fits perfectly with IT, but raises several issues: (1) How far can social understanding in general be described according to this model of joint activities, which is grounded in coordination?

(2) Furthermore, there is the worry that the process of coordination can only get off the ground if every agent already has a basic form of understanding of the other: and what is this basic form of understanding others? (Martens and Schlicht 2017). If it is based primarily on direct perception, we are back to criticism (1).

Concerning (1), many of the everyday cases of social understanding are not cases of joint motor activities to be established by motor coordination between an agent and an interpreter, but are cases of direct perception of the mental states of the other (like a perception of a basic emotion), or of understanding others by observing them in a conventionally regulated context (like sitting in the bus or in the classroom, or observing someone who is refueling her car at a gas station), or are cases of linguistic communication about another's mental state—where the latter example may and often does involve some coordination, e.g., in turn-taking, but this coordination does not (or does only minimally) constrain the linguistic content. Furthermore, the content of the linguistic communication may not concern the social relation between the agent and the interpreter but rather a third person they are both talking about. As I argued earlier, neither observation is accounted for by IT.

Let me introduce an everyday example: in aiming to understand his son, who is telling him about bad results attained in a math exam, a father may be strongly biased by their social relation, which is activated and also influenced by the actual situation in which the conversation takes place; this often leads to self-deception in the evaluation of the data. The linguistic information in the conversation does not help us to understand the specific interpretation the father may adopt—say, that the bad marks do not indicate that his son is incompetent at mathematics but rather that the teacher is deficient.[1] This example shows, first, that most everyday situations involve relevant social relations, and those social relations are not only those created in the actual online situations but also those linked to underlying long-term relationships (here between father and son). Although the theory of participatory sense-making correctly highlights the relevance of online social relations that are created in the actual interaction as one factor influencing social understanding (e.g., in the case of dancing or jointly carrying a table), social relations shape our understanding even more intensely when they are not created in the actual situation but are already long established and thus being simply reactivated. Furthermore, the relevant long-term social relation may not even be lived through in the relevant situation, e.g., if I am talking to a friend about my son without my son being present. In such a case, of course, my understanding of his remarks is strongly influenced by my intense long-term relationship with my son. This observation can be generalized to an overview of the merits and limitations of IT: IT has correctly described one important case of social understanding, which is *online understanding* of others in the sense of understanding the other during activities of joint (or coordinated) action in the present; but, as already demonstrated, there are other important cases for which IT

[1] Thus, social understanding is biased by self-deception, which can be analyzed as pseudo-rational reorganization of beliefs (Michel and Newen 2010), and one source of the bias in social understandings is involving social relations.

cannot adequately account. The latter can be summarized as offline understanding, i.e., understanding others without being involved in any interaction or coordination with them. Offline understanding includes two cases: (1) the obvious case of understanding others while the other is not present, e.g., by relying on memories, by getting information about the other, talking to further persons or getting information via media of all kinds; and (2) offline understanding while the other is present, e.g., a reporter listening to an official political discussion in order to write an article without interacting with the participants to the discussion. Thus, the difference between online understanding and offline understanding is best conceptualized as a different *stance*: online understanding is an understanding of the other from an interactor's point of view, while offline understanding is done from an observer's point of view (Schilbach 2014). The presence of the other is necessary but not sufficient for online understanding, since even while others are present one can remain in an offline, observational stance in understanding the other.

With these distinctions in hand, we can now summarize the merits and limits of IT: IT can account nicely for cases of basic online understanding that focus on coordination between agent and interpreter in an actual situation. In these cases, a new social relation is created that is clearly relevant in cases of social understanding of joint or coordinated actions, like starting a joint dance for the first time. IT correctly highlights the relevance of synchronic activity and imitation as well as complementary activities in these cases of basic social understanding. It may indeed be called understanding others by coordination, since adequate coordination pertains not only to the observed phenomenon but is at the same time the realization of a basic form of social understanding. But there are also more complex forms of online understanding, like the case in which the father seeks to understand his son's problems with mathematics: IT overlooks or strongly underestimates the difference between basic and complex forms of social understanding or it lacks an argument why the complex forms of social understanding still should be constituted by interaction. Furthermore, it overlooks the importance of the long-term social relationships that are habitual and reactivated in social interactions. This long-term, person-centered information can become strongly relevant in shaping an online interaction, much more so than any specific information about the son in the situation, and it is also relevant in offline understanding, such as when trying to understand the son while discussing him with a friend. This criticism can be condensed into one core difference for which IT cannot account: namely, the difference between the social understanding of a person's actions in one and the same situation type, where in one case the person is a complete stranger and in the other a well-known family member. This is especially the case since this difference is already implemented in early infancy.[2]

[2] Defenders of IT may reply that they can include the *memorized* interactions to account for this. But then they start to change the proposal by accepting memorized models of other persons. Another move would be to claim that the memorized information is available in form of narratives since those are an additional tool in IT. Although narratives are an important tool to enrich information about others unfolding from two years of age onwards (Hutto 2008; Newen 2015), they cannot account for the relevant sensitivity in early infancy.

Let us take stock. Looking back, this defect of IT can also be diagnosed for TT and ST. First of all, TT is well known for focusing on general folk psychological rules in understanding others, but has a gap in considering the idiosyncratic behavior of individuals we know very well, and of course we take our experience of these idiosyncrasies strongly into account. ST can only account for idiosyncratic features that one shares with the other and thus can activate to project onto them; but—as mentioned earlier—idiosyncratic behavior that is distinct from one's own behavioral repertoire is not accounted for in ST. Thus, all the relevant theoretical frameworks fail to account for the observation that in understanding others we rely strongly on all the background knowledge we have established in long-term relations, especially including knowledge about idiosyncratic behavior or attitudes of well-known persons. Is there a theoretical framework that can do justice to this while not running into the problems of the theories discussed? We now turn to the *person model theory* as an alternative, which can account for special information about a well-known individual, while also including the difference between basic and complex forms of social understanding. Thus it integrates the merits of IT while at the same time doing justice to multiple forms of offline understanding. In the next section, the person model will be introduced, its advantages explored and defended, and we will ask which kinds of social understanding can be seen as situated in the new framework and to what extent.

Person Model Theory

There are a very few proposals that developed in parallel sharing the idea that we need something less demanding than a *theory* as proposed in TT and that this may be something similar to *models* in understanding science in an early stage. These developments are discussed under the label "model theory" (Newen and Schlicht 2009; Maibom 2009; Godfrey-Smith 2005). Without being able to discuss the different versions in this chapter, I will develop my own proposal, which has unfolded over several years (Newen and Vogeley 2011, 2015; Newen 2014, 2015).

The *person model theory* (PMT) consists of two main claims, one concerning the epistemic strategy humans use to access the mental states of others, and one concerning the organization of the background information we almost always rely on to understand others. Concerning the *epistemic* strategy, PMT defends the multiplicity view: we do not rely on one epistemic strategy, as is suggested by most proposals in the literature (e.g., ST claims that simulation is the only or at least the absolute dominant strategy), but rather we rely on a multiplicity of strategies that, for the most part, are implicitly activated by contextual cues. These strategies include at least simulation strategies, theory-based inferences, and direct perception, as well as understanding based on social interaction and by relying on narratives.[3] Concerning the organization of the relevant background

[3] A plurality of social understanding was described by Andrews (2012), but she did not work out the important difference between epistemic strategies and the relevant background information that allows a systematic analysis of the rich and varying phenomena.

information, the central claim is that information about other humans as individuals or types of persons is stored and organized in *person models*, and that these are realized on two levels, i.e., the implicit level of person schemata and the explicit level of person images. It is further argued that philosophical theories so far have tended to ignore the fact that we usually understand others by relying on rich background information concerning them and their situation.[4] Thus, the relevant background knowledge involves both models of person and models of situations. The two central aspects of PMT, the multiplicity of epistemic strategies and the organization of relevant background information in person models and situation models, are explicated in what follows.

A Multiplicity of Strategies of Understanding Others

There are two main arguments employed to defend the multiplicity view concerning epistemic strategies: (1) the ontogenetic argument, where the ontogenetic development of understanding others can best be explained by describing the development of a multiplicity of strategies of understanding others such that no strategy is eliminated once acquired; (2) the pathology argument, which turns on the observation that some cases of mental disorder can best be described by demonstrating that some epistemic strategies are lacking and thus others—which are still available—are used as substitutions, even though they often cannot really compensate for the complete lack of the original strategies.

As for (1), the ontogenetic argument: quite early on, babies rely on *online understanding by coordinated interaction*. They are sensitive to others' facial expressions (neonate imitation) and develop an expectation of an interaction scheme as demonstrated by the still-face paradigm. *Direct perception* is very relevant starting from early infancy, as proven by face-based recognition of emotions based on direct perception (Zahavi 2011; Newen et al. 2015). During ontogeny we develop further important strategies for understanding others, which also include strategies of offline understanding. It will also be indicated that we cannot observe any general dominance of one of these strategies, but that the activation of a specific strategy is dependent on the context, while strategies are often activated in combination. At the age of 9 to 12 months, children learn *to understand others as participating in joint attention and joint action* (Tomasello 1999), where the latter is demonstrated, e.g., by understanding the other as following a plan like jointly constructing a Lego house (18 months) or similar plans (Metlzoff 1995). At 2.5 years children become sensitive to rules and norms such that they insist that group members follow rules. This involves an *understanding of others as rule-followers*, i.e., as members of the group governed by expectations concerning rule-following behavior in relevant situations (Rakoczy et al. 2008). Furthermore, there is the well-studied ability to *understand others by explicit false beliefs* (age four onwards) which enables *explicit theory-based inferences* or *explicit simulation strategies* to

[4] A possible exception within representatives of IT is Gallagher (2011).

understand others.[5] Let me finally mention *understanding by explicit second-order false beliefs* developed between ages seven and nine (Wimmer and Perner 1983), although I leave it open which additional epistemic strategies can be fruitfully distinguished as developing later in the process of growing up. There is consensus that these abilities come step by step, and that abilities acquired early remain intact and in use even when more sophisticated abilities are available. To illustrate: looking at the face of a person, I may directly perceive an expression of anger. But when I am informed that she is suffering from Parkinson's disease and therefore has severe limitations in controlling her facial expression, I will evaluate the same facial expression quite differently. Despite having a standard "reading" of the emotion expressed in the face, this new knowledge about Parkinson's disease helps me to override my spontaneous perception: although the direct perceptual impression is still in place, I will override it in this context and use a theory-based inference to reach a new evaluation of the person, relying on other cues including the person's linguistic utterances. This also illustrates the context-dependence of the preference of one strategy over the other.

Concerning (2), the pathology argument: some mental disorders essentially involve significant deficits in social understanding, and these cases can best be explained such that at least one strategy of the normal bundle of strategies is lacking. This can be illustrated by having a look at people with Asperger's syndrome: they lack an intuitive understanding of others. They are unable to directly perceive emotions on the basis of facial expressions, and they tend to avoid social interaction (Vogeley 2012). Thus intuitive understanding by primary interaction or direct perception is (almost) unavailable for them. Since they also tend to experience themselves as being different (Vogeley 2012), they do not use simulation as a strategy: so they are left principally with theory-based inferences (Kuzmanovic et al. 2011). Since they are often intelligent they can learn some theory-based inferences that are helpful in specific situations, but they lack an intuitive generalization of this knowledge. Thus in new or slightly modified situations, they again feel lost since they do not even have a theory on which basis to apply theory-based inferences. Since we have to deal with new or modified situations almost every day, autistic people notice their tendency to get lost and many of them avoid social encounters. This special situation is explained by the fact that in contrast to the usual availability of multiple strategies of understanding, they are left mainly with theory-based inferences alone, and need an explicit corpus of knowledge to apply this (since they lack intuitive generalization) (for further arguments concerning the multiplicity view, see Newen 2015; Fiebich and Coltheart 2015). Thus, social understanding usually relies on a multiplicity of epistemic strategies that are selected in a highly context-dependent manner (as demonstrated with the Parkinson case). Concerning the epistemic strategies of social

[5] It is an open debate how exactly explicit theory-of-mind abilities and understanding by narratives are related to each other. While Hutto (2008) claims that the latter is more primitive than the former, I presuppose here that *understanding* by narratives is based on a theory-of-mind ability and enriches it. Thus, I do not discuss its role in addition to theory-of-mind abilities.

understanding, we may indicate that social understanding is strongly dependent on the actual context.

Person Models as Unified Information Structures (Person Files)

Having argued for the multiplicity view of epistemic strategies for social understanding, let me now focus on the organization of the relevant background information. The central claim here is that relevant information about other humans as individuals or types of persons is stored and organized in *person models*, which are either implicitly available person schemata or explicitly available person images. A person schema remains implicit; it can typically be described as a unity of sensorimotor abilities and basic mental phenomena associated with one human being (or a group of humans), while the schema typically functions without any explicit considerations and is activated when directly seeing or interacting with another person. A person image is constituted by explicitly (i.e., typically consciously) available information concerning physical and mental phenomena associated with and unified to belong to one human being (or a group of humans). Thus, a person image is the unity of rather easily and explicitly available information about a person, including the person's mental setting. Both person schemata and person images can be developed for an individual, e.g., one's mother, brother, best friend, etc., as well as for groups of people, e.g., students, medical doctors, lawyers, etc. Furthermore, person models are created for other people but also for oneself. This results in the following 3×2 design of the relevant person models (Table 25.1)

The person model theory can account for several important aspects which are highlighted as defects of at least one of the competitors: (1) The person model theory can convincingly account for the difference between understanding a complete stranger by relying on a group model only, and understanding a well-known family member by relying on a rich explicit person image of the individual, which can contain very specific information. No other theory can account for the systematic understanding of individual idiosyncrasies of others that are different from one's own dispositions, but individual person models can do the job. (2) By appealing to the distinction between implicit

Table 25.1 Types of person models

Person models	Self	Other: Individuals	Other: Groups
Person schema	Self-schema	Individual person schema	Group person schema
Person image	Self-image	Individual person image	Group person image

and explicit person models, PMT can account for the difference between basic or intuitive understanding and complex or theory-based understanding of others, which is underdeveloped in TT. (3) With the difference between a person model of oneself and person models of others, PMT can account for an understanding of others that goes beyond one's own self-model as the sole source of understanding others, contrary to ST. (4) By including the multiplicity view concerning the epistemic strategies of understanding, PMT can account for the fact that we actually use different strategies of understanding that vary with the context and other constraints, which clearly distinguishes it from TT and ST. (5) PMT can be distinguished from its competitors: it is different from TT since TT cannot adequately account for the plurality and development of social understanding, which is attained in PMT by the multiplicity view of epistemic strategies, and it is different from TT because PMT can account for an early intuitive understanding by implicit person schemata. It is different from ST because it can account both for an understanding of others based on the self-model and an understanding of others based on the person model of other individuals or a group model, which can be radically different from the self-model. It is different from IT since it addresses not only basic online understanding but also offline social understanding. (6) Finally, there is recent empirical evidence from neuroscience that we actually construct and rely on person models (Hassabis et al. 2013; see Newen 2015, §5.3).[6] Thus, it seems correct to call PMT a new approach, not just a variant of an existing one.

An account of full-blown PMT needs to mention one further component, namely, situation models. We have the ability to understand others by completely abstracting from the individual, e.g., it can be sufficient to predict the behavior of a restaurant guest by expecting her to act according to the conventions of a high-level restaurant which includes the expectation of a certain sequence of events. This type of understanding is developed together with rule-based understanding of others at the age of 2.5 years (see earlier). In new contexts, especially in new cultural contexts, we begin by employing an understanding of others mainly on the basis of noticing rule-based behavior that we discern as being adequate in a situation—e.g., as a European one learns to understand other restaurant guests by learning the rule-based behavior characteristic of high-quality restaurant situations in Japan (special greetings, taking off shoes, sitting in a special way, etc.). Thus, we not only create person models but also situation models, and our understanding of others uses both types of model as input and selects the model most helpful for evaluating the other person's behavior. How do person models and situation models interact in this implicit evaluation process? Without being able to unfold this fully here, we can note that at this point the person model theory meets and aims to develop further the old psychological attribution theory (Kelley 1967; Weiner 1995),

[6] In the study (Hassabis et al. 2013), it is shown that there are neural correlates of imagining two central features of the Big Five in personality psychology, i.e., agreeable in contrast to antisocial personalities and introvert in contrast to extrovert personalities. Furthermore, it was shown that the combinations of personality types like an agreeable and extrovert personality is represented in a modulation of medial prefrontal cortex.

since this theory already has worked-out criteria for under which conditions an internal attribution is favored (which in PMT involves an attribution focusing on a person model), or an external attribution is chosen (which in PMT takes place when the situation model is dominant). This outline of the person model theory is sufficient as a basis for us to turn to the question of whether and to what extent social understanding is situated in the sense of implementing the 4E criteria.

The Situatedness of Person Model Theory

Embodiment

To what extent does PMT fit the 4E criteria? Let us start with the role of embodiment in PMT: embodied abilities usually remain implicit, and the place to account for implicit abilities in PMT is the implicit person schemata. Do we have to presuppose embodied features to be involved in an implicit person schema? Yes. First of all, there are many studies showing the embodiment of social cognition, which is accepted by all theories of social understanding, and thus this is not specific to PMT: these include the well-known observations of mirror neurons not only in the case of doing a basic intentional action but also in observing the same action in others (Rizzolatti et al. 2002). In the same line, Adolphs et al. (2000) offer evidence that the somatosensory cortex is activated in face-based recognition of emotion. These experiments demonstrate embodiment in the weak sense of involving sensorimotor activities in the brain as being involved in basic social understanding. Embodiment in the strong sense of involving the brain and body of the interpreter can be proven by online embodiment in registering emotions: mimicry of the facial expression of the other's emotion is intensely involved in emotion recognition, as was observed by Wallbott (1991) and then demonstrated in detail by Niedenthal et al. (2001), who showed that participants free to mimic an emotional expression detected a fine-grained change in emotional expression as it morphed into another expression earlier than did participants who were prevented from mimicking.

There are many more studies illustrating the weak and strong embodiment of social understanding in a way that does not allow one to distinguish between the theories discussed above (for an overview, see Niedenthal et al. 2005). Let me now illustrate one specific case of embodiment that provides further evidence for PMT and the interaction between person schema and person image in person impression formation: this is the case of a typical patient suffering from Capgras syndrome. This misidentification syndrome consists primarily in the patient's delusional belief that one of his closest relatives, e.g., his wife, has been replaced by an impostor (Davies et al. 2001). The misidentification is not due to the lack of any explicit knowledge about the relevant person, since the patient correctly identifies clothing, voice, face, the wedding ring she is wearing, etc. (in my terminology we would say that the person image is intact). What is lacking, according to the standard analysis, is a feeling of familiarity that normally comes with perceiving a well-known person. Since the resulting social understanding of the relevant person is radically different for the Capgras patient (who does not express greetings to his wife, but accuses her of being an imposter), this shows that the embodied feeling of familiarity is an essential part of social understanding. Thus, the feeling of familiarity is a constitutive component of the person schema that needs to be activated in person recognition. While the person image is intact, the disturbance of the person schema leads to the radically inadequate interpretation of the situation (Davies et al. 2001; Vogerau and Newen 2007). In all these cases the relevant embodied features remain inaccessible.

Can we offer a convincing case of embodiment in which the relevant feature is accessible and thus can be made explicit? In principle, all explicitly noticed body contacts with the other person are candidates, an obvious one being the embodied greeting often expressed in our culture by handshaking. What can we learn from explicitly being aware of a firm handshake by the other person? In a study by Chaplin et al. (2000), it was first shown concerning the expressive dimension that individuals whose handshakes are firmer (i.e., have a more complete grip, are stronger, more vigorous, longer in duration, and associated with more eye contact) are more extraverted and are less neurotic and shy. The important second step is that four professional interpreters of the handshakes who shook the hands with all participants were also asked to analyze their first impression according to nine standard personality features, and those were compared to a self-report of the test persons. The handshaking characteristics are related to both objective personality measures and the first impressions of the other formed by the trained interpreters.[7] Another example relying only on everyday experience is walking hand in hand, e.g., mother and child or two romantic partners, where this involves the mother's registering that her child wants to look at the toy shop they are passing, or one partner's registration that something is wrong with the other (she is nervous, impatient). There are quite intense forms of embodiment implemented in basic forms of social understanding, while it remains an open question whether this is true for complex forms like high-level mindreading based on an explicit theory of mental abilities (Goldman and de Vignemont 2009, p. 157).

Extended, Enacted, and Embedded? The Variability of Social Understanding

Let me first introduce my understanding of the central notions: the weakest claim concerning 4E features being made of a cognitive ability is the embeddedness of social understanding understood as the claim that a contextual feature *strongly* modulates social

[7] For a critical understanding of this study it is important to mention that the authors discuss that they have to include a top-down effect: the interpreter takes some information about the personality from the handshake but this information is modulated top down by the general person impression and is not completely independent (Chaplin et al., 2000, Discussion section).

understanding, without involving a claim about its being constitutive. For the other notions of embodied, extended, and enacted, I think they are especially useful if we take them to involve a claim about, respectively, the body, an external entity, or the action of the relevant person to play a *constitutive role* for a cognitive ability of the person (see chapter 1 for details). Can we find some clear cases of embodiment, extendedness, and enactment of social understanding?

Concerning embodiment, there are clear cases concerning emotion recognition that I already reported (see earlier): Niedenthal et al. (2001) showed that participants free to mimic an emotional expression detected a fine-grained change in emotional expression as it morphed into another expression earlier than did participants who were prevented from mimicking. In addition having a smiling facial expression compared to a neutral or frowning face makes you evaluate the same pictures as funnier (Niedenthal 2007).

One interesting feature is the early availability of interaction schemata, such as the mother–child interaction schema, which is used in the understanding of the famous triangle movements of Heider and Simmel (1944). One way to interpret the perception of characteristic movements of geometrical figures as mother–child interactions is to say that there is an implicit interaction schema that is activated to understand the situation. Now, this is not yet in any way to go beyond an internal representation of this schema. Only if the interaction schema is constitutively shaped in the actual situation by the other person or by the ongoing interaction with the other person would it be a candidate for extendedness or enactedness. According to the proponents of this view (e.g., De Jaegher et al. 2010; Froese 2018), the best candidate for enactment of social understanding—i.e., a constitutive role of interaction for social understanding—is the perceptual crossing experiment, especially the version realized by Auvray et al. (2009). A careful analysis (Herschbach 2012; Auvray and Rohde 2012) shows that the data are still compatible with much more modest interpretations, such as that "even in simple environments, embodied and embedded interaction can bring about coordination and/or synchronisation" (Auvray and Rohde 2012, p. 9). From my point of view, a modest interpretation could also be that their perceptual crossing experiment shows that even on the basis of sparse information (due to the settings), we are able to register the agency of another person. This would allow us to integrate the results of this interesting experiment with the well-known effects of registering social features on the basis of point-light displays of person behavior. The claim about a constitutive role for interaction in social understanding is still open to empirical tests, and it is accepted here only for one of the many kinds of social understanding, i.e., basic social online understanding as it is realized in joint or coordinated actions. This is an obvious constitutive claim of the same kind as "At one spatiotemporal location only one color can be instantiated." In our case this is the claim that to have a basic understanding of joint or coordinated action it is constitutive to be *adequately involved in the relevant coordination*. The latter is a typical way to characterize the standard of basic social understanding and thus it is strongly connected with it. But for other cases of social understanding, the debate is still open.

Is there is a candidate for extendedness of social understanding, i.e., that social understanding of others constitutively involves the other person (not just a representation of her)? The best cases are those of social understanding in early infancy, e.g., the cognitive abilities underlying the still-face paradigm (see earlier): the caregiver is a constituent of the expected social interaction since at least at the age of stranger anxiety the reaction of the infant is strongly dependent on the other person involved. Furthermore, social understanding is intuitively essentially influenced by our explicit self-image, since we create person models of others in relation to our self-model. Extreme cases make this everyday intuition obvious: in the case of narcissism the person always understands others in a way that is constitutively influenced by their own radically overestimated self-esteem. There is further empirical evidence that social understanding is strongly shaped by our self-esteem (Greenwald and Banaji 1995), and there is the well-known evidence that at least for Asian cultures (China and Japan) the mother is not only relevant for the self but constitutively involved in it: the so-called interdependent self in some Asian cultures is such that it involves not just one person (oneself) but also often a very important other, usually the mother (Markus and Kitayama 1991). Evidence for this Asian concept of an interdependent self (in contrast to a Western independent self) has also been delivered by a discovery of a respective difference in neural correlates (Sui and Han 2007). Now, if the self is essentially involved in the process of understanding others, and the self is—at least in some cultures—extended, then the process of social understanding can be extended too. But, again, this interesting observation can also be consistently described from a purely internalist perspective according to which we are just relying on rich internal representation of the self, which accounts for the cultural variation of the interdependent as well as the independent self. Within the person model theory this comes along with the introduction of a self-model in addition to person models for others. A self-model is a unity of information about oneself, where describing it in this terminology does not decide in favor of an internalist account or exclude externalism. Whether the role of a self in social understanding grounds extendedness is thus still an open question. But coming back to the *basic form* of social understanding in joint or coordinated action, there it is also rather clear that the self-esteem of a young child is strongly influenced by the other in the actual situation, and thus this basic form of social understanding is a good candidate for extended social understanding.

Concluding Remarks on the Situatedness of Social Understanding

To summarize these considerations, *basic social understanding* as realized in emotion recognition or in joint or coordinated action in early infancy (which are unified in person schemata) are good candidates for being embodied, extended, and/or being enacted. And this is an important discovery because these modes of social understanding remain intact when we become adults. But for more complex forms of online social understanding and for all forms of offline understanding, we do not have any convincing evidence on the basis of which we can generalize 4E features for the time being. Must we simply continue searching for better experiments? Although to find a decisive experiment would indeed be welcome (e.g., as suggested by Schilbach 2014), there are principled worries that we may have to reconsider this debate because the inability to generalize may not be just an accidental condition, but rather, for several reasons, we should expect there to be strong variability. (1) Thanks to recent memory research, the activation of a memorized relevant person model can no longer be seen as a reactivation of static information but needs to be accounted for as a reconstruction of person models on the spot in the actual situation on the basis of memory traces (Hassabis and Maguire 2007). Thus, person models are not static person files unifying information about a person, but are flexibly constructed templates resulting from memory traces that are selected and modified by contextual information and the self-model. (2) Social understanding is based on the activation of person models as background information in combination with epistemic strategies for using this information. Given the arguments for the multiplicity of epistemic strategies of social understanding (see earlier), it follows that we should expect further variation in social understanding. (3) If we accept that a blind person's stick is a good candidate for extended cognition, we would still need to qualify the usage: if it is used by a blind person for spatial orientation, then this ability is co-constituted by the stick, while the same stick could also be used by the same person to defend herself and then it would no longer be part of extended cognition. The implication is that social understanding is also strongly influenced by contextual and situational factors including such usage: this makes it difficult to *generalize* on whether a relevant feature of social understanding renders this process embodied, extended, or enacted. It follows that we can only work out very *specific type cases* (not rather general types), since we know that every ability humans develop undergoes a process of habituation such that the first instances of realizing the ability often involve intense (conscious) control and may essentially rely on contextual elements in the environment, and subsequently the ability is modified by habituation into a process that can be interpreted in both ways leading to an impasse: e.g., using a blind stick can either be interpreted as extended (essentially involving the stick) or as essentially relying on an internal representations (while the stick is only a causally relevant, contextual tool) since the habituation leads to a neural representation of the stick as part of the body schema. Since accounting for habituation leaves the interpretation open, the best candidates for situatedness in the sense of being extended or enacted are newly acquired as well as the spontaneously and creatively formed acts of social understanding. Should we better then not refrain from using the characterization of situatedness, since we cannot easily generalize from relevant features of social understanding to its situatedness? I do not think so, but here I can only outline an alternative perspective on situatedness in general.

We may accept that a mental phenomenon can only be described as situated when describing detailed and specific versions of social understanding. This can be made fruitful if we accept that social understanding is a strongly varying set of abilities that can be usefully described as being individuated as an integrated pattern of characteristic features[8] consisting of (1) typical input in the actual situation (seeing the other person, her appearance, activity in the actual situation), (2) typical bodily arousal and behavioral dispositions, (3) the involvement of at least one of the relevant epistemic strategies, (4) the intentional object of the act of social understanding, and (5) the activation of the relevant person models as well as (6) the self-model. The realization of a typical token pattern can vary a lot since it can be based on varying subgroups of features realizing these components, and the question whether a 4E feature (being embodied, extended, enacted, or embedded) is implemented can only be answered for a specific realized type of pattern (considering the process of habituation). Thus, in the case of highly variable mental phenomena like social understanding, we do not have the means to convincingly answer the question concerning situatedness in the form of a general metaphysical theory of social understanding. The best we could get is a pragmatic framework that enables us to describe typical *specific* cases that are part of a relevant determining integrated pattern of characteristic features, e.g., understanding a basic emotion on the basis of facial expression is usually embodied since it is strongly modified by mimicry of the observer (see earlier). If these final considerations are leading in the right direction, the metaphysical debate about the relevance of 4E features should no longer be discussed for highly variable mental phenomena in general, and we should be satisfied with a pragmatic framework that allows us to decide about specific cases concerning the 4E features. Social understanding is highly variable; it is creative, dynamic, and highly context dependent; but none of these criteria implies that it needs to be applied by implementing one of the 4E features. Potentially, it is only for a specific type of implementation that we will be able to decide which of the 4E features are realized.

References

Adolphs, R., Damasio, H., Tranel, D., Cooper, G., and Damasio, A.R. (2000). A role for somatosensory cortices in the visual recognition of emotion as revealed by three dimensional lesion mapping. *Journal of Neuroscience*, 20, 2683–90.

Andrews, K. (2012): *Do apes read minds? Toward a new folk psychology*. Cambridge, MA: MIT Press.

Atkinson, A.P., Dittrich, W.H., Gemmell, A.J., and Young, A.W. (2004). Emotion perception from dynamic and static body expressions in point-light and full-light displays. *Perception*, 33(6), 714–46.

Auvray, M., Lenay, C., and Stewart, J. (2009). Perceptual interactions in a minimalist environment. *New Ideas in Psychology*, 27, 79–97.

Auvray, M. and Rohde, M. (2012). Perceptual crossing: the simplest online paradigm. *Frontiers in Human Neuroscience*, 6, 191.

Baron-Cohen, S. (1995). *Mindblindness: an essay on autsim and theory of mind*. Cambridge, MA: MIT Press.

[8] This idea is developed in detail for emotions in Newen et al. 2015.

Bayne, T. (2009). Perception and the reach of phenomenal content. *Philosophical Quaterly*, 59(236), 385–404.
Becchio, C., Manera, V., Sartori, L., Cavallo, A. and Castiello, U. (2012). Grasping intentions: from thought experiments to empirical evidence. *Frontiers of Human Neuroscience*, 6(117).
Bertin, E. and Striano, T. (2006). The still-face response in newborn, 1.5-, and 3-month-old infants. *Infant Behavior and Development*, 29(2), 294–7. doi:10.1016/j.infbeh.2005.12.003
Butterfill, S.A. (2009). Seeing causings and hearing gestures. *Philosophical Quaterly*, 59(236), 405–28.
Carey, S. (2009). *The origin of concepts*. Oxford: Oxford University Press.
Chaplin, W.F., Phillips, J.B., Brown, J.D., Clanton, N.R., and Stein, J.L. (2000). Handshaking, gender, personality, and first impressions. *Journal of Personality*, 79(1), 110–17.
Damasio, A.R. (1999). *The feelings of what happens: body and emotion in the making of consciousness*. New York: Harcourt Brace.
Davies, M., Coltheart, M., Langdon, R., and Breen, N. (2001). Monothematic delusions: towards a two-factor account. *Philosophy, Psychiatry and Psychology*, 8(2/3), 133–58. doi:10.1353/ppp.2001.0007
de Bruin, L., van Elk, M., and Newen, A. (2012). Reconceptualizing second-person interaction. *Frontiers in Neuroscience*, 6, 151. doi:10.3389/fnhum.2012.00151
De Jaegher, H. and Di Paolo, E. (2007). Participatory sense-making: an enactive approach to social cognition. *Phenomenology and the Cognitive Sciences*, 6(4), 485–507.
De Jaegher, H., Di Paolo, E., and Gallagher, S. (2010). Can social interaction constitute social cognition? *Trends in Cognitive Sciences*, 14(10), 441–7.
Ekman, P. (1999). Basic emotions. In: T. Dalgleish and M.J. Power (eds.), *The handbook of cognition and emotion*. New York: Wiley, pp. 45–60.
Fiebich, A. and Coltheart, M. (2015). Various ways to understand other minds: toward a pluralistic approach to the explanation of social understanding. *Mind and Language*, 30(3), 235–58.
Froese, T. (2018). Searching for the conditions of genuine intersubjectivity: from agent-based models to perceptual crossing experiments. In: this volume, pp. 163–86.
Gallagher, S. (2001). The practice of mind: theory, simulation, or interaction? *Journal of Consciousness Studies*, 8(5–7), 83–107.
Gallagher, S. (2008). Direct perception in the intersubjective context. *Consciousness and Cognition*, 17(2), 535–43. doi:10.1016/j.concog.2008.03.003
Gallagher, S. (2011). Narrative competency and the massive hermeneutical background. In: P. Fairfield (ed.), *Hermeneutics in education*, New York: Continuum, pp. 21–38.
Gallagher, S. and Hutto, D. (2008). Understanding others through primary interaction and narrative practice. In: J. Zlatev, T.P. Racine, C. Sinha, and E. Itkonen (eds.), *the shared mind: perspectives on intersubjectivity*. Amsterdam: John Benjamins, pp. 17–38.
Gao, T., Newman, G.E., and Scholl, B.J. (2009). The psychophysics of chasing: a case study in the perception of animacy. *Cognitive Psychology*, 59(2), 154–79.
Godfrey-Smith, P. (2005). Folk psychology as a model. *Philosophers' Imprint*, 5(6), 1–16.
Goldman, A.I. (2006). *Simulating minds: the philosophy, psychology, and neuroscience of mindreading*. Oxford: Oxford University Press.
Goldman, A. and de Vignemont, F. (2009). Is social cognition embodied? *Trends in Cognitive Sciences*, 13(4), 154–9.
Gopnik, A. (1993). How we know our minds: the illusion of first-person knowledge of intentionality. *Behavioral and Brain Sciences*, 16(1), 1–15, 90–101. doi:10.1017/S0140525X00028636

Greenwald, A.G. and Banaji, M.R. (1995). Implicit social cognition: Attitudes, self-esteem, and stereotypes. *Psychological Review*, 102(1), 4–27.

Griffiths, P.E. (1997). *What emotions really are: the problem of psychological categories.* Chicago: University of Chicago Press.

Hassabis D. and Maguire E.A. (2007). Deconstructing episodic memory with construction. *Trends Cognitive Science*, 11, 299–306. doi:10.1016/j.tics.2007.05.001

Hassabis, D., Spreng, R.N., Rusu, A.A., Robbins, C.A., Mar, R.A., and Schacter, D.L. (2013). Imagine all the people: how the brain creates and uses personality models to predict behavior. *Cerebral Cortex*. doi:10.1093/cercor/bht042

Heider, F. and Simmel, M. (1944). An experimental study of apparent behavior. *American Journal of Psychology*, 57, 243–59.

Herschbach, M. (2012). On the role of social interaction in social cognition: a mechanistic alternative to enactivism. *Phenomenology and the Cognitive Sciences*, 11(4), 467–86.

Holodynski, M. and Friedlmeier, W.E. (1999). *Emotionale Entwicklung*. Heidelberg: Springer.

Hutto, D. (2008). *Folk psychological narratives*. Cambridge, MA: MIT Press.

Izard, C.E. (1993). Four systems for emotion activation: cognitive and noncognitive processes. *Psychological Review*, 100, 68–90.

Kelley, H.H. (1967). Attribution theory in social psychology. In: D. Levine (ed.), *Nebraska Symposium on Motivation* (vol. 15). Lincoln: University of Nebraska Press, pp. 192–238.

Kuzmanovic, B., Schlibach, L., Lehnhardt, F., and Vogeley, K. (2011). A matter of words: impact of verbal and nonverbal information on impression formation in high-functioning autism. *Research in Autism Spectrum Disorders*, 5(1), 604–13.

Lewis, M. (2000). The emergence of human emotions. In: M. Lewis, & J.M. Haviland-Jones (eds.), *Handbook of emotions* (2nd ed.). New York: Guilford, pp. 265–80.

Maibom, H.L. (2009). In defence of (model) theory theory. *Journal of Consciousness Studies*, 16(6–8), 360–78.

Markus, H.R. and Kitayama, S. (1991). Culture and the self: implications for cognition, emotion, and motivation. *Psychological Review*, 98(2), 224–53. doi:10.1037/0033-295X.98.2.224

Martens, J. and Schlicht, T. (2017). Individualism versus interactionism about social understanding. *Phenomenology and the Cognitive Sciences*, 16, 1–22.

Meltzoff, A.N. (1995). Understanding the intentions of others: re-enactment of intended acts by 18-month-old children. *Developmental Psychology*, 31 (5), 838–50.

Meltzoff, A.N. and Moore, M.K. (1977). Imitation of facial and manual gestures by human neonates. *Science*, 198(4312), 75–8. doi:10.1126/science.198.4312.75

Michel, C. and Newen, A. (2010). Self-deception as pseudo-rational regulation of belief. *Consciousness and Cognition*, 19(3), 731–44.

Nagy, E. (2008). Innate intersubjectivity: newborn's sensitivity to communication disturbance. *Developmental Psychology*, 44(6), 1779–84. doi:10.1037/a0012665

Newen, A. (2015). Understanding others—the person model theory. In: T. Metzinger, and J.M. Windt (eds.), *Open MIND*, 26. doi:10.15502/9783958570320, 1–28

Newen, A. (2014). Selbst- und Fremdverstehen: die Personenmodelltherie als Analyserahmen für mentale Störungen. In: R. Vogt (ed.), *Verleumdung und Verrat: Dissoziative Störungen bei schwer traumatisierten Menschen in Folge von Vertrauensbrüchen*. Kröning: Asanger, 209–18.

Newen, A. (2016). Defending the liberal-content view of perceptual experience: direct social perception of emotions and person impressions. *Synthese*. doi:10.1007/s11229-016-1030-3

Newen, A. and Schlicht, T. (2009). Understanding other minds: a criticism of Goldman's simulation theory and an outline of the person model theory. *Grazer Philosophische Studien*, 79(1), 209–42.
Newen, A. and Vogeley, K. (2011). Den anderen verstehen. *Spektrum der Wissenschaft*, 8.
Newen, A., Welpinghus, A., and Juckel, G. (2015): Emotion recognition as pattern recognition: the relevance of perception. *Mind & Language*, 30(2), 187–208.
Niedenthal, P.M. (2007). Embodying emotion. *Science*, 316, 1002–5.
Niedenthal, P.M., Barsalou, L.W., Winkielman, P., Krauth-Gruber, S., and Ric, F. (2005). Embodiment in attitudes, social perception and emotion. *Personality and Social Psychology Review*, 9(3), 184–211.
Niedenthal, P.M., Brauer, M., Halberstadt, J.B., and Innes-Ker, A.H. (2001). When did her smile drop? Facial mimicry and the influences of emotional state on the detection of change in emotional expression. *Cognition and Emotion*, 15, 853–64.
Pacherie, E. (2005). Perceiving intentions. In: J. Sàágua (ed.), *A explicação da interpretação humana*. Lisbon: Edições Colibri, pp. 401–14.
Perner, J. and Ruffman, T. (2005). Infants' insight into the mind: how deep? *Science*, 308(5719), 214–6.
Rakoczy, H., Warneken, F., and Tomasello, M. (2008). The sources of normativity: young children's awareness of the normative structure of games. *Developmental Psychology*, 44(3), 875–81.
Rizzolatti, G., Fadiga, L., Fogassi, L., and Gallese, V. (2002). From mirror neurons to imitation: facts and speculation. In: A.N. Meltzoff and W. Prinz (eds.), *The imitative mind: development, evolution and brain bases*. New York: Cambridge University Press, pp. 247–66.
Rutherford, M.D. and Kuhlmeier, V.A. (2013). *Social perception: detection and interpretation of animacy, agency, and intention*. Cambridge, MA: MIT Press.
Schaffner, H.R. (1974). Cognitive components of the infant's response to strangeness. In: M. Lewis and L. A. Rosenblum (eds.), *The origins of fear*. New York: Wiley, pp. 11–24.
Schilbach, L. (2014). On the relationship of online and offline social cognition. *Frontiers in Human Neuroscience*, 8, 278.
Siegel, S. (2009). The visual experience of causation. *Philosophical Quaterly*, 59(236), 519–40.
Stenberg, C., Campos, J., and Emde, R. (1983). The facial expression of anger in seven-month-old infants. *Child Development*, 54, 178–84.
Sui, J. and Han, S. (2007). Self-construal priming modulates neural substrates of self-awareness. *Psychological Science*, 18(10), 861–6. doi:10.1111/j.1467-9280.2007.01992.x
Tomasello, M. (1999). *The cultural origins of human cognition*. Cambridge, MA: Harvard University Press.
Vogeley, K. (2012). *Anders sein—Hochfunktionaler Autismus im Erwachsenenalter*. Weinheim: Beltz-Verlag.
Vosgerau, G. and Newen, A. (2007). Thoughts, motor actions and the self. *Mind and Language*, 22(1), 22–43. doi:10.1111/j.1468-0017.2006.00298.x
Wallbott, H.G. (1991). Recognition of emotion from facial expressions via imitation? Some indirect evidence for an old theory. *British Journal of Social Psychology*, 30, 207–19.
Weiner, B. (1995). *Judgments of responsibility: a foundation for a theory of social conduct*. New York: Guildford Press.
Wicker, B., Keysers, C., Plailly, J., Royet, J.-P., Gallese, V., and Rizzolatti, G. (2003). Both of us disgusted in my insula: the common neural basis of seeing and feeling disgust. *Neuron*, 40(3), 655–64. doi:10.1016/S0896-6273(03)00679-2

Wimmer, H. and Perner, J. (1983). Beliefs about beliefs: representation and constraining function of wrong beliefs in young children's understanding of deception. *Cognition*, 13(1), 103–28. doi:10.1016/0010-0277(83)90004-5

Zahavi, D. (2011). Empathy and direct social perception: a phenomenological proposal. *Review of Philosophy and Psychology*, 2(3), 541–58.

Zinck, A. and Newen, A. (2008). Classifying emotions: a developmental account. *Synthese*, 1, 1–25.

CHAPTER 26

FALSE-BELIEF UNDERSTANDING, 4E COGNITION, AND PREDICTIVE PROCESSING

LEON DE BRUIN

INTRODUCTION

MANY philosophers in the debate on social cognition assume that predicting or explaining another agent's behavior depends on a specialized capacity that allows us to track and "read" his or her mental states. According to these philosophers, our prediction that John will go shopping is based on (our representation of) his desire to restock his wardrobe and his belief that the end-of-season sale starts today. Similarly, our explanation of why Jill took the last piece of cake without asking the others is based on (our representation of) her fondness for cake and her belief that no one else was interested. Empirical support for the existence of such a capacity for "mindreading" has allegedly been found in psychological and neuroscientific research on theory of mind. In particular, developmental studies on false-belief understanding have received a lot of attention, because they seem to provide concrete evidence for the claim that, at some point in ontogeny, children come to understand the distinction between mind and world. That is, they learn to distinguish between what people may falsely believe about the world and how the world really is.

Mindreading accounts of social cognition have been increasingly challenged over the past couple of decades. Proponents of the 4E cognition movement have argued that social cognition is not (primarily) about reading another agent's mental states in order to predict or explain his or her behavior. A large part of their criticism has focused on the empirical findings on false-belief understanding, questioning both the philosophical interpretation of these findings and the experimental designs that give rise to them.

My aim in this chapter is twofold. First, I discuss the empirical findings on false-belief understanding in light of the criticism leveled by proponents of 4E cognition. Second, I propose an alternative interpretation of these findings and investigate to what extent this interpretation is compatible with the main insights of 4E cognition.

The chapter has the following structure. The second section provides an overview of the traditional "elicited-response" studies on false-belief understanding, in which children are asked to predict another agent's action on the basis of his or her (false) belief. It also discusses some objections to these studies that have been put forward by proponents of 4E cognition. The third section focuses on recent "spontaneous-response" studies on false-belief understanding, which measure children's understanding of false belief in terms of the behavior they spontaneously produce. These studies address some of the problems pointed out by proponents of 4E cognition, but also give rise to new questions about false-belief understanding. In the fourth section, I present several interpretations of the data from spontaneous-response studies on false-belief understanding, and examine the prospects of adjudicating between them on empirical grounds. The fifth and final section explores the viability of an alternative theoretical framework for understanding the experimental results on false-belief understanding, which is inspired by the predictive processing paradigm.

Elicited-Response Measures of False-Belief Understanding

Premack and Woodruff's paper "Does the Chimpanzee Have a Theory of Mind?" (1978) is commonly considered as the starting point for research on theory of mind. It reports an experiment in which an adult female chimpanzee called Sarah was shown videotapes of a human actor who struggled with a variety of problems (e.g., trying to obtain bananas out of reach). With each videotape, Sarah was given several photographs, one of which represented a solution to the problem (e.g., a moveable box that allowed the actor to reach the bananas). Premack and Woodruff argued that Sarah's consistent choice of the correct picture strongly suggested that she has a theory of mind.

However, as Bennett (1978), Dennett (1978), and Harman (1978) suggested in their commentaries on the article, a key element missing from Premack and Woodruff's experiment was a measure of false-belief understanding. This was necessary to rule out the possibility that Sarah chose on the basis of her own belief instead of the belief of the human actor.

These suggestions inspired Wimmer and Perner (1983) to develop an experimental design that could be used to investigate false-belief understanding in young children: the so-called false-belief test. Wimmer and Perner told children a story about a puppet called Maxi, who puts his chocolate in the kitchen cupboard and leaves the room to play. While he is away, his mother (whom he cannot see) moves the chocolate from the cupboard to a drawer. Maxi returns. At this point, children were asked the following question: "Where will Maxi look for his chocolate, in the drawer or in the cupboard?" In order to give the correct answer, children needed to realize that Maxi's action depends on his belief rather than how things really are in the world (according to the children).

In Wimmer and Perner's original version of the false-belief test, children are asked to (verbally) predict where another agent will look for the object. Other tests have been designed to investigate whether children also understand another agent's false beliefs about the identity (rather than the location) of the object. For example, in the so-called Smarties test, children are shown a Smarties tube, which is actually full of pencils (this is revealed to the children). Then they are asked what they think other people will think is in the box (Gopnik and Astington 1988).

The results of these and other false-belief tests show that three-year-old children consistently give wrong predictions about what the agent will think or do next and thus fail to take into account his false belief. Instead of answering randomly, they make a specific false-belief error—they assert that Maxi will look in the drawer, or that other people will think there are pencils in the Smarties tube. Four-year-olds, by contrast, correctly predict that Maxi will look in the cupboard, and that other people will think there are Smarties in the Smarties tube. On the basis of these results, researchers have concluded that the capacity to understand false-belief, which is at the core of our theory of mind, normally emerges around four years of age (see Wellman et al. 2001 for a meta-analysis).

In their criticism of the elicited-response false-belief test, proponents of 4E cognition such as Gallagher (2001, 2004), Hutto (2008, 2015a), Ratcliffe (2005, 2007), and Zahavi (2004, 2014) have primarily concerned themselves with the picture of social cognition that it presupposes. They have argued that the false-belief test focuses on a specialized capacity to predict and explain behavior from an observational, third-person point of view. However, they claim that this is not how we understand others in everyday life. First of all, social understanding emerges as the result of dynamic second-person interaction, rather than passive third-person observation. Second, our interactions with others are embodied: we are able to understand them on the basis of their facial expressions, goal-directed gestures, and bodily movements without having to read or represent their mental states. What they believe, desire, and intend is directly expressed in their behavior, rather than hidden in their mind. Third, understanding others is embedded or "situated": it happens in a broader social and pragmatic context, which constrains the social information that is relevant to our understanding of them, and it is based on a history of interactions, which shapes and structures our expectations about their behavior.

According to proponents of 4E cognition, these considerations suggest a non-mentalistic view of social understanding that is primary to the specialized capacity to explain and predict behavior on the basis of mental states, in the sense that it comes earlier in ontogenetic development.[1] However, as I will show in the next section, recent "spontaneous-response" studies on false-belief understanding seem to challenge this assumption.

[1] See Currie (2008) and Spaulding (2010) for a critical assessment.

Spontaneous-Response Measures of False-Belief Understanding

In contrast to the elicited-response false-belief tests described in the previous section, spontaneous-response studies no longer require children to verbally predict another agent's action on the basis of her mental states. Instead, their understanding of false-belief is inferred from the behavior they spontaneously produce in "violation-of-expectation," "anticipatory-looking," and "active-helping" paradigms.

The Violation-of-Expectation False-Belief Test

The violation-of-expectation paradigm tests whether children look reliably longer when another agent's action is inconsistent with his or her false belief. This longer looking is associated with a violated expectation; when children see what they expect their looking times are considerably shorter that when they see something unexpected. Thus, looking times are a means of measuring the expectations of infants.

Onishi and Baillargeon (2005) used this paradigm to examine whether 15-month-olds understood an agent's false belief about the location of an object. In the first familiarization trial, infants observed how the experimenter played with a toy standing between a yellow and a green box in front of her, and then hid the toy inside the green box. In the second and third familiarization trials, the experimenter reached inside the green box (as though to grasp her toy) and then paused. Next, the infants either watched how, in the experiment's absence, the toy was moved to the yellow box ("false-belief-yellow" condition), or they watched how, in the experimenter's presence, the toy was moved to the yellow box but then returned to the green one after the experimenter left the room ("false-belief-green" condition). Finally, during the test trial, the experimenter reached inside either the yellow or the green box and then paused. It was found that in the final test trial of the false-belief-green condition, infants who watched the agent reaching inside the yellow box looked reliably longer than those who watched the agent reaching into the green box; whereas they showed the reversed looking-pattern in the false-belief-yellow condition. Onishi and Baillargeon (2005) argued that this showed that infants expected the agent to reach where they falsely believed the toy to be.

Other experiments have focused on the question whether infants understand another agent's false beliefs about the identity of an object. In a study by Song and Baillargeon (2008), for example, 14.5-month-olds observed how the experimenter's hand placed a doll with blue hair and a stuffed skunk with a pink bow on placemats or in shallow containers. Afterwards, the experimenter showed her preference for either of the two toys by reaching for it. Then, in the experimenter's absence, the toys were put in two boxes, one of which had a tuft of blue hair on the lid, suggesting it contained the doll. The doll was placed in the plain box, and the skunk in the box with the tuft of blue hair on it. When the experimenter returned and reached for the box with the tuft of hair after having showed a preference for the skunk, the infants looked considerably longer than when she reached for the plain box—and conversely. According to the experimenters, this meant that the infants expected the agent to falsely conclude that the doll was hidden in the hair box.

The Anticipatory-Looking False-Belief Test

The anticipatory-looking false-belief test investigates whether children can visually anticipate where another person with a false belief about the location of an object will search for this object. In a study by Clements and Perner (1994), three-year-olds observed how mouse Sam placed a piece of cheese in a blue box and fell asleep. While Sam was sleeping, mouse Katie took the cheese and put it in a red box. When Sam returned, the children correctly anticipated his behavior by looking at the blue box where he falsely believed the cheese to be hidden—even when they gave the wrong prediction that Sam would search for the cheese in the red box. By contrast, younger children erroneously looked at the object's real location, which they gave for their answer. On the basis of these findings, Clements and Perner concluded that children under the age of three do not have an understanding of false-belief.

However, Southgate et al. (2007) pointed out that the anticipatory-looking condition in Clements and Perner's experiment still included a verbal element. In order to maximize the frequency of anticipatory looking at one of the object locations, the experimenter said aloud, "I wonder where Sam is going to look." According to Southgate et al. (2007), this primed younger children to look at the object's real location.

In their own study, Southgate et al. (2007) removed this verbal element. They familiarized 25-month-olds to a scene in which a bear puppet hid a ball in one of two boxes. The experimenter, whose head was visible above a panel with two small windows (one above each box), looked on and retrieved the ball by opening the correct window. In the test trial, a phone rang behind the experimenter after she watched how the ball was hidden in either the left or the right box. While the experimenter was facing away, the bear retrieved the ball and left. Then the phone stopped ringing and the agent turned toward the boxes. It was found that most infants correctly anticipated the experimenter's behavior by looking at the window above the box where she falsely believed the ball to be hidden. According to Southgate et al. (2007), this indicated that children younger than three years in fact do have some grasp of the experimenter's false belief.

The Active-Helping False-Belief Test

The active-helping false-belief test examines whether infants are able to actively assist the experimenter in a false-belief situation. For example, in an experiment by Buttelmann et al. (2009), 18-month-olds watched as a toy was transferred from Box A to Box B while an experimenter either witnessed the transfer (true-belief condition) or not (false-belief condition). Then the experimenter attempted unsuccessfully to open Box A—the empty box. In the true-belief condition, infants could follow their natural tendency to help the experimenter by opening Box A for him. In the false-belief condition, if infants understood the experimenter's false belief, they had to infer that he wanted the toy he thought was in there. In this case they should not simply go help him open Box A, but rather go to Box B and extract the toy for him. It was found that 18-month-olds were able to actively assist the experimenter in his search for the toy.

In another active-helping false-belief test by Southgate et al. (2010), 17-month-olds watched how an experimenter placed two objects into two boxes and left the room. In her absence, a second experimenter appeared and switched the locations of the objects. When the first experimenter returned, she pointed at one of the boxes and asked, "Can you pass me the sefo [a nonsense name]?" (in one condition) or "Can you pass me it?" (in another). Southgate et al. (2010) found that most infants moved toward the box that was not indicated by the experimenter, and thus seemed to understand that she had a false belief about the location of the object.

The Verbal Spontaneous-Response False-Belief Test

Active-helping false-belief tests show that speech comprehension by itself cannot explain why children have difficulties with versions of the false-belief test that include a verbal element. This is because children younger than two years are able to pass these tests, despite the fact that they involve verbal encouragements and requests. Thus, in the study by Buttelmann et al. (2009), infants are verbally encouraged to help the target agent unlock a box, and in the study by Southgate et al. (2010), they are asked to reach for a designated object.

Other studies show that young infants even succeed at spontaneous-response false-belief tasks that imposed more significant linguistic demands. Scott et al. (2012), for example, tested 2.5-year-olds with two verbal spontaneous-response false-belief tests that imposed significant linguistic demands: an anticipatory-looking test in which children listened to a story about a protagonist with a false belief while looking at a picture book (with matching and non-matching pictures), and a violation-of-expectation test in which children watched an agent answer (correctly or incorrectly) a standard false-belief question. Despite their linguistic demands, positive results were obtained with both tests.

The Perspective-Tracking False-Belief Test

Rubio-Fernández and Geurts (2013) have shown that three-year-olds can pass a streamlined version of the elicited-response false-belief test, the so-called Duplo test, which is designed in such a way as to support their ability to track the perspective of the agent. First, children are familiarized with a set of Duplo toys: a girl figure, a bunch of bananas, and two little cupboards ("fridges"). They are told that the girl loves bananas and had one for breakfast this morning. The experimenter then manipulates the girl to put the bananas inside one of the two fridges and tells the child that the girl now wants to go for a walk. The Duplo girl walks in the direction of the child and turns her back on the scene. Importantly, the child can still see the Duplo girl (in contrast to false-belief studies that make the protagonist disappear). The experimenter asks the child, in a secretive manner, "Can the girl see me from where she is?" Then, she moves the bananas from one fridge to the other and asks the child, pointing at the girl figure, "She hasn't seen what I did, has she?" These prompts are intended to help the child keep track of the girl's perspective. Likewise, rather than introducing a second character in the story, distracting the child from the protagonist's perspective, the experimenter is the one who moves the bananas. The experimenter then returns the girl figure back to the center of the scene, places the figure in front of the two fridges, and asks the child whether he or she would like to play with the girl now. The child is prompted with the question: "What happens next?. . . What is she going to do now?" The child then continues acting out the next event with the Duplo girl.

Rubio-Fernández and Geurts showed that 80 percent of the three-year-olds passed the Duplo test by making the Duplo girl look in the empty fridge for her bananas, compared to 23 percent on standard elicited-response false-belief tests. When the girl was made to disappear on her walk (as in the standard test), the success rate fell to 17.6 percent. When the child was not allowed to interact with the Duplo girl, the success rate fell to 22 percent.

Interpreting the Spontaneous-Response False-Belief Test

Spontaneous-response false-belief tests seem to address some of the objections to the elicited-response false-belief test (see the end of Section 2), insofar as they no longer focus solely on the capacity to verbally predict another agent's action on the basis of her mental states. For example, although the violation-of-expectation and the anticipatory-looking false-belief tests are still about third-person observation rather than second-person interaction, the active-helping false-belief tests by Buttelmann et al. (2009) and Southgate et al. (2010) and the perspective-tracking test by Rubio-Fernández and

Geurts (2013) *do* involve second-person interaction between the child and the protagonist and/or the experimenter.

At the same time, however, proponents of mindreading have used the results of the spontaneous-response false-belief tests to counter the claim that non-mentalistic forms of social understanding are primary to the specialized capacity to explain and predict behavior on the basis of mental states. They have argued that these results show that false-belief understanding is in fact a basic capacity that emerges early in development (Overgaard and Michael 2015). One of the main questions in the current debate on social cognition, therefore, is whether children's behavior in spontaneous-response false-belief tests should indeed be interpreted as mentalistic. Are looking times, anticipatory looking, and active helping representative of false-belief understanding?

In what follows I will discuss several accounts of the results of the spontaneous-response false-belief test, ranging from full-blown mentalistic to non-mentalistic. Most of these accounts also attempt to explain how these results are related to the results of the elicited-response test.

Dual-System Accounts

Apperly and Butterfill (2009) have proposed a dual-system account of the false-belief data (see also Butterfill and Apperly 2013). According to them, children's verbal predictions in the elicited-response false-belief test are facilitated by a full-blown propositional theory of mind, but their performance on the spontaneous-response false-belief tests is best explained in terms of a "minimal," non-propositional theory of mind. This minimal theory of mind is insensitive to the aspectual nature of belief and does not support the attribution of propositional attitudes. It allows infants to *track* (rather than represent) beliefs as the result of *registerings* of facts, such as the location of an object. Another agent's registering of the location of an object is said to be off-target when the object is not located where it is registered to be. According to Apperly and Butterfill (2009), spontaneous-response false-belief tests show that infants are able to track off-target registerings and use them to predict or anticipate another agent's action.

Baillargeon et al. (2010) have also put forward two systems to account for the false-belief data.[2] The first system ("subsystem 1") enables infants to attribute motivational and "reality-congruent" informational states to other agents, and is operational by the end of the first year. Motivational states are defined in terms of another agent's goals and dispositions, whereas reality-congruent informational states are defined as states that specify another agent's knowledge about the scene. The second system ("subsystem 2") becomes operational in the second year of life. It underlies the attribution

[2] Baillargeon et al. (2010) present their theory as a dual-system account. However, since only subsystem 2 is used to explain the false-belief data, the theory can also be classified as a single-system account of false-belief understanding (cf. Carruthers 2013).

of "reality-incongruent" informational states to other agents, which specify what they falsely believe about the scene.

According to Baillargeon et al. (2010), to explain the different results of the spontaneous-response and the elicited-response false-belief test, we need to consider the extent to which both tests recruit the second system. They suggest that the spontaneous-response false-belief test only involves the representation of another agent's false belief. By contrast, the elicited-response false-belief test also involves a selection process (infants have to access their representation of another agent's false belief) and an inhibition process (infants have to inhibit any tendency to answer the test on the basis of their own knowledge). The joint activation of these three processes is said to overwhelm the child's limited information-processing resources.

Single-System Accounts

Single-system accounts propose that the results of the spontaneous-response and the elicited-response false-belief test can be explained in terms of a single theory of mind system, which provides children with the concepts and core knowledge to represent another agent's mental states. Carruthers (2013, 2016), for example, argues that it is possible to make sense of the false-belief data by postulating a single theory of mind system that operates with various executive functions, depending on the task demands (see also Leslie 1994; Leslie et al., 2004). He claims that the spontaneous-response false-belief test is a "single-mindreading" task, which requires children to inhibit their own (true) belief, and select and maintain the (false) belief of another agent.[3] The elicited-response false-belief test, by contrast, is a "triple-mindreading" task: it requires (1) inhibiting a true belief, and selecting and representing a false belief, as well as (2) processing the speech of the experimenter and figuring out his or her communicative intentions, and (3) formulating an action that would serve to communicate the target agent's false belief to the experimenter. According to Carruthers, in order to engage in all three things at once children need to switch back and forth between the different perspectives (now the protagonist, now the experimenter, now one's own), bearing in mind the output of "one round of computation" while undertaking another, and then accessing the former at the appropriate time. The combination of multiple processing demands overwhelms the resources of their mindreading system, and makes them default to the next most relevant answer, which is what they themselves believe. Carruthers emphasizes the importance of language production (task element 3) in particular, which he takes to be the main disrupting factor in the elicited-response false-belief test.

[3] Note that this is how Baillargeon et al. (2010) describe the task requirements of the *elicited-response* false-belief test. They claim that the spontaneous-response false-belief test only involves representing another agent's false belief.

Behavior-Rule Accounts

Thus far I have discussed mentalistic accounts of the results of the spontaneous-response false-belief test. These accounts postulate domain-specific theory of mind systems that are dedicated to representing the mental states of other agents.

Behavior-rule accounts, by contrast, postulate domain-specific systems that do not represent another agent's mental states, but certain principles that specify what another agent will do under specified, observable conditions (Perner 2010; Perner and Ruffman 2005; Ruffman and Perner 2005). For example, a number of results (Onishi and Baillargeon 2005; Surian et al. 2007; Southgate et al. 2007; Buttelmann et al. 2009) could be explained in terms of behavioral rules, such as "people tend to look for an object where they last saw it and not necessarily where the object actually is." According to proponents of behavior-rule accounts, young infants could have a grasp of this rule without any notion of false belief as something that causes behavior. Another rule that has been proposed to explain the results of Onishi and Baillargeon (2005) and Surian et al. (2007) is "ignorance leads to error." This would lead infants to expect an agent who is ignorant of the location of a desired object to search in the wrong location, where the object is not.

In order to explain all the experimental results, many behavioral rules have to be postulated. However, Povinelli and Vonk (2004) have argued that it will always be possible to find some behavior rule that could account for any given item of data. They claim that whenever a mindreading proponent postulates that children link another agent's circumstances to his subsequent behavior via an internally attributed mental state, one could claim instead that these children employ a behavior-rule linking the circumstances to the behavior directly.[4]

Interactionist Accounts

The main idea behind interactionist accounts is that false-belief understanding is facilitated by second-person interactions, rather than a dedicated internal capacity to represent mental states or behavioral rules. Gallagher (2015), for example, proposes that children understand another agent's action as aimed at the world in ways that offer social affordances for interaction with them. Take the active-helping false-belief test by Buttelmann et al. (2009). According to Gallagher, the infant knows that the agent has seen the switch or not. This, in combination with the agent's behavior with respect to Box A (moving to A and attempting to open it), signals a difference in affordance, i.e., a difference in how the infant can interact with the agent. There is no need to represent the

[4] See also Träuble et al. (2010, pp. 442–3), who suggest that the behavior rules may be formulated more flexibly or more generally.

agent's mental states since all of the information needed for understanding and interaction is already available in the infant's perception of the situation.

Gallagher (2015) argues that the importance of second-person interaction becomes very clear in the experiment by Rubio-Fernández and Geurts (2013): three-year-olds pass the Duplo experiment not only because it facilitates the tracking of perspectives (result b), but also because it allows for a higher degree of second-person interaction with the experimenter and the protagonist. When the interactive elements are removed, the third-person observation design is re-established, and, on average, the child does worse (results a and c).

It is not only the quantity but also the *quality* of second-person interaction that makes the Duplo experiment easier for three-year-olds. Gallagher suggests that in the standard elicited-response false-belief test, the second-person interaction (with the experimenter) has a saliency that takes precedence over the third-person task and biases the child's answer. Both the child and the experimenter know where the toy really is, and this shared knowledge motivates the (wrong) answer to the third-person task. However, if the task is rearranged in a way that (a) allows the child to interact with the agent, and (b) makes the interaction with the experimenter support (rather than distract from) the child's ability to track the perspective of the agent, the child does much better.

Low-Level Novelty Accounts

Heyes (2014) argues that the results of the spontaneous-response false-belief studies are indicative of "low-level novelty" rather than false-belief understanding. In some studies this novelty depends on the observable properties of the test stimuli (e.g., test events that display new spatiotemporal relations among colors, shapes, and movements), whereas in other studies it depends on memory or imagination (e.g., test events that are remembered or imagined to have new spatiotemporal relations among colors, shapes, and movements).

On a low-level novelty account, the results of the anticipatory-looking false-belief test by Southgate et al. (2007), for example, could be explained as follows: infants looked at the correct location of the ball because they *themselves* imagined the ball to be there, not because they understood the experimenter's false belief. The bell ringing and head turning, which was supposed to signal to the infants that the experimenter could not see movements of the ball, may instead have distracted the infants so that they didn't see, or didn't remember, those movements. In that case, the infants might assume the ball to be at the location where they last saw it and, on the basis of their familiarization experience, expect the hand to appear above that location.

The low-level novelty account assumes that the results of the spontaneous-response false-belief tests can be fully explained in terms of domain-general processes (e.g., perception, attention, motivation, learning, and memory). There is no need to postulate domain-specific theory of mind systems or behavior rules.

How to Evaluate These Accounts?

Various objections have been raised against the accounts of false-belief understanding presented in the previous section. Some have argued for alternative interpretations of the data on the basis of *parsimony*. For example, Carruthers (2016) has proposed that the false-belief data can be explained more parsimoniously by postulating one rather than two theory of mind systems. Gallagher (2015) has gone even further by suggesting that the false-belief data can be explained more parsimoniously by postulating no theory of mind systems whatsoever. However, it is difficult to assess the strength of these proposals, since there is no general consensus about the criteria that should be used to adjudicate between the different accounts of false-belief understanding. There are many different kinds of parsimony and it is not clear how they can be weighed against each another.

Others have ruled out certain interpretations of the false-belief data on the basis of empirical evidence. For example, Scott and Baillargeon (2014) have argued that a low-level novelty explanation of the data is implausible, because there is a considerable amount of evidence suggesting that infants do not merely react to the perceptual novelty of colors, shapes, and movements, but rather respond to agents acting on objects. The appeal to empirical evidence is not without problems, however. Indeed, some have argued that there is an "insurmountable" gap between theory and evidence, thereby questioning the very idea of a straightforward empirical test to decide between the various alternatives (Buckner 2014; Hutto 2015a). To illustrate this last point, consider Heyes's (2014) proposal to test whether spontaneous-response false-belief tests require the attribution of mental states by means of an experimental design that involves a "self-informed belief induction variable," i.e., a variable that, if the infant is capable of mental state attribution, she knows "only through extrapolation from her own experience to be indicative of what an agent can or cannot see, and therefore does or does not believe" (p. 656). What is problematic here is that what needs to be empirically demonstrated, namely, mental state attribution, is already part of the definition of a self-informed belief induction variable (the step from "can or cannot see" to "does or does not believe").

False-Belief Understanding: An Alternative Framework?

Rather than attempting to decide the debate on false-belief understanding on empirical grounds or by appeal to parsimony, it might be more fruitful to look at the results of false-belief studies from a radically different theoretical perspective. In the remainder of this chapter I will explore the prospects of an alternative account of false-belief understanding, one that is based on an explanatory framework that has steadily been gaining ground throughout the cognitive sciences: the predictive processing (PP) framework.

The Predictive Processing Paradigm

The predictive processing framework conceives of the brain as a probabilistic inference system, which attempts to predict the input it receives by constructing models of the possible causes of this input. The main aim of the system is to minimize the "prediction error"—the discrepancy between the predicted and the actual input.

PP postulates that the models generated by the brain are not only evaluated according to how well they fit the evidence, i.e., how well they predict the input in question, but also according to how likely they are in the first place, i.e., their "prior probability." When making sense of new input the brain does not start from scratch, but rather updates the models with the highest prior probability in order to accommodate the new evidence.

Furthermore, PP assumes that these models are organized in a *hierarchy*. At the lowest level of the hierarchy, neural populations encode such features as surfaces, edges, and colors. At a hierarchically superordinate level, these low-level features are grouped together into objects, while even further up the hierarchy these objects are grouped together as components of larger scenes involving multiple objects. The key organizational principle of this hierarchy is that of time scale: features of the world that change over shorter time scales are represented at the bottom of the hierarchy, whereas features that change over longer time scales are represented at higher levels of the hierarchy.

Finally, PP proposes that the brain has two options for reducing prediction error. The first option is to revise its model of the world until the prediction error is satisfactorily diminished ("perceptual inference"). The second option is to change the world so that it matches the model ("active inference"). For example, if one expects to see a banana in the fridge, but it turns out not to be there, one might simply conclude that one was mistaken (i.e., change the model). But one might also adjust one's head or even one's bodily position until one does see the banana, e.g., behind the broccoli or sandwiched between two cartons of milk. In this case, one has changed the world in the sense of changing the position of one's body in the world. Together with the concept of perceptual inference, the concept of active inference provides a unifying framework for perception and action: both can be viewed as means of reducing prediction error.

The basic principles of PP have been discussed extensively by theorists such as Clark (2013, 2016), Friston and Stephan (2007), Friston (2010), and Hohwy (2013, 2014). In what follows, I will therefore concentrate on the implications of PP for our understanding of the false-belief data.

Predictive Processing and False-Belief Understanding

Let us first consider the violation-of-expectation false-belief test. From a PP perspective, what happens in this experiment can be explained as the result of a prediction error—infants form expectations about the behavior of another agent, which are violated at some point. However, this does not necessarily demonstrate that infants have an understanding of false belief, in the sense that they represent another agent's belief about the world (specifically, the location of an object). Rather, it suggests that infants are able to track features of the agent that remain stable over longer time scales (in this case the agent's preference), at a higher level of the hierarchy compared to other features of the agent (e.g., his or her bodily movement) or the world (e.g., the location of the object). Furthermore, the results of the violation-of-expectation false-belief test show that the infant's expectations about another agent's behavior are violated, but they do not show how the infant deals with this prediction error. From a PP perspective, this is actually a more important question than the question whether infants are able to represent false beliefs.

Similar considerations apply to the results of the anticipatory-looking false-belief test. On a PP interpretation, these results indicate that infants form certain expectations about the behavior of another agent, which are manifested in their looking behavior. However, the anticipatory-looking false-belief test does not address the question whether or how infants are able to reduce the prediction error that follows when their expectations are violated.

The results of the active-helping false-belief test are far more interesting in this respect. They suggest that children are not merely surprised when their expectations about another agent's behavior are violated, but actively try to reduce the resulting prediction error by assisting the agent in question. That is, children engage in *active inference* by changing the world so that it matches their model of it. For example, the experiment by Buttelmann et al. (2009) shows that 18-month-olds, when confronted with a mismatch between (their representation of) the preference of the agent and the new location of the object, reduce this prediction error by directing the agent to the actual location of the object and extracting the object for him.

An intriguing question is whether this also suggests that children engage in *perceptual inference*, i.e., reduce the prediction error by revising their model of the world. On the one hand, perceptual inference seems to play a crucial role in the active-helping false-belief test insofar as children have to update their model of the object and its changing location in the world—not only because this is what gives rise to a prediction error in the first place (the mismatch between the preference of the agent and the new location of the object), but also because it allows them to assist the agent in his search for the object. On the other hand, children do not have to update their model of the agent's preference for the toy or their model of the agent's mental states more in general. Now it might be objected that children need to update their model of the agent's belief about the location of the object and engage in perceptual inference once they have shown the agent the actual location of the object (Box B) by means of active inference. However, it is not clear that the active-helping false-belief test requires children to represent another agent's belief about the location of an object. Just like the violation-of-expectation and the anticipatory-looking false-belief test, children only have to track the agent's preference for the object, and this preference remains stable during the experiment (and therefore does not need to be updated by means of perceptual inference).

These considerations suggest a possible answer to the "developmental paradox," i.e., the finding that infants pass the spontaneous-response false-belief test but somehow consistently fail the (traditional) elicited-response false-belief test. On a PP interpretation, what makes the latter test more difficult is precisely the fact that children cannot engage in active inference. As a result, they have no means of reducing the prediction error that ensues when the agent leaves the scene and the object is transferred to another location.

Mindreading accounts such as those of Baillargeon et al. (2010) and Carruthers (2016) argue that, in order to correctly predict where the agent will look for the object, children have to inhibit their own true belief and select and represent the false belief of the agent. According to these accounts, young children fail the elicited-response false-belief test because they cannot handle the complexity of the task, which overwhelms the resources of their mindreading system, and therefore default to what they themselves believe to be the case. However, just like parsimony, there are many different kinds of complexity and it is not clear how they can be weighed against each another. Furthermore, as Gallagher (2015) points out, if it were merely complexity overloading the mindreading system, we might just as well expect confused or arbitrary answers.

PP offers a potential explanation of what goes wrong in the elicited-response false-belief test that goes beyond simply postulating "complexity" as the root of the problem. If we assume that children have a natural inclination to reduce prediction error, then the elicited-response false-belief test might be more difficult because young children need to inhibit their tendency to engage in perceptual inference and update their model of the agent's belief to get rid of the discrepancy between the preference of the agent and the new location of the object. In the active-helping false-belief test, children do not need to engage in perceptual inference because they can reduce the prediction error by means of active inference. However, in the elicited-response false-belief test, this option is not available. This is why the children resort to perceptual inference. However, what younger children fail to take into account is that perceptual inference is bound to certain constraints, and that the natural tendency to reduce prediction error has to be resisted when these constraints are not met. In the elicited-response false-belief test, the relevant constraint on perceptual inference is that one's model of another agent's belief about the location of an object should not be updated when this object is moved to another location while the agent is absent (and thus cannot see what happens). Older children, by contrast, do understand this and they pass the elicited-response false-belief test because they can resist the natural tendency to reduce prediction error when this constraint is not met.

Of course, this interpretation of the false-belief data needs to be fleshed out in much more detail. For example, we have to spell out the nature of the infant's expectations about the agent's behavior. Also, we need to explain how the constraints on perceptual inference should be articulated (e.g., as biologically innate or scaffolded by the social environment), and how the relationship between false-belief understanding and inhibition processes (Perner and Lang 1999; Cole and Mitchell 2000; Carlson and Moses 2001), working memory (Carlson et al. 2004; Hala et al. 2003; Perner et al. 2002), and language production (Astington and Jenkins 1999; Astington and Baird 2005) should be construed.

However, what is important here is that PP seems to have the resources to explain *why* younger infants systematically give incorrect answers in the elicited-response false-belief test. It is not merely because of the complexity of the task. It is because of their tendency to reduce prediction error.

Predictive Processing and 4E Cognition

To what extent is this account of false-belief understanding—in particular, the PP principles that motivate it—compatible with the main insights of 4E cognition?

PP agrees with the claim that understanding others is situated and based on a history of interactions (see the end of the second section), insofar as it maintains that the predictions made by our brains are contextualized and based on prior expectations (Clark 2013, 2016). It is also compatible with the assumption that this understanding emerges as the result of a dynamic process of second-person interaction. However, in the context of the false-belief test, PP suggests that the relevant contrast is not between second-person interaction and third-person observation, but rather between two ways of reducing prediction errors: active inference and perceptual inference.

To illustrate this, consider Rubio-Fernández and Geurts's (2013) proposal that the false-belief test basically measures whether children are able to track another agent's perspective despite sudden changes (prediction errors) that may disrupt this process. Rather than focusing on how these changes disrupt children's capacity for perspective tracking, the account of false-belief understanding presented in the previous section puts more emphasis on whether the experimental design offers them possibilities for dealing with prediction errors by means of active inference and/or perceptual inference. But it still acknowledges the importance of second-person interaction, insofar as active inference can be seen as an effective way to reduce prediction error.

4E cognition's emphasis on embodiment is harder to reconcile with the main principles underlying PP. Most advocates of PP believe that predictive processing depends on a notion of represention. For example, Howhy (2014) has argued that although the concept of active inference entails a central role for the body in reducing prediction error, this role is only significant in the sense that the body is *represented* in the model, as a parameter useful for minimizing prediction error. He concludes that PP is therefore incompatible with radical, nonrepresentationalist forms of 4E cognition (e.g., Radcliffe 2007; Gallagher 2008; De Jaegher et al. 2010; Hutto and Myin 2013). Clark (2016) also has difficulty seeing how the story of PP can be told in entirely nonrepresentational terms, without invoking the concept of a hierarchical probabilistic generative model. However, he sees no harm in talking about representations in the context of 4E cognition. Clark proposes that the notion of representation that is required for PP is "action-oriented" rather than "action-neutral," and that the generative model functions—just as a proponent of 4E cognition might insist—to enable and maintain structural couplings that "serve our needs." Furthermore, he claims that PP does not "seem to imply the presence of inner models or content-bearing states of the kind imagined in traditional cognitive science. Instead, what are picked out seem to be physical processes defined over states that do not bear contents at all" (2016, p. 18).

Nevertheless, the claim that PP cannot do without representations may be hard to swallow for those philosophers who push a more radical 4E cognition agenda. As I see it, there are (at least) three ways in which they could respond to this claim. First of all, they could emphasize that, pace Howhy (2014), prediction error minimization essentially depends on the body itself—and not just on the model's representation of the body. It is the embodied interaction of an organism with its local environment that is required for prediction error minimization, and not the secluded brain alone. Second, they could acknowledge the representational nature of PP, but propose to see it merely as a steppingstone toward a better explanatory model. Accordingly, the R-word would serve as "nothing other than a place-holder for an explanation that needs to be cast in dynamical terms of an embodied, environmentally embedded, and enactive model" (Gallagher 2008, p. 365). Third, they could actually try to provide such a model, and articulate an account of PP that is not committed to representations (see, e.g., Bruineberg et al. 2016 for a nonrepresentational reading of PP).

My own sympathies tend toward Clark's "reconciliation view" of representations and 4E cognition. Clark argues that PP leaves open a number of deep and important questions concerning the nature of representation, and suggests that a "deep (but satisfyingly natural) engagement with evolutionary, embodied, and situated approaches" (2013, p. 200) is required to address these questions. I realize that this is probably not enough for more radical proponents of 4E cognition. But I hope that, at the very least, it will not stop them from considering potential synergies between PP and 4E cognition in the explanation of important social capacities such as false-belief understanding.

References

Apperly, I.A. and Butterfill, S.A. (2009). Do humans have two systems to track beliefs and belief-like states? *Psychological Review*, 116(4), 953–70.

Astington, J.W. and Bair, J.A. (2005). *Why language matters for theory of mind.* New York: Oxford University Press.

Astington, J.W. and Jenkins, J.M. (1999). A longitudinal study of the relationship between language and theory-of-mind development. *Developmental Psychology*, 35, 1311–1320.

Baillargeon, R., Scott, R.M., and He, Z. (2010). False-belief understanding in infants. *Trends in Cognitive Sciences*, 14, 110–18.

Bennett, J. (1978) Some remarks about concepts. *Behavioral and Brain Sciences*, 1, 557–60.

Bruineberg, J., Kiverstein, J., and Rietveld, E. (2016). The anticipating brain is not a scientist: the free-energy principle from an ecological-enactive perspective. *Synthese.* doi:10.1007/s11229-016-1239-1

Buckner, C. (2014). The semantic problem(s) with research on animal mind-reading. *Mind and Language*, 29(5), 566–89.

Buttelmann, D., Carpenter, M., and Tomasello, M. (2009). Eighteen-month-old infants show false-belief understanding in an active helping paradigm. *Cognition*, 112(2), 337–42.

Butterfill, S. and I. Apperly. (2013). How to construct a minimal theory of mind. *Mind & Language*, 28, 606–37.
Carlson, S.M., Mandell, D.J., and Williams, L. (2004). Executive function and theory of mind: stability and prediction from ages 2 to 3. *Developmental Psychology*, 40, 1105–22.
Carlson, S.M. and Moses, L.J. (2001). Individual differences in inhibitory control and children's theory of mind. *Child Development*, 72, 1032–53.
Carruthers, P. (2013). Mindreading in infancy. *Mind & Language*, 28(2), 141–72.
Carruthers, P. (2016). Two systems for mindreading? *Review of Philosophy and Psychology*, 7(1), 141–62.
Clark, A. (2013). Whatever next? Predictive brains, situated agents, and the future of cognitive science. *Behavioral and Brain Sciences*, 36(3), 181–204.
Clark, A. (2016). *Surfing uncertainty: prediction, action, and the embodied mind*. New York: Oxford University Press.
Clements, W.A. and Perner, J. (1994). Implicit understanding of belief. *Cognitive Development*, 9(4), 377–95.
Cole, K. and Mitchell, P. (2000). Siblings in the development of executive control and a theory of mind. *British Journal of Developmental Psychology*, 18, 279–95.
Currie, G. (2008). Some ways to understand people. *Philosophical Explorations*, 11, 211–8.
De Jaegher, H., Di Paolo, E., and Gallagher, S. (2010). Can social interaction constitute social cognition? *Trends in Cognitive Sciences*, 14(10), 441–7.
Dennett, D. (1978). Beliefs about beliefs. *Behavioral and Brain Sciences*, 1, 568–70.
Friston, K.J. (2010). The free-energy principle: a unified brain theory? *Nature Neuroscience*, 11, 127–38.
Friston, K.J. and Stephan, K.E. (2007). Free-energy and the brain. *Synthese*, 159, 417–58.
Gallagher, S. (2001). The practice of mind: theory, simulation or primary interaction? *Journal of Consciousness Studies*, 8, 83–108.
Gallagher, S. (2004). Understanding interpersonal problems in autism: interaction theory as an alternative to theory of mind. *Philosophy, Psychiatry, and Psychology*, 11(3), 199–217.
Gallagher, S. (2008). Are minimal representations still representations? *International Journal of Philosophical Studies*, 16(3), 351–69.
Gallagher, S. (2015) The problem with 3-year-olds. *Journal of Consciousness Studies*, 22(1–2), 187–90.
Gopnik, A. and Aslington, J.W. (1988). Children's understanding of representational change and its relation to the understanding of false-belief and the appearance-reality distinction. *Child Development*, 59(1), 26–37.
Hala, S., Hug, S., and Henderson, A. (2003). Executive functioning and false-belief understanding in preschool children: two tasks are harder than one. *Journal of Cognition and* Development, 4, 275–98.
Harman, G. (1978) Studying the chimpanzee's theory of mind. *Behavioral and Brain Sciences*, 1, 515–26.
Heyes, C. (2014). False-belief in infancy: a fresh look. *Developmental Science*, 17(5), 647–59. doi:10.1111/desc.12148
Hohwy, J. (2013). *The predictive mind*. Oxford: Oxford University Press.
Hohwy, J. (2014). The self-evidencing brain. *Noûs*. doi:10.1111/nous.12062
Hutto. D. (2008). *Folk psychological narratives: the sociocultural basis of understanding reasons*. Cambridge, MA: MIT Press.
Hutto, D. (2015a). REC: revolution effected by clarification. *Topoi*. doi:10.1007/s11245-0159358-8

Hutto, D. (2015b). Basic social cognition without mindreading: minding minds without attributing contents. *Synthese*. doi:10.1007/s11229-015-0831-0

Hutto, D. and Myin, E. (2013). *Radicalizing enactivism. basic minds without content*. Cambridge, MA: MIT Press.

Leslie, A.M. (1994). Pretending and believing: issues in the theory of ToMM. *Cognition*, 50(1–3), 211–38.

Leslie, A.M., Friedman, O., and German, T.P. (2004) Core mechanisms in "theory of mind." *Trends in Cognitive Sciences*, 8(12), 528–33.

Onishi, K.H. and Baillargeon, R. (2005). Do 15-month-old infants understand false beliefs? *Science*, 308(5719), 255–8.

Overgaard, S. and Michael, J. (2015). The interactive turn in social cognition research: a critique. *Philosophical Psychology*, 28(2), 160–83.

Perner, J. (2010). Who took the cog out of cognitive science? *International Perspectives on Psychological Science*, 1, 241–62.

Perner, J. and Lang, B. (1999). Development of theory of mind and executive control. *Trends in Cognitive Sciences*, 3, 337–44.

Perner, J., Lang, B., and Kloo, D. (2002). Theory of mind and self-control: more than a common problem of inhibition. *Child Development*, 73, 752–67.

Perner, J. and Ruffman, T. (2005). Infants' insight into the mind: how deep? *Science*, 308, 214–16.

Povinelli, D.J. and Vonk, J. (2004). We don't need a microscope to explore the chimpanzee's mind. *Mind & Language*, 19(1), 1–28.

Premack, D. and Woodruff, G. (1978). Does the chimpanzee have a theory of mind? *Behavioral and Brain Sciences*, 1(4), 515–26.

Ratcliffe, M. (2005). Folk psychology and the biological basis of intersubjectivity. In: A. O'Hear (ed.), *Philosophy, biology and life*. Royal Institute of Philosophy Supplement 56. Cambridge: Cambridge University Press, pp. 211–33.

Ratcliffe, M. (2007). *Rethinking commonsense psychology: a critique of folk psychology, theory of mind and simulation*. New York: Palgrave Macmillan.

Rubio-Fernández, P. and Geurts, B. (2013). How to pass the false-belief task before your fourth birthday. *Psychological Science*, 24, 27–33.

Ruffman, T. and Perner, J. (2005). Do infants really understand false-belief? Response to Leslie. *Trends in Cognitive Sciences*, 9, 462–3.

Scott, R.M. and Baillargeon, R. (2014). How fresh a look? A reply to Heyes. *Developmental Science*, 17, 660–4.

Scott, R.M., He, Z., Baillargeon, R., and Cummins, D. (2012). False-belief understanding in 2.5-year-olds: evidence from two novel verbal spontaneous-response tasks. *Developmental Science*, 15, 181–93.

Song, H. and Baillargeon, R. (2008). Infants' reasoning about others' false perceptions. *Developmental Psychology*, 44(6), 1789–95.

Southgate, V., Chevallier, C., and Csibra, G. (2010). Seventeen-month-olds appeal to false beliefs to interpret others' referential communication. *Developmental Science*, 13(6), 907–12.

Southgate, V., Senju, A., and Csibra, G. (2007). Action anticipation through attribution of false belief by two-year-olds. *Psychological Science*, 18(7), 587–92.

Spaulding, S. (2010). Embodied cognition and mindreading. *Mind and Language*, 25(1), 119–40.

Surian, L., Caldi, S., and Sperber, D. (2007). Attribution of beliefs by 13-month-old infants. *Psychological Science*, 18(7), 580–6.

Träuble, B., Marinovic, V., and Pauen, S. (2010). Early theory of mind competencies: do infants understand others' beliefs? *Infancy*, 15, 434–44.
Wellman, H.M., Cross, D., and Watson, J. (2001). Meta-analysis of theory-of-mind development: the truth about false-belief. *Child Development*, 72, 655–84.
Wimmer, H. and Perner, J. (1983). Beliefs about beliefs: representation and constraining function of wrong beliefs in young children's understanding of deception. *Cognition*, 13, 103–28.
Zahavi, D. (2004). The embodied self-awareness of the infant: a challenge to the theory-theory of mind? In: D. Zahavi, T. Grünbaum, and J. Parnas (eds.), *The structure and development of self-consciousness: interdisciplinary perspectives*. Amsterdam: John Benjamins, pp. 35–63.
Zahavi, D. (2014). *Self and other: exploring subjectivity, empathy, and shame*. Oxford: University Press.

CHAPTER 27

CRITICAL NOTE

How Revisionary are 4E Accounts of Social Cognition?

MITCHELL HERSCHBACH

Consider this section's chapters in relation to what I'll call the "traditional mindreading account" of human social cognition. Since the late 1970s, this research program has characterized human social cognition as essentially involving an ability to understand other agents' minds (i.e., to "mindread") so as to explain and predict their behavior. In this tradition, mindreading is described and studied from an individualist, internalist, cognitivist perspective. Mindreading is characterized as a set of representational, cognitive capacities realized in the individual's brain. As Newen (2018) notes, the theory-theory (TT) and the simulation theory (ST) offer competing (but ultimately compatible) accounts of the cognitive processes used to represent others' mental states. As de Bruin (2018) describes, experiments in this tradition often use behavioral tasks to try to assess whether an individual agent possesses particular mindreading capacities—in particular, when human children develop an understanding of different types of mental states, especially false beliefs. Neuroscientists measure brain activity while participants are engaged in these sorts of behavioral tasks to determine how our mindreading capacities are physically realized in the brain (e.g., Saxe et al., 2004).

As several of the chapters review, researchers in the early 2000s started criticizing this mindreading account along several lines (e.g., Gallagher 2001, 2008; Hutto 2004, 2008; Hutto and Ratcliffe 2007; Ratcliffe 2007; Schilbach et al. 2013):

1. *Clarifying the sociocognitive phenomena to be explained*: The traditional mindreading account was criticized for too narrowly focusing on "third-person, detached observation" of other people's behavior, and not adequately characterizing the sociocognitive processes driving "second-person," unreflective, social interactions.

2. *Expanding the types of mindreading processes beyond TT and ST*: Direct perception was posited as an alternative, non-inferential psychological process for appreciating others' mental states, particularly for second-person interactions.
3. *Going beyond mindreading*: Some argued that non-mindreading sociocognitive processes shouldn't be overshadowed by mindreading; others more strongly argued that mindreading is rarely used in everyday life, where "second-person" social interactions guided by non-mindreading processes predominate.
4. *Emphasizing conscious experience in accounts of social cognition*: Traditional mindreading accounts were criticized for contradicting the facts about our conscious experiences of other people. In particular, TT and ST were criticized for (supposedly) claiming that much of our social lives are spent consciously reflecting on the mental causes of people's behavior. The direct perception view claimed to more accurately characterize our experience of others' mental states during second-person interactions, and other sociocognitive processes were thought to better explain the times when we interact without thinking about others' mental states at all.

These criticisms have often been characterized within a 4E framework that challenged the cognitivism of the traditional mindreading account. For example, the non-mindreading processes thought to drive social interaction are often referred to as instances of embodied cognition (Gallagher 2001). And De Jaegher (2018) pitches her entire account of social cognition within the enactivist framework.

But there is also plenty of recent research on social cognition that accepts these criticisms as *refinements* of the traditional mindreading approach, without explicitly adopting much or any of the 4E cognition approach. For example, Apperly and Butterfill's (Apperly 2011; Butterfill and Apperly 2013; Low et al. 2016) work on mindreading attempts to characterize the sociocognitive processes behind both fast, dynamic social interactions and slow, reflective thought about people's minds and behavior, all within a traditional cognitivist framework. Apperly adopts a dual process/systems approach, which at least in part is driven by the phenomenological contrast between these different kinds of social experiences. Leon de Bruin (2018) reviews how this sort of approach has revolutionized the kinds of behavioral tasks used to study mindreading in developmental psychology. At most, this work agrees with the embedded/situated cognition perspective that agents' interactions with their social environment are essential to accurately characterizing the internal cognitive resources used to guide that interaction.

In light of this, my commentary will focus on two core issues about the relevance of the 4E cognition framework for social cognition. One is the extent to which a 4E perspective informs and is supported by "pluralistic" accounts of social understanding that emphasize non-mindreading processes and go beyond TT's and ST's account of mindreading. I'll focus on this issue mostly with regard to the chapters by Gallese and Sinigaglia, Newen, and de Bruin. A second issue is whether there is anything theoretically special about *social interaction* itself—for example, whether there is a form of social understanding unique to social interaction. This will be my focus when commenting on Reddy's and De Jaegher's chapters.

Gallese and Sinigaglia

Gallese and Sinigaglia offer a rather modest challenge to traditional mindreading research. They do not question its basic cognitivist commitments, since their "embodied" approach only posits cognitive processes involving bodily formatted neural representations. What they implicitly challenge is the importance of mindreading to human social cognition, by articulating an understanding of others' action goals that does not involve mental state understanding. This goal understanding is driven by "embodied simulation" (ES), that is, "by reusing the processes and representations that would occur if the observer herself were planning to execute those actions" (p. 422). Gallese and Sinigaglia helpfully review neuroscientific and developmental evidence in support of their view.

Gallese and Sinigaglia treat ES-based action understanding as "basic" in being distinct from and perhaps a developmental precursor to mindreading. ES-driven understanding of others' goals isn't considered a form of mindreading since "goals" are conceived of non-mentalistically as the outcomes or endpoints of actions (a point defended further elsewhere, e.g., Sinigaglia and Butterfill 2015). They also suggest that ES might be a developmental precursor to mindreading, and in adults both feed into and be modulated by mindreading processes (p. 423). Both are interesting proposals, but Gallese and Sinigaglia don't focus on filling them out or defending them in this chapter.

This leaves one wondering what their exact commitments are regarding the relative *importance* of basic goal understanding and mindreading to human social cognition. Some 4E cognition advocates (e.g., Gallagher and Hutto 2008) contend that non-mentalistic forms of social cognition like ES are the *norm* and that mindreading is a relatively infrequent part of human social cognition. Gallese and Sinigaglia make no such claims here. They briefly discuss how ES could be used for social interactions, including joint actions (p. 424). But they don't make any claims about ES being *exclusive* to social interaction. ES appears to be just as useful for passive observation of the actions of people with whom one is not interacting. And they make no claims about how frequently ES-driven goal understanding is used in everyday life. They admit that mirror-driven ES need not be considered the *only* way to ascribe goals to others' actions, and do not question the idea that genuine mindreading provides a fuller understanding of the *reasons* behind their actions (pp. 423–5).

In conclusion, Gallese and Sinigaglia's account may well be plugged into a more radical 4E account of social cognition, de-emphasizing the role of mindreading. But they do not themselves highlight these stronger theses. As it is formulated here, their account appears largely compatible with the way the traditional mindreading account has been supplemented in recent years to acknowledge non-mindreading processes.

Newen

Newen more explicitly defends a pluralistic account of social cognition, integrating insights from traditional cognitivist and 4E accounts of social cognition. Newen starts with the traditional mindreading account, accepting that we use explicit, conscious, theory-based inferences and mental simulations to understand other people's mental states. But in line with advocates of 4E cognition, Newen believes TT and ST poorly capture intuitive, *implicit* forms of social understanding that are often (but not exclusively) used for social interaction; these contexts often use different *epistemic strategies* for social cognition, including the direct perception of others' mental states. But Newen's account also includes representations of conventional social situations or scripts, which he calls "situation models," and De Jaegher's (2018) form of participatory sense-making that does not involve mindreading at all—which may count as an example of extended social cognition. A novel feature of Newen's account is its emphasis on the use of specific, idiosyncratic knowledge of individual persons and groups with which we are familiar from a history of interactions.

Though I endorse Newen's commitment to pluralism, I have a few critical comments about how these pieces fit together. First, Newen moves rather quickly in characterizing and distinguishing epistemic strategies. For example, Newen argues that TT and ST poorly characterize intuitive, implicit social cognition, and posits direct perception as an alternative epistemic strategy supplying such intuitive understanding of other people. Many, however, will resist this, and treat TT and/or ST as accounts of not just explicit, reflective mindreading, but also the direct perception of others' mental states (Carruthers 2015; Herschbach 2008; Lavelle 2012). Thus, more must be said to distinguish direct perception as a unique epistemic strategy distinct from the traditional strategies described by TT and ST.

Further, Newen is not especially clear about the kinds of properties of other people that are represented and used by these different epistemic strategies. For example, are all of the epistemic strategies capable of representing the same information about other people? Or are the various epistemic strategies sensitive to different sorts of properties about people, and thus useful for different purposes? For example, Newen mostly talks about the direct perception of emotions. Can all mental states be directly perceived, or only some? If only some, what explains that limitation in this epistemic strategy?

This leads to a more fundamental question about the role of mental state understanding in Newen's account of "social understanding." Newen's acceptance of many of the non-mindreading processes advocated by 4E accounts suggests Newen might similarly downplay the importance of mindreading. But when defining his person model theory, he describes its first main claim as concerning the epistemic strategies "humans use to access *the mental states of others*" (p. 478, emphasis added). The other main part of his account concerns the background information about others that is accessed via these epistemic strategies. But Newen is vague about what information we possess about other people in our person models and situation models. It seems like "person models" always contain information about others' mental states, but Newen is most clear about this with regard to explicit "person images" and not implicit "person schemas" (p. 481). Are Newen's "situation models" (representations of routine, conventional behaviors in particular situations) devoid of mental state representations? Newen isn't exactly clear. Without clear answers to these questions, it is hard to say how Newen's pluralistic account stands on the priority of mental state understanding to human social cognition, and thus how strongly he endorses the 4E critics of mindreading.

I do wholeheartedly agree, however, with Newen's conclusion that the debate about 4E cognition in the sociocognitive domain (and probably all domains) should focus on narrowly defined types of cases. Gallese and Sinigaglia have provided us a nice example of one type of embodied cognition. Whether social cognition ever counts as extended is an issue I cover later in my commentary on De Jaegher's chapter.

de Bruin

Like Newen, de Bruin explicitly addresses the debate between traditional and 4E accounts of human social cognition, focusing on false-belief understanding. He reviews how traditional "elicited-response" false-belief tasks were criticized by researchers in the 4E camp in the ways mentioned earlier, but how newer "spontaneous-response" false-belief tasks sidestep these criticisms by using nonverbal and/or interactive behavioral measures. De Bruin surveys the various explanations that have been given of successful performance on spontaneous-response false-belief tasks, from full-blown mindreading accounts to completely non-mentalistic accounts. He helpfully summarizes the difficulties in evaluating these accounts in terms of empirical adequacy and theoretical virtues like parsimony. He offers a novel perspective to the debate by analyzing false-belief tasks from the perspective of a popular general theoretical framework in cognitive science: the predictive processing paradigm (PPP). Given the popularity of PPP, it is interesting to see how it could inform accounts of social cognition. Unfortunately, it is difficult to parse which parts of de Bruin's analysis depend on PPP and which do not. I worry that PPP does not offer any new insight into how children behave on false-belief tasks.

According to PPP, the brain consists of hierarchical, generative models of the world that try to predict sensory input, and uses two strategies to minimize prediction error: perceptual inference and active inference. De Bruin contends that the design of false-belief tasks constrains which error-minimization strategies can be used, affecting their difficulty. This is central to de Bruin's solution to the "developmental paradox" of why young children pass spontaneous-response false-belief tasks before elicited-response tasks.

De Bruin contends that one type of spontaneous-response task, active-helping tasks, allows children to reduce prediction error through active inference, specifically, interaction with the target agent. Elicited-response false-belief tasks, however, require the child to respond to the experimenter's questions about the target agent's behavior or mental states, and don't allow the child to interact with the agent. Thus, these tasks do not permit active inference as a strategy for reducing prediction error; they instead require the child to use perceptual inference to update their representation of the agent's beliefs, e.g., about the location of the preferred object (pp. 506–7). This can only be done accurately if a child recognizes that an agent's belief about an object's location does not change if the agent does not witness the object's moving location. Younger children lack this recognition, so they'll succeed on active-helping tasks while failing elicited-response tasks.

I have a few concerns about this interpretation of false-belief tasks. In his discussion of elicited-response tasks, de Bruin claims that young children (a) whose perceptual inference does not display a genuine understanding of the conditions that generate false beliefs, and (b) who can't engage in active inference, will engage in a problematic form of perceptual inference: they will update their model of the agent's belief to take into account its new location. That is, they will mistakenly represent the agent has having a *true* belief about the object's current location. But, as de Bruin reviews, this is *not* what we see in one- to two-year-olds' responses to violation-of-expectation and anticipatory-looking tasks: their looking behavior shows they expect the agent to look for the desired object where they *falsely* believe it to be located. Why does de Bruin think young children will, on elicited-response tasks, engage in inaccurate perceptual inference about the agent's beliefs, which leads to unsuccessful task performance, yet children of the same age will succeed on spontaneous-response tasks without forming representations of the agent's beliefs?

Further, while de Bruin's solution to the developmental paradox indeed appeals to PPP's contrast between perceptual inference and active inference, de Bruin's claim that belief representations are not needed for spontaneous-response tasks seems independent of the commitments of PPP. PPP's different strategies for minimizing prediction error can't be the basis for this claim about spontaneous-response tasks. That is because, according to de Bruin, only active-helping tasks even address how children deal with prediction errors generated during such tasks. So what leads to this claim? In PPP terms, it would be a claim about the kind of generative model needed to generate the behavioral responses measured in these tasks. But that's just PPP's way of talking about the psychological processes driving successful performance on these tasks. In the fourth section de Bruin surveys the various theoretical accounts of these psychological processes, with the alternatives to full-blown mentalistic accounts ranging from "minimal mindreading" to non-mentalistic accounts (e.g., behavior-reading accounts and low-level novelty accounts) that completely deny that success on these tasks requires representing others' mental states. Unfortunately, it isn't clear how de Bruin's account relates to these alternatives, or why we should prefer de Bruin's to these others. Most importantly, it is not at all obvious how PPP provides any reason for preferring de Bruin's account of the generative models used for different false-belief tasks.

Thus, it seems like de Bruin's conclusion about which false-belief tasks do or do not require participants' generative models to accurately model agents' false beliefs doesn't really hinge on any details of the PPP approach; its contribution to the larger debate is rather unclear. But let's just say it did help. What are the implications for 4E accounts of social cognition? As de Bruin admits, not all that many. PPP's emphasis on internal, generative models makes it tough to reconcile PPP with more radical forms of 4E cognition; PPP is most clearly compatible with embedded cognition and modest embodied cognition theses. And since the core of de Bruin's account really seems to hinge on the nature of these generative models, it's not clear how much these 4E cognition theses are *supported* by this analysis of false-belief tasks, rather than simply being *compatible* with them. While it's not a loss, this isn't much of a win for the 4E approach.

Reddy

Reddy's chapter focuses on the role of social interaction in the development of human social cognition. She argues that children's "second-person engagements" with adults are essential to the typical development of human social cognition. Reddy aims to first clarify her theoretical framework for social cognition, then to provide empirical evidence that second-person engagements are developmentally crucial to children's developing understanding of other agent's attention and intentions. I will mostly comment on Reddy's theoretical framework, before briefly addressing her developmental thesis.

Reddy aims for conceptual clarification of the role of the "social" in social cognition, focusing on the concepts of "engagement" and "second-person relations." Reddy says "engagement" is her preferred alternative for "interaction" because it better captures certain "emotional qualities" she says are essential to these sociocognitive phenomena (p. 434). What kind of emotional quality? Reddy initially describes it as an "acknowledgment of, or response to, the person-ness of the other" (p. 434). She doesn't offer a clear definition, but her examples suggest the following: *engagement is a social interaction involving an emotionally charged awareness of (and response to) a person's mental states* (since possessing mental states seems central to being a person and not a mere object).

But there are several problems with this interpretation. First, Reddy says interactions with *objects* can count as engagements: "relishing the warmth of a hot shower" and infants' attraction to grasping objects (p. 435) are given as examples of engagement. Second, Reddy does not even require engagement to involve *interaction*. She writes that "emotional involvement without action—as in intense feelings while watching a movie—is indeed engagement" (p. 436). Engagement can't be defined as a certain kind of social interaction *plus* some special stance toward those other person(s) if *interaction* with a *person* is not a necessary condition for engagement. This makes it seem like Reddy's concept of engagement just picks out our *affective* response to the world. But this would conflict with Reddy's claims about engagement described earlier. Reddy admits in her conclusion that "engagement is a vague and multifaceted term" (p. 447). But the reader is left with far too many questions and not enough clarity from Reddy's discussion.

A similar lack of clarify infects Reddy's discussion of the concepts of "second-person" and "third-person" relations between persons. She emphasizes that these concepts should be thought of as two different sociocognitive *stances* one can take toward other people, which come with distinct *experiences* (affective and otherwise). The second-person stance "involves the experience of being addressed by another, of being seen as a You by another person, and of the mutuality that is generated in seeing the other as a You in turn" while the third-person stance "involves a more detached, observational, stance in which one sees the other as a He or She rather than as a You" (p. 437). Reddy claims this distinction between stances can cross-cut the "structural"/causal distinction between mutual interaction and non-interactive observation: one can take a third-person stance toward people with whom you are interacting (e.g., by thinking about them "in detached analytic terms"), and can take a second-person stance toward people with whom you are not interacting (e.g., watching a movie "but still see[ing] the other(s) as if they were speaking to oneself and feel[ing] involved with responsive sympathy or hate," p. 437). This conflicts, however, with Reddy's references to *mutuality* and *reciprocity* when characterizing these stances. For example, Reddy writes that that these stances are distinguished by the "openness or closed-ness with which one faces (and is faced by) the other" (p. 437). But how can the other person recognize (or fail to recognize) you as a person *if they're not interacting with you*? How can mutuality be essential to the second-person stance if that stance can be taken in purely observational contexts? Does that mean actual responses by another person are not necessary to instantiate a second-person relation, and that *imagining* such responses is sufficient?

Overall, Reddy's notions of "second-person stance" and "engagement" appear to largely overlap. It is unclear if there is really much difference between the two, and why Reddy needs both in her theoretical framework. The only unique aspect in Reddy's explication of engagement seems to be the affective element—but it's not obvious that this alone justifies treating "engagement" and "second-person stance" as separate concepts for characterizing sociocognitive phenomena. Further, both concepts fail to adequately distinguish between sociocognitive stances of the individual and the nature of the causal relations between persons. As Schönherr and Westra (in press) have recently argued, research on social interaction has largely failed to identify a unique contribution of actual social interaction to social cognition, since in many studies the mere belief that one is being interacted with produces the same cognitive effects as actually being engaged in real social interaction.

Despite these conceptual concerns, Reddy roughly homes in on a type of sociocognitive phenomenon that seems to be of genuine theoretical interest, at least for the development of our understanding of mental states. Reddy argues that children's awareness and understanding of others' mental states develop in the context of "second-person engagements," where child and others mutually interact. Reddy contends that infants follow a distinctive trajectory in their understanding of the *content* of the other's attentional states and intentions: infants are first aware of and emotionally invested in the other's attention to themselves, then their actions, then to distal objects. In other words, one's awareness of others' minds begins with the way the other is mentally engaged with *oneself*, then expands to recognize the way others can be mentally directed toward other things, including other people. As Reddy acknowledges, this coheres with Henrike Moll and Michael Tomasello's thesis (e.g., Moll et al. 2007) that children's understanding of other persons' minds is largely or entirely restricted to people with whom they're reciprocally engaged, and that understanding the mental perspectives of people with whom they're not engaged only develops later in childhood.

If this developmental trajectory holds true, this would be an important defense of the thesis that social interaction plays an important role in at least the development of human social cognition, even if it doesn't establish the further claims that adults possess a form of social cognition unique to social interaction, or that social interaction itself can metaphysically constitute social cognition. There are important theoretical and empirical questions, however, about this phenomenon. For example, in what sense is early social cognition "mutualistic"? Can this sort of mutuality be captured without positing higher-order mental state representations (e.g., "I see that you see what I see")? Or does it establish an earlier existence of such complex representations? Further, can existing accounts of mindreading help to explain this developmental trajectory? Perhaps minds are first understood when engaged with oneself because they're easier to *simulate*: the other will often be attending to the same things as oneself, so simulating the other's mental states wouldn't require much adjustment from one's *own* mental perspective. These sorts of issues would need to be addressed in order to answer how much this developmental phenomenon requires modifying the traditional mindreading account.

De Jaegher

De Jaegher's chapter is devoted to articulating a comprehensive theoretical and methodological framework for studying intersubjectivity, based on the enactivist approach to 4E cognition. De Jaegher believes there are five necessary elements for such a comprehensive framework (pp. 454–61). I will only focus on the first two: that we should account for both subjectivity and interaction; and that we should adopt a multilevel conceptual and methodological framework, from the physiological parts of agents all the way up to the social groups made up of these agents. I wholeheartedly agree with both these general points, but I want to challenge a few more specific claims De Jaegher makes when explicating her perspective on them.

One concerns De Jaegher's endorsement of the extended-cognition thesis that social interaction can constitute social cognition. Here a "constitutive element" is thought of as a compositional part of the cognitive phenomenon at issue. As I've argued elsewhere (Herschbach 2012), De Jaegher and colleagues have been rather confusing about what phenomena they're trying to characterize the constitutive parts of, and seem at times to conflate phenomena at different compositional levels (e.g., social vs. individual). That confusion continues in this chapter. For example, De Jaegher mentions cases of "interpersonal synchrony," where "neurological and physiological processes" "coordinate across individuals" (pp. 457–8). When discussing constitutive explanations of these cases, she writes:

> Aspects of this self-organization or coordination may be explained by individual factors, where the interaction is considered no more than a context for the synchronizing effect. For instance, the interaction may be considered an input for a contingency perception system (Gergely and Watson 1999). But social interaction processes may themselves enable or even constitute the interpersonal synchrony, and, therefore, part of the social understanding that takes place (see, respectively, Di Paolo, Rohde, and Iizuka 2008; and De Jaegher et al. 2010). (pp. 457–8)

"Interpersonal synchrony" is a social-level phenomenon consisting of two or more individuals. So how could this phenomenon ever be explained by only factors internal to *one* individual? The activity of a neural mechanism like a contingency perception system could never alone explain the synchrony exhibited between individuals. It is uncontroversial to say the social interaction consisting of both individuals *constitutes* this social-level phenomenon of interpersonal synchrony. But should we call interpersonal synchrony a kind of "social understanding"? Or is "social understanding" referring to the individual-level cognition of each individual agent? Vagueness about what particular phenomena we're calling "social cognition" impedes the evaluation of De Jaegher's thesis that social interaction can constitute social cognition.

De Jaegher here and elsewhere (De Jaegher et al. 2010; De Jaegher et al. 2016) has identified perceptual crossing experiments as a primary example of social cognition's being partly constituted by social interaction. In these experiments, the collective dynamics of the interaction between the two agents simplify the sensory discrimination task that each agent needs to solve. I've previously argued (Herschbach 2012; see also Newen 2018) that these experiments can be plausibly analyzed as cases of *embedded* rather than *extended* cognition. Appealing to the mechanistic framework from philosophy of science, I argued that we can decompose the social interaction into the two individual agents, and constitutively explain the search behavior of each agent in terms of their internal neural mechanisms—while noting that the complexity of the cognitive task faced by each agent is simplified by the nature of the social environment in which they're embedded. Certainly an individual agent receives sensory inputs from its social environment, and performs actions that affect its environment (e.g., the behavior of the other agent). But the causal interactions between the individual and their environment seem straightforward enough to treat the individual as a unit whose behavior is primarily driven by its internal parts. If we're trying to understand why an individual receives those particular sensory inputs, we'll need to understand the dynamics of the social interaction between individuals. But it seems a confusion of levels to say the social interaction partly constitutes the internal processes driving the behavior of the individual agent, or to just relabel the entire social interaction as the sociocognitive phenomenon to be explained. On my analysis, we would in this case label as "social cognition" the internal processes that drive the individual's search behavior. Is that what enactivists want to call "social cognition"? Or do they want to call the social-level phenomenon "social cognition"? From De Jaegher's chapter, I can't say I'm any clearer about how enactivists want to answer these questions. Even if we remain open-minded about how to define cognition, I think all parties would agree that we should distinguish these individual- and social-level phenomena and what explains them.

Whether this embedded cognition interpretation is preferable to an extended cognition interpretation really comes down to how we identify constitutive vs. non-constitutive factors in cognitive explanations. This is something that De Jaegher does not address in any detail here, but which seems essential to evaluating her enactivist framework for social cognition. Work within the mechanistic approach in philosophy of science has focused on how to characterize the complex dynamical interactions found in such phenomena (Bechtel 2016; Bechtel and Abrahamsen 2010; Kaplan and Bechtel 2011). One account of constitutive parts appeals to "mutual manipulability" (Craver 2007), which Kaplan (2012) has explicitly applied to demarcate the constitutive elements of cognitive systems (see Kirchhoff 2017 for a recent development of the view). I believe De Jaegher's enactivist framework should further engage with these issues to help evaluate under what conditions, if any, social cognition would be constituted by parts of the individual's social environment. De Jaegher et al.'s (2016) discussion of "interaction-dominant systems" is the beginning of this sort of engagement.

I also want to challenge De Jaegher's view that enactivism's focus on social interaction and subjectivity means human social cognition is "not in the first place a question of figuring out each other's mental states" (p. 454). I don't see how accepting the importance of non-mindreading processes in a pluralistic account of social cognition *entails* rejecting the priority of mindreading to human social cognition (see Di Paolo and De Jaegher 2012). It has certainly been a fault of traditional mindreading research to give inadequate attention to non-mindreading processes, social interaction, and social experience. But recent work has certainly tried to address these faults in the ways as I've described earlier: by including non-mindreading processes, positing "minimal" mindreading systems for dynamic social interaction, and acknowledging the phenomenon of direct perception of mental states. Perhaps these modifications aren't enough and mindreading really is less important than the mindreading literature contends. My point here is that the importance of mindreading to the development of social cognition and how often it is deployed by adults are empirical issues that don't seem settled by identifying non-mindreading forms of social cognition. In my opinion, De Jaegher and other critics of the traditional mindreading account move too quickly in concluding that mindreading should be displaced from its traditional place at the core of human social cognition.

Conclusions for 4E Cognition

I believe these five chapters most firmly defend the importance of embedded cognition to investigations of human social cognition. Appreciating the ways in which social agents interact is essential to understanding the sociocognitive processes driving agents' behavior in those social contexts. The modest form of embodiment endorsed by Gallese and Sinigaglia seems well grounded empirically, but its importance to human social cognition is still an open question. Only De Jaegher pushed the most radical 4E extended cognition thesis, from the enactive perspective. But, as I argued, this position is quite controversial.

This, on the whole, constitutes a fairly modest challenge to the traditional mindreading approach, one that can plausibly be seen as enriching that perspective without rejecting its core claims about the importance of the "representation-hungry," internal capacity for mindreading.

References

Apperly, I.A. (2011). *Mindreaders: the cognitive basis of "theory of mind."* Hove, UK: Psychology Press.

Bechtel, W. (2016). Mechanists must be holists too! Perspectives from circadian biology. *Journal of the History of Biology*, 49(4), 705–31.

Bechtel, W. and Abrahamsen, A. (2010). Dynamic mechanistic explanation: computational modeling of circadian rhythms as an exemplar for cognitive science. *Studies in History and Philosophy of Science*, 41(3), 321–33.

Butterfill, S.A. and Apperly, I.A. (2013). How to construct a minimal theory of mind. *Mind & Language*, 28(5), 606–37.

Carruthers, P. (2015). Perceiving mental states. *Consciousness and Cognition*, 36, 498–507.

Craver, C.F. (2007). *Explaining the brain*. Oxford: Oxford University Press.

de Bruin, L. (2018). False-belief understanding, 4E cognition, and the predictive processing paradigm. In: this volume, pp. 493–512.

De Jaegher, H. (2018). The intersubjective turn. In: this volume, pp. 453–68.

De Jaegher, H., Di Paolo, E., and Adolphs, R. (2016). What does the interactive brain hypothesis mean for social neuroscience? A dialogue. *Philosophical transactions of the Royal Society of London. Series B, Biological sciences*, 371(1693), 20150379. doi:10.1098/rstb.2015.0379

De Jaegher, H., Di Paolo, E., and Gallagher, S. (2010). Can social interaction constitute social cognition? *Trends in Cognitive Sciences*, 14(10), 441–7.

Di Paolo, E.A. and De Jaegher, H. (2012). The interactive brain hypothesis. *Frontiers in Human Neuroscience*, 6(163), 1–16.

Di Paolo, E. A., Rohde, M., and Iizuka, H. (2008). Sensitivity to social contingency or stability of interaction? Modelling the dynamics of perceptual crossing. *New Ideas in Psychology*, 26(2), 278–94.

Gallagher, S. (2001). The practice of mind: theory, simulation or primary interaction? *Journal of Consciousness Studies*, 8(5–7), 83–108.

Gallagher, S. and Hutto, D.D. (2008). Understanding others through primary interaction and narrative practice. In: J. Zlatev et al. (eds.), *The shared mind: perspectives on intersubjectivity*. Amsterdam: John Benjamins, pp. 17–38.

Gallese, V. and Sinigaglia, C. (2018). Embodied resonance. In: this volume, pp. 417–32.

Gergely, G. and Watson, J. S. (1999). Early socio-emotional development: contingency perception and the social-biofeedback model. In: P. Rochat (ed.), *Early social cognition*. Hillsdale, NJ: Erlbaum, pp. 101–137.

Herschbach, M. (2008). Folk psychological and phenomenological accounts of social perception. *Philosophical Explorations*, 11(3), 223–35.

Herschbach, M. (2012). On the role of social interaction in social cognition: a mechanistic alternative to enactivism. *Phenomenology and the Cognitive Sciences*, 11(4), 467–86.

Hutto, D.D. (2004). The limits of spectatorial folk psychology. *Mind & Language*, 19(5), 548–73.

Hutto, D.D. (2008). *Folk psychological narratives*. Cambridge, MA: MIT Press.

Hutto, D.D. and Ratcliffe, M.M. (2007). *Folk psychology re-assessed*. Dordrecht: Springer.

Kaplan, D.M. (2012). How to demarcate the boundaries of cognition. *Biology and Philosophy*, 27(4), 545–70.

Kaplan, D.M. and Bechtel, W. (2011). Dynamical models: an alternative or complement to mechanistic explanations? *Topics in Cognitive Science*, 3(2), 438–44.

Kirchhoff, M.D. (2017). From mutual manipulation to cognitive extension: challenges and implications. *Phenomenology and the Cognitive Sciences*, 16(5), 863–78.

Lavelle, J.S. (2012). Theory-theory and the direct perception of mental states. *Review of Philosophy and Psychology*, 3(2), 213–30.

Low, J., Apperly, I.A., and Rakoczy, H. (2016). Cognitive architecture of belief reasoning in children and adults: a two-systems account primer. *Child Development Perspectives*, 10(3), 184–9.

Moll, H., Carpenter, M., and Tomasello, M. (2007). Fourteen-month-olds know what others experience only in joint engagement. *Developmental Science*, 10(6), 826–35.

Newen, A. (2018). The person model theory and the question of situatedness of social understanding. In: this volume, pp. 469–92.

Ratcliffe, M. (2007). *Rethinking commonsense psychology: a critique of folk psychology, theory of mind and simulation*. Basingstoke, UK: Palgrave Macmillan.

Reddy, V. (2018). Why engagement? A second-person take on social cognition. In: this volume, pp. 433–52.

Saxe, R., Carey, S., and Kanwisher, N. (2004). Understanding other minds: linking developmental psychology and functional neuroimaging. *Annual Review of Psychology*, 55(1), 87–124.

Schilbach, L., Timmermans, B., Reddy, V., Costall, A., Bente, G., Schlicht, T. et al. (2013). Toward a second-person neuroscience. *Behavioral and Brain Sciences*, 36(4), 393–414.

Schönherr, J. and Westra, E. (in press). Beyond "interaction": how to understand social effects on social cognition. *The British Journal for the Philosophy of Science*.

Sinigaglia, C. and Butterfill, S. (2015). Motor representation in goal ascription. In: M.H. Fischer. and Y. Coello (eds.), *Conceptual and interactive embodiment* (vol. 2). New York: Routledge, pp. 149–64.

PART VII

SITUATED AFFECTIVITY

CHAPTER 28

EMBODIMENT OF EMOTION AND ITS SITUATED NATURE

EVAN W. CARR, ANNE KEVER, AND PIOTR WINKIELMAN

INTRODUCTION

EMBODIMENT theories have now become a major conceptual framework for understanding the mind. The main idea behind these theories is that higher-level processing is at least partly based in one's perceptual, motor, and somatosensory systems—leading such frameworks to often be called *grounded cognition* theories (Barsalou 2008; Wilson 2002). The origin of modern embodiment theories can be traced back to classic research on mental imagery (which showed the involvement of perceptual systems in conceptual operations; Kosslyn 1994), cognitive linguistics (which highlighted the metaphorical grounding of many concepts; Gibbs 2003; Lakoff 1993), and work on motor affordances (which demonstrated tight links between perception and action; Tucker and Ellis 1998). However, one could argue that the recent exponential growth in embodiment research comes from its extension to social and emotional life (see Figure 1 in Mahon 2015). As such, embodiment theories have been applied to understanding the processing of facial expressions, emotional words and concepts, emotion regulation, social relations, and even moral concepts (Barrett et al. 2015; Meier et al. 2012; Niedenthal et al. 2005; Schubert and Semin 2009). In this chapter, we describe recent evidence for this account—particularly in the domain of emotion processing across social contexts. We also discuss potential limits of the framework and challenges for future research.

To do this, we start our chapter by briefly contrasting embodiment theories with other more "traditional" frameworks that emphasize an amodal and propositional nature of mental representations. We then review evidence for embodiment's role in emotional processing—focusing on responses to facial expressions and emotional language. Next, we move on to describing embodiment in an interpersonal domain with action *mirroring*—or the replication of others' actions, behaviors, gestures, and expressions. Throughout the chapter, we argue that in order to reach a comprehensive account of embodiment, one must consider its situated, flexible, and dynamic nature—especially within the social environment. We conclude that while embodiment theories likely cannot satisfactorily account for all aspects of cognition and emotion, they have led to fundamental empirical discoveries and theoretical developments. As such, embodiment theories offer substantive contributions to illuminating how the mind works.

"Traditional" Frameworks: Amodal Processing

The human conceptual system performs a vast array of cognitive operations—from simple stimulus recognition to complex decision-making. Major models of the conceptual system within cognitive psychology have traditionally been focused on associative networks (e.g., Anderson 1983; Collins and Loftus 1975). For instance, imagine that you perceive an individual in your social circle (such as a friend or family member). On this amodal view, inputs are initially encoded in the brain's modal systems, such as the visual and auditory regions (which are likely linked to affective systems). This information is then extracted into an abstract, language-like symbol (a *proposition*) and stored as a *node*. Within an associative network, in this example, the node might be something like the word "BROTHER." This symbol or node is stored with connections to other representative features like "SMART," "FUN," and "STRONG." Critically, these features were initially encoded in the brain's modal systems, but they now represent abstract conceptual nodes. Consequently, when thinking about one's brother at a later time, information is extracted in a language-like form (i.e., a concept label with associated features), which is then used for inference and responding.

Thus, nodes generally represent units of information in associative network models. These nodes are further interconnected by *associative links* (although this term can be slightly misleading, since links can have different structures, such as having certain properties, inheriting certain features, etc.). When a node is activated, linked nodes are also activated as a function of the associative strength between them, via *spreading activation* (i.e., the greater the interconnectedness among the nodes, the greater the probability that it will be activated by its neighbors). As a result, ideas that are already stored in memory impact online processing in accordance with their activation level—and, overall, the full set of nodes in the associative network constitutes a person's conceptual system. In theory, this conceptual system then serves as the basis for processes such as inference, categorization, and other higher-order cognitive operations.

Importantly, though, note that amodal and associative network theories require endorsing two key assumptions. First, on this view, advanced mental operations are basically symbol manipulations, which are defined by their functional roles. As such, they are not essentially constrained by the specific physical states and structure of the body and brain. That is, amodal theories subscribe to the principles of *functionalism*, which views the mind as software that operates on the brain's hardware, but could principally be ported to any other hardware that can realize the same functional roles (Block 1995). Second, amodal views assume that advanced cognition consists of operations on abstract symbols—"thoughts" that operate as amodal, propositional states, which are fully detached from the original analog form encoded in the perceptual system (e.g., vision or audition; Fodor 1975; Newell and Simon 1972; Pylyshyn 1984). Surely these symbols can be still be associatively connected to different perceptual and motor modalities, but these embodied components do not causally contribute to any "pure" cognitive activities such as categorization, comprehension, or inference (Mahon 2015). As we explain next, research in psychology, cognitive science, and neuroscience casts doubt on both assumptions (Barsalou 2008).

Embodiment Theories

On the most general level, embodied cognition theories propose that information processing is shaped by the brain, body, and its interactions with the external physical world (Barsalou 1999, 2008; Clark 1999; Wilson 2002). This claim by itself may be relatively uncontroversial, but specific branches of embodiment theories go much further, in different directions (see Wilson 2002 for a broad overview of different meanings for embodiment). Thus, some embodied theories focus on a much stronger claim that the body (or even the world, according to "extended cognition" theses) is literally a part of the physical realization of cognition. On this view, cognition is seen as emerging from the interactions of the brain with the body and the world (Clark 1999). Emotion can be viewed similarly—a thesis that is elaborated and extended by Colombetti (2014, 2018).

In contrast, our perspective is more modest and informed by Barsalou's perceptual symbol system view (Barsalou 1999). On this view, the generation of a thought involves partial reproduction or *simulation* of experiential and motor states that occur when the perceiver has actually encountered the object. For example, imagine you are trying to describe one of your favorite colleagues to a friend—you might recall and/or actively generate traces of direct perceptual experiences with that colleague (e.g., mimicking his or her movements and expressions, reconstructing the sound of his or her voice, etc.). In other words, embodied conceptual processing involves partial reactivations of the sensorimotor systems that are used during real-world interactions with the stimulus in question. Note here that the "simulation" assumption is not shared by some of the more radical, anti-representational views on embodiment within philosophy (but going into this debate is not essential for the current chapter).

Crucially for this chapter, an embodiment perspective applies particularly well to thinking about emotion (Niedenthal 2007; Niedenthal and Barsalou 2009; Niedenthal et al. 2005; Winkielman, Niedenthal, and Oberman 2008). William James (1890) observed that "every representation of a movement awakens in some degree the actual movement which is its object" (p. 526). Similarly, when our thoughts focus on some joyful personal moment, we also partially reproduce the joyful feeling. Embodiment theories hold that such re-enactments are far from incidental byproducts of recalling past states—instead, they can be crucial to social reasoning, processing emotional concepts, interpreting language, and relating to others' emotional states. These re-enactments also do not have to be conscious, full-blown physical episodes, but could involve only a sufficient level of re-instantiation for the original experience such that it is useful for social and conceptual processing. As we discuss later, embodied reactions do not only result from associative connections between percepts, concepts, and somatic states, but they can also be selectively produced when it is necessary to represent the conceptual content during information processing.

Emotional Embodiment

William James was one of the main predecessors of modern embodiment theories, and his canonical example of coming upon a bear in the woods is an apt starting point for this discussion. James (1896) roughly said the following—you see such a bear, and as a result, your autonomic nervous system is automatically activated (e.g., elevated heart rate and blood pressure, increased sweating, etc.). Upon noticing this altered bodily state, you recognize that you are afraid—simply put, we do not tremble because we are afraid; we are afraid because we tremble. More generally across emotional states, James proposed that an external exciting stimulus leads to a physiological reaction, with the interpretation of said reaction leading to a given emotion.

Obviously, modern emotion theories deem the emotional cascade of events as a much more complex, iterative process—with somatosensory and motor resources recruited at multiple times in the emotion perception, understanding, experience, and production process (Barrett et al. 2009; Cunningham et al. 2013; Scherer 2009; Wood et al. 2016). Further, most modern theories do not take actual changes in bodily states to be necessary to experience an emotion, instead focusing on brain representation of somatosensory and motor processes. For instance, the somatic marker hypothesis argues that the brain transforms physiological events into representations of the bodily state to constitute an emotional feeling—and together with past outcomes, these emotional feelings can guide decision-making strategy (Damasio 1999). Nevertheless, the basic Jamesian point remains, in that mental representations of the bodily state are integral parts of emotional processing and experience. This core insight has been supported by several lines of research, including the role of embodiment in processing of facial expressions, somatic involvement in valence processing (e.g., approach-avoidance effects), affective processing as a modality, and comprehension of emotional language.

Processing of Facial Expressions

Perhaps the most investigated contribution of emotional embodiment is with the processing of facial expressions (for a review, see Wood et al. 2016). Note that on many traditional models, expression recognition is basically a matter of detecting features and the configurations that are probabilistically associated with an expression (Haxby et al. 2000). For example, detecting a smile relies on the decoding of features such as curves at the corners of the mouth, lines in the corners of the eyes, and relative changes in mouth and eye positions (among others). In other words, recognition of a smile is very much like the recognition of any other stimulus, and thus can be potentially implemented in completely disembodied systems (Dailey et al. 2000). However, on the embodied perspective, recognition of a smile involves motor reproduction of the smile and a partial simulation of the presumed state of happiness (Goldman and Sripada 2005). This can bootstrap recognition and, via facial feedback, provide verification of a match between one's own state and the mood of the person we are imitating (Niedenthal et al. 2010).

There is quite a bit of evidence for correlational links between the perception of facial expressions and activation of spontaneous facial motor movements (e.g., Dimberg 1990), along with greater activity in somatosensory areas of brain (e.g., Carr et al. 2003). One classic interpretive objection, however, is that such findings do not reflect "simulations" (i.e., attempts to build a re-enactment of the social stimulus); instead, they are just simple associative reactions. These imitative reactions, while reliably produced, are essentially byproducts of the frequent association between smile perception and smile action (when we see smiles, we usually smile back).

Fortunately, though, there is some evidence suggesting a causal role of embodied resources in emotion recognition. As an example, temporarily preventing participants from engaging expression-relevant facial muscles impairs their ability to detect briefly presented or relatively ambiguous facial expressions that involve that muscle (Niedenthal et al. 2001; Oberman et al. 2007; Ponari et al. 2012). Lesion studies have also revealed emotion recognition impairments to occur after damage to somatosensory and motor feedback mechanisms (Adolphs et al. 2000; Pistoia et al. 2010). Further, temporary inactivation of the brain's face representation areas with repetitive transcranial magnetic stimulation (rTMS) tends to impair emotion recognition (Pitcher et al. 2008).

Of course, this does not mean that embodiment is *always* involved in the processing of facial expressions, nor is it always causally necessary. For instance, patients with facial paralysis (also called Moebius syndrome, primarily from the underdevelopment of cranial nerves VI and VII) can recognize facial expressions using non-embodied routes (Bogart and Matsumoto 2010). In another example, a recent study found that inhibiting the right primary motor (M1) and somatosensory (S1) cortices with TMS reduced spontaneous facial mimicry and delayed the perception of changes in facial expressions (Korb et al. 2015). However, note that these effects occurred only for female participants. TMS had no influence on males—perhaps reflecting greater reliance by females on the embodied simulation strategy, in order to enhance emotion understanding. These findings highlight our message of contextual flexibility of embodied simulation and its dependence on resource availability and situational goals.

Somatic Involvement in Valence Processing: Approach-Avoidance Effects

Another well-investigated connection between embodiment and emotion is the link between valence perception and specific action tendencies. One classic illustration comes from Darwin (1872), when he visited a snake exhibit in a London zoo. He writes the following:

> I put my face close to the thick glass-plate in front of a puff-adder in the Zoological Gardens, with the firm determination of not starting back if the snake struck at me; but, as soon as the blow was struck, my resolution went for nothing, and I jumped a yard or two backwards with astonishing rapidity. (p. 40)

Since then, psychological research has found plenty of evidence for emotion-action links using a variety of stimuli. This has led many modern emotion theories to posit that affective and motivational processes are organized along an *approach-avoidance* dimension, which are linked to specific action tendencies (Harmon-Jones et al. 2013). Thus, individuals are generally faster to pull a lever toward the body for positive stimuli and/or push a lever away from the body for negative stimuli (Chen and Bargh, 1999; Krieglmeyer et al. 2010). These experiments were motivated by the observation that we use the motor action of pushing to stimuli that we do not like (either as a practical action or as a communicative gesture), but we pull objects that we do like toward us (or indicate our liking of objects to others via some type of pulling gesture). Considering the notion that specific motor actions are tied to abstract valence representations, it follows that reactions to the task would be facilitated when the valence of the physical action is congruent with that of the concept being evaluated. Indeed, several studies have shown that positive stimuli (e.g., words, picture, and faces) trigger faster approach reactions, whereas negative stimuli trigger faster avoidance reactions.

Note, however, that the automaticity of approach-avoidance effects is still debated. The initial claims were that affective stimuli spontaneously and rapidly trigger adaptive actions, without much involvement of conceptual processes (Chen and Bargh 1999). In contrast, many of these effects require focusing participants on the evaluative nature of the task. For example, happy faces trigger approach movements but only when they are evaluated on emotion, not gender (Rotteveel and Phaf 2004; Krieglmeyer et al. 2010). Further, notice that the specific nature of the link between valence and action appears to be at least partially shaped by conceptual processes. For instance, originally, these approach-avoidance tendencies have been reported to follow body-centered direction (i.e., movements toward vs. away the subject's own body; Chen and Bargh 1999) and specific muscle activation (i.e., bicep flexion vs. tricep extension; Cacioppo et al. 1993). However, later experiments found such effects with participants' movements that brought the object closer or farther to the "self," even when the self is a mere symbolic representation on the screen by participants' names (e.g., Markman and Brendl 2005; but also see Van Dantzig et al. 2009). In short, along with other research presented in our chapter, this argues that while some embodied connections are relatively simple (and perhaps associative), others interact with sophisticated conceptual processes.

In this context, one intriguing line of research shows that rapid approach-avoidance actions are shaped by the ease of information processing. This is interesting because it suggests that rapid embodied responses can result from the mere dynamics of mental operations. Specifically, extensive research has shown that *fluency* (or fast and easy processing) augments positive affective responses (Winkielman and Cacioppo 2001), evaluative judgments like attractiveness and trust (Schwarz 1998; Winkielman et al. 2003; Winkielman et al. 2015), and real-world decisions like brand assessments and stock purchases (Alter and Oppenheimer 2006; Lee and Labroo 2004). These effects can occur with manipulations that target the ease of semantic understanding and categorization, such as making a stimulus more difficult to assign into categories (e.g., Owen et al. 2016). They can also arise even with manipulations that are purely perceptual (e.g., increased stimulus repetition, priming, clarity, contrast, and/or duration). According to the hedonic fluency model, fluency impacts ratings and behavior because easy processing elicits mild positive affect, which is then (mis)attributed to the target stimulus (Winkielman et al. 2003).

Given these fluency findings, Carr, Rotteveel, and Winkielman (2016) tested whether increasing the perceptual fluency of neutral pseudowords (i.e., pronounceable strings of letters with no actual meaning) would facilitate rapid approach movements. Participants were instructed to use their "gut" to evaluate different pseudowords on whether they were "good or bad" (i.e., affective judgment; Experiments 1 and 3) or "living or non-living" (i.e., non-affective judgment; Experiments 2 and 4). Participants made these decisions using a vertical button-stand, which mapped responses onto approach (bicep contraction with arm flexion to hit the top button) and avoidance (tricep contraction with arm extension to hit the bottom button). The findings showed that fluent (easy-to-read) pseudowords led to faster approach movements, consistent with the idea that fluency can infuse initially neutral stimuli with a positive valence. Interestingly, this fluency-action effect occurred only in the affective judgment context (i.e., experiments that involved good/bad judgments, not living/non-living judgments). This is intriguing when considering the EMG findings, which measured physiological muscle responses for smiling (*zygomaticus major*) and frowning (*corrugator supercilii*). Carr, Rotteveel, and Winkielman (2016) observed increased smiling EMG and reduced frowning EMG to fluent pseudowords across both contexts (good/bad and living/non-living). These findings are important for two main reasons. First, they suggest that perceptual fluency can imbue neutral stimuli with enough positivity to make them elicit embodied action responses that are similar to intrinsically valenced stimuli (e.g., emotional words or facial expressions). Second, the findings illustrate the flexible nature of these effects, where fluent stimuli increased smiling and reduced frowning across both contexts—but only led to approach actions in affective contexts. Therefore, fluency instantiates a low-level hedonic response across multiple domains, but this affective response is only selectively translated to action-tendency based on the relevance to the task at hand (i.e., ones that require emotionally based judgments). While these findings should certainly be extended, they raise the interesting possibility that embodied approach-avoidance responses go beyond highly valenced, intrinsically affective stimuli.

It is worth noting here that the context-dependency of approach-avoidance effects has also been observed in other experiments. In terms of task context, one review argues that approach-avoidance effects on action are only robustly observed when such action is embedded in affective classification tasks (even to strongly valenced stimuli; see Phaf et al. 2014 for a meta-analysis). For instance, positively valenced faces facilitate approach action when participants classify such emotional faces into affective categories, such as positive or negative, but not into non-affective categories, such as male or female (Rotteveel et al. 2015; Rotteveel and Phaf 2004). Moreover, the internal emotional context also matters. For example, in an investigation of approach- and avoidance-related physical responses to face images, van Peer et al. (2007) found faster responses to press a button using arm flexion (i.e., approach movement) when viewing happy faces (compared to angry faces), while the opposite was true when the button press required arm extension. Critically, this motor effect of viewing angry faces was heightened by the administration of cortisol vs. placebo in highly anxious participants, providing additional evidence that the effect is linked to emotional context. Thus, physical movements are not context-independent—they are enhanced or impeded by specific task and affective environments.

Harmon-Jones (Harmon-Jones et al. 2011; Harmon-Jones and Peterson 2009) and Price (Price et al. 2011) have applied this approach-avoidance framework to explore other emotional embodiment effects. Here, emotions associated with approach include anger and joy, in contrast to avoidant emotions, such as fear, anxiety, and disgust. Neurophysiological measures like asymmetries in frontal cortical electroencephalographic (EEG) activation can dissociate these motivational tendencies—with left-sided activation indicating approach, and right-sided activation indicating avoidance.

Aside from hand movements, body posture involved in approach and avoidance is also linked to affect. Harmon-Jones and Peterson (2009) showed that in the upright posture, hearing a negative, anger-inducing peer evaluation increased participants' left cortical EEG activity (associated with approach). However, no such response was observed in the reclining position. Furthermore, Price et al. (2012) found that forward-leaning body postures lead to diminished startle magnitude and larger late positive potential (LPP) responses to those pictures—but only when the pictures were appetitive, creating a context of potential reward. In contrast, this effect was completely absent when neutral pictures (with no appetitive value) were displayed.

In short, these approach-avoidance effects demonstrate somatic involvement in processing of emotional stimuli. That is, they demonstrate that people's bodies are adaptively engaged when processing emotional stimuli and that manipulating body engagement changes the response to incoming emotional information. At the same time, they highlight that the specific task and motivational contexts flexibly and dynamically shape the nature of these effects. As discussed later, similar conclusions follow from other lines of research on emotion concepts.

Affect as a Type of Modality

One of the classic lines of evidence for the grounded cognition approach comes from studies that ask participants to verify properties for concepts (i.e., judge whether an object or a concept does or does not have certain attributes). Importantly, when two subsequent concepts imply different modalities, this produces switching costs (Pecher et al. 2003). Thus, verifying a property that involves the auditory modality (e.g., BLENDER-loud) was slower after verifying a property that involves a gustatory modality (e.g., CRANBERRIES-tart) than after verifying a different property in the same auditory modality (e.g., LEAVES-rustling). Experiments using the switching paradigm also suggest that affect can be considered as a modality unto itself (Vermeulen et al. 2007). In those experiments, participants had to verify auditory (e.g., KEYS-jingling) and visual (e.g., TREASURE-bright) features of nouns that were either neutral or had strong affective value (either positive or negative). Each target pair was preceded by a priming pair (e.g., TRIUMPH-exhilarating followed by COUPLE-happy). The structure of these pairs was experimentally manipulated so that subjects had to consecutively verify properties either of the same or different modalities (visual, gustatory, auditory, affective) with either similar or different valences (positive or negative). For example, a same-modality/same-valence pair might be [TANK-khaki: WOUND-open]; a different-modality/same-valence pair might be [TANK-khaki: SOB-moaning]; and a different-modality/different-valence pair could be [TANK-khaki: VICTORY-sung]. Another example includes [BABY-babbling], where the same-modality/same-valence pair would be [LAUGHTER-heard] and different-modality/same-valence pair would be [ROSE-red].

The findings showed that verifying features of concepts from different modalities produced costs of longer reaction times and higher error rates than concepts from the same modality. Critically, this included switching between affective and other modalities. There were also costs of crossing modality of processing while keeping valence constant. These results are difficult to account for with amodal models of concept representation, which view affect as just another node in the semantic network.

Situated Simulation in the Embodiment of Emotional Language

Much of the research in the embodiment literature focuses on what specific features people simulate (and when). This focus reflects the growing appreciation that people do not simulate all properties but only those that are required by the current situation (hence, *situated* simulation). One of the earliest demonstrations comes from cognitive work (Wu and Barsalou 2009) showing that different perceptual features are simulated when participants are asked to list properties of different objects, like watermelon vs. half watermelon (e.g., green and striped vs. red with seeds). Presumably, the concept of "half watermelon" invites a different simulation (one dealing with internal properties, like red with seeds) than "watermelon" (one dealing with external properties, like green with stripes).

More recent experiments explored situated simulation processes in understanding abstract mental states. Note that these mental states can be affective (e.g., anger or happiness) or cognitive (e.g., thinking or remembering). Here, the idea is that the understanding of abstract concepts referring to mental states can vary, depending on what perceptual perspective is activated. Many mental states have a clear *internal* or *experiential* component—people "feel" a certain way when they are in these states (e.g., anger feels hot, memory retrieval feels effortful). These internal experiences may be simulated when people understand conceptual references to mental states. On the other hand, mental states can also be described from an *external* or *action* perspective—where simulation of visible outside features may be more relevant for understanding (e.g., anger makes the face red, memory retrieval involves head-scratching). In one study using a switching-costs paradigm, participants saw semantically unrelated sentences describing emotional and non-emotional mental states, while manipulating their internal or external focus (Oosterwijk et al. 2012). For example, the concept of disgust could be presented in an internal sentence like "She was sick with disgust" or an external sentence like "Her nose wrinkled with disgust." Results showed that switching costs (slower reaction times) also occurred when participants shifted between emotional sentences with an internal and external focus. Moreover, a follow-up study showed that reading sentences about emotional states that imply an internal vs. external perspective activates different brain areas related to different aspects of the emotion process (Oosterwijk et al. 2015). Specifically, emotion sentences written from the external perspective activate a brain region related to action representation (i.e., the inferior frontal gyrus), whereas sentences about the same emotion written from the internal perspective activate a brain region associated with the generation of experiential states (the ventromedial prefrontal cortex). These findings further emphasize that different forms of simulation contribute to understanding mental states from different points of view. This conclusion is important, as it shows that even abstract emotion concepts are grounded in modalities, but this grounding is flexible (and the specific properties that are revealed depends on the requirements of a specific "perspective").

Causal role of embodied processes in emotional language

Critically, there is also causal evidence for the role of embodiment in how people process emotional words and concepts (using their own facial muscles). In one study, authors first used subcutaneous injections of Botox to temporarily paralyze the facial muscle used in frowning, after which they had participants read emotional sentences (Havas et al. 2010). The results showed that participants were slower to understand emotional sentences that involved the use of the paralyzed muscle. Davis, Winkielman, and Coulson (2015) also demonstrated that facial blocking impedes comprehension of emotional words (e.g., via the N400 event-related potential component that indexes processing of semantic meaning).

Another important example comes from Niedenthal et al. (2009), who had participants view emotional words (e.g., concrete nouns, such as SUN or SLUG, or abstract words, such as FOUL and JOYFUL). Some participants were simply asked whether the words were capitalized or not (a shallow perceptual task), while others were asked whether the words were associated with an emotion (a deeper conceptual task). Crucially, during this task, the activation of participants' facial muscles was measured with EMG. Consistent with the idea of strategic and context-dependent modal processing, the results showed that facial muscles were subtly activated in emotion-specific patterns when participants were evaluating the meaning of the words—but not when they made judgments of letter case. These results clearly argue against a "simple emotional reaction account," where measured emotions are just reflexive reactions to reading the word. Furthermore, Niedenthal et al. (2009) conducted another experiment explicitly designed to manipulate the strategic need for emotion simulation. Participants generated features of emotional concepts (e.g., FRUSTRATION) by thinking of and listing associated properties, while facial EMG was recorded. Crucially, depending on condition, participants were informed that the features were being produced either for an audience that was interested in emotional, "hot" features of the concepts (like a good friend with whom they have a close relationship and can share anything) or for an audience interested in more factual, "cold" features of the emotion concepts (like a supervisor with whom they have a formal relationship). Interestingly, participants were able to produce normatively appropriate emotion features in both the "hot" and "cold" conditions. However, the physiological results showed that there was greater activation of the expected sets of facial muscles when participants were asked for features of emotion words in the "hot" condition, compared to the "cold" audience condition. This demonstrates that embodied simulations are recruited in concept understanding, but only if they are relevant for solving the task (Wu and Barsalou 2009). Again, this is important because it argues against the idea that embodiment reactions are passive byproducts of conceptual processing (i.e., sensorimotor reflexes that are just there "for the ride"). Note that these studies also somewhat qualify "strong" embodiment claims, which suggest that embodied reactions are *necessary* to understand such concepts. Given that participants in the "cold" condition were still able to successfully generate emotion features, this suggests that embodied responses may serve as one input that informs understanding (but there are likely alternative routes to understanding that do not require embodiment).

Finally, it is worth discussing recent evidence suggesting that the representation of emotional concepts is affected by externally induced changes in physiological arousal (Kever et al. 2015). Concretely, participants saw emotion words in the classic attentional blink (AB) paradigm, requiring detection of a target word in a stream of distractors (Raymond et al. 1992). Participants completed the AB task once after a short physical exercise session (increased arousal) and once after a relaxation session (decreased arousal). During the AB task, two target words (T1 and T2) were presented in close succession in a rapid serial visual presentation (RSVP) of distractor items. The AB effect refers to the reduced ability to report the second of two targets (T2) if it appears 200 to

500 milliseconds after the first to-be-detected target (T1). In the present study, T1 was always a neutral word, whereas T2 was either a high-arousal word (e.g., *terror, orgasm*) or low-arousal word (i.e., *distress, flower*). The results revealed that increased physiological arousal led to improved reports of high-arousal T2 words, while reduced physiological arousal led to improved reports of low-arousal T2 words. Neutral T2 words remained unaffected by the arousal conditions. These findings emphasize that experiencing a level of physiological arousal that matches the emotional arousal value of a concept promotes the awareness of the concept (e.g., being highly aroused while processing a highly arousing word such as *terror* or *orgasm*). In other words, congruence between an individual's level of bodily arousal and the emotional arousal of a to-be-processed stimulus influences the way attention is allocated (and, consequently, its access to awareness). One could even go further in this interpretation by suggesting that the arousal dimension of emotional concepts is (at least partially) grounded in our bodily systems of arousal.

Emotional language and memory for faces

If the perception of facial expressions and perception of emotion words both involve embodiment, one may expect interactive links between these processes. Specifically, motor-conceptual interactions might serve to support (as well as distort) memories of other people's facial expressions. This was investigated by Halberstadt et al. (2009), where participants were first asked to look at faces of several ambiguous facial expressions for different individuals and think about why each of these individuals might possibly feel "happy" or "angry" (concept labels were randomly paired with faces). Later, participants were asked to recall what exact expression was presented for each individual. The data showed that participants' memory for the expressions was biased in the direction of the earlier language concept (e.g., remembering a face as happier when it was earlier associated with a happy label). Critically, this memory distortion was related to the degree to which the conceptual label assigned to the expression (i.e., happy vs. angry) elicited a corresponding facial EMG response during the initial perception of the face. Presumably, this concept-driven motor representation got tied with the actual perceptual representation of the face and later served as a retrieval cue.

Emotional language and metaphor

One of the most fascinating questions about the embodiment of emotion is its relation to higher-order conceptual processes. This classic question, discussed since William James, is receiving new urgency in light of data showing the strong influence of conceptualization and categorization on basic processes of emotional recognition and experience (Barrett et al. 2015). Clearly, conceptual processes are informed by and grounded in bodily states. At the same time, conceptual processes shape how exactly the body is being used. One source and result of that linkage are linguistic and conceptual metaphors, which are presumably not only grounded in embodied processes but also invite us to think about emotions in embodied terms (Lakoff 1993). We will briefly review some of the key studies on embodied metaphor and emotion. Importantly, note that some studies make a direct claim about the role of somatosensory processes and embodied simulation, whereas others explore the body–emotion link in a more linguistic realm without committing to any role for somatosensory involvement (for a discussion of relations between linguistic and simulationist views, see Casasanto and Gijssels 2015; Landau et al. 2010).

Consider a restaurant where you feel most at home—it may have a "warm" waiter who talks to you in a familiar way. In contrast, the places you least like to go (e.g., government offices) may be described as feeling "cold." Similarly, an old friend can be described as being "close" to us, or perhaps we have gone different directions and have actually gotten quite "distant" from one another. Several lines of research suggest that thinking about emotion is metaphorically tied to physical distance and temperature. One set of studies suggests that manipulations of physical distance can increase feelings of emotional distance (Williams and Bargh 2008a; but also see Pashler et al. 2012). In these studies, subjects completed a priming manipulation where they were asked to plot two points in two-dimensional space—with some subjects plotting points that were very close together, and others plotting points that were quite far apart. Subsequently, subjects who plotted the far (as opposed to close) points perceived themselves as having weaker emotional attachment to their hometowns and family members, and were less affected by a story relating a distressing experience (presumably because they felt more distant from the situations presented). In this context, it is interesting that one way in which psychological distance (Trope and Lieberman 2010) can decrease emotional responses is by reducing reliance on embodied responses (Maglio and Trope 2012).

Another set of experiments used a similar logic, and primed participants with physical sensations that bear on the embodied metaphor for warmth (Williams and Bargh 2008b). These studies showed that participants find an interaction partner as a "warmer" person when they were holding a warm cup of coffee (rather than a cold one) in an unrelated task. Zhong and Leonardelli (2008) also asked some subjects to recall an experience of social rejection, while others had to recall an experience of social inclusion. Afterwards, both groups were asked to estimate room temperatures, and, interestingly, participants that recalled exclusion guessed significantly lower temperatures than those who recalled inclusion. Their second experiment involved the "cyberball" manipulation, where subjects played an online game. This game was supposedly multiplayer, but, in reality, all participants were playing with a computer program that either included or excluded them in a game of virtual catch. After the game, participants were given a marketing survey where they had to rate the desirability of various food and drink items, some warm and some cold (e.g., cold cola vs. hot coffee, cold apples vs. hot soup). Excluded participants (i.e., those who were tossed the virtual cyberball just two times out of 30) rated hot items as more desirable than included participants (i.e., tossed the cyberball an equal number of times as all other players). Finally, a provocative result suggests that psychological coldness (i.e., social rejection) may even literally reduce skin temperatures (IJzerman et al. 2012).

Other research suggests that bottom-up embodied cues influence which conceptual metaphor guides emotion understanding. Note that multiple metaphors are usually available for understanding an emotional state. One classic example for this is "love," which could be understood as a journey, a flower, a game, or a unity (Lakoff 1993). These framings are consequential for emotional reactions, as shown by a study in which participants were more distressed by relationship conflicts when they had a unity rather than journey frame in mind (Lee and Schwarz 2014). But when and why are specific metaphors preferred? And how do they guide our understanding of emotional material? These questions were addressed in an exploratory study by Tseng et al. (2007), where the researchers noticed that similar emotional words (e.g., *happiness* and *joy*) are differently associated with metaphorical frames. Thus, the expression "searching for happiness" is more common than "searching for joy," whereas the expression "full of joy" occurs more often than "full of happiness." In turn, subtly activating different metaphors through embodied cues should influence which emotional term, *joy* or *happiness*, is applied to ambiguous emotional material. To test this, the experimenters approached participants as they were either searching for something (like a book in a library) or drinking something from a container. Participants were then shown a photo of a person with a very positive facial expression and asked whether it is better described as happiness or joy. As expected, participants who were drinking (and presumably activated the container metaphor) were more likely to describe the picture as showing joy, whereas people who were searching were more likely to describe the expression as showing happiness.

The just-discussed work also fits well with studies on embodied metaphors of morality. These studies show that the physical act of washing seems to remove the negative feeling associated with a moral transgression. In an experiment by Zhong and Liljenquist (2006), participants were asked to write about a past moral transgression and then either cleaned their hands with wipes or did not. Next, they filled out an emotional state questionnaire and were later approached (without forewarning) about participating in another study for a desperate graduate student. In previous studies, the reliving of moral transgressions had been shown to increase propensity to engage in good deeds—this held in the current study, but only for participants who had not cleaned themselves. Participants who had already cleaned their hands were less likely to help the graduate student, and they also reported lower scores on measures of moral emotions (e.g., disgust, regret, guilt, shame, embarrassment, and anger) but not for amoral emotions (e.g., confidence, calm, excitement, and distress). In the same vein, Lee and Schwarz (2010) showed that the need for cleansing induced by immoral action was specific to the body part used for the "dirty" deed. Participants asked to type a lie showed a greater preference for hand wipes, whereas participants who spoke a lie preferred mouthwash. It is important to note that the connection between physical cleanliness and morality is not completely straightforward or consistent, with some studies reporting that cleanliness reduces the severity of moral judgments (Schnall et al. 2008), other studies reporting that cleanliness enhances moral harshness (Zhong et al. 2010), and yet other studies failing to obtain such effects altogether (Johnson et al. 2014). Clearly, whether and how specifically a bodily feeling translates into an abstract moral decision must depend on a variety of interpretational factors (e.g., Who is "clean" here—me or someone else? Is the feeling induced by the target of judgment? Is this a moral problem for which a feeling is relevant?).

Summary

The social environment is complex, flexible, and dynamic. As a result, we have to efficiently construct context-appropriate affective responses. The lines of research summarized earlier (as well as others not mentioned here) provide increasing evidence that situated and embodied aspects of emotion processing contribute to generating such adaptive responses.

Mirroring as Embodied Cognition

As mentioned, one prominent idea guiding embodiment research is that perceivers often replicate others' states using their own motor and somatosensory resources. Neurally, researchers often assign this function to the putative mirror neuron system (MNS), which re-enacts the observed action in the perceiver's motor system (Rizzolatti and Craighero 2004). However, there is no accepted psychological explanation to date for the function of mirroring and behavioral mimicry. Is it mostly an epiphenomenon, a simple byproduct of frequent perception-action links (Heyes 2011)? Does it reflect a common cognitive representational format for perception and action (Preston and de Waal 2002)? Or is it a processing strategy in service of better understanding (Goldman and Sripada 2005) or social-emotional regulation (Hess and Fischer 2013)?

One debate that bears on this issue concerns the relative role of representational and nonrepresentational processes in mirroring responses (see Carr and Winkielman 2014). Note that while definitions of "mental representations" do vary across different papers and fields, they are generally viewed as products of propositional encoding from incoming sensory information. This sensory information is translated into abstract mental symbols that can be cognitively manipulated, after which they are back-translated into motor responses (Gallese 2003). Several mirroring accounts appeal to higher-order representational processes (e.g., Goldman and Sripada 2005), and arguably some of the strongest pro-representational evidence comes from studies showing that mirroring responses dynamically adapt to environmental cues. As examples, motor imitation (e.g., finger-lifting, hand-opening and -closing) is modulated by prosocial attitudes (Leighton et al. 2010), affiliative drive (Lakin and Chartrand 2003), and eye contact (Wang et al. 2011). This also extends to spontaneous facial mimicry of emotional expressions (i.e., smile-to-a-smile, frown-to-a-frown), which can shift dramatically based on social factors, such as negative attitudes (Likowski et al. 2008), competition (Weyers et al. 2009), and social power (Carr, Winkielman, and Oveis 2014).

In contrast, others argue that mirroring can occur with *minimal* contribution from conceptual representations (i.e., mirroring does not necessitate the use of abstract symbolic representations to generate action from perception). For example, one major nonrepresentational framework proposes that "mirror" neurons are just reactions generated from perception-action links that are built throughout development, which pair contingent and contiguous stimulus and response (Heyes 2011). Here, the MNS does not code the goal-directed nature of observed actions, nor does it necessarily even strictly "mirror" those perceived behaviors (Cook et al. 2014). Other nonrepresentational accounts for mirroring echo this same general sentiment, but they also posit that action understanding is purely motoric in nature and based within the specific neural function of the MNS (Sinigaglia 2009). Much work in psychology and neuroscience on spontaneous imitation in humans seems to support this nonrepresentational perspective. These experiments have shown that mirroring is modifiable by low-level perceptual and motor factors, like simple training (Catmur et al. 2007), visual feedback (Cook, Johnston, and Heyes 2013), and testing environment (Cook, Dickinson, and Heyes 2012). In social situations, people also seem to imitate even against strong competitive incentives *not* to do so (Belotet al. 2013) and when they are clearly convinced that a model lacks any intentionality (e.g., androids that have no conscious will or agency for action; Hofree et al. 2014; Hofree et al. 2015). Finally, mirroring of some basic actions (e.g., intransitive finger movements) seem impervious to some higher-level manipulations of social context, such as status and power of the model (Farmer et al. 2016).

An advantage of the embodiment framework is that it could possibly integrate this rather confusing mirroring literature. One way is by deepening the understanding of the functionality of spontaneous mimicry (Kavanagh and Winkielman 2016). As an example, gestural and postural mimicry have been frequently linked to affiliation and rapport between the mimicked party (the model) and the person doing the imitating (the mimic). Individuals who like each other tend to mimic one another, and being mimicked by another person tends to increase one's feelings of affiliation toward that person (Chartrand and Bargh 1999). These tendencies have led to mimicry being labeled a form of "social glue" because it seems to foster cohesion between social groups (Lakin et al. 2003). Embodiment theories help to explain why this is the case—mimicry contributes to creation of the same somatically grounded emotional state, thus facilitating understanding (Kavanagh and Winkielman 2016).

Critically, just like with other forms of embodied information, the effects of dyadic mimicry are moderated by contextual information. One of most basic forms of socially relevant information is group membership. Indeed, recent studies show that gestural mimicry by an in-group member makes one feel socially and physically warmer, yet mimicry by an out-group member makes one feel colder (Leander et al. 2012). These data are consistent with previous work on facial mimicry, which found that negative attitudes toward the model are associated with counter-mimicry (Likowski et al. 2008), and subliminally priming the concept of competition reduces or even reverses facial mimicry (Weyers et al. 2009).

Keep in mind that mimicry's function is not limited to directly interacting parties. Observers can also make social judgments about people based on whom they mimic. In some situations, third-party observers will infer from the presence of mimicry that the members of the dyad are socially related and positively affiliated (Bernieri 1988).

Another example from Kavanagh et al. (2011) showed even more complex inferences. They found that if a target person mimics a model who is rude to the person, third-party observers of this interaction will judge the mimic as incompetent—even when observers fail notice the presence of mimicry. In fact, in that situation, the mimic was rated as *less* competent than the non-mimic. This suggests that if a person chooses to mimic a rude model, the person is nonselective or injudicious in using embodied responses (i.e., socially incompetent).

In short, inferences supported by embodied cues can be quite complex and context-dependent. Importantly, this does not challenge the value of embodiment theories, but instead suggests that such social reactions and inferences are based on a flexible and situated recruitment of embodied resources that are relevant to the current context.

Conclusion

In the current chapter, we discussed theory and evidence for the embodiment of emotional and affective information across social contexts. On this perspective, embodied conceptual processing involves the partial reuse and reinstatement of experiential and motor states that occur when the perceiver has actually encountered the object. In other words, reactivations of the sensorimotor systems can be used to inform understanding of emotional stimuli during real-world interactions.

We have illustrated that embodied resources are often routinely activated in information processing, including higher-order conceptual tasks. They can also play a causal role in emotion understanding, and perceivers can flexibly deploy them in order to facilitate mental processing. Note, however, that such embodied responses do not always appear to be necessary to perceive, understand, or generate affective information. Thus, while we embrace the strengths of the embodiment perspective, we nevertheless emphasize that a satisfactory account must also consider the interaction of embodied processes with context-sensitive, conceptual processes. As such, future work on the embodiment of emotion should further explore the interplay between amodal frameworks and modal embodiment theories, to better gauge when and how these responses adapt to a constantly changing social environment.

References

Adolphs, R., Damasio, H., Tranel, D., Cooper, G., and Damasio, A.R. (2000). A role for somatosensory cortices in the visual recognition of emotion as revealed by three-dimensional lesion mapping. *Journal of Neuroscience*, 20, 2683–90.

Alter, A.L. and Oppenheimer, D.M. (2006). Predicting short-term stock fluctuations by using processing fluency. *Proceedings of the National Academy of Sciences*, 103, 9369–72.

Anderson, J.R. (1983). A spreading activation theory of memory. *Journal of Verbal Learning and Verbal Behavior*, 22(3), 261–95.

Barrett, L.F. and Bliss-Moreau, E. (2009). Affect as a psychological primitive. *Advances in Experimental Social Psychology*, 41, 167–218.

Barrett, L.F., Wilson-Mendenhall, C.D., and Barsalou, L.W. (2015). The conceptual act theory: a road map. In: L.F. Barrett and J.A. Russell (eds.), *The psychological construction of emotion*. New York: Guilford, pp. 83–110.

Barsalou, L.W. (1999). Perceptual symbol system. *Behavioral and Brain Sciences*, 22, 577–660.

Barsalou, L.W. (2008). Grounded cognition. *Annual Review of Psychology*, 59, 617–45.

Belot, M., Crawford, V.P., and Heyes, C. (2013). Players of matching pennies automatically imitate opponents' gestures against strong incentives. *Proceedings of the National Academy of Sciences*, 110(8), 2763–8.

Bernieri, F. (1988). Coordinated movement and rapport in teacher-student interactions. *Journal of Nonverbal Behavior*, 12, 120–38.

Block, N. (1995). The mind as the software of the brain. In: E.E. Smith and D.N. Osherson (eds.), *Thinking*. Cambridge, MA: MIT Press, pp. 377–425.

Bogart, K.R. and Matsumoto, D. (2010). Facial expression recognition by people with Moebius syndrome. *Social Neuroscience*, 5, 241–51.

Cacioppo, J.T., Priester, J.R., and Bernston, G.G. (1993). Rudimentary determination of attitudes: II. Arm flexion and extension have differential effects on attitudes. *Journal of Personality and Social Psychology*, 65, 5–17.

Carr, E.W., Rotteveel, M., and Winkielman, P. (2016). Easy moves: perceptual fluency facilitates approach-related action. *Emotion*, 16, 540–52.

Carr, E.W. and Winkielman, P. (2014). When mirroring is both simple and "smart": how mimicry can be embodied, adaptive, and non-representational. *Frontiers in Human Neuroscience*, 8(1–7), 1.

Carr, E.W., Winkielman, P., and Oveis, C. (2014). Transforming the mirror: power fundamentally changes facial responding to emotional expressions. *Journal of Experimental Psychology: General*, 143(3), 997–1003.

Carr, L., Iacoboni, M., Dubeau, M.C., Mazziotta, J.C., and Lenzi G.L. (2003). Neural mechanisms of empathy in humans: a relay from neural systems for imitation to limbic areas. *Proceedings of the National Academy of Sciences*, 100, 5497–502.

Casasanto, D. and Gijssels, T. (2015). What makes a metaphor an embodied metaphor? *Linguistics Vanguard*, 1, 327–37.

Catmur, C., Walsh, V., and Heyes, C. (2007). Sensorimotor learning configures the human mirror system. *Current Biology*, 17(17), 1527–31.

Chartrand, T.L. and Bargh, J.A., (1999). The chameleon effect: the perception-behavior link and social interaction. *Journal of Personality and Social Psychology*, 76, 893–910.

Chen, M. and Bargh, J.A. (1999). Consequences of automatic evaluation: immediate behavioral predispositions to approach or avoid the stimulus. *Personality and Social Psychology Bulletin*, 25, 215–24.

Clark, A. (1999). An embodied cognitive science? *Trends in Cognitive Sciences*, 3(9), 345–51.

Collins, A.M. and Loftus, E.F. (1975). A spreading-activation theory of semantic processing. *Psychological Review*, 82(6), 407–28.

Colombetti, G. (2014). *The feeling body: affective science meets the enactive mind*. Cambridge, MA: MIT Press.

Cook, R., Bird, G., Catmur, C., Press, C., and Heyes, C. (2014). Mirror neurons: from origin to function. *Behavioral and Brain Sciences*, 37(2), 177–92.

Cook, R., Dickinson, A., and Heyes, C. (2012). Contextual modulation of mirror and countermirror sensorimotor associations. *Journal of Experimental Psychology: General*, 141(4), 774–87.

Cook, R., Johnston, A., and Heyes, C. (2013). Facial self-imitation: objective measurement reveals no improvement without visual feedback. *Psychological Science*, 24(1), 93–8.

Cunningham, W.A., Dunfield, K.A., and Stillman, P. (2013). Emotional states from affective dynamics. *Emotion Review*, 5, 344–55.

Dailey, M.N., Cottrell, G.W., and Adolphs, R. (2000). A six-unit network is all you need to discover happiness. In: L.R. Gleitman and A.K. Joshi (eds.), *Proceedings of the Twenty-Second Annual Conference of the Cognitive Science Society*, 22, 101–6.

Damasio, A.R. (1999). *The feeling of what happens: body and emotion in the making of consciousness*. Boston, MA: Houghton Mifflin Harcourt.

Darwin, C. (1872). *The expression of the emotions in man and animals with photographic and other illustrations*. London: J. Murray.

Davis, J.D., Winkielman, P., and Coulson, S. (2015). Facial action and emotional language: ERP evidence that blocking facial feedback selectively impairs sentence comprehension. *Journal of Cognitive Neuroscience*, 27, 2269–80.

Dimberg, U. (1990). Facial electromyography and emotional reactions. *Psychophysiology*, 27, 481–94.

Farmer, H., Carr, E.W., Svartdal, M., Winkielman, P., and Hamilton, A.F.C. (2016). Status and power do not modulate automatic imitation of finger movements. *PLoS ONE*, 11(4), e0151835.

Fodor, J. (1975). *The language of thought*. Cambridge, MA: Harvard University Press.

Gallese, V. (2003). The manifold nature of interpersonal relations: the quest for a common mechanism. *Philosophical transactions of the Royal Society of London. Series B, Biological sciences*, 358(1431), 517–28.

Gibbs, R.W. (2003). Embodied experience and linguistic meaning. *Brain and Language*, 84, 1–15.

Goldman, A.I. and Sripada, C.S. (2005). Simulationist models of face-based emotion recognition. *Cognition*, 94, 193–213

Halberstadt, J., Winkielman, P., Niedenthal, P.M., and Dalle, N. (2009). Emotional conception: how embodied emotion concepts guide perception and facial action. *Psychological Science*, 20, 1254–61.

Harmon-Jones, E., Gable, P.A., and Price, T.F. (2011). Leaning embodies desire: evidence that leaning forward increases relative left frontal cortical activation to appetitive stimuli. *Biological Psychology*, 87, 311–13.

Harmon-Jones, E., Harmon-Jones, C., and Price, T.F. (2013). What is approach motivation? *Emotion Review*, 5, 291–5.

Harmon-Jones, E. and Peterson, C.K. (2009). Supine body position reduces neural response to anger evocation. *Psychological Science*, 20, 1209–10.

Havas, D.A., Glenberg, A.M., Gutowski, K.A., Lucarelli, M.J., and Davidson, R.J. (2010). Cosmetic use of botulinum toxin-A affects processing of emotional language. *Psychological Science*, 21, 895–900.

Haxby, J.V., Hoffman E.A., and Gobbini M.I. (2000). The distributed human neural system for face perception. *Trends in Cognitive Sciences*, 4(6), 223–33.

Hess, U. and Fischer, A. (2013). Emotional mimicry as social regulation. *Personality and Social Psychology Review*, 17(2), 142–57.

Heyes, C.M. (2011). Automatic imitation. *Psychological Bulletin*, 137, 463–83.

Hofree, G., Ruvolo, P., Bartlett, M.S, and Winkielman, P. (2014). Bridging the mechanical and the human mind: spontaneous mimicry of a physically present android. *PLoS ONE*, 9(7), e99934.

Hofree, G., Urgen, B.A., Winkielman, P., and Saygin, A.P. (2015). Observation and imitation of actions performed by humans, androids, and robots: an EMG study. *Frontiers in Human Neuroscience*, 9(364). doi:10.3389/fnhum.2015.00364

IJzerman, H., Gallucci, M., Pouw, W.T., Weiβgerber, S.C., Van Doesum, N.J., and Williams, K.D. (2012). Cold-blooded loneliness: social exclusion leads to lower skin temperatures. *Acta Psychologica*, 140(3), 283–8.

James, W. (1890). *The principles of psychology* (vol. 1). New York: Holt.

James, W. (1896). The physical basis of emotion. *Psychological Review*, 101, 205–10.

Johnson, D.J., Cheung, F., and Donnellan, M.B. (2014). Does cleanliness influence moral judgments? A direct replication of Schnall, Benton, and Harvey (2008). *Social Psychology*, 45, 209–15.

Kavanagh, L., Suhler, C., Churchland, P., and Winkielman, P. (2011). When it's an error to mirror: the surprising reputational costs of mimicry. *Psychological Science*, 22, 1274–6.

Kavanagh, L.C. and Winkielman, P. (2016). The functionality of spontaneous mimicry and its influences on affiliation: an implicit socialization account. *Frontiers in Psychology*, 7, 458.

Kever, A., Grynberg, D., Eeckhout, C., Mermillod, M., Fantini, C., and Vermeulen, N. (2015). The body language: the spontaneous influence of congruent bodily arousal on the awareness of emotional words. *Journal of Experimental Psychology: Human Perception and Performance*, 41, 582–9.

Korb, S., Malsert, J., Rochas, V., Rihs, T.A., Rieger, S.W., Schwab, S., et al. (2015). Gender differences in the neural network of facial mimicry of smiles—an rTMS study. *Cortex*, 70, 101–14.

Kosslyn, S. (1994). *Image and brain: the resolution of the imagery debate*. Cambridge, MA: MIT Press.

Krieglmeyer, R., Deutsch, R., De Houwer, J., and De Raedt, R. (2010). Being moved: valence activates approach-avoidance behavior independently of evaluation and approach-avoidance intentions. *Psychological Science*, 21, 607–13.

Lakin, J.L. and Chartrand, T.L. (2003). Using nonconscious behavioral mimicry to create affiliation and rapport. *Psychological Science*, 14(4), 334–9.

Lakin, J.L., Jefferis, V.E., Cheng, C.M., and Chartrand, T.L. (2003). The chameleon effect as social glue: evidence for the evolutionary significance of nonconscious mimicry. *Journal of Nonverbal Behavior*, 27, 145–62.

Lakoff, G. (1993). The contemporary theory of metaphor. In: A. Ortony (ed.), *Metaphor and thought* (2nd ed.). Cambridge: Cambridge University Press, pp. 205–51.

Landau, M.J., Meier, B.P., and Keefer, L.A. (2010). A metaphor-enriched social cognition. *Psychological Bulletin*, 136, 1045–67.

Leander, N.P., Chartrand, T.L., and Bargh, J.A. (2012). You give me the chills: embodied reactions to inappropriate amounts of behavioral mimicry. *Psychological Science*, 23, 772–9.

Lee, A.Y. and Labroo, A.A. (2004). The effect of conceptual and perceptual fluency on brand evaluation. *Journal of Marketing Research*, 41, 151–65.

Lee, S.W.S. and Schwarz, N. (2010). Of dirty hands and dirty mouths: embodiment of the moral purity metaphor is specific to the motor modality involved in moral transgression. *Psychological Science*, 21, 1423–5.

Lee, S.W.S. and Schwarz, N. (2014). Framing love: when it hurts to think we were made for each other. *Journal of Experimental Social Psychology*, 54, 61–7.

Leighton, J., Bird, G., Orsini, C., and Heyes, C. (2010). Social attitudes modulate automatic imitation. *Journal of Experimental Social Psychology*, 46(6), 905–10.

Likowski, K.U., Mühlberger, A., Seibt, B., Pauli, P., and Weyers, P. (2008). Modulation of facial mimicry by attitudes. *Journal of Experimental Social Psychology*, 44, 1065–72.

Maglio, S.J. and Trope, Y. (2012). Disembodiment: abstract construal attenuates the influence of contextual bodily state in judgment. *Journal of Experimental Psychology: General*, 141, 211–6.

Mahon, B.Z. (2015). What is embodied about cognition? *Language, Cognition, and Neuroscience*, 30, 420–9.

Markman, A.B. and Brendl, C.M. (2005). Constraining theories of embodied cognition. *Psychological Science*, 16, 6–10.

Meier, B.P., Schnall, S., Schwarz, N., and Bargh, J.A. (2012). Embodiment in social psychology. *Topics in Cognitive Science*, 4(4), 705–16.

Newell, A. and Simon, H.A. (1972). *Human problem solving*. Englewood Cliffs, NJ: Prentice-Hall.

Niedenthal, P.M. (2007). Embodying emotion. *Science*, 316, 1002–5.

Niedenthal, P.M. and Barsalou, L.W. (2009). Embodiment. In: D. Sander and K.S. Scherer (eds.), *Oxford companion to emotion and the affective sciences*. London: Oxford University Press, p. 140.

Niedenthal, P.M., Barsalou, L., Winkielman, P., Krauth-Gruber, S., and Ric, F. (2005). Embodiment in attitudes, social perception, and emotion. *Personality and Social Psychology Review*, 9, 184–211.

Niedenthal, P.M., Brauer, M., Halberstadt, J.B., and Innes-Ker, Å (2001). When did her smile drop? Facial mimicry and the influences of emotional state on the detection of change in emotional expression. *Cognition and Emotion*, 15, 853–64.

Niedenthal, P.M., Mermillod, M., Maringer, M. and Hess, U. (2010). The simulation of smiles (sims) model: embodied simulation and the meaning of facial expression. *Behavioral and Brain Sciences*, 33, 417–80.

Niedenthal, P.M., Winkielman, P. Mondillon, L., and Vermeulen, N. (2009). Embodiment of emotional concepts: evidence from EMG measures. *Journal of Personality and Social Psychology*, 96, 1120–36.

Oberman, L.M., Winkielman, P., and Ramachandran, V.S. (2007). Face to face: blocking expression-specific muscles can selectively impair recognition of emotional faces. *Social Neuroscience*, 2, 167–78.

Oosterwijk, S., Mackey, S., Wilson-Mendenhall, C., Winkielman, P., and Paulus, M.P. (2015). Concepts in context: processing mental state concepts with internal or external focus involves different neural systems. *Social Neuroscience*, 10, 294–307.

Oosterwijk, S., Winkielman, P., Pecher, D., Zeelenberg, R., Rotteveel, M., and Fischer, A.H. (2012). Mental states inside out: processing sentences that differ in internal and external focus produces switching costs. *Memory and Cognition*, 40, 93–100.

Owen, H.E., Halberstadt, J., Carr, E.W., and Winkielman, P. (2016). Johnny Depp, reconsidered: how category-relative processing fluency determines the appeal of gender ambiguity. *PLoS ONE*, 11(2), e0146328.

Pashler, H., Coburn, N., and Harris, C.R. (2012). Priming of social distance? Failure to replicate effects on social and food judgments. *PLoS One*, 7(8), e42510.

Pecher, D., Zeelenberg, R., and Barsalou, L.W. (2003). Verifying different-modality properties for concepts produces switching costs. *Psychological Science*, 14, 119–24.

Phaf, R.H., Mohr, S.E., Rotteveel, M., and Wicherts, J.M. (2014). Approach, avoidance, and affect: a meta-analysis of approach-avoidance tendencies in manual reaction time tasks. *Frontiers in Psychology*, 5, 378

Pistoia, F., Conson, M., Trojano, L., Grossi, D., Ponari, M., Colonnese, C. et al. (2010). Impaired conscious recognition of negative facial expressions in patients with locked-in syndrome. *Journal of Neuroscience*, 30, 7838–44.

Pitcher, D., Garrido, L., Walsh, V., and Duchaine, B.C. (2008). Transcranial magnetic stimulation disrupts the perception and embodiment of facial expressions. *Journal of Neuroscience*, 28(36), 8929–33.

Ponari, M., Conson, M., D'Amico, N.P., Grossi, D. and Trojano, L. (2012). Mapping correspondence between facial mimicry and emotion recognition in healthy subjects. *Emotion*, 12(6), 1398–403.

Preston, S.D. and de Waal, F.B. (2002). Empathy: its ultimate and proximate bases. *Behavioral and Brain Sciences*, 25(1), 1–20.

Price, T.F., Dieckman, L.W., and Harmon-Jones, E. (2012). Embodying approach motivation: body posture influences startle eyeblink and event-related potential responses to appetitive stimuli. *Biological Psychology*, 90, 211–7.

Price, T.F., Peterson, C.K., and Harmon-Jones, E. (2012). The emotive neuroscience of embodiment. *Motivation and Emotion*, 36(1), 27–37.

Pylyshyn, Z.W. (1984). *Computation and cognition: towards a foundation for cognitive science.* Boston, MA: MIT Press.

Raymond, J.E., Shapiro, K.L., and Arnell, K.M. (1992). Temporary suppression of visual processing in an RSVP task: an attentional blink? *Journal of Experimental Psychology: Human Perception and Performance*, 18(3), 849–60.

Rizzolatti, G. and Craighero, L. (2004). The mirror-neuron system. *Annual Review of Neuroscience*, 27, 169–92.

Rotteveel, M., Gierholz, A., Koch, G., van Aalst, C., Pinto, Y., Matzke, D. et al. (2015). On the automatic link between affect and tendencies to approach and avoid: Chen and Bargh (1999) revisited. *Frontiers in Psychology*, 6, 335.

Rotteveel, M., and Phaf, R.H. (2004). Automatic affective evaluation does not automatically predispose for arm flexion and extension. *Emotion*, 4, 156–72.

Scherer, K.R. (2009). The dynamic architecture of emotion: evidence for the component process model. *Cognition and Emotion*, 23, 1307–51.

Schnall, S., Benton, J., and Harvey, S. (2008). With a clean conscience: cleanliness reduces the severity of moral judgments. *Psychological Science*, 19, 1219–22.

Schubert, T.W. and Semin, G. (2009). Embodiment as a unifying perspective in psychology. *European Journal of Social Psychology*, 39, 1135–41.

Schwarz, N. (1998). Accessible content and accessibility experiences: the interplay of declarative and experiential information in judgment. *Personality and Social Psychology Review*, 2, 87–99.

Sinigaglia, C. (2009). Mirror in action. *Journal of Consciousness Studies*, 16(6–8), 309–34.

Trope, Y. and Liberman, N. (2010). Construal-level theory of psychological distance. *Psychological Review*, 117, 440–63.

Tseng, M., Hu, Y., Han, W.-W., and Bergen, B. (2007). "Searching for happiness" or "full of joy"? Source domain activation matters. In: R.T. Cover and Y. Kim (eds.), *Proceedings of the Thirty-First Annual Meeting of the Berkeley Linguistics Society*. Berkeley, CA: Berkeley Linguistics Society, pp. 359–70.

Tucker, M. and Ellis R. (1998). On the relations between seen objects and components of potential actions. *Journal of Experimental Psychology: Human Perception and Performance*, 24, 830–46.

Van Dantzig, S., Zeelenberg, R., and Pecher, D. (2009). Unconstraining theories of embodied cognition. *Journal of Experimental Social Psychology*, 45, 345–51.

van Peer, J.M., Roelofs, K., Rotteveel, M., van Dijk, J.G., Spinhoven, P., and Ridderinkhof, K.R. (2007). The effects of cortisol administration on approach–avoidance behavior: an event-related potential study. *Biological Psychology*, 76(3), 135–46.

Vermeulen, N., Niedenthal, P.M., and Luminet, O. (2007). Switching between sensory and affective systems incurs processing costs. *Cognitive Science*, 31, 183–92.

Wang, Y., Ramsey, R., and Hamilton, A.F.D.C. (2011). The control of mimicry by eye contact is mediated by medial prefrontal cortex. *Journal of Neuroscience*, 31(33), 12001–10.

Weyers, P., Mühlberger, A., Kund, A., Hess, U., and Pauli, P. (2009). Modulation of facial reactions to avatar emotional faces by nonconscious competition priming. *Psychophysiology*, 46(2), 328–35.

Williams, L.E. and Bargh, J.A. (2008a). Keeping one's distance: the influence of spatial distance cues on affect and evaluation. *Psychological Science*, 19, 302–8.

Williams, L.E. and Bargh, J.A. (2008b). Experiencing physical warmth promotes interpersonal warmth. *Science*, 322, 606–7.

Wilson, M. (2002). Six views of embodied cognition. *Psychonomic Bulletin and Review*, 9, 625–36.

Winkielman, P. and Cacioppo, J.T. (2001). Mind at ease puts a smile on the face: psychophysiological evidence that processing facilitation elicits positive affect. *Journal of Personality and Social Psychology*, 81(6), 989.

Winkielman, P., Niedenthal, P., and Oberman, L. (2008). The embodied emotional mind. In: Semin, G.R. and Smith, E.R. (eds.), *Embodied grounding: social, cognitive, affective, and neuroscientific approaches*. New York: Cambridge University Press, pp. 263–88.

Winkielman, P., Olszanowski, M., and Gola, M. (2015). Faces in between: evaluative responses to faces reflect the interplay of features and task-dependent fluency. *Emotion*, 15, 232–42.

Winkielman, P., Schwarz, N., Fazendeiro, T., and Reber, R. (2003). The hedonic marking of processing fluency: implications for evaluative judgment. In: J. Musch and K.C. Klauer (eds.), *The psychology of evaluation: affective processes in cognition and emotion*. Mahwah, NJ: Lawrence Erlbaum, pp. 189–217.

Wood, A., Rychlowska, M., Korb, S., and Niedenthal, P. (2016). Fashioning the face: sensorimotor simulation contributes to facial expression recognition. *Trends in Cognitive Sciences*, 20(3), 227–40.

Wu, L.L. and Barsalou, L.W. (2009). Perceptual simulation in conceptual combination: evidence from property generation. *Acta Psychologica*, 132, 173–89.

Zhong, C.B. and Leonardelli, G.J. (2008). Cold and lonely: does social exclusion feel literally cold? *Psychological Science*, 19, 838–42.

Zhong, C.B. and Liljenquist, K. (2006). Washing away your sins: threatened morality and physical cleansing. *Science*, 313, 1451–2.

Zhong, C.B., Strejcek, B., and Sivanathan, N. (2010). A clean self can render harsh moral judgment. *Journal of Experimental Social Psychology*, 46(5), 859–62.

CHAPTER 29

THINKING AND FEELING

A Social-Developmental Perspective

R. PETER HOBSON

Introduction

In order to understand something, often it is helpful to understand how the something is made, or how it arises, or how it comes to take the shape and have the properties that it does. In addition, there may be much to learn from instances in which construction or development goes wrong. It is from these vantage-points I shall consider what thinking entails.

My aim is to trace how thinking—or at least, one important form of creative, flexible thinking—develops over the first two years of a human being's life. I shall emphasize how thinking develops through communication between expressive, embodied persons, where such communication requires infants to have affective engagement with their social and non-social environment. Through the case of autism, I shall explore how thinking is affected when children are limited in their capacity for social-affective relatedness.

A developmental approach of this kind promises to alter our conception of what it means to think, for example, in explicating the intimate relation between thinking and feeling. My thesis is that in order to think symbolically, a human being needs to share and coordinate experiences of the world on a non-inferential (affective) basis, drawing on a biologically based capacity to identify with the attitudes of others. I shall focus on communicative role-taking as the source of human-specific forms of flexible, creative thinking.

Developmental Principles

I begin with some developmental principles that (in my view) establish an appropriate conceptual framework for characterizing the emergence of symbolic thinking.

At the outset, we need to consider how the means to think differentiates out of infants' experienced relations with the world, because, as Buber (1937/1958, p. 18) insists, "In the beginning is relation." I take it that many forms of person–world relation, and specifically infants' relations toward other persons, have cognitive, conative, and affective *aspects* (Hobson 2008). Aspects are not components. We should be wary of assuming that the earliest modes of engagement between infants and the world are built up, Lego-like, from distinct sources of motivation, feeling, and thought. Just as we can consider the mass, form, color, and temperature of an object as aspects of that single thing—the object's color and temperature presuppose it has form and mass; these are not distinct components—so, too, we can consider the relations between infants and their environment as having a cognitive aspect (for instance, they register the difference between people and things), a motivational aspect (you may try in vain to induce infants to do something against their will), and an affective aspect (their behavior expresses their feeling states). Although in our own alternative conceptual takes on the phenomena of infant–world relations, we may introduce divisions among thought, feeling, and motivated action, these categories of mental functioning are not carving nature at its joints, at least not at this earliest stage of life. The *relative* separation of motivation, feeling, and thought is a developmental achievement yet to come.

Here I am trying to settle upon an appropriate unit of analysis for thinking about the early development of thought. Vygotsky (1962/1986) writes:

> [Rather than analyzing complex psychological wholes into elements,] the right course to follow is to use the other type of analysis, which may be called *analysis into units*. By *unit* we mean a product of analysis which, unlike elements, retains all the basic properties of the whole and which cannot be further divided without losing them . . . . Unit analysis . . . demonstrates the existence of a dynamic system of meaning in which the affective and the intellectual unite. It shows that every idea contains a transmuted affective attitude toward the bit of reality to which it refers. (Vygotsky 1962, pp. 4, 8)

In short, we should beware of pursuing analytic dissection to the point at which we have excised the meaning of what we are studying. I take "mode of relatedness" to be a fundamental unit of analysis for characterizing early cognitive as well as social development.

In the passage just cited, Vygotsky makes a second, more specific point about thinking. He claims that *every* idea is intimately and essentially linked with an affective attitude. That attitude has an intrinsic connection with the world toward which it is directed (compare Goldie's, 2000, notion of "feeling towards"). Vygotsky's phrase "transmuted affective attitude" implies that ideas are distilled out of relations with the surroundings that are affectively configured. If this is so, then it suggests that "symbolic mental representations" emerge from person–world relations that have affective aspects, and also, for this reason, inherent links with what they represent. Moreover, what in later life becomes thought (sometimes relatively emancipated from feeling) is embedded in what becomes feeling (sometimes relatively emancipated from thought).

Werner (1948) proposed that "the awareness of objects during early childhood depends essentially on the extent to which objects can be responded to in motor-affective behavior" (p. 66). In primitive or concrete abstraction, things may be grouped according to the objects' "equal affective value" for the subject (Werner 1948, p. 232), as well as by their perceptual similarity. In true conceptual abstraction, by contrast, "the quality (e.g., a color) common to all the elements is deliberately detached—mentally isolated, as it were—and the elements themselves appear only as visible exemplifications of the common quality" (Werner 1948, p. 243). At the conceptual level, a person can shift points of view in a purposeful grouping activity.

I should like to draw attention to two features of this account. Werner's point about the child understanding objects through the motor and affective attitude of the subject highlights the significance of attitudes and relatedness for knowledge about the world (cf. Piaget's emphasis on the sensorimotor origins of thinking, which I shall not dwell on here; e.g., Piaget 1972). Then there is the notion that applying concepts might involve a person in shifting "points of view." If higher forms of abstracting and conceptualizing involve flexibilty in categorizing the world according to *various* points of view (Weigl 1941), then we shall need to see how very young children come to take, and subsequently understand, points of view.

Abstract notions of thought, feeling, and points of view are just that—abstracted from the people whose thoughts, feelings, and points of view they are. Critical for the development of thinking is communication among embodied individuals. In an essay entitled "Internalization of Higher Psychological Functions" (reprinted in Vygotsky 1978), Vygotsky writes:

> Every function in the child's cultural development appears twice: first, on the social level, and later, on the individual level; first, *between* people (*interpsychological*), and then *inside* the child (*intrapsychological*). This applies equally to voluntary attention, to logical memory, and to the formation of concepts. All the higher functions originate as actual relations between human individuals. (Vygotsky 1978, pp. 56–7)

Goings-on *between* people become internalized, so they become features of individual intelligence. Werner and Kaplan (1963/1984) develop this line of thinking by exploring how infants' emerging capacity to symbolize has an intimate relation to evolving forms of infant-caregiver transaction:

> In the course of development there is a progressive distancing or polarization between person and object of reference, between person and symbolic vehicle, between symbolic vehicle and object, and between the persons in the communication situation, that is, addressor and addressee. . . . Initially, inter-individual interaction occurs in purely sensory-motor-affective terms. Sooner or later, however, a novel and typically human relationship emerges, that of "sharing" of experiences, which probably

> has its clearest early paradigm in the nonreflexive smile of the infant in response to its mother's smile. . . . Thus, the act of reference emerges not as an individual act, but as a social one: by exchanging things with the Other, by touching things and looking at them with the Other. Eventually, a special gestural device is formed, *pointing* at an object, by which the infant invites the Other to contemplate an object as he does himself. . . . Whereas pointing entails only reference, the indication or denotation of a concretely present object, symbolization involves differentiation and integration of two aspects: reference to an object and representation of that object. In reference by pointing, the referent (the object) remains "stuck" in the concrete situation; in reference by symbolization, the characteristic features of the object (its connotations) are lifted out, so to speak, and are realized in another material medium (an auditory, visual, gestural one, etc.). (Werner and Kaplan 1963/1984, pp. 42–3)

In due course, names (which are designators linked to a concrete-pragmatic sphere of action) become words located in a linguistic field with semantic contrasts, and two-word combinations make explicit the child's capacity to differentiate between referent and attitude in topic-comment utterances (e.g., "Ball gone").

In their rather different ways, then, Vygotsky and Werner and Kaplan (and also Piaget) depicted how, from primary modes of relatedness between infants and their physical and human environment, toddlers develop capacities for conceptual thinking *about* the world. The question is: how are the requisite forms of sharing with others, distancing between self and others, and "lifting out" of connotations effected?

Philosophical Considerations

I am moving toward the proposal that an individual acquires symbolic thought along with a self-reflective grasp that one's takes on the world (through attitudes) are both separate from, and coordinated with, the subjective takes of other people. In order for the explanation to work, this "grasp" cannot itself be derived from already-established forms of thinking it is supposed to explain, even if a *full* grasp is attained only once such thinking has developed. Therefore there needs to be a non-intellectual, non-inferential mode of pre-understanding others.

A variety of philosophers have pointed to the primacy of interpersonal engagement, rather than intellectual deliberation, for grounding our understanding of minds. For example, in *Philosophical Investigations*, Wittgenstein (1958, p. 178) argues for the epistemological primacy of relatedness in our dealings with people: "My attitude towards him is an attitude towards a soul. I am not of the *opinion* that he has a soul." Malcolm (1962) explicates how the bedrock for explaining our understanding of other people does not lie in what we believe, and concludes with a quotation from Wittgenstein (1958, p. 226). Malcolm writes:

> As philosophers we must not attempt to justify the forms of life, to give reasons for *them*—to argue, for example, that we pity the injured man because we believe, assume, presuppose, or know that in addition to the groans and writhing, there is pain. The fact is, we pity him! "What has to be accepted, the given, is—so one could say—*forms of life*." (Malcolm 1962, p. 92)

We apprehend feelings *in* the expressions of other people (Wittgenstein 1980, vol. 2, §570). Our engagement with other people is such that it provides direct access to other people with their own subjective lives, through responsiveness to their expressions of feeling. But Wittgenstein (and compare Merleau-Ponty 1964) goes further than this. Wittgenstein indicates there is something primitive about the human ability to assume the actions and expressions of others:

> "I see that the child wants to touch the dog, but doesn't dare." How can I see that? Is this description of what is seen on the same level as a description of moving shapes and colours? Is an interpretation in question? Well, remember that you may also *mimic* a human being who would like to touch something, but doesn't dare. (Wittgenstein 1980, vol. 1, §1066)

There is something basic about the ability to transpose what one perceives in the expression of someone else, to one's own bodily expression and activity. This fact is relevant for understanding what the expression means.

I suggest that Wittgenstein's example captures one implication of the human capacity to "identify with" another person's attitudes (Freud 1955/1921), namely, the potential to assimilate the stance of someone else so that it becomes a potential stance for oneself. This explains how one could mimic, even if one does not explicitly do so. Taking up Wittgenstein's approach, Hampshire (1976) explored how something like identification (he calls it "the primitive faculty of imitation") provides a psychological bridge between people, or as he expresses it, "a necessary background to the communication of feeling" (p. 73). I would stress that identifying-with is not a kind of merging, in that it involves the individual adopting an other-person-centered stance. At the same time, it creates the potential for the individual to shift (either partially or more fully, and either in the present or future) *into* what he or she registers of that stance.

The Origins of Symbolizing

Having framed an interpersonal-affective context for early development, I want to build upon the groundwork of Werner and Kaplan in explicating the emergence of symbolic thinking. For this, we need more precise specification of what it is we are trying to explain. In 1923, Ogden and Richards published their classic work, *The Meaning of Meaning* (reprinted in 1985), and wrote as follows:

> Between a thought and a symbol causal relations hold. When we speak, the symbolism we employ is caused partly by the reference we are making and partly by social and psychological factors—the purpose for which we are making the reference, the proposed effect of our symbols on other persons, and our own attitude. When we hear what is said, the symbols both cause us to perform an act of reference and to assume an attitude which will, according to circumstances, be more or less similar to the act and the attitude of the speaker. (Ogden and Richards 1923, pp. 10–11)

Although Ogden and Richards emphasize communication insofar as one person intends to have an effect on others, they also indicate that symbols may express a speaker's attitudes, and elsewhere refer to the "emotive aspects of language" (p. 10). They stress that it is only when a thinker makes use of a symbol that the symbol stands for anything. Moreover, a symbol carries a thinker's *way* of thinking about and/or attitude to the referent. As Langer (1942) expresses it:

> Symbols are not proxy for their objects, but are *vehicles for the conception of objects*. To conceive a thing or a situation is not the same thing as to "react toward it" overtly, or to be aware of its presence. In talking *about* things we have conceptions of them, not the things themselves; and *it is the conceptions, not the things, that symbols directly "mean."* (Langer 1942, pp. 60–1)

Elsewhere, Langer points out the way in which animals may use signs to indicate things, but humans are unique in using them to represent. Correspondingly, words "let us develop a characteristic attitude toward objects *in absentia*, which is called 'thinking of' or 'referring to' what is not here" (Langer 1942, p. 31). Thus a symbol is coupled, for a certain subject, with a conception that fits an object or event. The connotation of a word is the conception it conveys.

It is one thing for a child to make something "refer to" or even "stand for" another. It involves more than this when the child employs a symbol with an understanding that the symbol means the same to the child him/herself as it does to other people. It was Mead (1934) who expounded this point most vigorously, insisting as he did so that in order to employ "significant symbols," an individual requires a degree of self-reflective awareness. Mead wrote:

> It involves not only communication in the sense in which birds and animals communicate with each other, but also an arousal in the individual himself of the response he is calling out in the other individual, a taking of the role of the other, a tendency to act as the other person acts. (Mead 1934, p. 73)

From the perspective of the infant learning to speak with symbols, "The signals the infant learns to produce are the same signals he comprehends when produced by others, and therefore in using these symbols he anticipates how they will be interpreted" (Kaye 1982, p. 150). The position of the listener is that already described by Ogden and Richards (1923, p. 11): "When we hear what is said, the symbols both cause us to perform an act of reference and to assume an attitude which will, according to circumstances, be more or less similar to the act and the attitude of the speaker." Although a child might begin with but a partial understanding of what a symbol is, for example, by comprehending a symbolic meaning while as yet unable to use the symbol adequately or flexibly, this simply illustrates that the child has some way to go before she appreciates how symbols function.

The story about the development of linguistic reference and predication is also a story about an infant's growing capacities for abstraction and thought. "For a thought to be a genuine thought it must be about something, and something else must be thought about that something; so there is in that thought something corresponding to the distinction between subject and predicate that becomes explicit in language itself" (Hamlyn, 1978, p. 76). In what follows, I illustrate how infants come to realize (albeit not in so many words) how attitudes have aboutness, how people can intend to convey their attitudes about the world to others in subject-predicate communicative acts, and how all this has a profound effect on their ability to think.

Early Typical Development

An observer who watches a two-month-old infant relating to his or her caregiver is likely to be struck by the fluid interweaving of each participant's facial expressions, vocalizations, and gestures within coordinated interpersonal exchanges (Trevarthen 1979). Perhaps the most telling demonstration that a dyadic system is in play is when a perturbation is introduced to disrupt such mutual relatedness. Consider the following description of a typically developing two-month-old videotaped during a still-face procedure (Cohn and Tronick 1983) conducted by colleagues and myself. When the mother assumed the still face and unreactive posture as requested, the infant responded by becoming uneasy, restless, and jerky in her movements, and lost the infectious smiling and smooth tonguing movements that had been evident just moments before. Her bright, protracted gazes into her mother's eyes were transformed into brief, checking glances. After about 40 seconds her behavior changed again, and she began to give longer looks to her mother accompanied by forced smiles. There was a strong impression that she was making bids to re-establish contact with her mother, trying to elicit a resumption of the joyful interpersonal exchange that was now missing.

These observations illustrate how infants participate in experience with another person. Sharing experience with someone else is not merely like having one's own experience of the world, and then adding something. It is more like having one's own subjective state and registering something of the other's attitudes conjointly, in a qualitatively new form of experience. The other person has a physical location, and the person's bodily expressiveness has a marked impact on the dyadic exchanges.

Then, toward the end of the first year of life, there is a watershed in the emergence of the infant's capacity to identify with the attitudes of someone else. Now the infant engages in new forms of joint attention that appear to reflect awareness (not yet full understanding) of other points of view (e.g., Bretherton, McNew, and Beeghly-Smith 1981; Carpenter, Nagell, and Tomasello 1998). For example, colleagues and I videotaped a 13-month-old girl seated across a table from an adult tester. After a period playing together, the tester secured the child's gaze to her face and then looked to her right while extending her right arm and finger into a point and exclaiming, "Look at that." Initially, the child's gaze lingered on the tester's outstretched hand, then returned to the tester's face as if to ascertain what the latter might be trying to communicate. For a moment she dwelt on her face, then suddenly shifted her gaze to a target beyond the tester's still outstretched finger. This illustrates something about the means by which infants achieve shared reference by their psychological movement *through* the other. It is not simply that the other's point or gaze or other gesture serves as a signpost to objects and events in the world. It is also that the infant is drawn into alignment with the other's orientation toward a shared world.

Moreover, and importantly, an infant's attitude to objects and events in the world may be altered through the infant's appraisal of the attitudes of another person. When an infant is confronted with an emotionally ambiguous object or event, the infant is likely to look and then respond to the affective expression of a parent, as this has directedness to the object or event in question. In one well-known early study of such social referencing (Sorce et al. 1985), what seemed like a cliff prevented the child reaching a goal. The majority of 12-month-olds who perceived that their mothers were looking to the cliff with smiles tentatively proceeded toward their goal, whereas none of those who witnessed their mothers showing fear toward the cliff did so. The cliff became an object of fear, through the mother's attitude toward this specific part of the environment.

What all this indicates is that by the end of the first year of life, not only does the typically developing infant experience self and other as distinct and potentially separable sources of attitudes toward the world, but the infant can also link in with, and at times assume, such attitudes for himself or herself. I do not think this justifies the claim that infants of this age have a conceptual understanding of another person's subjective orientation. Rather, the infant is endowed with a non-inferential, affectively grounded responsiveness structured in terms of self–other relations embedded in a shared world. Such responsiveness provides the foundations for the acquisition of conceptual understanding over the coming months of life. It establishes the conditions for achieving not only interpersonal connectedness, but also distancing between self and other. Through the other, the infant can be moved from one person-anchored stance/attitude/perspective on a given object or event, to another and distinct person-anchored stance/attitude/perspective toward that very same object or event. It is a relatively small step for the infant to become aware that a single object or event can have multiple meanings for persons, and multiple meanings for him- or herself. This form of distancing is needed for reference to become representation.

As a child moves out of infancy in the middle of the second year, he or she comes to *think about* self and other as separate individuals who have their own takes on the world. At this age, toddlers come to make self-descriptive utterances such as "my book" or "Mary eat" (Kagan 1982, p. 371), they show abilities to comply and cooperate with others, and they engage in coordinated role-responsive interactions (e.g., Kaler and Kopp 1990). Hoffman (1984) tells the story of how Marcy, a girl of 20 months, drew her sister away from playing with a toy she wanted by climbing up on her sister's favorite horse and crying: "Nice horsey! Nice horsey!" (p. 109). The capacity to symbolize in play becomes manifest from around the middle of the second year of life, hand in hand with these more or less explicit forms of emotional understanding, role-taking, and self-reflective awareness (Hobson 2000; Zahn-Waxler et al. 1992).

An observation on the acquisition of personal pronouns is a reminder of the role of identifying-with in these developments. Charney (1980) recorded that among one- or two-year-olds, nearly all of the first uses of "my" were produced when the child was grabbing, acting on, or claiming an object, usually when the object was *not* the child's. In order to have learned the meaning of "my," the child would have heard someone else use the word when that *other person* was appropriating something. Through identifying-with *and* anchoring the linguistic term "my" in the attitude of the other, the child came to express his or her same attitude with the same linguistic term. This instance exemplifies a number of features of communication and symbolizing already highlighted, not least how symbols arise and function through reversible role-taking, and how they anchor and evoke person-centered attitudes.

Here, then, we discern the underpinnings and structure of an important species of symbolic thinking. The ability to apprehend, respond to, and be moved by the subjective states of other people in relation to the world is that such movements create mental space for negotiating attitudes and meanings. It is only once an infant has experience of shifting across person-anchored stances through assimilating another person's attitudes—but at the same time, registering the source of those attitudes as other—that it becomes possible for the infant to achieve understanding of what it means to have alternative perspectives.

Understanding what it means to have a perspective is part of achieving a conceptual grasp of the relation between selves and the world. It is the essence of the "lifting out" process by which symbolic meanings are generated and anchored in representations. The very idea of a self is that each person is a self with his or her own take on the world, a take that may be shared or challenged, aligned with or repudiated. As Mead (1934) emphasized, this not only involves taking the role of the other in communicative exchanges, but also is critical for the capacity knowingly to introduce new meanings/perspectives on the materials of symbolic play. If I make this box a pretend-bed, I knowingly apply bed meanings to something that yields up its identity for the purposes of play. So by the middle of the second year of life, already-established pre-reflective and intuitive forms of role-taking based on identifying-with are yielding both the means to conceptualize (symbols), and contents to conceptualize selves-with-attitudes in relation to a world that is the focus of attitudes and co-reference.

According to this account, it is in the propensity to dwell in the experience of the other, and to experience the world through the other—that is, to identify with the attitudes of another person from the stance of the other—that the specialness of human forms of sharing are grounded. The kinds of "distancing" described by Werner and Kaplan are achieved through an infant relating through others to a shared world, but one that can have new meanings according to the person-centered stance from which it is newly apprehended (Hobson 1990). Moreover, one-year-olds can come to achieve distance from their own attitudes, through identifying with the attitudes of others toward themselves.

Insights from Autism

If it is the case that interpersonal engagement and identifying-with the attitudes of others plays a pivotal role in the genesis of creative, flexible symbolic thinking, then it follows that individuals who are seriously compromised in identifying with others should manifest corresponding abnormalities and limitations in thinking.

Consider this condensed version of a classic description of a person with autism called L (Scheerer, Rothmann, and Goldstein 1945). L was first seen at the age of 11 with a history of severe learning difficulties. He was said never to have shown interest in his social surroundings. Although he had an IQ of only 50 on a standardized test of intelligence, L could also recount the day and date of his first visit to a place, and could usually give the names and birthdays of all the people he met there. He could spell forwards and backwards.

L's background history included the fact that in his fourth and fifth years, he rarely offered spontaneous observations or reasons for any actions or perceived event. Nor would he imitate an action of others spontaneously. He was unable to understand or create an imaginary situation. He did not play with toys, nor did he show any conception of make-believe games. He was unable to converse in give-and-take language. He barely noticed the presence of other children, and was said to have "little emotionality of normal depth and coherence."

Up to 15 years of age, L was unable to define the properties of objects except in terms relating to his own use of the objects or to specific situations. For example, he defined orange as "that I squeeze with," and an envelope was "something I put in with." He could neither grasp nor formulate similarities, differences, or absurdities, nor could he understand metaphor. At the age of 15 he defined the difference between an egg and a stone as "I eat an egg and I throw a stone." Once, when the doctor said, "Goodbye, my son," L replied, "I am not your son." When asked "What would happen if you shot a person?" L replied, "He goes to the hospital." He showed no shame in parading naked through the house.

I give this case description for two reasons. First, it captures something of the quality of rigid thinking that typifies autism—a "stuckness" that does not extend to all kinds of thought. Second, it sets L's cognitive limitations in a broader context (see also Kanner 1943). Not only do we see that there is specificity to L's impairments—indeed, that in some respects he has exceptional cognitive abilities as well as disabilities—but also our attention is drawn to aspects of L's communicative and social-emotional relations that might bear an intimate relation to the nature of his thinking disorder. For example, L would rarely see someone else's actions and make those his own through imitation; he seemed not to engage with others' feelings or perspectives in either verbal or nonverbal give-and-take exchanges; and he appeared to be unselfconscious.

It is well known that children with autism have limitations in joint attention and other kinds of person-with-person engagement well before they could be expected to conceptualize minds (e.g., Charman et al. 1997). They are atypical in their relative lack of engaging in "sharing" forms of joint attention and in social referencing (Sigman et al. 1992). All this makes it plausible that, as I argued some years ago (e.g., Hobson 1990, 1993a, 1993b), it is in the children's limited engagement with other people's attitudes, both in one-to-one mutual exchanges and in person–person–world interactions such as those of joint attention and social referencing, that we find the *source* of later deficits in interpersonal understanding (so-called theory of mind), communication, and thinking.

Identifying-with: The Case of Autism

More specifically, I suggest, the final common pathway to autism is a limitation in the children's propensity to identify with the attitudes of others toward a shared world (Hobson 2002, 2007). A part of what it means to identify with someone else is that one individual apprehends and responds to the other-person-centered source of attitudes. In other words, one is affectively engaged with and cares about the other person's feelings *as* the other's. In a program of research on social emotions among children and adolescents with autism (Hobson et al. 2006), we interviewed parents of matched children with and without autism. Parents felt they could recognize in their children with autism not only emotions such as anger and fear and emotional responsiveness to other people's mood states, but also pride and jealousy. Yet in contrast with parents of children without autism, rarely were they able to report that their children showed more person-focused emotions or qualities of relatedness, for example, guilt or empathic concern.

Here we can see that understanding minds involves more than thought. If to think were enough to cause a sea change in personal relations, then children's acquisition of seemingly coherent thoughts about people's feelings, wishes, and so on—and many children with autism do acquire such thoughts—would be the harbinger of radical changes in clinical features of autism. Yet such alterations are often modest. The fact is that feelings are constitutive of the kinds of thought about minds (or more fundamentally, persons) that really make a difference.

Now consider a study of communication that focuses upon the kind of immediate, unreflective (and unconceptualized) role-taking that characterizes the process of identifying with the attitudes of someone else. Hobson and Meyer (2005) presented a "sticker test" in which children needed to communicate to another person where on her body she should place her sticker-badge. The majority of children without autism pointed to a site on their own bodies to indicate the tester's body, that is, anticipating that the other person would identify with their act of identifying with her body. The children with autism rarely communicated in this way; instead, most pointed to the body of the investigator to indicate where the sticker should be placed. Although it is possible that flexible self–other communication depends upon thinking or understanding other people's minds, we consider it more likely that such seemingly effortless and natural stance-shifting among children without autism reflects a more basic form of self–other connectedness and differentiation that has a cognitive aspect—not as a cause, but as an essential property of such relations—and affective and motivational aspects, too. Since the "sticker test," a body of research has accumulated to confirm that, indeed, there is fragility in autistic individuals' propensity to identify with the attitudes of others (e.g., Hobson et al. 2012).

Symbolic Play in Autism

Symbolic play is an especially clear manifestation of children's ability knowingly to apply alternative meanings to materials that do not usually have these meanings. There is substantial evidence that children with autism are limited in their creative representational play, especially in their spontaneous play (e.g., Lewis and Boucher 1988; Wing et al. 1977). On the other hand, many children with autism do achieve some ability to make one thing stand for another. The important question here is whether the developmental underpinnings of the play they achieve, and therefore the qualities and creativity of this play, are the same as in the case of typically developing children.

Colleagues and I compared two matched groups of children, one with and one without autism, for their ability to engage in symbolic play (Hobson, Lee, and Hobson 2009). As it turned out, these two groups were similar in the mechanics of play, that is, on ratings of whether they were able to make one thing stand for another, or represent absent properties, or pretend that something was present when it was not. Yet there were significant group differences insofar as the children with autism were rated as showing less creativity and fun; they were less invested in the new meanings of the play; and they were less aware of themselves as initiators of the new meanings. Of course, these results might be interpreted as demonstrating that the children with autism were perfectly able to symbolize in play, but, as a separate matter, they showed little affective and motivational investment. In our view, by contrast, these very qualities of affective and motivational investment betrayed what it meant for the children *without* autism to symbolize.

The investment and fun that accompanied their play was intrinsic to their symbolic thinking, and reflected the interpersonal transactions from which this form of thinking was derived.

The Case of Congenital Blindness

We considered that children with congenital blindness might be at serious risk for developing autism. Our reasoning was as follows. Among young sighted children, vision affords not only perception of affective expressions as belonging to physically located, embodied individuals, but also experience of the *directedness of another person's attitudes* toward given objects and events in visually specified, shared surroundings. Blind children need to rely on senses that are less suited to navigate the geometry that configures emotionally structured child–adult–world transactions. Without vision, it must be very difficult (though not impossible) to register shifts from one "take" to another *through* movement to the stance of another embodied person vis-à-vis given objects or events that are perceived to be at the focus of the other person's attitudes.

It is not that touch, hearing, and perhaps other modes of perception are totally without value for such intentionally directed role-shifting. Indeed, these senses need to be recruited in appropriate ways to foster blind children's appreciation of joint attention and coordinated perspectives. But congenitally blind children are at risk of seriously impoverished experience of interpersonally co-referential attitudes to a visually specified world. Without vision, they are deprived of opportunities to experience first one set of attitudes to something, and then movement to different other-person-derived takes on that something. Such shifts lead a child to discover—first by registering, only later by understanding—how a given something can be construed in more than one way, or have more than one person-anchored description.

The empirical question was whether there is in fact substantial clinical overlap between a group of congenitally blind children and sighted children with autism. Colleagues and I documented that, indeed, clinical features like those of autism are highly prevalent among congenitally blind children. Brown et al. (1997) reported that approximately one half of such children between the ages of four and eight in special schools for the blind (this was not an epidemiological study) who were without obvious neurological impairment met the criteria for the full syndrome of autism, and many of those who did not have the diagnosis displayed some features of autism. Congenitally blind children showed atypicalities in communication (for instance, confusions over the use of personal pronouns), activity, and symbolic play very similar to those seen among sighted children with autism. Importantly, the more socially impaired blind children, and not necessarily those with the lowest IQs, were most limited in their creative symbolic play (Bishop, Hobson, and Lee 2005).

Conclusion

To conclude, it may be worth returning to the original passage in which Scheerer et al. (1945) summarized what they meant by L's impairment in abstract attitude in order to see whether one might encompass their formulation within a social-developmental account of creative symbolic thinking. These authors considered the abstract attitude to be a "common functional basis" for the following (among other things):

> To behold simultaneously different aspects of the same situation or object, to shift from one aspect to another; to understand a general frame of reference, a symbolic meaning as relation between a given specific percept and a general idea; to evolve common denominators, to reason in concepts, categories, principles; to assume different mental states . . . to plan ahead ideationally . . . to behave symbolically (e.g., demonstrating, make belief, etc); to reflect upon oneself, giving verbal account of acts; to detach one's ego from a given situation or inner experience; to think in terms of the "mere possible," to transcend the immediate reality and uniqueness of a given situation, a specific aspect or sense impression. (Scheerer et al. 1945, p. 37)

I believe that the forms of abstract attitude and detachment that are limited but usually not absent in autism are precisely those for which appropriately patterned intersubjective experience is necessary. Note how in the formulation by Scheerer et al., as well as in the case illustration of L with which we began, there is a close connection between aspects of self-reflective awareness and thinking. In particular, L had difficulty reflecting on himself, detaching himself from a situation or inner experience, and transcending immediate reality. The story I have sketched about young children's conceptual development in differentiating between self and other with their distinct and yet connected attitudes to a shared world is also a story about the growth in infants' understanding of the differentiation between people's takes and attitudes toward objects or events, and those objects or events as foci for such attitudes. In accordance with the ideas of Mead (1934) as well as Werner and Kaplan (1963/1984), the achievement of self-reflective awareness and the ability to think about self and other appears to be one side of a coin, of which the other face is the ability to grasp how symbols can function as the means to thought. One could not think about self and other without symbols, but one could not achieve the requisite ability to use symbols without grasping how symbolic vehicles and their referents are linked and yet differentiated—something that requires an understanding of what it means to take alternative person-anchored perspectives on a shared world.

Self/other awareness and role-shifting, with the movement among alternative perspectives that these entail, are intrinsic to a certain mode of creativity, flexibility, and context-sensitivity in thinking. I have suggested that these features of mental functioning are structured and motivated by the social-developmental process of identifying with other people's attitudes. If certain forms of difficulty in generating ideas are linked with communicative impairments in autism, this may be explained by the importance of communicative transactions for establishing the capacity to generate new perspectives within an individual's own mind.

In this developmental perspective on the origins of symbolic thinking, we have taken a fresh look at social vis-à-vis individual factors in the genesis of thought, and re-examined how relations among cognition, conation, and affect evolve over infancy and toddlerhood. In early life and beyond, it seems, the structure of emotional relations is far less dependent on cognition than many philosophers and psychologists have supposed (e.g., Hobson 2012), and the nature of thinking is far more dependent on feeling-imbued social and non-social relatedness than many have appreciated. Thinking is embedded in dynamic relations that transcend the individual, and draw upon more than cognitive dimensions of experience.

References

Bishop, M., Hobson, R.P., and Lee, A. (2005). Symbolic play in congenitally blind children. *Development and Psychopathology*, 17, 447–65.

Bretherton, I., McNew, S., and Beeghly-Smith, M. (1981). Early person knowledge as expressed in gestural and verbal communication: when do infants acquire a "theory of mind? In: M.E. Lamb and L.H. Sherrod (eds.), *Infant social cognition: empirical and theoretical considerations*. Hillsdale, NJ: Erlbaum, pp. 333–73.

Brown, R., Hobson, R.P., Lee, A., and Stevenson, J. (1997). Are there "autistic-like" features in congenitally blind children? *Journal of Child Psychology and Psychiatry*, 38, 693–703.

Buber, M. (1937/1958). *I and Thou* (2nd ed.; trans. R.G. Smith). Edinburgh: Clark.

Carpenter, M., Nagell, K., and Tomasello, M. (1998). Social cognition, joint attention, and communicative competence from 9 to 15 months of age. *Monographs of the Society for Research in Child Development*, 63, 1–174.

Charman, T., Swettenham, J., Baron-Cohen, S., Cox, A., Baird, G., and Drew, A. (1997). Infants with autism: an investigation of empathy, pretend play, joint attention, and imitation. *Developmental Psychology*, 33, 781–9.

Charney, R. (1980). Speech roles and the development of personal pronouns. *Journal of Child Language*, 7, 509–28.

Cohn, J.F. and Tronick, E.Z. (1983). Three-month-old infants' reaction to simulated maternal depression. *Child Development*, 54, 185–93.

Freud, S. (1955/1921). Identification. In: J. Strachey (ed.), *The standard edition of the complete psychological works of Sigmund Freud* (vol. 18). London: Hogarth, pp. 105–10.

Goldie, P. (2000). *The emotions: a philosophical exploration*. Oxford: Clarendon.

Hamlyn, D.W. (1978). *Experience and growth of understanding*. London: Routledge and Kegan Paul.

Hampshire, S. (1976). Feeling and expression. In: J. Glover (ed.), *The philosophy of mind*. Oxford: Oxford University Press, pp. 73–83.

Hobson, R.P. (1990). On acquiring knowledge about people and the capacity to pretend: response to Leslie. *Psychological* Review, 97, 114–21.

Hobson, R.P. (1993a). *Autism and the development of mind*. Hove, UK: Erlbaum.

Hobson, R.P. (1993b). The emotional origins of social understanding. *Philosophical Psychology*, 6, 227–49.
Hobson, R.P. (2000). The grounding of symbols: a social-developmental account. In: P. Mitchell, and K.J. Riggs (eds.), *Reasoning and the mind*. Hove, UK: Psychology Press, 11–35.
Hobson R.P. (2002). *The cradle of thought*. London: Macmillan (and 2004; New York: Oxford University Press).
Hobson, R.P. (2007). Communicative depth: soundings from developmental psychopathology. *Infant Behavior and Development*, 30, 267–77.
Hobson, R.P. (2008). Interpersonally situated cognition. *International Journal of Philosophical Studies*, 6, 377–97.
Hobson, R.P. (2012). Emotion as personal relatedness. *Emotion Review*, 4, 169–75.
Hobson, R.P., Chidambi, G., Lee, A., and Meyer, J. (2006). Foundations for self-awareness: an exploration through autism. *Monographs of the Society for Research in Child Development*, 71, 1–165.
Hobson, R.P., Hobson, J.A., Garcia-Perez, R., and Du Bois, J. (2012). Dialogic resonance and linkage in autism. *Journal of Autism and Developmental* Disorders, 42, 2718–28.
Hobson, R.P., Lee, A., and Hobson, J.A. (2009). Qualities of symbolic play among children with autism: a social-developmental perspective. *Journal of Autism and Developmental Disorders*, 39, 12–22.
Hobson, R.P. and Meyer, J.A. (2005). Foundations for self and other: a study in autism. *Developmental Science*, 8, 481–91.
Hoffman, M.L. (1984). Interaction of affect and cognition in empathy. In: C.E. Izard, J. Kagan, and R.B. Zajonc (eds.), *Emotions, cognition and behaviour*. Cambridge: Cambridge University Press, pp. 103–31.
Kagan, J. (1982). The emergence of self. *Journal of Child Psychology and* Psychiatry, 23, 363–81.
Kaler, S.R. and Kopp, C.B. (1990). Compliance and comprehension in very young toddlers. *Child Development*, 61, 1997–2003.
Kanner, L. (1943). Autistic disturbances of affective contact. *Nervous Child*, 2, 217–50.
Kaye, K. (1982). *The mental and social life of babies*. London: Methuen.
Langer, S.K. (1942). *Philosophy in a new key*. Cambridge, MA: Harvard University Press.
Lewis, V. and Boucher, J (1988). Spontaneous, instructed and elicited play in relatively able autistic children. *British Journal of Developmental Psychology*, 6, 325–39.
Malcolm, N. (1962). Wittgenstein's *Philosophical Investigations*. In: V.C. Chappell (ed.), *The philosophy of mind*. Englewood Cliffs, NJ: Prentice-Hall, pp. 74–100.
Mead, G.H. (1934). *Mind, self, and society* (ed. C.W. Morris). Chicago: University of Chicago Press.
Merleau-Ponty, M. (1964). The child's relations with others (trans. W. Cobb). In: M. Merleau-Ponty, *The primacy of perception*. Evanston, IL: Northwestern University Press, pp. 96–155.
Ogden C.K. and Richards, I.A. (1923/1985). *The meaning of meaning*. London: Routledge.
Piaget, J. (1972). *The principles of genetic epistemology*. London: Routledge and Kegan Paul.
Scheerer, M., Rothmann, E., and Goldstein, K. (1945). A case of "idiot savant": an experimental study of personality organization. *Psychological Monographs*, 58, 1–63.
Sigman, M.D., Kasari, C., Kwon, J.H., and Yirmiya, N. (1992). Responses to the negative emotions of others by autistic, mentally retarded, and normal children. *Child Development*, 63, 796–807.
Sorce, J.F., Emde, R.N., Campos, J., and Klinnert, M.D. (1985). Maternal emotional signaling: its effect on the visual cliff behavior of 1-year-olds. *Developmental Psychology*, 21, 195–200.

Trevarthen, C. (1979). Communication and cooperation in early infancy: a description of primary intersubjectivity. In: M. Bullova (ed.), *Before speech: the beginning of human communication*. Cambridge: Cambridge University Press, pp. 321–47.
Vygotsky, L.S. (1962/1968). *Thought and language* (trans. E. Hanfmann and G. Vaker). Cambridge, MA: MIT Press.
Vygotsky, L.S. (1978). Internalization of higher psychological functions. In: M. Cole, V. John-Steiner, S. Scribner, and E. Souberman (eds.), *Mind in society: the development of higher psychological processes*. Cambridge, MA: Harvard University Press, pp. 52–7.
Weigl, E. (1941). On the psychology of so-called processes of abstraction (trans. M.J. Rioch). *Journal of Abnormal and Social Psychology*, 36, 3–33.
Werner, H. (1948). *Comparative psychology of mental development*. Chicago: Follett.
Werner, H. and Kaplan, B. (1963/1984). *Symbol formation*. Hillsdale, NJ: Erlbaum.
Wing, L., Gould, J., Yeates, S.R., and Brierly, L.M. (1977). Symbolic play in severely mentally retarded and in autistic children. *Journal of Child Psychology and Psychiatry*, 18, 167–78.
Wittgenstein, L. (1958). *Philosophical investigations* (trans. G.E.M. Anscombe). Oxford: Blackwell.
Wittgenstein, L. (1980). *Remarks on the philosophy of psychology* (vols. 1–2; ed. G.H. von Wright and H. Nyman; trans. C.G. Luckhardt and M.A.E. Aue). Oxford: Blackwell.
Zahn-Waxler, C., Radke-Yarrow, M., Wagner, E., and Chapman, M. (1992). Development of concern for others. *Developmental Psychology*, 28, 126–36.

CHAPTER 30

ENACTING AFFECTIVITY

GIOVANNA COLOMBETTI

THE philosophical debate on the "4Es"—the embodied, embedded, enactive, and extended nature of the mind—has focused primarily on cognition, understood as distinct from affectivity. Yet there is a lot that recent debates on the 4Es can contribute to an analysis of affectivity, thereby deepening our understanding of this complex phenomenon.[1] In this chapter I focus on the "enactive" approach (or "enactivism") and provide an overview of the contribution that this framework can make to our understanding of affectivity. I shall favor breadth rather depth, as I take it that readers of a handbook are interested primarily in getting a general sense of the issues, questions, and possibilities that pertain to a certain domain of inquiry with which they would like to become more familiar. I will draw largely on my own work (past and current) on enactivism and affectivity, provide further references for readers who want to learn more, and also indicate questions that still remain unanswered and that, in my view, need to be addressed in the future.

WHAT IS ENACTIVISM?

Enactivism is arguably the most complex of the 4Es. It entails that cognition is both "embodied" (realized, enacted, or "brought forth" not just by the brain but by the whole

[1] The terms "affect," "affectivity," and "emotion" are not used in the same way across and even within disciplines. Some philosophers use the term "affect/affects" as synonymous with "emotion/emotions"; others explicitly distinguish between the two and single out "affect" as unconscious and nonspecific (e.g., Massumi 2002). Many psychologists use the term "affect" today to refer to feeling states characterized by "valence" (pain or pleasure) and "arousal" (the intensity of the feeling), and taken to characterize all emotional states (e.g., Russell 2003), but even this use is not universal—for example, Ekman's idea that emotions are "affect programs" (e.g., Ekman 2003). I do not use the term "affect" here. My preferred word is "affectivity," which, as I clarify below, I use broadly to refer to the capacity of organisms (not just human ones) to be sensitive to, and affected by, what matters to them. As such, affectivity includes emotions, moods, and motivational states (see main text).

organism) and "embedded" (realized by the organism in interaction with the environment). Whether it also entails that cognition is "extended" is a matter of current debate; as I will argue in the third section of the chapter, I think that enactivism does allow the mind to "extend" into parts of the world. But in addition to its overlap with the other three E's, enactivism is a particularly complex framework because it is in effect a synthesis of quite different but complementary approaches in cognitive science, phenomenology, and biology. It is also a developing framework, one that acknowledges its own incompleteness and the existence of a variety of challenges that still need to be addressed.

So, what is the enactive approach? Historically, the term "enaction" was made popular in cognitive science by Varela, Thompson, and Rosch (1991), who wrote:

> We propose as a name the term *enactive* to emphasize the growing conviction that cognition is not the representation of a given world by a pregiven mind but is rather the enactment of a world and a mind on the basis of a history of the variety of actions that a being in the world performs. (9)

Varela et al. (1991) in particular emphasized that "(1) perception consists in perceptually guided action and (2) cognitive structures emerge from the recurrent sensorimotor patterns that enable action to be perceptually guided" (p. 173). The main idea conveyed by these passages is that cognition emerges from the actions or performances of a sensorimotor being. The idea that perception, specifically, is a form of action, has been developed further more recently by Susan Hurley, Alva Noë, and Kevin O'Reagan (e.g., Hurley 1998; O'Reagan and Noë 2001; Noë 2004), and is today often regarded as a central claim of enactivism: the world that we perceive is "enacted" in the sense that action is constitutive of how we perceive the world.

Sensorimotor integration is, to be sure, a very important enactive theme, but it would be restrictive to reduce enactivism to the claim that perception is a form of action. Varela et al. introduced several other themes that are distinctive of enactivism in its widest sense. One of these is the centrality that *lived experience* should be assigned in the study of the mind. In fact, at the very beginning of their book, Varela et al. wrote that "The new sciences of mind need to enlarge their horizon to encompass both lived human experience and the possibilities of transformation inherent in human experience" (1991, p. xv). On that same first page, they presented their approach as a continuation of the philosophy of Merleau-Ponty, and emphasized the need for a *mutual circulation* between the analysis of our biological and phenomenological embodiment—i.e., between our existence as living organisms, and our experience of ourselves as lived bodies. Another theme that is central to the enactive approach is its emphasis on *biological autonomy* and the idea that we are cognitive systems because we are autonomous organisms (see Varela 1979). This theme is a development of Maturana and Varela's (1972/1980) theory of autopoiesis—a theory of living systems as cognitive in virtue of their self-maintaining and self-identifying nature. Although enactivists have distanced themselves from the original formulation of the theory of autopoiesis[2] (see Di Paolo and Thompson 2014), they are still interested in the organization of living systems and in clarifying how that organization makes cognition possible.

All these themes have been synthesized more recently by Evan Thompson (Thompson 2007, 2011) and others, and make up what I regard as the "canonical" version of enactivism. Note, however, that the term "enactivism" has recently been used by philosophers coming from other traditions as well, such as Hutto and Myin (2013). The latter have proposed what they regard as a "radicalized" version of enactivism, characterized by the rejection of all representational talk, and of related talk of the "content" and "vehicles" of mental states. Their work, however, engages neither with phenomenology nor with theories of biological autonomy. In this sense, it is quite far removed from the spirit of canonical enactivism. One pending question that needs to be addressed in the future is whether and how these two versions of enactivism relate to one another, and whether there is a set of core claims that can be associated with enactivism as a unitary framework, and that distinguishes it from other "embodied" and/or "embedded" approaches.

Addressing this question, however, will take me too far astray from my main goal for this chapter, so I will leave it aside. In this chapter I shall focus exclusively on canonical enactivism (which I call simply "enactivism") as the framework that, in my view, has the most suitable conceptual resources to address the phenomenon I am most interested in—namely, affectivity.

Enactive Themes and Affectivity

Because of its emphasis on lived experience and biological autonomy, and the related call for a mutual circulation between biological and phenomenological descriptions of the living organism, enactivism can be a very productive framework for thinking about a variety of affective phenomena. In this section I provide a broad overview of some of the ways in which enactivist ideas can be applied to affectivity. The aim of this overview is not so much to develop arguments in detail, but to give the reader a general sense of what can be said about affectivity from an enactive perspective.[3]

[2] It is therefore misleading to call this version of enactivism "autopoietic" (as, e.g., Hutto and Myin 2013 do).

[3] I provide references to further works along the way. The interested reader is directed to Colombetti (2014) for a more in-depth elaboration of the ideas presented in this section.

The Inherently Affective Character of Enactive Cognition

One important and distinctive implication of enactivism in the domain of affectivity is that the mind, including cognition, is at its roots inherently affective and, relatedly, that affectivity permeates the mind—namely, affectivity is not a distinct "part" of the mind that merely "interacts" with other, non-affective parts; if we take affectivity away from the mind, we do not have a mind any more.

To appreciate this claim, it is important to note that saying that the mind is affective is not the same as saying that the mind is "emotional." Affectivity is best understood as a broad capacity to be affected or "touched" by something. Affectivity in this broad sense entails a lack of indifference, and rather a sensibility, interest, and concern; it is required for anything to *matter* to an agent. The folk psychological concept of emotion, on the other hand, usually refers to the capacity to undergo various different emotions, such as anger, sadness, joy, envy, hope, and many others. These, on their part, are conceived primarily as intense or even overwhelming experiences. It seems correct to say that the mind is not emotional in the sense of always undergoing some emotional episode—if we understand the latter as intense experiences categorizable as "happiness," "fear," "surprise," etc. But if we consider that there is much more to affectivity than episodes of this kind, then the claim that the mind is inherently affective immediately appears more plausible. Non-emotional affective states may include at least moods (feeling cranky, bored, upbeat, up or down, having the blues, etc.), long-term dispositional sentiments (love, hate, etc.), and motivational states (desire, hunger, pleasure, pain, etc.); when we are in these states, things matter to us in one way of another; we are affected by them and our condition.

What is distinctive about enactivism is that it provides a theory of biological organization and of its relation to the mind that entails that not just emotions, moods, motivational states, etc., are affective, but that *cognition* is too. More precisely, as we are about to see, enactivism claims that the hallmark of cognition is "sense-making," and a close look at this notion reveals that sense-making is simultaneously a cognitive and an affective phenomenon.

So here is the story.[4] According to enactivism, all living systems are sense-making systems in virtue of their organization. This organization is one of *precarious adaptive autonomy*. Autonomy refers here to the interdependence of the constituent processes of the living system, such that they "(1) recursively depend on each other for their generation and their realization as a network, (2) constitute the system as a unity in whatever domain they exist, and (3) determine a domain of possible interactions with the environment" (Thompson 2007, p. 44). Living systems are also adaptive, that is, able to monitor and regulate themselves with respect to their conditions of viability, and to improve their situation when needed. Finally, living systems are precarious in the sense that their

[4] I can provide here only a succinct account. For more details, see Di Paolo (2005), Thompson (2007), Thompson and Stapleton (2009), Di Paolo and Thompson (2014).

component processes will run down or cease to exist in the absence of the enabling relations established by the rest of the system. A living system, in other words, is a network of interrelated processes that has to work continually in order to maintain itself, to counteract the unstable and decaying tendencies of its component processes. This precarious nature is also what makes the living system continually seek interactions with its surroundings, to get the necessary energetic and material resources.

Importantly, enactivism also provides a *phenomenological* characterization of life, inspired primarily by Hans Jonas (1966). On this characterization, living systems, in virtue of their organization, have a *perspective* or *point of view* from which their surroundings acquire a certain meaning. Varela initially used the term "world-making" (see Varela 1991) to refer to the transformation of the physiochemical environment of the organism into a world that is meaningful to it. Since Weber and Varela (2002), the preferred term has become "sense-making." Drawing on Uexküll's (1934/2010) terminology, we can say that enactive sense-making refers to the transformation of the organism's *Umgebung* (its physical and chemical context) into an *Umwelt* (the environment from the perspective of the organism). For enactivists, this process of enacting or creating a world of meaning is the most basic form of cognition.[5]

If we understand affectivity in the broad way suggested earlier, we can easily see that sense-making, as the constitution of an *Umwelt*, is inherently an affective phenomenon: the constitution of an *Umwelt* is by definition the constitution of a world that is relevant for the organism; the *Umwelt* is the world as it touches, strikes, or, indeed, "affects" the organism as significant relative to its organization and purposes. Jonas (1966, p. 99) aptly talked here of the "irritability" of life to refer to the sensitivity and receptivity of organisms to their environments, and wrote that "irritability is the germ, as it were the atom, of having a world."[6]

From an enactive perspective, then, affectivity turns out to be a primordial phenomenon, inherent in all forms of life, including the simplest ones.[7] It is not a psychological faculty distinct from (and at most interacting with) cognition, but cognitive capacities, as well as emotions, moods, sentiments, motivational states, etc., are all modes of functioning of this foundational affectivity. What this entails is that the mind remains inherently affective also as living systems become more complex, and the capacities

[5] The qualification "basic" here is meant to allow for the appearance of more complex forms of cognition in more complex living systems, including forms of human cognition dependent on organisms being situated in cultural symbolic contexts.

[6] For a more detailed discussion of the affective character of sense-making, see in particular Colombetti (2014, ch. 1, and 2015).

[7] The idea that affectivity is somehow primordial and foundational is not new. For example, Spinoza in the *Ethics* proposed that every aspect of existence (not only life) is driven by an effort or strive—a *conatus*—to persevere in its being. Another related account is Henry's (1965) suggestion that the most primordial level of subjectivity is a "latent tension," "interior quivering," or "affective tonality" that permeates all mental states. Similarly, Patočka (1995/1998) proposed that subjectivity (not only human, but also animal) is always in a mood and is characterized by a "primordial dynamism" (see Colombetti 2014, ch. 1).

to judge, reflect, criticize, and so on, appear. Human cognition, from this perspective, is still a modality of affectivity; most generally speaking, it is a way of engaging and making sense of oneself and the world. As Heidegger (1926/1996) put it, we are always "attuned" to the world through a mood (a *Stimmung*, which in German has the same root as *stimmen*, "to tune," as in to tune a piano). Even when we are engaged in what contemporary cognitive scientists call "higher cognitive capacities," we are so from an interested and motivated perspective, one from which we care about what we do.

Implications for the Notion of Cognitive Appraisal

Thinking of cognition as enacted by the whole organism and as inherently affective has various implications. One concerns the so-called cognitive approach to emotion and one of its central notions, the one of *appraisal*.[8] Cognitive theories of emotions generally claim that cognition is necessary for emotion, where "emotion" typically refers to, as mentioned earlier, the capacity to undergo distinct emotions or emotional episodes (fear, anger, etc.). Although cognitive theories come in different guises, they all emphasize that emotions have cognitive aspects or components. The cognitive component of appraisal is regarded as particularly important in emotion psychology, and much effort in this field has gone into characterizing the variety of appraisal processes taken to elicit emotional episodes and their aspects (e.g., Scherer 2009). In its most general characterization, appraisal refers to a cognitive process that evaluates the agent's situation in relation to the agent's goals and capacity to cope with the situation.

The enactive approach does not deny that emotions are cognitive. Yet because it claims that cognition is enactive, it entails that emotions are cognitive "in an enactive sense," i.e., a sense that does not draw a sharp wedge between cognition and the body,[9] but rather sees the cognitive aspects of emotions as themselves embodied. Compare, for a contrast, Robert Solomon's (1976/1993) early cognitive account of the emotions. He claimed that emotions *are* cognitions, and emphasized their world-constituting role and their deep personal significance. At the same time, he diminished their bodily aspects and characterized bodily changes occurring during emotions as mere contingent causal concomitants.[10] If, for example, you become jealous upon discovering that your partner has a lover, on this view your jealousy is entirely constituted by how you cognize the situation; the fact that you are also shaking, that your throat is constricted, or that you

[8] The psychologist Magda Arnold (1960) originally introduced the notion of appraisal to explain how emotions and associated bodily responses come about in the first place, including why different people respond with different emotions to the same stimuli.

[9] By "body" here I refer to the organism "minus" the brain. Cognitive approaches to emotion generally assume that cognition is minimally embodied in the sense of "embrained," but regard changes that happen in the rest of the organism as part of the emotion's "bodily arousal," distinct from the cognitive component of the emotion.

[10] Solomon (2004) later claimed that his earlier view on bodily processes was too radical and misguided.

are crying is not essential to your jealousy—it is just an empirical contingency. This is what I call a "disembodied" cognitive view of emotions. It is a view that assumes and reinforces a cognition-vs.-body dichotomy, by characterizing the cognitive appraisal as the smart faculty that evaluates, monitors, regulates, decides how to cope, etc.; and, relatedly, bodily changes as not-so-smart processes that merely "respond" to the appraisal and maybe even influence it, but have no cognitive-evaluative function.

Enactivism, on the other hand, advances an embodied view of emotions as cognitive episodes. It agrees with Solomon and others that emotions are world-disclosing and personally significant, but holds that they are so in virtue of being processes of a living, cognitive-affective organism (rather than "merely embrained," disembodied cognitions). On the enactive view, the bodily processes you undergo when in the grip of jealousy are not mere contingent responses to essential, brain-bound cognitive evaluations; they are, rather, bodily ways of making sense of the situation, and as such they are cognitive-evaluative. Applied to appraisal theories in psychology, enactivism thus entails that appraisal ought not to be seen as a component of emotion distinct from other, bodily components with no cognitive-evaluative character.[11] Rather, when we undergo an emotional episode, this is simultaneously bodily (it is a whole-organism process) and cognitive-evaluative (it is an evaluation of one's situation). Shaking, crying, etc., are bodily ways of evaluating the fact that one's partner has a lover as upsetting and threatening. Bodily changes are here parts of the material processes realizing a certain appraisal.[12]

In Colombetti (2014, ch. 4), I outlined an enactive view of appraisal in more detail, by following two complementary strategies: (1) by arguing that, at the subpersonal (i.e., neural and more generally physiological) level of description of the agent, it is not clear that we can draw a line around cognitive and bodily components of emotion; and (2) by developing a phenomenological account of the experience of undergoing an emotion and showing that, in experience, bodily feelings are part of the process of appraising (see also Colombetti 2007, 2010; Colombetti and Thompson 2008). The first strategy draws on recent neuroscientific accounts arguing that the distinctions between cognition and emotion, and appraisal and the rest of an emotion, do not map neatly onto the brain's functional organization (e.g., Lewis 2005; Pessoa 2008, 2013, 2014), and that the

[11] In Scherer's (2009) component process model, for example, the bodily components of emotion are physiological arousal (the label reserved for involuntary bodily changes in the heart and other organs and smooth muscles regulated by the autonomic nervous system), musculoskeletal activity (behavior, expressions), and bodily feelings, including feelings of "action tendencies" (urges to move in one way or the other; see Frijda 1986).

[12] This does not mean that there cannot be different *phases* within an emotional episode. Some emotions may involve an initial "ruminative" phase in which one is pondering the implications of a situation (such as discovering that one's long-term partner has a lover), and a subsequent "explosive" phase in which one lets oneself express the emotion in an outward way (by crying, shouting, or kicking). Rather than a sequence of a merely embrained, non-embodied cognitive-appraising phase, and a bodily-expressive response, on the enactive view we have here a continuous unfolding organismic process, where the ruminative phase has one cognitive-affective-bodily-experiential profile, and the explosive part a different one.

brain itself is deeply integrated with the rest of the organism (Thompson and Cosmelli 2012). Walter Freeman's (2000) dynamical neuroscientific account of how perception is always "emotion-laden" and influenced by the organism's state, intentions, and state of expectancy, together with Barrett and Bar's (2009) more recent predictive model of how the organism's state of arousal participates in the construction of affect-laden visual percepts, provide further support to the view that, at the subpersonal level, cognitive and bodily aspects of emotion overlap greatly.

Central to the second, phenomenological strategy is the consideration that bodily feelings are not restricted to feelings *of* the body, where the body is the intentional object of experience. If we think of bodily feelings this way, then we will have to attribute to some other "part" of the emotion experience (typically, the cognitive one) the capacity to be about objects or situations in the world, and we will end up with the familiar cognition/body split. But it is possible to conceive of bodily feelings in other ways, drawing on the philosophical tradition of phenomenology, in particular on the notion of the *lived* body as *that through which* we experience the world. In a nutshell, the proposal here is that bodily feelings in emotion experience are not just feelings of the body, but ways of feeling the world and of perceiving its affective qualities through how the body is experienced. For example, experiencing a pang in the stomach upon discovering that one's partner has a lover is not just a feeling of something happening *in* the stomach, but a feeling *of* the situation as painful *through* the pang in the stomach (for more detailed discussions, see Slaby 2008; Ratcliffe 2008; Colombetti 2011, 2014, ch. 4 and 5; Maiese 2014). In sum, then, phenomenological considerations appear to converge with considerations at the subpersonal level in undermining the traditional separation between the cognitive-evaluative and bodily components of affective episodes.[13]

A Dynamical Systems Approach to Emotions and Other Affective Phenomena

Enactivism, as we saw earlier, views cognitive systems as autonomous, and autonomous systems are, importantly, dynamical (made of mutually influencing processes) and self-organizing (their overall behavior is the result of the interactions of their various

[13] It might be useful at this point to comment on the relation between the considerations developed in this section and those advanced by Carr, Kever, and Winkielman (2018). Their discussion focuses primarily on how the bodily elements of emotion influence our perception and understanding of emotional states in other people. On their view, it seems, emotions necessarily come with changes in the body, and these changes are also necessary to read other people's emotions accurately. This view *could* be interpreted as claiming that, at least in the case of evaluating/appraising others' emotions, this process is not merely embrained but embodied. Note, however, that enactivism does not entail that we need to "reproduce" or "simulate" the other's emotional state in ourselves in order to perceive and/or understand it (see Colombetti 2015, ch. 7). In addition, it is not clear that Carr and colleagues regard the cognitive-evaluative aspects of emotion as always embodied. What they refer to as "the embodiment of emotion" in their chapter thus refers to quite a different view from the one I sketched in this section.

processes, which are not directed or determined by a dedicated system or process). Enactivism is thus sympathetic to the "dynamical systems" approach in cognitive science, which emphasizes the temporal nature of cognition and its emergence from the mutual interactivity of brain, body, and world (e.g., Port and van Gelder 1995; Beer 2000). According to Thompson (2007), indeed, "embodied dynamicism" is an integral part of the enactive approach.

We can then regard recent proposals to conceptualize emotions and other affective phenomena by drawing on dynamical systems theory as part of developing an enactive account of affectivity (existing accounts include Freeman 2000; Lewis 2000, 2005; Camras and Witherington 2005; Thompson 2007, ch. 12; Colombetti 2009, 2014, ch. 3). A major difference between such a dynamical approach and more mainstream frameworks is that the former characterizes affective events (emotions, moods, others) not as "outputs" of some internal determining factor, but as self-organizing episodes of mutually influencing and constraining brain and bodily processes. In this respect, a dynamical systems approach differs from the influential "basic emotions" framework, according to which at least some emotions are "affect programs," i.e., internal sets of (genetic and/or neural) instructions that determine relatively fixed and pancultural patterns of responses in the rest of organism (baring the teeth in anger, increasing adrenaline in fear, etc.).[14] The basic emotions framework has been criticized on various grounds, including the fact that it cannot easily account for the high variability of emotions over development and across different cultures. A dynamical systems approach, on its part, can naturally account for such variability by characterizing emotional episodes as highly context-sensitive bundles of self-organizing organismic processes, while still allowing for evolutionary influences on emotional development. Consider, for example, how human adults manifest anger in different cultures—such as, say, the UK and Italy (to pick the two cultures I am most familiar with).[15] There are commonalities attributable to the evolutionary history of anger—such as the tendency to attack the object of one's anger, together with a certain physiological activation supporting this tendency; and facial and vocal features signaling hostility (frowning, raised voice). Yet there are also differences that depend on local implicit norms of behavior. For example, Italians move the body (especially the upper body, arms and hands) more overtly and assertively than the British, who on their part perform subtler facial changes (and these gestures come, like accents and dialects, in further regional, and class- and gender-dependent variations).

[14] The notion of "affect programs" is due to Tomkins (1962) and is associated primarily with the work of Paul Ekman. See Scarantino and Griffiths (2011) for a clear overview of the notion of basic emotions and the debate around it.

[15] There is little research on differences in the bodily manifestations of emotion across European cultures, so I need to resort here to personal observation. Psychologist Jeanne Tsai has studied similarities and differences in emotions across cultures (in particular between European-Americans and Asian-Americans), looking at physiology and facial expressions (e.g., Tsai et al. 2002; Chentsova-Dutton, Ryder, and Tsai 2014), but not at bodily posture and gestures. Sociobiologist Desmond Morris has written various books on body language and gestures across cultures (e.g., Morris 1977/2002), which include a few examples of bodily emotion expressions.

The basic emotions framework claims that affect programs are the same across cultures, and so to explain these variations it posits so-called display rules: culture-specific rules that determine how to regulate one's expressions depending on context (Ekman 1971). Although this construct might explain cases in which, as Ekman (1971) first observed, people from different cultures exhibit the same facial expression just after presentation of an emotional stimulus, and only later "mask it" with a different expression, it seems ad hoc and overly complicated to posit display rules for every variation one can find across cultures and individuals. From an enactive, dynamical systems perspective, there is no need to appeal to "rules" here. Rather, emotions are conceptualized as patterns of organismic processes (from physiology to posture and behavior) that take the form they do in virtue of evolutionary and developmental forces that together funnel specific organismic trajectories. The organism self-organizes into specific patterns or configurations that constitute attractors carved by selective forces as well as developmental factors encountered in one's lifetime. Our organisms are not blank slates but come into existence with expressive tendencies and capacities shaped by evolution; as they progress through the life span, other forces (from caregivers to colleagues to explicit norms of behavior) contribute to shaping their expressive/gestural repertoire. Thus anger, to stick to our example, can be seen as a configuration of the organism that has a certain evolutionarily prepared form at birth, which is molded and refined as the organism grows and interacts with the environment. Variations are characterized here as culturally developed attractors rather than as "outputs" of acquired internal rules. On this view, all emotions are both evolutionarily prepared and culturally and developmentally shaped. Thus, a dynamical systems approach to emotions can help cut across the nature/nurture dichotomy still influential in much of affective science, and particularly apparent in the distinction between biologically basic, innate, and pancultural emotions on the one hand, and acquired, culture-dependent, and cognitive emotions on the other.

A dynamical systems approach also naturally allows us to conceptualize the relationship between emotions and other affective phenomena in a way that respects their phenomenology. Consider the relationship between emotional episodes (of anger, sadness, etc.) and moods (being cranky, cheerful, having the blues, etc.). The former are relatively short-lived episodes directed at specific objects or events (I am angry at my neighbor, sad at the loss of my favorite pet, and so on). Moods usually last longer, for hours, days, weeks, or even more; they are typically not about anything in particular, although they will make some emotional episodes more likely than others. For example, being in a grumpy mood will make it more likely to flip one's lid at a colleague. Generally, different moods increase the occurrence of different emotional episodes, even in the same person; when we are in a downward mood we are more likely to undergo episodes of sadness and fear than when we are in a cheerful mood. Moods have thus been characterized phenomenologically as pre-intentional (Ratcliffe 2008): they are not themselves about specific events, but they shape our experience of the world so as to make certain intentionally directed experiences possible. At the same time, our emotions can shift our moods. Undergoing repeated episodes of fear can induce a general state of anxiety, for example, and thus change the way in which we are pre-intentionally attuned to the world. The affect-program framework does not really address these differences and dynamical relations between emotions and moods, and tends to conceptualize emotions and moods in the same way, as outputs of internal programs that have been selected by evolution. On the other hand, the conceptual tools of dynamical systems theory make it possible to describe emotions and moods at the subpersonal level in a way that respects these phenomenological relations. Specifically, short-lasting emotional episodes can be modeled as attractor points in the organism's state space. On their part, moods, as longer-lasting affective states, can be seen as topologies of attractors, i.e., different distributions of attractors in the organism's state space. In other words, in a mood the overall configuration of the organism changes in such a way that some specific patterns of self-organization (corresponding to specific emotional episodes) become more likely than others (see Lewis 2000; Thompson 2007, ch. 12; Colombetti 2014, ch. 3). In turn, the reiteration of certain emotional episodes will create attractors that will lead to new topologies.

This dynamical framework can be applied also to the interdependencies between emotions, moods, and personality traits (see Lewis 2000). Things like temperament and character traits influence the kinds of mood and therefore the emotions one is disposed to have, and are in turn influenced by the moods and emotions one repeatedly undergoes. Developing an enactive approach to affectivity further would require providing a detailed account of the phenomenology of character and personality (e.g., their development and manifestation), together with a more detailed account, at the subpersonal level, of how organismic patterns constrain and influence one another at different time scales.

Extending Affectivity, the Enactive Way

So far I have discussed affectivity as something that occurs within individual organisms. Yet arguably enactivism does not imply that affectivity remains necessarily "inside the organism," so to speak. In other words, it does not entail what philosophers call an "internalist" view of affectivity, according to which the material processes underpinning affective states remain within the boundary of the organism. The charge that enactivism is internalist has been raised by Wheeler (2010), who argues that enactivism entails that mind and life are co-located. Relatedly, he concludes that enactivism is incompatible with one of the other 4Es, the "extended mind thesis," according to which the physical processes that realize the mind are, sometimes, best regarded as extending beyond the organism, to include environmental processes (e.g., Clark and Chalmers 1998; Clark 2008).

There are certainly important differences between enactivism and the extended mind thesis. A major one is that the latter assumes a computational-functionalist view of cognition that enactivism rejects. Yet, when it comes to the internalism/externalism debate, enactivists deny that enactivism is necessarily internalist (see Di Paolo 2009; Thompson and Stapleton 2009; Thompson 2011; Colombetti 2015). As I see it, enactivism importantly allows *hybrid* living systems composed by organic and non-organic processes to enact sense-making.[16] In this section I elaborate briefly on this point, with a specific emphasis on affectivity.

Consider first Di Paolo's (2009) response to Wheeler's charge of internalism. Whereas Wheeler takes it to be a "*highly* plausible claim" (2010, p. 35; italics in original) that living systems cannot extend, Di Paolo has responded that indeed they can. More precisely, his claim is that enactivism allows organisms to integrate "mediating structures" into their organization. To illustrate this point he offers the example of an aquatic insect that can breathe underwater by trapping air bubbles on its abdominal hairs. As the insect consumes the oxygen contained in these bubbles, a partial pressure deficit is created in them, which is then compensated by dissolved oxygen that diffuses in from the water, so that the process can continue indefinitely. Di Paolo characterizes this network of processes as a *new form of life* involving a structure (the air bubbles) that mediates the organism's regulatory activity of its coupling (i.e., reciprocal interactions) with the environment. As he notes, "The mediation in cases like this is so intimately connected with vital functions that the living system itself might be called extended" (2009, p. 17). "Intimate connection" refers here to the integration or assimilation of the mediating structure into a new precarious adaptive autonomous system—the system "insect-plus-air-bubbles." This is a system underdetermined by metabolism, but it is a (new) form of life because its structure exhibits the proper organization. This is tantamount to saying that the air bubbles are not external to the living form, merely allowing the insect to stay alive underwater; they are, rather, constitutive parts of the new form of life.

Although Di Paolo (2009) does not put it this way, in my view his argument importantly implies that enactivism also allows sense-making (including affectivity) to extend—in the specific sense that it allows the physical processes enabling sense-making to go beyond the physical boundaries of the organism. In other words, if living systems, enactively characterized, are cognitive-affective sense-making systems in virtue of their organization, then extended living systems will be as well—and the physical underpinnings of sense-making in these cases will correspond to the new, extended living form. This point applies not only to simple organisms like insects, but to more complex animals as well as humans. In all these cases, the logic will be the same: if it is possible to integrate an artifact into the living system in the right way, such that the resulting system shows the organization typical of living systems, then we have an extended living system.

I have recently developed this view in more detail elsewhere (see Colombetti 2015). Here I want to add to this point by linking it to another line of argument one finds

[16] This idea is prefigured in Maturana and Varela (1972/1980, p. 107–11), who already talked of "higher-order composite" autopoietic systems.

among existing enactive responses to Wheeler's charge of internalism, in particular to Thompson and Stapleton's (2009) suggestion that to determine whether a part of the world is part of a cognitive system, one needs to determine whether the part of the world in question is *incorporated*. The criterion they offer to determine whether objects are incorporated (i.e., assimilated or integrated within the body) is the one of *phenomenological transparency*. Phenomenologically transparent objects "are no longer experienced as objects: rather the world is experienced through them" (p. 29). The typical example of a phenomenologically incorporated object is the blind person's cane: the blind person does not experience the cane as just another physical object in the world, but as that through which she experiences objects in the world. As we saw earlier, this is how we often experience our body in our engagements with the world.

Thompson and Stapleton (2009) also suggest that this is not the only criterion for determining whether objects are part of one's mental processes. They claim that "some kind of intimate coupling with the body's autonomous dynamics is necessary" (p. 29). In particular, "external resources [need to] be subject to active regulation by the body.... The body has to be capable of leading the dance" (p. 29). The blind person's cane appears to satisfy this criterion as well, and thus should count as a constitutive part of the person's sensorimotor activity.

Following this logic, to determine when affectivity, from an enactive perspective, extends, one needs to establish (1) whether parts of the world can be incorporated into affective states, i.e., become phenomenologically transparent; and (2) whether those parts of the world are also intimately coupled with the body and subject to active regulation by it. Given these criteria, a plausible candidate for incorporation into an affective state in the human case is the musical instrument played by a skilled musician. Consider, for example, a skilled saxophone player who picks up his instrument and improvises a sad melody, and who in the course of this activity undergoes a specific affective episode (a mood, an emotion, perhaps a complex mixture of both) that he would otherwise not be able to achieve.[17] The instrument in this scenario appears to satisfy the second criterion for incorporation, for it appears to be "intimately coupled" with the musician's organism (his brain-plus-body), and subject to "active regulation" by it. The musician manipulates the instrument in specific ways, and the instrument accordingly produces sounds that influence what the musician will play next. Although there is a two-way direction of influence, the musician is "leading the dance," deciding what to play next (the decision can be rapid, pre-reflective, or even unconscious, and merely the result of habit). As for the first criterion, it seems plausible to say that the instrument is phenomenologically transparent: while performing, the musician arguably does not experience his instrument as an object in the world, but as something that enables him to "articulate" a certain affective state—to perform it in real time, to construct and give form to it. In other words, the musician experiences the saxophone as a means through

[17] I borrow the example of the saxophone improviser from Cochrane (2008), who, however, does not use it specifically to argue for the extension of an emotional episode.

which he can achieve a certain affective state, similarly perhaps to how professional actors and dancers experience their performing bodies (see Colombetti and Krueger 2015; Colombetti 2016 for a more detailed discussion of the phenomenon I call "affective object-incorporation").

Conclusion

In this chapter I have provided a broad overview of the implications of enactivism for our understanding of affectivity. There are, as always, many other questions that still need to be addressed in this area. One, as briefly mentioned earlier, regards the relation between what I called "canonical" enactivism and Hutto and Myin's (2013) "radical" enactivism. Clarifying this relation will have implications for how to understand affectivity from an enactive perspective. For example, Hutto and Myin want to ban talk of "vehicles" of mental states, but what I said in the previous section about the possibility of hybrid systems to enact extended forms of affectivity does rely on the existence of vehicles in the sense of identifiable physical processes underpinning or enacting sense-making. Thus, a question that needs to be addressed is exactly *which* notion of vehicles, if any, would need to be dropped from enactivism, and what such a move would imply for the proposal that affectivity can be enactively extended.

A related issue that needs clarification is the relationship between enactivism and what Hutto and Myin (2013, ch. 7) call the "extensive mind view," according to which "basic minds are fundamentally, constitutively already world-involving" (p. 137), as opposed to world-involving only under certain conditions. In one sense, enactive sense-making is "extensive," because it is always necessarily enacted by an organism that continually relates dynamically to an environment (no organism without environment). Yet Hutto and Myin's extensive mind notion refers to something different, namely, to the idea that basic minds need to be supported or "scaffolded" by social practices in order to become adult minds that can operate relatively independently from context (as when one performs a mathematical calculation without external props). Whereas this view is in line with Thompson's (2007, ch. 13) discussion of enculturation, it overlooks that, as we saw in the previous section, enactivists have posed criteria for establishing when the physical processes underpinning a certain mental activity should be seen as including parts of the world. What is needed, then, is a comprehensive account of the various senses in which enactivism entails that the mind, including affectivity, is world-involving: from the sense in which organisms require an environment to differentiate themselves from it and enact sense-making, to the sense that living organisms can come to integrate mediating structures into their organization, to the even further sense that organisms and their mental lives are scaffolded by highly diverse sociocultural practices. Clarifying how these different processes of world-involvement relate to one another and contribute to the emergence of ever more complex cognitive and affective phenomena is, I believe, one the most interesting and exciting challenges for enactivism for the near future.

References

Arnold, M.B. (1960). *Emotion & personality* (vols. 1–2). New York: Columbia University Press.

Barrett, L.F. and Bar, M. (2009). See it with feeling: affective predictions during object perception. *Philosophical transactions of the Royal Society of London. Series B, Biological sciences*, 364(1521), 1325–34.

Beer, R.D. (2000). Dynamical approaches to cognitive science. *Trends in Cognitive Sciences*, 4, 91–9.

Camras, L.A. and Witherington, D.C. (2005). Dynamical systems approaches in emotional development. *Developmental Review*, 25, 328–50.

Carr, E.W., Kever, A., and Winkielman, P. (2018). Embodiment of emotion and its situated nature. In: this volume, pp. 529–52.

Chentsova-Dutton, Y., Ryder, A., and Tsai, J.L. (2014). Understanding depression across cultural contexts. In: I. Gotlib and C. Hammen (eds.), *Handbook of depression* (3rd ed.). New York: Guilford Press, pp. 337–52.

Clark, A. (2008). *Supersizing the mind: embodiment, action, and cognitive extension*. Oxford: Oxford University Press.

Clark, A. and Chalmers, D. (1998). The extended mind. *Analysis*, 58, 7–19.

Cochrane, T. (2008). Expression and extended cognition. *The Journal of Aesthetics and Art Criticism*, 66(4), 329–40.

Colombetti, G. (2007). Enactive appraisal. *Phenomenology and the Cognitive Sciences*, 6, 527–46.

Colombetti, G. (2009). From affect programs to dynamical discrete emotions. *Philosophical Psychology*, 22, 407–25.

Colombetti, G. (2010). Enaction, sense-making and emotion. In: J. Stewart, O. Gapenne, and E. Di Paolo (eds.), *Enaction: toward a new paradigm for cognitive science*. Cambridge, MA: MIT Press, pp. 145–64.

Colombetti, G. (2011). Varieties of pre-reflective self-awareness: foreground and background bodily feelings in emotion experience. *Inquiry*, 54(3), 293–313.

Colombetti, G. (2014). *The feeling body: affective science meets the enactive mind*. Cambridge, MA: MIT Press.

Colombetti, G. (2015). Enactive affectivity, extended. *Topoi*, doi:10.1007/s11245-015-9335-2

Colombetti, G. (2016). Affective incorporation. In: J.A. Simmons and J.E. Hackett (eds.), *Phenomenology for the twenty-first century*. New York: Palgrave Macmillan.

Colombetti, G. and Krueger, J. (2015). Scaffoldings of the affective mind. *Philosophical Psychology*, 28(8), 1157–76.

Colombetti, G. and Thompson, E. (2008). The feeling body: towards an enactive approach to emotion. In: W.F. Overton, U. Müller, and J.L. Newman (eds.), *Developmental perspectives on embodiment and consciousness*. New York: Lawrence Erlbaum, pp. 45–68.

Ekman, P. (1971). Universals and cultural differences in facial expressions of emotion. In: J. Cole (ed.), *Nebraska symposium on motivation* (vol. 19). Lincoln: University of Nebraska Press, pp. 207–82.

Ekman, P. (2003). *Emotions revealed: understanding faces and feelings*. London: Weidenfeld & Nicolson.

Di Paolo, E.A. (2005). Autopoiesis, adaptivity, teleology, agency. *Phenomenology and the Cognitive Sciences*, 4, 429–52.

Di Paolo, E.A. (2009). Extended life. *Topoi*, 28, 9–21.

Di Paolo, E.A. and Thompson, E. (2014). The enactive approach. In: L.A. Shapiro (ed.), *The Routledge handbook of embodied cognition*. Oxford, UK: Routledge, pp. 69–77.

Freeman, W.J. (2000). Emotion is essential to all intentional behaviors. In: M.D. Lewis and I. Granic (eds.), *Emotion, development, and self-organization*. Cambridge: Cambridge University Press, pp. 209–35.

Frijda, N.H. (1986). *The emotions*. Cambridge: Cambridge University Press.

Heidegger, M. (1996). *Being and time* (trans. J. Stambaugh). Albany, NY: SUNY Press.

Henry, M. (1965). *Philosophie et phénomenologie du corps: essai sur l'ontologie biranienne*. Paris: Presses Universitaires de France.

Hurley, S.L. (1998). *Consciousness in action*. Cambridge, MA: Harvard University Press.

Hutto, D.D. and Myin, E. (2013). *Radicalizing enactivism: basic minds without content*. Cambridge, MA: MIT Press.

Jonas, H. (1966). *The phenomenon of life: toward a philosophical biology*. Evanston, IL: Northwestern University Press.

Lewis, M.D. (2000). Emotional organization at three time scales. In: M.D. Lewis and I. Granic (eds.), *Emotion, development, and self-organization: dynamic systems approaches to emotional development*. Cambridge: Cambridge University Press, pp. 37–69.

Lewis, M.D. (2005). Bridging emotion theory and neurobiology through dynamic systems modeling. *Behavioral and Brain Sciences*, 28(2), 169–93.

Maiese, M. (2014). How can emotions be both cognitive and bodily? *Phenomenology and the Cognitive Sciences*, 13(4), 513–31.

Massumi, B. (2002). *Parables for the virtual: movement, affect, sensation*. Durham, NC: Duke University Press.

Maturana, H.R. and Varela, F.J. (1972/1980). *Autopoiesis and cognition: the realization of the living*. Dordrecht: D. Reidel.

Merleau-Ponty, M. (1964). The child's relations with others. In: J. Edie (ed.), *The primacy of perception*. Evanston, IL: Northwestern University Press, pp. 96–155.

Morris, D. (1977/2002). *Peoplewatching: the Desmond Morris guide to body language*. London: Vintage.

Noë, A. (2004). *Action in perception*. Cambridge, MA: MIT Press.

O'Regan, J.K. and Noë, A. (2001). A sensorimotor account of vision and visual consciousness. *Behavioral and Brain Sciences*, 24(5), 939–73.

Patočka, J. (1995). *Body, community, language, world* (trans. E. Kohák). Chicago: Open Court.

Pessoa, L. (2008). On the relationship between emotion and cognition. *Nature Reviews Neuroscience*, 9, 148–58.

Pessoa, L. (2013). *The cognitive-emotional brain: from interactions to integration*. Cambridge, MA: MIT Press.

Pessoa, L. (2014). Précis on *The Cognitive-Emotional Brain*. *Behavioral and Brain Sciences*, 38, e71. doi:10.1017/S0140525X14000120

Port, R.F. and van Gelder, T.J. (eds.) (1995). *Mind as motion*. Cambridge, MA: MIT Press.

Ratcliffe, M. (2008). *Feelings of being: phenomenology, psychiatry and the sense of reality*. Oxford: Oxford University Press.

Russell, J.A. (2003). Core affect and the psychological construction of emotion. *Psychological Review*, 110, 145–72.

Scarantino, A. and Griffiths, P. (2011). Don't give up on basic emotions. *Emotion Review*, 3(4), 444–54.

Scherer, K.R. (2009). The dynamic architecture of emotion: evidence for the component process model. *Cognition and Emotion*, 23(7), 1307–51.
Slaby, J. (2008). Affective intentionality and the feeling body. *Phenomenology and the Cognitive Sciences*, 7(4), 429–44.
Solomon, R.C. (1976/1993). *The passions: emotions and the meaning of life* (2nd rev. ed.). Indianapolis: Hackett.
Solomon, R.C. (2004). Emotions, thoughts, and feelings. In: R.C. Solomon (ed.), *Thinking about feeling: contemporary philosophers on emotions*. Oxford: Oxford University Press, pp. 76–89.
Thompson, E. (2007). *Mind in life: biology, phenomenology, and the sciences of mind*. Cambridge, MA: Harvard University Press.
Thompson, E. (2011). Reply to commentaries. *Journal of Consciousness Studies*, 18(5–6), 176–223.
Thompson, E. and Cosmelli, D. (2012). Brain in a vat or body in a world? Brainbound versus enactive views of experience. *Philosophical Topics*, 39(1), 163–80.
Thompson, E. and Stapleton, M. (2009). Making sense of sense-making: reflections on enactive and extended mind theories. *Topoi*, 28(1), 23–30.
Tomkins, S.S. (1962). *Affect, imagery, consciousness* (2 vols.). Oxford: Springer.
Tsai, J.L., Chentsova-Dutton, Y., Freire-Bebeau, L., and Przymus, D.E. (2002). Emotional expression and physiology in European Americans and among Americans. *Emotion*, 2(4), 380–97.
Uexküll, J. von. (1934). *A foray into the worlds of animals and humans. With a theory of meaning* (trans. J.D. O'Neil). Minneapolis: University of Minnesota Press.
Varela, F.J. (1979). *Principles of biological autonomy*. New York: Elsevier.
Varela, F.J. (1991). Organism: A meshwork of selfless selves. In: A.I. Tauber (ed.): *Organism and the origin of the self*. Boston: Kluwer Academic Publishers, pp. 79–107.
Varela, F.J., Thompson, E., and Rosch, E. (1991). *The embodied mind: cognitive science and human experience*. Cambridge, MA: MIT Press.
Weber, A. and Varela, F.J. (2002). Life after Kant: natural purposes and the autopoietic foundations of biological individuality. *Phenomenology and the Cognitive Sciences*, 1, 97–125.
Wheeler, M. (2010). Minds, things and materiality. In: L. Malafouris and C. Renfrew (eds.), *The cognitive life of things: recasting the boundaries of the mind*. Cambridge: McDonald Institute for Archaeological Research, pp. 29–37.

CHAPTER 31

BEYOND MIRRORING

4E Perspectives on Empathy

DAN ZAHAVI AND JOHN MICHAEL

Introduction

The notion of empathy does not have a long history. The German term *Einfühlung* was introduced into the field of social cognition by the psychologist Theodor Lipps at the beginning of the twentieth century and used as a label for our basic understanding of others; an understanding that, according to Lipps, involved a combination of imitation and projection. It was Lipps's notion that Edward Titchener had in mind when he in 1909 translated *Einfühlung* as "empathy" (Titchener 1909).

When considering the current debate on empathy, it quickly becomes evident that a diversity of definitions of and approaches to the topic are available, and that no consensus seems forthcoming. A recent issue of the *Boston Review* entitled *(Against) Empathy* can serve as a good illustration of this.

In his target contribution, Paul Bloom concedes that the term "empathy" is used in many ways, but he maintains that it typically refers to a process whereby one comes to experience the world as others do, be it through imaginative perspective taking or by some kind of affective matching. Bloom further argues that empathy serves to dissolve the boundaries between one person and another, and that it can therefore be a force against selfishness and indifference. It is consequently not surprising that some have seen empathy as a moral virtue and have argued that we need to nurture and expand our empathic powers, since a high degree of empathy might be a requirement for being good and doing good (Bloom 2014, p. 15).

For a variety of reasons, however, Bloom is quite skeptical about this line of reasoning. This is in part because empathy concerns our relation to specific individuals. If we are striving for a better world, one that might involve an increase of humanitarian aid or a rethinking of the criminal justice system, i.e., policies affecting large groups of people, a keen sense of justice or moral obligation might be far more relevant than any empathic skill. This is all the more true since, according to Bloom, empathy is biased: we tend to empathize more with those whose needs are salient, who are similar to ourselves, and who are close by. If we want to promote impartiality and fairness, we should consequently put empathy aside.

In addition, Bloom is skeptical about the value of empathy even in relationships with specific other people. To empathize with another person in pain or distress is, according to Bloom, to feel what the other person is feeling. But if the empathizer suffers as a result of empathizing with your suffering, it is not obvious that this is to your advantage. Empathic distress can lead to egoistic drift, where the empathizer becomes more concerned with alleviating her own distress (for instance, by absenting himself) than with caring about you. You want the other to respond with care and concern and help, rather than to relive your pain and distress. You want the physician to be calm and confident when she is treating you, not to be overwhelmed by negative emotions (Bloom 2014, p. 16).

Given the definition of empathy that Bloom starts out with, this line of reasoning might seem compelling.[1] But, as pointed out by various commentators in the *Boston Review* issue, it is by no means obvious that Bloom's definition is the most appropriate one, i.e., that we should conceptualize empathy as involving an affective matching between empathizer and target. Moreover, how can I expect to receive care or help from another unless she understands my situation, and isn't that understanding exactly what empathy on some accounts is supposed to provide? Should we then distinguish among different kinds of empathy—say, an affective type and a cognitive type? If so, what is the relation between cognitive empathy and ordinary mindreading? Questions abound. People disagree about the role of affective matching, caring, understanding, and imagination in empathy, just as they disagree about the relation between empathy and social cognition in general, and about whether empathy is a natural kind or rather a heterogeneous construct (for an overview, see Zahavi 2014a).

In the following, we will not attempt to resolve these disputes or to argue in favor of any one particular way of conceptualizing empathy. Instead, our aim is to open up a new perspective by exploring the potential of applying embodied, extended, enactive, and embedded approaches to empathy research. As we shall see, these approaches provide useful resources for thinking about empathy, and in particular for going beyond the notion of affective matching. First, however, it will be useful to begin by articulating the notion of affective matching as clearly as possible, and by illustrating how it, too, can be enriched and sharpened by drawing upon ideas from embodied cognition approaches.[2]

[1] For a more extensive discussion of the relation between empathy and morality, see Maibom 2014.

[2] The following discussion partly draws on and expands on points made in Zahavi (2011, 2014a, 2014b), Michael and Fardo (2014), and Michael (2014).

Empathy and Affective Matching

Lipps

It is natural to start by taking a closer look at Lipps's contribution. Although the German term for empathy (*Einfühlung*) was first used in the domain of aesthetics, it was Lipps who started to use it in the context of interpersonal understanding. Whereas a number of contemporary empathy theorists have taken empathy to denote a particular prosocial attitude vis-à-vis others, the original discussion was to a much larger extent epistemologically oriented. In various of his writings, Lipps argues that there are three distinct domains of knowledge: (1) knowledge of external objects, (2) self-knowledge, and (3) knowledge of others—and he took these domains to have three distinct cognitive sources, namely, perception, introspection, and empathy (Lipps 1909, p. 222). Lipps consequently insisted that empathy, which he took to be a psychological and sociological core concept, qualified as a modality of knowledge sui generis (Lipps 1907, pp. 697–8, p. 710).

Sometimes Lipps refers to what he calls the "instinct" of empathy, and argues that it involves two components, a drive directed toward imitation and a drive directed toward expression (Lipps 1907, p. 713). In the past, I have been sad, and have experienced an instinctual tendency to express that sadness. The expression was not experienced as something next to or on top of the sadness but as an integral part of the feeling. Now, when I see the expression elsewhere, I have an instinctual tendency to imitate or reproduce it, and this tendency will then evoke the same feeling to which it was intimately connected in the past (Lipps 1909, pp. 229–30; 1907, p. 719). When I experience the feeling anew, it will remain linked to the expression I am currently perceiving and will be projected into or onto it. For example, when I see an angry face, I will reproduce the expression of anger, this will evoke a feeling of anger in me, and this felt anger that is co-given with the currently perceived facial expression will then be attributed to the other, thereby allowing for a form of interpersonal understanding (Lipps 1907, pp. 717–19).

One implication of Lipps's model is that there are rather strict limitations to what I can come to understand empathically about the other. The imitated expression can only evoke an affective state in myself that resembles the affective state of the other if I have had the affective state in question in the past (Lipps 1907, pp. 718–19). Consequently, I can only understand those of the other's experiences that I have already enjoyed myself, or, to put it differently, Lipps's account of empathy doesn't allow me to recognize anything in the other that is new, anything that I am not already familiar with, anything that I haven't put there myself.

Goldman and de Vignemont and Colleagues

Lipps's position has remained influential and has a number of modern heirs. Not surprisingly, it is in particular within the simulationist camp that the notion of empathy has resurfaced as a central category. Indeed, it has even been argued that simulationists are today's equivalents of empathy theorists (Stueber 2006, p. ix). Goldman has acknowledged that simulationist themes can be found in earlier theorists such as Lipps (Goldman 2006, p. 18), and, in *Simulating Minds*, Goldman explicitly equates empathy theory with simulation theory (Goldman 2006, p. 11), and states that mindreading is an extended form of empathy (Goldman 2006, p. 4).

In recent work, Goldman has emphasized that an account of mindreading should be able to cover the whole range of mental states, including sensations, feelings, and emotions—i.e., it shouldn't just address the issue of belief ascription (Goldman 2006, p. 20). This is precisely why Goldman now distinguishes what he calls "low-level" mindreading from "high-level" mindreading (Goldman 2006, p. 43), and argues that we need to recognize the existence of a simple, primitive, and automatic ability to attribute basic emotions such as fear, anger, and disgust to others on the basis of their facial expressions (Goldman and Sripada 2005).

How can we explain this kind of basic "mindreading," this ability to recognize someone's face as expressive of a certain emotion? In *Simulating Minds*, Goldman considers different models, and ultimately opts for one he calls the unmediated resonance model (Goldman 2006, p. 132), which avoids some of the limitations of Lipps's original proposal.

According to the unmediated resonance model, the same neural substrate is activated both when we experience an emotion ourselves and when we recognize the emotion in others. This occurs because the perception of a target's emotional expression directly triggers activation of the neural substrate of the same type of emotion in ourselves, thereby making the process a kind of unmediated matching, one that bypasses the need for and feedback from facial mimicry (Goldman 2006, p. 128). When compared to Lipps's model, this proposal has the distinct advantage that it does not require an agent to have had any particular past experience in order to empathize with the experience that some other agent is currently having. Rather, the coupling is hardwired. In principle, observing the facial expressions of others might give rise to new emotions in yourself, emotions you haven't felt before.

Goldman suggests that my observation of another's emotional expression automatically triggers the experience of that emotion in myself, and that this first-person experience then serves as the basis for my third-person ascription of the emotion to the other. As he writes—in the context of discussing disgust expressions—"the evidence points toward the use of one's disgust experience as the causal basis for third-person disgust attributions" (Goldman 2006, p. 137). It is consequently no coincidence that Goldman considers a more apt name for the whole process to be simulation-plus-projection (Goldman 2006, p. 40), thereby affirming the structural similarity between his own account and the one we found in Lipps.

Like Goldman, de Vignemont and colleagues (de Vignemont and Jacob 2012; see also de Vignemont and Singer 2006) have articulated a conception of empathy that builds upon the core idea found in Lipps. In contrast to Goldman, however, de Vignemont and her coauthors have insisted on the need for a quite narrow definition of empathy.

Rather than seeing empathy simply as another label for (basic) mindreading, they have stipulated necessary and sufficient conditions for empathy, which, when taken together, make it possible to distinguish empathy from related phenomena such as emotional contagion, sympathy, and mindreading.[3] In this vein, de Vignemont and Jacob (2012) have recently proposed the following conditions:

1. the affectivity condition;
2. the interpersonal similarity condition;
3. the causal path condition;
4. the ascription condition;
5. the caring condition.

Let us briefly examine each of these in turn. The *affectivity* condition demands that an empathizer and her target both experience an affective state. This condition rules out the possibility of empathizing with someone who is not experiencing an affective state of any kind. It also makes it possible to distinguish empathy from standard mindreading: merely identifying that somebody is in a particular affective state would be mindreading, whereas empathy involves the empathizer also coming to experience an affective state.

The *interpersonal similarity* condition demands that the affective states of the empathizer and the target be similar to each other (de Vignemont and Singer 2006 refer to an "isomorphism" between the affective states of the empathizer and her target). This is intended to make it possible to distinguish empathy from sympathy. A sympathizer also experiences an affective state, but it is different from the affective state of the target. In fact, according to de Vignemont and Jacob, sympathy is a "sui generis social emotion."

The *causal path* condition requires that the empathizer's affective state is caused by the target person's affective state. This is intended to rule out cases in which two people have similar affective experiences because of some common cause. The *ascription* condition requires that the empathizer be aware that the target person is in an affective state and that this is the source of her own affective state. This distinguishes empathy from contagion. The *caring* condition is based on the observation that when empathizing (as when sympathizing), one tends to be concerned about the target person's well-being, i.e., insofar as she is suffering, one is motivated to alleviate that suffering. It is worth noting that this condition is also assumed in psychological theories in which empathy is associated with prosocial motivation (Batson 1991; Hoffman 1982).

Taken together, these five conditions present a clear and concise conception that systematically relates empathy to mindreading, sympathy, and contagion (for a critical discussion, see Zahavi 2011; Zahavi and Overgaard 2012). It is also worth emphasizing that de Vignemont and Jacob (2012) have developed this conception in a manner that

[3] There is broad agreement that a conceptual model of empathy should make it possible to distinguish empathy from these related phenomena. In future research, it may also be useful to investigate how empathy relates to pity, compassion, and other associated phenomena.

incorporates impulses from recent embodied cognition approaches. To see how, consider their analysis of one particular type of empathy, namely, empathy for pain. Their account is based on the common view that the experience of pain derives from the processing and integration of nociceptive inputs and complex emotional and cognitive processes, implicating the participation of several pain-specific brain structures that may be functionally distinct. The neural network involved in pain processing is often referred to as the "pain matrix," the primary components of which are sometimes said to be a sensory-discriminative and an affective-motivational network (e.g., Singer et al. 2004; Aydede 2006). On this view, primary and secondary somatosensory and posterior insular cortices are thought to serve the processing of sensory-discriminative features of pain stimuli, such as location, duration, and stimulus intensity. In the affective-motivational domain, anterior cingulate and anterior insular cortices are thought to mediate these aspects of pain processing, for example, the unpleasantness of pain.

De Vignemont and Jacob (2012) suggest that these two components can be dissociated, and that this provides a basis for distinguishing pain empathy from the phenomenon of contagious pain. Specifically, they suggest that contagious pain is more likely to recruit the sensory-discriminative component, whereas empathy is more likely to recruit the affective component of the pain matrix. In accordance, they argue that empathy is "other-centered" insofar as it involves a concern for the other person's affective state, whereas contagion is "self-centered." In support of this position, they refer to research suggesting that in pain empathy the affective neural components are selectively activated (Singer et al. 2004; Botvinick et al. 2005). Thus, de Vignemont and colleagues' proposal is an embodied account insofar as it conceptualizes affective matching as a bodily response to the others' experiences, and insofar as it distinguishes empathy from related phenomena partly in virtue of the specific nature of this bodily response (for critical discussion, see Michael and Fardo 2014; Michael 2014).

For de Vignemont and colleagues, empathy is a special kind of third-person mindreading (de Vignemont and Jacob 2012, p. 310), which is more complex and less direct than standard mindreading (de Vignemont 2010, p. 292; de Vignemont and Singer 2006, p. 439). After all, whereas empathy must meet five requirements, the simpler and presumably more widespread standard mindreading only has to meet one requirement—that of attributing a mental state to another (de Vignemont and Jacob 2012, p. 307). Thus, and this is obviously quite significant, on their proposal, empathy is not what establishes an awareness of the other person's mental state in the first place. Rather, empathy requires a prior understanding of the other's mental life in order to get off the ground, and is then supposed to allow for an enhanced understanding of the other's feeling.

As mentioned before, one noteworthy difference between Goldman and de Vignemont is consequently that whereas the former takes (basic) empathy to be involved in simple mindreading, the latter operates with a far more restrictive use of the term. Despite this difference, it bears emphasizing that both Goldman and de Vignemont understand affective matching to be integrated with various other social-cognitive processes, from drawing inferences about others' situations and mental states to being motivated to alleviate others' suffering. This is important in connection with the concerns raised by Bloom about empathy (i.e., that affective matching may lead to biases and may be of little use to a person who is suffering). While Bloom may well be right that affective matching, taken alone, may be prone to such dangers and limitations, empathy may not be, because it typically involves a consideration of others' situations and mental states as well as a motivation to alleviate their suffering.

While Goldman and de Vignemont illustrate how approaches to empathy may include affective matching but also incorporate other resources that help to address Bloom's concerns, we shall see in the next section that some 4E approaches, also inspired in important ways by themes within the phenomenological tradition, depart more radically from the debate that structures Bloom's critique of empathy.

INSIGHTS FROM 4E APPROACHES

Embodied Simulation: Gallese and Iacoboni

Lipps's influence on the contemporary empathy debate is wide-reaching. Other authors inspired by him include Iacoboni and Gallese, who both endorse Lipps's idea that empathy involves a form of inner imitation (Gallese 2003, p. 519; Iacoboni 2007, p. 314). However, both have also been inspired and influenced by the post-Lippsian discussion of empathy found in the phenomenological tradition. Gallese, for instance, references Stein's account of empathy, and Husserl's and Merleau-Ponty's discussion of intersubjectivity and intercorporeity (Gallese 2001), and is quite explicit in arguing that his own notion of embodied simulation is akin to, and a further development of, the phenomenological proposal (Gallese et al. 2004, p. 397; cf. Iacoboni 2009).

Building upon similar findings to those discussed by Goldman and de Vignemont, embodied simulationists such as Gallese and Iacoboni draw more far-reaching conclusions. For them, empathy is not a relatively rare instance of an enhanced understanding of others' affective states. Rather, for them empathy constitutes a basic and important form of social understanding that renders further inferences or explicit attribution of mental states to others otiose (Gallese 2001, 2009; Gallese et al. 2004). More specifically, the discovery of what have been called mirror neuron systems or neuronal resonance mechanisms has been interpreted as lending support to the existence of a low-level, simulation-based form of empathy—one that explains the ease with which we "mirror" ourselves in the behavior of others and recognize them as similar to us. Indeed, it is the neural matching mechanism constituted by mirror neurons that allows for a direct, automatic, non-predicative, and non-inferential empathic link between different individuals (Gallese 2001, p. 42, 44).

Gallese ultimately claims that all kinds of interpersonal relations—including action understanding, the attribution of intentions, and the recognition of emotions and sensations in others—rely on automatic and unconscious embodied simulation routines (Gallese 2003, p. 517). The very same neural substrate, which is activated when we execute actions or subjectively experience emotions and sensations, is also activated when we observe somebody else act or experience emotions and sensations. So, when we encounter somebody, and observe their actions, or their displayed emotions or sensations, we don't just see them. In addition to the sensory information we receive from the other, internal representations of the body states associated with the other's actions, emotions, and sensations are evoked in us, and it is "as if" we are doing a similar action or experiencing a similar emotion or sensation. Because of this automatic, non-predicative, and non-inferential embodied simulation mechanism—because the activation of these neural mechanisms allows us to share actions, intentions, feelings, and emotions with others—we are able to understand others (Gallese 2001, pp. 44–45; 2009, p. 527).

We shall here not rehearse in any detail the debate concerning the precise contribution of the mirror neurons (see Michael 2011a). For our purpose, the more interesting aspect concerns the extent to which empathy on this model is taken to involve a rupture with standard accounts of social cognition, and ultimately point beyond the dichotomy of behavior-reading and mindreading (cf. Sinigaglia 2008). According to Iacoboni, mirror neuron activity links self and other in a way that questions traditional Cartesian as well as more recent cognitivist assumptions of how social understanding comes about. Indeed, as Iacoboni also writes, the functioning of mirror neurons only makes sense if we are dealing with agents that interact with other people in a shared environment, where the classical dichotomies (such as action–perception, subject–world, and inner–outer) have dissolved (Iacoboni 2007, 2009). This view is, according to Iacoboni, reminiscent of themes found in existential phenomenology, which is why he has labeled his own approach, "existential neuroscience" or "neurophysiologic phenomenology" (Iacoboni 2007, p. 319; 2009, p. 17). Indeed, for Iacoboni the discovery of mirror neurons has not only for the first time in history provided a plausible neurophysiological explanation for complex forms of social cognition and interaction (Iacoboni 2009, p. 5). Mirror neurons also seem to explain why, as he puts it, "existential phenomenologists were correct all along" (Iacoboni 2009, p. 262).

On this account, embodied social cognition might involve an attempt to replicate, imitate, or simulate the mental life of the other, but the simulation process in question is automatic, unconscious, prelinguistic, and non-metarepresentational. As Gallese puts it, intercorporeity is more fundamental than any explicit attribution of propositional attitudes to others, and remains the main source of knowledge we directly gather about others (Gallese 2009, p. 524). In other words, while de Vignemont and colleagues conceptualize embodied affective sharing as a component of empathy, Gallese goes a step further in drawing the conclusion that embodied simulation may suffice for empathic understanding and thus renders the attribution of mental states superfluous (i.e., he rejects de Vignemont and colleagues' ascription condition).

Gallese's work on embodied simulation is undoubtedly more in line with 4E approaches to cognition than Goldman's or de Vignemont's. However, this is not to say that there are no tensions within his conception of empathy. As we have seen,

Gallese has emphasized the affinities between his own position and that of both Lipps and the classical phenomenologists. This is in itself slightly surprising, since the latter were highly critical of Lipps's account (see Zahavi 2014a). Gallese's commitment to the idea that empathy is at bottom to be explained in terms of mirroring or matching mechanisms is also not unequivocal. In a 2009 publication, Gallese observes that the mirror metaphor itself might be misleading, since it suggests the presence of an exact match between object and observer, thereby disregarding individual differences (Gallese 2009, p. 531), and he has also conceded that imitation cannot really account for interpersonal understanding, since the latter calls for a preservation of difference and otherness (Gallese 2007, p. 11; 2009, p. 527).

The idea that empathy, rather than involving identity, similarity, and affective matching, might crucially preserve interpersonal difference, and indeed highlight other-centeredness, is something that other phenomenologically inclined empathy theorists have explored further. To see how, let us briefly return to the phenomenological reception of Lipps.

Other-Centeredness

In the wake of Lipps's investigation, a number of phenomenologists engaged in intensive discussions regarding the nature and structure of empathy. While they accepted the idea that empathy must be equated with (a basic form of) other-understanding, they were more critical of Lipps's suggestion that empathy involves a form of inner imitation, and rejected various attempts to explain empathy in terms of mirroring or mimicry.

In a number of publications, one of us has offered a systematic reconstruction of the phenomenological discussion of empathy (drawing especially on Husserl, Stein, and Scheler), and has defined empathy as "a distinctive form of other-directed intentionality, distinct from both self-awareness and ordinary object-intentionality, which allows foreign experiences to disclose themselves as foreign rather than as own" (Zahavi 2014b, p. 138). On this account, empathy is a perceptually based experience of another person's mental life, one that more complex and indirect forms of social cognition presuppose as well as rely on. To insist that the empathizer must have the same (kind of) state as the target, is on this account to miss what is distinctive about empathy, namely, the fact that it confronts you with the presence of an experience that you are not living through yourself. Rather than blurring the distinction between self and other, rather than leading to some sense of merged personal identities (Cialdini et al. 1997), the asymmetry between self-experience and other-experience is consequently crucial for empathy. One might say that empathy provides a special kind of knowledge by *acquaintance*. It is not first-person acquaintance, but rather a distinct other-acquaintance. Empathy denotes a special kind of epistemic access and should not be confounded with sympathy or compassion. Thus, for Zahavi, it is perfectly coherent to think that an expert torturer may rely on empathy in order to work out how best to push her victim's buttons (Zahavi 2011, 2014a, 2014b).

One benefit of this lean account is that it can elegantly distinguish between empathy and emotional contagion—empathy, according to Zahavi, does not require that the empathizer have an affective experience that is similar to that of the target. Moreover, this way of conceptualizing empathy also admits of a clear and straightforward distinction between empathy and sympathy: sympathy, but not empathy, involves concern for the well-being of the target person.

However, a challenge for this view is to articulate a distinction between mindreading and empathy. One option would be to regard empathy as a *kind* of mindreading—for example, as perceptual mindreading. It is no coincidence that Zahavi has occasionally presented his investigation of empathy in the framework of the direct social perception debate (Zahavi 2011). For some, it may seem odd to restrict empathy to perception. After all, why should we exclude cases in which one learns about someone's suffering through a third person and empathizes with them without ever perceiving them? But, as we have already noted, people's intuitions about borderline cases of empathy tend to diverge greatly, so one should be highly cautious about basing theoretical claims on them. Our evaluation of such a restriction would therefore have to be based on other factors, such as the overall coherence of the proposal, its empirical fruitfulness, or its explanatory power. For the moment, then, it must be regarded as perfectly legitimate to advance a conception of empathy that very much targets the face-to-face encounter. The guiding idea would be the following: just as we ought to consider the difference between thinking about a tiger, imagining a tiger, and seeing a tiger, we also ought to acknowledge the difference between referring to Berta's compassion or sadness, imagining in detail what it must be like for her to be compassionate or sad, and being empathically acquainted with her compassion or sadness in the direct face-to-face encounter. In the latter case, our acquaintance with Berta's experiential life has a directness and immediacy to it that is not possessed by whatever beliefs we might have about her in her absence.

But this is not the only option for Zahavi. Another option would be to say that empathy is more basic and fundamental than mindreading proper. The coherency of this proposal obviously depends on what one understands by mindreading. For some, the term suggests that we come to identify mental states on the basis of bodily behavior in a manner analogous to the way in which we grasp meaning on the basis of written inscriptions (see Heyes and Frith 2014; Apperly 2011). According to this usage, mindreading qua mental state attribution is a skill that has to be acquired just as we need to learn how to read texts (since there is no intrinsic or natural connection between the psychologically meaningful mental states and what is perceptually available). Given such a usage, empathy could be seen as an immediate and direct form of social understanding (involving sensitivity to the animacy, agency, and emotional expressivity of others) that is manifest from the outset and which any attempt to explain or predict the other's mental states and behaviors rely on and presuppose.

Even if children from birth onward might have the empathic ability to distinguish animate creatures from inanimate objects, the introduction of a developmental perspective on empathy complicates matters somewhat. It might, for instance, put pressure on an overly epistemic account of empathy by suggesting that the most basic form of social relatedness isn't emotionally neutral. As Hobson has observed, interpersonal understanding normally involves emotional responsivity, and this especially holds true for early infancy, where it is the infant's affective engagement with others that provides it with salient interpersonal experiences encompassing an interplay between similarity and difference, connectedness and differentiation (Hobson 2007; see Reddy 2008). Indeed, such emotionally structured interactions in early infancy might be crucial for the further development of social cognition. A more in-depth exploration of the relationship between interpersonal affectivity and social cognition is obviously of crucial importance, but beyond the scope of this chapter.

At any rate, Zahavi's conception is clearly leaner than that espoused by de Vignemont and colleagues. In order to see just how much leaner it is, let us examine which of de Vignemont and colleagues' conditions it incorporates and which it does not. Like de Vignemont and colleagues, Zahavi requires that there be a causal link between the mental state of the empathizer and that of the target person. Thus, he endorses the causal path condition—condition 3. In fact, since empathy in his sense is elicited by perception, it appears to exclude some other types of causal path that would satisfy de Vignemont and colleagues' condition 3—e.g., imagination, memory, or communication. In this sense, Zahavi's version of condition 3 is stricter than de Vignemont and colleagues'. As for de Vignemont and colleagues' requirement that the empathizer ascribes a mental state to the target person (condition 4), Zahavi certainly does maintain that empathy is other-centered or other-directed, but he also allows for cases of empathy where it doesn't involve any specific mental state ascription, but simply an experience of the presence of other-mindedness. Thus, Zahavi endorses the ascription condition in a loose sense. However, Zahavi dispenses with the affectivity condition (condition 1), since he rejects the claim that one can only empathize with affective states. We can see the other's elation or doubt, surprise or attentiveness in his or her face, we can hear the other's trepidation, impatience, or bewilderment in her voice, feel the other's enthusiasm in his handshake, grasp his mood in his posture, and see her determination and persistence in her actions. Moreover, Zahavi also rejects the interpersonal similarity condition (2). In fact, Zahavi is at pains to insist that empathy primarily confronts one with the presence of an experience that one is not living through oneself. Finally, as already noted, empathy in Zahavi's sense does not entail any prosocial motivation. Thus, Zahavi also rejects the requirement that the empathizer be concerned for the well-being of the target person (i.e., condition 5, the caring condition).

Intentional Alignment

Building upon the theme of other-centeredness, some theorists (within and outside of the phenomenological tradition) do make room for matching of a sort to play an important role in empathy—but it is of a very different sort than that envisaged by proponents of the Lippsian model of projective empathy.

To begin with, Merleau-Ponty maintains that when we perceive an angry gesture, we are perceiving the anger itself and not merely psychologically meaningless behavior. At the same time, though, he denies that the meaning of the gesture is perceived in the same way as the color of the carpet. The other's gestures point to an intentional object, and I understand the meaning of those gestures, not by looking behind them, but by attending to the part of the world that they highlight (Merleau-Ponty 2012, pp. 191–2). The idea here is that in perceiving another as a minded being whose experience is directed toward the same world as that toward which my experience is directed, the intentional content or structure of their experience may also become a focus of my experience.

Based upon this idea, Shaun Gallagher has put forth a proposal that explicitly incorporates elements of enactive and embedded cognition approaches. His starting point is to highlight that emotions at least usually involve not only a qualitative feel but also an intentional object or structure. Thus, while it is not at all necessary for an empathizer to enter into an affective state that matches that of her target with respect to its qualitative feel, it is necessary for her to attune to the same intentional object or state of affairs that is the focus of the target's experience. This, according to Gallagher, is what distinguishes empathy from sympathy:

> Empathy: A feels sad [and/or outrage] *about the injustice done to* B, knowing that B also feels sad [and perhaps outrage] about the injustice done to her (A's feeling has a similar intentional structure as B's affective state). (2012, p. 6)

In contrast, A may also feel sad for B without agreeing with B that an injustice really occurred:

> Sympathy: A feels sad *for B*, who is sad [and perhaps outraged] about an injustice done to B (dissimilar intentional structure). (2012, p. 6)

With this distinction in hand, Gallagher is able to distinguish empathy from contagion, since contagion clearly does involve matching with respect to affect states but not with respect to the intentional structures of experiences. Thus, Gallagher's account includes a kind of alignment between empathizer and target, but this alignment is specified in relation to intentional content rather than affective or qualitative experience. One possible objection to this view is that it appears to rule out the possibility of empathizing with someone when one does not know what the focus of their experience is. And yet, if one, for example, enters a room and finds someone sitting there weeping, it seems odd to suggest that one could not empathize with her. In response to this objection, one reasonable response for Gallagher would be to place the emphasis on the empathizer's motivation to identify the focus of the target person's experience rather than on the successful identification per se. In other words, empathy may be understood to involve the project of identifying and understanding a target's experience. (See Goldie 1999 for an account that is not inspired by phenomenology but that also conceptualizes empathy as a project of reconstructing the other's experience without matching its internal features.)

In this context, it is interesting to take a second look at the body of research on pain empathy, which, as mentioned earlier, has informed the approach of de Vignemont and colleagues. Specifically, some of this research suggests that the areas associated with the affective component of the pain matrix (anterior cingulate cortex and anterior insula) are in fact involved in modulating attention to reflect what is salient in the environment (Iannetti and Mouraux 2010; Legrain et al. 2011; Mouraux et al. 2011). In other words, whenever there is something salient, and attention needs to be directed to it, these areas are recruited; of course, pain is one example of a salient stimulus, but so are lots of other non-painful things. Crucially, registering salience would fit well with the idea of aligning the intentional structure of one's experience with that of the person with whom one is empathizing. Hence, this data raises the possibility that the findings that de Vignemont and colleagues take to provide evidence of embodied affective sharing in fact reveal the matching of intentional structures of experiences rather than of qualitative features of those experiences.

Complementarity and Reciprocity: The Importance of the Second-Person Perspective

In developing the core 4E themes of embodied, enactive, extended, and embedded cognition, some recent theorists have also homed in on the overlapping themes of complementarity and reciprocity in empathy—both of which themes are neglected by approaches focusing on affective sharing. Here, again, the writings of Merleau-Ponty have proven to be a fruitful starting point.

In characterizing the internal relation that obtains between my own body and that of the other, for example, Merleau-Ponty claims that the other appears as the completion of the system and that "the other's body and my own are a single whole, two sides of a single phenomenon" (Merleau-Ponty 2012, p. 370). Thus, to speak, as Merleau-Ponty does, of self and other as "collaborators in perfect reciprocity" (2012, p. 370) suggests an approach to social cognition where the encounter with the other's actions, rather than simply occasioning a mere replication or simulation of those actions, elicits a dynamic response that takes those actions as affordances for further complementary actions (cf. Gallagher and Miyahara 2012). In order to capture what Merleau-Ponty has in mind, it might consequently be better to liken social understanding to dancing than to mirroring. In any case, Merleau-Ponty's emphasis on complementarity provides one more reason why proponents of the embodied simulation approach should distance themselves more clearly from the Lippsian model of projective empathy if they wish to retain a link to the classical phenomenological account of empathy.

In the current research landscape, the themes of complementarity and reciprocity provide useful keys to appreciating ongoing debates about the importance of the second-person perspective. Over the past 15 years or so, interest in the second-person perspective has been stoked by dissatisfaction with the two hitherto dominant mainstream positions in the theory of mind debate, the theory-theory (in its different versions) and the simulation theory (in its different versions). It has occasionally been argued that a limitation of both of the traditional positions is that they privilege either the first-person perspective (this would be the simulation theory) or the third-person perspective (the theory-theory), and that an adequate account of social cognition should also explicitly target the second-person perspective.

However, there is still considerable disagreement about what exactly second-person perspective taking involves. One influential account can be found in a target article in *Behavioral and Brain Sciences*, written by Schilbach and colleagues. For them, the second-person perspective concerns the issue of directly interacting and emotionally engaging with others (rather than simply observing them from a distance). Thus, the second-person perspective is contrasted with what is called the spectatorial stance (Schilbach et al. 2013). Indeed, given that face-to-face interaction engages complementary affective, motoric, and higher cognitive processes that are not engaged in observational settings, an important new challenge for researchers investigating social cognition is to develop more interactive paradigms (Michael 2011b; Overgaard and Michael 2015).

One aspect that may not have been sufficiently highlighted in Schilbach and colleagues' article, however, is the role of reciprocity (de Bruin et al. 2012; Fuchs 2013). After all, the second-person perspective involves bidirectionality and reciprocation. In short, to adopt the second-person perspective is to engage in a subject–subject (you–me) relation where I not only respond to the other but am aware of the other as another, and, at the same time, implicitly aware of myself in the accusative, as attended to or addressed by the other (Husserl 1973, p. 211). This process has been described in much detail by developmental psychologists working on dyadic joint attention (Rochat 2001; Reddy 2008), but also by classical phenomenologists such as Walther, Stein, and Schutz who argued that reciprocal empathy is a key to experiential sharing and communal experiences (cf. Zahavi and Rochat 2015; León and Zahavi 2016; Zahavi and Salice 2017).

Conclusion

In sum, then, we have seen that impulses from 4E approaches have informed recent attempts to articulate the notion of affective matching and to use it as a basis for conceptualizing empathy, but also that 4E approaches have inspired other approaches to empathy that do not require affective matching at all. But most importantly, we have also seen that all of these approaches offer resources that help to address the concerns raised by Bloom in his target article. While Bloom is surely right that merely matching one's affective state with that of someone who is suffering is not necessarily much use to them or to anyone else, empathy need not be limited to such a matching relation, and in fact it need not involve such a matching relation at all.

For those who do include affective matching in their conception of empathy, it is crucial to bear in mind that empathy is integrated with various other social-cognitive processes, from drawing inferences about others' situations and mental states, to being motivated to alleviate others' suffering. Furthermore, as we have already seen, there are many who simply do not consider affective matching to be a constitutive feature of empathy at all. For them, empathy may in fact include a sort of matching insofar as the empathizer comes to focus on the intentional content or structure of the target's experience—and indeed this is just the sort of contextual feature that embedded and extended approaches would urge us to take into account. But it is difficult to see how this sort of matching would fall prey to the same dangers as the affective matching criticized by Bloom. Moreover, whatever one might think of this sort of matching as intentional alignment, the themes of other-centeredness, complementarity, and reciprocity—all of which are inspired by insights from embodied and enactive approaches—provide further insurance against the perils identified by Bloom. In sum, then, the outlook for empathy is not as bleak as Bloom envisions.

Of course, this still leaves us with the question of how to understand empathy. In the contemporary debate, one can encounter distinctions between mirror empathy, motor empathy, affective empathy, perceptually mediated empathy, re-enactive empathy, and cognitive empathy, to mention just a few of the options available. As should have become clear by now, one reason why it continues to be so difficult to reach a commonly accepted definition of empathy is that people have been using the notion to designate rather different phenomena. For the same reason, it is not obvious that it makes that much sense to try to determine once and for all what empathy really is. Although one might make the case that one ought to stick to the traditional use of the term—as already mentioned, it was introduced by Lipps as a general term for our understanding of others—instead of identifying it with, say, prosocial behavior or a very special kind of imaginative perspective taking, it is not evident that such a strategy would be particularly productive or illuminating. Thus, rather than promoting a specific account of empathy as the right account, a more reasonable verdict might be that the different analyses of empathy contain various insights that contemporary debates on social cognition and interpersonal understanding ought to incorporate.

References

Apperly, I. (2011). *Mindreaders: the cognitive basis of "theory of mind."* Hove, UK: Psychology Press.

Aydede M. (ed.) (2006). *Pain: new essays on its nature and the methodology of its study.* Cambridge, MA: MIT Press.

Batson, C.D. and Shaw, L.L. (1991). Evidence for altruism: toward a pluralism of prosocial motives. *Psychological Inquiry*, 2(2), 107–22.

Bloom, P. (2014). Against empathy. *Boston Review*, 39(5), 14–19.

Botvinick, M., Jhab, A., Bylsmaa, L., Fabian, S., Solomon, P., and Prkachin, K. (2005). Viewing facial expressions of pain engages cortical areas involved in the direct experience of pain. *NeuroImage*, 25(1), 312–19.

Cialdini, R.B., Brown, S.L., Lewis, B.P., Luce, C., and Neuberg, S.L. (1997). Reinterpreting the empathy-altruism relationship: when one into one equals oneness. *Journal of Personality and Social Psychology*, 73(3), 481–94.

de Bruin, L., van Elk, M., and Newen, A. (2012). Reconceptualizing second-person interaction. *Frontiers in Human Neuroscience*, 6, 1–14.

de Vignemont, F. (2010). Knowing other people's mental states as if they were one's own. In: S. Gallagher and D. Schmicking (eds.), *Handbook of phenomenology and cognitive science*. Dordrecht: Springer, pp. 283–99.

de Vignemont, F. and Jacob. P. (2012). What is it like to feel another's pain? *Philosophy of Science*, 79(2), 295–316.

de Vignemont, F. and Singer, T. (2006). The empathic brain: how, when and why? *Trends in Cognitive Sciences*, 10, 435–41.

Fuchs, T. (2013). The phenomenology and development of social perspectives. *Phenomenology and the Cognitive Sciences*, 12(4), 655–83.

Gallagher, S. (2012). Empathy, simulation, and narrative. *Science in Context*, 25(3), 355–81.

Gallagher, S. and Miyahara, K. (2012). Neo-pragmatism and enactive intentionality. In: J. Schulkin (ed.), *Action, perception and the brain: adaptation and cephalic expression*. Basingstoke, UK: Palgrave Macmillan, pp. 117–46.

Gallese, V. (2001). The "shared manifold" hypothesis: from mirror neurons to empathy. *Journal of Consciousness Studies*, 8(5–6), 33–50.

Gallese, V. (2003). The manifold nature of interpersonal relations: the quest for a common mechanism. *Philosophical transactions of the Royal Society of London. Series B, Biological sciences*, 358(1431), 517–28.

Gallese, V. (2007). Embodied simulation: from mirror neuron systems to interpersonal relations. *Empathy and Fairness: Novartis Foundation Symposium*, 278, 3–19.

Gallese, V. (2009). Mirror neurons, embodied simulation, and the neural basis of social identification. *Psychoanalytic Dialogues*, 19(5), 519–36.

Gallese, V., Keysers, C., and Rizzolatti, G. (2004). A unifying view of the basis of social cognition. *Trends in Cognitive Sciences*, 8(9), 396–403.

Goldie, P. (1999). How we think of others' emotions. *Mind and Language*, 14(4), 394–423.

Goldman, A.I. (2006). *Simulating minds*. New York: Oxford University Press.

Goldman, A.I. and Sripada, C.S. (2005). Simulationist models of face-based emotion recognition. *Cognition*, 94(3), 193–213.

Heyes, C. and Frith, C. (2014). The cultural evolution of mind reading. *Science*, 344(6190). doi:10.1126/science.1243091

Hobson, R.P. (2007). Communicative depth: soundings from developmental psychopathology. *Infant Behavior & Development*, 30, 267–77.

Hoffman, M.L. (1982). Development of prosocial motivation: empathy and guilt. In: N. Eisenberg (ed.), *The development of prosocial behavior*. New York: Academic Press, pp. 281–338.

Husserl, E. (1973). *Zur Phänomenologie der Intersubjektivität: Texte aus dem Nachlass. Zweiter Teil: 1921–28*. Husserliana 14. Den Haag: Martinus Nijhoff.

Iacoboni, M. (2007). Existential empathy: the intimacy of self and other. In: T.F.D. Farrow and P.W.R. Woodruff (eds.), *Empathy in mental illness*. Cambridge: Cambridge University Press, pp. 310–21.

Iacoboni, M. (2009). *Mirroring people: the science of empathy and how we connect with others*. New York: Picador.

Iannetti G.D. and Mouraux, A. (2010). From the neuromatrix to the pain matrix (and back). *Experimental Brain Research*, 205, 1–12.

Legrain, V., Iannetti G.D., Plaghki, L., and Mouraux, A. (2011). The pain matrix reloaded: a salience detection system for the body. *Progress in Neurobiology*, 93, 111–24.

León, F. and Zahavi, D. (2016). Phenomenology of experiential sharing: the contribution of Schutz and Walther. In: A. Salice and H.B. Schmid (eds.), *The phenomenological approach to social reality: history, concepts, problems*. Dordrecht: Springer, pp. 219–34.

Lipps, T. (1907). Das Wissen von fremden Ichen. In: T. Lipps (ed.), *Psychologische Untersuchungen I*. Leipzig: Verlag von Wilhelm Engelmann, pp. 694–722.

Lipps, T. (1909). *Leitfaden der Psychologie*. Leipzig: Verlag von Wilhelm Engelmann.

Maibom, H.L. (ed.) (2014). *Empathy and morality*. Oxford: Oxford University Press.

Merleau-Ponty, M. (2012). *Phenomenology of perception* (trans. D.A. Landes). London: Routledge.

Michael, J. (2011a). Four models of the functional contribution of the mirror neuron system. *Philosophical Explorations*, 14(2), 185–94.

Michael, J. (2011b). Interaction and mindreading. *Review of Philosophy and Psychology*, 2(3), 559–78.

Michael, J. (2014). Towards a consensus about the role of empathy in interpersonal understanding. *Topoi*, 33(1), 157–72.

Michael, J. and Fardo, F. (2014). What (if anything) is shared in pain empathy? A critical discussion of de Vignemont and Jacob's (2012) theory of the neural substrate of pain empathy. *Philosophy of Science*, 81, 154–60.

Mouraux, A., Diukova, A., Lee, M.C., Wise, R.G., and Iannetti, G.D. (2011). A multisensory investigation of the functional significance of the "pain matrix." *NeuroImage*, 54, 2237–49.

Overgaard, S. and Michael, J. (2015). The interactive turn in social cognition research: a critique. *Philosophical Psychology*, 28(2), 160–83.

Reddy, V. (2008). *How infants know minds*. Cambridge, MA: Harvard University Press.

Rochat, P. (2001). *The infant's world*. Cambridge, MA: Harvard University Press.

Schilbach, L., Timmermans, B., Reddy, V., Costall, A., Bente, G., Schlicht, T. et al. (2013). Toward a second-person neuroscience. *Behavioral and Brain Sciences*, 36(4), 393–414.

Singer, T., Seymour, B., O'Doherty, J., Kaube, H., Dolan, R., and Frith, C. (2004). Empathy for pain involves the affective but not sensory components of pain. *Science*, 303, 1157–62.

Sinigaglia, C. (2008). Mirror neurons: this is the question. *Journal of Consciousness Studies*, 15(10-11), 70–92.

Stueber, K.R. (2006). *Rediscovering empathy: agency, folk psychology, and the human sciences*. Cambridge, MA: MIT Press.

Titchener, E.B. (1909). *Lectures on the experimental psychology of thought-processes*. New York: Macmillan.

Zahavi, D. (2011). Empathy and direct social perception: a phenomenological proposal. *Review of Philosophy and Psychology*, 2(3), 541–58.

Zahavi, D. (2014a). *Self and other: exploring subjectivity, empathy, and sharing*. Oxford: Oxford University Press.

Zahavi, D. (2014b). Empathy and other-directed intentionality. *Topoi*, 33(1), 129–42.

Zahavi, D. and Overgaard, S. (2012). Empathy without isomorphism: a phenomenological account. In: J. Decety (ed.), *Empathy: from bench to bedside*. Cambridge, MA: MIT Press, pp. 3–20.

Zahavi, D. and Rochat, P. (2015). Empathy ≠ sharing: perspectives from phenomenology and developmental psychology. *Consciousness and Cognition*, 36, 543–53.

Zahavi, D. and Salice, A. (2017). Phenomenology of the we: Stein, Walther, Gurwitsch. In: J. Kiverstein (ed.), *The Routledge handbook of philosophy of the social mind*. London: Routledge, pp. 515–27.

CHAPTER 32

CRITICAL NOTE

3E's Are Sufficient, But Don't Forget the D

ACHIM STEPHAN

A purely disembodied human emotion is a nonentity.

(William James 1890, vol. 2, p. 452)

INTRODUCTION

The four chapters in this section contribute, in one way or another, to the study of situated affectivity and situated cognition. While the contributions of both Giovanna Colombetti and Dan Zahavi and John Michael aim at elucidating aspects of human affectivity from a situated perspective, the chapters by Peter Hobson and Evan Carr et al. mainly shed light on aspects of human cognition from a situated perspective on emotions, while adding less to a better understanding of affectivity proper.

Before turning to the four contributions in more detail, though, I will first introduce the debate on human affectivity from a more general perspective, in order to then locate to what particular aspects the articles contribute, what they could have considered in addition, and what important aspects are to be studied in the future.

The situated perspective on human affectivity is a response to several one-sided and unbalanced theories of emotions in the past. The most salient one—the cognitivist theory of emotions (or cognitivism)—had been pursued to counterbalance the alleged shortcomings of what is known as the feeling theory of emotions (or reflexive perceptualism),[1] a position generally thought of as ignoring the cognitive part of

[1] The theory was introduced and defended by William James who claimed "that *the bodily changes follow directly the* PERCEPTION *of the exciting fact, and that our feeling of the same changes as they occur* IS *the emotion*" (1884, pp. 189–90; emphasis by James).

emotions, particularly their intentionality and their evaluative function.[2] By emphasizing, in contrast, the judgmental and intentional aspects of emotions, cognitivists typically ignored the role of the body in emotions. In retrospect, even one of the protagonists of cognitivism, Robert Solomon, conceded that they had "veered too far in the other direction" (Solomon 2004, p. 85).

Already in the heyday of cognitivism, some emotion psychologists pointed to the misguided study of cognition and affectivity from a disembodied perspective (e.g., Zajonc and Marcus 1984), thereby bringing the body back to consideration. Subsequently and still continuing, a plethora of studies investigated the role of the body in emotional tasks. Carr et al. (2018) review and systematize many of these studies in their contribution to this volume.

We also know of attempts in the philosophy of emotions to reconsider the role of the body and to overcome the gulf between feeling theories and cognitivist theories. In what is now referred to as a neo-Jamesian approach, Jesse Prinz explicitly introduced so-called *embodied appraisals*, thereby enriching James's perceptualism with a cognitive component. His emotion theory, however, is still limited to one-shot encounters with emotion-eliciting events examined from the perspective of individual persons. Thus, even positions that merge cognitive and bodily aspects of affectivity turn out to be restricted and unbalanced in various ways, as they do not consider the dynamical unfolding of emotions in social encounters, or the scaffolding impact of the social and physical environment on facilitating certain emotional processes, or any collective forms of affectivity—though social psychologists such as, e.g., Brian Parkinson (1995), started to study these phenomena in the 1990s.

Only recently have philosophers of emotion widened the focus of their research to include the synchronic and diachronic scaffolding of emotions by environmental structures and their dynamic unfolding in social interactions. Pioneering work with regard to these topics comprise both Giovanna Colombetti and Evan Thompson's "The Feeling Body: Towards an Enactive Approach to Emotion" (2008) and Paul Griffiths and Andrea Scarantino's "Emotions in the Wild" (2009) that appeared nearly simultaneously, but independently of each other. With these approaches all pieces were in play to get the idea of situated affectivity in all its facets off the ground. What the debate lacked until recently, however, is a clear and shared understanding of the terms used. Among the first attempts to trace the geography of the involved concepts, including embodied, embedded, extended, enacted, as well as distributed emotions, are Wilutzky et al. (2011) and Stephan et al. (2014), who provided detailed suggestions for how to understand these terms. What follows is, more or less, structured along these concepts.

[2] But see Ratcliffe (2008, pp. 220–2).

Embodied Emotions

While it seems a fair issue to controversially discuss whether the human body is essentially involved in cognitive processes, the analogous dispute about the bodily involvement in affective processes seems odd. In contrast to cognitions, which in early cognitive science had been deemed to be disembodied entities, not to speak of their past non-bodily bearers such as the Aristotelian *nous poiêtikos* or the Cartesian *res cogitans*, emotions have always—except for the short period of radical cognitivism—been seen as hybrid phenomena including the body. This explicitly holds true for Aristotle and Descartes, but also for recent psychological theories of emotions such as, for example, the component process model of emotions as proposed by Klaus Scherer (2005), in which four of the five components are essentially bodily.

It is therefore no surprise that Evan W. Carr, Anne Kever, and Piotr Winkielman (2018) do not aim at showing that emotions are essentially embodied. They take this for granted, or so it seems. In their comprehensive review, they rather argue that *embodied* emotional processes strongly (though not indispensably) influence various cognitive and motivational tasks if affective issues are concerned. The unhindered use of one's own facial muscles, for example, assists both the recognition of facial expressions of others (pp. 532–4) and—if a task requires it—the smooth understanding of sentences and phrases with affective content (pp. 537–43). Analogously, if certain affectively construable stimuli are, in fact, conceptualized in *affective* terms, they facilitate corresponding motor actions tied to abstract valence representations along an *approach-avoidance* dimension (pp. 534–6).[3] Furthermore, the comprehension of sentences with affective content seems to activate different brain regions depending on whether the sentences are about an external observer's perspective or the experiential participant's perspective—suggesting that different forms of simulation are involved in the understanding of the detailed affective states (pp. 537–8); related studies report so-called switching costs in cognitive processing either when switching between an observer's and a participant's perspective or when switching between affective and sensory modalities—indicating that affective concepts establish a distinct modality analogous to auditory or gustatory modalities (pp. 537–8). Eventually, after having gone through an extensive amount of mirroring literature with quite confusing results, Carr et al. emphasize that the embodiment framework might help to account for some of the findings, particularly those about mirroring in diverse social settings. Since mimicry fosters the formation of congruent, bodily grounded emotional states, it is seen as a sort of "social glue"—liking one another supports mimicking one another, and being mimicked supports liking the mimicker, especially if she is an in-group member. In contrast, mimicry by out-group members often has negative effects (pp. 543–5).

[3] This fits nicely with Scarantino's motivational theory of emotion, which expands on Nico Frijda's idea of action readiness (see Scarantino 2014).

To sum up, the studies reported by Carr et al. do not question that emotional processes are embodied—they presuppose their bodily nature. Instead, the results support the much more controversial embodiment paradigm with regard to cognitive processing, though not without qualification. Although the studies overwhelmingly show that embodied components of emotions contribute in many ways to cognitive activities such as categorization and comprehension as well as to forming and sustaining social relationships, they nevertheless speak against a strong embodiment thesis according to which the bodily involvement is indispensable for these tasks. Many of the cognitive tasks can still be performed if embodied components of emotional processes are inhibited, though they slow down considerably (p. 531). Moreover, often a cognitive, context-sensitive, top-down component affects whether or not the emotional aspect of a particular stimulus will inform the cognitive processing—which shows that embodied emotions are not bound to contribute automatically, in a reflex-like way, to cognitive processes. The results show how strongly and context-sensitively affective and cognitive processes interact.

Environmental Scaffolds of Human Affectivity

The feature of context-sensitivity points to two other paradigms of the current debate about affective situatedness: the hypotheses that—at least some—emotions (or other affective phenomena) are embedded or extended, respectively. In metaphysically different formats, both hypotheses emphasize the huge impact the physical and social environment is supposed to have on emotional processes—though not in the trivial sense that most of our emotions are triggered by environmental stimuli in a certain context. In analogy to recent developments in the philosophy of cognition, it has become common to understand the hypothesis of embedded affectivity as the claim that affective processes are co-dependent upon or causally coupled to extrabodily processes, while, in contrast, the hypothesis of extended affectivity is understood as the claim that affective processes are co-constituted by extrabodily processes (Stephan et al. 2014, pp. 68–9). Though the distinction between causal coupling and constitution can precisely be drawn in philosophical terms, it is not promising to wait for an empirical answer whether some kind of affective process is embedded or extended. There will be no answer, as Baumgartner and Wilutzky (2017) have convincingly argued.[4] I will therefore

[4] They demonstrate that the borderline between constituents and causes of cognitive processes cannot be drawn experimentally since there is no way to generate decisive experimental evidence for constitution. Instead, we should acknowledge that the "inference to constitution is of *inherently pragmatic* nature, involving, for instance, measures of explanatory power" (Baumgartner and Wilutzky 2017, p. 17)—a diagnosis that might help to calm down the ongoing debate about the so-called coupling-constitution fallacy charged upon the hypothesis of extended cognition (cf. Adams and Aizawa 2001; 2009; also see Newen et al. 2018, pp. 7–8).

bracket the metaphysical debate about causal dependency and co-constitution, and instead turn to what is discussed under the heading of diachronic and synchronic affective scaffolding.[5]

The idea of environmental scaffolding, which can be traced back to the work of Lev Vygotsky, was reintroduced by Andy Clark to comprise all sorts of environmental structures that can be used for enabling or facilitating certain cognitive tasks (Clark 1997, pp. 45–7, 179–80, 194ff.). In this broad sense, it also applies to various affective processes: in order to affect their emotional condition, human beings both actively structure their environment or use given environmental resources as affective scaffolds. They seek spectacular natural sites to experience awe, joy, or admiration; they deliberately undergo psychotherapy to master their anxieties; or they down excessive amounts of alcohol to get a kick out of engaging in mass brawls. In addition to these synchronically scaffolded emotional engagements, people are also diachronically scaffolded in the acquisition of their emotional repertoire (Griffiths and Scarantino 2009, p. 443), for example, when children and adolescents grow into the affective regime of an honor culture.

It is this latter type of diachronic scaffolding to which Peter Hobson (2018) basically contributes. Yet, just as for Carr et al., his main target is not human affectivity, nor the ontogenesis of human emotionality, in particular. Rather, he argues for the essential impact of an infant's capacity for social-affective relatedness on her cognitive development. With reference to the results of the thoroughly tested and replicated still-face experiment and the visual cliff paradigm,[6] Hobson states that an infant "is endowed with a non-inferential, affectively grounded responsiveness structured in terms of self–other relations embedded in a shared world" and emphasizes that "such responsiveness provides the foundations for the acquisition of conceptual understanding over the coming months of life" (p. 560). Thinking, he claims, "develops through communication between expressive, embodied persons, where such communication requires infants to have affective engagement with their social and non-social environment" (p. 553). Adding further support from his rich clinical experience, Hobson regards the deficiencies in affectively grounded interactions with other persons—to be seen from early on in autistic children—as the main source of their subsequent weaknesses in interpersonal understanding, communication, and thinking (cf. p. 563).

Hobson might have liked to notice how strongly his insights and ideas correspond to the recent debate about environmentally scaffolded affectivity, to the hypotheses of extended cognition and emotion, and to further adjustments of these paradigms, particularly in the work of Jennifer Greenwood (2015), Joel Krueger (2014), and Somogy

[5] My move is also supported by Kim Sterelny's claim that the most plausible cases for the extended mind hypothesis are merely "limiting special cases of scaffolded minds," thus depriving the extended mind paradigm of its heuristic potential, since it "obscures rather than highlights both the continuities and the differences amongst external resources and their contributions to cognitive competence" (Sterelny 2010, p. 473).

[6] See Adamson and Frick (2003) and Adolph and Kretch (2012).

Varga (2016). All three focus on developmental aspects in early dyadic relationships to shed more light on extended cognition and affectivity. Thus, Krueger takes infant-caregiver interactions that establish shared emotions and provide mutual affect regulation as exemplars for collectively extended emotions (2014, pp. 545–8). Varga also focuses on the topic of emotion regulation in infants and emphasizes the essential role of successful synchronic interactions between infant and caregiver for both the regulation of an infant's current emotions and the maturation of her emotional self-regulatory abilities in an evolving diachronic process (2016, pp. 2475–6). Although he considers early dyadic interactions, in principle, as instances of socially extended cognition (cf., e.g., Gallagher 2013), he argues for augmenting the extended cognition paradigm by a thesis of extended emergent cognition (2016, p. 2489), which he expects to better account for the peculiarities of collective cognitive systems. I will come back to this issue under the heading of distributed affectivity in the fifth section.[7]

Greenwood offers a comprehensive study of the affective and cognitive development in early childhood. In examining both synchronic and diachronic scaffoldings provided by caregivers in specific sociocultural settings, she does not spend much time on fully controlled unidirectional support as seen, e.g., in assisting an infant's first steps. Instead, she explores in depth the synchronous interactions of infant and caregiver that diachronically shape and enhance the infant's behavioral repertoire by transforming the neural structure of the child's brain (2015, pp. 8–9, ch. 4–6). Regarding the 4E debate on cognition and affectivity, Greenwood claims the setup, in which human emotional ontogenesis materializes, to be a paradigm case for extended mind scenarios, though not for the "orthodox" first- and second-wave positions that adhere to the parity principle and the model of complementarity, respectively (2015, pp. 56–7).[8] Rather, she supports what has been called the third-wave model, according to which "human emotional ontogenesis extends from the brains of people in the external environment and the sociocultural products into which they are deeply integrated into other people's developing brains. What this clearly implies is that emotional ontogenesis is a world-to-brain transcranial achievement" (2015, p. 107)—a position she prefers to dub "contingent transcranialism" to emphasize her rejection of the unidirectional mind-to-world extension suggested by first- and second-wave externalism (2015, p. 10, ch. 3).

[7] To avoid confusion, we should distinguish between extended and distributed cognition (and affectivity). Distributed cognitive processes are typically instantiated by several individuals, no one of which can be singled out as the "center" of the processing. In contrast, extended cognitive processes are always attributable to an individual agent as the "heart" of the processing (see Stephan et al. 2014, pp. 69, 77–8).

[8] According to the parity principle, cognitive processes roughly count as co-constituted by extrabodily processes if, more or less, the latter play the same functional role as corresponding intracranial processes we would regard as constituents proper. In contrast, according to the principle of complementarity, extrabodily processes count as co-constituents if they complement intracranial processes to jointly instantiate features the internal processes alone could not exhibit (cf. Kirchhoff 2012, pp. 289–91; Stephan et al. 2014, p. 73).

How strongly Hobson's and Greenwood's ideas correspond with each other becomes apparent with regard to the pioneering work of Vygotsky both refer to (Greenwood 2015, pp. 45, 91). Vygotsky, cited by Hobson (p. 555), emphasizes that "every function in the child's cultural development appears twice: first, on the social level, and later, on the individual level; first, *between* people (*interpsychological*), and then *inside* the child (*intrapsychological*)" (Vygotsky 1978, p. 57)—an insight Hobson endorses when claiming that "goings-on between people become internalized, so they become features of individual intelligence" (p. 555). Though rather tacitly, Hobson's contribution relates in some measure to the current debate about environmental scaffolds of cognition and affectivity.

Enactive Human Affectivity

Besides the developmental perspective considered earlier, dynamical affective couplings between two or more emoters (or an emoter and her environment) have not been in the focus thus far. They are, however, a central topic for the enactive approach to human affectivity.

To begin with, enactivism presupposes that cognitive and affective processes are embodied and environmentally scaffolded. Furthermore, and in contrast to previous approaches, Giovanna Colombetti, being one of the key figures of an enactive approach to human affectivity,[9] argues for abandoning the clear-cut distinction between cognition and affectivity (Colombetti 2018). She instead emphasizes the inherently affective character of the mind, stipulating that we would not have a mind any more if we deprive it of affectivity (p. 575).[10] Her claim has its roots in a specific understanding of the life-mind continuity thesis (Thompson 2007, pp. 128–9). While enactivism usually confines itself to the thesis that *cognitive* processes are essentially acts of "sense-making" conducted by autonomous and adaptive living beings in interaction with their environment (Thompson 2007, pp. 158–9), Colombetti understands the implied claim that organisms have a capacity to be sensitive to their needs and concerns explicitly as an affirmation of their endowment with primordial affectivity (2014, p. 2). Accordingly, sense-making is always both a cognitive and an affective phenomenon (p. 574); even human cognition remains an essentially affective affair—being a special manifestation of our capacity *to give a damn*, as John Haugeland would have had it.[11]

[9] In her concise introduction to the canonical form of enactivism, Colombetti (2014) also sketches further versions of enactivism (pp. 571–3).

[10] We may speculate about how Colombetti might assess Mr. Spock cases. Would she—contrary to ordinary intuition—deprive such individuals of a mind, or would she grant them a form of basic though unnoticeable affectivity, or would she rather claim them to be nonentities in human terms?

[11] With reference to Vygotsky, Hobson emphasizes the unity of cognitive and affective processes as well (pp. 554–6).

With regard to widely acknowledged psychological theories of emotion, particularly appraisal theories, enactivism requires some modifications in our understanding of the cognitive component responsible for appraisals. In opposition to, e.g., Klaus Scherer's component process model of emotions, according to which each emotional episode consists of the synchronization of several subsystems primarily steered by the cognitive component that continuously runs through cycles of stimulus evaluation checks (cf. 2005, pp. 697–9), Colombetti claims that we should not subdivide an organism into principally separable components. In particular, appraisal should not be conceived of as an essentially intracranial (cognitive) process distinct from subsequent bodily processes that generate facial expression, bodily posture, action readiness, and the like. Instead, we should think of these bodily processes as "bodily ways of making sense of the situation" (p. 577). Hence, appraisals should be recognized as being simultaneously bodily and cognitive-evaluative.[12]

Though the holistic picture enactivism proffers might prima facie be quite appealing, it deliberately gives up on structuring emotional processes in distinct functional units, and thereby runs the risk of blurring what could be distinguished. If, as Scherer claims, a certain timeline in the recursive unfolding of evaluative and bodily processes can be vindicated within emotional episodes (Scherer 2001, pp. 100–2, 106–8), enactivists should provide further arguments to actually challenge the component process model of emotions. As in her disagreement with the basic emotion framework, Colombetti might refer here, too, however, to the dynamical systems approach according to which "emotions are conceptualized as patterns of organismic processes . . . that take the form they do in virtue of evolutionary and developmental forces that together funnel specific organismic trajectories" (p. 580).

But instead of moving deeper into internal quarrels among emotion researchers, I finally want to shed some light on how Colombetti comments on the relationship between enactivism and the extended mind paradigm. Moreover, this gives us the opportunity to see how the dynamical unfolding of emotional processes among several persons provides an important addition to the traditional enactivist agenda. Enactivists typically regard the internalism–externalism debate as pretty alien to enactivism if the options are merely internalism and first- or second-wave theories of the extended mind, since these approaches all share the assumption that intracranial processes provide paradigm cases for cognitive processes—an assumption enactivists reject. In contrast, they judge cognitive and affective processes as basically *relational* processes of sense-making that occur *between* an organism and its environment (Thompson and Stapleton 2009, pp. 25–6). Colombetti, however, makes greatest efforts to argue that organismic sense-making processes may extend beyond the boundaries of an organism—thereby responding to both orthodox enactivists and to those critics who claim that enactivism is in fact incompatible with the extended mind thesis (pp. 582–4).

[12] In a cross-reference to Carr et al. (2018), Colombetti states that it remains open whether they judge the cognitive-evaluative appraisal aspects of emotion as always embodied (p. 578, fn. 13).

Leaving this quarrel aside, I take it to be more promising to examine how enactivism relates to externalism within the framework of environmentally scaffolded affectivity. Recently, Giovanna Colombetti and Tom Roberts elaborated on a proposal of Andy Clark (2008, pp. 129–31) according to whom a system is best regarded as an extended cognitive system when (1) it is coupled to an environmental item (2) through which the system loops some kind of self-stimulating activity, (3) and this self-stimulating activity in particular has been set in place and maintained over time to achieve a certain cognitive result (Colombetti and Roberts 2015, p. 1248). With regard to affective processes, they consider situations where couplings are selected and maintained for the function they perform, and view the activity of the whole loop—in essentially enactivist terms—as responsible for the episode's overall affective character. Their paradigm case is a deeply mourning jazz musician, whose "playing sets up a mutually constraining cycle of affective responding and expression: the qualities of the music performed, and of the actions and gestures initiated, feed back into the character of the musician's emotional experience, which in turn governs what he plays next" (Colombetti and Roberts 2015, p. 1258; cf. Colombetti, p. 583).

Though being an instance of environmental coupling and self-stimulatory loops, the case of the musician still somewhat exemplifies the second-wave externalists' mind-to-world extension—the instrument does not by itself interact with the musician. In contrast, the coupling between infant and caregiver as examined by Greenwood should be a much more fitting type of environmentally scaffolded affectivity for enactivists. Here we clearly see a dynamically driven, mutual affective-cognitive process between two persons in relation to a specific sociocultural environment. The coupling of caregiver and child, moreover, is a compelling example for the relational and distributed process of joint recursive sense-making that eventually will enable the child to advance her own cognitive and affective resources. Therefore, it should be quite straightforward to integrate Greenwood's third-wave externalism into the enactive approach Colombetti favors.[13]

Distributed Affectivity

Dynamical affective couplings between two and more persons or even within larger groups, in general, need to move more into the focus of enactivism and paradigms of environmental scaffolding. In these social settings no individual can be singled out as the "hub" of the affective process (as is the case for first- and second-wave externalism). Instead, *distributed* affectivity or *collective* emotions are instantiated by a dyad or

[13] Greenwood is right, indeed, to strictly distinguish her account from the typical extended mind scenarios. The mutual interactions she studies are better regarded as exemplars of distributed cognition and affectivity than as instances of cognitive (or affective) extensions of one single cognizer (or emoter) into the environment.

collectives—a topic not captured by any of the contributions to this section under this heading (but cf. von Scheve and Salmela 2014).

Typical examples are, e.g., on a large scale, contagious euphoria or panic of a crowd, on a smaller scale, social interactions like starting to flirt with a stranger or a family row, in which emotions are dynamically unfolding between social agents, where the outcome is initially open, with many factors influencing the development of this process, such as the social setting and cultural conventions and practices (Parkinson 1995, ch. 7, 8; Wilutzky 2015, pp. 6–7). In such social encounters, we do not see just one individual responding emotionally to another person's behavior; rather we notice a continuous exchange between reciprocally interacting agents shaping the whole affective atmosphere hereby online (Griffiths and Scarantino 2009, pp. 438–40; Stephan et al. 2014, pp. 76–8).

Such examples of distributed affectivity—where groups of emoters and not individuals coupled to some technical instrument or natural item jointly instantiate an affective atmosphere—count among the best candidates for affective phenomena beyond an individual's skin. Although we still miss a fully fleshed out account of distributed affectivity, it should be obvious that the social couplings to be studied there are quite distinct from those ordinarily covered by the extended mind paradigm. Distributed affectivity spreads out over a collective; it does not extend from one particular individual into the social or natural environment.

Empathy in 4E Perspective

In their contribution to this volume, Dan Zahavi and John Michael (2018) promise to present new insights from the 4E perspective on empathy. To begin with, they introduce two neo-Lippsian approaches to empathy as proposed by Alvin Goldman and by Frédérique de Vignemont and colleagues, respectively. While according to Goldman, empathy is just a subtype of unmediated mindreading governed by a simulation-plus-projection process (pp. 591–5), de Vignemont and colleagues offer a more elaborated account that contrasts empathy with such related phenomena as, e.g., mindreading, sympathy, and emotional contagion. According to them, empathy (1) is an affective state, which (2) is similar to the affective state of the other, by which state (3) it is elicited; moreover, empathy (4) presupposes to differentiate between oneself and the other, and (5) implies the prosocial attitude of caring (pp. 591–5).

Before commenting on de Vignemont's list of features she supposes empathy to have, Zahavi and Michael carry on with surveying recent literature. They present two hybrid approaches to empathy, in which—inspired by the mirror neuron paradigm— both Gallese and Iacoboni fuse a neo-Lippsian take with the phenomenological tradition by endorsing the view that empathy involves a low-level, simulation-based form of inner imitation, thereby constituting a basic, non-inferential form of embodied social understanding within a shared environment (pp. 595–6). Gallese himself, however, seems to notice that uncritically overemphasizing the mirroring mechanism might produce a tension to an insight of phenomenologists that empathy essentially requires a preservation of self–other difference to enable genuine interpersonal understanding (p. 597).

For Zahavi, whose recent phenomenologically inspired take on empathy the authors recapitulate, it is of particular import to understand empathy as a specific form of intentionality that reveals and presents, while directed at others, their experience as foreign (p. 597). Re-examining de Vignemont's proposal, they point out that Zahavi approves of both the third feature, if empathy is elicited by perception, and the fourth feature, if empathy is understood as other-centeredness. He declines, however, the remaining three features, arguing instead that (1*) we can also empathize with other than affective states of another person, that (2*) an empathizer's state need not be similar to the other's state, and that (3*) an empathizer need not have prosocial attitudes toward the other. The latter claim emphasizes the epistemic understanding of empathy while keeping the feature of caring for sympathy.

On the whole, Zahavi's modifications are well founded. I have only some worries regarding the first feature. Renouncing the affectivity condition seems to entail that neither another's nor an empathizer's corresponding state need be an affective state. If so, then Zahavi is about to give up the distinction between what is called affective empathy on the one hand and cognitive empathy on the other (see Stephan 2015, pp. 111–13), a result hardly enticing for phenomenologists, since this could make artificial agents, who may be endowed with the capacity for cognitive empathy, be genuine empathizers.

Zahavi and Michael announce at the beginning and resume at the end of their contribution that 4E approaches provide useful resources for conceptualizing empathy and distinguishing it from related phenomena (pp. 590, 602). The actual results in-between, however, are somewhat meager; they all could have been achieved without reference to the 4E approaches, or so it seems. Most often, Zahavi and Michael mention the four positions all at once, not elucidating in what particular sense one of these paradigms could contribute to empathy proper or to a better account of empathy. To point to our bodily nature when thinking about disclosing another's affective state is not something deeply new. They could, for example, have examined in what way particular forms of environmental scaffolding could enhance or enable empathy,[14] how empathic capacities are diachronically shaped by recursive interactions between infant and caregiver, or whether there are any forms of collective empathy. They also could have discussed in more detail how the features of overlapping complementarity and reciprocity in empathy, as touched upon by Merleau-Ponty (p. 601), relate to a potential enactive account of empathy. If, however, it might be better to liken social understanding to dancing than to mirroring, as they say, then what we study might no longer be empathy.

[14] In Stephan et al. (2014, pp. 74–5), we introduced Arnold as a cousin of Otto, an autistic person hardly capable of directly recognizing the emotional states of others in social interactions. If Arnold's limited empathic capacities were compensated, say, by a headset camera connected to a computer running a program for decoding human emotional states providing online information about the affective states of others to him, we may claim this to be a case of orthodoxly extended empathy.

References

Adams, F. and Aizawa, K. (2001). The bounds of cognition. *Philosophical Psychology*, 14, 43–64.

Adams, F. and Aizawa, K. (2009). Why the mind is still in the head. In: P. Robbins and M. Aydede (eds.), *The Cambridge handbook of situated cognition*. Cambridge: Cambridge University Press, pp. 78–95.

Adamson, L.B. and Frick, J.E. (2003). The still-face: a history of a shared experimental paradigm. *Infancy*, 4, 451–73.

Adolph, K.E. and Kretch, K.S. (2012). Infants on the edge: beyond the visual cliff. In: A.M. Slater and P.C. Quinn (eds.), *Developmental psychology: revisiting the classic studies*. Los Angeles: Sage, pp. 36–55.

Baumgartner, M. and Wilutzky, W. (2017). Is it possible to experimentally determine the extension of cognition? *Philosophical Psychology*, 30, 1104–25.

Carr, E.W., Kever, A., and Winkielman, P. (2018). Embodiment of emotion and its situated nature. In: this volume, pp. 529–52.

Clark, A. (1997). *Being there: putting brain, body, and world together again*. Cambridge, MA: MIT Press.

Clark, A. (2008). *Supersizing the mind: embodiment, action, and cognitive extension*. Oxford: Oxford University Press.

Colombetti, G. (2007). Enactive appraisal. *Phenomenology and the Cognitive Sciences*, 6, 527–46.

Colombetti, G. (2009). From affect programs to dynamical discrete emotions. *Philosophical Psychology*, 22, 407–25.

Colombetti, G. (2010). Enaction, sense-making and emotion. In: J. Stewart, O. Gapenne, and E. Di Paolo (eds.), *Enaction: toward a new paradigm for cognitive science*. Cambridge, MA: MIT Press, pp. 145–64.

Colombetti, G. (2011). Varieties of pre-reflective self-awareness: foreground and background bodily feelings in emotion experience. *Inquiry*, 54(3), 293–313.

Colombetti, G. (2014). *The feeling body: affective science meets the enactive mind*. Cambridge, MA: MIT Press.

Colombetti, G. (2018). Enacting affectivity. In: this volume, pp. 571–80.

Colombetti, G. and Roberts, T. (2015). Extending the extended mind: the case for extended affectivity. *Philosophical Studies*, 172, 1243–63.

Colombetti, G. and Thompson, E. (2008). The feeling body: towards an enactive approach to emotion. In: W.F. Overton, U. Mueller, and J. Newman (eds.), *Body in mind, mind in body: developmental perspectives on embodiment and consciousness*. Mahwah, NJ: Erlbaum, pp. 45–68.

Gallagher, S. (2013). The socially extended mind. *Cognitive Systems Research*, 25(26), 4–12.

Greenwood, J. (2015). *Becoming human: the ontogenesis, metaphysics, and expression of human emotionality*. Cambridge, MA: MIT Press.

Griffiths, P. and Scarantino, A. (2009). Emotions in the wild. In: P. Robbins and M. Aydede (eds.), *The Cambridge handbook of situated cognition*. Cambridge: Cambridge University Press, pp. 437–53.

Hobson, P. (2018). Thinking and feeling: a social-developmental perspective. In: this volume, pp. 553–70.

James, W. (1884). What is an emotion? *Mind*, 9(34), 188–205.

James, W. (1890). *The principles of psychology* (2 vols.). New York: Henry Holt.

Kirchhoff, M.D. (2012). Extended cognition and fixed properties: steps to a third-wave version of extended cognition. *Phenomenology and the Cognitive Sciences*, 11, 287–308.
Krueger, J. (2014). Varieties of extended emotions. *Phenomenology and the Cognitive Sciences*, 13, 533–55.
Newen, A., de Bruin, L., and Gallagher, S. (2018). 4E cognition: historical roots, key concepts, and central issues. In: this volume, pp. 3–16.
Parkinson, B. (1995). *Ideas and realities of emotion*. London: Routledge.
Ratcliffe, M. (2008). *Feelings of being: phenomenology, psychiatry and the sense of reality*. Oxford: Oxford University Press.
Sacarantino, A. (2014). The motivational theory of emotions. In: J. D'Arms and D. Jacobson (eds.), *Moral psychology and human agency: philosophical essays on the science of ethics*. Oxford: Oxford University Press, pp. 156–85.
Scherer, K.R. (2001). Appraisal considered as a process of multi-level sequential checking. In: K.R. Scherer, A. Schorr, and T. Johnstone (eds.), *Appraisal processes in emotion: theory, methods, research*. Oxford: Oxford University Press, pp. 92–120.
Scherer, K. (2005). What are emotions? *Social Science Information*, 44, 695–729.
Solomon, R. (2004). Emotions, thoughts, and feelings. In: R. Solomon (ed.), *Thinking about feeling*. Oxford: Oxford University Press, pp. 76–88.
Stephan, A. (2015). Empathy for artificial agents. *International Journal of Social Robotics*, 7, 111–16.
Stephan, A., Walter, S., and Wilutzky, W. (2014). Emotions beyond brain and body. *Philosophical Psychology*, 27 (1), 65–81.
Sterelny, K. (2010). Minds: extended or scaffolded? *Phenomenology and the Cognitive Sciences*, 9, 465–81.
Thompson, E. (2007). *Mind in life: biology, phenomenology, and the sciences of mind*. Cambridge, MA: Belknap Press.
Thompson, E. and Stapleton, M. (2009). Making sense of sense-making. *Topoi*, 28, 23–30.
Varga, S. (2016). Interaction and extended cognition. *Synthese*, 193, 2469–96.
von Scheve, C. and Salmela, M. (eds.) (2014). *Collective emotions: perspectives from psychology, philosophy, and sociology*. Oxford: Oxford University Press.
Vygotsky, L.S. (1978). *Mind in society: the development of higher psychological processes* (ed. M. Cole, V. John-Steiner, S. Scribner, and E. Souberman). Cambridge, MA: Harvard University Press.
Wilutzky, W. (2015). Emotions as pragmatic and epistemic actions. *Frontiers in Psychology*, 6, 1593, 1–10.
Wilutzky, W., Stephan, A., and Walter, S. (2011). Situierte Affektivität [Situated affectivity]. In: J. Slaby, A. Stephan, H. Walter, and S. Walter (eds.), *Affektive Intentionalität* [*Affective Intentionality*]. Paderborn: Mentis, pp. 283–320.
Zahavi, D. and Michael, J. (2018). Beyond mirroring: 4E perspectives on empathy. In: this volume, pp. 589–606.
Zajonc, R.B. and Marcus, H. (1984). Affect and cognition: the hard interface. In: C. Izard, J. Kagan, and R.B. Zajonc (eds.), *Emotion, cognition, and behavior*. Cambridge: Cambridge University Press, pp. 73–102.

PART VIII

LANGUAGE AND LEARNING

CHAPTER 33

THE EMBODIMENT OF LANGUAGE

MARK JOHNSON

The new field that came to be known in the 1960s as "philosophy of language" had its origins in theories that either denied or minimized the role of embodiment in syntax, semantics, and pragmatics. Although up through the mid-twentieth century there was significant work on the bodily basis of meaning and language by phenomenologists like Maurice Merleau-Ponty (1962) and pragmatist philosophers like William James and John Dewey (1925/1981), analytic philosophers mostly ignored that work. Consequently, it is still the case today that the idea that linguistic phenomena are substantially shaped by the nature of our bodies has not been embraced in mainstream analytic philosophy of language.

However, over the past three decades, embodiment has become a central topic of attention, due mostly to the rapid growth of cognitive science research on mind, thought, and language. The role of the body and brain in shaping meaning, concepts, and reasoning is becoming more obvious as a result of research coming out of cognitive linguistics and cognitive neuroscience. We are now in a position to give fairly detailed explanations of how our bodily makeup and brain architecture profoundly shape thought and language.

As a prelude to a survey of some of the bodily processes of meaning-making that underlie all our forms of symbolic expression, both linguistic and nonlinguistic, it is useful to understand the general philosophical orientation that led so many to neglect the bodily sources of meaning.

Disembodied Views of Language

The perspectives known loosely as "analytic philosophy" and "linguistic philosophy" emerged in the first half of the last century in an atmosphere profoundly influenced by Gottlob Frege's (1892) views about meaning and reference and the logical empiricist theories that emerged within that framework. Frege distinguished sharply between (1) the sign (word or expression), (2) its reference (the object or state of affairs referred to), (3) its sense (the understanding, or the mode of presentation, of the reference), and (4) any subjective "associated ideas" that might be brought to an individual's mind by a sign. The *sense* was supposedly the public, shared meaning or understanding of the referred-to object, while the *associated idea* was an image or idea called up by a sign in the particular subjective mind of an individual. Frege summarizes:

> The reference and sense of a sign are to be distinguished from the associated idea. If the reference of a sign is an object perceivable by the senses, my idea of it is an internal image, arising from memories of sense impressions which I have had and acts, both internal and external, which I have performed. . . . The same sense is not always connected, even in the same man, with the same idea. The idea is subjective: one man's idea is not that of another. . . . This constitutes the essential distinction between the idea and the sign's sense, which may be the common property of many and therefore is not part or a mode of the individual mind. (Frege 1892, p. 59)

Notice that, in this famous passage, there is no mention of the body in relation to the sense of a sign. As presumably objective, senses supposedly cannot depend on the peculiarities of particular minds, let alone of particular bodies. They are universal and objective, in sharp contrast to associated "ideas," which depend on the body and experiences of those who have the ideas. Thus, Frege says "one need have no scruples in speaking simply of *the* sense, whereas in the case of an idea one must, strictly speaking, add to whom it belongs and at what time" (Frege 1892, p. 60). For example, the sense of the English word *mother* would allegedly be an abstract meaning or understanding "grasped" (to use Frege's term) by all who understand English. In addition, each of those individuals would have their own associated (and highly subjective) ideas that come to mind when he or she thinks about mothers, none of which is part of the objective sense of the term. Consequently, senses are supposedly not dependent on the particulars of the bodies and brains that grasp them, whereas associated ideas and images lay no claim to universality, precisely because they depend on our embodiment and experiences. Frege summarizes this distinction between objective senses and subjective ideas as follows: "The reference of a proper name is the object itself which we designate by its means; the idea, which we have in that case, is wholly subjective; in-between lies the sense, which is indeed no longer subjective like the idea, but is yet not the object itself" (Frege 1892, p. 60).

Frege fatefully went on to argue that the *proposition*, not the word, was the basic unit of meaning. Propositions have a subject-predicate structure. When the subject is specified and a concept is predicated of it, only then does the whole expression (i.e., the proposition) have a truth value (i.e., true or false). As a mathematician and logician, Frege was especially concerned with explaining how there could be shared, public meaning that provides a basis for objective knowledge. His answer was that to understand the thought (i.e., proposition) expressed in a sentence is to grasp its public, universal sense, which is "not the subjective performance of thinking but its objective content, which is capable of being the common property of several thinkers" (Frege 1892, p. 62n.).

In order to explain the objectivity of the senses of terms, Frege had proposed a somewhat odd ontology with three independent realms: the physical, the mental, and a third realm (to which he gave no name) that consists of abstract quasi-entities including senses, concepts, propositions, numbers, functions, and the strange objects "The True" and "The False." Because Frege believed that both physical (bodily) events and mental (psychological) processes were incapable of guaranteeing the objective and universal character of publicly shareable meaning and thought, he posited the third realm to house the objective contents (and objects) of thought. Consequently, on this view, a theory of language need not pay any special attention to our embodiment, other than to notice how perception might be shaped by our bodily capacities.

With Frege, the dice were fatefully cast. Few philosophers could fully embrace Frege's unusual ontological picture, but the vast majority of so-called "analytic" philosophers agreed with his basic assumption that meaning and thought are essentially linguistic (i.e., sentential, propositional, conceptual) in nature and are objective. Not surprisingly, one can find no serious account in Frege of the body's contributions to meaning and thought. This neglect of the body carried over into most of the major figures in the analytic tradition, such as Bertrand Russell, J.L. Austin, Rudolf Carnap, Carl Hempel, W.V.O. Quine, and Donald Davidson, who had nothing deep or extensive to say about the body's role in meaning and thought. Even Hilary Putnam, who is much celebrated for his brain-in-a-vat thought experiments (Putnam 1981), in which he emphasized that meaning requires a body interacting with a world, has never supplied any detailed account of how the body shapes our thought and communicative practices. This is not to deny that there may be some insightful comments on embodiment scattered throughout their writings (especially in Wittgenstein), but their perspective remains mostly disembodied in accounting for meaning, language, and thought. By a "disembodied" account of thought and language, I mean one that assumes that it is possible to explain the syntax, semantics, and pragmatics of natural languages without a detailed explanation of how grammatical forms and meaning are shaped by the nature of our bodies, brains, and the physical environments we inhabit. The overwhelming tendency in mainstream analytic philosophy of language is to begin with concepts more or less well formed, and then to analyze their relations to one another in propositions and to objects of reference in the world. This leads one to overlook the bodily origins of those concepts and patterns of thought that constitute our understanding of, and reasoning about, our world.

What is at issue here are the very origins of meaning and language. Anglophone philosophy of language developed mostly in this "disembodied" mode, not in the sense that we could engage in linguistic communication without a body, but rather in the sense that a theory of meaning, thought, and language is given without any serious study of the workings of the body and brain in how we make and communicate meaning. In the 70 or so years since the emergence of the field of philosophy of language, there has been remarkably little variance from these early ideas that (1) language is conceptual and propositional, and (2) concepts, propositions, and thoughts are not profoundly shaped by the nature of our bodily capacities and modes of engagement with our material environments, other than as a medium for supplying the content of perceptions. The first-generation cognitive science that developed within this framework was therefore a blending of analytic philosophy of language, generative linguistics, information-processing psychology, computer science, and budding artificial intelligence research—all of which were relatively disembodied perspectives.

Embodied Mind and Meaning

By the mid-1970s, the landscape was beginning to change, as a number of research programs in philosophy, psychology, linguistics, computer science, and neuroscience coalesced into a new, second-generation brand of cognitive science that recognized how important our embodiment is for everything we experience, think, mean, communicate, and do. The more researchers studied how people actually think—as opposed to how philosophers, logicians, and computer scientists said they ought to think—the clearer it became that our bodies impose the very conditions of our experience, thinking, and communicating with others. This new perspective has come to be known as *the cognitive science of the embodied mind*, or, more simply, *embodied cognitive science* (Lakoff and Johnson 1999; Gibbs 2006; Feldman 2006).

A theory of the embodiment of language has to begin with how it is possible for us to experience and make meaning through our bodily interactions with our environment, and it then has to show how abstract concepts are generated by recruiting body-based meaning. This account must include an explanation of how the syntactic patterns and relations so characteristic of natural languages are grounded in the patterns of our bodily engagement with our environment. What follows are the basic components of such a theory.

Meaning Arises from Organism–Environment Interactions

The origin of meaning and thought is the activity of a bounded, embodied organism as it engages its various environments in ways that allow it to maintain the basic conditions for life and growth. The more complex the organism is, the more ways it has by which it can meaningfully interact with the energy structures that make up its environments. Depending on the specific bodily makeup of the organism, particular situations will provide for the organism what James Gibson (1979) called affordances—patterns for meaningful perception and action relative to the nature of the organism, its needs, and its purposive activity in the world that it inhabits. For example, for human animals of our size, makeup, and interests, certain caves afford relations of containment (here, as space for habitation). Small caves do not afford access to large mammals, such as elephants, and so such enclosures do not have the same meaning to elephants as they do to humans.

Notice that, already at this basic level of animal–environment affordances and transactions, I have spoken of the "meaning" of certain environmental structures for a certain type of creature. I use "meaning" here for any experiences enacted or suggested by various affordances in our surroundings (Johnson 2007). Any aspect or quality of a situation means (for a specific type of creature) what it calls forth by way of experience. That includes past experiences, present experiences, and projected future experiences perceived to be possibilities developing out of one's current situation.

Meaning so described is obviously quite a bit more expansive than the conceptual/propositional accounts put forward in analytic philosophy of language. It is not just words and sentences that have meaning, but also any affordances within our environment. Conceptual/propositional meaning depends on, and emerges from, this much deeper and broader embodied process of meaning-making.

There are at least two very important consequences of this conception of meaning. (1) It acknowledges our evolutionary continuity with many other species, and therefore allows that certain nonhuman animals might be capable of various sorts of meaning-making. However, in species lacking capacities for abstraction, it will be a greatly attenuated range of meanings, relative to the richness of meaning available to humans, but it will be meaning nonetheless. (2) Conceiving of meaning in this embodied, experiential manner enables us to go beyond the narrower confines of language-based meaning to embrace the full range of human meaning-making in such practices as painting, sculpture, music, architecture, dance, spontaneous gesture, and ritual practices, in a way that no merely linguistically centered account of meaning can. No traditional understanding of signs as having meaning only through some conceptual/propositional content grounded in reference to states of affairs in the world could even begin to capture the richness of body-based meaning that is experienced in all of these varied forms of human meaning-making and communicative activity.

Body-Part Projections

One important way that the body undergirds languages the world over is the use of *body-part projections* for understanding objects, events, and scenes (Lakoff and Johnson 1999; Feldman 2006). We use our own body-part relations to make sense of objects and spatial relations in our surroundings. A good example of this is the way we experience our own bodies as having *fronts* and *backs*, and so it seems natural for us to project these *front/back* relations onto other objects, such as houses, TV screens, automobiles, signs, refrigerators, and lines of people, none of which have inherent fronts or backs. We experience computer screens as *facing* us, when we sit *in front of* them. We tend to project fronts onto moving objects (cars, buses, airplanes, ships), with the front defined relative to the canonical direction of motion for the object. Cars, buses, airplanes, and ships mostly move forward, and so their "front" is specified by that direction of motion. If they reverse direction, they are then said to "back up." We extend this *front/back* orientation even onto simple physical objects like bottles, balls, and rocks. For instance, if I rotate a plastic water bottle into a horizontal orientation, and then move the bottle in a line through space, you will project a front onto the bottle based on the direction of its motion.

The relation *in front of* is defined in most languages relative to the space between a viewer and some object in their field of vision. So, a dog that is located between me and a tree is experienced by me as in front of that tree, as if the tree faces me. Some languages, such as Hausa, reverse this, projecting the front of the tree as facing away from the viewer. Therefore, in Hausa, a dog that is located between the viewer and the tree would be "behind" or "in back of" that tree; and if the tree is between me and the dog, then the dog is described as "in front of" the tree. However, despite such orientational reversals between English and Hausa, in both cases the *in front of* relation is the result of a body-part or body-orientation projection. A number of extensive cross-linguistic studies suggest that, in languages around the world, we project body-part relations onto scenes and objects in our surroundings (Brugman 1983; Lakoff 1987 Gibbs 2006). It is also common to experience objects such as mountains, trees, towers, poles, and people as being oriented *up* and *down*, as having *tops* and *bottoms*, and often as having *heads* and *feet* (as in the *foot of* a mountain, tree, or tower). Moreover, as will be discussed later, we use body-part terms imagistically and metaphorically when we conceive of rivers as having *arms*, and when we attribute body parts like *eyes* and *hearts* to objects and events, such as the *eye* of a needle or a storm, or the *heart* of an artichoke or a problem.

Image-Schematic Affordances

Body-part projections are meaningful because they enact aspects of our fundamental ways of relating to, and acting within, our environment. The way our perceptual and motor systems get characteristically wired up (neuronally) as we grow and develop, through ongoing relations with energy patterns in our environment, establishes a large number of intrinsically meaningful patterns that George Lakoff (1987) and I (Johnson 1987) dubbed *image schemas*. The basic idea was that, given the nature of our bodies (how and what we perceive, how we move, what we value) and the general dimensions of our surroundings (stable structures in our environment), we will experience regular recurring patterns (such as up–down, right–left, front–back, containment, iteration, balance, loss of balance, source–path–goal, forced motion, locomotion, center-periphery, straight, curved, etc.) that afford us possibilities for meaningful interaction with our surroundings, both physical and social. For example, the fact that humans exist and operate within earth's gravitational field generates repeated experiences of *up/down* (i.e., *verticality*) relations. We understand objects as rising *up* and falling *down*, as *up*right or lying *down*, as *on top of* or *below/under*, relative to our own bodily orientation and experiences. The fact that we routinely, and crucially, experience *balance* or *lack of* *balance* gives rise to a *balance* schema that applies literally to balancing physical objects and metaphorically to our internal bodily states, to mathematical equations, and to notions of political fairness and justice (Johnson 1987). Our numerous daily experiences with containers and contained spaces generate a pervasive *container* schema in our conceptual system (Lakoff 1987). Our numerous daily encounters with moving objects and with moving our own bodies give rise to a *locomotion* schema (Dodge and Lakoff 2005). A list of common image schemas might run into the scores or even hundreds (Cienki 1997). Importantly, such schemas are typically multimodal, and so are not tied to any single sensory or motor area of the brain. This multimodality is evident when we experience containment both through vision and touch, or when we see something far off and also hear it as far away. Ellen Dodge and George Lakoff (2006) conclude that, although all languages do not have the same spatial relations concepts, nevertheless, they appear to build their particular spatial relations from "a limited inventory of basic primitive image schemas and frames of reference" (p. 71).

Perceptual Concepts

Lawrence Barsalou (1999, 2003) argues that our perceptual symbols for various concrete objects (e.g., cars, glasses, houses) are grounded in the sensory and motor experiences afforded us by those objects. The key idea is that the same sensory, motor, and affective neural processes involved in our interaction with such objects are activated when we understand, reason, and talk about those objects (see, e.g., van Elk and Bekkering 2018). There are not two different and independent systems, one for perception and another for conception; instead, to conceive some object is a matter of engaging in a simulation process of selective aspects of that object and our typical engagements with it. In "Percept and Concept" William James (1911) first presented this idea of a continuity between perception and conception, but Barsalou has the advantage of a century of subsequent psychological and neuroscience research to flesh out some of the details of the process. For example, understanding a concept like *chair* involves a sensory, motor, and affective simulation of possible experiences with chairs of all sorts. Such simulations will involve multiple modalities (such as vision, touch, audition, and proprioception), insofar as our interactions with chairs are multimodal. We see chairs from various points of view as we walk around them, we know what it feels like to sit on and touch various types of chairs made from different materials, and we know the types of motor programs required for sitting in and standing up from chairs. To know the meaning of *chair*, to understand what a chair is in a certain context, is to simulate experiences with chairs using all of the sensory, motor, and affective modalities available to us. Barsalou summarizes the six basic dimensions of his theory of body-based perceptual symbols as follows:

> Perceptual symbols are neural representations in sensory-motor areas of the brain; they represent schematic components of perceptual experience, not entire holistic experiences; they are multimodal, arising across the sensory modalities,

> proprioception, and introspection. Related perceptual symbols become integrated into a simulator that produces limitless simulations of a perceptual component (e.g., *red, lift, hungry*). Frames organize the perceptual symbols within a simulator, and words associated with simulators provide linguistic control over the construction of the simulation. (Barsalou 1999, p. 582)

Barsalou's use of the term *representation* might seem to support what is known as a representational theory of mind, in which thought proceeds via operations on internal mental representations that are somehow supposedly relatable to external, mind-independent realities. However, Barsalou's view could be made compatible with a nonrepresentational theory of mind, where having or entertaining a concept is merely running a neural simulation in which sensory, motor, and affective areas of the brain are activated, not as representations mediating between an inner and outer world, but rather as *the very understanding of the concept.* In other words, the neural activations involved in the simulations within a specific context *just are* what it is to grasp the meaning of the concept in question. For any given concept, such as *chair*, there will be a general, culturally shared meaning of the term, insofar as different people's simulations include embodied experiences arising from shared structures of perception, bodily interaction, and feeling response, but there are also likely to remain certain idiosyncratic aspects of the simulation process.

In several articles (e.g., Gallese 2003, 2007, 2008; Gallese and Lakoff 2005; Glenberg and Gallese 2012; Gallese and Cuccio 2015) neuroscientist Vittorio Gallese and colleagues have provided evidence of embodied simulation in our processing of action verbs. Gallese summarizes the results of those studies as follows:

> Language, when it refers to the body in action, brings into play the neural resources normally used to move that very same body. Seeing someone performing an action, like grabbing an object, and listening to or reading the linguistic description of that action lead to a similar motor simulation that activates some of the same regions of our cortical motor system, including those with mirror properties, normally activated when we actually perform that action. (Gallese and Cuccio 2015, p. 13)

Simulation Semantics: Language Understanding as Embodied Simulation

The notion of understanding as online neural simulation has now been expanded to a general theory of language understanding. Ben Bergen (2012) has surveyed a large number of neuroimaging studies of how such simulations work as we read or hear sentences. He proposes an *embodied simulation hypothesis*, namely, that "we understand language by simulating in our minds what it would be like to experience the things that the language describes" (Bergen 2012, p. 13). For example, one group of studies reveals that:

> the motor system is often used when people are understanding language about action, and that this is more likely when the language uses progressive than perfect aspect. We know that interfering with the perceptual system—by having people

> look at lines or spirals moving in one direction or another—affects how long it takes them to determine that a(n) (action) sentence makes sense, and so on. (Bergen 2012, p. 249)

In other words, research suggests that sentences with the progressive aspect, in which the action is currently ongoing, are more likely to activate motor and premotor cortical areas used in that specific kind of action than in a sentence in which the action is already completed (i.e., perfect aspect). Moreover, if we interfere with the normal direction or motor processes for the action specified in the sentence, it takes longer to understand the sentence.

It is too early to make any sweeping claims about the scope and adequacy of the *embodied simulation hypothesis*, but there is growing evidence that many parts of language understanding work in this fashion. It is an elegant and parsimonious hypothesis that meshes well with many neuroimaging studies that are revealing how our understanding of sentences is an ongoing temporal process in which sensory, motor, and affective areas of the brain are progressively activated as we read or hear parts of sentences. Moreover, it is a highly testable hypothesis, especially as we are developing more fine-grained and time-sensitive techniques for neuroimaging.

The Embodiment of Abstract Concepts

Disembodied theories of language often cite abstract concepts as evidence that thought and language cannot be accounted for solely in terms of bodily processes. Chairs and cars may be good candidates for an embodied simulation treatment, but what about abstract concepts like mind, freedom, love, knowledge, and property? The most well-researched and detailed accounts of such concepts to date, from an embodied perspective, come from the orientation known as Conceptual Metaphor Theory, which gives evidence that body-based metaphor is our principal means of abstract conceptualization and reasoning. This idea was first put forward in a detailed fashion by George Lakoff and Mark Johnson in *Metaphors We Live By* (1980). They observed that our abstract concepts are typically defined by multiple metaphors, by which we map entities, structures, and relations from a physical or social domain to structure our understanding of a more abstract domain of experience. For example, English, and most known languages, has a basic metaphor by which acts of thought are understood metaphorically as acts of perception. A much studied example would be the UNDERSTANDING IS SEEING metaphor, in which thinking is conceptualized as seeing something, which gives rise to expressions such as "I *see* what you mean," "Could you *shed* more *light* on that idea?" "What she said was incredibly *illuminating*," and "His concept of time is too *obscure*." Over the past three decades, hundreds of cross-linguistic studies have revealed scores of conceptual metaphors shared across disparate cultures (Gibbs 2008).

A second crucial claim of Conceptual Metaphor Theory is that the vast majority of metaphors is not based on similarities, but rather emerges from common experiential correlations occurring between the source and target domains. So, for example, it is not that vision and thought are "similar" or share literal similarities as the basis for metaphorical mapping, but instead we have certain vision metaphors in cultures the world over because people routinely experience the correlation between seeing something and thereby gaining an understanding of it. Joe Grady (1997) called these basic correlational phenomena "primary metaphors," and he showed how we could learn scores of them simply by having the kinds of bodies we have and interacting with the kinds of environments that we routinely inhabit. In this way, primary metaphors would arise without any conscious awareness, simply from the neural co-activations of the source and target domains. In this fashion, we would acquire the neural basis of primary metaphors like MORE IS UP, AFFECTION IS WARMTH, INTIMACY IS CLOSENESS, SIMILARITY IS CLOSENESS, IMPORTANT IS BIG, PURPOSES ARE DESTINATIONS, ACTIVITIES ARE MOTIONS, ORGANIZATION IS PHYSICAL STRUCTURE, and CAUSES ARE FORCES. There is a large literature surveying various types of cross-linguistic evidence for primary metaphor (Lakoff and Johnson 1999; Kövecses 2010; Dancygier and Sweetser 2014).

Combinations of two or more primary metaphors can give rise to more complex metaphor systems that constitute our abstract conceptual systems. The LIFE IS A JOURNEY metaphor, for example, is built from a number of primary metaphors that constitute the submappings of the metaphor. Conceptualizing life metaphorically as a journey involves primary metaphors like ACTIONS ARE MOTIONS, DIFFICULTIES ARE IMPEDIMENTS TO MOTION, CAUSES ARE PHYSICAL FORCES, RESPONSIBILITIES ARE BURDENS, PURPOSES ARE DESTINATIONS, etc. (Lakoff and Johnson 1999). The submappings of complex metaphors give rise to specific inferential patterns by which we are able to reason about the target domain given our knowledge of the source domain. For instance, in the source domain of physical motion, we experience the way a physical obstacle or impediment can temporarily or permanently stop our forward motion. Via the DIFFICULTIES ARE IMPEDIMENTS TO MOTION submapping, we draw the target-domain inference that difficulties can temporarily or permanently impact our ability to realize some purpose we are pursuing.

Embodied Construction Grammar

Embodiment is not just the source of semantic content that would then somehow be ordered by a pure, disembodied system of formal relations, manifested either as syntax or logical patterns of thought. Instead, even syntax is shaped and given meaning by the contours of our bodily experience.

Noam Chomsky famously argued that syntactic deep structures are innate universals, in no way dependent on our particular embodiment. On this generative linguistics view, syntax is a matter of purely formal relations—that is, of our innate ability to recognize and produce linguistic expressions according to specific structural patterns. As such, syntactic form was believed to operate separately from our capacities to process meaning (construed as semantic content of sentences), and it was supposedly not dependent on the uses (pragmatics) to which linguistic utterances are put.

As far back as the late 1960s, however, there was growing evidence for the intertwining of syntax, semantics, and pragmatics and for the importance of the body for defining the syntax of natural languages. On this view, linguistic forms are meaningful (hence, tied to semantics and pragmatics), and they are meaningful because they encode the structures of events, actions, agents, purposes, objects, etc., that we experience as embodied creatures in a world. Instead of being born with all of the syntactic knowledge we need, we learn the syntax of specific languages by exposure to conceptual and linguistic constructions that capture important aspects of our daily experiences and actions. We are exposed to grammatical constructions, which are "pairings of form with semantic or discourse function" (Goldberg 2003, p. 219). Grammatical patterns are thus the product of cognitive mechanisms for sensory and motor processing that shape our conceptualization and reasoning.

Over the past 40 years, a number of these cognitive structures have been studied as they operate in our conceptual systems and determine all the ways we have of making sense of a situation. These structures include body-part relations (Brugman 1983), image schemas (Johnson 1987; Lakoff 1987; Hampe and Grady 2005), semantic frames (Fillmore 1982), action schemas (Narayanan 1997); prototype effects (Rosch 1975), radial category structure (Lakoff 1987), force dynamics (Talmy 2000), conceptual metaphors (Lakoff and Johnson 1980, 1999), and blends (Fauconnier and Turner 2002).

A good example of this type of research is the grammar of actions. Showing how words are defined relative to semantic frames, Charles Fillmore (1982) argued that prototypical actions involve a conceptual frame with slots of the following sorts: action, actor, object acted with or upon, goal of action, manner of action, etc. Srini Narayanan (1997), in developing neurocomputational models of schemas for certain kinds of bodily actions, discovered a general control structure for actions, which he dubbed an *executing schema*, with the following temporal dimensions: preparatory state (getting into a state of readiness), starting the process for the event, main process (either instantaneous or prolonged), an option to stop, an option to resume, an option to continue or to reiterate the main process, a check to see if the goal has been met, the finishing process, and the final state. Narayanan was able to create neurocomputational models that could both recognize and carry out (via appropriate robotics) action events.

According to *embodied construction grammar*, people learn, in addition to action frames and executing schemas, a range of schemas for other dimensions of experience, simply by perceiving and having bodily experiences in the world. These embodied conceptual structures underlie grammatical constructions. We would thus expect any natural language to have some way of coding each of these general dimensions and parameters of actions, events, and objects, although different languages will often vary in the details of how they grammatically code these recurring dimensions of human experience. The relevant frames and schemas will depend on the nature of our bodies, our brain architecture, and the recurring dimensions of the environments we inhabit. Embodied construction grammar extrapolates from cases of this sort to propose that we learn *all* of our basic grammatical patterns in this experiential manner (Goldberg 2003), based on our shared sensory and motor capacities, other shared general cognitive mechanisms, and our exposure to natural languages.

The Neural Theory of Language

The cognitive processes and structures discussed earlier—image schemas, semantic frames, primary metaphors, complex metaphors, and grammatical constructions—are just part of the large array of cognitive mechanisms that give rise to thought and language. The approach that has come to be known as the neural theory of language (Feldman 2006) attempts to model the neural mechanisms that give rise to the structures and processes that make language possible. George Lakoff and Srini Narayanan (in press) have emphasized that an adequate theory of concepts, meaning, and language would need to involve multiple levels of explanation:

> Neuroscience is only one part of the answer, for a simple reason. Neuroscience does not study the detailed nature of thought and language. For that you also need the field of cognitive linguistics. . . . We also need to understand how neural circuitry functions to produce thought. For that we need neural computation to model that functioning. And we need to know how thought impacts behavior. For that you need experimental research on embodied cognition. In short, what is needed is a way to integrate all four sciences: Neuroscience, Cognitive Linguistics, Neural Computation, Experimental Embodied Cognition. (Lakoff and Narayanan in press)

The neural theory of language is thus a vastly ambitious attempt to understand how our brains and bodies give rise to our conceptual systems, meaning, thought, language, and other forms of symbolic interaction. It studies, from multiple perspectives, how conceptual development is tied to the body and its engagements with its environments to generate the marvelous accomplishments of natural languages. It models the various neural and brain architectures that make linguistic activity possible.

The neural theory of language is based on the hypothesis that, over the course of evolutionary history, humans developed the capacity to recruit areas of the brain originally evolved for perception and motor activity for the purpose of "higher"-level acts of conceptualization, reasoning, and linguistic communication. Exaptation—the technical term for this type of process—is used to explain how embodied creatures could possibly acquire abstract thought and expression by making use of the body-based syntax and semantics of perception and bodily action (Tucker and Luu 2012; Lakoff and Narayanan in press).

One key challenge is to supply bridges between the various levels of processing, starting with molecules and biological neural systems, which then need to be modeled neurocomputationally, guided by the cognitive linguistics of natural languages, which tells us precisely which constructions need to be modeled and explained. Jerome Feldman captures the challenge of this vast undertaking as follows:

> This integrated, multifaceted nature of language is hard to express in traditional theories, which focus on the separate levels and sometimes view each level as autonomous. But constructions can provide a natural description of the links between form and meaning that characterize the neural circuitry underlying real human language. They offer a high-level computational description of a neural theory of language (NTL). . . . In particular, it allows the embodied and neural character of thought and language to take center stage. (Feldman 2006, p. 9)

Emotion and Meaning

Neurocomputational theories of thought and language are typically not good at capturing the emotional and feeling dimensions of human meaning, since they do not include the hormonal processes so crucial to emotions. Traditional philosophy of language has fared no better. In fact, it has tended to either downplay or entirely dismiss the emotional dimensions of meaning. Ogden and Richards (1923) set the stage for this when they distinguished "descriptive" from "emotive" meaning, and then concluded that only the former has significant cognitive content relevant to understanding and knowing our world. The unfortunate result of this illegitimate bifurcation has been, until quite recently, the ignoring of emotions in mainstream accounts of language. William James (1890) long ago deplored this radical separation of thought and feeling, arguing that all thought has a feeling dimension that includes both a felt sense of the horizon or fringe of meaning surrounding a particular term, and also a feeling of the direction of our thinking. Instead of "thought" or "feeling," James preferred the hyphenated "thought-feeling" to capture the true embodied, affective character of our mental processes.

In a series of books, Antonio Damasio (1999, 2003, 2010) has developed a theory of emotions that places them at the center of human thought, meaning, and value. He argues that emotions are automated neurochemical response patterns to the body-mind's ongoing assessment of how it is being affected by its environment. In order to survive and flourish, organisms must establish a semi-permeable boundary within which they sustain the conditions of life by maintaining an allostasis (Schulkin 2011), or dynamic equilibrium (Luu and Tucker 2003). This, in turn, requires that the body continually monitor changes in its body state in response to its engagement with its environment. Emotional response patterns arise, therefore, when an "emotionally competent stimulus" (Damasio 2003, p. 53) causes the body to recover equilibrium by adjusting its internal milieu and instigating bodily changes that often result in overt action. Emotional response patterns typically run their course automatically, without need of conscious reflection. However, on those occasions when we become aware of changes in our body as it interacts with its environment and gives rise to an emotional response pattern, we then *feel* an emotion.

The connection between emotion and meaning is that emotions are our most elementary way of taking the measure of our current or anticipated situation and responding to it. As such, they can be said to indicate "how things are going for us" and "what's happening" (Damasio 1999). Johnson (2007) summarizes this emotion/meaning connection:

> Our emotional responses are based on both our non-conscious and conscious assessments of the possible harm, nurturance, or enhancement that a given situation may bring to our lives.... Emotional responses... are bodily processes (with neural and chemical components) that result from our appraisal of the meaning and significance of our situation and consequent changes in our body state, often initiating actions geared to our fluid functioning within our environment. (pp. 60–1)

Rather than opposing emotion to reason, emotions and feelings are an integral part of how we conceptualize and reason. They provide ongoing contact with our situation at the most primordial level where we feel ourselves in our environment. These affective dimensions are not lost when language comes on the stage; rather, they are taken up into the very processes of meaning-making and come to permeate our words, phrases, and sentences (Gendlin 1997).

Language and the Seven "E's"

To sum up, language is intimately shaped by all aspects of our bodily being in the world—from perception, to movement, to feeling. Empirical studies of language processing do not support a disembodied mind. On the contrary, they reveal that meaning, concepts, and understanding emerge from our physical interactions with things and events, and from our interpersonal interactions with other human and nonhuman animals. Cognitive neuroscience is beginning to provide us an elementary understanding of the neural architectures that give rise to thought and language. Through a multilevel dialogue between neurocomputational modeling, cognitive linguistic accounts of language understanding, and psychological experiments on thought and language, we are beginning to understand how the body lies at the heart of our ability to make, understand, and communicate meaning and thought.

In recent years, this general orientation toward the grounding of mind in organism-environment interactions has come to be known as "4E cognition," that is, cognition as *embodied, embedded, enactive,* and *extended.* Cognition is *embodied* in some of the ways surveyed earlier; it is *embedded* insofar as it arises from interactions with its environments (both physical and social); it is *enactive* in the way it creates meaning and thought in an ongoing fashion; and it is *extended* in the sense that we offload certain cognitive operations and contents onto (or into) aspects of our environment, such as books, computers, buildings, and signs.

However, the neural theory of language suggests that we should add at least three more E's to this list: *emotional, evolutionary*, and *exaptative*. First, I have stressed the importance of emotions in our ability to grasp the meaning and significance of our situation and of what is happening at any given moment as we engage language, through reading, writing, and speaking. The simulation theory of meaning recognizes the deep affective dimensions of our concepts. Second, since all meaning and thought arise through ongoing processes of organism–environment interactions, there is an evolutionary dimension to our conceptual systems and languages. The relative stability (not fixity) of language is the result of relative evolutionary stability even in the face of change, rather than being the result of innate, universal modules or architectures. Third, I have noted that the neural theory of language adopts a naturalistic orientation that attempts to explain how conceptual systems and languages can emerge in an evolutionary process in which "higher" functions (such as abstract conceptualization and reasoning) operate through the recruitment of the syntax, semantics, and logics of sensory and motor areas of the brain. As Lakoff and Narayanan (in press) put it: "With the development of larger forebrains, human beings have 'repurposed' the circuitry types already present in animals. The technical term is *exaptation*, the use of evolutionarily inherited traits for new purposes. We hypothesize that this evolutionary repurposing underlies much, if not all, of human thought and language" (in press, ch. 1, §1). As discussed earlier, our concrete concepts (i.e., concepts of physical objects, properties, events, and actions) involve activation of some of the same sensory and motor areas that are involved in our actual perception and manipulation of those objects, as well as our performance of those actions. These same sensorimotor activations are exapted for our understanding of abstract concepts, and for the reasoning we do with them, typically in the form of conceptual metaphor in which structures and inference patterns from concrete source domains are used to understand abstract concepts.

Language therefore appears to be very much a product of "E" cognition, whether we count four E's or seven. However long the list, embodiment comes first, because it is our bodily habitation of our world that gives rise to our capacity to create and use language. In just three decades, we have begun to move from relatively disembodied views of thought and language to perspectives that appreciate the ways our meaning, concepts, and language are deeply rooted, at the most visceral level, in our bodily engagements with our world.

References

Barsalou, L. (1999). Perceptual symbol systems. *Behavioral and Brain Science*, 22, 577–660.

Barsalou, L. (2003). Situated simulation in the human conceptual system. *Language and Cognitive Processes*, 18(5/6), 513–62.

Bergen, B. (2012). *Louder than words: the new science of how the mind makes meaning*. New York: Basic Books.

Brugman, C. (1983). *The use of body-part terms as locatives in Chalcatongo Mixtec*. Report No. 4 of the Survey of California and Other Indian Languages. Berkeley: University of California, pp. 235–90.

Cienki, A. (1997). Some properties and groupings of image schemas. In: M. Verspoor, K. Lee, and E. Sweetser (eds.), *Lexical and syntactical constructions and the construction of meaning*. Amsterdam: John Benjamins, pp. 3–15.

Damasio, A. (1999). *The feeling of what happens: body and emotion in the making of consciousness*. New York: Harcourt Brace.

Damasio, A. (2003). *Looking for Spinoza: joy, sorrow, and the feeling brain*. Orlando, FL: Harcourt.

Damasio, A. (2010). *Self comes to mind: constructing the conscious brain*. New York: Pantheon.

Dancygier, E. and Sweetser, E. (2014). *Figurative language*. Cambridge: Cambridge University Press.

Dewey, J. (1925/1981). *Experience and nature* (vol. 1 of *The Later Works, 1925–1953*; ed. J.A. Boydston). Carbondale: Southern Illinois University Press.

Dodge, E. and Lakoff, G. (2005). Image schemas: from linguistic analysis to neural ground. In: B. Hampe and J. Grady (eds.), *From perception to meaning: image schemas in cognitive linguistics*. Berlin: Mouton de Gruyter, pp. 57–91.

Fauconnier, G. and Turner, M. (2002). *The way we think: conceptual blending and the mind's complexities*. New York: Basic Books.

Feldman, J. (2006). *From molecule to metaphor: a neural theory of language*. Cambridge, MA: MIT Press.

Fillmore, C. (1982). Frame semantics. In: The Linguistic Society of Korea (ed.), *Linguistics in the morning calm*. Seoul: Hanshin Publishing Co., pp. 111–37.

Frege, G. (1892). On sense and reference. In: *Translations from the philosophical writings of Gottlob Frege*. Oxford: Basil Blackwell, pp. 56–78.

Gallese, V. (2003). A neuroscientific grasp of concepts: from control to representation. *Philosophical transactions of the Royal Society of London. Series B, Biological sciences*, 358(1435), 1231–40.

Gallese, V. (2007). Before and below theory of mind: embodied simulation and the neural correlates of social cognition. *Philosophical transactions of the Royal Society of London. Series B, Biological sciences*, 358(1480), 659–69.

Gallese, V. (2008). Mirror neurons and the social nature of language: the neural exploitation hypothesis. *Social Neuroscience*, 3, 317–33.

Gallese, V. and Cuccio, V. (2015). The paradigmatic body: embodied simulation, intersubjectivity, the bodily self, and language. In: T. Metzinger and J.M. Windt (eds.), *Open mind*. Frankfurt: MIND Group, pp. 1–23.

Gallese, V. and Lakoff, G. (2005). The brain's concepts: the role of the sensory-motor system in conceptual knowledge. *Cognitive Neuropsychology*, 22, 455–79.

Gendlin, E. (1997). How philosophy cannot appeal to experience, and how it can. In: D. Levin (ed.), *Language beyond postmodernism: saying and thinking in Gendlin's philosophy*. Evanston, IL: Northwestern University Press.

Gibson, J.J. (1979). *The ecological approach to visual perception*. Boston: Houghton Mifflin.

Glenberg, A.M. and Gallese, V. (2012). Action-based language: a theory of language acquisition, comprehension, and production. *Cortex*, 48(7), 905–22.

Gibbs, R. (2006). *Embodiment and cognitive science*. Cambridge: Cambridge University Press.

Gibbs, R. (2008). *The Cambridge handbook of metaphor and thought*. Cambridge: Cambridge University Press.

Goldberg, A. (2003). Constructions: a new theoretical approach to language. *Trends in Cognitive Sciences*, 7(5), 219–24.

Grady, J. (1997). *Foundations of meaning: primary metaphors and primary scenes* [PhD dissertation]. Berkeley: University of California.

Hampe, B. and Grady, J. (eds.) (2005). *From perception to meaning: image schemas in cognitive linguistics*. Berlin: Mouton de Gruyter.

James, W. (1890/1950). *The principles of psychology* (2 vols.). New York: Dover.

James, W. (1911). Percept and concept. In: *Some problems of philosophy*. Cambridge, MA: Harvard University Press, pp. 21–60.

Johnson, M. (1987). *The body in the mind: the bodily basis of meaning, imagination, and reason*. Chicago: University of Chicago Press.

Johnson, M. (2007). *The meaning of the body: aesthetics of human understanding*. Chicago: University of Chicago Press.

Kövecses, Z. (2010). *Metaphor: a practical introduction*. Oxford: Oxford University Press.

Lakoff, G. (1987). *Women, fire, and dangerous things: what categories reveal about the mind*. Chicago: University of Chicago Press.

Lakoff, G. (2008). The neural theory of metaphor. In: R. Gibbs (ed.), *The Cambridge handbook of metaphor and thought*. Cambridge: Cambridge University Press, pp. 17–38.

Lakoff, G. and Johnson, M. (1980). *Metaphors we live by*. Chicago: University of Chicago Press.

Lakoff, G. and Johnson, M. (1999). *Philosophy in the flesh: the embodied mind and its challenge to Western thought*. New York: Basic Books.

Lakoff, G. and Narayanan, S. (in press). Conceptual science: the embodiment of thought and language.

Luu, P. and Tucker, D. (2003). Self-regulation by the medial frontal cortex: limbic representation of motive set-points. In: M. Beauregard (ed.), *Consciousness, emotional self-regulation and the brain*. Amsterdam: Benjamins, pp. 123–61.

Tucker, D. and Luu, P. (2012). *Cognition and neural development*. Oxford: Oxford University Press.

Merleau-Ponty, M. (1962). *Phenomenology of perception* (trans. C. Smith). London: Routledge.

Narayanan, S. (1997). Embodiment in language understanding: sensory-motor representations for metaphoric reasoning about event descriptions [PhD dissertation]. Berkeley: University of California.

Putnam, H. (1981). Brains in a vat. In: *Reason, truth, and history*. Cambridge: Cambridge University Press, pp. 1–21.

Rosch, E. (1975). Cognitive representations of semantic categories. *Journal of Experimental Psychology, General*, 104, 192–233.

Schulkin, J. (2011). *Adaptation and well-being: social allostasis*. Cambridge: Cambridge University Press.

Talmy, L. (2000). *Toward a cognitive semantics* (2 vols.). Cambridge, MA: MIT Press.

Tucker, D. and Luu, P. (2012). *Cognition and neural development*. Oxford: Oxford University Press.

van Elk, M. and Bekkering, H. (2018). The embodiment of concepts: theoretical perspectives and the role of predictive processing. In: this volume, pp. 641–60.

CHAPTER 34

THE EMBODIMENT OF CONCEPTS

Theoretical Perspectives and the Role of Predictive Processing

MICHIEL VAN ELK AND HAROLD BEKKERING

INTRODUCTION

IMAGINE a Martian coming to planet Earth. What would his experience be like? In the Dutch countryside he might see slowly moving objects covered by patches of black and white and making "moo"-like sounds. After observing these objects for a while, he might classify these objects as belonging to the category of "living creatures." But now imagine that he sees a Scottish Highlander that looks very different, but makes similar sounds and moves in a similar way. Would he recognize that this exemplar belongs to the same category of "cows"? Or imagine that the alien returns to his home planet. How would he communicate about his experiences with cows to another Martian who did not visit Earth? He might describe a list of features that are characteristic of cows (i.e., having legs, having a specific color). Alternatively, he might try to draw a painting of a cow or try to imitate the sounds and movements made by cows.

This example raises two questions that are central to any theory of concepts, namely (1) what does it take to learn a concept? and (2) how are concepts represented? Regarding *the acquisition of concepts*, one could argue that an alien has no other option than to rely on his sensory experiences to acquire concepts related to creatures living on planet Earth. Alternatively, it could be that the alien uses relevant background knowledge about similar exemplar categories from his own planet to make inferences about what he sees. Regarding the *representation of concepts*, one could argue that a concept consists of an abstract list of properties or that concepts are primarily sensory-based, entailing the same format as used for actually perceiving the real-world referents of concepts (see also Johnson 2018).

The role of experience in concept formation dates back to philosophical discussions about whether knowledge is innate or derived from experience (Prinz 2002) and has received renewed attention in recent discussions on evolutionary psychology and the notion of cognitive modules (Barrett and Kurzban 2006). Discussions on the representational format of concepts have a long philosophical history as well and this topic has recently re-emerged in cognitive theories of language understanding and neuroscientific debates on the modality-specific nature of concepts (Barsalou 2010). In the present chapter we primarily focus on the topic of concept acquisition and representation from an embodied cognition perspective. We will argue that a theory of concepts qualifies as embodied depending on its stance on a number of issues, namely: (1) the nature of concepts, (2) the relation between language and concepts, (3) the function of concepts, (4) the acquisition of concepts, (5) the representation of concepts, and (6) the role of context (see Figure 34.1). We propose that theoretical positions regarding these different

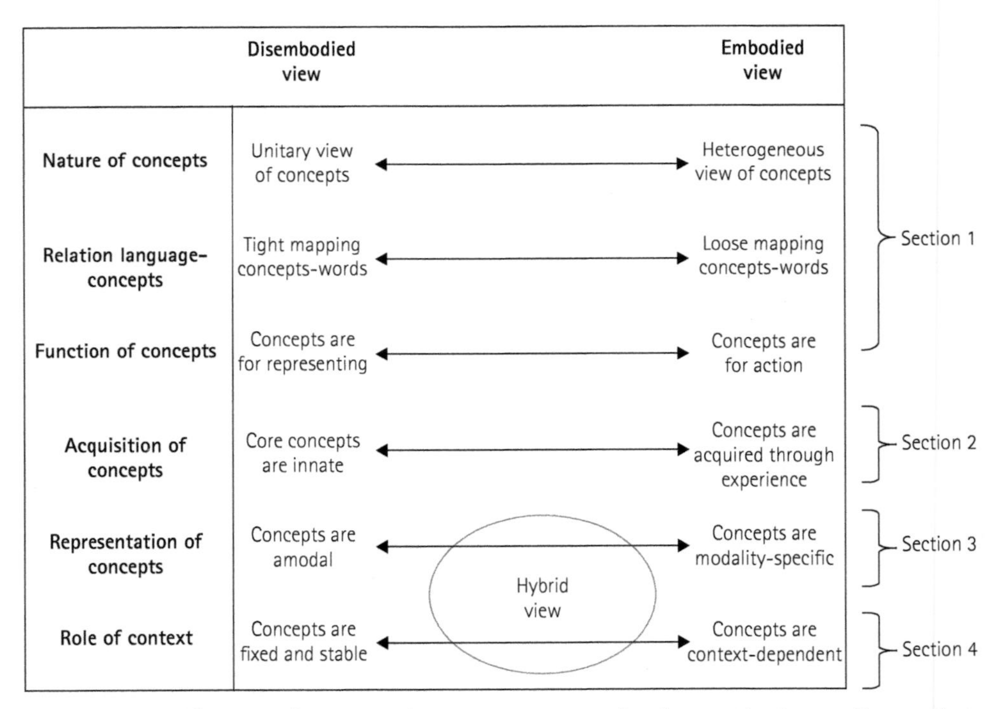

FIGURE 34.1. Theories of conceptual representation can be characterized according to their view on the nature of concepts, the relation between language and concepts, the function of concepts, the acquisition of concepts, the representation of concepts, and the role of context. Theoretical positions tend to covary in their view on the different axes (e.g., the view that concepts are innate often goes hand in hand with the view that there is a tight mapping between words and concepts), but in principle the axes could be conceived of as a multidimensional space along which different theories can be characterized.

dimensions often tend to covary, such that one's position on the nature of concepts is strongly linked to one's view of how concepts are acquired and represented. However, in principle the different axes could also be conceived of as being independent, resulting in a multidimensional complex space along which different theories can be characterized.

In the first section, we will argue that theories of concepts start from (implicit) assumptions about the nature of concepts (i.e., are concepts homogeneous or heterogeneous?), the relation between concepts and language (i.e., is there a one-to-one mapping between words and concepts?), and the function of concepts (i.e., are concepts for representing the world or for action?). In the second section, we will focus on current debates in the field of cognitive science and psychology, arguing on the one hand for the role of innate cognitive biases in concept acquisition and for the role of embodied interactions and experiences on the other hand. In the third section, we will discuss theoretical views about the representational format of concepts (i.e., as being modality-specific or amodal), and in the fourth section we will focus on the role of context on conceptual representations.

In the final section, we propose to extend an embodied view of concepts through the integration of a predictive processing framework (Clark 2013). Starting from the free energy principle and the notion that the brain aims to minimize prediction error signals (Friston 2010), we argue that concepts are used as prior models to generate multimodal expectations, thereby enhancing predictability and precision in the perception of exemplars. At the same time, concepts are dynamically acquired and updated, based on recurrent processing of prediction error signals in a hierarchically structured network. We argue that this extended-embodied view of concepts provides a more complete account of concept acquisition and representation that takes into account the role of learning and the relevance of concepts for making predictions.

Concepts: Definition, Relation with Language, and Function

Before starting our discussion on the acquisition and representation of concepts, an important question is how a concept can actually be defined. Typically, in cognitive science concepts are considered the basic informational units of higher cognitive competences, such as categorization, reasoning, and learning (Machery 2005). Fodor even argued that a theory of concepts should be at the heart of cognitive science, as concepts are truly central to human cognition (Fodor 1998). It should be noted that the definition of concepts as used by psychologists and cognitive scientists differs from a philosophical definition of concepts. Philosophers refer to concepts as that which enables people to have propositional attitudes, such as beliefs and desires about objects, and consequently they focus on the boundary conditions that enable people to have propositional attitudes (Machery 2009). Instead, psychologists are primarily interested in how concepts enable cognitive processes like categorization or induction. In this section we will shortly highlight three central issues regarding (1) the question of whether concepts are a single construct, (2) the relation between concepts and language, and (3) the functional role of concepts (see Figure 34.1).

First, according to a dominant view in cognitive science, concepts are characterized by a common set of properties that enable higher cognitive processes (Medin, Lynch, and Solomon 2000). For instance, the concept of a "cow" is the body of information that is used to categorize something as a cow, to draw an analogy between cows and other species, and to understand sentences containing the word "cow." However, an obvious challenge for this view is the question how one common set of properties can be used to support such diverse processes as categorization, inductive and deductive reasoning, and language understanding. Therefore others have argued for the *heterogeneity hypothesis* of concepts, according to which different processes (e.g., categorization, induction) involve different types of concepts, such as prototypes, exemplars, and theories (Machery 2009). A *prototype* can be defined as the statistical or paradigmatic information about a specific set of properties that is characteristic of an object (e.g., cows have legs, produce milk, eat grass, etc.). An *exemplar* is a concrete representation of a single object (e.g., this is a Dutch Belted) and a set of exemplars may constitute a specific concept (e.g., Dutch Belted, Dutch Friesian, and Groningen-Whiteheaded are all instances of "cows"). A *theory* can be defined as descriptive information (e.g., nomological, causal, or functional) about an object (e.g., knowing that a cow needs four stomachs to digest grass). Thus, on the heterogeneity hypothesis, "concepts" are not a unitary construct and psychologists should instead focus on the role that prototypes, exemplars, and theories play in higher cognitive processes. An alternative hybrid account has been proposed, according to which a class of exemplars is represented by a single concept, consisting of different parts (Vicente and Manrique 2014). The hybrid view assumes that the different aspects are in fact part of a common representation, which constitutes a concept and which is stored in long-term memory.[1] Thus, theories of concepts may start from the assumption that concepts are a unitary construct, a heterogeneous construct, or a unitary construct with different properties.

Second, any theory of concepts either implicitly or explicitly entails a view about the relation between concepts and language. On Fodor's account, concepts are like words that function as arbitrary symbols in a language of thought (Fodor 2004). The main reasons for postulating this strong relation between words and concepts are the requirements that concepts should combine compositionally (which explains our ability to generate new thoughts) and that concepts should be stable over places and times (i.e., to enable communication and the exchange of meaning). Some proponents of the view that words directly relate to concepts, which in turn are constituent of our thought, have argued that prelinguistic concepts may provide the basis for word learning (Bloom 2000;

[1] We note that despite these philosophical considerations, many researchers in cognitive science and psychology continue to treat the notion of "concept" as a unitary construct.

Carey 2001). Vice versa: others have argued that language and linguistic input determine the acquisition of concepts, which can be considered as a version of the Whorfian hypothesis that language shapes thought (Boroditsky 2001; Choi and Bowerman 1991). An alternative view proposes a looser link between language and concepts, based on findings indicating that—in at least some domains—people from different cultures tend to diverge in the naming and grouping of different objects (Malt, Gennari, and Imai 2010). Thus, theories of concepts may entail a tight or a loose coupling between words and concepts.

Finally, all theories about concepts start from an implicit assumption about what concepts are actually for. As our example with the alien illustrates, a concept may help in categorizing particular examples as belonging to a specific category. Moreover, it is often pointed out that concepts can be used in making inferences about the world through a process of inductive or deductive reasoning, by drawing analogies, and by enabling communication (Machery, 2009). One's theoretical stance on how concepts are acquired and represented is directly related to the function of concepts (Prinz 2005). Rationalists can be characterized by the assumption that concepts are innate and amodal: on this account, concepts are primarily used for representing things or state of affairs in the world (Fodor 2004). In contrast, (neo-)empiricists argue that concepts are acquired through experience and represented in a modality-specific format: this view entails that concepts are primarily used for acting and interacting with the surrounding world. According to Prinz (2005),

The difference between Fodor and the empiricists is captured by a difference in the nature of the mental representations that they postulate. Fodor thinks concepts are like words. They are arbitrary symbols in a language of thought. There is little one can do with an arbitrary symbol. A symbol does not include any instructions for how to interact with the category that it represents. For the empiricist, concepts are more like mental images, or inner models. Representations of that kind can be used to guide action. We can read features of a category off of our concepts if empiricism is right. (p. 680)

The notion that concepts primarily serve the guidance of action is central to many theories of embodied cognition (Glenberg, Witt, and Metcalfe 2013; see also later).[2]

Thus, theories about concepts can be characterized by their view of the nature of concepts (i.e., as being a homogeneous or heterogeneous category), the relation between concepts and language (i.e., implicating a direct or a loose mapping between concepts and words), and the role of concepts in cognition and action. We will return to these distinctions toward the end of the chapter. We believe these philosophical distinctions are important because they directly bear relevance to the two topics that are

[2] Besides the rationalist and the empiricist notion of concepts, there is also a theoretical position inspired by Gibson's ecological psychology that tries to get rid of the notion of concepts in explaining cognition altogether (Chemero 2009). Starting from a dynamical systems perspective, on this account it is argued that cognition arises from the complex interplay between an organism and the environment. However, in the remainder of this chapter we will primarily focus on the rationalist and the empiricist accounts of concepts.

central to this chapter, namely: (1) how concepts are acquired and (2) how concepts are represented.

The Acquisition of Concepts: Nativism vs. Empiricism

A dominant view in developmental psychology is the idea that infants are born with innate cognitive mechanisms that enable the acquisition of knowledge, concepts, and language (Carey 2009, 2011; Keil 1989; Kinzler and Spelke 2007; Spelke and Kinzler 2007). At least five core systems have been identified that enable the representation of inanimate objects and their interactions: agents and goal-directed actions, sets and numerical relationships, places and spatial layout, and reasoning about social partners and group members (Spelke and Kinzler 2007). Based on developmental studies in children, cross-cultural research, and studies with nonhuman primates, it is argued that these core systems are deeply embedded in human phylogeny and ontogeny and that these mechanisms determine cognitive development. For instance, with respect to the core system of object representation, it has been found that young infants already show a basic understanding of the spatiotemporal principles of cohesion (objects move as connected wholes), continuity (objects move on unobstructed paths), and contact (objects do not interact at a distance; Spelke 1990). These principles have been identified already in newborn infants and newly hatched chickens (Regolin and Vallortigara 1995; Valenza et al. 2006) and in adult nonhuman primates (Santos 2004). It is argued that core systems for objects, agents, numbers, and places underlie the acquisition of concepts, such as the concept of "number" or "map" and the learning of words (Bloom 2000). More specifically, it is proposed that children's core systems enable the acquisition of world knowledge through a process of bootstrapping, whereby core systems provide initial expectations and representations that are updated in response to incoming data (Carey 2011).

Although still widely prevalent, the core systems view has been criticized by several authors (Haith 1998; Heyes 2003; Prinz 2005).[3] It has been suggested, for instance, that alternative theoretical accounts, such as the theory of associative sequence learning (Heyes 2003) or ideomotor theory (Paulus 2014) could explain the available data equally well or even in a more parsimonious fashion. Thus, in contrast to the nativist view, other researchers (also referred to as "neo-empiricists") have stressed the role of experience in concept acquisition (Prinz 2002, 2005). This view dates back to Piaget, who stressed the importance of basic sensorimotor processes for cognitive development (Piaget 1962).

[3] It has been argued, for instance, that the empirical data on which the core systems view is based has been over-interpreted and that several low-level confounds in the design of the experiments and the stimuli could actually explain the effects observed (Prinz 2005).

Piaget proposed that two central processes govern cognitive development, namely, assimilation (i.e., the integration of information into an existing mental representation or schema) and accommodation (i.e., the adaptation of an existing schema to novel information). According to Piaget, children are born with a basic set of innate reflexes that in the first year of life are released by different external stimuli, which in turn results in the assimilation of objects in their schemas. With increased age, through experience and interaction with the surrounding physical world, children acquire the ability for symbolic thinking and more abstract thought.

A prime example showing the importance of physical experience can be found in developmental studies on visual cliff avoidance. Rather than having an innate bias for perceiving depth, several studies have shown that children need to rely on locomotor experience (e.g., crawling, walking) to develop a bias for avoiding visual cliffs (Adolph 2000). Other studies have shown the importance of experience for acquiring a goal-directed representation of other agents (Paulus et al. 2011) and for imitation (Ray and Heyes 2011). It has been found, for instance, that frequency information related to previous experiences dominates action perception (Paulus et al. 2011), rather than innate mechanisms that would predispose infants to infer the most efficient means for attaining a specific goal (Gergely and Csibra 2003). An ongoing controversy concerns the relative importance of innate processes compared to the role of experience on the development of language. Based on the "poverty-of-stimulus" argument, for instance, it has been argued that language acquisition can only be accounted for by specific innate language modules in the brain that enable recursive processing (Hauser, Chomsky, and Fitch 2002). In contrast to this dominant view, others have argued that children learn language by picking up the statistical regularities of speech through basic perceptual biases (Kuhl 2000). The linguistic input in turn also shapes perceptual processes, thereby further enabling the acquisition of language. The poverty-of-stimulus argument has also been criticized, for instance, by the observation that adults typically adjust their speech in order to facilitate language processing in the infant (i.e., "motherese"); an alternative account has been proposed according to which grammar is acquired through a process of Bayesian learning (Perfors, Tenenbaum, and Regier 2006).

With respect to the development of concepts, the empiricist view argues that concepts are learned based on experience with category instances (i.e., exemplars). One basic idea is that children acquire concepts by starting to note perceptual similarities between specific exemplars. In support of this account, it has been found, for instance, that concepts in young children refer to shapes (e.g., round, rectangular, etc.) rather than natural kinds (e.g., animals vs. artifacts; cf. Landau, Smith, and Jones 1988). In addition, it has been found that children learn through experience the distinction between living things and artifacts: younger children displayed a quite imperfect understanding of this distinction, while older children had much a more accurate representation of the different categories (Simons and Keil 1995).[4] Thus, these studies show that with increased age

[4] These studies have in turn been criticized by proponents of the nativist view (e.g., Soja, Carey, and Spelke 1991), arguing for an alternative interpretation of the data that is more in line with a nativist account.

children acquire more accurate and detailed conceptual representations, and experience likely plays a central role in this process. In the next section we will argue that the empiricist view of concepts is also compatible with findings regarding the representation of concepts and the role of context.

Concept Representation: Amodal vs. Modality-Specific Views

According to an embodied view of cognition, concepts are represented in modality-specific brain systems, such as those involved in action, vision, audition, and touch. Proponents of the embodied view of concepts often start by discussing the so-called classical cognitivist account of representation (Barsalou 2010; Fischer and Zwaan 2008; Glenberg et al. 2013; Zwaan 2014). According to the classical view, concepts are conceived of as nodes in an amodal semantic network that derive meaning from the connections with other nodes. The classical view has been criticized because of the so-called grounding problem: if nodes are recurrently connected, it remains unclear how basic concepts derive their meaning in the first place (Barsalou 1999). The notion that concepts are grounded in modality-specific brain areas is proposed as a solution to this problem: concepts are based on our sensorimotor experiences in the concrete world and the retrieval of a concept entails the re-enactment of these sensorimotor experiences.

Evidence for the notion that concepts are embodied can be found in behavioral experiments, research with neuropsychological patients, and in neuroimaging studies. These findings have been extensively reviewed elsewhere (Barsalou 2010; Fischer and Zwaan 2008; Glenberg et al. 2013; Zwaan 2014), so here we will only briefly point to some of the evidence that is typically cited in support of an embodied account of concepts. Behavioral evidence comes, for instance, from studies in which participants were required to conduct a property verification task (Pecher, Zeelenberg, and Barsalou 2003; Van Dantzig et al. 2008): a modality-switch cost was observed, reflected in slower reaction times when participants switched to identifying properties from a different modality (e.g., visual–auditory) compared to the same modality (visual–visual) as the previous trial. These findings suggest that the retrieval of conceptual information (e.g., to decide whether leaves are green) recruits modality-specific processes. Studies with neuropsychological patients have furthermore shown that damage to visual or motor areas can result in selective impairments of conceptual categories for which visual or motor information is a defining feature (for a review, see Caramazza and Mahon 2003). It has been found, for instance, that patients with frontal lesions show an impaired performance on a lexical decision task in their response to action verbs, whereas patients with occipital lesions were impaired in their response to nouns that had a strong visual association (Pulvermuller and Fadiga 2010). Findings like these are also compatible with the view that concept retrieval necessarily involves activation of modality-specific brain areas. Finally, neuroimaging studies have directly shown that the processing of words referring to actions or nouns is accompanied by the activation of sensory brain areas (for a review, see Pulvermuller and Fadiga 2010). For instance, the processing of action verbs referring to different effectors (e.g., kick, lick, pick) has been associated with a somatotopic activation of corresponding motor areas (Hauk, Johnsrude, and Pulvermuller 2004). Interestingly, in more recent studies it has been found that different brain areas are involved in representing action concepts at different levels of abstraction, with more frontal regions being involved in concrete representations, while posterior areas code for actions at a more abstract level (Wurm et al. 2015), suggesting a concrete-to-abstract gradient in conceptual representation at the brain level.

The classical view that concepts are represented in an amodal format rather than according to modality-specific features is typically attributed to cognitivists, like Fodor (Fodor 1975). More recently, it has been argued that the data from studies with neuropsychological patients with semantic dementia is indeed more compatible with the view that concepts are not represented in the same brain areas as involved in perceptual processes (McCaffrey 2015). Furthermore, some authors have pointed out that the available evidence that is typically interpreted in favor of an embodied account of concepts can be equally well explained by a disembodied account of cognition (Caramazza et al. 2014; Mahon and Caramazza 2008). However, these authors do not necessarily argue that concepts are represented in an amodal format—rather that the observed modality-specific effects can be accounted for by spreading activation or post-lexical imagery as well. Others have argued that the notion of modality-specific brain areas being involved in the retrieval of conceptual knowledge is overly simplistic, and have proposed dynamic accounts of neural networks underlying meaning construction and unification in language processing instead (Hagoort 2014; Hagoort and Indefrey 2014). Thus, rather than proposing an amodal or disembodied account of conceptual representation, it might be more accurate to state that some researchers tend to remain agnostic about the representational format of concepts. Still, proponents of an embodied view of concepts have argued that a major strength of the embodied cognition hypothesis is that it allows making specific predictions that have been confirmed by empirical research (Prinz 2002, 2005). The findings discussed earlier could be predicted a priori based on an embodied view of concepts, whereas a disembodied view would not have led to these specific predictions.

It is often pointed out that a potential problem for embodied theories of concepts concerns the representation of abstract concepts. That is, if concepts are constituted by modality-specific representations, it remains unclear how more abstract notions like "democracy" or "liberty" could be represented. Several responses to this concern have been proposed, for instance, by proposing that ultimately abstract concepts are also derived from concrete perceptual experiences (Barsalou and Wiemer-Hastings 2005). Furthermore, as we will discuss later, hybrid theories have been proposed that could account for the representation of both concrete and abstract concepts.

Hybrid Views, Context, and the Semantic Hub Hypothesis

In recent years several authors have proposed a hybrid view of conceptual representation, according to which conceptual processing involves the use of both embodied and amodal representations (Dove 2009; Louwerse 2011; Zwaan 2014). For instance, it has been argued that amodal abstract symbols may serve as a heuristic to support fast and online processing of language, whereas embodied representations could be primarily involved in offline processing, providing a more in-depth level of understanding (Louwerse 2011). Zwaan (2014) proposed that the relative importance of embodied processing depends on the level of embeddedness in the environment: embodied representations are primarily recruited in concrete interactions with the environment, but with increased levels of abstraction one needs to rely more strongly on abstract representations from long-term memory. Both accounts argue that the important next step for theories of embodied cognition is to determine the boundary conditions for "embodied" effects to occur (see also Willems and Francken 2012). Thus, rather than using specific findings to argue for or against the notion that concepts are represented in a modality-specific or amodal fashion, future studies should determine if and when (i.e., in which context) cognitive processing involves the recruitment of embodied conceptual representations.

First evidence for the role of context on conceptual processing can be found in the early work of Barsalou (1982), showing that the properties that are retrieved in association with a concept depend on the context. For instance, the concept of a cow in the countryside vs. a cow at the butcher evokes a different representation (Barsalou 1987). Examples such as these provide an important argument for the notion that there are no core concepts, but that conceptual representations are flexible instead (Prinz 2002, 2005). In addition, neuroimaging studies also underline the importance of context for conceptual processing (Schuil, Smits, and Zwaan 2013; van Dam et al. 2012). It has been found, for instance, that task context (i.e., verifying the action associated with an object compared to verifying the color of an object) determines the activation of modality-specific brain areas (i.e., activation of motor-related regions; cf. van Dam et al. 2012). Similarly it was found that when action verbs were presented in a literal context, the motor system was activated to a stronger extent compared to when the sentences were presented in a symbolic context (Schuil et al. 2013). These findings also argue against the notion that conceptual processing is automatic and bottom up, but rather suggest that the retrieval of concepts is a top-down process, involving the activation of modality-specific features that are relevant to the current task or context.

A related view to the hybrid account of conceptual representation can be found in the semantic hub hypothesis (Patterson, Nestor, and Rogers 2007; Pobric, Jefferies, and Ralph 2010), according to which modality-specific representations converge in higher-order convergence zones (such as the anterior temporal lobes; ATL). Semantic hubs may support a supra-modal or multimodal representation of concepts that is recruited irrespective of the sensory modality through which the concept is accessed (auditory, visual, tactile, etc.). Conceptual retrieval through activation of the semantic hub, in turn, may co-activate modality-specific brain areas that support more detailed retrieval about specific properties of concepts. We note that the specific claim regarding the involvement of the semantic hubs, such as the ATL, in multimodal conceptual processing has been criticized based on a close inspection of the available neuroimaging and neuropsychological data (Gainotti 2011; Martin et al. 2014). Still, an important argument for the involvement of higher-order multimodal representations in conceptual processing is that for many cognitive tasks one needs to abstract away from concrete features of exemplars, and multimodal representations could serve as a proxy to provide direct access to a specific concept without entailing the need to first go through a process of mental simulation.

Theoretical Integration

We have discussed different theoretical views regarding the nature, acquisition, and representation of concepts. In Figure 34.1 we presented an integrated perspective on how theories of conceptual representation can be characterized. As discussed in the preceding sections, theories of concepts can be classified according to their view on the nature of concepts, the relation between language and concepts, the function of concepts, the acquisition of concepts, the representation of concepts, and the role of context in conceptual processing. Disembodied views of concepts can be broadly placed on the left side of the spectrum, whereas embodied views can be placed on the right side. Hybrid accounts of conceptual representation propose that conceptual processing entails both modality-specific representations as well as multimodal or supra-modal representations (Dove 2009; Louwerse 2011; Zwaan 2014). Hybrid accounts account well for the finding that conceptual processing is flexible and context-dependent and therefore these are placed on the middle of the spectrum in Figure 34.1.

In our own previous empirical and theoretical work we have advocated an embodied account of cognition, and in a more recent review article we proposed a theoretical model according to which action concepts are represented in both modality-specific and supra-modal brain areas (van Elk, van Schie, and Bekkering 2014a, 2014b). Here we would like to extend this view by integrating an embodied notion of concepts with the theory of predictive processing (Clark 2013). On the predictive processing account, concepts are dynamically acquired and updated, based on the recurrent processing of prediction error signals in a hierarchically structured network. In turn, concepts are used as prior models to generate multimodal expectations, thereby enabling greater precision in the perception of exemplars. We believe that the extension of an embodied theory of concepts with a predictive processing account provides a more complete and unifying account of concept acquisition and representation that takes into account the role of learning and the relevance of concepts for making predictions.

Predictive Coding and Concepts

On the one hand nativists have argued that core knowledge systems enable the acquisition of concepts, while on the other hand empiricists have stressed that concepts are primarily acquired through experience. However, even empiricists need to start from (implicit) basic assumptions regarding the innate capabilities of the brain for learning. That is, some empiricists have argued that concept acquisition starts by detecting the overlap between perceptual experiences, which requires at the minimum a basic process of associative learning (Heyes 2012). As discussed earlier, Piaget proposed that the two general processes of accommodation and assimilation underlie cognitive development and learning (Piaget 1962). More recently, similar notions have been integrated in a unifying theory of brain function and cognition, called the predictive coding framework (Clark 2013).

The theoretical background of the predictive coding framework can be found in the free energy principle, according to which self-organizing systems that are in equilibrium with their environment must minimize their amount of free energy (Friston 2010). From an information-theoretical perspective, the equivalent of minimizing states of high entropy amounts to the avoidance of surprise (Applebaum 2008). At the brain level, surprise minimization is accomplished by minimizing the prediction error between anticipated and actual sensory experiences. Thus, according to the theory of predictive coding, we rely on generative models of the surrounding world that generate top-down predictions, which are in turn fed back to our perceptual systems. In case of a mismatch between the sensory input and top-down predictions, a prediction error signal is generated that cascades up in the processing hierarchy, thereby resulting in an updating of the internal model (i.e., the equivalent of Piaget's notion of accommodation).

On the predictive coding account, concepts can be described as high-level internal models that are derived from sensory experiences through a process of recurrent error processing. Throughout development, every time an infant encounters a specific exemplar, sensory information is fed forward to higher levels to construct and update an internal model (i.e., that can be considered a "prototype"). Different conceptual models (i.e., of different concepts) are constructed and selected through a process of Bayesian learning and error minimization: whenever a specific exemplar does not match prior models, a new model is constructed, which can in turn be updated through additional sensory experiences (for a more elaborate description of model updating/revising, see later). The conceptual model in turn generates top-down predictions to lower-level sensory regions that are used to minimize prediction error signals. Thus, according to this account the primary function of concepts is to reduce surprise and to enable the prediction of sensory experiences. For instance, seeing a cow results in the bottom-up activation of the prior model of a "cow," which in turn facilitates top-down predictions to lower sensory areas (e.g., anticipating that a cow will make specific sounds, will move in a specific way, etc.). At the same time, it is easy to see that prior conceptual models allow enhanced precision in the detection and categorization of specific exemplars. For instance, the prior model of a "cow" generates top-down predictions to sensory systems and small deviations in the experience of a "cow" result in prediction error signals to higher-level systems, allowing one to distinguish one cow from another (e.g., based on the specific pattern of black and white patches). We note that although the predictive processing account places a strong emphasis on learning for updating one's model of the world, in principle the framework does not preclude the importance of innate mechanisms for acquiring concepts. On the Bayesian account, priors shape the learning process and are updated based on new incoming information, but initial priors could be acquired through innate and/or evolved mechanisms as well (Barrett 2014). Thus the predictive processing framework takes an intermediate position with respect to the role of innate mechanisms and experience for concept acquisition.

The predictive processing account differs from the previously discussed hub or multimodal view of conceptual representation. Both views hold that an important prerequisite for the acquisition of concepts is the ability to form multimodal associations through a process of associative learning (Connolly 2014). Examples of this approach can be found in the perceptual symbols theory as proposed by Barsalou (1999) and in the conceptual empiricism proposed by Prinz (2002, 2005). However, the predictive processing account provides a more elaborate account of the *acquisition* of concepts based on perceptual experiences: rather than simply relying on a process of associative learning, concepts are acquired through Bayesian learning and error minimization.

For instance, if concepts were solely based on perceptual experience and passive associative learning, it remains unclear how children learn to distinguish between concepts (e.g., living vs. non-living) that are based on perceptually similar experiences (e.g., a living bird vs. a stuffed bird).

The predictive processing model of concepts proposes that prior conceptual models are used to make predictions regarding the multisensory experiences associated with a specific concept. Thus, upon seeing both a living and a stuffed bird, multisensory expectations are initially generated (i.e., that the bird will fly, whistle, etc.). In case of a living bird, the predictions are confirmed and the exemplar is categorized as a bird. In contrast, in case of a stuffed bird, the multimodal predictions are violated and a subsequent prediction error signal is generated, resulting in the generation of a novel conceptual model of "stuffed birds." Thereby the predictive coding theory provides an account of the apparent ease whereby infants acquire conceptual knowledge and update their conceptual models throughout development.

The notion of a hierarchically structured network in which different brain areas are involved in different levels of abstraction fits well with the *heterogeneity view* of concepts: lower-level sensory regions may be involved in representing specific exemplars, higher-level areas in representing prototypes, and at the highest level in the hierarchy theoretical knowledge may be represented (e.g., knowing that a cow needs four stomachs to digest grass). The predictive coding account also offers a different perspective on the *function* of concepts: whereas embodied theories of concepts have argued that concepts are primarily used for action (Prinz 2005), the predictive coding account argues that concepts have a more general function, namely, enhancing predictability and minimizing free energy (Friston 2010). With respect to the relation between language and concepts, most embodied theories have argued for a loose mapping between words and concepts: the modality-specific associations that are activated in association with word processing are highly context-dependent (Barsalou 1982; Prinz 2002). The predictive coding account makes similar predictions regarding the role of context on conceptual processing but also specifies *how* contexts affect conceptual processing: specific contexts will active prior models that in turn generate top-down predictions regarding the specific exemplars that are to be expected in that context.

Recently, we have begun to test these ideas in social situations (Hartstra, Vijayakumar, and Bekkering, in preparation). For example, we manipulate the knowledge you have about group preferences to make predictions about individuals that belong to the group; we investigate how the brain generates predictions on the basis of acquired knowledge, and how it incorporates prediction errors to improve the model about group and individual preferences. In addition, prior expectations about the expertise of an observed actor determine our perception of the action that he/she performs: if we see an expert bowler throw a ball we make different predictions regarding the outcomes of his throw than when watching a novice player, and through a process of learning and experience we come to associate specific players with specific outcomes. By using these paradigms—and by focusing specifically on the gamma-frequency band as reflective of bottom-up sensory signals and the beta-frequency band as related to feedback predictions (Bastos et al. 2012)—direct insight can be obtained in how predictions and prediction error signals unfold over time and support the making of behavioral inferences.

Questions that are currently debated in relation to the predictive coding account are, for example, how the precision of the prediction error is weighted so that it can be used to bring prediction and observation closer to each other. Think about the example of how to acquire the category of birds. They share visual features (e.g., they have a beak, feathers), they share auditory features (e.g., birdsongs), and we might have knowledge about some of their behavior (e.g., they fly in the sky, lay eggs, etc). Now suppose we meet a penguin, and your friend tells you this is a bird. Basically, the original hypothesis can be revised with an updated hypothesis (typically implemented by some form of gradient descent; cf. Friston 2002), for example: flying may no longer be part of your category of a bird. Alternatively, you can revise the parameters of the model that generated the predictions that birds fly (Friston 2003) such that the beak feature gets more important. Alternatively, you could try to obtain additional information, i.e., by sampling the world (Friston et al. 2012); in this case by asking your friend whether penguins lay eggs. Finally, you could manipulate the state of the world, i.e., by active inference (Brown, Friston, and Bestmann 2011) and you might approach a penguin to see if they have feathers, a beak, and eggs. Which strategy is employed depends on various aspects, for example, the amount of reducible versus irreducible uncertainty in the environment

(Yu and Dayan 2005). On a longer time scale, prediction errors (or the relative absence thereof) shape the generative models, to improve future predictions (Montague et al. 2012), and concepts may be conceived of as such high-level generative models.

Conclusions

In this chapter we have discussed how theories of concepts can be characterized according to their view on the nature of concepts, the relation between concepts and language, the function of concepts, the acquisition of concepts, the representation of concepts, and the role of context on conceptual representations. We have shown how the theory of predictive coding can provide a complementary perspective on embodied theories of concepts by underlining the central role of concepts for prediction and perception. An important next step for integrating these two perspectives would be to use formal modeling and experimentation to elucidate how Bayesian inferential processes can indeed be used to form and use new conceptual categories.

Acknowledgments

The writing of this chapter was supported by a VENI grant no. 016.135.135 to the first author from the Netherlands Organization for Scientific Research (NWO).

References

Adolph, K.E. (2000). Specificity of learning: why infants fall over a veritable cliff. *Psychological Science*, 11(4), 290–95.

Applebaum, D. (2008). *Probability and information: an integrated approach*. Cambridge, UK: Cambridge University Press.

Barrett, H.C. (2014). *The shape of thought: how mental adaptations evolve*. Oxford: Oxford University Press.

Barrett, H.C. and Kurzban, R. (2006). Modularity in cognition: framing the debate. *Psychological Review*, 113(3), 628–47.

Barsalou, L.W. (1982). Context-independent and context-dependent information in concepts. *Memory and Cognition*, 10(1), 82–93.

Barsalou, L.W. (1987). The instability of graded structure: implications for the nature of concepts. In: U. Neisser (ed.), *Concepts and conceptual development: ecological and intellectual factors in categorization*. Cambridge, MA: Cambridge University Press, pp. 101–40.

Barsalou, L.W. (1999). Perceptual symbol systems. *Behavioral and Brain Sciences*, 22(4), 577–660.

Barsalou, L.W. (2010). Grounded cognition: past, present, and future. *Topics in Cognitive Science*, 2(4), 716–24.

Barsalou, L.W. and Wiemer-Hastings, K. (2005). Situating abstract concepts. In: D. Pecher and R.A. Zwaan (eds.), *Grounding cognition: the role of perception and action in memory, language, and thought*. Cambridge: Cambridge University Press, pp. 129–63.

Bastos, A.M., Usrey, W.M., Adams, R.A., Mangun, G R., Fries, P., and Friston, K.J. (2012). Canonical microcircuits for predictive coding. *Neuron*, 76(4), 695–711.

Bloom, P. (2000). *How children learn the meaning of words*. Cambridge, MA: MIT Press.

Boroditsky, L. (2001). Does language shape thought? Mandarin and English speakers' conceptions of time. *Cognitive Psychology*, 43(1), 1–22.

Brown, H., Friston, K.J., and Bestmann, S. (2011). Active inference, attention, and motor preparation. *Frontiers in Psychology*, 2, 218.

Caramazza, A., Anzellotti, S., Strnad, L., and Lingnau, A. (2014). Embodied cognition and mirror neurons: a critical assessment. *Annual Review of Neuroscience*, 37, 1–15.

Caramazza, A. and Mahon, B.Z. (2003). The organization of conceptual knowledge: the evidence from category-specific semantic deficits. *Trends in Cognitive Sciences*, 7(8), 354–61.

Carey, S. (2001). Whorf versus continuity theorists: Bringing data to bear on the debate. In: M. Bowerman and S.C. Levinson (eds.), *Language acquisition and conceptual development*. Cambridge, UK: Cambridge University Press, pp. 185–214.

Carey, S. (2009). *The origin of concepts*. Oxford: Oxford University Press.

Carey, S. (2011). Précis of *The Origin of Concepts*. *Behavioral and Brain Sciences*, 34(3), 113–7.

Chemero, A. (2009). *Radical embodied cognitive science*. Harvard, MA: MIT Press.

Choi, S. and Bowerman, M. (1991). Learning to express motion events in English and Korean—the influence of language-specific lexicalization patterns. *Cognition*, 41(1-3), 83–121.

Clark, A. (2013). Whatever next? Predictive brains, situated agents, and the future of cognitive science. *Behavioral and Brain Sciences*, 36(3), 181–204.

Connolly, K. (2014). Multisensory perception as an associative learning process. *Frontiers in Psychology*, 5, 1095.

Dove, G.O. (2009). Beyond perceptual symbols: a call for representational pluralism. *Cognition*, 110, 412–31.

Fischer, M.H. and Zwaan, R.A. (2008). Embodied language: a review of the role of the motor system in language comprehension. *Quarterly Journal of Experimental Psychology*, 61(6), 825–50.

Fodor, J.A. (1975). *The language of thought*. Harvard, MA: Harvard University Press.

Fodor, J.A. (1998). *Concepts: where cognitive science went wrong*. Oxford: Oxford University Press.

Fodor, J.A. (2004). Having concepts: a brief refutation of the 20th century. *Mind and Language*, 19, 29–47.

Friston, K.J. (2002). Beyond phrenology: what can neuroimaging tell us about distributed circuitry? *Annual Review of Neuroscience*, 25, 221–50.

Friston, K.J. (2003). Learning and inference in the brain. *Neural Networks*, 16(9), 1325–52.

Friston, K.J. (2010). The free-energy principle: a unified brain theory? *Nature Reviews Neuroscience*, 11(2), 127–38.

Friston, K.J., Adams, R.A., Perrinet, L., and Breakspear, M. (2012). Perceptions as hypotheses: saccades as experiments. *Frontiers in Psychology*, 28(3), 151.

Gainotti, G. (2011). The organization and dissolution of semantic-conceptual knowledge: is the "amodal hub" the only plausible model? *Brain and Cognition*, 75(3), 299–309.

Gergely, G. and Csibra, G. (2003). Teleological reasoning in infancy: the naive theory of rational action. *Trends in Cognitive Sciences*, 7(7), 287–92.

Glenberg, A.M., Witt, J.K., and Metcalfe, J. (2013). From the revolution to embodiment: 25 years of cognitive psychology. *Perspectives on Psychological Science*, 8(5), 573–85.

Hagoort, P. (2014). Nodes and networks in the neural architecture for language: Broca's region and beyond. *Current Opinion in Neurobiology*, 28, 136–41.

Hagoort, P. and Indefrey, P. (2014). The neurobiology of language beyond single words. *Annual Review of Neuroscience*, 37, 347–62.

Haith, M. (1998). Who put the cog in infant cognition? Is rich interpretation too costly? *Infant Behavior & Development*, 21(2), 167–79.

Hauk, O., Johnsrude, I., and Pulvermuller, F. (2004). Somatotopic representation of action words in human motor and premotor cortex. *Neuron*, 41(2), 301–7.

Hauser, M.D., Chomsky, N., and Fitch, W.T. (2002). The faculty of language: what is it, who has it, and how did it evolve? *Science*, 298(5598), 1569–79.

Heyes, C. (2003). Four routes of cognitive evolution. *Psychological Review*, 110(4), 713–27.

Heyes, C. (2012). Simple minds: a qualified defence of associative learning. *Philosophical transactions of the Royal Society of London. Series B, Biological sciences*, 367(1603), 2695–703.

Johnson, M. (2018). The embodiment of language. In: this volume, pp. 623–40.

Keil, F.C. (1989). *Concepts, kinds, and cognitive development*. Cambridge, MA: MIT Press.

Kinzler, K.D. and Spelke, E.S. (2007). Core systems in human cognition. *Progress in Brain Research*, 164, 257–64.

Kuhl, P.K. (2000). A new view of language acquisition. *Proceedings of the National Academy of Sciences*, 97(22), 11850–7.

Landau, B., Smith, L.B., and Jones, S.S. (1988). The importance of shape in early lexical learning. *Cognitive Development*, 3(3), 299–321.

Louwerse, M.M. (2011). Symbol interdependency in symbolic and embodied cognition. *Topics in Cognitive Science*, 3, 273–302.

Machery, E. (2005). Concepts Are Not a Natural Kind. *Philosophy of Science*, 72(3), 444–67.

Machery, E. (2009). *Doing without concepts*. Oxford: Oxford University Press.

Mahon, B.Z. and Caramazza, A. (2008). A critical look at the embodied cognition hypothesis and a new proposal for grounding conceptual content. *Journal of Physiology (Paris)*, 102, 59–70.

Malt, B.C., Gennari, S., and Imai, M. (2010). Lexicalization patterns and the world-to-words mapping. In: B. Malt and P. Wolff (eds.), *Words and the mind: how words capture human experience*. Oxford: Oxford University Press, pp. 29–57.

Martin, A., Simmons, W.K., Beauchamp, M.S., and Gotts, S.J. (2014). Is a single "hub," with lots of spokes, an accurate description of the neural architecture of action semantics? Comment on "Action semantics: a unifying conceptual framework for the selective use of multimodal and modality-specific object knowledge" by van Elk, van Schie, and Bekkering. *Physics of Life Reviews*, 11(2), 261–2.

McCaffrey, J. (2015). Reconceiving conceptual vehicles: lessons from semantic dementia. *Philosophical Psychology*, 28(3), 337–54.

Medin, D.L., Lynch, E.B., and Solomon, K.O. (2000). Are there kinds of concepts? *Annual review of psychology*, 51(1), 121–47.

Montague, P.R., Dolan, R.J., Friston, K.J., and Dayan, P. (2012). Computational psychiatry. *Trends in Cognitive Sciences*, 16(1), 72–80.

Patterson, K., Nestor, P.J., and Rogers, T.T. (2007). Where do you know what you know? The representation of semantic knowledge in the human brain. *Nature Reviews Neuroscience*, 8(12), 976–87.

Paulus, M. (2014). How and why do infants imitate? An ideomotor approach to social and imitative learning in infancy (and beyond). *Psychonomic Bulletin & Review*, 21(5), 1139–56.

Paulus, M., Hunnius, S., van Wijngaarden, C., Vrins, S., van Rooij, I., and Bekkering, H. (2011). The role of frequency information and teleological reasoning in infants' and adults' action prediction. *Developmental Psychology*, 47(4), 976–83.

Pecher, D., Zeelenberg, R., and Barsalou, L.W. (2003). Verifying different-modality properties for concepts produces switching costs. *Psychological Science*, 14(2), 119–24.

Perfors, A.F., Tenenbaum, J.B., and Regier, T. (2006). Poverty of the stimulus? A rational approach. In: R. Sun and N. Miyake (eds.), *Proceedings of the 28th Annual Conference of the Cognitive Science Society*, 28, 663–8.

Piaget, J. (1962). *Play, dreams and imitation in childhood*. New York: Norton.

Pobric, G., Jefferies, E., and Ralph, M.A.L. (2010). Amodal semantic representations depend on both anterior temporal lobes: evidence from repetitive transcranial magnetic stimulation. *Neuropsychologia*, 48(5), 1336–42.

Prinz, J.J. (2002). *Furnishing the mind: concepts and their perceptual basis*. Cambridge, MA: MIT Press.

Prinz, J.J. (2005). The return of concept empiricism. In: H. Cohen and C. Lefebvre (eds.), *Handbook of categorization in cognitive science*. Amsterdam: Elsevier, pp. 679–95.

Pulvermuller, F. and Fadiga, L. (2010). Active perception: sensorimotor circuits as a cortical basis for language. *Nature Reviews Neuroscience*, 11(5), 351–60.

Ray, E. and Heyes, C. (2011). Imitation in infancy: the wealth of the stimulus. *Developmental Science*, 14(1), 92–105.

Regolin, L. and Vallortigara, G. (1995). Perception of partly occluded objects by young chicks. *Perception & Psychophysics*, 57(7), 971–6.

Santos, L.R. (2004). "Core knowledges": a dissociation between spatiotemporal knowledge and contact-mechanics in a non-human primate? *Developmental Science*, 7(2), 167–74.

Schuil, K.D., Smits, M. and Zwaan, R.A. (2013). Sentential context modulates the involvement of the motor cortex in action language processing: an fMRI study. *Frontiers in Human Neuroscience*, 7, 100.

Simons, D.J. and Keil, F.C. (1995). An abstract to concrete shift in the development of biological thought—the *insides* story. *Cognition*, 56(2), 129–63.

Soja, N.N., Carey, S., and Spelke, E.S. (1991). Ontological categories guide young children's inductions of word meaning: object terms and substance terms. *Cognition*, 38(2), 179–211.

Spelke, E.S. (1990). Principles of object perception. *Cognitive Science*, 14(1), 29–56.

Spelke, E.S. and Kinzler, K.D. (2007). Core knowledge. *Developmental Science*, 10(1), 89–96.

Valenza, E., Leo, I., Gava, L., and Simion, F. (2006). Perceptual completion in newborn human infants. *Child Development*, 77(6), 1810–21.

van Dam, W.O., van Dijk, M., Bekkering, H., and Rueschemeyer, S.A. (2012). Flexibility in embodied lexical-semantic representations. *Human Brain Mapping*, 33(10), 2322–33.

Van Dantzig, S., Pecher, D., Zeelenberg, R., and Barsalou, L.W. (2008). Perceptual processing affects conceptual processing. *Cognitive Science*, 32(3), 579–90.

van Elk, M., van Schie, H., and Bekkering, H. (2014a). Action semantics: a unifying conceptual framework for the selective use of multimodal and modality-specific object knowledge. *Physics of Life Reviews*, 11(2), 220–50.

van Elk, M., van Schie, H., and Bekkering, H. (2014b). The scope and limits of action semantics: reply to comments on "Action semantics: a unifying conceptual framework for the selective use of multimodal and modality-specific object knowledge." *Physics of Life Reviews*, 11(2), 273–9.

Vicente, A. and Manrique, F.M. (2014). The big concepts paper: a defence of hybridism. *The British Journal for the Philosophy of Science*, 67(1), 1–30.

Willems, R.M. and Francken, J.C. (2012). Embodied cognition: taking the next step. *Frontiers in Psychology*, 3, 582.

Wurm, M.F., Ariani, G., Greenlee, M.W., and Lingnau, A. (2015). Decoding concrete and abstract action representations during explicit and implicit conceptual processing. *Cerebral Cortex*, 26(8), 3390–3401.

Yu, A.J. and Dayan, P. (2005). Uncertainty, neuromodulation, and attention. *Neuron*, 46(4), 681–92.

Zwaan, R.A. (2014). Embodiment and language comprehension: reframing the discussion. *Trends in Cognitive Sciences*, 18(5), 229–34.

CHAPTER 35

ORIGINS AND COMPLEXITIES OF INFANT COMMUNICATION AND SOCIAL COGNITION

ULF LISZKOWSKI

THE current chapter investigates the intertwined emergence of human communication and social cognition in infancy. I review a body of recent and new evidence on how infants communicate before they speak, and to what extent this entails a theory of mind. I end by discussing the origins of infant communication from an evolutionary and social-cultural perspective.

Humans engage with each other in unique ways. Together, they create and transmit new knowledge and practices, which is at the heart of cultural group formation and transgenerational learning. What constitutes this ability, and where does it come from? Developmentally, one hypothesis is that before they communicate through language, infants already engage with others nonlinguistically, and these communicative exchanges are meaningful because they entail a social-cognitive understanding of what others intend to bring across and make of one's communication (Tomasello 2008). Alternative views have suggested that children's communication begins to entail an understanding of others' minds only later in development when children already use language (e.g., Shatz and O'Reilly 1990) and that infants initially lack an understanding of others' minds (Carpendale and Lewis 2004). On the latter view, social activity is the developmental motor and starting point of becoming a competent interactant. This contrasts with the former view, which assumes a social-cognitive basis that enables the developing organism to enter into meaningful social activity in the first place. There is merit in both approaches. Logically, there must be some cognitive basis down the line enabling one to engage in, and benefit from, social interaction. However, interactional experiences gained from social activity certainly also shape one's understanding of others. Just how complex this cognitive basis is and how early in life social experiences might shape cognitive development is a matter of empirical investigation. Scrutiny is required in delineating what competencies emerge at what ages, and what the driving forces of their emergence are. Infancy is a test bed to find out about the origins of human communication before language has emerged, and to find out about the origins of understanding others before an explicit verbal theory-like reasoning system about others' minds has been formed.

The ontogenetic origins of mutual engagement and understanding reach into infancy, and their development is intertwined from early on. However, research into the origins of these skills has often followed different strands. On the cognitive side, the main question has been about the cognitive basis, or precursors, of a theory of mind (ToM). Experiments on infants' processing of social information, often based on visual looking paradigms, have revealed early social-cognitive sensitivity to others' behaviors in terms of goals and intentions, and to others' perceptions in terms of attention and knowledge (see Wellman 2010). However, this line of research has less often related infants' social-cognitive abilities to their actual interactional skills. Therefore it has remained less clear whether early social-cognitive abilities are prerequisites, or at all related, to infant communication. On the social side, the main question has been whether and how nonlinguistic means of communication are related to later skills of linguistic communication (Bruner 1975). Initial observations of infants' preverbal interactional skills revealed early language-independent communication strategies that were suggested to carry the seeds of mature forms of human communication (Bates, Camaioni, and Volterra 1975). However, that line of research has less often related infants' prelinguistic communication skills to their social-cognitive understanding of others. Therefore, it has remained less clear what the cognitive underpinnings of infant communication are. To get at the origins of the human ability to connect with and understand each other, one has to consider both sides. How do infants communicate before they have acquired a language? What social-cognitive abilities are involved in infant communication? How early in life does social interaction shape infants' communicative and social-cognitive abilities?

How Do Infants Communicate?

Pragmatic theories of language development have argued for continuity in the development from prelinguistic to linguistic communication (Bruner 1975; Bates 1979; Werner and Kaplan 1963). From this perspective it is not the structure of communication that changes but the communicative means. This change involves a shift from manual attention-directing gestures to symbolic spoken or signed words to meaningful linguistic constructions (Tomasello 2003). Infants begin to stick out their finger around 12 months of age, and parents typically react by commenting about the things to which the finger appears directed (with some variations in object- and action-related comments depending on the situation; Puccini et al. 2010). Because this pointing behavior emerges early in ontogeny, before infants communicate linguistically, it is of theoretical importance to know about its underlying communicative complexities.

Initial empirical studies employed free observations of a few focal infants (Bates et al. 1975; Bruner 1983) and established that infants begin to point around their first birthdays, before they have acquired language. It remained controversial, however, to what extent this behavior is communicative. Do infants indeed intend to communicate when they point, and if so, at what age? Or is it only the caregivers who interpret infants' pointing as communicative? Another debate concerns the reasons *why* infants point. Does pointing serve a single purpose or function? Do infants mostly point to get things or to get praised and rewarded? Or do infants have various motives to point? Connected to these questions is the debate on whether infants point at all to anything specific in the environment. For adults, infants' stuck-out index finger implies a target to which the finger is directed. But do infants intend to direct others attention to something specific when they stick out their index finger?

Because none of these questions are easily settled by observations alone, my colleagues and I ran a series of controlled experiments to test whether 12-months-olds intend to communicate when they point, and if so, about what and why. These questions were motivated by theories of human communication (Grice 1957; Searle 1969) which posit that adult communication involves communicative intentions, reference to content, and reasons for making the communicative act ("illocutionary forces" like statements, requests, etc.). From a social-pragmatic perspective we assumed language-independent roots of human communication and predicted that infants point with communicative intent, with referential intent, and for various flexible reasons, or motives, with cooperative assumptions of being together in a shared activity. An alternative view was that infants point only for themselves (Bates et al. 1975; Desrochers et al. 1995; Carpendale and Carpendale 2010), have no intention to refer someone else's attention to a specific thing in the environment, and point egocentrically to obtain attention to the self (Moore and D'Entremont 2001). Table 35.1 provides an overview of key findings of our research program and recent studies that have followed in the by now industrious research field on infant pointing. These studies have looked at both infants' production and comprehension of pointing in social interaction, to probe a unitary competence of prelinguistic communication. The findings converge to show that when they point around 12 months of age:

- Infants have and understand *communicative intentions*: they point to address another person in expectation of a response.
- Infants have and understand *referential intentions*: they point to refer a recipient to specific things or aspects in expectation of a response about these specific things.
- Infants have and understand *social intentions*: they point for social reasons in expectation of a cooperative activity with the recipient, for example, to request help, to share attitudes, or to provide help by informing others (a.k.a. "*motive*" (Liszkowski 2005), "*performative*" (Bates et al. 1975), "*illocutionary force*" (Searle 1969)).

Table 35.1 Evidence for infants' intentions to communicate with others

Intentionality	Mode	Evidence
Communicative Intention	Production	• Repeat point when partner sees point but doesn't react. Cease pointing for a partner who never looks at or comments on referents.[1] • Point less when partner cannot see point.[2] • Point less when alone.[3]
	Comprehension	• Comprehend point only if being addressed.[4]
Referential Intention	Production	• Repeat point when partner doesn't look at or looks at wrong referent.[1,5] • Point to location of ceased event/occluded referent.[6,7,8] • Point to empty location to communicate about object that was previously there.[8]
	Comprehension	• Uncover occluded non-perceivable object at indicated location.[7,9] • Choose object that is absent from indicated location (looking measure).[10]
Social Intention	Production	*Cooperative requests* • Repeat point until receive successful assistance.[11] • Point even when able to achieve goal alone.[12] • Distinct prosody compared to other social intentions.[13] *Expressives* • Cease pointing for partner who looks at but doesn't comment on referent or comments without interest.[1,5] • Point as part of a shared activity.[14,15] • Point preferentially to a previously shared referent.[10,16] • Distinct prosody compared to requestive social intention.[13] *Informatives* • Partner searches; infant points to specific object the partner needs[17] and is ignorant about.[18] • Unsolicited pointing to hidden object if relevant to both partner's intention and knowledge/belief.[19,20,21] • Use distinct prosody compared to requestive social intention.[13]
	Comprehension	*Cooperative requests* • Offer part of two-part toy that partner points to (partner has the other part).[22] • Clean up or continue to play with toy that partner points at, depending on previous activity with the partner.[23] • Offer object when partner points at it with requestive accompaniments/prosody.[24] *Expressives* • Point to object that partner has pointed to and commented on.[15] • Attend to partner's referent and sometimes smile.[22] • Attend to referent when partner points at it with expressive accompaniments/prosody.[24] *Informatives* • Use point as cue in hiding game.[7,9] • Inspect/search occluder when partner points at it with informative accompaniments/prosody.[24]

[1]Liszkowski et al. 2004; [2]Liszkowski et al. 2008a; [3]Franco et al. 2009; [4]Behne et al. 2005; [5] Liszkowski et al. 2007a; [6]Liszkowski et al. 2007b; [7]Behne et al. 2012; [8]Liszkowski et al. 2009; [9]Liszkowski and Tomasello 2011; [10]Liszkowski and Ramenzoni 2015; [11]Carpenter et al. 1998; [12]van der Goot et al. 2014; [13]Grünloh and Liszkowski 2015; [14]Murphy 1978; [15]Liszkowski et al. 2012; [16]Liebal et al. 2010; [17]Liszkowski et al. 2006; [18]Liszkowski et al. 2008b; [19–21]Knudsen and Liszkowski 2012a,2012b,2013; [22] Camaioni et al. 2004; [23]Liebal et al. 2009; [24]Esteve-Gibert et al. 2017.

These three layers of intentions are interwoven and together they form the basis for meaningful interactions through which infants learn about the physical and cultural world, about what and how persons see things, and about how one communicates about all this with language.

Communicative Intention

A simple conception of communicative intent, suitable to investigate nonverbal communication, operationalizes a given behavior as flexibly and intentionally performed *for* someone else in order to instigate in the other an appropriate reaction without physically coercing the reaction. I will deal with a more subtle definition of communicative intentions derived from philosophy in a moment. The top row of Table 35.1 summarizes evidence for infants' intentions to communicate with others and their expectations that the recipient responds as a consequence. The findings demonstrate unequivocally that infants do not point randomly or solely in response to arousing events. Instead, infants tailor their pointing toward a recipient: they point less when a recipient cannot see their point (Liszkowski et al. 2008a), and they repeat and augment their communication if the recipient sees their point but does not react (Liszkowski et al. 2004). Further, infants often accompany their index finger points with vocalizations (more so than their hand points; Liszkowski and Tomasello 2011), and they increase their point-accompanying vocalizing when they are misunderstood (Liszkowski et al. 2008a). The opposing view—that infants point only for themselves—mostly bears on anecdotal observations that infants often do not alternate gaze with a recipient, which has been taken to mean that they do not address anyone. However, gaze alternation is not the only, and perhaps not even the best, criterion to define communicative intent (Liszkowski et al. 2008a). For example, the referent might simply capture infants' attention more than an anticipated response from the recipient; and infants might mostly look at a recipient when he does *not* respond as expected (Liszkowski et al. 2004). The absence of gaze alternation is thus not evidence for the absence of communicative intent. Further, the communicativeness is often apparent through preceding interactions (and perhaps the default in infants' "primordial sharing situation"; Werner and Kaplan 1963), rendering additional indications of communicative intent unnecessary. Indeed, one of the earliest contexts in which pointing can be reliably elicited is when infants are in their caregiver's arms and already share a visual perspective (Rüther and Liszkowski 2016)—and often in these cases neither the infants nor caregivers alternate gaze.

Beyond the established experimental evidence that 12-month-olds' pointing entails communicative intent, the paramount debate about communicative versus non-communicative pointing has led to new research directions. One direction explores earlier roots of communicative intentionality. Are there ontogenetic earlier behaviors in younger, pre-pointing infants that reveal communicative intent? We will come back to this question in the third section. The idea is that if pointing is communicative, then the developmental story of communicative intent must begin before, not after, pointing. Another direction has explored the roots of non-communicative pointing. Is there any positive evidence that infants point non-communicatively? And does non-communicative pointing disappear in favor of communicative pointing, as Bates et al. (1975) suggested, or does it coexist with communicative pointing? Delgado, Gómez, and Sarriá (2011) have suggested a dual account of pointing. In analogy to Vygotsky's private speech account, these authors have argued and provided evidence that five- and three-year-old children point to orient their own attention in demanding problem-solving situations. The current challenge is thus to test whether, and if so *why*, younger infants point non-communicatively. Do they have "tasks" and "problems" that require them to focus their own attention through pointing gestures? Some preliminary results with 18-month-olds may suggest so, as they have been shown to point alone in a room (Delgado Gómez, and Sarriá, 2009) and perhaps use pointing as a mnemonic strategy (DeLoache Cassidy, and Brown, 1985). However, other studies find that 12-month-olds point less when alone in a room (Franco et al. 2009). In analogy to private speech development, the dual account of pointing would suggest that the communicative function of pointing is developmentally primary and becomes internalized into a private non-communicative function later.

I now turn to the more subtle definition of intentionality in human communication. In conceptual accounts (Grice 1957; Sperber and Wilson 1995), the sender intends to change the cognitive environment of the recipient (the "informative intention") and intends the recipient to understand this informative intention (the "communicative intention"). One can express an informative intention without the higher-order communicative intention. For example, one can push an empty wine glass to the edge of the table so the waiter will see it and refill it (the empty glass at the edge of the table changed his cognitive environment); but one can do so without letting the waiter know that this was one's intention (thus hiding one's communicative intention). Returning to infant pointing, the conceptual question is whether it entails a higher-level communicative intention to overtly manifest an informative intention. While the conceptual distinction is clear, the independent existence of the two intentions is typically demonstrated with regard to hiding authorship. Hiding authorship appears beyond infants' communicative skills and motivation for quite some time (infants rather emphasize their social contributions rather than hide them). In search of an empirical distinction between the two forms of intentions one could thus ask whether there is positive evidence for *emphasizing* authorship. Indeed, 12-month-olds' communicative acts appear fueled with signs of overt expressions of communicativeness. The canonical form of the extended index finger pointing gesture has no other instrumental function than communication, and it is accompanied by vocalizations with prosodic characteristics of communication (Esteve-Gibert and Prieto, 2012), which all appears to overtly mark and manifest the informative intention. An ultimate test would need to measure infants' reactions to selective violations of their communicative, but not informative, intentions. Positive evidence for a distinction has been shown at 2.5 years for verbal communication (Shwe and Markman 1997) and for gestural communication at 18 months (Grosse et al. 2010). In those studies, an experimenter misunderstood the communicative intention while reacting appropriately to the informative intention. Although infants achieved what they wanted (the informative intention), they still augmented their communication in an attempt to have their communicative intention understood, too. From the comprehension side, Csibra (2010) has argued that infants comprehend communicative intentions already in the first year of life, based on a set of codes (e.g., mutual gaze, contingency, motherese), and that the comprehension of the communicative intention occurs prior to inferences about a nested informative intention. From this perspective it is conceivable that infants' face-to-face proto-conversations in the first year of life manifest their communicative intentions without carrying informative intentions, that is, without being about something or without referring the recipient to something.

Referential Intention

Unlike communicative intentions, which aim at producing an effect in the addressee, referential intentions specify the content of communication. A simple operationalization is that a given behavior is done for the other (i.e., communicatively) to relate the other to some aspect in the world; in prelinguistic creatures often by directing others' visual attention to something relevant. The middle row of Table 35.1 summarizes the evidence for infants' comprehension and production of referential intentions when pointing. Of course it is possible to communicate without referential intentions, for example, in emotional proto-conversations or rituals, and this has been one reading of Bates's interpretation of pointing, perpetuated by Moore and D'Entremeont (2001). In their experiments, Moore and D'Entremont (2001) found that 12- (but not 24-) month-olds pointed equally often to a toy to which a recipient was or was not attending. They concluded that young infants do not intend to direct the experimenter's attention but only want to elicit attention to the self. Logically, however, one can refer to something to which another person is already attending, for example, when the pointer wants to expand on the topic or express his or her own referential relation to the referent. More convincing cases are repairs of referential misunderstandings (Liszkowski et al. 2004, 2007a). We found that when infants point and the recipient attends only to the infant, or to an incorrect referent, infants repeat their pointing to the referent in an attempt to redirect attention to it. This provides evidence that infants want to communicate a specific referential relation. Further lines of evidence show that infants even communicate about non-perceptible and absent entities by pointing to occluded sites or empty but habitual locations of absent referents (Liszkowski, Carpenter, and Tomasello 2007b; Liszkowski et al. 2009; Behne et al. 2012). Similarly, they comprehend others' referential intentions when following a point as revealed in scenarios in which the entity about which is communicated is perceptually absent because it is occluded or displaced (Liszkowski and Tomasello 2011; Behne et al. 2012; Liszkowski and Ramenzoni 2015).[1]

[1] A theoretical debate is whether attention-directing communicative acts convey propositional content (i.e., reference *plus* predicate). One reading of visual attention-directing acts is that the deictic gesture directs attention to a site (the referent) under the ascription of a predicate (e.g., an object, a color, and so forth), which would make pointing indeed propositional.

Social Intention

Table 35.1 provides an overview of evidence that infants have and understand different social intentions. Bates et al.'s (1975) initial insight was that infants point for different reasons, in analogy to different illocutionary forces of speech acts. On their account, infants point imperatively by using the adult as a tool to obtain an object, and declaratively by using the object as tool to obtain adult attention. Subsequent accounts have either attempted to reduce all pointing to an imperative motive ("infants want to get something") or grabbled with how to best characterize the motive of declarative pointing.

Deflationary accounts have suggested that imperative pointing is an egocentrically motivated act that constitutes a rather primitive immature form of communication. For example, Camaioni et al. (2004) suggested that imperative pointing emerges before declarative pointing. However, their claim was based on higher rates of imperative compared to declarative pointing, not on differences in the age of emergence. A longitudinal study (Carpenter et al. 1998) found instead that declarative and imperative motives emerge together. Nevertheless, imperative pointing is often conceptualized as a form of instrumental reaching. For example, in many instances of imperative pointing the reward is too distant for infants to retrieve it by themselves (e.g., on a shelf), or infants are strapped in a seat or carried in the arms and are thus physically constrained from retrieving an object. Therefore, imperative pointing could simply be a consequence of physical constraints rather than a communicative expression of one's social intention to get help to obtain a desired thing.

We tested whether infants indeed simply use instrumental action attempts to signal their goal when they cannot execute the action successfully due to physical constraints (van der Goot et al. 2014). We created a situation in which infants needed some additional objects to continue playing with an apparatus they were sitting next to. The objects were freely accessible either at a distance of just under a meter or a bit beyond a meter. At least half of the infants simply stayed next to the apparatus and pointed to request the objects from there, although no constraints hindered them from getting the objects themselves. A locomotion check confirmed that all infants were able to crawl there to get more objects. Thus, infants chose a communicative strategy over an instrumental one in the apparent cooperative expectation that the experimenter would interpret their point as a request for help to get more objects. These findings suggest also that imperatives have the primary function of directing attention to something relevant, and are better understood as cooperative-communicative requests than signals of instrumental action attempts.

Declarative pointing was first mainly characterized by the absence of an imperative motive: infants were observed to point at birds, airplanes, the location of a non-visible clock emanating a sound, to persons, in picture-book reading routines, etc. All these cases lack an apparent desire to obtain an object (the classic definition of an imperative) and have been interpreted as "comments" rather than requests (e.g., Baron-Cohen 1993). In support of a distinction, Grünloh and Liszkowski (2015) established that infants' imperative points differ in prosody and hand shape from infants' declarative points. In contrast,

Moore and D'Entremont (2001) interpreted declarative pointing as another instance of a direct imperative request for attention to the self. However, one piece of evidence against declarative pointing as a direct request for attention to the self is that when an adult comments on infants' pointing by expressing a lack of interest in the referent, infants do not persist in their pointing to obtain a more rewarding emotional response (as if to use a tool to bring about a desired effect; Liszkowski, Carpenter, and Tomasello 2007a). Rather, they cease pointing for her on further occasions, although they may continue to engage with the experimenter about other things and do not opt out of the session altogether. Presumably, infants cease pointing because the experimenter is not interested in the specific referent. In our view, one of the motives underlying the "comment" function of declarative pointing is to share and align referential attitudes ("expressive pointing"; Tomasello, Carpenter, and Liszkowski 2007). For example, when an adult has looked at and expressed interest in an event, infants subsequently point for the adult more to the location of the event than when the adult has previously expressed lack of interest in it—a selective effect of sharing interest (Liszkowski, Carpenter, and Tomasello 2007b; see also Liebal, Carpenter, and Tomasello 2010). Notably, the less enthusiastic emotion of non-interest itself does not hamper infants' pointing: when the adult has not seen the event and expresses non-interest, infants point as much as when she expresses interest—a general effect of attention-directing (Liszkowski, Carpenter, and Tomasello 2007b). This shows that declarative pointing serves a primary attention-directing function (the referential intention, see earlier section), which is then complemented by a motive for doing so, in this case the apparent desire to express and align interests.

Subsequent studies have found further motives underlying declarative pointing. One such sub-motive of declarative pointing is to provide information for others, to help others find something they need ("informative pointing"). For example, in Liszkowski et al. (2008b, Exp. 2) an experimenter needed to clean a table and sorted items into various drawers. One item fell down unbeknownst to her, and when she searched for it, infants readily pointed out its new location. The acts had no requestive accompaniments or overt signs of emotional joy, and the item itself had not been valued positively before or after and was subsequently simply sorted into a drawer. This makes it unlikely that infants wanted to get the object, or wanted the experimenter to help them operate the object, or wanted her to proceed with an important next step in a sequence, or share interest in it. Further, pointing ceased significantly when the experimenter had seen the object fall down and knew its location. Instead, infants simply wanted the experimenter to find what he was looking for. Another study showed that infants point to warn others (Knudsen and Liszkowski 2013). In the experiment, the experimenter accidentally touched material that she had not seen lying in her reaching trajectory for a desired object. She then emoted either negatively or, in another condition, she emoted neutrally surprised. Subsequently she removed the material and pursued her reach for the desired object. When the material was later surreptitiously placed back into the reaching path unbeknownst to the experimenter, infants selectively pointed it out for her when it had bothered her before and she wanted to avoid it. They volunteered this information even though the experimenter had not searched or asked for any information or help.

None of these findings are easily compatible with a deflationary account that infants point only to directly elicit positive emotional reactions. Instead, the current findings show that infants have various motives to point, which all coexist from early on. Like adults, infants point to express and share their attitudes, to cooperatively request help form others, and to cooperatively provide help for others. Just as speakers use linguistic utterances flexibly, infants are flexible in using one and the same gesture for various reasons.

Recent accounts have attempted to reduce declarative pointing to the sole function of requesting information, excluding expressive and informative motivations (Southgate, Maanen, and Csibra 2007). In the framework of natural pedagogy (Csibra and Gergeley 2009), pointing is seen as an evolutionary adaptive strategy to learn about opaque cultural knowledge. Our framework and current evidence on infant communication contrast with "one motive" views of declarative pointing, and I maintain that infants direct attention for various social intentions. Notably, however, our framework does not exclude a motive to request information and, in fact, I have speculated before that infants might point interrogatively (Liszkowski 2005; see also Baldwin and Moses 1996). There is no direct evidence to my knowledge, however, that would support the natural pedagogy hypothesis that infants point only, or primarily, to request culturally relevant opaque knowledge. Less stringent accounts have acknowledged that interrogative pointing may coexist with other motives (Kovács et al. 2014; Begus and Southgate 2012; Begus, Gliga, and Southgate 2014), although they have maintained claims about an ontogenetic primacy of interrogative pointing despite a lack of longitudinal comparisons, and although most these studies involve infants well beyond the onset age of pointing (e.g., at 16 months).

Begus et al. (2014) have argued that pointing reflects a readiness to acquire information, a facilitative state for social learning. Communication most certainly presents an optimal situation for social learning. The central question is what information infants learn when they point, and what part of this learning is directly intended and not just a consequence of joint engagement. Infants want to obtain a comment about the things they point to (Liszkowski et al. 2004). Natural pedagogy holds that the information should be object-based and opaque, that is, it should be about the valence, cultural function, or label of an object. Available evidence to date is equally compatible with the idea that infants want to learn about others' relations to objects, e.g., their preferences and attitudes. In contrast to natural pedagogy's narrow conceptualization of pointing as a tool to obtain opaque generalizable information, pointing is perhaps better understood as primarily enabling social contact, providing opportunities for learning about persons, and building up familiarity with others and trust in their information, which are necessary components for successful cooperation. From this perspective, the primary motivation is then to share and align with others, a motive that presumably originates in earlier forms of non-referential proto-conversational exchanges early in the first year of life.

The Social Cognition of Infant Pointing

Given that infants make choices about when to communicate with whom, why, and about what, these choices must be guided by some cognitive system. In none of what follows do I intend to equate infants' information processing with adults' explicit, verbal thought; nor do I claim that infants balance reasons for what they do in a deep reflective way. However, as we will see, experimental findings show that infants' communication is predicated on others' relations to the world, that is, their intentions, attention, knowledge, and beliefs, which must mean that infants somehow process information about others' mental states. Importantly, beyond typical processing paradigms like visual looking, interaction-based studies reveal that infants' social information processing is linked to acting appropriately and meaningfully in the social world.

Point Production

When infants point they have several expectations about the recipient's behaviors. They know their act will only be efficient when the partner attends to it. Infants do not just discern whether they are looked at or not—a processing bias working from few months of life—they understand it to be a condition for the recipient to react to them and accordingly point only then (Liszkowski et al. 2008a). Further, infants know that their act will only be efficient if the partner attends to the referent. Again, infants do not just discriminate whether the partner's eyes are on them or turned away; and their pointing is not explained by their documented preference for gaze turned to them. Instead, they expect the partner's gaze to turn away: they understand whether the partner has turned to attend to the same thing as them, and if not redirect her attention to it (Liszkowski, Carpenter, and Tomasello 2007a). Similarly, when an adult is already attending to a new event, infants will point to it less than when the adult attends to another site, considering in their communication what others do and do not attend to (Liszkowski, Carpenter, and Tomasello 2007b). However, infants do not just react to perceptually available behavioral cues about others' relation to the world. When an adult has seen one object fall down, but not another; or has seen an object but did not witness its secret relocation, and she then searches around, infants recover her information state—they understand what part she does not know and point it out for her appropriately (Liszkowski, Carpenter, and Tomasello 2008b). In fact, infants do not even need to see the adult search to inform her about something relevant she does not know. Unprompted, they keep track of the information state of their interaction partner such that when she re-enters a room in which her object of desire has been relocated, infants will spontaneously point to its new location if she has not witnessed the location change and if it is relevant to her acting (Knudsen and Liszkowski 2012a, 2013). Infants do not only represent others' "gaps" of information (i.e., their "ignorance") but specific contents that may no longer correspond to reality (i.e., "false beliefs"). For example, when 18-month-olds have seen that an experimenter wants to avoid aversive material, and this material has surreptitiously been placed into two boxes, one of which the experimenter thinks contains her toy, infants will selectively warn her about the material in the box that the adult believes contains her toy, even when in reality the toy is no longer in either of the two boxes (Knudsen and Liszkowski 2012b).

Thus, beyond simple processing biases to the scleral white of the eye, infants understand that others have mental relations to the world (i.e., have epistemic states). Infants track the contents of these representations; and they expect to influence and change these with their communicative attempts. Importantly, infants' expectations about others' epistemic states are linked to their expectations about others' conative states, like their preferences, intentions, and desires. When infants point at interesting events, they consider not only whether the adult attends to them correctly, but also whether the adult is interested in these. And when they inform an adult about a changed state of affairs they consider whether the change is relevant to her intention. For example, when the adult re-enters a room, and the toy she last saw in one box is now in another box, infants will selectively inform her about this change if she previously searched for the object, but less if she previously found it accidentally in the course of another action (Knudsen and Liszkowski 2012a). This is not only the case for positive desires but also for negative desires, when the adult wants to avoid something disturbing to her course of action (Knudsen and Liszkowski 2013). Again, infants do not just react to perceptually available cues, like eight-month-olds who can visually anticipate a goal by extrapolating an ongoing reaching trajectory of an arm to its end point (Brandone et al. 2014). Instead, their pointing involves an understanding of the recipient's intentions based on her previous behaviors and interactions.

All these findings then show that infants do not blatantly point any time anywhere, or follow a rule based on specific perceptual cues or triggers. Instead, infants take others' mental relations to the world into account. They understand conative and epistemic states as related and connected to behavior, and even in the absence of immediately perceivable behavioral cues they can react to and modify others' mental relations with their communicative acts in order to be understood.

Point Comprehension

Comprehension studies, too, show that infants do not just follow a point like a perceptual discriminative cue. Instead they expect a point to be about something for a reason. This suggests they are recovering the referential and social intentions of a communicative act. In a seeking-finding game, when 12-month-olds do not know where a desired toy is hidden, and an adult points to the hiding location, infants do not just look at the hiding spot. Instead, they comprehend the point directed at the hiding cloth to mean that the toy is hidden under the cloth, and appropriately retrieve it from that location (Behne et al. 2012; Liszkowski and Tomasello 2011).

Arguably, infants might simply extrapolate the vectorial direction of the point to then search for something interesting at that location. But sometimes a point directs attention to a site that is different from the one containing the intended referent, as in cases of false-belief or absent reference. For example, in a study (Southgate et al. 2010), an adult and infant both knew that in each of the two boxes there was an object. The adult pointed to one of the two boxes to request the infant to retrieve the object for her. However, unbeknownst to the adult, the objects had been swapped. Thus, the adult thought her intended referent was in the box she pointed to, when it really was in the other box. Seventeen-month-old infants understood the mismatch of the adult's information state and current reality (her false belief) and retrieved for her the object from the other box. Instead, when she had seen the swap, infants retrieved the object from the box she pointed to. In another paradigm (Liszkowski and Ramenzoni 2015), we tracked 18-month-olds' eye movements as they watched an adult ostensibly place two objects in distinct but perceptually equal looking locations. These objects then disappeared and the adult pointed to one of the two empty locations. We found that the point to the empty location led infants in a subsequent new scene, in which the two objects were displayed at spatially distinct new locations on the center of the screen, to look more to the object belonging to the previously pointed to, empty place. Ongoing work in my lab reveals that this effect is confined to pointing cues and absent for endogenous light flashes. The studies thus show that infants' comprehension of pointing is not confined to an extrapolation of the vectorial direction of the finger to a perceptually available object or location. Instead it is better explained as an inference about the intended referent, which need not be apparent in the perceptual scene.

Just as they point for different reasons, infants also draw distinct inferences about why someone pointed for them (i.e., the social intentions). Camaioni et al. (2004) suggested that infants distinguish between imperative and declarative motives. In their study, 12-month-old infants typically followed a declarative point to a distal target, sometimes accompanied by joyful expressions; and they typically followed an imperative point to a proximal toy by handing it over to the pointer. In our lab we found that a crucial piece of information that infants use is the distance between pointer and target. In the study (Thorgrimsson 2014), an object was placed equidistant to the infant on one of two sides. The object was either closer to or farther from an experimenter who sat perpendicular to the infant. When the experimenter, who was previously unengaged, then pointed to the object, 12-month-old infants offered the object more often when it was distal to him compared to when it was proximal. Presumably, infants reasoned that the experimenter could have taken the object himself in the proximal condition, and so his point must have meant something else than a request to obtain it (for example, to share interest in it or explore it). This distance effect was absent when the experimenter used unambiguous gestures, like a palm-up beg, ruling out the explanation that infants selectively offer objects that are more distal.

Distance alone is of course not always decisive in recovering the meaning of a point, and not the only available information. In a lot of cases a point is situated within a meaningful context of a shared activity, joint attentional scenes, or follows other preceding actions. In an experiment (Liebal et al. 2009), 14- and 18-month-olds engaged with the experimenter in an activity—for example, they cleaned away toys. One object was surreptitiously left on the floor. At test, either the first experimenter with whom they had shared the activity pointed to the toy, or another experimenter with whom they had not engaged in the clean-up game. The distances to the toy were the same for both pointers. Infants were significantly more likely to understand the point as a directive to clean away the toy in the former case, presumably because it was immediately meaningful in the context of the preceding shared activity.

Sometimes, however, helpful context from preceding shared activities is less available to disambiguate the meaning of a point. This is especially the case when shared activity is yet to be initiated, or when contexts or activities are new, a situation that novices and infants often face. When information from the distances or shared activities is not available, act-accompanying features provide another source of information. For example, requests typically take on different prosodic characteristics than offers. In recent studies we found that parents accompany their pointing with distinct prosody and use distinct hand shapes to express different social intentions (Esteve-Gibert, Liszkowski, and Prieto 2016, 2017): expressive pointing was accompanied by vocalizations with wide pitch range and long syllables; informative pointing had a narrow pitch range and intermediate syllable duration; and imperative pointing had an intermediate pitch range and short syllables. Imperative requests were mostly realized with palm-up open hands; the former two types with the extended index finger. In an experimental setting, we then tested whether infants would react appropriately on the basis of these different accompanying characteristics (Esteve-Gibert et al. 2017). In three conditions, adults were instructed to direct attention to an appearing upside-down cupcake, and either share interest in it, request it, or inform about a hidden item under the cupcake. We controlled for disambiguating information stemming from the distance to the referent, preceding action sequences, and lexical cues. Infants made appropriate inferences about the distinct intentions based on the accompanying characteristics of the acts and, respectively, mostly shared attention to the referent, offered it, or searched for it.

The studies reveal that infants recover the social intentions of pointing from spatial information, social activity, and the form of the act (i.e., prosody and hand shape). They all converge to show that infants understand pointing not just as a spatial directional cue, or object-directed behavior (Woodward and Guajardo 2002). They understand it as an intentional communicative-referential act with distinct illocutionary forces.

Does the infant then have a theory of mind? Yes—if you define it that way. Certainly, they have practical knowledge of the workings of interaction, of how their own behavior can influence others and vice versa, and how this is mediated by the relation of conative and epistemic states. Infants use this knowledge to flexibly predict others' behaviors, beyond perceptually available behavioral cues, and they use these predictions in the course of acting together with others. Their understanding is more practical in the sense that they presumably do not reason about others' thoughts and desires in hypothetical scenarios, perhaps because they lack abstract language to represent these scenarios. Third-party paradigms, in which infants are not directly addressed but observe others interact, reveal that during the second year of life, infants' expectations become less dependent on direct second-person engagement, which suggests a somewhat abstracted, internalized understanding of the structure and meanings of social interactions (Gräfenhain et al. 2009; Fawcett and Liszkowski 2012a, 2012b; Martin et al. 2012; Thorgrimsson, Fawcett, and Liszkowski 2014; Thorgrimsson, Fawcett, and Liszkowski 2015).

Likely, infants' theory of mind emerges for social interaction and therefore initially takes on a form of practical knowledge, perhaps better described as a theory of *action*. Expectations about actions may of course be very simple initially (Reddy et al. 2013), but in the second year of life the evidence speaks against a shallow reactive understanding of others based on a few behavioral cues. Instead, it reveals a flexible understanding of others that infants use to interact meaningfully.

Limits of Infant Communication

While the origins of human communication clearly reach into infancy, this is not to say that infant and adult communication are just the same. The obvious difference is that infants lack symbolically structured language. A closer look reveals that infants are limited in their ability to communicate representationally. Bates (1979) viewed representational communication as the "second dawn" in communicative development (the "first dawn" was the emergence of deictic communication, like pointing). Representational communication requires using a sign that stands for a referent, often in an arbitrary (symbolic) way, or in an iconic way, in which the sign somehow resembles the referent or aspects of the referent. Creatively produced iconic gestures are a good indicator that the communicator intends to refer a recipient to a referent by invoking a representation of the referent. Initial observations had suggested that infants show a nascent competence around 13 to 14 months of age, when they play with pretend props (e.g., putting a doll shoe on a doll foot; Bates 1979) and use non-deictic gestures in a circumscribed context (Acredolo and Goodwyn 1988). These observations, however, did not convincingly show that the behaviors were intended to stand for something. They remained amenable to an alternative interpretation in which the behaviors rather reflect a participatory activity in routinized rituals or play, but do not stand for something other than the activity itself (Liszkowski 2010). Recent longitudinal corpus data suggest that a spontaneous communicative use of iconic gestures emerges around two years of age (Özçalişkan and Goldin-Meadow 2011).

An experimental study shows that 27- and to some extent 21-month-olds will invent iconic gestures to communicate a solution of a problem (e.g., how to open a vessel) to someone who doesn't know the solution (Behne, Carpenter, and Tomasello 2014). In a similar study conducted in my lab with Reyhan Furman, we had a hand puppet violate conventional uses of tools to probe whether toddlers would protest and correct her. Twenty-four- but not 20-month-olds spontaneously created iconic gestures depicting how to use the tools. We also tried to elicit such forms of iconic gesturing in younger infants at 18 months of age using a simplified paradigm (Puccini 2013). On several trials we demonstrated how to operate toys with objects in a very basic way. Infants had no problem imitating either horizontal ("stroking") or vertical ("banging") movements with the tool, for example, to make a xylophone sound or produce patterns in sand, etc. At test, another experimenter entered, pretending not to know what to do. Most infants pointed to the tool or object, presumably to communicate to the adult something about the previous activity with the object. The adult then took the tool and went on with quizzical looks asking for clarification what to do with it. However, infants did not make any horizontal or vertical movements to somehow communicate to the adult *what* to do with the object. Thus, 18-month-old infants communicated with deictic gestures about the relevant objects, but not with representational gestures about the relevant actions. Together, the experimental findings confirm that before their second birthdays, creative representational communication through gestures is rare, if not absent. These findings concur with several comprehension studies which show that infants comprehend iconic gestures by 26 months of age, or a bit earlier depending on the conservativeness of the analyses, but not before 18 months of age (Namy 2008).

Origins of Infant Communication

Where do the seemingly sophisticated communication skills that emerge relatively early in infancy come from? One possibility is that these communication skills have deep evolutionary roots and are shared with nonhuman primates. Species comparisons with our closest living ancestors, the great apes, however, reveal that their communication differs from that of human infants on referential and social levels. Apes do not naturally point for each other. In experiments, zoo-housed apes can indicate choices through gestural behaviors to make a keeper give them food items they want to obtain, or tools necessary to obtain these (Call and Tomasello 1994). On most accounts and evidence, they do this behavior for human keepers, not conspecifics.

A closer look suggests that apes use a different communicative strategy that derives from their instrumental actions. They use instrumental action attempts to signal their action intentions, that is, they make an action intention overt through an abbreviated instrumental action. This strategy has two consequences. First, this form of communication requires expressing (making overt) the intention to act on something (e.g., to obtain it), and this should require a marked decrease in the distance between actor and object, usually through clear bodily approach toward the object. Thus, apes should approach a desired object as close as possible, or *necessary*, so that a human keeper can understand the goal to obtain it from their bodily directions and action toward the object. Second, this form of communication is bound to things that are present, because instrumental actions are directed at objects. Thus, apes should not be able to communicate about items that are not in the "here and now." We tested these predictions in two studies.

In Study 1, to test whether apes signal their goal to others through their approach behaviors, we assessed whether they can in favorable cases also communicate from a distance without exhibiting their goal through effortful approach behaviors, i.e., without a marked decrease in distance between them and the referent object. In our study (van der Goot et al. 2014), we centered apes at one end of the enclosure by delivering food to them just at that place. Whenever they ran out of food, a second experimenter delivered a refill from the far edge of the enclosure. At some point, however, the delivery stopped and the food remained at the far edge of the enclosure. We measured whether and how apes would request more items. Distance was the key variable to discern strategies: most efficient would be to stay at the delivery place and simply point from a distance to the food. This is what many human 12-month-olds did in a comparable situation (see earlier). Apes, in contrast, always moved across the distance and protruded their fingers through the mesh in front of the food to indicate their intention to get more items. Thus, they decreased the distance between them and the object to express their action intention to obtain the objects. Leavens et al.'s (2015) rebuttal of these findings failed because in their experiment they decreased the distance to the object, which was key to distinguishing the two strategies (in their study the "distal" reward was so close to the ape that extending the arm was an appropriate and sufficient action to express the instrumental intention to get the object). To further argue for apes' ability to point, Leavens et al. (2015) published decontextualized photographs of apes extending their arms, presumably to request items far out of reach. However, the items were located somewhere above them outside their enclosure so that the apes simply could not get any closer to the referent objects. The depicted behaviors support the interpretation that apes use instrumental action attempts, like approaches or reaches, to signal their action intentions. There is little doubt that apes would execute their instrumental actions successfully were they able to get the items themselves—unlike human 12-month-olds who communicated their request with the conventional pointing gesture even when the objects were fully accessible to them, and relatively close by (van der Goot et al. 2014).

In Study 2 (Liszkowski et al. 2009), we tested whether the strategy to exhibit their instrumental intentions through instrumental action attempts would limit apes' communication to act on objects in the "here and now." Apes watched as two experimenters requested and offered each other either food (desirable) or cage bedding (neutral) that was always located in the same of two distinct locations. After intensive familiarization, the locations were empty, but in one condition the items were placed inside the locations (these were platforms that could be opened), and apes saw this. Apes had no problem approaching the platform that contained the food and often extended their finger through the mesh to convey their intention to get the object. Thus, apes did not need to see the object in order to generate and exhibit the intention to act on the object.

However, in another condition, the locations were empty and remained empty. In these cases, apes did not selectively approach the empty location or signal specific action attempts—presumably because there were no objects to act on, and so no way to signal their intention through instrumental action attempts. Apes did not cease requesting altogether. They banged on the cage bars, and stuck their fingers unsystematically through the bars at various places. These requests were "unspecific," best glossed as "more" (perhaps analogous to newborns' requests). This is in stark contrast to human 12-month-olds, who simply pointed to the empty location to specifically request more of the items that usually belong in the location.

A rebuttal by Lyn et al. (2014) failed due to methodological confounds. In their study, apes had not watched the contingencies of placing and offering from a third-person perspective. Instead, they had obtained training in directly approaching the location with food items to obtain food until the location was empty. The approach to the empty location at test was thus most likely driven by the previous contingency of getting food when exhibiting the action of approaching or acting toward the location. In other words, their action of approaching the location had been sufficiently often interpreted as a request for the items that this approach behavior was then sufficient to make their intention overt. The same problem applies to a study by Bohn et al. (2015), although their positive findings with apes are fragile (16 of 337 points to an empty site from which the apes had previously obtained food).

Another difference to ape communication pertains to the underlying motives. Apes communicate to request but not to inform a keeper helpfully of what he needs to know to fulfill his own desires, or to share interest with him in new or liked events, like human 12-month-olds do (Bullinger et al. 2011). This is in line with the broader picture that great apes and humans differ in the motivation to align and cooperate with others (Tomasello, 2009). The difference in the motivation underlying their communication goes hand in hand with their instrumental way of communicating: communicating by signaling action attempts enables them to communicate about things they want to act on, but not about things another individual would want to see or act on. The picture is further complemented by comprehension studies which show that apes infer the presence of a hidden item when they see a human performing an instrumental action attempt toward it (e.g., reaching; Hare and Tomasello 2004). However, unlike human 12-month-olds, apes fail to recover the communicative-referential intention when they see a human cooperatively directing attention to the hiding place with a distal index finger gesture that is not instrumental in acting on it.

Turning to the ontogenetic origins of infant communication, it must be stressed that before they turn one year of age and begin to point, infants' communication skills are still limited in several ways. They do not communicate from the beginning with different layers of referential and social intentionality. Rather, pointing and its communicative infrastructure are a developmental outcome of the first year of life. Precursors of the key components emerge before infants begin to point.

Infants clearly interact with others in the first year of life, but it is debatable whether they communicate intentionally. Interactive behaviors like smiles, cries, mutual looks, or re-engagement behaviors are perhaps initially driven by their perlocutionary effects (unintended consequences) and turn into instrumental actions to elicit these effects, as illustrated by cases in which infants suddenly "switch on and off" their crying to obtain a change in the environment. There is also little doubt that communication in the first year of life involves various reasons, or motives. Most notably, infants do not only whine and request—they also smile and look into the eyes to share and regulate their emotional states with others, which is perhaps the origin of the expressive motive underlying later pointing acts. Around four months of age, infants vocalize in a functionally flexible way, accompanying similar vocal categories with different facial expressions of affect, which suggests that a vocal signal can take on various functions (Oller et al. 2013). Thus, requestive and affiliative social motives are part of infants' communication from early on.

The perhaps clearest limitation of infant communication in the first half year of life pertains to the referential level—infants do not seem to specify *what* they communicate about. Precursor behaviors to reference emerge a few months before infants begin to point, for example, when infants reach for or hold out things. In an experiment (Ramenzoni and Liszkowski 2016), we found that eight-month-olds will refrain from reaching for objects out of reach—that is, they know about their action boundaries. However, when someone else is sitting close by to help, infants will again reach for these unobtainable objects, in an apparent expectation that their unsuccessful reach will signal their intention to obtain the object. This strategy resembles that of apes in that eight-month-old infants signal their intention through an instrumental action attempt. Another study (Cameron-Faulkner et al. 2015) found that the frequency of holding out and giving objects at 10 to 12 months is related to the frequency of subsequently emerging pointing gestures, suggesting a developmental—and, by inference, conceptual—relation between these morphologically distinct behaviors.

How then do these "proto"-referential behaviors emerge and develop into communicative-referential pointing at the end of the first year? Regarding the development to communicative pointing, empirical findings suggest that cognitive and social development interact. A longitudinal training study (Matthews et al. 2012) found that point-following predicts point production, revealing the importance of social-cognitive understanding in the emergence of pointing production. Ongoing longitudinal work in our lab finds that parents point for their eight-month-old infants before they begin to point, and that the frequency of parents' pointing is predictive of the age of emergence of index finger pointing (Rüther and Liszkowski 2016). Our further findings reveal that parents point significantly less for six- compared to eight-month-olds, suggesting that parents gauge their communicative behaviors to infants' cognitive processing abilities. In a cross-cultural study (Salomo and Liszkowski 2013) we found differences in the natural amount of triadic joint engagement and deictic gestures directed at 8- to 15-month-old infants during home visits with Yucatec Mayans in Mexico, Dutch in the Netherlands, and Chinese in Shanghai. These natural differences across cultural settings were positively correlated with infants' gesturing and predictive of the emergence of the index finger pointing gesture. The findings thus suggest that beyond necessary social-cognitive components, social interactional context exerts an early influence on the emergence of communicative behaviors.

The likely departure point of communicative development then is the basic social motivation to belong to and engage with others, connected to the ability to detect and establish direct face-to-face engagement. The element of reference then emerges through both cognitive and social changes: the cognitive achievement of instrumental action production and understanding lends itself in social situations to signal an instrumental goal to others, provided there is a fundamental trust in, and orientation to, others' cooperativeness and helpfulness. Because of the unidirectional cooperative relation between apes and human keepers, and because of their understanding of instrumental actions, captive apes too achieve this level of communication with human keepers. In the case of humans, however, infants' cognitive skills of action production and understanding get affected by their deep interest in others, and others' deep interests in them in ways much stronger than for great apes. And so, human infants begin to attend to others' objects and activities as "objects-of-regard" (Werner and Kaplan 1963) and "demonstrations," and caregivers mark their objects and actions *for* infants. Infants in turn show their objects and activities to others, and responsive caregivers do the same. Together they will then take off down the road to reference, showing each other the world, the things they act on, they like, and find relevant.

Acknowledgments

This research was supported by the German Federal Ministry of Education and Research (BMBF, grant number 01DL14007) and the German Science Foundation (DFG, grant number FOR2253, LI 1998/3-1).

References

Acredolo, L.P. and Goodwyn, S.W. (1988). Symbolic gesturing in normal infants. *Child Development*, 59, 450–66.

Baldwin, D.A. and Moses, L.J. (1996). The ontogeny of social information gathering. *Child Development*, 67, 1915–39.

Bates, E. (1979). *The emergence of symbols: cognition and communication in infancy*. New York: Academic Press.

Bates, E., Camaioni, L., and Volterra, V. (1975). The acquisition of performatives prior to speech. *Merrill-Palmer Quarterly*, 21, 205–26.

Begus, K., Gliga, T., and Southgate, V. (2014). Infants learn what they want to learn: responding to infant pointing leads to superior learning. *PLoS One*, 9(10), e108817.

Begus, K. and Southgate, V. (2012). Infant pointing serves an interrogative function. *Developmental Science*, 15(5), 611–7.

Behne, T., Liszkowski, U., Carpenter, M., and Tomasello, M. (2012). Twelve-month-olds' comprehension and production of pointing. *British Journal of Developmental Psychology*, 30(3), 359–75.
Behne, T., Carpenter, M., and Tomasello, M. (2005). One-year-olds comprehend the communicative intentions behind gestures in a hiding game. *Developmental Science*, 8, 492–9.
Behne, T., Carpenter, M., and Tomasello, M. (2014). Young children create iconic gestures to inform others. *Developmental Psychology*, 50(8), 2049–60.
Bohn, M., Call, J., and Tomasello, M. (2015). Communication about absent entities in great apes and human infants. *Cognition*, 145, 63–72.
Brandone, A.C., Horwitz, S., Wellman, H.M., and Aslin, R.N. (2014). Infants' goal anticipations during failed and successful reaching actions. *Developmental Science*, 17, 23–34.
Bruner, J.S. (1975). The ontogenesis of speech acts. *Journal of Child Language*, 2, 1–19.
Bruner, J.S. (1983). *Child's talk*. New York: Norton.
Bullinger, A.F., Zimmermann, F., Kaminski, J., and Tomasello, M. (2011). Different social motives in the gestural communication of chimpanzees and human children. *Developmental Science*, 14(1), 58–68.
Call, J. and Tomasello, M. (1994). Production and comprehension of referential pointing by orangutans (*Pongo pygmaeus*). *Journal of Comparative Psychology*, 108(4), 307–17.
Camaioni, L., Perucchini, P., Bellagamba, F., and Colonnesi, C. (2004). The role of declarative pointing in developing a theory of mind. *Infancy*, 5(3), 291–308.
Cameron-Faulkner, T., Theakston, A., Lieven, E., and Tomasello, M. (2015). The relationship between infant holdout and gives, and pointing. *Infancy*, 20(5), 576–86. doi:10.1111/infa.12085
Carpendale, J.E.M. and Lewis, C. (2004). Constructing an understanding of mind: the development of children's understanding of mind within social interaction. *Behavioral and Brain Sciences*, 27, 79–150.
Carpendale, J.I.M. and Carpendale, A.B. (2010). The development of pointing: from personal directedness to interpersonal direction. *Human Development*, 53, 110–26.
Carpenter, M., Nagell, K., and Tomasello, M. (1998). Social cognition, joint attention, and communicative competence from 9 to 15 months of age. *Monographs of the Society of Research in Child Development*, 63(4), 1–143.
Csibra, G. (2010). Recognizing communicative intentions in infancy. *Mind & Language*, 25, 141–68.
Csibra, G. and Gergely, G. (2009). Natural pedagogy. *Trends in Cognitive Sciences*, 13, 148–53.
Delgado, B., Gómez, J.C., and Sarriá, E. (2009). Private pointing and private speech: developing parallelisms. In: A.Winsler, C. Fernyhough, and I. Montero (eds.), *Private speech, executive function and the development of verbal self-regulation*. Cambridge: Cambridge University Press, pp. 153–62.
Delgado, B., Gómez, J.C., and Sarriá, E. (2011). Pointing gestures as a cognitive tool in young children: experimental evidence. *Journal of Experimental Child Psychology*, 110, 299–312. doi:10.1016/j.jecp. 2011.04.010
DeLoache, J.S., Cassidy. D.J., and Brown, A.L. (1985). Precursors of mnemonic strategies in very young children. *Child Development*, 56, 125–37.
Desrochers, S., Morissette, P., and Ricard, M. (1995). Two perspectives on pointing in infancy. In: C. Moore and P. Dunham (eds.), *Joint attention: Its origin and role in development*. Hillsdale, NJ: Erlbaum, pp. 85–101.
Esteve-Gibert, N., Liszkowski, U., and Prieto, P. (2016). Prosodic and gestural features distinguish the intention of pointing gestures in child-directed communication. In: M.E.

Armstrong, N. Henriksen, and M.M. Vanrell (eds.), *Interdisciplinary approaches to intonational grammar in Ibero-Romance*. Amsterdam: John Benjamins, pp. 251–76.

Esteve-Gibert, N., Prieto, P., and Liszkowski, U. (2017). Twelve-month-olds understand social intentions based on prosody and gesture shape. *Infancy*, 22(1), 109–29. (First published April 1, 2016.)

Fawcett, C. and Liszkowski, U. (2012a). Observation and initiation of joint action in infants. *Child Development*, 83, 434–41.

Fawcett, C. and Liszkowski, U. (2012b). Infants anticipate others' social preferences. *Infant and Child Development*, 21, 239–49.

Franco, F., Perucchini, P., and March, B. (2009). Is infant initiation of joint attention by pointing affected by type of interaction? *Social Development*, 18(1), 51–76.

Gräfenhain, M., Behne, T., Carpenter, M., and Tomasello, M. (2009): One-year-olds' understanding of nonverbal gestures directed to a third person. *Cognitive Development*, 24(1), 23–33.

Grice, H.P. (1957). Meaning. *Philosophical Review*, 66(3), 377–88. (Reprinted as ch. 14 on Grice (1989) *Studies in the way of words*. Cambridge, MA: Harvard University Press, pp. 213–23.)

Grosse, G., Behne, T., Carpenter, M., and Tomasello, M. (2010). Infants communicate in order to be understood. *Developmental Psychology*, 46(6), 1710–22.

Grünloh, T. and Liszkowski, U. (2015). Prelinguistic vocalizations distinguish pointing acts. *Journal of Child Language*, 42(6), 1312–36. doi:10.1017/S0305000914000816

Hare, B. and Tomasello, M. (2004). Chimpanzees are more skilful in competitive than in cooperative cognitive tasks. *Animal Behaviour*, 68(3), 571–81.

Knudsen, B. and Liszkowski, U. (2012a). Eighteen- and 24-month-old infants correct others in anticipation of action mistakes. *Developmental Science*, 15, 113–22.

Knudsen, B. and Liszkowski, U. (2012b). 18-month-olds predict specific action mistakes through attribution of false belief, not ignorance, and intervene accordingly. *Infancy*, 17, 672–91.

Knudsen, B. and Liszkowski, U. (2013). One-year-olds warn others about negative action outcomes. *Journal of Cognition and Development*, 14(3), 424–36.

Kovács, Á.M., Tauzin, T., Téglás, E., Gergely, G., and Csibra, G. (2014). Pointing as epistemic request: 12-month-olds point to receive new information. *Infancy*, 19(6), 543–57.

Leavens, D.A., Reamer, L.A., Mareno, M.C., Russell, J.L., Wilson, D., Schapiro, S.J. et al. (2015). Distal communication by chimpanzees (*Pan troglodytes*): evidence for common ground? *Child Development*, 86(5), 1623–38.

Liebal, K., Behne, T., Carpenter, M., and Tomasello, M. (2009). Infants use shared experience to interpret pointing gestures. *Developmental Science*, 12(2), 264–71.

Liebal, K., Carpenter, M., and Tomasello, M. (2010). Infants' use of shared experience in declarative pointing. *Infancy*, 15(5), 545–56.

Liszkowski, U. (2005). Human twelve-month-olds point cooperatively to share interest with and helpfully provide information for a communicative partner. *Gesture*, 5(1–2), 135–54. doi:10.1075/gest.5.1.11lis

Liszkowski, U. (2010). Before L1: a differentiated perspective on infant gestures. In: M. Gullberg and K. De Bot (eds.), *Gestures in language development*. Amsterdam: Benjamins, pp. 35–51.

Liszkowski, U., Albrecht, K., Carpenter, M., and Tomasello, M. (2008a). Twelve- and 18-month-olds' visual and auditory communication when a partner is or is not visually attending. *Infant Behavior and Development*, 31(2), 157–67. doi:10.1016/j.infbeh.2007.10.011

Liszkowski, U., Brown, P., Callaghan, T., Takada, A., and De Vos, C. (2012). A prelinguistic gestural universal of human communication. *Cognitive Science*, 36, 698–713.

Liszkowski, U., Carpenter, M., Henning, A., Striano, T., and Tomasello, M. (2004). Twelve-month-olds point to share attention and interest. *Developmental Science*, 7(3), 297–307.

Liszkowski, U., Carpenter, M., Striano, T., and Tomasello, M. (2006). Twelve- and 18-month-olds point to provide information for others. *Journal of Cognition and Development*, 7(2), 173–87.

Liszkowski, U., Carpenter, M., and Tomasello, M. (2007a). Reference and attitude in infant pointing. *Journal of Child Language*, 34(1), 1–20.

Liszkowski, U., Carpenter, M., and Tomasello, M. (2007b). Pointing out new news, old news, and absent referents at 12 months of age. *Developmental Science*, 10(2), F1–F7.

Liszkowski, U., Carpenter, M., and Tomasello, M. (2008b). Twelve-month-olds communicate helpfully and appropriately for knowledgeable and ignorant partners. *Cognition*, 108(3), 732–9. doi:10.1016/j.cognition.2008.06.013

Liszkowski, U. and Ramenzoni, V. (2015). Pointing to nothing? Empty places prime infants' attention to absent objects. *Infancy*, 20(4), 433–44.

Liszkowski, U., Schäfer, M., Carpenter, M., and Tomasello, M. (2009). Prelinguistic infants, but not chimpanzees, communicate about absent entities. *Psychological Science*, 20, 654–60.

Liszkowski, U. and Tomasello, M. (2011). Individual differences in social, cognitive, and morphological aspects of infant pointing. *Cognitive Development*, 26, 16–29.

Lyn, H., Russell, J.L., Leavens, D.A., Bard, K.A., Boysen, S.T., Schaeffer, J. et al. (2014). Apes communicate about absent and displaced objects: methodology matters. *Animal Cognition*, 17, 85–94.

Martin, A., Onishi, K.H., and Vouloumanos, A. (2012). Understanding the abstract role of speech in communication at 12 months. *Cognition*, 123, 50–60.

Matthews, D., Behne, T., Lieven, E., and Tomasello, M. (2012). Origins of the human pointing gesture: a training study. *Developmental Science*, 15(6), 817–29.

Moore, C. and D'Entremont, B. (2001). Developmental changes in pointing as a function of attentional focus. *Journal of Cognition & Development*, 2, 109–29.

Murphy, C.M. (1978). Pointing in the context of a shared activity. *Child Development*, 49(2), 371–80.

Namy, L.L. (2008). Recognition of iconicity doesn't come for free. *Developmental Science*, 11, 841–6.

Oller, D.K., Buder, E.H., Ramsdell, H.L., Warlaumont, A.S., Chorna, L., and Bakeman, R. (2013). Functional flexibility of infant vocalization and the emergence of language. *Proceedings of the National Academy of Sciences*, 110(16), 6318–23.

Özçalişkan, S. and Goldin-Meadow, S. (2011). Is there an iconic gesture spurt at 26 months? In: G. Stam and M. Ishino (eds.), *Integrating gestures: the interdisciplinary nature of gesture*. Amsterdam: Benjamins, pp. 163–74.

Puccini, D. (2013). *The use of deictic versus representational gestures in infancy* [PhD thesis]. Nijmegen: Radboud University.

Puccini, D., Hassemer, M., Salomo, D., and Liszkowski, U. (2010). The type of shared activity shapes caregiver and infant communication. *Gesture*, 10(2/3), 279–97. doi:10.1075/gest.10.2-3.08puc

Ramenzoni, V. and Liszkowski, U. (2016). The social reach: 8-month-olds reach for unobtainable objects in the presence of another person. *Psychological Science*, 27(9), 1278–85.

Reddy, V., Markova, G., and Wallot, S. (2013). Anticipatory adjustments to being picked up in infancy. *PLoS ONE*, 8, e65289. doi:10.1371/journal.pone.0065289

Rüther, J.N. and Liszkowski, U. (2016). *Social and cognitive origins of infant pointing*. Poster presented at the International Conference on Infant Studies, New Orleans, May 26–28.

Salomo, D. and Liszkowski, U. (2013). Sociocultural settings influence the emergence of prelinguistic deictic gestures. *Child Development*, 84(4), 1296–307. doi:10.1111/cdev.12026

Searle, J. (1969). *Speech acts: an essay in the philosophy of language*. Cambridge: Cambridge University Press.

Shatz, M. and O'Reilly, A.W. (1990). Conversational or communicative skill? A reassessment of two-year-olds' behaviour in miscommunication episodes. *Journal of Child Language*, 17, 131–46.

Shwe, H.I. and Markman, E.M. (1997). Young children's appreciation of the mental impact of their communicative signals. *Developmental Psychology*, 33(4), 630–6.

Southgate, V., Chevallier, C., and Csibra, G. (2010). Seventeen-month-olds appeal to false beliefs to interpret others' referential communication. *Developmental Science*, 16, 907–12.

Southgate, V., van Maanen, C., and Csibra, G. (2007). Infant pointing: communication to co-operate or communication to learn? *Child Development*, 78(3), 735–40.

Sperber, D. and Wilson, D. (1995): *Relevance: communication and cognition* (2nd ed.), Oxford/Cambridge: Blackwell Publishers.

Thorgrimsson, G. (2014). *Infants' understanding of communication as participants and observers* [PhD thesis]. Nijmegen: Radboud University.

Thorgrimsson, G., Fawcett, C., and Liszkowski, U. (2014). Infants' expectations about gestures and actions in third-party interactions. *Frontiers in Psychology*, 5, 321.

Thorgrimsson, G., Fawcett, C., and Liszkowski, U. (2015). 1-and 2-year-olds' expectations about third-party communicative actions. *Infant Behavior and Development*, 39, 53–66.

Tomasello, M. (2003). *Constructing a language: a usage-based theory of language acquisition*. London: Harvard University Press.

Tomasello, M. (2008). *Origins of human communication*. Cambridge, MA: MIT Press.

Tomasello, M. (2009). *Why we cooperate*. Cambridge, MA: MIT Press.

Tomasello, M., Carpenter, M., and Lizskowski, U., (2007). A new look at infant pointing. *Child Development*, 78, 705–22.

van der Goot, M.H., Tomasello, M., and Liszkowski, U. (2014). Differences in the nonverbal requests of great apes and human infants. *Child Development*, 85(2), 444–55. Epub July 31, 2013. doi:10.1111/cdev.12141

Wellman, H. (2010). Developing a theory of mind. In: U. Goswami (ed.), *The Blackwell handbook of cognitive development* (2nd ed.). Oxford: Blackwell.

Werner, H. and Kaplan, B. (1963). *Symbol formation: an organismic-developmental approach to language and the expression of thought*. New York: Wiley.

Woodward, A.L. and Guajardo, J.J. (2002). Infants' understanding of the point gesture as an object-directed action. *Cognitive Development*, 83, 1–24.

CHAPTER 36

DEVELOPING AN UNDERSTANDING OF NORMATIVITY

MARCO F.H. SCHMIDT AND HANNES RAKOCZY

Introduction

From Prediction and Causes to Prescription and Reasons

The capacity for *cognition* allows human and nonhuman animals to navigate the physical world effectively and adaptively. For instance, animals can estimate distances, memorize events, track objects in space, detect regularities, discriminate between small sets of objects exactly and between large sets approximately, and make causal inferences. Thus, over the last decades, developmental and comparative research have gained more and more insights into the development of human and nonhuman thinking about the natural world including its entities, regularities, and causal structure (Baillargeon and Carey 2012; Call and Tomasello 2005; Rakoczy 2014; Tomasello 2014).

But many animals, including humans, also evolved a form of cognition that does not serve to deal with the physical world per se, but rather with the observable (behavioral) and unobservable (mental) states of conspecifics (and other species)—typically called *social cognition.* The study of human and nonhuman social-cognitive capacities has mostly been concerned with issues of mindreading (i.e., understanding individual psychological states, such as beliefs and desires). Based on research conducted over the last couple of years in particular, scholars have suggested that nonhuman animals (e.g., chimpanzees) understand simple perceptual states (e.g., seeing), but potentially not propositional mental states, such as beliefs proper (Andrews 2012; Call and Tomasello 2008; but see Carruthers 2013 for arguments in favor of continuity between animals and humans, and Penn and Povinelli 2007 for arguments in favor of more fundamental discontinuity). Regarding social cognition and mindreading in human ontogeny (and adulthood), two-process accounts have been advocated lately. One process (for tracking "belief-like" states) is thought to be fast and efficient, but inflexible, while the other (for explicit reasoning about beliefs and propositional attitudes in general) is thought to be slow but flexible (Apperly and Butterfill 2009; Butterfill and Apperly 2013; Rakoczy 2015).

Human (and perhaps even nonhuman) social cognition, however, not only deals with the prediction and explanation of others' behavior and individual mental states in a causal-descriptive sense—that is, humans not only understand themselves to live in a world of social regularities and causes for action. Human social cognition also gives rise to the phenomenon of *normativity* in thought and action, a fundamental notion in philosophy, but somewhat less noted in psychological research. For instance, humans understand themselves to have such things as obligations, commitments, rights, entitlements, social institutions, cultural knowledge, traditions, customs, mores, and rules. In other words, humans not only have (and think about) causes for belief and action. They also have and recognize *reasons* to believe certain things and reasons to act in certain ways (Raz 1999; Scanlon 1998; Searle 2001); and in psychological terms, this reason responsiveness might be based on the human-specific ability to take normative attitudes toward their own and others' thought and action (Schmidt and Rakoczy 2016). Normativity thus poses a problem not only conceptually and ontologically (e.g., What are normative facts and how do they relate to empirical facts? Brandom 1994; Sellars 1963), but also psychologically and empirically: how do humans, psychologically, come to integrate the two "worlds"—the realm of predictions and causes on the one hand and the realm of prescriptions and reasons on the other—in their everyday reasoning and acting? (Hitchcock and Knobe 2009; Kalish 2006). Despite the pervasiveness of normativity, there is little research on the developmental origins of understanding normativity. Here we review developmental research conducted over the last couple of years suggesting that even young children have some basic gasp of a variety of different normative phenomena. But two things should be said and clarified in advance. First, not all kinds of normativity are alike. Thus, we first provide a brief overview of different types of norms infants and children need to develop an understanding of. And second, not all kinds of verbal or nonverbal behaviors (e.g., imitation) are indicative of an understanding of normativity, nor practical if we are to investigate the roots of normativity in early human ontogeny. Hence, we briefly discuss the methodological question of how to measure whether a creature understands something about normativity.

Features and Types of Normativity

When talking about the normative, we mean something distinct from how the world *is*. In a general sense, we mean some ideal state in the world that can be attained or not. That is, there are conditions of success and failure regarding some state of affairs (Brandom

1994; Kripke 1982; McDowell 1984). Examples are mental states (which can successfully represent reality) and linguistic expressions (which can successfully be applied). Here, we are interested in normativity in a narrow sense, that is, in norms that set standards of correctness, come with normative force and authority, are valid both in general (agent-independent) and context-relative ways:

- *Standards of correctness:* a given human action in a given social interaction can be assessed as right or wrong according to some *standard* accepted by a given group of people (Hechter and Opp 2001; Popitz 2006). What this implies is that for an agent to be granted an understanding of norms, the agent must be capable of comparing (not necessarily in propositional terms) an observed action with an ideal "standard," an ideal act.
- *Generality:* norms entail some abstractness and general applicability, such that they are valid for any agent (including oneself) in equivalent circumstances—they are valid in *general, agent-independent* ways (Nagel 1986).
- *Normative force:* norms are peculiar phenomena in that they do not have brute physical force (e.g., the law of gravity makes us fall, but laws of logic do not knock us down) and they are distinct from mere coercion (e.g., performing an action because someone is holding a gun to your head)—rather, norms have *binding force* and authority over us and there is typically the possibility to violate them (Korsgaard 1996; Rousseau 1762/1997, pp. 43–4). Thus, we have normative expectations about what we (oneself and others) "ought" to do in a particular situation (Chudek and Henrich 2011; Gloor 2014). Crucially, we could do otherwise, but we think we should adhere to the norm. Normative expectations are to be kept distinct from descriptive expectations about how people "will" behave. Normative expectations come with motivational force and are about how people "should" behave. Therefore, descriptive expectations are typically said to have a mind-to-world direction of fit (analogous to epistemic states), whereas normative expectations are construed to have a world-to-mind direction of fit (analogous to volitional states; Christen and Glock 2012; Schmid 2011; Searle 1983).[1]
- *Context-relativity:* norms, such as standing in line in a grocery store, typically apply in one context but not in another. Norms are thus usually context-relative. Even more non-arbitrary norms (such as moral norms we discuss later) can be relative to context: for instance, it is somewhat fine to harm someone in a boxing match (opponents have a legitimate reason to harm each other), although it is usually forbidden to harm someone (without any reason). Note that the generality feature of norms is not opposed to context relativity, since for a norm to apply in general (i.e., in all contexts of a certain category) does not preclude that it applies only in certain contexts (i.e., not in other categories of contexts).

[1] Alternatively, if one follows accounts that stress the role of beliefs in explaining norms (e.g., Bicchieri 2006; Lewis 1969), normative expectations could be considered having a double direction of fit.

Having these key features of normativity at hand, let us look how one can categorize normative phenomena. Surely, not all norms are created equal. One way to delineate normative phenomena is to talk about *practical* and *theoretical* (or *epistemic*) normativity (Engel 2011; Littlejohn and Turri 2014). Practical norms pertain to human actions; they give reasons to act in certain ways, and are thus part and parcel of human cultural practices and values. Epistemic norms pertain to human beliefs; they give reasons to believe certain things, and are thus fundamental to our theoretical reasoning, cultural knowledge, and understanding of truth.[2] Our focus here is on practical norms.

Many subscribe to the view that there are different types of practical norms. Perhaps the most famous contrast is that between *conventional* norms and *moral* norms (Korsgaard 1996; Lewis 1969; Scanlon 1998; Turiel 1983, 2006). Conventional norms regulate, organize, and constitute social practices and are typically arbitrary (i.e., another form of behavior could have become the norm or the "equilibrium"). A common further distinction is often made between conventional norms that are constitutive of some (social) behavior and conventional norms that merely regulate pre-existing (social) behavior (e.g., greeting conventions, etiquette rules, traffic rules). Constitutive norms create new social and institutional facts by the formula "X counts as Y in context C" (Rawls 1955; Searle 1995, 2010)—and if collectively accepted, they have normative consequences and prescribe or proscribe certain actions for agents in certain roles (Searle 1995, 2010). All kinds of social institutions, such as money, marriage, and games, are constituted by constitutive norms. Moral norms (at least prototypical ones), however, are considered non-arbitrary, as they are about issues of well-being, justice, and rights (Turiel 1983, 2006). And perhaps some moral norms spread more easily because they capitalize on something prior to the norm like a predisposition to feel averse to harming others (Nichols 2004). By contrast, without etiquette rules that regulate ways of eating, people might simply use their hands. *Norms of instrumental rationality* are different from both conventional and moral norms, because here the focus is on the efficiency (or rationality) of a means-end relation: an agent ought to adopt the most efficient means to reach his or her end (Korsgaard 1997).[3] Interestingly, however, they can be considered wide in scope—similar to moral norms—in that they apply to any rational agent.

How to Measure Norm Sensitivity

A major question is how we can assess whether infants and children understand something about normativity. One strategy could be to investigate whether the young learner's behavior conforms to certain norms. But mere (accidental) acting in accordance with a

[2] However, the practical-epistemic distinction need not be understood categorically (Graham 2015; Littlejohn and Turri 2014). For some epistemic norms might be considered practical (perhaps even moral), such as the norm to give true and relevant information (Graham 2015; Rescorla 2007).

[3] Norms of instrumental rationality can be construed as governing both practical and theoretical reasoning and thus not only prescribe certain actions, but also certain beliefs (Kelly 2003).

norm is not indicative of truly following a norm based on an understanding that one's action is subject to a norm (Brandom 1994; Wittgenstein 1953/2001). That is, the acting in accordance with a norm does not reveal whether the child understands the important features of normativity outlined earlier (in particular, standards of correctness, normative force, and generality). One can act in accordance with a norm for many different reasons, such as preferring a course of action or being afraid of sanctions. Then, might it be better to just ask children whether a given action is right or wrong? This is of course an important approach. Interview studies based on Elliot Turiel and colleagues' social domain theory on children's judgment and reasoning about norm transgressions in hypothetical scenarios have revealed that even preschoolers make subtle distinctions between moral and conventional norms (and that with age, children are able to justify their judgments), such as that they consider prototypical moral transgressions as more severe and wrong independent of an authority's opinion (Killen and Smetana 2014; Smetana 2006; Turiel 1983, 2006; Turiel and Dahl 2016). Two caveats, however, need to be raised. First, interview techniques have their limits when it comes to investigating younger children's, infants', and nonhuman animals' understanding of norms. And second, they focus on children's knowledge about (prevailing) norms and do not directly assess children's understanding of the normative force of norms; for the normative force of normativity essentially reveals itself most clearly in social interactions when actual norm transgressions occur.[4] More specifically, an understanding of normativity with its main features can be assessed most convincingly by confronting an individual with an actual norm violation and testing whether the individual—as an unaffected observer—enforces the norm via critique, sanctioning, and the like. Precursors to such spontaneous third-party norm enforcement have been reported in a few studies, for instance, regarding infants spontaneous reactions to malfunctioning artifacts (Kagan 1981), or regarding two- and three-year-olds' spontaneous rejections of assertions that do not match reality (Pea 1982). Over the last couple of years, researchers have begun to systematically investigate young children's understanding of normativity in different domains and contexts using the method of spontaneous third-party norm enforcement. Overall, this research suggests between two and three years of age, children begin to show a robust understanding of different types of norms in a variety of contexts, and that they not only understand them cognitively, but also care about them motivationally (e.g., by upholding norms in cases of violation). In what follows, we will take a closer look at this research.

[4] This does not mean that statements and judgments in an interview have nothing to do with an individual's understanding of norms and actual behavior in social interactions. Turiel (2008), for instance, found that children make similar distinctions between moral and conventional transgressions for both real and hypothetical violations, and that their behavior in observed social interactions also corresponds to the moral-conventional distinction. This, however, does not obviate the need for a direct assessment of children's understanding of the normative force and generality of norms (see also Blasi 1983 for a discussion of the complex relation between moral cognition and behavior).

Children's Developing Understanding of Normativity

Conventional Normativity

The first experimental research on children's understanding of normativity focused on simple solitary cooperative rule games (i.e., each player is supposed to perform the same action) based on conventional (constitutive) norms. The first study tested two- and three-year-old children's understanding of such simple game rules by giving them the opportunity to spontaneously criticize a third party (a hand puppet) that violated the rules of the game (Rakoczy, Warneken, and Tomasello 2008). An adult model introduced the game by using normative language and novel words ("This game is called Daxing!"), and then the third-party puppet said that she was going to play the game (Daxing), too. The puppet, however, showed a different action, which was explicitly introduced as wrong before by the adult—thus, this act was against the constitutive norms of the game. Three-year-olds protested and criticized the puppet, often by using normative language (e.g., "This is wrong. One must do it like this."). They intervened less when the puppet said that she would show the child something (i.e., when not playing the game) and performed the same action. Two-year-olds showed the same general pattern of protest behavior, but used less normative language. This study provided the first evidence that young children understand that established constitutive norms have normative consequences for parties who engage in activities that are subject to these norms. In addition, children's disinterested enforcement of these norms suggests that they have some grasp of the three features of normativity, since they applied the norms to other participants of the social practice.

Rakoczy and colleagues (2009) followed up on this first study and looked more closely at young children's understanding of the context-relativity of conventional norms. They had young children (ages two and three) again play a simple game, but this time the action was prescribed at one location (Table A), not at another (Table B). Three-year-olds (but not two-year-olds) took into account the context-relativity of these game rules and intervened against third-party transgressions only when the action was wrong in a given context.

Pretense can be considered another paradigmatic case of conventional constitutive norms (Currie 1998; Rakoczy 2008a), since in a pretend game, players act as if a certain object were another object (e.g., using a banana as a telephone), and thereby treat object X as Y in game context C (Rakoczy 2008a). Rakoczy (2008b) had two- and three-year-old children play a simple pretend game. An adult demonstrated, for instance, that an object is to be treated as a knife in a pretend game. When a puppet pretended to eat the object (i.e., the knife), children protested. They did not protest, however, when the puppet pretended to eat an object that was designated as a carrot. In a subsequent study, it was found that three-year-olds, but not two-year-olds, are able to switch between different pretend identities in two game contexts. For instance, a yellow stick may count as a toothbrush in one game at one location and as a carrot in another game at a different location (Wyman, Rakoczy, and Tomasello 2009). Hence, by three years of age, children understand something about the context-relative bindingness of conventional norms.

The scope of conventional norms is not only relative in spatial, but also in sociocultural terms. That is, many conventional norms are group-relative (e.g., etiquette rules, currency). Schmidt, Rakoczy, and Tomasello (2012) investigated how three-year-old children understand the social scope of prototypical conventional norms (simple solitary game-like actions), moral norms (destroying someone's property), and norms of instrumental rationality (failing to use the necessary means to an end). Children did not treat all norm transgressions by all transgressors alike: for conventional norm transgressions, they criticized an in-group member more than an out-group individual, but for moral and instrumental norm transgressions, children protested equally against in-group and out-group violators. This suggests that children recognize that conventional norms are limited in scope to members of their own group, whereas they understand moral and instrumental norms to have a much wider scope (see later for further research on moral norms).

Norm Psychology, Intentionality, and Rationality

Are norm psychology and other forms of social cognition related from early in development? In their normative evaluation of an action, adults take into account the agent's intentionality, and they do so differentially for different types of norms (Giffin and Lombrozo 2015). Current research suggests that even children reason in such ways. One dimension on which conventional and moral norms differ is how much adults take into account an agent's freedom to act and other aspects of intentionality in normative assessment: If a soccer player is unable to reach the ball with his or her head and uses the hand instead (be it a reflex action or intentional), the referee will blow the whistle in any case (according to the constitutive norms of the game). Whether, however, a soccer player caused a severe injury to another player unintentionally or intentionally (e.g., because he or she is angry due to a prior foul), the referee would blow the whistle in both cases, but we would evaluate the situation differently in moral terms (i.e., less blame in case of unintentional harm). In a recent study, Josephs and colleagues (2016) found that young children (four-year-olds more so than three-year-olds) make this distinction between moral and conventional norms. They did not blame a third party for committing a moral violation when this agent was physically constrained, but they still criticized a violator of conventional norms who was under physical constraint (although less than a violator under no constraint; see also Tunçgenç, Hohenberger, and Rakoczy 2015, for similar findings with Turkish children).

All studies on children's understanding of conventional norms reported so far were concerned with (solitary) cooperative games, that is, with activities in which all participants of a social practice have the same goal and are supposed to do something in the same way (without any need for simultaneous coordination). Many human institutionalized practices, however, are characterized by a friendly juxtaposition of cooperation and competition. In a competitive game, for instance, players jointly intend to compete within a cooperative framework, that is, a set of constitutive norms. And, importantly, opponents in a competitive game expect each other to try to win and thus to employ rational game-playing strategy. This, however, means that a player has to coordinate normative expectations about her opponent's rational game-playing, the constitutive norms of the game, and her own goal to win. A purely egocentric player should actually applaud an opponent who plays irrationally, since this is beneficial to the egocentric player's goal attainment. Schmidt, Hardecker, and Tomasello (2016) investigated whether preschoolers (three- and five-year-olds) form such normative expectations about rational game-playing in a simple two-player competitive game. Children played against a puppet, and sometimes the puppet helped children to get closer to winning the game. Five-year-olds protested irrational play regardless of whether their opponent adhered to the constitutive norms of the game or not. Three-year-olds showed a more ambiguous protest pattern. This study thus suggests that even preschoolers understand something about the bindingness of cooperatively structured competition.

Second, and less intuitively, children's ascription of intentionality to an agent is influenced by their normative assessment of her behavior. In particular, recent research has shown that children from age four, much like adults, are subject to the so-called "side-effect effect" (Knobe 2003), interpreting the bringing about of foreseen but unintended side effects as more intentional when they are negative than when they are positive (Leslie, Knobe, and Cohen 2006; Pellizzoni, Siegal, and Surian 2009; Rakoczy et al. 2015). Overall, these studies suggest that from early in development, children's understanding of normativity and other forms of social cognition are intimately related and well integrated (see also Smetana, Jambon et al. 2012, for reciprocal relations between children's moral judgment and theory of mind).

Norm Learning Mechanisms

Besides investigating young children's understanding of different types of norms and interrelations between normativity and theory of mind, researchers have begun looking at mechanisms of norm learning. That is, the young learner needs to solve an epistemological problem: on which basis shall he or she infer that a single observed action is subject to norms (and thus generalizable) as opposed to an idiosyncratic action (and thus not generalizable)? In real life, infants and children observe many actions that are not accompanied by explicit language and instruction (e.g., that "this is the way we do it"), even more so in non-Western cultures (Lancy 1996; Rogoff 2003). We note that there is a rich literature on children's sociocultural learning (e.g., assessing children's imitation of others' actions) that is beyond the scope of this chapter (see, e.g., Legare and Harris 2016; Legare and Nielsen 2015; Tomasello 2016).

When it comes to learning norms, reliability and competence, for instance, are important social-epistemic cues. Rakoczy, Warneken, and Tomasello (2009) found that four-year-olds selectively learn rule games from reliable models (e.g., who previously labeled objects correctly) over unreliable models and that children formed normative expectations about the way the games were played. Thus, when a third party violated the rules of the game (as demonstrated by the reliable model), children protested and corrected the deviator. Competence might also be expressed in mere age differences. In a different study, three- and four-year-olds watched as an adult and a peer model performed two game-like actions in different ways (Rakoczy et al. 2010). Children at both ages preferred to imitate the action performed by the adult, and, crucially, they also attributed normativity to the adult's action: when a third-party puppet deviated from the demonstrated action (the puppet performed the action the peer had demonstrated), they criticized the puppet, but they did not protest when the puppet performed the adult's action.

But what if children incidentally observe an adult who performs a new action on some artifacts, but does not tell the child that this action is the right way to do things? Schmidt, Rakoczy, and Tomasello (2011) explored this question and found that children at age three attribute normativity and generality to novel game-like acts when observing an adult who intentionally and confidently performed these actions. Importantly, the adult did not explicitly teach children anything or address them. Children nevertheless attributed normativity to the action and later protested against a third-party puppet that performed a deviating action. In a control condition, children inferred significantly less normativity when the adult performed the action as if she invented it on the spot, although even in this context, some children protested against the puppet. It is possible that children have a natural tendency to "promiscuously" impute normativity to others' intentional actions similar to their propensity to attribute purpose to objects and others' actions and minds more generally (Kelemen 1999, 2004). A recent study provides evidence for such promiscuous normativity in young children (Schmidt et al. 2016): three-year-olds incidentally witnessed an unknown adult who, in one experiment, spontaneously took some junk objects out of a trash bag. The person then performed a brief idiosyncratic, arbitrary, and intentional action without obvious purpose (e.g., taking a damaged snail shell and pushing it a bit forward with a piece of wood). In another condition, the adult used pedagogical cues ("Look!") before performing the action. Thus, the evidence spoke against the possibility that this act was subject to any norms. Nonetheless, children even normalized such singular and individual behavior (both in the incidental observation and in the pedagogical context) unless it was marked as an accident (as in a control condition). That is, they protested and intervened against a puppet that performed a slightly different action (reaching a similar goal, but in a different way) with the junk objects. Hence, it seems that young children have a strong tendency to violate Hume's law, that is, to go from "is" to "ought" (Hume 1739/2000) and to construct social rules out of the blue. Although pedagogy did not make a difference in the studies discussed, it might still be a catalyst for normative learning in other situations—for instance, with respect to the strength of normative learning. An experiment found that young children are more resistant to counter-evidence when they have learned conventional norms pedagogically for their benefit than when they merely incidentally observed an adult performing an action that is subject to conventional norms (Butler et al. 2015).

Promiscuous normativity may be a mechanism important for explaining children's tendency to overimitate, that is, to imitate adult actions that are not necessary to reach a goal. Recent studies suggest that young children's overimitation is at least partly normatively motivated in that they think that even unnecessary actions are supposed to be performed. Kenward (2012), for instance, found that three- and five-year-old children who learned instrumental actions (necessary to achieve a goal) and some unnecessary actions (not necessary to achieve a goal) protested against a third-party puppet that omitted the unnecessary acts. In a further study (Keupp, Behne, and Rakoczy, 2013), three- and five-year-olds criticized a puppet more that omitted irrelevant actions when they had learned the actions in a conventional context (e.g., "This game is called Daxing!") than when they had learned the actions in a means-end context (i.e., the adult emphasized the goal of the action sequence, such as ringing some bells). This suggests that children attributed normativity to these irrelevant actions, presumably because they inferred that they were also part of the conventional activity. Furthermore, recent studies found that children's overimitation is not automatic, but rather flexible and rational: children criticize a third-party puppet less when she does not perform irrelevant actions in a novel context (Keupp et al. 2015), but more when the irrelevant actions cause harm (destruction of an adult's belongings; Keupp et al. 2016).

In sum, these findings suggest that young children's norm learning is far from being a passive process—children actively seek out norms, are highly motivated to identify actions that are valid beyond the here and now, and use social-pragmatic and epistemic cues in rational and selective ways to make the inductive leap that some behavior is subject to norms.

Ontology of Norms

A thorough understanding of normativity not only requires children to follow and enforce norms in rational and context-relative ways, but also to learn that norms are essentially human-made social facts that can be changed or brought into existence under certain conditions (e.g., by collectively aligning our beliefs, desires, and intentions). Thus, children face the developmental task to learn about the social nature of norms. One mechanism by which norms can come into existence is agreement among a local group of people. Schmidt and colleagues (2016b) investigated under which conditions three-year-old children understand arbitrary game rules as established and valid. If all participants (several puppets and the child) agreed upon a game rule, children enforced this rule on deviators. If, however, there was dissent during the norm-setting process, children failed to see a norm as established for anyone at all, not even for people who had agreed—even a majority of 90 percent would not create a norm. This suggests that even young children understand something of the role of agreement in creating norms, but that their early grasp of the ontology of norms is confined to conditions of unanimity.

Another study looked at spontaneous norm creation in five-year-old peers (Göckeritz, Schmidt, and Tomasello 2014). Five-year-old children worked together on an apparatus in order to achieve a shared goal (getting some rewards). Children co-constructed their own norms for coordination (including assignment of roles) and thereby regulated their interaction. When paired with novice peers, children transmitted their created norms as objective facts using generic normative language (e.g., "One should do it like this!") instead of renegotiating how to coordinate, suggesting that they reified the co-constructed norms as if they had discovered them (see also Köymen et al. 2014, 2015, for children's use of generic normative language and tendency to objectify norms).

Norms in Language Use

Our everyday use of language is governed by norms, too. And language is an especially interesting case regarding the world of causes and the world of reasons, since some types of speech acts (assertions) are used to merely describe the causal world, while others (imperatives) are used to change the causal world (Searle 1969, 1983). Both types of speech acts, however, are assessed by human speakers within the normative world of reasons. How do young children, then, make sense of different types of speech act (i.e., different directions of fit)? Rakoczy and Tomasello (2009) assessed whether three-year-olds understand this structural difference between assertions and imperatives and found that children protested against a commentator who asserted that an actor was performing a certain action (although this was not the case), but that they protested against the actor if she was not doing what the commentator told her to do. In another study, four-year-olds showed an understanding of the normativity of future-directed speech acts (Lohse et al. 2014), such that they recognized that a speaker made a mistake when her prediction ("A will do X") did not come true, but that an actor made a mistake when she did not follow an imperative with the same content that had been given earlier by a speaker.

Moral Normativity

Perhaps the most famous kind of normativity is moral normativity. A proper treatment of moral norms would go beyond the scope of this chapter. We therefore confine ourselves to briefly discussing current work on children's understanding of the binding force and generality of moral norms. But before doing so, we should note that over the last couple of years, researchers have accumulated evidence that even infants have prosocial preferences (e.g., for helping over hindering agents; see Hamlin 2013 for a review), descriptive expectations about equality (third-party fairness) in resource allocation (Geraci and Surian 2011; Schmidt and Sommerville 2011; Sloane, Baillargeon, and Premack 2012), and empathic tendencies toward others in distress (Svetlova, Nichols, and Brownell 2010; Vaish, Carpenter, and Tomasello 2009)—a suite of cognitive and motivational tendencies that may plausibly ground morality proper (Jensen, Vaish, and Schmidt 2014; Roughley 2016). It is not clear yet, however, how exactly these early capacities relate to young children's developing understanding of moral norms (in particular, regarding the features' normative force and generality). One finding worth noting is that infants' early descriptive expectations about third-party fairness are related to their own prosocial sharing behavior (Schmidt and Sommerville 2011; Sommerville et al. 2013; see also Dahl, Schuck, and Campos 2013): that is, infants who engage in costly sharing (giving away a toy they like) are more concerned about third-party fairness than infants who engage in non-costly sharing (giving away a toy they do not prefer). This interrelation opens the possibility—to be investigated further—that the development of normative expectations (beyond purely descriptive ones) about morally relevant actions is fostered by other-regard and sympathy.

Regarding a more mature understanding of moral normativity, researchers have found that young children at age three protest violations of moral norms, for instance, when a third party harms someone by destroying or throwing away her property (Rossano, Rakoczy, and Tomasello 2011; Schmidt, Rakoczy, and Tomasello 2012; Vaish, Missana, and Tomasello 2011). And with regard to distributive justice, preschoolers start to enforce the norm of equality (Rakoczy, Kaufmann, and Lohse, submitted), and at early school age, children begin to understand that sometimes inequality is normatively justified, such as when one individual is needier or more meritorious than another (Schmidt et al. 2016c). Moreover, young children's understanding of moral normativity goes beyond the notion of obligation and extends to issues of rights and entitlements. For instance, Schmidt, Rakoczy, and Tomasello (2013) found that three-year-olds defend an actor's entitlement (e.g., to play with a toy) against someone else who threatens the actor's entitlement (i.e., children actively intervene and show some early form of moral courage; Baumert, Halmburger, and Schmitt 2013).

In sum, the research reported here suggests that even young children have some basic grasp of a variety of normative phenomena, apply norms in context-specific ways, and, with age, become more flexible in their understanding of norms, including their social ontology. Young children's selective and rational third-party enforcement of norms provides evidence that they understand the main features of normativity (standards of correctness, normative force, generality, context relativity) and also care about normativity.

Conclusion and Outlook

Humans share with many other species, notably primates, basic capacities for representing the natural and social world around them in terms of enduring objects governed by natural regularities. But human cognition seems unique in being "fraught with ought" (Sellars 1963): it is concerned not only with what is the case, tends to happen, or occurs regularly, but with normative questions of what is appropriate or correct and what ought to be done. The developmental research reviewed in this chapter suggests that basic forms of normative cognition emerge early in human ontogeny: from as early as two to three years of age, children understand and enforce simple social norms governing conventional activities such as games and language use. They learn such norms swiftly, sometimes overeagerly, from observing others, and their normative assumptions themselves play important roles in imitation and other forms of social learning. And generally, normative cognition and other forms of social cognition seem to be intimately related from early in development.

From this developmental research, we have thus learned about our early developing norm psychology and its relation to other cognitive capacities. But fundamental questions remain open for future inquiry: How does this norm psychology emerge ontogenetically? What are its phylogenetic origins, and what its cognitive foundations? One promising avenue, in our view, will be to investigate different forms of norm psychology in relation to different forms of intentionality. Arguably, every form of individual intentionality as such already brings with it basic forms of normativity mentioned at the outset of this chapter: intentional beings are subject to correctness conditions of belief, for example, and to success conditions of action (e.g., Burge 2009; Hurley 2003). But the more complex forms of norm psychology under review in this chapter—norm psychology in which the agent herself is not only subject to norms, but has some grasp of them and some stake in enforcing them—quite plausibly are the upshot of more complex and more social forms of intentionality, in particular, shared or collective intentionality (Rakoczy and Tomasello 2007; Schmidt and Rakoczy 2016; Schmidt and Tomasello 2012; Tomasello 2014; Tomasello and Rakoczy 2003). Intuitively, the most basic forms of shared or collective intentionality involve two or more agents acting in ways that transcend purely individual intentions and actions, intending "that we . . ." (e.g., dance tango, take a walk together, lift a table together, play ball together; Bratman 1992; Gilbert 1989; Searle 1990; Tuomela and Miller 1988). Even the most basic and mundane forms of shared intentionality such as taking a walk together establish new and more social forms of normativity: individual agents are now not only subject to normative assessment in terms of success or failure vis-à-vis their own goals, but subjects of (and subject to) normative expectations toward each other: when we have committed ourselves to taking a walk together, each of us is now committed to fulfilling her part in this project and subject to critique in case of deviation (e.g., "Hey, you can't just go shopping without any explanation—we are taking a walk together!"). More complex forms of shared intentionality involve not only coordination of shared activities, but the conventional creation of new, so-called "institutional" facts (e.g., Searle 1995). Shared intentional practices of playing games or speaking languages, for example, create such new facts as "this is checkmate," "this figure is a king," "'dog' means dog"—observer-dependent facts that are not out there as facts about the natural world, but only hold because we as participants of the practice take them to hold. And such facts have inherent normative implications: when something is a king in chess, it licenses certain movements; when some sound pattern refers to dogs in a given language, it licenses and requires certain usage.

From an ontogenetic point of view, basic forms of shared intentionality seem to develop from the second year of life: from 12 to 18 months, children begin to engage in simple cooperative activities, both instrumental and playful, with others involving preverbal indicators of true shared intentionality such as coordination, communication, division of labor, and role reversal. More complex forms of shared intentionality with conventional fact-creation emerge from the end of the second year, in particular, in the form of joint pretense and other games. From this time on, children also show the first signs of actively tracking and enforcing the socially constituted norms of such practices (Rakoczy 2008a; Rakoczy, Warneken, and Tomasello 2008; Schmidt, Rakoczy, and Tomasello 2011).

From a comparative point of view, while social coordination is, of course, widespread in the animal kingdom, to date there is no clear and convincing evidence that nonhuman animals, even great apes, engage in anything like proper cooperation involving joint goals and coordinated roles (Tomasello 2014; Tomasello et al. 2012). Similarly, although chimpanzee groups do have something like behavioral traditions and culture sensu lato (Boesch 2012), to date there is no evidence that chimpanzees understand certain behaviors as enforceable generic types (Rudolf von Rohr, Burkart, and van Schaik 2011; Schmidt and Rakoczy 2016), and it is possible that these behavioral regularities are based on individual learning strategies or genetic variability (Langergraber et al. 2011; Tennie, Call, and Tomasello and 2009). In general, more complex forms of shared intentionality involving shared conventional practices and fact creation seem quite clearly to be a unique human capacity (Rudolf von Rohr et al. 2011; Schmidt and Rakoczy 2016).

So, one picture that is worth being explored more systematically in future research is that while humans and other species, notably primates, share basic forms of individual intentionality (and the corresponding natural norms of correctness and success), uniquely human forms of norm psychology and uniquely human forms of shared intentionality develop in close tandem in early ontogeny, the former building on and growing out of the latter.

References

Andrews, K. (2012). *Do apes read minds? Toward a new folk psychology*. Cambridge, MA: MIT Press.

Apperly, I.A. and Butterfill, S.A. (2009). Do humans have two systems to track beliefs and belief-like states? *Psychological Review*, 116(4), 953–70. doi:10.1037/a0016923

Baillargeon, R. and Carey, S. (2012). Core cognition and beyond: the acquisition of physical and numerical knowledge. In: S. Pauen (ed.), *Early childhood development and later outcome.* New York: Cambridge University Press, pp. 33–65.

Baumert, A., Halmburger, A., and Schmitt, M. (2013). Interventions against norm violations: dispositional determinants of self-reported and real moral courage. *Personality and Social Psychology Bulletin*, 39(8), 1053–68. doi:10.1177/0146167213490032

Bicchieri, C. (2006). *The grammar of society: the nature and dynamics of social norms.* Cambridge: Cambridge University Press.

Blasi, A. (1983). Moral cognition and moral action: a theoretical perspective. *Developmental Review*, 3(2), 178–210. doi:10.1016/0273-2297(83)90029-1

Boesch, C. (2012). *Wild cultures: a comparison between chimpanzee and human cultures.* Cambridge, MA: Cambridge University Press.

Brandom, R.B. (1994). *Making it explicit.* Cambridge, MA: Harvard University Press.

Bratman, M.E. (1992). Shared cooperative activity. *Philosophical Review*, 101(2), 327–41.

Burge, T. (2009). Primitive agency and natural norms. *Philosophy and Phenomenological Research*, 79(2), 251–78.

Butler, L.P., Schmidt, M.F.H., Bürgel, J., and Tomasello, M. (2015). Young children use pedagogical cues to modulate the strength of normative inferences. *British Journal of Developmental Psychology*, 33(4), 476–88. doi:10.1111/bjdp.12108

Butterfill, S.A. and Apperly, I.A. (2013). How to construct a minimal theory of mind. *Mind & Language*, 28(5), 606–37. doi:10.1111/mila.12036

Call, J. and Tomasello, M. (2005). Reasoning and thinking in nonhuman primates. In: K. Holyoak and B. Morrison (eds.), *The Cambridge handbook of thinking and reasoning.* Cambridge University Press, pp. 607–32.

Call, J. and Tomasello, M. (2008). Does the chimpanzee have a theory of mind? 30 years later. *Trends in Cognitive Sciences*, 12(5), 187–92.

Carruthers, P. (2013). Animal minds are real, (distinctively) human minds are not. *American Philosophical Quarterly*, 50(3), 233–48.

Christen, M. and Glock, H.-J. (2012). The (limited) space for justice in social animals. *Social Justice Research*, 25(3), 298–326. doi:10.1007/s11211-012-0163-x

Chudek, M. and Henrich, J. (2011). Culture–gene coevolution, norm-psychology and the emergence of human prosociality. *Trends in Cognitive Sciences*, 15(5), 218–26. doi:10.1016/j.tics.2011.03.003

Currie, G. (1998). Pretence, pretending and metarepresenting. *Mind & Language*, 13(1), 35–55. doi:10.1111/1468-0017.00064

Dahl, A., Schuck, R.K., and Campos, J.J. (2013). Do young toddlers act on their social preferences? *Developmental Psychology*, 49(10), 1964–70. doi:10.1037/a0031460

Engel, P. (2011). Epistemic norms. In: S. Bernecker and D. Pritchard (eds.), *The Routledge companion to epistemology*. New York: Routledge, pp. 47–57.

Geraci, A. and Surian, L. (2011). The developmental roots of fairness: infants' reactions to equal and unequal distributions of resources. *Developmental Science*, 14(5), 1012–20. doi:10.1111/j.1467-7687.2011.01048.x

Giffin, C. and Lombrozo, T. (2015). Mental states are more important in evaluating moral than conventional violations. In: R. Dale, C. Jennings, P. Maglio, T. Matlock, D. Noelle, A. Warlaumont et al. (eds.), *Proceedings of the 37th annual conference of the Cognitive Science Society*. Austin, TX: Cognitive Science Society, pp. 800–805.

Gilbert, M. (1989). *On social facts.* London: Routledge.

Gloor, J. (2014). Collective intentionality and practical reason. In: A. Konzelmann and H.B. Schmid (eds.), *Institutions, emotions, and group agents: contributions to social ontology*. Dordrecht, The Netherlands: Springer Verlag, pp. 297–312.

Göckeritz, S., Schmidt, M.F.H., and Tomasello, M. (2014). Young children's creation and transmission of social norms. *Cognitive Development*, 30, 81–95. doi:10.1016/j.cogdev.2014.01.003

Graham, P.J. (2015). Epistemic normativity and social norms. In: D. Henderson and J. Greco (eds.), *Epistemic evaluation: purposeful epistemology*. New York: Oxford University Press, pp. 247–73.

Hamlin, J.K. (2013). Moral judgment and action in preverbal infants and toddlers: evidence for an innate moral core. *Current Directions in Psychological Science*, 22(3), 186–93. doi:10.1177/0963721412470687

Hechter, M. and Opp, K.D. (2001). *Social norms*. New York: Russell Sage Foundation.

Hitchcock, C. and Knobe, J. (2009). Cause and norm. *Journal of Philosophy*, 11, 587–612.

Hume, D. (1739/2000). *A treatise of human nature* (ed. D.F. Norton and M.J. Norton). Oxford: Oxford University Press.

Hurley, S.L. (2003). Animal action in the space of reasons. *Mind and Language*, 18(3), 231–56.

Jensen, K., Vaish, A., and Schmidt, M.F.H. (2014). The emergence of human prosociality: aligning with others through feelings, concerns, and norms. *Frontiers in Psychology*, 5, 822. doi:10.3389/fpsyg.2014.00822

Josephs, M., Kushnir, T., Gräfenhain, M., and Rakoczy, H. (2016). Children protest moral and conventional violations more when they believe actions are freely chosen. *Journal of Experimental Child Psychology*, 141, 247–55. doi:10.1016/j.jecp. 2015.08.002

Kagan, J. (1981). *The second year: the emergence of self-awareness*. Cambridge, MA: Harvard University Press.

Kalish, C.W. (2006). Integrating normative and psychological knowledge: what should we be thinking about? *Journal of Cognition & Culture*, 6(1/2), 191–208. doi:10.1163/156853706776931277

Kelemen, D. (1999). The scope of teleological thinking in preschool children. *Cognition*, 70(3), 241–72.

Kelemen, D. (2004). Are children "intuitive theists"?: Reasoning about purpose and design in nature. *Psychological Science*, 15(5), 295–301. doi:10.1111/j.0956-7976.2004.00672.x

Kelly, T. (2003). Epistemic rationality as instrumental rationality: a critique. *Philosophy and Phenomenological Research*, 66(3), 612–40. doi:10.1111/j.1933-1592.2003.tb00281.x

Kenward, B. (2012). Over-imitating preschoolers believe unnecessary actions are normative and enforce their performance by a third party. *Journal of Experimental Child Psychology*, 112(2), 195–207. doi:10.1016/j.jecp. 2012.02.006

Keupp, S., Bancken, C., Schillmöller, J., Rakoczy, H., and Behne, T. (2016). Rational over-imitation: preschoolers consider material costs and copy causally irrelevant actions selectively. *Cognition*, 147, 85–92. doi:10.1016/j.cognition.2015.11.007

Keupp, S., Behne, T., and Rakoczy, H. (2013). Why do children overimitate? Normativity is crucial. *Journal of Experimental Child Psychology*, 116(2), 392–406. doi:10.1016/j.jecp. 2013.07.002

Keupp, S., Behne, T., Zachow, J., Kasbohm, A., and Rakoczy, H. (2015). Over-imitation is not automatic: context sensitivity in children's overimitation and action interpretation of causally irrelevant actions. *Journal of Experimental Child Psychology*, 130, 163–75. doi:10.1016/j.jecp. 2014.10.005

Killen, M. and Smetana, J.G. (eds.) (2014). *Handbook of moral development* (2nd ed.). New York: Psychology Press.

Knobe, J. (2003). Intentional action and side effects in ordinary language. *Analysis*, 63(279), 190–4. doi:10.1111/1467-8284.00419

Korsgaard, C.M. (1996). *The sources of normativity*. Cambridge, UK: Cambridge University Press.

Korsgaard, C.M. (1997). The normativity of instrumental reason. In: G. Cullity and B. Gaut (eds.), *Ethics and practical reason*. Oxford: Clarendon Press, pp. 215–54.

Köymen, B., Lieven, E., Engemann, D.A., Rakoczy, H., Warneken, F., and Tomasello, M. (2014). Children's norm enforcement in their interactions with peers. *Child Development*, 85(3), 1108–22. doi:10.1111/cdev.12178

Köymen, B., Schmidt, M.F.H., Rost, L., Lieven, E., and Tomasello, M. (2015). Teaching versus enforcing game rules in preschoolers' peer interactions. *Journal of Experimental Child Psychology*, 135, 93–101. doi:10.1016/j.jecp. 2015.02.005

Kripke, S.A. (1982). *Wittgenstein on rules and private language. an elementary exposition*. Cambridge, MA: Harvard University Press.

Lancy, D.F. (1996). *Playing on the mother-ground: cultural routines for children's development*. New York: Guilford Press.

Langergraber, K.E., Boesch, C., Inoue, E., Inoue-Murayama, M., Mitani, J.C., Nishida, T. et al. (2011). Genetic and "cultural" similarity in wild chimpanzees. *Proceedings of the Royal Society B: Biological Sciences*, 278(1704), 408–16. doi:10.1098/rspb.2010.1112

Legare, C.H. and Harris, P.L. (2016). The ontogeny of cultural learning. *Child Development*, 87(3), 633–42. doi:10.1111/cdev.12542

Legare, C.H. and Nielsen, M. (2015). Imitation and innovation: The dual engines of cultural learning. *Trends in Cognitive Sciences*, 19(11), 688–99. doi:10.1016/j.tics.2015.08.005

Leslie, A.M., Knobe, J., and Cohen, A. (2006). Acting intentionally and the side-effect effect: theory of mind and moral judgment. *Psychological Science*, 17(5), 421–7. doi:10.1111/j.1467-9280.2006.01722.x

Lewis, D.K. (1969). *Convention: a philosophical study*. Cambridge, MA: Harvard University Press.

Littlejohn, C. and Turri, J. (eds.) (2014). *Epistemic norms: new essays on action, belief and assertion*. Oxford, UK: Oxford University Press.

Lohse, K., Gräfenhain, M., Behne, T., and Rakoczy, H. (2014). Young children understand the normative implications of future-directed speech acts. *PLOS ONE*, 9(1), e86958. doi:10.1371/journal.pone.0086958

McDowell, J. (1984). Wittgenstein on following a rule. *Synthese*, 58(3), 325–63. doi:10.1007/BF00485246

Nagel, T. (1986). *The view from nowhere*. New York: Oxford University Press.

Nichols, S. (2004). *Sentimental rules: on the natural foundations of moral judgment*. Oxford: Oxford University Press.

Pea, R.D. (1982). Origins of verbal logic: spontaneous denials by two- and three-year olds. *Journal of Child Language*, 9(3), 597–626.

Pellizzoni, S., Siegal, M., and Surian, L. (2009). Foreknowledge, caring, and the side-effect effect in young children. *Developmental Psychology*, 45(1), 289–95. doi:10.1037/a0014165

Penn, D.C. and Povinelli, D.J. (2007). On the lack of evidence that non-human animals possess anything remotely resembling a "theory of mind." *Philosophical transactions of the Royal Society of London. Series B, Biological sciences*, 362(1480), 731–44.

Popitz, H. (2006). *Soziale Normen* (ed. F. Pohlmann and W. Essbach). Frankfurt am Main: Suhrkamp.

Rakoczy, H. (2008a). Pretence as individual and collective intentionality. *Mind & Language*, 23(5), 499–517. doi:10.1111/j.1468-0017.2008.00357.x

Rakoczy, H. (2008b). Taking fiction seriously: young children understand the normative structure of joint pretence games. *Developmental Psychology*, 44(4), 1195–1201. doi:10.1037/0012-1649.44.4.1195

Rakoczy, H. (2014). Comparative metaphysics: the development of representing natural and normative regularities in human and non-human primates. *Phenomenology and the Cognitive Sciences*, 14(4), 683–97. doi:10.1007/s11097-014-9406-7

Rakoczy, H. (2015). In defense of a developmental dogma: children acquire propositional attitude folk psychology around age 4. *Synthese*, 1–19. doi:10.1007/s11229-015-0860-8

Rakoczy, H., Behne, T., Clüver, A., Dallmann, S., Weidner, S., and Waldmann, M.R. (2015). The side-effect effect in children is robust and not specific to the moral status of action effects. *PLoS ONE*, 10(7), e0132933. doi:10.1371/journal.pone.0132933

Rakoczy, H., Brosche, N., Warneken, F., and Tomasello, M. (2009). Young children's understanding of the context relativity of normative rules in conventional games. *British Journal of Developmental Psychology*, 27, 445–56. doi:10.1348/026151008X337752

Rakoczy, H., Hamann, K., Warneken, F., and Tomasello, M. (2010). Bigger knows better: young children selectively learn rule games from adults rather than from peers. *British Journal of Developmental Psychology*, 28(4), 785–98. doi:10.1348/026151009X479178

Rakoczy, H., Kaufmann, M., and Lohse, K. (submitted). Young children understand the normative force of standards of equal resource distribution.

Rakoczy, H. and Tomasello, M. (2007). The ontogeny of social ontology: steps to shared intentionality and status functions. In: S.L. Tsohatzidis (ed.), *Intentional acts and institutional facts: essays on John Searle's* Social Ontology. Berlin: Springer Verlag, pp. 113–37.

Rakoczy, H. and Tomasello, M. (2009). Done wrong or said wrong? Young children understand the normative directions of fit of different speech acts. *Cognition*, 113(2), 205–12. doi:10.1016/j.cognition.2009.07.013

Rakoczy, H., Warneken, F., and Tomasello, M. (2008). The sources of normativity: young children's awareness of the normative structure of games. *Developmental Psychology*, 44(3), 875–81. doi:10.1037/0012-1649.44.3.875

Rakoczy, H., Warneken, F., and Tomasello, M. (2009). Young children's selective learning of rule games from reliable and unreliable models. *Cognitive Development*, 24, 61–9. doi:10.1016/j.cogdev.2008.07.004

Rawls, J. (1955). Two concepts of rules. *The Philosophical Review*, 64(1), 3–32. doi:10.2307/2182230

Raz, J. (1999). *Practical reason and norms* (2nd ed.). Oxford: Oxford University Press.

Rescorla, M. (2007). A linguistic reason for truthfulness. In: D. Greimann and G. Siegwart (eds.), *Truth and speech acts: studies in the philosophy of language* (vol. 69). New York: Routledge, pp. 250–79.

Rogoff, B. (2003). *The cultural nature of human development*. Oxford: Oxford University Press.

Rossano, F., Rakoczy, H., and Tomasello, M. (2011). Young children's understanding of violations of property rights. *Cognition*, 121(2), 219–27. doi:10.1016/j.cognition.2011.06.007

Roughley, N. (2016). Moral normativity from the outside in. In: K. Bayertz and N. Roughley (eds.), *The normative animal? On the anthropological significance of social, moral and linguistic norms*. New York: Oxford University Press.

Rousseau, J.-J. (1762/1997). *The social contract and other later political writings* (ed. V. Gourevitch). Cambridge, UK: Cambridge University Press.

Rudolf von Rohr, C., Burkart, J.M., and van Schaik, C.P. (2011). Evolutionary precursors of social norms in chimpanzees: a new approach. *Biology and Philosophy*, 26(1), 1–30.

Scanlon, T.M. (1998). *What we owe to each other*. Cambridge, MA: Harvard University Press.

Schmid, H.B. (2011). The idiocy of strategic reasoning: towards an account of consensual action. *Analyse & Kritik*, 33(1), 35–56.

Schmidt, M.F.H., Butler, L.P., Heinz, J., and Tomasello, M. (2016a). Young children see a single action and infer a social norm: promiscuous normativity in 3-year-olds. *Psychological Science*, 27(10), 1360–1370. doi:10.1177/0956797616661182

Schmidt, M.F.H., Hardecker, S., and Tomasello, M. (2016). Preschoolers understand the normativity of cooperatively structured competition. *Journal of Experimental Child Psychology*, 143, 34–47. doi:10.1016/j.jecp. 2015.10.014

Schmidt, M.F.H. and Rakoczy, H. (2016). On the uniqueness of human normative attitudes. In: K. Bayertz and N. Roughley (eds.), *The normative animal? On the anthropological significance of social, moral and linguistic norms*. New York: Oxford University Press.

Schmidt, M.F.H., Rakoczy, H., Mietzsch, T., and Tomasello, M. (2016b). Young children understand the role of agreement in establishing arbitrary norms—but unanimity is key. *Child Development*, 87(2), 612-626. doi:10.1111/cdev.12510

Schmidt, M.F.H., Rakoczy, H., and Tomasello, M. (2011). Young children attribute normativity to novel actions without pedagogy or normative language. *Developmental Science*, 14(3), 530–9. doi:10.1111/j.1467-7687.2010.01000.x

Schmidt, M.F.H., Rakoczy, H., and Tomasello, M. (2012). Young children enforce social norms selectively depending on the violator's group affiliation. *Cognition*, 124(3), 325–33. doi:10.1016/j.cognition.2012.06.004

Schmidt, M.F.H., Rakoczy, H., and Tomasello, M. (2013). Young children understand and defend the entitlements of others. *Journal of Experimental Child Psychology*, 116(4), 930–44. doi:10.1016/j.jecp. 2013.06.013

Schmidt, M.F.H. and Sommerville, J.A. (2011). Fairness expectations and altruistic sharing in 15-month-old human infants. *PLoS ONE*, 6(10), e23223. doi:10.1371/journal.pone.0023223

Schmidt, M.F.H., Svetlova, M., Johe, J., and Tomasello, M. (2016c). Children's developing understanding of legitimate reasons for allocating resources unequally. *Cognitive Development*, 37, 42–52. doi:10.1016/j.cogdev.2015.11.001

Schmidt, M.F.H. and Tomasello, M. (2012). Young children enforce social norms. *Current Directions in Psychological Science*, 21(4), 232–6. doi:10.1177/0963721412448659

Searle, J.R. (1969). *Speech acts: an essay in the philosophy of language*. Cambridge: Cambridge University Press.

Searle, J.R. (1983). *Intentionality: an essay in the philosophy of mind*. Cambridge: Cambridge University Press.

Searle, J.R. (1990). Collective intentions and actions. In: P. Cohen, J. Morgan, and M.E. Pollack (eds.), *Intentions in communication*. Cambridge, MA: MIT Press, pp. 401–15.

Searle, J.R. (1995). *The construction of social reality*. New York: Free Press.

Searle, J.R. (2001). *Rationality in action*. Cambridge, MA: MIT Press.

Searle, J.R. (2010). *Making the social world*. Oxford: Oxford University Press.

Sellars, W. (1963). *Science, perception and reality*. London: Routledge & Kegan Paul.

Sloane, S., Baillargeon, R., and Premack, D. (2012). Do infants have a sense of fairness? *Psychological Science*, 23(2), 196–204. doi:10.1177/0956797611422072

Smetana, J.G. (2006). Social-cognitive domain theory: consistencies and variations in children's moral and social judgments. In: M. Killen and J.G. Smetana (eds.), *Handbook of moral development*. Mahwah, NJ: Erlbaum, pp. 119–54.

Smetana, J.G., Jambon, M., Conry-Murray, C., and Sturge-Apple, M.L. (2012). Reciprocal associations between young children's developing moral judgments and theory of mind. *Developmental Psychology*, 48(4), 1144–55. doi:10.1037/a0025891

Sommerville, J.A., Schmidt, M.F.H., Yun, J., and Burns, M. (2013). The development of fairness expectations and prosocial behavior in the second year of life. *Infancy*, 18(1), 40–66. doi:10.1111/j.1532-7078.2012.00129.x

Svetlova, M., Nichols, S.R., and Brownell, C.A. (2010). Toddlers' prosocial behavior: from instrumental to empathic to altruistic helping. *Child Development*, 81(6), 1814–27. doi:10.1111/j.1467-8624.2010.01512.x

Tennie, C., Call, J., and Tomasello, M. (2009). Ratcheting up the ratchet: on the evolution of cumulative culture. *Philosophical transactions of the Royal Society of London. Series B, Biological sciences*, 364(1528), 2405–15. doi:10.1098/rstb.2009.0052

Tomasello, M. (2014). *A natural history of human thinking*. Cambridge, MA: Harvard University Press.

Tomasello, M. (2016). Cultural learning redux. *Child Development*, 87(3), 643–53. doi:10.1111/cdev.12499

Tomasello, M., Melis, A.P., Tennie, C., Wyman, E., and Herrmann, E. (2012). Two key steps in the evolution of human cooperation: the interdependence hypothesis. *Current Anthropology*, 53(6), 673–92. doi:10.1086/668207

Tomasello, M. and Rakoczy, H. (2003). What makes human cognition unique? From individual to shared to collective intentionality. *Mind & Language*, 18(2), 121–47.

Tunçgenç, B., Hohenberger, A., and Rakoczy, H. (2015). Early understanding of normativity and freedom to act in Turkish toddlers. *Journal of Cognition and Development*, 16(1), 44–54. doi:10.1080/15248372.2013.815622

Tuomela, R. and Miller, K. (1988). We-intentions. *Philosophical Studies*, 53(3), 367–89.

Turiel, E. (1983). *The development of social knowledge: morality and convention*. Cambridge: Cambridge University Press.

Turiel, E. (2006). The development of morality. In: W. Damon, R.M. Lerner, and N. Eisenberg (eds.), *Handbook of child psychology, vol. 3: social, emotional, and personality development* (6th ed.). Hoboken, NJ: John Wiley & Sons, pp. 789–857.

Turiel, E. (2008). Thought about actions in social domains: morality, social conventions, and social interactions. *Cognitive Development*, 23(1), 136–54. doi:10.1016/j.cogdev.2007.04.001

Turiel, E. and Dahl, A. (2016). The development of domains of moral and conventional norms, coordination in decision-making, and the implications of social opposition. In: K. Bayertz and N. Roughley (eds.), *The normative animal? On the anthropological significance of social, moral and linguistic norms*. New York: Oxford University Press.

Vaish, A., Carpenter, M., and Tomasello, M. (2009). Sympathy through affective perspective taking and its relation to prosocial behavior in toddlers. *Developmental Psychology*, 45(2), 534–43.

Vaish, A., Missana, M., and Tomasello, M. (2011). Three-year-old children intervene in third-party moral transgressions. *British Journal of Developmental Psychology*, 29(1), 124–30. doi:10.1348/026151010X532888

Wittgenstein, L. (1953/2001). *Philosophical investigations: the German text, with a revised English translation* (3rd ed.; trans. G.E.M. Anscombe). Malden, MA: Blackwell.

Wyman, E., Rakoczy, H., and Tomasello, M. (2009). Normativity and context in young children's pretend play. *Cognitive Development*, 24(2), 146–55. doi:10.1016/j.cogdev.2009.01.003

CHAPTER 37

CRITICAL NOTE

Language and Learning from the 4E Perspective

HANS-JOHANN GLOCK

I shall comment on the chapters in a sequence taking us from concepts through language to social cognition and normativity. My perspective will be philosophical in stressing conceptual questions, yet in a way that acknowledges their dynamic interconnections with empirical and methodological issues. What constitutes the phenomena investigated by 4E cognitive science, such as concepts, meaning, interaction, and normativity? How are they understood in different scientific paradigms, research programs, and theories?

"The Embodiment of Language," M. Johnson

Johnson argues that language is pervasively and profoundly shaped by our bodies. Meaning, concepts, and understanding emerge from our bodily interactions both with inanimate things and events, and with other human and nonhuman animals. There is much to applaud in this case. Johnson recognizes that embodied accounts of language like that of Barsalou are representationalist only in the minimal sense that neural processes enable subjects to grasp concepts (pp. 629–30). Defying Chomskian prejudices, he also acknowledges the connections between syntax, semantics, and pragmatics (p. 632). Last but not least, he skillfully summarizes the research—pioneered by Lakoff and Johnson—into the way in which features of our bodies find expressions in metaphors that can be found in many, and in some cases perhaps all, natural languages.

On the downside, Johnson's attack on "disembodied views of language" is unfair and uncompelling. He uses this tendentious label indiscriminately for all analytic philosophers of language from Frege to the present. It suggests that the analytic philosophers he attacks either *deny* that languages are spoken by embodied agents and depend causally on bodily processes, or that they are at least *incompatible* with this insight. But although *some* of his bogeys may be committed to Platonic or Cartesian conceptions that preclude taking seriously the bodily aspects of language, this certainly does not hold of the loosely speaking pragmatist strand, which treats language primarily as an intersubjective practice. All members of this tradition can allow for this bodily dimension, and some of them have actively explored it. This holds not just for Wittgenstein, to whom Johnson gives a passing nod, but also, e.g., for Strawson (1959), who explored the ways in which the semantic operation of reference presupposes the need of speakers to locate their bodies within a spatiotemporal framework (an aspect subsequently explored by Evans 1982 and Campbell 1994).

At the same time, Johnson alerts us to a curious fact: even among the pragmatists, many (e.g., Austin and Brandom) who go on at length about the practical nature of language have nothing to say about the fact that practices are undertaken by embodied agents. But this is a lacuna rather than a principled incompatibility. That much holds even for Frege, whom Johnson singles out for special censure. His portrayal features several historical and conceptual inaccuracies (pp. 624–5). Frege assigned an *argument-function* rather than the traditional "subject-predicate structure" to "propositions" (in a single paragraph Johnson first inaccurately treats Fregean propositions—*Gedanken*—as sentences to be contrasted with words and then, accurately, characterizes them as something expressed by sentences). Worse still, Johnson lumbers Frege with the howler of thinking that the sense of a word is "an abstract meaning or understanding 'grasped'"; fortunately, Frege was clear that while "senses," a.k.a. meanings, are *things grasped*, the understanding is *the grasping*.

The root cause of Johnson's misunderstandings of and subsequent animosity toward analytic philosophy of language lies in his failure to distinguish two separate projects. He defines a "disembodied" account of thought and language as "one that assumes that it is possible to explain the syntax, semantics, and pragmatics of natural languages without a detailed explanation of how grammatical forms and meaning are shaped by the nature of our bodies, brains, and the physical environments we inhabit" (p. 625). But it all depends what kind of explanation is at issue. On the one hand, there is the loosely speaking causal explanation, be it of the evolutionary emergence of thought and language, of the diachronic development of individual natural languages, of language acquisition, or of the way in which physical vehicles enable individual speakers to speak. On the other hand, there is an explanation of what language, meaning, understanding *consists* in, i.e., a determination of what counts as a language, an expression having a meaning, a speaker understanding what it means, etc. Such a constitutive explanation can either be regarded as an ontological one of the essence of these phenomena in all possible worlds, or as a conceptual analysis or explication of what the relevant terms mean. In either case, however, they determine the *phenomena*, the emergence, development, and prerequisites of which are explained by causal explanations.[1] For instance, Johnson's discussion of body-part projections and image-schematic affordances promises insights into how—through what neural and computational means—we conceptualize; yet they do not tell us *what it is* to conceptualize, and how conceptual thinking and speaking differ from less advanced cognitive achievements.

Johnson's own account is occasionally marred by dubious conceptions of the phenomena he rightly identifies as important. Thus he employs "meaning" "for any experiences enacted or suggested by various affordances in our surroundings. Any aspect or quality of a situation means (for a specific type of creature) what it calls forth by way of experience" (p. 627). This conception equates meaning with the causal consequences of a subject perceiving something. It is not just wider than the established notion of linguistic meaning in philosophy, logic, linguistics, and everyday parlance (a fact Johnson acknowledges); this widening stands in tension with his own talk of "meaning-making" and his stress on the adaptive evolutionary function of cognition. For even creatures capable of intentional action (only a small fraction of those for which things "mean" something in Johnson's undiscriminating sense) do not "make" their experiences actively, they *undergo* them.

Johnson's neo-empiricist proposal about how abstract concepts can directly be constructed from purely sensory experience by way of "body-based metaphor" is equally contentious. First, it leaves it utterly mysterious why only humans command abstract concepts. Second, his assurances notwithstanding, the "inferential patterns" that he correctly regards as part and parcel of abstract concepts are simply not captured by the "primary metaphors" that are supposed to explain them (see pp. 632–4). Otherwise it would be impossible to say, for instance, that small things can be important, that one can do more by way of digging deeper, that a brake failure is a difficulty in spite of not impeding motion, etc.

Johnson ends his contribution with the suggestion that "4E cognition," the idea that cognition is *embodied, embedded, enactive,* and *extended,* should be expanded by three additional "E's," i.e., that it is *emotional, evolutionary,* and *exaptative*. This proposal is intriguing. However, Johnson's account of cognition and language intermittently underestimates the social, cultural, and normative aspects of both, e.g., when he explains our sense of fairness by reference to our sense of bodily balance (rather than the requirements of cooperative primates), or in ignoring the interplay between biological and cultural evolution. So, make my seventh E *enculturated* rather than *exaptative*. At least some crucial factors in the development of both the language faculty and specific languages are not side effects of purely biological adaptations; instead they are directly functional for cooperative and language-wielding primates, and they may be the result

[1] I much prefer the conceptual gloss, not least since it allows that the clarification of what counts as "language," "meaning," "syntax," etc., can be vague, provisional, and relative to speakers and research programs, thereby facilitating the interaction between philosophy and empirical disciplines (see Glock 2016).

of cumulative cultural development or even of (admittedly complex and messy) intentional innovations.

"The Embodiment of Concepts: Theoretical Perspectives and the Role of Predictive Processing," M. van Elk and H. Bekkering

Their contribution revolves around a trichotomy of theories of conceptual representation into embodied, disembodied, and hybrid. They develop an embodied view, but one that is hybrid in taking into account the importance of multi- and supra-modal associations. Furthermore, they enhance their embodied view through the "theory of predictive processing." This approach underlines the role of concepts for perception and prediction. Concepts are dynamically acquired and updated through Bayesian learning, based on the recurrent processing of prediction error signals in a hierarchically structured network. Concepts serve as prior models that generate multimodal expectations, thereby enabling greater precision in the perception of exemplars.

The article provides an illuminating taxonomy of theories of concepts, illustrated in Figure 34.1. Theories are distinguished along the following parameters: the nature of concepts, their relation to language, their function, their acquisition, their sensory representation, and the role of context for conceptual cognition. They rightly point out that theories tend to covary along these parameters: embodied theories often distance concepts from language, stress their role in action, are empiricist rather than nativist, and regard concepts as multimodal and context-dependent. But their discussion of these parameters is uneven. As regards the nature of concepts, they note the differences between psychological and philosophical definitions. Philosophers, they contend, refer to a concept as that which enables "propositional attitudes" such as beliefs, and consequently focus on "boundary conditions" for such attitudes. By contrast, psychologists are interested in how concepts enable cognitive processes like categorization and induction.

This suggests two distinct types of definition: philosophers define concepts as that which enable "propositional attitudes," less presumptuously called intentional states; psychologists define them as that which enable cognitive processes. But that would be misleading. For the cognitive process of categorization results in intentional states, like believing that this *x* is *F*; and both deduction and induction presuppose and yield such states, as in believing that all *x* that are *F* so far encountered have also been *G*, and inferring that the next *x* that is *F* will be *G* as well.

More seriously, van Elk and Bekkering distinguish different theories of the nature of concepts exclusively according to whether they are "unitary" or "heterogeneous," that is, according to how many types of concepts they detect. This is to put the cart before the horse. Prior to the question of how many species of concepts there are, is a question concerning the genus, namely of what concepts are to begin with. Regarding the latter, van Elk and Bekkering simply take for granted that they are mental representations, apparently in the Fodorian sense of particulars in the minds or brains of individuals. One problem with this orthodoxy is that this fails to do justice to the fact that concepts (just like intentional states) can be *shared* between different subjects. Fodor purports to resolve this difficulty, yet only by oscillating between representation *types*, which can be shared but are *universals*, and representation *tokens*, which are particulars but *cannot be shared*. Alternative views that regard concepts as abstract entities (notably Fregean senses) fare better in this respect. On the other hand, Platonist accounts have difficulties accounting for the role of concepts in cognition and action, rightly stressed by embodied accounts. But that shortcoming is rectified by relating concepts to abilities (without identifying them with the latter), e.g., by defining them as principles or rules guiding higher cognitive operations, more specifically operations of classification and inference (see Glock 2013). Like the hybrid view, such an account avoids the mistake of tying concepts too closely to activity. It is a distinguishing feature of conceptual cognition involving belief and inference that it is decoupled both from conative states like desires and from immediate reactions to the environment. To be sure, conceptual knowledge provides reasons not just for holding other beliefs but also for behavior (otherwise the "grounding problem" diagnosed by hard-line embodiment theorists would arise); it is "available" not just for belief-formation but also for behavior. That downright entails, however, that conceptual knowledge does not *reduce* to the behavior it enables.

This perspective can also accommodate the empirical findings judiciously assessed by van Elk and Bekkering. Concepts can be more or less abstract, independent of perceptual experience, depending on the *properties* according to which they classify or categorize objects and events. At the most fundamental level, as empiricists rightly stress, these are properties ascertained through sensory perception. Moreover, regarding the findings favoring modality-specific theories, it is not particularly puzzling that the facility with which subjects can classify phenomena according to such properties depends on the kind and number of sense modalities involved. At the same time, however, concepts also feature in more advanced cognitive capacities, e.g., in analogical and metaphorical thinking and various types of inference (Newen and Bartels 2007). And they are defined not by the sense modalities on which we draw in employing them, but by the properties according to which we classify and infer (albeit in a manner that is by and large more vague, complex, and dynamic than the classic "definitional" view of concepts as a set of defining features had it).[2]

[2] Another quibble: van Elk and Bekkering write, "On Fodor's account, concepts are like words that function as *arbitrary* symbols in a language of thought" (p. 644, my emphasis). But unlike the connection between concepts—the words of a language of thought—and words of public languages, the one between the world and our perception of it and (type-) words of the language of thought is precisely not arbitrary, a matter of convention, but a causal one, albeit described at a computational rather than neurophysiological level. When a subject engages in conceptual thinking, Mother Nature inscribes patterns of neural firing into her brain.

On the one hand nativists have argued that core knowledge systems enable the acquisition of concepts, while on the other hand empiricists have stressed that concepts are primarily acquired through experience. However, even empiricists need to start from (implicit) basic assumptions regarding the innate capabilities of the brain for learning. (p. 652).

As described, this Homeric struggle should never have started. There is not even a tension between acknowledging that core capacities and dispositions must be in place for concepts to be acquired and modified, while at the same time recognizing that the acquisition itself also requires experience and learning. That process may well conform to the predictive processing account. But to settle that matter is a task for empirical research.

"Origins and Complexities of Infant Communication and Social Cognition," U. Liszkowski

The acquisition of concepts in ontogenesis receives attention in the final two articles. Liszkowski scrutinizes the interaction between the ontogenetic development of linguistic communication on the one hand and social cognition on the other. He adjudicates between two general perspectives. According to the first, linguistic communication is an outgrowth of prior, nonlinguistic communication, which in turn is based on a "social-cognitive understanding of what others intend to bring across and make of one's communication" (p. 661). According to the second, the initial social activity of infants does not yet evince possession of a "theory of mind," but underlies its acquisition along with that of language. Liszkowski prudently detects merit in both views. Social interaction provokes experiences that shape the way infants understand others. Regarding the most fundamental level, however, he sides with the theory of mind first approach. "Logically, there must be some cognitive basis down the line enabling one to engage in, and benefit from, social interaction" (p. 661). Now, it is granted that any learning presupposes some innate capacities of a cognitive kind. But insofar as Liszkowski's point *is* logical, a.k.a. conceptual, it would seem to presuppose a cognitively loaded understanding of what it is to engage in and benefit from interaction. This leaves open the option of explaining the emergence of a theory of mind and meaningful communication as consequences of social interactions that rely only on cognitive capacities shy of a theory of mind.

On the other hand, Liszkowski marshals impressive findings, many from his own research group, indicating that as a *matter of empirical fact*, children understand and influence the mental states of others through meaningful communication before attaining full linguistic competence. His chief piece of evidence is infant pointing. He makes a convincing case for the thesis that infants have and understand *communicative, referential*, and *social intentions*: they point to address another person, to refer to a specific feature, and in expectation of a cooperative activity with the recipient.[3]

Another respect in which Liszkowski steers a middle course between the theory-of-mind-first and the interaction-first approach is this: he regards the theory of mind that is both presupposed and in turn propelled by social interaction as a "form of practical knowledge, perhaps better described as a theory of *action*" (p. 675). The first concession raises the question of whether the term "theory" is appropriate. After all, we are dealing with a capacity that enables skillful performance. Now, that capacity may amount to "knowing how" to communicate. And there is a case for holding that such knowing how, as opposed to a mere being able to, implies knowledge that—namely, that the way to communicate successfully in such-and-such respects and situations is to do this-and-that. Even then, however, it is far from clear that such knowing must or, given the early developmental stages concerned, could amount to a systematic theory. It is therefore preferable to speak of "mindreading." The second concession suggests that even this label needs to be questioned. Do infants read minds or action? At the same time, that contrast requires qualification. It is plausible to insist that both infants and, e.g., chimpanzees read intentions by reading actions; they register "intentions in action," to use Searle's phrase.

Liszkowski ends with a discussion of infant communication from an evolutionary and social-cultural perspective. That discussion would have been even more illuminating if he had not simply taken for granted that ontogeny recapitulates phylogeny. Leaving that lacuna aside, Liszkowski argues powerfully for the anti-Chomskian and anti-representationalist view that the "departure point" of the development of communication is neither syntax nor reference, but the infant's urge to engage and belong with others. This bears additional witness to the promise of embodied theories for developmental psychology and psycholinguistics.

"Developing an Understanding of Normativity" M. Schmidt and H. Rakoczy

This contribution takes us from social cognition to the more specific and perhaps uniquely human phenomenon of normativity. It reports impressive recent research into

[3] Liszkowski illustrates the last point by the fact that pointing serves to request help, share attitudes, and to provide help by informing others. He describes these as cases of "*illocutionary force*." But illocutionary force attaches to particular expressions (in context), as a matter of general linguistic rules. What Liszkowski has in mind is closer to the intention of achieving a "perlocutionary" effect, something which is specific to a situation, does not presuppose language, and is not guaranteed by conventions. Perhaps Liszkowski does not consider this label because he believes that perlocutionary effects are "unintended (p. 678). But this is not part of the meaning of that technical term as introduced by Austin (1975, p. 101).

the ontogenesis of norms and norm psychology (much of it by the authors). Normative cognition and other forms of social cognition seem to be intimately related from early on in development. Even young children have a strikingly robust grasp of normative phenomena. Their selective and rational third-party enforcement of norms indicates that they understand central features of normativity (standards of correctness, normative force, generality, context-relativity). Children also *care* about normativity, to the point of actively seeking out norms and "promiscuously" imputing normativity to others' intentional actions (pp. 688–9). The chapter ends with the plausible conjecture that our singular norm psychology develops in close tandem with our equally unique forms of shared intentionality. In both, one should add, we find an intertwinement of cognitive and conative/affective dimensions that characterizes the new paradigm of 4E cognition more generally, a point noted in Johnson's proposal of "emotional" as a fifth E.

Schmidt and Rakoczy are exemplary in attending to the methodological challenges posed by diagnosing various types and stages of normative comprehension (pp. 685–96). They point out the importance of considering norm transgressions and the attending reactions. These methodological considerations are in turn based on sophisticated philosophical reflections concerning conceptual and ontological aspects. Schmidt and Rakoczy are exercised by the phenomenon of "normative force." They rightly distinguish the latter from physical influence and coercion, yet they stick their necks out unnecessarily in their attempt to capture normativity's distinctive force: "Crucially, we could do otherwise, but we think we should adhere to the norm" (p. 687). However, normativity does not require the ability to do otherwise, which is known as "liberty of indifference." It only requires "liberty of spontaneity," the power to act in accordance with one's wants. If these wants are themselves causally predetermined, as determinism has it, a subject can follow a norm because it wants to, yet without the possibility of an alternative outcome.

Schmidt and Rakoczy diagnose that "normativity in a narrow sense, that is, in norms that set standards of correctness, come with normative force and authority, are valid both in general (agent-independent) and context-relative ways" (p. 687). But while the idea of a standard of correctness is crucial, it at best demarcates normativity sensu lato, not in more demanding senses requiring rules or, more specifically, prescriptions and prohibitions. Normativity exists only where there is a standard against which either beliefs or actions (the distinction between practical and theoretical or epistemic normativity appears to me to be less important than to Schmidt and Rakoczy, p. 688) are to be adjusted, in most if not all cases pro tanto, i.e., subject to the interference of other normative standards. Anything short of a standard requiring or licensing correction is at most *evaluative*. Values constitute an intermediate category, neglected by most studies, between norms and facts. They give rise to *assessment*, yet fall short of any *demand* to bring beliefs or actions in line. Because values do furnish reasons without being normative in this sense, I am also skeptical of the widespread conviction that reasons are per se normative (Schmidt and Rakoczy prudently do not subscribe to this view, but they do link "reason responsiveness" to "normative attitudes," p. 686).

According to an attractive "objectivist" account, reasons are not inner mental causes of behavior and beliefs, but facts or states of affairs that normally concern the world rather than the subject's mental state. Now, any facts—including purely "descriptive" ones—can furnish reasons to do or believe something. What is special about normative *and* evaluative facts is that they provide *intrinsic*, albeit defeasible, reasons, independently of empirical facts. The fact that I am wearing a striped tie over a plaid shirt per se does not license a negative assessment, yet the fact that I am dressed *badly* does. By the same token, it gives me some reason to alter my attire. By contrast to being dressed incorrectly, however, it does not license *correction*, let alone coercion. Similarly, that I am without cap and gown does not provide per se provide a reason for changing my attire; but the fact that I am dressed incorrectly—given the context of a graduation ceremony—does. Indeed, unlike being dressed badly, it warrants a pro tanto demand for correction.

That the idea that a standard of correctness is normative only in a minimal sense is also evident from the fact that it applies not just to "normative expectations" but also to "descriptive expectations" and beliefs more generally. Although the latter have a "mind-to-world" rather than "world-to-mind direction of fit" (p. 687), the very idea of *fitting* indicates a standard of correctness—in line with the etymology of Latin *regula* and *norma*. The difference lies in what violating the standard amounts to and what kind of correction it requires. This reinforces, however, another guiding theme of Schmidt and Rakoczy. The key to an understanding of normativity in both more or less restrictive senses lies in understanding *infringements*.

There is a more general moral for the topics revolving around language and learning. While it is central to look at standard cases of things going right, it may also be indispensable to investigate what happens when they go wrong.

References

Austin, J.L. (1975). *How to do things with words* (2nd ed.). Oxford: Oxford University Press.

Campbell, J. (1994). *Past, space and self.* London: MIT Press.

Evans, G. (1982). *The varieties of reference.* Oxford: Oxford University Press.

Glock, H.J. (2013). Animal minds: a non-representationalist approach. *American Philosophical Quarterly*, 50, 213–32.

Glock, H.J. (2016). Impure conceptual analysis. In: S. Overgaard and G. d'Oro (eds.), *The Cambridge companion to philosophical methodology*. Cambridge: Cambridge University Press, pp. 83–107.

Johnson, M. (2018). The embodiment of language. In: this volume, pp. 623–40.

Liszkowski, U. (2018). Origins and complexities of infant communication and social cognition. In: this volume, pp. 661–84.

Newen, A. and Bartels, A. (2007). Animal minds and the possession of concepts. *Philosophical Psychology*, 20, 283–308.

Schmidt, M.F.H. and Rakoczy, H. (2018). Developing an understanding of normativity. In: this volume, pp. 685–706.

Strawson, P.F. (1959). *Individuals*. London: Methuen.
van Elk, M. and Bekkering, H. (2018). The embodiment of concepts: theoretical perspectives and the role of predictive processing. In: this volume, pp. 641–60.
Wittgenstein, L. (1953/2009). *Philosophische Untersuchungen/Philosophical investigations* (trans. G.E.M. Anscombe, P.M.S. Hacker, and J. Schulte). Chichester, UK: Wiley-Blackwell.

PART IX

EVOLUTION AND CULTURE

CHAPTER 38

THE EVOLUTION OF COGNITION

A 4E Perspective

LOUISE BARRETT

What is Cognition Anyway?

How and why cognition evolved depends on what one thinks cognition is. The classic definition by Neisser (1967) identified cognition as the processes by which sensory inputs are transformed, manipulated, augmented, and used to give rise to motor outputs, with the implicit assumption that these processes took place solely in the brain. There is a distinctly anthropocentric tinge to this definition, grounded as it is in the cognitive revolution, which aimed to model (or even recreate) human intelligence via the use of computers. Consequently, the processes usually considered to be cognitive include concept formation, reasoning, and problem-solving abilities, theory of mind, natural language, memory, prospective planning, and the ability to represent objects in their absence. This view of cognition often results in what Lyons (2006) terms an "anthropogenic" approach to its evolution, in which we "assume, to a greater or lesser extent, that human psychological attributes are the hallmarks of cognition and ask what sort of biological or evolutionary story might account for them" (Lyons 2006, p. 12).

Accordingly, one of the tasks of comparative psychology is to identify which, if any, cognitive traits are shared by other species. Basically, we engage in a sorting process that distinguishes creatures capable of flexible "human-like" thought and behavior from those capable only of fixed, "instinctive" noncognitive responses to environmental stimuli. Via this sorting process, one can identify the conditions under which cognition was selected, and how it contributes to organisms' fitness. These views fit with Darwin's dictum that organisms should differ only by degree, rather than in kind, such that we should expect to see the precursors of our own abilities in related species. Even in the context of a more "ecological" approach to cognition, where the demands of the animal's ecological niche are argued to drive the evolution of particular cognitive capacities, cognitive processes are conceptualized in the same representational, information-processing terms that characterize the anthropogenic approach (Shettleworth 1999; Penn et al. 2008).

So far, so good. The problem is that, as Rowlands (2003, p. 167) has pointed out, there is very little consensus on what entitles something to be called a cognitive process; often, he says, we simply point toward the set of abilities we consider to be cognitive, like reasoning and language, and leave it at that, without specifying exactly why such processes count and others don't. Deciding what counts as cognitive therefore becomes rather fraught at times, with a continual battle over whether a particular behavior is the result of a cognitive process or "merely" the outcome of associative learning (e.g., Buckner 2011; Byrne and Bates 2006). This state of affairs is made even more confusing by the fact that the leading textbook on comparative cognition undercuts this distinction by defining cognition as "any process by which animals acquire, process, store, and act on information from the environment" (Shettleworth 1999), which would seem to render debates over whether a particular result supports an associative or cognitive account redundant, if not incoherent.

Despite these various efforts, then, it seems that, as van Duijn et al. (2006) suggest, we still lack a means of navigating between overly "mindless" accounts of behavior versus "mindful" accounts that involve the use of mentalistic terms, where the latter, as Penn (2011) points out, fail to identify any cognitive process at all, but simply fall back on folk psychological descriptors (see also Ramsey 2007). In particular, we seem to lack the appropriate vocabulary to discuss the abilities of simple creatures in ways that capture the flexibility of their behavior, but that do not generate an inappropriately "mentalistic" view of matters (van Duijn et al. 2006).

Determining when or why an organism can be deemed cognitive is, therefore, not clear-cut, and tracing the evolutionary origins of cognition accordingly is more difficult. Perhaps this is why accounts of cognitive evolution are sometimes couched in terms of brain size (Roth and Dicke 2005; Dunbar and Shultz 2007). Brain size is, of course, more easily and unambiguously measured than cognitive capacity, and has the advantage that it can be obtained from species that are long dead and fossilized, as well as from extant species. Although measuring brains avoids the difficulty of defining cognitive processes, this approach rests on the assumption that brain capacity is an accurate proxy for cognitive abilities. As van Duijn et al. (2006, p. 163) point out, however, it appears that our best reason for linking brains and cognition is that it "seems so obviously true": those animals we characterize as obviously cognitive, like humans and apes, have very large brains for their body size, making the connection seem entirely logical. Creatures like bacteria, which lack nervous tissue cannot, by definition, engage in a cognitive process, and this also seems reassuringly logical. Equating brains (and nervous systems) with cognition thus provides another (simple) means by which different creatures can be sorted into those that are cognitive and those that are not.

The problem with this solution is that bacteria have revealed themselves capable of a remarkable flexibility in behavior, despite lacking the only structure, i.e., a nervous system, allegedly capable of producing such flexibility (e.g., Shapiro 2007). The brain = cognition equation is also undermined by creatures like salticid spiders, which possess brains that are absolutely tiny, no bigger than a poppy seed, yet are capable of remarkably flexible hunting tactics, including the ability to engage in deceptive mimicry, create diversions to distract prey, and take long, complex detours in order to better position themselves for prey capture (Tarsitano and Jackson 1994; Tarsitano 2006; Barrett 2011). If a certain degree of behavioral flexibility is possible without brains, or with only very small ones, then the idea that more brain means more cognition seems rather shaky, as is our ability to distinguish cognitive from noncognitive creatures using simple metrics.

The situation is made even worse by the fact that, among creatures in possession of fully fledged brains, the manner in which cognitive capacity can be mapped onto brain structure is a highly vexed issue—Healy and Rowe (2007, 2013), for example, identify a number of problematic assumptions within the literature and dispute the idea that smart behavior maps onto brain anatomy in the way some have argued. Even within the context of this debate, however, the essential assumption that brains are the organs of cognition remains intact: elsewhere, these same authors have argued for better controlled experiments that can eliminate the influence of "noncognitive" factors on behavioral performance in order to achieve a "pure" estimate of brain-based cognitive capacity (Rowe and Healy 2014).

The rise of 4E cognition (embodied, embedded, enactive, and extended) has raised awareness of the inconsistencies and problems of the traditional cognitivist view, making it clear that brains cannot be divorced from their bodily and environmental contexts (Clark 1997; Gallagher 2005; Rosch, Thompson, and Varela 1992; Chemero 2009; Hutto and Myin 2012; Barrett 2011). Whatever cognition might be, it is not best captured by a purely brain-based computational perspective that places brains at a remove from the rest of the body and the worlds they inhabit. This is true of humans, but even more so for other creatures, given that the computational views on offer take some quite peculiarly human abilities as a benchmark for cognitive processes in general; these might not apply to other animals, or at least they might not operate in quite the same way. 4E cognition thus offers an alternative to the anthropogenic approach, by providing a bottom-up or "biogenic" perspective on cognitive evolution, where the principles of biological organization, and their links to fitness, are taken to be the most productive means by which we can understand what cognition is, what it does, and how and why it evolved (Lyon 2006). A 4E biogenic approach thus seems worth considering, given that—although psychologists readily accept that cognition occurs in a biological context—the inherent anthropocentrism of the representational, computational view of cognition is less widely recognized, even among authors who attempt a more ecological approach. Hence, many authors clearly consider representational and computational theories of cognition to be "species neutral," when in fact these are highly human-oriented and hence much less "biogenic" than they imagine. In what follows, I first consider some ideas relating to the minimal criteria for cognition, before going on to discuss how cognitive evolution can be conceived within a 4E framework. I conclude with a short consideration of why more radical forms of embodied cognitive science may represent the way forward.

Cognition as Coordination

If we move away from thinking of cognition as brain-bound computational processes, and move toward a more functional definition, the line between cognitive and noncognitive creatures becomes increasingly blurred. Lyon (2006), for example, defines cognition as the "capacity to infer relations between external circumstance and internal need to facilitate agency" whose function is to "enable successful action and interaction in a niche." Anderson (2003) offers an even simpler definition of cognition as "situated activity." It is not immediately apparent that, under these definitions, cognition depends on the possession of a particular kind of brain or nervous system organization, or that there are particular kinds of representational or computational processes that qualify as cognitive while others don't.

Brooks (1991, 1999), for example, famously rejected the idea that representation and reason were necessary for intelligent behavior, and sought instead to study the adaptive behavior of whole agents operating in an uncontrolled real-world environment (see also Beer 1996; Pfeifer and Scheier 2001; Pfeifer and Bongard 2006). Brooks's work demonstrated that perception and action could often be so closely intertwined that there was no need for "cognition": the discrete action-producing systems of his robots resulted in adaptive, robust behavior in the complete absence of a central processing unit that could integrate inputs, plan, and produce coordinated action. His robots did not need a complete internal representation of the world in order to produce coherent behavior. Instead, they were "situated": sensitive to relevant information in their surroundings, with tightly coupled perception-action mechanisms that were, in turn, directly coupled to an environment that they dynamically and frequently sampled. As Chemero (2013) suggests, Brooks's work was thus a logical extension of Gibson's (1979) anti-representational ecological theory of perception (where psychological phenomena are seen as relations between an organism's physical constitution and its environment) along with Barwise and Perry's (1983) notion of situation semantics (the idea that the meaning of thoughts and utterances has nothing to do with mental representations, but with the relationship between thinkers/speakers and their environment). For Brooks, like his predecessors, intelligence was a relational property: it came about via possessing a particular physical constitution operating within a particular environment.

This last point is key, as it helps illustrate how the rationale for Brooks's research program was genuinely evolutionary. As he pointed out, most of the four billion years of life on earth has been spent refining the perception and action mechanisms that permit organisms to cope with a dynamic, ever-changing environment (Brooks 1999). These seemingly more "primitive" nonrepresentational forms of intelligence therefore must have been the more difficult forms to evolve and implement, whereas the highly representational processes that we humans deem intelligent—language, mathematical reasoning, symbolic logic—must have been relatively easy to set in place. Following this reasoning, Brooks decided to focus on insect-level intelligence, arguing it was essential to understand the 97 percent of behavior that was controlled by nonrepresentational processes, in order to understand the remaining 3 percent that could be deemed representational (where these values refer to humans). As such, Brooks's work represented an early attempt to rid artificial intelligence of its anthropocentric orientation, grounding cognition in biological processes, and not human artifacts.

This proposal was met with skepticism by some, however, who considered it unlikely that insect intelligence would scale up to human intelligence. As Kirsh (1991, p. 162) put it: "Insect ethologists are not cognitive scientists." This latter pronouncement seems increasingly insecure in the face of scientific findings that make a strong case for insect cognition (e.g., Leadbeater and Chittka 2007; Lihoreau et al. 2012). One could argue, though, that the decision to pursue insect-level intelligence as the basis of intelligent "cognitive" behavior was somewhat arbitrary, given that insects themselves appear pretty late in the day, evolutionarily speaking.

Indeed, some in the classic enactivist camp suggest that any attempt to distinguish cognition from other kinds of processes is, essentially, meaningless, and consider all life to be inherently cognitive (Rosch et al. 1992). More specifically, Thompson (2007) offers us the position that "life and mind share a set of basic organizational properties, and the organizational properties of mind are an enriched version of those fundamental to life. Mind is life-like, and life is mind-like" (p. 128). The problem with this position is that, if everything is cognitive, then nothing is (van Duijn et al. 2006): we are left no means of coming to grips with why a sea urchin differs from a shark or a slow worm from a squirrel monkey. If we want to understand differences between minimally cognitive systems and more advanced systems, van Duijn et al. (2006, p. 159) argue, then "it seems important to differentiate between the cognitive processes of the rabbit and those of the carrot": we need to distinguish the subset of living systems that can be classified as cognitive systems, and make clear the basis on which we make such a classification.

As noted earlier, one means of doing so is to assume minimal cognition requires the possession of a nervous system. Moreno and colleagues, for example, argue that the sensorimotor behavior of creatures without a nervous system is simply an extension of their metabolic processes whereas, on their account, minimal cognition requires "meta-metabolic functions" that can transcend these processes, generating an autonomous sensorimotor domain that can sustain an "independent domain of patterns" (Moreno et al. 1997; Moreno and Etxeberria 2005). As van Duijn et al. (2006) point out, however, several authors have now questioned whether a nervous system is essential for cognitive (i.e., meta-metabolic) behaviors, arguing that the evolution of nervous systems represents the augmentation of abilities that already exist in unicellular organisms, rather than nervous systems marking the crossing of some "cognitive Rubicon" (Lyon 2006; Taylor 2004).

To make this point more forcefully, van Duijn et al. (2006) give the specific example of the sensorimotor abilities of the bacterium, *Escherichia coli*, and its two-component signal transduction (TCST) system. In brief, and to oversimplify a little, chemotaxis in *E. coli* relies on two interacting signaling pathways, a fast one (of the order of milliseconds) that mediates perception, and a slower one (of the order of seconds) that mediates adaptation to the signals in the environment (and therefore constitutes a form of memory). The perception pathway consists of surface receptors for chemical attractants or repellents, the occupation of which triggers phosphorylation of other molecules inside the bacterium, and influences the direction in which the cells' flagella rotate (so controlling whether the organism "runs" or "tumbles"). The adaptation process involves the methylation of occupied receptors. As this process is slower than the perceptual process, the number of methylated receptors registers the concentration of chemicals in the environment a few seconds previously. The receptors are continuously methylated and demethylated in response to the occupation of receptor sites by attractants and repellant chemicals as the organism moves through the environment while, at the same time, receptor occupation and levels of methylation influence chemical processes inside the cell which, in turn, influence how the organism moves. The interaction and feedback between the two systems thus produces a dynamic form of molecular memory. The difference between the current state of the receptors and the internal level of methylation means that the organism becomes sensitive to the relative concentration of chemicals in the environment, rather than their absolute value, and hence can register any large-scale abrupt changes in chemical concentration but will adapt to a constant level of signal. This, in turn, means that organisms can move toward high concentrations of attractants and away from repellents, but will remain in an environment with an optimum concentration of attractants.

Van Duijn et al. (2006, p. 163) use this example to undercut the intuitively appealing idea that minimally cognitive behavior requires a nervous system. Chemotaxis is more than just an aspect of the cell's metabolism because it involves physical changes of the organism in relation to the environment that are distinct from metabolic processes. Metabolic processes benefit from the changes in the environment the organism produces, but the changes produced (moving toward a food source or away from a noxious stimulus) are not themselves part of metabolism. Thus, they fulfill Moreno and colleagues' requirement of autonomy and "meta-metabolic functions." Given this, van Duijn et al. (2006) suggest that sensorimotor coordination offers a conceptually clearer and more effective starting point for minimal cognition.

Grounding cognition in sensorimotor coordination emphasizes that organisms are always fully situated (or embedded) in their environments, and it naturally incorporates physical embodiment. For example, *E. coli* uses a temporal system (memory) to act adaptively in the environment because it is too small to employ any kind of spatial mechanism (e.g., it cannot position its sensors/receptors in such a way that maintaining equal stimulation at each sensor or receptor would drive the organism toward a target). Its rod-like shape improves its ability to engage in chemotactic behavior, and it possesses flagella that are activated by particular kinds of proteins, as well as the TCST system itself, which provides an internal connection between sensors and effectors. This approach therefore not only specifies what organisms are able to do, but also how they are able to do it, something that any adequate theory of cognition must provide. This fits with a biogenic perspective, as it identifies the need to optimize the conditions needed to maintain metabolic processes within acceptable limits as the fundamental problem to be solved. This, van Duijn et al. (2006) suggest, may lie at the origin of many basic forms of cognition. This being so, it is crucial to understand the variety of solutions that exist if we are to understand the nature of cognition, and how more complex forms of cognition arise. In effect, then, van Duijn et al. (2006) reiterate Brooks's (1999) earlier views, but provide a more principled argument for where we should begin our exploration of the origins and evolution of cognition, rather than taking insect-level cognition as a convenient, but arbitrary, starting point.

The Skin-Brain Thesis

This view has further implications for how to think about nervous system evolution. Keijzer et al. (2013), for example, suggest that nervous systems may have evolved not to enable more intelligent behavior (after all, bacteria, it seems, are pretty smart), but to enable muscular behavior—what they refer to as the "skin-brain thesis" (SBT). Among mobile organisms, the evolution of multicellular forms gave rise to selection for movement via muscular contraction, rather than by the use of cilia or flagellae (the slender hair-like or whip-like protuberances that single-cell organisms use to propel themselves) in order to coordinate and control the movement of the body in adaptive ways (Keijzer et al. 2013). (Other multicellular organisms, like plants, followed a different evolutionary trajectory, and there is fascinating work on the nature of plant learning and intelligence, see, e.g., Gagliano et al. 2014.) A simple analogy helps make clear why this should be: building a pyramid from Lego is much easier than building it from massive stone blocks—it is obvious that the greater size of the latter entails a completely different means of moving things around. Thus, the rise of muscle-based motility is seen as a key transition in animal evolution, with muscular contractions central to understanding the evolution of nervous systems themselves: "while cilia—literally—only allow your skin to crawl, internal sheets of contractile cells make it possible to use the body itself to accomplish motility" (Keijzer et al. 2013, p. 12). According to this hypothesis, then, the earliest nervous systems were coordination systems—diffuse nerve nets that spread out over a large portion of the animal's body (as seen in species like hydra and jellyfish)—and not input-output systems (like the classic notion of a reflex arc, where a sensor is linked to an effector by neural connections).

Keijzer et al. (2013) put forward a two-phase evolutionary process. The first (hypothetical) phase involved the evolution of excitable myoepithelia (epithelia with contractile properties that can conduct electrical activity across their surface) that was capable of chemical transmission between cells. This myoepithelia is thought to have comprised an unbroken sheet of protonervous and muscle tissue, with each cell signaling to its immediate neighbors. Keizer et al. (2013) refer to these muscle surfaces as "Pantin surfaces" (in honor of Carl Pantin, who first had the idea that the need for mobility drove nervous system evolution; Pantin 1956), each of which has a specific size, shape, and extension for each species (and indeed each individual, given that these will not be identical to each other). They refer to the self-organized waves of coordinated contraction that the surface produces as "patterning." This aspect of the hypothesis is therefore of particular relevance to an embodied perspective, because such patterning is argued to be particular to the Pantin surface that produces it, i.e., it inherently reflects the structure of an animal's body. Patterning cannot, therefore, be interpreted as some form of generic motor output function (e.g., feeding, mating), but should be viewed more like the action of an organ: as a single coordinated system of movement that is characteristic of a particular species and individual. As such, there is no controller that stands in contrast to the element being controlled.

The same goes for sensory capacities (Keijzer 2015). In order for patterning over the Pantin surface to remain functional, the organism must be sensitive to disturbances in such patterning (e.g., due to growth, short-term changes in activity, temperature, chemical concentrations), as this will allow it to compensate and restore functionality. The animal must therefore be sensitive to the ongoing dynamical changes in contraction and extension across its Pantin surface, so that it can influence the processes that produce and maintain these same contractions. This sensitivity to its own bodily dynamics naturally incorporates sensitivity to any external disturbances that hinder or enhance bodily contractions and extensions. Thus, the animal can sense the environment via the skin-brain despite the absence of any dedicated external receptors. In this view, the "animal body itself becomes a sensing device through its use of contractile tissues and the environmental feedback this generates, both within and external to the body." (Keijzer 2015, p. 14).

In the second phase of evolution, Keijzer et al. (2013) argue that a comparatively small number of more specialized signaling cells evolved, with elongated axodendritic processes (i.e., neurons) that enabled them to signal to non-adjacent cells. This would give rise to a diffusely connected nerve net spread over the entire body. The advantage of such nerve nets is that they would add to the self-organizing properties of existing epithelial and muscular tissues. Specifically, Keijzer et al. (2013) argue that the addition of long-range neural processes could thus enable wave-like patterning across larger Pantin surfaces, releasing a constraint on size, as well as offering greater flexibility: in contrast to sheet-like myoepithelia, nerve nets can take many forms, so removing constraints on body form. Rather than providing specific, more precise connections from sensors to effectors, as an input-output view of nervous system evolution would argue, the function of nerve nets was to control, modify, and extend self-organized patterning across a Pantin surface. Much like the memory system of *E. coli*, then, diffuse nerve nets offered organisms an extra loop of control over behavior.

With diffuse neural nets in place, Keijzer et al. (2013) suggest that, evolutionarily, the contractile apparatus could be cut loose from its placement on the animal's skin and repositioned as internal muscle. The general differentiation of muscle from skin would then give rise to more complex nervous systems, and the kinds of animal bodies that we are familiar with today, which fit well with the input-output conduction view of nervous systems (see also Jekely et al. 2015). Thus, there is nothing in the SBT that suggests we should abandon the conduction view. Rather the SBT offers a different, more embodied standpoint from which to consider possible hypotheses about the origins of nervous systems, and how we can think about cognitive processes. The SBT does, however, present a major contrast to the one-size-fits-all view of classical cognitivism, which "casts the operation of basic nervous systems in terms of abstract computational functions." Instead, "the SBT draws it as a producer of complex but concrete dynamical patterns of activation" (Keijzer et al. 2013, p. 15).

Behavioral Flexibility, the Elaboration of Nervous Systems, and the *Umwelt*

I have explained the SBT in some detail because it offers a view of nervous system evolution that fits perfectly with an argument I have offered elsewhere, which considers cognitive evolution to be a process by which organisms became better equipped to track and deal with unpredictable contingencies in real time in changeable environments (Barrett 2011). In brief, what has taken place over the course of evolutionary history is the progressive elaboration of animals' central nervous systems, with larger, long-lived species tending to possess more complex brains and more elaborate nervous systems than small, short-lived species. Longevity is a crucial factor because long-lived creatures are likely to encounter a wide variety of conditions across their lifetimes, and higher levels of behavioral flexibility enable animals to cope with environmental change more effectively (although, of course, this depends crucially on the rate of change: climate change and habitat loss are exposing the limits of species' flexibility).

It is important to emphasize the nervous system as a whole, and not just the brain, because species with larger brains also tend to have more elaborate sensory apparatus. Although we often speak of humans having a poorer sense of smell, or less sensitive hearing, than other creatures, our senses of smell and hearing are actually pretty good, and we also have good vision, along with a sense of touch, taste, and balance. There is often the assumption that more brain tissue allows animals to engage in more complex "thinking" processes, but the control of complex sensorimotor systems also requires a lot of neural tissue. So, in order to employ our receptors effectively, they are incorporated into perceptual systems (e.g., we have two eyes in our head, that in turn is positioned on a moveable neck, which is attached to a body that can move in a variety of ways) that allow us to actively explore and sample the environment, and not passively receive it (Gibson 1979). Part of the reason why human brains are so big, then, is to help control and coordinate our sensory and motor apparatus. This, in turn, is partly why humans show such high levels of behavioral flexibility: our greater sensitivity to many different aspects of our environments expands our *Umwelt*—the term used by Uexküll (2014) to capture the notion that organisms are sensitive only to those aspects of the environment that hold significance for their survival and reproduction. This also helps us recognize why other animals, with different kinds of sensorimotor systems, which have developed to differing degrees, vary in their flexibility and problem-solving capacities.

An embodied perspective on comparative cognition therefore includes the idea that other features of an animal's anatomy can contribute to successfully navigating the world, including how morphology itself can contribute to successful problem-solving. For example, New Caledonian crows are large-brained birds, and have exceptionally good tool-using abilities compared to other members of the crow family, despite the fact that these are similarly large-brained. This is partly because they have much greater convergence (binocular vision) and straighter beaks (Troscianko et al. 2012). It is much easier for New Caledonian crows to visually guide the stick-like tools they use to obtain food than it is for other crow species, and this helps explain why they perform so much better on tool-using problems than other corvids in the laboratory, but show comparable performance on non-tool-using tasks.

Once one is attuned to the idea that behavioral flexibility is a property of bodies as well as brains, and that the exploitation of environmental structure can serve a similar function, examples appear all over the place—from the plant hoppers that have a gearing mechanism in their legs that synchronize their jumping movements more precisely than neural control can achieve (Burrows and Sutton 2013), to the seahorses that have square rather than cylindrical tails (Porter et al. 2015), which improve their ability to maintain a controlled grasp onto coral reefs, to the way that mudskippers (fish with the ability to feed on land) form a "hydrodynamic tongue" to help capture food: the fish emerge onto land and eject a mouthful of water, using a pattern of motion that shows a clear resemblance to the way newts protrude their sticky tongues to capture prey (Michel et al. 2015). From an evolutionary comparative perspective, a 4E approach allows us to appreciate the capacities of other species on their own terms: it brings the organism's body and environment into clear focus, in a way that allows the anthropocentric biases of the cognitivist position to be circumvented, and (hopefully) overcome.

Going Radical and Getting Our Continuity Right

Originally, the argument presented here was couched in terms of "conservative" embodied cognitive science (CEC) (Hutto and Myin 2012), which brings mind, body, and world together by reconfiguring representations as controllers of action ("action-oriented representations") rather than the abstract, action-neutral "mirrors" of the world favored by the classical cognitivist position (e.g., Clark 1997, 2008). More recently,

I have become persuaded by radical embodied and enactivist positions (REC), which embrace an anti-representational stance (Chemero 2009; Hutto and Myin 2012).

One reason is because, as Chemero (2013, p. 147) notes, CEC maintains a commitment to a computational theory of mind, "embracing some of the ideas of Gibson, Barwise and Perry and Brooks, but backpedaling on the strongest claims these authors made." Given this, REC should not be seen as the radicalization of Clark-style embodied cognitive science (as is often assumed); rather, Clark-style CEC should be seen as a watered-down version of REC. Hence, in Chemero's (2013) view, it is the computational elements of CEC that actually requires justification, and not the radical anti-representational stance, which is inherent to a functionalist embodied approach. Given the anthropocentric origins of the computational theory of mind (Barrett 2015a), ridding embodied theories of a computational stance allows us to begin thinking of cognition in a truly biogenic, bottom-up fashion.

Rejecting the notion of contentful representation for "basic minds" (i.e., those that are non- or prelinguistic) thus seems to offer the same horizon-expanding promise as the SBT. REC presents a new way of thinking about and conceiving of cognition, of understanding what minds are and how they become more elaborate. A representational stance, in contrast, seems to lead inexorably to an intellectualist view of other animals (and indeed our own species; Barrett 2011), in ways that may hinder rather than help understand cognitive evolution.

As I have argued elsewhere, the search for the precursors of our own cognitive and linguistic abilities in other animals reflects the same "explanatory need" that plagues studies of human cognition (Hutto 2013a): contentful representations are deemed to be present because a need for them has been established by a logical argument that simply insists this must be the case (Barrett 2015b). Specifically, the logic of evolutionary continuity is used to insist that, as human contentful representational knowledge could not have sprung from nowhere, then its precursors must be found in other species.

For example, Dorothy Cheney and Robert Seyfarth, pioneers and leaders in the field of comparative cognition, have argued that baboon social knowledge "reveal[s] a hierarchical, rule-governed structure with many properties similar to those found in human language" (Seyfarth et al. 2005, p. 264), as well as suggesting that vervet alarm calls are "best described as a proposition" and that "non-human primates certainly act as if they are capable of thinking (as it were) in sentences" (Cheney and Seyfarth 2005, pp. 150–1). But it is only the assumption that nonhuman minds *must* contain content-bearing representations that makes it necessary to characterize monkeys as possessing this kind of knowledge, when many aspects of their behavior actually speak against this view. Cheney and Seyfarth (2005) freely admit that their monkeys seem "unaware of their effects on their audiences' knowledge and beliefs" (p. 139), and do not "know that" others have beliefs and knowledge, nor do they show any evidence of "knowing that" they themselves possess knowledge and beliefs (Cheney and Seyfarth 1992; Cheney and Seyfarth 2008). Indeed, it is this mismatch that requires certain concepts either to be modified or watered down (e.g., a proposition is simply "a single utterance or thought that simultaneously incorporates a subject and a predicate," with no conditions of satisfaction or truth-preserving component; Cheney and Seyfarth 2005, pp. 150–1). As a result, it is unclear the extent to which these concepts can contribute to an understanding of either monkey or human cognitive evolution.

A commitment to both evolutionary continuity and the computational theory of mind forces comparative psychologists to ground human knowledge and language in more basic forms of knowledge of our closest relatives—to accept that at the basis of our knowledge is yet more knowledge. But this premise is shaky. As Wittgenstein (regarded by some as the first enactivist; Moyal-Sharrock 2013) argues, our most basic beliefs, our "hinge certainties," are non-epistemic and non-propositional: it is action that lies at the base of our knowledge (Wittgenstein 1974, p. 204). If this is correct, then it makes no sense to look for the precursors of human knowledge and linguistic thought in the (putative) thought processes of the living primates (or other organisms). For RECers, there are no precursors of this nature, and whatever we share in common with other species is to be found by seeking similarity in the nonrepresentational ways they act and control behavior, as these will form the common ground between humans and other living beings. In other words, we have to get our continuity the right way around, and the dominant computational view of mind seems to get it wrong.

REC is therefore attractive because it is explicitly pluralist in its approach. It does not argue against all forms of representational thought—only the notion that basic minds must possess content (Chemero 2009; Hutto and Myin 2012; Hutto 2013b). Linguistic, socioculturally scaffolded minds like ours clearly deal in rules and representations (else you would not be able to read and comprehend the words on this page), and it is quite clear that some of the interesting issues of human psychology are best understood using a representational framework. It is a mistake, however, to assume that, given this state of affairs pertains in the human species, that contentful thought must therefore characterize cognition across the animal kingdom.

Interestingly enough, this pluralist stance is seen as problematic precisely because it seems to deny evolutionary continuity. Menary (2015), for example, suggests that, if representational content is a feature only of cognitive systems scaffolded by language and sociocultural practices, REC is left with the problem of bridging the gap between basic cognitive processes and enculturated ones. That is, if we are to explain human abilities in natural terms we must ensure a seamless psychological continuity across all forms of life, and content-bearing representations provide the necessary link. In response to this, Hutto and Satne (2017) ask: does evolutionary continuity logically require psychological continuity? After all, as already discussed, this assumption raises its own problems. The insistence that there be no discontinuity between humans and other species also seems to deny the fact that humans occupy a unique sociocultural cognitive niche. If, as REC argues, content arises with the mastery of sociocultural practices and artifacts in this particular kind of niche—that is, if representations exist publicly externally first and are then "internalized," as Vygotsky (1997) argued—then the evolution and maintenance of these public sociocultural practices can, in fact, explain both the initial emergence and continued maintenance of content-involving minds (Hutto and Satne, 2017). It may be that public practices like the high-fidelity transmission of cultural behaviors, along with teaching (i.e., an elaboration of other skills also found in nonhuman species), are crucial to generating human minds (Laland, 2017). Here, then, we come full circle, returning to Brooks's (1999) argument that, as human cognitive abilities appeared in a mere blink of evolution's eye, they must have been relatively simple to implement. "Basic minds" were the hard part. REC therefore seems to get its continuity just right: it allows these abilities to emerge late, and builds them on top of basic minds, rather than extending content-dependent processes far back into the evolutionary past.

Acknowledgments

Thanks to Jessica Parker, Gert Stulp, and an anonymous reviewer for comments on an earlier draft. This work was supported by the Canada Research Chairs and NSERC Discovery Grant Programs.

References

Anderson, M. L. (2003). Embodied cognition: a field guide. *Artificial Intelligence*, 149, 91–130.
Barrett, L. (2011). *Beyond the brain: how body and environment shape animal and human minds*. Princeton, NJ: Princeton University Press.
Barrett, L. (2015a). A better kind of continuity. *The Southern Journal of Philosophy*, 53, 28–49.
Barrett, L. (2015b). Back to the rough ground and into the hurly-burly. In: D. Moyal-Sharrock, V. Munz, and A. Coliva (eds.), *Mind, Language and Action: Proceedings of the 36th Wittgenstein Symposium*. Berlin: De Gruyter, pp. 299–316.
Barwise, J., and Perry, J. (1983). *Situations and attitudes*. Cambridge, MA: MIT Press.
Beer, R.D. (1996). Toward the evolution of dynamical neural networks for minimally cognitive behavior. *From Animals to Animats*, 4, 421–9.
Brooks, R.A. (1991). Intelligence without representation. *Artificial Intelligence*, 47, 139–59.
Brooks, R.A. (1999). *Cambrian intelligence: the early history of the new AI*. Cambridge, MA: MIT Press.
Buckner, C. (2011). Two approaches to the distinction between cognition and "mere association." *International Journal of Comparative Psychology*, 24, 314–48.
Burrows, M., and Sutton, G. (2013). Interacting gears synchronize propulsive leg movements in a jumping insect." *Science*, 341, 1254–6.
Byrne, R.W. and Bates, L.A. (2006). Why are animals cognitive? *Current Biology*, 16, R445–48.
Chemero, A. (2009). *Radical embodied cognitive science*. Cambridge, MA: MIT Press.
Chemero, A. (2013). Radical embodied cognitive science. *Review of General Psychology*, 17, 145–50.
Cheney, D.L., and Seyfarth, R.M. (1992). *How monkeys see the world: inside the mind of another species*. Chicago: University of Chicago Press.
Cheney, D.L. and Seyfarth, R.M. (2005). Constraints and preadaptations in the earliest stages of language evolution. *The Linguistic Review*, 22, 135–59.
Cheney, D.L., and Seyfarth, R.M. (2008). *Baboon metaphysics: the evolution of a social mind*. Chicago: University of Chicago Press.

Clark, A. (1997). *Being there: putting brain, body, and world together again*. Cambridge, MA: MIT Press.

Clark, A. (2008). *Supersizing the mind: embodiment, action, and cognitive extension: embodiment, action, and cognitive extension*. Oxford: Oxford University Press.

Dunbar, R.I.M. and Shultz, S. (2007). Evolution in the social brain. *Science*, 317, 1344–7.

Gagliano, M., Renton, R., Depczynski, M., and Mancuso, S. (2014). Experience teaches plants to learn faster and forget slower in environments where it matters. *Oecologia*, 175, 63–72.

Gallagher, S. (2005). *How the body shapes the mind*. Cambridge: Cambridge University Press.

Gibson, J.J. (1979). *The ecological approach to visual perception*. Brighton: Psychology Press.

Healy, S.D. and Rowe, C. (2007). A critique of comparative studies of brain size. *Proceedings of the Royal Society B: Biological Sciences*, 274, 453–64.

Healy, S.D. and Rowe, C. (2013). Costs and benefits of evolving a larger brain: doubts over the evidence that large brains lead to better cognition. *Animal Behaviour*, 86, e1–e3.

Hutto, D.D. (2013a). Enactivism, from a Wittgensteinian point of view. *American Philosophical Quarterly*, 50, 281–302.

Hutto, D.D. (2013b). Psychology unified: from folk psychology to radical enactivism. *Review of General Psychology*, 17, 174–8.

Hutto, D.D., and Myin, E. (2012). *Radicalizing enactivism*. Cambridge, MA: MIT Press.

Hutto, D.D., and Satne, G. (2017). Continuity skepticism in doubt: a radically enactive take. In: C. Durt, T. Fuchs, and C. Tewes (eds.), *Embodiment, enaction, and culture: investigating the constitution of the shared world*. Cambridge, MA: MIT Press, pp. 107–28.

Jekely, G., Keijzer, F., and Godfrey-Smith. P. (2015). An option space for early neural evolution. *Philosophical transactions of the Royal Society of London. Series B, Biological sciences*, 370(1684), 20150181.

Keijzer, F. (2015). Moving and sensing without input and output: early nervous systems and the origins of the animal sensorimotor organization. *Biology and Philosophy*, 30, 311–31.

Keijzer, F., van Duijn, M., and Lyon, P. (2013). What nervous systems do: early evolution, input-output, and the skin brain thesis. *Adaptive Behavior*, 21, 67–85.

Kirsh, D. (1991). Today the earwig, tomorrow man? *Artificial Intelligence*, 47, 161–84.

Laland, K. (2017). *On the evolution of culture and the origins of human mind and intelligence*. Princeton, NJ: Princeton University Press.

Leadbeater, E. and Chittka, L. (2007). Social learning in insects—from miniature brains to consensus building. *Current Biology*, 17, R703–13.

Lihoreau, M., Latty, T., and Chittka, L. (2012). An exploration of the social brain hypothesis in insects. *Frontiers in Physiology*, 3, 442.

Lyon, P. (2006). The biogenic approach to cognition. *Cognitive Processing*, 7, 11–29.

Menary, R. (2015). Mathematical cognition: a case of enculturation. *Open MIND*, 25, 1–20.

Michel, K.B, Heiss, E. Aerts, P., and Van Wassenbergh, S. (2015). A fish that uses its hydrodynamic tongue to feed on land. *Proceedings of the Royal Society B, Biological Sciences*, 282. doi:10.1098/rspb.2015.0057

Moreno, A. and Etxeberria, A. (2005). Agency in natural and artificial systems. *Artificial Life*, 11, 161–75.

Moreno, A., Umerez, J., and Ibañez, J. (1997). Cognition and life: the autonomy of cognition. *Brain and Cognition*, 34, 107–29.

Moyal-Sharrock, D. (2013). Wittgenstein's razor: the cutting edge of enactivism. *American Philosophical Quarterly*, 50, 265–79.

Neisser, U. (1967). *Cognitive psychology*. Brighton: Psychology Press.

Pantin, C.F.A. (1956). The origin of the nervous system. *Pubblicazioni Della Stazione Zoologica Di Napoli*, 28, 171–81.

Penn, D.C. (2011). How folk psychology ruined comparative psychology: and how scrub jays can save it. In: J. Fischer (ed.), *Animal thinking: contemporary issues in comparative cognition*. Cambridge, MA: MIT Press, pp. 253–65.

Penn, D.C., Holyoak, K.J., and Povinelli, D.J. (2008). Darwin's mistake: explaining the discontinuity between human and nonhuman minds. *Behavioral and Brain Sciences*, 31, 109–78.

Pfeifer, R. and Bongard, J. (2006). *How the body shapes the way we think: a new view of intelligence*. Cambridge, MA: MIT Press.

Pfeifer, R. and Scheier, C. (2001). *Understanding intelligence*. Cambridge, MA: MIT Press.

Porter, M.M., Adriaens, D., Hatton, R.L., Meyers, M.A., and McKittrick, J. (2015). Why the seahorse tail is square. *Science*, 349, 6243.

Ramsey, W.M. (2007). *Representation reconsidered*. Cambridge: Cambridge University Press.

Rosch, E., Thompson, E., and Varela, F.J. (1992). *The embodied mind: cognitive science and human experience*. Cambridge, MA: MIT Press.

Roth, G. and Dicke, U. (2005). Evolution of the brain and intelligence. *Trends in Cognitive Sciences*, 9, 250–7.

Rowe, C. and Healy, S.D. (2014). Measuring variation in cognition. *Behavioral Ecology*, 25, 1287–92.

Rowlands, M. (2003). *Externalism: putting mind and world back together again*. Montreal: McGill-Queen's Press.

Seyfarth, R.M., Cheney, D.L., and Bergman, T.J. (2005). Primate social cognition and the origins of language. *Trends in Cognitive Sciences*, 9, 264–6.

Shapiro, J.A. (2007). Bacteria are small but not stupid: cognition, natural genetic engineering and socio-bacteriology. *Studies in History and Philosophy of Science, Part C: Studies in History and Philosophy of Biological and Biomedical Sciences*, 38, 807–19.

Shettleworth, S.J. (1999). *Cognition, evolution, and behavior*. Oxford: Oxford University Press.

Tarsitano, M. (2006). Route selection by a jumping spider (*Portia labiata*) during the locomotory phase of a detour. *Animal Behaviour*, 72, 1437–42.

Tarsitano, M.S., and Jackson, R.R. (1994). Jumping spiders make predatory detours requiring movement away from prey. *Behaviour*, 131, 65–73.

Taylor, B.L. (2004). An alternative strategy for adaptation in bacterial behavior. *Journal of Bacteriology*, 186, 3671–3.

Thompson, E. (2007). *Mind in life: biology, phenomenology, and the sciences of mind*. Cambridge, MA: Harvard University Press.

Troscianko, J., von Bayern, A.M.P., Chappell, J., Rutz, C., and Martin, G.R. (2012). Extreme binocular vision and a straight bill facilitate tool use in New Caledonian crows. *Nature Communications*, 3, 1110.

Uexküll, Von J. (2014). *Umwelt Und Innenwelt Der Tiere*. Berlin: Springer-Verlag.

van Duijn, M., Keijzer, F., and Franken, D. (2006). Principles of minimal cognition: casting cognition as sensorimotor coordination. *Adaptive Behavior*, 14, 157–70.

Vygotsky, L.V. (1997). *The collected works of L.S. Vygotsky: problems of the theory and history of psychology* (vol. 3). Berlin: Springer Science & Business Media.

Wittgenstein, L. (1974). *On certainty* (ed. G.E.M. Anscombe and G.H. Von Wright; trans. D. Paul and G.E.M. Anscombe). Oxford: Basil Blackwell.

CHAPTER 39

MINDSHAPING

TADEUSZ WIESŁAW ZAWIDZKI

Introduction

Mindshaping is a hypothesis about what makes human social cognition distinctive.[1] There are persuasive reasons for holding that distinctively human social cognition is key to our evolutionary success, i.e., the fact that our species dominates the planet, while our closest extant cousin-species teeter on the brink of extinction. Our technological feats rely heavily on our superb abilities at social organization. Unlike any other mammalian species, we routinely coordinate on complex, cooperative projects with complete strangers, often comprising groups of thousands of individuals, many of whom are often even unaware of each other (Seabright 2004; Sterelny 2012).[2] Our dazzling technological capacities would be impossible without traditions of social learning, enabling the preservation and gradual improvement, over historical time, of techniques of resource extraction, processing, and distribution, and of social communication and interaction. This "cumulative cultural evolution" (Boyd and Richerson 1996) or "ratchet effect" (Tomasello 1999) is the only known mechanism for generating the complex and sophisticated toolkits on which the biological success of human populations depends.[3] The

[1] Mameli (2001) introduces the concept of mindshaping as a potential function of mental state attribution. McGeer (1996, 2007) defends a similar notion: the regulative dimension of folk psychology. Zawidzki (2008, 2013) argues that it is the key to understanding the phylogeny of distinctively human social cognition.

[2] Indeed, there is evidence that demographic changes leading to increased contact among unfamiliar individuals triggered the so-called "broad spectrum revolution" (Sterelny 2012) in human culture and technology in the late Pleistocene (Mellars 2005; Powell et al. 2009). And there is evidence that military success in contemporary nomadic societies relies on cooperation among unfamiliar individuals (Matthew and Boyd 2011).

[3] Morin (2015) has challenged the consensus on this view of human cultural evolution. He argues that disruptive change and foresighted invention are much more important than faithful copying to human cultural evolution. And capacities not requiring faithful copying, like argumentation and communication, are much more central to such phenomena. This is an important corrective to excessive

social capacities enabling this "human cooperation syndrome" (Sterelny 2012) thus distinguish us from other species, and are among the most important pieces of the puzzle of human evolution.

There is a widely held and persuasive theory of the social capacities that make all of this possible (Humphrey 1980; Tooby and Cosmides 1995; Baron-Cohen 1999; Leslie 2000; Mithen 2000; Sperber 2000; Dunbar 2000, 2003, 2009; Siegal 2008). According to this received view, our species is distinct in its capacity to correctly represent mental states. It is because we have a neurally implemented capacity to correctly ascertain each other's beliefs, desires, and other propositional attitudes that we are able to coordinate and cooperate with, and learn from each other so well. There are many disagreements within this broad paradigm. Some concern the form that this neurally implemented cognitive capacity takes: is it something akin to a scientific theory (re)discovered during human ontogeny? (Wellman 2014). Or is it instead an innately specified, domain-specific computational module that gradually comes online during human ontogeny, as domain-general capacities for attention and working memory improve? (Baillargeon et al. 2010). Or is it more akin to a skill at simulating the perspectives of others, gradually acquired during human ontogeny? (Goldman 2006). There are also lively debates about how and where the human brain implements our capacities to read one another's minds (Saxe 2010). But all parties to these debates share a basic assumption: what sets us apart from other species, and explains the complex social capacities on which our evolutionary success depends is a neurally implemented, individual capacity to correctly ascertain each other's mental states, especially propositional attitudes, like beliefs and desires.

The mindshaping hypothesis rejects this assumption, and proposes an alternative. According to this alternative, our social accomplishments are not due to an individual, neurally implemented capacity to correctly represent each other's mental states. Rather, they rely on less intellectualized and more embodied capacities to shape each other's minds, e.g., imitation, pedagogy, and norm enforcement.[4] We are much better mindshapers, and we spend much more of our time and energy engaged in mindshaping than any other species. Our skill at mindshaping enables us to insure that *we come to have* the complementary mental states required for successful, complex coordination, without requiring us to solve the intractable problem of *correctly inferring* the independently constituted mental states of our fellows.

focus on conformism; however, it in no way challenges the thesis that conformism is necessary for cumulative cultural evolution. It is mysterious how communication, argumentation, foresighted invention, and disruptive change could solve the problem of reinventing the wheel: a population starting from scratch with each generation, no matter how innovative, would not soon achieve technological sophistication. And the kinds of abilities on which effective communication and argumentation rely presuppose faithful copying of communicative conventions, symbol systems, lexicons, etc., by new generations.

[4] I address the obvious rejoinder to this claim in the third section of the chapter: contra the consensus, these capacities do not require intellectualized, mindreading abilities.

Of course no champion of the received view would deny the importance of human mindshaping. Instead, they would claim that mindreading and mindshaping are complementary components of distinctively human social cognition. However, there remains a difference in emphasis. On the received view, mindreading is the key innovation on which the rest of the distinctively human sociocognitive syndrome depends. Without a capacity to correctly represent independently constituted beliefs and desires we could not shape each other's minds as effectively as we do. The mindshaping hypothesis reverses this priority: without appropriate mindshaping, attributing propositional attitudes is pointless and intractable. Furthermore, sophisticated versions of the kinds of *behavior*-reading available to our closest primate cousins are sufficient to support the sophisticated mindshaping practices necessary for successful mutual interpretation and coordination in human populations.

The mindshaping hypothesis is a natural ally of 4E approaches to human social cognition. Rather than conceptualize distinctively human social cognition as the accomplishment of computational processes implemented in the brains of individuals, involving the correct representation of mental states, the mindshaping hypothesis conceptualizes it as emerging from embodied and embedded practices of tracking and molding behavioral dispositions in situated, sociohistorically, and culturally specific human populations. Our sociocognitive success depends essentially on social and hence extended facts, e.g., social models we shape each other to emulate, both concrete ones, e.g., high status individuals, and "virtual" ones, e.g., mythical ideals encoded in external symbol systems. And social cognition, according to the mindshaping hypothesis, is in a very literal sense enactive: we succeed in our sociocognitive endeavors by cooperatively enacting roles in social structures.

The mindshaping hypothesis is also an ambitious attempt at theoretical integration. It seeks to reconcile insights about human social life from traditions often thought to be unrelated or even antithetical. Most dramatically, it rests on a neo-Darwinian justification for a broadly Nietzschean understanding of human sociality. Rather than conceive of human-specific biological adaptations to social life as neurally implemented computational systems aiming to correctly represent unobservable mental states, we should think of such adaptations as capacities to institute and enact social roles in social structures, and otherwise shape each other in ways that make coordination possible. Put another way, according to the mindshaping hypothesis, culturally specific ideologies to which members of human populations try to conform are the most adaptive way to solve the coordination problems that characterize distinctively human socio-ecology.

Theoretical stances in the sciences are often motivated by background metaphors that are seldom defended explicitly. For example, the idea that the universe is a blind mechanism replaced the idea that it is a collection of agencies hierarchically organized by a supernatural power during the scientific revolution of the seventeenth and eighteenth centuries. Since the publication of Wilfrid Sellars's *Empiricism and the Philosophy of Mind* (1956/1997), the philosophy and psychology of social cognition have accepted, largely uncritically, the metaphor that forms the centerpiece of that work: social cognition is conceptualized on the model of theoretical inference in science. On this metaphor, the success of our quotidian interactions depends largely on our ability to infer concrete, unobservable mental states that are causally responsible for observable behavior. The mindshaping hypothesis is above all an attempt to introduce, articulate, and defend an alternative metaphor. Rather than conceive of successful human social agents as scientific psychologists, it proposes that we conceive of them as engineers, teachers, pupils, actors, and advocates. Our social success depends on capacities to engineer social environments by teaching and learning roles to play in social structures, and defending our status within such structures through reasoned advocacy. As with all such highly abstract and metaphorical construals, it is difficult to identify empirical tests that vindicate this metaphor over the older one. However, science thrives when different paradigms are brought to bear on the same phenomena, and mindshaping should be understood as a new and viable alternative to the metaphor of human interpreters as scientific psychologists.

In what follows, I put some flesh on this skeletal outline. In the first section, I make clearer what mindshaping is supposed to be, focusing on how it can be independent of sophisticated mindreading, and illustrating its different varieties. In the process, I identify some ways in which human mindshaping is distinctive. In the second section, I motivate the mindshaping hypothesis by identifying some puzzles about human social cognition that it seems better suited to address than the received, mindreading view. The third section responds to the most persuasive criticism of the mindshaping hypothesis: our best theories of the various capacities on which sophisticated mindshaping relies claim these capacities presuppose the ability to accurately represent mental states. I conclude in the fourth section.

What is Mindshaping?

Any attempt to define mindshaping in a way that coheres with the spirit of the hypothesis articulated earlier immediately runs into a problem. It is unclear how social agents can purposefully and intelligently shape each other's minds without first accurately representing them. Surely, to intelligently shape a mind, whether one's own or another's, one must, at the very least, represent the current state of this mind, the state one desires for it, and some means of minimizing the difference between these. But this way of conceptualizing mindshaping makes it parasitic on sophisticated mindreading, and hence immediately jeopardizes its role in formulating an alternative to the received theory of what makes human social cognition distinctive. Fortunately, Darwin's theory of natural selection provides the resources necessary to make sense of intelligent behavior, including mindshaping, without assuming sophisticated representation, like the attribution of mental states.

Following a philosophical framework articulated and defended by Ruth Millikan (1984), I define mindshaping by appeal to the proper functions of cognitive mechanisms and the normal conditions on their operation. This definition is neutral on the precise means by which mindshaping occurs, and hence leaves room for the possibility of mindshaping mechanisms that do not rely on the representation of mental states. On Millikan's theory, evolved mechanisms have proper functions, i.e., effects that explain their selection in evolution. So, for example, the proper function of the heart is to pump blood. There are also normal conditions on the execution of such proper functions, e.g., hearts can perform their proper functions only if certain arteries are unobstructed. Millikan applies this framework to cognitive states in order to specify naturalistic conditions for individuating them in terms of their contents. For example, desires are individuated in terms of their proper functions: they aim to get organisms to bring about certain states of affairs; this is what explains their selection in evolution.[5] Mechanisms giving rise to desires to ingest food were selected because they led to the ingestion of food. Beliefs also have proper functions: they aim to combine with desires in order to give rise to practically rational behavior. Mechanisms giving rise to beliefs were selected because they produced beliefs that were accurate enough to guide organisms in ways that led to the satisfaction of their desires, e.g., by representing where food was.

Although Millikan's framework is unabashedly representationalist, when it is applied to the case of mindshaping, it need not be *metarepresentationalist*. That is, it provides the resources required to define mindshaping without presupposing a capacity to represent mental states. I define mindshaping as a relation between a target mind (the mind being shaped), a cognitive mechanism (the proper function of which involves shaping that mind), and a model that the mindshaping mechanism works to make the target mind match. Thus, mindshaping occurs when a cognitive mechanism selected in evolution for making target minds match models performs its proper function, in Millikan's sense.[6] Clearly, normal conditions on this must include representing the model accurately, but this need not involve the attribution of mental states. The reason is that mindshaping can occur simply in virtue of making the target mind disposed to match a pattern of behavior. Thus, all that needs to be represented is the model's behavior.[7] If there are mechanisms that use such representations to alter the dispositions of target minds in ways that make them more likely to match model behavior, they constitute mindshaping mechanisms that require no representation of mental states.

[5] Some radical versions of 4E cognitive science reject any appeals to contentful states, on the grounds that naturalization projects like Millikan's fail (Hutto and Myin 2013). However, even Hutto (2008) embraces Millikan's framework as an account of what he calls "biosemiotics"—the biological basis for determinate, intentional, though contentless directedness in natural cognitive systems. For my purposes here, it is unimportant whether the cognitive phenomena that constitute the explananda of Millikan's framework are understood as content-bearing states or as mere manifestations of intentional directedness.

[6] As I argue later, the precise selection pressures include advantages accruing to individuals capable of feats of coordination enabled by mindshaping.

[7] Again, note that the use of "represented" here is intended to be maximally neutral with regard to any controversies over the nature of representation. For those skeptical of full-blown, contentful representation, this can be rephrased in terms of Hutto's notion of "intentional directedness": target minds need only be intentionally directed at the model's behavior, not her mental states.

Let us make this more concrete by applying it to a specific example. A human infant observes an adult model turn on a light panel resting on a table by leaning over and touching it with her forehead (Meltzoff 1988). After seeing this, the infant is disposed to do the same when put in similar circumstances. This early form of infantile imitation clearly fits the definition of mindshaping. There is some cognitive mechanism in the infant that treats the behavior of the adult as a model to be matched, and disposes the infant to match it. However, there appears to be no reason to assume that the infant need represent the adult's intentions or other mental states in order to shape its mind in this way.[8] On this definition, mindshaping is widespread among nonhuman animals. For example, it applies to baby rats learning which foods to favor based on odors they smell on their mother's breath (Galef et al. 1983). In all such cases, it is arguable that there are cognitive mechanisms involved that alter behavioral dispositions to approximate behavioral patterns observed in social models.

Such a minimalist understanding of mindshaping raises another problem, however. If mindshaping is so widespread among nonhuman animals, how can it be used to explain what is distinctive about human social cognition? Here, there is again a temptation to collapse the distinction between the mindshaping hypothesis and the received view that human social cognition is distinctive in its reliance on sophisticated mindreading. How else can human-specific mindshaping be distinguished from other varieties? A brief survey of recent empirical work on human social learning shows that there are actually at least four ways of distinguishing human-specific mindshaping from other varieties, without assuming that it relies on sophisticated mindreading.

First, the developmental and comparative literature on imitation provides overwhelming evidence of a clear distinction in the *scope* of human vs. nonhuman imitation. Most nonhuman species are limited to acquiring *new goals* from observing the behavior of others, while selecting their *own methods* of accomplishing those goals. For example, many bird species can learn from observing conspecifics that food can be extracted from a particular location, but then go on to discover their own method of extracting it, ignoring the method used by their model (Zentall 2006). The one nonhuman exception to this appears to be chimpanzees (Horner and Whiten 2005). They can sometimes acquire both goal and method from a model, but only when there is no alternative method available to them. If they come to discover a different, more efficient method to accomplish the goal, chimpanzees immediately switch to it, ignoring the model's method. Surprisingly, this is *not* the case with human children. When shown a method to accomplish some goal by an adult model, human children persist in using that method, even after they are made aware of a more efficient method, through demonstrations that components of the modeled method are superfluous or irrelevant to accomplishing the goal. They persist in the modeled method, even when the adult model is not present and

[8] This case, however, is much more complicated than I suggest here. Infants respond very differently to subtle variations in such scenarios, and this leads many researchers to conclude that even such apparently simple imitation relies on sophisticated mindreading. I respond to this claim later, in the third section.

they think they are alone and unobserved, so fear of contradicting an adult cannot explain this phenomenon. Such "overimitation" (Lyons et al. 2007; Nielsen and Tomaselli 2010) is a distinctively human form of mindshaping. Yet, it does not appear to require sophisticated mindreading, like the attribution of propositional attitudes. Human children need only represent the goal of an adult model's behavior and the precise sequence of behavioral steps used by her in accomplishing the goal.

A second distinctive feature of human mindshaping is a plausible explanation of phenomena like overimitation. Matching model behavior, for humans but not nonhumans, appears to be its own reward. Nonhumans will imitate a target to the extent that it helps accomplish some further goal, like extracting food from a novel location (Zentall 2006). Humans, on the other hand, seem to find matching a model's goal intrinsically rewarding. This explains overimitation: children imitate the precise means of accomplishing a desired goal even if they are aware of more efficient means of accomplishing the same goal. It is plausible that this is due to the fact that they experience some kind of reward signal for matching model behavior precisely that outweighs the value of accomplishing the goal as efficiently as possible. There are other forms of mindshaping that also appear to show intrinsic motivation. For example, the costly punishment of norm flouters appears widespread in human populations (Henrich et al. 2006). Since this involves incurring a cost in order to punish counter-normative behavior, it suggests that shaping minds to respect norms is intrinsically motivating (Sripada and Stich 2006). Thus, the fact that human mindshaping appears intrinsically motivating is another feature that sets it apart from nonhuman varieties.

A third distinctive feature of human mindshaping is the socially extended nature of many human mindshaping mechanisms. For example, although there are some limited examples of pedagogy among nonhuman species (Thornton and McAuliffe 2006), none come close to the sophistication and pervasiveness of pedagogy in human populations. Unlike imitation, pedagogy relies on extra-mental components, e.g., active guidance by a teacher. Perhaps the most pervasive form of pedagogy in human populations takes the master-apprentice form, in which experts provide subtle behavioral guidance to novices (Sterelny 2012). This kind of pedagogy is possible without a sophisticated language or even, arguably, a sophisticated theory of mind. It seems to involve the gradual molding of novice behavioral dispositions via skilled expert demonstrations and interventions. There is evidence that human infants are, from a very young age, extremely receptive to this style of "natural pedagogy" (Csibra and Gergely 2011). For example, from a very young age they interpret certain stereoptyped adult communicative behaviors, such as eye contact, as overtures to demonstrating novel, generalizable information about referential objects specified by subsequent stereotyped behaviors, such as eye saccades. As human civilization grew more complex, socially extended mindshaping mechanisms became more sophisticated. We now have institutions of formal education and sanctioning to shape group members to play highly specific roles in very complex social structures.

A fourth distinctive feature of human mindshaping concerns its use of abstract, fictional models. All nonhuman mindshaping involves matching some aspect of the observable behavior of another, actual, concrete individual. But many of the most sophisticated forms of human mindshaping involve matching the behavior of fictional models like protagonists of myths or morally ideal agents. This is possible due to the representational power of public language. We can formulate public representations of non-actual states of affairs, including non-actual patterns of behavior by fictional agents that we go on to imitate. This form of mindshaping does not obviously require sophisticated mindreading, only a public language for representing the *behavior* of fictional models.[9]

Thus, it is possible to describe forms of mindshaping unique to humans without assuming that they rely on sophisticated mindreading, like propositional attitude attribution. Distinctively human mindshaping is (1) intrinsically motivating, (2) maximally flexible in the aspects of model behavior that it seeks to match (as in overimitation), (3) often reliant on external components (e.g., expert guidance and sophisticated pedagogical practices and institutions), and (4) often seeks to match the behavior of fictional models. Of course, it is empirically possible that these varieties of mindshaping presuppose, in practice, sophisticated mindreading, including the attribution of propositional attitudes. I respond to variants on this objection later, in the third section. However, there are good empirical reasons to doubt this. First, as I argue next, in the second section, it is puzzling how full-blown propositional attitude attribution can be accurate, timely, and computationally tractable at the same time; so it is not clear that it can support the mindshaping practices described earlier. Second, there is already empirical evidence of low-level mindshaping mechanisms in humans that appear independent of propositional attitude attribution. For example, an fMRI study identified low-level mechanisms of social conformism that make use of basic reward circuits involved in behavioral conditioning (Klucharev et al. 2009).[10] Signals that one's behavior fails to conform to group behavior can play the same role as prediction errors in individual learning (i.e., when a planned behavior does not have its intended effect), driving individuals to conform to their groups. There is no evidence that such mechanisms require the representation of propositional attitudes.

I now turn to a discussion of various explanatory advantages of the mindshaping hypothesis over the received mindreading hypothesis. It turns out that there are a number of deep problems about distinctively human social cognition that arise for the received view that can be solved or avoided on the mindshaping hypothesis.

[9] This may seem far-fetched: how can sophisticated uses of language, such as the production of fictional narratives, *not* require sophisticated mindreading? Later, in the third section, I address this worry, arguing that, on the typical understanding of "sophisticated mindreading," there are good reasons to doubt that it is necessary for sophisticated language use, like the production of fictional narratives.

[10] Although there are good reasons for general skepticism about the neurocognitive/functional interpretation of fMRI data (Eklund et al. 2016), in conjunction with other evidence from a variety of fields applying different methodologies, it can strengthen the case for hypotheses about cognitive function. As the study cited here notes, there are many kinds of converging evidence that human beings are prone to social conformism in judgment, and that the relevant brain circuits are involved in behavioral conditioning.

The Advantages of Mindshaping

The basic story motivating the received view that human social cognition is distinctive in its reliance on sophisticated mindreading is well known and persuasive. The idea is that our prehistoric ancestors faced strong selection pressures for Machiavellian intelligence (Humphrey 1980). Due to unusually large and complex groups, they had to learn both how to take advantage of others and how to prevent others from taking advantage of them. This triggered an evolutionary "arms race," the result of which was an advanced theory of mind, supporting the reliable attribution of propositional attitudes like belief and desire. It is easier to take advantage of others, e.g., by deceiving them, if one can correctly represent their mental states, especially false beliefs. Once this capacity is prevalent in a population, it pays to detect deception, so the capacity to attribute more complex mental states, like intentions to deceive or (higher-order) beliefs about false beliefs, is incentivized. Once the capacity to detect deception is widespread in a population, a new incentive arises: knowing when one's deception is likely to be detected. This requires an even more sophisticated theory of mind. In this way, on the received view, social complexity in human prehistory triggered Machiavellian adaptations in the form of capacities to attribute increasingly complex types of mental states.

Although the social brain hypothesis is increasingly taking forms other than this Machiavellian variant (Dunbar and Shultz 2007), the notion that the capacity to attribute mental states was the central function driving the expansion of the social brain in human prehistory continues to wield enormous influence. Consider, for example, this preface to a recent report of empirical evidence that even seven-month-old human infants attribute mental states:

> Humans are guided by internal states such as goals and beliefs. Without an ability to infer others' mental states, society would be hardly imaginable. Social interactions, from collective hunting to playing soccer to criminal justice, critically depend on the ability to infer others' intentions and beliefs. Such abilities are also at the foundation of major evolutionary conundra. For example, the human aptitude at inferring mental states might be one of the crucial preconditions for the evolution of the cooperative social structure in human societies. (Kovács et al. 2010, p. 1830)

However, there are a number of serious problems with this picture. First, it appears that our closest nonhuman cousins, chimpanzees, live in groups large and complex enough to incentivize deception and deception detection. The Machiavellian nature of "chimpanzee politics" is well known (de Waal 2000). Yet the consensus in experimental, comparative psychology is that chimpanzees are incapable of attributing full-blown propositional attitudes, like beliefs and sophisticated desires or intentions (Call and Tomasello 2008). They can certainly interpret conspecific behavior in terms of its goals, and the perceptions or knowledge by which it is guided. But there is little evidence that they conceive of their conspecifics as animated by unobservable states of mind with complex relations to each other and observable behavior (Povinelli and Vonk 2003, 2004; Penn and Povinelli 2007). Given that chimpanzees and other intelligent nonhuman animals living in complex social groups appear to manage Machiavellian intelligence without the attribution of full-blown propositional attitudes, the evolutionary story behind the received view that human social cognition is distinctive in this capacity seems unmotivated. It seems that sophisticated mindreading, such as the accurate attribution of propositional attitudes, is unnecessary for Machiavellian success in socially complex groups.

A second problem with the received mindreading view concerns the computational tractability of accurate mindreading. The attribution of full-blown propositional attitudes is holistically constrained. There is no simple, one-to-one mapping between the observable behavior and circumstances of a target of attribution and the propositional attitudes that she instantiates in her mind. It is a familiar philosophical point that any finite set of beliefs and desires is compatible with any observable behavior or circumstance, given appropriate adjustments to background beliefs and desires (Morton 1996, 2003; Bermúdez 2003, 2009). An agent may want to stay dry and believe that it is raining, while standing in the rain with an unopened umbrella, due to certain background beliefs, such as the belief that opening the umbrella will trigger a bomb, or that the umbrella does not work, etc. Given this holism, it is hard to see how interpreters can come to accurate attributions and predictions in time to react appropriately to rapidly evolving, dynamic, real-world social situations.

This is particularly problematic if, as seems to be the case with nonhuman Machiavellian intelligence, successful social cognition can make do with far less complicated cognitive capacities. If tracking the goals and information access of conspecifics is enough to support adaptive behavior in real-world social contexts, then why would natural selection support a further capacity to correctly represent the actual psychological causes of conspecific behavior, together with the complex "ceteris paribus" laws that link them (Gauker 2003, p. 240), when this appears to be too computationally demanding to make a difference in real time? This point is often lost on empirical researchers because they equate the attribution of beliefs and desires with far simpler capacities like tracking goals and information access. However, it is certainly possible to perceive a bout of behavior as aiming at a goal, and informed by a (potentially non-actual) worldly situation, without conceiving of it as caused by an unobservable mental state with complex connections to behavior encoded in "ceteris paribus" laws. Such an embodied, perceptual attunement to relational properties of bouts of behavior is more likely to support timely and accurate behavioral anticipation than inference over hidden mental states with tenuous connections to behavior. And there is increasing consensus that the quotidian social cognition of nonhumans, human infants, and even human adults in most circumstances takes something like this less intellectualized form (Hutto 2008; Gallagher and Hutto 2008; Apperly and Butterfill 2009; Apperly 2011; Butterfill and Apperly 2013).

A final set of problems with the received view of distinctively human social cognition as dependent on reliable propositional attitude attribution concerns the kinds of social situations likely faced by our prehistoric ancestors, especially opportunities to deceive and otherwise free ride on the cooperative dispositions of group mates. Given the holism problem and the relatively long time course for developing the capacity to attribute higher-order propositional attitudes in ontogeny (Perner and Wimmer 1985), it is unlikely that deception in prehistory was checked through more sophisticated mindreading. But it had to be checked somehow. After all, our capacity to coordinate on extremely complex cooperative projects with large numbers of individuals, of whom we have little personal knowledge, is one of the most important distinguishing marks of human sociality.

Furthermore, even assuming largely cooperative dispositions among early human populations, it is not obvious how sophisticated mindreading could help them solve *coordination problems*. You might think that knowing what your partner in a coordination problem is thinking would help you select behavior that leads to successful coordination. However, things are not so simple. The problem is that your partner is in exactly the same situation as you: she must know what you are thinking. But if the mindreading is accurate, she will learn only that you are thinking about what she is thinking, just as you will learn only that she is thinking about what you are thinking (Gilbert 1996; Bacharach 2006). Trying to reconnect a disconnected telephone conversation is a good example of this. If both parties call back at the same time then they will not reconnect. If both parties wait for the other to call, then they will not reconnect. They must somehow figure out who is to call back and who is to wait. But mindreading appears to be of no help here, since A's accurate mindreading of B reveals only that B is trying to read A's mind, and vice versa. Thus, it is not even clear that sophisticated, accurate mindreading is sufficient for solving simple coordination problems of the kind likely faced by our prehistoric ancestors.

The mindshaping hypothesis, coupled with some plausible conjectures about the role of "cultural group selection" (Henrich 2004) in human prehistory, offers a way of avoiding these problems. The basic idea is simple: if members of human populations are shaped via the kinds of mechanisms discussed earlier to routinely adopt similar or complementary mental states and behavioral dispositions when coordinating on cooperative tasks, then the holism problem should never arise, and deception and other forms of free-riding should be rare. If there are mindshaping practices in human populations that insure that most of one's potential interactants react in familiar, coordination- and cooperation-enhancing ways—e.g., by conforming to norms promulgated via the behavior of well-known real and fictional models—then our social cognition should succeed even if it relies on relatively unsophisticated sociocognitive mechanisms. We should manage to track the behavior of our group mates simply by attributing to the behavior goals by which we think people ought to be motivated in such circumstances, and assuming it is guided by information we think people ought to find relevant to such goals. This should work because, as a matter of fact, most people with whom we interact are products of similar mindshaping regimes as we are; such mindshaping prevents the radical cognitive heterogeneity that might thwart such simple heuristics. Furthermore, groups composed of members shaped to favor cooperation over free-riding and to respect coordination norms should outcompete groups composed of members not so shaped, and this could explain, via cultural group selection, the evolution of human capacities for coordination on cooperative projects (Henrich 2004).

On the face of it, tracing distinctively human social cognition to virtuosity at the sorts of mindshaping practices I described in the second section appears to avoid the major shortcomings of the received view that it depends on sophisticated mindreading, like the attribution of propositional attitudes. However, many will find this unconvincing. It is not obvious that the mindshaping mechanisms described in the second section, e.g., overimitation, natural pedagogy, and the emulation of fictional agents encoded in public language, require no sophisticated mindreading, no attribution of propositional attitudes. In fact, many theories of such mechanisms posit precisely such capacities. This is the most serious problem for the mindshaping account, considered as an alternative to the received view. I address it next.

Sophisticated Mindshaping without Sophisticated Mindreading

The received view is not just a hypothesis about the phylogenetically most important component of distinctively human social cognition. It is also central to most current explanations of most sophisticated, human social capacities. The distinctively human mindshaping mechanisms and practices I discussed in the second section are no exception. Consider overimitation. The capacity of human infants to imitate adult models who switch on light panels lying on tables with their foreheads is a classic example of overimitation: they learn an inefficient method of accomplishing a goal which they could accomplish much more easily, i.e., by switching the light on by hand. Subsequent experiments show that this is not mere blind copying (Gergely et al. 2002). If the adult model's hands are otherwise occupied or out of view when she switches on the light panel with her forehead, infants learn to switch on the light panel using the most efficient method available to them: with their hands. A natural interpretation of this is that infant imitators rely on the attribution of intentions to adult models. When an adult model switches on the light panel with her forehead while her hands are free, she must intend specifically to use her forehead, since she could more easily switch it on with one of her hands. But when an adult model switches on the light panel with her forehead while her hands are occupied, she must intend to switch it on by the most efficient method available to her.

Natural pedagogy is also typically explained in terms of sophisticated infant mindreading. For example, Csibra (2010) argues that natural pedagogy relies on the capacity to attribute higher-order intentions. On this explanation, infants interpret eye contact as expressing the communicative intention that immediately ensuing behavior be interpreted as intending to inform the infant of novel information concerning some salient object. On this view, natural pedagogy relies on infant capacities to attribute second-order propositional attitudes.

Finally, in the second section I suggested that our capacity to copy non-actual patterns of behavior by fictional agents encoded in public language is one of the most sophisticated forms of distinctively human mindshaping. But mastering a public language is routinely explained in terms of capacities to attribute complex propositional attitudes. For example, according to Sperber and Wilson (2002), all linguistic communication presupposes the capacity to attribute nested intentions and beliefs. And, according to Bloom (2002), word learning requires the capacity to attribute referential intentions to adult models. Thus, it would seem that any mindshaping reliant on the representation of model behavior in public language presupposes sophisticated mindreading. If these theories of overimitation, natural pedagogy, and language use are correct, then the distinctively human mindshaping practices and mechanisms discussed in the second section presuppose sophisticated mindreading, and hence cannot constitute an alternative to the received view of what makes human social cognition distinctive.

This whole question turns on what we mean by "sophisticated mindreading" and "propositional attitude attribution." Most philosophers of psychology follow Wilfrid Sellars (1956/1997) when interpreting these concepts. Propositional attitudes are treated as states of an unobservable causal nexus responsible for an agent's behavior: the agent's mind. Furthermore, as I noted earlier, their relations to observable circumstances and behavior are holistically constrained: what one does in specific circumstances depends on indefinitely broad networks of propositional attitudes; hence, it should be difficult to determine an agent's propositional attitudes based on observations of finite bouts of behavior, and an agent's future behavior based on attributions of finite sets of propositional attitudes. Finally, if we take the Sellarsian picture seriously, and think of propositional attitude attribution on the model of scientific hypotheses about unobservable causal factors, then propositional attitude attribution should involve a strong appearance/reality distinction. Think of medical diagnosis here. Because the causes of symptoms, e.g., bacteria, are unobservable factors independent of the symptoms, it is always possible that two qualitatively similar sets of observable symptoms are products of radically different unobservable factors. Appearance does not determine reality. If propositional attitude attribution is supposed to be like this, then it requires an appreciation of the possibility that two qualitatively indistinguishable patterns of observable behavior are caused by radically different sets of propositional attitudes.

If we conceive of propositional attitude attribution along these lines, there is good reason to doubt that sophisticated human mindshaping, like overimitation, natural pedagogy, and language-assisted mindshaping presuppose propositional attitude attribution. For one thing, the speed and fluency with which infants overimitate, interpret pedagogical interactions, and engage in linguistic interactions suggest that they are not engaging in scientific reasoning about unobservable causes with tenuous connections to observable behavior. Second, it is very unlikely that such mindshaping capacities rely on an appreciation of a strong behavioral appearance/mental reality distinction. There is no evidence that human infants can conceptualize the possibility that qualitatively indistinguishable patterns of behavior might be products of radically different sets of propositional attitudes. Typically, when tested for capacities to interpret behavior, infants and children show no hesitation: they see behavior as unambiguously directed at specific goals and informed by specific situations. Thus, if we think of sophisticated mindreading and propositional attitude attribution along Sellarsian lines, there is no reason to suppose that distinctively human mindshaping depends on them.

How else might we conceive of the sociocognitive capacities underlying distinctively human mindshaping? One possibility is to think of human mindshapers and "mindshapees" as operating with an ontology of informed, goal-directed bouts of behavior. To be goal-directed, a bout of behavior must be predictable on the assumption that it constitutes the most efficient of observable means to some observable end state. To be informed by some (possibly non-actual) situation, a bout of behavior must count as the most efficient of observable means to some observable end state *relative to that situation*. One can perceive bouts of behavior as goal-directed and informed in these ways, *without* thinking of them as caused by representations of goals and information within the unobservable minds of agents. A number of theorists have defended the hypothesis that human infant and even most human adult social cognition relies on such minimalist assumptions about behavior (Gergely and Csibra 2003; Apperly and Butterfill 2009; Apperly 2011; Butterfill and Apperly 2013). Although these are still early days, this is a viable hypothesis about the sociocognitive capacities underlying sophisticated human mindshaping that does not presuppose a capacity for sophisticated mindreading, at least not in a Sellarsian sense.

Furthermore, there is evidence that our closest nonhuman cousins, chimpanzees, also sometimes rely on the assumption that their conspecifics engage in goal-directed bouts of behavior that are informed by (possibly non-actual) worldly states (Crockford et al. 2011). Thus, if sophisticated human mindshaping relies exclusively on similar assumptions, then distinctively human mindshaping does not presuppose *distinctively* human means of interpreting behavior. Motivations to treat conspecific behaviors as models for one's own seem more important to distinguishing human from nonhuman social cognition than assumptions about goal-directedness or informedness. Of course, it is true that humans deploy such assumptions about behavior in far more subtle, complex, and diverse ways than chimpanzees or any other nonhuman species. But we can conceive of such differences as products of the gradual evolutionary accumulation of tweaks to a basic sociocognitive capacity we share with other social primates, aimed at improving mindshaping practices, i.e., making us better overimitators, pupils, teachers, and conformers to linguistically encoded fictional models. On this view, though there are differences between humans and nonhumans in the scope and sophistication of our means of interpreting behavior, these are products of evolution for better mindshaping, which was necessary in our distinctive, cooperative socio-ecology. Mindshaping remains the source of our sociocognitive distinctiveness.

This perspective raises another worry, however. If our most important sociocognitive feats are products of motivations to use others as models for our own behavior, guided by enhanced versions of ancient primate capacities to interpret behavior as goal-directed and informed by (possibly non-actual) worldly states, then why do we engage in sophisticated mindreading at all? What function is left for the attribution of full-blown propositional attitudes? On the mindshaping hypothesis, this is a late-arriving capacity involved in the *justification* rather than the prediction of behavior (cf. Bruner 1990; Andrews 2009). Given our highly interdependent, cooperative socio-ecology, one's social status and, hence, ultimate biological success is heavily dependent on being perceived as a competent and reliable potential partner in coordination on complex, cooperative projects. Therefore, it is unsurprising that anomalous behavior jeopardizing one's reputation for such competence and reliability, e.g., misinforming someone or reneging on an explicit commitment, immediately puts one's social status at risk. In fact our status may be even more precarious. Assuming that our potential cooperation partners see behavior that resembles their own, is familiar, and, in general, respects prevalent norms as a signal of general trustworthiness,[11] any kind of deviant behavior might put status at risk. In such circumstances, it is useful to have a practice of rehabilitating status, of explaining away apparent deviance by showing the behavior to be reasonable in the light of propositional attitudes of which witnesses might be unaware. On this view, the attribution of full-blown propositional attitudes first gains traction as a means of normalizing apparently deviant behavior (Bruner 1990). This proposal even has some empirical support (Malle et al. 2007).

If full-blown propositional attitude attribution functions to mitigate the social fallout from apparently deviant behavior, many of the properties that make it unsuitable as a prediction device start to look adaptive. Holism seems tailor-made for this function, as any behavior can be made to accord with any set of propositional attitudes, given appropriate adjustments to background propositional attitudes. A behavioral appearance/mental reality distinction also begins to make sense: the whole point of rationalization is that behavior seemingly caused by one set of mental states might actually be caused by a different set. Once such a practice prevails in a population, one would expect members to start interpreting all behavior, both their own and others', in terms of potential rationalizations. There would also be incentives to police behavior to insure that it stays rationalizable in terms of widely tolerated propositional attitudes. In such populations, individuals would actively shape themselves and each other to conform to expectations generated by propositional attitude attributions. On the mindshaping hypothesis, this dynamic characterizes many modern human populations. The idea that propositional attitude attribution is the most important component of human social cognition is an illusion born of these relatively recent, sociohistorical circumstances.

[11] An assumption for which there is some empirical evidence (Wiltermuth and Heath 2009).

Conclusion

Of necessity, this has been a relatively superficial exploration of the mindshaping hypothesis. But the broad contours, I hope, are clear. On this hypothesis, what sets human social cognition apart from other varieties is our capacity to shape each other and ourselves into the kinds of agents that can coordinate successfully on cooperative projects. This capacity does not presuppose sophisticated mindreading, like propositional attitude attribution. Instead, it relies on sophisticated versions of behavior-reading strategies also present in nonhuman species, coupled with unusually strong motivations to copy the behavior of others. This variant of the basic primate sociocognitive toolkit was adaptive in the distinctively cooperative socio-ecologies of prehistoric hominins. Eventually it gave rise to such human-specific phenomena as overimitation, pervasive pedagogy, and the imitation of fictional agents. The contrast with the received view is clear: we are not scientific psychologists first, and only later social engineers, teachers, pupils, actors, and advocates. In fact, this gets things almost exactly backward. It is only relative to a social environment engineered via sophisticated practices of mindshaping that the practice of justifying behavior in terms of full-blown propositional attitudes makes sense.

Also clear from the foregoing are the rich affinities between the mindshaping hypothesis and 4E approaches to social cognition. With many champions of such approaches, the mindshaping hypothesis rejects the common assumption that solving the "other minds" problem via inference to unobservable mental states is the basis for human sociocognitive competence (Hutto 2008; Gallagher and Hutto 2008). The kinds of low-level behavior tracking and shaping assumed in my characterization of mindshaping mechanisms plausibly qualify as examples of embodied and embedded cognition. They involve attunement to low-level, bodily dispositions, contextualized to specific sociocultural embeddings. Furthermore, external or extended structures are central to many human-specific forms of mindshaping, including the roles of teachers and pedagogical institutions, as well as both concrete and abstract (fictional) models, the latter encoded in public systems of representation. It is true that, in assuming Millikan's teleosemantics in the definition of mindshaping, I accept a kind of representationalism that some champions of 4E approaches to cognition might find anathema. However, there are moderate varieties of 4E approaches that allow for the possibility of some forms of representation (Clark 1997), and at least one prominent defender of 4E approaches endorses Millikan's teleosemantics, though as a theory of biosemiotics rather than biosemantics (Hutto 2008). Thus, there is much potential for a productive mutualism between the mindshaping hypothesis and 4E approaches to human social cognition.

References

Andrews, K. (2009). Understanding norms without a theory of mind. *Inquiry*, 52(5), 433–48.
Apperly, I.A. (2011). *Mindreaders*. Hove, UK: Psychology Press.

Apperly, I.A. and Butterfill, S.A. (2009). Do humans have two systems to track beliefs and belief-like states? *Psychological Review*, 116(4), 953–70.
Bacharach, M. (2006). *Beyond individual choice.* Princeton: Princeton University Press.
Baillargeon, R., Scott, R., and He, Z. (2010). False-belief understanding in infants. *Trends in Cognitive Sciences*, 14(3), 110–8.
Baron-Cohen, S. (1999). The evolution of a theory of mind. In: M.C. Corballis and S.E.G. Lea (eds.), *The descent of mind.* New York: Oxford University Press, pp. 261–77.
Bermúdez, J.L. (2003). The domain of folk psychology. In: A. O'Hear (ed.), *Minds and persons.* Cambridge: Cambridge University Press, pp. 25–48.
Bermúdez, J.L. (2009). Mindreading in the animal kingdom. In: R. Lurz (ed.), *The philosophy of animal minds.* Cambridge: Cambridge University Press, pp. 145–64.
Bloom, P. (2002). Mindreading, communication, and the learning of names for things. *Mind and Language*, 17(1), 37–54.
Boyd, R. and Richerson, P.J. (1996). Why culture is common but cultural evolution is rare. *Proceedings of the British Academy*, 88, 73–93.
Bruner, J. (1990). *Acts of meaning.* Cambridge, MA: Harvard University Press.
Butterfill, S. and Apperly I.A. (2013). How to construct a minimal theory of mind. *Mind and Language*, 28(2), 606–37.
Call, J. and Tomasello, M. (2008). Does the chimpanzee have a theory of mind? 30 years later. *Trends in Cognitive Sciences*, 12, 187–92.
Clark, A. (1997). *Being there: putting brain, body and world together again.* Cambridge, MA: MIT Press.
Crockford, C., Wittig, R.M., Mundry, R., and Zuberbühler, K. (2011). Wild chimpanzees inform ignorant group members of danger. *Current Biology*, 22(2), 142–6.
Csibra, G. (2010). Recognizing communicative intentions in infancy. *Mind and Language*, 25, 141–68.
Csibra, G. and Gergely, G. (2011). Natural pedagogy as evolutionary adaptation. *Philosophical transactions of the Royal Society of London. Series B, Biological sciences*, 366, 1149–57.
de Waal, F.B.M. (2000). *Chimpanzee politics.* Baltimore: Johns Hopkins University Press.
Dunbar, R. (2000). On the origin of the human mind. In: P. Carruthers and A. Chamberlain (eds.), *Evolution and the human mind: modularity, language, and meta-cognition.* Cambridge: Cambridge University Press, pp. 238–53.
Dunbar, R. (2003). The social brain: mind, language, and society in evolutionary perspective. *Annual Review of Anthropology*, 32, 163–81.
Dunbar, R. (2009). Why only humans have language. In: R. Botha and C. Knight (eds.), *The prehistory of language.* Oxford: Oxford University Press, pp. 12–35.
Dunbar, R. and Shultz, S. (2007). Evolution in the social brain. *Science*, 317(5843), 1344–7.
Eklund, A., Nichols, T., and Knutsson, H. (2016). Cluster failure: why fMRI inferences for spatial extent have inflated false-positive rates. *Proceedings of the National Academy of Sciences*, 113(28), 7900–5.
Galef, B.G., Wigmore, S.W., and Kennett, D.J. (1983). A failure to find socially mediated taste aversion learning in Norway rats (*R. norvegicus*). *Journal of Comparative Psychology*, 97(4), 358–63.
Gallagher, S. and Hutto, D. (2008). Understanding others through primary interaction and narrative practice. In: J. Zlatev et al. (eds.), *The shared mind: perspectives on intersubjectivity.* Amsterdam: John Benjamins, pp. 17–38.
Gauker, C. (2003). *Words without meaning.* Cambridge, MA: MIT Press.

Gergely, G., Bekkering, H., and Király, I. (2002). Rational imitation in preverbal infants. *Nature*, 415, 755–6.
Gergely, G. and Csibra, G. (2003). Teleological reasoning in infancy: the naive theory of rational action. *Trends in Cognitive Sciences*, 7(7), 287–92.
Gilbert, M. (1996). *Living together: rationality, sociality, and obligation*. Lanham, MD: Rowman & Littlefield.
Goldman, A.I. (2006). *Simulating minds: the philosophy, psychology, and neuroscience of mindreading*. Oxford: Oxford University Press.
Henrich, J. (2004). Cultural group selection, coevolutionary processes, and large-scale cooperation. *Journal of Economic Behavior and Organization*, 53, 3–35.
Henrich, J., McElreath, R., Barr, A., Ensminger, J., Barrett, C., Bolyanatz, A. et al. (2006). Costly punishment across human societies. *Science*, 312, 1767–9.
Horner, V. and Whiten, A. (2005). Causal knowledge and imitation/emulation switching in chimpanzees (*Pan troglodytes*) and children (*Homo sapiens*). *Animal Cognition*, 8, 164–81.
Humphrey, N. (1980). Nature's psychologists. In: B.D. Josephson and V.S. Ramachandran (eds.), *Consciousness and the physical world*. Oxford: Pergamon Press, pp. 57–80.
Hutto, D.D. (2008). *Folk psychological narratives: the sociocultural basis of understanding reasons*. Cambridge, MA: MIT Press.
Hutto, D.D. and Myin, E. (2013). *Radicalizing enactivism*. Cambridge, MA: MIT Press.
Klucharev, V., Hytönen, K., Rijpkema, M., Smidts, A., and Fernández, G. (2009). Reinforcement learning signal predicts social conformity. *Neuron*, 61, 140–51.
Kovács, Á.M., Téglás, E., and Endress, A.D. (2010). The social sense: susceptibility to others' beliefs in human infants and adults. *Science*, 330, 1830–4.
Leslie, A.M. (2000). How to acquire a "representational theory of mind." In: D. Sperber (ed.), *Metarepresentations: a multidisciplinary perspective*. Oxford: Oxford University Press, pp. 197–223.
Lyons, D.E., Young, A.G., and Keil, F.C. (2007). The hidden structure of overimitation. *Proceedings of the National Academy of Sciences*, 104(50): 19751–6.
Malle, B.F., Knobe, J., and Nelson, S.E. (2007). Actor-observer asymmetries in behavior explanations: new answers to an old question. *Journal of Personality and Social Psychology*, 93, 491–514.
Mameli, M. (2001). Mindreading, mindshaping, and evolution. *Biology and Philosophy*, 16, 597–628.
Matthew, S. and Boyd, R. (2011). Punishment sustains large-scale cooperation in prestate warfare. *Proceedings of the National Academy of Sciences*, 108(28), 11375–80.
McGeer, V. (1996). Is "self-knowledge" an empirical problem? Renegotiating the space of philosophical explanation. *Journal of Philosophy*, 93(10), 483–515.
McGeer, V. (2007). The regulative dimension of folk psychology. In: D.D. Hutto and M. Ratcliffe (eds.), *Folk psychology re-assessed*. Dordrecht: Springer, pp. 137–56.
Mellars, P. (2005). The impossible coincidence: a single-species model for the origins of modern human behavior in Europe. *Evolutionary Anthropology*, 14, 12–27.
Meltzoff, A.N. (1988). Infant imitation after a 1-week delay: long-term memory for novel acts and multiple stimuli. *Developmental Psychology*, 24, 470–6.
Millikan, R.G. (1984). *Language, thought, and other biological categories: new foundations for realism*. Cambridge, MA: MIT Press.

Mithen, S. (2000). Palaeoanthropological perspectives on the theory of mind. In: S. Baron-Cohen, H. Tager-Flusberg, and D.J. Cohen (eds.), *Understanding other minds*. Oxford: Oxford University Press.

Morin, O. (2015). *How traditions live and die*. Oxford: Oxford University Press.

Morton, A. (1996). Folk psychology is not a predictive device. *Mind*, 105(417), 119–37.

Morton, A. (2003). *The importance of being understood: folk psychology as ethics*. London: Routledge.

Nielsen, M. and Tomaselli, K. (2010). Overimitation in Kalahari Bushman children and the origins of human cultural cognition. *Psychological Science*, 21(5), 729–36.

Penn, D.C. and Povinelli, D.J. (2007). On the lack of evidence that non-human animals possess anything remotely resembling a "theory of mind." *Philosophical transactions of the Royal Society of London. Series B, Biological sciences*, 362, 731–44.

Perner, J. and Wimmer, H. (1985). "John *thinks* that Mary *thinks* that . . .": attribution of second-order beliefs by 5- to 10-year-old children. *Journal of Experimental Child Psychology*, 39(3), 437–71.

Povinelli, D.J. and Vonk, J. (2003). Chimpanzee minds: suspiciously human? *Trends in Cognitive Sciences*, 7(4), 157–60.

Povinelli, D.J. and Vonk, J. (2004). We don't need a microscope to explore the chimpanzee's mind. *Mind and Language*, 19(1), 1–28.

Powell, A., Shennan, S., and Thomas, M.G. (2009). Late Pleistocene demography and the appearance of modern human behavior. *Science*, 324, 1298–301.

Saxe, Rebecca. (2010). The right temporo-parietal junction: a specific brain region for thinking about thoughts. In: A. Leslie and T. German (eds.), *Handbook of theory of mind*. Hove, UK: Psychology Press.

Seabright, P. (2010). *The company of strangers*. Princeton: Princeton University Press.

Sellars, W. (1956/1997). *Empiricism and the philosophy of mind*. Cambridge, MA: Harvard University Press.

Siegal, M. (2008). *Marvelous minds: the discovery of what children know*. Oxford: Oxford University Press.

Sperber, D. (ed.) (2000). *Metarepresentations*. New York: Oxford University Press.

Sperber, D. and Wilson, D. (2002). Pragmatics, modularity, and mind-reading. *Mind and Language*, 17(1–2), 3–23.

Sripada, C. and Stich, S. (2006). A framework for the psychology of norms. In: P. Carruthers, S. Laurence, and S. Stich (eds.), *The innate mind: culture and cognition*. New York: Oxford University Press, pp. 280–301.

Sterelny, K. (2012). *The evolved apprentice*. Cambridge, MA: MIT Press.

Thornton, A. and McAuliffe, K. (2006). Teaching in wild meerkats. *Science*, 313, 227–9.

Tomasello, M. (1999). *The cultural origins of human cognition*. Cambridge, MA: Harvard University Press.

Tooby, J. and Cosmides, L. (1995). The language of the eyes as an evolved language of mind. Foreword to: S. Baron-Cohen (ed.), *Mindblindness: an essay on autism and theory of mind*. Cambridge, MA: MIT Press.

Wellman, H. (2014). *Making minds*. New York: Oxford University Press.

Wiltermuth, S.S. and Heath, C. (2009). Synchrony and cooperation. *Psychological Science*, 20(1), 1–5.

Zawidzki, T.W. (2008). The function of folk psychology: mind reading or mind shaping? *Philosophical Explorations*, 11(3), 193–210.
Zawidzki, T.W. (2013). *Mindshaping*. Cambridge, MA: MIT Press.
Zentall, T.R. (2006). Imitation: definitions, evidence, and mechanisms. *Animal Cognition*, 9, 335–53.

CHAPTER 40

BRINGING THINGS TO MIND

4Es and Material Engagement

LAMBROS MALAFOURIS

Introduction: Understanding the Changing Material Ecologies of Mind

Things. Human life, society, and evolution would have been very different without them. The human mind has always been inextricably coupled with the material forms people make. But exactly what do things do for the mind? The answer I propose in this chapter is that human beings are not merely embedded in a rich and changing universe of things. Rather, I argue that human cognitive and social life is a process genuinely mediated and often constituted by them. Think of development. Experiencing things, the child learns what it feels and means to be tangible and perceptible. By touching it learns to objectify. I do not just mean that by touching and seeing the things that surround us we get to know them and learn about their physical properties. Instead, what I mean is that by engaging the things that surround us we get to know about touching and seeing and learn about their phenomenal properties. Although we tend to think that we come to feel and know the world through the body's senses, it is also the case that we come to feel and know our bodily senses by engaging the world. Material engagement is the basic process by which we discover the feel and functions of our senses and through them the capacities, limits, and boundaries of our bodies. This is also how we discover affordances, enact possibilities of action, and appreciate the varieties of consciousness by which we apprehend and come to know "reality." Those processes are of course inseparably social, cognitive, and material. There is a lost sense of "we" in the so-called we-intentionality (Tomasello et al. 2005); a neglected stage of material engagement precedes, allows, scaffolds, and supports human intersubjective development and social interaction (see also Gallagher and Ransom 2016; Gallagher 2017).

Turning to a longer time scale, human evolution is by no means an exception. A simple look in the deep time history of our species can reveal that far from being immune to the affective power and agency of things, the nature of change in human beings incorporates, from the very start, the enactive logic of creative material engagement (Malafouris 2004, 2008, 2010, 2013, 2014, 2015, 2016a, 2016b; Renfrew, Frith, and Malafouris 2008). From discovering or understanding material properties of form (as in the case of tool making) to changing or influencing material behavior (as in the case of compound manufacture and pyrotechnology), human thinking can be seen as a craft patiently absorbed in the manufacture of complex surfaces of sorts (plastic or rigid). Many other animals, of course, develop ways to engage with their relevant material worlds, and there is no place for old-fashioned discontinuities here (Barrett 2015). Still, no other form of animal intelligence has been or can be defined on the basis of its relationship with material culture and its form-making abilities (objects, materials, artifacts, techniques, tools). Human intelligence is precisely of this rather peculiar sort. Attachment, ownership, causality, intentionality, selfhood—they are all emergent properties of what I call the gray zone of material engagement where brains, bodies, and things conflate, mutually constituting each other (Malafouris 2004, 2008a, 2008b, 2008c, 2013).

This central feature of the engagement of mind with the material world—I call it *metaplasticity* (Malafouris 2008b, 2010, 2015)—has yet to receive the systematic attention it deserves in philosophy and cognitive science (even within the 4Es camp). I would make the case in this chapter for the need to add a strong material culture dimension of research in the area of 4Es. I propose material engagement theory (MET) (Malafouris 2013) as a framework suitable for the cross-disciplinary study of the interaction between cognition and material culture. MET is less of a theory in the conventional sense of the term and more of a research program aiming to extend the epistemic domain of the cognitive sciences beyond what was considered until now as the "cognitive" in order to grasp the material domain. Material engagement incorporates three major working hypotheses, each one targeting a different but complementary aspect of human becoming—i.e., cognition, signification, and agency (Figure 40.1). Taken together, these hypotheses provide a unit of analysis that allows us to view the mind as situated within and constituted by the material world rather than merely being *about* it. For MET the question "What are things?" and the question "What are minds?" are inseparable. Material engagement collapses the unhelpful antinomies of nature/culture and people/things by focusing on the ways materiality becomes entangled with our everyday life and thinking. Things play an important part in the integration and coordination of processes that operate on radically different time scales (e.g., neural, bodily, cultural, and evolutionary) (Malafouris 2013, 2015, 2016). Through their physical persistence, they help us to move across the scales of time and to construct bridges between temporal phenomena that operate at different experiential levels. Things also work best over the long term, accumulating biographies through joint participation in cultural practices, in ways that often escape the temporal limits and rhythms of individual experience.

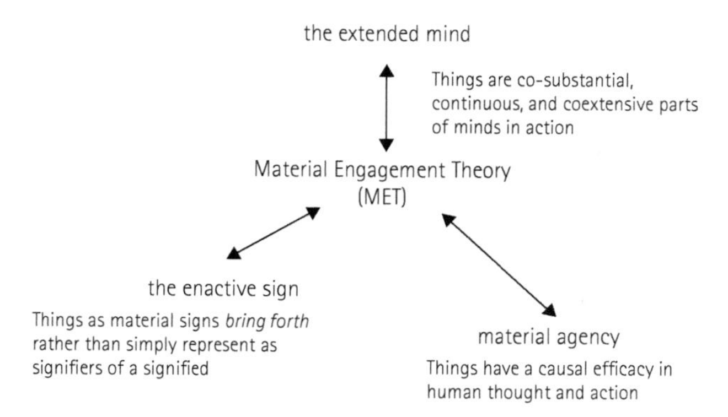

FIGURE 40.1. Material engagement theory.

Heidegger's Jug

"Near to us are what we usually call things," the philosopher Martin Heidegger writes in his famous essay "Das Ding" ("The Thing") (1975, p. 166). Yet, "despite all conquest of distances," which in our days goes well beyond the media of radio and film that Heidegger had in mind when delivering his lecture to the Bayerichen Akademie der Schönen Künste in 1950, "the nearness of things remains absent" (p. 166). It is worth noting that what seems to motivate Heidegger's preoccupation with the "nearness," "remoteness," or "absence" of things, is not his more famous concern with the ontology of hands and tools that we see in his approach to the question of technology.[1] Neither it is a concern to understand the different ways in which things (tools or objects) can be argued to stretch the space of our body and modify our bodily awareness (for a good review of those questions, see Frédérique de Vignemont 2018), or even the transforming powers of new media and technologies to become, in the famous words of Marshall McLuhan (1964/1994), an extension of ourselves. Rather, Heidegger in this essay is after something more basic and elusive; he calls it the "thingness" or "thinghood" of things. He builds his case using the example of a jug.

This simple example of Heidegger's jug is a good starting point for my purpose in this chapter, which, as mentioned, is to argue that material things play an active role in human becoming by immanently extending our bodily, perceptual, and cognitive faculties. However, contrary to what the name "jug" seems to imply in Heidegger's example, no two jugs are the same. Rather, they instantiate different ontologies, meanings, and life histories. Of course, the kind of things we call "jugs" have similar form and

[1] As Graham Harman discuss in his article "Technology, Objects and Things in Heidegger" (2010), the tool analysis was first published in 1927 in his masterwork *Being and Time* (Heidegger 1962) but dates back to his 1919 lecture course during the so-called Freiburg War Emergency Semester. Heidegger's views on technology are well known through his famous lecture "The Question Concerning Technology," delivered in Munich in 1953 and published the following year (Heidegger 1982).

share a number of functional properties and affordances for action. Still, the affordances of things, in the ecological sense of this term (Knappett 2004, 2005; Chemero 2009; Rietveld and Kiverstein 2014), is the contextual and relational product of an organism's cognitive life as situated action. So, to make my argument more effective, I will abandon Heidegger's jug and I will use instead a ceramic vase (unknown to Heidegger) that is "near to me" and thus I happen to know better (Figure 40.2). Let's try to think about its "thingness." What makes this vase a thing really?

One way we could answer that question and come to know the "thingness" of the vase is by experiencing some of its phenomenal qualities. We could perceive its color, follow the shape of its body, or the ways the light is reflected on its exterior surface. If we want, we could reach out and grasp it. We could feel its temperature, weight, or maybe—depending on how experienced we are—also sense something of the texture and physical qualities of the clay that it is made of. Another way to think about the "thingness" of this artifact is to try to identify its possible function. We could, for instance, classify it as a container that can be used to hold something within it and name it as a vase. We could also perhaps think about its origin, about the processes responsible for making it the object it is. We could think of the people, the actions, the materials, the tools, and the places involved in its manufacture, as well as think about it in terms of style, texture, value, aesthetic, skill, time, and technique. We could attempt to draw it on a piece of paper. Of course, there is a personal or subjective element in all these. I, for instance, just by looking at this vase, can remember persons, events, and places associated with it in the past. The vase has a life history of which I am a part. As a result, this object is "mine" in a sense that it can never be "yours." By the same token, the vase belongs to the potter who made it—on the wheel—or to the person(s) who use(s) it in a way that is different from my remote sense of "myness."

FIGURE 40.2. A ceramic vase.

I can go on writing about this vase for hours. But even then we would have merely scratched the surface of the varieties of our sensuous material engagement. Have we made any progress with our initial query about what is the mode of existence of what we call "thingness" or "thinghood"? Not really. It seems that the answer to that question is more difficult than common sense and our use of language in everyday experience might have it. Heidegger, for instance, maintained that things are not what we usually call "objects." Things become "objects" when they lose their "thingness." Reflecting on how a thing retains or loses its "thingness," Heidegger comes up with the following articulation: "The vessel's thingness does not lie at all in the material of which it consists, but in the void that holds" (Heidegger 1975, p. 169). Understanding "thingness" then, is less about producing a list of primary and secondary material or phenomenal qualities and more about penetrating the ontological power of this vessel to "gather" space and time. But what does it mean to say that a thing "gathers"? Heidegger introduces the term "thinging" as a means of expressing the "gathering" or tying together of the vessel's ontological constituents that make up not just the totality of the form-making processes brought together in the making of this particular thing, but also the material conditions and relations that will sustain the vessel's social, cognitive, and emotional life (actual and possible).

This idea of "thinging" as "gathering" provides an interesting metaphor to think about the "nearness" of things. However, my own use of the term *thinging* (Malafouris 2014, 2016a), although rooted in this idea of gathering, diverges from the Heideggerian phenomenological path. Instead, I am seeking those aspects of "thingness" related to the vitality and agency of things in human thinking (Malafouris 2008c, 2015; Bennett 2009; Latour 1990), or else the cognitive life of things (Malafouris and Renfrew 2010; Sutton 2002, 2008). Such a cognitive life is inextricably sensorial and affective; it is also transactional and inherently transformational, thus, it should not be understood in the restricted computational sense associated with cognitive artifacts. The processes I am trying to understand with the notion of *thinging* are not merely those we see in the special case of external representations, maps, or other old or new media of exographic storage, distributed memory, and problem-solving. The usual notions of tools or of external scaffolding represent only a small part of the processes I seek to articulate with the notion of *thinging*. Moreover, as I discuss in the last part of the chapter, the process of *thinging* is not only *spatially* but also *temporally* distributed over brain, body, and world. The ontological messiness of this ongoing dialectic of co-constitution of minds and things can be the cause of cognitive dissonance and epistemic embarrassment, but also the source of inspiration. For now I want to return to the question I raised at the beginning: what do things do for the mind?

Unthinking Cognitivism

Paradoxically, the answer, mostly implicit, to the question of what things do for the mind has been, and for a big part of cognitive science continues to be, nothing. That is, things do nothing that *really matters* to the mind. In stark opposition to the argument for the primacy of material engagement I present in this chapter, the main assumption has been—at least to those committed to the cognitivist vision of mind as consisting of formal computational reasoning processes acting on syntactic algorithms and internal representations (e.g., Fodor 1975)—that thinking is never *thinging*. Or else, the major underlying assumption seems to be that we can never really think *with* or *through* things; rather, we can only think *about* them. There is no place for things inside the mind, the argument goes, because the mind's location is supposedly inside the head and by implication things have no access to that area, but only by means of representation (mental or neural). Of course, nowadays, you can have an implant inserted in your brain. Or recall the famous case of Phineas Gage, a railway construction worker injured[2] from a massive 6-kg iron bar that was propelled through his left cheek and skull by an accidental explosion in 1848 (Ratiu et al. 2004, p. 638). Putting aside those remarkable curiosities, the general consensus can be described as follows: the human mind can be aware of things, or perceive things, but the mental actions that enable us to think about things are seen as different *in kind* from the physical actions we use to touch, smell, create, and manipulate them. Whatever it is that things do is no concern for cognitive science, because the process of mind is not "really"—or in any sense, directly—affected by their doings. Minds and things seem to pass each other by.[3] Even in those well-controlled special cases where things, objects, toys, and material structures might be employed to form an integral part of the experimental setting and the experimental procedure for the study of mind, the assumption is that those material objects are essentially passive or in any case neutral. That is, they *do not matter* for the cognitive phenomenon under investigation. Rather, the role of these objects is merely instrumental; they act as cognitive and behavioral prompts—scaffolds at best. Even when things are conspicuously present, cognitive science prefers to treat them as absent.

Though this epistemic inattentional blindness[4] about things is unfortunate, it is hardly surprising. One of the major objectives in the study of mind has been to demarcate and insulate the mental realm of cognitive action from the physical realm of bodily, social, and material action. It is only when the different realms of life can be separated and compartmentalized that a science of mind can emerge. To accomplish that, material culture, the world of material objects and artifacts, had to be left out of the cognitive system proper. They had to be left out because, as we said, things do not get inside the head like food gets inside the stomach or air in the lungs.

[2] See Damasio et al. (1994) and Macmillan (2002).

[3] As Turkle observes: "We find it familiar to consider objects as useful or aesthetic, as necessities or vain indulgences. We are on less familiar ground when we consider objects as companions to our emotional lives or as provocations to thought" (Turkle 2007, p. 5).

[4] The notion inattentional blindness comes from cognitive neuroscience and refers to the failure to notice a salient object or visible feature in a scene owing to misdirected attention or attention that is not engaged at a level sufficient to achieve awareness of the object.

"Near to us" things might be, but a long tradition of cognitivism has managed to keep them at a safe distance. The legacy of cognitivism in the study of mind seems to have left us with an unbridgable ontological gap between the "internal" world of mental action and the "external" world of physical action (bodily, technical, or social). The solution to that problem came by in the form of mental or more recently neural "representations." In the place of real objects and artifacts out there, in the world, various neural and mental substitutes were created and placed inside the cranium. Of course, from a methodological perspective, one could think of several reasons why sometimes it pays to keep things at a distance (even if localized close to the body in peripersonal space). The problem starts when people confuse this analytical separation of mind and matter for an ontological one. Maybe what is even worse than the tyranny of representation is the implicit assumption that if a phenomenon has no neural representation, it has no cognitive life.

How to Think about Things: Situating the 4E's

Now I turn to the 4E (embodied–embedded–extended–enactive) mind (Chemero 2009; Clark 1997, 2008; Gallagher 2017; Hutto and Myin 2013; Varela, Thompson, and Rosch 1991). No doubt, in spite of their many differences, the proponents of the 4Es seem to share a commitment to breaking with reductionistic explanations and the cognitivist tradition. What about things? What changes in the way you understand and make sense of the human–thing entanglement once you adopt the 4E perspective? One would think that the moment you start recognizing the tight connections between perception and action, shifting the boundaries of mind beyond skin and skull, *everything changes*. Things are no longer just "near to us"—passive objects to be perceived, thought about, and represented. Rather, things now become Us. That is, they emerge as genuine parts of the extensive physical machinery of human thought.

Unfortunately, what in theory sounds a like a small revolution[5] that could transform forever how we think about minds and things has yet to be followed to its paradigm-shifting results and conclusions (see also Wheeler 2015). One noticeable obstacle here (at least from the perspective of cognitive archaeology) is that although proponents of the 4E approach seem to agree that they can now provide a new type of cognitive explanation of natural and cultural phenomena, they seem unwilling or they simply fail to realize that this theoretical shift can only succeed if they also admit that, in many cases, it will be a new type of sociomaterial practice-based explanation of cognitive phenomena that is needed. I would argue then with Edwin Hutchins (2011) that accounting for the

[5] As Michael Wheeler observes (2015, p. 1): "Optimising the 4E (embodied–embedded–extended–enactive) revolution in cognitive science arguably requires the rejection of two guiding commitments made by orthodox thinking in the field, namely that the material realisers of cognitive states and processes are located entirely inside the head (internalism), and that intelligent thought and action are to be explained in terms of the building and manipulation of content-bearing representations (representationalism)."

varieties of material mediations and the temporality of ecological assemblies is the central and unsolved problem in the 4E approach. Put simply, Andy Clark's imperative of *Being There* (1997) will be more useful if translated to, and contextualized through, a series of micro-ethnographies about "being where, when, and how."[6]

Unfortunately, material culture—the practices, techniques and the social skills it embodies—seems to introduce a level of complexity, and thus of uncertainty, that can be perceived as unhelpful, at least from the perspective of method. The convenient ontological boundaries and the familiar representational logic of the computational metaphor that we have made the yardstick of anything "truly" cognitive—at this moment in the history of our species—is hard to overcome. It is hard to overcome not because it allows us to understand better the complexity of the mind, but because it allows us to bypass and to forget it by getting rid of all the supposedly "peripheral" distractions that live on the "outside." As a consequence, any perspective that disrespects representationalism is often perceived more of a threat than a help to the scientific study of mind. Of course, a lot can be said about the meaning of "science" in the sciences of mind, about the split with society, history, and culture, that it often promotes, as well as about the assumptions upon which it operates. But that has to be the topic of a different paper. Suffice it to say here that the complexity of mind must be respected and should not be as easily dismissed or explained away for the sake of disciplinary integrity or scientific purity—if such a thing exists in the first place.[7]

In any case, the fact remains that in most cases philosophers' and cognitive scientists' engagement with material culture stops with the usual examples of pens and paper, notebooks, laptops, PDAs, and iPhones, that were briefly introduced to illustrate the varieties of cognitive extension. However, there is little substantive engagement with the specific ecologies and biographies of these objects. Fortunately, the great analytical potential of the interaction between 4E's and 4A's (anthropology, art, architecture, archaeology)[8] is becoming increasingly recognized not just within archaeology and anthropology (Malafouris and Renfrew 2010; Malafouris 2013; Hutchins 2011, 2014; Barrett 2011; Baber, Chemero, and Hall 2017; Durt, Fuchs, and Tewes 2017; Froese 2017; Garofoli 2015, 2017; Garofoli and Iliopoulos 2017; Iliopoulos 2017; Iliopoulos and Garofoli 2016; Knappett 2004, 2005; Overmann 2017a, 2017b; Poulsgaard and Malafouris 2017; Poulsgaard 2017; Walls and Malafouris 2016; Walls 2018) but also in other fields. (Good examples of such interaction can be seen in the work by Erik Rietveld on affordance-based design in architecture (Rietveld and Brouwers 2017; Rietveld, and Martens 2017), or by Paul March (2017), Ricardo Manzotti (2011), and Simon Penny

[6] To what extent his more recent preoccupation with the "predictive mind/brain" can help us to answer those questions remains to be seen. Here my concern is with material engagement.

[7] One may ask here: Is not the brain complex enough on its own? Of course it is, and there are many questions and analytical levels that it can be in itself the object of scientific study. But there are limits to what can be asked, beyond which a brain in a vat is not just meaningless as a unit of analysis but categorically misleading.

[8] I borrow the term from the anthropologist Tim Ingold (see his book Making (2013))

(2017) on art and enactive cognitive science.) More collaborative and cross-disciplinary work of that kind is needed for making progress toward a new ecology of mind. Many years have passed since Andy Clark and Chalmers introduced Inga and Otto to the world (Clark and Chalmers 1998). We have learned a great deal about "parity" but nothing really about the "notebook," which became just a placeholder for whatever material vehicles of memory stands for in common language. It seems that even in this philosophical plot that would have never been possible in the absence of things—there is no sense you can talk about parity without Otto's notebook—things are forgotten. Paradoxically, finding his way to the New York Museum of Modern Art (MOMA), it was Otto's close partnership with his notebook rather than with Inga that made the difference so far as the parity principle is concerned. A more careful look would have revealed something that escaped our attention: the cognitive processes that enable Otto and Inga to find their way to the MOMA are not as different as Clark and Chalmers would have us believe. They are similar not because thinking (mediated or not) always happens on the "inside." Quite the contrary, they are similar because no internal process on its own would have ever allowed Otto or Inga to reach their destination. *Knowing that* the museum is on 53rd Street is no guarantee that you are going to find it. *Knowing how* to get there, by contrast, is. A migratory seabird could find its way returning to nest and breed in the same place often with little external support; we humans rarely, if ever, do so—thus the constitutive role that prosthesis has in human cognitive life.

Naturally there is more to this cognitive life than conventional categories like "tools," "technology," and "cognitive artifacts" allow comprehending. The devil is in the details. Mental extensiveness in the form of material engagement is constantly changing and is varying not just through history and evolution but even within a single lifetime. If words like extension, enaction, embodiment, and embeddedness are to have any real anthropological significance, then cognitive science must develop ways to understand the specificity and varieties of human cognitive becoming. New things, technologies, and material environments can stretch and enhance our minds, but they can also shrink them, blind them, or deprive them of their creative abilities and what Michel Foucault referred to as critical consciousness (Foucault 1988).

How to Think about Things: Toward a Theory of MET

What can be done? As I see it, the moment you adopt the general principles of a 4E approach (independent of discipline and specialization), you are immediately faced with two main available options: either to remain soft, avoiding getting your hands dirty by messing with the real mind-stuff of the material world; or on the other hand, to seriously engage things, to "go native," as we say in anthropology, and seek to follow the stuff of mind "in the wild." The danger with the former option is that you might well end up with an ahistorical empty concept of mind that adds very little to the previous cognitivist establishment (embodied cognitivism and most "soft" versions of the 4E approach are no exception). If you follow the second option, however, I suspect there is a good chance you will discover the meaning of "extensiveness" (Hutto and Myin 2013) in human life providing a new ontological basis for rethinking the usual debates around the so-called mark of the cognitive, the role of parity, and the limits of functionalism. Needless to say, this brave new insight comes at a price. Entering into the field of material engagement and taking material culture seriously opens a new window onto the mind but at the same time makes the project of cognitive science much more difficult, albeit far more variable and maybe also more interesting. Nevertheless, my contention is that it is the only way for a truly "locationally uncommitted account of the cognitive" (Wheeler 2010, 2014, 2015).

Here is the challenge: explaining the role of things in human cognition entails more than a thorough analytical re-examination of the basic ontological ingredients of human thought. It also demands a detailed understanding of the embodied cultural practices that turn those ingredients into cognitive processes in different contexts and across the scales of time. The challenge for the 4E approach demands reconnecting the brain with the body and beyond, with proper attention to specific activities and varieties of material practices in specific contexts. One possible strategy to that end is to adopt the perspective of the material engagement approach and follow the enactive thread that it embodies. Material engagement theory wants to refocus the study of mind by recognizing the impact of materiality (the world of things, objects, materials, and artifacts) not simply in the shaping of the emergent dynamical patterns of brain activity, but importantly in the constitution of the conditions for their extension and reorganization. In that sense, the material engagement approach suggests that cognitive processes—far from being brain-bound representational events—cut across the brain–body–world divisions. To investigate those processes we require an adequate form of description and possibly a new conceptual vocabulary. Hence, I devised the neologism "thinging" to articulate and draw attention to the kind of cognitive life instantiated in acts of thinking and feeling *with, through*, and *about* things (Malafouris 2014). The latter aspect of thinging, *aboutness*, has received a lot of attention in philosophy of mind due to its association with the contents of human intentionality, which enable us to be conscious of things in the world. On the contrary, *withness* and *throughness* remain little understood due to their association with the nonrepresentational performative aspects of embodied intentionality, or what is known as intention *in* action.

In particular, the notion of *thinging* refers to the process of thinking incorporating things. This kind of thinking blends with feeling and affect. It is primarily through feeling and affect that things are presented *to* us through acts of material engagement. This is very different from the usual process by which things are seen as represented *in* us by way of internalization and mental substitution. It follows that with *thinging* the focus falls on a variety of material assemblages (Deleuze and

Guattari 2004) and ecologies (Ingold 2012; Hutchins 2010) rather than just on specific objects, tools, and external representations. *Thinging* embodies culture-specific bodily techniques; it also extends to sensory and cognitive prostheses and interfaces of any kind. The analytic philosopher may protest here that this description of *thinging* is too general to serve as the basis for useful taxonomic considerations. But the anthropological value of the notion of *thinging* lies not in the way it can help us to differentiate and classify brains, bodies, and things; rather, it is found in the ways it can help us bring them back together. What may initially appear as an inherent analytical limitation of our notion is actually the product of our persistent framing of the question of what mind consists of within the classical antinomies of nature/culture and mind/matter. Once that old cognitivist frame is removed, the analytical value of the notion *thinging* becomes clearer in helping us to understand how brains, bodies, and things come to be, that is, how they come to possess ontological specificity or multiplicity in the course of their life history. This process will inevitably vary in different times and places.

Take, for instance, the example of the clay vase we discussed at the start. What kind of relationship best describes the creative process by which such a thing can be formed? No doubt, if there is a sense in which what we call mind or intelligence can be captured or embodied in a single material form, then objects like that vase offer good examples. They constitute mind-traps. Clay vases are not grown by themselves; they are made. Acts of making are both mental and physical in that they presuppose the blending of various creative intelligences and agencies (human and nonhuman). They also demand a great deal of energy exchange and advanced bodily skills. In this *hylonoetic* field the material properties of clay are also important (maybe as important as the neural properties of the potter's brain). This is not a universal statement about mind in general. There is no such thing as mind in general. There are only specific people, situated in time and space, engaged in specific cognitive processes, according to a given cognitive ecology. In the case of our example, that situated process is the making of a vase from clay. So far as the cognitive ecology of this process is concerned the plasticity of clay matters as much as the plasticity of the neural networks that the potter's body sets in motion or activates while making the vase.

Cognitivism, of course, has no problem describing all that by employing a language that speaks of mental intentions and representations. But this convenient vocabulary and the logic of unidirectional action it imposes on the phenomenon we try to understand seriously distorts exactly those aspects of embodied action and skillful copying that make this process unique (no two vases are made the same), creative (the making of a vase is a process of enactive discovery), and experientially rich (a lived experience of creative material engagement). To explain, cognitivism would see the vase as the product of intelligent action or creativity that happened because the potter's body followed and carefully executed (consciously or tacitly) the instructions, plans, and intentions of the potter's internal mind or brain. By contrast, the material engagement approach has a decentralized view of the process, according to which the vase has used the potter's muscles and skills to bring about its final form as much as the potter has used the material affordances of the clay to create the form of the vase. What cognitivism seeks to explain away by means of mental intentions and representations, the material engagement approach places at the center of experience, that is, where brain, body, clay, and wheel conflate. I call this the feeling *of* and *for* clay (Malafouris 2014).,

The feeling *of* and *for* clay designates, on the one hand, the experience of absorption in and submission to the material, and, on the other, the parallel active exploration of and ongoing improvisation with the material. Notions of "material agency" (Malafouris 2008; Knappett and Malafouris 2008), "vital materiality" (Wheeler 2010), and "vibrant matter" (Bennett 2009) can help us to understand better the phenomenological significance of this transactive ensemble between the affordances of the potter's body and the affordances of wet spinning clay. This is also where the traditional phenomenological insistence on the ontological purity and priority of the first-person perspective breaks down. There is no autonomous subjectivity but a flow of energies within and between varieties of heterogeneous materials (organic and inorganic). This Whiteheadean "event" (1929/1978) instantiates the process ontology of decentered becoming that characterizes the phenomenon of material engagement. It also exemplifies the constitutive intertwining of mind with matter and shows how energies are being transformed into agencies. Agencies, when embodied in living bodies, can also acquire experiential content and sometimes develop awareness, i.e., a "sense" of agency. But this awareness of agency, characteristic of human bodies, is largely an illusion. There is no agent apart from the action. Agency is not a permanent feature or property that someone (human or nonhuman) has independently of situated action, but the emergent product of material engagement seen in our image of the clay vase as a creative tension of mind and matter or of flow and form (Gosden and Malafouris 2015). The classical mistake is to perceive the clay as inanimate and passive when in fact it is the source of the potter's agency and a psychoactive path of self-identification. By contrast the notion of *thinging* suggests that only by looking at this performative transactional environment that permits and constrains movement (bodily and neural) can we ever understand how the potter's intention to act comes to life. The idea of the isolated human agent that acts upon the inert world must be abandoned. Of course, the feeling or sense of agency might be something the potter would still admit to. Nevertheless, no aspect of the potter's creative intelligence and agency can be accurately delimited and restricted before and outside the act of making. With *thinging*, the purity of action, and of the bodies that carry those actions, gives way to situated action. The potter's body, in its capacity as the body of a potter, is more than a body—it is a *situated body*. That means that this body embodies a unique developmental life history largely shaped by the kinesthetic experiences and skills of pottery making. This is embodiment in the radical sense of the word. The point is not just to recognize the bodily basis of the human body but to expand that basis beyond the organismic boundaries of the skin (Malafouris 2016; Malafouris and Koukouti 2017).

Conclusion

As we saw, according to mainstream cognitive sciences, to say that things have a cognitive dimension is at best to show that they have some kind of "representational" or "symbolic" value within an information-processing system implemented by the brain. What that means really is that things do *not matter* for what they are, or for what they do, that is, for their distinctive sociomaterial properties and affordances. Instead, if they have a cognitive dimension it is because they can be vehicles for representing "externally," in the material world, something that has originated "internally," inside the head. I argue that things don't simply have a cognitive dimension by executing the orders of a central executive or as passive placeholders for cognitive content to be realized somewhere else, but, instead, a full-blown cognitive life of their own. The latter might seem strange within the limited time scales that cognitive science usually operates, but from the angle of a process archaeology of mind, it is commonplace (Gosden and Malafouris 2015). This is where things start to become really messy by revealing to us on the one hand that often the material world has no need of representation and, on the other hand, that *every representation is almost by definition a misrepresentation*. To illustrate the latter premise, I will borrow and reinterpret the example of the desktop of a Windows interface discussed by Hoffman and Prakash (2014). Think of the small rectangle icon in my desktop as the *representation* of the text file I use to write this essay. I can click on that icon to open the file; I can drag that icon to the trash if I want to delete the file; or I can drag it to the icon of a different folder if I want to create a copy. Computer scientists design those representations in this way, that is, they make them resemble the physical form and action of real objects, because they found that it simplifies the cognitive task involved. It can be argued then that the icons on my desktop are both real—they have shape and color and I can physically manipulate their position indirectly—and useful—they make the task of writing easier. So far, so good. The problem begins, however, if you are misled to think that those visible properties of the icon that help you carrying out the task of writing and editing an essay can actually "stand for" the real processes responsible for the task and thus that they can be used to explain the properties of the file and the operations involved. To commit this category mistake is to completely misunderstand the purpose of the interface and the function of the computer. The use and meaning of representation in computational cognitive science suffers from this representational fallacy.

The material engagement approach allows for a dynamic reciprocal relationship between brains, bodies, and things. Human cognition is made of action and for action. Instead of thinking in terms of neural "representations" as the way the brain represents reality, we should be thinking of neural activations by which the brain and the body "engages" the world. MET proposes we replace the passive language of representations with the active language of material engagement. An important difference to note here is that while representations are seen as complete and localized on the "inside," material engagement is always partial, transactive, and incomplete, and thus "in-between."

A mind is what a mind does. We can all agree—proponents and opponents of the 4E approach—that what a mind does is thinking. We can also agree that there are certain processes we can clearly identify as forms of thinking. Where most people seem to disagree is on what constitutes or counts as thinking. I suggest that to ask and to debate that question is both productive and useful. I say that for two main reasons. First, it avoids essentialism about location (it assumes nothing about the location of thinking). Second, it views thinking not as fixed entity with set pre-specified characteristics, i.e., the mind in the head, but instead as a process open to change and transformation. The important question is now: *How and where does the mind do its thinking*? The contribution of MET to answering that question lies in understanding thinking as *thinging*. As mentioned, *thinging* is not going to provide a definitive answer to the questions of what things and minds are. Nonetheless, it can help us understand what it is that minds and things *do* together, shifting our attention away from the sphere of isolated and fixed categories (objects, artifacts, etc.) to the sphere of the fluid *interactions* and relational *transactions* between people and things, that is, to the realm of material engagement.

Acknowledgment

The writing of this chapter was assisted by the John Templeton Foundation Grant, *Self-Bound: The Making of Human Consciousness* (ID 60652).

References

Aston, A. (2017). Cognition and the city: cognitive ecology and the Paris commune of 1871. In: S.J. Cowley and F. Vallée-Tourangeau (eds.), *Cognition beyond the brain*. Berlin: Springer, pp. 215–31.

Baber, C., Chemero, T., and Hall, J. (2017). What the jeweller's hand tells the jeweller's brain: tool use, creativity and embodied cognition. *Philosophy & Technology*, 1–20.

Barrett, L. (2011). *Beyond the brain: how body and environment shape animal and human minds*. Princeton, NJ: Princeton University Press.

Barrett, L. (2015). A better kind of continuity. *The Southern Journal of Philosophy*, 53(S1), 28–49.

Bennett, J. (2009). *Vibrant matter: a political ecology of things*. Durham, NC: Duke University Press.

Chemero, A. (2009). *Radical embodied cognitive science*. Cambridge, MA: MIT Press.

Clark, A. (1997). *Being there: putting brain, body and world together again*. Cambridge, MA: MIT Press.

Clark, A. (2008). *Supersizing the mind*. New York: Oxford University Press.

Clark, A. and Chalmers, D. (1998). The extended mind. *Analysis*, 58(1), 7–19.

Damasio, H., Grabowski, T., Frank, R., Galaburda, A.M., and Damasio, A.R. (1994). The return of Phineas Gage: clues about the brain from the skull of a famous patient. *Science*, 264(5162), 1102–5.

Deleuze, G. and Guattari, F. (2004). *A thousand plateaus* (trans. B. Massumi). London: Continuum.

de Vignemont, F. (2018). The extended body hypothesis: referred sensations from tools to peripersonal space. In: this volume, pp. 389–404.

Durt, C., Fuchs, T., and Tewes, C. (eds.) (2017). *Embodiment, enaction, culture: investigating the constitution of the shared world*. Cambridge, MA: MIT Press.

Fodor, J. A. (1975). *The language of thought*. Cambridge, MA: Harvard University Press.

Foucault, M. (1988). Technologies of the self. In: L.H. Martin, H. Gutman, and P.H. Hutton (eds.), *Technologies of the self: a seminar with Michel Foucault*. Amherst: University of Massachusetts Press, pp. 16–49.

Froese, T. (2017). Making sense of the chronology of Paleolithic cave painting from the perspective of material engagement theory. *Phenomenology and the Cognitive Sciences*, 1–22. doi:10.1007/s11097-017-9537-8

Gallagher, S. (2017). *Enactivist interventions: rethinking the mind*. Oxford: Oxford University Press.

Gallagher, S. and Tailer, G.R. (2016). Artifacting minds: material engagement theory and joint action. In: G. Etzelmüller and C. Tewes (eds.), *Embodiment in evolution and culture*. Heidelberg: Mohr Siebeck, pp. 337–53.

Garofoli, D. (2017). RECkoning with representational apriorism in evolutionary cognitive archaeology. *Phenomenology and the Cognitive Sciences*, 1–23. doi:10.1007/s11097-017-9549-4

Garofoli, D. and Iliopoulos, A. (2017). Replacing epiphenomenalism: a pluralistic enactive take on the metaplasticity of early body ornamentation. *Philosophy & Technology*, 1–28. doi:10.1007/s13347-017-0296-9

Gosden, C. and Malafouris, L. (2015). Process archaeology (P-Arch). *World Archaeology*, 47(5), 1–17.

Harman, G. (2010). Technology, objects and things in Heidegger. *Cambridge Journal of Economics*, 34(1), 17–25.

Heidegger, M. (1962). *Being and time* (trans. J. Macquarrie and E. Robinson). New York: Harper & Row.

Heidegger, M. (1975). The thing. In: *Poetry, language, thought* (trans. A. Hofstadter). New York: Harper & Row, pp. 161–84.

Heidegger, M. (1982). *The question concerning technology* (trans. W. Lovitt). New York: Harper.

Hoffman, D.D. and Prakash, C. (2014). Objects of consciousness. *Frontiers in Psychology*, 5, 577.

Hutchins, E. (1995). *Cognition in the wild*. Cambridge, MA: MIT Press.

Hutchins, E. (2008). The role of cultural practices in the emergence of modern human intelligence. *Philosophical transactions of the Royal Society of London. Series B, Biological sciences*, 363, 2011–9.

Hutchins, E. (2011). Enculturating the supersized mind. *Philosophical Studies*, 152(3), 437–46.

Hutto, D.D. and Myin, E. (2013). *Radicalizing enactivism: basic minds without content*. Cambridge, MA: MIT Press.

Iliopoulos, A. (2016). The evolution of material signification: tracing the origins of symbolic body ornamentation through a pragmatic and enactive theory of cognitive semiotics. *Signs and Society*, 4(2), 244–77.

Iliopoulos, A. and Garofoli, D. (2016). The material dimensions of cognition: re-examining the nature and emergence of the human mind. *Quaternary International*, 405, Part A (The material dimensions of cognition), 1–7.

Ingold, T. (2013). *Making*. London & New York: Routledge.
Ingold, T. (2015). *The life of lines*. New York: Routledge.
Kirsh, D. (1995). The intelligent use of space. *Artificial Intelligence*, 73, 31–68.
Knappett, C. (2004). The affordances of things: a post-Gibsonian perspective on the relationality of mind and matter. In: E. DeMarrais, C. Gosden, and C. Renfrew (eds.), *Rethinking materiality: the engagement of mind with the material* world. Cambridge: McDonald Institute for Archaeological Research, pp. 43–51.
Knappett, C. (2005). *Thinking through material culture: an interdisciplinary perspective*. Philadelphia: University of Pennsylvania Press.
Knappett, C. and Malafouris, L. (eds.) (2008). *Material agency: towards a non-anthropocentric approach*. New York: Springer.
Macmillan, M. (2002). *An odd kind of fame: stories of Phineas Gage*. Cambridge, MA: MIT Press.
Malafouris, L. (2004). The cognitive basis of material engagement: where brain, body and culture conflate. In: E. DeMarrais, C. Gosden, and C. Renfrew (eds.), *Rethinking materiality: the engagement of mind with the material world*. Cambridge: McDonald Institute for Archaeological Research, pp. 53–62.
Malafouris, L. (2008a). Between brains, bodies and things: tectonoetic awareness and the extended self. *Philosophical transactions of the Royal Society of London. Series B, Biological sciences*, 363, 1993–2002.
Malafouris, L. (2008b). Beads for a plastic mind: The "blind man's stick" (BMS) hypothesis and the active nature of material culture. *Cambridge Archaeological Journal*, 18, 401–14.
Malafouris, L. (2008c). At the potter's wheel: an argument for material agency. In: C. Knappett and L. Malafouris (eds.), *Material agency: towards a non-anthropocentric perspective*. New York: Springer, pp. 9–36.
Malafouris, L. (2010). Metaplasticity and the human becoming: principles of neuroarchaeology. *Journal of Anthropological Sciences*, 88, 49–72.
Malafouris, L. (2012). Prosthetic gestures: how the tool shapes the mind. *Behavioral and Brain Sciences*, 35(4), 230–1.
Malafouris, L. (2013). *How things shape the mind: a theory of material engagement*. Cambridge, MA: MIT Press.
Malafouris, L. (2014). Creative thinging: the feeling of and for clay. *Pragmatics and Cognition*, 22(1), 140–58.
Malafouris, L. (2015). Metaplasticity and the primacy of material engagement. *Time and Mind*, 8(4), 351–71.
Malafouris, L. (2016a). On human becoming and incompleteness: a material engagement approach to the study of embodiment in evolution and culture. In: G. Etzelmüller and C. Tewes (eds.), *Embodiment in evolution and culture*. Heidelberg: Mohr Siebeck, pp. 289–306.
Malafouris, L. (2016b). Material engagement and the embodied mind. In: T. Wynn and F.L. Coolidge (eds.), *Cognitive models in palaeolithic archaeology*. Oxford: Oxford University Press, pp. 69–82.
Malafouris, L. and Koukouti, M.D. (2017). More than a body. In: C. Meyer, J. Streeck, and J.S. Jordan (eds.), *Intercorporeality: emerging socialities in interaction*. Oxford: Oxford University Press, pp. 289–303.
Malafouris, L. and Renfrew, C. (eds.) (2010). *The cognitive life of things: recasting the boundaries of the mind*. Cambridge: McDonald Institute for Archaeological Research.
Manzotti, R. (2011). (ed.) *Situated aesthetics: art beyond the skin*. Exeter: Imprint.

March, P. L. (2017). Playing with clay and the uncertainty of agency: a material engagement theory perspective. *Phenomenology and the Cognitive Sciences*, 1–19. doi:10.1007/s11097-017-9552-9

McLuhan, M. (1964/1994). *Understanding media: the extensions of man*. Cambridge, MA: MIT Press.

Overmann, K.A. (2017a). Concepts and how they get that way. *Phenomenology and the Cognitive Sciences*, 1–16. doi:10.1007/s11097-017-9545-8

Overmann, K.A. (2017b). Thinking materially: cognition as extended and enacted. *Journal of Cognition and Culture*, 17(3–4), 354–73.

Penny, S. (2017). *Making sense: cognition, computing, art, and embodiment*. Cambridge, MA: MIT Press.

Poulsgaard, K.S. (2017). Enactive individuation: technics, temporality and affect in digital design and fabrication. *Phenomenology and the Cognitive Sciences*, 1–18. doi:10.1007/s11097-017-9539-6

Poulsgaard, K.S. and Malafouris, L., (2017). Models, mathematics and materials in digital architecture. In: S.J. Cowley and F. Vallée-Tourangeau (eds.), *Cognition beyond the brain*. Berlin: Springer, pp. 283–304.

Ratiu, P., Talos, I.F., Haker, S., Lieberman, D., and Everett, P. (2004). The tale of Phineas Gage, digitally remastered. *Journal of Neurotrauma*, 21(5), 637–43.

Renfrew, C., Frith, C., and Malafouris, L. (2008). Introduction: the sapient mind. *Philosophical transactions of the Royal Society of London. Series B, Biological sciences*, 363, 1935–8.

Rietveld, E. and Brouwers, A.A. (2016). Optimal grip on affordances in architectural design practices: an ethnography. *Phenomenology and the Cognitive Sciences*, 1–20.

Rietveld, E., Rietveld, R., and Martens, J. (2017). Trusted strangers: social affordances for social cohesion. *Phenomenology and the Cognitive Sciences*, 1–18. doi:10.1007/s11097-017-9554-7

Sutton, J. (2002). Porous memory and the cognitive life of things. In: D. Tofts, A. Jonson, and A. Cavallaro (eds.), *Prefiguring cyberculture: an intellectual history*. Cambridge, MA and Sydney: MIT Press and Power Publications, pp. 130–41.

Sutton, J. (2008). Material agency, skills, and history: distributed cognition and the archaeology of memory. In: L. Malafouris and C. Knappett (eds.), *Material agency: towards a non-anthropocentric approach*. Berlin: Springer, pp. 37–55.

Tomasello, M., Carpenter, M., Call, J., Behne, T., and Moll, H. (2005). Understanding and sharing intentions: the origins of cultural cognition. *Behavioral and Brain Sciences*, 28, 675–735.

Turkle, S. (ed.) (2007). *Evocative objects: things we think with*. Cambridge, MA: MIT Press.

Walls, M. and Malafouris, L. (2016). Creativity as a developmental ecology. In: V.P. Glăveanu (ed.), *The Palgrave handbook of creativity and culture research*. London: Palgrave Macmillan, pp. 623–38.

Wheeler, M. (2010). Minds, things, and materiality. In: L. Malafouris and C. Renfrew (eds.), *The cognitive life of things: recasting the boundaries of the mind*. Cambridge: McDonald Institute Monographs, pp. 29–38.

Wheeler, M. (2014). Revolution, reform, or business as usual? The future prospects for embodied cognition. In: L. Shapiro (ed.), *The Routledge handbook of embodied cognition*. Abingdon and New York: Routledge, pp. 374–83.

Wheeler, M. (2015). The revolution will not be optimised: radical enactivism, extended functionalism and the extensive mind. *Topoi*, 36(3), 457–72.

CHAPTER 41

CULTURE AND THE EXTENDED PHENOTYPE

Cognition and Material Culture in Deep Time

KIM STERELNY

The Extended Phenotype

This chapter explores through deep time the connections between human cognition and the external resources that shape and amplify our cognitive powers. The chapter largely focuses on material resources—physical tools that support thought—but in the final section, the scope expands to include social resources. Human cognition is enhanced by the ways we rely on one another as well as by our material culture. The aim of the chapter is to compare and evaluate the different frameworks that have been developed to capture the role of these external resources. I begin with the extended phenotype, and then discuss ideas from the extended mind, distributed cognition, and niche construction literatures. The choice between these frameworks is largely heuristic, as all recognize the most central phenomena. But I shall suggest that the niche construction framework is the most general and systematic.

In his 1982 classic, *The Extended Phenotype*, Richard Dawkins tried to reshape biological thinking about adaptive design. The standard model of evolutionary biology took adaptive design to be a feature of organisms: as organisms in a population compete for the essentials of life and for opportunities to reproduce, we should expect to see a robust tendency for organisms to trend toward optimal design for their specific circumstances. Of course, optimal design is *constrained* optimal design: organisms face trade-offs between differing demands; for example, attempting to reproduce often lowers the chances of survival. Moreover, evolutionary adjustment to local circumstances depends on appropriate variants being fed into the engine of selection; luck, good or bad, can have a significant effect on historical trajectories, especially in small populations. Nonetheless, the framework of many evolutionary biologists involves identifying local optima and expecting populations to be at those optima. Luck, limits on variation, and/or subtle trade-offs will be revealed by departures from this expectation (Godfrey-Smith 2001).

Dawkins wanted to shake up this thinking in two deep ways (Dawkins 1982). First, in step with the main thrust of his evolutionary thinking, he reminded the evolution community that the phenotype of an organism is the evolutionary result of an alliance between the genes—the replicators—that shape that organism's ontogeny. Moreover, this alliance is often unstable and partial; the genetic consensus on what organism to build is not always complete. Coincidence of genetic interest is partial rather than complete, in part because different genes have different kin in other bodies. As David Haig (especially) has made clear, the genes from your father have a stake in your father's relatives' prospects for reproduction, but not in those of your mother's relatives, and vice versa for the genes from her (Haig 2014). Second, and this is the dominant theme of *The Extended Phenotype*, to the extent that your genes are a cooperative alliance leveraging copies of themselves into the next generation, the levers whose construction they guide include you and your phenotypic features, but not *only* you and your features. Thornbill genes build thornbill nests in exactly the same sense in which they build thornbills. That is, there are interventions on thornbill genes that would change thornbill phenotypes in adaptively relevant ways. Likewise, there are interventions on those genes that would change thornbill nests in adaptively relevant ways. That is why there can be, and are, guides to nests as well as to the birds themselves. Nests are not a one-off. Gene alliances engineer the physical and biological environment of the bodies they ride around in, and do so in systematic, patterned, and multi-generational ways. Termite mounds are feats of collective physical engineering, but they are feats of multi-species biological engineering too, as termites and termite genes depend on symbiotic alliances with bacteria and fungi (Turner 2000). Thus Dawkins suggests that we should think of the genotype as a set of genes in provisional and partial alliance, and the phenotype they build is often an organism adaptively coupled with, and/or embedded in, a set of environment resources that those genes also shape. This extended phenotype reliably changes and develops over ontogeny, and in ways shaped by evolution.

A few years later, Andy Clark (initially in partnership with David Chalmers), Rob Wilson, and a few others began to explore the idea that human cognition, including many of our distinctive cognitive powers, depends on, and is tuned to, out-of-the-body resources (Wilson 1994; Clark and Chambers 1998; Dennett 2000; Clark 2001, 2008). Clark suggested that we should not think of cognition as brain-bound. The classic arguments for extended cognition, beginning with Otto and his notebook, do not depend on the theoretical apparatus developed in *The Extended Phenotype*; they were not framed in the context of evolutionary hypotheses about human cognition. But, as Clark notes, there is an obvious resemblance between the two frameworks (e.g., Clark 2007, p. 176). As with many of Dawkins's examples, the extended mind literature details the smooth and powerful integration of material resources into our cognitive and behavioral repertoire. Moreover, as with Dawkins's examples, hominin interaction with the material resources of their environment has been shaped by selection. Our hands are adapted to make and use tools, as we can grip with both power and precision (Ambrose

2001). Our arms and shoulders are adapted for hard, accurate throwing (Henrich 2016). We are adapted to the tools we have produced. Very likely, the same is true of our cognitive organization: our minds, not just our hands, are probably adapted to the cultural tools we make and use (Tomasello 1999b; Alvard 2003; Tomasello 2014).

The Hominin Extended Phenotype in Deep Time

As noted earlier, the main aim of this chapter is to explore the emergence and importance of our cognitive dependence on material culture, and I shall begin this exploration with three examples, which will serve to illustrate the importance and temporal depth of the connection between how we think and what we make. Three salient examples are:

Fire. Fire is most obviously important for cooking (see later), for warmth in cold periods, and perhaps for protection. But one other important product of fire is light, for firelight has affected human circadian rhythms and hence human time budgets. Most day-active animals move into the sleep cycle for most of the hours of darkness, minimizing energy expenditure and reducing exposure to nocturnal predators. Our sleep rhythms are more truncated. While there is obviously much personal and cultural variation, hominins have over the darker months four more hours or so a day to eat, to tinker, to socialize, to learn from one another (Dunbar and Gowlett 2014; Gowlett 2016). To the extent that fire is shared (and that is a much more efficient use of resources), fire trained and selected our ancestors for social tolerance. When fire is shared, agents must function for extended periods in close proximity to one another. So fire reshaped our social environment, not just our nutrition.

Food processing. Hominins have coevolved with their food processing techniques and technologies. That is ongoing: lactose tolerance has coevolved with the establishment of dairy farming as an important human lifeway. The relevant gene is common only in those cultures with a historical commitment to dairy consumption (O'Brien and Laland 2012). But Richard Wrangham (with various co-workers) has argued that this is a very deep feature of hominin history; beginning over two million years ago (Wrangham 2009; Wrangham and Carmody 2010; Gowlett and Wrangham 2013). Hominin teeth, jaw muscles, and guts all shrank despite increasing energy demands. The evolution of larger brains (in both absolute and relative terms), and the evolution of lifeways involving much larger territories and longer daily journeys through those territories, both increased energy demands. Wrangham himself argues that cooking was an early hominin innovation: perhaps as early as 2 mya. There is no direct trace evidence of controlled fire for a very early date. That said, very ancient sites, especially in the open, are most unlikely to survive well enough to preserve a clear signal of domestic fire. There are a few ambiguous traces in Africa earlier than 1 mya; a convincing site at about 800 kya; regular, undeniable evidence from around 400 kya. However, even if cooking is not an early Pleistocene innovation, the controlled application of heat is not the only processing technology. Ingestion and digestion is eased by chopping, grinding, pounding, soaking. Hominins probably coevolved with their food processing techniques through most of the Pleistocene, even if cooking is a mid-to-late Pleistocene innovation.

Endurance hunting. Joseph Henrich makes a surprisingly strong case for the claim that we are born to run. Humans are adapted for endurance running—through our lower-body skeletal and muscle structure; through our efficient gait; through being able to shed heat; and through our technology (Henrich 2016). We cannot out-sprint large- and medium-size herbivores, but hominins of a certain physical type common in southern Africa—gracile, wiry—can grind them down. Henrich's case, importantly, is that endurance running is an evolved complex of morphological traits, physiological traits (full body sweating), and cognitive traits—especially tracking, but also teamwork—and technology. Endurance hunters sweat heavily to shed heat, and this water must be replaced. So endurance hunting depended on water-carrying technology.

We adapt to external resources that amplify our capacities, though that amplification depends on the smooth integration of our embodied and our external resources, including cognitive resources. The effective use of technology depends not just on skin, muscle, and bone. It depends as well on skill, and hence on the investment of time, energy, effort, and forethought in acquiring skill. For humans acquire skill by focused practice, not just trial-and-error experimentation. Technology must be mastered, and usually mastered well enough so that its use is both appropriate and unreflective. For technology is often used as part of a suite of individual and collective activities, in situations of high demand on attention. So the integration of technology depends on know-how; on skills that are typically acquired over many hours of learning and practice. These skills thus depend on neural hyper-plasticity and social support (Malafouris 2010). At the same time, these technologies *supported* the cognitive skills needed to retain and transmit them. The human world is now richly endowed with specialist tools for thinking. Maps and compasses, for example, are powerful aids for navigation, but are not much use for anything else. That is much less true of the earliest technologies; these often have both physical and informational value. As Stephen Mithen and Dan Dennett have pointed out, directly utilitarian early technology is also very information-rich (Mithen 1998; Dennett 2000; Mithen 2000).

The fishing spear in Figure 41.1 (the photo is of a spear in the author's collection) carries a lot of information about how to make a spear. An artisan who picks it up can judge its weight and balance. It can be used as a template for others of the same length, shape, and barb pattern. Someone with a reasonable knowledge of the local woods could probably identify the timber. Much early technology can be wholly or partially reverse-engineered. That is true even of early-generation composite tools: hafted spears and axes, bows. The composition and production of adhesives cannot be reverse-engineered. But an artisan could reverse-engineer a tool's bindings, attachment points, and the shape and placing of its different components. Peter Hiscock points out that stone tools can carry information over centuries and longer. There are known examples that have been picked up, reworked, and used centuries after their initial manufacture.

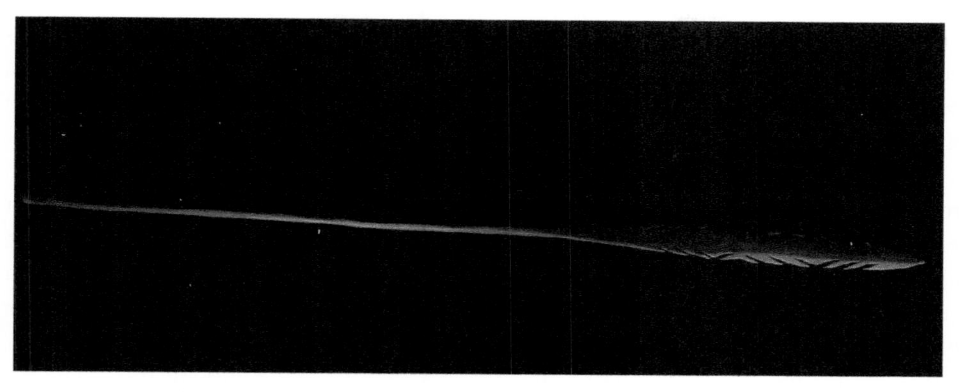

FIGURE 41.1. Fishing spear.

This gap in time between first manufacture and later use is archaeologically visible, because a patina forms and changes color on the scar formed as a flake is struck from the core. So a core with different patinas on its scars has been retouched centuries after its first shaping. The example in Figure 41.2 is one of his images (for which I thank him), showing the phenomenon very clearly.

Terrence Twomey has argued, convincingly, that domesticated fire is a cognitively demanding technology (Twomey 2013). Ignition at will is obviously demanding: fire drills require technical skills similar to those needed for making bows. But he points out that fire demands planning. Fuel must be collected well in advance of need. Careful

FIGURE 41.2. Core showing different patinas on scars.

placement and control is important, especially in those areas where wildfire is a serious and dangerous risk. But while that is true, to the extent that fire is important, and lives become organized around the expectation of light and warmth, the regular use of fire alters the local campsite, reminding those agents of the need to organize fuel and to protect the site from wind and rain. Fire use helps create a social and informational environment that supports the very capacities on which fire depends (Dunbar and Gowlett 2014). Hearths, fireplaces, and fuel dumps remind agents of what they will need, and set up adaptive habits. The traces of old fires serve as reminders, cuing agents of the need to collect fuel and keep the fire alive, and as prompts for paths of least resistance for regular patterns of use.

In these respects, the world has changed completely. There is now a much sharper divide between physical and informational technology. Many contemporary devices are hopelessly ill-suited to serve as templates for their own construction. Raw materials are profoundly transformed. The devices themselves are often complex, causally opaque, and error intolerant. Many are operational only as part of an industrial network. They need fuel, batteries, power sources, lubricants. They are often sealed, so working parts can be reached only destructively or with special tools. There are many devices on my desk: pens, pencils, a phone, a computer, a stapler, knives, scissors, a magnifying glass, paper, duct tape. Probably only the knife could be reverse-engineered by a reasonably competent handyman, and even then only one with a considerable toolkit.

Obviously, there is much we do not know about early hominin technology. Soft materials rarely preserve. The oldest wooden artifacts are spears, which may be about 400,000 years old. Moreover, in contrast with genetically built adaptations, gaps in the material record—for example, the ambiguous record of fire between about 1.5 mya and 400 kya—might record genuine technological loss rather than the failure to preserve material traces of early hominin activities. Even so, the record is rich enough to reveal five salient features of early hominin life: (1) We have been dependent on technology for millions of years; quite likely for over two million years. (2) Our bodies have coevolved with that technology; we are physically shaped by our own tools. (3) Our technologies have long been cognitively demanding to both make and use. Very likely, the skills needed have long depended on socially supported social learning, perhaps including active teaching. (4) While cognitively demanding, those technologies were informationally rich in ways that industrial technology is not. Those technologies were exploitable informational resources. (5) We have no smoking-gun proof that they *were* used informationally as well as materially. Using a technology informationally is itself a skill that has to be learned and transmitted. But it is very likely that they were.

Niche Construction

The extended mind research tradition is undoubtedly right to emphasize the immense and positive impact of material culture on cognition. Likewise, that tradition is right to place the ways material culture scaffold and power cognition in an evolutionary context. Materially supported cognition is not a recent human innovation. These factors do indeed connect extended minds to extended phenotypes. That said, Dawkins's extended phenotype framework is not the ideal theoretical lens through which to study this phenomenon. That framework is *genetic*: extended phenotypes are phenotypes of individual organisms (or sometimes kin groups), and these extended phenotypes are shaped genetically. They evolve (Dawkins supposes) by selection among alternative genetic variants; variants with differential effects on the extended trait. In particular, neither phenotypic plasticity nor individual and social learning play much role in Dawkins's discussion. Yet individual developmental plasticity, social learning and cultural transmission, and gene-culture coevolution will be central to any evolutionary account of extended mind phenomena. It is also true, especially in more recent times, that much of our material culture is produced and used collectively, perhaps even over generations. This probably dates back at least to Pleistocene cave art. In contrast, the extended phenotype framework focuses on individual adaptation.

It is true that some of our cognitive tools—and tools with cognitive effect—have long been part of the way we experience and interact with the world. That is certainly true of important stone-working traditions, but I would bet that hominins have been marking trails to make them easier to see and recall for hundreds of thousands of years. Indeed, the famous Laotoli footprints show that we have been able to use the information trails carry for millions of years (Shaw-Williams 2014). For one set of tracks has been used as a guide by a second hominin following and stepping into them. But many of our cognitive tools are recent innovations, too recent to have elicited an evolved response. That is true even taking into account the fact that evolutionary responses mediated by cultural inheritance appear much more rapidly than those mediated by genetic inheritance. For example, I am a habitual user of Post-it Notes to annotate and keep track of key ideas in those books relevant to my research. Likewise, I use computational cut-and-paste and multiple drafts extensively in my academic writing. These ways of organizing my workspace are novel: they were certainly not inherited, culturally, from either my biological family or my academic mentors. Our reliance on cognitive tools depends on a tripod of (1) human hyper-plasticity, coupled to (2) highly structured and enriched learning environments, and (3) material support for skill acquisition long into adolescence and early adulthood (Gurven et al. 2006). This combination makes it possible for novel tools to be integrated within an agent's repertoire, often in months rather than generations.

For this reason, I have argued that niche construction is a more natural framework for investigating the ways hominins are embedded in, and interact with, their environment. Niche construction recognizes that fact that agents change their environment; in favorable cases, adapt their environment (Scott-Phillips et al. 2014; Laland, Matthews, Feldman 2016). Often, these effects are minor, ephemeral, or unsystematic, and evolutionary and developmental biology can safely treat the environment as fixed or autonomous. Nature chooses the environment; the agents make their best play in that environment. But often is not always: many agents select and/or modify their environments. Hominins, of course, are extreme niche constructors. These days most of us live in environments that are largely human constructions. But we do not live in environments that we have built ourselves, individually, as the caddis fly larva has built its own house. One central difference between the niche construction framework and that of the extended phenotype is that, as noted earlier, the latter is *individualist*. Extended phenotypes are the expression of an individual genome and exist to benefit that genome. The niche construction picture offers a more general framework because niche construction theory recognizes explicitly that niche construction is sometimes collective rather than individual. Moreover, niche construction effects are often transgenerational, not just in affecting the fitness of the next generation (gene selection can readily accommodate selection for transgenerational fitness benefits), but in structuring the developmental environment of the next generation. Niche construction theorists sometimes call this "ecological inheritance." For example, many Australia trees and shrubs rely on fire to initiate seed germination in favorable conditions (rain into ash-fertilized soil) and to suppress competition. As a consequence, many are adapted to make their local habitat more fire-prone. Some species of eucalypt have highly flammable streams of bark hanging down from the trunk and major branches; others have leaves primed with volatile oils. But this is a collective effect. A single scribbly gum will not make an environment fire-prone. Moreover, as noted earlier, niche construction effects are often effects on the ontogeny of the next generation. Nests shelter and conceal adult birds while they are brooding (and hence when escape has high fitness costs). But they have dramatic effects on the ontogeny, not just the fitness, of the eggs and developing nestlings. The nest is warm, dry, concealed.

The same is true of much of the material support of hominin cognition. It is collective and multi-generational, and hence does not fit naturally into an extended phenotype framework. Almost certainly, hominins' most important cognitive tools are collective, multi-generational products. These include languages with their specialist vocabularies and taxonomies that mark and make salient important similarities and differences; databases (like forager herbals or aboriginal star codes; Norris and Harney 2000); scripts; notation systems; and systems of depiction. Many of these are not just multi-generation, multi-user constructions. They have been modified through repeated cycles of transmission and learning to make them more learnable, more usable. Language itself might well be the product of repeated cycles of use and transmission, structured by trade-offs between efficiency of use and ease of acquisition (Tamariz and Kirby 2015). For the same reason, I have been skeptical of the extended mind framework as the most insightful way of exploring the ways material culture powers cognition, for it too is individualist. I have argued that we should see agent–environment support as the environment scaffolding hominin cognition, rather than as agents having extended cognitive systems (Sterelny 2010). It is true that some examples are intuitively compelling: Otto's notebook and a blind person's stick are cases in which it is very natural to think of external resources as a genuine part of a specific agent's cognitive system. Otto's notebook really does have a memory-like role in his life. But these intuitive examples have special features in common: these external resources (1) are trusted (and trustworthy); (2) are resources to which the agent has reliable and easy access; (3) are customized to that one user. The examples are striking, but they are not the most important examples of the environmental support of cognition. Otto did not invent, and probably did not customize, the alphabet he uses in recording his plans, nor the vocabulary he uses to express them. Moreover, while these standard extended mind examples are intuitively persuasive, there is no qualitative break between these and environmental supports that are less individualized, less automatically trusted, and less reliably accessible. In my view, the extended phenotype phenomena are special cases of niche construction and can be accommodated within that more general framework. Likewise, we should prefer a more general framework in which the extended mind examples emerge as special cases of cognitive niche construction or (more or less equivalently) the construction of scaffolds that support and enhance hominin cognition.

A second reason for preferring a more general framework is that not all of the tools that culture provides are material or even external. Consider, for example, Arabic numerals and positional notation. Those with the good luck to be exposed to, and soaked in, that notation system acquire skills in quantitative reasoning that are unavailable to others (or available only at much greater effort and with higher risks of error). As these skills are first acquired, they depend on external, and typically material, support. Children learn arithmetic skills with the help of inscriptions and diagrams (like number lines), rather than just verbal rehearsal. Acquisition requires practice in writing rows and columns of numerals, manipulating them, and annotating these arithmetical scripts as the manipulation proceeds. But, at least in simple cases, many agents learn to perform some of these operations silently, "in their head." These are still culturally built tools for thinking, installed internally via the use of external tools in structuring the developmental environment. These skills and their transmission depend on material scaffolds in the developmental environment (hence on developmental niche construction). But adults do not always continue to depend on external scaffolds. Likewise, the transmission of natural taxonomies through language sensitizes agents to relevant similarities and differences in thought and decision, before the development of scripts, and even in non-social circumstances. Spoken language is a mind-transforming cultural tool, but before scripts, it is not an enduring chunk of material culture.

External Support from Other Minds: Distributed Cognition

Hominin cognition is supported by material resources and by the ways hominins organize their workspace and environment. Built environments, for instance, do not just provide shelter, protection, and storage: they also carry information about the social environment and about the norms and regularities of social living, by dividing settlements into households and organizing life within those households, and by screening off some activities fully or partially from public view (Watkins 2008; Sterelny and Watkins 2015). But we are also resources for one another. It has never been denied that social learning is central to human life, though the idea that it is the most distinctive feature of hominin social life is more recent (Tomasello 1999a). The recognition that the division of cognitive labor is an *ancient* and *essential* feature of hominin social life is more recent still. Until around a decade ago, it was received opinion in the archaeological world that there was a major and abrupt change in the human material record somewhere between 50 kya and 40 kya; a transition known as the Upper Paleolithic Revolution or, less dramatically, the Upper Paleolithic Transition (Mellars 1989, 2005; Mellars and Stringer 1989). On the view then current, human material, economic, and social lives changed in important and linked ways. The transition saw (1) the establishment of improved stone toolmaking traditions, with soft hammer percussion. This in turn made possible more precise control, enabling the knapper to remove smaller flakes. (2) This improved control led to more blade-like tools (i.e., longer and thinner cutting implements). (3) The range of materials expanded to include bone, ivory, and antler. Moreover, some of these tools seemed designed to make fitted clothes. (4) There was evidence of an expanded trade network, which in turn was the signal of a more complex social world. (5) Ornaments, art, and music appeared in the physical record. For example, the first clearly identifiable musical instruments are flutes from about 42 kya. In the eyes of many, ornaments and art—material symbols—were especially important. They were evidence of new forms of cognitive sophistication—the capacity to represent yourself as others see you; the capacity to represent non-literally (Henshilwood and Marean 2003, 2006).

The prevailing wisdom at the time was that this abrupt transition was seen as a sign that the *sapiens* body had been joined with a fully human mind, a mind whose capacities and idiosyncrasies fell within the range of human variation known from the ethnographic record. Humans had become "behaviorally modern," not just anatomically modern. Moreover, it was the result of a genetic change that upgraded the then-existing level of cognitive competence. Perhaps this upgrade was to full language; perhaps to a fully integrated, domain-general intelligence; perhaps it was an enhancement of executive control and planning capacities. Furthermore, since the change was genetically triggered, it took place in *sapiens* populations, but not in populations of our still-living sister species, the Neanderthals. As a consequence, in regions where they both lived, these changes made *sapiens* populations competitively superior to Neanderthals, with their simpler technologies and less sophisticated social organization. Finally, that competitive superiority was the probable cause of Neanderthal extinction (for a clear and accessible history of this supposed revolution and its relevance to Neanderthal extinction, see Papagianni and Morse 2015).

This picture of a genetically triggered shift to more complex cognition and culture still has a few defenders (Klein and Steele 2013; Tattersall 2016). But beginning with a landmark paper in 2000, it has been undermined in two ways (McBrearty and Brooks 2000). First, the appearance of the new technology was not coordinated in space and time. The various signals of "behavioral modernity" came on stream in Africa, over perhaps 100,000 years, and these signals are spatially scattered as well. Of course, much is lost from the material record, and the first material traces of a capacity are vanishingly unlikely to be due to the first expression of that capacity. Even so, the record does not suggest an innovation package appearing at a point of origin and then radiating through space and time. Second, the record does not suggest a clean shift to a new threshold of cultural and cognitive complexity. The record seems to show that these supposed "signature innovations" (material symbols, advanced composite tools) appear as part of the material repertoire, and then disappear. Jewelry and material symbols appear in Africa sometime a little less than 100 kya, but disappear for about 30,000 years (between around 60 kya and 30 kya). Likewise, bow and arrow technology seems to disappear from the same regions it was established in (Lombard and Parsons 2011). This should not have been surprising. History surely tells us that the material complexity of a culture is strongly influenced by economic and organizational factors, not just the informational and cognitive capacities of the individuals in that culture. The genetically triggered Upper Paleolithic Revolution framework was tacitly nativist and individualist: it presupposed that the material culture of a community is sharply constrained by the individual cognitive capacities of the members of that culture, and those capacities in turn are sharply constrained by their genetic makeup.

It is unlikely then that there was a sharp, genetically triggered upward shift in human cognitive capacity and, as a consequence, technical and social complexity, in southern Africa shortly before one *sapiens* lineage expanded out of Africa. That said, it is still widely agreed that there is a phenomenon to be explained. Hominin technical, social, and ecological takeoff seems to have had a very long, slow-burning fuse. Moreover, changes in technology and lifeway do not seem to covary with the appearance of new hominin species, or with physical changes (for example, changes in brain size) of existing species. Simple stone tools appear in the hominin record at about 3.2 mya; more complex Acheulian ones from about 1.7 mya; perhaps at about 250 kya we see the first indications of "prepared core" techniques (which allow knappers to make lighter, more easily carried, and more blade-like tools). At very roughly the same time, we see the first signs of hafted tools (Barham 2013). Hominin morphology shows signs of dietary improvement over this same period, presumably through some combination of food preparation techniques and improved access to meat (perhaps from initial scrounging and opportunistic small-game hunting to power scavenging and medium- and large-game hunting). But as with technology, the pace of change seems very slow, and those changes that we do see do not seem clearly correlated with speciation events[1] or morphological transitions in the hominin lineages.[2] For example, the emergence of *Homo heidelbergensis*, the very large-brained presumptive common ancestor to *sapiens* and Neanderthals, does not seem to be associated with any technological innovation.

[1] An exception may be the evolution of *erectus* coinciding with the introduction of Acheulian technology.

[2] As always, we must be very cautious about these claims, given the extremely fragmentary material record. We are very unlikely to discover the origin points in space and time of specific hominin species or specific hominin technologies.

The picture seems to be that until approximately 250–200 kya, hominins had the ability to retain a relatively small set of core skills (some of which were demanding), but found it difficult to add to this skill set. Indeed, the rate of innovations that established and spread (so becoming archaeologically visible) is so low that it has been seriously suggested that hominin lithic skills were as genetically canalized and inflexible as birdsong (Corbey et al. 2016). Then, from about 250 kya, this constraint on innovation gradually weakened. Why? Perhaps the most influential suggestion turns on the power of *distributed cognition*, a suggestion developed especially, but not only, by Joseph Henrich.[3] On his view, recognizably human lives depend on our capacity to use one another as cognitive resources. We innovate, recognize our own successful innovations, discard unsuccessful ones, pick up useful innovations from others, and transmit to others collectively. Yet individual innovation, retention, learning, and passing on skills to others often depend on quite coarse, rule-of-thumb heuristics. Despite the fact that at this individual level, these learning and decision procedures are noisy and not especially efficient, collectively, at the level of groups and populations of groups, cultural selection biases uptake in favor of practices and customs that are associated with success. Successful innovations often involve lucky accidents, but they tend to be preserved (Richerson and Boyd 2005). The secret of our success is collective. We succeed by collective information pooling and filtering.

Collective intelligence (the analysis continues) rests on demographic foundations. Groups have to be either reasonably large, or small but in a well-connected network, for cognitive capital to be preserved and enhanced by the distribution of information and skill among the local community. The size of the community—both the size of the core foraging band and the other bands with which there is regular, friendly interaction—really matters. Under those circumstances, information is protected by redundancy. Information in only a few heads is easily lost by unlucky accident. If there is only one woman in the band who knows how to find, recognize, and use medicinal herbs, this skill could be very easily lost. Moreover, social learning is powered up in well-connected groups. If a particular skill is difficult to acquire, it helps to have more models and more opportunities for a novice to see a skill expressed. If, say, kayaks are made only very occasionally on a per capita basis, in very small social worlds a novice might see one made only a few times in a life. In larger groups, a naive subject will see this skill expressed more often. Size buffers a group against the loss of cognitive capital through unlucky accident. In addition, larger or well-networked groups are more likely to support specialization (Ofek 2001). A group of ten is unlikely to be able to allow a particularly skilled knapper to focus on making arrowheads; perhaps a group of 50 could. Specialization makes it economically viable to expand the range and quality of technology. It cannot pay a forager to make fishing nets if those nets are only used occasionally. If specialists are more likely to innovate in the areas of their skill, specialization will also increase the innovation rate (though perhaps at the cost of reducing redundancy).

[3] See especially Henrich (2004), Richerson and Boyd (2013), Henrich (2016), Muthukrishna and Henrich (2016).

The idea, then, is that the gradual easing of the constraints on innovation beginning around 250 kya is a signal of the power of distributed intelligence gradually coming on stream, as the human population began to cross a demographic threshold, with groups becoming larger enough or (more likely) well-enough connected for the generation, selection, and retention of innovation to become more reliable (Sterelny 2011). For 150,000 years or more, this crossed threshold was very fragile. Local, regional, and global disturbances had the power to force local populations through population bottlenecks, or force them to disperse, disrupting the flow of information between local bands. Indeed, the last glacial maximum, around 22 kya, may well have forced many regional populations through such a bottleneck. On this analysis, given this demographic fragility, it is no surprise that earlier innovations in this period of more relaxed constraint do nonetheless disappear. Notice that the explanation for the social learning bottleneck of (say) 250 kya is environmental. Our ancestors were, as individuals, cognitively poised to take advantage of one another. But populations were too small and dispersed for collective intelligence to stabilize.

This distributed cognition framework for explaining behavioral modernity is very plausible, and it is quite well supported by formal models (Henrich 2004; Powell, Shennan, and Thomas 2009). In addition, there is some ethnographic support for the idea. For example, there seems to be a positive correlation between population size and technical complexity of islands in Oceania (Kline and Boyd 2010). Likewise, there are reports of skill attrition among the Inuit from accident and disease (Henrich 2016). However, there is no direct evidence of increasing human population size in southern Africa, where the most consistent early signs of increased innovation are found. Indeed, Klein[4] and Steele argue, on the basis of shellfish data, that there is evidence *against* population growth in the relevant parts of southern Africa. When shellfish are intensively harvested, middens show a decline in shell size as larger individuals become rarer. While we see this in shellfish middens later in the Pleistocene, we do not see it in these early African sites of increasing innovation (Klein and Steele 2013). So while the model has both intuitive plausibility and ethnographic support, it may not have direct archaeological support.

This fact makes Holocene Tasmania a central case. For Henrich has argued that the Tasmanian archaeological record does show the interaction of demography with distributed cognition. He reports that when Tasmanian aboriginal communities became isolated from the main Australian population as Holocene sea levels rose and Bass Strait formed, their technology became depauperate (Henrich 2004, 2016; critiqued by Read 2006). Peter Hiscock is very skeptical of this view of technological decline in Holocene Tasmania, and even more skeptical of a demographic explanation of that decline (Hiscock 2008). As he points out, we have little credible information about late

[4] Klein continues to defend a genetic trigger model, though recognizing that this is now a minority view (Klein 2009; Klein and Steele 2013).

Holocene population sizes. In particular, estimates made from colonial records cannot be trusted, as these were estimates of a persecuted hence cryptic population, in deep decline as a result of European disease impacts and persecution.[5] Most saliently, there is apparently no good evidence that Pleistocene aboriginal Tasmanians had ground-edge axes, boomerangs, spear throwers, or fitted clothing made with bone tools. As far as we can tell, ancient Tasmanians never had those technologies. If so, their absence from the Holocene material record is not caused by a loss of demographic connection. The one clear case is the abandonment of fishing. Hiscock suggests that this might have an economic explanation: a shift to specialization on woodland resources as wallaby populations rebounded. So the poster case for the demography-distributed cognition model of human takeoff is open to question.

Moreover, the changes taking place in hominin life in the latter Pleistocene were not just demographic (if Klein is right, perhaps not demographic at all). Life in the latter Pleistocene was becoming more economically complex, with a shift from immediate return mutualism to more reciprocation-mediated forms of cooperation. Around this time (the exact dates are controversial) high-speed projectiles appear in the record,[6] and these projectiles make small hunting parties possible. A greater range of resources was harvested as well (projectiles might make this possible, as hunting and power scavenging no longer depends on the united power of the band as a whole). Small foraging parties and a greater range of foraging targets will increase the total resource take and reduce daily variation in success for the band as a whole. These more complex economies select for more diverse toolkits. Riverine and seashore resources can be harvested without specialist equipment. Shellfish can be gathered by hand; fish can be caught in tidal traps or speared with hunting spears. But much more can be gathered much more efficiently with specialist equipment. Once made, this equipment is part of the informational environment. To some extent, technological diversification can be self-sustaining, once agents learn to exploit the information available from their subsistence technology. Information is encoded in the structure of more causally transparent forms of technology, not just in the heads of those who make and use that technology.

These more complex economies select for cultural innovations too. While increasing the resource portfolio of the group reduces temporal variation, individuals and small groups will have days of failure and success; days in which they give, and others in which they receive. Reciprocation-based cooperation imposes greater delays between contribution and return; contribution and return are less easily weighted (how much of an impala were two fish last week worth?); return is sometimes from third parties, and so reputation is increasingly important. Since different resources are found in different places, groups were more dispersed; more fission, less fusion, and so cooperation and social bonds were not as well mediated by face-to-face interaction. The more complex

[5] Indeed, Hiscock points out that the archaeological evidence, if anything, suggests late Holocene population growth, with increases in the number of midden sites and the re-exploitation of Bass Strait islands.

[6] Spear-thrower-assisted javelins; arrows.

economy increases the cognitive and motivational burdens of cooperation. In other work, I have argued that the role of explicit norms in human social life originated, or massively expanded, to manage the increased cognitive and motivational demands of reciprocation-based cooperation and the conflicts that could easily emerge from those demands. If a group as a whole drives hyenas from a carcass, or ambushes a horse at a waterhole, the profits of cooperation are immediate, and immediately divided. Conflicts are still possible, but there is no problem of temporal discounting; no weighing of the value of differing benefits (everyone gets a chunk of horse); no need to track and aggregate contributions and benefits over time. At about the time in hominin evolutionary history that technology becomes more complex, and the hominin resource envelope expands, we also see the appearance of material symbols in the record. The absence of jewelry and of structured disposal of the dead earlier than around 100,000 years ago is unlikely to be the result of demographic constraints on the transmission of the appropriate know-how between the generations. While the production of some material symbols is technically demanding (e.g., cave art, figures, ivory flutes), much is not. Incised shell jewelry was (very likely) not beyond the technical skills of Heidelbergensians. Nor was burying their dead. Yet there is no sign of either. The appearance of material symbols in the record is a sign of much greater investment in ritual, and that in turn is a signal of a social life much more regulated by explicit norms (Sterelny 2014). To some extent, then, the increased material complexity and diversity of the late Pleistocene is probably a reflection of changes in forager economic lives, rather than either a relaxation of constraints on social learning or a change in the innate cognitive powers of those foragers.

There are (somewhat fragile) positive feedback relations between demography and these other factors selecting for a more diverse toolkit. A group that searches its territory more efficiently (by splitting into small parties, on the condition, of course, that these are still effective) and that harvests a greater range of resources can support larger numbers.[7] Larger numbers store information and skill more robustly. A third factor may be important as well. As Celia Heyes has noted, those working within the cultural evolution research tradition have tended to assume that the cognitive mechanisms of cultural learning are not themselves the products of (or deeply influenced by) cultural learning (Heyes 2012). We do not learn from our cultures how to learn from our cultures. Rather, the cognitive mechanisms of cultural learning are special-purpose devices that have evolved genetically (Tomasello 1999a, 2014). Heyes points out that there is no a priori reason to accept this framework: genetic evolution might give us nothing more than very general learning capacities (association; abstract, non-sensory categorization of environmental phenomena; the capacity to recognize hidden causal relationships). These general mechanisms are tuned in ontogeny in ways that vary due to variation in the environment. She puts flesh on this suggestion by showing how important learning

[7] There is some evidence that Neanderthals exploited a narrower range of resources and lived at significantly lower population densities than *sapiens* populations.

mechanisms (imitation, theory of mind) could themselves be built in ontogeny through learning (Heyes 2012; Heyes and Frith 2014; Heyes 2016).

In developing this analysis of the transition to behaviorally modern humans, I do not intend to downplay the importance of distributed cognition. We are critical resources for one another. Human intelligence is collective as well as individual. But overcoming the social learning bottleneck through richer and more stable social networks was probably not the only factor in the transition to the more socially and materially complex lives of the later Pleistocene foragers. Social and material support for hominin cognition has very deep roots indeed. It probably extends back to the erectines and the establishment of hand-ax technology about 1.7 mya, perhaps even earlier. But Heyes's program opens up an intriguing possibility though one very difficult to test. Perhaps the relaxation of constraints on the volume and fidelity of cultural learning beginning around 250 kya might at least in part be a signal of improved social learning capacities interacting with an environment of gradually increasing material scaffolds for social learning, and with richer and more stable social networks. A richer material culture scaffolds its own transmission to the next generation (to the extent that it is not causally opaque), but only if agents are primed to exploit the information available from these artifacts; only if the artifacts are readily accessible to novices; and novices have the time, the freedom, and the support to experiment and play. My best guess is that (1) distributed cognition, (2) the material scaffolding of skill acquisition enhanced by selection for a more diverse technology, and (3) improved learning strategies were all important in the acceleration of technical, ecological, and cultural change beginning about 250 kya. Part of that improvement may be just more time to learn: there is some dental evidence that *sapiens* childhood and adolescence was longer than that of our Neanderthal sister species[8] (Smith et al. 2007). But identifying the material traces of these cognitive and social factors remains a major and open challenge.

References

Alvard, M. (2003). The adaptive nature of culture. *Evolutionary Anthropology*, 12, 136–49.

Ambrose, S. (2001). Paleolithic technology and human evolution. *Science*, 291, 1748–53.

Barham, L. (2013). *From hand to handle: the first industrial revolution*. Oxford: Oxford University Press.

Clark, A. (2001). Reasons, robots and the extended mind. *Mind and Language*, 16(2), 121–45.

Clark, A. (2007). Curing cognitive hiccups: a defense of the extended mind. *Journal of Philosophy*, 104(4), 163–92.

[8] The evidence is from growth patterns on Neanderthal teeth—evidence that Neanderthals grew to maturity significantly more rapidly than did *sapiens* children (Smith et al. 2007). If this data holds up, and if the Heidelbergensian common ancestor to *sapiens* and Neanderthals had the same life history pattern as Neanderthals, then a (presumably) genetic change gave *sapiens* children and adolescents much longer to learn adult life skills than the children of the sibling and ancestor species.

Clark, A. (2008). *Supersizing the mind: embodiment, action, and cognitive extension*. Oxford: Oxford University Press.

Clark, A. and Chambers, D. (1998). The extended mind. *Analysis*, 58(1), 7–19.

Corbey, R., Jagich, A., Vaesen, K., and Collard, M. (2016). The Acheulean handaxe: more like a bird's song than a Beatles' tune? *Evolutionary Archaeology*, 25, 6–19.

Dawkins, R. (1982). *The extended phenotype*. Oxford: Oxford University Press.

Dennett, D.C. (2000). Making tools for thinking. In: D. Sperber (ed.), *Metarepresentations: a multidisciplinary perspective*. Oxford: Oxford University Press, pp. 17–29.

Dunbar, R. and Gowlett, J. (2014). Fireside chat: the impact of fire on hominin socioecology. In: R. Dunbar, C. Gamble, and J. Gowlett (eds.), *Lucy to language: the benchmark papers*. Oxford: Oxford University Press, pp. 277–96.

Godfrey-Smith, P. (2001). Three kinds of adaptationism. In: S. Orzack and E. Sober (eds.), *Adaptation and optimality*. Cambridge: Cambridge University Press, pp. 335–57.

Gowlett, J. (2016). The discovery of fire by humans: a long and convoluted process. *Philosophical transactions of the Royal Society of London. Series B, Biological sciences*, 371(1696), 20150164.

Gowlett, J. and R. Wrangham (2013). Earliest fire in Africa: towards the convergence of archaeological evidence and the cooking hypothesis. *Azania: Archaeological Research in Africa*, 48(1), 5–30.

Gurven, M., Kaplan, H., and Gutierrez, M. (2006). How long does it take to become a proficient hunter? Implications for the evolution of extended development and long life span. *Journal of Human Evolution*, 51, 454–70.

Haig, D. (2014). Coadaptation and conflict, misconception and muddle, in the evolution of genomic imprinting. *Heredity* 113, 96–103.

Henrich, J. (2004). Demography and cultural evolution: why adaptive cultural processes produced maladaptive losses in Tasmania. *American Antiquity*, 69(2), 197–221.

Henrich, J. (2016). *The secret of our success: how culture is driving human evolution, domesticating our species and making us smarter*. Princeton: Princeton University Press.

Henshilwood, C. and Marean, C. (2003). The origin of modern human behavior. *Current Anthropology*, 44, 627–51.

Henshilwood, C. and Marean, C. (2006). Remodelling the origins of modern human behaviour. In: H. Soodyall (ed.), *The human genome and Africa: tracing the lineage of modern man*. Cape Town: Jonathan Ball, pp. 31–46.

Heyes, C. (2012). Grist and mills: on the cultural origins of cultural learning. *Philosophical transactions of the Royal Society of London, Series B, Biological sciences*, 367, 2181–91.

Heyes, C. (2016). Blackboxing: social learning strategies and cultural evolution. *Philosophical transactions of the Royal Society of London. Series B, Biological sciences*, 371(1693), 20150369.

Heyes, C. and Frith, C. (2014). The cultural evolution of mind reading. *Science*, 344, 6190.

Hiscock, P. (2008). *Archaeology of ancient Australia*. London: Routledge.

Klein, R. (2009). Darwin and the recent African origin of modern humans. *Proceedings of the National Academy of Sciences*, 106(38), 16007–9.

Klein, R. and Steele, T. (2013). Archaeological shellfish size and later human evolution in Africa. *Proceedings of the National Academy of Sciences*, 110(27), 10910–5.

Kline, M. and Boyd, R. (2010). Population size predicts technological complexity in Oceania. *Proceedings of the Royal Society B: Biological Sciences*, 277, 2559–64.

Laland, K., Matthews, B., and Feldman, M. (2016). An introduction to niche construction theory. *Evolutionary Ecology*, 30(2), 191–202.

Lombard, M. and Parsons, I. (2011). What happened to the human mind after the Howiesons Poort? *Antiquity*, 85(330), 1433–43.

Malafouris, L. (2010). Metaplasticity and the human becoming: principles of neuroarchaeology. *Journal of Anthropological Sciences*, 88, 49–72.

McBrearty, S. and Brooks, A. (2000). The revolution that wasn't: a new interpretation of the origin of modern human behavior. *Journal of Human Evolution*, 39(5), 453–563.

Mellars, P. (1989). Major issues in the emergence of modern humans. *Current Anthropology*, 30(3), 348–85.

Mellars, P. (2005). The impossible coincidence: a single-species model for the origins of modern human behavior in Europe. *Evolutionary Anthropology*, 14, 12–27.

Mellars, P. and Stringer, C. (eds.) (1989). *The human revolution*. Edinburgh: Edinburgh University Press.

Mithen, S. (1998). A creative explosion? Theory of the mind, language and the disembodied mind of the Upper Palaeolithic. In: S. Mithen (ed.), *Creativity in human evolution and prehistory*. New York: Routledge, pp. 165–86.

Mithen, S. (2000). Mind, brain and material culture: an archaeological perspective. In: P. Carruthers and A. Chamberlain (eds.), *Evolution and the human mind: modularity, language and metacognition*. Cambridge: Cambridge University Press: pp. 207–17.

Muthukrishna, M. and Henrich, J. (2016). Innovation in the collective brain. *Philosophical transactions of the Royal Society of London. Series B, Biological sciences*, 371(1690), 20150192.

Norris, R.P. and Harney, B.Y. (2000). Songlines and navigation in Wardaman and other Australian Aboriginal cultures. *Journal of Astronomical History and Heritage*, 17(2), 1–15.

O'Brien, G. and Opie, J. (2015) Intentionality lite or analog content? *Philosophia*, 43(3), 723–30.

O'Brien, M. and Laland, K. (2012). Genes, culture, and agriculture. *Current Anthropology*, 53(4), 434–70.

Ofek, H. (2001). *Second nature: economic origins of human evolution*. Cambridge: Cambridge University Press.

Papagianni, D. and Morse, M. (2015). *The Neanderthals rediscovered*. London: Thames and Hudson.

Powell, A., Shennan, S., and Thomas, M. (2009). Late Pleistocene demography and the appearance of modern human behavior. *Science*, 324, 1298–1301.

Read, D. (2006). Tasmanian knowledge and skill: maladaptive imitation or adequate technology. *American Antiquity*, 71(1), 164–84.

Richerson, P. and Boyd, R. (2013). Rethinking paleoanthropology: a world queerer than we had supposed. In: G. Hatfield and H. Pittman (eds.), *The evolution of mind, brain and culture*. Philadelphia: University of Pennsylvania Press, pp. 263–302.

Richerson, P.J. and Boyd, R. (2005). *Not by genes alone: how culture transformed human evolution*. Chicago: University of Chicago Press.

Scott-Phillips, T., Laland, K., Dickins, T., and West, S. (2014). The niche construction perspective: a critical appraisal. *Evolution*, 68, 1231–43.

Shaw-Williams, K. (2014). The social trackways theory of the evolution of human cognition. *Biological Theory*, 9(1), 16–26.

Smith, T.M., Toussaint, M., Reid, D.J., Olejniczak, A.J., and Hublin, J. (2007). Rapid dental development in a middle Paleolithic Belgian Neanderthal. *Proceedings of the National Academy of Sciences*, 104(51), 20220–5.

Sterelny, K. (2010). Minds—extended or scaffolded? *Phenomenology and the Cognitive Sciences*, 9(4), 465–81.

Sterelny, K. (2011). From hominins to humans: how *sapiens* became behaviourally modern. *Philosophical transactions of the Royal Society of London. Series B, Biological sciences*, 366(1566), 809–22.

Sterelny, K. (2014). A paleolithic reciprocation crisis: symbols, signals, and norms. *Biological Theory*, 9(1), 65–77.

Sterelny, K. and Watkins, T. (2015). Neolithization in southwest Asia in a context of niche construction theory. *Cambridge Archaeological Journal*, 25(3), 673–91.

Tamariz, M. and Kirby, S. (2015). The cultural evolution of language. *Current Opinion in Psychology*, 8, 37–43.

Tattersall, I. (2016). Language Origins: An Evolutionary Framework. *Topoi*. doi:10.1007/s11245-016-9368-1

Tomasello, M. (1999a). *The cultural origins of human cognition*. Cambridge: Harvard University Press.

Tomasello, M. (1999b). The human adaptation for culture. *Annual Review of Anthropology*, 28, 509–29.

Tomasello, M. (2014). *A natural history of human thinking*. Cambridge: Harvard University Press.

Turner, J.S. (2000). *The extended organism: the physiology of animal-built structures*. Cambridge: Harvard University Press.

Twomey, T. (2013). The cognitive implications of controlled fire use by early humans. *Cambridge Archaeological Journal*, 23(1), 113–28.

Watkins, T. (2008). Natural environment versus cultural environment: the implications of creating a built environment. In: J.C. Zoilo (ed.), *Proceedings of the 5th International Congress on the Archaeology of the Ancient Near East*. Madrid: Universidad Autónoma de Madrid, pp. 427–37.

Wilson, R.A. (1994). Wide computationalism. *Mind*, 103, 351–72.

Wrangham, R. (2009). *Catching fire: how cooking made us human*. London: Profile Books.

Wrangham, R. and Carmody, R. (2010). Human adaptation to the control of fire. *Evolutionary Anthropology*, 19(5), 187–99.

CHAPTER 42

CRITICAL NOTE

Evolution of Human Cognition. Temporal Dynamics at Biological and Historical Time Scales

TOBIAS STARZAK AND ANDREAS ROEPSTORFF

THE four chapters in this section explore from different angles how approaching the topic of 4E cognition from an evolutionary perspective can shed light on aspects of cultural and biological evolution, the nature of cognition in general, and of human cognition in particular. Each from their own perspective, they offer a temporal analysis of human cognition that puts cognition into a context of larger processes that span historical and indeed evolutionary time, and they embed the dynamics of cognition beyond brains and individuals, into groups and even species.

Cognition as Coordination in Unpredictable Environments

In her chapter "The Evolution of Cognition: A 4E Perspective," the biologist Louise Barrett elegantly applies evolutionary thinking to explore the questions: *What is cognition essentially? How can we explain cognitive evolution? What is the role of brains and nervous systems for cognition? How should we approach animal cognition?* and *What is specific about human cognition?* The most straightforward evolutionary question concerning cognition is how and why it evolved. How and why cognition evolved depends on what cognition is, but, as Barrett points out, the cognitivist attempt to link cognition to brains and nervous systems is deeply problematic. We neither know how brains are related to cognitive processes nor why they should be necessary: we cannot map cognitive abilities on brain anatomy, brain size (absolute or relative) is a bad indicator of cognitive abilities, and we find remarkably flexible behavior in tiny-brained and even brainless organisms.

Here, it seems that Barrett somewhat downplays what we know about the importance of brains for cognition. While it is true that brain size is a bad indicator of cognitive abilities, it has recently been argued that the absolute number of neurons is robustly correlated with cognitive abilities (Herculano-Houzel 2016, 2017). Furthermore, for all we know our failure to map cognitive abilities on brain architecture could simply reflect a lack of knowledge on our part rather than the impossibility of such mapping. Still, even if it turns out that brains support cognition in a specific and determinable way, it should be an open question whether brains are necessary for cognition. Thus Barrett is right to argue that we should not define cognition in terms of brains or nervous systems.

As a promising alternative she proposes a *biogenic* bottom-up approach: rather than starting with paradigmatic cases of representational human cognition (i.e., processes involving the modification of representations) and work our way back (the *anthropogenic* top-down approach), we should take an evolutionary perspective to look at the function of minimal cognition in order to understand how to best think about more advanced cases and about cognition in general. Thus, Barrett turns the tables and claims that what cognition is depends on how and why it evolved.

If we reject brain-based definitions, we need a functional criterion, something in virtue of which a process is cognitive. It needs to be wide enough to capture all kinds of flexible behavior, but specific enough to let us draw a line between cognitive and noncognitive processes. Barrett suggests that processes underlying *sensorimotor coordination* fulfill these requirements. Since we find minimally cognitive behavior in this sense even in bacteria, neither brains nor nervous systems are necessary for minimal cognition. But these processes can be distinguished from noncognitive, merely metabolic processes of an organism.

A worry here is that Barrett simply presupposes that there *must* be a criterion that captures all cases. Against this view, Colin Allen (2017) argues that cognition may be too general a term to find a defining criterion. There may be other, more pragmatic reasons that explain why we consider some processes as cognitive and others not. If Allen is right and there is no feature all cases share and in virtue of which we consider them cognitive, it would be a mistake to identify a lowest common denominator with what's essential for cognition. By the same reasoning, it would be a mistake to classify as cognitive every process that fulfills this criterion.

Barrett goes on to argue that a good criterion should also have the potential to explain differences between minimally and more advanced cognitive systems. This is where nervous systems and brains enter her picture. Following Keijzer and colleagues (2013), she argues that the original and more fundamental function of nervous systems is to internally coordinate action, or, as Godfrey-Smith puts it, to work out *how* to do something rather than *what* to do (Godfrey-Smith 2016). Barrett argues that in this evolutionary picture, brains and nervous systems are best understood as a means to solve the problem of successful sensorimotor coordination of situated and embodied organisms, once increasing body size and longevity make these problems more pressing. Thus, brains and nervous systems do figure in explaining the difference between minimally cognitive systems and more advanced systems, though in a very different way than envisaged by classical cognitivism. They function to enable situated and embodied action rather than to compute passively received input. In this picture, cognitive evolution is best understood as a process by which organisms became better equipped to track and deal with unpredictable contingencies in real time in changeable environments (Barrett 2011).

Barrett's conclusion has further implications for how we should think about typical cases of human representational cognition and their evolution, closing the circle of the questions mentioned at the beginning. This is where it becomes obvious that she endorses a radical rather than a conservative enactivist and embodied approach to both the question what cognition is and how we should study it. While some human processes are certainly representational—language, mathematical reasoning, symbolic logic—Barrett argues that these are very special cases indeed. They only arise with the mastery of sociocultural practices and are the result of internalized public representations (Vygotsky 1997). This makes them special for two reasons: First, they will not be shared by other animals, because nonlinguistic animals don't share the sociocultural practices needed to build contentful representations. While Sterelny's contribution in this volume explores the profound ways in which niche construction shaped human cognition, Barrett's claim that representational cognition itself is a product of cultural processes seems far more radical. Second, humans are situated and embodied organisms as well. Hence, Barrett claims that what we take to be *paradigmatic* cases for human cognition turn out to be merely a small part of their cognitive lives, which for the most part is as embodied, enactive, and nonrepresentational as those of nonlinguistic *basic minds*. This, she claims, is where we should locate evolutionary continuity. Finally, she agrees with Brooks (1999) that even in humans, explaining the nonrepresentational part is more fundamental; understanding the nonrepresentational part of cognition is essential to understand the small fraction of representational cognition in humans.

Barrett makes a strong case for putting embodied and enactive aspects of agency into focus when studying flexible behavior, and she demonstrates how an evolutionary perspective can shed light on the nature of cognition. Concerning her claim that representational cognition is unique to humans we are more skeptical. According to Barrett, it is representationalism that is in need of justification. But shifting the burden of proof strongly depends on the plausibility of Vygotsky's claim that human representational cognition really is the product of internalized public sociocultural practices. Given its radical nature, this claim needs further support to convince us to go radical as well.

From Mindreading to Mindshaping

In its own right, the philosopher Tadeusz Zawidzki's contribution continues Barrett's analysis at the level of the human species. He demonstrates how most current models of human social cognition stress the importance of representations and contentful thought for figuring out what other people may be thinking and intending. This line of analysis poses that "it is because we have a neurally implemented capacity to correctly ascertain each other's beliefs, desires, and other propositional attitudes, that we are able to coordinate and cooperate with, and learn from each other so well" (p. 736). Zawidzki fundamentally questions the foundational importance of this idea of *mindreading*. Instead, he proposes that humans have a special ability to shape each other's minds so that coordination emerges, apparently spontaneously. The key claim is that this is enabled by specific low-level mechanisms, none of which require a detailed representation of other people's minds. Crucially, humans appear to pay thorough attention to the specific process of an action, the *how* of doing, and not just the end result, the *what* of doing. One may see signs of this attentional sensitivity already in infants and small children, who will meticulously imitate and copy very specific, embodied ways of doing things. This attention to processes appears rare among animal species, with only traces of it seen among chimpanzees, but not in other primates. Thus human infants and adults seem to have a particular attention to the *style* of an action that allows them to establish particular normative templates for actions.

Zawidzki suggests that this attention is tied to reward mechanisms, so that mimicking or even overimitating behavior to templates is intrinsically motivating, while the apparently costly punishment of those, who don't follow norms is outweighed by internal rewards elicited by the very attempt of matching others' actions to these stylistic templates. Importantly, many of these mindshaping mechanisms are embedded beyond the individual into formal and informal practices and institutions that support learning and development. Furthermore, these templates for actions do not only emerge from concrete, embodied patterns, mediated by publicly shared language. They may equally well derive from fictional agents, like characters in myths and narratives.

Zawidzki proposes that mindshaping mechanisms ensure the development and stabilization of internal cohesion within a group. Over time, this may drive cultural group selection where in a competitive environment, groups dominated by cooperation can outcompete groups composed of members not so shaped. Thus, according to Zawidzki, the key mechanisms to look for in the evolution of human sociality are those that support *mindshaping*.

This suggestion invites a quite radical rethinking of the role of social and cultural processes in providing minds with specific patterns and priors. However, the approach may be seen as somewhat one-directional: in Zawidzki's account, mindshaping emanates from figures of authority, be they real or fictional, and it spreads out to people who take it in more or less passively. Processes of influence, from pedagogy to propaganda, seldom work like that. There seems to be a need to embed in the model a possibility for agency in the recipients, be that in the form of resistance, interpretation, or exploration. Both in evolution and ontogeny, a critical feature for that may be the extended period of immaturity in human childhood, which not only allows them to receive mindshaping inputs from figures of authority, but also to actively explore fields of possibilities and inference (e.g., Buchsbaum et al. 2012). Thus, there is a need to account not just for the emergence of intergroup difference through shared mindshaping mechanisms within groups, but also of intragroup diversity and creativity. Here, culturally supported active exploration as in play and playfulness may be important mechanisms for mindshaping along with those more passive modes of transmission proposed by Zawidzki.

In interesting ways, Zawidzki's argument parallels and extends Barrett's analysis of early animal cognition. She argued that intraorganismic coordination with respect to a changing environment is at the very heart of the evolution of animal cognition. Zawidzki's central claim is that in humans, low-level, socially embedded mechanisms underlie processes of mindshaping, which enable and ensure coordination within *and* between individuals. Thus, human behavior is embedded in a socially mediated shaping of mind that works at a group level. Neither Barrett nor Zawidzki deny the existence and indeed importance of a human ability to represent abstract content, be that of the world or of other people's minds. However, rather than making that faculty a gold standard for universal cognition, and the representation of other minds a hypothetical foundation of human uniqueness and sociality, they demonstrate how potentials for representations may integrate with, and indeed emerge from, much more basic mechanisms of coordination and attention.

Thinging: Human Cognition and Material Engagement

In his contribution, "Bringing Things to Mind," the cognitive archaeologist Lambros Malafouris stresses the critical importance of objects and materials in human culture and interaction. The claim is not just stating the obvious fact that humans use objects to a degree not seen in other species. Rather, Malafouris argues that for humans the mental and the material are so intimately intertwined that "human thinking can be seen as a craft patiently absorbed in the manufacture of complex surfaces of sorts (plastic or rigid)" (p. 756). This intertwinement lies at the root of Malafouris' material entanglement hypothesis. It claims that our active involvement with materiality is at the root of *agency*, in that it shapes the ability to act, of *signification*, in that it brings forth meaning, and of *cognition*, in that it shapes the very process of thinking. At the same time, the stability of things allows for a coordination of human action across time. They shape and structure processes of learning and transmission, and they offer support and anchoring for shared cultural and ritual practices. The stability and physical persistence of things and objects thus allow humans to construct bridges between temporal phenomena that operate at different time scales, from neural and cellular processes over individual experiential life, to historical and evolutionary dynamics.

It is not surprising that an archaeological perspective provides a sensitivity to the importance of things. Indeed, archaeology is at its heart concerned with objects and materials as they are preserved through the ages, and with deducing what they may tell us about distant lives, cultures, and worlds. However, the perspective offered by

Malafouris is more radical than a simple reminder about the historical importance of things. Malafouris proposes the neologism *thinging* to capture this intimate interrelation between mind and matter. *Thinging* is about processes of thinking and feeling *with, through*, and *about* things. As an example, he analyzes the making of a clay vase. It does not come about simply by the potter projecting a representation onto inert matter. Rather, the vase emerges in an experiential process, "where brain, body, clay and wheel conflate" (p. 766).

Obviously, this analysis challenges representational accounts of human cognition. However, as Malafouris suggests, it also highlights an important dimension that 4E cognition needs to take into account: the world of things, objects, materials, and artifacts does not just shape patterns of brain activity. More fundamentally, these are constitutive of the plasticity and stability that is at the heart of the extension and reorganization of neural, embodied, and cultural processes. Cognition as *thinging* thus cuts across brain–body–world divisions, with agency at all levels.

Suggesting strong links between materiality and agency offers an important perspective on human cognition, but it leaves fundamental evolutionary questions open. It is becoming increasingly clear that many nonhuman animals engage extensively with materials, tools, and technologies (Seed and Byrne 2010). Even in "lowly" species such as spiders, objects are used in ways that remind us of gift-giving (Ghislandi et al. 2015). If this is the case, what, if anything, is then special about human engagement with materiality? Is it an attention to style and form, and not just use, as suggested by Zawidzki? Is it flexibility and plasticity of use? If there is something specific about the human way of engaging with objects, does it, then, have to do with a particular relation to the material, in how things and meanings, words and objects, are intertwined in material symbols, that are simultaneously inside and outside? (Clark 2006). Thus, we seem to need key mechanisms that are not solely in the world but importantly also in the human mind, or brain, and in particular we need to know how the mind or brain relates to objects and materials (Roepstorff 2008).

Like Zawidzki, Malafouris identifies processes of shaping and being shaped as a fundamental feature of human cognition. Human cognition, in groups and individuals, is at the same time provided with a plasticity enabled through interaction with other minds and materials, and a stability offered by norms and narratives, things, and objects. Concerning mindshaping and material engagement, Zawidzki and Malafouris jointly draw a contemporary picture of the human being akin to the philosopher Friederich Nietzsche's notion of *das noch nicht festgestellte Tier*, the animal whose nature is not yet fixed. However, it is given form and stability in cultural practices and creative interactions with subjects and objects, people and things.

Evolution and Niche Construction

As do the other contributions in this section, the philosopher Kim Sterelny approaches the subject of cognition from a temporal perspective. However, whereas Barrett uses a temporal perspective to defend a claim about the general nature of cognition, and Zawidzki and Malafouris analyze particular constituents of human cognition, Sterelny's focus is on how it came about that human cognition became so different from nonhuman animal cognition in the transition to behaviorally modern humans. Sterelny's analysis demonstrates that a key component of human cognition is that it relies heavily on external support—on tools, material or cognitive, that humans have created themselves. This expands classical 4E cognition in a fundamental way. The frame of analysis extends not only in space beyond the skull, taking into account the body and external resources. It critically extends in time, too. Thus, one must explore the interplay and resulting feedback mechanisms between biological evolution, cultural dynamics, and environmental changes. Importantly, the importance of external tools for human cognition is not exhausted by their role for scaffolding cognitive processing. Sterelny argues that one must also understand the fundamental role of external resources like physical and social learning environments, group size, material culture, and biological and cultural transmission processes in shaping and developing minds like ours.

Sterelny's main theoretical project is to explore which framework best captures "the ways material culture powers cognition" (p. 780). For this he compares Dawkins' *extended phenotype* (1982), Clark's *extended cognition* (2007, 2008), and Henrich's *distributed cognition* (Henrich 2016) with the niche construction framework (Laland 2016), arguing that the latter is best suited to account for the complex relations between human cognition and external resources. Briefly, the *extended phenotype* is a gene-based model that focuses on individuals. However, it fails to capture the influence of tools "too recent to have elicited an evolved responses" (p. 779), and of nongenetic transmission (like environmental engineering and culture). Furthermore, it cannot deal with the collectively embedded dimension of human cognition and transmission. Being too individualistic is also a problem the *extended cognition* framework faces. Furthermore, its application is restricted to very special cases in which humans make use of external resources (like a notebook or a personal computer), thus leaving the majority of cases, where external resources influence human cognition, unaccounted for. Some of these cases concern material scaffolding of cognitive processes. But Sterelny also draws attention to internalized *cognitive* tools, like language or linguistic concepts, that influence the way we categorize and think about the world. Although being internalized in individual agents, they have a critical external and collective dimension. They are to some degree the product of cultural evolution (they have been created by others) and their acquisition during development is mediated through material tools and other minds, like script symbols and active teaching. But because their use ceases to depend on external scaffolding once agents have mastered using them, theories that focus too narrowly on mind–world interaction miss the external nature of these tools.

Henrich's *distributed cognition* framework captures this aspect. However, Sterelny convincingly argues that this framework somewhat overemphasizes the importance of demographic features for structuring our social environment in ways that cross natural thresholds for innovation and cultural evolution. By contrast, the *niche construction* framework recognizes the fact that agents change and, in favorable cases, adapt their environment. This is particularly the case for modern humans, who live embedded in environments that are to a large extent structured by humans, but these are not environments that each individual has built on their own. To a large degree they are inherited with all the material resources and cultural practices they contain. Thus, Sterelny argues, only the niche construction framework proves sufficiently resourceful to account for the whole range of relations between individual cognition and material, external, and collective support that characterizes human development and existence, while at the same time allowing for an evolutionary analysis that also integrates with nonhuman species.

Although Sterelny makes a convincing case for the general framework, here are some critical remarks. First, in defending niche construction Sterelny argues that "the choice between [the different] frameworks is largely heuristic" (p. 773). This is a fairly modest claim. If what we are after is a general account of the role external resources play for human cognition, the fact that niche construction *can* account for all (or at least a great number of) relevant cases, and alternative accounts cannot, should be viewed as a strong argument in favor of niche construction.

On the other hand, there is a worry that this should *not* be our evaluative criterion, and as a consequence this comparison does not do justice to these alternative accounts. Clark suggests that instead of picking a *best* one, we should "practice the art of flipping between the different perspectives. The cure for cognitive hiccups[1] is to stop worrying and enjoy the ride" (Clark 2008, p. 139). Thus, Clark argues that the different frameworks are *all* useful in their own right by focusing our attention on different aspects of the relation between cognition and external resources. So, given that we want a general account of how material culture and cognition interact, Sterelny's niche construction framework may very well be the theory of choice. But if we want to answer whether in some cases we should describe the use of external resources rather as scaffolding than as a literal extension of our cognitive processes beyond the skull and the body, we may better look at the embedded or extended cognition frameworks. As a consequence, proponents of these alternative approaches may simply deny being in the business of explaining the same thing that a niche construction account does; their accounts are not, as Sterelny says, "developed to capture the role of . . . external resources" (p. 773) in the same general way than the niche construction framework is.

The upshot is that Sterelny makes a strong case for niche construction being the most promising framework to capture the various ways in which material culture powers cognition, but we should not interpret this as a strong argument against the alternative frameworks in general. A strength of the general niche construction approach is its flexibility: it leaves a lot of room to tweak the details to adjust to new

[1] The *hiccups* refer to the question which we should prefer, HEC (hypothesis of extended cognition), HEMC (hypothesis of embedded cognition), or HOC (hypothesis of organism-centered cognition).

empirical evidence. However, this implies that the explanatory power of any specific niche construction account depends on *how* the details will be spelled out. Sterelny puts some flesh on the bones, suggesting that distributed cognition, the material scaffolding of skill acquisition, and improved learning strategies each play a central role in accounts of human niche construction. Relating his framework to extended phenotype, extended cognition, and distributed cognition illuminates how difficult it is to keep them neatly apart. But the main challenge for Sterelny is to compare and defend his account against alternative niche construction accounts that spell out the details differently.

Conclusion: The Specificity of the Human Niche

In line with Sterelny's analysis, the niche construction framework indeed suggests a tentative synthesis across the four papers in this section. Evolutionarily speaking, humans appear to have emerged to be the niche constructors par excellence; they inhabit niches of a very particular sort that allow them to solve problems of interindividual coordination through particular processes, two of which may be mindshaping and material engagement. As in other species, human cognition has a fundamentally embodied anchoring, which allows for coordination within an organism and for adaptability to changing environments. However, in humans these processes expand beyond the individual in time and space, in ways that support coordination and forms of collaboration (at least within groups), and that may have even given rise to representational cognition. This appears to generate particular patterns of intergroup diversity, temporally and spatially. Moreover, this may be a key mechanism in the human potential for rapid cultural evolution and diversification, which appears unique in the animal kingdom. Perhaps the tightly intertwined processes of mindshaping and material engagement are key to this. Cultural practices of shaping minds and matter may be at the heart of the dynamic evolution of that particular human niche, which according to some theoreticians of the Anthropocene now affects and even encompasses many if not most other biological niches on the planet.

A 4E-based analysis of the evolution of human cognition does not deny the importance of abstract representations for human cognition. However, all four contributions suggest that in focusing too narrowly on representational cognition, we lose sight of the basic mechanisms supporting and driving human cognition. To understand these, we may need to understand not only how these processes are embodied, embedded, enactive, and extended, but also how they are shaped, transmitted, and diversified in processes of group formation. This may also be critical for understanding how to solve some of the major problems of intergroup coordination and collaboration that we as a species are currently facing.

References

Allen, C. (2017). On not defining cognition. *Synthese*. doi:10.1007/s11229-017-1454-4

Barrett, L. (2011). *Beyond the brain: how body and environment shape animal and human minds*. New Jersey: Princeton University Press.

Barrett, L. (2018). The evolution of cognition: a 4E perspective. In: this volume, pp. 719–34.

Brooks, R.A. (1999). *Cambrian intelligence: the early history of the new AI*. Cambridge, MA: MIT Press.

Buchsbaum, D., Bridgers, S., Weisberg, D.S., and Gopnik, A. (2012). The power of possibility: causal learning, counterfactual reasoning, and pretend play. *Philosophical transactions of the Royal Society of London. Series B, Biological sciences*, 367(1599), 2202–12.

Clark, A. (2006). Material symbols. *Philosophical Psychology*, 19(3), 291–307.

Clark, A. (2007). Curing cognitive hiccups: a defense of the extended mind. *Journal of Philosophy*, 104(4), 163–92.

Clark, A. (2008). *Supersizing the mind: embodiment, action, and cognitive extension*. Oxford: Oxford University Press.

Dawkins, R. (1982). *The extended phenotype*. Oxford: Oxford University Press.

Godfrey-Smith, P. (2016). *Other minds: the octopus and the evolution of intelligent life*. London: HarperCollins UK.

Henrich, J. (2016). *The secret of our success: how culture is driving human evolution, domesticating our species and making us smarter*. Princeton: Princeton University Press.

Herculano-Houzel, S. (2016). *The human advantage: a new understanding of how our brains became remarkable*. Cambridge, MA: MIT Press.

Herculano-Houzel, S. (2017). Numbers of neurons as biological correlates of cognitive capabilities. *Current Opinion in Behavioral Sciences*, 16, 1–7.

Ghislandi, P., Bilde, T., and Tuni, C. (2015). Extreme male mating behaviours: anecdotes in a nuptial gift-giving spider. *Arachnology*, 16(8), 273–5.

Keijzer, F., van Duijn, M., and Lyon, P. (2013). What nervous systems do: early evolution, input-output, and the skin brain thesis. *Adaptive Behavior*, 21, 67–85.

Laland, K., Matthews, B., and Feldman, M. (2016). An introduction to niche construction theory. *Evolutionary Ecology*, 30(2), 191–202.

Malafouris, L. (2018). Bringing things to mind: 4Es and material engagement. In: this volume, pp. 755–72.

Roepstorff, A. (2008). Things to think with: words and objects as material symbols. *Philosophical transactions of the Royal Society of London. Series B, Biological sciences*, 363, 2049–54.

Seed, A. and Byrne, R. (2010). Animal tool-use. *Current Biology*, 20(23), R1032–R1039.

Sterelny, K. (2018). Culture and the extended phenotype: cognition and material culture in deep time. In: this volume, pp. 773–92.

Vygotsky, L.S. (1997). *The collected works of L.S. Vygotsky: problems of the theory and history of psychology* (vol. 3). Berlin: Springer Science & Business Media.

Zawidzki, T.W. (2018). Mindshaping. In: this volume, pp. 735–54.

PART X

APPLICATIONS

CHAPTER 43

COMMUNICATION AS FUNDAMENTAL PARADIGM FOR PSYCHOPATHOLOGY

KAI VOGELEY

INTRODUCTION

COGNITIVE sciences in general are either based on theories that emphasize "the role of internal representations—paradigmatically internal models—of the agent's body and environment in explaining an agent's behavior" or are based on accounts that focus on "the role of high-bandwidth agent–environment interactions in producing adaptive behavior without much or any representation on the part of the agent" (Grush 2005, p. 209). The latter viewpoint has been recently explicated and developed further in a number of ways under the umbrella term of 4E cognition summarizing the multiple facets of this integrative view of cognition covering extended, embodied, enactive, and embedded aspects of cognition.

Cognition is "extended" as it is in a continuous exchange with its environment: we need to understand the "intelligent system as a spatio-temporally extended process not limited by the tenuous envelope of skin and skull" (Clark 1997, p. 221). This aspect is closely related to the feature of embodiment: cognition cannot be understood without the integration of our cognitive systems and our brains within our body allowing locomotion and action (Wilson 2002). Based on the concept of affordances (Gibson 1979), "enactive cognition" is based on the understanding of "perceiving as a way of acting" (Noë 2004, p. 1). Cognition should not be perceived as abstract and amodal information processing in detached brains or observers, but as being "grounded in multiple ways, including simulations, situated action, and, on occasion, bodily states" (Barsalou 2008, p. 619). Brain and environment are closely interconnected: on the one hand, "the environment, situations, the body, and simulations in the brain's modal systems ground" our cognitive capacities, whereas, on the other hand, "the cognitive system utilizes the environment and the body as external informational structures" (Barsalou 2010, p. 717).

Following classical accounts of psychopathology, it is the individual suffering from a mental disturbance that can be diagnosed objectively and needs therapy or support provided by the medical system. This view is challenged by communicative accounts that propose that psychopathological conditions are constituted in dyadic interactions and is supported by recent debates on views that extend cognition from individual to social interactions with others (Sullivan 1953; Ruesch 1957; Ruesch and Bateson 1951; Watzlawick et al. 1967; Glatzel 1977, 1978, 1981).

Persons and Things

Successful interactions between two partners crucially depend on the adequate mutual understanding of both interactants. During social encounters we seemingly effortless ascribe mental states to other persons. This ability to non-inferentially and nonverbally interact with other persons, even in the absence of any kind of previous knowledge of the other persons, presupposes that we immediately experience the interaction partner as a person—and not as a physical object.

As Fritz Heider points out in his canonical work, *Psychology of Interpersonal Relations* (1958), understanding the inner experience of others, for instance, on the basis of a stimulus like directed gaze, relies on the fundamental difference between "thing perception" or "nonsocial perception" on the one hand and "person perception" or "social perception" on the other (Heider 1958, p. 21). Although both persons and things are "real, solid objects with properties of shape and color . . . that occupy certain positions in the environment," there are decisive differences: (1) persons are "rarely mere manipulanda," but they are "action centers and as such can do something to us"; (2) persons have "abilities, wishes and sentiments"; "they can act purposefully, and can perceive or watch us" (p. 21); and (3) "there is a peculiar functional closeness and interaction" in social encounters, for instance, "in a mutual glance" (p. 77). This is, of course, associated with a high degree of unpredictability: "Probably the constancy in social perception, however, is less perfect than the constancy in thing perception" (p. 29) (Table 43.1).

With the background of this fundamental distinction and following the so-called attribution theory in social psychology, we can understand the inner mental states of other persons on the basis of the following three different types of data: (1) the stimuli provided, e.g., a nonverbal signal like a smile or directed gaze of another person; (2) the situational context of the social encounter, which either allows to apply generally accepted rules for one's own behavior (formal encounters) or not (informal encounters); and (3) previously acquired knowledge about the interaction partner (e.g., Kelley 1967). However, we cannot fully understand the other person as our understanding of others is often enough "unformulated or only vaguely conceived" and "may not be directly evident" (Heider 1958, p. 2). This leads to an inherent ambiguity and uncertainty (Heider

Table 43.1 Fundamental differences between persons and things

Criteria	Persons	Things
Inner Experience	Existent and relevant for behavior	Nonexistent
Behavior	Internal intentions to act ("action centers"); external physical forces	External physical forces ("mere manipulanda")
Predictability	Low ("less perfect constancy")	High ("perfect constancy")

Data From Fritz Heider, *The psychology of interpersonal relations*, John Wiley & Sons, 1958.

1958, p. 29) which substantially limits our capacity to predict the behavior of other persons (p. 2). Persons behave on the basis of "inner" mental states such as experiences, motivations, and intentions that allow us to predict their behavior. In contrast, things or physical objects rely on natural laws valid in the framework of Newtonian mechanics. For instance, things reproducibly fall to the ground if we let them go because of physical forces such as gravity. In contrast, if we look at another person, we cannot always expect that the other person will respond to this nonverbal signal.

Closing the Communicative Loop

The capacity to communicate with others belongs to the basic cognitive functions of humans that are necessary for our survival, our navigation in the social world, and the full participation in culture and society (Heinz 2014; Tomasello 2008). This is especially important from the perspective of psychopathology. "Man requires interpersonal relationships" (Sullivan 1953, p. 32), and this is because "the developmental history of personality . . . is actually the developmental history of possibilities of interpersonal relations" (Sullivan 1953, p. 30). Social interactions constitute our personality, the "relatively enduring pattern of recurrent interpersonal situations which characterize a human life" (Sullivan 1953, p. 111). Interpersonal relations are vital: an infant who starts life under disadvantageous circumstances of communication and cooperation will finally perish (Sullivan 1953, p. 61). Communicative capacities are acquired during ontogeny; the moving force of this developmental process is the "human being's need for social action" (Ruesch and Bateson 1951, p. 38); it is learned over decades and plausibly evolves as a function of biological maturation (Ruesch 1957, p. 59).

In essence, communication can be considered a closed loop constituted by three different elements: (1) the signal sent out by the first interaction partner ("sender"), (2) the adequate processing in the addressee or second interaction partner "(recipient"), and (3) her/his reaction to the signal of the sender that demonstrates that the sent signal was perceived and understood by the recipient (Ruesch and Bateson 1951, p. 15; Ruesch

1957, p. 34, p. 189). By combining these three elements, a "system of communication" (Ruesch and Bateson 1951, p. 21) or a "social situation" (Ruesch and Bateson 1951, pp. 23, 28) is constituted. Obviously, the feedback signal in the third step of the communicative loop can itself stimulate a reaction of the original sender who was asked to respond to the feedback signal of the recipient. This can launch a series of communicative events in the sense of turn-taking that might end up in a conversation of two persons (Figure 43.1).

This is the general, socially embedded architecture for all different kinds of interactions, "interpersonal systems—stranger groups, marital couples, families, psychotherapeutic, or even international relationships, etc.—may be viewed as feedback loops, since the behavior of each person affects and is affected by the behavior of each other person" (Watzlawick et al. 1967, p. 31). Of course, it is important to note that this definition of a basic communicative event does not imply that communication is always simple and uniform—the variance of the individual configuration or design of communicative encounters is virtually infinite. Even unintended signals sent out to other persons can be perceived and are usually interpreted as communicative signals. In other words, "all actions and events have communicative aspects, as soon as they are perceived by a human being" (Ruesch and Bateson 1951, pp. 6, 31). Communication is the "matrix in which all human activities are embedded"; it "links object to person and person to person" (Ruesch and Bateson 1951, p. 13). It is simply impossible to stop from communicating: "one cannot not communicate" (Watzlawick et al. 1967, p. 51).

Communicative encounters always comprise a relational component and a content that are not completely independent from each other: "Relationship patterns exist independently from content although, of course, in actual life they are always manifested by and through content" (Watzlawick et al. 1967, p. 153; similarly Ruesch 1957, p. 9). The different relationships can vary, which is documented by the fact that we can experience the relation between the interaction partners in different ways: "To be understood is a pleasure; to reach an agreement is expedient and pleasant; to be understood and to reach agreements is deeply gratifying." (Ruesch 1957, p. 36). And it is not only the relationship but also the content of the communication that differs, for instance, due to different signal channels or due to individual and cultural variations of expressing one's own inner experiences. This motivated Ruesch to develop a taxonomy of different types of personalities, namely, the "person of action," the "demonstrative," the "logical," the "withdrawn, nonparticipating," and the "anxious and fearful" person (Ruesch 1957, pp. 120ff.; similarly Glatzel 1977, p. 45).

Nonverbal and Verbal Communication

Nonverbal communication is probably much more powerful as opposed to verbally conveyed information. It was estimated that approximately two thirds of our communicative signals are based on nonverbal signals (Burgoon 1994). On the basis of nonverbal signals we can infer the inner experience of a person in a given situation: "A person

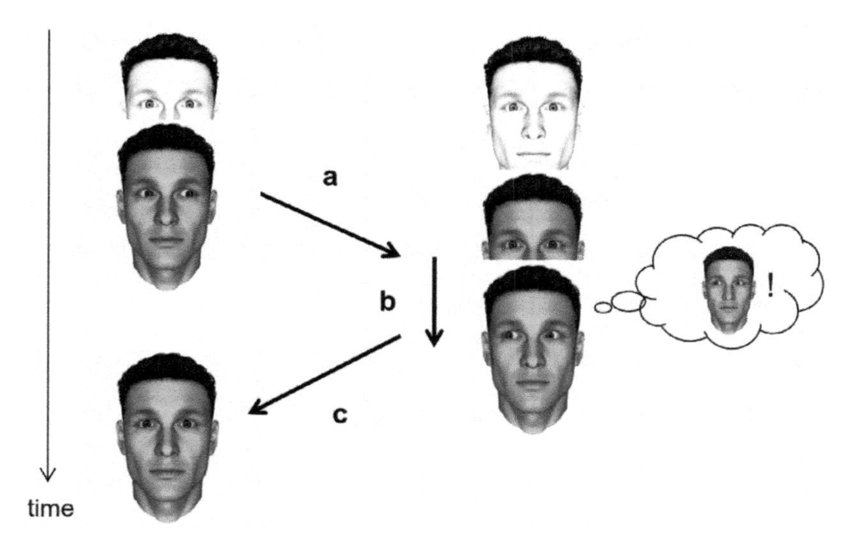

FIGURE 43.1. Two persons establish a "social situation" as soon as the "loop is closed," this is constituted on the basis of three different components: (a) one person sends a signal to another person (e.g., eye gaze contact), (b) the addressee processes this information and, subsequently, (c) prepares an answer (e.g., responds to direct gaze), thereby closing the loop.

Data from Jurgen Ruesch and Gregory Bateson, *Communication. The Social Matrix of Psychiatry*, Norton & Company, Inc., 1968.

reacts to what he thinks the other person is perceiving, feeling, and thinking, in addition to what the other person may be doing" (Heider 1958, p. 1). Facial expressions, gaze behavior, gestures, postures, and body movements deeply influence our communications (Argyle et al. 1970; Mehrabian and Wiener 1967), and they do so very early during the process of a social encounter (Willis and Todorov 2006). Nonverbal behavior serves at least four different purposes, namely, (1) modeling and coordination functions, (2) discourse functions, (3) dialogue functions, and (4) socio-emotional functions (Vogeley and Bente 2010). Nonverbal communication can be opposed to verbal communication, corresponding to "digital language" and "analogical language" (Watzlawick et al. 1967, pp. 66–7; Table 43.2).

One of the most relevant nonverbal cues to prepare for imitation and/or coordination before the onset of action is the observed gaze direction of others: "The complexity of feelings and actions that can be understood at a glance is surprisingly great" (Heider 1958, p. 2). However, gaze can also be used as a deictic cue, to influence the attention of another person and direct it to an object; "even the direction of a glance may provide a strong hint as to what the person is thinking, feeling, and wishing" (Heider 1958, p. 43). Gaze helps to regulate dyadic encounters, and its coordination can help to establish three-way relations between self, other, and the object world (Argyle and Cook 1976). A particularly interesting phenomenon in this respect is the experience of "joint attention" that is established as soon as a given person follows another individual's gaze

Table 43.2 Digital and analogical communication

Criterion	Digital Communication	Analogical Communication
Syntax	Complex, logical	Inadequate or lacking
Semantics	Inadequate or lacking	Adequate
Application	Nature of relationship	Content of relationship

Data From Paul Watzlawick, Janet Beavin Bavelas, and Don D. Jackson, *Menschliche Kommunikation—Formen, Störungen, Paradoxien* (Verlag Hans Huber, 1967); Eng. trans. as *Pragmatics of Human Communication: A Study of Interactional Patterns, Pathologies, and Paradoxes* (Norton & Company, Inc., 2011).

to a novel focus of visual attention. The total duration of gaze contact between persons appears to be universal—approximately three seconds—independent of different possible influential factors such as sex, personality traits, perceived attractiveness, and cultural background. However, persons preferring longer durations of gaze contact showed faster increases in pupillary size, suggesting that the dynamics of pupil dilation might have a modulatory influence on gaze contact (Binetti et al. 2016). Within this gaze contact window, sex-dependent gaze behavior can be identified, with female persons being more explorative than males (Coutrot et al. 2016).

Experimental studies on gaze-based interactions in real time have already been widely used employing both live video feeds of real interaction partners and virtual reality setups. They allow test persons to have the impression that they are involved in an actual ongoing social encounter with another person, even though the interaction is only conveyed by gaze-contingent behavior, simply based on the phenomenon of joint attention (Schilbach et al. 2010; Pfeiffer et al. 2013, 2014).

Understanding and Explaining in Psychopathology

The aim of psychopathology as an independent scientific endeavor is to understand and describe the inner experience of persons, including their deviations. It is not surprising that we find a corresponding distinction between persons and things (Vogeley 2013), namely, "understanding" ("Verstehen") and "explaining" ("Erklären") (Jaspers 1913/1997, pp. 301–2). Whereas understanding refers to the empathic appreciation of conflicts, hopes, and desires of an individual person, explaining relates to the attempt to consider mental disorders including their neurobiological and genetic prerequisites. Explaining as the only accepted approach in natural sciences is not available if we refer to the inner experience of persons. This led to the proposal of the "psychology of meaning"

("Verstehende Psychologie") (Jaspers 1913/1997, p. 301ff.). The inner experience of persons needs to be understood "from within" (Jaspers 1913/1997, p. 28). This capacity of understanding employing the so-called genetic mode of understanding is limited, for instance, by cultural influences (Jaspers 1913/1997, pp. 307–8), and "must itself be inconclusive . . . a final 'terra firma' is never reached" (Jaspers 1913/1997, p. 357). As our understanding of others is limited, we are forced to "interpret" (Jaspers 1913/1997, p. 305).

It is important to note that in contrast to the model of disturbed communication (Ruesch 1957) or "interactional psychopathology" (Glatzel 1978, 1981), Jaspers still follows in essence a biomedical concept of disease that focuses on the individual organism that is suffering from different pathological states, disturbances, and diseases. The biomedical disease model is the leading concept behind the international classification systems of mental disorders. The *Diagnostic and Statistical Manual of Mental Disorders (DSM-5)* (American Psychiatric Association 2013) integrates scientific findings from latest research in cognitive neuroscience, neuroimaging, epidemiology, and genetics in order to take into account the complex etiology of mental disorders in a "disorder spectrum based on common neurocircuitry, genetic vulnerability, and environmental exposures" (DSM-5, xiii). Interestingly, culture is recognized as a dynamical system (Vogeley and Roepstorff 2009) that is "transmitted, revised, and recreated within the family and other social systems and institutions." Mental disorders, including their symptoms and etiology, but also their adaptation capacities and required support must be defined in "relation to cultural, social, and familial norms and values." It is also acknowledged that the boundaries of what counts as disorder may vary from culture to culture (DSM-5, pp. 14–15). However, even with a more fluid concept of mental disorders—no longer understood as rigid categorical entities but as spectra without well-defined boundaries between one another or between themselves and the condition of mental health—it is still the individual who carries the symptoms and suffers from the disorder that can be diagnosed independently of the observer (DSM-5, p. 12ff.), even if diagnostic criteria may change according to cultural background.

Psychopathology of Self-Consciousness and Intersubjectivity

Psychopathological phenomena refer to norm-deviant disturbances of subjective experiences that are related essentially to one or more of the following domains: (1) changes in interactive and communicative behavior, (2) inadequate emotional experiences or changes in sharing emotional experiences with others, (3) inconsistency of subjective experiences or incongruence with experiences of others leading to a loss of a sense of reality that can be shared by the majority of other persons within the same cultural and traditional background (Vogeley and Newen 2009). Remarkably, all three domains refer to a social community as a background against which the subjective experience of the person suffering from a psychopathological condition stands in contrast. It seems to be constitutive for mental disorders that they are defined on the basis of norms that are generated or constituted by groups of persons, populations, or social systems.

Following this line of thought, it therefore seems that the communicative account as sketched earlier (Ruesch and Bateson 1951; Sullivan 1953; Ruesch 1957; Watzlawick et al. 1967; Glatzel 1977, 1978, 1981) should be the central and fundamental concept in psychopathology and psychiatry that transforms the dyadic interaction into the fundamental unit of analysis. In addition, how we approach others and how we are perceived by them crucially depends on the situational context in which the social encounter is embedded. This context provides a set of gestures, symbols, and meanings that need to be shared in the given situation in order to establish a common ground between both partners (Glatzel 1977, pp. 97, 129). As a consequence, communication can only be studied in the "context in which it occurs," this is especially true under conditions of psychopathology; accordingly, "the terms 'sanity' and 'insanity' practically lose their meanings as attributes of individuals" (Watzlawick et al. 1967, p. 46). This leads to the insight that psychopathological phenomena are substantially determined by cultural influences: "Sickness is culturally defined" (Ruesch and Bateson 1951, p. 73); "the notions of health and sickness . . . are a function of cultural values" (Ruesch and Bateson 1951, p. 92; similarly Heinz 2014, pp. 27, 102). This requires us to understand "the cultural matrix" as the basis for any "understanding of the nature of interaction between persons" (Ruesch and Bateson 1951, p. 168). The adequate interpretation of the communicative behavior "thus requires knowledge of the social matrix or the culture in which the exchange of messages takes place" (Ruesch 1957, p. 180).

Successful and Disturbed Communication

Based on the aforementioned explorations about the fundamental significance of communication as a key function of human cognitive equipment (Heinz 2014, p. 170), it is reasonable to defend the thesis that "psychopathology is defined in terms of disturbances of communication" (Ruesch and Bateson 1951, p. 79). Communication needs to be conceptualized as the key to understanding psychopathology and mental disorders. In a very similar way, this idea of mental disorders as disorders of communication or interpersonal relations was brought forward by Harry Stack Sullivan: "If the term, mental disorder, is to be meaningful, it must cover like a tent the whole field of inadequate or inappropriate performance in interpersonal relations" (Sullivan 1953, p. 313).

Along these lines, Ruesch developed a formal catalog of criteria for successful communications (Ruesch 1957, p. 34ff.; Table 43.3). Taken together, successful communication "takes place when correspondence of information between the two persons has been established" (Ruesch 1957, p. 37). "The ability to mutually correct the meaning

Table 43.3 Criteria for successful and disturbed communication

Criterion	Successful Communication	Disturbed Communication
Flow	Feedback circuits	Interference with free flow of messages
Fit	Mutual fit of overall patterns	Inappropriateness of reply
Efficiency	Efficient	Inefficient
Flexibility	Flexible	Lack of flexibility
Experience	Gratification	Frustration

Data From Jurgen Ruesch, *Disturbed communication: the clinical assessment of normal and pathological communicative behavior* (Norton & Company, Inc., 1957).

of messages and to mutually influence each other's behavior to each other's satisfaction is the result of successful communication" (Ruesch and Bateson 1951, p. 87). This is fundamentally different from an instance of disturbed communication (Ruesch 1957, p. 43ff.; Table 43.3).

Obviously, not every communication is successful throughout; an essential aspect refers to the goals that the interaction partners have in mind. If they both want to succeed with their communication, misunderstandings can be analyzed and corrected. "Understanding and nonunderstanding, agreement and disagreement, thus are phases which occur in the process of healthy, normal, and potentially successful communication. But if misunderstanding and disagreement become goals in themselves, then we deal with a pathological process of communication" (Ruesch 1957, pp. 29–30). Either quantitative deviations or inappropriate patterns of communication are the mechanisms by which a communication becomes unsuccessful: "Too much, too little, too early, too late, at the wrong place, is the disturbed message's fate" (Ruesch 1957, p. 41). Interaction partners are also expected to share with each other the reference frames or let the partner know the "context in which they are operating" (Ruesch 1957, p. 52). The transition between successful and disturbed communication is a graded phenomenon: "Disturbed and undisturbed communication has been outlined as one continuous function: the disturbances are considered either as quantitative exaggerations of normal behavior or as behavior that does not fit a given situation." (Ruesch 1957, p. 190).

Autism as Nonverbal Communication Deficit

The most impressive show case of disturbed or unsuccessful communication is the case of autism spectrum disorder (ASD). That nonverbal communication is a very prominent and important communicative signal system is not only true for the field of psychopathological phenomena as a whole in the sense of an "interactional psychopathology" (Glatzel 1977, p. 103), but it is explicitly true for ASD. Persons with ASD suffer from "mindblindness" (Baron-Cohen 1995). ASD is characterized by lifelong and stable deficits in nonverbal communication and social interaction, while verbal and general learning and memory abilities are independently developed and often fully preserved. It is interesting to note that one of the person types that Ruesch had proposed, namely, the "logical person," demonstrates close resemblance with characteristic features of persons with autism: "The perfection of logical thinking and expression in word or number is matched by an inadequate appreciation of feeling and emotions.... [The person with ASD] overcontrols and underadapts.... He does not care to acknowledge the intent of the other person (Ruesch 1957, p. 128).

Individuals with ASD have difficulties in the adequate processing and integration of nonverbal communication cues into their person judgments (Kuzmanovic et al. 2011) while the general verbal intelligence is preserved. They do not spontaneously attend to social information, and are thus less able to intuitively interact in social contexts (Klin et al. 2003). When confronted with nonverbal signals such as eye gaze, facial expressions, or gestures, individuals with ASD have shown atypical detection (Dratsch et al. 2013; Senju et al. 2008) and interpretation of such cues (Uljarevic and Hamilton 2013). Generally speaking, they seem to be less affected by nonverbal cues when processing a task, as compared with typically developed control persons (Schwartz et al. 2010), and/or they seem to use atypical strategies for social processing (Walsh et al. 2014; Kuzmanovic et al. 2014). On a neural level, the processing of socially relevant information recruits significantly less key regions of the "social neural network" (comprising essentially anterior medial prefrontal cortex, posterior cingulate cortex, and temporoparietal junction), both in a study of gaze detection and evaluation (Georgescu et al. 2013) and in a study on animacy experience (Kuzmanovic et al. 2014). As a general result, these research findings show atypical processing of nonverbally presented, socially relevant information in ASD both on the behavioral and neural level. This can be interpreted as a decrease in the salience of nonverbal information for individuals with ASD compared to control persons. While the mere perception of nonverbal cues is often comparable to that of control persons, ASD individuals seem to employ different strategies (Georgescu et al. 2014).

A very interesting field is the study of social gaze behavior. Autistic children appear to be able to follow someone's gaze, but tend to spend less attention on congruent objects in a gaze-following task (Bedford et al. 2012). Autistic persons avoid the eye region during the visual inspection of faces (Pelphrey et al. 2002) and spend significantly less time fixating the eye region of people in passive viewing studies involving social scenes (Klin et al. 2002). Furthermore, they have difficulties with interpreting gaze as a nonverbal cue supporting the disambiguation of social scenes, thereby suggesting a more general problem in using gaze as a tool to infer the mental states of others (Boraston and Blakemore 2007). This suggests that while core processes of social gaze can be functional, they might be driven by different motives than in non-autistic individuals, possibly indicating changes in the functional connectivity of autistic brains (e.g., Murphy et al. 2012).

Taken together, the study of nonverbal communication in ASD in general and of social gaze in particular is likely to foster our understanding of the underlying communicative deficits in ASD, while simultaneously providing valuable information about the cognitive mechanisms underlying the processing of dynamic social interactions. Thus, nonverbal information may influence social perception substantially. In addition, nonverbal behavior exhibits specifically high levels of complexity that are closely related to intuitive cognitive and affective processing. For this reason, investigating nonverbal behavior processing in ASD helps to understand (1) social cognition and possibly its underlying neural mechanisms, but also (2) the specific cognitive style characteristic of ASD. These insights, in turn, are most valuable for improving supportive therapy and training options (Ruesch 1961), which may improve the lives of affected individuals and their families.

Conclusions and Outlook

The communicative situation in the dyadic encounter provides the fundamental database for the psychiatrist: "The only thing that the psychiatrist really can rely upon is the state of communication as it is observed at a given moment, within a given context, and involving specific people" (Ruesch and Bateson 1951, p. 91). The essential information is the deviation from the normal unaffected population: "The patient is different—or feels himself to be different—from the remainder of the population" (Ruesch and Bateson 1951, p. 232). Accordingly, what has to be studied is the "interaction of personalities in particular recurrent situations or fields which 'include' the observer" (Sullivan 1953, p. 368). If communication "is the only scientific model which enables us to explain physical, intrapersonal, interpersonal, and cultural aspects of events within one system" (Ruesch and Bateson 1951, p. 5), then the systematic study of communication is obviously the only valid and useful paradigm to study psychopathological phenomena adequately. Studies on the nonverbal communication capacities in persons with ASD show a clear circumscribed deficit in embodied nonverbal but not verbal communication. These findings can be taken to suggest a focused disturbance of "primary intersubjectivity" (Gallagher 2013; Fuchs 2015) or a "lack of embodied common sense" (Fuchs 2015).

This conceptual view of psychopathology is in line with developments in social cognition research that focus on nonverbal communication. This research domain makes use of methodological advancements in virtual reality research that allows "transformed social interaction" (Bailenson et al. 2004; Bente et al. 2008). In this approach, motion is captured and rendered on a virtual character. Not only the appearance of the virtual character, which can contain information on sex, identity, ethnicity, or attractiveness, but also their nonverbal behavior can be manipulated, by blending particular channels, or by modifying specific nonverbal cues (Bente and Krämer 2011). By doing this in a systematic manner, it can be determined which aspects of nonverbal behavior are necessary and/or most efficient with regard to various social contexts. This makes it possible to systematically explore how manipulations of appearance and/or behavior of one agent or a dyad affect the experience and the course of social interactions. "Investigators can take apart the very fabric of social interaction using IVET [i.e., immersive virtual environment technology], disabling or altering the operation of its components and thereby reverse engineering social interaction. With this approach, social psychologists could systematically determine the critical aspects of successful and unsuccessful social interactions" (Blascovich et al. 2002, p. 121). The transformed social interaction approach has been used to study the effects of experimentally manipulated gaze behavior in ongoing interactions. Bente and colleagues (2007), used eye-tracking and motion capture to control two avatars representing two interactants during an open conversation. While gestures and movements were conveyed in real time, the display of gaze direction was manipulated. The authors could show that longer periods of directed gaze fostered the positive evaluation of the partner. The study demonstrated how experimental control of nonverbal cues can be implemented within a rich and fluent social interaction.

Although the study of the neural mechanisms and the underlying neurobiology of communication is beyond the scope of this chapter, it is worth mentioning that a full description and scientific understanding includes the determination of both cognitive and possibly neural correlates of persons interacting with others. On the basis of the insights of this chapter, this cannot be adequately studied in a detached-observers mode "offline," but only "online" during truly ongoing social interactions (Schilbach et al. 2010; Pfeiffer et al. 2013, 2014). We proposed to pursue this line of research further under the heading of so-called second-person neuroscience (Schilbach et al. 2013). Developments in social neuroscience tried to operationalize the experience of "true interaction" between two persons involved in a social encounter (Schilbach et al. 2010; Pfeiffer et al. 2014). Making use of virtual characters and mediated environments, it could be demonstrated that virtual characters not only convey social information to human observers, but are in turn also perceived as social agents thus exerting social influence on human interactants. Results demonstrated that not only the medial prefrontal cortex as part of the social neural network was recruited but, interestingly, also the ventral striatum as part of the brain's reward system associated with hedonic aspects of sharing attention (Schilbach et al. 2010; Pfeiffer et al. 2014). Future prospects should focus on the admittedly ambitious approach "hyperscanning," namely, the simultaneous study of two persons participating in an ongoing social encounter. This approach allows the simultaneous and time-locked measurement of two persons being analyzed in parallel with respect to neural and psychological synchronization measures during ongoing social interactions (Montague et al. 2002; Hasson et al. 2012).

In conclusion, the conceptual approach of socially embedded and—via nonverbal communication—embodied cognition allows us to substantially modify our understanding of mental disorders: we consider the dyadic interaction and its context as the fundamental unit of analysis. Therefore, no longer do we take mental disorders to be symptoms of isolated, detached individuals. We consider instead that such disorders are constituted by norm deviations in the dyadic interaction.

References

American Psychiatric Association (APA). (2013). *Diagnostic and statistical manual of mental disorders (DSM-5)*. Washington, DC: American Psychiatric Association.

Argyle, M. and Cook, M. (1976). *Gaze and mutual gaze*. Oxford: Cambridge University Press.

Argyle, M., Salter, V., Nicholson, H., Wiliams, M., and Burgess, P. (1970). The communication of inferior and superior attitudes by verbal and non-verbal signals. *British Journal of Social and Clinical Psychology*, 9(3), 222–31.

Bailenson, J.N., Beall, A.C., Loomis, J., Blascovich, J., and Turk, M. (2004). Transformed social interaction: decoupling representation from behavior and form in collaborative virtual environments. *Presence*, 13(4), 428–41.

Baron-Cohen, S. (1995). *Mindblindness*. Cambridge, MA: MIT Press.

Barsalou, L.W. (2008). Grounded cognition. *Annual Review of Psychology*, 59(1), 617–45. doi:10.1146/annurev.psych.59.103006.093639

Barsalou, L.W. (2010). Grounded cognition: past, present, and future. *Topics in Cognitive Science*, 2, 716–24.

Bedford, R., Elsabbagh, M., Gliga, T., Pickles, A., Senju, A., Charman, T. et al. (2012). Precursors to social and communication difficulties in infants at-risk for autism: gaze following and attentional engagement. *Journal of Autism and Developmental Disorders*, 42, 2208–18.

Bente, G., Eschenburg, F., and Aelker, L. (2007). Effects of simulated gaze on social presence, person perception and personality attribution in avatar-mediated communication. In: L. Moreno (ed.), *Proceedings of the 10th Annual International Workshop on Presence*. Barcelona: Starlab Barcelona S.L., pp. 207–14.

Bente, G. and Krämer, N.C. (2011). Virtual gestures: embodiment and nonverbal behavior in computer-mediated communication. In: A. Kappas and N.C. Krämer (eds.), *Face-to-face communication over the internet: issues, research, challenges*. Cambridge: Cambridge University Press, pp. 176–209.

Bente, G., Senokozlieva, M., Pennig, S., Al Issa, A., and Fischer, O. (2008). Deciphering the secret code: a new methodology for the cross-cultural analysis of nonverbal behavior. *Behavior Research Methods*, 40, 269–77.

Binetti, N., Harrison, C., Coutrot, A., Johnston, A., and Mareschal, I. (2016). Pupil dilation as an index of preferred mutual gaze duration. *Royal Society Open Science*, 3, 160086.

Blascovich, J., Loomis, J., Beall, A.C., Swinth, K.R., Hoyt, C.L., and Bailenson, J N. (2002). Immersive virtual environment technology as a methodological tool for social psychology. *Psychological Inquiry*, 13(2), 103–24.

Boraston, Z. and Blakemore, S.-J. (2007). The application of eye-tracking technology in the study of autism. *The Journal of Physiology*, 581, 893–8.

Burgoon, J.K. (1994). Nonverbal signals. In: M.L. Knapp and G.F. Miller (eds.), *Handbook of interpersonal communication*. Thousand Oaks, CA: Sage, pp. 450–507.

Clark, A. (1997). *Being there: putting brain, body, and world together again*. Cambridge, MA: MIT Press.

Coutrot, A., Binetti, N., Harrison, C., Mareschal, I., and Johnston, A. (2016). Face exploration dynamics differentiate men and women. *Journal of Vision*, 16(14), 1–19. doi:10.1167/16.14.16

Dratsch, T., Schwartz, C., Yanev, K., Schilbach, L., Vogeley, K., and Bente, G. (2013). Getting a grip on social gaze: control over others' gaze helps gaze detection in high-functioning autism. *Journal of Autism and Developmental Disorders*, 43(2), 286–300.

Fuchs, T. (2015). Pathologies of intersubjectivity in autism and schizophrenia. *Journal of Consciousness Studies*, 22(1–2), 191–214.

Gallagher, S. (2013). Intersubjectivity and psychopathology. In: B. Fulford, M. Davies, G. Graham, J. Sadler and G. Stanghellini (eds.), *Oxford handbook of philosophy of psychiatry*. Oxford: Oxford University Press, pp. 258–74.

Georgescu, A.L., Kuzmanovic, B., Roth, D., Bente, G., and Vogeley, K. (2014). The use of virtual characters to assess and train nonverbal communication in high-functioning autism. *Frontiers in Human Neuroscience*, 8, 807.

Georgescu, A.L., Kuzmanovic, B., Schilbach, L., Tepest, R., Kulbida, R., Bente, G. et al. (2013). Neural correlates of "social gaze" processing in high-functioning autism under systematic variation of gaze duration. *NeuroImage: Clinical*, 3, 340–51.

Gibson, J.J. (1979). *The ecological approach to visual perception*. Hillsdale, NJ: Lawrence Erlbaum.

Glatzel, J. (1977). *Das psychisch Abnorme. Kritische Ansätze zu einer Psychopathologie*. München, Wien, Baltimore: Urban & Schwarzenberg.

Glatzel, J. (1978). *Allgemeine Psychopathologie*. Stuttgart: Enke Verlag.

Glatzel, J. (1981). *Spezielle Psychopathologie*. Stuttgart: Enke Verlag.

Grush, R. (2005). Internal models and the construction of time: generalizing from state estimation to trajectory estimation to address temporal features of perception, including temporal illusions. *Journal of Neural Engineering*, 2(3), S209–218.

Hasson, U., Ghazanfar, A.A., Galantucci, B., Garrod, S., and Keysers, C. (2012). Brain-to-brain coupling: a mechanism for creating and sharing a social world. *Trends in Cognitive Sciences*, 16(2), 114–21.

Heider, F. (1958). *The psychology of interpersonal relations*. New York: John Wiley and Sons.

Heinz, A. (2014). *Der Begriff der psychischen Krankheit*. Berlin: Suhrkamp.

Jaspers, K. (1913/1997). *General psychopathology* (2 vols.; trans. J. Hoenig and M.W. Hamilton). London and Baltimore: The Johns Hopkins University Press.

Kelley, H.H. (1967). Attribution theory in social psychology. *Nebraska Symposium on Motivation*, 15, 192–238

Klin, A., Jones, W., Schultz, R., and Volkmar, F. (2003). The enactive mind, or from actions to cognition: lessons from autism. *Philosophical transactions of the Royal Society of London. Series B: Biological scieneces*, 358, 345–60.

Klin, A., Jones, W., Schultz, R., Volkmar, F., and Cohen, D. (2002). Defining and quantifying the social phenotype in autism. *American Journal of Psychiatry*, 159, 895–908.

Kuzmanovic, B., Schilbach, L., Georgescu, A., Kockler, H., Santos, N., Shah, N.J. et al. (2014). Dissociating animacy processing in high-functioning autism: neural correlates of stimulus properties and subjective ratings. *Social Neuroscience*, 9, 309–25.

Kuzmanovic, B., Schilbach, L., Lehnhardt, F.G., Bente, G., and Vogeley, K. (2011). A matter of words: impression formation in complex situations relies on verbal more than on nonverbal information in high-functioning autism. *Research in Autism Spectrum Disorders*, 5, 604–13.

Mehrabian, A. and Wiener, M. (1967). Decoding of inconsistent communications. *Journal of Personality and Social Psychology*, 6(1), 109–14.

Montague, P.R., Berns, G.S., Cohen, J.D., McClure, S.M., Pagnoni, G., Dhamala, M. et al. (2002). Hyperscanning: simultaneous fMRI during linked social interactions. *NeuroImage*, 16, 1159–64.

Murphy, E.R., Foss-Feig, J., Kenworthy, L., Gaillard, W.D., and Vaidya, C.J. (2012). Atypical functional connectivity of the amygdala in childhood autism spectrum disorders during spontaneous attention to eye-gaze. *Autism Research and Treatment*, 652408.

Noë, A. (2004). *Action in perception*. Cambridge, MA: MIT Press.

Pelphrey, K.A., Sasson, N.J., Reznick, J.S., Paul, G., Goldman, B.D., and Piven, J. (2002). Visual scanning of faces in autism. *Journal of Autism and Developmental Disorders*, 32, 249–61.

Pfeiffer, U., Schilbach, L., Timmermans, B., Kuzmanovic, B., Georgescu, A., Bente, G. et al. (2014). Why we interact: on the functional role of the striatum in the subjective experience of social interaction. *NeuroImage* 101, 124–37.

Pfeiffer, U.J., Vogeley, K., and Schilbach, L. (2013). From gaze cueing to dual eye-tracking: novel methods to investigate the neural correlates of gaze in social interaction. *Neuroscience and Biobehavioral Reviews*, 37, 2516–28.

Ruesch, J. (1957). *Disturbed communication: the clinical assessment of normal and pathological communicative behaviour*. New York: Norton & Company.

Ruesch, J. (1961). *Therapeutic communication*. New York: Norton & Company.

Ruesch, J. and Bateson, G. (1968, 1951). *Communication: the social matrix of psychiatry*. New York: Norton & Company.

Schilbach, L., Timmermans, B., Reddy, V., Costall, A., Bente, G., Schlicht, T., and Vogeley, K. (2013). A second-person neuroscience in interaction. *Behavioral and Brain Sciences*, 36, 441–62.

Schilbach, L., Wilms, M., Eickhoff, S.B., Romanzetti, S., Tepest, R., Bente, G. et al. (2010). Minds made for sharing: initiating joint attention recruits reward-related neurocircuitry. *Journal of Cognitive Neuroscience*, 22, 2702–15.

Schwartz, C., Bente, G., Gawronski, A., Schilbach, L., and Vogeley, K. (2010). Responses to nonverbal behaviour of dynamic virtual characters in high-functioning autism. *Journal of Autism and Developmental Disorders*, 40(1), 100–11.

Senju, A., Kikuchi, Y., Hasegawa, T., Tojo, Y., and Osanai, H. (2008). Is anyone looking at me? Direct gaze detection in children with and without autism. *Brain and Cognition*, 67, 127–39.

Sullivan, H.S. (1953). *The interpersonal theory of psychiatry*. New York, London: Norton & Company.

Tomasello, M. (2008). *Why we cooperate*. Cambridge, MA: MIT Press.

Uljarevic, M. and Hamilton, A. (2013). Recognition of emotions in autism: a formal meta-analysis. *Journal of Autism and Developmental Disorders*, 43, 1517–26.

Vogeley, K. (2013). A social cognitive perspective on "understanding" and "explaining." *Psychopathology*, 46, 295–300.

Vogeley, K. and Bente, G. (2010). "Artificial humans": psychology and neuroscience perspectives on embodiment and nonverbal communication. *Neural Networks*, 23, 1077–90.

Vogeley, K. and Newen, A. (2009). Consciousness of oneself and others in relation to mental disorders. In: S. Wood, N. Allen, and C. Pantelis (eds.), *The neuropsychology of mental illness*. Cambridge: Cambridge University Press, pp. 420–9.

Vogeley, K. and Roepstorff, A. (2009). Contextualising culture and social cognition. *Trends in Cognitive Sciences*, 13, 511–6.

Walsh, J.A., Vida, M.D., and Rutherford, M.D. (2014). Strategies for perceiving facial expressions in adults with autism spectrum disorder. *Journal of Autism and Developmental Disorders*, 44, 1018–26.

Watzlawick, P.B., Bavelas, J.B., and Jackson, D.D. (1967/2011). *Menschliche Kommunikation—Formen, Störungen, Paradoxien* (*Pragmatics of human communication: a study of interactional patterns, pathologies, and paradoxes*) (12th ed.). Bern: Verlag Hans Huber (New York: Norton & Company).

Willis, J. and Todorov, A. (2006). First impressions: making up your mind after a 100-ms exposure to a face. *Psychological Science*, 17, 592–8.

Wilson, M. (2002). Six views of embodied cognition. *Psychonomic Bulletin & Review*, 9, 625–36.

CHAPTER 44

SCAFFOLDING INTUITIVE RATIONALITY

CAMERON BUCKNER

The Problem of Intuitive Rationality

There is much about the mind that we understand comparatively well. Explicit, effortful, language-dependent cognition in adult humans—deliberate planning, problem-solving, mathematical cognition, rule-based categorization, and the like—have been extensively and successfully studied by over 60 years of cognitive psychology. Science has accumulated hundreds of well-confirmed, broadly computational models of these forms of cognition and such cognition is often evaluated according to the internal coherence norms of classical rationality (such as avoid contradictions, follow the axioms of probability theory, and choose so as to maximize expected utility). From behaviorism, ethology, and comparative psychology, we also have a decent understanding of the most basic inherited and learned causes of behavior in nonlinguistic animals—reflexes, imprinting, innate releasing mechanisms, and fixed action patterns, and the basic mechanisms of classical and instrumental conditioning. Science has accumulated hundreds of well-confirmed, broadly stimulus-driven models of these forms of behavioral causation; and these psychological processes are often evaluated according to the externalist and ecological norms of biological rationality (such as how well they maximize fitness in the organism's natural environment—see Kacelnik 2006). These are extremely rough characterizations and much remains much to be learned about both groups of psychological processes, to be sure; but we should also not minimize the remarkable progress that has been made over the last century in understanding their course and nature.

However, much intelligent behavior—perhaps even most intelligent behavior in both humans and animals—is produced instead by processes that fall into what Nagel (2012) has called the "murky zone" between these two better understood forms of processing. This is the kind of processing you perform when you effortlessly read another's emotional state from subtle facial and behavioral cues; attempt to guess whether sharks or mosquitoes are more dangerous to humans; drive to work along a familiar route "on autopilot" (though responding flexibly to dynamic traffic conditions and obeying complex traffic laws); or guide the performance of a skilled motor activity, such as quickly choosing whether to serve a tennis ball hard down the outside line or go for a safer spin serve. Implicit, tacit, nonlinguistic: the labels we use to describe these forms of psychological processing tell us more about what they are not than what they actually are.

I will broadly refer to these forms of decision-making as "intuitive judgment." By calling such judgment "intuitive," I mean minimally that even if these judgments allow agents to reliably succeed in typical environments through sophisticated information processing, much of the structure and course of this processing is opaque to our introspection. When we are aware of the outcomes of such processing, we would often report them not as taking the form of explicit judgments with introspectible grounds, but rather as "seemings" of one form or another—often described in folk terms as "intuitions," "hunches," or "gut feelings." The widely shared assumption—for which I offer no novel defense here—is that this kind of cognition serves as the primary form of flexible cognition in nonlinguistic animals and preverbal infants, who lack the language-dependent capacity for explicit, deductive thought.[1]

Many aspects of study of intuitive judgment remain mired in foundational controversy. Fifty years after Tversky and Kahneman made the case for the study of intuitive judgment as a subdiscipline in psychology, we still cannot decide whether these forms of decision-making should be explained by appeal to qualitative labels (such as "representativeness," "availability," and "anchoring and adjustment"; Kahneman and Frederick 2002; Kahneman and Tversky 1996; Tversky and Kahneman 1974); connectionist networks built of links and nodes (Clark 1989); fast and frugal algorithms that emphasize the cues processed and their orders (Gigerenzer 1996; Gigerenzer, Todd, and ABC Research Group 1999); or associative learning models that stress how such strategies are learned from complex spatiotemporal contiguities between cues (Heyes 2012; Papineau and Heyes 2006; Pearce 2002). In epistemology, the "rationality wars" (Samuels, Stich, and Bishop 2012) were fought to decide whether the prevalence of such nonclassical procedures indicate that we are fundamentally irrational (because the processes routinely violate the axioms of probability theory) or rational (because they are efficient and reliably succeed in familiar environments). And ethics is currently embroiled in disputes as to whether we can be held responsible for behaviors generated by such implicit processing, dividing especially over whether we possess the ability to bring their course more in line with our endorsed norms (for diverse positions in this debate, see the articles in Brownstein and Saul 2016a, 2016b). It is perhaps unsurprising

[1] We should also accept that the addition of language to the system might fundamentally "rewire" even these intuitive forms of judgment (Anderson 2010; Bermúdez 2007b); and I shall mention some ways this might happen later.

that we lack a folk familiarity with these forms of processing given their combination of complexity and opacity; but it remains a challenge for science and philosophy that we understand so little about perhaps the primary cause of intelligent action in humans and animals.

Two broad strategies have been popularly deployed to plumb these murky depths. First, we might begin with a classical, formal model of explicit rational thought—for example, the rules of deductive logic and decision theory—and "revise down" to fit the behavioral patterns of nonlinguistic thought. This kind of approach often begins with what I will call a "neo-Fregean" conception of thought and reasoning—which takes on board Evans's views on the generality constraint, Davidson's views on rationality, and/or Fodor's views on the language of thought—and proceeds to ask whether nonlinguistic agents are capable of anything remotely resembling systematic use of concepts and the truth-functional thought transitions characteristic of post-Fregean formal logic (for critical explorations of this terrain, see Camp 2009 and Cussins 1992). However, it is difficult to see how agents could master the rules of classical logic and decision theory in a fully domain-general, abstract way without the use of correspondingly domain-general and linguistic representations, either deriving from an external natural language or an internal language of thought (Bermúdez 2003; Hurley 2003; Watanabe and Huber 2006). A prominent example of this "revise-down" strategy can be found in Bermúdez's protological approach, which attempts to find analogs of deductive operators that could modify non-sentential representations (Bermúdez 2006, 2007a). For example, Bermúdez describes a "protoconditional" reasoning based in instrumental beliefs about means to achieve desired ends, and "protonegation"-like relationships holding between contrary concepts that might enable an agent to reason by exclusion.

In this chapter, I will offer and defend a version of the alternative "build-up" strategy, which begins instead with more basic associative learning capacities to see what must be added to achieve the behavioral flexibility found in intuitive judgment (other prominent examples of build-up include Allen 1999; Beisecker 1999; Bennett and Bennett 1989; Dennett 2008; Hurley 2003). First, I will sketch a new model of intuitive inferences that describes how they could be practically rational. This model involves a novel hybrid of internalism and externalism, which I argue is the key to understanding the role of psychological explanation in comparative psychology and ethology. A major advantage of the build-up approach is that if we begin with learning mechanisms that we know animals possess, and ask what must be minimally added to achieve the increasingly flexible behavior found in the murky zone, our explanations will more fully recognize the rich dependence of these strategies on ecological scaffolding and developmental shaping, which will in turn increase the explanatory power of those models. Finally, I will suggest some ways to answer the deeper question that faces much of this literature: how domain-specific, ecological processes could ever achieve the domain-general, validity-preserving transitions characteristic of the classical approach. The answer may be disappointing: the role of such thought is vanishingly small even in adult humans; and while it is extremely powerful and may explain much of our mastery over the natural world, it need only capture the most symbolically and socially scaffolded human activities—explicit scientific and mathematical reasoning—an ideal our cognition rarely actually achieves.

Intuitive Practical Inference

On the neo-Fregean picture, inference consists in a systematic transition between thoughts with propositional contents. Some initial thoughts are taken to be true, and in virtue of formal connections between those thoughts' truth-functional structure (fit to a rule of classical logic or decision theory), those premises are taken to provide support for an appropriately structured conclusion (Boghossian 2014). In theoretical inference, the conclusion will be a further belief; in practical inference, the conclusion is an action (or intention to perform an action; I remain agnostic on this issue here). I will here focus on practical inference, since we know even less about nonlinguistic theoretical inferences.

One of the key arguments offered in favor of this package is that it satisfies what Boghossian has called the "taking condition" (often thought to be inspired by a passage from Frege that we will examine in the final section): that inferring necessarily involves thinkers taking their premises to support their conclusions and drawing their conclusions because of that fact (Boghossian 2014; Valaris 2016). A model of practical inference needs to show how inference rationalizes actions if reasons are to explain them. For a rationalizing explanation of behavior to identify its actual causes, it cannot be phrased in terms of what the agent could or should have recognized. The agent must have actually performed the action because of those particular contents, as those contents are presented in its own psychology. This is often taken to be something of a conceptual point—a necessary condition on inference as such—but the point could also be bolstered empirically given a real architectural difference between cases of behavioral causation that are a-rational in the relevant sense (such as reflexes, stimulus-response conditioning, or lexical priming) and those that depend upon a process of reasoning.

If intuitive inferences are to be interpreted as rational, then we indeed need to satisfy something like the taking condition. However, this not because we need to satisfy some conceptual condition on thought or inference as such, but rather because these judgments are typically explained in terms of representations in the empirical literatures that study them, and representations are individuated in terms of their semantic contents. The appeal to a concept, memory, transitive ordering, cognitive map, or other representational state cannot causally explain some behavior unless the representational content of that state is causally relevant to the behavior it goes on to produce. Instead of formality, this makes intensionality—in the sense that "creature with a heart" and "creature with a kidney" are extensionally equivalent but intensionally distinct—the key to understanding practical inference. I thus minimally interpret the taking condition as requiring that inferences be causally sensitive to the intensions of the representations that govern them. Formal relations are one way for the course of an inference to be intensionally sensitive—through the old computationalist mantra that "syntax tracks semantics"—but similarity relations can satisfy this requirement as well.

Though similarity has been much maligned in philosophy and psychology of the twentieth century (Goodman 1972), most of these well-known criticisms are now obsolete. Since Goodman's day, psychology and philosophy have developed a variety of empirically supported, formal models of similarity, including Tversky's influential feature-matching model, Gärdenfors's new geometric model, transformational models, and popular mathematical models of prototype and exemplar-based categorization (Ashby and Maddox 1993; Decock and Douven 2011; Gärdenfors 2004; Larkey and Markman 2005; Tversky 1977). An organizing principle unifying these models is that similarity judgments are causally sensitive to the structure of the target representations on which they operate. My decision to enter a particular restaurant on vacation will be determined by how similar it is to my prototype for a *good restaurant*—for example, I may score options based on how many of the following features they possess: clean tables, a dining room that is reasonably full of non-tourists, and pleasant smells wafting from the doorway. The presence or absence of such features causally governs my choice, and had I associated different features with good restaurants—say, modernist decor, high Yelp reviews, the necessity for reservations, and a smartly dressed wait staff—I would predictably make different choices about where to eat. Of course, if the similarity judgment is intuitive, then the particular features matched can be opaque to the agent's introspection; and this opacity will have consequences for the kind of metacognitive awareness and control we have over such judgments. Nevertheless, this weakened kind of governance does not challenge the fact that our judgments are intensionally sensitive in a relevant sense.

In brief, such similarity-based judgments satisfy a more modest but adequate version of the taking condition: they tell us what agents saw in their actions, namely, that agents judged the option to be a sufficiently similar to their idea of the typical or ideal member of the desired target category.[2] A target category is desired because it is associated with certain consequences (say, a pleasant meal that is not too expensive), and a particular option is chosen due to the individual's intension for that target category. This story is enough to rationalize the agent's action in the "explanatory" sense of rationality—it tells us why agents acted as they did, what they saw in their choices. However, this story does not (and need not) tell us whether those reasons are ultimately good ones, whether the course of inference weighed the evidence appropriately, or whether the decision was likely to succeed.

These questions concern instead rationality in the so-called justificatory sense. Remarkably, the classical inferentialist picture can give the same answer to the explanatory and justificatory questions about explicit deductive inferences; we acted as we did because we followed the rules, and since we followed the rules—which ensure that we are led from true premises to true conclusions, update our likelihood estimates in

[2] For a full defense of this claim and other ideas offered in this section, see Buckner (2017).

a consistent manner, optimize expected utility according to our current assumptions, and so on—our action was rationally justified. This convenient simplicity, however, is just that—a contingent feature of the classical view. There is no deeper reason why our answer to questions about the two senses of rationality must coincide in this way. In fact, their coincidence on the classical picture has become something of a liability, as recent literature on judgment and decision-making has demonstrated that truth-preserving and utility-optimizing approaches are routinely outcompeted by more ecological and frugal strategies that do not aspire to infallibility (Gigerenzer et al. 1999). In most cases, it actually does not pay to follow such rules—being classically rational usually turns out to be instrumentally ill-advised.

The liabilities are due to many factors often acknowledged in models of "bounded rationality," such as the lack of unlimited time, attention, and computational power; but a deeper problem arises from the fact that agents often act in noisy and unpredictable environments. There are many environments in which fast and frugal strategies will reliably outperform classical strategies regardless of how much time and processing power is available, but perhaps the most dramatic cases involve the phenomena of overfitting (Gigerenzer and Brighton 2009). Overfitting occurs when the stimuli channels that provide evidence about the structure of the environment are noisy, or the environmental structure itself is partly stochastic or dynamic—the agent receives some information about environmental structure but also some noise along the same channel, and has no privileged way to distinguish signal from noise. In their attempt to take into account all relevant evidence, optimizing approaches can treat the noise as information, using it to build structured expectations about environmental regularities. As a result, these strategies are routinely outcompeted in such environments by more frugal strategies that simply make a bet about the structure of the environment, with no attempt to optimize on all available evidence. Overfitting demonstrates that the problem with classical approaches is not just that we lack unlimited cognitive resources—even a demon with unlimited time and processing power could be beaten in such environments by an appropriately lazy dog with the right epistemic bet.

Thus, the right justificatory norms for this class of judgment are externalist and ecological. The researchers who study these forms of decision-making highlight the concept of ecological validity, where a strategy is ecologically valid if it has the right kind of fit with the informational structure of the environment in which it is deployed. Agents never have a god's-eye view of that environmental structure, so being rational in this sense always involves a gamble. However, this form of ecological rationality should not be confused with a raw, case-by-case consequentialism; ecologically rational strategies will outcompete other strategies in those environments over the long run, largely through effective combinations of ecologically valid cues, where cue validity is the conditional probability that an exemplar falls under a target category given the cue in that environment. By combining these cues in the right way (as with a prototype match in the representativeness heuristic, or a cue search tree as with the take the best heuristic; Gigerenzer and Goldstein 1999; Kahneman and Frederick 2002), they can make the most instrumentally rational use of available evidence over the long term.

To summarize, this model of intuitive judgment involves a novel hybrid of internalism and externalism. Intuitive judgments are rational in the explanatory sense when they operate on similarity judgments that are intensionally sensitive to the modes of presentation of the agent's internal category representations, but they are rational in the justificatory sense when they achieve the right kind of fit with their external environment.[3] By giving different answers to the explanatory and justificatory questions, we can understand both how such judgments are governed by the agent's distinctive assumptions about the world and how they can routinely outcompete classical strategies that attempt to make optimal use of available evidence.

Building Up

One worry we might have about such similarity judgments is that they could not be bootstrapped or scaffolded into something more recognizable as reasoning of the deliberate, rule-governed variety—thereby rendering the evolution or development of this kind of reasoning more mysterious. Granted, similarity judgments allow agents to generalize flexibly to at least some novel cases that they have never before encountered; I had never before seen that particular restaurant I went to on vacation, but because my intension for *good restaurant* directed me toward a dining room reasonably full of locals, I was non-accidentally led to an excellent choice. This ability to answer correctly in novel cases, however, depends on relevant similarities between the new cases and those that have already shaped my decision-making through evolution or learning. As a result, similarity-based categorization judgments have problems accommodating perceptually atypical cases, such as that penguins are birds and whales are mammals. Since penguins and whales fail to score highly on surface similarity matches to most birds and mammals, respectively, similarity-based strategies will often struggle to categorize such exemplars correctly; yet this is something that every schoolchild eventually learns to do. The general solution to this worry is to understand how representational similarity spaces can be reshaped and transformed to focus on deeper invariances that are conserved across differences in surface properties—but this idea requires significant elaboration.

Many theorists may only be interested in animal and infant judgments insofar as they are relevant to the kind of adult, deductive inference that accounts for our uniquely human cultural and scientific achievements—which characteristically aim at the deepest causal structure or principles found in the explicitly theoretical discourse of human science and culture. A terribly interesting question is how we could ever

[3] This is not to say that all judgments in the murky zone are similarity-based; in fact, this is not the case. This is merely one way that murky judgments can count as rational in the explanatory sense; there may be yet other ways not explored here. However, the full taxonomy of murky zone judgments requires a complex discussion that is best saved for another day.

progress—both in evolution and development—from the kind of basic intuitive similarity judgments I have discussed so far, to the kind of explicit, rule-based inferences found in this theoretical discourse.[4]

I will answer this worry in three parts; first by pointing out how similarity-based inferences can become more sophisticated by matching increasingly abstract and higher-order features, second by noting how forms of reasoning more recognizable as planning might emerge by chaining together similarity assessments to simulated future possibilities, and finally—in the next section—by suggesting how intuitive inferences, with the aid of language and social collaboration, might be bootstrapped into something evaluable by the abstract, validity-preserving coherence norms familiar to us from deductive logic and decision theory. Deductive logic finally comes onto the scene as an abstraction upon the end products of these most advanced consequence judgments themselves—formal logic serving as a technological artifact designed to help extend our native powers of reasoning. None of these ideas is entirely novel and each deserves its own paper-length treatment; I here intend only to sketch a plausible way forward from the model of similarity judgment offered earlier.

First, we should note that from an uncritical perspective, it is natural to array the kinds of features assessed by a similarity-based judgment along a spectrum, from the most "stimulus-bound" and perceptually basic on one end (simple colors, sounds, or tastes), to the most abstract, complex, and higher-order on the other. For example, contrast one pigeon which has been trained to peck at stimuli that are red with another which has been trained to peck at triangles (Watanabe 1991), from a third which has been trained to peck at Impressionist art (Watanabe 2011). In a straightforward sense, the features learned by the second pigeon, whatever they might be, are more complex and abstract than those learned by the first pigeon, and those learned by the third are more complex and abstract still.

Things get somewhat trickier if we are called upon to justify this natural-seeming continuum from basic to abstract. We might note that the first pigeon's responding will be governed by a stimulus generalization curve centered on a single color value—responding will peak if presented with a centrally "red" stimulus, and gradually drop off as the reflectance profile of the stimulus grades away from peak redness to other colors (Watanabe 2006). The pigeon's learned behavior is in this sense comparatively inflexible. The responding of the triangle-identifying pigeon, however, is less brittle. We can vary several of the stimulus' features—colors, lines, sizes, angular rotation—and the pigeon will still reliably peck at "triangular" shapes and avoid non-triangular ones.[5] In this sense, it is more stimulus-independent, since a much wider range of perceptually disparate stimulus situations are linked to a common response. Such abstract features are probably constructed by clustering and configuring several more basic perceptual features, as has been presumed by the perceptual processing cascade in classical

[4] For an empirical attempt to do just that directly, see Vigo and Allen (2009).

[5] Note that, obviously, the pigeons do not learn the same triangle concept that we possess (Watanabe 2006).

cognitive theories of perception and vindicated by more recent hierarchical theories of Bayesian perception or predictive inference (Marr et al. 1991; Tenenbaum et al. 2011). And the art-classifying pigeon has learned to pick up on something more abstract and configural yet. As the features become more abstract, complex, and configural, clustering together more complex and diverse perceptual situations, the agent's thinking becomes even more flexible, in the sense that it can recognize opportunities for action and desired outcomes across a wider range of different perceptual situations, and is better able to generalize its behavioral strategies to perceptually novel circumstances.

Granted, there is some reason to worry about the artificiality of this sense of basic and abstract. What deeper metaphysical or epistemological fact, we might wonder, renders the color red simpler and more basic than whatever holistic, statistical property the pigeons use to identify Impressionist artwork? Would it be somehow impossible to design an agent for which the latter property were more basic, and individual features like red could only be extracted with difficulty? In fact, might this be a plausible developmental story about how infants start out with respect to at least some features that adults regard as basic—that they begin with a more holistic "blooming, buzzing confusion," and only through a significant restructuring of their conceptual scheme manage to localize individual features in that perceptual gestalt?

For present purposes, it is fine to concede that this distinction lacks any deeper metaphysical justification, and exists only relative to the processing architecture or epistemology of some particular agent. In this sense, features which are more "basic" are prior—either temporally, in some processing cascade or developmental trajectory, epistemically, in some process of justification, or both—to features that are more abstract. But especially in mature processing frameworks that are capable of learning to discriminate what we would typically regard as abstract categories from the kind of information the environment typically presents to the sensory organs, such processing hierarchies frequently appear. And there may be deeper mathematical reasons why they do reliably appear that can shed a bit more light on the role of abstraction in cognition.

Consider, for example, the most successful image classifying systems in artificial intelligence today, the deep learning "convnets" used to automate the process of image labeling. These neural networks involve dozens of hierarchical layers of processing, where each layer in the hierarchy consists of a paired linear and a nonlinear classifier layer, with pixel-by-pixel luminance and color channel information presented to the inputs at the bottom layer and high-level classification decisions received from outputs at the topmost layer. Through exposure to massive training sets, each layer of the network is eventually tuned to respond to increasingly abstract configurations of output from the previous layer in the hierarchy. A mathematical analysis of these networks reveals that they work so well because they are uniquely well suited to conquering what researchers call "nuisance variables" in perceptual processing tasks (Patel, Nguyen, and Baraniuk 2015). These are variables that frequently produce a high degree of variation in the input signal without providing diagnostic information about the classification task, variables such as position, scale, and rotation in visual processing tasks, or pitch, tone, and volume in auditory processing tasks. It is in the nature of an embodied agent's sensory interactions with three-dimensional space that these variables will frequently pose challenges to any classification task seeking deeper scientific or taxonomic principles, as regularities in nuisance variables are frequently irrelevant to distinctions in these deeper features. The system thus must develop ways of transforming perceptual input into representational formats that are increasingly insensitive to nuisance variation. In this sense, we might give a slightly more technical definition of abstraction: a representational format that is more resistant to nuisance variation for some classification task.

Once a system can transform perceptual input so as to recognize deeper perceptual regularities across nuisance variation—as one comes to see all cubes as cubes despite different positions, sizes, or angular rotations—one also now has the architecture to discriminate deeper regularities along other dimensions as well. The first component we might want to add to this kind of hierarchical discrimination system is the ability to further reshape similarity spaces on the basis of succeeding or failing to predict future contingencies. Some of the more significant advances in cognitive science over the last 15 years concern the phenomena of prediction error learning, where animals learn to group exemplars into categories not just by surface similarity but also by the future contingencies they predict (Clark 2013, 2015). There are a variety of models of such learning, many of which are mathematically based, but also empirically confirmed and neuroanatomically plausible. For example, the connectionist model of the mammalian hippocampal system offered by Gluck and Myers (1993) combines similarity-based configural learning and contingency-based predictive learning into a neuroanatomically plausible package that can reshape representational similarity space in just these ways.[6] When two perceptually disparate outcomes predict the same desired consequence, this can allow the organism to reshape the similarity space until those two previously dissimilar situations are located closer to each other in the transformed feature space.

The combination of abstraction with predictive learning is sufficient to generate striking examples of what appears from the outside to be advanced reasoning and insight. For example, consider an experiment on theory of mind in ravens conducted by Bugnyar, Reber, and Buckner (2016). To provide some background, ravens spontaneously cache food items as part of their normal foraging behavior, and, like chimpanzees, previous experiments have shown that they behave differently when they are being watched by a competitor (who might pilfer their caches). In the experiment, ravens trained to use peepholes to pilfer caches made by an experimenter later inferred that when they cache in the presence of an open peephole—even if they had no prior experience caching in the presence of peepholes—that unseen competitors might be able to watch them through the peephole, too. As a result, they later guarded their own caches against observation in the presence of the open peephole, even if they could not see any competitors (or their gaze) at the time (Figure 44.1).

[6] Since this is a connectionist network, the way in which this network reshapes similarity spaces is by increasing or decreasing the distance between hidden-layer activation vectors for two stimuli. For a thorough discussion of the way this is achieved in the network, see Gluck and Myers (2001).

FIGURE 44.1. Sketch of the experimental setup of Bugnyar et al. (2016). (a) Observed (Obs) condition: The cover of the window is open (white bar) and the focal subject (storer, st) caches food in the visual presence of a conspecific observer. (b) Non-observed (Non) condition: The cover of the window is closed (gray bar) and the focal subject caches food in visual isolation of a conspecific (non-observer). Both observers and non-observers make sounds in the experimental chamber, which are audible to the storer. (c) Peephole (Peep) condition: The cover of the window is closed (gray bar) but one of the two peepholes (small white square) is open; the focal subject caches food in the absence of any behavioral cues, whereas the presence of conspecifics is simulated via playback of sounds recorded from non-observed trials (symbolized by loudspeaker). In this experiment, the storer (raven) was first trained to use peepholes to pilfer caches it observes being made by an experimenter. The Obs and Non conditions were used to see how the raven behaved when it knew it was and was not being observed, respectively. Caching behavior in the peep condition—the first time the raven had ever cached in the presence of the peephole—was then found to match that in the Obs condition and to differ from that in the Non condition, despite the fact that no gaze cues were present.

Reproduced from Thomas Bugnyar, Stephan A. Reber, and Cameron Buckner, Ravens attribute visual access to unseen competitors, *Nature Communications*, 7 (10506), Figure 1, doi:10.1038/ncomms10506 © 2016 Thomas Bugnyar, Stephan A. Reber, and Cameron Buckner. This work is licensed under the Creative Commons Attribution 4.0 International License. It is attributed to the authors Thomas Bugnyar, Stephan A. Reber, and Cameron Buckner.

A plausible explanation of this result is that their own experience using the peepholes allowed them to learn that two perceptually dissimilar situations[7]—<A sees B out in the open> and <A sees B through a peephole>—afforded them the same pilfering opportunities, and so they compressed their representations of these two situations into an "abstract equivalence class" that predicts the same opportunities for others as well (Whiten 1996). This abstract encoding in turn allowed them to draw the "insightful" generalization in the test condition that competitors might be able to pilfer their caches by watching them through the peephole, too.

One question we might pose at this point is what this kind of abstraction could become at its limit. Suppose we had an animal whose representational revision was optimized to predict a highly desired outcome to the exclusion of other forms of learning by even abstract perceptual similarity. Could the most abstract features of that similarity space eventually collapse into something like a singularity—a representational "slot" much like the holes in a tintamarresque, those endearing "head in a hole" photo-op cardboard cutouts of comical scenes that one finds at fairs and amusement parks? In other words, could abstraction delimit a region in similarity space that ascribes no particular features, but into which some component must fit for the representation to be complete? Less metaphorically, could the placeholders for "A" and "B" in the specification of the equivalence classes above function like slots into which any arbitrary individual actors could fit, but some such individual must fit in order for the consequences to be predicted? If so, this might render the category representation "unsaturated" in the old Fregean notion of the term, a kind of representational function that must be supplemented with additional particulars in order to perform its representational role.

Granted, this would not yet be a real predicate slot into which one could quantify and thereby draw further generalized inferences. An agent outfitted with the sort of tintamarresque representations I introduced here could not yet reason in existential or universal terms about what must be true if some or every individual could fit into that slot. For that, one would need other cognitive resources corresponding to variables and quantifiers, likely of a more explicitly metarepresentational or linguistic nature. But ravens equipped with the kind of abstract equivalence class I proposed earlier could recognize the ability of other actors to play a role in the scene when confronted with them, and generalize behavioral abilities they observed themselves to possess when occupying a certain position in that scene to others who occupy it later on. This behavior begins to look from the outside very much as though it were governed by abstract rules.

A significant concern remains about this kind of inference even supplemented with abstraction, prediction, and tintamarresque slots, however: that agents with these representations might be stimulus-bound in another sense, in that they could only think about goals and categories triggered by current drives and perceptions. Such creatures may still fail to satisfy the generality constraint in a second sense—perhaps increasing

[7] It may be that ravens as highly social songbirds had a variety of other sources of evidence that had already prepared them to form self–other mappings in such situations, such as the constant variation provided by self and other vocalizations (Derégnaucourt and Bovet 2016).

abstractness can allow them to deploy the same conceptual resources across a wider range of perceptually disparate situations, but their active goals are still controlled by exposure to an occurrent triggering stimulus, whether internal (like hunger or proprioception) or external (such as a secondary reinforcer, like money or social approval). If this were the whole story, animals and children would still not be capable of "mental time travel," the ability to think about goals and means to achieve them that are not present at hand (Suddendorf and Corballis 2008). Perhaps this is not fatal to their ability to draw insightful inferences in some significant sense, but it would block them from many forms of foresight and planning for future scenarios that have long been considered the paradigm cases of deductive reasoning.

It is an open question though whether humans are really capable of this kind of trigger-free thought in their own cognition. The most likely way that humans transcend these difficulties is by using words as perceptual proxies for their associated triggers, whether through internal rehearsal (as in episodes of inner speech) or external reminders (such as a Google calendar pop-up or "to do" list). The dependence of human mental time travel on such symbolic scaffolding is not to be underestimated; but since its importance has been richly elaborated elsewhere (Clark 2006), I will not belabor this point here.

Even in the absence of such linguistic triggers to transcend the here and now, however, animals might achieve a more limited ability to respond to more temporally distant motivational factors merely by chaining together abstract associations along more and more iterations. For example, in an apparent demonstration of mental time travel, Clayton and colleagues observed scrub jays caching a form of food on which they were currently sated in another chamber where they had learned they would be deprived of it in the future (Clayton, Russell, and Dickinson 2009). Their behavior has been thought evidence of mental time travel, as they could not have currently desired that type of food. However, their environment was rich with cues that might have also been associated with the absence of that food type in the other chamber. If the scrub jays were capable of predictive inference of the sort we have been discussing earlier, they could then simulate the absence of this food type at breakfast tomorrow, which could then serve as an independent associative trigger for a desire for that food type tomorrow. It should be expected that this form of projective inference would be more fragile in general than mental time travel guided by linguistic tags and cues, as associations are fraught with ambiguities that could multiply at every iteration. For example, it would be much more difficult to locate absolute temporal coordinates without the aid of time words; upon glancing at this enclosure—should the scrub jay think about tomorrow morning in this space, tomorrow afternoon, or last week?—and upon simulating an experience in that location, should the jay focus on the caching trays, the food sources, or the presence or absence of onlooking competitors? Such powerful forms of associative learning would, however, grant a limited form of transcendence of the here and now that is clearly relevant to intelligent planning and action.

Much of this is admittedly very speculative, but it is offered only as a proof of concept that basic forms of associative and predictive learning can be bootstrapped into the subtler forms of cognition that increasingly approximate the representational and inferential power of classical inference. Each, however, falls well short of the true domain generality and formality of classical deductive and decision-theoretic reasoning. Where, then, does this kind of reasoning finally fit into the picture?

Relearning Old Lessons about the Border Between Logic and Psychology

As mentioned earlier, revise-down theorists often take inspiration from Frege when discussing the potential for nonlinguistic inference. The assumption is that Frege's powerful insights into the nature of truth-conducive thought and validity-preserving inference—filtered through a long tradition of Evans, Davidson, Fodor, and perhaps others—ground the subject matter in a definitive way. We can only understand nonlinguistic agents as having thoughts and performing inferences worthy of the name, the thought goes, if we can somehow understand them as having thoughts involving Fregean logical form, and performing inferences based on something like that form. The task of the revise-down camp is then to see how little we must tweak in the Fregean picture to plausibly account for the flexibility observed in nonlinguistic behavior.

My claim here is that it is time to remove the more modern filters and re-fortify the Fregean boundary between the disciplines of psychology and logic. Though Frege was perhaps solely concerned with guarding one side of this boundary—in preventing the infirmities of the mind from infecting the pristine laws of truth—we should perhaps now be equally invested in preventing the normative idealizations of logic from scaling the boundary in the other direction and warping our descriptive science of reasoning. I aim here to give researchers philosophical cover to reject the assumption that they should be asking whether animals *really* or merely *as-if* reason by exclusion, or *really* or merely *as-if* optimize utility. These questions reflect the height of a bias that I have called anthropofabulation—a combination of anthropocentrism and confabulation that ties the criteria for the possession of some cognitive capacity to an artificially inflated sense of what humans can or routinely do (Buckner 2013). I will thus end by simply rehearsing some reminders about the (lack of) relationship that Frege saw between psychology and logic, which, despite a brief stint during which we can be forgiven for supposing computationalism to have made the language of thought empirically plausible, should have always been the correct attitude to adopt toward nonlinguistic thought.

To review some history, Frege is throughout his writings adamant that logic must be strictly separated from the psychology of reasoning, and indeed held that even adult humans typically do not reason in the way recommended by his *Begriffshrift*. Logic, for Frege, is not the scientific study of reasoning as it occurs in actual minds, but rather the most general science, the science aimed at discovering the laws of truth itself. "Law" here can be interpreted in either a descriptive or prescriptive sense—as describing the most general truths about the structure of the world itself (i.e., that no contradictory situations exist), or how we ought to think if we want to achieve the truth. However, Frege in no way assumes that humans or animals typically or even very often think in this manner, and asserts that psychology should make no such presupposition, for "error and superstition have causes" too (Frege 1956, p. 290), causes that would surely fall under the explanatory purview of descriptive psychology.

Frege further distinguishes the mental process of arriving at an assertion and a logical proof of the assertion, noting that though the former might sometimes accord with the rules of logic, logic concerns the latter, and drawing any necessary connections between the two would court "misunderstanding" and involve a "blurring of the boundary between psychology and logic" (p. 290) that should always be preserved. Given the conclusions explored earlier—that animals appear to be capable only of "ratiomorphic" (Brunswik 1955) thinking rather than true deductive proof; that we can satisfy something like the taking condition with similarity-based inference; and that we can come quite far in understanding the flexibility of such inference through abstraction, prediction, and iteration—it may perhaps be worth considering a re-establishment of the Fregean boundary between the psychology of reasoning and the discipline of logic.

Granted, Frege might have wanted to reserve the word "inference" for the latter, logical investigation. The quotation often deployed to support the imposition of Fregean constraints on the psychology of inference (for example, by Boghossian 2014 and Valaris 2016) comes from his posthumous writings on logic, where he seems to offer a definition of inference:

> To make a judgment because we are cognizant of other truths as providing a justification for it is known as inferring. There are laws governing this kind of justification, and to set up these laws of valid inference is the goal of logic. (Frege 1979, p. 3)

Several things are peculiar about deploying the paragraph this way, however, as Boghossian himself notes. First, Frege here seems to presume that we infer only from truths, whereas in practice surely we often reason from falsehoods as well. Second, "cognizant" seems to have what Boghossian calls a "success grammar" built into it, implying the impossibility of incorrect inference. Essentially, these two restrictions would conflate "inference" with "sound inference," making it conceptually impossible to explain any inferences based on untrue assumptions or invalid reasoning. In concert with the points noted earlier, this would render the study of inference entirely the province of logic rather than psychology, with any attempt to engage in a descriptive psychology of unsound inference impossible and confused. For these reasons, Boghossian rejects these constraints and offers the taking condition as the minimal constraint to extract from Frege's remarks—and this weakened interpretation of the constraint is the one I have argued earlier that similarity-based inference can satisfy.

Frege would have perhaps been content to leave the study of inference as the exclusive province of logic; but leaving Frege behind, we might still wonder whether any additional constraints should be imposed on the descriptive psychology of reasoning from the fact that humans at least sometimes "grasp" Fregean deductive entailment relations. In other words, sound inferential transitions must be something of which we are at least sometimes capable in practice—lest the laws of logic could not be discovered by logicians, or taught to undergraduates in logic courses—so shouldn't some more minimal logical constraint on the psychology of reasoning remain? Granting all this for the sake of argument, it is worth noting that humans would only need to approach this ideal in our most scaffolded and deliberate cognitive activities—when we are engaged in explicit, collaborative discourse about theoretical scientific principles. The laws of logic could then be induced not by collecting myriad examples of actual reasoning and performing an empirical induction, as Mill, Erdmann, and other purportedly psychologistic logicians are accused of suggesting, but rather by collecting examples of sound reasoning from the theoretical sciences when we take things to have gone well, and let logic proceed as the (normative) attempt to understand and codify how things went well in those cases.

These exemplary instances of scientific reasoning—whether from physics, chemistry, economics, or psychology—will reflect our most laborious epistemic achievements, often the product of weeks or months of linguistically and socially scaffolded labor. They will typically involve writing, reading, revision, effortful reflection upon criticism from peers, and so on. The laws of logic, on this way of understanding matters, are the thought patterns that remain brilliant and robust in the face of all this critical consideration, the diamonds that have emerged once the coal of the scientist's workaday cognition has enjoyed the limit of criticism and reflection. Only in this sense does human deductive behavior need to satisfy the laws of truth, which could then be induced and systematized from observation of the inferential relationships that tend to arise repeatedly in these particular laudable inferences (for a further defense of such a "particularist" way of conceiving of the relationship between human reasoning and logic, see Lowe 1993). And even then, logic must typically "round off some corners" in these collected thought patterns in the interests of simplicity and systematic considerations—as we find, for example, in the infamous differences between implication as it occurs in natural language and the material conditional deployed in Fregean logic (Stalnaker 1968).

From this perspective, it can seem obviously inappropriate to begin with the Fregean model of classical logic as a descriptive model of nonlinguistic thought, even if one intends to put it through heavy revision. Classical logic and decision theory are the pinnacles of scientific technology, themselves the result of painstaking reflection on idealized scientific reasoning, which can be used perhaps to evaluate scientific cognition (further strengthening the revision process, if only further in a particular, socially scaffolded direction) but not to predict or psychologically explain it. The laws of logic or decision theory here stand in the same relationship to even adult theoretical cognition as the laws of quantum mechanics stand to adult physical cognition. The patterns of thought are achieved by domain experts only with great effort and practice, and through frequent use of decision aids that might include not only words, but other tools such as calculating devices, truth tables, lists of inference rules, and so on.

In short, the only constraint from logic and decision theory that a psychological theory of inference need satisfy is that it must allow for the possibility that effortful, scaffolded scientific cognition in language-using adult humans sometimes satisfies logical norms in its end products—that is, in the linguistic artifacts (research articles, books, and other defenses of explicit theory) that professional scientists produce. That provides a vanishingly weak restriction on the psychology of animal and infant thought, one that would rule out none of the more modest forms of inference I suggested earlier.

Conclusion

In this chapter, I have sketched a similarity-based model of intuitive inference, suggested how such a model might be bootstrapped to cover more flexible forms of reasoning, and defended the model against the charge that it is too deflationary to meet conceptual constraints on inference. These constraints are often motivated by Fregean considerations, but I argued that they are not in the spirit of the relationship Frege originally proposed between psychology and logic. In short, one must reject nearly everything Frege says about this relationship to regard their motivations as plausible.

As I have noted earlier, there are other philosophical reasons not discussed here for linking inference and logic in this way—emerging perhaps from Evansian considerations about the criteria for concept possession, Davidsonian considerations about rational interpretation, and/or Fodorian computationalism about cognitive architecture. Though none of these packages today enjoy the pride of place they once held in philosophy of cognitive science, the revise-down project is still regarded as a default choice in the philosophy of nonlinguistic thought; and when the platform is thought to require additional support, its proponents often reach for Fregean buttresses that actually weaken this connection when closely examined. Philosophical and empirical researchers of animal and infant cognition should thus no longer be expected to answer the question as to whether the patterns of thought they study merely approximate those of classical logic and decision theory—and indeed, the whole debate would benefit from the rejection of the assumptions behind this skeptical challenge.

REFERENCES

Allen, C. (1999). Animal concepts revisited: the use of self-monitoring as an empirical approach. *Erkenntnis*, 51(1), 537–44. doi:10.1023/A:1005545425672

Anderson, M.L. (2010). Neural reuse: a fundamental organizational principle of the brain. *The Behavioral and Brain Sciences*, 33(4), 245–66. doi:10.1017/S0140525X10000853

Ashby, F.G. and Maddox, W.T. (1993). Relations between prototype, exemplar, and decision bound models of categorization. *Journal of Mathematical Psychology*, 37(3), 372–400.

Beisecker, D. (1999). The importance of being erroneous: prospects for animal intentionality. *Philosophical Topics*, 27(1), 281–308.

Bennett, J. and Bennett, J.F. (1989). *Rationality: an essay towards an analysis*. Indianapolis, IN: Hackett Publishing.

Bermúdez, J.L. (2003). *Thinking without words*. New York: Oxford University Press.

Bermúdez, J.L. (2006). Animal reasoning and proto-logic. In: S. Hurley and M. Nudds (eds.), *Rational animals*. Oxford: Oxford University Press, pp. 127–37.

Bermúdez, J.L. (2007a). Negatlion, contrariety, and practical reasoning: comments on Millikan's varieties of meaning. *Philosophy and Phenomenological Research*, 75(3), 663–9.

Bermúdez, J.L. (2007b). *Philosophy of psychology: contemporary readings*. New York: Routledge.

Boghossian, P. (2014). What is inference? *Philosophical Studies*, 169(1), 1–18. doi:10.1007/s11098-012-9903-x

Brownstein, M. and Saul, J. (2016a). *Implicit bias and philosophy, vol. 1: metaphysics and epistemology*. New York: Oxford University Press.

Brownstein, M. and Saul, J. (2016b). *Implicit bias and philosophy, vol. 2: moral responsibility, structural injustice, and ethics*. New York: Oxford University Press.

Brunswik, E. (1955). "Ratiomorphic" models of perception and thinking. *Acta Psychologica*, 11, 108–9. doi:10.1016/S0001-6918(55)80069-8

Buckner, C. (2013). Morgan's canon, meet Hume's dictum: avoiding anthropofabulation in cross-species comparisons. *Biology & Philosophy*, 28(5), 853–71.

Buckner, C. (2017). Rational inference: the lowest bounds. *Philosophy and Phenomenological Research*. doi:10.1111/phpr.12455

Bugnyar, T., Reber, S.A., and Buckner, C. (2016). Ravens attribute visual access to unseen competitors. *Nature Communications*, 7. doi:10.1038/ncomms10506.

Camp, E. (2009). Putting thoughts to work: concepts, systematicity, and stimulus-independence. *Philosophy and Phenomenological Research*, 78(2), 275–311.

Clark, A. (1989). *Microcognition: philosophy, cognitive science, and parallel distributed processing* (vol. 6). Cambridge, MA: MIT Press. Retrieved from http://uclibs.org/PID/8692

Clark, A. (2006). Language, embodiment, and the cognitive niche. *Trends in Cognitive Sciences*, 10(8), 370–4.

Clark, A. (2013). Whatever next? Predictive brains, situated agents, and the future of cognitive science. *Behavioral and Brain Sciences*, 36(3), 181–204.

Clark, A. (2015). *Surfing uncertainty: prediction, action, and the embodied mind*. New York: Oxford University Press.

Clayton, N.S., Russell, J., and Dickinson, A. (2009). Are animals stuck in time or are they chronesthetic creatures? *Topics in Cognitive Science*, 1(1), 59–71. doi:10.1111/j.1756-8765.2008.01004.x

Cussins, A. (1992). Content, embodiment and objectivity: the theory of cognitive trails. *Mind*, 101(404), 651–88.

Decock, L. and Douven, I. (2011). Similarity after Goodman. *Review of Philosophy and Psychology*, 2(1), 61–75.

Dennett, D.C. (2008). *Kinds of minds: toward an understanding of consciousness*. New York: Basic Books.

Derégnaucourt, S. and Bovet, D. (2016). The perception of self in birds. *Neuroscience & Biobehavioral Reviews*, 69, 1–14. doi:10.1016/j.neubiorev.2016.06.039

Frege, G. (1956). The thought: a logical inquiry. *Mind*, 65(259), 289–311.

Frege, G. (1979). Logic. In: *Posthumous writings*. Chicago: University of Chicago Press.

Gärdenfors, P. (2004). *Conceptual spaces: the geometry of thought*. Cambridge, MA: MIT Press.

Gigerenzer, G. (1996). On narrow norms and vague heuristics: a reply to Kahneman and Tversky (1996). *Psychological Review*, 103(3), 592–6.

Gigerenzer, G. and Brighton, H. (2009). Homo heuristicus: why biased minds make better inferences. *Cognitive Science*, 1, 107–43. doi:10.1111/j.1756-8765.2008.01006.x

Gigerenzer, G. and Goldstein, D.G. (1999). Betting on one good reason: the take the best heuristic. In: G. Gigerenzer, P.M. Todd, and ABC Research Group (eds.), *Simple heuristics that make us smart*. New York: Oxford University Press, pp. 75–95.

Gigerenzer, G., Todd, P.M., and ABC Research Group. (1999). *Simple heuristics that make us smart*. New York: Oxford University Press.

Gluck, M.A. and Myers, C.E. (1993). Hippocampal mediation of stimulus representation: a computational theory. *Hippocampus*, 3(4), 491–516.

Gluck, M.A. and Myers, C.E. (2001). *Gateway to memory: an introduction to neural network modeling of the hippocampus*. Cambridge, MA: MIT Press.

Goodman, N. (1972). Seven strictures on similarity. In: N. Goodman (ed.), *Problems and projects*. New York: Bobbs-Merrill, pp. 437–46.

Heyes, C. (2012). Simple minds: a qualified defence of associative learning. *Philosophical transactions of the Royal Society of London. Series B, Biological sciences*, 367(1603), 2695–703.

Hurley, S. (2003). Animal action in the space of reasons. *Mind & Language*, 18(3), 231–57. doi:10.1111/1468-0017.00223

Kacelnik, A. (2006) Meanings of rationality. In: S. Hurley and M. Nudds (eds.), *Rational animals?* Oxford: Oxford University Press, Oxford, pp. 87–106.

Kahneman, D. and Frederick, S. (2002). Representativeness revisited: attribute substitution in intuitive judgment. *Heuristics and Biases: The Psychology of Intuitive Judgment*, 49, 49–81.

Kahneman, D. and Tversky, A. (1996). On the reality of cognitive illusions. *Psychological Review*, 103(3), 582–91.

Larkey, L.B. and Markman, A.B. (2005). Processes of similarity judgment. *Cognitive Science*, 29(6), 1061–76.

Lowe, E.J. (1993). Rationality, deduction and mental models. In: K.I. Manktelow and D.E. Over (eds.), *Rationality: psychological and philosophical perspectives*. Florence: KY: Taylor & Frances/Routledge, pp. 211–30.

Marr, D., Poggio, T., Hildreth, E.C., and Grimson, W.E.L. (1991). A computational theory of human stereo vision. In: Vaina, L. (ed.), *From the retina to the neocortex: selected papers of David Marr*. New York: Springer, pp. 263–95.

Nagel, J. (2012). Gendler on alief. *Analysis*, 72(4), 774–88.

Papineau, D. and Heyes, C. (2006). Rational or associative? Imitation in Japanese quail. In: S. Hurley and M. Nudds (eds.), *Compare: a journal of comparative education*. New York: Oxford University Press, pp. 187–96.

Patel, A.B., Nguyen, T., and Baraniuk, R.G. (2015). A probabilistic theory of deep learning. *arXiv:1504.00641* [stat.ML]. Retrieved from http://arxiv.org/abs/1504.00641

Pearce, J.M. (2002). Evaluation and development of a connectionist theory of configural learning. *Animal Learning Behavior*, 30(2), 73–95.

Samuels, R., Stich, S., and Bishop, M. (2012). Ending the rationality wars. In: S. Stich (ed.), *Collected papers, vol. 2: knowledge, rationality, and morality, 1978–2010*. Oxford: Oxford University Press, pp. 191–223.

Stalnaker, R.C. (1968). A theory of conditionals. In: Rescher, N. (ed.), *Studies in logical theory. American Philosophical Quarterly, Monograph 2*. Oxford: Blackwell, pp. 98–112.

Suddendorf, T. and Corballis, M. (2008). New evidence for animal foresight? *Animal Behaviour*, 75(5), e1–e3. doi:10.1016/j.anbehav.2008.01.006

Tenenbaum, J.B., Kemp, C., Griffiths, T.L., and Goodman, N.D. (2011). How to grow a mind: statistics, structure, and abstraction. *Science*, 331(6022), 1279–85.

Tversky, A. (1977). Features of similarity. *Psychological Review*, 84(4), 327.

Tversky, A. and Kahneman, D. (1974). Judgment under uncertainty: heuristics and biases. *Science*, 185(4157), 1124–31.

Valaris, M. (2016). What reasoning might be. Update to:194(6), pp 2007–2024. doi:10.1007/s11229-016-1034-z

Vigo, R. and Allen, C. (2009). How to reason without words: inference as categorization. *Cognitive Processing*, 10(1), 77–88.

Watanabe, S. (1991). Effects of ectostriatal lesions on natural concept, pseudoconcept, and artificial pattern discrimination in pigeons. *Visual Neuroscience*, 6(5), 497–506.

Watanabe, S. (2006). The neural basis of cognitive flexibility in birds. In: E.A. Wasserman and T.R. Zentall (eds.), *Comparative cognition: experimental explorations of animal intelligence*. New York: Oxford University Press, pp. 619–39.

Watanabe, S. (2011). Discrimination of painting style and quality: pigeons use different strategies for different tasks. *Animal Cognition*, 14(6), 797–808.

Watanabe, S. and Huber, L. (2006). Animal logics: decisions in the absence of human language. *Animal Cognition*, 9(4), 235–45. doi:10.1007/s10071-006-0043-6

Whiten, A. (1996). When does smart behaviour-reading become mind-reading? In: P. Carruthers and P. Smith (eds.), *Theories of theories of mind*. New York: Cambridge University Press, pp. 277–92.

CHAPTER 45

ROBOTS AS POWERFUL ALLIES FOR THE STUDY OF EMBODIED COGNITION FROM THE BOTTOM UP

MATEJ HOFFMANN AND ROLF PFEIFER

INTRODUCTION

THE study of human cognition—and human intelligence—has a long history and has kept scientists from various disciplines—philosophy, psychology, linguistics, neuroscience, artificial intelligence, and robotics—busy for many years. While there is no agreement on its definition, there is wide consensus that it is a highly complex subject matter that will require, depending on the particular position or stance, a multiplicity of methods for its investigation. Whereas, for example, psychology and neuroscience favor empirical studies on humans, artificial intelligence has proposed computational approaches, viewing cognition as information processing, as algorithms over representations. Over the last few decades, overwhelming evidence has been accumulated showing that the pure computational view is severely limited and that it must be extended to incorporate embodiment, i.e., the agent's somatic setup and its interaction with the real world, and, because they are real physical systems, robots became the tools of choice to study cognition. There have been a plethora of pertinent studies, but they all have their own intrinsic limitations. In this chapter, we demonstrate that a robotic approach, combined with information theory and a developmental perspective, promises insights into the nature of cognition that would be hard to obtain otherwise.

We start by introducing "low-level" behaviors that function without control in the traditional sense; we then move to sensorimotor processes that incorporate reflex-based loops (involving neural processing). We discuss "minimal cognition" and show how the role of embodiment can be quantified using information theory, and we introduce the so-called SMCs, or sensorimotor contingencies, which can be viewed as the very basic building blocks of cognition. Finally, we expand on how humanoid robots can be productively exploited to make inroads in the study of human cognition.

Behavior Through Interaction

What cognitive scientists are regularly forgetting is that complex coordinated behaviors—for example, walking, running over uneven terrain, swimming, avoiding obstacles—can often be realized with no or minimal involvement of cognition/representation/computation. This is possible because of the properties of the body and the interaction with the environment, that is, the embodied and embedded nature of the agent. Robotics is well suited for providing existence proofs of this kind and then to further analyze these phenomena. We will only briefly present some of the most notable case studies.

Low-Level Behavior: Mechanical Feedback Loops

A classical illustration of behavior in complete absence of a "brain" is the passive dynamic walker (McGeer 1990): a minimal robot that can walk without any sensors, motors, or control electronics. It loosely resembles a human, with two legs, no torso, and two arms attached to the "hips," but its ability to walk is exclusively due to the downward slope of the incline on which it walks and the mechanical parameters of the walker (mainly leg segment lengths, mass distribution, foot shape, and frictional characteristics). The walking movement is entirely the result of finely tuned mechanics on the right kind of surface. A motivation for this research is also to show how human walking is possible with minimal energy use and only limited central control. However, most of the problems that animals or robots are faced with in the real world cannot be solved solely by passive interaction of the physical body with the environment. Typically, active involvement by means of muscles/motors is required. Furthermore, the actuation pattern needs to be specified by the agent,[1] and hence a controller of some sort is required. Nevertheless, it turns out that if the physical interaction of the body with the environment is exploited, the control program can be very simple. For example, the passive dynamic walker can be modified by adding a couple of actuators and sensors and a reflex-based controller, resulting in the expansion of its niche to level ground while keeping the control effort and energy expenditure to a minimum (Collins et al. 2005).

However, in the real world, the ground is often not level and frequent corrective action needs to be taken. It turns out that often the very same mechanical system can

[1] In this chapter, we will use "agent" to describe humans, animals, or robots.

generate this corrective response. This phenomenon is known as *self-stabilization* and is a result of a mechanical feedback loop. To use dynamical systems terminology, certain trajectories (such as walking with a particular gait) have attracting properties and small perturbations are automatically corrected, without control—or one could say that "control" is inherent in the mechanical system.[2] Blickhan et al. (2007) review self-stabilizing properties of biological muscles in a paper entitled "Intelligence by Mechanics"; Koditschek et al. (2004) analyze walking insects and derive inspiration for the design of a hexapod robot with unprecedented mobility (RHex—e.g., Saranli et al. 2001).

Sensorimotor Intelligence

Mechanical feedback loops constitute the most basic illustration of the contribution of embodiment and embeddedness to behavior. The immediate next level can be probably attributed to direct, reflex-like, sensorimotor loops. Again, robots can serve to study the mechanisms of "reactive" intelligence. Grey Walter (Walter 1953), the pioneer of this approach, built electronic machines with a minimal "brain" that displayed phototactic-like behavior. This was picked up by Valentino Braitenberg (Braitenberg 1986) who designed a whole series of two-wheeled vehicles of increasing complexity. Even the most primitive ones, in which sensors are directly connected to motors (exciting or inhibiting them), display sophisticated behaviors. Although the driving mechanisms are simple and entirely deterministic, the interaction with the real world, which brings in noise, gives rise to complex behavioral patterns that are hard to predict.

This line was picked up by Rodney Brooks, who added an explicit anti-representationalist perspective in response to the in-the-meantime-firmly-established cognitivistic paradigm (e.g., Fodor 1975; Pylyshyn 1984) and "good old-fashioned artificial intelligence" (GOFAI) (Haugeland 1985). Brooks openly attacked the GOFAI position in the seminal articles "Intelligence without Reason" (Brooks 1991a) and "Intelligence without Representation" (Brooks 1991b), and proposed *behavior-based robotics* instead. Through building robots that interact with the real world, such as insect robots (Brooks 1989), he realized that "when we examine very simple level intelligence we find that explicit representations and models of the world simply get in the way. It turns out to be better to use the world as its own model" (Brooks 1991b). Inspired by biological evolution, Brooks created a decentralized control architecture consisting of different layers; every layer is a more or less simple coupling of sensors to motors. The levels operate in parallel but are built in a hierarchy (hence the term *subsumption architecture*; Brooks 1986). The individual modules in the architecture may have internal states (the agents are thus not purely reactive any more); however, Brooks argued against calling the internal states representations (Brooks 1991b).

[2] The description is idealized—in reality, a walking machine would fall into the category of "hybrid dynamical systems," where the notions of attractivity and stability are more complicated.

Minimal Embodied Cognition

In the case studies described in the previous section, the agents were either mere physical machines or they relied on simple direct sensorimotor loops only—resembling reflex arcs of the biological realm. They were reactive agents constrained to the "here-and-now" time scale, with no capacity for learning from experience and also no possibility of predicting the future course of events. Although remarkable behaviors were sometimes demonstrated, there are intrinsic limitations.

The introduction of first instances of internal simulation, which goes beyond the "here-and-now" time scale, is considered the hallmark of cognition by some (e.g., Clark and Grush 1999). This could be a simple forward model (as present already in insects—see Webb 2004) that provides the prediction of a future sensory state given the current state and a motor command (efference copy). Forward models could provide a possible explanation of the evolutionary origin of first simulation/emulation circuitry[3] and of environmentally decoupled thought—the agent employing primitive "models" before or instead of directly operating on the world.

> Early emulating agents would then constitute the most minimal case of what Dennett calls a Popperian creature—a creature capable of some degree of off-line reasoning and hence able (in Karl Popper's memorable phrase) to "let its hypotheses die in its stead" (Dennett 1995, p. 375). (Clark and Grush 1999, p. 7)

Importantly, we are still far from any abstract models or symbolic reasoning. Instead, we are dealing with the sensorimotor space and the possibility for the agent to extract regularities in it and later exploit this experience in accordance with its goals. For example, the agent can learn that given a certain visual stimulation, say, from a cup, a particular motor action (reach and grasp) will lead to a pattern of sensory stimulation (in humans: we can feel the cup in the hand). The sensorimotor space plays a key part here and it is critically shaped by the embodiment of the agent and its embedding in the environment: a specific motor signal only leads to a distinct result if embedded in the proper physical setup. If you change the shape and muscles of the arm, the motor signal will not result in a successful grasp.

Quantifying the Effect of Embodiment Using Information Theory

For cognitive development of an agent, the "quality" of the sensorimotor space determines what can be learned. First, the type of sensory receptors—their mechanism

[3] See Grush (2004) for the similarities and differences between emulation theory (Grush 2004) and simulation theory (Jeannerod 2001).

of transduction—determines what kind of signals the agent's brain or controller will be receiving from the environment. Furthermore, the shape and placement of these sensors will perform an additional transformation of the information that is available in the environment.

For example, different species of insects have evolved different non-homogeneous arrangements of the light-sensitive cells in their eyes, providing an advantageous non-linear transformation of the input for a particular task. One example is exploiting ego-motion together with motion parallax to gauge distance to objects in the environment and eventually facilitate obstacle avoidance. Using a robot modeled after the facet eye of a housefly, Franceschini et al. (1992) showed that the nonlinear arrangement of the facets—more dense in the front than on the side—compensates for the motion parallax and allows uniform motion detection circuitry to be used in the entire eye, which makes it easy for the robot to avoid obstacles with little computation. These findings were confirmed in experiments with artificial evolution on real robots (Lichtensteiger 2004). Artificial eyes with designs inspired by arthropods include Song et al. (2013) and Floreano et al. (2013).

It is not always possible to pinpoint the specific transformation of sensory signals that is facilitated by the morphology as in the previous case. A more general tool is provided by the methods of information theory. Information is used in the Shannon sense here—to quantify statistical patterns in observed variables. The structure or amount of information induced by particular sensor morphology could be captured by different measures, for example, entropy. However, information (structure) in the sensory variables tells only half of the story (a "passive perception" one in this case), because organisms interact with their environments in a closed-loop fashion: sensory inputs are transformed into motor outputs, which in turn determine what is sensed next. Therefore, the "raw material" for cognition is constituted by the sensorimotor variables and it is thus crucial to study relationships between sensors and motors, as illustrated by the sensorimotor contingencies (see next section). Furthermore, time is no less important a variable. Lungarella and Sporns (2006) provide an excellent example of the use of information theoretic measures in this context. In a series of experiments with a movable camera system, they could show that, for example, the entropy in the visual field is decreased if the camera is tracking a moving visual target (a red ball) compared to the condition where the movement of the ball and the camera were uncorrelated. This is intuitively plausible, because if the object is kept in the center of the visual field, there is more "order," i.e., less entropy. A collection of case studies on information-theoretic implications of embodiment in locomotion, grasping, and visual perception is presented by Hoffmann and Pfeifer (2011).

Sensorimotor Contingencies

Sensorimotor contingencies (SMCs) were originally presented in the influential article by O'Regan and Noë (2001) as the structure of the rules governing sensory changes produced by various motor actions. The SMCs, according to O'Regan and Noë, are the key "raw material" upon which perception, cognition, and eventually consciousness operates. Furthermore, they sketch a possible hierarchy ranging from modality-related (or apparatus-related) SMCs to object-related SMCs. The former, the modality-related SMCs, would capture the immediate effect that certain actions (or movements) have on sensory stimulation. Clearly, these would be sensory modality specific (e.g., head movement will induce a different change in the SMCs of the visual and auditory modalities—turning the head will change the visual stimulation almost entirely, whereas changes in the acoustic system will be minimal) and would strongly depend on the sensory morphology. Therefore, this concept is strongly related to what we have discussed in the previous sections: (1) different sensory morphology importantly affects the information flow induced in the sensory receptors and hence also the corresponding SMCs; (2) the effect of action is already constitutively included in the SMC notion itself.

Although conceptually very powerful, the notion of SMCs was not articulated concretely enough in O'Regan and Noë (2001) such that it could be expressed mathematically or directly transferred into a robot implementation, for example. Bührmann et al. (2013) have proposed a formal dynamical systems account of SMCs. They devised a dynamical system description for the environment and the agent, which is in turn split into body, internal state (such as neural activity), motor, and sensory dynamics. Bührmann et al. are making a distinction between sensorimotor (SM) environment, SM habitat, SM coordination, and SM strategy. The SM environment is the relation between motor actions and changes in sensory states, independent of the agent's internal (neural) dynamics. The other notions—from SM habitat to SM strategies—add internal dynamics to the picture. SM habitat refers to trajectories in the sensorimotor space, but subject to constraints given by the internal dynamics that are responsible for generating motor commands, which may depend on previous sensory states as well—an example of closed-loop control. SM coordination then further reduces the set of possible SM trajectories to those "that contribute functionally to a task." For example, specific patterns of squeezing an object in order to assess its hardness would be SM coordination patterns serving object discrimination. Finally, SM strategies take, in addition, "reward" or "value" for the agent into account.

As wonderfully illustrated by Beer and Williams (2015), the dynamical systems and information theory are two complementary mathematical lenses through which brain–body–environment systems can be studied. While acknowledging the merits of both frameworks as "intuition, theory, and experimental pumps" (Beer and Williams 2015), it is probably fair to say that compared to dynamical systems, information theory has been thus far more successfully applied to the analysis of real systems of higher dimensionality. This is true for both natural systems—in particular, brains (Garofalo et al. 2009; Quiroga and Panzeri 2009)—and artificial systems. Thus, to study sensorimotor contingencies in a real robot beyond the simple simulated agents of Bührmann et al. (2013) and Beer and Williams (2015), we chose to use the lens of information theory. Following up on related studies of e.g., Olsson et al. (2004), we conducted a series of studies in a real quadrupedal robot with rich nonlinear dynamics and a collection of sensors from different modalities (Hoffmann et al. 2012; Hoffmann et al. 2014; Schmidt et al. 2013) (see Box 45.1). We have applied the notion of "transfer entropy"

Box 45.1 Sensorimotor contingencies in a quadruped robot

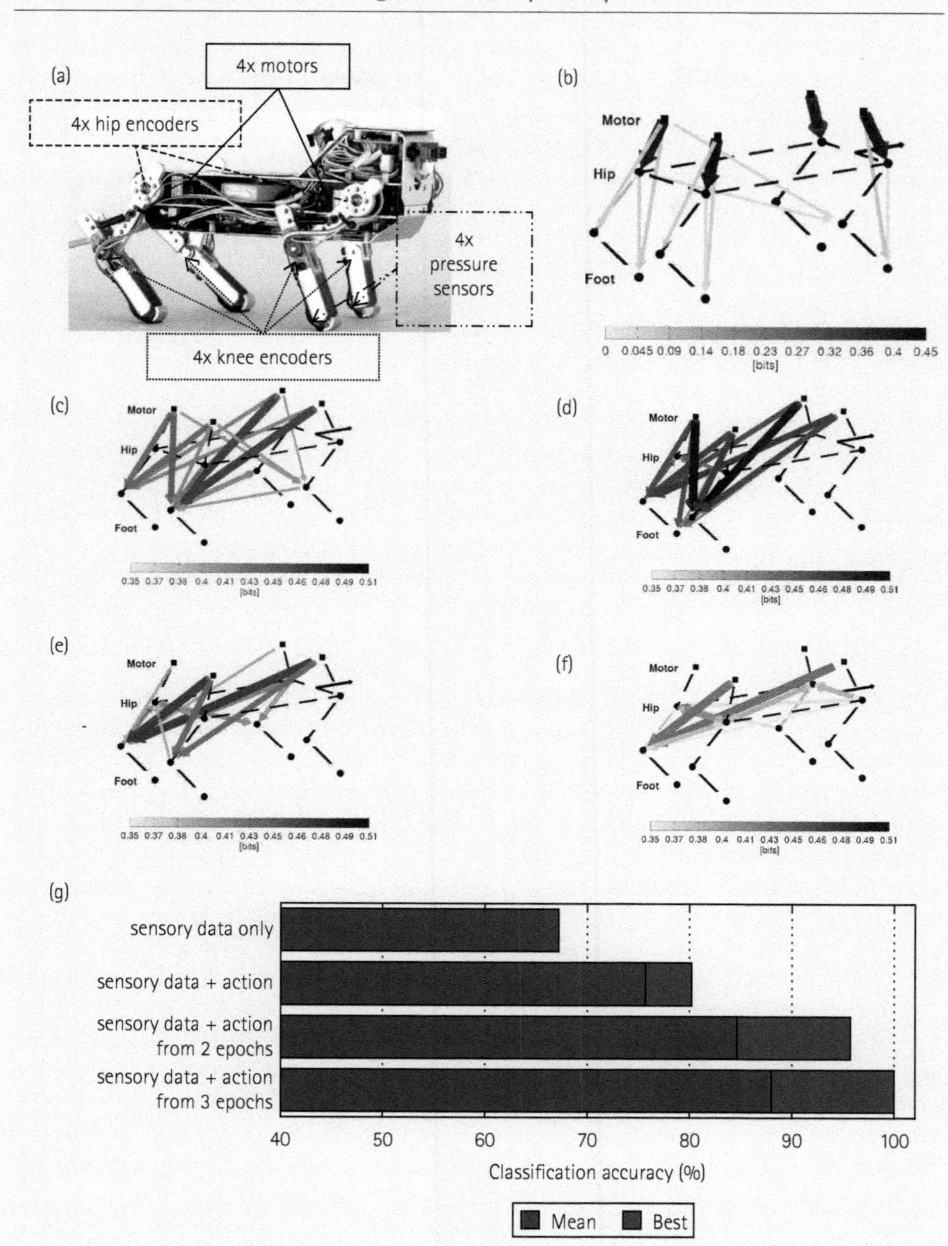

FIGURE 45.1. Robot "Puppy" and sensorimotor contingencies.

Experiments were conducted on the quadrupedal robot Puppy (Figure 45.1a), which has four servomotors in the hips together with encoders measuring the angle at the joint, four encoders in the passive compliant knees, and four pressure sensors on the feet. We used the notion of "transfer entropy" from information theory, which can be used to measure directed information flows between time series. In our case, the time series were collected from individual motor and sensory channels and the information transfer was calculated for every pair of channels two times, once in every direction (say, from hind right motor to front right knee encoder and also in the opposite direction). Loosely speaking, transfer entropy from channel A to channel B measures how well the future state of channel B can be predicted knowing the current state of channel A (see Schmidt et al. 2013 for details).

First, we wanted to investigate the "sensorimotor structure," i.e., the relative strengths of relationships between different sensors and motors, which is intrinsic to the robot's embodiment (body + sensor morphology only). To this end, random motor commands were applied and the relationships between motor and sensory variables were studied, closely resembling the notion of SM environment (Bührmann et al. 2013). The strongest information flows between pairs of channels were extracted and are shown overlaid over the schematic of the Puppy robot (dashed lines) in panel B. The transfer entropy is encoded as thickness and gray level of the arrows. The strongest flow occurs from the motor signals to their respective hip joint angles, which is clear because the motors directly drive the respective hip joints. The motors have a smaller influence on the knee angles (stronger in the hind legs) and on the feet pressure sensors—on the respective legs where the motor is mounted, thus illustrating that body topology was successfully extracted (at the same time, the flows from the hind leg motors and hips to the front knees highlight that the functional relationships are different than the static body structure; see also Schatz and Oudeyer 2009). These patterns are analogous to the modality-related SMCs; just as we can predict what will be the sensory changes induced by moving the head, the robot can predict the effects of moving the hind leg, say.

In a second step, we studied the relationships in the sensorimotor space when the robot was running with specific coordinated periodic movement patterns or gaits. The results for two selected gaits—turn left and bound right*—are shown in panels C and D, respectively. The flows from motors to the hip joints, which would again dominate, were left out of the visualization. The plots clearly demonstrate the important effect of specific action patterns in two ways. First, they markedly differ from the random motor command situation: the dominant flows are different and, in addition, the magnitude of the information flows is bigger (the number of bits—note the different range of the color bar compared to B), illustrating how much information structure is induced by the "neural pattern generator." Second, they also significantly differ between themselves. The "turn left" gait in panel C reveals the dominant action of the right leg and in particular the knee joint. In the "bound right" gait in D, the motor signals are predictive of the sensory stimulation in the hind knees and also the left foot. The gaits were obtained by optimizing the robot's performance for speed or for turning and thus correspond to patterns that are functionally relevant for the robot and can even be said to carry "value." Thus, in the perspective of Bührmann et al. (2013), our findings about the sensorimotor space using the gaits can be interpreted as studying the SM coordination or even SM strategy of the quadruped robot.

Finally, next to the embodiment or morphology (shape of the body and limbs, type and placement of sensors and effectors, etc.) and the brain (the neural dynamics responsible for generating the coordinated motor command sequences), the SMCs are co-determined by the environment as well. All the results thus far came from sensorimotor data collected from the robot running on a plastic foil ground (low friction). Panels E and F depict how the information flows for the bound right gait are modulated when the robot runs on a different ground (E—Styrofoam, F—rubber). The overall pattern is similar to D, but the flows to the left foot disappear, and eventually flows to the left knee joint become dominant. This is because the posture of the robot changed: the left foot contacts the ground at a different angle now, inducing less stimulation in the pressure sensor. Also, as the friction increases (from the foil over Styrofoam to rubber), the push-off during stance of the left hind leg becomes stronger, resulting in more pronounced bending of the knee. Finally, since the high-friction ground poses more resistance to the robot's movements, the trajectories are less smooth and the overall information flow drops.

While all the components (body, brain, environment) have a profound effect on the overall sensorimotor space, our analysis reveals that in this case, the gait used (as prescribed primarily by the "neural/brain" dynamics) is a more important factor than the environment (the ground)—the latter seems to modulate the basic structure of information flows induced by the gait. This has important consequences for the agent when it is to learn something about its environment and perform perceptual categorization, for example. In order to investigate this quantitatively, we have presented the robot with a terrain (the surface/ground it was running on) classification task. Relying on sensory information alone leads to significantly worse terrain classification results than when the gait is explicitly taken into account in the classification process (Hoffmann, Stepanova, and Reinstein 2014). Furthermore, in line with the predictions of the sensorimotor contingency theory, longer sensorimotor sequences are necessary for object perception (Maye and Engel 2012). That is, while in short sequences (motor command, sensory consequence), modality-related SMCs (panel B) will be dominant, longer interactions will allow objects the agent is interacting with to stand out. Using data from our robot, this is convincingly demonstrated in panel G. The first row shows classification results when using data from one sensory epoch (two seconds of locomotion) collapsed across all gaits, i.e., without the action context. Subsequent rows report results where classification was performed separately for each gait and increasingly longer interaction histories were available. "Mean" values represent the mean performance; "best" are classification results from the gait that facilitated perception the most (see Hoffmann et al. 2012 for details).

* "Turn left" was a movement pattern dominated by the action of the right hind leg that was pushing the robot forward and left. Regarding "bound right," bounding gait is a running gait used by small mammals. It is similar to gallop, and features a flight phase, but is characterized by synchronous action of every pair of legs. However, in this study, we used lower speeds without an aerial phase. In addition, the symmetry of the motor signals was slightly disrupted, resulting in a right-turning motion.

from information theory, which can be used to characterize sensorimotor flows in the robot—for example, how strongly sensors are affected by motor commands—and we tried to isolate the effects of the body, motor programs (gaits), and environment in the agent's sensorimotor space. Finally, we tested the predictions of SMC theory regarding object discrimination. In our investigations, we have chosen the situated perspective—analyzing only the relationships between sensory and motor variables that would also be available to the agent itself. However, information-theoretic methods can also be productively applied to study relationships between internal and external variables, such as between sensory or neuronal states and some properties of an external object (e.g., its size, Beer and Williams 2015; or any other property that can be expressed numerically). Using this approach, one can obtain important insights into the operation and temporal evolution of categorization, for example. Performing this in the ground discrimination scenario on the quadrupedal robot constitutes our future work.

While the studies on "minimally cognitive agents" are of fundamental importance and lead to valuable insights for our understanding of intelligent behavior, the ultimate target is, of course, human cognition. Toward this end, one may want to resort to more sophisticated tools, for example, humanoid robots.

Human-like Cognition in Robots

In the previous section, we showed how robots can be beneficial in operationalizing, formalizing, and quantifying ideas, concepts, and theories that are important for understanding cognition but that are often not articulated in sufficient detail. An obvious implication of this analysis is that the kind of cognition that emerges will be highly dependent on the body of the agent, its sensorimotor apparatus, and the environment it is interacting with. Thus, to target human cognition, the robot's morphology—shape, type of sensors, and their distribution, materials, actuators—should resemble that of humans as closely as possible. Now we have to be realistic: approximating humans very closely would imply mimicking their physiology, the sensors in the body, and the inner organs, the muscles with comparable biological instantiation, and the bloodstream that supplies the body with energy and oxygen. Only then could the robot experience the true concept, e.g., of being thirsty or out of breath, hearing the heart pumping, blushing, or the feeling of quenching the thirst while drinking a cold beer in the summer. So, even if, on the surface, a robot might be almost indistinguishable from a human (like, for example, Hiroshi Ishiguro's recent humanoid "Erica"), we have to be aware of the fundamental differences: comparatively very few muscles and tendons, no actuators that can get sore when overused, no sensors for pain, only low-density haptic sensors, no sweat glands in the skin, and so on and so forth. Thus, "Erica" will have a very impoverished concept of drinking or feeling hot. In other words, we have to make substantial abstractions.

Just as an aside, making abstractions is nothing bad—in fact, it is one of the most crucial ingredients of any scientific explanation because it forces us to focus on the essentials, ignoring whatever is considered irrelevant (the latter most likely being the majority of things that we could potentially take into account). Thus, the specifics of the robot's cognition—its concepts, its body schema—will clearly diverge from that of humans, but the underlying principles will, at a certain level of abstraction, be the same. For example, it will have its own sensorimotor contingencies, it will form cross-modal associations through Hebbian learning, and it will explore its environment using its sensorimotor setup. So if the robot says "glass," this will relate to very different specific sensorimotor experiences, but if the robot can recognize, fill, and hand a "glass" to a human for drinking, it makes sense to say that the robot has acquired the concept of "glass."

Because the acquisition of concepts is based on sensorimotor contingencies, which in turn require actions on the part of the agent, and because the patterns of sensory stimulation are associated with the respective motor signals, the robot platforms of choice will ideally be tendon-driven—just like humans who use muscles and tendons for movements. Given our discussion on abstraction earlier, we can also study concept acquisition in robots that have motors in the joints—we just have to be aware of the concrete differences. Still, the principles governing the robot's cognition can be very similar to that of humans (see Box 45.2 for examples of different types of humanoid robots).

BOX 45.2 Humanoid embodiment for modeling cognition

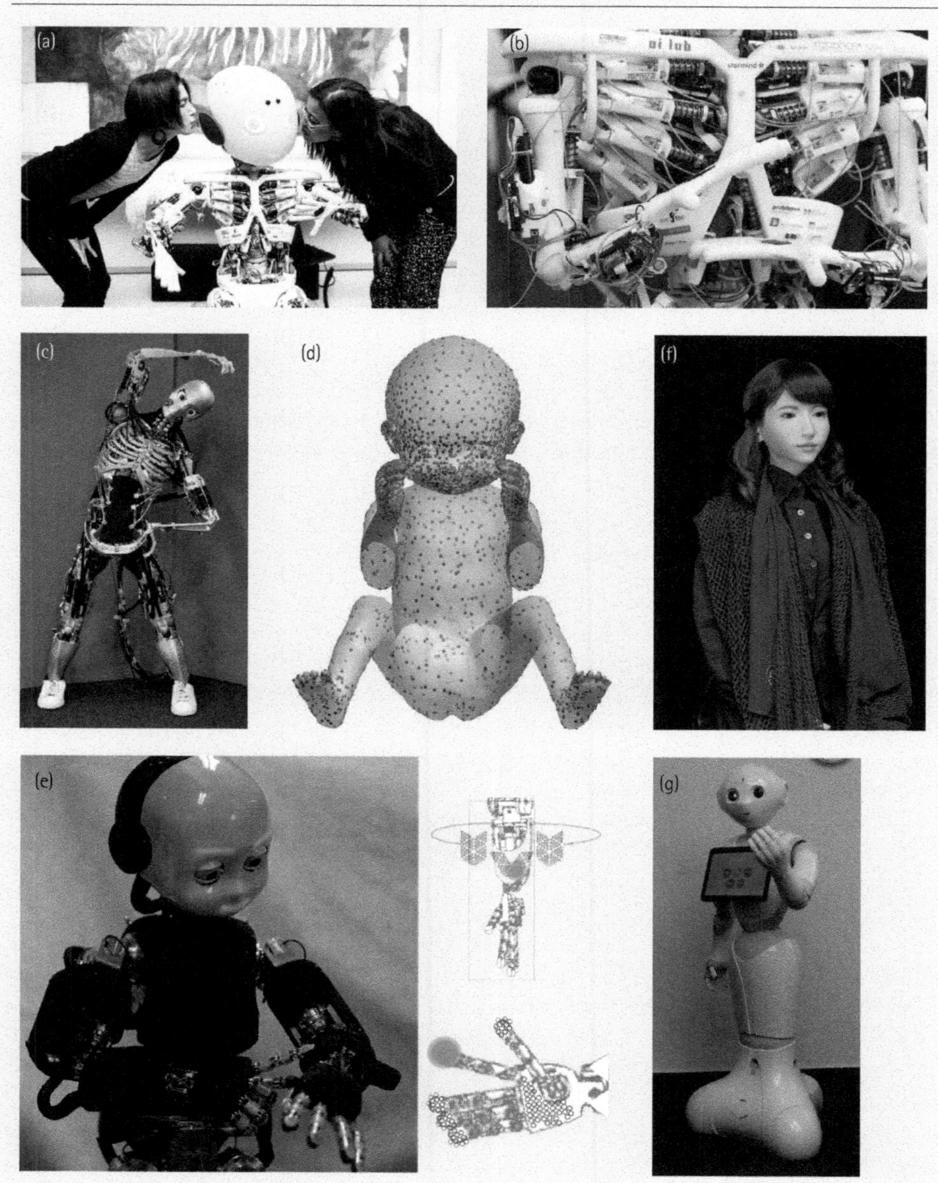

FIGURE 45.2. Humanoid robots.

A large number of humanoid robots have been developed over the last decades and many of them can, one way or other, be used to study human cognition. Given that all of them to date are very different from real humans—each of them, implicitly or explicitly, embodies certain types of abstractions—there is no universal platform, but they have all been developed with specific goals in mind. Here we present a few examples and discuss the ways in which they are employed in trying to ferret out the principles of human cognition. The categories shown in Figure 45.2 are musculoskeletal robots (Roboy and Kenshiro), "baby" robots with sensorized skins (iCub and fetus simulators), and social interaction robots (Erica and Pepper).

In order to use the robots for learning their own complex dynamics and for building up a body schema, both Roboy and Kenshiro (Nakanishi et al. 2012) need to be equipped with many sensors so that they can "experience" the effect of a particular actuation pattern. Given rich sensory feedback, using the principle that every action leads to sensory stimulation, both these robots can, in principle, employ motor babbling in order to learn how to move. Especially for Kenshiro, with his very large number of muscles, learning is a must. A very important step in this direction is the work of Richter et al. (2016), who have combined a musculoskeletal robotics toolkit (Myorobotics) with a scalable neuromorphic computing platform (SpiNNaker) and demonstrated control of a musculoskeletal joint with a simulated cerebellum.

Finally, if the interest is social interaction, it might be more productive to use robots like Erica or Pepper. Both Erica and Pepper are somewhat limited in their sensorimotor abilities (especially haptics), but are endowed with speech understanding and generation facilities; they can recognize faces and emotions; and they can realistically display any kind of facial expression.

Musculoskeletal robots: Roboy and Kenshiro

Figure 45.2a. Roboy overview: The musculoskeletal design can be clearly observed. At this point, Roboy has 48 "muscles." Eight are dedicated to each of the shoulder joints. This can no longer be sensibly programmed: learning is a necessity. Currently, Roboy serves as a research platform for the EU/FET Human Brain Project to study, among other things, the effect of brain lesions on the musculoskeletal system. Because it has the ability to express a vast spectrum of emotions, it can also be employed to investigate human–robot interaction, and as an entertainment platform.

Credit: © Embassy of Switzerland in the United States of America.

Figure 45.2b. Close-up of the muscle-tendon system. Although the shoulder joint is distinctly dissimilar to a human one—for example, it doesn't have a shoulder blade—it is controlled by eight muscles, which require substantial skills in order to move properly: which muscles have to be actuated to what extent in order to achieve a desired movement?

Credit: © Erik Tham/Corbis Documentary/Getty Images.

Figure 45.2c. Kenshiro's musculoskeletal setup. The musculoskeletal design is clearly visible. At this point, Kenshiro has 160 "muscles"—50 in the legs, 76 in the trunk, 12 in the shoulder, and 22 in the neck. In terms of musculoskeletal system, it is the one robot that most closely resembles the human. So, if learning of the dynamics in this system is the goal, Kenshiro will be the robot of choice. Note that although Kenshiro is "closest" to a human in this respect, it is still subject to enormous abstractions. Currently, Kenshiro serves as a research platform at the University of Tokyo to investigate tendon-controlled systems with very many degrees of freedom (Nakanishi et al. 2012).

Credit: Photo courtesy Yuki Asano.

"Baby" robots with sensitive skins

Figure 45.2d. Fetus simulator. A musculoskeletal model of human fetus at 32 weeks of gestation has been constructed and coupled with a brain model comprising 2.6 million spiking neurons (Yamada et al. 2016). The figure shows the tactile sensor distribution, which was based on human two-point discrimination data.

Reproduced from Yasunori Yamada, Hoshinori Kanazawa, Sho Iwasaki, Yuki Tsukahara, Osuke Iwata, Shigehito Yamada, and Yasuo Kuniyoshi, An Embodied Brain Model of the Human Foetus, *Scientific Reports*, 6 (27893), Figure 1d, doi:10.1038/srep27893 © 2016 Yasunori Yamada, Hoshinori Kanazawa, Sho Iwasaki, Yuki Tsukahara, Osuke Iwata, Shigehito Yamada, and Yasuo Kuniyoshi. This work is licensed under the Creative Commons Attribution 4.0 International License (CC BY 4.0). It is attributed to the authors Yasunori Yamada, Hoshinori Kanazawa, Sho Iwasaki, Yuki Tsukahara, Osuke Iwata, Shigehito Yamada, and Yasuo Kuniyoshi.

Figure 45.2e. The iCub baby humanoid robot. The iCub (Metta et al. 2010) has the size of a roughly four-year-old child and corresponding sensorimotor capacities: 53 degrees of freedom (electrical motors), two stereo cameras in a biomimetic arrangement, and over 4,000 tactile sensors covering its body. The panel shows the robot performing self-touch and corresponding activations in the tactile arrays of the left forearm and right index finger.

Social interaction robots: Erica and Pepper

Figure 45.2f. Erica, the latest creation of Prof. Hiroshi Ishiguro, was designed specifically with the goal of imitating human speech and body language patterns, in order to have "highly natural" conversations. It also serves as a tool to study human–robot interaction, and social interaction in general. Moreover, because of its close resemblance to humans, the "uncanny valley"—the fact that people get uneasy when the robots are too human-like—hypothesis can be further explored and analyzed (see, e.g., Rosenthal-von der Pütten, Marieke, and Weiss 2014, where the Geminoid HI-1 modeled after Prof. Ishiguro was used).

Credit: Photo courtesy of Hiroshi Ishiguro Laboratory, ATR and Osaka University.

Figure 45.2g. Pepper, a robot developed by Aldebaran (now Softbank Robotics), although much simpler (and much cheaper!) than Erica, is used successfully on the one hand to study social interaction, for entertainment, and to perform certain tasks (such as selling Nespresso machines to customers in Japan).

The Role of Development

A very powerful approach to deepen our understanding of cognition, and one that has been around for a long time in psychology and neuroscience, is to study ontogenetic development. During the past two decades or so, this idea has been adopted by the robotics community and has led to a thriving research field dubbed "developmental robotics." Now, a crucial part of ontogenesis takes place in the uterus. There, tactile sense is the first to develop (Bernhardt 1987) and may thus play a key role in the organism's learning about first sensorimotor contingencies, in particular, those pertaining to its own body (e.g., hand-to-mouth behaviors). Motivated by this fact, Mori and Kuniyoshi (2010) developed a musculoskeletal fetal simulator with over 1,500 tactile receptors, and studied the effect of their distribution on the emergence of sensorimotor behaviors. Importantly, with a natural (non-homogeneous) distribution, the fetus developed "normal" kicking and jerking movements (i.e., similar to those observed in a human fetus), whereas with a homogeneous allocation it did not develop any of these behaviors. Yamada et al. (2016), using a similar fetal simulator and a large spiking neural network brain model, have further studied the effects of intrauterine (vs. extrauterine) sensorimotor experiences on cortical learning of body representations. A physical version—the fetusoid—is currently under development (Mori et al. 2015). Somatosensory (tactile and proprioceptive) inputs continue to be of key importance also in early infancy when "infants engage in exploration of their own body as it moves and acts in the environment. They babble and touch their own body, attracted and actively involved in investigating the rich intermodal redundancies, temporal contingencies, and spatial congruence of self-perception" (Rochat 1998, p. 102). The iCub baby humanoid robot (Metta et al. 2010) (Box 45.2E), equipped with a whole-body tactile array (Maiolino et al. 2013) comprising over 4,000 elements, is an ideal platform to study these processes. The study of Roncone et al. (2014) on self-calibration using self-touch is a first step in this direction.

Applications Of Human-Like Robots

Finally, this research strand—employing humanoid robots to study human cognition—has also important applications. In traditional domains and conventional tasks—such as pick-and-place operations in an industrial environment—current factory automation robots are doing just fine. However, robots are starting to leave these constrained domains, entering environments that are far less structured and are starting to share their living space with humans. As a consequence, they need to dynamically adapt to unpredictable interactions and guarantee their own as well as others' safety at every moment. In such cases, more human-like characteristics—both physical and "mental"—are desirable. Box 45.3 illustrates how more brain-like body representations can help robots to become more autonomous, robust, and safe. The possibilities for future applications of robots with cognitive capacities are enormous, especially in the rapidly

BOX 45.3 Body schema in humans vs. robots

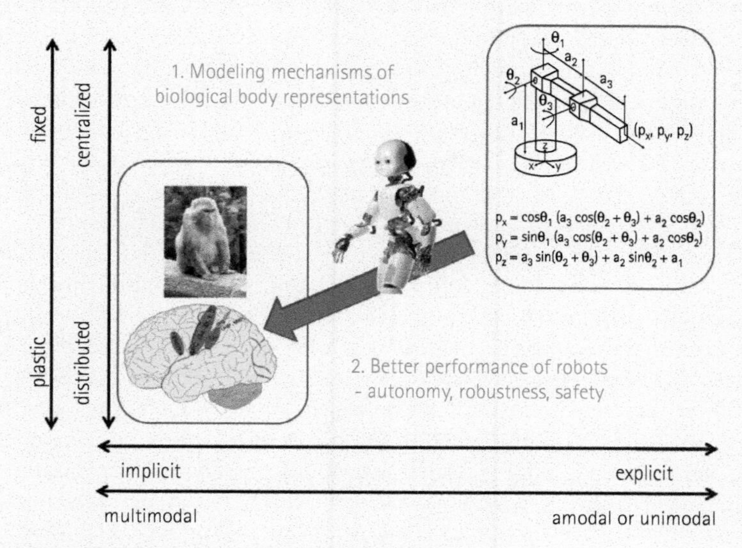

FIGURE 45.3. Characteristics of body representations.

Credit: Monkey photo source: Einar Fredriksen/Flickr/Attribution-ShareAlike 4.0 International (CC BY-SA 4.0)
Credit: Brain image source: Hugh Guiney/Attribution-ShareAlike 3.0 Unported (CC BY-SA 3.0)
Credit: Line drawing and equations source: Reproduced with the permission of Dr. Hugh Jack from http://www.engineeronadisk.com
Credit: iCub Robot source: © iCub Facility—IIT, 2017

A typical example of a traditional robot and its mathematical model is depicted in the upper right of Figure 45.3. The robot is an arm consisting of three segments with three joints between the base and the final part—the end-effector. Its model is below the robot—the forward kinematics equations that relate configuration of the robot (joint positions θ_1, θ_2, θ_3) to the Cartesian position of the end-effector (p_x, p_y, p_z). The model has the following characteristics: (1) it is explicit—there is a one-to-one correspondence between its body and the model (a_1 in the model is the length of the first arm segment, for example); (2) it is unimodal—the equations directly describe physical reality; one sensory modality (proprioception—joint angle values) is needed to get the correct mapping in the current robot state; (3) it is centralized—there is only one model that describes the whole robot; (4) it is fixed—normally, this mapping is set and does not change during the robot operation. Other models/mappings are typically needed for robot operation, such as inverse kinematics, differential kinematics, or models of dynamics (dealing with forces and torques), but they would all share the above-mentioned characteristics (see Hoffmann et al. 2010 for a survey).

As pointed out earlier, animals and humans have different bodies than robots; they also have very different ways of representing them in their brains. The panel in the lower left shows the rhesus macaque and below some of the key areas of its brain that deal with body representations (see, e.g., Graziano and Botvinick 2002). There is ample evidence that these representations differ widely from the ones traditionally used in robotics—namely, "the body in the brain" would be (1) implicitly represented—there would hardly be a "place" or a "circuit" encoding, say, the length of a forearm; such information is most likely only indirectly available and possibly in relation to other variables; (2) multimodal—drawing mainly from somatosensory (tactile and proprioceptive) and visual, but also vestibular (inertial) and closely coupled to motor information; (3) distributed—there are numerous distinct, but partially overlapping and interacting representations that are dynamically recruited depending on context and task; (4) plastic—adapting over both long (ontogenesis) and short time scales, as adaptation to tool use (e.g., Iriki et al. 1996) or various body illusions testify (e.g., humans start feeling ownership over a rubber hand after minutes of synchronous tactile stimulations of the hand replica and their real hand under a table; Botvinick and Cohen 1998).

The iCub robot "walking" from the top right to the bottom left in the figure is illustrating two things. First, in order to be able to model the mechanisms of biological body representations, the traditional robotic models are of little use—a radically different approach needs to be taken. Second, by making the robot models more brain-like, we hope to inherit some of the desirable properties typical of how humans and animals master their highly complex bodies. Autonomy and robustness or resilience are one such case. It is not realistic to think that conditions, including the body, will stay constant over time and a model given to the robot by the manufacturer will always work. Inaccuracies will creep in due to wear and tear and possibly even more dramatic changes can occur (e.g., a joint becomes blocked). Humans and animals display a remarkable capacity for dealing with such changes: their models dynamically adapt to muscle fatigue, for example, or temporarily incorporate objects like tools after working with them, or reallocate "brain territory" to different body parts in case of amputation of a limb. Robots thus also need to perform continuous self-modeling (Bongard et al. 2006) in order to cope with such changes. Finally, unlike factory robots that blindly execute their trajectories and thus need to operate in cages, humans and animals use multimodal information to extend the representation of their bodies to the space immediately surrounding them (also called peripersonal space). They construct a "margin of safety," a virtual "bubble" around their bodies that allows them to respond to potential threats such as looming objects, warranting safety for them and also their surroundings (e.g., Graziano and Cooke 2006). This is highly desirable in robots as well, and can transform them from dangerous machines to collaborators possessing whole-body awareness like we do. First steps along these lines in the iCub were presented by Roncone et al. (2016).

growing area of service robotics, where robots perform tasks in human environments. Rather than accomplishing them autonomously, they often do it in cooperation with humans, which constitutes a big trend in the field. In cooperative tasks, it is of course crucial that the robots understand the common goals and the intentions of the humans in order to be successful. In other words, they require substantial cognitive skills. We have barely started exploiting the vast potential of these types of cognitive machines.

Conclusion

Our analysis so far has demonstrated that robots fit squarely into the embodied and pragmatic (action-oriented) turn in cognitive sciences (e.g., Engel et al. 2013), which implies that whole behaving systems rather than passive subjects in brain scanners need to be studied. Robots provide the necessary grounding to computational models of the brain by incorporating the indispensable brain–body–environment coupling (Pezzulo et al. 2011). The advantage of synthetic methodology, or "understanding by building" (Pfeifer and Bongard 2007), is that one learns a lot in the process of building the robot and instantiating the behavior of interest. The theory one wants to test thus automatically becomes explicit, detailed, and complete. Robots become virtual experimental laboratories retaining all the virtues of "theories expressed as simulations" (Cangelosi and Parisi 2002), but bring the additional advantage that there is no "reality gap": there is real physics and real sensory stimulation, which lends more credibility to the analysis if embodiment is at center stage.

We are convinced that robots are the right tools to help us understand the embodied, embedded, and extended nature of cognition because their makeup—physical artifacts with sensors and actuators interacting with their environment—automatically warrants the necessary ingredients. It seems that they are particularly suited for investigations of cognition from bottom up (Pfeifer et al. 2014), where development under particular constraints in brain–body–environment coupling is crucial (e.g., Thelen and Smith 1994). It also becomes possible to simulate conditions that one would not be able to test in humans or animals—think of the simulation of fetal ontogenesis while manipulating the distribution of tactile receptors (Mori and Kuniyoshi 2010). Furthermore, many additional variables (such as internal states of the robot) become easily accessible and lend themselves to quantitative analysis, such as using methods from information theory. Therefore, the combination of a robot with sensorimotor capacities akin to humans, the possibility of emulating the robot's growth and development, and finally the ease of access to all internal variables that can be subject to rigorous quantitative investigations create a very powerful tool to help us understand cognition.

We want to close with some thoughts on whether it is possible to realize—next to embodied, embedded, and extended—enactive robots as well. Most researchers in embodied AI/cognitive robotics automatically adopt the perspective of extended functionalism (Clark 2008; Wheeler 2011), whereby the boundaries of cognitive systems can be extended beyond the agent's brain and even skin—including the body and environment. However, it has been pointed out by the proponents of enactive cognitive science (Di Paolo 2010; Froese and Ziemke 2009) that in order to fully understand cognition in its entirety, embedding the agent in a closed-loop sensorimotor interaction with the environment is necessary, yet may not be sufficient in order to induce important properties of biological agents such as intentional agency. In other words, one should not only study instances of individual closed sensorimotor loops as models of biological agents—that would be the recommendation of Webb (2009)—but one should also try to endow the models (robots in this case) with similar properties and constraints that biological organisms are facing. In particular, it has been argued that life and cognition are tightly interconnected (Maturana 1980; Thompson 2007), and a particular organization of living systems—which can be characterized by autopoiesis (Maturana 1980) or metabolism, for example—is crucial for the agent to truly acquire meaning in its interactions with the world. While these requirements are very hard to satisfy with the artificial systems of today, Di Paolo (2010) proposes a way out: robots need not metabolize, but they should be subject to so-called precarious conditions. That is, the success of a particular instantiation of sensorimotor loops or neural vehicles in the agent is to be measured against some viability criterion that is intrinsic to the organization of the agent (e.g., loss of battery charge, overheating leading to electronic board problems resulting in loss of mobility, etc.). The control structure may develop over time, but the viability constraint needs to be satisfied, otherwise the agent "dies" (McFarland and Boesser 1993). In a similar vein, in order to move from embodied to enactive AI, Froese and Ziemke (2009) propose to extend the design principles for autonomous agents of Pfeifer and Scheier (2001), requiring the agents to generate their own systemic identity and regulate their sensorimotor interaction with the environment in relation to a viability constraint. The unfortunate implication, however, is that research along these lines will in the short term most likely not produce useful artifacts. On the other hand, this approach may eventually give rise to truly autonomous robots with unimaginable application potential.

Acknowledgments

M.H. was supported by a Marie Curie Intra European Fellowship (iCub Body Schema 625727) within the 7th European Community Framework Programme and the Czech Science Foundation under Project GA17-15697Y.

References

Beer, R.D. and Williams, P.L. (2015). Information processing and dynamics in minimally cognitive agents. *Cognitive Science*, 39, 1–38.

Bernhardt, J. (1987). Sensory capabilities of the fetus. *MCN: The American Journal of Maternal/Child Nursing*, 12(1), 44–7.

Blickhan, R., Seyfarth, A., Geyer, H., Grimmer, S., Wagner, H., and Günther, M. et al. (2007). Intelligence by mechanics. *Philosophical transactions. Series A*, 365, 199–220.

Bongard, J., Zykov, V., and Lipson, H. (2006). Resilient machines through continuous self-modeling. *Science*, 314, 1118–21.

Botvinick, M. and Cohen, J. (1998). Rubber hands "feel" touch that eyes see. *Nature*, 391(6669), 756.

Braitenberg, V. (1986). *Vehicles—experiments in synthetic psychology*. Cambridge, MA: MIT Press.

Brooks, R. (1986). A robust layered control system for a mobile robot. *IEEE Journal of Robotics and Automation*, 2(1), 14–23.

Brooks, R.A. (1989). A robot that walks: emergent behaviors from a carefully evolved network. *Neural Computation*, 1, 153–62.

Brooks, R.A. (1991a). Intelligence without reason. In: J. Myopoulos (ed.), Proceedings of the Twelfth International Joint Conference on Artificial Intelligence (vol. 1). San Francisco, USA: Morgan Kaufmann, pp. 569–95.

Brooks, R.A. (1991b). Intelligence without representation. *Artificial Intelligence*, 47, 139–59.
Bührmann, T., Di Paolo, E., and Barandiaran, X. (2013). A dynamical systems account of sensorimotor contingencies. *Frontiers in Psychology*, 4, 285.
Cangelosi, A. and Parisi, D. (2002). Computer simulation: a new scientific approach to the study of language evolution. In: *Simulating the evolution of language*. London: Springer Science & Business Media, pp. 3–28.
Clark, A. (2008). *Supersizing the mind: embodiment, action, and cognitive extension*. New York: Oxford University Press.
Clark, A. and Grush, R. (1999). Towards cognitive robotics. *Adaptive Behaviour*, 7(1), 5–16.
Collins, S., Ruina, A., Tedrake, R., and Wisse, M. (2005). Efficient bipedal robots based on passive dynamic walkers. *Science*, 307, 1082–5.
Dennett, D. (1995). *Darwin's dangerous idea*. New York: Simon & Schuster.
Di Paolo, E. (2010). Robotics inspired in the organism. *Intellectica*, 53–54, 129–62.
Engel, A.K., Maye, A., Kurthen, M., and König, P. (2013). Where's the action? The pragmatic turn in cognitive science. *Trends in Cognitive Sciences*, 17(5), 202–9.
Floreano, D., Pericet-Camara, R., Viollet, S., Ruffier, F., Brückner, A., Leitel, R. et al. (2013). Miniature curved artificial compound eyes. *Proceedings of the National Academy of Sciences*, 110(23), 9267–72.
Fodor, J. (1975). *The language of thought*. Cambridge, MA: Harvard University Press.
Franceschini, N., Pichon, J., and Blanes, C. (1992). From insect vision to robot vision. *Philosophical transactions of the Royal Society of London. Series B, Biological sciences*, 337, 283–94.
Froese, T. and Ziemke, T. (2009). Enactive artificial intelligence: investigating the systemic organization of life and mind. *Artificial Intelligence*, 173(3), 466–500.
Garofalo, M., Nieus, T., Massobrio, P., and Martinoia, S. (2009). Evaluation of the performance of information theory-based methods and cross-correlation to estimate the functional connectivity in cortical networks. *PLoS ONE*, 4(8), e6482.
Graziano, M. and Botvinick, M. (2002). How the brain represents the body: insights from neurophysiology and psychology. In: W. Prinz and B. Hommel (eds.), *Common mechanisms in perception and action: attention and performance*. New York: Oxford University Press, 136–57.
Graziano, M. and Cooke, D. (2006). Parieto-frontal interactions, personal space, and defensive behavior. *Neuropsychologia*, 44(6), 845–59.
Grush, R. (2004). The emulation theory of representation—motor control, imagery, and perception. *Behavioral and Brain* Sciences, 27, 377–442.
Haugeland, J. (1985). *Artificial intelligence: the very idea*. Cambridge, MA: MIT Press.
Hoffmann, M., Marques, H., Arieta, A., Sumioka, H., Lungarella, M., and Pfeifer, R. (2010). Body schema in robotics: a review. *IEEE Transactions on Autonomous Mental Development*, 2(4), 304–24.
Hoffmann, M. and Pfeifer, R. (2011). The implications of embodiment for behavior and cognition: animal and robotic case studies. In: W. Tschacher and C. Bergomi (eds.), *The implications of embodiment: cognition and communication*. Exeter: Imprint Academic, pp. 31–58.
Hoffmann, M., Schmidt, N.M., Pfeifer, R., Engel, A.K., and Maye, A. (2012). *Using sensorimotor contingencies for terrain discrimination and adaptive walking behavior in the quadruped robot Puppy*. In: T. Ziemke, C. Balkenius, and J. Hallam (eds.), *From animals to animats 12*. SAB 2012. Lecture Notes in Computer Science (vol. 7426). Berlin, Heidelberg: Springer, pp. 54–64.

Hoffmann, M., Stepanova, K., and Reinstein, M. (2014). The effect of motor action and different sensory modalities on terrain classification in a quadruped robot running with multiple gaits. *Robotics and Autonomous Systems*, 62(12), 1790–8.

Iriki, A., Tanaka, M., and Iwamura, Y. (1996). Coding of modified body schema during tool use by macaque postcentral neurones. *Neuroreport*, 7, 2325–30.

Jeannerod, M. (2001). Neural simulation of action: a unifying mechanism for motor cognition. *NeuroImage*, 14, 103–9.

Koditschek, D.E., Full, R.J., and Buehler, M. (2004). Mechanical aspects of legged locomotion control. *Arthropod Structure and Development*, 33, 251–72.

Lichtensteiger, L. (2004). *On the interdependence of morphology and control for intelligent behavior* [PhD dissertation]. Zurich: University of Zurich.

Lungarella, M. and Sporns, O. (2006). Mapping information flow in sensorimotor networks. *PLoS Computational Biology*, 2, 1301–12.

Maiolino, P., Maggiali, M., Cannata, G., Metta, G., and Natale, L. (2013). A flexible and robust large scale capacitive tactile system for robots. *Sensors Journal, IEEE*, 13(10), 3910–7.

Maturana, H.a.V.F. (1980). *Autopoiesis and cognition: the realization of the living*. Dordrecht: D. Reidel Publishing.

Maye, A. and Engel, A.K. (2012). Time scales of sensorimotor contingencies. In: H. Zhang, A. Hussain, D. Liu, and Z. Wang (eds.), *Advances in brain inspired cognitive systems*. BICS 2012. Lecture Notes in Computer Science (vol. 7366). Berlin, Heidelberg: Springer, 240–9.

McGeer, T. (1990). Passive dynamic walking. *The International Journal of Robotics Research*, 9(2), 62–82.

Metta, G., Natale, L., Noei, F., Sandini, G., Vernon, D., Fadiga L. et al. (2010). The iCub humanoid robot: an open-systems platform for research in cognitive development. *Neural Networks*, 23(8–9), 1125–34.

Mori, H., Akutsu, D., and Asada, M. (2015). Fetusoid35: a robot research platform for neural development of both fetuses and preterm infants and for developmental care. In: A. Duff, N.F. Lepora, A. Mura, T.J. Prescott, and P.F.M.J. Verschure (eds.), *Biomimetic and biohybrid systems*. Living Machines 2014. Lecture Notes in Computer Science (vol. 8608). New York: Springer International Publishing, pp. 411–13.

Mori, H. and Kuniyoshi, Y. (2010). A human fetus development simulation: self-organization of behaviors through tactile sensation. In: *2010 IEEE 9th International Conference on Development and Learning*. doi:10.1109/DEVLRN.2010.5578860

Nakanishi, Y., Asano, Y., Kozuki, T., Mizoguchi, H., Motegi, Y., Osada, M. et al. (2012). Design concept of detail musculoskeletal humanoid "Kenshiro"—toward a real human body musculoskeletal simulator. In: *2012 12th IEEE-RAS International Conference on Humanoid Robots (Humanoids)*. doi:10.1109/HUMANOIDS.2012.6651491

Olsson, L., Nehaniv, C.L., and Polani, D. (2004). Sensory channel grouping and structure from uninterpreted sensory data. In: *Proceedings. 2004 NASA/DoD Conference on Evolvable Hardware, 2004*. doi:10.1109/EH.2004.1310825

O'Regan, J.K. and Noë, A. (2001). A sensorimotor account of vision and visual consciousness. *Behavioral and Brain Sciences*, 24, 939–1031.

Pezzulo, G., Barsalou, L.W., Cangelosi, A., Fischer, M.H., McRae, K., and Spivey, M.J. (2011). The mechanics of embodiment: a dialog on embodiment and computational modeling. *Frontiers in Psychology*, 2, 5.

Pfeifer, R. and Bongard, J.C. (2007). *How the body shapes the way we think: a new view of intelligence*. Cambridge, MA: MIT Press.

Pfeifer, R., Iida, F., and Lungarella, M. (2014). Cognition from the bottom up: on biological inspiration, body morphology, and soft materials. *Trends in Congnitive Sciences*, 18(8), 404–13.

Pfeifer, R. and Scheier, C. (2001). *Understanding intelligence*. Cambridge, MA: MIT Press.

Pylyshyn, Z. (1984). *Computation and cognition: toward a foundation for cognitive science*. Cambridge, MA: MIT Press.

Quiroga, R.Q., and Panzeri, S. (2009). Extracting information from neuronal populations: information theory and decoding approaches. *Nature Reviews Neuroscience*, 10(3), 173–85.

Richter, C., Jentzsch, S., Hostettler, R., Garrido, J.A., Ros, E., Knoll, A. et al. (2016). Musculoskeletal robots: scalability in neural control. *IEEE Robotics & Automation Magazine*, 23(4), 128–37. doi:10.1109/MRA.2016.2535081

Rochat, P. (1998). Self-perception and action in infancy. *Experimental Brain Research*, 123, 102–9.

Roncone, A., Hoffmann, M., Pattacini, U., Fadiga, L., and Metta, G. (2016). Peripersonal space and margin of safety around the body: learning tactile-visual associations in a humanoid robot with artificial skin. *PLoS ONE*, 11(10), e0163713.

Roncone, A., Hoffmann, M., Pattacini, U., and Metta, G. (2014). Automatic kinematic chain calibration using artificial skin: self-touch in the iCub humanoid robot. In: *2014 IEEE International Conference on Robotics and Automation (ICRA)*. doi:10.1109/ICRA.2014.6907178

Rosenthal-von der Pütten, A.M., Marieke, A., and Weiss, A. (2014). The uncanny in the wild: analysis of unscripted human–android interaction in the field. *International Journal of Social Robotics*, 6(1), 67–83.

Saranli, U., Buehler, M., and Koditschek, D. (2001). RHex: a simple and highly mobile hexapod robot. The *International Journal of Robotics Research*, 20, 616–31.

Schatz, T. and Oudeyer, P.Y. (2009). Learning motor dependent Crutchfield's information distance to anticipate changes in the topology of sensory body maps. *2009 IEEE 8th International Conference on Development and Learning*. doi:10.1109/DEVLRN.2009.5175526

Schmidt, N., Hoffmann, M., Nakajima, K., and Pfeifer, R. (2013). Bootstrapping perception using information theory: case studies in a quadruped robot running on different grounds. *Advances in Complex Systems*, 16(2-3), 1250078.

Song, Y.M. et al. (2013). Digital cameras with designs inspired by the arthropod eye. *Nature*, 497(7447), 95–9.

Thelen, E. and Smith, L. (1994). *A dynamic systems approach to the development of cognition and action*. Cambridge, MA: MIT Press.

Thompson, E. (2007). *Mind in life: biology, phenomenology, and the sciences of mind*. Cambridge, MA: MIT Press.

Walter, G.W. (1953). *The living brain*. New York: Norton & Co.

Webb, B. (2004). Neural mechanisms for prediction: do insects have forward models? *Trends in Neurosciences*, 27(5), 278–82.

Webb, B. (2009). Animals versus animats: or why not model the real iguana? *Adaptive Behavior*, 17, 269–86.

Wheeler, M. (2011). Embodied cogntion and the extended mind. In: J. Garvey (ed.), *The Continuum companion to philosophy of mind*. London: Continuum, pp. 220–36.

Yamada, Y., Kanazawa, H., Iwasaki, S., Tsukahara, Y., Iwata, O., Yamada, S. et al. (2016). An embodied brain model of the human foetus. *Scientific Reports*, 6. doi:10.1038/srep27893

CHAPTER 46

INTERPERSONAL JUDGMENTS, EMBODIED REASONING, AND JURIDICAL LEGITIMACY

SOMOGY VARGA

MANY have noted that in the Western philosophical tradition, the body has not been regarded as the subject of serious philosophical inquiry and has, at best, been deemed of marginal importance to our mental lives. A particularly influential strand of this view, often labeled "Cartesian," subscribes to what we could call the "body neutrality view," holding that we can explain how the organism's mind contributes to the creation of intelligent behavior without having to take into consideration aspects of the organism's physical embodiment. Although it is debatable to what extent this view really maps completely onto Descartes's own views (see Wheeler 2005), it has been influential in traditional approaches to understanding cognition. Crudely put, on this view, the brain is the sole locus of cognitive processing, which consists in the manipulation of representations in conformity with particular formal rules (see Haugeland 1995; Shapiro 2004). But if computations are simply formal operations on internal representations, as researchers like Fodor (1979, 1981) believe, then it seems right to think that the organism's body and sensorimotor systems merely deliver sensory input and enable behavioral output, but do not actively participate in carrying out cognitive activity (Shapiro 2012, p. 120).

The relatively recent embodied cognition (EC) research program positions itself against this view and provides an approach that places significant emphasis on the role that the organism's sensorimotor functions play in cognitive processes. Instead of comprehending the domain of cognition as functionally separate from sensorimotor processes, proponents of EC think that cognition is "grounded" in the sensorimotor system and can, in many cases, best be explicated in terms of a tight interaction between neural and non-neural processes (Foglia and Wilson 2013).

EC involves several relatively heterogeneous fields of research that in some cases operate with slightly dissimilar concepts, commitments, and research foci (Wilson 2002). Nonetheless, the past decade has witnessed a veritable explosion of research in several disciplines (for reviews, see Meier et al. 2012; Barsalou 2010), providing preliminary evidence that higher-level cognition is founded on modal systems (Martin 2007; Pulvermüller 2005), that understanding social situations involves executing motor system simulations (Rizzolatti and Craighero 2004), that influencing parts of the body causally affects higher cognitive processes (Niedenthal et al. 2005; Barsalou 2003), and that bodily movements and gestures can facilitate calculation (Andres et al. 2008; Goldin-Meadow 2003; Goldin-Meadow and Wagner 2004) and memory tasks (Wilson 2001; Niedenthal et al. 2005; Dijkstra et al. 2007). Some of this research has been interpreted as supporting the idea that, at least in some cases, the body acts as a partial realizer of cognitive processes, easing the cognitive burden.[1]

In this chapter, the focus will be on a particular approach that elucidates abstract concepts as grounded in particular bodily experiences. This is followed by a detailed examination of recent behavioral and neuroscientific research that explores a number of surprisingly substantial influences that bodily experiences and states have on judgment and behavior. Finally, further developing Adam Benforado's (2009) discussion of potential implications of embodied cognition for courtroom interactions, the chapter addresses what might be seen as challenges to juridical legitimacy.

Embodiment and Conceptual Metaphors

One area of EC research focuses on the manner in which concepts are "scaffolded onto" particular domains of bodily experience. Some of this work concentrates on the role metaphors play in developing and comprehending concepts (Lakoff and Johnson 1980, 1999; Lakoff 2014). For instance, the concept "argument" is structured by and comprehended in terms of the metaphor "argument is war." But if understanding a concept like "argument" requires metaphoric connections to other concepts, then avoiding circularity requires that at least some concepts be independently grounded. Lakoff and Johnson argue that certain concepts and metaphors are grounded in bodily experience, while they also function as the foundation for more abstract concepts and complex conceptual metaphors. These can typically be broken down into arrangements of more basic, cross-culturally occurring "primary metaphors," which arise from cross-domain mapping through regular correlations between embodied experiences. For example, infants who are frequently embraced tenderly by their caretakers experience the stable

[1] According to a more radical approach, the realizers of cognitive processes extend beyond the boundaries of the body (Clark and Chalmers 1998; Clark 2008).

correlation of affection and warmth. They "conflate" these experiences such that "for a time children do not distinguish between the two when they occur together" (Lakoff and Johnson 1999, p. 50). During this period, firm associations are involuntarily created between the domains, leading to conceptual metaphors "that will lead the same infant, later in life, to speak of a warm smile, a big problem, and a close friend" (Lakoff and Johnson 1999, p. 50).

Representative primary metaphors include "important is big," "change is motion," "intimacy is closeness," "more is up," and "understanding is grasping." For example, abstract judgments like "prices rose" or "prices fell" are conceptualized in terms of the basic sensorimotor experience linked to the vertical orientation of human bodies: "more is up" and "less is down." But many emotion-related metaphors also arise from primary metaphors intimately linked to the bodily correlates of affective experiences. The fact that anger is closely linked to "heat" and "pressure" is consistent with the fact that bodily correlates of anger include elevated temperature and blood pressure. "Boiling with anger" exemplifies the conceptual metaphor that "anger is a hot fluid," while "burning with love" exemplifies the conceptual metaphor that "love is fire" (Kövecses 2000, pp. 4–5). Such examples also help demonstrate that metaphors for emotions can be constitutive of our understanding of these emotions. For instance, the concept of love would probably be significantly different without the link to "electricity," "madness," "closeness," etc. This is no different for basic moral concepts, which arise from primary metaphors that connect embodied experiences of well-being (feeling health, strength, wealth, purity, control, nurturance, and empathy) to particular embodied experiences (Lakoff 2012, 2014). As we shall see in the course of this chapter, this also means that moral reasoning can be constrained by the logic of these metaphors.

Drawing on different sources, Williams, Huang, and Bargh (2009) argue that sensorimotor experiences function as the basis for subsequently developing a grasp of abstract concepts. The idea is that abstract concepts are difficult to process for the child, while other concepts linked to weight, height, closeness, etc., are relatively effortless, because they are directly related to the child's physical experience of the environment. These concepts provide a "scaffolding" for abstract concepts in a way that makes it possible for sensorimotor experiences to influence higher cognitive processes. Such scaffolding is often taken to reflect a central feature of cognition, namely, neural "reuse." This term describes processes in which sensorimotor neural mechanisms adapt to serve new cognitive roles, while simultaneously preserving their original functions (Gallese and Lakoff 2005, p. 456). This view also appears plausible from an evolutionary perspective. Natural selection favors shortcuts, exploiting existing adaptive structures instead of constructing completely new ones (Dawkins 1976; Dennett 1995). In the same way that biological adaptations build on and retain elements of design from previous adaptations, more complex evolved cognitive structures also build on structures originally designed for other functions. Simply put, we evolved from beings with neural resources principally dedicated to sensorimotor processing, and these simpler structures now also serve as the foundation for more abstract cognitive functions (Anderson 2010).

Before going further, it is worthwhile to pause and briefly situate conceptual metaphor theory (CMT) in the EC landscape. Although the various approaches within EC share reservations about the body neutrality view, we may distinguish between more conservative and more radical versions of EC theories. We may thus distinguish between simple embodiment, which interprets facts about embodiment as constraining cognitive processing and organization, and radical embodiment, which interprets such facts as significantly changing the theoretical framework of cognitive science (Clark 1999, p. 248). CMT is located on the "simple" side of the continuum and is not necessarily irreconcilable with the computational theory of mind: while it may be right that algorithms of disembodied computational approaches fail due to their omission of the peculiarities of the body, there is no reason to assume that there could be no computational approach that represents the relevant bodily properties (Shapiro 2012, pp. 92–3). CMT certainly opposes the view that concepts are disembodied and the sensorimotor system is modular, maintaining instead that the sensorimotor system structures and influences concepts and their semantic content. However, the fact that the particular structures of the human body enable the acquisition of a distinctive range of basic concepts does not necessarily support the stronger thesis that lacking very human-like bodies, organisms would not be capable of acquiring the kind of basic concepts that human beings operate with. It is difficult to exclude the possibility that analogous concepts could be developed through different processes.

Bottom-Up Effects

CMT thus emphasizes the groundedness of mental processes in the bodily experience of the world, the expansion of cognitive processes through scaffolding, and the intimate intertwining of bodily and abstract domains via conceptual metaphors. Behavioral and neuroscientific research helps support and demonstrate the full depth and consequences of CMT. Numerous empirical studies explore the extent to which the activation of certain concepts and metaphors affect emotions, behavior, and cognitive processing. Experiments using priming methodologies expose individuals to certain sensory and perceptual experiences in order to activate concepts or goals and explore the extent to which this exposure has *bottom-up* effects on higher-order judgments. It should be mentioned that replication failures have haunted priming research in social psychology, leading to debates about the epistemic status of certain findings. However, it is helpful to point out that even if some individual studies fail to replicate, there is a large number of studies that appear to support bottom-up influences. Moreover, some argue that it is unsurprising that priming effects exhibit a wide range of variations and that absent robust theories for identifying the relevant variables, replication failures only allow for modest conclusions (Cesario 2014). Others point to the impressive number of studies over the last few years that appear to confirm the effects, or to meta-analyses of priming effects in hundreds of studies concluding that priming is a valid tool (Cameron, Brown-Iannuzzi, and Payne 2012; DeCoster and Claypool 2004).

With these preliminary considerations in mind, let us now turn to a representative sample of the available empirical material (for a more comprehensive review, see Varga 2018). The idea is that simple judgments based on bodily experience (warm vs. cold; clean vs. dirty) ground complicated social categorization and judgment (kind vs. unkind; moral vs. immoral), such that the activation of related embodied concepts influence moral judgment and behavior in a bottom-up fashion. Take for instance "affection is warmth," which is a common conceptual metaphor, expressed for instance in ordinary sentences like "She was cold to me" or "She is a warm person." In such sentences, we use concepts linked to experiences of physical temperature to describe other persons and social interactions. When we say "She is a warm person," warmth refers to a collection of traits that includes, for instance, pleasantness and kindness, while coldness refers to opposite traits, and coldness-related expressions (like "give the cold shoulder") describe social rejection or exclusion. Importantly for our purposes, the intertwinement of the relevant bodily and abstract domains leads to interesting bottom-up effects. Let us see how.

A number of studies have investigated the influence of temperature on interpersonal judgments. Gockel and colleagues (2014) tested the effect of ambient temperature on judgments of criminals (related to both attributing criminal intent and inferring criminality). Participants were asked to inspect pictures of individuals who had committed a crime while they were exposed to different room temperatures within a comfort zone (corresponding to upper, medium, or lower levels). Consistent with the metaphorical expression "hot-headed," participants exposed to a higher temperature were more likely to ascribe impulsive crimes than participants exposed to a lower temperature. Consistent with the expression "cold-hearted," participants exposed to a lower temperature were more likely to ascribe premeditated crimes to the criminals as well as crimes resulting in a more severe penalty. While ambient temperature noticeably influenced the judgment of criminals, changes in affective states could not explain the results. In a somewhat similar study (see Nakamura et al. 2014), participants exposed to a cold room temperature were more likely to opt for a utilitarian judgment, possibly because coldness facilitates the decrease of empathetic involvement, rendering participants more "cold-hearted" in their assessment of moral problems.

Similar effects can be found in studies exploring the effects of tactile experiences of physical warmth or cold. In two studies by Williams and Bargh (2008), participants held a hot or a cold cup of coffee for a short time and subsequently rated a target person. Participants holding the hot beverage rated the target person as substantially warmer, but for both groups, the experience did not influence judgments about traits unrelated to warmth and coldness. The second study extended the reach from interpersonal judgment to behavior and showed that such priming also impacts the participants' behavior. Participants primed in the cold condition tended to choose a reward for themselves, while participants in the warm condition preferred the option of rewarding a friend.

Much like "warm" and "cold," "clean" and "pure" designate both moral and physical states, while commonplace expressions like "dirty hands" and "clean conscience" show

how morality is often expressed in terms of cleanliness and purity. The strong association between moral and bodily purity gives rise to interesting effects, such that physical cleansing can actually work as a substitute for moral purification. A couple of intriguing studies (Zhong, Strejcek, and Sivanathan, 2010; Schnall, Benton, and Harvey, 2008; Lobel et al. 2014) show that physical cleansing can compensate for moral impurity and assist restoring a sense of moral integrity. For example, when participants physically cleansed after recalling an unethical act, they reported reduced moral emotions, and their direct compensatory behavior for unethical acts dropped by almost half, indicating that physical cleansing can actually reduce the intensity of emotional states associated with moral wrongdoing. This is consistent with the finding that individuals make larger donations to charity prior to, rather than after, bathing for religious purification.

Also, a sense of cleanliness improves one's moral self-assessment, which, in turn, results in stricter judgments on matters that have potential moral implications (Zhong, Strejcek, and Sivanathan 2010). But the mere perception of physical cleanliness in the environment also appears to have an effect on behavior. In a study using an anonymous trust game, participants could exploit the sender by keeping an entire sum or honor the trust by returning some of the money. Compared to those in a baseline condition, participants placed in a clean-scented room returned considerably more money, chose to reciprocate trusting behavior, and articulated greater motivation to volunteer for a charitable nonprofit organization and to donate money (Liljenquist et al. 2010). Moreover, consistent with the metaphorical expression "smells fishy," exposure to fishy smells stimulated suspicion in participants and undercut their willingness to engage in cooperation and in trust-based economic exchanges. Also, participants who engaged in a two-investor public goods game and were exposed to fishy smells were less inclined to trust others, to act prosocially, and to contribute to shared resources (Lee and Schwarz 2012; for a cross-cultural study of moral purity, see Lee et al. 2015).

In all, a range of rather unremarkable bodily states influence judgments and behavior. Consistent with the basic idea of CMT, embodied metaphors firmly associate bodily experiences with complex concepts, such that metaphor-specific bottom-up influence runs from sensory perception to cognition and behavior.

CMT and Juridical Legitimacy

CMT and the research examined in this chapter demonstrate that individuals are susceptible to relatively minor influences. The fact that these influences occur in a way that is insensitive to their reflectively endorsed norms gives rise to a number of concerns. One might worry that susceptibility of cognition and behavior to influences undermines certain conceptions of character or perhaps even personal autonomy (see Varga 2018). Another worry concerns the influence on important judgments, for instance, in the juridical arena. If it turns out that the experience of, for example, ambient temperature can influence the judgments of judges, then it is likely that the public confidence in the fairness of judicial procedures will decrease (see Benforado 2009). In the following, the focus is on what this might mean for the perceived legitimacy of legal procedures. Let me explain.

In most democratic Western societies, individuals usually recognize particular legal institutions as possessing a *legitimate* right to exercise authority. Such legitimacy is an important social good that renders social regulation less taxing and gives rise to forms of voluntary compliance that are absolutely crucial for the functioning of the legal system. While compliance is vital for criminal courts to accomplish their societal task, it is less likely to occur if individuals regard laws, procedures, or legal actors as lacking legitimacy (Tyler and Fagan 2008). As Max Weber (1978, p. 37) has famously argued, in contemporary societies, the main source of legitimacy is "the belief in legality, the compliance with enactments which are formally correct and which have been made in the accustomed manner." Weber has proposed a descriptive account of legitimacy that distinguishes between three main sources. Individuals may regard a legal or political institution as legitimate (1) because it has a long tradition or history in that particular society, (2) because of their confidence in those holding power, or (3) because they trust its legality. The key to such legitimacy is not that the participants have agreed to specific terms the law, but that the laws have emerged and are administered according to procedures previously accepted as requirements to establishing laws (Hermann 1983).

Judicial legitimacy thus partly descends from the confidence that judges and other actors appointed to administer the law apply the relevant legal norms to concrete facts in a "rational" and coherent manner, and arrive at decisions in an impartial way that is firmly grounded in law, through processes that are demonstrably rational and fair. In fact, one important factor that distinguishes legal institutions from political ones is that they are perceived as legitimate by their constituents, because their decisions are viewed as being impartial and principled rather than motivated by external considerations and interests. However, such perceptions are changing as higher courts, including the US Supreme Court, are to an increasing extent making politically relevant decisions ("judicialization of politics"), and on important and contentious cases, the opinions of Supreme Court justices can be fairly reliably predicted based on the views of the party of their appointing president. Accordingly, what could potentially undermine the legitimacy of a legal institution is what Weber called the "irrational" administration of the law, which occurs when legal decisions are made under the influence of external evaluative standards, for instance, based on emotionally, ethically, or politically determined responses to the details of a concrete case.[2] Such "irrational" administration might result in weakening the legitimacy of the institution, which may lead individuals to regard these institutions' exercise of power as unjustified and the decisions they produce as not worthy of respect or obedience.

[2] Note that this is a simplified version of Weber's account, which introduces further distinctions, for instance, between "formal" and "substantial" forms of irrationality.

While the legitimacy of legal institutions is a crucial matter, there is currently so little public trust in the fairness of the criminal justice system that some experts diagnose a "modern crisis of legitimacy in American criminal justice" (Fagan and Meares 2008; Hough and Roberts 2005).[3] This crisis has several sources, including concerns about procedural fairness, the stigmatization of mainly minority communities, the relative neglect of resource-weak individuals by legal actors, and skepticism about the consistency of legal responses to criminal acts (Fagan and Meares 2008; Ewick and Silbey 1998).

The available studies do not warrant far-reaching conclusions, in particular as we currently lack studies that examine the robustness of bottom-up influences in legal settings. However, assuming that the experimental findings extend to such situations, we have preliminary reasons for thinking that it might present a challenge to juridical legitimacy by increasing the skepticism about the consistency of legal responses to criminal acts. The effects described in the previous sections could have a decisive impact at several stages in a criminal case, for instance, when the prosecutor charges the suspect with a crime, when the judge makes decisions about bail and pretrial detention, when the jury decides whether the defendant is guilty, and when the defendant is sentenced. Clearly, skepticism about the juridical system based on perceptions of political, ethnical, racial, sexual, and other forms of bias is very different from the kind of influences that this chapter explored. One might argue that instead of the latter increasing the former, they lead to two different forms of skepticism about legal rationality. But if it turns out that judgments of judges and jurors at some or all of these stages are influenced by ambient temperature, by the surface of the chair they sit on, by their feeling of cleanliness, or by the temperature of the beverages they consume, then it will likely fuel several forms of skepticism about the legitimacy, consistency, and fairness of, for instance, legal responses to criminal acts.

These considerations raise a number of important issues that need more careful and sustained study. But if the experimental findings exhibit ecological validity and can be confirmed outside of the lab, then they raise important questions about the permissibility of interventions. An anti-interventionist might argue that the effects are less grave than injustice born of negligence or failure to counterbalance, for example, racial biases, as they operate arbitrarily and "blindly" in the sense that they do not target individuals from certain racial, ethnic, or social groups. Also, the anti-interventionist might argue that given that bidirectional effects are automatic and remain beyond the agent's control, it is unfair to hold people responsible for biased judgments. However, such arguments are not entirely convincing: while agents informed about bottom-up effects will probably still lack willful control over the relevant automatic processes as they occur, one might still argue that they are under the obligation to try to control the circumstances that trigger these processes.

[3] The Harvard IOP Spring 2015 poll shows that 49 percent of "millennials" lack confidence in the fairness of the justice system. Available at: http://iop.harvard.edu/sites/default/files_new/150424_Harvard%20IOP%20Spring%202015%20Report_FINAL_WEB.pdf

But while growing evidence about the effects might warrant taking steps toward some kind of prophylactic intervention, in order to design concrete measures, further knowledge is needed on the circumstances in which the effects are most acute, the prevalence of the effects in specific juridical settings, and the extent to which the effects can be mitigated. Moreover, thinking about possible interventions opens up a host of new questions. For instance, should redesigned guidelines and architecture merely prevent certain bodily experiences, or should they also increase the occurrence of bodily experiences that are favorable for equal consideration? Whatever the answer will be, comprehending the influences of cognitive processes and behavior within the theoretical framework of CMT offers some preliminary insight into how we might go about minimizing its effects.

Conclusion

Following a brief introduction to the main tenets of the embodied cognition research program, this chapter focused on conceptual metaphor theory, which elucidates abstract concepts and metaphors as built on specific domains of bodily experience. The chapter offered an examination of recent behavioral and neuroscientific research that appears to support CMT, exposing a number of surprisingly strong influences on judgment and behavior. The chapter closed by offering some reflections on a particular concern in the realm of law that such findings may lead to.

References

Anderson, M.L. (2010). Neural reuse: a fundamental organizational principle of the brain. *Behavioral and Brain Sciences*, 33, 245–66.

Andres, M, Olivier, E, and Badets, A. (2008). Actions, words, and numbers: a motor contribution to semantic processing? Current direction in psychology. *Science*, 17, 313–17.

Bargh, J.A. (2012). Priming effects replicate just fine, thanks. Available at: https://ST.psychologytoday.com/blog/the-natural-unconscious/201205/priming-effects-replicate-just-fine-thanks

Barsalou, L.W. (2003). Abstraction in perceptual symbol systems. *Philosophical transactions of the Royal Society of London. Series B, Biological sciences*, 358, 1177–87.

Barsalou, L.W. (2010). Grounded cognition: past, present, and future. *Topics in Cognitive Science*, 2, 716–24.

Benforado, A. (2009). The body of the mind: embodied cognition, law, and justice. *St. Louis University Law Journal*, 54, 1185.

Cameron, C.D., Brown-Iannuzzi, J.L., and Payne, B.K. (2012). Sequential priming measures of implicit social cognition: a meta-analysis of associations with behavior and explicit attitudes. *Personality and Social Psychology Review*, 16(4), 330–50.

Casasanto, D. and Boroditsky, L. (2008). Time in the mind: using space to think about time. *Cognition*, 106, 579–93.

Cesario, J. (2014). Priming, replication, and the hardest science. *Perspectives on Psychological Science*, 9(1), 40–8

Clark, A. (1999). An embodied cognitive science? *Trends in Cognitive Sciences*, 3, 345–51.

Clark, A. (2008). *Supersizing the mind: embodiment, action, and cognitive extension*. New York: Oxford University Press.

Clark, A. and Chalmers, D. (1998). The extended mind. *Analysis*, 1998, 58, 10–23.

Dawkins, R. (1976). *The selfish gene*. New York: Oxford University Press.

DeCoster, J. and Claypool, H.M. (2004). A meta-analysis of priming effects on impression formation supporting a general model of informational biases. *Personality and Social Psychology Review*, 8, 2–27.

Dennett, D.C. (1995). *Darwin's dangerous idea: evolution and the meanings of life*. New York: Simon and Schuster.

Dijkstra, K., Kaschak, M.P., and Zwaan, R.A. (2007). Body posture facilitates retrieval of autobiographical memories. *Cognition*, 102, 139–49.

Ewick, P. and Silbey, S.S. (1998). *The common place of law: stories from everyday life*. Chicago: University of Chicago Press.

Fagan, J. and Meares, T.L. (2008). Punishment, deterrence and social control: the paradox of punishment in minority communities. *Ohio State Journal of Criminal Law*, 6, 173.

Fiske, S.T., Cuddy, A., and Glick, P. (2007). Universal dimensions of social cognition: warmth, then competence. *Trends in Cognitive Sciences*, 11, 77–83.

Fodor, J. (1979). *The language of thought*. Cambridge, MA: Harvard University Press.

Fodor, J. (1981). *Representations*. Cambridge, MA: MIT Press.

Foglia, L. and Wilson, R.A. (2013). Embodied cognition. *WIREs Cognitive Science*, 4, 319–25.

Gallese, V. (2005). Embodied simulation: from neurons to phenomenal experience. *Phenomenology and the Cognitive Sciences*, 4, 23–48.

Gallese, V. and Lakoff, G. (2005). The brain's concepts: the role of the sensory-motor system in conceptual knowledge. *Cognitive Neuropsychology*, 22, 455–79.

Gockel, C., Kolb, P.M., and Werth, L. (2014). Murder or not? Cold temperature makes criminals appear to be cold-blooded and warm temperature to be hot-headed. *PLoS ONE*, 9(4), e96231.

Goldin-Meadow, S. (2003). *Hearing gesture: how our hands help us think*. Cambridge, MA: Harvard University Press.

Goldin-Meadow, S. and Wagner. S. (2004). How our hands help us learn. *Trends in Cognitive Sciences*, 9(5), 234–41.

Häfner, M. (2013). When body and mind are talking: interoception moderates embodied cognition. *Experimental Psychology*, 60(4), 255–9.

Haugeland, J. (1995). Mind embodied and embedded. In: L. Haaparanta and S. Heinämaa (eds.), *Mind and cognition: philosophical perspectives on cognitive science and artificial intelligence*. Acta Philosophical Fennica (vol. 58). Helsinki, Finland: Philosophical Society of Finland, pp. 233–67.

Hermann, D.H. (1983). Max Weber and the concept of legitimacy in contemporary jurisprudence. *DePaul Law Review*, 33, 1.

Hough, M. and Roberts, J. (2005). *Understanding public attitudes to criminal justice*. London: McGraw-Hill Education (UK).

Kaspar, K., Krapp, V., and König, P. (2015). Hand washing induces a clean slate effect in moral judgments: a pupillometry and eye-tracking study. *Scientific Reports*, 5, 10471.

Kövecses, Z. (2000). *Metaphor and emotion*. New York: Cambridge University Press.
Lakoff, G. (2012). Explaining embodied cognition results. *Topics in Cognitive Science* 4(4), 773–85.
Lakoff, G. (2014). Mapping the brain's metaphor circuitry: metaphorical thought in everyday reason. *Frontiers in Human Neuroscience*, 8, 958. doi:10.3389/fnhum.2014.00958
Lakoff, G. and Johnson, M. (1980). *Metaphors we live by*. Chicago: University of Chicago Press.
Lakoff, G. and Johnson, M. (1999). *Philosophy in the flesh: the embodied mind and its challenge to Western thought*. New York: Basic Books.
Lee, S.H.S. and Schwarz, N. (2012). Bidirectionality, mediation, and moderation of metaphorical effects: the embodiment of social suspicion and fishy smells. *Journal of Personality and Social Psychology*, 103, 737–49.
Lee, S.H.S., Tang, H., Wan, J., Mai, X., and Liu, C. (2015). A cultural look at moral purity: wiping the face clean. *Frontiers in Psychology*, 6, 577. doi:10.3389/fpsyg.2015.00577
Liljenquist, K., Zhong, C.B., and Galinsky, A.D. (2010). The smell of virtue: clean scents promote reciprocity and charity. *Psychological Science*, 21(3), 381–3.
Lobel, T.E., Cohen, A., Kalay Shahin, L., Malov, S., Golan, H., and Busnach, S. (2014). Being clean and acting dirty: the paradoxical effect of self-cleansing. *Ethics & Behavior*. doi:10.1080/10508422.2014.931230
Martin, A. (2007). The representation of object concepts in the brain. *Annual Review of Psychology*, 58, 25–45.
Meier, B.P., Schnall, S. Schwarz, N., and Bargh J.A. (2012) Embodiment in social psychology. *Topics in Cognitive Science* 4(4), 705–16.
Nakamura, H., Ito, H., Honma, H., Mori, T., and Kawaguchi, J. (2014). Cold-hearted or cool-headed: physical coldness promotes utilitarian moral judgment. *Frontiers in Psychology*, 5, 1086.
Niedenthal, P.M., Barsalou, L.W., Winkielman, P., Krauth-Gruber, S., and Ric, F. (2005). Embodiment in attitudes, social perception, and emotion. *Personality and Social Psychology Review*, 9, 184–211.
Pulvermüller, F. (2005). Brain mechanisms linking language and action. *Nature Reviews Neuroscience*, 6, 576–82.
Rizzolatti, G. and Craighero, L. (2004). The mirror-neuron system. *Annual Review of Neuroscience*, 27, 169–92.
Schnall, S., Benton, J., and Harvey, S. (2008). With a clean conscience: cleanliness reduces the severity of moral judgments. *Psychological Science*, 19, 1219–22.
Shapiro, L. (2012). *Embodied cognition*. New York: Routledge.
Shapiro, L. (2004). *The mind incarnate*. Cambridge, MA: MIT Press.
Tyler, T.R. and Fagan, J. (2008). Legitimacy and cooperation: why do people help the police fight crime in their communities. *Ohio State Journal of Criminal Law*, 6, 231.
Varga, S. (2018). Embodied concepts and mental health. *Journal of Medicine and Philosophy*, 43(2), 241–60.
Weber, M. (1978) *Economy and society: an outline of interpretive sociology* (2 vols.; ed. G. Roth and C. Wittich). Berkeley: University of California Press.
Wheeler, M. (2005). *Reconstructing the cognitive world: the next step*. Cambridge, MA: MIT Press.

Williams, L.E. and Bargh, J.A. (2008). Experiencing physical warmth promotes interpersonal warmth. *Science*, 322, 606–7.

Williams, L.E., Huang, J.Y., and Bargh, J.A. (2009). The scaffolded mind: higher mental processes are grounded in early experience of the physical world. *European Journal of Social Psychology*, 39, 1257–67.

Wilson M. (2001). The case for sensorimotor coding in working memory. *Psychonomic Bulletin & Review*, 9, 49–57.

Wilson M. (2002). Six views of embodied cognition. *Psychonomic Bulletin & Review*, 9, 625–36.

Zhong, C.B., Strejcek, B., and Sivanathan, N. (2010). A clean self can render harsh moral judgment. *Journal of Experimental Social Psychology*, 46, 859–62.

CHAPTER 47

4E COGNITION AND THE HUMANITIES

AMY COOK

INTERPLAY

THE arts and humanities move us and change our minds—not just metaphorically but literally. Making sense of how we are moved and changed and what it takes to do this to another person requires a convergence of methods, evidence, lenses, and insights. In the last 20 years or so, scholars in the arts and humanities have integrated findings from the cognitive sciences into their research. Some of the early work feels disembodied now—its interest being the impact of literature, film, or theater on the brain. Yet a growing group of scholars are thinking through the implications of embodied, embedded, extended, and enacted cognition on critical questions within the humanities. How does art function as a tool? Is there a difference between the impact of a book read in bed versus one read on the subway? Does it matter how we hold the book we are reading—or "turn" the pages? How does the ecosystem of a theater produce moving experiences? What does it mean to pretend to be someone else? Why come together with conspecifics to watch others make up stories and pretend? What is the impact of misreadings on our experience of literature? How are experimental works of fiction or theater reflecting and shaping new ways of understanding cognition? Interdisciplinary scholarship is opening up new questions, new approaches, and new answers to challenges all of us face.

Speaking across disciplines means knowing your audience; translating the right words in the right rooms. Speaking in a room of theater scholars, I begin with articulating the ways in which embodied cognition upends traditionally understood conceptions of cognitivism. I give examples of how a particular cognitive event is embodied, embedded, extended, and enacted. Situated within this handbook, I know only that I do not need to explain 4E cognition or cognitivism; this is not the same, however, as believing the meanings of those four lovely E's are stable and clear. There are real and critical questions to the field presented by differences and conflicts between the E's or even various interpretations of each E. Here, however, I wish to speak about the value of the research within the cognitive sciences to the arts and humanities. I will set aside the relationship between enactivism and extended mind theories or whether the body/mind contains anything that might be thought of as representations. The growing consensus within the cognitive sciences is that thinking is not computing in the brain but action with the body in the world. While the details are crucial, the research thus far seems to support at least three radical claims for work in the arts and humanities:

1. reading is not just about "meaning"
2. the staging of the audience is critical to theatrical experience
3. engaging with art is tool use

These are not small changes to the general received wisdom about literature, theater, and art. If thinking is "world-making," rather than processing stimuli into meaning, then the hermeneutic tradition of literary and art scholarship must adapt; we do not read to attain meaning; we read to enact worlds within which to experience anew. If cognition is embedded in a given environment, always affective, and extended outside of skin and skull, then a spectator's experience, emotion, and attention depends less on what is happening "inside" the actors onstage—as the acting style of Constantin Stanislavsky and others and the playwriting of Tennessee Williams and others seem to suggest—and more on a staged, enacted experience of their own. If thinking means using objects in our environment in order to make changes to our own and extended ecosystem, then an interaction with a work of art can be aesthetic, poetic, and autopoietic.

Interdisciplinary scholars are taking up the challenge posed by 4E cognition and working to spread the cry for new research, new thinking, and new models of training. I will provide a brief history of the work at the intersection of the fields, focusing particular attention on work that most deeply engages with embodied cognition. This reads, I'm afraid, like a list of "meanings" gleaned from a string of influential scholars; the experience I hope this provides is one of respect for how work in the humanities may not be sequential as it can be in the sciences but rather cumulative. We are also seeing that artists are generating work that calls attention to the embodied, embedded, extended, and enacted cognition of our engagement. These works create new models for non-scientists to think about thinking in a new way. This work, both artifact and evidence of cognition, puts pressure on the sciences to ask and answer new questions.

My own work started with a desire to understand how complicated poetry affected an audience. I wanted to know, for instance, why an audience sits forward at a particular piece of poetry. Why does everyone remember some lines ("hold the mirror up to nature") but not others ("'twas caviar to the general")? Cognitive linguistics, with its insistence on embodiment and scales of evidence similar to literary studies, has been a powerful tool for these kinds of inquiries. Using conceptual integration and metaphor theory, I unravel the "mirror held up to nature" at the center of Shakespeare's play and argue that what Hamlet depicts is a conceptual blend of three different ideas of "mirror" historically available at the time and that this mirror conception goes on to structure his attempts to understand the relationship between seeing and knowing and cause and effect.[1] Through examining *how* audiences have understood this creative image, rather than what they have understood, I found a web of evoked sources (convex mirror, political tracts, the use of glass in scientific instruments, and the small flat glass mirrors newly available from Italy) tied together. Understanding "the purpose of playing" evokes a number of tools for vision that do different things.

This is only helpful to see, of course, if seeing it allows us to ask new questions about the poetry or to perceive new angles in the performance. It does not prove that Shakespeare "anticipated" current theories in cognitive science and it does not prove that these theories are accurate because we see evidence in Shakespeare. Cognitive linguists have received many calls to find a way to empirically verify their theories and the status of some of the work that has been drawn on in theater and performance studies is unstable. This caveat must be reiterated and care must be taken at the start of any project like this: new research is happening every day, challenging and stabilizing interpretations and assumptions of the past. While I cannot use Shakespeare's poetry to prove a question in cognitive linguistics and cognitive linguistics cannot prove the value of Shakespeare, integrating the two enriches both, and ignoring the knowledge and research across the disciplines imperils the work in our own. My means are interdisciplinary but my goals are disciplinary.

Cognitive linguistics has proved to be tremendously influential in analyses of Shakespeare and other classical texts. Donald Freeman (1995) views *Macbeth* as being tightly constructed around the image-schemata of PATH (e.g., "LIFE IS A JOURNEY") and CONTAINER (e.g., the castle, Duncan's body). This moves from a smart close reading to an influential argument about the impact of poetry, however, when he notes that the metaphoric structure of the play then becomes the metaphoric structure for the critics who write about the play; the play's linguistic scaffolding becomes contagious.[2] Eve Sweetser (2006) finds the rhyming in *Cyrano de Bergerac* a critical part of the verbal jousting: the winner is the one able to "control the relations between meaning and poetic form."[3] Mary Thomas Crane's *Shakespeare's Brain* (2001) examines how the language of the early modern period reflects and illuminates the brain that created the work. By

[1] Amy Cook, *Shakespearean Neuroplay: Reinvigorating the Study of Dramatic Texts and Performance through Cognitive Science* (New York: Palgrave Macmillan, 2010). Conceptual integration, or blending, is from Gilles Fauconnier and Mark Turner, *The Way We Think: Conceptual Blending and the Mind's Hidden Complexities* (New York: Basic Books, 2002).

[2] Freeman, "'Catch[ing] the Nearest Way': *Macbeth* and Cognitive Metaphor," *Journal of Pragmatics* 24 (1995), 689–708.

[3] Sweetser, "Whose Rhyme Is Whose Reason? Sound and Sense in *Cyrano De Bergerac*," in *Language and Literature* (London: Sage, 2006), p. 30. Reuven Tsur has a similar, though slightly conflicting theory of rhythm in poetry; see, for example, Tsur, "Delivery Style and Listener Response in the Rhythmical Performance of Shakespeare's Sonnets," *College Literature*, 33(1), 170–96; *Cognitive Shakespeare: Criticism and Theory in the Age of Neuroscience* (Winter, 2006).

focusing on how Shakespeare's famous wordplay relates to a shifting subjectivity, Crane finds a feedback between brain, performance, language, and culture:

> It seems possible that the process of creating fictional characters to exist in a three-dimensional stage space brought out the spatial structures of language to an unusual degree. Perhaps it is enough to say that these effects "emerge" through Shakespeare's almost uniquely rich use of language. Shakespeare (i.e., Shakespeare's language-processing functions) causes us to notice these connections—which in turn reveal information about the culture and also about the organizational tendencies of the brain.[4]

Here she shares important contributions to "neural historicism" with Alan Richardson (Crane and Richardson, 1999; Richardson and Spolsky, 2004).[5] For those of us writing about text, it is important to see our work in the context of (at this point) a fairly long tradition.

If the terrain is cognitive scientific approaches to the arts and humanities, the work is vast and varied. Though they have different takes on "cognitive poetics," Peter Stockwell and Reuven Tsur have embraced a more embodied approach to thinking about reading and poetry. Tsur (2012) has importantly integrated the physical impact of rhythm in poetry, but he generally isolates the experience of poetry as fundamentally different from everyday language use.[6] Stockwell (2013) seeks to generate a framework for the "analysis of ethics as an interaction between readerly disposition and textual imposition, to produce a sense of a 'positioned reader' of a literary work."[7] His "position" here is not embedded, extended, or enacted; however, it is more a representational positioning of perspective taking conducted by the story telling. Lisa Zunshine's (2008) work on literature and theory of mind,[8] Frederick Luis Aldama's (2010) approach to narrative theory,[9] and Paul Armstrong's (2013) phenomenological examination of the work of literature on the brain,[10] all think about the readers' *mental* experience almost completely separated from the body. Although they probably would not agree that their work is "disembodied" (nor are they likely to agree that they even belong in the same category as

[4] Mary Thomas Crane, *Shakespeare's Brain: Reading with Cognitive Theory* (Princeton, NJ: Princeton University Press, 2001), pp. 24–5.

[5] See, for example, Mary Crane and Alan Richardson's "Literary Studies and Cognitive Science: Toward a New Interdisciplinarity," *Mosaic: A Journal for the Interdisciplinary Study of Literature* 32(2) (June 1999), 123–40, as well as *The Work of Fiction: Cognition, Culture, and Complexity*, ed. Alan Richardson and Ellen Spolsky (Hampshire, UK: Ashgate, 2004).

[6] Reuven Tsur, *Poetic Rhythm: An Empirical Study in Cognitive Poetics* (2nd ed.). (Eastbourne, UK: Sussex Academic Press, 2012).

[7] Peter Stockwell, "The Positioned Reader," *Language and Literature*, 22(3) (2013), 263.

[8] Lisa Zunshine, *Strange Concepts and the Stories They Make Possible* (Baltimore: Johns Hopkins University Press, 2008).

[9] Frederick Luis Aldama, *Toward a Cognitive Theory of Narrative Acts* (Austin: University of Texas Press, 2010).

[10] Paul B. Armstrong, *How Literature Plays with the Brain: The Neuroscience of Reading and Art* (Baltimore: Johns Hopkins University Press, 2013).

each other), their focus is on reading as an internal mental activity and thus seemingly independent of the body that holds the book. Although the terrain is shifting to stronger claims of embodiment, those of us pushing for a more thorough integration of 4E cognition in the arts and humanities often find ourselves querying work that speaks of "embodied cognition" as if it were a special *kind* of cognition and "body/mind" as if the connection were a special union and not a clumsy linguistic attempt to right Cartesian wrongs.

Shaun Gallagher (2015) calls this the work of the "body snatchers" and he notes this very challenge, raised by the circulation of the term "embodied cognition," in the humanities. Gallagher rightly points out that there are those who label their work "embodied" but who "devise a version of embodied cognition that leaves the body out of it."[11] Although I have encountered theater and dance scholars who talk about the "wisdom of the body" in a way that seems to leave the *brain* out of it, Gallagher's description is apt. One of the major challenges of interdisciplinary research is clarifying terminology, which is particularly challenging when the terminology is not set or stable. Freeman's and Crane's scholarship, for example, is persuasive but qualifies as "theories that discount or eliminate the role of the body or organism-environment per se in cognition" (Gallagher, 2015, p. 101). They are discussing the processes at work once the information/text is somehow *inside* the reader; there is no discussion of the hand that holds the book, the ecology that contains the hand.

The humanities have historically been interested in hermeneutics: what does this poem, this play, this historical event, this painting, *mean*. We discover this *meaning* through a *reading*. The metaphor of reading presumes a linearity of information intake and meaning output: decoding. The reader's eye, disembodied, remains chained to the string of words, and they go into the processing system in the brain and generate knowledge. We know through studies of eye saccades that readers go backward as well as forward through the words to accrete meaning, just as listeners must recall and anticipate as they hear the present word in the sentence string. Still, the metaphor of reading as meaning-making highlights key elements of the experience of reading or listening while hiding others. Havas and colleagues (2007) asked people to hold a pencil in their mouths, either with only their teeth or only their lips, and then to answer questions about emotionally positive or negative events. Though this awkward posturing of subjects' faces into smiles or frowns was entirely artificial, and independent of endogenous emotional state, subjects responded more quickly to sentences relating emotions congruent with their expression.[12] The state of the body is not only an input into language interpretation, it is also an output. When we read "open the drawer" we are much quicker to perform a movement moving our hands toward our bodies than away, for

[11] Shaun Gallagher, "Invasion of the Body Snatchers: How Embodied Cognition Is Being Disembodied." *Philosophers' Magazine*, April (2015), 97.

[12] David A. Havas, Arthur M. Glenberg, and Mike Rinck, "Emotion Simulation during Language Comprehension." *Psychonomic Bulletin & Review* 14(3) (2007), 436–41, cited by Arthur M. Glenberg, "Embodiment as a Unifying Perspective for Psychology." *WIREs: Cognitive Science* 1 (2010), 586–96.

example, suggesting that the comprehension of the sentence accessed the motor cortex sufficiently to prime one physical action (movement toward) rather than another (movement away) (Bergen 2012).[13] Others have extended this kind of result to show that the hand muscles are primed even by sentences that describe metaphorical exchanges ("You delegate the responsibilities to Anna.") (Glenberg et al. 2008).[14] We cannot rely on readings that disembody language and talk about meaning as a kind of semiotic code. We require a new look at the literature that moves us.

There are exciting examples of scholars demonstrating the potential of this interdisciplinary turn. A recent special of the journal *Style* focused on the "second generation" of cognitive literary approaches "foregrounding the embodiment of mental processes and their extension into the world through material artifacts and sociocultural practices" (Kukkonen and Caracciolo 2014, p. 261).[15] Barbara Dancygier, a literary scholar and cognitive linguist, focuses on the multimodality of language and argues that complicated literary stories can "stretch the cognitive abilities of 'making sense' to their limits" (Dancygier 2012, p. 11).[16] In an essay, Dancygier turns to the multimodality of stage language, arguing that props can work as dramatic anchors, holding parts of the story or meaning in their presence:

> Dramatic anchors have a dual function, and they are both material anchors (though only to the ongoing conceptualization of the play's meaning) and narrative anchors, as they contribute to the understanding of the story. They do materially participate in the performance, but they also guide the viewer in constructing the meaning of the play. (Dancygier, 2016, p. 32)[17]

Caesar's mantle, held up by Marc Antony at the funeral oration ("I come to bury Caesar not to praise him"), stands in for the victories of Caesar and Rome's glorious past. Antony then points out holes of the murderers' daggers that now mark the mantle and uses it as a "material metonymic device" through which Antony places the murder in front of the eyes of the mourners/spectators while he describes the cruelty and betrayal Caesar must have felt. Of course, Antony's whole speech could be taken as a challenge to literal meaning theory or cognitivism: the meaning of what he is saying is nothing; how

[13] Benjamin Bergen, *Louder Than Words: The New Science about How the Mind Makes Meaning* (New York: Basic, 2006), pp. 79–80.

[14] Arthur M. Glenberg, M. Sato, L. Cattaneo, L. Riggio, D. Palumbo, et al. "Processing Abstract Language Modulates Motor System Activity." *Quarterly Journal of Experimental Psychology*, 61(6) (2008), 905–19.

[15] Karin Kukkonen and Marco Caracciolo, "Introduction: What Is the 'Second Generation'?" *Style* 48(3) (2014), 261–74. See also Daniel Irving, "Presence, Kinesic Description, and Literary Reading" *CounterText* 2.3 (2016), 322–37. Irving argues that novels in which nothing happens generate an experience of "presence."

[16] Barbara Dancygier, *The Language of Stories: A Cognitive Approach* (Cambridge: Cambridge University Press, 2012).

[17] B. Dancygier, "Multimodality and Theatre: Material Objects, Bodies and Language." In: *Theatre, Performance and Cognition: Languages, Bodies and Ecologies* (London: Methuen, 2016), pp. 21–39.

he is saying it initiates a riot. The murder becomes present for and through the bodies of the spectators as Antony holds up the bloody cloak and the funeral oration inspires the organism of the crowd to want to make change, to get revenge. The bloody cloak, lying at his feet, inspires Antony to perform with it. It affords a response while it anchors the murder.

As the actors and spectators work with the props provided by the playwright, humans interact with the objects and technology in their world to see and think differently. Ellen Spolsky has made extraordinary contributions to the field, from *Gaps in Nature* (1993) to *Contracts of Fiction* (2015). Although her work occasionally snatches or eclipses the body, it is often focused on the interaction between art and the individual. In her 2015 book, Spolsky maintains that it is through the "impurities of fiction"—and of art, in general—that we work through difficult social/cognitive problems. Spolsky examines the history of religious art, architecture, and sacred relics, insisting that it "has for centuries been an 'external technology' expanding the cognitive power of an individual or a community as a whole by expanding the communicative reach of the church."[18] The church works with and through its art. While many scientists and philosophers have made claims about the value of literature and art, Spolsky's historically engaged and deeply humanistic understanding of these works of art—from a Shakespeare play to a painting by Manet or the Stonyhurst Salt Peter—suggests that the works of art are not (just) better understood through reference to cognitive science, but are in fact sources of evidence for the cognitive sciences. This does not mean that a painting can be evidence in and of itself, though; only that a conversation across the disciplines might lead to a controlled experiment wherein the artifacts that remain are examined systematically for the performances they invite, entail, and require of the humans for whom they are made.

Matt Hayler (2016) argues that humans don't just use art as part of their thinking but are rather "mediated and co-constituted by our engagement with material artefacts."[19] He posits a visual grammar that controls the sense we make out of what we see and that can be expanded or shifted, depending on what and how we see: "I'm interested, here, in those interactions with artefacts that alter our abilities by manipulating the grammars through which we meaningfully conceive of our potential for action" (p. 162). His book, which begins with an investigation into the impact of e-readers and moves toward a theory of technology and human interaction, demonstrates that "the expert use of artefacts is intimately a part of what it means to be human, that we use equipment in order to apprehend the world, to define our place within it, as extensions of our bodies and the cognition they are entwined with."[20] Hayler, a scholar at the intersection

[18] Ellen Spolsky, *Contracts of Fiction: Cognition, Culture, Community* (Oxford: Oxford University Press, 2015), p. 100.

[19] Matt Hayler, "Another Way of Looking: Reflexive Technologies and How They Change the World." In: *Theatre, Performance and Cognition: Languages, Bodies and Ecologies* (London: Methuen, 2016), pp. 159–73.

[20] M Hayler, *Challenging the Phenomena of Technology: Embodiment, Expertise, and Evolved Knowledge* (London: Palgrave, 2015), p. 68.

of digital humanities, philosophy, English, and cognitive science, exemplifies the power and influence that the move toward 4E cognition has had on the humanities: it cannot be isolated.

When most cognitive scientists stopped thinking of the brain as a computer, it became easier for artists to think about science. The deployment of cognitive science in theater and performance studies has potential because they are not very far apart. The trick is to tie them together in the right places in the right ways. In the last 15 years, theater and performance scholars have been asking and answering many great questions given the research within cognitive science. How might cognitive science enable us to rethink questions in theater history?[21] What are the different kinds of questions we might ask about audience perception and the experience of going to the theater?[22] How might we reimagine the situation of the performer/environment interaction?[23] In what ways might theater and performance provide new ways of treating people with autism[24] or memory loss[25] or Parkinson's?[26] How can it improve our methods of actor training?[27]

Rhonda Blair insists traditional methods of American acting training—based mostly on Constantin Stanislavsky—already presume an embodied mind. They "were about facilitating the actor's dynamic engagement with her environment through working with and through the body—the senses—as the source of imagination, in a way that acknowledged the actor's experience as being simultaneously, inseparably imaginary and real."[28] She calls on actors to attend to the way richly specific images and a deep

[21] Bruce McConachie, *American Theater in the Culture of the Cold War* (Iowa City: University of Iowa Press, 2003); Tobin Nellhaus, *Theatre, Communication, Critical Realism (What Is Theatre?)* (London: Palgrave Macmillan, 2010); Marla Carlson, *Performing Bodies in Pain* (London: Palgrave Macmillan, 2010); Jill Stevenson, *Performance, Cognitive Theory, and Devotional Culture: Sensual Piety in Late Medieval York*, (London: Palgrave Macmillan, 2010); Evelyn B. Tribble, *Cognition in the Globe: Attention and Memory in Shakespeare's Theatre* (New York: Palgrave Macmillan, 2011); Naomi Rokotnitz, *Trusting Performance: A Cognitive Approach to Embodiment in Drama*, (London: Palgrave Macmillan, 2011).

[22] Bruce McConachie and F. Elizabeth Hart, *Performance and Cognition: Performance Studies and the Cognitive Turn* (London: Routledge, 2007); McConachie, *Engaging Audiences: A Cognitive Approach to Spectating in the Theatre* (New York, Basingstoke, and London: Palgrave Macmillan, 2008); Amy Cook, *Shakespearean Neuroplay: Reinvigorating the Study of Dramatic Texts and Performance through Cognitive Science* (London: Palgrave Macmillan, 2010); Naomi Rokotnitz, *Trusting Performance: A Cognitive Approach to Embodiment in Drama* (London: Palgrave Macmillan, 2011).

[23] Teemu Paavolainen, *Theatre/Ecology/Cognition: Theorizing Performer-Object Interaction in Grotowski, Kantor, and Meyerhold* (London: Palgrave Macmillan, 2012).

[24] Nicola Shaughnessy, "Imagining Otherwise: Autism, Neuroaesthetics and Contemporary Performance." *Interdisciplinary Science Reviews*, 38(4) (2013), 321–34.

[25] Helga Noice and T. Noice, "What Studies of Actors and Acting Can Tell Us About Memory and Cognitive Functioning." *Current Directions in Psychological Science* 15(1) (2006), 14–18.

[26] Nicola Modugno, Sara Iaconelli, Mariagrazia Fiorilli, Francesco Lena, Imogen Kusch, and Giovanni Mirabella, "Active Theater as a Complementary Therapy for Parkinson's Disease Rehabilitation: A Pilot Study." *The Scientific World* 10 (2010), 2301–13.

[27] Rhonda Blair, *The Actor, Image, and Action* (New York: Routledge, 2008); John Lutterbie, *Toward a General Theory of Acting: Cognitive Science and Performance* (New York: Palgrave Macmillan, 2011); Rick Kemp, *Embodied Acting: What Neuroscience Tells Us About Performance* (London: Routledge, 2012).

[28] Blair, *The Actor, Image, and Action* (New York: Routledge, 2008), p. 45.

investment in the given circumstances shift their reactions in and to the scene. In his *Toward a General Theory of Acting*, John Lutterbie sees the work of the actor through the lens of dynamic systems theory (DST); despite many different methods of creating a performance, the art of acting is about being present and responsive to perturbations in a system. In his discussion of Katie Mitchell's *The Waves* at The National Theatre, he presents a process and a product that serves as a kind of staging of DST. The production uses video and live action to take apart, tell, and show the story of Virginia Woolf's *The Waves*. The meaning and emotional impact of the play builds: "By the end, the feelings experienced are quite powerful and unexpected. The emotions were always present, but there was no time to appreciate them fully. They are implicit in the action, but it is only through their accretion over time that the full impact is felt."[29] For Lutterbie, the active process of constructing the story and staging that construction as well as the story, made the performance a kind of demonstration of boundary conditions, perturbations, and phase synchrony.

Imagining Shakespeare's Globe as an environmental tool used by the players to facilitate the heavy cognitive task of performing five to six different plays per week, Evelyn Tribble finds that the players and environment of the Globe can be thought of as a system that creates and perpetuates cognition. Her book *Cognition in the Globe: Attention and Memory in Shakespeare's Theatre*, is an excellent example of the potential for cross-disciplinary work: she applies the paradigm shift coming out of the sciences about how cognition works to one of history's most amazing cognitive events—Shakespeare's plays—and weaves a story impossible without both fields lending their best insight to the mystery at hand.[30] Seeking to answer the question of how playwrights and players came together and, in such a short time and night after night, put on a play for a demanding audience, Tribble looks for an answer not in the individuals, but in the system. As Tribble importantly points out, "distributed" does not mean parceled out; it means spread over. If the cognitive environment of a theater includes its constitutive elements—the audience, the props, the plots backstage, the verse structure of the poetry, the conventions of staging, the girl selling oranges to the groundlings—it is not just individual cognition that needs to be questioned but our investment in the idea of individuals.

Writing with cognitive scientist John Sutton, Tribble thinks about performances as cognitive ecologies:

> Cognitive ecologies are the multidimensional contexts in which we remember, feel, think, sense, communicate, imagine, and act, often collaboratively, on the fly, and in rich ongoing interaction with our environments. . . . The idea is not that the isolated, unsullied individual first provides us with the gold standard for a cognitive agent, and that mind is then projected outward into the ecological system: but that from

[29] J. Lutterbie, *Toward a General Theory of Acting* (New York: Palgrave Macmillan, 2011), p. 287.

[30] Evelyn B. Tribble, *Cognition in the Globe: Attention and Memory in Shakespeare's Theatre* (New York: Palgrave Macmillan, 2011).

> the start (historically and developmentally) remembering, attending, intending, and acting are distributed, co-constructed, system-level activities. (Tribble and Sutton, 2011, pp. 94, 96)[31]

A focus on cognitive ecologies stops extracting individual "thinking" agents as figures from the "ground" within which they stand and work; cognitive ecologies examine the situated and distributed system of cognition. The "ecological approach to performance and cognition" is gaining influence and importance. Teemu Paavolainen (2012) suggests that:

> the "ecology" of a stage performance involves Gibsonian "affordances" well beyond such fixed typologies of theatrical objects as props, scenery, and costume: just as the objects on stage always enable and constrain forms of action available for the performers, the interplay of actors and objects will also enable and constrain the range of interpretations the audience is liable to come up with. (p. 19)[32]

Paavolainen's ecology of the stage performance aims at extending the interpretive field but does not decenter the importance of interpretation. The audience is still *out* there. I want to point to contemporary performance ecologies wherein one is brought to feel *in* a cognitive system.

Work in Action

A growing number of theatrical experiences are not validating an understanding of cognition as some internal processing of information—secret goes in, gets computed, evokes memories from filmic Rolodex, causes emotions to come up from subconscious, evokes emotions that alter the output. Theater is staging embodied cognition. I do not just mean that we can see evidence of embodied cognition in the theater—as the prop table organizes what the actor needs between scenes so he can offload the cognitive task in what David Kirsh and Paul Maglio call "epistemic action"—I mean that the theater is providing an experience that enables spectators to shift their conception of cognition. There are many theatrical experiences giving us a new pair of "calipers" to interact with our world. Alva Noë (2012) offers this term as a way to suggest the work done by conceptual categories: "Don't think of a concept as a label you can slap on a thing; think of it as a pair of calipers with which you can pick the thing up. If there is a difference between seeing something and thinking about it, it is because of differences in our calipers"

[31] Evelyn Tribble and John Sutton, "Cognitive Ecology as a Framework for Shakespearean Studies." *Shakespeare Studies* 39 (2011), 94–103.

[32] Teemu Paavolainen, *Theatre/Ecology/Cognition: Theorizing Performer-Object Interaction in Grotowski, Kantor, and Meyerhold* (London: Palgrave Macmillan, 2012), p. 19.

(p. 36).[33] The arts and humanities are always more than manifestations of theories of cognition and the self; they provide the calipers necessary for the world to show up differently for us. We can see this in how the theater stages the audience to make use of a strange new tool.

Human artifacts afford and invite certain engagements and dissuade or resist others. This is as true for Heidegger's hammer as it is for an epic poem, a novel, a realist drama, or an immersive theater piece. Each creation presumes the figure on the other end of the engagement: one does not tell stories to creatures who cannot envision a "What if?" One doesn't write stories down for a community that does not read. Of course, stories, both in form and content, also shape those for whom they are meant. An artist has a folk theory of how her audience will consume or make sense of her work and the work enacts the subject in its use. Monet's "Water Lilies" moves the viewer to the position necessary to take the large painting in. She must stand back; she must squint. Bringing the flowers into being, the viewer becomes aware of the operation of her visual system—the work necessary to turn color and shape into an image; it wasn't just there, out there in the world for the artist to capture, it had to be actively composed by the viewer. As opposed to much of the art that came before, where the skill of the artist was assessed based on mimetic ability—to capture the flowers as they *really* are—Impressionism foregrounded the perceiving subject. This experience presumes and enacts the modern fascination with self, a self similarly enacted and presumed by the works of Virginia Woolf and Anton Chekhov. What we see as interesting to the modernist artist, through his/her art, is the individual differences in perception. As Marjorie Garber (2008) said of much criticism of *Hamlet*: it "holds the mirror up to nature and finds the critic reflected there" (p. 201).[34] This may challenge critics' claims about the universality of certain works of art—or at least the assumption that the themes or meanings remain the same—but suggests a richer potential in the work the art does within and to a particular observer/participant. Art gives us an opportunity to stage the impossible to comprehend: a shifting conception of who we are and how we operate.

When most people think of theater, they imagine sitting in a dark room and watching actors pretend to be characters acting out a story as if the audience isn't there. Maybe there's a kitchen table, a window with a curtain and light coming through it. The audience may know, when the woman comes through the door at the center of the stage, that she has not just come in from outside but from backstage, but they "willingly suspend disbelief" and pretend she's really carrying groceries and that, when she fights about a family secret with the man who comes on next, that she really is mad, that the man and woman do not know what is going to happen next.[35] Spectators of varying levels of

[33] Alva Noë, *Varieties of Presence* (Cambridge, MA: Harvard University Press, 2012).

[34] Marjorie Garber, *Shakespeare and Modern Culture* (New York: Anchor Books, 2008). See also Terence Hawkes, "Telmah" in Patricia Parker and Geoffrey Hartman (eds.), *Shakespeare & the Question of Theory* (New York: Methuen, 1985), pp. 310–32.

[35] For a critique of the "willing suspension of disbelief," see Richard Gerrig and G. Egidi, "*The Bushwhacked Piano* and the Bushwhacked Reader: The Willing Construction of Disbelief," *Style* 44

fluency in the conventions of theater may very well know what kind of gobo makes the light look like a tree on the floor of the stage or simply that when the man threatens to hurt the woman, they need not run onstage to protect her. Spectators expect to stay quiet and to *read* the story being acted out—to find the *meaning* of the story.

This theatrical experience would fall into what Noë (2012) calls a "distinctively modern way of understanding theater": actors are behind the fourth wall, "they reside in a symbolic space.... We read them, or interpret them, or try to understand the story. We don't witness anything" (p. 5).[36] Because, for Noë, this "modern theater" does not allow the spectators to participate, to be active eyewitnesses, as they are with a sporting event, "Modern theater denies real presence" (p. 5). Presence, for Noë, is the way in which things "show up" for us; it is not strictly meaning or understanding but rather a "skillful engagement" with the world. What he is calling "modern" theater is what most of us imagine when we think about going to the theater: try not to cough, watch what's happening in the symbolic space onstage, look past the falseness of the living room wall or prop gun, read the scene, explore what the meaning of the scene is a symptom of. Perhaps the secret the family is working so hard to keep in—the lost child, the drug addiction, the infidelity—represents the history of racism, or sexism, or the ways the modern family unit squeezes the life out of its individuals. Our job, as audience, is to read the meaning and then probe the illness through the symptomatic behavior of the characters. While I agree with parts of the distinction Noë makes, I want to push against this categorization of modern theater.

Although critics in the humanities are slow to challenge the long tradition of finding meaning, artists in the last 10 years are more and more interested in exploring presence or—though I do not suggest they are conscious of this—experimenting with their art form to stage or enact embodied cognition. I would go further and suggest that what Noë calls the presence-denying "modern theater" was simply staging a kind of cognition that was committed to individuality, internal life, a subconscious, and a Cartesian split between body and mind. When I saw *Long Day's Journey into Night* by Eugene O'Neill in 11th grade, I was moved—though the *meaning* was slow to reveal itself. My teenage self felt that the play provided me with a way of understanding the self as having an internal psychology messed up by one's parents, containing memories that haunt until purged. This genre of theater stages a Freudian self and a Cartesian mind, which was tremendously valuable when this was the dominant scientific paradigm. This theater may have denied "presence," but that's only because so too did the science of the time. More and more, though, I find myself with spectators who are engaged in the environment of the theatrical experience: not asked to make meaning but given opportunities to enact an experience.

(2010), 189–206. See also Amy Cook, "Staging Nothing: *Hamlet* and Cognitive Science," *SubStance*, 35(2) (2006), 83–99.

[36] Alva Noë, *Varieties of Presence* (Cambridge, MA: Harvard University Press).

Jet-lagged and cold, I enter Shakespeare's Globe in London to meet a colleague to stand for what will turn out to be over three hours to watch a production of *As You Like It*. The space is packed and, over the course of the afternoon, there will be rain falling on those of us standing in the open yard in front of the stage, helicopter interruptions, a growing late-summer chill, and at least one crying baby. We are all there to listen to poetry performed by people in costumes. The actors involve us, including us in their story, with a gesture or a look. I move around the yard a bit to shift my perspective of the action. Sometimes I just watch the spectators. They laugh, adjust, respond with "ahhs" to romantic couplings, clap, and cheer. Why are we here? I'm not waiting to find out what happens—it's a comedy so it ends in marriage—and the play makes very little sense from a psychological perspective, so it isn't that I feel some "internal" truth churning. All these people, held together in this wooden O, come together to hear a story told in relatively complicated language. The audience moves and flows, bumping into each other, spreading out, closing in, ebbing and flowing like a murmuration of starlings. The experience is one of embodied, embedded, enacted, and extended cognition. This is not to say that when I attended *Long Day's Journey into Night* that I processed it in a less embodied way than I did *As You Like It*—cognition *is* embodied and one experience cannot be more embodied than another—but rather that the theatrical experience renders salient different conceptions of the self and cognition.

I walk by it. That's the Shoreditch Town Hall *Hotel* so maybe the Shoreditch Town Hall is around here somewhere—that's where the performance is supposed to be, but without a poster out front, I assume I'm in the wrong place. I'm embarrassed by my stupidity now, my reliance on certain conventions and frames that announce a theatrical experience. I go back when I realize how wrong I must be. Inside, they ask me if I'd like to "check in." They are staging me, casting me as a hotel guest, and I try to adjust. Because I am early, they ask if I want to sit at the bar to wait. There are copies of the day's *Evening Standard* at the bar. When I'm called in, I wait on a chair in the lobby. There's the paper again. I decide I should take a look. Buried in the paper is a story about a duchess who has been evicted from the Shoreditch Town Hall Hotel, about the entrepreneur/building investor who has plans to modularize all the rooms, updating it and making a killing through removing any signs of where you are. There's also a video playing a press conference or promotional video of the same man explaining this business opportunity. I am dislocated. Then I am directed to follow the hotel manager. Is *Absent*, the play I have come to see by Tristan Sharps of dreamthinkspeak, starting now? There's no curtain, no program, no seats, and I'm unsettled but alive to the interaction of people and architecture.

Two of us are led back outside and under the main steps into a small empty hotel room, like the ones shown in the video played while we waited. It could be a single dorm room: single bed, small dresser on one side, night stand on the other, large rectangular frames on three walls. The doors close and the frames come to life: two of them show sepia-toned silent videos of the duchess arriving at the hotel, waving at paparazzi. Her clothes are maybe 1950s, she's then at a party . . . it seems like documentary footage. The rectangle above the bed comes to life and it is an actor, on the other side of the wall, the mirror image of the room we are in. She is an older woman, clearly drunk, drinking whiskey from the bottle and packing her suitcase. She lifts lovely outfits from the 50s and 60s up to her as she puts them in. There's a teddy bear she seems to coddle and talk to. She seems to be getting more and more hysterical. She then notices us or starts to look in the rectangle above her bed. Maybe she sees a mirror, maybe a movie of her past, maybe us. The lights go dark.

Someone opens the door and says, "This way." We are in a long hallway. There are doors with little peepholes. One of them has a step stool in front of it to facilitate looking. We are being invited to peep. There are slightly differently sized videos in each hole. In one we see someone looking back at us: the younger woman from the video in the first room. We see the same woman, a younger version of the woman we saw packing, unpacking her suitcase in another room. Then at the end of the long hall we see the woman with her suitcase and whiskey bottle. It's a video projected where the end of the hall should be, but it looks like we are watching her walk down the hall, drunkenly. She's weaving; her lack of balance makes me feel dizzy. I hold on to the wall as I watch her make her way down the hall and (presumably) out of the hotel.

A door opens and we're invited in. We are in a room like the one before, but instead of video screens, there are replicas of the rooms on all sides, like rooms in a dollhouse. We look through glass into a miniature replica of the room we are in. But each has a slightly different scale or perspective. The bed is longer in one and in another it's farther away from us. Again our sense of balance, of proportion is confused and we are made unsure of our own size and shape. I feel almost drugged.

We go through a door and we are in an unfinished part of a construction site. There's a sign that says that the new hotel will be ready in 2025—are we in the future? I realize we've been in the same contemporary room the whole time, the modular capitalist hell, and we've been looking back on time. We have lost a sense of our size and our time and our place. We are more than disoriented or mis-placed: we are suspended. Time is layered in here like the excavated walls that reveal paint, wallpaper, and cement: beautiful, geologic layers of time.

Near the end, after going through different time periods of the same hotel room—and sometimes two time periods at once, with the new barely covering the old, like the room where tiles on the floor only partially obscure the unfinished cement below, with discarded cigarettes, lost pearls, and business cards—we enter the large ballroom we've seen in the video. It's destroyed as in the film—chairs and tables knocked over—but the lights are still on. The plants are wilted and dying, but not thoroughly dead—the way they should be if the room hadn't been touched since the destruction that occurred in the early 60s, the time period depicted in the film. There's still the smell of perfume in the air and there's music playing. Violence has been done but it's incomplete. It is unclear what or when the absent destruction occurred. It's missing. We are coroners, trying to figure out time of death. I want to know what happened in the room and when it happened. This desire is frustrated: the "meaning" of what happened in the room does not follow the narrative frame. The failure to find a story, with a moral or a causal chain of events, leaves me with the "presence" that Noë describes when an artwork opens itself up to you.

Going to the theater is most often about watching the ballet of intention, cause, and effect. Usually visible in the story or characters—Claudius wants to be king, Hamlet causes Polonius's death without intending it, Ophelia's madness and suicide is the effect of Polonius's death—it is present even in postdramatic performances, dance, ritual, and performance art. We watch agents do things, causing things, and we imagine this was intended. This is how we organize the world around us, into intentional agents taking actions, causing effects. This may be a very useful way of understanding the world, but that does not mean it is always accurate. The destruction of the ballroom is not an event and no one person caused it. It does not belong in a narrative world that requires such things and therefore it allows us to see the fallibility of our reliance on causal chains. In 1955, Hanson wrote an essay challenging our assumption that effects have causes and link up in clean causal chains.[37] As Hanson points out: "I hear that John is indisposed and ask 'Why?' . . . But on any ordinary Tuesday morning I would not ask after the cause of John's moderate good health" (p. 309). The attempt to find a cause is motivated by the assessment of something as an effect: "The main reason for referring to the cause of X at all is to explain X. Of course, there are as many causes of X as there are explanations of X" (p. 309). How we decide that there is an event, what the boundaries are, and what the question is will change our cause-and-effect chain. How delightful if life could be understood as a series of billiard balls striking one another and thus I could stop one and avoid the promised end—but it isn't this way. Our systems are dynamic and always in flux and that which counts as an event, or an agent for that matter, is only visible post hoc—which renders it shaky ground on which to carry too much empirical weight. Contemporary theater that calls attention to the staging of the audience manifests new ways of thinking about the self, the group, and cognition.

This kind of theater does not ask spectators to diagnose internal emotional duress or psychological problems, as most twentieth-century theater does. It is not teaching us something, like a moral, or attempting to evoke emotions, like a medieval morality play or a melodrama. It gives us something to do and frustrates our desire to find meaning. We are rewarded for engaging, for patiently awaiting the opening up of presence. We make things up—"she must be exiting the hotel" or "that must be her lipstick"—and when things are not confirmed for us, we must become aware of our compulsion to find agents, causes, effects, narrative. We find ourselves sensing space and time differently and even ourselves. We navigate the experience with others: now two, then six—there's a different tension because we are not carefully placed in protective seats. We are standing, free-floating, and the darkness calls for a kind of renegotiation of our collective agreement to manage individuality, to behave, to blend in. Untethered to my individual seat, I become aware of the collective, of being one, none, all. I leave the performance, and the area that was unfamiliar before is now clear, and yet my way of experiencing it has been made strange.

[37] N.R. Hanson, "Causal Chains," *Mind* 64(255) (1955), 289–311.

Conclusion

The cognitive turn in the humanities has infused literary and theatrical scholarship with the kind of fresh jolt that new historicism or psychoanalysis once did, but there remain skeptics from the sciences who see the work as lacking rigor and from the humanities who view the work as universalizing or positivist. The problem is partially language: despite the rejection of Cartesian dualism in the sciences, for example, our language still separates the body from the mind, emotions from thinking. We have to stop talking about embodied cognition if there is no such thing as disembodied cognition. Research in cognitive science is profoundly reimagining who we are and how we come to be, but if even theater and literature scholars still think of cognitive science as threatening, the sciences are not performing their scholarship sufficiently. Further, now that the brain is no longer a black box or a computer, separated out from what we feel, what we want, what we imagine, those of us in the humanities have expertise in the very areas scientists are studying. Interdiscourse might allow for progress on both sides, particularly if there was an area of research critical to both cognitive science and theater that relied on evidence of the same scalar level. Scientists, I would argue, also do well to attend to the humanities. We need the humanities because we need the language. The paradigm shift that keeps trying to arrive is about recategorizing what we are talking about when we talk about thinking, or meaning, or touching, or feeling, or embodiment, or performance. This requires close reading and creativity and a facility with metaphor. All the money in the sciences cannot do what those of us in the arts and humanities can do: give us new ways to see and stage—and thus to understand and to use and to develop—the implications of embodied, embedded, enacted, and extended minds. Contemporary art, literature, music, and performance may suggest a new language—and a richer perspective on old language—for 4E cognition.

CHAPTER 48

EMBODIED AESTHETICS

BARBARA GAIL MONTERO

Introduction and the Traditional View

Aestheticians have always revered the body. Artworks, such as Michelangelo's *David* or Modigliani's *Red Nude* are lavished with aesthetic honorifics: they can be beautiful, graceful, powerful, effortless, fierce, evocative. *Other* bodies are aesthetically valuable. But what about one's own? Is it possible to derive *aesthetic* pleasure from the experience of one's own body?

Traditionally, aestheticians have said "no." Or rather, aestheticians have excluded a particular kind of bodily experience from the aesthetic realm: experiences of the body via the so-called "lower" or "bodily" senses. These are the senses, according to George Santayana, that "call our attention to some part of our own body, and which make no object so conspicuous to us as the organ in which they arise" (1955, p. 24). Typically, only vision and audition—the "higher" or "intellectual" senses—escape this demotion, as they are seen as the only senses that direct our attention primarily to external objects rather than to the sensations themselves. On the traditional view, as Prall explains it, "experience is genuinely and characteristically aesthetic only as it occurs in transactions with external objects of sense" (1929, pp. 28, 56). Hegel upheld this view, telling us that "art is related only to the two theoretical senses of sight and hearing, while smell, taste and touch remain excluded from the enjoyment of art" (1835/1975, p. 38); similarly, the ancient Greeks, according to Francis Hutcheson, "observe[ed] a peculiar dignity of the senses of seeing and hearing that in their objects we discern the *kalon* [beautiful], which we do not ascribe to the objects of the other senses" (1725/1973, p. 47). Aesthetic experience, according to these thinkers, while sensuous (depending on sense experience), is not supposed to be sensual pleasure, not pleasure in our own bodily sensations. Or, in Santayana's words, it is an experience in which, "the soul . . . is glad to forget its connection with the body" (1955, p. 24).

Although the confluence of great minds often points toward the truth, the traditional exclusion of the bodily senses from the aesthetic realm is in tension with the experience of expert dancers who find intense aesthetic pleasure in the experience of their own bodily movements. How to resolve this tension and reconnect the soul to the body in aesthetic experience is the goal of this chapter. More specifically, in contrast to the traditional view that denigrates the bodily even while elevating the body, I aim to make sense of dancers' embodied aesthetic experience of their own movements, as well as audience members' embodied aesthetic experience of watching dance.[1]

The Dancer's Experience

Traditionally, aestheticians have rejected what I shall refer to as the thesis of embodied aesthetics:

Embodied aesthetics: one can have an aesthetic experience of one's own body as perceived through senses other than vision and audition.

Dancers, however, implicitly accept this thesis. Ask a dancer why he changed a certain movement to make it cover less space, or decided to bend his wrist just so, and the answer will sometimes be, "I can feel that this particular way of movement is better than the other way," or, more simply, "It feels better that way." And the means by which the dancer feels this is proprioception, our nonvisual sense of where our limbs are in space via receptors in the joints, skin, muscles, and tendons. It is via proprioception that you can know in the dark that your arms are raised, or that your knees are bent. And when dancers are talking about "feeling" one movement as better than another, they mean that they can proprioceptively feel this, that is, that via proprioception, they are aware of their dancing as exciting, or graceful, or brilliant, or any other number of aesthetic qualities that bodily movements can manifest. Not only is this a common sentiment among dancers, but, I submit, it is a universally accepted one, and I take it as my central piece of data that proprioceptive aesthetic experiences occur.

Thomas Aquinas, a proponent of the traditional view, pointed out that "we do not speak of beautiful tastes and beautiful odors" (1485/1960, p. 27). But even if one thinks that this is a stumbling block for a gustatory or olfactory aesthetics, it is not an impediment to a proprioceptive aesthetics since it is natural, at least for dancers, to talk of experiencing beauty and other aesthetic qualities proprioceptively. A dancer, during a rehearsal onstage—a situation in which there are no mirrors from which to glean visual feedback—may claim that a certain movement or position is beautiful or, since dancers tend to be a self-critical lot, complain that the beauty, or whatever other aesthetic

[1] Though the tradition goes against me, I am certainly not alone in breaking it. See, for example, Richard Shusterman's (2008, 2012) work on what he refers to as "somaesthetics," a practice that encompasses both theorizing about the body and engaging in various bodily practices that aim to heighten our first-person awareness of our bodies (see Montero 2015 for a review); Carolyn Korsmeyer's (1999) defense of the aesthetics of gustatory taste; Dominic Lopes's (2002) defense of the aesthetics of touch; and Barry Smith's (2015) argument against the higher/lower hierarchy of the senses.

quality he or she is aiming at, is lacking: "The movement is too abrupt," "The line is ugly," "I'm not feeling the connections," are all phrases that roll naturally off a dancer's tongue.

How do I know that dancers claim to experience aesthetic properties of their movements via proprioception? Before entering academia, I spent a number of years as a professional ballet dancer, and it is something I remember. It is also something I still experience since, after a long hiatus, I have been performing again. Furthermore, I have heard this sentiment expressed by many other dancers: in the rehearsal room a valid complaint to a choreographer or rehearsal director is, "It doesn't feel right." This is a valid complaint for a dancer to make not only because a movement that feels right often looks right, but also because dancers deserve to experience the grace, power, precision, or humor (or whatever aesthetic quality the movement aims to capture) of their own dancing; it makes all the pain worth it. It is not an inviolable prerogative—some movements might never feel right, but might be called for since they look right nonetheless. However, many choreographers, in creating a work, consider dancers' opinions about how a movement feels.

To say that dancers make claims that suggest that they experience aesthetic properties via proprioception is not to say that they never talk about how they look. They talk about this too. Endlessly. Yet sometimes vision is not a practical means of evaluating one's own movement. For example, turning one's head to look in the mirror can lead to performing a movement inaccurately since head directions are components of dancers' movements, and a correct head position might preclude even taking a peek. Moreover, onstage, productive visual evaluation of one's own movements falls by the wayside. Even ignoring these practical impediments, a trained dancer may at times trust proprioception more than vision. Sometimes this is because proprioception allows one to see in the mind's eye what a movement looks like. However, although proprioception can provide a platform upon which visual imagination works, sometimes proprioception itself is the vehicle of aesthetic experience; at these times, one appreciates the sheer proprioceptive quality of movement—or so I shall argue. In fact, a style of dance called "gaga" not only takes place in studios with covered mirrors so as to remove the temptation to look, but explicitly encourages dancers to forget their visual appearance and focus on ideas, the space around them, and their proprioceptive experience of movement (Katan-Schmid 2016).

Although gaga deliberately eschews the visual, in many if not most forms of dance, the visual and the proprioceptive feed each other: dancers take their own bodily movements to be *proprioceptively* graceful (for example), in part because they judge that if seen, these movements would look graceful, and they take certain bodily movements of others to be *visually* graceful, in part because they judge that if they were to move like so, such movements would feel graceful.

Proprioception, according to dancers' accounts, thus seems to be a means by which dancers experience the aesthetic properties of their own bodily movements. And it's not just dance aesthetics that is proprioceptively informed. For example, the pianist and writer Charles Rosen highlights the idea of proprioceptively appreciating the aesthetic qualities of a piece of music in explaining what he calls "Chopin's ruthlessness" (Rosen

1987). Chopin, he tells us, makes no concession to any technical limitations of pianist. In his words:

> The Etudes generally begin easily enough—at least the opening bars fit the hand extremely well. With the increase of tension and dissonance, the figuration quickly becomes almost unbearably awkward to play. The positions into which the hands are forced are like a gesture of exasperated despair. . . . The performer literally feels the sentiment in the muscles of his hand. This is another reason why Chopin often wanted the most delicate passages played with the fifth finger alone, the most powerful cantabile with the thumbs. There is in his music an identity of physical realization and emotional content. (Rosen 1987)

Rosen is illustrating the proprioceptive aesthetic experience of playing; through an awareness of his body, the pianist experiences the emotional content of the music: in the unbearably awkward figuration of the hands, one finds an aesthetics of exasperated despair. Here proprioceptive awareness does not interfere with performance: if anything, it seems conducive to it. Or at least it does if pianists do not get so proprioceptively immersed they fail to pay attention to other relevant aspects of their performance, such as how they sound. Yet, according to Rosen, pianists frequently forget to listen to themselves (2002, p. 36). Surprisingly, conductors, Rosen tells us, may also fail to take in aural information, not realizing, for example, that their orchestra has fallen back into a habitual way of playing rather than following the conductor's lead. Rosen's explanation: "So much of the sentiment that rightly belongs to the sound of music is embodied, for pianists and for conductors in the physical effort, in gesture" (2002, p. 36).

Other forms of artistic creation may also be guided, in part, by proprioception. For example, could there be an embodied aesthetics of singing, according to which singers garner aesthetic information about the quality of their voice via proprioceptive awareness of their vocal cords, mouth, and throat? Or could there be an embodied aesthetics of sculpture, according to which the sculptor's aesthetic decisions are guided not only by how her sculpture will look, but also by how it feels (via proprioception and touch) in her hands, arms, shoulders, and back as she is working?[2] Or an embodied aesthetics of painting, according to which painters experience some of the aesthetic qualities of their artworks through the movements of their bodies? Indeed, could it even be that writers sometimes make aesthetic choices about the cadence of words in literature and poetry based on a felt rhythm and flow that such words produce in their own bodies?

Outside of the arts, one might propose that athletic performance occasionally involves proprioceiving aesthetic properties of one's own movements—and not merely in sports such as gymnastics, which have a dancelike elements, but even in sports such as baseball, soccer, and hockey (see Cohen 1991). Perhaps the beauty or graceful feeling of a

[2] Though not explicitly about proprioception, Herder's (1778/2002) discussion of sculpture (especially pp. 41 and 91) highlights a role for touch (as broadly construed to include proprioception) in the appreciation of sculpture. See also Zuckert (2009).

movement can even be a guide to what works in sports. Rather than thinking about the movement on a muscular level, perhaps sometimes the best way to assure a slam dunk is to focus on the aesthetic qualities of the movement. However, rather than presenting further speculations about the reach of embodied aesthetics, let me return to my focus, which is dance, and which, from at least my perspective as a former dancer, provides us with most palpable example.

Are First-Person Reports Reliable?

Is what dancers say a reliable guide to what they are experiencing? It is sometimes pointed out that we are frequently mistaken about the contents of our own minds and therefore neither introspection nor first-person reports can be trusted.[3] However, although I assume that sometimes we are mistaken about what is going on in our own minds, especially with respect to the psychological instigators of our actions, I also think that, in general, as long as there are no good grounds to question any particular report about what someone says he or she is experiencing, if someone claims to be experiencing p, this is defeasible evidence for the view that he or she is experiencing *p*. For example, if you have good reason to believe that Sally is lying or that she is likely self-deceived (based on, say, personal interactions), or that she is responding to leading questions, or that she has been reading about a theory that may be coloring her own experience, or that the results of a psychology experiment or a philosophical argument show that the type of introspection she claims to employ is more likely wrong than right, then there may be reason to doubt Sally's reports about her experiential life. However, barring any good reasons for doubt, we ought to take Sally's and others' first-person accounts of their own mental processes at face value.

This methodological principle, which I rely on, follows roughly from what Tyler Burge calls the "acceptance principle," which holds that "a person is entitled to accept as true something that is presented as true and that is intelligible to him, unless there are stronger reasons not to do so" (1993, p. 467).[4] I have claimed that dancers talk in ways that indicate that they make judgments about how to move based on proprioceptive information. What they say indicates that they uphold the thesis of embodied aesthetics with respect to proprioception. The question, then, is whether there are good reasons to doubt what they say is correct.

[3] See, for example, Nisbett and Wilson (1977) and Schwitzgebel (2011), as well as Petitmengin et al. (2013) for a critical analysis of the former.

[4] Whether the acceptance principle is a fundamental principle that we know a priori, as Burge thought, or is justified through our experience of interacting with people, is a question I leave to the side.

Is Proprioception Sensory?

Ought we to believe dancers and others who claim to experience the beauty and grace of their own movements via proprioception? Let's put away the possibility that such individuals are flat out lying. Could they be mistaken? One reason to think that they are might seem to flow out of Elizabeth Anscombe's arguments against the idea that proprioception provides us with sensory information (1957/2000, 1962). When we come to learn (nonvisually) of the movements and positions of our limbs, we do so, Anscombe tells us, without having any sensory awareness of where they are, without perceiving them, without what she refers to as "observation." If she is correct we have reason to doubt dancers' claims, since if there are no proprioceptive sensations, it would not be possible for such sensations to ground aesthetic experience.[5]

Why does Anscombe think that our knowledge of our movements and positions is not observational? She argues that if we were to have observational or proprioceptive knowledge of our bodily positions and movements, it would need to be based on what she calls "separately describable" sensations. A separately describable sensation, as I understand her, is merely the sensation of one part of an entire movement. In having our legs crossed, Ansombe (1962) tells us, we might have separately describable sensations of tingles, pressure, and touch. But this, she thinks, does not provide us with enough information to ground our knowledge that our legs are crossed. And thus she concludes that proprioception can't ground knowledge of the movements and positions of our own bodies (cf. Bermúdez 2010).

If not via proprioception, how, then, does Anscombe think that we come to have knowledge of our bodily movements and positions? On her view, we know the positions and movements of our body because we have directed our body to move or assume a position; we know that our legs are crossed, for example, because we have directed them to cross (1957/2000, p. 14). As she sees it, just as an architect might know what a completed building looks like without seeing it (because she has directed the building to be built in a certain way), we can know where our limbs are without having sensations of them. Our knowledge of our bodily movements and positions, according to Anscombe, is not based on sensory experiences. And if this is correct, dancers cannot actually experience the beauty, grace, and so forth of their movements via proprioceptive sensations.

I would like to suggest, however, that although a dancer may know where her limbs are because she has directed them to go there, I think this is not the only way dancers and others come to understand their bodily positions and movements. When Anscombe attempts to identify the sensations associated with having crossed legs, she mentions that tingles, pressure, and touch are not normally sufficient for us to know that our legs are crossed. However, just because we do not typically arrive at knowledge of our bodily

[5] Or at least it wouldn't be possible to ground aesthetic experience, if one thinks that aesthetic experience must be grounded in conscious sensations.

positions via those sensations does not mean that we do not typically arrive at this knowledge via the sense of proprioception itself, for proprioception is what it is and not something else like the sense of pressure or the experience of a tingle. Moreover, given that we can have knowledge of entirely passive movements as well as bodily movements that fail to match our intended movements, directors' knowledge cannot be the entire story.[6]

It seems, then, that Anscombe's argument that proprioception is not based on sensory experiences does not disembody aesthetics, for it does not provide us with a reason to think that dancers do not experience the aesthetic qualities of their movements. Are there other reasons to think that proprioception has no sensory component? Some claim there are no such things as sensations at all; neurological activity exists, but not sensations (Churchland 1981; Churchland 1986). This position—which in philosophy of mind is referred to as "eliminativism"—I shall ignore in what follows, save for pointing out that if eliminativism is true, then at least all types of sensory experiences—the lower and the higher, those directed to oneself and those directed toward others—are on the same plane: all equally nonexistent. I shall also put to the side any other arguments for the nonexistence of proprioceptive sensations, if there are any, and in what follows assume that proprioception provides not only sensory, but, at times, conscious sensory information about bodily movements and positions (for a further defense of the idea that proprioception is a veritable sense, see Ritchie and Carruthers 2015). The question I turn to now is whether it is possible to reap information about aesthetic properties via proprioceptive sensations.

Distancing (and other) Requirements for Aesthetic Experience

The search for necessary and sufficient conditions for aesthetic experience has led, rather than to a set of criteria that all agree upon, to a tangled web of sometimes contradictory views.[7] Dancers' embodied experience seems to readily fit under the umbrella of some of these. For example, it has been claimed by Monroe Beardsley (1981/1958), John Dewey (1934/2005), and others that aesthetic experience must be unified. Although what this comes to is itself debated, in at least one good sense of the term, dancers' embodied aesthetic experience meets this criterion since the dancer herself, if her claims are ultimately accepted, is aware of a distinct kind of experience when she evaluates the aesthetic

[6] Whether or not proprioception produces conscious sensations, studying individuals who have suffered from a loss of proprioception indicates how important this sense is to our knowledge of our bodily movements and positions (see, e.g., Cole 1995).

[7] See Shusterman (2006) for an explanation of how the concepts of aesthetics and experience are both "vague, variable, and contested"; I agree with this as well as with his claim that these concepts can still be useful even if they are not definable in a univocal way (pp. 217–18).

qualities of her movements proprioceptively. It has also been thought that, in Bernard Bosanquet's words, "The aesthetic want is not a perishable want, which ceases in proportion as it is gratified" (1915, p. 4). But although one may get physically exhausted in moving, as one may get physically exhausted traipsing through a museum, one seems to never tire of the experience of moving in aesthetically valuable ways.

However, rather than systematically plowing through each proposed criterion and judging whether it permits dancers' proprioceptive experience of movement to count as aesthetic, let me focus on addressing what I see as the most challenging criteria for the thesis of embodied aesthetics: the related criteria of aesthetic distance and shareability. Many have thought that the aesthetic senses must, in some sense, distance the observer from the observed (Shaftesbury 1711/1900; Kant 1790; Dickie 1964; Beardsley 1981/1958; Bullough 1912; Iseminger 2003.) Yet, if proprioception is, as Oliver Sacks puts it, "the inner sense by which the body is aware of itself" (Sacks 1995, p. x), what is observed and the observed seem to be about as close together as you can get. Thus, on at least a first pass, the distancing requirement would seem to preclude proprioceptive embodied aesthetic experience.

But what exactly does it mean to distance an observer from the observed? The concept of distance takes on difference guises: there is physical distance, practical distance, psychical distance, and what I like to refer to as "metaphysical distance." Fortunately, the first three of these are readily addressed. Attaining physical distance requires that the object of aesthetic appreciation is not in direct contact with the observer. But contact is pervasive. The light waves that bounce off a painting must come in contact with one's eyes, no less than the molecules wafting from the perfume bottle must come in contact with one's nose. And so, let us simply reject this as a criterion for an experience to be aesthetic. The practical distancing requirement mandates that no practical need or compulsion is satisfied in aesthetic experience. However, although one might dance to pay the bills, one need not dance for practical rewards, and barring such afflictions as St. Vitus dance—a neurological disorder characterized by rapid, jerking movements—one is not physically compelled to dance. Finally, attaining what Edward Bullough (1912) refers to as "psychical distance" implies that in aesthetic experience one is not in the same psychological state as one would be in if the represented act were really occurring (for example, one ought not to climb up on the stage and prevent Hamlet from slaying Polonius). But whether or not this is an appropriate requirement for having an aesthetic experience of art, it would seem that the dancer's proprioceptive experience of dancing easily satisfies it since dancing seems to naturally distance the dancers from what they are portraying—there is too much technique to think about to ever forget entirely that you are a person with a body rather than, for example, ethereal sylph. Diderot (1830) emphasized the importance not losing the self in acting, arguing that the consummate actor deliberately controls her expressions of emotion, else she would not be able to perform as consistently as she does. The same could be said of dance: to perform well, dancers need to distance themselves from their roles in this way.

The metaphysical distancing requirement, however, seems to pose a greater obstacle for an embodied aesthetics of dance. Satisfying it requires us to distinguish in aesthetic experience the object one senses from the bodily sensation. In other words, it requires aesthetic experience to be about an object that is distinct from the experience of that object. In still other words, it mandates that in aesthetic experience we must find both a subject of experience and an object of experience. Can proprioception provide this?

Metaphysical distancing is important because it is thought to set aesthetic experience apart from mere sensory experience, such as the sensation of pain. The aesthetic senses, in contrast to our sense of pain, are supposed to direct us to objects in the world, which, in turn, is important since it is thought to permit what Kant thought of as the aesthetic virtue of shareability. According to Kant (1790/2007), aesthetic judgments (judgments based on aesthetic experiences) must be "shareable" and capable of grounding genuine disagreement. "A pinprick hurts" is not a shareable judgment; if you say it does and I say it doesn't, we are just expressing our own opinions about this. In contrast, "this table is marble" is shareable; barring vagueness, if we disagree, one of us is right and the other wrong. Aesthetic judgments are not the same as judgments about the table, on Kant's view. If correct, they are not objectively valid, since, according to Kant, they depend on our experience of aesthetic pleasure, but they must be *intersubjectively* valid—that is, though based on a feeling, they are judgments everyone ought to agree about. Metaphysical distancing would seem to be a necessary requirement for such shareablity.

The Subject and Object of Proprioception

Does proprioception permit metaphysical distancing? We can distinguish the object one senses from the bodily sensation in the realm of sight, smell, taste, touch, and hearing. For example, when I see an apple, the experience is distinct from the apple. Proprioception, although it is a type of self-perception, also distinguishes subject from object since the self in question is not merely sensory. As Merleau-Ponty (1945/2005) points out, we perceive our bodies as both subject and object, both as the locus of sensory awareness and as the object of such awareness. He uses touch as an example to illustrate this point: when one hand touches the other, he explains, it is possible to move back and forth between noticing the tactile experience of touch and what one is having an experience of (pp. 130–1). To bring this idea into focus, consider an artist who creates beautiful shapes with her hands; she sculpts not only with her hands but also sculpts her own hands. If she created a representation of a bird in flight with just her left hand, she could, by representing her body both as subject and as object, experience the soft curves and strong lines of her creation by touching and exploring her left hand with her right; such an action would effect both a subjective tactile experience (the tactile experience in her right hand) and, intimately intertwined with this, an experience of an object: her left hand.

Perceiving the body as object, rather than as subject only, allows for misrepresentation. Proprioception does as well. One way to misrepresent the world is to represent p as q when p is not q. When I look at a field and see it as covered with snow when it is actually covered with clover in bloom, my visual experience misrepresents the field. Proprioception can similarly misrepresent the world. Choreographers, rehearsal directors, ballet masters and mistresses, for example, often see dancers make mistakes based on such misrepresentations: a dancer might proprioceptively experience his knee as perfectly straight, when it is in fact bent, or his leg as directly behind him, when it off to the side. More dramatically, one can misrepresent p as q when p fails to exist. Proprioceptive mistakes occur with amputees who have phantom limbs; in this case, they may represent their right leg as bent when they no longer have a right leg. Of course, even pain judgments can be mistaken in this more dramatic sense: with phantom limb pain, one can feel foot pain without having a foot. Yet, arguably, judgments of pain are not mistaken in the former sense: if a pain appears sharp, then it is sharp. However, like vision, which represents actual objects in the world as being a certain way and is capable of misrepresenting them, proprioception seems to represent one's body as being in a certain way and is capable of misrepresenting it: one's limb may proprioceptively appear straight when it is bent.

Does proprioception, then, like vision, represent something in the word? Although there is a sense in which one's own body is not part of the world (when the world is considered as that which exists apart from oneself), this is not the relevant sense here; rather, the relevant contrast here is between one's body and one's bodily sensations, between the positions and movements of one's limbs and the sensations one has of these positions and movements, between body as subject and body as object—proprioception admits such a contrast.[8] Proprioception may be a type of self-perception, but the self in question is not merely sensory. Thus, proprioceptive experience, it seems, need not be doomed to exist solely in the realm of the mere agreeable, which, according to Kant (1790/2007), is entirely subjective and appeals only to the senses and not, in addition, to our aesthetic sensibilities (§7–8).

The Need for an Extended-Embodied Aesthetics

Though proprioceptive judgments can be about objects, the analogy to Merleau-Ponty's example of touch is not complete. The artist feels her left hand with her right, but another person could also reach out and touch the artist's hand. When we disagree about what is covering a field (snow or clover), we are disagreeing about properties in the

[8] This is consonant with the view put forth by José Bermúdez (1998), *The Paradox of Self-Consciousness*. See also Wittgenstein's (1953) distinction between the body as "object" and the body as "subject."

world that are visually presented to us both. However, the body I proprioceive is the one you see. Thus, unlike touch and vision, the object of my proprioception would seem to be shareable with another person only insofar as the other person employs touch or vision (or, indeed, hearing or olfaction). You and I can both touch your hand. However, although I can proprioceive my hand, you, it seems, can see it or touch it (or hear it or smell it), but cannot proprioceive it. [9] Does this mean that in proprioceiving one's own body, we cannot satisfy Kant's requirement that the object of aesthetic experience needs to be shareable?

Proprioception, then, seems private in a way that the traditional aesthetic senses are not. What does this mean for the possibility of an embodied proprioceptive aesthetics? One might question the inviolability of Kant's shareability requirement.[10] However, let me instead ask: is the object of your proprioceptive experience perceived only by you? Or is there a sense in which you can proprioceive someone else's movements? I would like to suggest that our bodies are a window not only into the aesthetic properties of our own movements, but also into the aesthetic properties of the movements of others; in other words, that via "kinesthetic sympathy" or "motor perception" we may come to appreciate various aesthetic qualities of the movements of others. Strange as this may sound, observers, I want to suggest, can "proprioceive" a dancer's movements; sitting motionless in a darkened theater, audience members can sense in their own bodies—or have a "motor perception" of—the movements of the dancers behind the proscenium arch.

More generally, in addition to the thesis of embodied aesthetics, I would like to propose the thesis of extended-embodied aesthetics:

Extended-embodied aesthetics: one can have an aesthetic experience of another's body as perceived through senses other than vision and audition.

Perhaps with the right kind of training, the thesis of embodied aesthetics can be extended even further: I proffer that not only are we able to attribute aesthetics properties to others' bodily movement via kinesthetic sympathy, but, via kinesthetic sympathy, we are also able to attribute certain bodily based aesthetic properties to non-bodily artworks created by bodies. This view was upheld by Johann Gottfried Herder (1778/2002), who argued for an extended-embodied experience of sculpture: "The eye that gathers impressions is no longer the eye that sees a depiction on a surface; it becomes a hand, the ray of light becomes a finger, and the imagination becomes a form of immediate touching" (quoted in Gaiger's introduction to Herder 1778/2002; see Zuckert 2009 for a discussion). Visually observing paintings can also have a bodily component. When

[9] To be sure, conjoined twins who share an arm, for example, each proprioceive the same object, that is, the movement of the shared arm. Virtual reality might be another way an individual can proprioceive another's movements, since virtual reality seems to create situations where we speak of proprioceiving a movement that is not a movement of our own body. That is, when watching an image of your arm reach out across the Hudson over to New Jersey, it seems that you are proprioceiving a virtual arm, which, in one sense, does not count as part of your body; though in another sense—an "extended body" sense—it does.

[10] I do this in Montero (2006a, 2006b).

you see a graceful curve on a Kandinsky painting, are you perhaps, in part, identifying it as graceful because in looking at the painting, you have a bodily experience of your arm moving in a graceful arc? When you identify a Jackson Pollock painting as energetic, is it in part because you can feel the energetic movement needed to create such a painting? Why is the Mona Lisa's smile so captivating? Certainly, it is visually captivating, but might it also be proprioceptively captivating: when you observe the smile, do you in part appreciate what it is like to smile in that way? What is it about live music that makes it so captivating? A number of reasons come to mind: we enjoy the unpredictability, we appreciate being part of a group, we like seeing which instruments produce which sounds. But could it also be that you are appreciating the sound as embodied in the movements you see the musicians make? And the experience of kinesthetic sympathy might not be limited to action that is visually observed. When listening to a great recording of Chopin's Etudes, are you in part appreciating the ruthlessness of the music that Rosen talks about in the muscles of your hand? The thesis of extended-embodied aesthetics answers such questions in the affirmative.

Support for the Thesis of Extended-Embodied Aesthetics

That there are neural underpinnings for the subpersonal components of an extended-embodied aesthetics is suggested by a large body of neuroscientific research that indicates that certain areas of the brain—areas that have been variously referred to as the "mirror system," the "action-observation network," and the "action resonance circuit"—exhibit increased activity when one executes a movement or sees that same movement performed by others (the literature here is vast; see Kilner and Lemon 2013 for a review). Behavioral studies have also supported the idea that visual impressions of movement in some sense resonate in the observer's body. For example, it has been shown that subjects in perceiving static photographs of an individual in motion are more likely to mistake the position of the individual as being further along in the action than in being in a position that is prior to the one depicted in the photograph, indicating that even in perceiving static images, we can represent dynamic information (Freyd 1983). Moreover, subjects tend to perceive geometrical figures in a way that is consistent with how that figure is naturally drawn. For example, if subjects see a circle being traced by a point of light which speeds up along the top and bottom of the circle and slows down along the sides, subjects tend to perceive an ellipse rather than a circle (Viviani and Stucchi 1989; Thoret et al. 2016a). Since we slow down in drawing sharp curves, one explanation of what is going on here is that our understanding of the circle is not entirely visual; we also understand its shape by feeling what it is like to draw it. When we watch a shape being drawn, the motion of drawing resonates in our own bodies. And, in relation to Rosen's comment about feeling the sentiment of the music in his hands, there are even studies that support the existence of a kind of audio motor perception, in which listening to the sound that a pencil on paper would make when drawing an ellipse flattens out one's drawings of circles (Thoret et al. 2016b).

But more pertinent to the question of not only whether proprioception is shareable but also whether it is capable of grounding aesthetic experience comes from observing how dance critics talk about dance. For example, though he prefers to use the term "kinesthesia" rather than "proprioception," dance critic John Martin tells us that in order to appreciate dance fully one must make use of "kinesthetic sympathy" (1972, p. 15). In his words, "Not only does the dancer employ movement to express his ideas, but, strange as it may seem, the spectator must also employ movement in order to respond to the dancer's intention and understand what he is trying to convey" (p. 15). "The irreducible minimum of equipment demanded of a spectator," Martin tells us, "is a kinesthetic sense in working condition" (p. 17). Accordingly, his reviews of dance were rife with references to the qualities of dance appreciated via kinesthetic sympathy, or motor perception. For example, he speaks of the dynamic variation in a dancer's movement that gave it a "rare beauty and a powerful kinesthetic transfer," or "[a] gesture which sets up all kinds of *kinesthetic* reactions," or even a dancer who "leaves you limp with vicarious *kinesthetic* experience" (p. 17). If proprioception grounds not only judgments about the aesthetic qualities of one's own movement but also about the movement of others, the objects of proprioception would then easily meet the aesthetic bar.

Edwin Denby is another critic who frequently emphasizes the relevance of motor perception in his writings. For example, in a review of a performance of *Afternoon of a Faun,* he mentions the bodily feeling that results from imitating the depictions of people on Greek vases and bas-reliefs: "The fact is that when the body imitates these poses, the kind of tension resulting expresses exactly the emotion Nijinsky wants to express" (1998, p. 34–5). He continues: "Both their actual tension and their apparent remoteness, both their plastic clarity and their emphasis by negation on the center of the body (it is always strained between the feet in profile and the shoulders *en face*)—all these qualities lead up to the complete realization of the faun's last gesture" (pp. 34–5). Denby is characterizing something kinesthetic: something that we come to understand not through our visual experience alone. Through motor perception, Denby feels the tension of the dancer's bodily torque.

Denby illustrates how kinesthetic qualities provide insight into what are commonly called the "expressive qualities" of the work. Expressive qualities reveal the emotion represented in a work of art, and Denby's contention in the review cited earlier is that Nijinsky expresses discomfiting emotions via the strained and twisted comportment of the faun. In a piece such as this, the dancer's movements, among other things, also represent some of the expressive qualities of the Debussy score. As such, it is in part via motor perception that audience members—especially those with dance training—experience both the expressive qualities of a dancer's movements and, more indirectly, the expressive qualities of the music. Yet the role of motor perception in aesthetic judgment, as I see it, is not limited to the judgment of expressive qualities, since part of the value of watching dance has to do with the motor-perceptual experience of beauty, precision, fluidity, and grace—qualities that are not emotions. In watching a dancer, we not only visually experience the beauty of her movements, but we may motor-perceptually experience it as well.

There are numerous other critics who also understand motor perception as a means by which we come to understand and experience aesthetic qualities. Alastair Macaulay, for instance, frequently alludes to motor perception—in one review telling us that Fredrick Ashton's choreography is "more kinesthetically affecting than any other ballet choreographer's," and that in "watching [it], you feel the movement so powerfully through your torso that it is often hard to sit still in your seat" (Macaulay 2008). Similarly, Louis Horst describes the "lyric beauty" of a dance choreographed by Anna Sokolow as having "a direct appeal to kinesthetic response" (Horst 1954, as quoted by Warren 1991, p. 114). And Michael Wade Simpson describes the finale in a piece by Helgi Tomasson as "satisfying musically, kinesthetically and emotionally" (Simpson 2005).

How do we know that such critics really have the sorts of experiences they claim to have? Again, as long as there are no good grounds to question such first-person knowledge, it seems that these dance critics' claims to experiencing a kinesthetic response upon watching dance should be taken as defeasible evidence for the truth of such claims.

Are there good reasons to think that these dance critics are not experiencing what they write about? Perhaps the strongest argument against kinesthetic sympathy comes from those who claim to not understand what in the world such a thing could be. Because such objectors have never experienced kinesthetic sympathy upon watching dance, they believe that dance critics must simply be making it up. One might try to respond to such doubters by saying, "I've experienced this myself so I know it exists," yet such an approach typically does not take one very far. However, this response combined with an explanation for why such doubters may have never experienced kinesthetic sympathy does, I think, carry some weight. Do we have such an explanation?

The experiments on the mirroring system in neuroscience suggest that the mirror response is most pronounced when one observes movements that are in one's motor repertoire (Buccino et al. 2004). Thus, it could be that those who doubt the existence of kinesthetic sympathy do not have the relevant motor skills. This is not to say that dance experience is a necessary condition for the ability to feel limp with vicarious kinesthetic sympathy, for example. Other physical activities might suffice for grounding the experience. But it does seem that having a significant amount of dance experience suffices for the ability to feel dancers' movements in one's body upon watching dance. Thus, those who doubt the existence of kinesthetic sympathy might want to consider the possibility that a world of embodied aesthetics could open up to them too, if only they had some training.[11]

[11] Does this mean that some individuals are barred from certain forms of aesthetic experience? And might this not be contrary to Kant's idea that correct aesthetic judgments are intersubjectively valid? To be sure, I am suggesting that having some dance or movement experience is relevant to one's ability to appreciate the kinesthetic resonance of dance. However, just as appreciating great literature requires training, it seems that appreciating great dance could require training as well. For further discussion, see Montero (2012, 2013a, 2013b).

Still, one might doubt that a critic's kinesthetic sympathy upon watching a dancer move is relevant to the aesthetic qualities of the dancer's movement (as does McFee 1992). But such doubts, I would like to suggest, are unfounded. Here's an analogy. It seems reasonable to understand scientific knowledge as providing the best picture of what sorts of things exist in the world. Or at least, it seems reasonable to say that if something is theorized by scientists to exist, such as atoms and cells, or if something is classified by scientists in a certain way, such as a whale being classified as a mammal, it is reasonable to think that these things do exist and that these classifications are accurate, unless we have very strong arguments to the contrary. (For example, some might take Bas van Fraassen's 1980 argument for constructive empiricism to show that some of the posits of science actually do not exist.) I think that our attitude toward the posits of art critics, when they are writing in their area of expertise, should be analogous to our attitude toward the posits of scientists when they are writing in their area of expertise. That is, if art critics generally accept something as a work of art (such as a Duchamp ready-made), it is reasonable to accept it as such unless one has good arguments to the contrary. Correlatively, if dance critics generally accept the aesthetic relevance of kinesthetic sympathy, it is reasonable for us to accept it too, unless there are good arguments to the contrary. And, as I have indicated, at least a good number of prominent ones do. Thus, we have more support for the idea that the objects perceived via proprioception are not private in the way that would preclude proprioception from being an aesthetic sense. We might not want to go as far as Alastair Macaulay and claim that watching dance may be "less visual than kinesthetic" (2012), but we seem to have good grounds for the claim that conscious motor perception is one pathway to the aesthetic experience of dance. It is not just the bodies of others as represented in painting or statuary, or as visually perceived on stage, that are objects of aesthetic evaluation; one's own body, via proprioception, can also bestow aesthetic pleasure.

There are, of course, arguments to the contrary, but it is not clear that there are any good arguments to the contrary. For example, according to Graham McFee (1992), if kinesthetic sympathy exists, it is irrelevant to one's appreciation of dance because "the focus on the performer is not appropriate to an art form such as dance" (p. 273). Rather, for McFee, the choreographer, and not the dancer, is the artist. However, in observing dance, it seems more accurate to say that one is observing the work of a number of artists, including the choreographer, the composer, the musicians, the lighting designer, the set designer, the costume designer, and the dancers. With many dances, some of these aspects should not be the focus of attention. For example, lightning, especially in classical ballet, is for the purpose of highlighting the dancer and should not be noticed by the audience. However, it seems that the dancer's movements are not merely for the purpose of highlighting the choreography, as the typical dance review, which spends a significant amount of time discussing the qualities of individual dancers' movements, illustrates. Perhaps McFee sees reviews of this sort misguided. However, it would seem to be incumbent on him to explain why.

Another argument against the view that kinesthetic sympathy is aesthetically relevant might proceed like this: although proprioception may provide insight into how one is actually moving, the aesthetically relevant qualities of movement depend on the visual illusion—the appearance of floating on the stage during a *bourrée*, of suspending oneself in the air in *grand jeté*, and so forth—produced by the movement. In other words, what is aesthetically relevant is how a movement (deceptively) looks, not how it actually feels, and proprioception only gives us the latter.

This argument would be powerful if it were the case that proprioception only provides information about how one is actually moving, for it is indeed true that the aesthetics of dance, particularly ballet, is sometimes based on the creation of visual illusions: leaps that appear to defy gravity, limbs that appear elongated, ease of movement in the presence of enormous effort. However, I would like to suggest that one can proprioceive an illusory movement. When one performs a "gravity defying" leap by further extending one's limbs at its height, one has a proprioceptive sensation of being suspended mid-air, and the same goes for watching such a leap: what in part makes watching a dancer leap aesthetically satisfying is that we feel the flight. Indeed, I would claim that one of the wonders of dancing—one of the reasons why dancers will put up with the pain it often involves—is that dance allows one to experience the impossible.

If one can proprioceptively experience a gravity-defying leap, this means that there are proprioceptive illusions. That there are such illusions is widely accepted: pilots in flight and in-orbit astronauts may experience proprioceptive illusions related to their position in space, and artificial muscle vibration can create a proprioceptive illusion that one's limb is bent at a certain angle when it is not (Goodwin et al. 1972; Previc and Ercoline 2004). These sorts of illusions may be more robust than the proprioceptive illusion one experiences in dance, but at the same time, the Müller-Lyer illusion, for example, is more robust than the visual illusion one has of seeing a dancer defy gravity. So, while the illusory element of the aesthetics of bodily movements cannot be overlooked, we can proprioceive, as well as see, illusory movement. We have, thus, found no reason to doubt that watching dance can deliver a powerful kinesthetic punch.

Conclusion and a Lingering Question

Proprioception, then, if I am correct, enables one to perceive aesthetic qualities of one's own bodily movements. There is a prima facie case to be made in favor of the claim, and the theoretical considerations that might tell against the possibility of proprioception enabling us to perceive aesthetic qualities of our own movements do not stand. Beyond sensing certain aesthetic qualities of our own movements, audience members may base certain aesthetic judgments about dancers in part on the internal experience of movement they have while watching dance. As I argued, neuroscience and behavioral studies suggest that in watching others move in familiar ways, we represent their movement in our bodies, and the words of dance critics lend support to the view that such motor resonance can be experienced consciously and is aesthetically valuable. In proprioceiving the aesthetic qualities of one's own movements or in having a kinesthetic response to the aesthetic qualities of the movements of others, the soul, contrary to Santayana's contention, is intimately connected to the body. Thus, aesthetics is embodied. It is not just the bodies that artists represent that are to be revered, but sometimes it is also, from their own point of view, the bodies of the artists themselves.

That said, one might have a lingering doubt about my whole approach, for it is sometimes claimed that a deliberate focus on how one is moving hinders one's performance of that movement. I claim that in focusing on her own movement, a dancer can aesthetically enjoy her own body. But doesn't deliberate focus on how one is moving stymie one's performance of that movement? Doesn't a dancer need to perform automatically and without any thought or awareness at all?

Dewey (1934/2005) had the right response to this question: "Because the artist is controlled in the process of his work by his grasp of the connection between what he has already done and what he is to do next, the idea that the artist does not think as intently and penetratingly as a scientific inquirer is absurd" (p. 360). (I have a much longer version of this response in Montero 2016; see also Montero 2013b, 2010, and Cole and Montero 2007.)

What does taking embodiment into account mean for aesthetics? Following Friedrich Nietzsche, who suggested that traditional aesthetics suffers from its exclusive focus on the reception rather than the creation of art, Richard Shusterman (2006) states that "philosophical accounts of aesthetic experience would be enriched by more attention to artists' experience" (p. 84). I have aimed to uncover one aspect of the artist's creative process—the dancer's embodied aesthetic experience of movement. Whether this has enriched our perception of aesthetics, I leave it to you, my readers, to judge.

References

Anscombe, G.E.M. (1957/2000). *Intention* (2nd ed.). Cambridge, MA: Harvard University Press.

Anscombe, G. E. M. (1962). On sensations of position. *Analysis*, 22(3), 55–8.

Aquinas, T. (1485/1960). *Summa theologiae* (vol 1). New York: McGraw-Hill.

Beardsley, M. (1981/1958). *Aesthetics*. Indianapolis, IN: Hackett Publishing Company, Inc

Bermúdez, J. (1998). *The paradox of self-consciousness*. Cambridge, MA: MIT Press.

Bermúdez, J.L. (2010). Action and awareness of agency. *Pragmatics and Cognition*, 18(3), 584–96.

Burge, T. (1993). Content preservation. *Philosophical Review* 102(4), 457–88.

Buccino, G., Lui, F., Canessa, N., Patteri, I., Lagravinese, G., Benuzzi, F. et al. (2004). Neural circuits involved in the recognition of actions performed by nonconspecifics: an fMRI study. *Journal of Cognitive Neuroscience*, 16, 114–26.

Bullough, E. (1912). "Psychical distance" as a factor in art and as an aesthetic principle. *British Journal of Psychology*, 5, 87–117.

Churchland, P.M. (1981). Eliminative materialism and the propositional attitudes. *Journal of Philosophy*, 78, 67–90.

Churchland, P.S. (1986). *Neurophilosophy: toward a unified science of the mind/brain*. Cambridge, MA: MIT Press.

Cohen, T. (1991). Sports and art: beginning questions. In: J. Andre and D.N. James (eds.), *Rethinking college athletics*. Philadelphia: Temple University Press, pp. 258–304.

Cole, J. (1995). *Pride and the daily marathon*. Cambridge, MA: MIT Press.

Cole, J. and Montero, B. (2007). Affective proprioception. In: S. Gallagher (ed.), *Special issue of Janus Head: Journal of Interdisciplinary Studies in Literature, Continental Philosophy, Phenomenological Psychology, and the Arts*, 299–317.

Denby, E. (1998). *Dance writings and poetry*. New Haven, CT: Yale University Press

Dewey, J. (1934/2005). *Art as experience*. New York: TarcherPerigee.

Dickie, G. (1964). The myth of the aesthetic attitude. *American Philosophical Quarterly*, 1, 56–65.

Diderot, D. (1830). *The paradox of acting* (trans. W.H. Pollock). Full text available: https://archive.org/details/cu31924027175961

Frede, J. (1983). The mental representation of movement when static stimuli are viewed. *Perception and Psychophysics*, 33(6), 575–81.

Goodwin G.M., McCloskey, D.I., and Matthews, P.B. (1972). The contribution of muscle afferents to kinaesthesia shown by vibration induced illusions of movement and by the effects of paralysing joint afferents. *Brain*, 95, 705–48.

Hegel, G.W.F. (1834/1975). *Aesthetics: lectures on fine art* (2 vols.; trans. T.M. Knox). Oxford: Clarendon Press.

Herder, J.G. (1778/2002). *Jason Gaiger translation of Herder's Sculpture: Some Observations on Shape and Form from Pygmalion's Creative Dream*. Chicago, IL: University of Chicago Press.

Hutcheson, F. (1725/1973). *An inquiry concerning beauty, order, harmony, design* (ed. P. Kivy). The Hague, Netherlands: Martinus Nijhoff.

Iseminger, G. (2003). Aesthetic experience. In: J. Levinson (ed.), *The Oxford handbook of aesthetics*. Oxford: Oxford University Press, pp. 99–116.

Kant, I. (1790/2007). *Critique of judgment* (ed. N. Walker; trans. J.C. Meredith). New York: Oxford University Press.

Katan-Schmid, E. (2016). *Embodied philosophy in dance: Gaga and Ohad Naharin's movement research*. Basingstoke, UK: Palgrave Macmillan UK.

Kilner, J.M. and Lemon, R.N. (2013). What we know currently about mirror neurons. *Current Biology*, 23, R1057–62.

Korsmeyer, C. (1999). *Making sense of taste: food and philosophy*. Ithaca, NY: Cornell University Press.

Lopes, D.M. (2002). Vision, touch and the value of pictures. *The British Journal of Aesthetics*, 42, 87–97.

Macaulay, Alastair (2008, December 22). A spinning, twisting tribute to Ashton, with skaters and pigeons. *New York Times*. Retrieved from http://www.nytimes.com/2008/12/23/arts/dance/23asht.html?pagewanted=all&_r=0

Macaulay, A. (2012, May 9). Works that are longer on style than on choreography [review]. *New York Times*. Retrieved from http://www.nytimes.com/2012/05/10/arts/dance/alonzo-king-lines-ballet-at-the-joyce-theater.html

Martin, J. (1972). *Introduction to the dance*. New York: Dance Horizons Inc.

McFee, G. (1992). *Understanding dance*. New York: Routledge.

Merleau-Ponty, M. (1945/2005). *Phenomenology of perception*. London: Taylor and Francis Books Ltd.

Montero, B. (2006a). Proprioceiving someone else's movement. *Philosophical Explorations*, 9(2), 149–61.
Montero, B. (2006b). Proprioception as an aesthetic sense. *Aesthetics and Art Criticism*, 64(2), 231–42.
Montero, B.G. (2010). Does bodily awareness interfere with highly skilled movement? *Inquiry*, 53, 105–22.
Montero, B.G. (2012). Practice makes perfect: the effect of dance training on the aesthetic judge. *Phenomenology and Cognitive* Science, 11(1), 59–68.
Montero, B.G. (2013a). The artist as critic: dance training, neuroscience and aesthetic evaluation. *Journal of* Aesthetics *and Art Criticism*, 71(2), 169–75.
Montero, B.G. (2013b). A dancer reflects: deliberation in action. In: J. Schear (ed.), *Mind, reason and being-in-the-world: the McDowell-Dreyfus debate*. London: Routledge.
Montero, B.G. (2015). Review of *Body Consciousness: A Philosophy of Mindfulness and Somaesthetics* by R. Shusterman and *Thinking through the Body: Essays in Somaesthetics* by R. Shusterman, *Mind*, 124(495), 975–9.
Montero, B.G. (2016). *Thought in action: expertise and the conscious mind*. Oxford: Oxford University Press.
Nisbett, R.E. and Wilson, T.D. (1977). Telling more than we can know: verbal reports on mental processes. *Psychological Review*, 84(3), 231–59.
Petitmengin, C., Remillieux, A., Cahour, B., and Carter-Thomas, S. (2013). A gap in Nisbett and Wilson's findings? A first-person access to our cognitive processes. *Consciousness and Cognition*, 22(2), 654–69.
Prall, D.W. (1929). *Aesthetic judgment*. New York: Thomas Y. Crowell.
Previc, F.H. and Ercoline, W.R. (2004). Spatial disorientation in aviation. Reston, VA: American Institute of Astronautics and Aeronautics.
Ritchie, J.B. and Carruthers, P. (2015). The bodily senses. In: M. Matthen (ed.), *The Oxford handbook of philosophy of perception*. Oxford: Oxford University Press, pp. 353–70.
Rizzolatti, G. and Craighero, L. (2004). The mirror-neuron system. *Annual Review of Neuroscience*, 27, 169–92.
Rosen, C. (1987/2002). *Piano notes: the world of the pianist*. New York: Free Press.
Sacks, O. (1995). Introduction. In: J. Cole, *Pride and the daily marathon*. Cambridge, MA: MIT Press.
Santayana, G. (1955). *The sense of beauty*. New York: Modern Library.
Schwitzgebel, E. (2011). *Perplexities of consciousness*. Cambridge, MA: MIT Press.
Shaftesbury, A.E. (1711/1900). *Characteristics of men, manners, opinions, times, etc.* (2 vols.; ed. J.M. Robertson). London: Grant Richards.
Shusterman, R. (2006). Aesthetic experience: from analysis to eros. *The Journal of Aesthetics and Art Criticism*, 64(2), 217–29.
Shusterman, R. (2008). *Body consciousness: a philosophy of mindfulness and somaesthetics*. New York: Cambridge University Press.
Shusterman, R. (2012). *Thinking through the body: essays in somaesthetics*. New York: Cambridge University Press.
Simpson, M.W. (2005, April 9). Dancers relay Robbins' gentler, reflective side in Chopin piece. *SFGate/San Francisco Chronicle*. Retrieved from http://www.sfgate.com/entertainment/article/Dancers-relay-Robbins-gentler-reflective-side-2687074.php
Smith, B. (2015). The chemical senses. In: M. Matthen (ed.), *The Oxford handbook of philosophy of perception*. Oxford: Oxford University Press, pp. 314–51.

Thoret, E., Aramaki, M., Bringoux, L., Ystad, S., and Kronland-Martinet, R. (2016a). When eyes drive hand: influence of non-biological motion on visuo-motor coupling. *Neuroscience Letters*, 612, 225–30.

Thoret, E., Aramaki, M., Bringoux, L., Ystad, S., and Kronland-Martinet, R. (2016b). Seeing circles and drawing ellipses: when sound biases reproduction of visual motion. *PLoS One*, 11(4): e0154475.

Toner, J., Montero, B., and Moran, A. (2015). The perils of automaticity. *Review of General Psychology*, 19(4), 431–42.

van Fraassen, B. (1980). *The scientific image*. New York: Oxford University Press.

Vecsey, G. (2010, September 11). A tournament filled with hope and grace. Retrieved from http://www.nytimes.com/2010/09/12/sports/tennis/12vecsey.html

Viviani, P. and Stucchi, N. (1989). The effect of movement velocity on form perception: geometric illusions in dynamic displays. *Perception and Psychophysics*, 46(3), 266–74.

Warren, L. (1991). *Anna Sokolow: the rebellious spirit*. Princeton, NJ: Princeton Book Co.

Wilson, J. (2014). No work for a theory of grounding. *Inquiry*, 57, 535–79.

Wittgenstein, L. (1953). *Philosophical investigations*. New York: Macmillan.

Zuckert, R. (2009). Sculpture and touch: Herder's aesthetics of sculpture. *The Journal of Aesthetics and Art Criticism*, 67(3), 285–99.

Name Index

Note: page numbers followed by *n* indicate footnotes.

G

H

I

J

K

Y

Z

Subject Index

Note: page numbers followed by *n* indicate footnotes.

F

G

H

M

N

O

R

S